PESTICIDES - FORMULATIONS, EFFECTS, FATE

Edited by **Margarita Stoytcheva**

INTECHOPEN.COM

Pesticides - Formulations, Effects, Fate
Edited by Margarita Stoytcheva

Published by InTech

Janeza Trdine 9, 51000 Rijeka, Croatia

Edition 2014

Publishing Process Manager Iva Lipovic

Technical Editor Teodora Smiljanic

Cover Designer InTech Design Team

Image Copyright B Brown, 2010. Used under license from Shutterstock.com

Additional hard copies can be obtained from orders@intechopen.com

Pesticides - Formulations, Effects, Fate, Edited by Margarita Stoytcheva
p. cm.
ISBN 978-953-307-532-7

INTECH

open science | open minds

Contents

Preface

A wide range of substances and mixtures of substances, highly effective in preventing, destroying, repelling or mitigating pests: insects, mice, unwanted plants, fungi, microorganisms such as bacteria and viruses, and prions are classified as pesticides. User benefit from pest control is obvious, in terms of increased agricultural production and elimination of pest-induced health problems. However, there are now evidence that pesticides do create risk to man and his environment. Therefore, the agricultural use of some persistent organochlorine insecticides, namely DDT, was banned after the 1960s. Other synthetic chemicals, organophosphates (1960s), carbamates (1970s) and pyrethroids (1980s) as well the emerging biopesticides (2000s) came on to substitute them.

This book comments on a large spectrum of pesticides issues, organized in three sections: new pesticides, pesticides formulations characteristics and pesticides application systems, assessment of pesticides pollution and exposure, and pesticides degradation and disposal.

Chapters 1-3, included in the first book section, highlight the growing interest on the development and use of pesticides derived from natural materials: plants and bacteria, as an environmental friendly alternative to the synthetic pesticides.

The design and the optimization of pesticides microemulsions, pesticides mixtures and controlled pesticides release formulations as well as the techniques for pesticides application are discussed in Chapters 4-9.

The second book section is dedicated to the following topics of great public concern: (i) the monitoring of the pesticides residues in agricultural products, the entry routes identification, the effect of food processing techniques to reduce the pesticides levels, the prevention of food contamination and the pesticides risk assessment (Chapters 10-15); (ii) the pesticides contamination of the environment: air, surface- and groundwater, soil, sediments and biota, the pesticides transport and distribution in the polluted area, the impacts of the pesticides presence on the ecosystems, the effects of pesticides exposure and the strategies for pesticides pollution surveillance (Chapters 16-26).

The third book section presents current investigations of the naturally occurring pesticides degradation phenomena, the environmental effects of the break down products and different approaches to pesticides residues treatment (Chapters 27-39).

The book covers selected noteworthy studies authored by leading international experts. A helpful feature of the volume is the extensive list of references at the end of each chapter, intended to reveal the current state of the art. The book is highly recommended to the professionals interested in pesticides issues.

The publication was made possible due to the efforts and the expertise of the contributing authors. They are gratefully acknowledged.

Margarita Stoytcheva
Mexicali, Baja California
Mexico

Part 1

Pesticides and Pesticides Formulations

1

Pesticides of Botanical Origin: A Promising Tool in Plant Protection

Nikoletta G. Ntalli and Urania Menkissoglu-Spiroudi*
Pesticide Science Laboratory, Faculty of Agriculture, Aristotle University of Thessaloniki
Greece

1. Introduction

Future agricultural and rural development is, to a large extent, influenced by the rapidly increasing food demand of 2.5 billion people expected to swell the world population by 2020. Achieving food sufficiency in a sustainable manner is a major challenge for farmers, agro-industries, researchers and governments (Schillhorn van Veen, 1999). The intensification of agriculture to fulfil food needs has increased the number of insect pest species attacking different crops and as a result the annual production losses of the standing crops. In the past, synthetic pesticides have played a major role in crop protection programmes and have immensely benefited mankind. Nevertheless, their indiscriminate use has resulted in the development of resistance by pests (insects,weeds, etc), resurgence and outbreak of new pests, toxicity to non-target organisms and hazardous effects on the environment endangering the sustainability of ecosystems (Jeyasankar & Jesudasan, 2005). In the recent years the EU has employed a fundamental reform of the Common Agricultural Policy (CAP) highlighting the respect to the environmental, food safety and animal welfare standards, imposing farmlands' cross-compliance with good agricultural and environmental conditions (Schillhorn van Veen, 1999). Due to environmental side effects and health concerns, many synthetic carbamate, organophosphate, and organophthalide pesticides have been banned (Council Directive 91/414/EEC) or are being under evaluation (Regulation 2009/1107/EC OL & Directive 2009/128/EC). On the other hand, industry does not equally sustain the economic cost of research and registration, of all pesticides' chemical classes. The development of nematicides is rarely supported, even though in some cases, such as in the Netherlands, they represent more than 60% of the total pesticides used in agriculture (Chitwood, 2002). This is due to the fact that nematodes are a rather difficult target and the economic cost of research and registration is an enormous hurdle for a prospective new synthetic nematicide to overcome (Chitwood, 2002). As a result, currently there are only few nematicides left in use, and their limited number makes the repeated applications of the same formulation, inevitable. This fact has led to the enhancement nematicides biodegradation in soil (Qui et al., 2004 , Karpouzas et al., 2004, Arbeli & Fuentes, 2007) and the development of resistance in pests. (Meher et al., 2009) These two phenomena are expressed in field as lack of efficacy of the applied pesticides. All the above facts necessitate the urge for new and alternative pest control methods (Chitwood, 2002).
An interesting way of searching for biorational pesticides is screening naturally occurring compounds in plants (Isman, 2006; 2008). Plants, as long-lived stationary organisms, must

Monoterpenes							
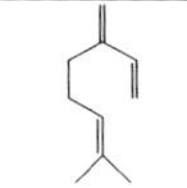 $C_{10}H_{16}$ Exact Mass: 136,13 Myrcene	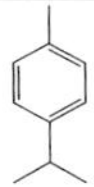 $C_{10}H_{14}$ Exact Mass: 134,11 p-Cymene	 $C_{10}H_{16}$ Exact Mass: 136,13 3-δ-Carene	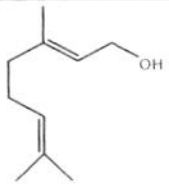 $C_{10}H_{18}O$ Exact Mass: 154,14 Geraniol	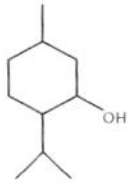 $C_{10}H_{20}O$ Exact Mass: 156,15 Menthol	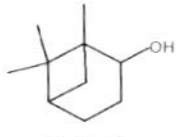 $C_{10}H_{18}O$ Exact Mass: 154,14 Borneol	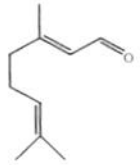 $C_{10}H_{16}O$ Exact Mass: 152,12 Geranial	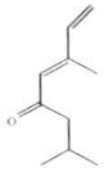 $C_{10}H_{16}O$ Exact Mass: 152,12 Tegetone
Hydrocarbon/acyclic	**Hydrocarbon/monocyclic**	**Hydrocarbon/bicyclic**	**Alcohol/acyclic**	**Alcohol/monocyclic**	**Alcohol/bicyclic**	**Aldehyde/acyclic**	**Ketone/acyclic**
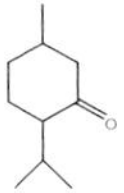 $C_{10}H_{18}O$ Exact Mass: 154,14 Menthone	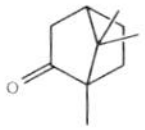 $C_{10}H_{16}O$ Exact Mass: 152,12 Camphor	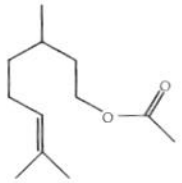 $C_{12}H_{22}O_2$ Exact Mass: 198,16 Citronellyl acetate	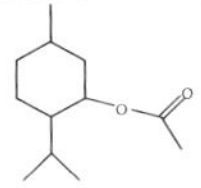 $C_{12}H_{22}O_2$ Exact Mass: 198,16 Menthyl acetate	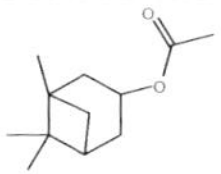 $C_{12}H_{20}O_2$ Exact Mass: 196,15 Isobornyl acetate	 $C_{10}H_{18}O$ Exact Mass: 154,14 Cineole 1,8	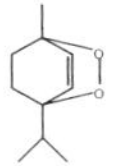 $C_{10}H_{16}O_2$ Exact Mass: 168,12 Ascaridole	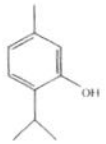 $C_{10}H_{14}O$ Exact Mass: 150,10 Thymol
Ketone/monocyclic	**Ketone/bicyclic**	**Ester/acyclic**	**Ester/monocyclic**	**Ester/bicyclic**	**Ethers**	**Peroxydes**	**Phenols**

Sesquiterpenes				Aromatic compounds				
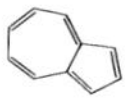 $C_{10}H_8$ Exact Mass: 128,06 Azulene	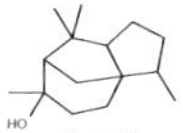 $C_{15}H_{26}O$ Exact Mass: 222,20 Centrol	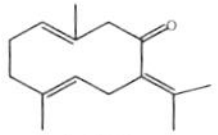 $C_{15}H_{22}O$ Exact Mass: 218,17 Germacrone	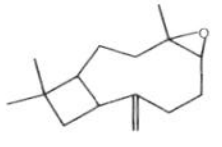 $C_{15}H_{24}O$ Exact Mass: 220,18 Caryophyllene oxide	 $C_{10}H_{10}O$ Exact Mass: 146,07 Cinnamaldehyde	 $C_{10}H_{12}O$ Exact Mass: 148,09 Cinnamyl alcohol	 $C_9H_{10}O$ Exact Mass: 134,07 Chavicol	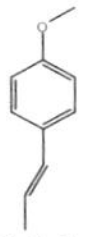 $C_{10}H_{12}O$ Exact Mass: 148,09 Anethole	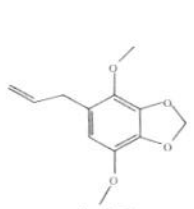 $C_{12}H_{14}O_4$ Exact Mass: 222,09 Apiole
Hydrocarbons	**Alcohol**	**Ketones**	**Epoxide**	**Aldehyde**	**Alcohol**	**Phenol**	**Methoxy derivative**	**Methylene dioxy**

Table 1. Chemical classes of essential oils' components

prototypical structure either containing or deriving from a precursor with a 4,4,8-trimethyl-17-furanylsteroid skeleton. Although hundreds of limonoids have been isolated from various plants, their occurrence in the plant kingdom is confined to plant families of the order Rutales and more abundantly in the families Meliaceae and Rutaceae, and less frequently in Cneoraceae and Harrisonia sp. of Simaroubaceae (Manners, 2007). Of the 300 limonoids known today, about one third is obtained from Meliaceae species (*Azadirachta indica* and *Melia azedarach*), also known as meliacins. The structural variations of limonoids found in Rutaceae are less than in Meliaceae and are generally limited to the modification of A and B rings. The limonoids of Meliaceae are more complex with very high degree of oxidation and rearrangement exhibited in the parent limonoid structure (Roy & Saraf, 2006; Connolly & Hill, 2008). Most work has been focused on azadirachtin, a limonoid PSM ($C_{35}H_{44}0_{16}$, a tetranortriterpenoid) of the Indian Neem tree (*Azadirachta indica* L., *Meliaceae*). The technical grade material of azadirachtin is used for the production of a wide range of commercial formulations exhibiting good efficacy against more than 400 insect species (Akhtar et al., 2008; Lee et al., 1991), mites (Flamini, 2003) and nematodes (Akhtar, 2000; Oka et al., 2007). In India the use of neem (*Azadirachta indica*) extracts in pest management is a part of the traditional practices. Neem is a mixture of more than 100 limonoid compounds, including azadirachtin, salannin, and nimbin and their analogues provoking repellence, feeding deterrence and insect growth inhibition (Schmutterer, 1990). Similar to *A. indica, M. azedarach* extracts possesses insecticidal, acaricidal and fungicidal properties and some of the limonoids isolated are 21-β-acetoxymelianol (Ntalli et al., 2010d), meliantriol, melianone, melianol (Lavie & Jain, 1967), meliacin (1-cinnamoyl melianone), meliacarpin (Li et al., 1999) and meliartenin (Carpinella et al., 2002), azedarachin B (Fukuyama et al., 2006) (Table 2). Limonoids do not have direct negative effects on beneficial insects (Charleston et al., 2005; Sengottayan & Sehoon, 2006), a fact that indicates their potential to be combined in biological pest control programmes. Azadirachtin is the mostly ever studied tetranortriterpenoid, which chemical structure required 18 years to solve and its total synthesis took almost 22 years (Morgan, 2009). The mode of action of azadirachtin lays on (i) deterrent effects on chemoreceptors resulting in antifeedancy (ii) effects on ecdysteroid and juvenile hormone titres through a blockage of morphogenetic peptide hormone release (e.g. PTTH; allatotropins) and (iii) direct effects on tissues resulting in an overall loss of fitness of the insect (Mordue & Blackwell, 1993). Within the azadirachtin molecule, the decalin fragment is responsible for the insect growth regulation and development effects observed, while the hydroxyl furan fragment causes the antifeedant effects more widely observed among target species (Table 2) (Aldhous, 1992). Recently it was proved that azadirachtin provokes a rapid increase in the mitotic index of insect cells, induces the appearance of many aberrant mitotic figures and prevents to some extend the polymerisatrion in vitro of mammalian tubulin (Salehzadeh et al., 2003). Interestingly the EC_{50} values for various cultured insect cell lines vary from 10^{-10} to 10^{-9} M, by which it is classed as highly toxic, whereas for all mammalian cell lines, values of EC_{50} are 10^{-5}–10^{-3} M, which places it in the mildly toxic to non-toxic class, and gives a margin of safety in excess of 100-fold between insect and mammalian cells. There is evidence that the difference in toxicity may be due to the ability of mammalian cells to remove azadirachtin from their body (Morgan, 2009). Azadiractin acute oral LD_{50} in rat is above 5000 mg kg^{-1} and this classifies it by the U.S. Environnemental Protection Agency (EPA) in class IV (no mammal toxicity). Additionally it has no effects on skin sensitization, eye irritation, and is not mutagenic (Isman, 1997). Under field conditions azadirachtin and other neem constituents, e.g., salannin, nimbin, deacetylnimbin, and deacetylsalannin, are not persistent. Three days post field application at five times the dose recommended by the manufacturer, residues of azadirachtin A and B were 0.03 and 0.01 mg/kg, respectively, while residues of salannin (LOQ 0.01 mg/kg) and nimbin (LOQ 0.5 mg/kg) were not detectable. Sunlight photodegradation is the main factor influencing the rate of its disappearance after greenhouse treatment while tomato epicuticular waxes double the photodegradation rate of a commercial formulation (Caboni et al., 2006; 2009). Besides limonoids, also the quassinoids and saponins fall in the PSMs'category of triterpenoids, being though much less studied (Table 2). Quassinoids, the bitter principles of the Simaroubaceae family (*Quassia amara, Cassia camara* and *Picrasma exelca*), are a group of structurally complex and highly oxygenated degraded triterpenes (Sarais et al., 2010). They are divided into five groups according to their basic skeleton, C-18, C-19, C-20, C-22 and C-25. In recent years, attention has been focused on quassinoids because several of them have shown promising biological activities phytotoxic, antifeedant, insecticidal (Almeida et al., 2007). Quassinoids act against insects, nematodes and weeds (Koul, 2008; Powell et al., 1998; Leskinen et al., 1984; Chitwod, 2002; Lin et al., 1995). In nematodes quassinoids acts as noncompetitive antagonists of the ionotropic GABARs to stabilize the closed conformation of the channel, resulting in the inhibition of the action of GABA (Kuriyama et al., 2005). No data are currently available concerning quassinoids fate in the environment. The plant-derived saponins are triterpene glycosides obtained from the soap bark (or soapbark) tree, *Quillaja saponaria* (Quillajaceae) as well as various other plant species of the families Alliaceae, Asteraceae, Polygalaceae and Agavaceae. Their side chains

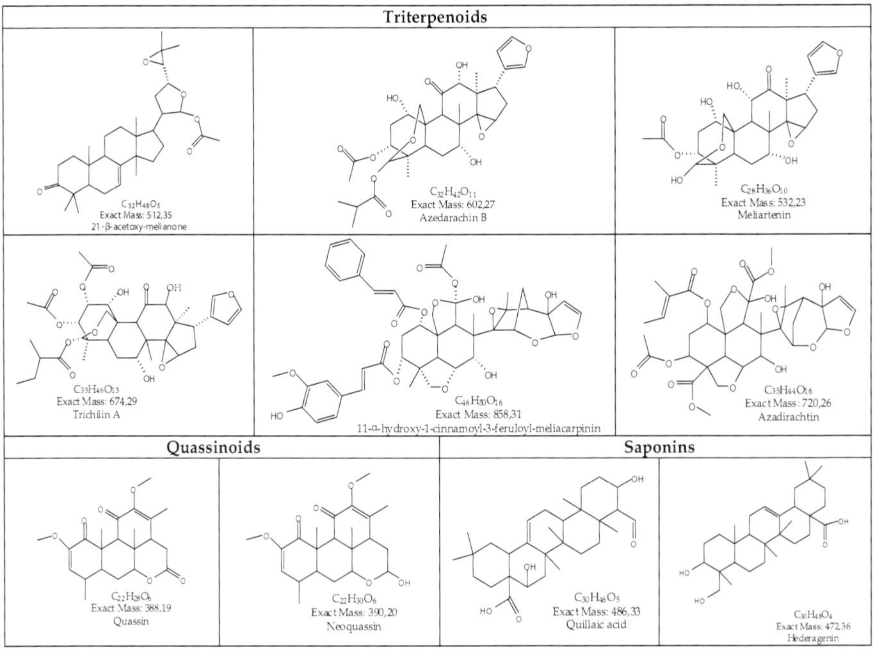

Table 2. Structures of intact and degraded triterpenoids as well as triterpenoid saponins.

of hydrophilic carbohydrates provide them with surfactant properties, but they possess also significant antifeedant, fungicidal and nematicidal properties (Chitwood, 2002; Koul, 2008; Duke et al., 2003; D'Addabbo et al., 2006; 2010; Ribera et al., 2008; Martin & Magunacelaya, 2005). Saponins disrupt also membranes (Majak, 1992).

Glucosinolates and Isothiocyanates (Brassicaceae)

Glucosinolates (GLSs) are sulphur and nitrogen containing PSMs produced by "mustards" (Brassica & Sinapis sp.) as well as other genus of the Capparales order. Glucosinolates are an important and unique class of secondary plant products containing b-D-thioglucose and sulphonated oxime moieties. These include thioglucosides, characterized by side chain with varying aliphatic, aromatic and heteroaromatic carbon skeletons. Glucosinolates get inverted into various degradation products (isothiocyanates, thiocyanates, indoles etc.), when ve getables containing them are cut or chewed, because during this process they come in contact with the enzyme myrosinase that hydrolyses them. By incorporating glucosinolate-containing plant material in soil their bioactive hydrolysis products, named isothiocyanates (ITCs) are released. These products can be used to control soil pests and weeds - a practice known as biofumigation (**Figure** 1). This practise is considered an ecological substitution of the soil fumigation with toxic fumigants such as MeBr, used in the past to suppress soil fungus, bacteria, nematodes and weeds, since it is considered fully biodegradable and less toxic (Vig et al., 2009). ITCs trigger the plant's defence mechanism, produce toxins that kill the target organisms, and produce defensive barriers around the roots of the host plant thus preventing the harmful fungi to enter the host. In fungus ITCs inhibit the oxygen uptake through the uncoupler action of oxidative phosphorylation in mitochondria, they inhibit the coupling between the electron transport and phosphorylation reactions and eventually hinder the ATP synthesis. In bacteria ITCs inactivate various intracellular enzymes by oxidative breakdown of-S-S- bridges and they obstruct ATP synthesis in cells through uncoupler action of oxidative phosphorylation in mitochondria. In insects ITCs inactivate the thiol group of essential enzymes, alkylate the nucleophilic groups of biopolymers like DNA and act as uncouplers accelerating the respiration, which needs more ATP as source of energy, while at the same time ATP production is blocked. This causes exhaustion of stored energy sources which finally leads to death of the pest. In weeds they inhibit seed germination by interfering with protein synthesis and formation of phosphorylated sugars, or inhibit plant enzyme activity (Vig et al., 2009). The ITCs' sorption, degradation or loss from soil mechanisms is fundamental for developing effective, but environmentally benign biofumigation strategies. Effective biofumigation relies on maximum hydrolysis of the glucosinolates in the plant tissue to generate high isothiocyanate concentrations in the soil after incorporation. This is favoured by maximum cell disruption, by addition of water, and a high soil temperature. Residual glucosinolates are very weakly sorbed, readily leached and are microbially degraded and mineralised in soil. In contrast, isothiocyanates are strongly sorbed by the organic matter in soil, react strongly with nucleophilic groups present in soil, and are prone to volatilization losses and microbial degradation and mineralisation minimizing the risks of persistence in the environment or leaching (Gimsing & Kirkegaard, 2009). During the recent years extensive reviews have concerned the chemical ecology of various Brassica towards parasitoids, predators, herbivores and nematodes emphasizing on GLSs and ITCs, their potential of integration in insect-pest management, the physiological and biochemical implications

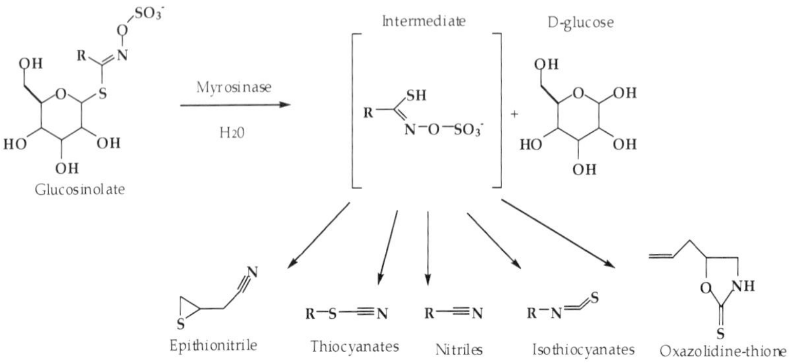

Fig. 1. A hydrolysis scheme of sinigrin under the action of myrosinase and respective degradation productions

underlying hydrolysis mechanisms (Ahuja, et al., 2010; Monfort et al., 2007; Kissen et al., 2009; Agerbirk et al., 2009).

Cyanogenic glycosides

Cyanogenic glucosides constitute a limited number of amino acid derived PSMs known to be present in more than 2500 plant species. This group of compounds is considered to play an important role in plant defence against herbivores due to their bitter taste and release of toxic hydrogen cyanide. Upon tissue disruption (e.g. chewing insects) the cyanogenic glucosides are released from the vacuoles and hydrolyzed by specific b-glucosidases to yield glucose, a ketone or an aldehyde and toxic HCN. This process is known as cyanogenesis and serves to facilitate a rapid HCN release (Figure 2) that suppress insects, fungus, nematodes and weeds (Zagrobelny et al., 2004; Morant et al., 2007; Bjarnholt et al., 2008; Carlsen & Fomsgaard, 2008). Cyanogenic glycosides, through the action of cyanide, prevent oxygen utilization by the inhibition of cytochrome oxidase (Majak, 1992).

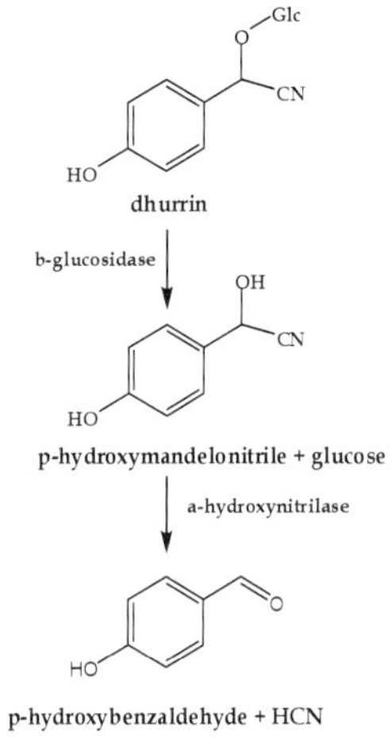

Fig. 2. Cyanogenesis (adapted by Morant et al, 2007)

Alkaloids

Alkaloids are PSMs containing nitrogen atoms, and derive from various botanical families amongst which the Solacaneae. Nicotine is undoubtedly the oldest alkaloid used in agriculture as well as the one of the first molecules used as insecticide (Table 3). It is an acetylcholine mimic binding to postsynaptic receptors and interfering with the transmission of signals in nerves, leading to a continuous firing of the neuroreceptor. This overstimulation leads to depression the central nervous system. It acts predominately through the vapour phase and to a less degree through stomach and contact. Nicotine's high toxicity to humans limited its use as a pesticide (Regnault-Roger & Philogéne, 2008). Bio-transformations of nicotine, involving activation reactions and detoxification mechanisms, have led to neonicotinoids, representing the current major class of insecticides of outstanding potency, systemic action and low toxicity to mammals (Tomizawa & Casida, 2008). Other alkaloids falling in the same category are veratrine and cevatrine, the major components of sabadilla (*Schoenocaulon officinale* Grey) seeds, which are mainly used to control thrips, but recently resistance issues have broken up (Humeres & Morse, 2006). Sabadilla alkaloids possess, like pyrethrins, a neurotoxic activity by slowing the shutting of Na+ channels and disturbing membrane depolarization. They cause paralysis before death. They are contact and nonsystemic insecticides, readily degraded in air and sunlight and are not considered hazardous to non target organisms (Copping, 2004). Ryanodine and its derivative, the dehydro-ryanodine, are extracted from *Ryania speciosa* (Liliaceae) naturalizing the Amazonian basin. Ryanodine acts against insects by interfering with the nerve impulse at the Ca^{2+} channel level and provoking a sustained contraction of the muscles and paralysis. This mode of action has inspired synthetic chemistry, and ryanodine receptors currently represent molecular targets for novel pest control chemicals (Sattelle et al., 2008). The toxicity of *Ryania* extracts towards mammals and fish has precluded its continuing use. Finally 2,5-dihydroxymethyl-3,4-dihydroxypyrrolidine (DMDP) is a sugar analogue, pyrrolidine alkaloid contained in the genera *Loncocarpus* and *Derris*, exhibiting nematicidal activity. It is downwardly mobile in plant phloem, applications on plant foliar decrease galling in roots, but its mode of action is under investigation (Chitwood, 2003).

Phenolics – Flavonoids

Phenolics are toxic to insects, fungi, bacteria, nematodes and weeds (Koul, 2008; Carlsen & Fomsgaard, 2008; Simmonds & Stevenson, 2001; Popa et al., 2008; Wu et al., 2001; Simmonds, 2003; Chitwood, 2002). Flavonoids, a major class of phenolic compounds, are distributed widely in vascular plants and Bryophytes, and ca. 5,000 kinds have been reported to possess feeding attractant and deterrent properties (Iwashina, 2003). Rotenone (Table 4), a flavonoid, present in plants of the genus *Derris* or *Lonchocarpus* (Leguminosae) is, with the alkaloid nicotine, one of the oldest insecticides used all over the world. The principal commercial product of the botanical insecticide rotenone comes from Cube resin, a root extract of *Lonchocarpus utilis* and *Lonchocarpus urucu*. Although rotenone is the primary major constituent in insecticides containing these preparations, a second isoflavone, deguelin, also possesses similar biological properties (Table 4) (Dayan et al., 2009; Caboni et al., 2004). Rotenone inhibits cellular respiration and energetic metabolism at the level of the mitochondrial respiratory chain. It is easily biodegradable and its half life under field conditions is 5 to 7 h (Cavoski et al., 2007). Initially it has been characterised moderately toxic to mammals, but it eventually links to Parkinson's disease (Giasson & Lee, 2000) and recently rotenone has not been included in Annex I of Council Directive 91/414 EEC

(2008/317/EC) and authorizations are withdrawn according to EU legislation. Finally, karanjin (3-methoxy-2-phenylfuro[2,3-h]chromen-4-one) is a furanoflavonol obtained from *Derris* that acts as acaricide and insecticide (Pavela, 2009).

Alkaloids	
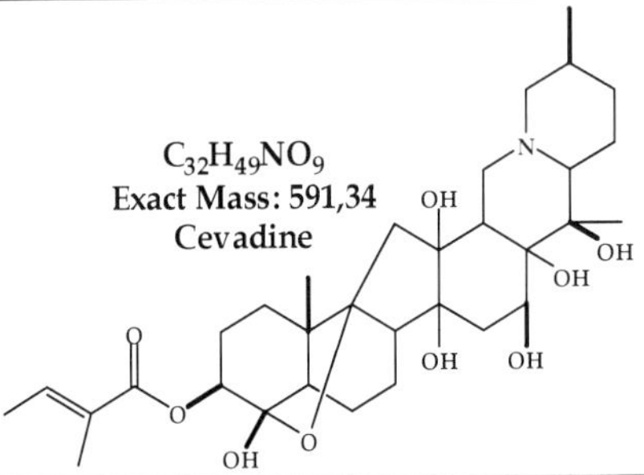 $C_{32}H_{49}NO_9$ **Exact Mass: 591,34** **Cevadine**	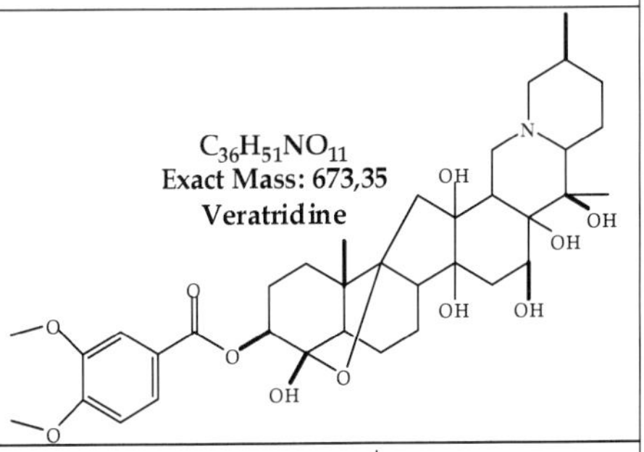 $C_{36}H_{51}NO_{11}$ **Exact Mass: 673,35** **Veratridine**
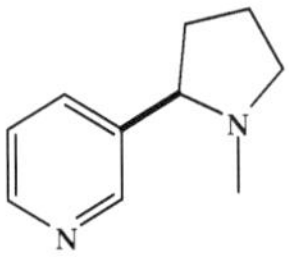 $C_{10}H_{14}N_2$ **Exact Mass: 162,12** **Nicotine**	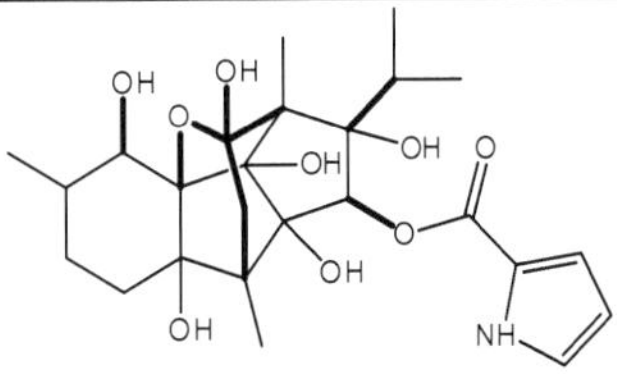 $C_{25}H_{35}NO_9$ **Exact Mass: 493,23** **Ryanodine**

Table 3. Chemical structures of nicotine and other alkaloids extracted from *Ryania* and Sabadilla species

Flavonoids		
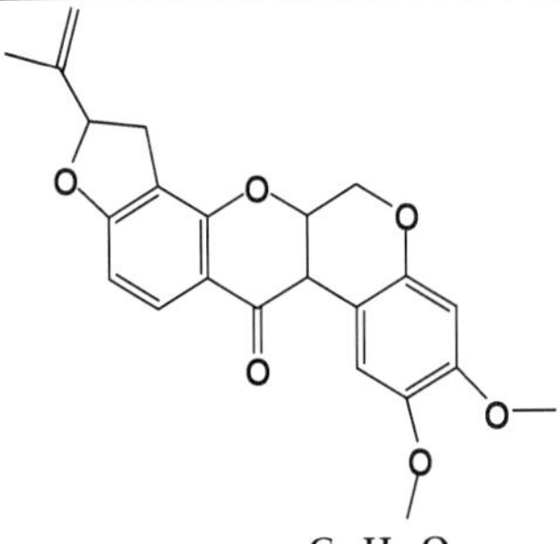 $C_{23}H_{22}O_6$ **Exact Mass: 394,14** **Rotenone**	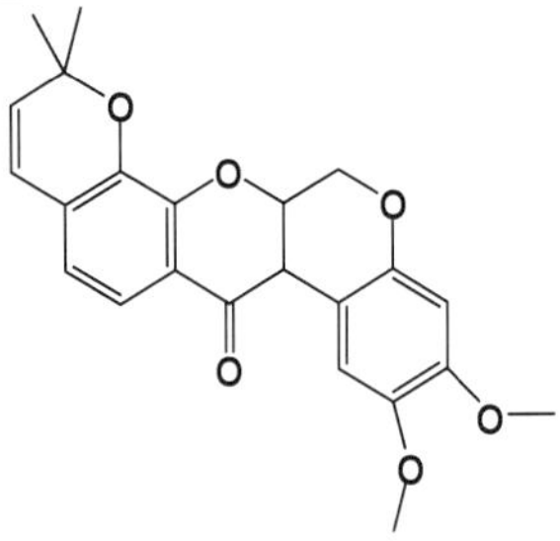 $C_{23}H_{22}O_6$ **Exact Mass: 394,14** **Dequelin**	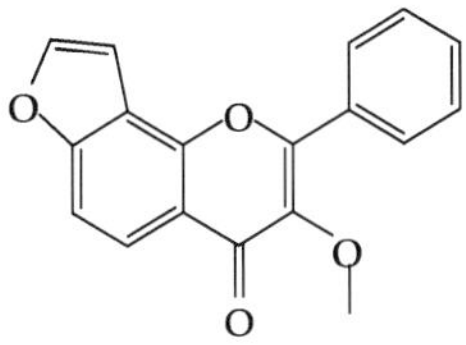 $C_{18}H_{12}O_4$ **Exact Mass: 292,07** **Karanjin**

Table 4. The major flavonoids contained in *Derris* or *Lonchocarpus*

Polyacetylenes & Polythienyls

They are substances present in *Tagetes* species, commonly "marigolds", of the botanical family Asteraceae. Polyacetylenes and Polythienyls possess insecticidal and nematicidal properties (Chitwood, 2002; Wat et al., 1981)

Pyrethrum

Pyrethrum is a powder obtained by crushing dried flowers of daisies belonging to the family of Asteraceae such as *Chrysanthemum.* spp., *Pyrethrum.* spp., and *Tanacetum.* spp. Pyrethrum is a mixture of six esters, pyrethrins I (**Figure 3**) and II, (the most abundant), cinerin I and II, and jasmoline I and II. Pyrethrins control a wide range of insects and mites binding to Na^+ channels and prolonging their opening. The insect presents hyperactivity followed by convulsions and finally it dies. The rapid action of pyrethrins is called knockdown effect. Pyrethrins have a relatively low toxicity toward mammals but toxicity is mentioned for nontargeted species, especially fish and bees. However, their great instability to light, air, and moisture reduce considerably the risks related to its use. Currently, pyrethrum is limited and costs have risen in recent years, making it inaccessible to less affluent societies (Isman, 2008).

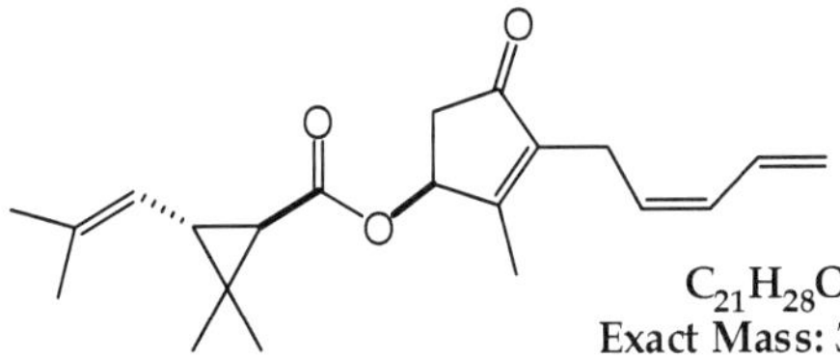

Fig. 3. Chemical structure of Pyrethrin I

Organic Acids

Vegetable oils contain large and heterogeneous quantities of fatty acids, saturated or unsaturated, with medium to long esterified carbon chains, and esters of fatty acids with high molecular weight. They develop toxicity by inhalation and contact suffocating the insect by forming an impermeable film upon the cuticle. Some compounds penetrate through the cuticle, disrupt cellular membrane, and uncouple oxidative phosphorylation. Some fatty acids, such as oleic (C18), have their own insecticidal activities, whereas undecylenic (C11) acid has a lower toxicity, but increases the activity of other insecticidal compounds by potentiation (Regnault-Roger & Philogene, 2008).

Others

The PSMs produced by many species in the genus *Piper* are called piperamides and they are characterised by insecticidal activity. Piperamides provoke contact toxicity, repellent and antifeedant activities and in a biochemical level act as neurotoxins. They quickly degrade under full sunlight (Scott et al., 2008). Capsaicin is obtained from the genus *Capsicum* such as chili peppers (*Capsicum frutescens*, Mill.) and is characterised by nematicidal, insecticidal and insect repellent properties (Neves et al., 2009; Dayan et al., 2009; Edelson et al., 2002). As an emerging biocide very little data is available on the environmental fate of capsaicin but initial assessment suggests it will bind to sediments (Tomas & Brooks, 2010).

3. Greek plants as a source of botanical pesticides: *Results of our ongoing research*

Botanicals have been in use for a long time for pest control offering an alternative approach to crop protection. Mediterranean area reserves abundance in plants exhibiting bioactive properties. Improvement in our understanding of plant allelochemical mechanisms of activity offer new prospects for using these substances in crop protection. In the frame of our research interests aiming at the plant chemical defense and the investigation for alternative chemical agents to control plant diseases and pests, we evaluated the activity of various indigenous plants, grown in the Mediterranean region against bacteria, insect, fungi and nematodes. The chemical composition of plants was investigated and determined in order to further be correlated with the biological activity of the tested plants.

Bacteria, fungi & terpenes

Studies on the activity of plant EOs and their terpenoid constituents against phytopathogenic or foodborne bacteria and fungi are receiving increased interest in scientific annals, as alternatives to synthetic pesticides and as key components of IPM. In our first attempts for eco-friendly methods to control phytopathogenic bacteria we investigated the role of PSMs obtained from plants indigenous or introduced, wild or cultivated in Greece. As a first trigger, we found that bacterial population on plant phyllosphere differ significantly from species to species and also between the same species. We also found a correlation between bacterial colonization of plant phyllosphere and presence of secondary metabolites (Karamanoli et al., 2000; 2005). Results of this work provided evidence of the allelopathic effect of PSMs in the field against epithytic bacteria and specifically Labiatae aromatic plants are protected against them by both EOs and leaf surface phenolics. Interestingly, the highest colonized lavender had the lowest EO content, compared to other aromatics (oregano, rosemary and sage) and this oil exhibited weak antibacterial activity. Particularly for EOs, another important factor is the amount of isoprenoid compounds that they contain, with phenols and oxygenated terpenoids generally, being the most effective (Karamanoli et al., 2000). Building on our team's previous work, the antibacterial activity of secondary metabolites (EOs, surface phenolics and leaf tissue extracts) from 19 Mediterranean species were evaluated and correlated with their chemical class and composition. Proper analytical techniques (hydrodistillation and LL extraction) were used for their isolation, followed by GC-MS analysis for chemical composition identification and bioassays or greenhouse experiments for antibacterial activity (MIC and MBC) estimation (Karamanoli et al., 2005). All aromatic plants were found active, with PSMs of *Oreganum* spp and *Thymus* spp being the most toxic. Furthermore, the activity of their main constituents was evaluated. Results showed carvacrol as the most potent compound (MBC=0.4mg mL^{-1}) against all tested eleven bacteria strains. Generally cyclic alcohols and ketones exhibited high activity, followed by long chain alcohols and hydrocarbons with less or no activity. Oregano oil, rich in carvacrol and thymol, also proved effective as an antibacterial biocontrol agent in field experiments. Consequently, the antibacterial activity of aromatic plants may have applications in agriculture for developing efficient biocontrol agents from these plants, with the prerequisite to overcome their limited persistence and formulation requirements. .

Similarly, we found a correlation between chemical composition of EO, isolated from *Pistachia* spp, and septoriosis caused by the pathogenic fungi *Septoria pistaciarum*. Terpinen-

4-ol was identified as the most active constituent of the e.o. against *S. pistaciarum* and determined at high concentration only in the EO of the tolerant plants (Douka et al., 2005).

Insects & terpens

Insecticidal effects of plant EOs and individual terpenes against disease vectors and insect pests consists a well studied case (Isman, 2000). Specifically, contact and fumigant activity of plant EOs against stored product pests have been reported, but the relationship between their chemical composition and their activity is always needed to be determined, in order the results to find further practical applications. Studies were performed aiming to evaluate the insecticidal activity (LC_{50}) of EOs obtained from the aromatic plants lavender (*Lavandula hybrida* Rev, Lamiaceae), rosemary (*Rosmarinus officinalis* L, Lamiaceae) and eucalyptus (*Eucalyptus globulus* Labill, Myrtaceae) and their main constituents against *Acanthoscelides obtectus*, Say (Papachristos et al., 2004). Strong insecticidal activity was found, and oxygenated monoterpenes were predominated over hydrocarbons. All EOs tested exhibited strong activity against *A. obtectus* adults and variability in their activity and chemical composition due to different plant part, season and insect sex was substantiated. Among 16 of the principal components of the EOs tested, the most active were teprinen-4-ol, camphor, 1,8-cineol and verbenone, followed by linalool (LC_{50}=0.8–7.1mg L^{-1} air), while linalyl and terpinyl acetate were active only against adult males. Overall, intact EOs are more potent in controlling *A. obtectus* than are their main constituents, the monoterpenoids.

Nematodes & terpenes

Phytonematodes are among the most notoriously difficult crop pests and their control is achieved mainly with cultural practices, crop rotation, and resistant cultivars, combined with a few available chemical nematicides that are still authorized. The need for discovering less toxic and environmentally acceptable substitutes for commercial nematicides is amplified, creating a significant market opportunity for alternative products such as biorationals. Essential oils and their components such as thymol, carvacrol, pulegone, limonene, anethole, geranial, and Artemisia ketone have been identified with nematicidal activity (Oka, 2000). As part of our ongoing effort towards the study of natural substances with biologically interesting properties, we have also focused on various plant species of Greek flora as sources of nematicidal compounds. We tested the paralysis activity of the EOs obtained with hydrodistillation from 15 botanical species on root knot nematodes (RKN), as well as the individual and paired activity of 23 pure terpenes, identified as components of the tested EOs (Ntalli et al., 2010b; 2010c). The activity of EOs against *M. incognita* was found to decrease in the order *O. vulgare, O. dictamnus, M. pulegium, M. officinalis, F. vulgare, P. anisum, E. meliodora*, and *P. terebinthus*. Among terpenes, the oxygenated compounds (alcohols and ketones) exhibited in general higher activity than hydrocarbons. The activity of the nematicidal terpenes was found to decrease in the order L-carvone, pulegone, *trans*-anethole, geraniol, eugenol, carvacrol, thymol, terpinen-4-ol, estragole and γ-eudesmol. Furthermore, the aromatic aldehyde benzaldehyde, found as a component of the EO from *E. meliodora*, exhibited the highest activity (EC_{50}=9 μg mL^{-1}). Our results, as well as results from similar studies (Lahlou, 2004; Jiang, 2009), cohere to the fact that the contribution of each ingredient compound to the overall activity of an EO is a complicated pattern of interactions. It is possible that they may act together synergistically or antagonistically to contribute to the toxicity of the totality of the tested oil. Notably, it was confirmed that "inactive" constituents may have some synergic effect on the "active" constituents and that, although not active individually, their presence is necessary to achieve full toxicity. Generally according to our findings, the EC_{50} values of the individual terpenes, measured at the same concentrations as expected in the EO at their EC_{50} value, was not as high as the activity of the corresponding EO indicating apparently, evidence of components interactions within the oil. To corroborate the role of individual constituents toward the synergistic and antagonistic actions, among each other, artificial blends were further tested for their activity. The terpenes' pairs exhibiting high binary action (synergism), in decreasing order were: *trans*-anethole/geraniol, *trans*- anethole/eugenol, carvacrol/eugenol and geraniol/carvacrol (Ntalli et al., 2010c). Understanding the interactions among the individual constituents of a natural extract/mixture provive significant information for further development of new plant protection products of environmental friendly approach. Active synergistic mixtures exhibiting high nematicidal activity could be used in artificial blends constituting promising agents of pest management. For this, further studies, concerning the interactions dependence on terpenes concentration and their ratios in mixtures, as well as the underling responsible biological mechanisms of terpenes provoking J2 paralysis, are in process. Additionally, further investigation is required in order to evaluate interaction effects between nematicidal and non nematicidal terpenes.

Nematodes & limonoids

The plant family *Meliaceae* (mahogany family) has received much attention especially because of the presence of limonoid triterpenes many of which are known for their insecticidal properties (Akhtar *et al.*, 2008). The majority of research has been focused on azadirachtin, a limonoid secondary metabolite ($C_{35}H_{44}0_{16}$, a tetranortriterpenoid) of the Indian Neem tree (*Azadirachta indica* L., *Meliaceae*). The technical grade material is used for the production of a wide range of commercial formulations exhibiting good efficacy against more than 400 insect species. Although several reports suggest neem products, such as seed powder, seed kernel powder, seed cake powder, dry leaf powder and aqueous neem extracts, to exhibit good efficacy against root-knot nematodes the results were rather contradictory. We proved that azadirachtin (Neemazal® 1%EC, Intrachem Hellas) acts against RKN at very high concentrations both concerning paralysis effects and biological cycle arrest (12.8 mg a.i. L^{-1} and 30.72 μg a.i. g^{-1}) and that the recommended dose for nematodes' control in field does not provide adequate control (Ntalli et al., 2009). In the frame of our research on limonoids against RKN we then explored the nematicidal activity of the botanical species *Melia azedarach* L. (Meliaceae), which is naturalized in Greece. Chinaberry demonstrated biofumigant properties when incorporated as pulverized fruits in *M. incognita* infested soil to be tested for its effect on nematode life cycle (EC_{50}=0.34 % w/w). First we distinguished for paralysis activity amongst the polar and non-polar fractions of the fruits extracts, and we established proper extraction procedures for maximum yields' obtainment. The activity of *M. azedarach* against RKN was attributed to the defatted methanol extract (polar fragment) of the fruit and this was used in extends in pot experiments to study its effect on the RKN biological cycle arrest. The EC_{50} values were calculated for all experiments. Four days (96 h) of juveniles immersion in a 0.03 % w/v Melia Methanol Extract (MME) solution paralysed half the population tested (EC_{50}=0.03 % w/v), while the EC_{50} value calculated for the pot experiments was 0.916 % w/w (Ntalli et al,. 2010a). Further studies on chemical characterization and biological activity of the defatted methanol extract *M. azedarach* L revealed that the nematicidal activity does not lay in its limonoids' contents (3-α-tigloylmelianol, melianone, 21-β-acetoxy-melianone, methyl kulonate) but in its organic acids, aldehydes and alcohols (Ntalli et al., 2010e). The aldehyde furfurale was the foremost nematicidal principle, exhibiting activity similar to that of the commercial nematicide fosthiazate, both after J2 immersion in test solutions and after exposure to its vapours (fumigant activity). Furfurale is already known to possess high nematicidal fumigant activity against *M. incognita*, tested in greenhouse and microplot conditions (Rodriges-Kabana et al., 1993) but no correlation had been ever made so far with Chinaberry. The nematicidal activity of *M. azedarach* as well as its contents in furfurale, reported by our group for the first time, reveals this species' nematicidal-biofumigant toxicity and its potency of incorporation into IPM programs. Because phytonematodes live in soil or within plant roots, the target of any chemical nematicide often resides at a fair distance away from the site of application. The fumigant activity of a nematicide is a rather important property enhancing its activity in the non treated soil layers.

These results combined with literature cited reports on activity of natural products of plant origin against various pests show their suitability to be considered as potential biopesticides. Our current studies are focused on the mode of action of these natural products specifically within nematodes' body, while further studies on the formulation and on the environmental fate are needed.

4. Current trend and future prospective

In this review we are aiming to explore the ability of the nature and the abundant resources for chemicals available for plant defence and suitable in pest management for crop protection. BCAs have long been touted as attractive alternatives to synthetic chemical pesticides for pest management because botanicals reputedly pose little threat to the environment and to human health. The body of scientific literature documenting bioactivity of plant derivatives to pests continues to expand rapidly, yet only a handful of botanicals are currently used in agriculture in the industrialized world, and there are few prospects for commercial development of new botanical products. Several factors appear to limit the success of botanicals, most notably regulatory barriers and the availability of competing products (newer synthetics, fermentation products, microbials) that are cost-effective and relatively safe compared with their predecessors (Isman, 2006). Botanical pesticides presently play only a minor role in crop protection; increasingly stringent regulatory requirements in many jurisdictions have prevented all but a handful of botanical products from reaching the marketplace in North America and Europe in the past 20 years (Isman, 2008).

Thus the regulation of natural products as crop-protection agents may have to undergo the same procedure as for a conventional chemical product. Each country or region has a different approach; for example, the USA EPA separates pesticides into two general categories: conventional chemical pesticides and biochemical/microbial pesticides, while Europe follows the OECD definition of biopesticides (rather than natural products), which includes pheromones, insect and plant growth regulators, plant extracts, transgenic plants and macro-organisms. The requirements, risk assessments and encouragement for such products differ from region to region (Neale, 2000). Concerted efforts are hoped to remove limitations in biopesticides raw material availability, potency variations, standardization of extraction methods, quality control, shelf life and improved bioefficacy (Mendki et al., 2001).

5. References

Aharoni, A.; Jongsma, M.A. & Bouwmeester, H. J. (2005). Volatile science? Metabolic engineering of terpenoids in plants. *Trends in Plant Science,* 10(12): 594-602, ISSN 1360-1385.

Ahuja, I.; Rohloff, J. & Bones, A.M. (2010). Defence mechanisms of Brassicaceae: implications for plant-insect interactions and potential for integrated pest management. A review. *Agron. Sustain. Dev.* 30: 311–348, ISSN: 1774-0746, EISSN: 1773-0155.

Agerbirk, N.; De Vos, M.; Kim, J.H. & Jander, G. (2009). Indole glucosinolate breakdown and its biological effects. *Phytochem. Rev.* 8:101–120, ISSN: 1568-7767, EISSN: 1572-980X.

Akhtar, M. & Mahmood, I. (1994). Potentiality of phytochemicals in nematode control: A review. *Bioresource Technol.*, 48: 189-201, ISSN: 0960-8524.

Akhtar, M. (2000). Nematicidal potential of the neem tree Azadirachta indica (A. Juss). *Integ. Pest Manag. Rev.*, 5: 57–66, ISSN: 1353-5226, EISSN: 1572-9745.

Akhtar, Y.; Yeoung, Y.-R. & Isman, M.B. (2008). Comparative bioactivity of selected extracts from Meliaceae and some commercial botanical insecticides against two noctuid caterpillars, *Trichoplusia ni* and *Pseudaletia unipuncta. Phytochem. Rev.*, 7: 77–88, ISSN: 1568-7767, EISSN: 1572-980X.

Aldhous, P. (1992). Neem chemical: the pieces fall in to place. *Science,* 258 (5084): 893, ISSN: 0036-8075.

Almeida, M.; Arriaga, M.B.; Dos Santos, Â.M.C.; Lemos, A.K.L.; Braz-Filho, T.L.G.; Curcino, R., & Vieira, I.J. (2007). Occurrence and biological activity of quassinoids in the last decade [Ocorrência e atividade biológica de quassinóides da última década] *Quimica Nova,* 30: 935-951, ISSN 0100-4042.

Arbeli, Z. & Fuentes, C.L. (2007). Accelerated biodegradation of pesticides: An overview of the phenomenon, its basis and possible solutions; and a discussion on the tropical dimension. *Crop Prot.* 26: 1733–1746, ISSN: 0261-2194.

Bakkali, F.; Averbeck, S.; Averbeck, D. & Idaomar, M. (2008). Biological effects of essential oils-A review. *Food Chem. Toxicol.*, 46: 446-475, ISSN: 0278-6915.

Bjarnholt, N.; Laegdsmand, M.; Hansen, H.C.B.; Jacobsend, O.H. & Møller, B.L. (2008). Leaching of cyanogenic glucosides and cyanide from white clover green manure. *Chemosphere,* 72(6): 897-904, ISSN: 0045-6535.

Caboni, P.; Sarais, G.; Angioni, A.; Lai, F.; Dedola, F. & Cabras, P. (2009). Fate of azadirachtin A and related azadirachtoids on tomatoes after greenhouse treatment. *J. Environ. Sci. Health, Part B: Pestic., Food Contam., Agric. Wastes,* 44(6, August): 598-605, ISSN: 0360-1234.

Caboni, P.; Sarais, G.; Angioni, A.; Garcia, A.J.; Lai, F.; Dedola, F. & Cabras, P. (2006). Residues and persistence of neem formulations on strawberry after field treatment. *J. Agric. Food Chem.*, 54: 10026-10032, ISSN: 0021-8561, EISSN: 1520-5118.

Caboni, P.; Sherer, T.B.; Zhang, N.; Taylor, G.; Na, H.M.; Greenamyre, J.T. & Casida, J.E. (2004). Rotenone, deguelin, their metabolites, and the rat model of Parkinson's disease. *Chem. Res. Toxicol.* 17 (11):, 1540-1548, ISSN: 0893-228X.

Carlsen, S.C.K. & Fomsgaard, I.S. (2008). Biologically active secondary metabolites in white clover (*Trifolium repens* L.) - a review focusing on contents in the plant, plant-pest interactions and transformation, *Chemoecology,* 18: 129–170, ISSN: 0937-7409, EISSN: 1423-0445.

Carpinella, M.C.; Giorda L.M.; Ferrayoli, C. G. & Palacios, S.M. (2003). Antifungal Effects of Different Organic Extracts from *Melia azedarach* L. on Phytopathogenic Fungi and Their Isolated Active Components. *J. Agric. Food Chem.*, 51(9): 2506–2511, ISSN: 0021-8561, EISSN: 1520-5118.

Carpinella, M.C.; Ferravoli, C.; Valladares, G.; Defago, M.; & Palacios, S.M. (2002). Potent limonoid insect antifeedant from *Melia azedarach. Biosc.Biotech. Biochem.*, 60: 1731–1736, ISSN: 0916-8451, EISSN: 1347-6947.

Cavoski, I.; Caboni, P.; Sarais, G.; Cabras, P.; & Miano, T. (2007). Photodegradation of rotenone in soils under environmental conditions. *J. Agric. Food Chem.*, 55(17): 7069-7074, ISSN: 0021-8561, EISSN: 1520-5118.

Charleston D.S.; Kfir R.; Dicke M. & Vet, L.E.M. (2005). Impact of botanical pesticides derived from *Melia azedarach* and *Azadirachta indica* on the biology of two parasitoid species of the diamondback moth. *Biol. Control*, 33: 131–142, ISSN: 1049-9644, EISSN: 1090-2112.

Chitwood, D.J. (2002). Phytochemical based strategies for nematode control. *Annu. Rev. Phytopathol.*, 40: 221-249, ISSN: 0066-4286.

Chitwood, D.J. (2003). Research on plant-parasitic nematode biology conducted by the United States Department of Agriculture-Agricultural Research Service. *Pest Manag. Sci.* 59: 748-53, ISSN 1526-498X, EISSN: 1526-4998.

Connolly, J.D. & Hill, R.A. (2008). Triterpenoids. *Nat. Prod. Rep.*, 25: 794-830, ISSN: 0265-0568, ISSN: E1460-4752.

Copping, L.G. (2004). The manual of biocontrol agents. Third edition of the biopesticide manual. BCPC publications. UK, pp. 250-252.

Dayan, F. E.; Cantrell, C.L. & Duke, S.O. (2009). Natural products in crop protection. *Bioorg. Med. Chem.*, 17: 4022–4034, ISSN: 0968-0896.

D'Addabbo, T.; Curto, G.; Santi, R. & Carella, A. (2006). Control of root-knot nematode Meloidogyne incognita by *Quillaja saponaria* extracts. *Giornate Fitopatologiche* Riccione (RN), 27-29 marzo 2006. Atti, volume primo.

D'Addabbo, T.; Carbonara, T.; Leonetti, P.; Radicci, V.; Tava, A. & Avato P. (2010). Control of plant parasitic nematodes with active saponins and biomass from Medicago sativa. *Phytochem. Rev.* DOI 10.1007/s11101-010-9180-2, ISSN: 1568-7767, EISSN: 1572-980X.

Douka, E.; Karamanoli, K.; Menkissoglu Spiroudi, U. & Tzavella-Klonari K. (2006). Antifungal activity of secondary metabolites of *Pistacia vera* L. against *Septoria pistaciarum*: Correlation with pathogenesis. Proceedings of the 12th Congress of the Mediterranean Phytopathological Union, Rhodes Island, Greece 10-15 June 2006, p. 310-312..

Duke, S.O.; Baerson, S.R., Dayan; F.E., Rimando, A.M.; Scheffler, B.E.; Tellez, M.R.; Wedge, D.E. Schrader, K.K.; Akey, D.H.; Arthur, F.H.; De Lucca, A.J.; Gibson, D.M.; Harrison Jr., H.F.; Peterson, J.K.; Gealy, D.R.; Tworkoski, T.; Wilson, C.L. & Brad, M. (2003). United States Department of Agriculture-Agricultural Research Service research on natural products for pest management. *Pest Manag. Sci.* 59 (6-7): 708-717, ISSN: 1526-498X, EISSN: 1526-4998.

Edelson, J.V.; Duthie, J. & Roberts, W. (2002). Toxicity of biorational insecticides: Activity against the green peach aphid, *Myzus persicae* (Sulzer). *Pest Manag. Sci.* 58 (3): 255-260, ISSN 1526-498X, EISSN: 1526-4998.

EEC, (1991). Council Directive 91/414/EEC of 15 July 1991 concerning the placing of plant protection products on the market. Office for Official Publications of the European Communities, *OJ No L 230*, 19.8.1991, p. 194.

European Commission (2008). Commission Decision 2008/317/EC of 10 April 2008 concerning the non-inclusion of rotenone, extract from equisetum and chinin-hydrochlorid in Annex I to Council Directive 91/414/EEC and the withdrawal of authorisations for plant protection products containing these substances. *OJ No L 108*, 18.04.2008, p. 30.

European Commission (2008). Commission Directive 2009/128/EC of 21 October 2009 establishing a framework for Community action to achieve the sustainable use of pesticides. *OJ No* L 309, 24.11.2009, p.71.

European Commission (2009). Council Regulation (EC) no. 1107/2009 of 21 October 2009 concerning the placing of plant protection products on the market and repealing Council Directives 79/117/EEC and 91/414/EEC. *OJ No L 309*, 24-11-2009, pp 50.

Flamini, G. (2003). Acaricides of natural origin, personal experiences and review of literature (1900-2001). *Studies in Natural Products Chemistry*, 28 (9): 381-451, Bioactive Natural Products (Part I), ISBN: 9780444514158.

Fukuyama, Y.; Nakaoka, M.; Yamamoto, T.; Takahashi, H. & Minami, H. (2006). Degraded and oxetane-bearing limonoids from the roots of *Melia azedarach. Chem Pharm Bull* (Tokyo). 54 (8): 1219-1222.

Giasson, B.I. & Lee, V.M.-Y. (2000). A new link between pesticides and Parkinson's disease. *Nature Neuroscience*. 3 (12): 1227-1228, ISSN: 10976256, EISSN: 1546-1726.

Gimsing, A.L. & Kirkegaard, J.A. (2009). Glucosinolates and biofumigation: fate of glucosinolates and their hydrolysis products in soil. *Phytochem. Rev.*, 8: 299–310 ISSN: 1568-7767, EISSN: 1572-980X.

Humeres, E. & Morse, J.C. (2006). Resistance of avocado thrips (Thysanoptera: Thripidae) to sabadilla, a botanically derived bait. *Pest Manag. Sci.*, 62: 886–889, ISSN 1526-498X, EISSN: 1526-4998.

Jeyasankar, A. & Jesudasan, R.W.A. (2005). Insecticidal properties of novel botanicals against a few lepidopteran pests. *Pestology*, 29: 42-44, ISSN: 0970-3012.

Jiang, Z.; Akhtar, Y.; Bradbury, R.; Zhang, X. & Isman, M. B. (2009). Comparative toxicity of essential oils of *Litsea pungens* and *Litsea cubeba* and blends of their major constituents against the cabbage looper, *Trichoplusia ni. J. Agric. Food Chem.*, 57: 4833–4837, ISSN: 0021-8561, EISSN: 1520-5118.

Iacobellis, N.S.; Lo, C.P.; Capasso, F. & Senatore, F. (2005). Antibacterial activity of Cuminum cyminum L. and Carum carvi L. essential oils. *J. Agric. Food Chem.* 53 (1): 57-61, ISSN: 0021-8561, EISSN: 1520-5118.

Isman, M.B. (1997). Neem and other botanical insecticides: Barriers to commercialization. *Phytoparasitica*. 25: 339-344, ISSN: 0334-2123.

Isman, M.B. (2000). Plant essential oils for pest and disease management. *Crop Prot.* 19: 603-608, ISSN: 0261-2194.

Isman, M.B. (2006). Botanical insecticides, deterrents, and repellents in modern agriculture and an increasingly regulated world. *Annu. Rev. Entomol.*, 51: 45-66, ISSN: 0066-4170.

Isman, M.B. (2008). Botanical insecticides: for richer, for poorer. *Pest Manag. Sci.*, 64: 8-11, 1207, ISSN 1526-498X, EISSN: 1526-4998.

Iwashina, T. (2003). Flavonoid function and activity to plants and other organisms. *Biol. Sci. Space*, 17: 24-44.

Karamanoli, K.; Vokou, D.; Menkissoglu-Spiroudi, U. & Constantinidou, H.-I.A. (2000). Bacterial colonization of phyllosphere of Mediterranean aromatic plants. *J.Chem.Ecol.*, 26 (9): 2035-2048, ISSN: 0098-0331.

Karamanoli, K. (2002). Secondary metabolites as allelochemicals in plant defence against microorganisms of the phyllosphere. Pp 277–288 in Reigosa M, Pedrol N (eds). Allelopathy: from molecules to ecosystems. USA-Enfield (NH): Science Publishers Inc.

Karamanoli, K.; Menkissoglu-Spiroudi, U.; Bosabalidis, A.M.; Vokou, D. & Constantinidou, H.-I.A. (2005). Bacterial colonization of the phyllosphere of nineteen plant species and antimicrobial activity of their leaf secondary metabolites against leaf associated bacteria *Chemoecology* 15: 59–67, ISSN: 0937-7409, EISSN: 1423-0445.

Karpouzas, D.G.; Karanasios, E. & Menkissoglu-Spiroudi, U. (2004). Enhanced microbial degradation of cadusafos in soils from potato monoculture: demonstration and characterization. *Chemosphere* 56 (4): 549–559, ISSN: 0045-6535.

Kissen, R.; Rossiter, J.T. & Bones, A.M. (2009). The 'mustard oil bomb': not so easy to assemble! Localization, expression and distribution of the components of the myrosinase enzyme system. *Phytochem. Rev.*, 8: 69–86, ISSN: 1568-7767, EISSN: 1572-980X.

Koul, O. (2008). Phytochemicals and insect control: An antifeedant approach. *Crit.Rev.Plant Sci.*, 27 (1): 1-24, ISSN: 0735-2689, EISSN: 1549-7836.

Kuriyama, T.; Ju, X.-L.; Fusazaki, S.; Hishinuma, H.; Satou, T.; Koike, K.; Nikaido, T. & Ozoe, Y. (2005). Nematocidal quassinoids and bicyclophosphorothionates: A possible common mode of action on the GABA receptor. *Pest. Biochem. Physiol.* 81 (3): 176-187, ISSN: 0048-3575.

Lahlou, M. (2004) Methods to study the phytochemistry and bioactivity of essential oils. *Phytother. Res.*, 18 (6): 435-448. EISSN: 1099-1573.

Lavie, D. & Jain, M.K. (1967). Tetranortriterpenoids from *Melia azadirachta* L. *Chemical Communications (London)*, 6: 278–280, ISSN: 0022-4936.

Lee, S.M.; Klocke, J.A.; Barnby, M.A.; Yamasaki, R.B. & Balandrin, M.F. (1991). Insecticidal constituents of *Azadirachta indica* and *Melia azedarach* (Meliaceae). In *Naturally Occurring Pest Bioregulators*, Volume 449, *American Chemical Society Symposium Series*, pp. 293–304. Ed. P.A. Hedin. Washington, DC: ACS.

Leskinen, V.; Polonsky, J. & Bhatnagar, S. (1984). Antifeedant activity of quassinoids. *J.Chem. Ecol.*, 10 (10): 1497-1507, ISSN: 0098-0331.

Li, X. (1999). Recent studies on insecticidal activities of limonoids from meliaceaous plants. Entomologia sinica. 6: 283-288. ISSN: 1672-9609, EISSN: 1744-7917.

Lin, L.-J.; Peiser, G.; Ying, B.-P.; Mathias, K.; Karasina, F.; Wang, Z.; Itatani, J.; Green, L. & Hwang, Y.-S. (1995). Identification of plant growth inhibitory principles in Ailanthus altissima and Castela tortuosa. *J. Agric. Food Chem.*, 43(6): 1708-1711, ISSN: 0021-8561, EISSN: 1520-5118.

Manners, G.D. (2007). Citrus limonoids: Analysis, Bioactivity and Biomedical prospects. *J. Agric. Food Chem.* 55: 8285-8294, ISSN: 0021-8561, EISSN: 1520-5118.

Majak, W. (1992). Mammalian metabolism of toxic glycosides from plants. *J. Toxicol.: Toxin Rev.*, 11 (1): 1-40, ISSN: 0731-3837.

Sattelle, D.B.; Cordova, D. & Cheek, T.R. (2008). Insect ryanodine receptors: molecular targets for novel pest control chemicals *Invertebr. Neurosci.*, 8: 107–119, ISSN: 1354-2516.

Schillhorn van Veen, T. W. (1999). Agricultural policy and sustainable livestock development. *Int. J. Parasitol.*, 29: 7-15, ISSN: 0020-7519.

Schmutterer, H., (1990). Properties and Potential of natural Pesticides from the neem tree, Azadirachta inidica. *Annu. Rev. Entomol.*, 35: 271–97, ISSN: 0066-4170.

Scott, I.M.; Jensen, H.R.; Philogène, B.J.R. & Arnason, J.T. (2008). A review of *Piper* spp. (Piperaceae) phytochemistry, insecticidal activity and mode of action *Phytochem. Rev.*, 7: 65–75, ISSN: 1568-7767, EISSN: 1572-980X.

Sengottayan, S.N. & Sehoon, K. (2006). Effects of *Melia azedarach* L. extract on the teak defoliator *Hyblaea puera* Cramer (Lepidoptera: Hyblaeidae). *Crop Prot.*, 25: 287–291, ISSN: 0261-2194.

Simmonds, M.S.J. & Stevenson, P.C. (2001). Effects of isoflavonoids from *Cicer* on larvae of *Heliocoverpa armigera*. *J. Chem. Ecol.*, 27 (5): 965-977, ISSN: 0098-0331.

Simmonds, M.S. (2003). Flavonoid-insect interactions: recent advances in our knowledge. *Phytochemistry*, 64 (1): 21-30, ISSN: 0031-9422.

Thomas, K.V. & Brooks, S. (2010). The environmental fate and effects of antifouling paint biocides *Biofouling* 26 (1): 73–88, ISSN: 0892-7014.

Tomizawa, M. & Casida, J.E. (2005). Neonicotinoid insecticide toxicology: Mechanisms of selective action. *Annu. Rev. Pharmacol. Toxicol.*, 45: 247-268, ISSN: 0362-1642.

Vig, A.P.; Rampal, G.; Thind, T.S. & Arora, S. (2009). Bio-protective effects of glucosinolates - A review. *LWT - Food Science and Technology* 42: 1561–1572, ISSN: 0023-6438.

Wat, C.-K.; Prasad, S.K.; Graham, E.A.; Partington, S.; Arnason, T.; Towers, G.H.N. & Lam, J. (1981). Photosensitization of invertebrates by natural polyacetylenes *Biochem. Syst. Ecol*,.9 (1): 59-62, ISSN: 0305-1978.

Wu, H., J.; Pratley, D. L. & Haig, T. (2001). Allelopathy in wheat (*Triticum aestivum*) *Ann. Appl. Biol.*, 139: 1-9, ISSN: 0003-4746.

Yadav, R.K.P.; Papatheodorou, E.M.; Karamanoli, K.; Constantinidou, H.-I.A. & Vokou, D. (2008). Abundance and diversity of the phyllosphere bacterial communities of Mediterranean perennial plants that differ in leaf chemistry. *Chemoecology*, 18: 217-226, ISSN: 0937-7409.

Zagrobelny, M.; Bak, S.; Rasmussen, A.V.; Jørgensen, B., Naumann, C.M. & Møller, B.L. (2004). Cyanogenic glucosides and plant-insect interactions. *Phytochemistry*, 65 (3): 293-306, ISSN: 0031-9422.

2

Baculovirus Biopesticides

Boguslaw Szewczyk[1*], Marlinda Lobo de Souza[2], Maria Elita Batista de Castro[2], Mauricio Lara Moscardi[3] and Flavio Moscardi[3]
[1]*Department of Molecular Virology, University of Gdansk and Medical University of Gdansk,*
[2]*Embrapa Recursos Genéticos e Biotecnologia, Brasília, DF,*
[3]*Dept. of Agronomy, Universidade Estadual de Londrina, Londrina, PR,*
[1]*Poland*
[2,3]*Brazil*

1. Introduction

Baculoviruses are a large group of double-stranded DNA viruses (almost 1000 species have been described); the majority have been isolated from a few insect orders: Lepidoptera, Diptera, Hymenoptera and Coleoptera. Viral genome ranges in size from 80 to 200 kb. Individual baculoviruses usually have a narrow host range limited to a few closely related species. The most widely studied baculovirus is the *Autographa californica* nuclepolyhedrovirus (AcMNPV).

Baculoviruses are arthropod viruses well known due to their potential as agents of biological control of pests in agriculture and forestry. They are also widely used as expression vectors in biotechnology. The family *Baculoviridae* contains diverse members and in the past the classification was based on virus morphology. It was divided into two genera: the *Nucleopolyhedrovirus* (NPVs) and the *Granulovirus* (GVs). Recently, this division was challenged (Jehle et al., 2006) because the comparison of 29 fully sequenced baculoviral genomes indicated that virus phylogeny followed more closely the classification of the hosts than the virion morphological traits, but the traditional division into two genera is still widely used.

Baculoviruses infect arthropods and they do not replicate in vertebrates, plants and microorganisms. Although they do not replicate, they may, under special conditions, enter animal cells. This unexpected property made them a valuable tool in the last few years for studies of transient expression of foreign genes under vertebrate promoters introduced into baculovirus genome (Boyce and Bucher, 1996; Kost et al., 2005).

The circular DNA genome of AcMNPV is surrounded by a small basic protein which neutralizes the negative charge of the DNA. This structure is protected by proteins forming a nucleocapsid. Virions consist of one or more nucleocapsids embedded in a membranous envelope. Two morphologically distinct, but genetically identical, viral forms are produced at different times post-infection:

- Budded virus particles (BV) which serve for the transmission of the virus to other tissues of the caterpillar body.
- Occlusion bodies (OB) which are responsible for the survival of the virus in the environment and the spread of the virus from insect to insect.

The occlusion bodies (polyhedra) of Nucleopolyhedrovirus contain many occlusion-derived virions (ODV) surrounded by a matrix composed mainly of polyhedrin, a major structural protein (Braunagel et al., 2003). Polyhedrin is produced in large quantities (around 30% of total protein mass at the time of host death) but it is not needed for the transmission of the virus from cell to cell. Polyhedra are relatively stable and the protected virions in the favourable conditions can survive in the environment for more than twenty years. Under magnification of around 1000x, polyhedra resemble clear, irregular crystals of salt so they are big enough to be seen in a light microscope.

Baculoviruses have gained great attention in molecular biology laboratories because they are very versatile genetic engineering tools (for a review see van Oers, 2006). In fact our current knowledge about the biology of AcMNPV is to a large extent a consequence of the developments of baculovirus-based expression vectors. Baculovirus system of expression of foreign genes has many advantages over other systems because high level of foreign gene expression is usually achieved compared to other eukaryotic expression systems. Baculovirus genome can accommodate large pieces (up to 50 kbp) of foreign DNA, so it is possible to express more than one foreign gene. Additionally, the insertion of specific signal sequences in front of a foreign gene leads very often to the export of the gene product outside of the infected cell.

Recombinant baculoviruses are usually constructed in two steps. Initially, a heterologous gene is introduced into a baculovirus transfer vector. It consists of a bacterial replicon of a multicopy plasmid, a selection marker gene, promoter and terminator regions along with flanking baculovirus sequences from a non-essential locus, and a multiple cloning site (or a single unique restriction site) downstream from a viral promoter. Most often the promoters and the flanking DNA originate from one of the late genes: polyhedrin or p10 gene. The latter is another viral gene coding for a protein which is produced in large quantities late in the infection. It is the main component of the fibrillar structures which accumulate in the nucleus and in the cytoplasm of infected cells. For some purposes weaker early promoters, such as basic protein promoter (p6.9), may be preferred.

Around 400 insect cell lines are known which potentially can be used for *in vitro* propagation of baculoviruses. Only a few of them support the growth of AcMNPV. These lines were obtained from two parental organisms: *Spodoptera frugiperda* and *Trichoplusia ni* (Lepidoptera: Noctuidae) The most widely used line is Sf9 which grows well in suspension. BTI-Tn5B1-4 derived from *T. ni*, known as High Five cells, has been also largely used for viral growth (Granados *et al.*, 1994). Cell lines which can be used for the propagation of *Lymantria dispar* nucleopolyhedrovirus (LdMNPV), *Heliothis zea* nucleopolyhedrovirus (HzSNPV), *Bombyx mori* nucleopolyhedrovirus (BmSNPV), *Anticarsia gemmatalis* nucleopolyhedrovirus (AgMNPV) and a few other baculoviruses are also currently available.

2. Baculovirus production technology

At present, commercial production of baculoviruses has been carried out only *in vivo*, either by applying the virus against the host insect in the field and collecting diseased or dead larvae, or by producing the target insect in the laboratory on an artificial diet. The latter is the most commonly used method for producing baculoviruses in many countries but both methods have been used successfully for commercial production of the *Anticarsia gemmatalis* baculovirus (AgMNPV) in Brazil (Moscardi, 1999, 2007). For some insects there are no available artificial diets and, therefore, commercial production of baculoviruses of these insects is generally too difficult or impossible under laboratory conditions. In such cases, field production of baculovirus stocks may be sometimes a method of choice, also from financial point of view (Moscardi, 1999). This approach is, however, difficult when liquefaction of the insect body is very intense, as, for instance, in larvae infected by *Spodoptera* spp. baculoviruses. In this case, live larvae must be collected close to death when the body has not yet ruptured.

Baculovirus production in insect cell cultures offers advantages over *in vivo* multiplication for being a controllable, sterile, highly pure product yield process. *In vitro* process of baculovirus production for agricultural pest control needs to be efficient, with competitive costs, leading to a final product which is highly pathogenic to the target pest. There is a strong limitation for *in vitro* production, however, since successive passages of the virus in cell culture result in genetic alterations leading to loss of virulence (Krell, 1996; Rhodes, 1996). In laboratory culture, production of occlusion derived virions (ODV) is not necessary for survival of the virus. The budded virus (BV) particle is the form used for cell-to-cell transmission in cell culture. The main protein of the BV particle is the GP64 (Blissard, 1996), essential for virus budding and responsible for entrance of the virus into the next host cell. Various culture conditions are known to influence infection of lepidopteran cells by baculoviruses and include temperature, pH, dissolved oxygen concentration, osmolality and nutrient composition of the culture medium. The investigation of factors associated with loss of genetic stability and the use of new strategies such as isolation of more stable variants, as well as the reduction of costs of cell culture medium components, are important requirements for process optimization of *in vitro* baculovirus production.

The requirements for productive insect cell lines (Jem et al., 1997) and for highly productive culture media (Chakraborty et al., 1999) are other challenges for *in vitro* production of baculovirus. Many cell lines are available for production purposes and are derived from various sources, thus exhibiting a wide variety of growth and production characteristics. Careful screening or formulation of media must be performed for a particular virus isolate-cell line combination, as different media can greatly affect polyhedra yields (Pedrini et al., 2006). Recently, a new strategy for *in vitro* production was proposed based on Many Polyhedra (MP) variants. These are clones selected using the plaque assay technique after several passages of the virus in cell culture. MPs maintain the wild type features such as formation of many polyhedra in the cell nucleus and Budded Virus high titer (Slavicek et al., 2001; Pedrini et al., 2005) which allow them, in principle, to compete with the population of Few Polyhedra mutants accumulated in cell culture.

3. Baculovirus pesticides in the past

Two strategies of pest management with baculovirus pesticides are usually employed (Fuxa, 2004):

- infested areas are sprayed with highly concentrated baculovirus to suppress the pest as quickly as possible,
- infested areas are sprayed with lower concentration of baculovirus and this results in establishment of the virus for more than one generation.

At present the number of registered pesticides based on baculovirus exceeds fifty formulations, some of them being the same baculovirus preparations distributed under different trade names in different countries. Both NPVs and GVs are used as pesticides but the former group is larger.

The first viral insecticide Elcar™ was introduced by Sandoz Inc. in 1975 (Ignoffo and Couch, 1981). Elcar™ was a preparation of *Heliothis zea* NPV which is relatively broad-range baculovirus and infects many species belonging to genera *Helicoverpa* and *Heliothis*. HzSNPV provided control of not only cotton bollworm, but also of pests belonging to these genera attacking soybean, sorghum, maize, tomato and beans. In 1982 the production of this biopesticide was discontinued. The resistance to many chemical insecticides including pyrethroids revived the interest in HzSNPV and the same virus was registered under the name GemStar™. HzSNPV is a product of choice for biocontrol of *Helicoverpa armigera* (Mettenmeyer, 2002). Countries with large areas of such crops like cotton, pigeonpea, tomato, pepper and maize, e.g. India and China, introduced special programs for the reduction of this pest by biological means. In Central India, *H. armigera* in the past was usually removed by shaking pigeonpea plants until caterpillars fell from the plants onto cotton sheets. This technique is now used to obtain caterpillars which are fed on virus-infected seeds. Baculovirus preparations obtained in this way are used by farmers to prepare a bioinsecticide spray applied on pigeonpea fields. Another baculovirus, HaSNPV is almost identical to HzSNPV. It was registered in China as a pesticide in 1993 (Zhang et al., 1995). It has been used for large scale biopesticide production and has been extensively used on cotton fields (over 100 000 ha of cotton in the last decade). Broad spectrum biopesticide based on HaNPV is also used in India (Srinivasa et al., 2008).

The forests of temperate regions are very often attacked and defoliated by larvae of Lepidoptera (most common pest species are: *Lymantria dispar, Lymantria monacha, Orgiya pseudotsugata* and *Panolis flammea*) and some Hymenoptera species (mainly *Neodiprion sertifer and Diprion pini*). *L. dispar* MNPV formulations marketed under trade names: Gypchek, Disparivirus, Virin-ENSH, and *O. pseudotsugata* MNPV under trade names: TM BioControl-1 and Virtuss (Reardon et al., 1996) are sometimes used for forest protection. Forest ecosystems tend to be more stable than agricultural systems, allowing for natural or applied baculoviruses to remain in the environment for long periods of time increasing the chance of natural epizootics by these agents.

Caterpillars of moths belonging to *Spodoptera* genus are of primary concern for agricultural industry in many countries of the world. Two commercial preparations based on *Spodoptera* NPV have been available. These are SPOD-X™ containing *Spodoptera exigua* NPV to control insects on vegetable crops and Spodopterin™ containing *Spodoptera littolaris* NPV which is used to protect cotton, corn and tomatoes. About 20 000 hectares of maize annually were controlled with *Spodoptera frugiperda* NPV in Brazil (Moscardi, 1999), but at present it has not been used due to technical problems in the virus production under laboratory conditions. Use of *Spodoptera litura* NPV has been tested on cabbage crops in India (Kumari et al., 2009). Many other species belonging to the *Noctuidae* family are economically important pests of sugarcane, legume, rice and others. *Autographa californica* and *Anagrapha falcifera* NPVs were registered in the USA and were field-tested at a limited scale. These two NPVs have relatively broad host spectrum and potentially can be used on a variety of crops infested with pests belonging to a number of genera, including *Spodoptera* and *Helicoverpa*.

Granulovirus CpGV is the active component of a number of biopesticides used for protection of apple and pear orchards against the coddling moth, *Cydia pomonella*. Some of the trade marks of GpGV-based products are following: Granusal™ in Germany, Carpovirusine™ in France, Madex™ and Granupom™ in Switzerland, Virin-CyAP in Russia. Annually up to 250 000 hectares of orchards have been protected with Madex™ in different European countries (Vincent et al., 2007). Considering application of all trade names of the CpGV, this may be the most important worldwide viral insecticide currently applied in terms of treated area.

Another granulovirus, *Erinnyis ello* (cassava hornworm) granulovirus, was found to be very efficient for protection of cassava plantations (Bellotti, 1999). This GV has been used for spraying cassava crops in some South American countries. In Brazil a successful program for cassava pest control was carried out in the eighties based on recovering the virus that were multiplied in the field larval population. However, due to *Erinnyis ello* cyclical behaviour and the difficulty in the insect mass production in laboratory conditions, the program was discontinued.

Other important viruses that are currently employed to control insects include the tea tortricids *Adoxophyes honmai* and *Homona magnanima* granuloviruses (GV) in Japan. The area sprayed with GVs comprised 5,850 ha in Kagoshima in 1995, equivalent to 80 % of all the tea fields in the prefecture (Nishi and Nonaka 1996). The GVs of *H. magnanima* and *A. honmai* were registered in 2003, however, the use of GVs has recently declined. One reason for the reduction in use of GVs in Japanese tea fields is the changing pattern of occurrence of other pests. Mulberry scale, for example, has been increasing recently and chemical treatment is required to control this insect at the same time GVs are sprayed. However, the spray also kills *H. magnanima* and *A. honmai*. Furthermore, GVs have been applied in Kagoshima for more than ten years and the populations of *H. magnanima* and *A. honmai* have been reduced (Nakamura 2003). In China twelve baculoviruses have been authorized as commercial insecticides (Sun and Peng 2007), including *H. armigera* NPV (the most widely used virus in China for cotton, pepper and tobacco protection), *S. litura* NPV (vegetables), *S. exigua* NPV (vegetables), *Buzura suppressaria* NPV (tea), *Pieris rapae* GV and *Plutella xylostella* GV (vegetables). Use of baculoviruses in China is the greatest worldwide, regarding the number of viruses being registered for insect control. Sun and Peng (2007) also reported a Cypovirus (CPV) produced in China for control of *Dendrolimus punctatus*, an insect pest of pine forests.

The well-known success of employing baculovirus as a biopesticide is the case of *Anticarsia gemmatalis* nucleopolyhedrovirus (AgMNPV) used to control the velvetbeen caterpillar in soybean (Moscardi, 1999). This program was implemented in Brazil in the early eighties, and came up to over 2,000,000 ha of soybean treated annually with the virus. Recently this number dropped down, mainly due to new emerging pests in the soybean complex. The use of AgMNPV in Brazil brought about many economical, ecological and social benefits. At the soybean grower level, the financial savings from the use of the virus may reach up to ca. U$ 7/ha/season, including product cost and application cost. The annual savings at the grower level, in the total area sprayed with the virus was over US$ 11,000.000. Since the beginning of the program more than 17 million liters of chemical insecticides were not sprayed in the environment. The protection of soybean fields in Brazil has proven that baculoviral control agents can be effectively produced on a large scale and they may be an alternative to broad-spectrum chemical insecticides. On the basis of this spectacular success of a baculovirus pesticide, it is needless to say that the advantages of biopesticides over chemical pesticides are numerous. Safety for humans and non-target organisms, preservation of biodiversity in the environment, reduction of toxic residues in agricultural end-products are just the examples of potential benefits. However, the cost of biopesticide production has been usually higher than the cost of conventional pesticides. So, paradoxically, countries where the cost of human labour is low are more open towards the use of baculoviral pesticides than highly-developed countries which claim that environmental protection is one of their priorities in the development.

Genomic variability has been described for many wild type virus including *A. californica* MNPV, *S. frugiperda* MNPV, *S. litura* MNPV, *P. flammea* MNPV and *Mamestra configurata* NPV. Genotypic variants can be recognized by the presence of submolar fragments in the electrophoretic patterns of restriction endonuclease digestion products of a viral genome. Genotypic variation in baculovirus genomes can include point mutations, both small and large deletions and insertions (Krell, 1996). Though mutations can occur in any place of the genome, the presence of some hot spots was observed for certain genomic alterations such as insertions due to transposable elements or deletions in the hypervariable DA26 gene region (Kamita et al., 2003). AgMNPV genomic variability has been also carefully studied because the selection pressure due to the application of AgMNPV in the field during subsequent years could lead to alterations in virus stability. The method of choice was the technique of restriction endonuclease analysis. Viral DNA were initially purified from diseased larvae collected during several crop seasons and compared to AgMNPV-79, a wild-type virus that was used originally and subsequently in this program (Souza et al, 2001). These results indicated that the virus maintains considerable stability, even with the existence of some genetic changes shown in the DNA restriction profiles. It has been also observed that the virus retains its virulence to the host insect throughout the years of its application.

4. Prognoses for future use of baculovirus pesticides

Large-scale application of AgMNPV in Brazil has proven that the baculovirus protection can be done at relatively low cost. It is very likely that the growing awareness of the benefits of the environment-friendly pesticides will result in the reevaluation of the prospects for biological protection with baculovirus preparations. However, until today, baculovirus insecticides have not met their full potential to control pest insects worldwide. In his review, Moscardi (1999) previewed the following: the expansion of baculovirus use, in the following five years, i.e., up to 2004, would depend on new developments in the areas of recombinant baculoviruses and in the *in vitro* commercial production of these agents. The development of recombinant baculovirus was efficiently completed by researchers in several countries, but the *in vitro* commercial technology still lags behind today, due to technical problems. Future development of baculovirus pesticides will probably depend on the attitude towards genetically modified organisms. In countries where use of genetically modified organisms is restricted, only naturally occurring baculoviruses will be used for protection of crops. In this case the improvements will be at the level of diagnostics of infection, development of the *in vitro* cultures and changes in the formulations of the biopesticide. In countries which favour the introduction of genetically modified organisms, the improvements will be achieved by introduction of exogenous genes into baculovirus genome, thus greatly enhancing the killing activity of bioinsecticide formulations.

Reliable assays for the progress of infection with baculovirus are necessary because the major problem in using biopesticide for crop protection is their slow action and lack of morphological changes in larvae in first stages of baculovirus propagation. Lack of such assays may incline agricultural services to use subsequent chemical means of protection which, from the ecological point of view, may be redundant. Fast and sensitive methods in diagnostics based on baculovirus genome detection will probably play a predominant role in future. They are relatively simple analytical methods giving precise information about occurrence and spread of the virus. Using specific primers, not only target larvae, but also vectors for baculovirus transfer - invertebrate and bird predators can be quickly analysed. For strictly quantitative assays, real-time PCR is a method of choice. The equipment required – light cyclers, are relatively expensive, but their prices decrease very quickly.

The *in vitro* production is still a strong requirement on a commercial perspective of baculoviruses use as insecticides. However the accumulation of genotypic variations by serial passage in cell culture prevents its large scale production. One of the most important effects of the viral passage is the change from the parental, many polyhedra per cell (MP) phenotype, to the few polyhedra per cell (FP) phenotype. The major problem of the passage effect is the reduced occlusion and loss of virulence of the occluded virus (Krell, 1996). Frequent mutations have been identified within a specific region in the Few Polyhedra mutants (FP) that contains the 25k *fp* locus (Harrison and Summers, 1995; Lua et al, 2002). This gene encodes a 25KDa protein that is essential for virion occlusion and polyhedron formation. Another type of mutants generated during serial passage of baculovirus is the formation of Defective Interfering Particles (DIs). These mutants have lost the ability to be replicated in the host cell without the aid of a helper virus and large sizes of their genome are usually deleted (Pijlman et al., 2001). These particles replicate faster because they are smaller, and inhibit the replication of a standard virus. The challenge to make *in vitro* commercial production of baculoviruses a viable initiative depends on development of new techniques to sustain MP production through passages in cell cultures from small flasks to large scale commercial fermentors.

The stability of baculoviruses is influenced by temperature, pH, humidity, presence of additives but ultraviolet light is probably the most detrimental factor to viral survival. Under field conditions little activity is left when the virus is not shaded by plant canopy, therefore much effort has been devoted to the development of UV protectants (Shapiro and Dougherty, 1994; Zou and Young, 1994, Morales et al., 2001). The best results were obtained for stilbene fluorescent brighteners which are marketed under many trade names (e.g. Phorwite AR, Blankophor and others). Future developments in the formulations of brighteners may lead to the reduction of cost of baculovirus production. Inactivation of baculoviruses may be also caused by plant metabolites such as peroxidases which generate free radicals (Hoover et al., 1998). The inactivation can be reduced by addition of free radical scavengers such as mannitol or enzyme superoxide dismutase to baculovirus preparations (Zhou et al., 2004).

The activity of baculoviruses against their natural hosts may be enhanced by introduction of insect-specific toxins or by interference with insect physiology (Bonning and Hammock, 1996; Inceoglu *et al.*, 2001). Baculovirus genome modifications by introduction of exogenous toxin genes were extensively studied in many laboratories. Most of the research was devoted to the studies of arthropod toxin genes isolated from the scorpion or spiders (Bonning and Hammock, 1996; Inceoglu et al., 2007). The most potent insect-specific toxin gene used for construction of baculovirus recombinants was the gene coding for a toxin from scorpion *Androctonus australis*. The feeding damage caused by larvae infected with this modified baculovirus was reduced by about 60% in comparison to a wild type baculovirus (Inceoglu et al., 2001). Toxin genes isolated from other scorpions, e.g. *Leiurus quinquestriatus hebraeus* (Froy et al., 2000), straw itch mite *Pyemotes tritici* (Burden et al., 2000), ants

(Szolajska et al., 2004) or spiders (Hughes et al., 1997) have been intensively studied as potential enhancers of baculovirus activity. Arthropod toxins usually attack insect sodium channels producing final effect similar to the chemical insecticides of the pyrethroid group. However, the specific target in sodium channels is different, so there is a potential possibility to produce synergistic effect by biopesticide/chemical pesticide application (McCutchen et al., 1997).

Baculovirus recombinants that produced occlusion bodies incorporating *Bacillus thuringiensis* toxin were constructed by making a fusion protein consisting of polyhedrin and Bt toxin (Chang et al., 2003). The pathogenicity of the recombinant was remarkably increased compared to wild-type virus. These studies proved that it is possible to construct a biopesticide which combines the advantages of the virus and the bacterial toxin.

The changes to host physiology were done by introducing genes coding for some insect hormones or hormone-modifying enzymes into baculovirus genome, or by deletion of the baculovirus-encoded ecdysteroid glucosyltransferase (*egt*) gene. The former approach was employed by cloning juvenile hormone esterase gene into baculovirus genome which overexpressed decreases the concentration of the juvenile hormone which is a signal for a caterpillar to stop feeding and pupate. This line of research is being pursued in some laboratories (Hammock et al., 1990; Inceoglu et al., 2001). The deletion of the baculovirus-encoded *egt* gene was used first by O'Reilly and Miller, 1991. The product of the *egt* gene interacts with larval moulting and indirectly increases the time of feeding of infected caterpillars. The *egt*-deletion from baculovirus genome resulted in 30% faster killing of caterpillars. Another advantage of this genomic modification is the fact that the *egt* gene is not essential for viral replication and can be replaced with an exogenous gene; the product of which may enhance the insecticidal activity of the recombinant virus (Sun et al., 2004).

In the future, genetically modified baculoviruses will contribute to the expansion of baculovirus use worldwide, as these GMOs are considered safe through extensive research conducted over many years. The scientific data indicate that baculoviruses pose no hazard to other animals than their hosts and this was documented by a number of studies from different laboratories. Recombinant baculoviruses were not pathogenic to bees and all vertebrate species (Sun et al., 2004) as well as to the natural enemies of larvae such as parasitoids and predators (Boughton et al., 2003). However, in spite of this sound evidence, preliminary field trials of genetically modified baculoviruses raised massive public protests which put on hold further trials for a long time. The slow progress in application of genetically modified baculoviruses as pesticides may be in part due to the choice of toxin genes used for modifications of the baculovirus genome which were isolated from highly dangerous invertebrates. Taking into account the origin of these social conflicts, the choice of toxin genes used for genome modifications should be restricted to genes coding for ecologically natural insect toxins, e.g. the genes coding for toxic polypeptides of parasitoid wasps occurring in regions infested by a particular pest. The more rational approach is also needed in the social perception of dangers associated with genetically modified baculoviruses by educating the public on risks and benefits of recombinant baculovirus pesticides.

5. References

Bellotti, A. C. 1999. Recent advances in cassava pest management. Ann. Rev. Entomol. 44:345-370.

Blissard, G.W. 1996. Baculovirus–insect cell interactions. Cytotechnology 20: 73–93.

Bonning, B. C., and Hammock, B. D. 1996. Development of recombinant baculoviruses for insect control. Ann. Rev. Entomol. 41:191-210.

Boughton, A. J., Obrycki J. J., and Bonning B. C. 2003. Effects of a protease-expressing recombinant baculovirus on nontarget insect predators of *Heliothis virescens*. Biol. Control 28:101-110.

Boyce, F. M., and Bucher, N. L. R. 1996. Baculovirus-mediated gene transfer into mammalian cells. Proc. Natl. Acad.Sci.USA 93:2348-2352.

Braunagel, S. C., Russell, W. K., Rosas-Acosta, G., Russell, D. H., and Summers, M. D. 2003. Determination of protein composition of the occlusion-derived virus of *Autographa californica* nucleopolyhedrovirus. Proc. Natl. Acad. Sci. USA 100: 9797-9802.

Burden J. P., Hails, R. S., Windass J. D., Suner M. M., and Cory, J. S. 2000. Infectivity, speed of kill, and productivity of a baculovirus expressing the itch mite toxin txp-1 in second and fourth instar larvae of *Trichoplusia ni*. J. Invertebr. Pathol.: 75:226-236.

Chakraborty, S., Monsour, C., Teakle, R., and Reid, S.1999. Yield,biologicalactivity,and field performance of a wild-type Helicoverpa nucleopolyhedrovirus produced in *H. zea* cell cultures. J. Invertebr. Pathol. 73:199–205.

Chang, J. H., Choi, J. Y., Jin, B. R., Roh, J. Y., Olszewski, J. A., Seo, S. J., et al., 2003. An improved baculovirus insecticide producing occlusion bodies that contain *Bacillus thuringiensis* insect toxin. J. Invertebr. Pathol. 84:30-37.

Froy, O, Zilberberg, N, Chejanovsky, N, Anglister, J, Loret, E, Shaanan, B, et al. 2000. Scorpion neurotoxins: structure/function relationship and application in agriculture. Pest Manag. Sci. 56:472-474.

Fuxa, J.R., 2004. Ecology of insect nucleopolyhedroviruses. Agriculture, Ecosystems & Environment., 103: 27-43.

Granados, R. R., Guoxun, L., Dersksen, A. C. G., and McKenna, K. A. 1994. A new insect cell line from *Trichoplusia ni* (BTI-Tn-5B1-4) susceptible to *Trichoplusia ni* single *nuclear* polyhedrosis virus. J. Invertebr. Pathol. 64:260-266.

Hammock, B. D., Bonning, B. C., Possee, R. D., Hanzlik, T. N., and Maeda, S. 1990. Expression and effects of the juvenile hormon esterase in a baculovirus vector. Nature 344: 458-461.

Harrison, R. L., and Summers, M. D. 1995. Mutations in the Autographa californica Multinucleocapsid Nuclear Polyhedrosis Virus 25 KDa protein gene result in reduced virion occlusion, altered intranuclear envelopment and enhanced virus production. J. Gen. Virol. 76:1451-1459.

Hoover, K, Kishida, K.T., DiGiorgio, L.A., Workman, J, Alaniz, S.A., Hammock, B.D., Duffey, S.S. 1998. Inhibition of baculoviral disease by plant-mediated peroxidase activity and free radical generation. J.Chem.Ecol., 24, 1949-2001.

Hughes, P. R., Wood, H. A., Breen, J. P., Simpson, S. F., Duggan, A. J., and Dybas, J. A. 1997. Enhanced bioactivity of recombinant baculoviruses expressing insect-specific spider toxins in lepidopteran crop pests. J. Invertebr. Pathol. 69: 112-118.

Ignoffo, C. M., and Couch, T. L. 1981. The nucleopolyhedrosis virus of *Heliothis* species as a microbial pesticide. In Microbial Control of Pests and Plant Diseases, ed. H. D. Burges, pp. 329-362. London: Academic Press.

Inceoglu, A. B., Kamita, S. G., Hinton, A. C., Huang, Q., Severson, T. F., Kang, K.D., and Hammock, B. D. 2001. Recombinant baculoviruses for insect control. Pest Manag. Sci. 57: 981-987.

Inceoglu, A. B., Kamita, S. G., and Hammock, B. D. 2007. Genetically modified baculoviruses: a historical overview and future outlook. Adv. Virus Res. 68:323-360.

Jem, K. J., Gong, T., Mullen, J., and Georgis, R. 1997. Development of an industrial insect cell culture process for large scale production of baculovirus biopesticides. In: Invertebrate Cell Culture: Novel Directions and Biotechnology Applications, eds. K. Maramorosch, and J. Mitsuhashi, pp. 173–180. New Hampshire: Science Publishers.

Jehle, J. A., Blissard, G. W., Bonning, B. C., Cory, J. S., Herniou, E. A., Rohrmann, G. F., Theilmann, D. A., Thiem, S. M., and Vlak, J. M. 2006. On the classification and nomenclature of baculoviruses: a proposal for revision. Arch. Virol. 151:1257–1266.

Kamita, G.K., Maeda, S., Hammock, B.D. 2003. High-frequency homologous recombination between baculoviruses involves DNA replication. J.Virol. 77: 13053-13061.

Kost, T. A., Condreay, J. P., and Jarvis, D. L. 2005. Baculovirus as versatile vectors for protein expression in insect and mammalian cells. Nature Biotech. 23:567-575.

Krell, P. J. 1996. Passage effect of virus infection in insect cells. Cytotechnology 20:125-137.

Kumari, V., and Singh, N. P. 2009. *Spodoptera litura* nuclear polyhedrosis virus (NPV-S) as a component in Integrated Pest Management (IPM) of *Spodoptera litura* (Fab.) on cabbage. J. Biopesticides 2: 84-86.

Lua, L. H. L., Pedrini, M. R. S., Reid, S., Robertson, A., and Tribe, D. E. 2002. Phenotypic and genotypic analysis of Helicoverpa armigera nucleopolyhedrovirus serially passed in cell culture. J. Gen. Virol.83: 945-955.

McCutchen, B. F., Hoover, K., Preisler, H. K., Betana, M. D., Herrmann, R., Robertson, J. L., and Hammock, B. D. 1997. Interaction of recombinant and wild-type baculoviruses with classical insecticides and pyrethroid-resistant tobacco budworm (Lepidoptera: Noctuidae). J. Econ. Entomol. 90:1170-1180.

Mettenmeyer, A., 2002. Viral insecticides hold promise for bio-control. *Farming Ahead.*, 124: 50-51.

Morales, L., Moscardi, F., Sosa-Gomez, D. R., Paro, F. E., and Soldorio, I. L. 2001. Fluorescent brighteners improve *Anticarsia gemmatalis* (Lepidoptera: Noctuidae) nucleopolyhedrovirus (AgMNPV) activity on AgMNPV susceptible and resistant strains of the insect. Biol. Control: 20: 247-253.

Moscardi, F., 1999. Assessment of the application of baculoviruses for control of Lepidoptera. Ann. Rev. Entomol. 44: 257–289.

Moscardi, F. 2007. A Nucleopolyhedrovirus for control of the velvetbean caterpillar in Brazilian Soybeans. In Biological Control: A Global Perspective, eds. C. Vincent, M. S. Goethel, and G. Lazarovits, pp. 344-352. Oxfordshire, UK, and Cambridge, USA: CAB International.

Nakamura T. 2003. Control of leafrollers in tea fields using "Hamaki tenteki". http://www.agrofrontier.com/guide/f_105.html

Nishi, Y., and Nonaka, T. 1996. Biological control of the tea tortrix using granulosis virus in the tea field. Agrochem. Jpn. 69: 7-10.

O'Reilly, D. R., and Miller, L. K. 1991. Improvement of a baculovirus pesticide by deletion of the *egt* gene. Biotechnology 9: 1086-1089.

Pedrini, M. R. S., Nielsen, L. K., Reid, S., and Chan, L. C. L. 2005. Properties of a unique mutant of Helicoverpa armigera single-nucleocapsid nucleopolyhedrovirus that exhibits a partial *Many Polyhedra* and *Few Polyhedra* phenotype on extended serial passaging in suspension cell cultures. In Vitro Cell. Devel. Biol. 41:289-297.

Pedrini, M. R. S., Christian, P., Nielsen, L. K., Reid, S., and Chan, L. C. L. 2006. Importance of virus-medium interactions on the biological activity of wild-type Heliothine nucleopolyhedroviruses propagated via suspension insect cell cultures. J. Virol. Meth. 136: 267-272.

Pijlman, G. P., van den Born, E., Martens, D. E., and Vlak, J. M. 2001. *Autographa californica* baculoviruses with large genomic deletions are rapidly generated in infected cells. Virology 283:132138.

Reardon, R., Podgwaite, J. P., and Zerillo, R. T. 1996. GYPCHECK - the gypsy moth nucleopolyhedrosis virus product. USDA Forest Service Publication FHTET-96 - 16.

Rhodes, D. J. 1996. Economics of baculovirus-insect cell production systems. Cytotechnology 20: 291-297.

Shapiro, M., Daugherty, E.M.1994. Enhancement in activity of homologous and heterologous viruses against gypsy moth (Lepidoptera, Lymantridae) by an optical brightener. J.Econ.Entomol. 87: 361-365.

Slavicek, J. M.. Hayes-Plazolles, N., and Kelly, M. E. 2001. Identification of a Lymantria dispar nucleopolyhedrovirus isolate that does not accumulate few-polyhedra mutants during extended serial passage in cell culture. Biol. Control 22:159-168.

Souza, M.L., Castro, M.E.B., Barros, A.M.L., Sihler, W., Moscardi, F., 2001. Análise de DNA de isolados de nucleopolyhedrovirus de *Anticarsia gemmatalis* utilizados no controle da lagarta da soja no Brasil. Boletim de Pesquisa e Desenvolvimento. Embrapa Recursos Genéticos e Biotecnologia.

Srinivasa, M., Jagadeesh Babu, C. S., Anitha, C. N., and Girish, G. 2008. Laboratory evaluation of available commercial formulations of HaNPV against *Helicoverpa armigera* (Hub.). J. Biopesticides. 1:138-139.

Sun, X., Wang, H., Sun, X., Chen, X., Peng, C., Pan, D., et al. 2004. Biological activity and field efficacy of a genetically modified *Helicoverpa armigera* SNPV expressing an insect-selective toxin from a chimeric promoter. Biol. Control 29:124-137.

Sun, X. L., and Peng, H. 2007. Recent advances in biological pest insects by using viruses in China. Virol. Sinica 22:158-162.

Szolajska, E., Poznanski, J., Ferber, M. L., Michalik, J., Gout, E., Fender, P., Bailly, I., Dublet, B., and Chroboczek, J. 2004. Poneratoxin, a neurotoxin from ant venom. Structure and expression in insect cells and bonstruction of a bio-insecticide. Eur. J. Biochem. 271: 2127-2136.

van Oers, M. M. 2006. Vaccines for viral and parasitic diseases produced with baculovirus vectors. Adv. Vir. Res. 68:193-253.

Vincent, C., Andermatt, M., Valéro, J. 2007. Madex® and Virosoft$^{CP4®}$, viral biopesticides for coddling moth control. In Biological Control: A Global Perspective, eds. C. Vincent,

M. S. Goethel, and G. Lazarovits, pp. 336-343. Oxfordshire, UK, and Cambridge, USA: CAB International.

Zhang, G. Y., Sun, X. L., Zhang, Z. X., Zhang, Z. F., and Wan, F. F. 1995. Production and effectiveness of the new formulation of *Helicoverpa* virus pesticide-emulsifiable suspension. Virol. Sinica 10:242-247.

Zou, Y., Young, S.Y., 1994. Enhancement of nuclear polyhedrosis virus activity in larval pests of Lepidoptera by a stilbene fluorescent brightener. *J. Entomol. Sci.,* 29: 130-133.

Zhou, M.Z., Sun, H.C., Hu, Z.H., Sun, X.L., 2004. SOD enhances infectivity of *Helicoverpa armigera* single nucleocapsid nucleopolyhedrosis against *H.armigera* larvae. *Virologica Sinica.,* 18: 506-507.

3

Acaricides – Biological Profiles, Effects and Uses in Modern Crop Protection

Dejan Marčić[1], Pantelija Perić[1] and Slobodan Milenković[2]
[1]*Institute of Pesticides and Environmental Protection, Belgrade*
[2]*Megatrend University, Graduate School of Biofarming, Bačka Topola*
Serbia

1. Introduction

Acaricide is a pesticide designed to control harmful species of mites (Acari)[1]. In crop protection practices, acaricides are used against phytophagous mites, pests causing economic injuries to agricultural crops and ornamental plants. Until mid-twentieth century, in agroecosystems of low-level productivity, phytophagous mite populations usually stayed below economic injury levels, due to natural regulation by predatory mites and insects, their natural enemies. The concept of secondary pest outbreak was introduced on spider mites (Tetranychidae), the most important plant-feeding mites, as a paradigm. Advances in agricultural production after World War II, based on the extensive use of pesticides and fertilizers, irrigation and other cultural practices, induced increase in spider mite populations far above economic threshold (Huffaker et al., 1970; McMurtry et al., 1970; Jeppson et al., 1975; Metcalf, 1980). Grown under favourable conditions, host plants became high quality food sources for the mite pests, which gave rise to outbreaks of their populations and made it possible to compensate for the losses caused by predators' activity. Moreover, widespread use of neuroactive insecticides (synthetic organic compounds used against insects as target pests, but toxic to other non-target insect and mite species as well) destroyed spider mite predators, generally more susceptible than their prey; on the other hand, heavy selection pressure by neuroactive insecticides caused emergence of tetranychid mite populations resistant to these compounds. Besides the resistance of spider mites and the elimination of their predators, as the primary causes, outbreaks are influenced by sublethal effects of pesticides on behaviour and physiology of pests and/or predators (Metcalf, 1980; Hardin et al., 1995; Dutcher, 2007).

Spider mites, mostly polyphagous species, are common pests in modern agroecosystems worldwide, and some of them are among the most important crop pests. After Tetranychidae,

[1] Mites (subclassis Acari), morphologically and ecologically very diverse assemblage of tiny invertebrates, belongs to class Arachnida (together with spiders and scorpions), subphylum Chelicerata and phylum Arthropoda. The arthropods also include insects, from which mites differ, beside being eight-legged animals (insects are hexapods) by the lack of true head and conspicuous body segmentation. There are some 50.000 mite species known today, but it is estimated that the true number is 20 times higher. Besides agricultural pests and their natural enemies (predators), mites include species of medical and veterinary importance (house dust mites, scabies mites, ticks), while the species living in soil and water are important environmental indicators.

the second most important mite pests are gall and rust mites (Eriophyoidea), while the other economically harmful species can be found among false spider mites (Tenuipalpidae), tarsonemid mites (Tarsonemidae) and acarid mites (Acaridae). Phytophagous mites feed on the liquid content of plant cells, thus disrupting the physiology of a host plant and causing various damages to plant tissues and organs, while some of the species can also act as vectors of plant viruses. In spite of relatively small size (100-400 µm), plant-feeding mites can cause considerable crop yield and quality losses, because they have short life span and under favourable conditions their populations quickly reach high abundance (Helle & Sabelis, 1985a,b; Lindquist et al., 1996; Zhang, 2003; van Leeuwen et al., 2010).

The use of acaricides has increased substantially over the past half of the 20th century. Since the first serious and widespread outbreaks of spider mites populations, during the 1950s, organophosphorous and other neuroactive insecticides were replaced by specific acaricides i.e. compounds exclusively or primarily effective against mites. Several generations of structurally diverse synthetic acaricides, directed against various biochemical and physiological targets, have been commercialized until now. Besides specific acaricides, a number of insecticides with considerable acaricidal activity (pyrethroids, avermectins, benzoylureas) have also been used, while some older neuroactive compounds are still available for the control of phytophagous mites (Jeppson et al., 1975; Knowles, 1997; Dekeyser, 2005; van Leeuwen et al., 2010). Most of the modern acaricides exert their effects through disruption of respiratory processes. Another approach in the development of synthetic acaricides launched compounds that act on growth and development (Dekeyser, 2005; Krämer & Schirmer, 2007). On the other hand, various natural bioactive products with acaricidal activity (botanical and microbial pesticides, essential oils, horticultural spray oils, mycopesticides) have become important alternatives to synthetic acaricides (Beattie et al., 2002; Copping & Duke, 2007; Faria & Wraight, 2007).

Acaricide resistance in phytophagous mites is a seriously increasing phenomenon, especially in spider mites which have a remarkable intrinsic potential for rapid evolution of resistance (Croft & van de Baan, 1988; van Leeuwen et al., 2009). Their populations have often developed a very high degree of resistance to a newly introduced compound after few years of use, with cross-resistance to other compounds with the same mode of action. According to APRD (*Arthropod Pesticide Resistance Database*) more than 700 cases of acaricide resistance in phytophagous mites have been reported. About 93% of these reports refer to spider mites resistance, and almost a half of spider mite resistance cases is related to the twospotted spider mite (*Tetranychus urticae*), highly polyphagous species, one of the most important pests in greenhouses throughout the world (Whalon et al., 2008, 2010). Therefore, there is a continual need for development and application of new acaricides with novel biochemical modes of action, but also for optimization of their use in order to prevent or delay the evolution of resistence and prolong their life span (Dekeyser, 2005). Considering biorational pest control as key approach to modern crop protection (Horowitz et al., 2009) new acaricides should be selective, that is, effective against the target pests and compatible with their natural enemies. Moreover, these compounds must be safe products with respect to human health, beneficial and non-target organisms (mammals, birds, earthworms, bees, aquatic organisms) and the environment in order to meet the regulatory requirements.

This chapter focuses on biological profiles of acaricides that have been commercialized at the end of the 20th and beginning of the 21st century, acaricide resistance in phytophagous mites, bioactive products of natural origin as alternatives to synthetic acaricides, compatibility of acaricides with the biological control agents, and other current issues related to acaricide uses in modern crop protection.

2. Summary of acaricide development

As already remarked, phytophagous mites became important pests of cultivated plants in the mid-20th century, during the „golden age of insecticide discovery" (Casida & Quistad, 1998) that was marked by intensive use of **organochlorines**, **organophosphates** and **carbamates,** broad spectrum insecticides which, as it was later discovered, included many acaricidal compounds. Those were the neuroactive compounds which disrupt the transmission of impulses between nerve cells of an insect by blocking the action of the enzyme acetylcholinesterase (organophosphates, carbamates) or interferring with ion channels in the nerve membrane (organochlorines) (Ishaaya, 2001). Figure 1 shows chlorpyrifos, probably the most common commercialized organophosphate today, carbaryl, the first synthesized carbamate, and endosulfan, one of the rare organochlorine compounds still in use.

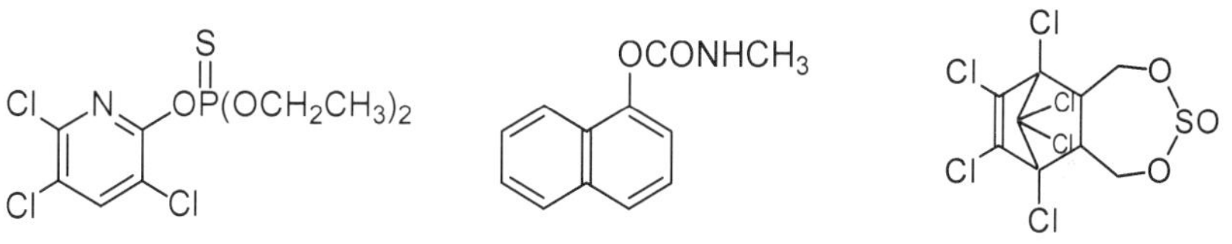

chlorpyrifos carbaryl endosulfan

Fig. 1. Neuroactive insecticides with acaricidal activity: chlorpyrifos (organophosphate), carbaryl (carbamate) and endosulfan (organochlorine)

The first serious and widespread spider mite outbreaks following applications of neuroactive insecticides, observed at the end of 1940s and beginning of 1950s, initiated the research and development of specific acaricides. These compounds, exclusively or primarily effective against mites, were gradually taking over the organochlorines, organophosphates and carbamates. **Bridged diphenyls** (bromopropylate, chloropropylate, chlorobenzilate, chlorfenethol, dicofol, tetradifon), the first specific acaricides, established themselves on the market in the 1950s. During the 1960s and early 1970s, the second generation of structurally rather different specific acaricides emerged, the most important of which were **propargite**, **organotins** (cyhexatin, fenbutatin-oxide) and **formamidines** (amitraz, chlordimeform). Most of first and second generation acaricides are not used any longer. Specific acaricides of the third generation are represented by **mite growth inhibitors** (clofentezine, hexythiazox), commercialized in the first half of the 1980s (Fig. 2) In addition to specific acaricides, several structurally diverse synthetic acaro-fungicides (dinocap, dinobuton, chinomethionate, dichlofluanid) were introduced; on the other hand, the use of sulfur products (that had been exploited as acaro-fungicides since 19th century) was largely displaced by novel synthetic compounds.

Introduction of specific acaricides reduced the adverse impact on beneficial insects (predators of insect and mite pests, polinators) to the minimum; at the same time, many specific acaricides proved to be selective, i.e. less toxic to predaceous mites than phytophagous mites. These acaricides effectively control populations of phytophagous mites resistant to neuroactive compounds, since they are compounds having different biochemical modes of action, with targets mostly being outside the nervous system (March, 1976; Knowles, 1976, 1997; Ishaaya, 2001; Krämer & Schirmer, 2007). Moreover, specific acaricides are far more safer for humans, non-target organisms and the environment in

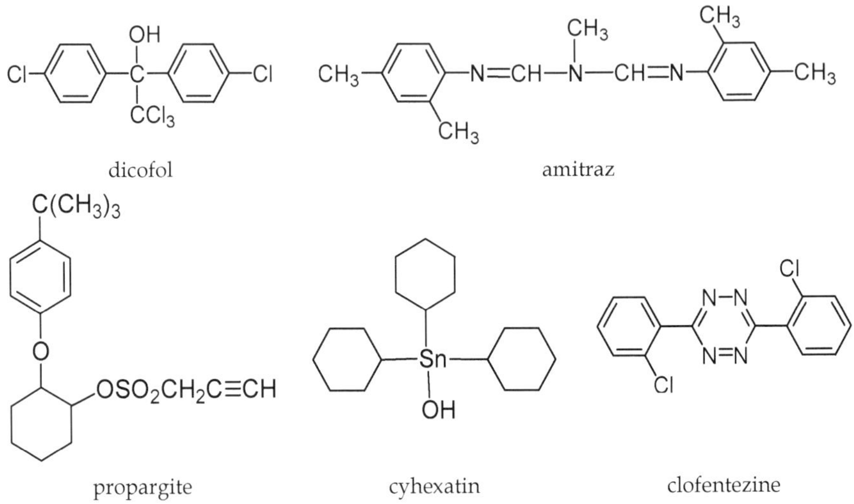

Fig. 2. Representatives of the first (dicofol), second (amitraz, propargite, cyhexatin) and third (clofentezine) generation of specific acaricides

comparison to neuroactive compounds, in particular organochlorines that were almost all severely restricted or banned in developed countries in the 1970s. Organophosphates and carbamates, however, remain to be the predominant group of insecticides by accounting for 35% of the global market (van Leeuwen et al., 2009).

In addition to specific acaricides, two new groups of synthetic insecto-acaricides were placed on the market in the 1970s and 1980s: **pyrethroids** (neuroactive compounds, sodium channel modulators) and **benzoylureas** (compounds acting on growth and development by inhibition of biosynthesis of chitin, a biopolymer present in the cuticle of arthropods). Another new commercial product was **abamectin**, neuroactive insecto-acaricide (chloride channel activator), a mixture of macrocyclic lactones avermectin B_{1a} and avermectin B_{1b}, natural products isolated from the fermentation of *Streptomyces avermitilis*, a soil Actinomycete (Fig. 3), (Ishaaya, 2001; Krämer & Schirmer, 2007). These compounds increased the biochemical diversity of acaricides and insecto-acaricides, but beside the partly expected resistance, some other problems emerged, such as the pyrethroid-induced spider mite outbreaks (Gerson & Cohen, 1989; van Leeuwen et al., 2009).

In the last two decades, a considerable number of non-neuroactive synthetic acaricides and insecto-acaricides emerged on the global market, but there is also a growing interest to find new and reinstate the already known acaricidal compounds of natural origin (Dekeyser, 2005; Copping & Duke, 2007; Krämer & Schirmer, 2007). The search for new chemistries that act on novel target sites and determination of an efficient strategy for use of acaricides that have different biochemical modes of action is presently the only sustainable solution that can prevent or delay the evolution of resistance and prolong the life span of acaricides (Dekeyser, 2005).

Nowadays, acaricides are developed under conditions marked by growing demand of the public opinion for safer, „greener" pesticides, and increasingly stricter toxicological and

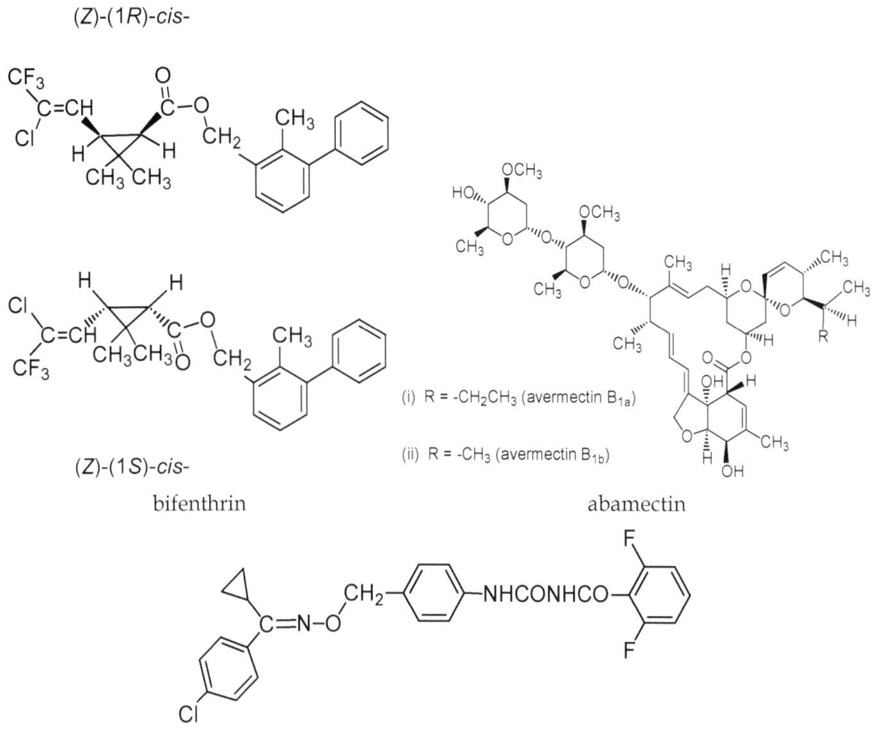

flucycloxuron

Fig. 3. Structural formulas of bifenthrin (pyrethroid), flucycloxuron (benzoylurea) and abamectin (natural product)

eco-toxicological criteria for market circulation of the existing pesticides and registration of the new ones imposed by the regulatory agencies and issues (Casida & Quistad, 1998; Dekeyser, 2005). In the USA, the passage of the Food Quality Protection Act (FQPA) of 1996 brought about significant changes in the way in which pesticides are registered by the U.S. EPA (Environmental Protection Agency). Besides re-evaluation of registered pesticides, priority in registration program has been given to "reduced-risk ***pesticides***", i.e. pesticides with reduced risk to human health, non-target organisms and environment as a replacement for older and potentially riskier chemicals. The list of reduced-risk pesticides includes several new acaricides and insecto-acaricides (EPA, 2009, 2010). In the European Union, implementation of Directive 91/414 that requires science-based assessment of pesticide risk to human health and the environment, has seriously impacted the EU acaricide portfolio. Nevertheless, new Regulation (EC) 1107/2009, revises the Directive and introduces hazard-based cut-off criteria, thus increasing the safety level (Balderacchi & Trevisan, 2010; van Leeuwen et al., 2010). When looking at the acaricides, from 103 substances, only 26 are currently included in a 'positive' list of compounds (Annex I) and the status of another four is pending (EU, 2010).

3. New synthetic acaricides

3.1 Acaricides acting on respiration targets

Similar to nervous system of insects, nervous system of mites has also long been the target for most chemicals used for their control (Casida & Quistad, 1998). The situation has somewhat changed during the last two decades due to commercialization of large number of acaricidal compounds acting on mitochondrial respiration process, that produces most of the energy in cells. This process includes two coupled parts: mitochondrial electron transport (MET) and oxidative phosphorylation. Although some of the older acaricides were known to inhibit respiration, the real exploitation of this target started no sooner than after the 1990s, with the prospects for expanding and developing new, more effective and safer products (Dekeyser, 2005; Lűmmen, 2007; Krämer & Schirmer, 2007).

Throughout the mitochondrial electron transport chain there are various potential sites for inhibition, but only three have been used so far as target sites of acaricidal activity, at transmembrane enzyme complexes. In the period 1991-93, four compounds from different chemical classes were successively commercialized: **fenpyroximate** (pyrazole), **pyridaben** (pyridazinone), **fenazaquin** (quinazoline) and **tebufenpyrad** (pyrazolecarboxamide) (Fig. 4), whose mode of action was inhibition of MET at complex I. These compounds, also known as METI acaricides, quickly gained the popularity worldwide owing to the high efficacy against both tetranychid and eriophyoid mites, quick knockdown effect and long-lasting impact. In addition, these substances have low to moderate mammalian toxicity and short to moderate environmental persistence (Dekeyser, 2005; Krämer & Schirmer, 2007, van Leewen et al., 2010). Fenpyroximate and tebufenpyrad are included in Annex I, while fenazaquin and pyridaben applications have been resubmitted for inclusion (EU, 2010). Fenpyroximate is also on the list of reduced risk and organophosphorus alternative pesticides (EPA, 2009). Complex I inhibitors also include **pyrimidifen** (pyrimidinamine), commercialized in 1995, as well as insecto-acaricide **tolfenpyrad**, another pyrazolecarboxamide, commercialized in 2002, and **flufenerim**, more recent derivative of pyrimidifen (Krämer & Schirmer, 2007).

The only known complex II inhibitor is the recently introduced insecto-acaricide **cyenopyrafen**, a compound from the acrylonitrile class of chemistry (Lűmmen, 2007). Complex III inhibition is mode of action of acequinocyl, fluacrypyrim and bifenazate. **Acequinocyl**, a naphthoquinone compound (Fig. 5) commercialized in 1999, is a pro-acaricide which is bioactivated via deacetylation. It is effective against all stages of spider mites, with low mammalian toxicity and short environmental persistence (Dekeyser, 2005). It is included in the EPA list of reduced risk pesticides, while the decision on its status under Directive 91/414 is pending (EPA, 2009; EU, 2010). **Bifenazate**, a carbazate compound (Fig. 5) is highly effective against immatures and adults of spider mites, with rapid knockdown effect (Ochiai et al., 2007). Although it was first considered to be a neurotoxin, more recent experimental results indicate complex III as target site (van Nieuwenhuyse et al., 2009). Bifenazate is a pro-acaricide which is bioactivated via hydrolysis of ester bonds, so the organophosphorous compounds, as inhibitors of esterase hydrolitic activity, can antagonize the toxicity of this acaricide (van Leeuwen et al., 2007). Bifenazate, introduced in 1999, is a compound of low mammalian toxicity and short environmental persistence; it is classified as a reduced risk and organophosphate alternative pesticide, and it is included in Annex I (Dekeyser, 2005; EPA, 2009; EU, 2010). **Fluacrypyrim**, introduced in 2002, shows acaricidal effect against all stages of tetranychids. This is the first strobilurin not commercialized as a fungicide (Dekeyser, 2005; Krämer & Schirmer, 2007), and more compounds with acaricidal effect from this group are anticipated (Li et al., 2010).

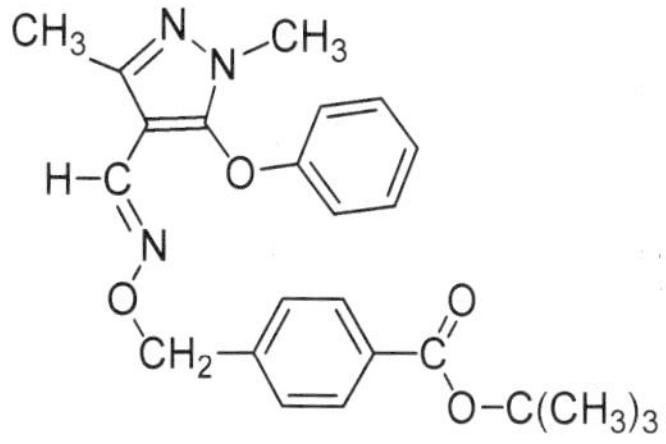

fenpyroximate

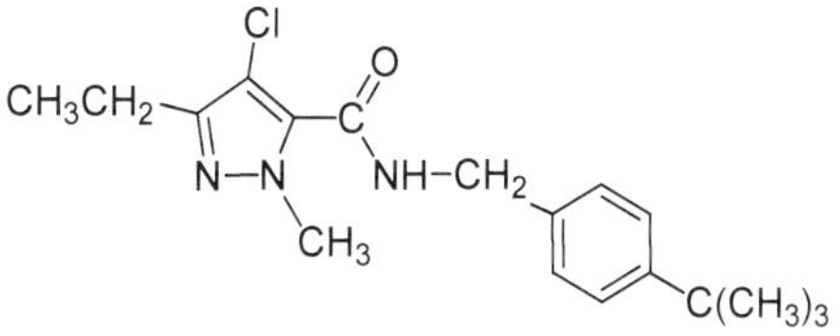

tebufenpyrad

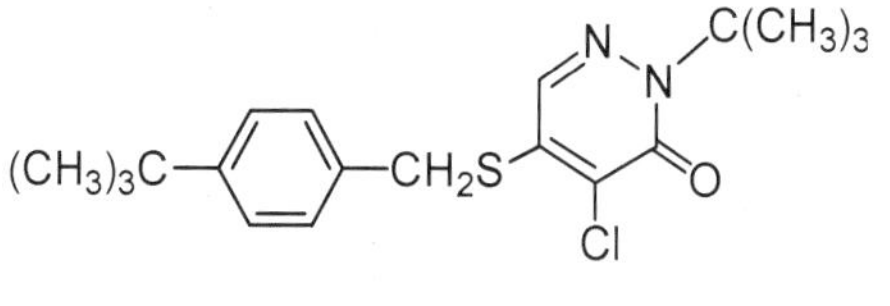

pyridaben

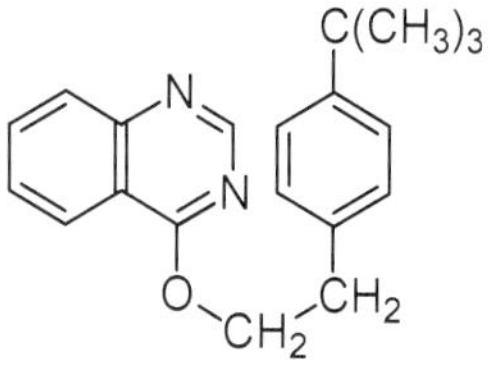

fenazaquin

Fig. 4. Structural formulas of acaricides acting on respiration targets: METI acaricides

Insecto-acaricide **diafenthiuron**, a novel thiourea compound (Fig. 5) launched in 1991, is the only modern representative of compounds that disrupt oxidative phosphorylation by inhibition of the mitochondrial ATP synthase, an enzyme with essential role in cellular bioenergetics (this mode of action has been recognized in propargites, tetradifons and organotin compounds). Diafenthiuron is a pro-acaricide, its carbodiimide metabolite inhibits the enzyme. It is effective against motile stages of spider mites and also provides good eriophyoid control. Diafenthiuron has low mammalian toxicity and short environmental persistence (Krämer & Schirmer, 2007; van Leewen et al., 2010).

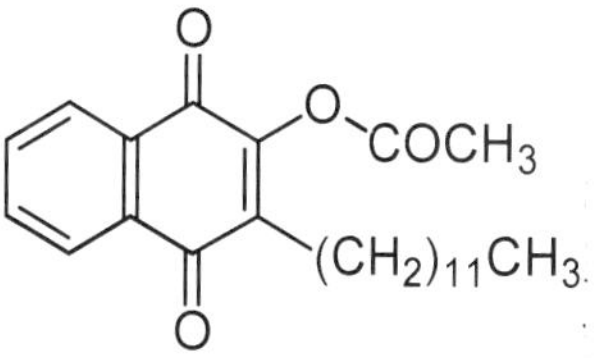

acequinocyl

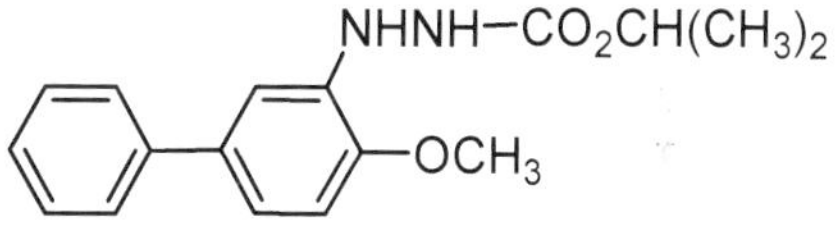

bifenazate

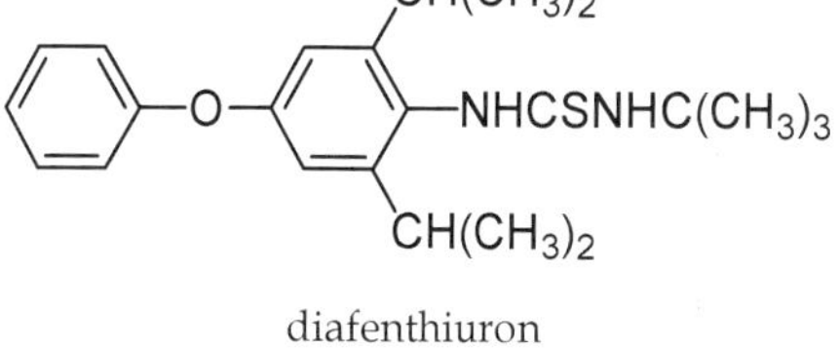

diafenthiuron

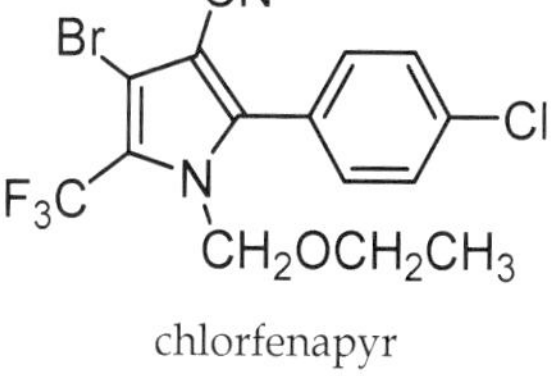

chlorfenapyr

Fig. 5. Structural formulas of acaricides acting on respiration targets: complex III inhibitors (acequinocyl, bifenazate) and inhibitors of oxidative phosphorylation (diafenthiuron, chlorfenapyr)

Another insecto-acaricide, **chlorfenapyr**, a pyrrole compound (Fig. 5) commercialized in 1995, at biochemical level acts as uncoupler of oxidative phosphorylation via disruption of the proton gradient. Chlorfenapyr is effective against all stages of spider mites and eriophyoid mites. This compound is a pro-acaricide activated by N-dealkylation. Chlorfenapyr is a compound of moderate mammalian toxicity, but long environmental persistence (Krämer & Schirmer, 2007; Van Leeuwen et al., 2010). It is included in the EPA list as an alternative to organophosphorus compounds (EPA, 2009).

3.2 Acaricides acting on growth and development targets

Another direction in research and development of synthetic acaricides is directed towards compounds affecting developmental processes. **Etoxazole,** a oxazoline compound (Fig. 6), is acaricide highly effective against eggs and immatures of spider mites, non-toxic to adults, but it considerably reduces fertility of treated females (Kim & Yoo, 2002; Dekeyser, 2005). This acaricide, launched in 1998, is usually classified among mite growth inhibitors, together with clofentezine and hexythiazox, older acaricides that cause similar symptoms (Marčić, 2003; Krämer & Schirmer, 2007), but whose exact mode of action is unknown. On the other hand, Nauen & Smagghe (2006) provided experimental evidence that etoxazole acts as a chitin synthesis inhibitor similar to benzoylureas. Etoxazole is on the EPA list of reduced risk and organophosphorus alternative pesticides (EPA, 2009).
Discovery of **spirodiclofen** and **spiromesifen,** tetronic acid derivatives (Fig. 6) launched in 2002-2004, broadened the biochemical diversity of acaricides by introducing a completely new mode of action. These compounds act as inhibitors of acetyl-CoA-carboxylase, a key enzyme in fatty acid biosynthesis. Spirodiclofen and spiromesifen are highly toxic to eggs and immatures of spider mites, while their effects on adult females are slower with fecundity and fertility reduction; their acaricidal effect is long-lasting and stable (Krämer &

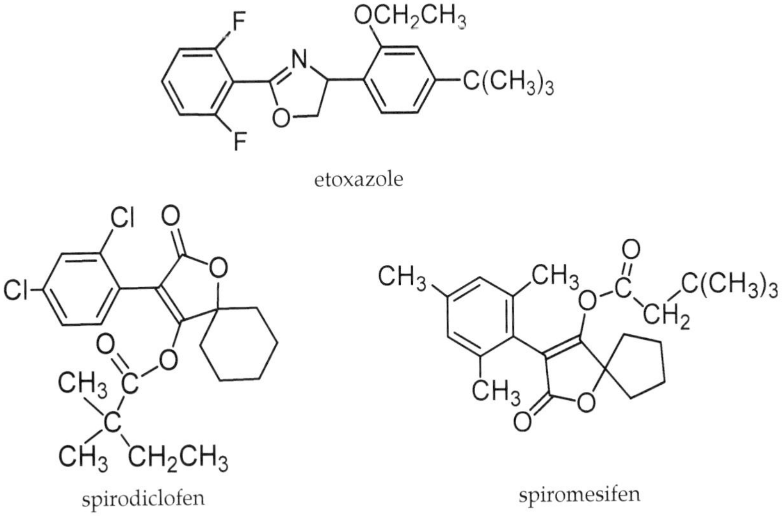

Fig. 6. Structural formulas of acaricides acting on growth and development

Schirmer, 2007; Marčić, 2007; Marčić et al., 2007; Van Pottelberge et al., 2009; Marčić et al., 2010). These two acaricides are the only new compounds used for control of eriophyoid mites as well (van Leeuwen et al., 2010). In addition to acaricidal effect, spirodiclofen has also shown considerable insecticidal activity against eggs and larvae of pear psylla and scales (Krämer & Schirmer, 2007; Marčić et al., 2007), while spiromesifen provides effective control of whiteflies (Krämer & Schirmer, 2007; Kontsedalov et al., 2008). Both compounds have low mammalian toxicity and short environmental persistence. Spirodiclofen is included in Annex I, and evaluation of spiromesifen is in progress (EU, 2010). **Spirotetramat**, a tetramic acid derivate recently introduced, belongs to inhibitors of acetyl-CoA-carboxylase. Although initially developed for control of whiteflies and aphids (Brück et al., 2009), the studies of its effects on *T. urticae* (Marčić et al., unpublished data) indicate that spirotetramat could potentially be an effective acaricide as well.

4. Natural acaricides and other alternative solutions

The use of natural products for plant and crop protection dates back to times long before the introduction of synthetic pesticides which imposed themselves as the main means for suppression of harmful organisms. In recent times, the significance of natural pesticides is constantly growing, primarily in organic agriculture, but also in the framework of biorational pest control programs which insist on use of environmentally-friendly pesticides and exploatation of novel biochemical modes of action (Isman, 2006; Isman & Akhtar, 2007; Copping & Duke, 2007; Horowitz et al., 2009). Some of the natural products are substances that have significant acaricidal effect.

Probably the most studied botanical insecticide in the last twenty years is a triterpenoid **azadirachtin** (Fig. 7), the major active ingredient of extracts, oils and other products derived from the seeds of the Indian neem tree (*Azadirachta indica*). Neem-products are registered in over 40 countries as products for suppression of arthropod pests important in growing of fruit, vegetables and ornamental plants (Kleeberg, 2004; Milenković et al., 2005). The effects of azadirachtin on treated insects manifest slowly and they include complete or partial antifeedant response, delayed and/or disrupted moulting, inhibited reproduction (Copping & Duke, 2007; Isman & Akhtar, 2007). The studies on spider mites (Sundaram & Sloane, 1995; Mansour et al., 1997; Martinez-Villar et al., 2005) indicate that azadirachtin, in addition to being toxic to various development stages, acts as antifeedant, reduces fecundity and fertility and shortens the life span of adult insects. Beside on spider mites, azadirachtin also exhibits acaricidal effect on some acarid and tarsonemid mites (Collins, 2006; Venzon et al., 2008). Azadirachtin is considered to be non-toxic to mammals and is not expected to have any adverse effects on the environment (Copping & Duke, 2007); its Annex I application is resubmitted (EU, 2010). Many neem/azadirachtin-based products are approved for use in organic crop production (Zehnder et al., 2007; EU, 2008; Dayan et al., 2009).

Products isolated from soil actinomycetes are an important source for deriving natural insecticides and acaricides. In early 1990s, several years after introduction of abamectin, another fermentation product, **milbemectin,** was commercialized. Milbemectin is a mixture of milbemycin A_3 and milbemycin A_4, natural products isolated from the fermentations of *Streptomyces hygroscopicus* subsp. *aureolacrimosus* (Fig. 7). Milbemectin is a neuroactive acaricide (chloride channel activator), effective against tetranychid and eriophyoid mites, relatively safe compound owing to the rapid uptake into treated plants combined with fast degradation of surface residues (Copping & Duke, 2007; Krämer & Schirmer, 2007). Like abamectin, milbemectin is also included in Annex I (EU, 2010).

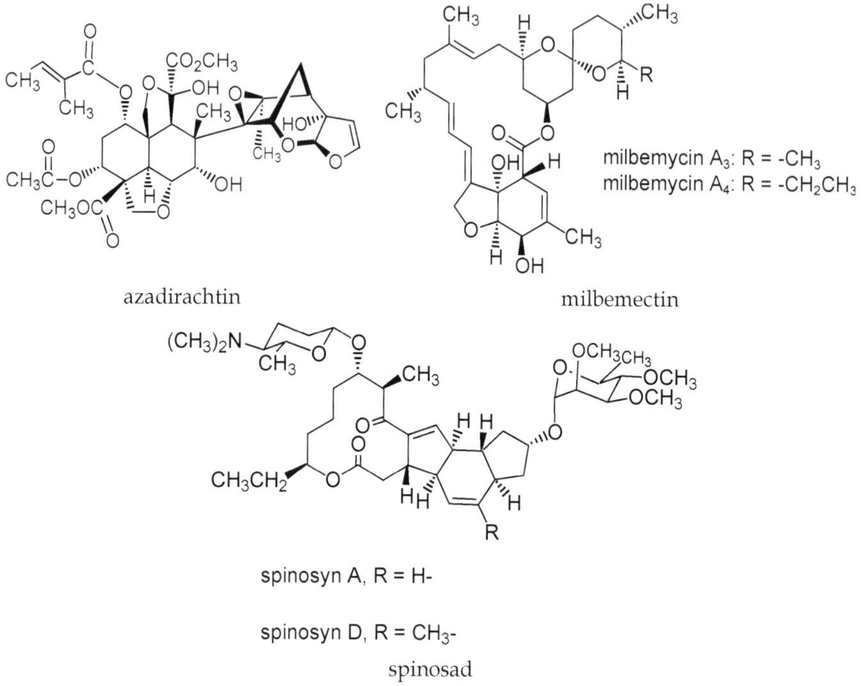

Fig. 7. Natural products with considerable acaricidal activity

The more recent example is **spinosad**, a mixture of spinosyn A and spinosyn D, secondary metabolites of *Saccharopolyspora spinosa*, (Fig. 7), introduced in 1997 as neuroactive insecticide, nicotinic acetylcholine receptor agonist (Copping & Duke, 2007; Krämer & Schirmer, 2007). This insecticide exerts significant acaricidal effect. Van Leeuwen et al. (2005) found out that the residual toxicity of spinosad to female *T. urticae* is equal to the level of toxicity resulting from application of dicofol, bromopropylate or fenbutatin oxide, while Villanueva & Walgenbach (2006) demonstrated that spinosad affects larvae and adults of this tetranychid, but relatively slowly, with the assumption that negative results of the previous testing of acaricidal properties were based on experiments that did not provide enough time for response. Spinosad shows systemic acaricidal effect, if used for substrate watering, such as rockwool, where the absorption level is reduced to the minimum (Van Leeuwen et al. (2005). This is a compound of very low mammalian toxicity and highly favourable environmental profile; it is included in the EPA list of organophosphorus alternative pesticides, and also in Annex I (EPA, 2009; EU, 2010). Spinosad is approved for use as an organic insecticide (EU, 2008; Dayan et al., 2009).

Essential oils, secondary metabolites abundant is some aromatic plants from families Lamiaceae, Apiaceae, Rutaceae, Myrtaceae and others, have been suggested as alternative sources for pest control products. Predominant bioactive ingredients of essential oils are monoterpenes and sesquiterpenes. Besides exerting acute toxicity to insects and mites, essential oils show sublethal effect as repellents, antifeedants and reproduction inhibitors. Lethal and sublethal effects of essential oils are the consequence of direct contact and/or uptake of gas-phase via respiratory system. Insect octopaminergic nervous system is considered to be the target site of action of some essential oil constituents, but this may not be the case considering their acaricidal activity. Moreover, there is a possibility that essential oils, as complex mixtures, act at multiple target sites (Isman, 2006; Miresmailli et al., 2006; Isman & Akhtar, 2007; Shaaya & Rafaeli, 2007). Essential oils extracted from caraway seeds, eucalyptus, mint, rosemary, basil, oregano, thyme, and other plants have shown a significant acaricidal activity (Aslan et al., 2004; Choi et al., 2004; Miresmailli et al., 2006). These oils could be useful as fumigants in the control of phytophagous mites in greenhouses; however, for improving acaricidal activity, their commercial formulations need to be developed (Choi et al., 2004; Han et al., 2010). Essential oils are mostly nontoxic to mammals; being volatile products, they have limited environmental persistence (Isman, 2006). Rosemary oil, thyme oil and some other essential oils are available for pest control in organic farming (Dayan et al., 2009).

Petroleum oils have been used for more than a century to control a wide range of crop pests, including spider mites. Because of their high phytotoxicity, the use of petroleum oils was limited to dormant or delayed dormant application against overwintering pest stages, to avoid injury to green plant tissue. Advances in petroleum chemistry considerably reduced phytotoxicity in newer, highly-refined petroleum-derived spray oils (PDSO), which are recognized today as an important alternative to synthetic pesticides. PDSO are environmentally-friendly products with negligible impact on human health and the environment. The most widely accepted theory on their mode of action is that PDSO primarily act physically by blocking the spiracles in insects (or the stigmata in mites) and thus causing suffocation, but it can not be presumed as the only mode of action (Taverner, 2002). At least some modern oils cause a range of cellular disruption leading to rapid insect death (Najar-Rodriguez et al., 2008). PDSO are highly effective against spider mites and eriophyoid mites in various field and greenhouse crops (Agnello et al., 1994; Nicetic et al., 2001; Marčić et al., 2009; Chueca et al., 2010). Beside mineral, **plant oils** proved to be effective acaricides as well, such as cottonseed oil (Rock & Crabtree, 1987) soybean oil (Lancaster et al., 2002; Moran et al., 2003) and rapeseed oil (Kiss et al., 1996; Marčić et al., 2009). PDSO and plant spray oils are considered compatible with organic farming (Zehnder et al., 2007; EU, 2008). Rapeseed oil is included in Annex I (EU, 2010).

Numerous studies indicate that entomopathogenic fungi, especially ascomycetes, can play an important role in regulation of harmful arthropod populations if used in biological control (Hajek & Delalibera, 2010), or applied as **mycoinsecticides** and/or **mycoacaricides** (Maniania et al., 2008; Jackson et al., 2010). Among the entomopathogenic fungi, the most potent pathogens of tetranychids and other pest mite species are *Beauveria basssiana, Hirsutella thompsonii, Lecanicillium* sp., *Metharizium anisopliae, Isaria fumosorosea, Neozygites floridana* (Chandler et al., 2000; Maniania et al., 2008), whose conidia and blastospores are used for formulation of fungal-based biopesticides. At the beginning of the 1980s, only one mycoacaricide was available (Mycar), formulated from conidia of *H. thompsoni* and intended for supression of citrus rust mite. Quarter of century later, there are some 30 commercial products acting against tetranychid, eriophyoid, and tarsonemid mites, mostly formulated as wettable powder or oil dispersion, and one third of which is made from conidia of *B. bassiana* (Faria & Wraight, 2007).

5. Acaricide resistance in phytophagous mites

As a result of exceptional intrinsic potential of mites for rapid development of resistance (Cranham & Helle, 1985; Croft & van de Baan, 1988; van Leeuwen et al., 2009) and often not so rational actions of humans, the acaricide resistance in mites, in particular the species from Tetranychidae family, has become a global phenomenon. Arthropod Pesticide Resistance Database (APRD) - managed by scientist from Michigan State University and supported by Insecticide Resistance Action Committee (IRAC), a specialist technical group of the industry association CropLife - contains published data on resistance in insects and mites important for agriculture, veterinary medicine and public health, from 1914 to date (Whalon et al., 2008, 2010). This database, which involves a large number of scientists and experts from around the world who work on its administration and upgrading, is useful for comprehension of acaricide resistance in mites on a global level.

In the mid-2010, APRD contained 9394 reports on resistance developed in 572 species of arthropods, of which 1130 reports refer to 82 species from Acari subclass. Out of this number, 745 reports concern 39 species belonging to four families of phytophagous mites: Tetranychidae, Acaridae, Eriophyidae and Tenuipalpidae. Approximately 93% of reports deal with the resistance of spider mites, with two predominant species two-spotted spider mite, *Tetranychus urticae* (53% of spider mite reports) and European red mite, *Panonychus ulmi* (26% of spider mite reports) (Tab. 1). The authors of the APRD created the list of the "top 20" resistant arthropod pests in the world, ranked by number of compounds with reported resistance. On this list, *T. urticae* and *P. ulmi* rank first and ninth, respectively, by data for 92 and 42 compounds for which the information about resistant populations exist (Whalon et al., 2008, 2010).

For both species, the majority of reports refer to resistance to organophosphates documented during the 1950s, 1960s, and 1970s. Together with carbamates, organophosphate compounds account nowadays for more than 35% of global insecticide market, so that the reports on resistant tetranychid populations/strains are still coming (Herron et al., 1998; Stumpf et al., 2001; Tsakaragkou et al., 2002; Kumral et al., 2009). The important part of the APRD database concerns pyrethroids resistance in tetranychids. Today, this class of compounds accounts for 20% of the market, but the increasing number of cases of resistance to bifenthrin and other pyrethroids has been registered in the recent past (Herron et al., 2001; Ay and Gürkan, 2005; Kumral et al., 2009; Tsakaragkou et al., 2009). As for other specific acaricides and insecto-acaricides, there is practically no active substance without documented cases of resistance, but there is an obvious difference in the scope of phenomenon between certain acaricides or groups of acaricides. For instance, global popularity of METI-acaricides contributed to relatively fast development of resistant spider mite populations in Japan, South Korea, Australia, Brazil, California and some European countries. On the other hand, there are only few reports on resistance to fenbutatin-oxide and other organotin compounds which have been used for four decades now (Stumpf & Nauen, 2001; Auger et al., 2004; van Leeuwen et al., 2009; Stavrinides et al., 2010).

Another phytophagous mite on the list of „top 20", bulb mite *Rhizoglyphus robini* (Acaridae), ranks 19th with 22 reports on resistance to almost exclusively organophosphate compounds. Cases of resistance to organophosphates are also registered for other species of acarids listed in the APRD. Among eriophyoid mites, resistance in citrus rust mite, *Phyllocoptruta oleivora*, to dicofol has been best documented and most studied (Omoto et al., 1994); other cases of resistance in Eriophyoidea also refer to organophosphorous compounds.

In addition to comprehensive documenting of acaricidal resistance in mites, the factors affecting this phenomenon of microevolution were also studied, as well as its physiological, biochemical ang genetic mechanisms. The results of these studies were summarized by Cranham & Helle (1985), Croft & van de Baan (1988), Messing & Croft (1996), Knowles

Mite species	No. of cases	No. of compounds
Tetranychus urticae	367	92
Panonychus ulmi	181	42
Panonychus citri	26	20
Tetranychus cinnabarinus	26	16
Tetranychus mcdanieli	19	13
Tetranychus kanzawai	12	12
Tetranychus viennensis	7	7
Tetranychus atlanticus	7	5
Tetranychus pacificus	7	5
Oligonychus pratensis	6	6
Tetranychus turkestani	5	5
Tetranychus hydrangaea	5	4
Tetranychus arabicus	3	3
Tetranychus crataegi	3	3
Tetranychus desertorum	3	3
Tetranychus ludeni	3	3
Tetranychus bimaculatus	2	2
Tetranychus cucurbitacearum	2	2
Tetranychus schoenei	2	2
Tetranychus tumidus	2	2
Eotetranychus hicoriae	1	1
Tetranychus althaeae	1	1
Tetranychus canadensis	1	1
Tetranychidae	691	
Rhizoglyphus robini	22	22
Rhizoglyphus echinopus	6	5
Acarus siro	4	3
Acarus chaetoxysilus	2	2
Acarus farris	1	1
Tyrophagus palmarum	1	1
Tyrophagus putrescentiae	1	1
Acaridae	37	
Phyllocoptruta oleivora	3	2
Aculus cornutus	3	3
Aculus pelekassi	3	3
Aculus fockeui	1	1
Aculus lycopersici	1	1
Aculus malivagrans	1	1
Aculus schlechtendali	1	1
Eriophyidae	13	
Brevipalpus chilensis	3	3
Brevipalpus phoenicis	1	1
Tenuipalpidae	4	
Total	745	

Tab. 1. Reported cases of acaricide resistance in phytophagous mites (Whalon et al., 2010)

(1997), van Leeuwen et al. (2009), and the largest number of data refers to populations/strains of *T. urticae*. As in other arthropods, the resistance in mites is caused by a less sensitive target site (**target site resistance**) and/or enhanced detoxification (**metabolic resistance**). The insensivity of acetylcholinesterase is the most common type of organophosphorous resistance in *T. urticae* (Cranham & Helle, 1985; Stumpf et al., 2001, Tsagkarakou et al., 2002; van Leeuwen et al., 2009; Khajehali et al., 2010; Kwon et al., 2010a). Metabolic resistance mediated by carboxylesterases was found in majority of cases of resistance development to pyrethroids in this species (Ay & Gűrkan, 2005; van Leeuwen et al., 2005b, van Leeuwen & Tirry, 2007), while the oxidative metabolism appears to play a major role in resistance to METI-acaricides (Stumpf & Nauen, 2001; Kim et al., 2004, 2006; van Pottelberge et al., 2009).

The results of numerous conventional genetic studies indicate that in most cases single major gene controls inheritance of resistance in spider mites (Cranham & Helle, 1985; van Leeuwen et al., 2009). Although the monogenic and dominant resistance has been expected due to intense selection pressure to which the populations under the open field or greenhouse conditions are exposed (Roush & McKenzie, 1987), some major exceptions occur, such as the monogenic-recessive resistance to dicofol (Rizzieri et al., 1988), propargite (Keena & Granett, 1990), pyridaben (Goka, 1998) and etoxazole (Uesugi et al., 2002), and poligenic resistance to cyhexatin (Mizutani et al., 1988). Lately, several studies dealing with molecular basis of the target site resistance to pyrethroids (Tsagkarakou et al., 2009; Kwon et al., 2010b), organophosphates (Khajehali et al., 2010; Kwon et al., 2010a) and bifenazate (van Leeuwen et al., 2008, van Nieuwenhuyse et al., 2009) have been published. Especially interesting discovery is that the bifenazate resistance in *T. urticae* is inherited only maternally, which is the first occurrence of non-Mendelian inheritance since the beginning of genetic studies on pesticide resistance in arthropods (van Leeuwen et al., 2008, van Nieuwenhuyse et al., 2009).

Biological, biochemical and genetic characterization of resistance is one of the essential elements in defining the strategy for management of acaricide resistance in phytophagous mites. An effective acaricide resistance management program could be based on general resistance management principles endorsed by IRAC (Krämer & Schirmer, 2007). The key recommendation is reduction of the selection for resistance which is possible to attain if there were available as many as possible acaricides with different modes of action. The history of resistance in *T. urticae* best illustrates the importance of the above: the first resistant populations can emerge as soon as after two or three years from the start of a new acaricide application, causing an obvious pest control failure (Cranham & Helle, 1985; Knowles, 1997; van Leeuwen et al., 2009).

In the European Union, the implementation of Directive 91/414 reduced the EU acaricide portfolio by more than 70% (EU, 2010; van Leeuwen et al., 2010). On the other hand, it is Directive 91/414 that requires pesticide registrants to address the risk of resistance development as part of dossiers submitted for EU registration (Thompson et al., 2008). In „Declaration of Ljubljana" (Bielza et al., 2008) a group of leading resistance management experts expressed strong concern that further loss of active ingredients resulting from the implementation of Directive 91/414 (and its revision) could endanger the sustainability of European farming, increasing the risk of developing resistance to the relatively few remaining substances. The scientists concluded that the resistance management requires access to a diversity of chemistries with different modes of action. Considering the fact that every year only few new active substances are registered in the EU, it is clear that the pesticide industry is unable to offer enough replacements for the products which are being withdrawn from the market (Thompson et al., 2008).

6. Acaricides and integrated control of phytophagous mites

Biological control of phytophagous mites by predatory mites (Phytoseiidae) and other predators proved to be a successful alternative to conventional chemical control, especially on greenhouse crops (Gerson & Weintraub, 2007). In spite of undoubtful advantages, biological control includes significant limitations as well (Gerson et al., 2003), which makes the use of acaricides still indispensable. In modern crop protection, these acaricides should be biorational compounds: highly effective against mite pests and relatively safe to their predators (i.e. selective), with low risk to human health and the environment. Biorational acaricides are important element of integrated control of phytophagous mites which is based on combination of chemical, biological and other control measures.

Therefore, it is very important to study the effects of acaricides and other pesticides on phytoseiid mites, other predatory mites and insect predators of phytophagous mites. Predators come into contact with pesticides if treated with them directly or exposed to their residues, if they feed on contaminated prey or pollen. Beside lethal effects (mortality), pesticides also cause a variety of sublethal effects, by changing the biological parameters and/or behaviour of survivors (Blűmel et al., 1999; Desneux et al., 2007). International Organization for Biological Control/Western Palearctic Regional Section (IOBC/WPRS) offered one of the most comprehensive programs to test lethal and sublethal side-effects of pesticides on beneficial organism, which comprise laboratory, semi-field and field trials. IOBC/WPRS working group „Pesticides and beneficial organisms" organized and carried out several joint testing programs for most of the predators, parazitoids and other beneficial organisms, including phytoseiid mites (Blűmel et al., 1999; Blűmel & Hausdorf, 2002). However, some methodological solutions within the IOBC procedures (way of exposure, choice of doses/concentrations, evaluation criteria) have been criticized as insufficiently realistic (Bakker & Jacas, 1995; Amano & Haseeb, 2001). In order to acquire an in-depth knowledge on sublethal effects of pesticides on biological control agents, the population level-toxicity approach was proposed; it is based on creation of life tables and calculation of population growth parameters, and/or projection of population growth rate based on matrix model (Stark & Banks, 2003; Stark et al., 2007).

The most frequently encountered on the lists of non-selective active substances are organochlorines, organophosphates, carbamates, pyrethroids and other broad-spectrum insecto-acaricides, which are *per definitionem* toxic to large number of insect and mite species, including Phytoseiidae and majority of other arthropods, predators of phytophagous mites (Croft & Brown, 1975; Knowles, 1997; Blűmel et al., 1999; Gerson et al., 2003). On the other side, abamectin and milbemectin, which are also broad-spectrum insecto-acaricides, are considered safe to beneficial arthropods under field conditions due to their short environmental persistance, rapid uptake into treated plants and fast degradation of surface residues (Krämer & Schirmer, 2007). Although beneficials may be killed when treated directly by spray oils or exposed to the vapor phase of essential oils, their short- term residual activity does not severely affect populations of phytoseiid mites and other predators (Chueca et al., 2010; Han et al., 2010).

It should be noted that certain fungicides (benomyl, dithocarbamates) are partly harmful to predatory mites (Blűmel et al., 1999; Gerson et al., 2003; Alston & Thomson, 2004). Sulfur, acaro-fungicide approved in organic farming (Zehnder et al., 2007; EU, 2008; Milenkovic et al. 2010) have been identified as disruptive to integrated mite control (Beers et al., 2009). Also, there are records of adverse effects of neonicotinoids (new class of neuroactive insecticides which is in great expansion in the last two decades), on survival and/or fecundity (James, 2003; Duso et al., 2008), predator activity (Poletti et al., 2007) and population growth (Stavrinides & Mills, 2009) of phytoseiid mites.
Specific acaricides are considered harmless to majority of predatory insects, while their toxicity to various development stages of the same mite species, and to different mite species, varies to a certain extent. From the standpoint of selectivity, it is essential to be aware of the comparative toxicity of acaricides to phytophagous mites and predatory mites (Knowles, 1997). Compounds, such as organotins, mite growth inhibitors and regulators, acequinocyl, diafenthiuron, some METI acaricides, bifenazate, spirodiclofen, spiromesifen, are usually graded as selective acaricides, much more toxic to phytophagous mites than to phytoseiid and other predatory mites (Blűmel et al., 1999; Knowles, 1976, Dekeyser, 2005; Krämer & Schirmer, 2007). Spinosad and azadirachtin appear to be compatible with predatory mites (Spollen & Isman, 1996; Williams et al., 2003; Raguraman et al., 2004).
Both positive and negative evaluation results are based on smaller or larger number of experimental data, but they should not be taken as general and final conclusions on (non)selectivity. Besides expected intrinsic differences among predatory species in susceptibility to the same pesticide, the literature provides different, and sometimes even contrasting results on compatibility for the same active substance and the same predatory species, due to different test procedures (applied doses/concentrations, way of treatment and exposure of test organisms, observed parameters, laboratory or field experiments); on the other hand, the results obtained by standardized methods are affected by the product formulation type, origin of test organism (autochthonous population or commercialized strain) and other factors (Blűmel et al., 1993; Duso et al., 2008).
Physiological selectivity, i.e. reduced susceptibility due to pesticide metabolism is the most desired testing result of pesticide effects to phytophagous mite predators. But, the non-selective compounds can be made safer for use by special application technology (Blűmel et al., 1999; van Leeuwen et al., 2005a), by reducing the doses/concentrations (Rhodes et al., 2006), by releasing the predators so that they would be exposed to older residues (Lilly & Campbell, 1999), by using the strain of predators with developed resistance to acaricides and other pesticides (Sato et al., 2007).
Application of selective acaricides (synthetic or natural) with releases of commercialized strains of phytoseiid mites and other predators is a sustainable alternative to an approach based on chemical measures only (Lilly & Campbell, 1999; Rhodes et al., 2006; Sato et al., 2007). According to Kogan (1998), a pest control program reaches the level at which it can be qualified as an integrated pest management (IPM) program only when biorational pesticides (acaricides) and release of predators are integrated with other control tactics, preventive and remedial (crop rotation, host plant resistance, cultural practices, mechanical and physical control measures etc). Higher IPM levels entail transfer from species/population level integration (the control of single species or species complexes), via community level integration (multiple pest categories e.g. insects, mites, pathogens, weeds and their control) to ecosystem level integration (the control of multiple pest impacts within the context of the total cropping system).

World-wide, IPM has become the accepted model for crop protection over the past decades, but the adoption of IPM programs has been generally slow in both the developed and the developing countries (Peshin et al., 2009). In the European Union, Directive 91/414 encourages Member States to take the principles of IPM into account, but the implementattion is voluntary (Freier & Boller, 2009). Success of IPM has often been measured by the reduction in pesticide usage, which is not necessarily a reliable indicator (Kogan, 1998). Transition from conventional pest control to IPM actually changes the role of pesticides (acaricides) in modern crop protection: within the principles of IPM, pesticides are applied highly rationally and in interaction with other control tactics.

7. References

Alston, D.G. & Thomson, S.V. (2004). Effects of fungicide residues on the survival, fecundity and predation of the mites *Tetranychus urticae* (Acari: Tetranychidae) and *Galendromus occidentalis* (Acari: Phytoseiidae). *Journal of Economic Entomology*, 97, 3, 950-956, ISSN 0022-1493

Amano, H & Haseeb, M. (2001). Recently-proposed methods anc concepts of testing the effects of pesticides on the beneficial mite and insect species: study limitations and implications in IPM. *Applied Entomology and Zoology*, 36, 1, 1-11, ISSN 0003-6862

Agnello, A.; Reissig, W.H. & Harris, T. (1994). Management of summer populations of European red mite (Acari: Tetranychidae) on apple with horticultural oil. *Journal of Economic Entomology*, 87, 1 148-161, ISSN 0022-1493

Aslan, I.; Özbek, H.; Çalmaşur, Ö. & Şahin, F. (2004). Toxicity of essential oil vapours to two greenhouse pests, *Tetranychus urticae* Koch and *Bemisia tabaci* Genn. *Industrial Crops and Products*, 19, 2, 167-173, ISSN 0296-6690

Auger, P.; Bonafos, R.; Guichou, S. & Kreiter, S. (2004). Resistance to fenazaquin and tebufenpyrad in *Panonychus ulmi* Koch (Acari: Tetranychidae) populations from South of France apple orchards. *Crop Protection*, 22, 8, 1039-1044, ISSN 0261-2194

Ay, R. & Gűrkan, M.O. (2005). Resistance to bifenthrin and resistance mechanisms of different strains of the two-spotted spider mite (*Tetranychus urticae*) from Turkey. *Phytoparasitica*, 33, br, 237-244, ISSN 0334-2123

Bakker, F.M. & Jacas, J.A. (1995). Pesticides and phytoseiid mites - strategies for risk assessment. *Ecotoxicology and Environmental Safety*, 32, 1, 58-67, ISSN 0147-6513

Balderacchi, M. & Trevisan, M. (2010). Comments on pesticide risk assessment by the revision of Directive EU 91/414. *Environmental Science and Polllution Research*, 17, 3, 523-528, ISSN 0944-1344

Beers, E.H.; Martinez-Rocha, L.; Talley, R.R. & Dunley, J.E. (2009). Lethal, sublethal, and behavioral effects of sulfur-containing products in bioassays of three species of orchard mites. *Journal of Economic Entomology*, 102, 1, 324-335, ISSN 0022-1493

Bielza, P.; Denholm, I.; Ioannidis, P.; Sterk, G.; Leadbeater, A.; Leonard, P. & Jørgensen, L.N. (2008). Declaration of Ljubljana - the impact of a declining European pesticide Portfolio on resistance management. *Outlooks on Pest Management*, 19, 6, 246-248, ISSN 1743-1026

Blümel, S. & Hausdorf, H. (2002). Results of the 8th and 9th IOBC joint pesticide testing programme: Persistance test with *Phytoseiulus persimilis* Athias Henriot (Acari: Phytoseiidae). *IOBC/WPRS Bulletin* 25, 11 43-51, ISSN 92-9067-148-X

Blümel, S.; Baker, F. & Grove, A. (1993). Evaluation of different methods to assess the side-effects of pesticides on *Phytoseiulus persimilis* A.-H. *Experimental and Applied Acarology*, 17, 3, 161-169, ISSN 1572-9702

Blümel, S.; Matthews, G.A.; Grinstein, A. & Elad, Y.(1999). Pesticides in IPM: selectivity, side-effects, application and resistance problems. In: *Integrated Pest and Disease Management in Greenhouse Crops,* Albajes, R.; Gullino, M.A.; van Lenteren, J.C. & Elad, Y. (ed.), 150-167, Kluwer Academic Publishers, ISBN 0-7923-5631-4, Dordrecht, the Netherlands

Brück, E.; Elbert, A.; Fischer, R.; Krueger, S.; Kühnhold, J.; Klueken A.M.; Nauen, R.; Niebes, J.F.; Reckman, U.; Schnorbach, J.J.; Steffens, R. & van Waetermeulen, X. (2009). Movento®, an innovative ambimobile insecticide for sucking insect pest control in agriculture: biological profile and field performance. *Crop Protection*, 28, 10, 838-844, ISSN 0261-2194

Casida, J.E. & Quistad, G.B. (1998). Golden age of insectiicde research: past, present, or future? *Annual Review of Entomology*, 43, 1-16, ISSN 0066-4170

Chandler, D.; Davidson, G.; Pell, J.K.; Ball, B.V.; Shaw, K. & Sunderland, K.D. (2000). Fungal biocontrol of Acari. *Biocontrol Science and Technology*, 10, 3, 357-384, ISSN 0958-3157

Chueca, P.; Garcera, C.; Molto, E.; Jacas, J.A.; Urbaneja, A. & Pina, T. (2010). Spray deposition and efficacy of four petroleum-derived oils used against *Tetranychus urticae*. *Journal of Economic Entomology*, 103, 2, 386-393, ISSN 0022-1493

Choi, W.I.; Lee, S.G.; Park, H.M. & Ahn, Y.J. (2004). Toxicity of plant essential oils to *Tetranychus urticae* (Acari: Tetranychidae) and *Phytoseiulus persimilis* (Acari: Phytoseiidae). *Journal of Economic Entomology*, 97, 2, 553-558, ISSN 0022-1493

Collins, D.A. (2006). A review of alternatives to organophosphorus compounds for the control of storage mites. *Journal of Stored Products Research*, 42, 4, 395-426, ISSN 0022-474X

Copping, L.G. & Duke, S.O. (2007). Natural products that have been used commercially as crop protection agents - a review. *Pest Management Science*, 63, 6, 524-554, ISSN 1526-498X

Cranham, J.E. & Helle W. (1985). Pesticide Resistance in Tetranychidae. In: *Spider Mites: Their Biology, Natural Enemies and Control*, Vol. 1B, Helle, W. & Sabelis, M.W. (ed.), 405-421, Elsevier, ISBN 0-444-42374-5, Amsterdam, the Netherlands

Croft, B.A. & van de Baan, H.E. (1988). Ecological and genetic factors influencing evolution of pesticide resistance in tetranychid and phytoseiid mites. *Experimental and Applied Acarology*, 4, 3, 277-300, ISSN 1572-9702

Croft, B.A. & Brown, A.W.A. (1975). Responses of arthropod natural enemies to insecticides. *Annual Review of Entomology*, 20, 285-335, ISSN 0066-4170

Dayan, F.E.; Cantrell, C.L. & Duke, S.O. (2009). Natural products in crop protection. *Bioorganic and Medicinal Chemistry*, 17, 12, 4022-4034, ISSN 0968-0896

Dekeyser, M.A. (2005). Acaricide mode of action. *Pest Management Science*, 61, 2, 103-110, ISSN 1526-498X

Desneux, N.; Decourtye, A. & Delpuech, J.M. (2007). The sublethal effects of pesticides on beneficial arthropods. *Annual Review of Entomology*, 52, 81-106, ISSN 0066-4170

Duso, C.; Malagnini, V.; Pozzebon, A.; Castagnoli, M.; Liguori, M. & Simoni, S. (2008). Comparative toxicity of botanical and reduced-risk insecticides to Mediterranean populations of *Tetranychus urticae* and *Phytoseiulus persimilis* (Acari: Tetranychidae, Phytoseiidae). *Biological Control* 47, 1, 16-21, ISSN 1049-9644

Dutcher, J.D. (2007). A review of resurgence and replacement causing pest outbreaks in IPM, In: *General Concepts in Integrated Pest and Disease Management*, Ciancio, A. & Mukerji, K.G. (ed.), 27-43, Springer, ISBN 978-1-4020-6060-1, Dordrecht, the Netherlands

EPA (2009). Reduced Risk/Organophosphate Alternative Decisions for Conventional Pesticides. www.epa.gov/opprd001/workplan/completionsportrait.pdf

EPA (2010). Implementation of Requirements under the FQPA. www.epa.gov/pesticides/regulating/laws/fqpa/fqpa_implementation.htm (Accessed July 31, 2010)

EU (2008). Comission Regulation (EC) No 889/2008. *Official Journal of the European Union*, 51, L 250, ISSN 1725-2555

EU (2010). EU Pesticide Database: http://ec.europa.eu/sanco_pesticides/public/index.cfm (Accessed July 31, 2010)

Faria, M.R. & Wraight, S.P. (2007). Mycoinsecticides and mycoacaricides: a comprehensive list with worldwide coverage and international classification of formulation types. *Biological Control*, 43, 3, 237-256, ISSN 1049-9644

Freier, B. & Boller, E.F. (2009). Integrated Pest Management in Europe - History, Policy, Achievements and Implementattion. In: *Integrated Pest Management: Dissemination and Impact*, Peshin, R. & Dhawan, A.K. (ed.), 435- 454, Springer Science+Business Media B.V. ISBN 978-1-4020-8989-3, Dordrecht, the Netherlands

Gerson, U. & Cohen, E. (1989). Resurgences of spider mites (Acari: Tetranychidae) induced by synthetic pyretrhroids. *Experimental and Applied Acarology*, 6, 1, 29-46, ISSN 1572-9702

Gerson, U.; Smiley, R.L. & Ochoa, R. (2003). Mites (Acari) for Pest Control. Blackwell Science Ltd., ISBN 978-0-632-05658-3, Oxford, UK

Gerson, U. & Weintraub, P.G. (2007). Mites for the control of pests in protected cultivation. *Pest Management Science*, 63, 7, 658-676, ISSN 1526-498X

Goka, K. (1998). Mode of inheritance of resistance to three new acaricides in the Kanzawa spider mite, *Tetranychus kanzawai* Kishida (Acari: Tetranychidae). *Experimental and Applied Acarology*, 22, 12, 699-708, ISSN 1572-9702

Hajek, A.E. & Delalibera Jr., I. (2010). Fungal pathogens as classical biological control agents against arthropods. *BioControl*, 55, 1, 147-158, ISSN 1386-6141

Han, J.; Choi, B.R.; Lee, S.G.; Kim, S.I. & Ahn, Y.J. (2010). Toxicity of plant essential oils to acaricide-susceptible and resistant *Tetranychus urticae* (Acari: Tetranychidae) and *Neoseiulus californicus* /Acari: Phytoseiidae). *Journal of Economic Entomology*, 103, 4, 1293-1298, ISSN 0022-0493

Hardin, M.R.; Benrey, B.; Coll, M.; Lamp, W.O.; Roderick, G.K. & Barbosa, P. (1995). Arthropod pest resurgence: an overview of potential mechanisms. *Crop Protection*, 14, 1, 3-18, ISSN 0261-2194

Helle, W. & Sabelis, M.W. (1985a). *Spider Mites: Their Biology, Natural Enemies and Control, Vol. 1A*, Elsevier, ISBN 0-444-42372-9, Amsterdam, the Netherlands

Helle, W. & Sabelis, M.W. (1985b). *Spider Mites: Their Biology, Natural Enemies and Control, Vol. 1B*, Elsevier, ISBN 0-444-42374-5, Amsterdam, the Netherlands

Herron, G.A.; Edge, V.E.; Wilson, L.J. & Rophail, J. (1998). Organophosphate resistance in spider mites (Acari: Tetranychidae) from cotton in Australia. *Experimental and Applied Acarology*, 22, 1, 17-30, ISSN 1572-9702

Herron, G.A.; Rophail, J. & Wilson, L.J. (2001). The development of bifenthrin resistance in two-spotted spider mite (Acari: Tetranychidae) from Australina cotton. *Experimental and Applied Acarology*, 25, 4, 301-310, ISSN 1572-9702

Horowitz, A.R.; Ellsworth, P.C. & Ishaaya, I. (2009). Biorational Pest Control - An Overview, In: *Biorational Control of Arthropod Pests*, Ishaaya, I. & Horowitz, A.R. (ed.), 1-20, Springer, ISBN 978-90-481-2315-5, Dordrecht, the Netherlands

Huffaker, C.B.; van de Vrie, M. & McMurtry, J.A. (1970). Ecology of tetranychid mites and their natural enemies: a review. II. Tetranychid populations and their possible control by predators: an evaluation. *Hilgardia*, 40, 391-458, ISSN 0073-2230

Ishaaya, I. (2001). Biochemical Sites of Insecticide Action and Resistance. Springer-Verlag, ISBN 3-540-67625-2, Berlin, Heidelberg and New York

Isman, M.B. (2006). Botanical insecticides, deterrents, and repellents in modern agriculture and an increasingly regulated world. *Annual Review of Entomology*, 51, 45-66, ISSN 0066-4170

Isman, M.B. & Akhtar, Y. (2007). Plant Natural Products as a Source for Developing Environmentaly Acceptable Insecticides. In: *Insecticide Design Using Advanced Technologies*, Ishaaya I.; Nauen R. & Horowitz R. (ed.), 235-248, Springer-Verlag, ISBN 10: 3-540-46904-4, Berlin, Heidelberg

Jackson, M.A.; Dunlap, C.A. & Jaronski, S.T. (2010). Ecological considerations in producing and formulating fungal entomopathogens for use in insect biocontrol. *BioControl*, 55, 1, 129-145, ISSN 1386-6141

James, D.G. (2003). Toxicity of imidacloprid to *Galendromus occidentalis*, *Neoseiulus fallacis* and *Amblyseius andersoni* (Acari: Phytoseiidae) from hops in Washington State, USA. *Experimental and Applied Acarology*, 31, 3-4, 275-281, ISSN 1572-9702

Jeppson, L.R.; Keifer, H.H. & Baker, E.W. (1975). *Mites Injurious to Economic Plants*, University of California Press, ISBN 0-520-02381-1, Berkeley and Los Angeles.

Keena, M.A. & Granett, J. (1990). Genetic analysis of propargite resistance in Pacific spider mites and twospotted spider mites (Acari: Tetranychidae). *Journal of Economic Entomology*, 83, 3, 655-661, ISSN 0022-1493

Kim, Y.J.; Lee, S.H.; Lee, S.W. & Ahn, Y.J. (2004). Fenpyroximate resistance in *Tetranychus urticae* (Acari: Tetranychidae): cross resistance and biochemical resistance mechanisms. *Pest Management Science*, 60, 10, 1001-1006, 1526-498X

Kim, Y.J.; Park, H.M.; Cho, J.R. & Ahn, Y.J. (2006). Multiple resistance and biochemical mechanisms of pyridaben resistance in *Tetranychus urticae* (Acari: Tetranychidae). *Journal of Economic Entomology*, 99, 3, 954-958, ISSN 0022-1493

Kim, S.S. & Yoo, S. (2002). Comparative toxicity of some acaricides to the predatory mite, *Phytoseiulus persimilis* and the twospotted spider mite, *Tetranychus urticae*. *BioControl*, 47, 5, 563-573, ISSN 1386-6141

Kiss, J.; Szendrey, L.; Schlosser, E. & Kotlar, I. (1996). Application of natural oil in IPM of grapevine with special regard to predatory mites. *Journal of Environmental Science and Health, Part B*, 31, 5, 421-425, ISSN 0360-1234

Khajehali, J.; van Leeuwen, T.; Grispou, M.; Morou, E.; Alout, H.; Weill, M.; Tirry, L.; Vontas, J. & Tsagkarakou, A. (2010), Acetylcholinesterase point mutations in European strains of *Tetranychus urticae* (Acari: Tetranychidae) resistant to organophosphates. *Pest Management Science*, 66, 2, 220-228, 1526-498X

Kleeberg, H. (2004). Neem based products - registration requirements, regulatory processes and global implications. In: *Neem, Today and in The New Millenium*, Koul O. & Wahab S. (ed.), 109-123, Kluwer Academic Publishers, ISBN-10: 1402012292, Dordrecht, the Netherlands

Knowles, C.O. (1976). Chemistry and toxicology of quinoxaline, organotin, organofluorine, and formamidine acaricides. *Environmental Health Perspectives*, 14, 93-102, ISSN 0091-6765

Knowles, C.O. (1997). Mechamisms of resistance to acaricides, In: *Molecular Mechanisms of Resistance to Agrochemicals*, Sjut, V. & Butters, J.A. (ed.), 58-78, Springer Verlag, ISBN 3-540-62461-9, Berlin - Heidelberg

Kogan, M. (1998). Integrated pest management - historical perspectives and contemporary developments. *Annual Review of Entomology*, 43, 243-270, ISSN 0066-4170

Kontsedalov, S.; Gottlieb, Y.; Ishaaya, I.; Nauen, R.; Horowitz, R. & Ghanim, M. (2008). Toxicity of spiromesifen to the developmental stages of *Bemisia tabaci* biotype B. *Pest Management Science*, 65, 1, 5-13, ISSN 1526-498X

Krämer, W. & Schirmer, U. (2007). *Modern Crop Protection Compounds, Vol. 3,* WILEY - VCH Verlag GmbH & Co., ISBN 978-3-527-31496-6, Weinheim, Germany

Kumral, N.A.; Susurluk, H.; Gencer N.S. & Gűrkan, M.O. (2009). Resistance to chlorpyrifos and lambda- cyhalothrin along with detoxifying enzyme activities in field collected female populations of European red mite. *Phytoparasitica*, 37, 1, 1-15, ISSN 0334-2123

Kwon, D.H.; Im, J.S.; Ahn, J.J.; Lee, J.H.; Clark, J.M. & Lee, S.H. (2010a). Acetylcholinesterase point mutations putatively associated with monocrotophos resistance in the two-spotted spider mite. *Pesticide Biochemistry and Physiology*, 96, 1, 36-42, ISSN 0048-3575

Kwon, D.H.; Clark, J.M. & Lee, S.H. (2010b). Cloning of a sodimu channel gene and identification of mutations putatively associated with fenpropathrin resistance in *Tetranychus urticae. Pesticide Biochemistry and Physiology*, 97, 2, 93-100, ISSN 0048-3575

Lancaster, A.L.; Deyton, D.E.; Sams, C.E.; Cummins, J.C.; Pless, C.D. & Fare, D.C. (2002). Soybean oil controls two-spotted spider mites on burning bush. *Journal of Environmental Horticulture*, 20, 2, 86-92, ISSN 0738-2898

Li, M.; Liu, C.L.; Li, L.; Yang, H.; Li, Z.N.; Zhang, H. & Li, Z.M. (2010). Design, synthesis and biological activities of new strobilurin derrivatives containing substituted pyrazoles. *Pest Management Science*, 66, 1, 107-112, ISSN 1526-498X

Lilly, R. & Campbell, C.A.M. (1999). Biological, chamical and integrated control of two-spotted spider mite *Tetranychus urticae* on dwarf hops. *Biocontrol Science and Technology*, 9, 4, 467-473, ISSN 0958-3157

Lindquist, E.E.; Sabelis, M.W. & Bruin, J. (1996). *Eriophyoid Mites: Their Biology, Natural Enemies and Control*, Elsevier, ISBN 0-444-88628-1, Amsterdam, the Netherlands

Lűmmen, P. (2007). Mitochondrial electron transport complexes as biochemical target sites for insecticides and acaricides, In: *Insecticide Design Using Advanced Technologies,* Ishaaya, I.; Nauen, R. & Horowitz, R. (ed.), 197-215, Springer-Verlag, ISBN 10: 3-540-46904-4, Berlin - Heidelberg

Maniania, N.K.; Bugeme, D.M.; Wekesa, V.W.; Delalibera, I.Jr. & Knapp, M. (2008). Role of entomopathogenic fungi in the control of *Tetranychus evansi* and *Tetranychus urticae* (Acari:Tetranychidae), pests of horticultural crops. *Experimental and Applied Acarology*, 46, 1-4, 259-274, ISSN 1572-9702

Mansour, F.A.; Ascher, K.R.S. & Abo-Moch, F. (1997). Effects of Neemgard on phytophagous and predacious mites and on spiders. *Phytoparasitica*, 25, 4, 333-336, ISSN 0334-2123

Marčić, D. (2003). The effects of clofentezine on life-table parameters in two-spotted spider mite *Tetranychus urticae*. *Experimental and Applied Acarology*, 30, 3, 249-263, ISSN 1572-9702

Marčić, D. (2007). Sublethal effects of spirodiclofen on life history and life-table parameters of two-spotted spider mite (*Tetranychus urticae*). *Experimental and Applied Acarology*, 42, 2, 121-129, ISSN 1572-9702

Marčić, D.; Ogurlić, I.; Mutavdžić, S. & Perić, P. (2010). The effects of spiromesifen on life history traits and population growth of two-spotted spider mite (Acari: Tetranychidae). *Experimental and Applied Acarology*, 50, 3, 255-267, ISSN 1572-9702

Marčić, D.; Perić, P.; Prijović, M.; Ogurlić, I. & Andrić, G. (2007). Effectiveness of spirodiclofen in the control of European red mite (*Panonychus ulmi*) on apple and pear psylla (*Cacopsylla pyri*). *Pesticides and Phytomedicine*, 22, 3, 301-309, ISSN 1820-3949

Marčić, D.; Perić, P.; Prijović, M. & Ogurlić, I. (2009). Field and greenhouse evaluation of rapeseed spray oil against spider mites (Acari: Tetranychidae), green peach aphid (Homoptera: Aphididae) and pear psylla (Hemiptera: Psyllidae) in Serbia. *Bulletin of Insectology*, 62, 2, 159-167, ISSN 1721-8861.

March, R.B. (1976). Properties and actions of bridged diphenyl acaricides. *Environmental Health Perspectives*, 14, 83-91, ISSN 0091-6765

Martinez-Villar, E.; Saenz-de-Cabezon, F.J.; Moreno-Grijalba, F.; Marco, V. & Perez-Moreno, I. (2005). Effects of azadirachtin on the two-spotted spider mite, *Tetranychus urticae* (Acari: Tetranychidae). *Experimental and Applied Acarology*, 35, 3, 215-222, ISSN 1572-9702

McMurtry, J.A.; Huffaker, C.B. & van de Vrie, M. (1970). Ecology of tetranychid mites and their natural enemies: a review. I. Tetranychid enemies, their biological characters and the impact of spray practices. *Hilgardia*, 40, 331-370, ISSN 0073-2230

Messing, R.H. & Croft, B.A. (1996). Pesticide Resistance in Eriophyoid Mites, Their Competitors and Predators. In: *Eriophyoid Mites - Their Biology, Natural Enemies and Control*, Lindquist, E.E.; Sabelis, M.W. & Bruin, J. (ed.), 695-726, Elsevier, ISBN 0-444-88628-1, Amsterdam, the Netherlands

Metcalf, R.L. (1980). Changing role of insecticides in crop protection. *Annual Review of Entomology*, 25, 219-256, ISSN 0066-4170

Milenković, S.; Tanasković, S. & Lazić, T. (2005). [Azadirachtin - the possibilities of use in plant protection]. *Voćarstvo*, 39, 1, 61-69, ISSN 1820-5054 (in Serbian)

Milenković, S.; Tanasković, S. & Sretenović, D. (2010). Organic raspberry production in Serbia, Proceedings of VII International Conference on Integrated Fruit Production, (IOBC/WPRS Bulletin, Vol. 54), pp. 413-416, ISBN 978-92-9067-213-5, Avignon, France, October 2008, IOBC/WPRS

Miresmailli, S.; Bradbury, R. & Isman, M.B. (2006). Comparative toxicity of *Rosmarinus officinalis* L. essential oil and blends of its major constituents against *Tetranychus urticae* Koch (Acari: Tetranychidae) on two different host plants. *Pest Management Science*, 62, 4, 366-371, ISSN 1526-498X

Mizutani, A.; Kumayama, F.; Ohra, K.; Ishiguro, T. & Hayashi, Y. (1988). Inheritance of resistance to cyhexatin in the Kanzawa spider mite, *Tetranychus kanzawai* Kishida (Acari: Tetranychidae). *Applied Entomology and Zoology*, 23, 3, 251-255, ISSN 0003-6862

Moran, R.E.; Deyton, D.E.; Sams, C.E.; Pless, C.D. & Cummins, J.C. (2003). Soybean oil as a summer spray for apple: European red mite control, net CO_2 assimilation and phytotoxicity. *HortEcience*, 38, 2, 234-238, ISSN 0018-5345

Najar-Rodriguez, A.J.; Lavidis, N.A.; Mensah, R.K.; Choy, P.T. & Walter, G.H. (2008). The toxicological effects of petroleum spray oils on insects - evidence for an alternative mode of action and possible new control options. *Food and Chemical Toxicology*, 46, 9, 3003-3014, ISSN 0278-6915

Nauen, R. & Smagghe, G. (2006). Mode of action of etoxazole. *Pest Management Science*, 62, 5, 379-382, ISSN 1526-498X

Nicetic, O.; Watson, D.M.; Beattie, G.A.C.; Meats, A. & Zheng, J. (2001). Integrated pest management of two-spotted mite *Tetranychus urticae* on greenhouse roses using petroleum spray oil and the predatory mite *Phytoseiulus persimilis*. *Experimental and Applied Acarology*, 25, 1, 37-53, ISSN 1572-9702

Ochiai, N.; Mizuno, M.; Mimori, N.; Miyake, T.; Dekeyser, M.; Canlas, L.J. & Takeda, M. (2007). Toxicity of bifenazate and its principal active metabolite, diazene, to *Tetranychus urticae* and *Panonychus citri* and their relative toxicity to the predaceous mites, *Phytoseiulus persimilis* and *Neoseiulus californicus*. *Experimental and Applied Acarology*, 43, 3, 181-197, ISSN 1572-9702

Omoto, C.; Dennehy, T.J.; McCoy, C.W.; Crane, S.E. & Long, J.W. (1994). Detection and characterisation of the interpopulation variation of citrus rust mite resistance to dicofol in Florida citrus. *Journal of Economic Entomology*, 87, 3, 566-572, ISSN 0022-1493

Peshin, R.; Bandral, R.S.; Zhang, W.J.; Wilson, L. & Dhawan, A. (2009). Integrated Pest Management: a Global Overview of History, Programs and Adoption. In: *Integrated Pest Management: Innnovation-Development Process* , Peshin, R & Dhawan, A.K. (ed.), 1-49, , Springer, ISBN 978-1-4020-8991-6, Dordrecht, the Netherlands

Poletti, M.; Maia, A.H.N. & Omoto, C. (2007). Toxicity of neonicotinoid insecticides to *Neoseiulus californicus* and *Phytoseiulus macropilis* (Acari: Phytoseiidae) and their impact on functional response to *Tetranychus urticae* (Acari: Tetranychidae). *Biological Control* 40, 1, 30-36, ISSN 1049-9644

Raguraman, S.; Ganapathy, N. & Venkatesan, T. (2004). Neem versus entomopathogens and natural enemies of crop pests: the potential impact and strategies. In: *Neem, Today and in The New Millenium*, Koul O. & Wahab S. (ed.), 109-123, Kluwer Academic Publishers, ISBN-10: 1402012292, Dordrecht, the Netherlands

Rhodes, E.M.; Liburd, O.E.; Kelts, C.; Rondon, S.I. & Francis, R.R. (2006). Comparison of single and combination treatments of *Phytoseiulus persimilis, Neoseiulus californicus* and Acramite (bifenazate) fro control of two-spotted spider mites in strawberries. *Experimental and Applied Acarology*, 39, 3-4, 213-225, ISSN 1572-9702

Rizzieri, D.A.; Dennehy, T.J. & Glover, T.J. (1988). Genetic analysis of dicofol resistance in two populations of twospotted spider mite (Acari: Tetranychidae) from New York apple orchards. *Journal of Economic Entomology*, 81, 5, 1271-1276, ISSN 0022-1493

Rock, G.C. & Crabtree, K.W. (1987), Biological activity of petroleum and cottonseed oils against two tetranychid mite species and tortricid insect species found on apple. *Journal of Agricultural Entomology*, 4, 3, 247-253, ISSN 0735-939X

Roush, R.T. & McKenzie, J.A. (1987). Ecological genetics of insecticide and acaricide resistance. *Annual Review of Entomology*, 32: 361-380, ISSN 0066-4170

Sato, M.E.; da Silva, M.Z.; de Souza Filho, M.F.; Matioli, A.L. & Raga, A. (2007). Management of *Tetranychus urticae* (Acari: Tetranychidae) in strawberry fields with *Neoseiulus californicus* (Acari: Phytoseiidae) and acaricides. *Experimental and Applied Acarology*, 42, 2, 107-120, ISSN 1572-9702

Shaaya, E. & Rafaeli, A. (2007). Essential Oils as Biorational Insecticides - Potency and Mode of Action. In: *Insecticide Design Using Advanced Technologies*, Ishaaya, I.; Nauen, R. & Horowitz, R. (ed.), 249-261, Springer, ISBN 10: 3-540-46904-4, Berlin - Heidelberg

Spollen, K.M. & Isman, M.B. (1996). Acute and sublethal effects of a neem insecticide on the commercial biological control agents *Phytoseiulus persimilis* and *Amblyseius cucumeris* (Acari: Phytoseiidae) and *Aphidoletes aphidimyza* (Diptera: Cecidomyiidae). *Journal of Economic Entomology*, 89, 6, 1379-1386, ISSN 0022-1493

Stark, J.D. & Banks, J.E. (2003). Population-level effects of pesticides and other toxicants on arthropods. *Annual Review of Entomology*, 48, 505-519, ISSN 0066-4170

Stark, J.D.; Sugayama, R.L. & Kovaleski, A. (2007). Why demographic and modeling approaches should be adopted for estimating the effects of pesticides on biocontrol agents. *BioControl*, 52, 3, 365-374, ISSN 1386-6141

Stavrinides, M.C. & Mills, N.J. (2009). Demographic effects of pesticides on biological control of Pacific spider mite (*Tetranychus pacificus*) by the western predatory mite (*Galendromus occidentalis*). *Biological Control*, 48, 3, 267-273, ISSN 1049-9644

Stavrinides, M.C.; van Niewenhuyse, P.; van Leeuwen, T. & Mills, N.J. (2010). Development of acaricide resistance in Pacific spider mite (*Tetranychus pacificus*) from California vineyards. *Experimental and Applied Acarology*, 50, 3, 243-254, ISSN 1572-9702

Stumpf, N. & Nauen, R. (2001). Cross-resistance, inheritance and biochemistry of mitochondrial electron transport inhibitor-acaricide resistance in *Tetranychus urticae* (Acari: Tetranychidae). *Journal of Economic Entomology*, 94, 6, 1577-1583, ISSN 0022-1493

Stumpf, N.; Zebitz, C.P.W.; Kraus, W.; Moores, G. & Nauen, R. (2001). Resistance to organophosphates and biochemical genotyping of acetylcholinesterases in *Tetranychus urticae* (Acari: Tetranychidae). *Pesticide Biochemistry and Physiology*, 69, 2, 131-142, ISSN 0048-3575

Sundaram, K.M.S. & Sloane, L. (1995). Effects of pure and formulated azadirachtin, a neem-based biopesticide, on the phytophagous spider mite, *Tetranychus urticae* Koch. *Journal of Environmental Science and Health (B)*, 30, 6, 801-814, ISSN 0360-1234

Taverner, P. (2002). Drowning or just waving? A perspective on the ways petroleum-derrved oils kill arthropod pests of plants. In: *Spray Oils beyond 2000 - Sustainable Pests and Diseases Management*, Beattie, G.A.C.; Watson, D.M.; Stevens, M.L.; Rae, D.J. & Spooner-Hart, R.N. (ed.)., 78-88, University of Western Sydney, ISBN 0-7341-0205-4, Sydney.

Thompson, G.D.; Matten, S.; Denholm, I.; Whalon, M.E. & Leonard, P. (2008). The Politics of Resistance Management: Working Towards Pesticide Resistance Management Globally. In: *Global Pesticide Resistance in Arthropods*, Whalon, M.E.; Mota-Sanchez,

D. & Hollingworth, R.M. (ed.), 146-165, CAB International, ISBN 978-1-84593-353-1, Wallingford, UK

Tsagkarakou, A.; Navajas, M.; Cuany, A.; Chevillon, C. & Pasteur, N. (2002). Mechanisms of resistance to organophosphates in *Tetranychus urticae* (Acari: Tetranychidae) from Greece. *Insect Biochemistry and Molecular Biology*, 32, 4, 417-424, ISSN 0965-1748

Tsagkarakou, A.; van Leeuwen, T.; Khajehali, J.; Ilias, A.; Grispou, M.; Williamson, M.S.; Tirry, L. & Vontas, J. (2009). Identification of pyrethroid resistance mutations in the *para* sodium channel of the twospotted spider mite *Tetranychus urticae* (Acari: Tetranychidae). *Insect Molecular Biology*, 18, 5, 583-593, ISSN 0962-1075

Uesugi, R.; Goka, K. & Osakabe M. (2002). Genetic basis of resistance to chlorfenapyr and etoxazole in the two-spotted spider mite (Acari: Tetranychidae). *Journal of Economic Entomology*, 95, 6, 1267-1274, ISSN 0022-1493

van Leeuwen, T. & Tirry. L. (2007). Esterase-mediated bifenthrin resistance in a multi-resistant strain of the two-spotted spider mite, *Tetranychus urticae*. *Pest Management Science*, 63, 2, 150-156, ISSN 1526-498X

van Leeuwen, T.; Dermauw, W.; van de Veire, M. & Tirry, L. (2005a). Systemic use of spinosad to control the two-spotted spider mite (Acari: Tetranychidae) on tomatoes grown in rockwool. *Experimental and Applied Acarology*, 37, 1, 93-105, ISSN 1572-9702

van Leeuwen, T.; van Pottelberge, S. & Tirry. L. (2005b). Comparative acaricide susceptibility and detoxifying enzyme activities in field-collected resistant and susceptible strains of *Tetranychus urticae*. *Pest Management Science*, 61, 5, 499-507, 1526-498X

van Leeuwen, T.; Vanholme, B.; van Pottelberge, S.; van Nieuwenhuyse, P.; Nauen, R.; Tirry, L. & Denholm, I. (2008). Mitochondrial heteroplasmy and the evolution of insecticide resistance: non-Mendelian inheritance in action. *Proceedings of the National Academy of Science*, 105, 16, 5980-5985, ISSN 0027-8424

van Leeuwen, T.; van Pottelberge, S. & Tirry, L. (2007). Organophosphate insecticides and acaricides antagonise bifenazate toxicity through esterase inhibition in *Tetranychus urticae*. *Pest Management Science*, 63, 12, 1172-1177, ISSN 1526-498X

van Leeuwen, T.; Vontas, J.; Tsagkarakou, A. & Tirry. L. (2009). Mechanisms of Acaricide Resistance in the Two-Spotted Spider Mite *Tetranychus urticae*, In: *Biorational Control of Arthropod Pests*, Ishaaya, I. & Horowitz, A.R. (ed.), 347-393, Springer, ISBN 978-90-481-2315-5, Dordrecht, the Netherlands

van Leeuwen, T.; Witters, J.; Nauen, R.; Duso, C. and Tirry, L. (2010). The control of eriophyoid mites: state of the art and future challenges. *Experimental and Applied Acarology*, 51, 1-3, 205-224, ISSN 1572-9702

van Nieuwenhuyse, P.; van Leeuwen, T.; Khajehali, J.; Vanholme, B. & Tirry, L. (2009). Mutations in the mitochondrial cytochrome b of *Tetranychus urticae* Koch (Acari: Tetranychidae) confer cross-resistance between bifenazate and acequinocyl. *Pest Management Science*, 65,4, 404-412, ISSN 1526-498X

van Pottelberge, S.; van Leeuwen, T.; Nauen, R. & Tirry, L. (2009). Resistance mechanisms to mitochondrial electron transport inhibitors in a field-collected strain of *Tetranychus urticae* Koch (Acari: Tetranychidae). *Bulletin of Entomological Research*, 99, 1, 23-31, ISSN 0007-4853

Venzon, M.; Rosado, M.C.; Molina-Rugama, A.J.; Duarte, V.S.; Dias, R. & Pallini, A. (2008). Acaricidal efficacy of neem against *Polyphagotarsonemus latus* (Banks) (Acari: Tarsonemidae). *Crop Protection*, 27, 3-5, 869-872, ISSN 0261-2194

Villanueva, R.T. & Walgenbach, J.F. (2006). Acaricidal properties of spinosad against *Tetranychus urticae* and *Panonychus ulmi* (Acari: Tetranychidae). *Journal of Economic Entomology*, 99, 3, 843-849, ISSN 0022-1493

Whalon, M.E.; Mota-Sanchez, D. & Hollingworth, R.M. (2008). *Global Pesticide Resistance in Arthropods*, CAB International, ISBN 978-1-84593-353-1, Wallingford, UK

Whalon, M.E.; Mota-Sanchez, D.; Hollingworth, R.M. & Duynslager, L. (2010). *Arthropod Pesticide Resistance Database*. www.pesticideresistance.com (Accessed: July 21, 2010)

Williams, T.; Valle, J. & Viñuela, E. (2003). Is the naturally derived insecticide spinosad compatible with insect natural enemies? *Biocontrol Science and Technology*, 13, 5, 459-475, ISSN 0958-3157

Zehnder, G.; Gurr, G.M.; Kũhne, S.; Wade, M.R.; Wartten, S.D. & Wyas, E. (2007). Arthropod pest management in organic crops. *Annual Review of Entomology*, 52, 57-80 (supplemental material), ISSN 0066-4170

Zhang, Z.O. (2003). *Mites of Greenhouses, Identification, Biology and Control*, CAB International, ISBN 0-85199-590-X, Wallingford, UK

4

Formula Optimization Design of Pesticide Microemulsion

Zhao Feng and Wu Yaqian
Jiangxi Key Laboratory of Organic Chemistry, Jiangxi Science & Technology Normal University Nanchang
China

1. Introduction

Water-based pesticide microemulsions have become one of the most potential pesticide formulation instead of the conventional pesticide formulation in recent years, which indeed have been a good insecticidal performance and a minimal impact on the environment. But the commonly organic solvents and cosurfactants such as toluene, xylene and methanol were widely applied in the procession of preparing the formulation. They are detrimental to environment and health of human beings all the same (Knowles., 2008). Therefore, controversy over the nocuous additives added excessively, really did put the production and application of pesticide microemulsion into trouble. But the traditional pesticide formulation can't be compared with the pesticide microemulsion based on excellent properties of microemulsion in some respects. We should prohibit the addition of toxic substances into microemulsion, rather than forbid the pesticide microemulsion formulation itself. The development of pesticide microemulsions need providing more equitable space and rational platform.

The formulation of pesticide microemulsion is an inexact science and eludes prediction for the most part, and largely dependent upon specific and incompletely understood interaction between the molecules of oil, emulsifiers, and water. So that, studies of pesticide microemulsions are limited to the combination of various components(Pratap, A. P.& Bhowmick, D. N.,2008; Narayanan, K. S., 1994; Narayanan, K. S., 1994; Jon, D. I. et al,1999). A considerable number of studies are processes of trial and error, similar to the proverbial "needle in a haystack". Very few works have studied the methodology of pesticide microemulsion preparation (Skelton, P. R. et al, 1988; Hiromoto, B., 2007). Therefore, investigations on the formula design of pesticide microemulsions are worthwhile for more detailed theoretical and technological studies.

In this paper, we reported our methods based on the pseudo-ternary phase diagram and orthogonal design. Although, the pseudo-ternary phase method has been employed to preparation of microemulsion, the combination method of the pseudo-ternary phase diagram and the orthogonal design has rarely been reported. The nearly non-toxic solvent and cosurfactants were employed to prepare pesticide microemulsion, and the optimization formulation of green environmentally friendly chlorpyrifos microemulsion was achieved. The experimental results were expected to provide new ideas and technical methods for the development of microemulsion.

2. Experimental section

2.1 Materials

Chlorpyrifos pesticide (96%) was provided by Jiangxi Agricultural University Plant Protection Chemical industry Co., Ltd. China; Nonionic surfactant (agricultural emulsifier.600) was obtained from local market in jiangxi province, China; Various grease compounds (natural carboxylate) was made present by Shandong University in China; Ethyl acetate, ethanol and ethylene glycol, analytical grade, were purchased from Tianjing Chemistry factory, China; Distilled water was used throughout the experiments.

2.2 Construction of pseudo-ternary phase diagram

The pseudo-ternary phase diagram can be drafted with the water titration method as described in many previous papers (Watnasirichaiku S. et al,2000; Chen H B. et al,2004; Zhang, Q. Z. et al,2004; Boonme, P. et al,2006). Chlorpyrifos pesticide was dissolved in Ethyl acetate as an oil phase (O); natural carboxylate, agricultural emulsifier.600 and mixed alcohol were blended at certain weight ratio to obtain the surfactant mixture (S), then mixture at the given weight ratios as an surfactant phase, the mixture of the oil phase and the surfactant phase at the weight ratios of 1:9, 2:8, 3:7, 4:6, 5:5, 6:4, 7:3, 8:2, and 9:1 were diluted with distilled water which remarked water phase (W) next by the microburst, at various concentrations, respectively. The dosage of water which made the mixture from the clarification to muddle was recorded, and the quality score of each phase point was analyzed. Phase diagrams were drawn based on visual inspection at ambient temperature.

2.3 Arrangement of orthogonal experiment

A standard orthogonal array matrix (L25; 3^5) was constructed with three factors and five levels (Table 1) to select optimum formation conditions in order to obtain the infinite dilute region(M) of pesticide microemulsion in the phase diagrams. The content of pesticide in solvent, the weight ratios of natural carboxylate/AE600 and mixed alcohol quality/S phase quality in total were selected as three impact factors.

Levels	Factors		
	Content of pesticide in solvent A (w/w)	Natural carboxylate/AE600 B (w/w)	Mixed alcohol C (w/w)
1	A_1 (50%)	B_1 (1:1)	C_1 (95%)
2	A_1 (40%)	B_2 (1:2)	C_2 (90%)
3	A_1 (30%)	B_3 (2:1)	C_3 (85%)
4	A_2 (20%)	B_1 (1:1)	C_2 (85%)
5	A_2 (10%)	B_2 (1:2)	C_3 (85%)

Table 1. A standard L25 (3^5) matrix

3. Results and discussion

3.1 Selection of formula components

Firstly, ester is a sort of lowly toxic matter, which has less impact on the environment. The experimental results show that the chlorpyrifos pesticide can be effectively solubilized in ethyl esters analogy to the most solubility of chlorpyrifos pesticide in toluene or xylene.

There was more 90%(weight ratios) chlorpyrifos pesticide to be able to be dissolved in the ethyl esters. So the ethyl ester was regarded as the substitute of tolunene or xylene. Secondly, the nonylphenols and alkylphenol phenoxy poly(ethyleneoxy)ethanol have been employed comprehensively in the commonly microemulsion so far. But they had even been limited to production and application with endangering the ecological environment in many countries. Then the nature carboxylate and AE600 at the weight ratios of 1:3, 1:2, 1:1, 2:1, and 3:1 were mixed and prepared to reach the demands of experiment. And the natural carboxylate is a kind of cheap, efficiency and safe anionic surfactant, because it derived from the oil scraps in green plants. Finally, the glycol and ethanol were regarded as the additives in the procession of preparing the microemulsion and their effectivity were predominant. All those could try to do minimum damage to the environment.

3.2 Choice of the best phase diagram

L_{25} (3^5) orthogonal table and orthogonal experimental data were listed in Table 2. Due to the pesticide microemulsion should be diluted with 200-fold water, the infinite dilute region's area (M) had been chosen as criterion in the pseudo-ternary phase diagram.
Table 2 shows that the order of the three factors' effect on the infinite dilution area with water is $R_A > R_B > R_C$ and the optimal conditions were found to be $A_5B_1C_5$ with the maximum infinite dilution area with water. To confirm the result of the orthogonal experiments,the pseudo-ternary phase diagram determined by $A_5B_1C_5$ (10% original drug phase content in the ethyl esters, natural carboxylate/AE600 = 1:3, Mixed alcohol quality/S phase quality in total = 85%) was illustrated in Fig.1. However, it is considered that the content of organic phase was only 10% and the content of chlorpyrifos also was 3%, the another optimal conditions was the $A_4B_2C_4$(20% original drug phase content, natural carboxylate/agricultural emulsifier.600 = 1:2, Mixed alcohol quality/S phase quality in total = 85%) (Fig.2) which was reasonable as compared with the area of the infinte dilution region (M=30.012) in the 25 group phase diagrams. The most significant reason is that the content of chlorpyrifos pesticide can reach 4%.

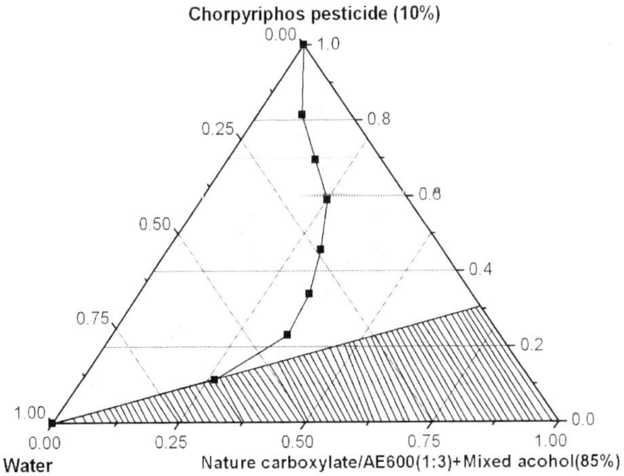

Fig. 1. Pesudo-ternary phase diagram defined by $A_5B_1C_5$[

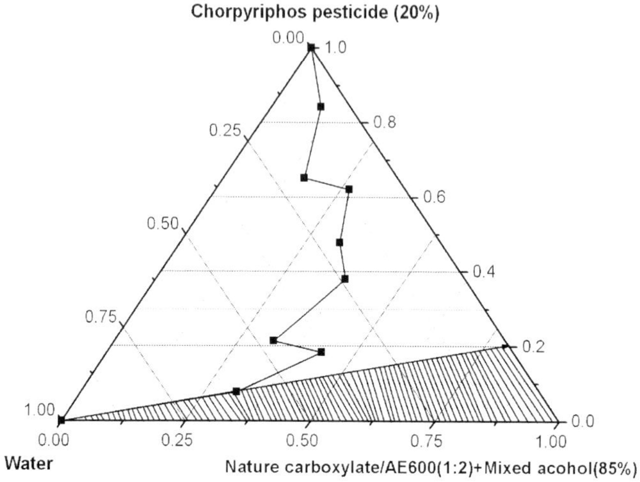

Fig. 2. Pesudo-ternary phase diagram defined by $A_4B_2C_4$

Level	Content of pesticide in solvent	Natural carboxylate/AE600	Mixed alcohol	Level	Content of pesticide in solvent	Natural carboxylate/AE600	Mixed alcohol
1	A1 (50%)	B1 (1:3)	C1 (95%)	14	A3 (30%)	B4 (2:1)	C5 (85%)
2	A1 (50%)	B2 (1:2)	C2 (90%)	15	A3 (30%)	B5 (3:1)	C1 (95%)
3	A1 (50%)	B3 (1:1)	C3 (85%)	16	A4 (20%)	B1 (1:3)	C5 (85%)
4	A1 (50%)	B4 (2:1)	C4 (85%)	17	A4 (20%)	B2 (1:2)	C1 (95%)
5	A1 (50%)	B5 (3:1)	C5 (85%)	18	A4 (20%)	B3 (1:1)	C2 (90%)
6	A2 (40%)	B1 (1:3)	C4 (85%)	19	A4 (20%)	B4 (2:1)	C3 (85%)
7	A2 (40%)	B2 (1:2)	C5 (85%)	20	A4 (20%)	B5 (3:1)	C4 (85%)
8	A2 (40%)	B3 (1:1)	C1 (95%)	21	A5 (10%)	B1 (1:3)	C3 (85%)
9	A2 (40%)	B4 (2:1)	C2 (90%)	22	A5 (10%)	B2 (1:2)	C4 (85%)
10	A2 (40%)	B5 (3:1)	C3 (85%)	23	A5 (10%)	B3 (1:1)	C5 (85%)
11	A3 (30%)	B1 (1:3)	C2 (90%)	24	A5 (10%)	B4 (2:1)	C1 (95%)
12	A3 (30%)	B2 (1:2)	C3 (85%)	25	A5 (10%)	B5 (3:1)	C2 (90%)
13	A3 (30%)	B3 (1:1)	C4 (85%)				
K_1	10.174	16.426	12.544				
K_2	10.022	14.118	10.096				
K_3	11.948	11.674	13.604				
K_4	14.408	10.01	12.024				
K_5	17.906	12.25	14.332				
R	7.884	6.416	5.236				

Table 2. Three factors and five levels orthogonal table and the analysis of experimental results

label	Surfactant phase /%	Oil phase /%	Water phase /%	Surfactant phase/Water phase(m/m)	Exterior	Dispersion into the water	cold storage	Heat storage
1	40.53	9.95	49.52	0.82	Transparent	Filamentous material sank, then mixture became transparent after shaking	Demixing in the 2nd day	Muddy
2	37.32	7.42	55.26	0.68	Transparent	Ibid	Demixing in the 3rd day	Muddy
3	35.15	5.03	59.82	0.59	Transparent	Ibid	Demixing in the 6th day	Muddy
4	32.27	2.41	65.31	0.49	Transparent	Ibid	Demixing in the 3rd day	Muddy
5	42.57	7.42	50.01	0.85	Transparent	Ibid	Demixing in the 2nd day	Transparent
6	44.84	5.07	50.08	0.90	Transparent	Ibid	Demixing in the 6th day	Transparent
7	51.48	2.76	45.77	1.12	Transparent	Ibid	Demixing in the 5th day	Transparent
8	48.02	11.96	40.02	1.20	Transparent	Ibid	Demixing in the 4th day	Transparent
9	48.24	10.10	41.65	1.16	Transparent	Ibid	Demixing in the 4th day	Transparent
10	48.46	7.55	44.00	1.10	Transparent	Ibid	Demixing in the 3rd day	Transparent
11	47.72	5.21	47.07	1.01	Transparent	Ibid	Demixing in the 2nd day	Transparent
12	49.63	2.61	47.76	1.04	Transparent	Ibid	Demixing in the 3rd day	Transparent
13	49.72	10.18	40.09	1.24	Transparent	Ibid	Stable over 7 days	Transparent
14	52.45	7.42	40.13	1.31	Transparent	Ibid	Stable over 7 days	Transparent
15	54.89	5.07	40.04	1.37	Transparent	Ibid	Stable over 7 days	Transparent
16	57.49	2.53	39.98	1.44	Transparent	Ibid	Demixing in the 2nd day	Transparent

Table 3. The composition of the points in the microemulsion and the result of stability text

3.3 The research of physical stability

The 16 points were selected in the best phase diagram($A_4B_2C_4$) and according to concentrations of the 16 points, the 16 experimental samples were prepared and then each experimental sample was divided into two parts. The one was examined in heat storage condition [(55 ± 2) degrees, 15d], and the other was examined in cold storage condition [(0 ± 1) degree, 7d]. Experimental results were listed in Table 3.

It is showed that the thirteenth, fourteenth and fifteenth group was raised above the sixteen groups experiment after heat treatment and cold storage in Table 3. Because the consumption of the surfactant was the lowest in the 13th group, and the concentration of the chorpyriphos pesticide content was the highest in the 13th group. So the composition of the

13th group was the best formula, which was listed as follows: chlorpyrifos (2.036%), ethyl acetate (8.154%), agricultural emulsifier.600(4.972%), natural carboxylate(2.486%, glycol (21.131%), ethanol (21.131%), water (40.090%).

4. Conclusions

The organic solvent, surfactant and kinds of additives were researched in the procession of preparing the chlorpyrifos microemulsion based on the method of pseudo-ternary phase diagram and orthogonal experiment in detail. The more green environment-friendly and extremely development potential chlorpyrifos microemulsion formulation was obtained. The optimized formula was composed of chlorpyrifos (2.036%), ethyl acetate (8.154%), agricultural emulsifier 600(4.972%), natural carboxylate(2.486%), glycol (21.131%), ethanol (21.131%), water (40.090%). Experimental results supported the foundation of the development in search of more green environmental protection chlorpyrifos microemulsion, and had a strong practical value on the principal of green environmental protection of chlorpyrifos microemulsion.

5. References

Knowles, A. (2008). Recent developments of safer formulations of agrochemicals. *Environmentalist*, 28., 1., (March 2008) 35-44, 0251-1088

Pratap, A. P. & Bhowmick, D. N. (2008). Pesticides as microemulsion formulations. Journal of Dispersion Science and Technology, 29., 9., (September 2008) 1325-1330, 0193-2691.

Narayanan, K. S. (1994). Water-based microemulsion formulations. *United States Patent*, 1994, Patent No: 5317042.

Narayanan, K. S. (1994). Water-based microemulsions of a triazole fungicide. *UnitedStates Patent*, (1994), Patent No: 5326789.

Jon, D. I.; Prettypaul, D. I.; Benning, M. J.; Narayanan, K. S. & Ianniello, R. M. (1999). Water-dilutable, microemulsion concentrate and pour-on formulations thereof. *UnitedStates Patent*, (1999), Patent No: 5968990.

Skelton, P. R.; Munk, B. H. & Collins, H. M. (1988). Formulation of pesticide In *Pesticide formulations and application systems*, Hovde D. A., Beestman.G..B. (Ed.), 36-45, American Society for Testing and Materials, 0-8031-0979-2, Philadelphia.

Hiromoto, B. (2007). Pesticide microemulsions and dispersant/penetrant formulations. *United States Patent*, (2007), Patent No: US 7297351.

Watnasirichaikus, S.; Davies, N. M.; Rades, T, *et al.* (2000). Preparation of Biodegradable Insulin Nanocapsules from Biocompatible Microemulsions. *Pharmaceutical research*, 17., 6.,(June 2000) 684-689, 0724-8741.

Chen, H. B.; Chang, X. L.; Weng, T. *et al.*(2004) .A study of Microemulsion Systems for Transdermal Delivery of Triptolide. *J Control Release*, 98., 3.,(August 2004) 427-436, 0168-3659.

Zhang, Q. Z.; Jiang, X. G.; Jiang, W. M.; Lu, W.; Su, L. N. & Shi, Z.Q. (2004). Preparation of nimodipine-loaded microemulsion for intranasal delivery and evaluation on the targeting efficiency to the brain. *International journal of pharmaceutics*, 275., 2., (May 2004) 85-96,0378-5173.

Boonme, P.; Krauel, K.; Graf, A.; Rades, T. & Junyaprasert, V. B. (2006). Characterization of microemulsion structures in the pseudoternary phase diagram of isopropyl palmitate/water/Brij 97:1-butanol. *AAPS PharmSciTech*,7., 2.,(March 2006) E99-E104, 1530-9932.

5

Pesticide Mixtures

Dr. Raymond A. Cloyd
Professor and Extension Specialist in Ornamental Entomology/Integrated Pest Management
Kansas State University, Department of Entomology, 123 Waters Hall, Manhattan, KS 66506-4004 USA

1. Introduction

A pesticide mixture is when two or more pesticides (in this case, insecticides and/or miticides) are combined into a single spray solution (Cloyd 2001a). A pesticide mixture entails exposing individuals in an arthropod (insect and/or mite) pest population to each pesticide simultaneously (Tabashnik 1989; Hoy 1998). Pesticide mixtures may be more effective against certain life stages including eggs, larvae, nymphs, and adults of arthropod pests than individual applications (Blümel and Gross 2001) although this may vary depending on the rates used and formulation of the pesticides mixed together (Blümel and Gross 2001).

There is already wide-spread use of pesticide mixtures associated with greenhouse and nursery operations world-wide, partly because combinations of selective pesticides may be required in order to deal with the arthropod pest population complex present in the crop (Tabashnik 1989; Bynum et al. 1997; Helyer 2002; Ahmad 2004; Warnock and Cloyd 2005; Cloyd 2009; Khajehali et al. 2009). Typically, two pesticides are mixed together; however, it has been demonstrated that three or more pesticides may be combined into a spray solution to target different insect and/or mite pests (Cloyd 2009). This book chapter discusses the benefits and concerns associated with pesticide mixtures, how pesticide mixtures may mitigate resistance, and the impact of pesticide mixtures on natural enemies.

2. Benefits associated with pesticide mixtures

Pesticide mixtures may enhance arthropod pest population suppression due to either synergistic interaction or potentiation between or among pesticides that are mixed together (All et al. 1977; Curtis 1985; Comins 1986; Ware and Whitacre 2004; Warnock and Cloyd 2005; Cloyd et al. 2007). Synergism refers to the toxicity of a given pesticide being enhanced by the addition of a less or non-toxic pesticide, or other compound such as a synergist (Chapman and Penman 1980; Ware and Whitacre 2004; Ahmad 2004). Potentiation involves an increased toxic effect on an arthropod pest population when mixing two compounds together, which by themselves are harmful to arthropod pests (Chapman and Penman 1980; Marer 1988; Ahmad 2004; Ahmad 2009).

The primary benefit of mixing pesticides together is a reduction in the number of applications required, which decreases labor costs (Cabello and Canero 1994; Blackshaw et al. 1995). Furthermore, pesticide mixtures may result in higher mortality of arthropod pest populations than if either pesticide were applied separately (Warnock and Cloyd 2005). Studies have demonstrated that pesticide mixtures increase efficacy against insect pests such as the western flower thrips, *Frankliniella occidentalis* Pergande (Cloyd 2003) and whiteflies (Brownbridge et al. 2000) compared to separate applications of each pesticide. For example, when permethrin (pyrethroid) is mixed with chlorpyrifos or methyl parathion (organophosphates), toxicity increases against certain insect pests (All et al. 1977; Koziol and Witkowski 1982). Pesticide mixtures associated with pyrethroid-based insecticides have been shown to potentiate the activity of the microbial, *Bacillus thuringiensis* Berliner subsp. *galleriae* against the cotton leafworm, *Spodoptera littoralis* (Boisduval) (Salma et al. 1984), and *B. thuringiensis* subsp. *kurstaki* against the fall armyworm, *S. frugiperda* (J. E. Smith) (Habib and Garcia 1981). Pesticide mixtures containing the botanical insecticide, pyrethrum appear to increase the efficacy of *B. thuringiensis* subsp. *kurstaki* against the fall webworm, *Hyphantria cunea* (Drury) (Morris 1972), and a combination of spinosad (spinosyn) and chlorpyrifos provided the best control of four species of *Liposcelis* (psocids) (Nayak and Daglish 2007).

Many studies have evaluated the effects of pesticide mixtures in suppressing populations of agricultural insect pests (All et al. 1977; Koziol and Witkowski 1982; Salma et al. 1984; Moar and Trumble 1987; Nayak and Daglish 2007) whereas there is less information associated with pesticides mixtures, and insect and mite pests of ornamental crops (Warnock and Cloyd 2005). However, Warnock and Cloyd (2005) demonstrated that all two, three, and four-way pesticide mixtures involving abamectin (macrocyclic lactone), bifenazate (carbazate), azadirachtin (limonoid insect growth regulator), and imidacloprid (neonicotinoid) along with spinosad did not affect suppression (based on percent mortality) of western flower thrips populations. This indicated that antagonism was not an issue in any of the pesticide mixtures. Cloyd et al. (2007) found that nearly all the two and three-way combinations associated with the pesticides acetamiprid (neonicotinoid), bifenazate, buprofezin (thiadiazine), and chlorfenapyr (pyrrole) exhibited no antagonistic activity with all the pesticide mixtures efficacious (based on percent mortality) against populations of the sweet potato whitefly B-biotype (*Bemisia tabaci* Gennadius) and the twospotted spider mite, *Tetranychus urticae* Koch. Mixtures of the insecticide/miticide abamectin and the fungicide triforine provided 95% control of twospotted spider mite adults, larvae, and eggs (Wang and Taashiu 1994). Improved control of the twospotted spider mite was obtained with a mixture of the miticides fenpyroximate (pyrazole) and propargite (organosulfur) compared to when both miticides were applied separately (Herron et al. 2003).

3. Concerns associated with pesticide mixtures

Although there are benefits associated with pesticide mixtures, potential problems need to be considered when two or more pesticides are mixed together. These include plant injury (=phytotoxicity), pesticide incompatibility (Cloyd 2001b), and antagonism (Lindquist 2002). Antagonism occurs when mixing two or more pesticides together results in reduced efficacy (based on percent mortality) compared to separate applications of each pesticide or when the combined toxicity of two materials when applied together is less than the sum of the toxicities of the materials when applied separately (Lindquist 2002). Antagonism may compromise the efficacy of insecticides and/or miticides under field conditions (Khajehali et al. 2009). For example, mixing together the miticide bifenazate with the organophosphate insecticide chlorpyrifos, and carbamate insecticides carbaryl, methomyl, and oxamyl decreased the efficacy of bifenazate against the twospotted spider mite indicating the occurrence of antagonism (Van Leeuwen et al. 2007; Khajehali et al. 2009). However, these effects may vary depending on the insect or mite strain (or strains), physiology, and resistance mechanisms present in the population (Ahmad 2004).

Incompatibility is a physical condition by which pesticides do not mix properly to form a homogenous solution or suspension. Instead, flakes, crystals, or oily clumps form or there is a noticeable separation. Incompatibility may be due to the chemical and/or physical properties of the pesticides, impurities in the water, or the types of pesticide formulations being mixed together (Marer 1988). In order to determine incompatibility (or compatibility) of a pesticide mixture, a 'jar test' should be conducted in which a representative sample of a pesticide mixture solution is collected in a glass jar and then allowed to remain stationary for approximately 15 minutes. If the solution is uniform or homogenous, then the pesticides are compatible; however, if there is clumping or separation, then the pesticides are not compatible with each other (Marer 1988).

4. Pesticide mixtures and resistance mitigation

It has been proposed that pesticide mixtures may delay the onset of resistance developing in arthropod pest populations (Skylakakis 1981; Mani 1985; Mallet 1989; Bielza et al. 2009). The implementation of pesticide resistance mitigating strategies is important for preserving the effectiveness of currently available pesticides (Hoy 1998). However, there is minimal evidence to suggest that pesticide mixtures may actually mitigate the onset of resistance (Immaraju et al. 1990).

Mixing pesticides with different modes of action may delay resistance developing within arthropod pest populations because the mechanism(s) required to resist each pesticide in the mixture may not be wide-spread or exist in arthropod pest populations (Georghiou 1980; Curtis 1985; Mani 1985; Mallet 1989; Ahmad 2004). As such, it may be difficult for individuals in the arthropod pest population to develop resistance to several modes of action simultaneously (Brattsten et al. 1986; Mallet 1989; Stenersen 2004; Yu 2008). Those arthropods present in the population resistant to one or more pesticides would likely succumb to the other pesticide in the mixture as long as pesticides with different modes of action are mixed together (Georghiou 1980; Mallet 1989; Yu 2008). For example, Crowder et al. (1984) reported that a mixture of chlordimeform (formamidine) with permethrin, delayed resistance development in populations of the tobacco budworm, *Heliothis virescens* (F.). However, pesticide mixtures may not always delay resistance (Burden et al. 1960). Attique et al. (2006) indicated that pesticide mixtures were less effective in delaying resistance associated with diamondback moth, *Plutella xylostella* (L.) populations than when applying insecticides separately. Furthermore, this approach may risk selecting for a detoxification mechanism that could allow survival to both pesticides (Stenersen 2004), and may actually enhance overall "selection pressure," thus accelerating the evolution of resistance (Curtis 1985; Via 1986; Brattsten et al. 1986).

The effect of pesticide mixtures is, however, unpredictable because differences in the mode of action do not necessarily insure a lack of common resistance mechanisms and may only reflect the specificity associated with enzymes responsible for detoxification (Sawicki 1981; Yu 2008). Moreover, the effects of pesticide mixtures may differ depending on the arthropod pest population as a result of peculiarities associated with species, strain, and even biotype

(Sawicki 1981; Georghiou and Taylor 1986; Ishaaya 1993). These differences could be related to physiology and the resistance mechanisms present in the population (Georghiou and Taylor 1977a; Brattsen et al. 1986). Also, resistance mechanisms typically don't respond to "selection pressure" or frequency of pesticide applications the same way based on the pesticide being applied. In fact, some resistance mechanisms may negate the advantages of pesticide mixtures (Tabashnik 1989; Stenersen 2004).

One aspect of pesticide mixtures is the opportunity for complex interactions including synergism or antagonism. Two active ingredients may compete for or inhibit the same enzyme (e.g., esterase), which could increase the toxicity of the pesticide mixture (Kulkrani and Hodgson 1980). Synergism may occur when one pesticide interferes with the metabolic detoxification of another pesticide (Corbett 1974; Kulkrani and Hodgson 1980). Certain organophosphate insecticides bind to the active site associated with esterase enzymes responsible for detoxification of pyrethroid-based insecticides (Kulkarni and Hodgson 1980; Ascher et al. 1986; Ishaaya et al. 1987; Bynum et al. 1997; Gunning et al. 1999; Ahmad 2004; Zalom et al. 2005; Ahmad et al. 2008; Ahmad 2009), and so organophosphate insecticides may be considered useful synergists for pyrethroids (Chapman and Penman 1980; Brattsten et al. 1986; Ishaaya et al. 1987; Gunning et al. 1999; Martin et al. 2003; Zalom et al. 2005; Attique et al. 2006). This is one of the main reasons why manufacturing companies formulate organophosphate and pyrethroid-based insecticide mixtures to manage arthropod pest complexes and counteract resistance (Ahmad 2004). Examples of commercially available products for use in greenhouse and/or nursery production systems include Tame/Orthene TR [fenpropathrin (pyrethroid) and acephate (organophosphate); Whitmire Micro-Gen Research Laboratories, Inc., St. Louis, MO] and Duraplex® TR [chlorpyrifos (organophosphate) and cyfluthrin (pyrethroid); Whitmire Micro-Gen Research Laboratories, Inc., St. Louis, MO]. Certain carbamate insecticides have also been reported to synergize the effects of pyrethroid-based insecticides. The carbamate insecticides methiocarb, pirimicarb and oxamyl, and even the fungicide propamocarb have been shown to synergize the efficacy (based on percent mortality) of the pyrethroid-based insecticide acrinathrin against the western flower thrips (Bielza et al. 2007; Bielza et al. 2009).

However, continued use of these types of pesticide mixtures may result in resistance to both modes of activity by arthropod pest populations, especially those that have the capacity of developing multiple resistance, which refers to an arthropod pest population resistant to pesticides with discrete modes of action or across chemical classes affiliated with the expression of different resistance mechanisms (Forgash 1984; Comins 1986; Georghiou 1986; Brattsten et al. 1986; Metcalf 1989; Attique et al. 2006; Ahmad et al. 2008).

As with applications of individual pesticides, it is important to only mix together pesticides with different modes of action or those that affect different biochemical processes in order to mitigate resistance developing in arthropod pest populations (Cranham and Helle 1985; Cloyd 2009). For example, acephate and methiocarb should not be mixed together because despite being in different chemical classes (organophosphate and carbamate) both have identical modes of action. Acephate and methiocarb block the action of acetylcholinesterase, an enzyme that deactivates acetylcholine, which is responsible for activating acetylcholine receptors. This then allows nerve signals to migrate through the central nervous system. Both acephate and methiocarb inhibit the action of acetylcholinesterase by attaching to the enzyme (Ware and Whitacre 2004; Yu 2008). Similarly, although the active ingredients acequinocyl, pyridaben, and fenpyroximate are in different chemical classes; napththoquinone, pyridazinone, and phenoxypyrazole, respectively all three are classified as mitochondrial electron transport inhibitors (METI). These active ingredients either inhibit nicotinamide adenine dinucleotide hydride (NADH) dehydrogenase (complex I) associated with electron transport, acting on the NADH CoQ reductase, or bind to the quinone oxidizing (Q_o) center or cytochrome bc_1 (complex III) of the mitochondria respiratory pathway. This reduces energy production by preventing the formation of adenosine triphosphate or ATP (Hollingworth and Ahammadsahib 1995; Yu 2008).

Pesticide mixtures may mitigate the onset of resistance under the following assumptions: 1) resistance associated with each pesticide in a mixture is monogenic (resistance resulting from the expression of a single gene) and independently genetically controlled (Curtis 1985; Tabashnik 1989). In addition, there is no cross resistance among individuals in the arthropod pest population to the pesticides used in the mixture (Mani 1985; Comins 1986; Tabashnik 1989; Tabashnik 1990). Cross resistance refers to a condition by which resistance to one pesticide confers resistance to another pesticide, even though the arthropod pest population was never exposed to the second pesticide; and insensitivity to pesticides with similar modes of action or in the same chemical class due to a common resistance mechanism or detoxification pathway associated with different pesticides (Cranham and Helle 1985; Georghiou and Taylor 1986; Roush 1993; Pedigo 2002). These conditions occur when there are different target sites and detoxification enzymes affiliated with resistance to the two pesticides. It is possible that under these given circumstances, individuals simultaneously possessing resistance mechanisms to both pesticides will be extremely rare (Curtis 1985; Brattsten et al. 1986; Mallet 1989; Roush 1993); 2) individuals in the arthropod pest population possess resistance genes (alleles) that are exclusively recessive and/or individuals that are doubly-resistant are very rare. Evolution of resistance will be instantaneous if any survivors possess doubly-resistant genes or multiple resistance mechanisms (Curtis 1985; Comins 1986; Tabashnik 1989; Mallet 1989); 3) some individuals in the arthropod pest population are not treated or exposed to the pesticide mixture primarily due to the presence of refugia (Georghiou and Taylor 1977b; Brattsten et al. 1986; Tabashnik 1989; Tabashnik 1990), or there is immigration of and mating with susceptible individuals, which reduces the frequency or proportion of resistant individuals (or resistant genes) in the arthropod pest population (Comins 1977; Georghiou and Taylor 1977b; Tabashnik and Croft 1982; Comins 1986; Georghiou and Taylor 1986; Mallet 1989; Jensen 2000; Stenersen 2004); 4) the pesticides mixed together have equal persistence so that any individuals in the arthropod pest population are not exposed to just one pesticide for an extended length of time (Forgash 1984; Curtis 1985; Tabashnik 1989; Tabashnik 1990; Roush 1993); and 5) resistance mechanisms to each pesticide are present at such low frequencies that they may not occur together in any individuals in an arthropod pest population (Yu 2008).

The assumptions presented above, in nearly all instances, are not realistic. For example, pesticide mixtures may, in fact, promote the expression of multiple resistance, which could extend across other chemical classes resulting in specific arthropod pest populations being very difficult to manage (Forgash 1984; Brattsten et al. 1986; Ahmad 2004; Attique et al. 2006). Furthermore, multiple evolutionary pathways may exist that eventually result in a pesticide-resistant arthropod pest population (Metcalf 1980; Georghiou 1983; Brattsten et al. 1986; Ishaaya 1993). Although pesticide mixtures may delay resistance due to target site insensitivity, which is usually specific to a certain class of pesticides, the use of pesticide mixtures enhances the selection for increased expression of metabolic enzymes that can simultaneously detoxify both pesticides (Roush and McKenzie 1987; Roush and Daly 1990; Roush and Tabashnik 1990; Stenersen 2004). Also, cross and multiple resistance may occur among some pesticides with similar modes of action (Stenersen 2004). Therefore, selecting for high levels of detoxification enzyme expression jeopardizes the usefulness of all pesticides, even those with new modes of action to which the arthropod pest population has never been previously exposed (Tabashnik 1989; Soderlund and Bloomquist 1990).

Additional problems associated with the assumptions of using pesticide mixtures to mitigate resistance are that the frequency of doubly-resistant individuals or those with multiple resistance mechanisms in the arthropod pest population may be extensive (Tabashnik 1989). This may be due to a history of pesticide exposure associated with selection for resistance in previous arthropod pest generations, which could imply that there may be some background levels of resistant traits or mechanisms in the arthropod pest population for each pesticide used in the mixture (Georghiou and Taylor 1977a). Also, there is usually no refuge to preserve susceptible individuals (Georghiou and Taylor 1986; Tabashnik 1989), particularly in enclosed ornamental production systems.

Is the use of pesticide mixtures the most appropriate way to extend their usefulness, or is it preferable to apply them individually? Pesticide mixtures, in fact, may be expensive, especially if the pesticides that are mixed together are used at the highest recommended label rate (Curtis 1985; Comins 1986; Mallet 1989; Attique et al. 2006). As such, a common practice is too use reduced rates of each pesticide in the mixture although this may not actually mitigate resistance developing in arthropod pest populations (Suthert and Comins 1979). More sophisticated uses of pesticide mixtures will require a thorough understanding of their interactions in order to optimize the dosage at below label rates when the components (active and inert ingredients) act synergistically (Tabashnik 1989; Attique et al. 2006). Pesticide mixtures may be an effective means of mitigating resistance as long as there is a high level of dominance in the arthropod pest population and immigration of susceptible individuals is prevalent (Mani 1985; Georghiou and Taylor 1986). Based on population genetic models, pesticide mixtures may effectively suppress resistance genes that are recessive and accord resistance to only one pesticide. However, it is possible that pesticide mixtures will select for dominant genes, which confer cross resistance (Tabashnik 1989).

The rate of resistance development in an arthropod pest population to two or more pesticides in a mixture may take longer than when the pesticides are applied separately (National Research Council 1986) although resistance to a pesticide mixture may occur at a similar rate as when the pesticides are applied individually (Kable and Jeffery 1980). The advantages of a pesticide mixture will only be sustained as long as resistance is not fully-dominant (Curtis 1985). Because the reliability of the pesticide mixture strategy depends on several assumptions, applying pesticides individually, or rotating those with different modes of action or that act on different target sites may be a more appropriate strategy (Roush 1993).

5. Pesticide mixtures and natural enemies

The use of pesticide mixtures with a broad-spectrum of arthropod pest activity and multiple modes of action may negatively impact biological control agents or natural enemies more so than separate applications of pesticides (Ahmad et al. 2004). However, only a few studies have evaluated the direct and indirect effects of pesticide mixtures on natural enemies and these primarily involve predatory mites. Lash et al. (2007) found that *Neoseiulus cucumeris* (Oudemans) deutonymphs were more sensitive to certain pesticide mixtures involving the insecticide spinosad, the insecticide/miticide abamectin, and the fungicides fenhexamid and thiophanate-methyl than adults. Predatory mite mortality, in general, associated with the pesticide mixtures was not significantly different from mortality when the pesticides were applied separately (Lash et al. 2007).

Field studies conducted with the predatory mite, *Typhlodromus pyri* Scheuten found that mixtures of the fungicides mancozeb or thiophanate-methyl with the insecticide chlorpyrifos were more harmful to the predatory mite than if the pesticides were applied by themselves (Cross and Berrie 1996). Sterk et al. (1994) determined that the fungicides maneb and mancozeb were moderately toxic to *T. pyri* when applied separately but their effects were diminished when both fungicides were mixed together. Blumel and Gross (2001) indicated no significant differences in the mortality rate or fecundity associated with *Phytoseiulus persimilis* Athias-Henriot females following exposure to the miticide (acaricide) hexythiazox (carboxamide), the fungicide triadimefon, and the insecticide heptenophos (organophosphate) when applied either individually or in mixtures.

Boomathi et al. (2005) evaluated the effects of pesticide mixtures on the parasitoid, *Trichogramma chilonis* Ishii and found that combinations of spinosad with *Bacillus thuringiensis* var. *galleriae* were toxic to adults (based on percent mortality) and inhibited adult emergence. Based on the studies presented above, pesticide mixtures may differentially directly (e.g., immediate mortality) or indirectly (e.g., delay female oviposition) impact natural enemies.

6. Summary

Pesticide mixtures involve combinations of two or more pesticides into a single spray solution. Pesticide mixtures are widely used to deal with the array of arthropod pests encountered in greenhouse and nursery production systems due to the savings in labor costs. Furthermore, the use of pesticide mixtures may result in synergism or potentiation (enhanced efficacy) and the mitigation of resistance (Ahmad 2009). However, antagonism (reduction in efficacy) may also occur due to mixing two (or more) pesticides together. Judicious use of pesticide mixtures or those that may be integrated with biological control agents is especially important because parasitoids and predators (and even microbials such as beneficial bacteria and fungi) can suppress arthropod pest populations irrespective of the arthropod pests' resistance traits or mechanisms (Tabashnik 1986). The use of pesticide mixtures to mitigate resistance must not divert attention from the implementation of alternative pest management strategies including cultural, sanitation, and biological control that can reduce reliance on pesticide mixtures and mitigate pesticide resistance (Georghiou 1983; Metcalf 1983; Tabashnik 1989; Roush 1989; Roush and Tabashnik 1990; Hoy 1998; Denholm and Jespersen 1998). Pesticide mixtures will continue to be an integral component of pest management programs due to the continual need to deal with a multitude of arthropod pests associated with ornamental cropping systems.

7. References

Ahmad, M. 2009. Observed potentiation between pyrethroid and organophosphate insecticides for the management of *Spodoptera litura* (Lepidoptera: Noctuidae). Crop Prot. 28: 264-268.

Ahmad, M. M. Ahmed Saleem, and A. H. Sayyed. 2008. Efficacy of insecticide mixtures against pyrethroid-and organophosphate-resistant populations of *Spodoptera litura* (Lepidoptera: Noctuidae). Pest Manag. Sci. 65: 266-274.

Ahmad, M. 2004. Potentiation/antagonism of deltamethrin and cypermethrins with organophosphate insecticides in the cotton bollworm, *Helicoverpa armigera* (Lepidoptera: Noctuidae). Pestic. Biochem. Physiol. 80: 31-42.

All, J. N., M. Ali, E. P. Hornyak, and J. B. Weaver. 1977. Joint action of two pyrethroids with methyl-parathion, methomyl, and chlorpyrifos on *Heliothis zea* and *H. virescens* in the laboratory and in cotton and sweetcorn. J. Econ. Entomol. 70: 813-817.

Ascher, K. R. S., M. Eliyahu, I. Ishaaya, M. Zur, and E. Ben-Moshe. 1986. Synergism of pyrethroid-organophosphorus insecticide mixtures in insects and their toxicity against *Spodoptera littoralis* larvae. Phytoparasitica 14: 101-110.

Attique, M. N. R., A. Khaliq, and A. H. Sayyed. 2006. Could resistance to insecticides in *Plutella xylostella* (Lep., Plutellidae) be overcome by insecticide mixtures? J. Appl. Entomol.130: 122-127.

Bielza, P., E. Fernández, C. Grávalos, and J. Albellán. 2009. Carbamates synergize the toxicity of acrinathrin in resistant western flower thrips (Thysanoptera: Thripidae). J. Econ. Entomol. 102: 393-397.

Bielza, P., P. J. Espinosa, V. Quinto, J. Abellán, and J. Contreras. 2007. Synergism studies with binary mixtures of pyrethroid, carbamate and organophosphate insecticides on *Frankliniella occidentalis* (Pergande). Pest Manag. Sci. 63: 84-89.

Blackshaw, R. P., A. D. Baylis, and P. F. Chapman. 1995. Effect of tank-mixing on economic threshold for pest control. Aspects Appl. Biol. 41: 41-49.

Blümel, S., and M. Gross. 2001. Effect of pesticide mixtures on the predatory mite *Phytoseiulus persimilis* A.H. (Acarina: Phytoseiidae) in the laboratory. J. Appl. Entomol. 125: 201-205.

Boomathi, N., P. Sivasubramanian, S. Raguraman, and K. Premalatha. 2005. Safety of botanical and microbial pesticide mixtures to *Trichogramma chilonis* Ishii (Hymenoptera: Trichogrammatidae). J. Biol. Control 19: 23-29.

Brattsten, L. B., C. W. Holyoke, J. R. Leeper, and K. F. Raffa. 1986. Insecticide resistance: challenge to pest management and basic research. Science 231: 1255-1260.

Brownbridge, M., M. Skinner, and B. L. Parker. 2000. Enhancing the activity of insect-killing fungi for floral IPM. Ohio Flor. Assoc. Bull. 842: 14-16.

Burden, G. S., C. S. Lofgren, and C. N. Smith. 1960. Development of chlordane and malathion resistance in the German cockroach. J. Econ. Entomol. 53: 1138-1139.

Bynum E. D., T. L. Archer, and F. W. Plapp. 1997. Comparison of banks grass mite and two-spotted spider mite (Acari: Tetranychidae): response to insecticide alone and in synergistic combinations. J. Econ. Entomol. 90: 1125-1130.

Cabello, T., and R. Canero. 1994. Pesticide mixtures used on garden crops in greenhouses in southeast Spain: cost analysis. Boletin Sanidad Vegetal, Plagus 20: 429-436.

Chapman, B. R., and D. R. Penman. 1980. The toxicity of mixtures of a pyrethroid with organophosphorus insecticides to *Tetranychus urticae* Koch. Pest. Sci. 11: 600-604.

Cloyd, R. A. 2009. Getting mixed-up: are greenhouse producers adopting appropriate pesticide mixtures to manage arthropod pests? HortTech. 19: 638-646.

Cloyd, R. A., C. L. Galle, and S. R. Keith. 2007. Greenhouse pesticide mixtures for control of silverleaf whitefly (Homoptera: Aleyrodidae) and twospotted spider mite (Acari: Tetranychidae). J. Entomol. Sci. 42: 375-382.
Cloyd, R. A. 2003. Control of western flower thrips on transvaal daisy. Arthr. Manag. Test 28: G23.
Cloyd, R. A. 2001a. The dilemma of tank mixing. Greenhouse Manag. Prod. 21(11): 66-67.
Cloyd, R. A. 2001b. Efficacy of selected insecticides for control of western flower thrips. Arthr. Manag. Test 26: G41.
Comins, H. N. 1986. Tactics for resistance management using multiple pesticides. Agri. Ecosyst. Environ. 16: 129-148.
Comins, H. N. 1977. The development of insecticide resistance in the presence of migration. J. Theor. Biol. 64: 177-197.
Corbett, J. R. 1974. The biochemical mode of action of pesticides. Academic Press, New York, NY.
Cranham, J. E., and W. Helle. 1985. Pesticide resistance in Tetranychidae, pp. 405-421. In: Helle W., and M. W. Sabelis [eds.]. Spider Mites, Their Biology, Natural Enemies and Control, Vol. 1. World Crop Pests. Elsevier, Amsterdam-New York, NY.
Cross, J. V., and A. M. Berrie. 1996. Further field evaluation of the effects of repeated foliar sprays of insecticides or fungicides alone and in admixture on an organophosphate-resistant strain of the orchard predatory mite *Typhlodromus pyri* on apple. Crop Prot. 15: 637-639.
Crowder, L. A., M. P. Jensen, and T. F. Watson. 1984. Permethrin resistance in the tobacco budworm, *Heliothis virescens*, pp. 223-224. In: Brown J. M. [ed.]. Proceedings of the Beltwide Cotton Conference, Atlanta, GA., January 9-12, 1984. National Cotton Council of America, Memphis, TN.
Curtis, C. F. 1985. Theoretical models of the use of insecticide mixtures for the management of resistance. Bull. Entomol. Res. 75: 259-265.
Denholm, I., and J. B. Jespersen. 1998. Insecticide resistance management in Europe: Recent developments and prospects. Pest. Sci. 52: 193-195.
Forgash, A. J. 1984. History, evolution, and consequences of insecticide resistance. Pest. Biochem. Physiol. 22: 178-186.
Georghiou, G. P., and C. E. Taylor. 1986. Factors influencing the evolution of resistance, pp. 157-169. In: Pesticide Resistance: Strategies and Tactics for Management. National Research Council. National Academy Press, Washington, D. C.
Georghiou, G. P., and C. E. Taylor. 1977a. Operational influences in the evolution of insecticide resistance. J. Econ. Entomol. 70: 653-658.
Georghiou, G. P., and C. E. Taylor. 1977b. Genetic and biological influences in the evolution of insecticide resistance. J. Econ. Entomol. 70: 319-323.
Georghiou, G. P. 1986. The magnitude of the resistance problem, pp. 14-43. In: Pesticide Resistance: Strategies and Tactics for Management. National Academy Press, Washington, D. C.
Georghiou, G. P. 1983. Management of resistance in arthropods, pp. 769-792. In: Georghiou G. P., and T. Saito [eds.]. Pest Resistance to Pesticides. Plenum Press, New York, NY.
Georghiou, G. P. 1980. Implications of the development of resistance to pesticides: basic principles and consideration of countermeasures, pp. 116-129. In: Gooding E. G. B

[ed]. Pest and Pesticide Management in the Caribbean. Proceedings Seminar and Workshop, Berkeley, CA. Consortium of International Crop Protection.

Gunning, R. V., G. D. Moores, and A. L. Devonshire. 1999. Esterase inhibitors synergise the toxicity of pyrethroids in Australian *Helicoverpa armigera* (Hubner) (Lepidoptera: Noctuidae). Pest. Biochem. Physiol. 63: 50-62.

Habib, M. E. M., and M. A. Garcia. 1981. Compatibility and synergism between *Bacillus thuringiensis* var. *kurstaki* and two chemical insecticides. Zietschrift fur Angewandte Entomologie 91: 7-14.

Helyer, N. 2002. Abamectin plus pymetrozine; an extremely useful addition to the IPM armoury. IOBC/WPRS Bull. 25: 101-102.

Herron, G. A., J. Rophail, J. Holloway, and J. Barchia. 2003. Potentiation of a propargite and fenpyroximate mixture against two-spotted spider mite, *Tetranchus urticae* (Acari: Tetranychidae). Exp. Appl. Acarol. 29: 115-119.

Hollingworth, R. M., and K. I. Ahammadsahib. 1995. Inhibitors of respiratory complex I: Mechanisms, pesticidal actions and toxicology. Rev. Pestic. Toxicol. 3: 277-302.

Hoy, M. A. 1998. Myths, models and mitigation of resistance to pesticides. Phil. Trans. Royal Soc. London Biol. Sci. 353: 1787-1795.

Immaraju, J. A., J. G. Morse, and R. F. Hobza. 1990. Field evaluation of insecticide rotation and mixtures as strategies for citrus thrips (Thysanoptera: Thripidae) resistance management in California. J. Econ. Entomol. 83: 306-314.

Ishaaya, I. 1993. Insect detoxifying enzymes: their importance in pesticide synergism and resistance. Arch. Insect Biochem. Physiol. 22: 263-276.

Ishaaya, I., Z. Mendelson, K. R. S. Ascher, and J. E. Casida. 1987. Cypermethrin synergism by pyrethroid esterase inhibitors in adults of the whitefly, *Bemisa tabaci*. Pest. Biochem. Physiol. 28: 155-162.

Jensen, S. E. 2000. Insecticide resistance in the western flower thrips, *Frankliniella occidentalis*. Integ. Pest Manag. Rev. 5: 131-146.

Kable, P. F., and H. Jeffery. 1980. Selection for tolerance in organisms exposed to sprays of biocide mixtures: a theoretical model. Phytopathol. 70: 8-12.

Khajehali, J., T. Van Leeuwen, and L. Tirry. 2009. Susceptibility of an organophosphate resistant strain of the two-spotted spider mite (*Tetranychus urticae*) to mixtures of bifenazate with organophosphate and carbamate insecticides. Exp. Appl. Acarol. 49: 185-192.

Koziol, F. S., and J. F. Witkowski. 1982. Synergism studies with binary mixtures of permethrin plus methyl parathion, chlorpyrifos and malathion on European corn borer larvae. J. Econ. Entomol. 75: 28-30.

Kulkrani, A. P., and E. Hodgson. 1980. Metabolism of insecticides by mixed function oxidase systems. Pharmacolog. Ther. 8: 379-475.

Lash, H. E., D. F. Warnock, and R. A. Cloyd. 2007. Effect of pesticide mixtures on the survival of the predatory mite *Neoseiulus cucumeris* (Acarina: Phytoseiidae). J. Entomol. Sci. 42: 311-319.

Lindquist, R. 2002. Tank-mixed pesticides offer benefits, disadvantages. Greenhouse Bus. 8(1): 25-29

Mallet, J. 1989. The evolution of insecticide resistance: have the insects won? Tree 4: 336-340.

Mani, G. S. 1985. Evolution of resistance in the presence of two insecticides. Genetics 109: 761-783.

Marer, P. J. 1988. The Safe and Effective Use of Pesticides. University of California Statewide Integrated Pest Management Project Division of Agriculture and Natural Resources. Publication 3324.

Martin, T., O. G. Ochou, M. Vaissayre, and D. Fournier. 2003. Organophosphorus insecticides synergize pyrethroids in the resistant strain of cotton bollworm, *Helicoverpa armigera* (Hübner) (Lepidoptera: Noctuidae) from West Africa. J. Econ. Entomol. 96: 468-474.

Metcalf, R. L. 1989. Insect resistance to insecticides. Pest. Sci. 26: 333-358.

Metcalf, R. L. 1983. Implications and prognosis of resistance to insecticides, pp. 703-734. In: Georghiou G. P., and T. Saito [eds.]. Pest Resistance to Pesticides. Plenum Press, New York, NY.

Metcalf, R. L. 1980. Changing roles of insecticides in crop protection. Ann. Rev.Entomol. 25: 219-256.

Moar, W. J., and J. T. Trumble. 1987. Toxicity, joint action and mean time of mortality of Dipel 2X, avermectin B1, neem and thuringiensin against beet armyworms (Lepidoptera: Noctuidae). J. Econ. Entomol. 80: 588-592.

Morris, O. N. 1972. Susceptibility of some forest insects to mixtures of commercial *Bacillus thuringiensis* and chemical insecticides, and sensitivity of the pathogen to the insecticides. Can. Entomol. 104: 1419-1425.

National Research Council. 1986. Tactics for prevention and management, pp. 313-326. In: Pesticide Resistance: Strategies and Tactics for Management. National Academy Press, Washington, D. C.

Nayak, M. K., and G. J. Daglish. 2007. Combined treatments of spinosad and chlorpyrifos-methyl for management of resistant psocid pests (Psocoptera: Liposcelididae) of stored grain. Pest Manag. Sci. 63: 104-109.

Pedigo, L. P. 2002. Entomology and Pest Management, 4th edition. Prentice Hall, Upper Saddle River, NJ.

Roush, R. T. 1993. Occurrence, genetics and management of insecticide resistance. Parasitol. Today 9: 174-179.

Roush, R. T. 1989. Designing pesticide management programs: how can you choose? Pest. Sci. 26: 423-441.

Roush, R. T., and J. C. Daly. 1990. The role of population genetics in resistance research and management, pp. 97-152. In: Roush R. T., and B. E. Tabashnik [eds.]. Pesticide Resistance to Arthropods, Chapman & Hall, New York, NY.

Roush, R. T., and B. E. Tabashnik. 1990. Pesticide resistance in arthropods. Chapman & Hall, New York, NY.

Roush, R. T., and J. E. McKenzie. 1987. Ecological genetics of insecticide and acaricide resistance. Ann. Rev. Entomol. 32: 361-380.

Salma, H. S., M. S. Foda, F. N. Zaki, and S. Moawad. 1984. Potency of combinations of *Bacillus thuringiensis* and chemical insecticides on *Spodoptera littoralis* (Lepidoptera: Noctuidae). J. Econ. Entomol. 77: 885-890.

Sawicki. R. M. 1981. Problems in countering resistance. Phil. Trans. Royal Soc. London. Biol. Sci. 295: 143-151.

Skylakakis. 1981. Effects of alternating and mixing pesticides on the build-up of fungal resistance. Phytopathol. 71: 1119-1121.

Soderlund, D. M., and J. R. Bloomquist. 1990. Molecular and biochemical mechanisms of insecticide resistance, pp. 58-96. In: Roush R. T., and B. E. Tabashnik [eds.]. Pesticide Resistance in Arthropods, Chapman & Hall, New York, NY.

Stenersen, J. 2004. Chemical Pesticides: Mode of Action and Toxicology. CRC Press LLC, Boca Raton, FL

Sterk, G., P. L. Creemers, and K. Merckx. 1994. Testing the side effects of pesticides on the predatory mite *Typhlodromus pyri* Scheuten (Acari: Phytoseiidae) in field trials. IOBC/WPRS Bull. 17(10): 27-40.

Suthert, R. W., and H. N. Comins. 1979. The management of acaricide resistance in the cattle tick, *Boophilis microplus* (Canestrini) (Acari: Ixodidae). Austral. Bull. Entomol. Res. 69: 519-537.

Tabashnik, B. E., and B. A. Croft. 1982. Managing pesticide resistance in crop-arthropod complexes: interactions between biological and operational factors. Environ. Entomol. 11: 1137-1144.

Tabashnik B. E. 1990. Modeling and evaluation of resistance management tactics, pp. 153-182. In: Roush R. T., and B. E. Tabashnik [eds.]. Pesticide Resistance in Arthropods, Chapman and Hall, New York, NY.

Tabashnik B. E. 1989. Managing resistance with multiple pesticide tactics: theory, evidence, and recommendations. J. Econ. Entomol. 82: 1263-1269.

Tabashnik B. E. 1986. Evolution of pesticide resistance in predator/prey systems. Bull. Entomol. Soc. Amer. 32: 156-161.

Van Leeuwen, T., S. Van Pottelberge, R. Nauen, and L. Tirry. 2007. Organophosphate insecticides and acaricides antagonize bifenazate toxicity through esterase inhibition in *Tetranychus urticae*. Pest Manag. Sci. 63: 1172-1177.

Via, S. 1986. Quantitative genetic models and the evolution of pesticide resistance, pp. 222-235. In: Pesticide Resistance: Strategies and Tactics for Management. National Academy of Science, Washington, D. C.

Wang, W., and L. Taashiu. 1994. Toxicity of mixtures of several miticides with the fungicide triforine against the two-spotted mite on Roses. Bull. Taichung Dist. Agric. Improv. Stn. 44: 1-11.

Ware, G. W., and D. M. Whitacre. 2004. The Pesticide Book. MeisterPro Information Resources, Willoughby, OH

Warnock, D. F., and R. A. Cloyd. 2005. Effect of pesticide mixtures in controlling western flower thrips (Thysanoptera: Thripidae). J. Entomol. Sci. 40: 54-66.

Yu S. J. 2008.The Toxicology and Biochemistry of Insecticides. CRC Press, Taylor & Francis Group, Boca Raton, FL

Zalom, F. G., N. C. Toscano, and F. J. Byrne. 2005. Managing resistance is critical to future use of pyrethroids and neonicotinoids. Calif. Agri. 59: 11-15.

6

Biodegradable Hydrogel as Delivery Vehicle for the Controlled Release of Pesticide

Fauze Ahmad Aouada[1,3], Márcia Regina de Moura[2,3],
Luiz Henrique Capparelli Mattoso[3]
[1]*IQ, Universidade Estadual Paulista - UNESP, Araraquara, SP*
[2]*IFSC, Universidade de São Paulo - USP, São Carlos, SP*
[3]*Laboratório Nacional de Nanotecnologia para o Agronegócio - LNNA, Embrapa Instrumentação Agropecuária - CNPDIA, São Carlos, SP*
Brazil

1. Introduction

1.1 Agrochemical controlled release goals

Controlled release is a chemical activation method, which is provided to specific plant species at preset rates and times. Different polymers are largely used to control the delivery rates, mobilities, and the chemicals period of effectiveness. The main benefit of the controlled release method is that if fewer chemicals are used for the protected plants over the predetermined period, then there is a lesser effect on the other plant species, while reducing leaching, volatilization, and degradation. The macromolecular nature of polymers is the key to chemical loss reduction throughout the production. Controlled release polymer systems can be divided into two categories. In the first, the active agent is dissolved, dispersed, or encapsulated within the polymeric matrix or coating. Its release takes place through diffusion or after biological or chemical breakdown of the polymer. In the second category, the active agent either constitutes a part of the macromolecular backbone or is attached to it. Here its release is the result of biological or chemical cleavage of the bond between the polymer and the bioactive agent [Mitrus et al., 2009].

The main problem with conventional agrochemical applications is using greater amounts of agrochemicals, over a long period of time, than what is actually needed, possibly leading to crop damage and environmental contamination [Bajpai & Giri, 2003]. Controlled release polymer matrix systems offer numerous advantages, not only to avoid treating excess amounts of active substances, but also to offer the most suitable technical solution in special fields of application [Wang et al., 2007]. The objective of controlled release systems is to protect the supply of the agent to allow the automatic release of the agent to the target at a controlled rate and to maintain its concentration in the system within the optimum limits over a specified period of time, thereby providing great specificity and persistence without diminishing efficiency. Controlled release of agrochemicals (pesticides, herbicides, nutrients) is used to maintain the local concentration of active ingredients in the soil and to reduce losses due to run off.

Controlled release systems for pesticides involve advanced pesticide delivery technologies, highlighting new means to reduce toxicity, increase efficacy, lessen the environmental

number of elastically effective chains, completely included in a perfect network, per unit volume (V_e); and controlled release of drugs [Koutroumanis et al., 2010; Liu & Lin, 2010; Sajeesh et al., 2010; Tanigo et al., 2010]; and agrochemicals [Saraydin et al., 2000; Bajpai & Giri, 2003; Bajpai et al., 2006; Wang et al., 2007; Pourjavadi et al., 2009].

3. Preparation of hydrogels and their application in pesticide controlled release

Our research group has recently focused on the preparation and characterization of polyacrylamide (PAAm) and methylcellulose (MC) biodegradable hydrogels, as potential delivery vehicles for the controlled release of paraquat pesticide, since they play an essential role in the use of hydrogels in controlled release technology.

3.1 Mechanism to form and prepare the PAAm and MC hydrogels

In a simplified preparation process of acrylamide hydrogel by the free radical co-polymerization of acrylamide (AAm) and a divinyl crosslinker, e.g. N,N'-methylene-bis-acrylamide (MBAAm), linear polymers are first formed in the solution during the fast propagation step, and later crosslinked with other molecules through their pendent double bonds and additional monomer units [Stepto, 1998]. According to Karadag et al., 2005 the polymerization of vinyl monomers, such as AAm and MBAAm in the presence of ammonium persulfate and N,N,N',N'- tetramethylethylene-diamine (TEMED), is first initiated by the reaction between ammonium persulfate and TEMED, in which the TEMED molecule is left with an unpaired valance electron. The activated TEMED molecule can combine with an AAm and/or crosslinker molecule, in which the unpaired electron is transferred to the monomeric units so that they then become reactive. Thus, another monomer or co-monomer can be attached and activated in the same way. The poly(AAm) or other copolymer hydrogel can continue growing indefinitely, with the active centre continually shifted to the free end of the chain.

The synthesis of PAAm-MC hydrogels was reported in the literature [Aouada et al., 2009*a*; Aouada et al., 2010]. AAm (3.6 - 21.7 in w:v%), MC (0 - 1.0 in w:v%), MBAAm, and TEMED were placed in a bottle and homogenized by stirred mixing. TEMED concentration was fixed at 3.21 μmol mL^{-1}. After the mixture was prepared, it was deoxygenated by N_2 bubbling for 25 min. Then, aqueous sodium persulfate (final conc. of 3.38 μmol mL^{-1}), also deoxygenated, was added to initiate the polymerization reaction. The resulting solution was quickly placed between two glass plates separated by a rubber gasket and kept at room temperature. The system was kept closed by means of metallic straps for 24 h at ambient temperature (ca. 25 °C). At this stage, the complete polymerization/cross-linking of AAm occurred. After 24 h, the hydrogels, in a membrane form (Scheme 1), were removed from the plates. These membranes (final thickness ≈ 9 - 10 mm) were then freed from the unreacted chemicals by dialysis with distilled/deionized water for 10 days. The polymeric network PAAm-MC was used to study the hydrophilic properties of the hydrogels and pesticide paraquat sorption from the aqueous solution. Polymeric networks were made by chemically induced polymerization through free radical mechanism, in which SP radical species generates the reactive sites on the MC, AAm and MBAAm. Due to the polyfunctionality of the crosslinker MBAAm, it has four reactive sites which can be linked to the radical on the methylcellulose and to the poly(acrylamide). Scheme 2 presents the formation of crosslinked network structures based on poly(acrylamide) and methylcellulose.

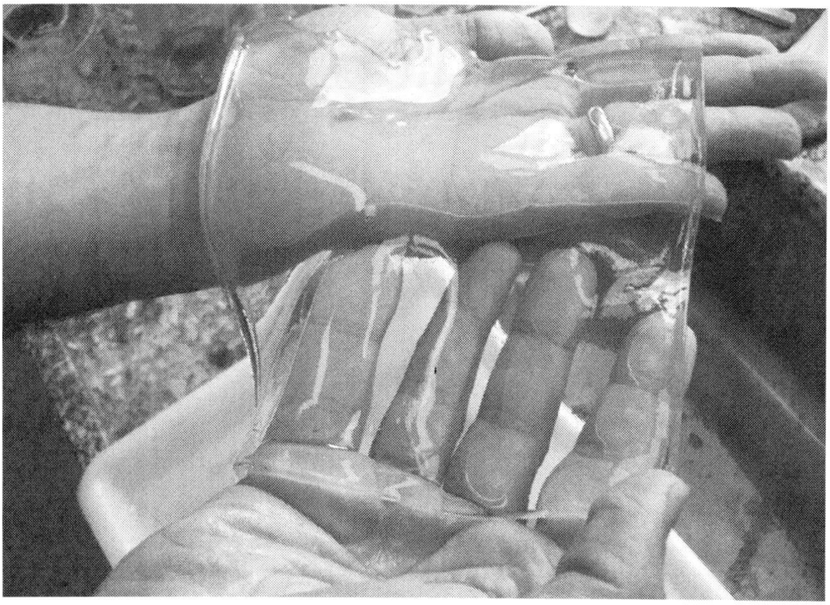

Scheme 1. Photo of hydrogel composed of PAAm and MC in membrane form after dialysis process: [AAm] = 6.0 in w:v%; [MC] = 1.0 in w:v% [Aouada et al., 2010].

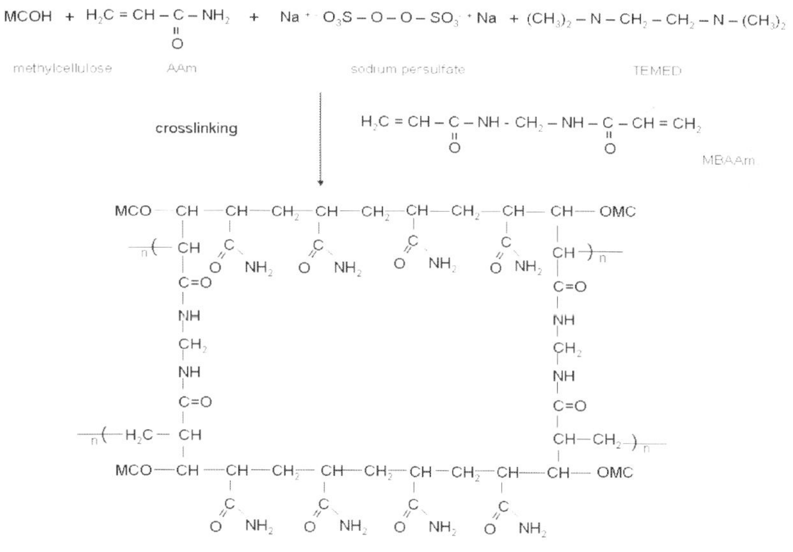

Scheme 2. Formation of crosslinked network structure based on poly(acrylamide) and methylcellulose [Aouada et al., 2009*a*].

3.2 Some physical-chemistry properties of PAAm-MC hydrogels

3.2.1 Hydrophilic properties

The hydrophilic properties of PAAm-MC hydrogels were investigated by measuring their water uptake (WU). For the water uptake studies, the swollen hydrogels in membrane form were cut into cylindrical shapes of 13 mm and the average of dry hydrogels used was of approximately 150 mg. WU values were obtained by the mass ratio of the swollen hydrogel to dried hydrogel. Measurements were performed in replicate at 25.0 °C to check reproducibility and the error bars indicate the standard deviation (n = 3).

Figure 1 shows the dependences of WU as a function of the immersion time of the hydrogels swelled in distilled water.

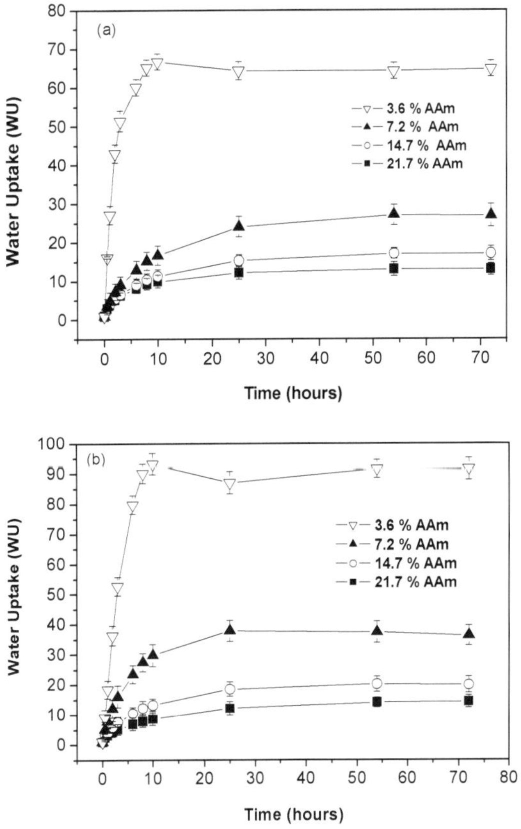

Fig. 1. Dependence of water uptake as a function of time for: (a) PAAm-MC0.5 and (b) PAAm-MC1.0, in distilled water (pH = 6.7), at 25.0 °C. Different concentrations of AAm were tested as indicated. Error bars represent standard deviations for the three experiments.

Changes on equilibrium in WU values as a function of the AAm concentration in the feed solution are shown in Figure 2. It can be pointed out that the value of WU decreases abruptly when the concentration of AAm in the gel-forming solution increases. This reduction is related to the increase of network rigidity, where the flexibility of a hydrogel network is directly related with the amount of total water absorbed by the hydrogel [Aouada et al., 2006]. The highest WU value obtained for 3.6 % AAm and 1.0 % MC, was of around 90 g/g. Also, the WU values abruptly increased when the concentration of MC in feed solution was increased. This trend is attributed to the increase in the hydrogel hydrophilicity (thus the increase in water absorption capacity) due to the incorporation of hydroxyl groups from MC segments. This tendency was also observed in the PAAm/poly(γ-glutamic acid) hydrogels studied by Rodríguez et al., 2006.

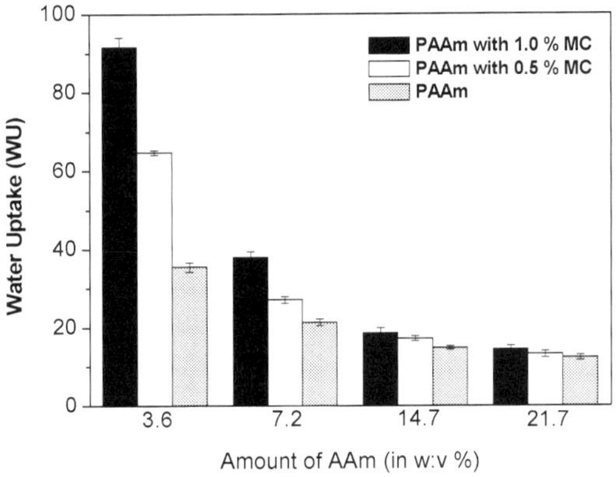

Fig. 2. Dependence of equilibrium water uptake as a function of acrylamide for PAAm, PAAm-MC0.5 and PAAmMC1.0 hydrogels, in distilled water (pH = 6.7), at 25.0 °C. Error bars represent standard deviations for the three experiments.

3.2.2 Mechanical properties

Uniaxial compression measurements were performed on equilibrium swollen hydrogels after their preparation. Compression tests were performed using a universal testing machine (Instron, Model 5500R, Canton, MA). Hydrogel compression was measured using a 1.27 cm diameter cylindrical probe. The probe was attached to the upper jaw of the Instron machine. The crosshead speed was of 12.0 mm min^{-1} with a 100 N load.

The measurements were conducted up to 30% compression of hydrogel. In this case, the maximum load (σ_{max}) of hydrogels was recorded. The modulus of elasticity (E) was calculated by Eq. (1), where F is the force and A is the cross-sectional area of the strained specimen. The relative strain (λ) was calculated from Eq. (2), where ΔL is the change in thickness of the compressed hydrogel and L_0 is the initial thickness. Six tests were run for each gel.

$$\sigma = \frac{F}{A} = E(\lambda - \lambda^{-2}) \quad (1)$$

$$\lambda = \frac{\Delta L}{L_0} \quad (2)$$

The effective (or apparent) cross-linking density, ν_e, was obtained from the slope of linear dependence of σ versus (λ - λ^{-2}), Eq. (3), where R is the universal gas constant, T is the temperature in absolute scale, $\phi_{g,0}$ and ϕ_g are the polymer volume fractions of the hydrogel in the relaxed state and in the swollen state, respectively.

$$\sigma = RT(\frac{\phi_{g,0}}{\phi_g})^{2/3}\phi_g \upsilon_e(\lambda - \lambda^{-2}) \tag{3}$$

The length of the effective chains between crosslinking points (N) is related to the effective cross-linking density ν_e by Eq. (4):

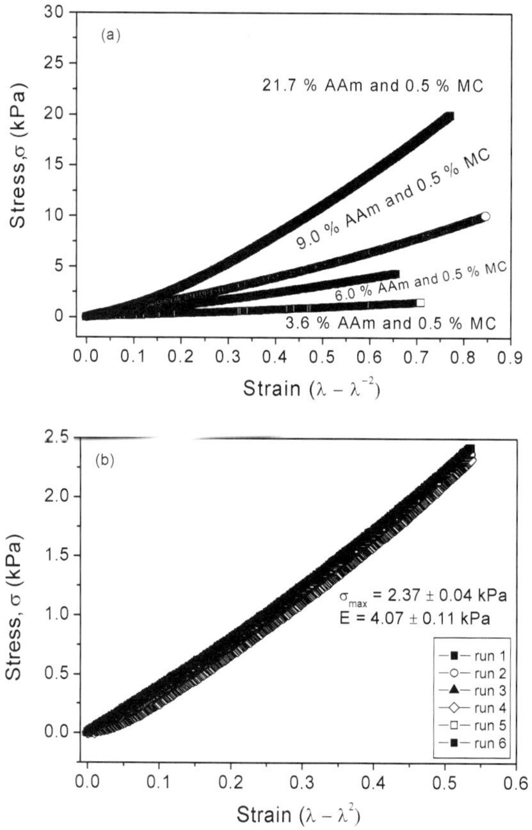

Fig. 3. Measured force and stress as a function of Strain (λ - λ^{-2}) at 25 °C for hydrogels synthesized with (a) [MC] = 0.5 in w:v%, [MBAAm] = 8.6 μmol mL^{-1} and different AAm concentrations; (b) [AAm] = 6.0 in w:v%, [MBAAm] = 10.0 μmol mL^{-1}, [MC] = 0.75 in w:v% [Aouada et al., 2009*b*].

$$N = (\upsilon_e V_l)^{-1} \quad (4)$$

where V_l is the molar volume of the segment, which is taken as the molar volume of water (18 cm^3 mol^{-1}).

To evaluate the mechanical properties of the PAAm and PAAm-MC hydrogel, the maximum load (σ_{max}) and modulus of elasticity (E) of the hydrogels were measured. Representative stress–strain curves for the hydrogels tested with uniaxial compression are shown in Fig. 3, where the linearity between force and strain can be observed. The reproducibility of the stress-strain experiments is shown in Fig. 3*b*.

The linear correlation indicates that elastic deformation occurred, i.e. the strain is recoverable after removing the applied stress. In the most elementary form, recoverable strain means that if the hydrogel is under an applied load, the polymer chains are rearranged to accommodate the deformation. At the same time, retractive elastic force develops in the polymer networks because of their tendency to return to their original formation [Buchholz & Graham, 1997].

The dependence of maximum load as a function of acrylamide concentration for hydrogels with different methylcellulose concentrations is shown in Fig. 4.

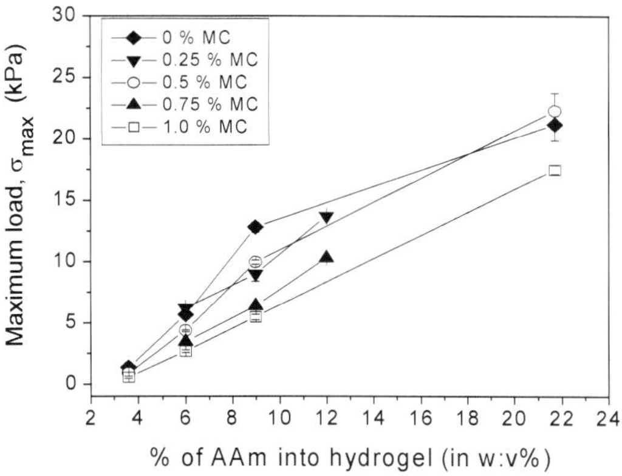

Fig. 4. Measured force and stress as a function of Strain (λ - λ^{-2}) at 25 °C for hydrogels synthesized with (a) [MC] = 0.5 in w:v%, [MBAAm] = 8.6 μmol mL^{-1} and different AAm concentrations; (b) [AAm] = 6.0 in w:v%, [MBAAm] = 10.0 μmol mL^{-1}, [MC] = 0.75 in w:v% [Aouada et al., 2009*b*].

The increase in mechanical property values was observed when the amount of acrylamide in the feed solution was increased. These results corroborate with the swelling degree results (see Table 1), where increasing AAm concentration and consequently, the rigidity of the networks, results in decreasing water-uptake. Maximum load and modulus of elasticity properties decrease with increasing MC concentration. The maximum load of the (3.6-8.6-MC) hydrogels, where MC is the methylcellulose concentration, were 1.35 ± 0.14, 0.89 ± 0.05 and 0.55 ± 0.09 kPa for M = 0, 0.5 and 1.0 (in w:v%), respectively. For the same hydrogel, the modulus of elasticity values were 1.85 ± 0.08, 1.43 ± 0.06 and 1.06 ± 0.15 kPa. Such a decrease is attributed to the increase of network hydrophilicity from an increase of hydroxyl groups entrapped in the PAAm network. Additionally, when the MC concentration was increased from 0 to 1.0 (in

Scanning electron microscopy technique was used to analyze the morphology of PAAm and PAAm-MC hydrogels. Average pore size values were estimated by considering at least 20 individual pore size values [Tang et al., 2007]. The SEM image of PAAm3.6 MC0.0 (3.6 % AAm and 0 % MC), shown in Fig. 5*a*, indicates the formation of homogeneous and highly porous material with a mean pore size of 90 (± 20) µm. The addition of MC into the solution-forming hydrogel caused morphological changes, mainly in the size and shape of the pores. From the SEM micrographs shown in Fig. 5*b* and 5*c*, it was possible to see that hydrogel pores are more foliaceous, larger, and highly heterogeneous than those shown in Fig. 5*a*. Due to the pore formation with high heterogeneity, it is not possible to accurately estimate the pore size of these hydrogels.

3.3 Pesticide controlled release from PAAm-MC hydrogels

3.3.1 Controlled release principles

The release of chemicals entrapped in a hydrogel occurs only after water penetrates the network to swell the polymer and dissolve the chemicals, followed by diffusion along the aqueous pathways to the surface of the device. The release of chemicals is closely related to the swelling characteristics of the hydrogels, which in turn is a key function for the chemical architecture of the hydrogels [Singh et al., 2008]. Scheme 3 shows the schematic representations of loading and the paraquat release process, which are directly correlated with the swelling capacity of the hydrogels. For instance, in the loading process, there is water and paraquat sorption. For the release case, the water sorption contributes to the pesticide desorption due to two main factors: (1) difference in chemical potential [Shang et al., 2008] (Eq. 5), and (2) osmotic pressure defined by the Donnan equilibrium theory [Liang et al., 2009] (Eq. 6).

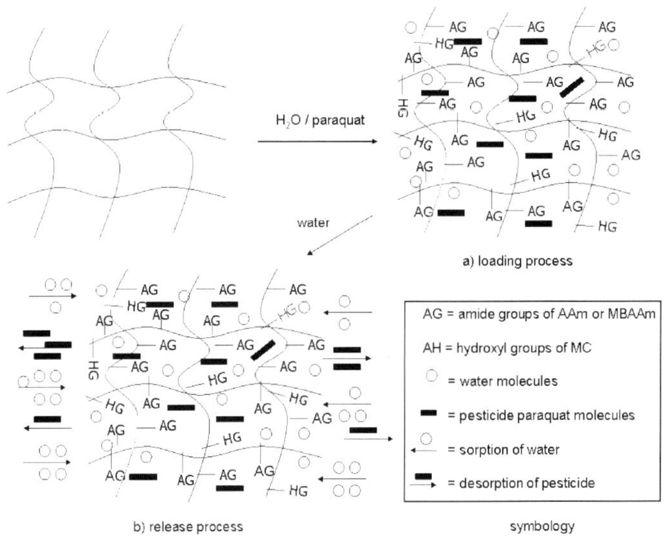

Scheme 3. Schematic representations of (a) the loading process showing the sorption of water and paraquat; and (b) the release process showing the sorption of water and desorption of paraquat.

$$\mu_i = \left(\frac{\partial G}{\partial n_i} \right)_{T,P,n_j} \tag{5}$$

where G is Gibb's free energy, n_i is the amount of component i, V is volume and P is pressure. The subscripts indicate that temperature, pressure and the amount of all other components are maintained constant.

$$\pi_{ion} = RT \sum_i (C_i^g - C_i^s) \tag{6}$$

where C_i is the mobile ion concentration of species i, and superscripts 'g' and 's' represent the gel and solution phase, respectively. R is the universal gas constant and T is the absolute temperature.

3.3.2 Effects of AAm and MC concentration on paraquat pesticide release

Hydrogels presented high loading capacity for paraquat pesticide. The pesticide was not chemically attached to the polymeric chain and the only likely interactions were ionic attractions. The hydrogels were loaded up to 82 % of paraquat, in relation to the amount of paraquat available in the loading solution. The maximum paraquat adsorption (q_{eq}) in hydrogels without MC was low, when compared with hydrogels containing MC, which was of around 0.7 mg g^{-1}. The low adsorption could be attributed to the absence of hydroxyl groups entrapped in PAAm chains. The paraquat molecules were absorbed into the hydrogels by an interaction with amide groups proceeding from PAAm chains. The general trend indicated that an increase in q_{eq} resulted from an increased MC concentration, due to the greater number of hydroxyl groups inherent in the MC. In these conditions, the adsorption was probably due to paraquat-MC interactions. It was also observed that an increased AAm concentration provoked a decrease in the q_{eq} values [Aouada et al., 2009*a*].

The varying effects of AAm and MC contents on releasing paraquat from PAAm-MC hydrogels were investigated in details and their results will be now discussed. Fig. 6 shows the amount of paraquat released as a function of time for PAAm-MC hydrogels prepared with 6.0 % AAm using different MC contents.

In general, the initial rate of paraquat release was fast, and after several days it decreased. This fact indicates that paraquat on the hydrogels surface (or close to) diffused rapidly from the initial swelling of the gel. Later, paraquat was released slowly from the hydrogels, up to 45 days. The content of methylcellulose significantly affects the amount of paraquat released, where the maximum release, close to 23 mgL^{-1}, was observed when an intermediate content of MC (0.5 %) was used.

Fig. 7 shows the effect of methylcellulose percentage on the kinetic behaviour of cumulative paraquat release. It is possible to see in Fig. 7*a* that the paraquat release from the hydrogel constituted of 6.0 % AAm is 100 % after 1 day. This fast release is attributed to the hydrophobic weak interactions between the cationic groups (from the paraquat) and amide groups from the PAAm chains. The Figure also reveals that the cumulative release occurred in a very controlled and sustained manner, in which the concentration of paraquat after 15 days was maintained constant up to 46 days. It was also observed that the quantity of paraquat release increases from 41.3 ± 5.6 % to 72.6 ± 6.1 % when the amount of MC in the gel-forming solution increases in the range of 0.25–0.5 (in w:v%), Fig. 7*b*. By increasing the

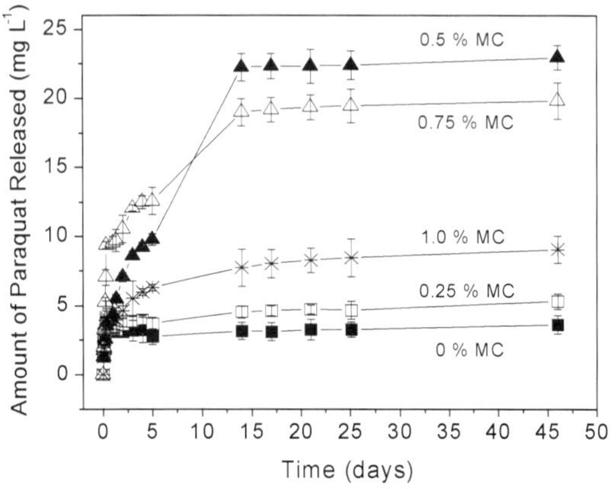

Fig. 6. Profiles of the amount of paraquat released as a function of time for PAAm-MC hydrogels with different MC concentrations: [MC] = 0; 0.25; 0.5; 0.75 and 1.0 in w:v%, [AAm] = 6.0 in w:v%, and C_0 = 37.48 mg L^{-1}. Error bars represent standard deviations for the three measurements (mean ± S.D., n = 3) [Aouada et al., 2010].

MC content of the matrix, the swelling of the matrix also increased due to the more hydrophilic nature of MC, leading to the percentage increase of the released paraquat. Similar observations have been noticed by Rokhade et al., 2007. The release profiles indicate that the amounts of paraquat released decreased in the hydrogel prepared with MC concentration above 0.5 %. At higher concentrations of MC (beyond 0.5 g), the density of network chains increases so much that both the diffusion of solvent molecules and relaxation of macromolecular chains are reduced. Similar behaviours have been observed in other studies on the characterizations of hydrogel hydrophilicity [Graiver et al., 1995; Bajpai & Giri, 2003]. This explains the drop in the hydrogels release capacity. Moreover, one of the primary factors in the application of hydrogels as a delivery vehicle for the controlled release of pesticide is the loading percentage effect on the solute release rate, because a larger hydrogel loading can facilitate the relaxation of macromolecular chains. In addition, the results of paraquat removal from aqueous solutions using PAAm and MC hydrogels, recently published by our group [Aouada et al., 2009*a*], indicated that paraquat adsorption is more favourable in hydrogels prepared with an MC concentration of around 0.5 %.

In general, the hydrogels did not release the total loaded paraquat because of the strong interaction of the paraquat-hydrogel matrix, specifically between the hydroxyl and amide groups (from MC and PAAm, respectively) with cationic regions from the paraquat. Controlled release systems studied by Alemzadeh & Vossoughi, 2002 and Sing et al., 2008 presented similar behaviours.

Fig. 8 shows the effects of acrylamide concentration on the cumulative paraquat release from PAAm-MC hydrogels prepared with different AAm and MC combinations. The releasing kinetic and the released quantity can be controlled up to 40-45 days and up to 75 % by adjusting the PAAm and MC contents in the gel-forming solution. In both cases, it was observed that as the polymeric matrix becomes rigid due to the increase in the concentration of acrylamide in the hydrogels, from 6.0 to 9.0 % (Fig. 8*a*) and from 6.0 to 12.0 % (Fig. 8*b*), the

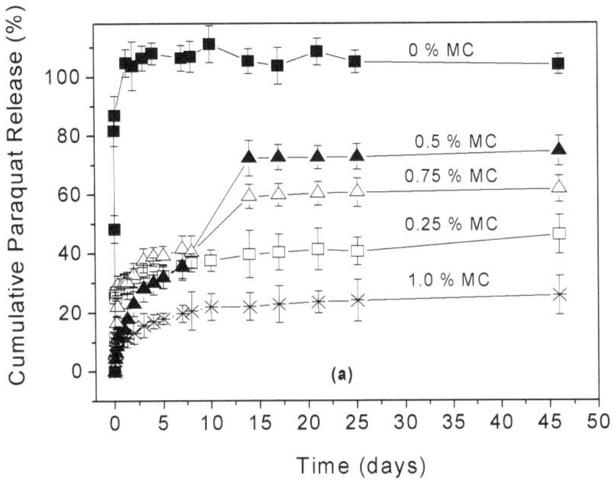

(a)

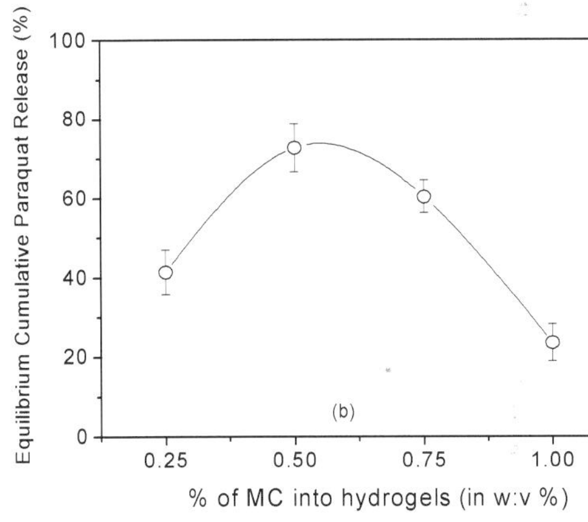

(b)

Fig. 7. (a) profiles of paraquat release as a function of time, and (b) dependence of equilibrium cumulative paraquat release as a function of methylcellulose concentration for PAAm and MC hydrogels: C_0 = 37.48 mg L^{-1}, and [AAm] = 6.0 in w:v%. Error bars represent standard deviations for the three measurements (mean ± S.D., n = 3) [Aouada et al., 2010].

cumulative paraquat release decreased. This tendency was also reported by Işiklan, 2007, where the author explained that the decreases in the cumulative release are due to the increasing of the monomer concentration, which gives rise to a compact network of the polymer, hence the free volume reduces and the penetration of water molecules and diffusion of pesticide molecules become difficult.

In accordance with Singh et al., 2009, the primary requisites for using agrochemicals to control the environment and health hazards are by means of controlled release and sustained manner. Also, PAAm-type hydrogels must act as carriers for herbicidal agents and hydrogels, such as in water preservation systems (soil conditioning), hence inducing aggregation, diminishing water evaporation and promoting plant growth [Siyam, 1994]. Moreover, acrylamide was selected due to its industrial importance and its better known properties [Kenawy, 1998]. Consequently, the hydrogels studied in this work have enormous potential to be applied in agriculture fields.

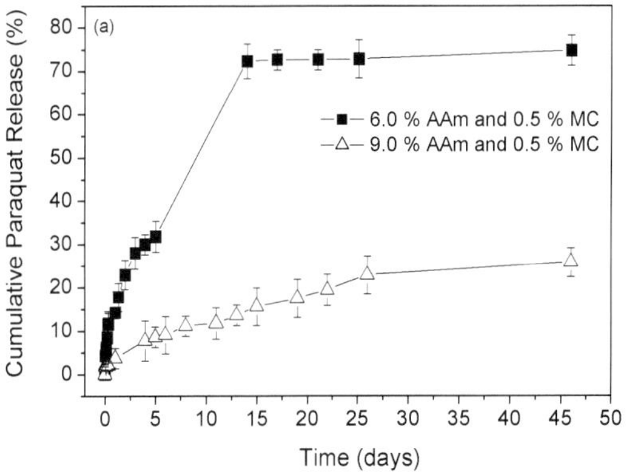

(a)

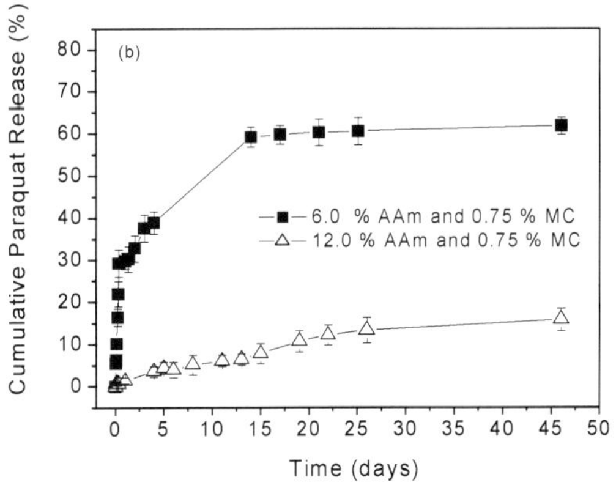

(b)

Fig. 8. Profiles of paraquat release from hydrogels constituted by PAAm and MC as a function of time in different conditions: (a) 6.0 % AAm and 0.5 % MC, 9.0 % AAm and 0.5 % MC; (b) 6.0 % AAm and 0.75 % MC, 12.0 % AAm and 0.75 % MC. C_0 = 37.48 mg L^{-1}. Error bars represent standard deviations for the three measurements (mean ± S.D., n = 3) [Aouada et al., 2010].

3.3.3 Mathematical modeling of paraquat release from PAAm and MC hydrogels

Hydrogels have a unique combination of characteristics that make them useful in controlled delivery applications. Due to their hydrophilicity, hydrogels can imbibe large amounts of water (> 90 in-wt %). Therefore, the molecule's release mechanisms from hydrogels are very different from hydrophobic polymers. Both simple and sophisticated models have been developed to predict the release of an active agent from a hydrogel device as a function of time. The most widely applicable mechanism for describing solute release from hydrogels is the diffusion-controlled release [Lin & Metters, 2006]. Fick's law of diffusion with either constant or variable diffusion coefficients is commonly used in modeling diffusion-controlled release [Andreopoulos & Tarantili, 2001]. Although there are a number of reports dealing with the mathematical modeling through swelling controlled release polymeric systems, no single model successfully predicts all the experimental observations [Singh et al., 2009].

The values of release exponent "*n*" and gel characteristic constant "*k*" calculated using Eq. 7 for the release dynamics of pesticide from the PAAm-MC hydrogels are in Table 2.

$$\frac{M_t}{M_\infty} = k\,t^n \tag{7}$$

where the M_t/M_∞ is the fractional release, k is a constant incorporating structural and geometric characteristics of the macromolecular polymeric system and the pesticide, and n is designated as the release exponent representing the release mechanism.

The curves obtained from Eq. 3 presented good linearity (regression coefficient, $R^2 \geq 0.999$), indicating that the Peppas model can be applicable to analyze the systems. The values of n remained in a range corresponding to Fickian diffusion (n = 0.45 - 0.5) until MC = 0.5 % for AAm concentration equal to 6.0 % (in w:v%). After this concentration, the paraquat release occurred through the non-Fickian diffusion. Non-Fickian or anomalous diffusion occurs when the diffusion and relaxation rates are comparable. Thus, the paraquat release depends on two simultaneous rate processes, water migration into the beads and diffusion through continuously swelling hydrogels [Ritger & Peppas, 1987]. The values of k showed that the release of paraquat becomes slower when the MC and AAm concentration increases.

Hydrogel	k (h^{-1})	n	Mechanism
(6.0-0)*	0.529 ± 0.0308	0.44 ± 0.02	Fickian
(6.0-0.25)	0.0678 ± 0.0008	0.44 ± 0.03	Fickian
(6.0-0.5)	0.0404 ± 0.0010	0.50 ± 0.02	Fickian
(6.0-0.75)	0.0541 ± 0.0021	0.63 ± 0.01	Anomalous
(6.0-1.0)	0.0375 ± 0.0010	0.58 ± 0.04	Anomalous
(9.0-0.5)	0.0147 ± 0.0302	0.34 ± 0.08	More-Fickian
(12.0-0.75)	0.00533 ± 0.00010	0.38 ± 0.09	More-Fickian

* [AAm] = 6.0 in w:v% and [MC] = 0 in w:v%.

Table 2. Parameters k and n obtained for paraquat pesticide release from hydrogels synthesized with various AAm and MC concentrations at 25.0 °C: C_0 = 37.48 mg L^{-1}. [Aouada et al., 2010].

Işıklan, N. (2007). Controlled release study of carbaryl insecticide from calcium alginate and nickel alginate hydrogel beads. *Journal of Applied Polymer Science,* 105, 2, (July 2007), 718-725, 0021-8995.

Jayakumar, R.; Menon, D.; Manzoor, K.; Nair, S. V. & Tamura, H. (2010). Biomedical applications of chitin and chitosan based nanomaterials— A short review. *Carbohydrate Polymers,* 82, 2, (September 2010), 227-232, 0144-8617.

Jiang, G.; Liu, C.; Liu, X.; Chen, Q.; Zhang, G.; Yang, M. & Liu, F. (2010). Network structure and compositional effects on tensile mechanical properties of hydrophobic association hydrogels with high mechanical strength. *Polymer,* 51, 6, (March 2010), 1507-1515, 0032-3861.

Karadag, E.; Uzum, O. B. & Saraydin, D. (2005). Water uptake in chemically crosslinked poly(acrylamide-co-crotonic acid) hydrogels. *Materials & Design,* 26, 4, (June 2005), 265-270, 0261-3069.

Kenawy, E. -R. (1998). Biologically active polymers: controlled-release formulations based on crosslinked acrylamide gel derivatives. *Reactive and Functional Polymers,* 36, 1, (February 1998), 31-39, 1381-5148.

Kim, K. -S. & Park, S. -J. (2010). Effect of porous silica on sustained release behaviors of pH sensitive Pluronic F127/poly(acrylic acid) hydrogels containing tulobuterol. *Colloids and Surfaces B: Biointerfaces,* 80, 2, (October 2010), 240-246, 0927-7765.

Koutroumanis, K. P.; Avgoustakis, K. Bikiaris, D. (2010). Synthesis of cross-linked N-(2-carboxybenzyl) chitosan pH sensitive polyelectrolyte and its use for drug controlled delivery. *Carbohydrate Polymers,* 82, 1, (August 2010), 181-188, 0144-8617.

Kulkarni, A. R.; Soppimath, K. S.; Aminabhavi, T. M.; Dave, A. M. & Mehta, M. H. (2000). Glutaraldehyde crosslinked sodium alginate beads containing liquid pesticide for soil application. *Journal of Controlled Release,* 63, 1-2, (January 2000), 97-105, 0168-3659.

Lee, E.; Kim, K. & Kim, B. (2009). Effect of P(MAA-co-PEGMA) hydrogel compositions and loading pH on the loading efficiency of cosmetic active agents. *Polymer-Korea,* 33, 5, (September 2009), 441-445, 0379-153X.

Li, C. A.; Han, K. N.; Bui, M. -P. N.; Pham, X. -H. & Seong, G. H. (2010). Development of hydrogel microstructures on single-walled carbon nanotube films. *Applied Surface Science,* 256, 24, (October 2010), 7428-7433, 0169-4332.

Liang, R.; Yuan, H.; Xi, G. & Zhou, Q. (2009). Synthesis of wheat straw-g-poly(acrylic acid) superabsorbent composites and release of urea from it. *Carbohydrate Polymers,* 77, 2, (June 2009), 181-187, 0144-8617.

Lin, C. C. & Metters, A. T. (2006). Hydrogels in controlled release formulations: Network design and mathematical modeling. *Advanced Drug Delivery Reviews,* 58, 12-13, (November 2006), 1379-1408, 0169-409X.

Liu, T. Y. & Lin, Y. L. (2010). Novel pH-sensitive chitosan-based hydrogel for encapsulating poorly water-soluble drugs. *Acta Biomaterialia,* 6, 4, (April 2010), 1423-1429, 1742-7061.

Lohakan, M.; Jamnongkan, T.; Pintavirooj, C.; Kaewpirom, S. & Boonsang, S. (2010). A numerical model for ultrasonic measurements of swelling and mechanical properties of a swollen PVA hydrogel. *Ultrasonics,* 50, 8, (August 2010), 782-789, 0041-624X.

Melchels, F. P. W.; Feijen, J. & Grijpma, D. W. (2010). A review on stereolithography and its applications in biomedical engineering. *Biomaterials,* 31, 4, (August 2010), 6121-6130, 0142-9612.

Mitrus, M.; Wojtowicz, A. & Moscicki, L. (2009). *Biodegradable polymers and their practical utility, In: Thermoplastic starch: a green material for various industries*, Leon Janssen, Leszek Moscicki, 1-33, WILEY-VCH Verlag GmbH & Co. KGaA, Weinheim - Germany.

Moura, M. R. de; Aouada, F. A.; Guilherme, M. R.; Radovanovic, E.; Rubira, A. F. & Muniz, E. C. (2006). Thermo-sensitive IPN hydrogels composed of PNIPAAm gels supported on alginate-Ca^{2+} with LCST tailored close to human body temperature. *Polymer Testing*, 25, 7, (October 2006), 961-969, 0142-9418.

Moura, M. R. de; Aouada, F. A.; Favaro, S. L.; Radovanovic, E.; Rubira, A. F. & Muniz, E. C. (2009). Release of BSA from porous matrices constituted of alginate-Ca^{2+} and PNIPAAm-interpenetrated networks. *Materials Science and Engineering: C*, 29, 8, (October 2009), 2319-2325, 0928-4931.

Panic, V.; Adnadjevic, B.; Velickovic, S. & Jovanovic, J. (2010). The effects of the synthesis parameters on the xerogels structures and on the swelling parameters of the poly(methacrylic acid) hydrogels. *Chemical Engineering Journal*, 156, 1, (January 2010), 206-214, 1385-8947.

Park, K.; Shalaby, W. S. W. & Park, H. (1993). *Biodegradable Hydrogels for Drug Delivery*, Technomic Publishers, 9781566760041, Lancaster.

Poujavadi, A.; Farhadpour, B. & Seidi, F. (2009). Synthesis and investigation of swelling behavior of new agar based superabsorbent hydrogel as a candidate for agrochemical delivery. *Journal of Polymer Research*, 16, 6, (November 2009), 655-665, 1022-9760.

Rao, K. S. V. K.; Rao, K. M.; Kumar, P. V. N. & Chung, I. D. (2010). Novel chitosan-based pH sensitive micro-networks for the controlled release of 5-fluorouracil. *Iranian Polymer Journal*, 19, 4, (April 2010), 265-276, 1026-1265.

Ritger, P. L. & Peppas, N. A. (1987). A simple equation for description of solute release II. Fickian and anomalous release from swellable devices. *Journal of Controlled Release*, 5, 1, (June 1987), 37-42, 0168-3659.

Rodkate, N.; Wichai, U.; Boontha, B. & Rutnakornpituk, M. (2010). Semi-interpenetrating polymer network hydrogels between polydimethylsiloxane/polyethylene glycol and chitosan. *Carbohydrate Polymers*, 81, 3, (July 2010), 617-625, 0144-8617.

Rodríguez, D. E.; Romero-Garcia, J.; Ramirez-Vargas, E.; Ledezma-Perez, A. S. & Arias-Marin, E. (2006). Synthesis and swelling characteristics of semi-interpenetrating polymer network hydrogels composed of poly(acrylamide) and poly(gamma-glutamic acid). *Materials Letters*, 60, 11, (May 2006), 1390-1393, 0167-577X.

Rokhade, A. P.; Shelke, N. B.; Patil, S. A. & Aminabhavi, T. M. (2007). Novel interpenetrating polymer network microspheres of chitosan and methylcellulose for controlled release of theophylline. *Carbohydrate Polymers*, 69, 4, (July 2007), 678-687.

Roy, A.; Bajpai, J. & Bajpai, A. K. (2009). Dynamics of controlled release of chlorpyrifos from swelling and eroding biopolymeric microspheres of calcium alginate and starch. *Carbohydrate Polymers*, 76, 2, (March 2009), 222-231, 0144-8617.

Sajeesh, S.; Vauthier, C.; Gueutin, C.; Ponchel, G. & Sharma, C. P. (2010). Thiol functionalized polymethacrylic acid-based hydrogel microparticles for oral insulin delivery. *Acta Biomaterialia*, 6, 8, (August 2010), 3072-3080, 1742-7061.

Saraydin, D.; Karadag, E. & Guven, O. (2000). Relationship between the swelling process and the releases of water soluble agrochemicals from radiation crosslinked

2. Diffusion model development

Figure 1 represents an idealized spherical capsule where pesticide is encapsulated by a polymer membrane. Pesticide must first diffuse across the membrane before becoming available for biological impact or subject to environmental dissipation mechanisms. The pesticide concentration within the capsule is greatest immediately following application and decreases with time as mass diffuses across the membrane. Characteristic length scales for diffusion include the membrane thickness and the environmental length scale associated with the pesticide transport distance once released. Numerous historical and mathematical descriptions for pesticide loss from microcapsules can be found elsewhere (Carslow and Jaeger, 1959; Collins and Doglia, 1973; Collins, 1974; Kydenieus, 1980, Coswar, 1981; Crank, 1993; Mogul et al, 1996)

C = concentration of pesticide within the capsule (assumed uniform throughout capsule).
C_o = initial concentration of pesticide within the capsule for $t \leq 0$.
C_s = concentration of pesticide at the membrane-environmental interface.
C_∞ = concentration of pesticide within the environment that is not impacted by the contribution from the capsule (typically equal to zero).
L = characteristic length scale for resistance to mass transfer within the environmental matrix where $C_{ext} \sim C_\infty$.
C_{ext}= concentration of pesticide outside the membrane wall and within the environment
a = capsule radius.
h = membrane thickness.
t = time.

What follows is a rudimentary, classical first-principles (conservation equations) approach for determining release rates from capsule release devices (Collins, 1974). The concentration of the pesticide at the membrane/environment interface (C_s) is approximately zero if mass transport away from the capsule is sufficiently rapid, leaving diffusion of pesticide across the membrane under these conditions as the rate-limiting step for environmental release.

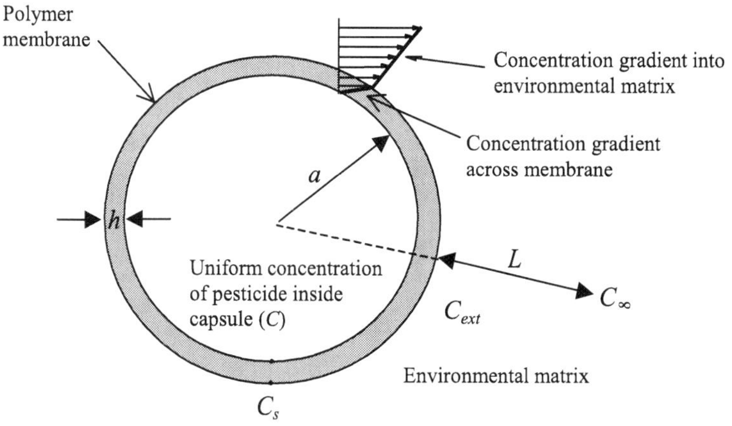

Fig. 1. Idealized cross-section of a spherical microcapsule.

The appropriate boundary condition at the membrane-environment interface is given by equating the mass flux loss from the membrane to the flux gain into the environment (Eq. 1);

$$D_m \left[\frac{\partial C}{\partial r}\right]_{m,\ r=a+h} = -D_{env} \left[\frac{\partial C}{\partial r}\right]_{env,\ r=a+h} \tag{1}$$

where:

D_m = diffusion coefficient of pesticide through membrane [$cm^2\ s^{-1}$]

D_{env} = diffusion coefficient of pesticide through environmental matrix (air, soil, water) [$cm^2\ s^{-1}$]

$\left[\frac{\partial C}{\partial r}\right]_{m,\ r=a+h}$ = concentration gradient of pesticide through membrane evaluated at the membrane-environment interface ($r=a+h$)

$\left[\frac{\partial C}{\partial r}\right]_{env,\ r=a+h}$ = concentration gradient of pesticide through environmental matrix evaluated at the membrane-environment interface ($r=a+h$).

The pesticide concentration gradients in the boundary condition of Eq. 1 are approximated as linear which, upon simplification, yields the concentration of pesticide at the membrane-environmental matrix interface (C_s).

$$C_s = \frac{C_\infty + \{\frac{D_m}{D_{env}}\frac{L}{h}\}C}{1+\{\frac{D_m}{D_{env}}\frac{L}{h}\}} = \frac{C_\infty + \alpha C}{1+\alpha} \tag{2}$$

where

$\alpha = \frac{D_m}{D_{env}}\frac{L}{h}$.

The parameter α is a measure of the relative diffusion magnitudes across the membrane and the environment. If the resistance to mass transfer within the environment is small ($D_{env} >> D_m$) then $\alpha = 0$ and Eq. 2 reduces to $C_s = C_\infty$.

The rate of diffusion (F) from the capsule across the membrane is given by Fick's first law

$$F = -D_m \frac{\partial C}{\partial r} = D_m \frac{(C - C_s)}{h} \tag{3}$$

where

F = Rate (flux) at which pesticide passes through the membrane [$g\ cm^{-2}\ s^{-1}$]

$\frac{\partial C}{\partial r}$ = Concentration gradient in radial direction across the membrane.

If one assumes the capsule keeps its spherical shape as pesticide is released through the membrane, the diffusion flux can also be represented as the rate of mass transport ($g\ s^{-1}$) across the surface area of the spherical capsule (A). The mass of pesticide inside the membrane is the volume (V) multiplied by the concentration (C). Thus,

$$F = -\frac{d}{dt}\left(\frac{VC}{A}\right) = \frac{d}{dt}\left(\frac{\frac{4}{3}\pi a^3 C}{4\pi a^2}\right) = -\frac{a}{3}\frac{dC}{dt} \tag{4}$$

where the volume and surface area of a sphere have been substituted. Equating Eq. 3 and Eq. 4 (and assuming $C_\infty = 0$), yields, upon simplification, a first order, separable, homogeneous ordinary differential equation, which simplifies to

$$\frac{dC}{dt} = -\frac{3\,D_m C}{a\,h(1+\alpha)} . \tag{5}$$

The solution to Equation 5 is

$$\frac{C(t)}{C_0} = \frac{M(t)}{M_0} = e^{-(\frac{3\,D_m}{a\,h})\frac{1}{(1+\alpha)}t} = e^{-0.693\frac{t}{\tau}} \tag{6}$$

where

$$\tau = 0.693\,\frac{a\,h(1+\alpha)}{3D_m} . \tag{7}$$

Equation 6 can be multiplied by the capsule volume to yield the mass remaining within the capsule. Therefore the dimensionless concentration ($\frac{C(t)}{C_0}$) or mass ($\frac{M(t)}{M_0}$) can be used interchangeably (since the capsule is assumed to remain spherical and of constant size over time). $M(t)$ is the mass of pesticide remaining within the capsule, and M_0 is the initial pesticide mass inside the capsule before diffusion losses begin. Here, τ is the half-life for release and is a function of the diffusion coefficient (D_m), membrane thickness (h), and capsule radius (a). τ will increase as the capsule radius, membrane thickness, and/or the parameter α increases and as the diffusion coefficient across the membrane decreases (Eq. 7). The diffusion loss from the capsule decreases as the mass transfer resistance in the environmental matrix increases ($\alpha\uparrow$, Figure 2). The actual amount of pesticide released from the capsule [$M_r(t)$] is determined by difference since the initial starting mass (M_0) is known and the mass at any time "t" [$M(t)$] is given by Eq. 6.

$$M_r(t) = M_0 - M_0\, e^{-0.693\frac{t}{\tau}} = M_0\,(1 - e^{-0.693\frac{t}{\tau}}). \tag{8}$$

Pesticide is often exposed to degradation and/or dissipation processes once outside of the capsule. Mechanisms for release and degradation/dissipation can be represented using a box model (Figure 3). Here, Q and R are the mass of pesticide inside and outside the capsule, respectively, and S represents the breakdown products of the pesticide. The mass of pesticide in the environment (i.e., R) is required from an efficacy standpoint. Thus, the material balance for R is written as

$$\frac{dR}{dt} = r_Q - r_R \tag{9}$$

where
r_Q = rate of mass transport of pesticide inside capsule to capsule surface/environment
r_R = rate of degradation of pesticide once outside capsule and in the environment.

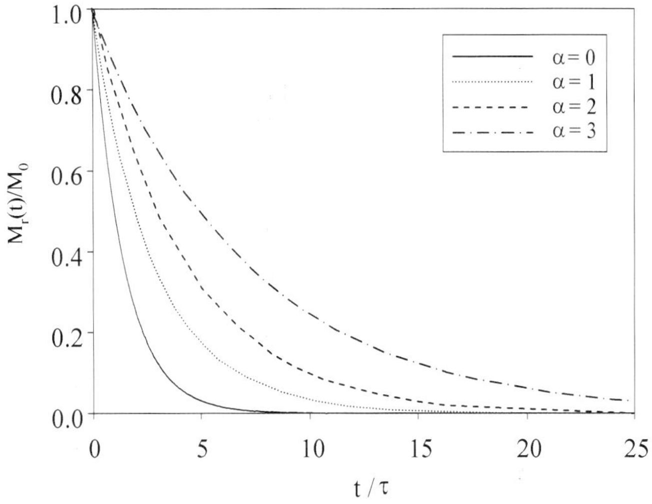

Fig. 2. Dimensionless mass of pesticide remaining in an idealized capsule as a function of time (dimensionless).

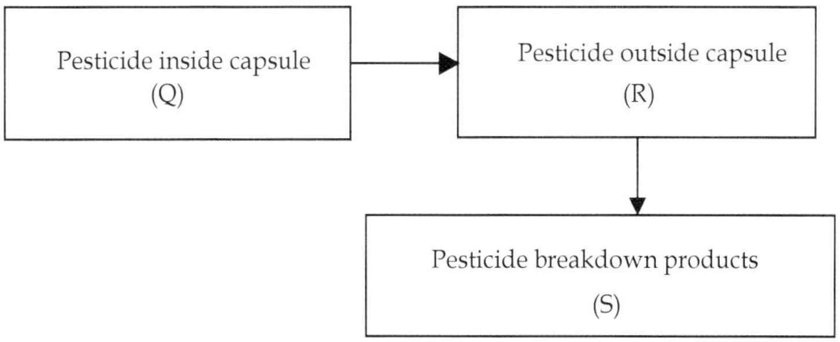

Fig. 3. Simple box model describing release rate and degradation of pesticide in the environment.

The rate of change of mass within the capsule (mass release rate, $\dot{M}_t$) is obtained by differentiating Equation 6 with respect to time.

$$\dot{M}_t = \frac{0.693 M_0}{\tau} e^{-0.693\frac{t}{\tau}} = r_Q \tag{10}$$

which is the rate of change represented by r_Q in the mass balance equation (Eq. 9). If the rate of degradation for pesticide once released into the environment is assumed to follow first order kinetics, then Eq. 9 can be written as

$$\frac{dR}{dt} + kR = \frac{0.693 M_0}{\tau} e^{-0.693\frac{t}{\tau}}. \tag{11}$$

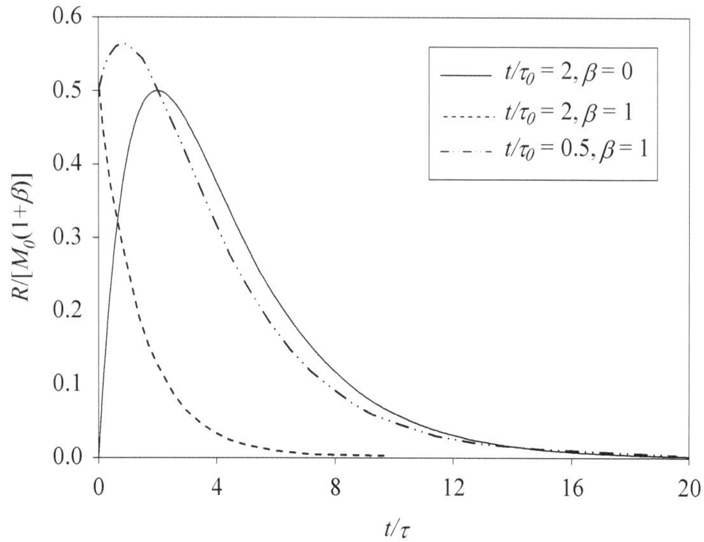

Fig. 5. Example of mass of pesticide in the environment for various ratios of capsule to degradation half-lives and conventional formulation addition ($\beta > 0$).

where

g_i = frequency (volume weighting) for capsules of size a_i (and membrane thickness h_i) within the distribution

f_i = mass fraction of capsule "i" size = (mass of capsules of size a_i)/ (total mass of all sizes).

The total amount of pesticide mass remaining within the capsule, assuming each capsule size (a_i, h_i) emits pesticide characterized by Equation 12, is the sum remaining for each unique, discrete capsule size distributions. The dimensionless scaling term γ has been added to account for different application rates for the pesticide. For γ = 1, then the scaled application rate is used. For γ = 2, then twice the scaled application rate is used, and so on. Here, M_{t0} is now defined as the total pesticide mass from all capsule sizes within the system at the time of application.

$$\frac{R}{\gamma M_{t0}} = \frac{1}{(1+\beta)}\{\sum_{i=1}^{n} f_i([\beta - \frac{1}{(\frac{\tau_i}{\tau_o}-1)}]e^{-0.693\frac{\tau_i}{\tau_o}(\frac{t}{\tau_i})} + \frac{1}{(\frac{\tau_i}{\tau_o}-1)}e^{-0.693(\frac{t}{\tau_i})}])\} \tag{17}$$

4. Application of single capsule diffusion equation

Characterizing pesticide release rates from microcapsules and relating them to biological observations provides an avenue to use mathematical models to design optimal release rate characteristics that yield maximum impact on the target pest (e.g., efficacy). To illustrate this principle, biological information and microcapsule release characteristics for a commercial herbicide for both a target (weed) and non-target plant (crop) species are determined via appropriate experimental observations, Figure 6. Linear regression lines have been drawn through each data set. Clearly, as the capsule release half-life increases, both target and non-target species injury decreases, and ideally, injury to non-target species should be minimal. However, efficacy will suffer as reductions in injury to non-target species are sought. The slope found in Figure 6 for the target species (e.g., weed) is greater than the slope for the non-target species. Thus, a change in τ has a more significant impact on weed species than the non-target species, and a conceptual optimal microcapsule release profile is anticipated that can maximize efficacy against the weed while minimizing injury to the non-target species.

A release half-life of approximately 400 hours is appropriate for this experimental data set if the goal is to have at least 90% control of the weed species (e.g., Figure 6). However, experiments used to generate Figure 6 were performed in controlled environments within a greenhouse where the dominant mode of dissipation for this herbicide (aqueous photolysis) did not occur. A laboratory aerobic aquatic degradation study yielded a half-life of 21 days, and an average soil dissipation half-life of 3.5 days for this herbicide. Thus, degradation is more rapid under field conditions then in the greenhouse where efficacy information was collected.

The modeling algorithm previously outlined can be useful to aid in formulation development. For this example, $\tau = 400$ hours, but the anticipated degradation half-life once released is $\tau_0 = 504$ hours (21 days) under aerobic, aquatic conditions. Assume five different capsule size distributions are available, as characterized by different size radius, to construct the optimal release profile, Figure 7. All distributions are log-normal, characterized by different means and standard deviations for capsule diameters. Capsule distribution mean radius values range from 2-200 μm, respectively. The diffusion coefficient across the membrane used was 3.6×10^{-13} cm^2 hr^{-1} and the membrane thickness (h_i) was approximated as 0.001 times the capsule radius (a_i). Thus, larger microcapsules had a thicker polymer membrane than smaller microcapsules.

The goal is to maximize efficacy while minimizing injury to the non-target crop, which simplifies to a mixing scenario where the unknowns are the volumes or mass fractions for each microcapsule size component (e.g., distributions 1-5, Figure 7). Mass fraction combinations of the different formulations were determined by a Levenburg-Marquardt procedure to minimize the sum of the squared residuals (ψ) between the greenhouse-measured optimal release profile (e.g. τ = 400 hours) and the predicted release patterns (Eq. 15 with $\beta = 0$) at selected time intervals 'i".

$$\psi = \sum_{i=1}^{n} (\text{Biological Observation}_i - \text{Predicted}_i)^2 \tag{18}$$

n = number of discrete time intervals over the mass release interval.

The greenhouse-determined optimal pesticide release pattern that maximized efficacy but minimized non-target plant injury is provided in Figure 8 (i.e., τ = 400 hrs, τ_0 = 21 days), along with two different "optimal" combinations of capsule mass fractions for the test problem. Here, the actual environmental degradation half-life is set equal to that

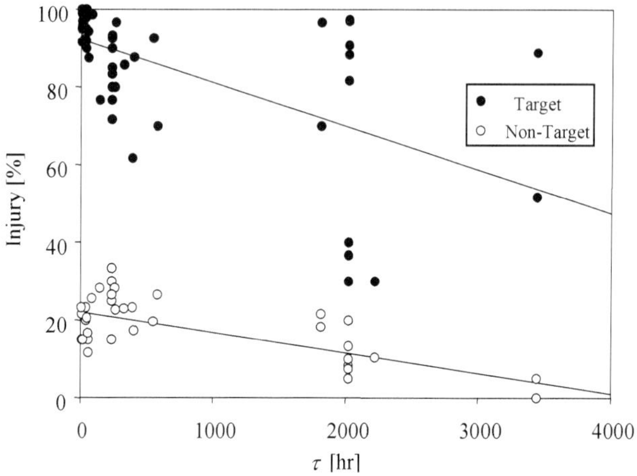

Fig. 6. Summary of greenhouse biological and laboratory determined release rate (τ= release half life) observations of plant injury for a commercial herbicide.

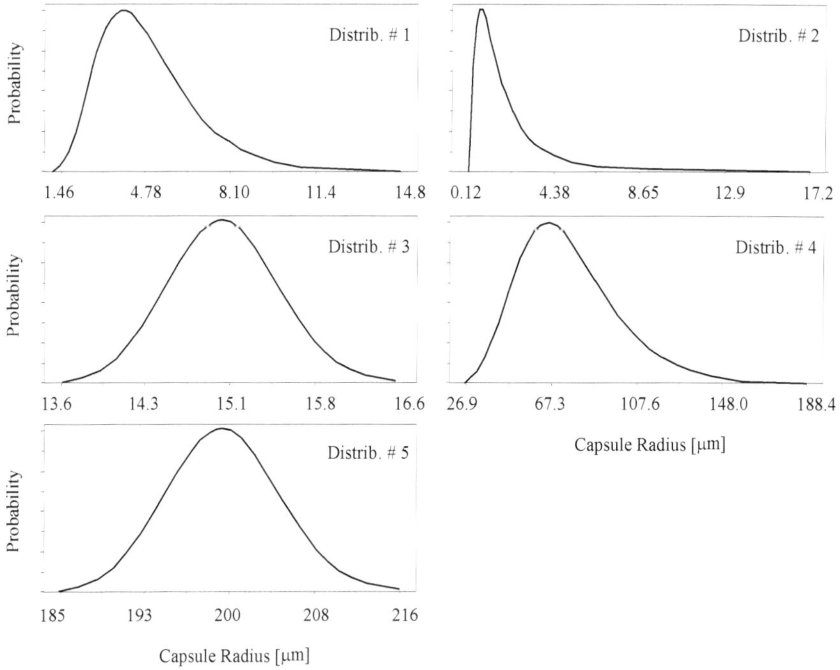

Fig. 7. Five capsule size distribution [radius (μm)] used in text example. The mean (μ) and standard deviation (σ) for all five distributions, given in increasing order are (5, 2), (2,2), (15, 0.5), 75, 25), (200, 5) for distributions 1-5, respectively.

approximated in greenhouse trials [e.g., τ_0 = 504 h = 21d]. The optimal concentration functions seen in Figure 8 represent predictions when γ and β are allowed to vary, and the application rate for field predictions is equal to that of greenhouse trials. The Solver routine of Microsoft excel (Branch and Bound solution technique) was employed to estimate the "optimal" parameter combinations for capsule mass fraction (f_i), γ, and β such that the objective function (Eq. 18) was minimized.

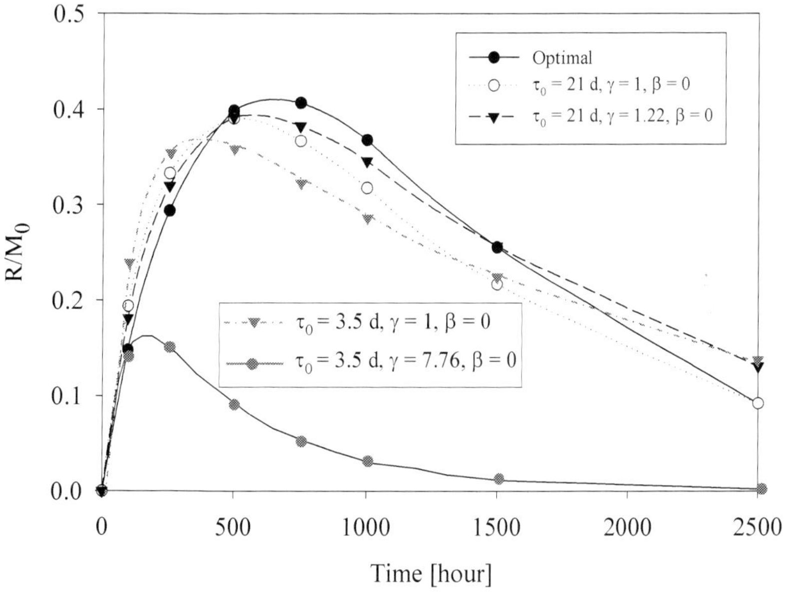

Fig. 8. Greenhouse determined optimal environmental matrix concentration with predicted "optimal" capsule formulation combinations when environmental degradation half-life (τ_0) is 21 or 3.5 days. γ and β parameters from Eq. 17.

Different mass fractions for capsule distributions of various sizes (e.g. Figure 7) ensue for each optimal calculation. Optimization results are summarized in Table 1 with optimal selection of mass fractions for the five different size formulations also provided. However, optimal results suggest the amount of conventional formulation added to the capsule formulations should be zero (β =0), and the resulting predicted concentration profile is similar to the biological optimal. Thus, one would expect field behavior for the first two entries found in Table 1 to yield similar results to greenhouse observations for this choice of inputs (assuming field dissipation occurs at a half-life of 21 days).

For the commercial herbicide used, the field dissipation half-life is approximately 3.5 days, while the green house aerobic aquatic half-life was estimated at 21 days. Therefore, one should *a priori* expect dramatically different behavior will ensue under field conditions. To appropriately mimic field behavior, the dissipation half-life was set to 3.5 days and the procedure for selecting mass-fractions of capsule distributions was repeated. Results of this exercise are also represented in Figure 8 and Table 1. What is evident in Figure 8 is the

Tarasevich, Y.Y. 2005. Simple analytical model of capillary flow in an evaporating sessile drop. Phys. Rev. E 71, 027301.

Widjaja, E., M.T. Harris. 2008. Particle Deposition Study During Sessile Drop Evaporation. AIChE J. 54:9, 2250-2260.

8

A New Technique for Safe Pesticide Spraying in Greenhouses

Prof. Guido Belforte, Ing. Gabriella Eula and Prof. Terenziano Raparelli
Department of Mechanics - Politecnico di Torino
C.so Duca degli Abruzzi, 24 - 10129 Torino
Italy

1. Introduction

Protection from parasites is an important factor in agricultural operations, and calls for continual monitoring and prompt action when needed. In many cases, the equipment, pesticides and manpower required for this purpose account for the majority of production expenses.

Though the use of chemicals has had a major influence on the development of agriculture in the twentieth century, bringing significant benefits, it has also had many side effects on human health, animals and the environment.

The ease with which these substances can be used, their initial low cost and the lack of knowledge on the part of growers has led to an overuse of pesticides, with dangerous consequences. Only in the last few years have agricultural techniques brought about improvements in the pesticides used (Hewitt, 2000).

There can be no doubt that advances in this field have provided a more effective range of choices, for growers and the environment in particular.

Parasites must thus be combated by producing a climate that is unfavourable for them, as well as by using a forceful, accurate and incisive spray technique.

To reduce pesticides by using more effective treatments, recent studies have investigated different spray techniques capable of reducing the pesticide dose with very low waste and outflow (Austerweil & Grinstein, 1997).

Efficacy of crop spraying depends on two main factors: coverage density and uniformity, and droplet size.

For the first factor, it should be emphasised that droplets should reach leaves without any overlapping.

The second factor is important because many studies have shown that coarse droplets reduce spraying treatment efficacy. In fact, smaller droplets penetrate the canopy better and, transported by air, can reach each part of the crop without dispersion. This is especially true in the greenhouse, where wind is not a problem.

These considerations have spurred interest in the idea of spraying pesticides in a defined volume, i.e., a confined area (Moltò et al, 2001) (Ebert et al, 2003).

The first step in designing an innovative pesticide sprayer is to study various spraying techniques.

In particular, investigating correct pesticide distribution entails considering the environment where the treatment is carried out, the crop growth rate and the characteristics of the chemical product used (Gil & Sinfort, 2005).
Standard atomizers are the most widely used equipment for crop treatments. They are based on three different designs: pneumatic atomizers, mechanical atomizers and mix atomizers (Cerreto & Failla, 2003) (Braekman et al, 2009).
These machines are employed both in the open field and in greenhouses, as they are highly reliable and easy to use, though their disadvantage is that the operator is directly exposed to chemicals. Consequently, the operator is forced to use eye protection, rubber gloves, and filter masks. All of this equipment is essential in order to avoid contact between pesticide and human skin or airways (Methner & Fenske, 1996) (Paul & Illing, 1997) (Nuyttens et al, 2004) (Nuyttens et al, 2009).
Another limitation of these machines consists in the absolute lack of control over the sprayed dose, which can be influenced by the nozzles' height from the ground and their orientation.
In addition, droplet size depends on operating pressure and nozzle type. Though a large number of models offering various levels of performance are available on the market, many growers base their choices more on economic considerations than on any knowledge of the nozzles' actual technical characteristics (Derksen & Sanderson, 1996) (Briand et al, 2002) (Ade et al, 2003).

2. Greenhouse spraying requirements

Spraying techniques used by many growers involve a number of significant problems.
Environment conditions in the greenhouse increase the incidence of plant diseases affecting common crops, because of intensive cultivation in an artificial ambient with heavy use of plant food. Consequently, the high value of the crops often grown in greenhouses can justify the use of more expensive protection treatments (Kondo et al, 1996) (Christensen et al, 2008).
Moreover, pesticides are used weekly throughout the entire production cycle (Maertens et al, 2005) to prevent aphids, mites and fungi.
In such cases, pesticides are generally atomized by means of mechanical sprayers.

2.1 Requirements for the new greenhouse pesticide spraying system

An effective system for treating crops in greenhouses should satisfy the following requirements (Naoki et al, 1995) (Mandow et al, 1996) (Acaccia et al, 1998) (Tian et al, 1999) (Nuyttens et al, 2003) (Singh et al, 2005) (Solanelles et al, 2006) (Piccarolo, 2008).

- **Quantity and quality of spraying control:** the crop must receive the correct dose of pesticide, as excessive doses increase costs and can cause damage to the environment and the crop. Conversely, insufficient doses cannot provide protection against pathogens, especially on the underside of the leaves. Delivering the correct dose is a question of controlling droplet size.
- **Efficient treatment with low pesticide doses:** it is important to avoid excessive pesticide doses, both for the operator and for the environment. This can be achieved by reducing pesticide loss by confining the spray within a defined volume where crops can be treated.
- **Operator safety**: the danger involved in using pesticides is underscored by the fact that operators are required to pass an exam before handling these chemicals. Operator safety could be guaranteed by a higher level of automation in pesticide spraying.

- **Ease of use and reliability**: the sprayer must guarantee that the operator's work can be carried on correctly and without difficulty, avoiding overly complex technical procedures. As such systems are currently required, it is necessary to develop a new device capable of satisfying all of these characteristics.
- **Flexibility:** the crop's volumes, geometries, leaf density, etc., all influence pesticide distribution, as do the grower's specific needs. Accordingly, the new machine must feature a wide range of regulations (nozzle orientation and height from ground, coverage area, etc...) so that it can be adapted to various cultivars.
- **Economy:** using an automatic pneumatic system makes it possible to obtain easy, durable and economical technical solutions. Though the market now provides a wide range of choices, few spraying techniques can satisfy all of these requirements together.

3. Design of the new system

Development of a new defined-volume sprayer involved the following stages:
- theoretical and experimental study of very fine pesticide fog generation to cover all parts of the canopy;
- study of a textile cover sheet capable of enclosing the spray area;
- construction of preliminary prototypes;
- design and testing of the system used to move the textile cover sheet;
- final testing in both the laboratory and the greenhouse.

All of these steps will now be analyzed.

4. Fog generation

The first step in developing an innovative pesticide spraying technique is an efficient and reliable fog generation system (Ade & Fabbri, 2000). To this end, many nozzle models were tested. All tests used a mixture of water and air, because the viscosity and the surface tension properties of this mix are similar to those of pesticide solutions in water (Singh et al, 2006) (Singh & Kumar, 2007) (Singh, 2007).

4.1 Atomizer nozzles

A pneumatic atomizer nozzle is a small mechanical component capable of generating a fine fog of droplets using a compressed gas. It has two inlet ports, one for gas and the other for liquid, as shown in Figure 1, and an outlet port where the mixture is produced (Tecsi, 2006). There are many models of atomizer nozzle.

a. *Internal mix model*

In this case, compressed air and liquid are mixed inside the component. It is suitable for fine fog generation (Figure 2a).

b. *External mix model*

In this model, compressed air and liquid are mixed outside the nozzle chamber (Figure 2b).

c. *Jet impact model*

An air stream mixes with the liquid outside the nozzle chamber. Using two nozzles makes it possible to reduce droplet dimensions by the impact of the two jets (Figure 2c).

Each nozzle is then characterized by its spray jet pattern. This pattern is the jet section at a plane orthogonal to the jet axis, in a defined outlet position. There are, generally, conical spray jets (full or hollow cone) and flat fan spray jets. The pattern which is best for the

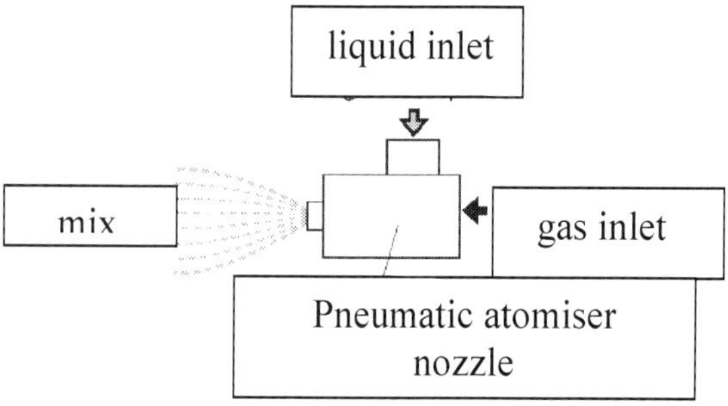

Fig. 1. Functional schematics of an atomizer nozzle

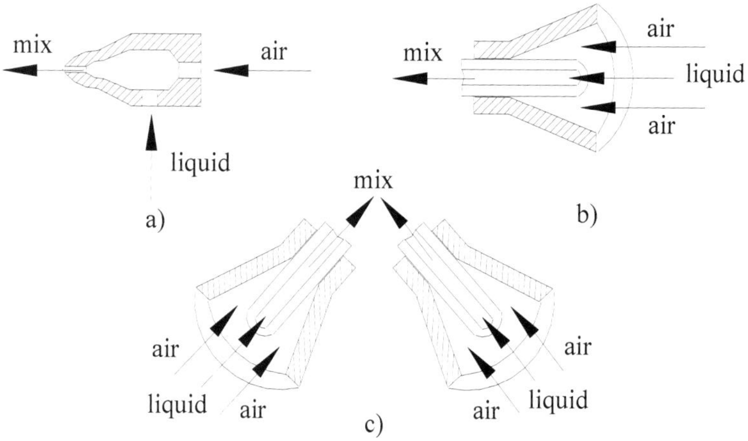

Fig. 2. a) internal mix; b) external mix; c) jet impact mix

specific application can be chosen. The nozzle spray pattern changes significantly with the distance between the target surface and the nozzle: increasing this distance also increases spray pattern diameter. In general, full cone jets have a smaller aperture angle than hollow cone jets. Atomizer nozzles are then classified by the operating pressure used for both air and water. Nozzle selection is also a question of droplet size, which is not an easy parameter to measure.

The basic thing is, obviously, to use the same method to compare droplets from different nozzles (Bouse, 1994) (Paice et al, 1995) (Nuyttens et al, 2007).

Droplets can be classified in three size groups. Droplets less than 10 μm in diameter generate what is called "dry fog", droplets whose diameter is between 10 and 100 μm form "fine fog", while diameters between 100 and 300 μm form "semi-fine fog" .

In this investigation, several full cone nozzles were tested, finally choosing the nozzle shown in Figure 3. This internal mix model was assembled together with another nozzle to obtain a jet impact mix (Figure 4) generating very fine fog (less than 50 μm).

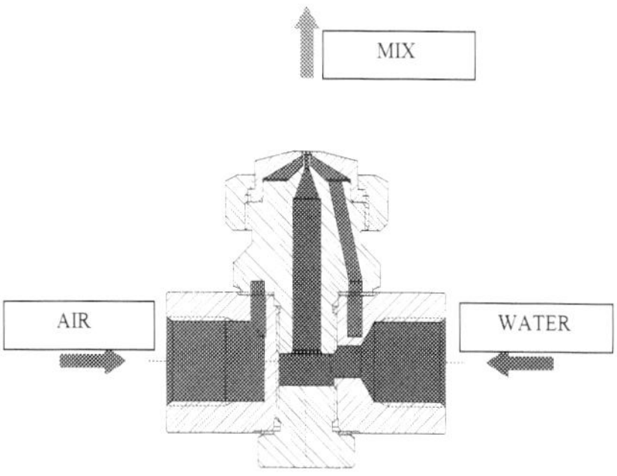

Fig. 3. Selected atomizer nozzle section

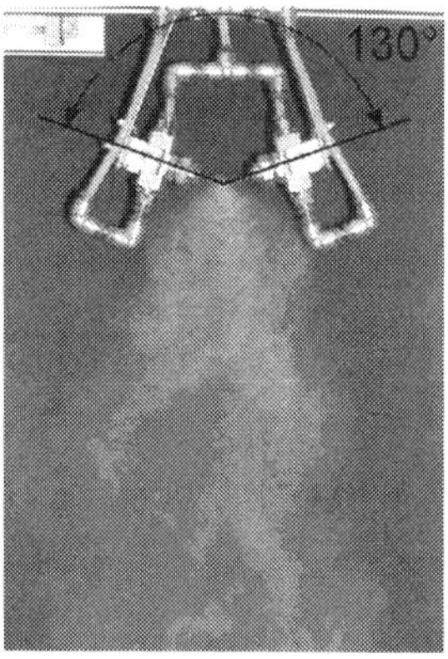

Fig. 4. Spraying technique with jet impact method

4.2 Experimental tests on nozzles

On the basis of a method developed at the Politecnico di Torino Department of Mechanics to measure atomized oil droplets in pneumatic circuits (Belforte et al, 1996), a special test-bench was designed for measuring nozzle droplet diameters. The method entails projecting a water-air spray against water-sensitive cards (Salvarani, 2006). They have a coated, yellow surface that is stained blue through a chemical reaction when contacted by water droplets or moisture. The cards are attached to a fixed wall, while a perforated moveable plate is placed between the wall and the nozzle in order to regulate spraying time and the surface exposed to the spray. The exposed cards are then examined under a fiberoptic microscope at x200 magnification. To improve analysis, enlarged specimens were also analyzed using imaging software (Cruvinel et al, 1996) (Kashden et al, 2006) (Qing et al, 2006).
The test bench is shown in Figure 5. It consists of a metal frame supporting the nozzles and a receiving screen to which the cards to be exposed to the spray are attached.
A rodless pneumatic cylinder moves the interrupter plate with a central hole measuring 120-40-26 mm in diameter which cuts the spray jet and establishes the time period for which the cards are exposed to the spray.

Tests were carried out with different types of nozzle, projecting droplets both horizontally and vertically. In this case, spraying conditions are adversely affected by gravity, because the highest droplets reach all of the leaves.
Cards were analyzed to construct graphs as shown in Figure 6, which refer to two nozzles assembled as in Figure 4. This graph shows percentage droplet dimensions in five consecutive tests: as can be seen, most of the droplets are less 50 μm in diameter, with a supply pressure of $1.6*10^5$ Pa for air and $0.5*10^5$ Pa for water.

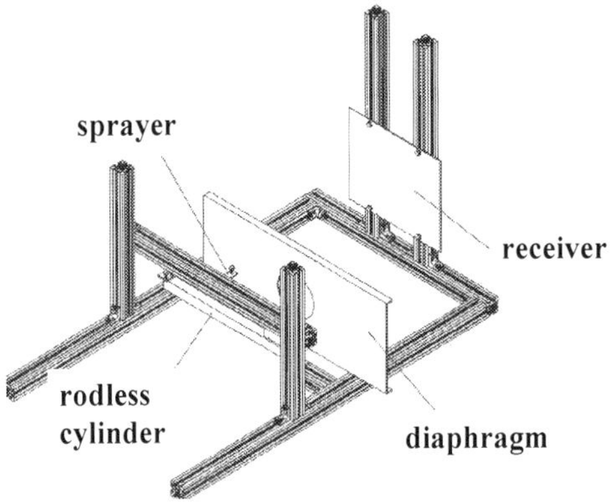

Fig. 5. Test-bench for experimental nozzle validation

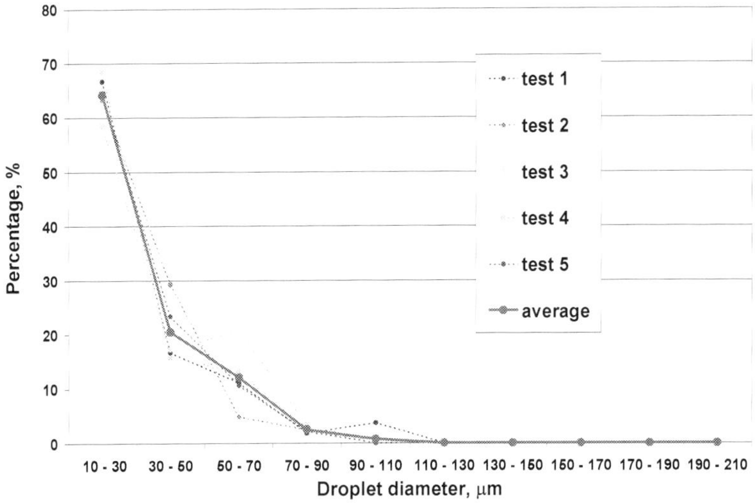

Fig. 6. Percentage dimensions of droplets

A Pitot tube was connected parallel to and in front of the spray jet (x axis) to measure spray profile and droplet velocity close to the leaf (Figure 7) (Belforte et al, 2009).

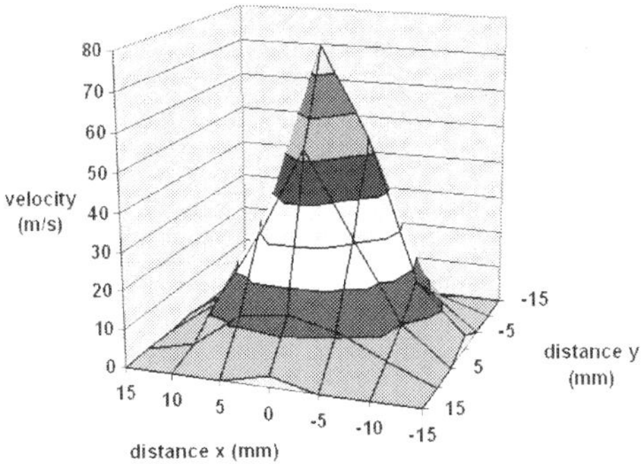

Fig. 7. Pitot tube results

4.3 Numerical simulation of leaf spraying

To assess the efficacy of spraying crops with these nozzle configurations, a numerical simulation was conducted to investigate droplet trajectories close to the leaves, in the presence of gravity (Lebeau, 2004).

In addition, an axially symmetric scale model was constructed which reproduces the nozzle, a leaf and the confined volume around it which represents the chamber where spraying takes place. In this model the leaf is simulated by a rigid 20 mm radius disk placed facing the spray jet at a distance of 280 mm from the nozzle outlet port.

Simulation was carried out by establishing the mass flowrates and the supply pressures measured experimentally at the nozzle input port, viz., 0.00018 kg/s and $1.6*10^5$ Pa for air, 0.00066 kg/s and $0.5*10^5$ Pa for water, temperature 300 K, steady-state flow.

In particular, simulation used a bi-phasic air-water mixture as the operating fluid to provide a better approximation of the experimental results. Droplet velocities near nozzle and leaf are shown in Figure 8.

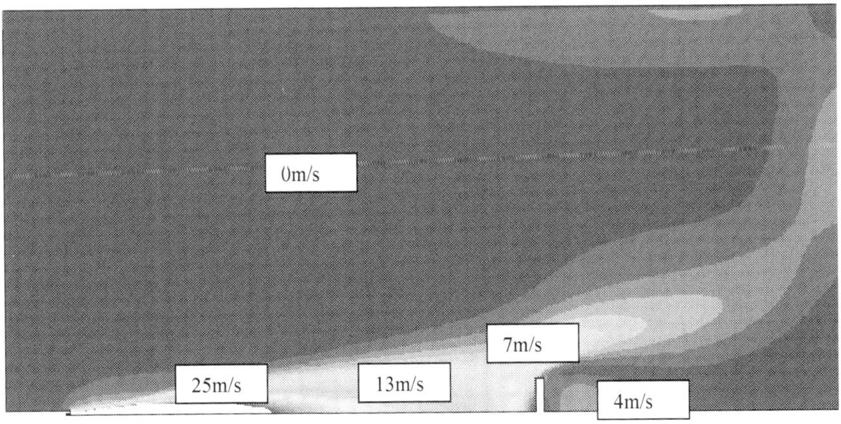

Fig. 8. Droplet velocities near leaf, with a bi-phasic air-water mixture

4.4 Experimental tests with various crops

Experimental spraying tests were carried out to assess droplet rebound and the resulting pesticide deposition on the top surface and underside of the leaf.
The test bench used for this purpose is shown in Figure 9. It consists of a rigid chamber of defined volume in which both nozzles and leaves to be sprayed are placed.
Initially, one leaf suspended from a small bar parallel to the ground was placed in the chamber and exposed to spray. The spray contained a yellow UV phosphorescent dye.
After treatment, the leaf was viewed under a UV lamp, where areas covered by the spray appear yellow and those that remain uncovered appear violet.
Surfaces were photographed using a digital camera to compare different crops in various test conditions.
Test parameters included crop type, type of ground surface, distance between leaf and ground, exposure time, and spray jet orientation. Tests were carried on using the following three types of leaf: flat, oily leaf (Cyclamen); smooth, irregularly shaped leaf (Pelargomium domesticum - geranium); flat, velvety leaf (Saintpaulia jonatha - African violet).
Crops were chosen on the basis of observations of the leaf surface.

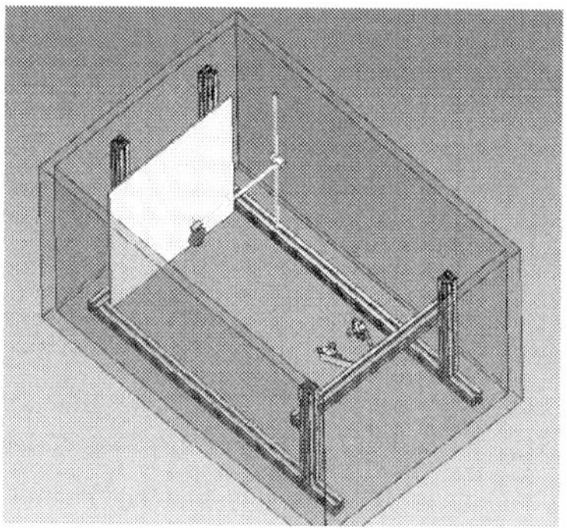

Fig. 9. Test bench for experimental tests on various crops

4.5 Test parameters and results

As the type of material used to cover the greenhouse benches on which potted plants are generally placed can influence droplet rebound towards the leaves, four types of surface in common use were considered, viz., stainless steel, linoleum, kraft paper and clay.
The distance between leaf and rebound surface was then varied, with all other parameters remaining constant.
This was done in order to simulate actual exposure conditions, as plants may stand at various distances between nozzles and ground. Tests were carried out at distances of h = 120-180-240 mm.
As an excessive dose of pesticide can damage leaves (causing spots and drying), it is necessary to stop spraying at the right time, before pooling and dripping take place. Spraying times used during test were approximately 5 and 10 s.
Test were carried out both horizontally and vertically. The test bench is shown in Figure 9.
Experimental results show that:

- the optimal distance between plant and ground is influenced by plant type, by exposure time and by the ground material;

- the underside of the leaf is difficult to reach unless vortexes around the leaf can be taken advantage of;
- droplets rebounding off the ground surface can reach the underside of the leaf more readily.

Figure 10 illustrates the results obtained with horizontal spray on various kinds of plant. Whether or not leaves are wrinkled affects spray coverage, while droplet rebound is affected by the type of ground surface.

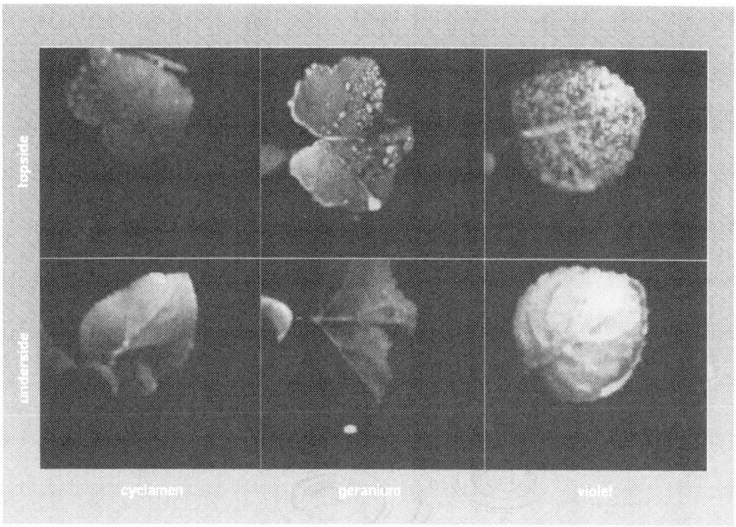

Fig. 10. Spraying tests on various plants

5. Defined volume

To spray pesticide in a confined area (Panneton et al, 2000) (Planas et al, 2002), a closed chamber with a metal top cover, pneumatic side walls, and a bottom surface consisting of soil or greenhouse bench covering was initially analyzed. Two atomizer nozzles were placed inside this chambers. Pneumatic walls are supplied with compressed air, thus producing a laminar fluidic layer. A large number of experimental tests were carried out to evaluate the behaviour of these pneumatic walls, especially as regards their interaction with the fine fog generated by the nozzles. Supplying pneumatic walls at a pressure of $6{*}10^5$ Pa proved unsuccessful, as vortexes generated on the ground interacted with the fog, spreading it all over the chamber. This is illustrated in Figure 11, where the arrows show the pesticide fog limited only in the upper part of the defined-volume chamber.

In addition, it was found necessary to supply pneumatic walls around the chamber at $5{*}10^5$ Pa for them to be effective. At this pressure, however, the pesticide fog, though limited to the chamber, caused too much ground and crop wetting for coalescent droplets to be produced.

Experimental tests thus indicated that pneumatic walls are not a good solution for this kind of fog application.

Consequently, a new chamber was designed with an rigid upper plate and retractable side walls consisting of an appropriate textile material: in this way, the chamber can be retracted when it is moved from one work position to the next, or when it is not in use.
The main requirement for this textile covering is absolute impermeability to air and to pesticide vapours, though it must also be lightweight and capable of resisting corrosive chemicals. For these reasons, various textile materials impermeable to liquid and to air were investigated, also attempting to find ways of guaranteeing these properties at the seams. Accordingly, seams were thermally bonded to prevent pesticide loss through needle holes. Results of air and water tightness tests on the chosen textile material were good.

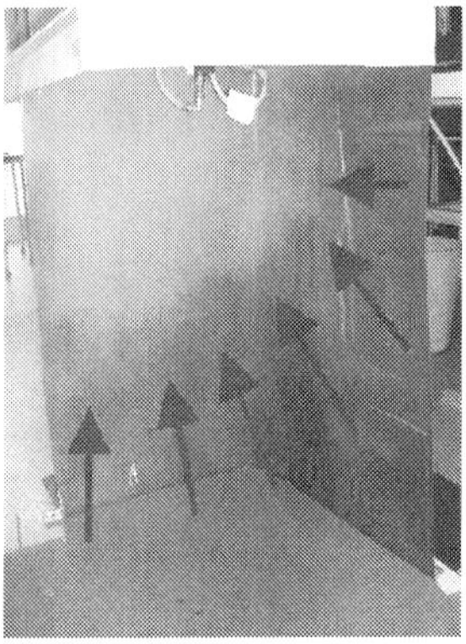

Fig. 11. Pneumatic walls supplied at $6*10^5$ Pa and 700 mm from ground

6. Fixed covering prototypes

Two preliminary fixed covering prototypes were constructed to assess spraying efficacy with the selected nozzles.

6.1 First fixed-covering prototype

An initial test bench for assessing nozzle efficacy was assembled using a metal frame for the textile walls, two pneumatic nozzles, a pneumatic control circuit, water-sensitive cards, flowmeters and manometers (class 1). The first step in experimental testing was to perform a three-dimensional evaluation of the fog generated in the chamber. A metal tree structure with a central trunk and six lateral branches was constructed to support the water-sensitive cards. To check plant coverage by the sprayed pesticide, twelve card positions were established, at 300-200-100 mm from the ground, at the center and edges of the tree. Cards were also positioned to simulate leaf top surfaces and undersides as shown in Figure 12a. Card placement on a tree is illustrated in Figure 12b. Using this metal tree, several preliminary tests were carried out with different nozzle orientations, varying supply pressure and exposure time. In this way, the best spraying conditions were identified whereby a very fine and concentrated fog can be produced. Further experimental tests were performed using a yellow UV phosphorescent dye in the spray. In this case, tests were conducted using real flowers. Results are indicated in Figures 13 and 14, using two nozzles assembled as shown in figure 4.

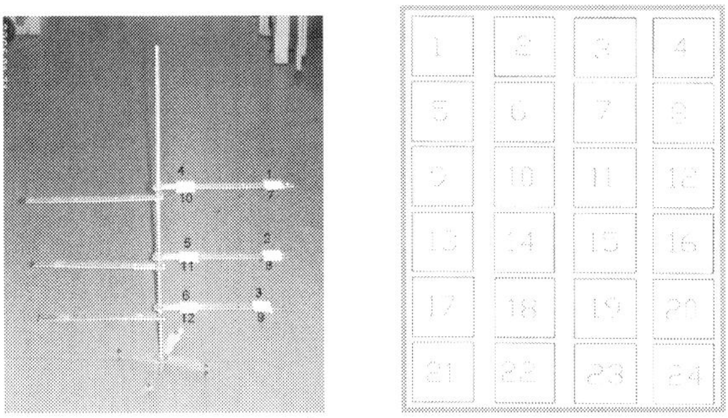

a) b)

Fig. 12. a) Card placement on metal tree b) Card numbering on the metal tree

Test parameters are illustrated in Table 1, including air and water supply pressure and flowrate, exposure time t_s, and time following treatment t_e (time between spraying and off-target pesticide recovery).

Fig. 13. Actual flower sprayed with phosphorescent dye

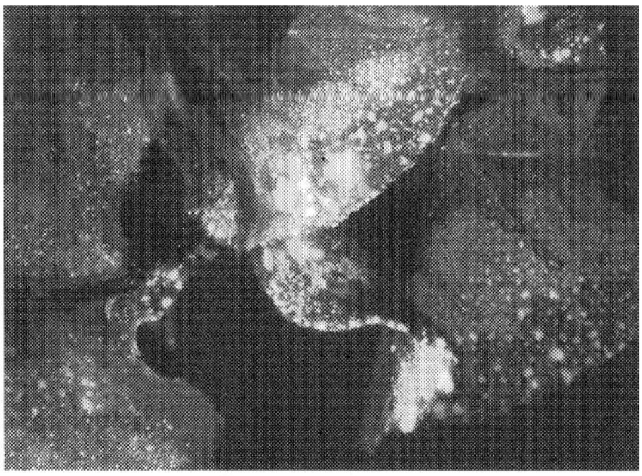

Fig. 14. Enlarged view of a treated crop

8. DeVoPeS work cycle

The DeVoPeS work cycle is as follows.

- The DeVoPeS is positioned over the area to be sprayed.
- The pantograph cylinder is actuated to spread the cover sheet.
- The corrugated tubing is extended to lower the side curtains.
- Pneumatic chambers are inflated at top and bottom. Top chambers stiffen the structure, while bottom chambers both stiffen the structure and create a seal.
- Air and pesticide are supplied to the atomizers.
- The rodless cylinder moves the atomizers to distribute pesticide uniformly on the plants.
- The atomizers are automatically shut off after a certain number of passes over the plants.
- The rodless cylinder stops.
- The aspiration system is activated to remove off-target pesticide, deflating the bottom sealing chambers.
- The corrugated tubing is retracted to raise the side curtains.
- The pantograph mechanism closes so that the system can be stowed or moved for a further spray cycle in another location.

The pneumatic control circuit used for DeVoPeS is shown in Figure 26. It consists of two pneumatic cylinders, one for the pantograph and the other for moving the nozzles over the plants; two ejectors, one for the corrugating tubes and the other for the pesticide recovery system; a number of solenoid valves for supplying actuators, ejectors, lower sealing chambers and nozzles for both air and water.

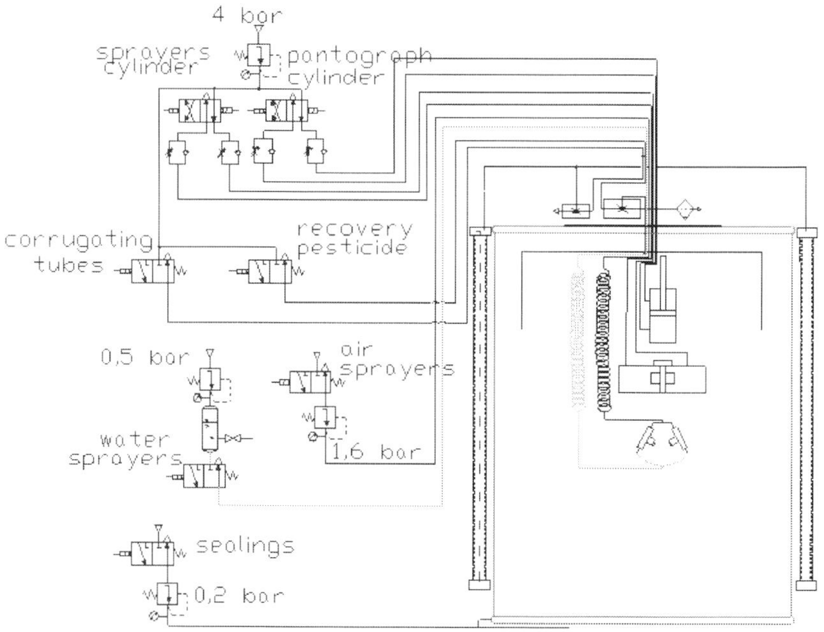

Fig. 26. DeVoPeS pneumatic control circuit

9. Laboratory testing

A large number of laboratory tests were conducted on the entire system to assess cycle times, air and water consumption, and retractable cover sheet movement.
The whole work cycle is accomplished in about 1.5 minutes.
Figure 27 shows overall DeVoPeS consumption during a work cycle. It should be emphasized that the largest consumption is due to the ejectors which, however, never work together. These ejectors generate a vacuum level of about $-0.6*10^5$ Pa when supplied at $4*10^5$ Pa as supply pressure, with a flowrate of 0.0015 m^3/s (ANR).
The cycle is automated by means of a PLC (Programmable Logic Controller) that receives a signal from end stroke actuators and sends commands to valves using timers. This PLC has 23 inputs and 15 outputs.

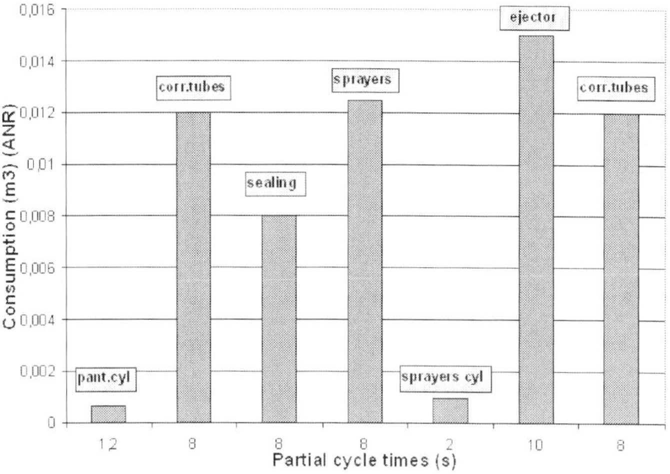

Fig. 27. DeVoPeS consumption in a work cycle

10. Greehouse testing

For greenhouse testing, DeVoPeS was connected to a robot capable of moving it onto plants. The DeVoPeS work area on a greenhouse bench is shown in Figure 28, together with the robot and the bench dimensions. It should be noted that the DeVoPeS was constructed as a half-scale prototype for demonstration purposes. Figure 29 shows the DeVoPeS connected to the robot. The robot is controlled by special position control software with NI PXI electronic cards and has a maximum acceleration of about 4 m/s^2 (Belforte et al, 2006) (Belforte et al, 2007) (Belforte et al, 2008). The DeVoPeS flow chart can be divided in two cycles:

- the first is used to move DeVoPeS by robot to the treatment area on the bench (the cover sheet is retracted in this phase);
- the second is used to carry out the treatment on plants with the work cycle described earlier.

Figure 30 shows DeVoPeS working in a greenhouse. Experimental tests yielded good results, demonstrating the usefulness of this new pesticide spraying technique.

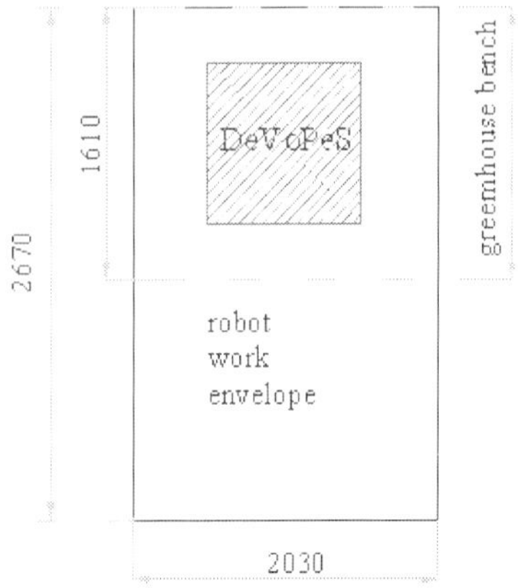

Fig. 28. DeVoPeS work area

Fig. 29. DeVoPeS on robot

Fig. 30. DeVoPeS working in greenhouse

11. Conclusion

A new spraying technique for safely distributing pesticides was investigated.
The spraying technique was studied using both numerical and experimental methods.
Overall results are good and provide an understanding of the interaction between leaf and droplets.
In particular, experimental tests on atomizer nozzles indicated that nozzles must be moved over the crop in order to achieve uniform pesticide distribution.
A new prototype called the DeVoPeS which can spray pesticide inside an enclosed, airtight chamber was designed, constructed and tested.
This machine offers a number of advantages: treatment is fully confined so that it does not affect the outside environment; the operator can remain in the greenhouse during spraying; pesticide losses are sharply reduced, increasing safety for both growers and the environment; the off-target pesticide recovery system provides economic benefits. DeVoPeS is currently a half-scale prototype, though its dimensions could readily be increased in the future to cover an entire greenhouse bench.

12. Acknowledgments

The investigation was funded by the Italian Ministry of Education, Universities and Research as part of the program *"Mechatronic and Pneumatic Systems for Safe, Environmentally Friendly and Competitive Automation in Floriculture"*.
The authors would like to thank the CeRSAA regional agricultural experimental station of Albenga (Savona - Italy) for its assistance. The authors would also like to thank MetalWork and Rockwell Automation for their cooperation and for supplying materials.

13. References

Acaccia G.M.; Callegari M.; Michelini R.C.; Molfino R.M. & Razzoli R.P. (1998). Greenhouse's Automation: Produce Tillage and Phytopatologies Treatments. *Proc. Of The 4th Ecpd Intl. Conference On Advanced Robotics, Intelligent Automation And Active Systems*, Moscow, 339-343.

Ade G. & Fabbri A. (2000). Theoretical Analisys on Bonds Between Covering Plants, Numerousness and Distribution of Droplets on Hydro-Sensible Cards (In Italian). *Riv. di Ing.Agr.*, 2, 104-108, ISSN: 1974-7071

Ade G; Rondelli V & Venturi P. (2003). Effects of Regulation in a Sprayer for Vineyard Studying Performance and Distribution in Some Test-Benches (In Italian). *Rivista Di Ing. Agr.*, 4, 17-25, ISSN: 1974-7071

Austerweil M. & Grinstein A. (1997). Automatic Pesticide Application in Greenhouses. *Phytoparasitica*, 718-808, ISSN: 0557-1413

Belforte Guido; Ferraresi C. & Raparelli T. (1986). Testing the Performance of Oil Fog Lubrificators. *Int.Fluid Power Symposium*, Bath, U.K., 359-368

Belforte Gustavo, Deboli R., Gay P., Piccarolo P. & Ricauda Aimonino D. (2006). Robot Design and Testing for Greenhouse Applications. *Biosystems Engineering*, 95, (3), 309-321, ISSN: 1537-5110

Belforte Gustavo, Gay P., Piccarolo P. & Ricauda Aimonino D. (2007). Robots for Greenhouses: Challenges and Opportunities. *Landwards*, 62, (1), 6-9, ISSN: 1363-8300

Belforte Gustavo, Gay P., Piccarolo P., Ricauda Aimonino D. & Deboli R. (2008). Horticultural Robotics: Seven Years of Experimentation. *Acta Horticulturae*, 801(1), 573-578, ISSN: 1369-8869

Belforte Guido; Eula G. & Raparelli T. (2008). Innovative Prototype for Automatic Pesticide Spraying in Greenhouses. *Proc. Of The Raad 2008*, 17th Int. Workshop On Robotics in Alpe-Adria-Danube Region, Ancona, Italy

Belforte Guido; Eula G. & Raparelli T. (2009). Analysis of Pneumatic Techniques for Pesticide Spraying in Greenhouses. *Experimental Techniques*, 33, Vi, 21-29, ISSN : 1369-8869

Belforte Guido; Eula G. & Raparelli T. Devopes: Defined Volume Pesticide Sprayer: Design and Testing. *Experimental Technique*, doi: 10.1111/J.1747-1567.2010.00642.X.

Bouse L.F. (1994). Effect Of Nozzle Type and Operation on Spray Droplet Size. *Transaction of Asae*, 37, 1389-1400, ISSN: 0001-2351

Braekman P.; Foquè D.; Van Labeke M.C.; Piteres J.G. & Nuyttens D. (2009). Influence of Spray Application Technique on Spray Deposition in Greenhouse Ivy Pot Plants Grown On Hanging Shelves, *Hort Science*, 44(7), 1921-1927, ISSN: 1592-1573

Briand O.; Bertrand F.; Seux R. & Millet M. (2002). Comparison of Different Sampling Techniques for the Evaluation of Pesticide Spray Drift in Apple Orchards. *The Science Of The Total Environment*, 288, 199-213, ISSN: 0048-9697

Cerreto E. & Failla S. (2003). Tests of Droplets Distribution in Vineyards by Sprayer with Mechanical Pulverization (In Italian). *Rivista Di Ingegneria Agraria*, 3, 13-22, ISSN: 1974-7071

Christensen S.; Lund I. & Jolsen H. (2008). Spraying Robots. *Aspects of Applied Biology 84 – Int.Advances in Pesticide Application*, 167-170, ISSN: 0265-1461

Cruvinel P.E.; Minatel E.R.; Mucheroni M.L.; Vieira S.R. & Crestona S. (1996). An Automated Method Based on Image Processing for Measurements of Drop Size Distribution from Agricultural Sprinklers. *Anais Do Ix Sibigrapi*, 39-46, ISSN: 0717-5000

Derksen R.C. & Sanderson J.P. (1996). Volume, Speed and Distribution Technique Effects on Poinsettia Foliar Deposits. *Transactions of The Asae*, 30 (1), 5-9, ISSN: 0001-2351

Ebert T.A.; Derksen R.C.; Downer R.A. & Krause C.R. (2003). Comparing Greenhouse Sprayers: the Dose-Transfer Process. *Pest Management Scienc*, 60 (5), 507-513, ISSN : 1526-4998

Gil Y. & Sinfort C. (2005). Emission of Pesticides to the Air During Sprayer Application: A Bibliographic Review, *Atmospheric Environment*, 39, 5183-5193, ISSN: 1352-2310

Hewitt A.J. (2000). Spray Drift: Impact of Requirements to Protect the Environment. *Crop Protection*, 19, 623-627, ISSN: 0261-2194

Kashden J.T.; Shrimpton J.S. & Whybrew A. (2006). A Digital Image Analysis Technique For Quantitative Characterization Of High Speed Sprays. *Optis And Lasers In Enginnering*, 45, 106-115, ISSN: 0030-3992

Kondo N.; Monta M. & Fujiura T. (1996). Basic Constitution of a Robot for Agricultural Use. *Advanced Robotics*, 10 (4), 339-353, ISSN: 1568-5535

Lebeau F. (2004). Modelling the Dynamic Distribution of Spray Deposits. *Biosystems Engineering*, 89 (3), 255-265, ISSN: 1537-5110

Maertens W.; Nuyttens D. & Sonck B. (2005). Optimisation of Sequence and Orientation for Used Nozzles Based on Few, Full Boom Distribution Measurements. *Commun.Agric.Appl.Biol.Sci.*, 70 (4), 989-995, ISSN:1379-1176

Mandow A.; Gomez De Gabriel J.M.; M.Rodriguez J.L.; Munoz V.F.; Ollero A. & Garcia-Cerezo A. (1996). The Autonomous Mobile Robot AURORA for Greenhouse Operation, *IEEE Robotics & Automation Magazin*, 18-28, ISSN: 0009-9147

Matthews G.A. (2000). Pesticide Application Methods. *Ed.Longman Scientific & Technical*, New York.

Methner M.M. & Fenske R.A. (1996). Pesticide Exposure During Greenhouse Applications. III. Variable Exposure Due To Ventilation Conditions And Spray Pressure. *Appl. Occup. Environ. Hyg.*, 11 (3), 174-180, ISSN: 1053-4245

Moltò E.; Martin B. & Gutierrez A. (2001). Pesticide Loss Reduction by Automatic Adaptation of Spraying on Globular Trees, *J.Agric.Eng.Res*, 78 (1), 35-41, ISSN: 0021-8634

Naoki S.; Yukio O.; Junichi A. & Nori H. (1995). Fully-Automatic Spraying System Employing Spray Car, *U.S.Patent* 5397056.

Nuyttens D.; Windey S.; Braekman P.; De Morr A. & Sonck B. (2003). Optimisation of a Vertical Spray Boom for Greenhouse Spraying Applications, *Comm.Agric.Appl.Biol.Sci.*, 68 (4), 905-921, ISSN: 1379-1176

Nuyttens D.; Widey S. & Sonck B. (2004). Comparison of Operator Exposure for Five Different Greenhouse Spraying Applications, *J. of Agricultural Safety and Health*, 10 (3), 187-195, ISSN: 1074-7583

Nuyttens D.; Baetens K.; De Schampheleire M. & Sonk B. (2007). Effect of Nozzle Type, Size And Pressure on Spray Droplet Characteristics, *Biosystems Engineering*, 97, 333-345, ISSN: 1537-5110

Nuyttens D.; De Schampheleire M.; Verboven P.; Brusselman E. & Dekeyser D. (2009). Droplet Size and Velocity Characteristics of Agricultural Sprays, *Transaction of The Asabe*, 52(5), 1471-1480, ISSN: 0001-2351

Nuyttens D.; Braekman P.; Windey S. & Sonck B. (2009). Potential Dermal Pesticide Exposure Affected By Greenhouse Spray Application Technique, *Pest Manag Sci*, 65, 781-790, ISSN: 1526-4998

Paice M.E.R.; Miller P.C. & Bodle J.D. (1995). An Experimental Sprayer for the Spatially Selective Application of Herbicides, *J.Agric.Eng.Res.*, 60, 107-116, ISSN: 0021-8634

Panneton B.; Philion H.; Theriault R.& Khelifi M. (2000). Spray Chamber Evaluation of Air-Assistend Spraying on Potato Plants. *Transaction of the Asae*, 43, 529-534, ISSN: 0001-2351

Paul H. & Illing A. (1997). The Management of Pesticide Exposure in Greenhouses, *Indorr And Built Environment*, 6 (5), 254-263, ISSN: 1423-0070

Planas S.; Solanelles F.; Fillat A. (2002). Assessment of Recycling Tunnel Sprayers in Mediterranean Vineyards and Apple Orchards, *Biosystem Engineering.*, 82, 45-52, ISSN: 1537-5110

Piccarolo P. (2008). Machinery for Protected Crops, *Machinery World*, Ed.Unacoma Service, 6, 64-73, ISSN: 1125-422X

Qing L.; Cooper S.E.; Qi L.; Fu Z. (2006). Experimental Study of Droplets Transport Time Between Nozzles and Target. *Biosystems Engineering*, 95 (2), 151-157, ISSN: 1537-5110

Salvarani (2006). http://www.salvarani.it. Accessed October 2006

Sammons P.J.; Furukawa T. & Bulgin A. (2005). Autonomous Pesticide Spraying Robot for Use in a Greenhouse, *Australasian Conference On Robotics And Automation*, 1-9, ISSN: 0921-8890

Singh S.; Burks T.F. & Lee W.S. (2005). Autonomous Robotic Vehicle Development for Greenhouse Spraying, *Trans. of The Asae*, 48 (6), 2355-2361, ISSN: 0001-2351

Singh M. (2006). Survismeter - Type I And II for Surface Tension, Viscosity Measurements of Liquids for Academic And Research and Developed Studies. *J. of Biochemical and Biophysical Methods*, 67 (2-3), 151-161, ISSN: 0165-022X

Singh M. & Kumar V. (2007). Effects of the Biomolecules: Vitamis, Proteins, Amino Acids, and Surfactants: Dtab, Mtoac, Tmsoi, Orcinol On Upper Critical Solution Temperatures. *Int. of Thermodynamics*, 10(3), 121-133, ISSN: 1301-9724

Singh M. (2007). Upper Critical Solution Temperatures for Immiscible Solvent Systems with Halide Salts, Carboxylic Acids, Surfactants and Polynuclear Aromatic Compounds and Benzene Derivatives. *J.Chem.Thermodynamics*, 39, 240-246, ISSN: 0021-9614

Solanelles F.; Escolà A.; Planas S.; Rosell J.R.; Camp F. & Gracia F. (2006). An Electronic Control System for Pesticide Application Proportional to the Canopy width of Tree Crops. *Biosystems Enginnering*, 95 (4), 473-481, ISSN: 1537-5110

Tecsi (2006). http://tecsi.it. Accessed October 2006

Tian L.; Reid J.F. & Hummel J.W. (1999). Development of a Precision Sprayer for Site-Specific Weed Management, *Transaction of Asae*, 42, 893-900, ISSN: 0001-2351

Distribution of Chemical and Microbial Pesticides Delivered Through Drip Irrigation Systems

Heping Zhu[1], Xiaochan Wang[2], Michael E. Reding[1] and James C. Locke[1]
[1]*USDA-ARS, Application Technology Research Unit, Wooster, OH 44691,*
[2]*Nanjing Agricultural University, Nanjing, Jiangsu 210031,*
[1]*USA*
[2]*China*

1. Introduction

The current lack of scientific data for effective delivery of pest-control agents for soil pests leaves the nursery or greenhouse industries vulnerable to possible crop loss or damage, rendering the crop unmarketable either due to appearance or quarantine. Spray is the most used method to apply agrichemicals and bio-products because spray application is timely and convenient to control pests, but there are several impediments with spraying pesticides. For example, pesticides may not be efficiently applied to intended targets because of large gaps between trees where pesticides are not needed; spray drift may affect residential areas nearby nurseries; plants physically disturbed or damaged by spraying may have lower market values; and extensive skills are required for applicators to properly operate sprayers. Also, applying pesticides uniformly and sufficiently to target pests under the soil or container substrates is a challenging constraint of current spray technologies.

Drip irrigation, noted for its highly efficient water distribution capability to increase crop yields (Grabow et al., 2006; Plaut et al., 1996; Bryla et al., 2003; Lamm and Trooien, 2003; Zhu et al., 2004), offers an alternative strategy to carry the label-allowed pest control agents to the target areas in the soil for effective insect or disease control (Felsot et al., 1998). Uniform chemical distribution pattern in the soil plays an important role to achieve high pest control efficiency and a sustainable safe environment. Chemigation includes applications of insecticides, fungicides, herbicides, fertilizers, microbial biopesticides and nematicides by means of irrigation systems. In production nurseries with drip irrigation systems in place, injection of pesticides into irrigation lines (or chemigation) offers an alternative strategy for efficient and economical application of pesticides to targeted zones in soil or container substrates The drift problem caused by spraying pesticides and costs associated with sprayers can be eliminated by using chemigation. This method has been shown to increase crop yields and reduce chemical leaching (Leib et al., 2000).

Injection of pesticides into drip irrigation lines takes advantage of the fact that active ingredients can be carried by water into root zones (Lamm et al., 2007). The ingredient distribution pattern in the soil plays an important role in pest control efficacy. Drip irrigation was demonstrated as an effective technique to apply water-soluble fumigants to the target soil (Ajwa et al., 2002). Leib and Jarrett (2003) reported limited leaching of Imidacloprid with 70% of the pesticide still located in the root zone at the end of the 40-day evaluation period when efficacy ended.

Because leaching potential is a major concern for using drip chemigation to apply agrichemicals (Jaynes et al., 1992), only few chemicals such as Imidacloprid are registered for drip applications. Delivery of alternative fumigants through drip irrigation system to treat soil to kill nematodes, fungi, and weeds greatly reduced use of fumigants and emission of fumigants from soil (Schneider et al., 2006; Ajwa et al., 2002). Considerable research has been conducted to investigate the efficacy and potential damage to the environments by applying Imidacloprid through irrigation systems (Fleischer et al., 1998; Van Iersel et al., 2000; Felsot et al., 2000; Byrne and Toscano, 2006) after it was released into the market. A multi-year field study reported that the level of Imidacloprid aphicide leaching after chemigation through a subsurface drip system could be significantly reduced if irrigation schedule matched the crop needs (Foslot et al., 2003). This is own to the mobility of Imidacloprid in the soil was very low (Leib et al., 2000).

Due to concerns about health and environmental hazards with traditional pesticides, the use of biopesticides to control pests and diseases has been dramatically increased for past decades (Fuxa and Richter, 1999; Gopal et al., 2001; Hajek et al., 2007; Dorner and Lamb, 2006; Grewal et al., 2001). Lumsden and Locke (1989) reported a very promising control of diseases caused by fungal root rot organisms in the greenhouse production of bedding plants by adding a microbial pesticide *Gliocladium virens* into the soilless substrate before planting seeds.

Since the fact that bio pest control agents are friendly to the environment with no leaching concerns, drip irrigation has become a convenient method to deliver them in the root zone. Entomopathogenic nematodes delivered through drip irrigation had very promising control of different soil pests (Wennemann et al, 2003; Ellsbury et al., 1996; Becker et al., 1989; Reed et al., 1986; Kramer and Grunder, 1998). Wennemann et al. (2003) applied entomopathogenic nematodes through drip irrigation with 2.0 L/h emitters in a vineyard and found the recovery rate of nematodes from drip emitters in 51 m long drip lines ranged from 42 to 92%. Reding et al. (2008) reported that Imidacloprid, clothianidin and entomopathogenic nematodes applied through drip irrigation effectively controlled white grubs in the root zones of various ornamental nursery trees. They also reported that nematodes applied through drip irrigation, injected into the soil, and surface drenched at a curative timing all significantly reduced numbers of grubs compared to untreated trees. These data illustrate drip irrigation is a viable delivery system for control of white grubs in nursery crop production.

Improving water distribution uniformity of drip irrigation systems has been studied extensively (Wu et al., 1979; Lamm et al., 1997; Camp et al., 2003; Clark et al., 2005; Grabow et al., 2006). However, the specific evaluation of a designated pest control agent's uniformity throughout drip lines is lacking, especially for the microbial bio-pesticides before they are used for field trials. Uniform distribution of the pest control agents throughout drip lines and in targeted areas is essential to assure drip chemigation can achieve both efficient pest/disease control and environmental safety. Physical properties of pest control agents, especially bio-pesticides, are quite variable. There are questions whether drip chemigation can uniformly distribute them throughout drip lines and within targeted areas. Efficacy of chemical and biopesticides is dependent on the amounts of water applied to facilitate movement of the chemical into the root zone. Deliverability and uniformity of many these materials have not been evaluated comprehensively under controlled conditions before they are released for field uses. Biopesticides are usually granular compounds or living organisms and are normally suspended in water. Their movement and mixture uniformity inside drip lines are influenced by fluid motion in either turbulent or laminar flow status, which varies with flow rate. Also, water use efficiency and irrigation practice are greatly affected by emitter flow capacities. Little information is available on distribution patterns of different water soluble or insoluble materials discharged from different flow-capacity (or flow-rate) emitters throughout drip lines. A quantitative relationship among emitter size, flow rate and chemical type may be helpful to develop strategies to apply suspendible granular biopesticides through drip irrigation systems.

Studies on water soluble agro-chemicals and nematodes injected into drip lines under field conditions have been reported, but there is no comprehensive study comparing distribution profiles of agro-chemicals and bio-compounds in the soil after they are injected into a system with different emitter flow capacities under controlled irrigation conditions. Little information is available on distribution patterns of different water soluble or insoluble materials in the soil under different emitter sizes and flow capacities. Another important consideration is that the chemical and microbial control agents may be less effective when surface applied due to active agent becoming bound in organic or other material at the surface. Consequently, excessive levels of surface application are required to achieve efficacy, at greater expense. Scientific information on distribution patterns of active agents in the soil with different emitter sizes and flow rate is essential to help develop strategies to apply suspendible granular biopesticides uniformly by using drip irrigation systems.

The objectives of this research were: (1) to investigate the capability of drip irrigation systems for delivering water soluble chemicals, suspendible microbial bio-insecticides and bio-fungicides, and entomopathogenic nematodes; (2) to investigate distribution patterns of these chemical and microbial pesticides in the soil. To achieve these objectives, the distribution uniformity and recovery rate of these materials throughout drip lines and in the soil were evaluated under controlled conditions as they were discharged from emitters of three flow capacities, and thus to verify whether increasing emitter flow capacity would significantly change distribution patterns of these materials along the drip lines and in the soil.

2. Materials and methods

2.1 Drip irrigation system design

A drip irrigation system was developed to test the application uniformity of agrochemicals and microbial bio-pesticides throughout drip lines (Wang et al., 2009). Variables including emitter flow, amount of injected materials and injection time could be individually controlled with the system. The system included three, 79 m long drip lines, a portable chemical injection unit, a shutoff valve for pressure control, a pressure sensor (Model 242PC60G, Micro Switch, Freeport, IL), a flow meter (Model DFS-2, DGH Corporation, Manchester, NH), and a backflow prevention check valve (Model T-413, Nibco Inc., Elkhart, IN). The portable chemical injection unit was installed at the beginning (upstream end) of each drip line.

The injection unit (fig. 1) included an injection valve assembled with a 1.27 cm (nominal ½ inch) thread PVC tee, a nominal 1.27 cm (nominal ½ inch) NPT electric wire connector (Kleinhuis North America, Inc., Worthington, OH), a bladder valve removed from a 40 cm diameter plastic toy ball (Item# 3314903313, Ball, Bounce, and Sport Inc. Ashland, OH), and a modified 50 ml Pro-Pistol™ pistol grip syringe (Model 1005, Neogen Corporation, Lexington, KY). The bladder valve performed as a one-way check valve for chemicals injected into the drip line with the syringe. After the syringe was removed, the valve prevented leakage of the pressurized liquid at the injection point. The backflow prevention valve was installed in the drip line upstream of the injection valve, to prevent chemicals flowing upstream to the main water line. The pressure and flow rate near the injection point were measured with the pressure sensor and flow meter which were connected to a micro data logger (Model CR23X, Campbell Scientific, Logan, UT). The data logger was programmed to acquire these data at 1-second intervals during the experiment.

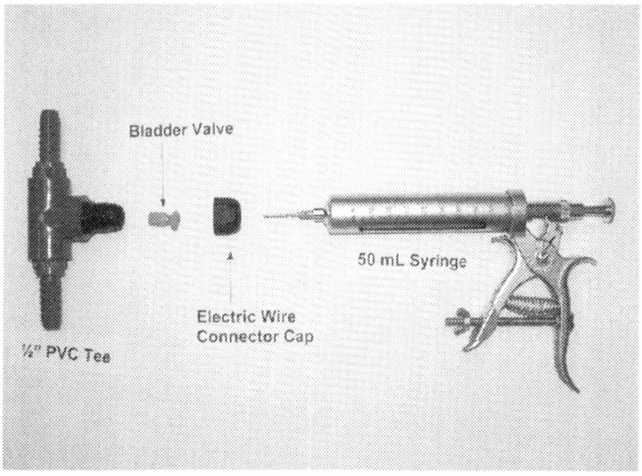

Fig. 1. Chemical injection unit assembly used to inject pesticide into drip lines.

The drip lines were three polyethylene tubes with external pressure-compensating emitters (Model WPC, Netafim USA, Fresno, CA) of three different flow capacities. The nominal flow capacity of emitters on the three drip lines was 1.9 L/h (line 1), 3.8 L/h (line 2), and 7.6 L/h (line 3), which covers the flow capacity ranges normally used for drip irrigation systems in ornamental nursery applications. The flow path of pressurized liquids within each emitter was controlled with a flexible diaphragm in the center donut-shaped chamber that reduces the flow path dimension with increasing pressure and with a series of baffles projecting from inside and outside the walls of the chamber. The diaphragm functioned as a variable-flow storage compartment as well as for a pressure compensation function. The inside and outside baffles alternating within the emitter formed a resistive flow path to discharge the pressurized liquid. The length, depth and width of the flow path in the chamber of the emitters for line 1 were 61, 1.07 and 1.17; for line 2 they were 1, 60, 1.30 and 1.40 mm; and for line 3 they were 17, 1.60 and 1.60 mm, respectively.

Each drip line contained a polyethylene tubing with 13.2 mm nominal inside diameter and 1.27 mm nominal wall thickness. The total number of emitters on each 79 m drip line was 87, spaced at 0.9 m intervals. In nurseries, it is common to grow crops in a row with less than 79 m length. The barb of emitters was inserted 4.2 mm inside the drip line tubing. Distance from the injection point to the first emitter was 0.45 m. For each replication of the treatment, only one line was used while the other two lines were disconnected.

To determine the flow rate and pressure to be used, water flow rate uniformity from emitters throughout each drip line was examined at four pressures of 69, 103, 138 and 276 kPa, respectively. The amount of water from 7 emitters at 4.1, 17.6, 31.1, 44.6, 58.1, 71.6 and

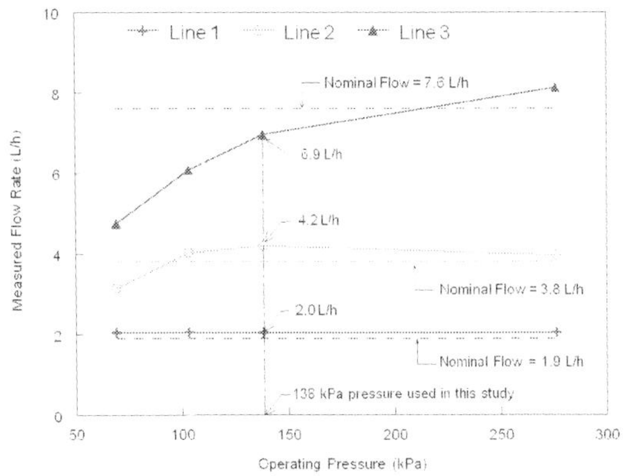

Fig. 2. Measured flow rate vs. pressure of the three different emitter capacities used in this study.

77.9 m (or 13.5 m apart) from the injection point was measured for 10 minutes and repeated three times. Figure 2 shows the relationship between the average water flow rate of the 7 emitters and the operating pressure for the three lines of different nominal flow capacities tested. The emitter flow rates on line 1 (nominal flow capacity of 1.9 L/h) and line 2 (nominal flow capacity of 3.8 L/h) were almost constant over pressure ranging from 69 kPa to 276 kPa, but on line 3 (nominal flow capacity of 7.6 L/h) the emitter flow rate increased as the pressure increased from 69 kPa to 276 kPa. The measured flow rate at 138 kPa on line 1 and line 2 was 2.0 and 4.2 L/h respectively which were very close to the nominal values, but on line 3 it was 6.9 L/h which was 15% (or 1.14 L/h) lower than the nominal value. Because of these preliminary test results, the operating pressure of 138 kPa was chosen for all treatments in this research. Also, the pressure 138 kPa was within the pressure range between 50 and 310 kPa recommended by the drip line manufacturer.

2.2 Materials tested

Tests were conducted with five different materials with different types and formulations (table 1): Brilliant Sulfaflavine (BSF, MP Biomedicals, Inc., Aurora, OH), Imidacloprid (Marathon II, OHP, Inc., Mainland, PA), an entomopathogenic fungus (EPF) (Novozymes Biologicals, Inc., Salem, VA), a microbial soil fungicide (SF) (SoilGuard 12G, Certis USA, LLC., Columbia, MD), and entomopathogenic nematodes (EPN, cultured by The Ohio State University, Entomology Department, Wooster, OH).

BSF is a water soluble, non sun-light degradable fluorescent tracer normally used to track pesticide deposition (Zhu et al., 2005). Also, the BSF solution has a nearly constant intensity over the pH range of 6.9-10.4. It was selected for this test because results of BSF could provide a reference to compare performances of other materials discharged through drip emitters.

Imidacloprid is the active ingredient of Marathon II which is a flowable formulation (suspension concentrate dispersible in water) with a viscosity of 84 mPa·s. It is a systemic chloronicotinyl insecticide normally applied to the soil for control of root feeding beetles such as scarab larvae (white grubs) or foliar feeding insects such as scale and leafhoppers in ornamental trees. The concentration of Imidacloprid in Marathon II is 0.24 kg/L.

Treatment	CV (%)*			DU*		
	2.0 L/h emitter	4.2 L/h emitter	6.9 L/h emitter	2.0 L/h emitter	4.2 L/h emitter	6.9 L/h emitter
BSF	2.3	1.7	1.8	0.95	0.97	0.96
Imidacloprid	43	36	54	0.76	0.80	0.76
EPF	90	104	119	0.56	0.26	0.33
SF	98	72	51	0.44	0.62	0.68
Nematodes	8.8	8.0	4.9	0.80	0.83	0.91

* Each value of CV or DU is the mean from 7 emitters in each drip line with three replications. CV and DU were calculated with equations (1) and (2), respectively.

Table 4. Mean coefficient of variation (CV) and distribution uniformity (DU) of BSF, Imidacloprid, EPF, SF and nematodes discharged from emitters with three different flow capacities (2.0, 4.2 and 6.9 L/h) at 138 kPa pressure.

EPF is a suspendible conidial powder microbial insecticide. Its active ingredient is Met52G and Tick EX EC with 2.0% by weight *Metarhizium anisopliae* Sorokin strain F52. The conidial powder contains approximately 5×10^{10} conidia per gram. The product has very uniform size. The sizes of conidia are typically 3-4 um wide and 7-9 um long while the average length of the powder granule is about 0.039 mm. The material is insoluble in water and remains on the top of water after standing for 10 minutes. Therefore, for this study, a suspension of conidia was made in 0.05% non-ionic surfactant Tween 80 (Sigma-Aldrich, St Louis, MO) to improve dispersion. The EPF was a fungus normally used to control insects, primarily beetle larvae and ticks on non-food use greenhouse and nursery crops.

SF is a suspendible, granular formulation of a naturally occurring soil fungus (*Gliocladium virens* strain GL-21), containing 12% by weight fungal fermentor biomass with 2×10^7 viable propagules per gram. The compound was formulated to be applied through irrigation lines or drenched into growing media to control diseases caused by fungal root rot pathogens such as *Pythium* and *Rhizoctonia*. The suspendible granules had irregular shapes with a considerably wide range of equivalent diameter: 10% volume less than 0.136 mm, 50% volume less than 0.32 mm, and 90% volume less than 1.06 mm. The maximal granule equivalent diameter was 1.47 mm, and the average diameter was about 0.345 mm. The granules suspend in the water when injected into the drip line.

Lastly, nematodes used in this study were *Heterorhabditis bacteriophora* Poinar strain GPS11 with a concentration of 2.0×10^6 nematodes per 100 ml of solution, estimated by counting with a microscope (Woodring and Kaya, 1988). They are typically 500 to 1000 μm long and 18 to 50 μm wide, and are normally used to carry and introduce symbiotic bacteria (*Xenorhabdus* spp.) into the body cavities of insects that eventually kill them within 48 hours. The species and strain of nematodes has shown efficacy for controlling scarab larvae in ornamental nurseries (Reding et al. 2008). The nematodes can be suspended in water but normally settled out within a few minutes if not agitated.

2.3 Experiments through drip lines

Drip lines were attached horizontally to three 2.5-mm diameter high-tensile electric fence wires suspended over 30 cm above the ground (fig. 3). Only one drip line was connected to the water source for each application. In each replication, the tested line was filled with water at a stabilized pressure (138 kPa) before injection and water sample collection.

Fig. 3. Three drip lines with three different capacity emitters suspended 30 cm above the ground used in this study.

The 50 mL syringe with a 0.9 mm inside diameter needle was modified to inject a fixed amount of materials into the drip line through the chemical injection unit (fig. 1). Each of the five materials was injected over a 1-minute period for each replication. All materials were mixed with water before they were injected into the drip line. Water samples mixed with each injected material were collected with 3.8 L plastic bottles from 7 emitters at 4.1, 17.6, 31.1, 44.6, 58.1, 71.6 and 77.9 m, respectively. The 7 emitters were the same ones at the same locations used for the water uniformity distribution test as mentioned above. The water flow test verified that the drip lines were able to uniformly distribute water flow through the drip lines as discussed in Results and Discussion section. The sampling began one minute before the start of injections. The collection time for samples from line 1 was 30 minutes, and 15 minutes from line 2 and 3. An estimated time for materials to flow from the injection point to the last emitter was 16.3 minutes in line 1, 7.8 minutes in line 2, and 4.7 minutes in line 3, respectively. These estimated times were calculated with a plug-flow equation similar to the flow equation used for boom sprayers (Zhu et al., 1998). The volume of collected sample from each emitter in line 1 was 1.0 L, 1.05 L in line 2, and 1.72 L in line 3. Each sample was shaken and sub-samples were decanted into glass bottles, and then taken to the laboratory for analysis. After samples were collected, the drip line was flushed by opening the end of the line for 10 minutes and a sample of the flush water was collected from the end of the drip line at the beginning of the flushing cycle. These flushing water samples were analyzed for observation only but not for quantitative comparison because it was very hard to control the amount of water collected at the end of lines during the flushing process. The above process that included the injection of a material into the drip line, the collection of samples and the flushing of drip line was repeated for three times representing three replications for each material and each drip line.

The starting concentration of the BSF solution was 3 g BSF per liter of water which was selected based on pre-trial tests for fluorescent intensity, and fell within the detection range of the spectrometer (Perkin-Elmer Limited, Beaconsfield, Buckinghamshire, England) used in this research. The viscosity of the solution was 0.887 mPa·s, and the amount of BSF solution injected into the drip line for each replication was 50 mL, equaling 0.06 g of BSF. Water samples collected from the emitters were taken to the laboratory where the BSF concentration was measured with the spectrometer calibrated to detect an emission wavelength of 500 nm (Zhu et al., 2005).

The Imidacloprid mixture for each replication was 13 mL of Marathon II and 25 mL of water. This rate was calculated based on the label rate of 50 mL Marathon II per 305 m row. After sample collection, the concentration of Imidacloprid was measured by filtering samples and adjusting the pH to 2. Aliquots of 2 mL were subsequently analyzed using liquid chromatography tandem mass spectrometry (LC-MS/MS). The LC-MS/MS system consists of a ProStar® 210 solvent delivery module with a ProStar 430 autosampler and a 1200L triple-stage quadrupole mass spectrometer with a dual off-axis ESI interface (Varian Inc., Walnut Creek, CA). A standard concentration of 5 mg/L prepared in methanol and water was used to optimize the instrument and attain precursor and transition ions using argon as the collision gas. A molecular mass to charge ratio of 256 in positive ion mode was used with transition ions (collision energy voltages in parenthesis) 208.8 (12.5), 175.1 (12.0) and 212.0 (8.5). Optimized parameters were 350 °C for the drying gas; ion transfer capillary, nebulizer needle and shield voltages were 40, 4500 and 200, respectively. Injection volume was 20 uL and scan time was 0.1 seconds. A Nova-Pak® C18 column (4 µm, 150 mm × 3.9 mm) and packing-matched guard were used for retention of the analyte (Waters Corporation, Milford, MA). A gradient elution using 0.1% formic acid (A) and 0.1% formic acid in acetonitrile (B) was used at a flow rate of 0.4 mL/min. Solvent B was held for one minute at 5%, and ramped to 95% over four minutes, held for three minutes, then gradually returned to initial conditions over one minute and held for three additional minutes for equilibration. A standard calibration curve, using matrix-matched standards, used the most abundant transition ion for quantitative analysis. The remaining ions, along with retention time as compared to the spiked matrix samples, were used for further confirmation in the sample matrix. The system was calibrated with Imidacloprid at known concentrations ranging from 0.036 to 3600 mg/L.

In the EPF trials, 5.5 g of the EPF formulation was mixed with 50 mL of water and 25 uL of Tween 80 for each replication. The mixture was stored at 5 °C for 24 h and then shaken well before the injection. After water samples were collected at seven locations, they were shaken and sub-samples were poured into 15 mL plastic vials, then sent overnight with an ice pack to a laboratory for analysis. In the laboratory, the water samples were placed in a sonicator for 20 minutes, and 1 mL of each sample was added to 99 mL of phosphate buffer, then 100 uL of this solution was spread on selective (Veen's) media for CFU's incubation (Veen and Ferron, 1966). The media plates were incubated for 4 to 5 days at 27°C and then fungal colonies were counted.

The SF mixture injected into the drip line for each replication was 10 g SF mixed with 100 mL water, and was shaken well before injecting into the pressurized drip line system. After samples were collected from emitters for each replication, they were transferred to the laboratory. Each collected sample was diluted two or three times and then 1 mL aliquoits from the diluted sample were deposited onto the surface of a semi-selective medium containing antibiotics to suppress bacterial growth (3 plates/dilution/sample). Following a period of incubation, colonies were counted and calculations made to determine the number of units (CFU) of active ingredient that were dispensed through each emitter.

The nematode mixture for each replication was 100 mL water and 2.0×10^6 nematodes. It was stored at 5°C for 24 h before the application. After nematode samples were collected, each plastic bottle was held at room temperature 20°C for 24 h to allow the nematodes to settle at the bottom of the bottles, and then most of the water was poured off. The solution remaining in the bottle was then poured into a glass bottle and allowed to set for another 24 h until the nematodes again settled at the bottom. A pipette was then used to remove most of the water until 15 mL of nematode suspension remained. Three 10 µL drops containing nematodes were taken from this suspension and spread on glass microscope slides, then the all the nematodes on a slide were counted under a stereoscopic microscope (Model SZX12, Olympus, Japan) at 50× magnification. The mean number of nematodes in three drops was reported.

After all samples were analyzed, the amount of materials discharged from emitters for each test was normalized for the specific emitter flow rate tested. To determine the effect of emitter flow capacity on the amount of materials discharged throughout the drip line, each group of data for the specific material treatments was first analyzed by one-way ANOVA to test the null hypothesis that all treatments had equal means of the material quantity with Duncan's methods using ProStat version 3.8 (Poly Software International, Inc., Pearl River, NY). If the null hypothesis was rejected, the multiple comparison procedure was used to determine differences among means of the material distributed throughout the drip line. Multiple comparisons for recovery rates across the drip line were also conducted among the five materials and three flow capacities. All differences were determined at the 0.05 level of significance.

Coefficient of variation (CV) and distribution uniformity (DU) were used to quantify the uniformity of distribution of each of the five materials throughout the drip line. CV and DU were calculated by replacing the flow rate with the amount of materials discharged from each emitter defined by ASAE Standards (ASAE EP405.1, 2008), Kruse (1978), and Keller and Bliesner (2000),

$$CV = \frac{s}{q_{ave}} \times 100 \quad (1)$$

$$DU = \frac{q}{q_{ave}} \quad (2)$$

where s is the standard deviation of the amount of materials discharged from emitters, q_{ave} is the mean amount of materials discharged from emitters, and q is the mean of the lowest one-fourth of the amount of materials discharged from each emitter sampled. These equations have also been used for evaluation of fertigation and chemigation performances such as the effect of injection methods and injection rates on fertigation uniformity (Bracy et al, 2003; Li et al., 2007). Since there were 7 samples of each material collected for each replication, the value of q was calculated by the following equation with expansion of the 7 data to 28 data,

$$q = \frac{4q_{L1} + 3q_{L2}}{7} \quad (3)$$

where, q_{L1} and q_{L2} are the lowest and second lowest amounts of the material among the 7 samples, respectively.

2.4 Experiments in the soil

To test distribution pattern of chemical and microbial pesticides in the soil after they were discharged from emitters, four materials were selected for the trials. They were BSF, Imidacloprid, EPF, and EPN. SF was not included in the soil test because previous tests demonstrated its uniformity and recovery rate through drip lines were very low.

The distribution patterns of the four materials were determined in a cultivated, 33.5 m long and 12.2 m wide field. No insecticides or other chemicals had been applied to this field for at least three years. Before the experiment, the plot was plowed and rototilled. Soil samples were collected from three depths and bulk density, water holding capacity, porosity, EC (electric conductivity), and pH were determined (Table 2). Water holding capacity was determined with the method used by Cassel and Nielsen (1986).

Soil Depth	Dry bulk density (g/ml)	Water holding capacity (%)	Total porosity (%)	EC (mS/cm)	pH
Top (7.6-10.2 cm)	1.16	29.8	49.3	601	5.6
Middle (15.2-20.3 cm)	1.20	30.2	51.8	202	5.8
Bottom (22.9-30.5 cm)	1.24	28.8	50.2	121	5.6

* All the values are the mean of three samples.

Table 2. Properties of the soil used for chemical distribution pattern tests*

The same drip irrigation system reported above was used to test the distribution pattern of agrichemical and microbial materials in the soil. The three drip lines were placed on the surface of the soil with the lines 5 m apart. The drip lines were longer than the soil plot so the excess tubing was coiled at the downstream border of the plot. The rates of BSF, Imidacloprid, EPF, and nematodes used for each trial were 150 mg, 2.8 mL, 5.5 g, and 2,000,000, respectively (Table 1).

A modified 50 mL syringe was used to inject a fixed amount of materials into the drip lines through a bladder valve. Each material was injected, one material at a time, into each line. Only one line at a time was connected to the water source. Irrigation was turned on and pressure was allowed to stabilize at 138 kPa before injection of each material. Injection of each material into the drip line took approximately 1 minute, and then the irrigation continued for 31 minutes for line 1, 14 minutes for lines 2, and 8 minutes for line 3. This irrigation schedule was repeated every 8 hours for 24 hours before soil samples were collected. A total 3.0 L of water was applied through each emitter with all the three flow capacities.

Soil samples of each material were collected 24 hours after last irrigation at 28 different locations near the emitter. A hollow auger with an inner diameter of 25 mm was used to take soil cores. Figure 4 shows the soil auger sampling geometry for each test. Soil samples were taken directly under emitters and 15, 30, 45 cm upstream and downstream from the emitters along the drip line laterally, and from the surface (0 cm depth) and at depths of 10, 20 and 30 cm vertically, with about 15 cm^3 sample from each depth. Thus, for each test, total 28 soil samples were taken from 28 locations evenly spread in the 30 x 91.2 cm section area under the drip line.

During the test period from injection of materials into drip lines to completion of soil sample collection, the ambient temperature ranged from 18° to 29 °C, relative humidity ranged from 45 to 86%, and soil moisture content at the non-irrigation point ranged from 36 to 41%. The

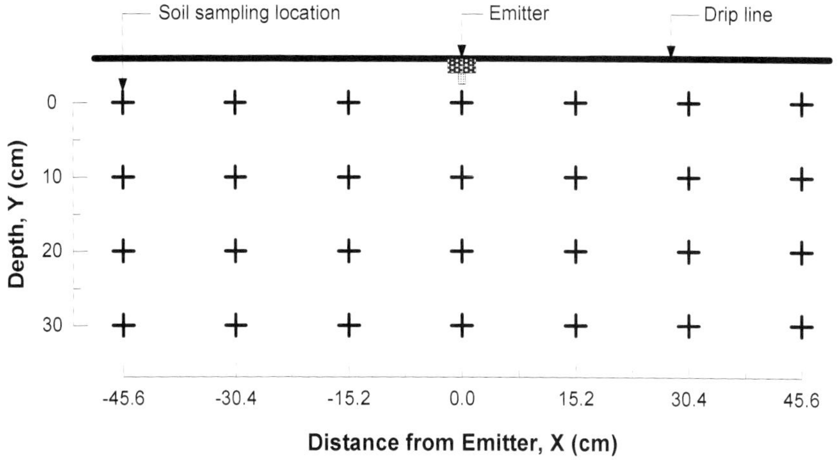

Fig. 4. Locations of soil samples collected for evaluation of chemical distribution patterns in the soil.

moisture content was measured with a Theta probe soil moisture sensor (Delta-T Devices Ltd, Cambridge, England) placed 10 cm in the soil at a point 2.5 m away from the drip lines. There was no precipitation during the experiments.

Amount of BSF in soil samples was determined by a method similar to that used by Barber and Parkin (2003) for a fluorescent tracer CBS-X. A pre-test was conducted to ensure that the concentrations of BSF soil samples obtained from spectrometer were linear with the actual BSF concentrations on the soil. An amount of 10 g soil was weighed into six 100 mL glass jars with six stock solutions (4 μL, 8 μL, 12 μL, 20 μL, 40 μL and 60 μL) at concentration of 3 g BSF per liter water and 50 mL distilled water was added, respectively. Samples were mixed in a rotating drum for 10 minutes under 500 rpm. To clear the samples after mixing, 0.5 g gypsum was added to each sample jar, shaken by hand for 10 s, and then left in the refrigerator until the supernatant had cleared. Once the supernatant was clear, fluorescence concentration readings were taken with the same method as the BSF-water samples mentioned above. The linear corresponding curve between readings and actual BSF concentrations had been obtained and used as a standard. After all soil samples were collected from the field, they were treated using the same methods as used for the pre-test, and the actual BSF concentration in the soil was then measured by the spectrometer.

Soil samples containing Imidacloprid were placed in glass jars, transported to the laboratory, and stored in a freezer at -40 °C until analysis. Methanol was used to extract Imidacloprid from the soil samples (Felsot et al., 1998) and ELISA kits (Envirologix, Inc., Portland, ME, USA) for Imidacloprid were used to determine the amount of Imidacloprid in the soil (Castle et al. 2005).

Soil samples containing EPF were placed in glass jars, and sent overnight with an ice pack to a laboratory for analysis. In the laboratory, the one gram of soil samples were placed in 99 mL of phosphate buffer, then 100 uL of this solution was spread on selective (Veen's) media for CFU's (Veen & Ferron, 1966). The media plates were incubated for 4 to 5 days at 27°C and the fungal colonies were counted.

Presence of nematodes in the soil samples was detected with larvae of the greater wax moth (*Galleria mellonella* L.), an available test for presence of nematodes in soil (Bedding and Akhurst, 1975). Wax moth larvae turned brick red when infected by the strain of nematodes used, and thus nematodes were considered present if larvae changed to that color. Soil samples were placed in metal containers (54 mm in diameter, 37 mm in height), and 3 wax moth larvae were carefully placed on top of the soil, the lid was placed on each container. The containers were incubated in the dark at room temperature for 5 days. During this time, color change and death of the larvae were monitored daily. Samples of untreated soil from the same field were collected, treated as above, and used for comparison.
To summarize soil experimental results, the mean concentration was calculated from three replications at each sampling point for BSF, Imidacloprid and EPF samples. For nematodes, the presence of nematodes was chosen as "Yes" if two or three replications showed the presence of nematodes at each sampling point. The presence was chosen as "No" if there was no or one replication showed the presence of nematodes.

3. Results and discussion

3.1 Water flow rate distribution throughout drip lines

Figure 5 shows the amounts of water collected from emitters at seven different distances from the injection point on line 1 for 30 minutes and lines 2 and 3 for 15 minutes, respectively. Data in the figure illustrates that there was little variation in the amount of water discharged from emitters along the drip lines 1 and 2 while there was 6% variation in line 3. That is, water distribution throughout the 79 m drip line varied very little with the 2.0 and 4.2 L/h flow rate emitters and varied considerably with the 6.9 L/h flow rate emitters. It is understandable that the 6.9 L/h emitters could not maintain a constant flow rate at various pressures because their short flow path and limitation of the flexible diaphragm limited the pressure compensation capability. Based on Bernoulli's equation in hydraulics,

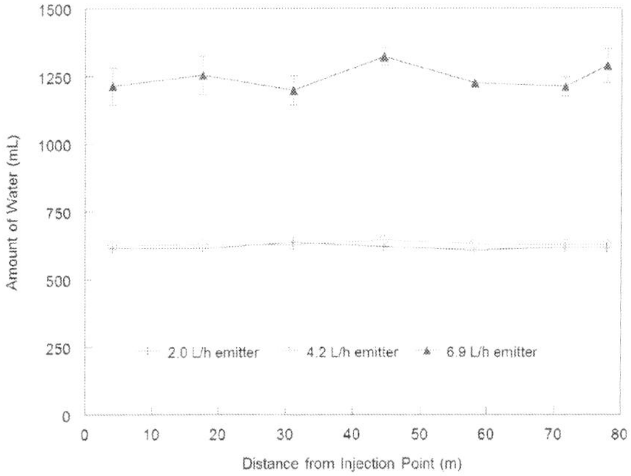

Fig. 5. Actual amounts of water samples collected from 7 locations throughout the 79 m long drip lines for each replication of the tests for the 2.0 L/h (line 1), 4.2 L/h (line 2) and 6.9 L/h (line 3) emitters. Error bars represent standard deviations around means.

flow rates from any hydraulic components are proportional to the multiplication of a flow constant, the cross section area of the flow path, and the squared root of pressure. For pressure compensated emitters with low flow capacity, the flexible diaphragm automatically maintain the flow constant and cross section area very well, and pressure changes within a small range will not produce noticeable changes in the flow rate.

3.2 Distribution of tested materials through drip lines

Amounts of measured BSF, Imidacloprid, EPF bio-compound insecticide, SF bio-compound fungicide, and nematodes throughout the three drip lines are shown in Figures 6 through 10. Table 3 shows the predicted and measured amounts of the five materials discharged from three different capacity emitters. The predicted amount was calculated by the amount of material injected into the drip line divided by 87 (the number of emitters in the drip line). The recovery rate shown in Table 3 is the percentage of the average measured amount of materials divided by the predicted amount of materials. Table 4 reports the mean CV and DU of the five materials throughout the entire drip lines with three different capacity emitters, respectively.

There was no significant difference in the amount of BSF discharged from different flow capacity emitters on three drip lines (fig. 6). The recovery rate of BSF was from 86% to 93% for the three drip lines (table 3). The average CV and DU of BSF throughout the 79 m drip line for all three drip lines was 1.9% and 0.96, respectively (table 4). The flow velocity near

Material	Emitter flow (L/h)	Mean quantity of material per emitter			Recovery rate (%)**
		Predicted	Measured*	Unit	
BSF	2.0	1724	1586 (99)A	μg	92a
BSF	4.2	1724	1486 (47)A	μg	86a
BSF	6.9	1724	1608 (98)A	μg	93a
Imidacloprid	2.0	35.9	18.0 (7.5)B	mg	50b
Imidacloprid	4.2	35.9	28.1 (10.7)A	mg	78a
Imidacloprid	6.9	35.9	12.3 (6.6)B	mg	34bcd
EPF	2.0	$5.69x10^7$	$5.1 x10^6$ ($2.8 x10^6$) A	CFU	9.0de
EPF	4.2	$5.69x10^7$	$3.4 x10^6$ ($2.9 x10^6$) A	CFU	6.0e
EPF	6.9	$5.69x10^7$	$5.4 x10^6$ ($6.6 x10^6$) A	CFU	9.5de
SF	2.0	$2.3 x10^6$	$3.89x10^5$ ($3.52x10^5$)A	CFU	17cde
SF	4.2	$2.3x10^6$	$4.75x10^5$ ($3.31 x10^5$)A	CFU	21cde
SF	6.9	$2.3x10^6$	$2.74x10^5$ ($1.45x10^5$)A	CFU	12de
Nematode	2.0	$2.3 x 10^4$	9387 (826)A	number	41bc
Nematode	4.2	$2.3 x 10^4$	9679 (774)A	number	42bc
Nematode	6.9	$2.3 x 10^4$	10754 (528)A	number	47b

* Means for the measured quantity of the same material in a column followed by a different uppercase letter are significantly different ($p<0.05$), but not for the comparison between materials.

** Recovery rate (%) = Measured quantity x 100 / Predicted quantity. Means for the recovery rate in a column followed by a different lowercase letter are significantly different among all the materials ($p<0.05$).

Table 3. Comparison of predicted and measured amounts of five materials discharged from individual emitters throughout drip lines with three different size emitters. Standard deviation is presented in parenthesis.

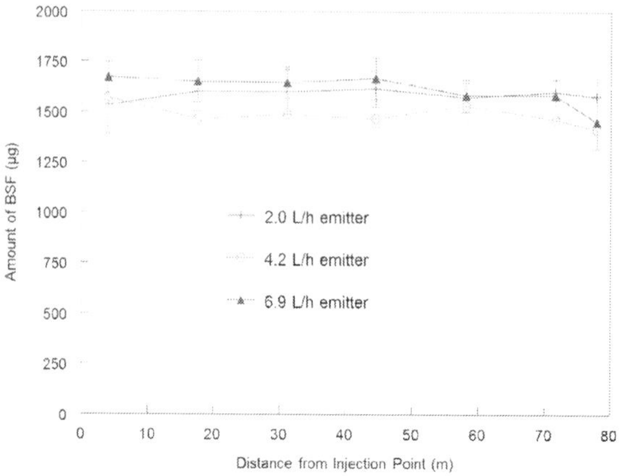

Fig. 6. Amounts of BSF discharged from individual 2.0, 4.2 and 6.9 L/h emitters throughout three different drip lines. Error bars represent standard deviations around means.

Treatment	CV (%)*			DU*		
	2.0 L/h emitter	4.2 L/h emitter	6.9 L/h emitter	2.0 L/h emitter	4.2 L/h emitter	6.9 L/h emitter
BSF	2.3	1.7	1.8	0.95	0.97	0.96
Imidacloprid	43	36	54	0.76	0.80	0.76
EPF	90	104	119	0.56	0.26	0.33
SF	98	72	51	0.44	0.62	0.68
Nematodes	8.8	8.0	4.9	0.80	0.83	0.91

* Each value of CV or DU is the mean from 7 emitters in each drip line with three replications. CV and DU were calculated with equations (1) and (2), respectively.

Table 4. Mean coefficient of variation (CV) and distribution uniformity (DU) of BSF, Imidacloprid, EPF, SF and nematodes discharged from emitters with three different flow capacities (2.0, 4.2 and 6.9 L/h) at 138 kPa pressure.

the injection point was 0.38 m/s in Line 1, 0.80 m/s in Line 2, and 1.32 m/s in Line 3. The Reynolds' number was 4807, 10120, and 16697 for the three lines, respectively. The flow near the injection point in all three lines was a turbulent flow. The flow rate in drip lines had little influence on the amount of BSF discharged from emitters. Therefore, the water soluble material could be well delivered throughout drip lines regardless of the emitter capacity.

The amount of Imidacloprid discharged from individual emitters varied with the emitter flow capacity (fig. 7). Emitter capacity also significantly influenced the distribution uniformity of Imidacloprid discharged throughout the drip line. The amount of Imidacloprid discharged from all emitters throughout drip line 3 (6.9 L/h emitters) was significantly lower than the other two drip lines. High concentrations of Imidacloprid were found in the flushing water samples collected from drip line 3. A large portion of the chemical injected into line 3 might have been carried by the high-speed water to the end of the line and then trapped. Drip line 2 (4.2 L/h emitters) had the highest amount of Imidacloprid discharged from emitters and highest recovery rate among the three lines (table 3). The average CV and DU of Imidacloprid throughout the 79 m drip line for all three drip lines was 44% and 0.78, respectively (table 4). Compared to BSF, Imidacloprid had a considerable high variation with the emitter flow capacity and the emitter location throughout the drip line.

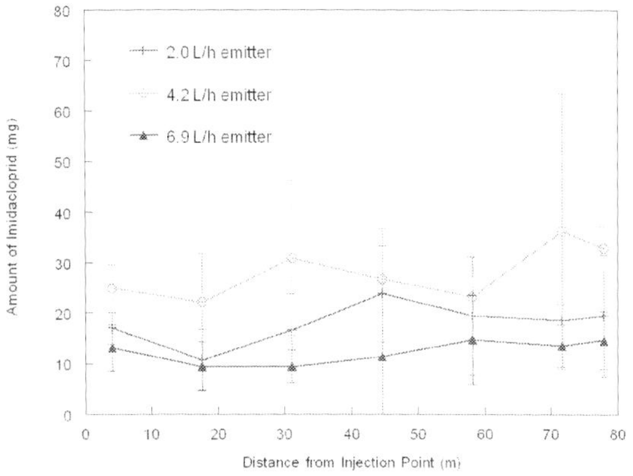

Fig. 7. Amounts of Imidacloprid discharged from individual 2.0, 4.2 and 6.9 L/h emitters throughout three different drip lines. Error bars represent standard deviations around means.

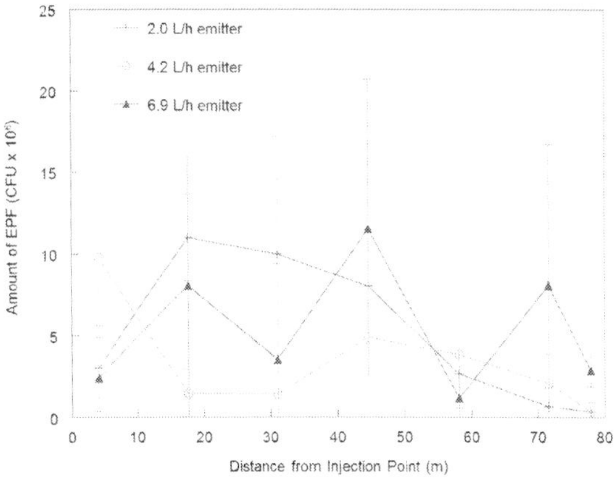

Fig. 8. Amounts of EPF discharged from individual 2.0, 4.2 and 6.9 L/h emitters throughout three different drip lines. Error bars represent standard deviations around means.

The emitter flow capacity did not significantly affect the amount of EPF discharged from individual emitters throughout the drip lines (fig. 8), which might be because the recovery rate of EPF was very low. Less than 10% EPF was recovered from all three drip lines (table 3). The low recovery rate might be caused by the adherence of EPF to the wall of drip lines. During preparation of the mixture of EPF and water, some suspensions were observed to adhere to the wall of plastic cups after standing for several minutes. The hydrophobic nature of *M. anisopliae* conidia combined with an emulsifiable concentrate formulation might require greater agitation to maintain suspension in water than that could be provided by a drip irrigation system. The CV for the amount of EPF throughout the entire drip line increased as the capacity of emitters increased, while the DU tended to decrease as the emitter capacity increased (table 4). The average CV and DU of EPF throughout the 79 m drip line for all three drip lines was 104% and 0.39, respectively (table 4).
Similar to the Imidacloprid, the amount of SF discharged from emitters was also affected by the emitter flow capacity (fig. 9). The amount of SF discharged from all emitters throughout drip line 3 was significantly lower than the other two drip lines while line 2 with 4.2 L/h emitter flow capacity had the highest amount of SF discharged (table 3). The recovery rate of SF ranged from 12 to 21% for all three emitter capacities. During three replications, there were three emitters clogged by materials in line 1, and 3 emitters clogged in line 2, but no emitters were clogged in line 3. The clogging problem was caused by particles with sizes larger than the depth of emitter flow path. Very high concentrations of SF remained in flushed water samples, which indicated that most SF remained in the drip line. Unlike EPF, the CV for the amount of SF throughout the entire drip line decreased and DU increased as the emitter capacity increased (table 4). The average CV and DU of SF throughout the 79 m drip line for all three drip lines was 74% and 0.58, respectively. The lowest and highest amounts of SF among all the emitters investigated in this study were 1.22×10^6 CFU and 6.67×10^6 CFU, respectively, both of which occurred on line 1, but the difference was not significant at the 0.05 probability level.

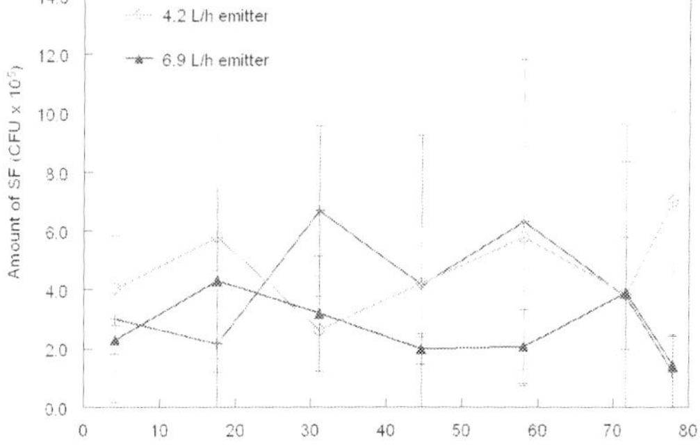

Fig. 9. Amounts of SF discharged from individual 2.0, 4.2 and 6.9 L/h emitters throughout three different drip lines. Error bars represent standard deviations around means.

The average number of nematodes discharged from individual emitters slightly increased as the emitter capacity increased, but the differences were not significant (table 3). The recovery rate of nematodes discharged from the 6.9 L/h emitters on line 3 was higher than that from the 2.0 and 4.2 L/h emitters. The emitter flow capacity influenced the distribution uniformity of nematodes throughout the drip lines (fig. 10). The value of CV decreased from 9.1% to 5.0% and DU increased from 0.80 to 0.91 when the emitter flow rate changed from 2.0 to 6.9 L/h (table 4).

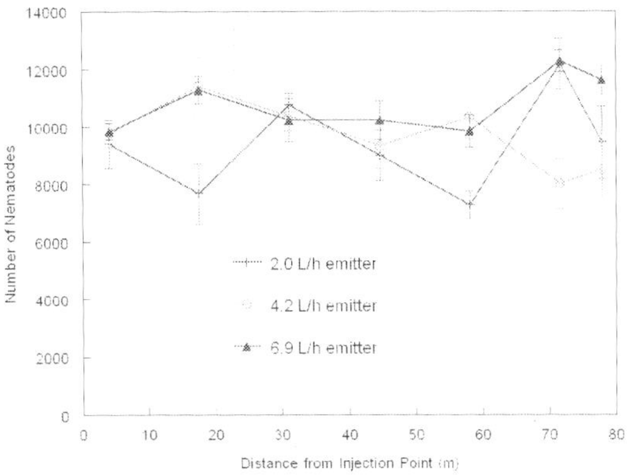

Fig. 10. Number of nematodes discharged from individual 2.0, 4.2 and 6.9 L/h emitters throughout three different drip lines. Error bars represent standard deviations around means.

Among the five materials tested, BSF had the highest recovery rate, the lowest CV and the highest DU across the three 79 m drip lines, followed by nematodes, Imidacloprid, and SF, while EPF had the highest CV and lowest DU and recovery rate for all three lines (tables 3, 4). The amount of BSF and number of nematodes discharged throughout all three drip lines had excellent distribution patterns (DU>0.80). The Imidacloprid DU was greater than 0.76 for all three drip lines. For the suspendible SF and EPF, their DU was less than 0.70, which was possibly caused by their particles not easily mixing with water. To compensate for the non-uniform delivery and low recovery rate of EPF and SF throughout the drip lines, the rate of EPF and SF required for effective insect/disease control must be determined before they are used in the field.

Both BSF and Imidacloprid solutions were soluble in water, but the viscosity of the Imidacloprid solution was much higher than the BSF solution. The Imidacloprid did not mix as well with the water in the drip line as the BSF after it was injected, which might be the reason that the Imidacloprid had higher CV and lower DU throughout all drip lines than the BSF. For nematodes and the bio-compound suspensions, the former were tiny worms suspended in water, while the latter were suspended organic particles with foam presumably from the wetting agent. Although the nematodes behaved as suspendible particles in solution, due to their small size and easy flow with water, their recovery rate and distribution uniformity throughout the entire drip lines performed much better than the two bio-compound suspensions EPF and SF.

The movement of chemical and microbial pesticides throughout the drip line is a complicated two-phase flow. Many factors influence the recovery rate and distribution uniformity of those materials throughout drip lines. The findings in this paper demonstrate the importance of evaluating these materials under controlled conditions before they are applied in the field. Future studies should further discover the influence of specific physical properties of these materials on the chemigation performances, and develop methods to improve the recovery rate and distribution uniformity of the suspended powder formulation of the microbial insecticide EPF and the suspended granular formulation of the microbial fungicide SF.

3.3 Distribution of tested materials in the soil

As mentioned before, BSF was used to track water movement in the soil and determine mobility of Imidacloprid, EPF and nematodes. Data in Table 5 illustrate lateral and vertical distribution patterns of BSF concentration in the soil for three different emitter capacities. BSF was detected at all 28 locations in the soil within the range from -45.6 to 45.6 cm laterally and from 0 to 30 cm vertically near each emitter. Within the 30 cm by 91.2 cm area, BSF presented the highest concentration at soil depths between 0 and 10 cm directly under the emitter. Concentration tended to decrease with the distance from the emitter for all three emitters, but the degree of such decrease was not very strong at soil depths of 20 and 30 cm. The BSF concentration diffused more evenly as soil depth increased. Within the area tested, the coefficient of variation of BSF concentration was 29 % for the 2.0 L/h emitters in line 1, 33% for the 4.2 L/h emitters in line 2, and 16% for the 6.9 L/h emitters in line 3, respectively. Thus, water discharged from the emitters travelled to the entire 91.2 cm by 30 cm area under the emitters with 2.0, 4.2 and 6.9 L/h flow capacities, with more water remaining near the emitters.

Soil Depth (cm)	Emitter flow (L/h)	Distance from emitter (cm)						
		-45.6	-30.4	-15.2	0	15.2	30.4	45.6
0	2.0	0.215	0.258	0.337	0.492	0.323	0.268	0.230
10	2.0	0.228	0.241	0.313	0.462	0.380	0.289	0.297
20	2.0	0.237	0.192	0.316	0.359	0.330	0.253	0.284
30	2.0	0.199	0.172	0.181	0.212	0.249	0.215	0.171
0	4.2	0.339	0.318	0.309	0.431	0.310	0.289	0.319
10	4.2	0.301	0.325	0.424	0.585	0.423	0.330	0.283
20	4.2	0.252	0.284	0.468	0.408	0.433	0.368	0.267
30	4.2	0.195	0.162	0.198	0.174	0.184	0.229	0.141
0	6.9	0.376	0.349	0.370	0.430	0.404	0.419	0.393
10	6.9	0.332	0.315	0.410	0.415	0.404	0.405	0.321
20	6.9	0.323	0.279	0.395	0.344	0.339	0.329	0.337
30	6.9	0.297	0.244	0.288	0.292	0.264	0.253	0.354

Table 5. Mean BSF concentration (µg/g) in the soil at various depths and lateral distances from the emitter with three flow capacities.

Unlike the distribution pattern of BSF in the soil, most Imidacloprid was distributed within a very narrow zone under the emitter (Table 6). Very little Imidacloprid was found at any depth when lateral distance from the emitter was greater than 30.4 cm. This was because the distribution coefficient of Imidacloprid in the soil was very small Cox et al. 1997). For the soil area 30 cm deep, 15.2 cm to the left and 15.2 cm to the right of the emitter, the CV of Imidacloprid concentration was 173 % for the 2.0 L/h emitters in line 1, 117% for the 4.2 L/h emitters in line 2, and 110% for the 6.9 L/h emitters in line 3, respectively. That is, higher flow provided better Imidacloprid distribution in the section close to the emitter, but the lateral distribution was very poor. The contour of Imidacloprid concentration in the soil also supported this statement (fig. 11).

Soil Depth (cm)	Emitter flow (L/h)	Distance from emitter (cm)						
		-45.6	-30.4	-15.2	0	15.2	30.4	45.6
0	2.0	0.26	0.64	12.1	2581	3.7	2.62	0.53
10	2.0	0.24	0.21	1.2	3099	3.1	0.85	0.30
20	2.0	0.28	0.16	9.5	2379	2.7	0.78	0.41
30	2.0	0.21	0.17	1.2	296	6.0	0.83	0.33
0	4.2	0.45	0.45	131.1	1170	957.4	4.05	0.49
10	4.2	0.25	0.18	4.6	1276	961.1	5.94	0.73
20	4.2	0.31	0.26	0.5	38	390.9	2.61	0.24
30	4.2	0.27	0.15	0.3	158	79.9	6.71	0.31
0	6.9	0.27	0.27	226.5	775	465.8	4.33	0.17
10	6.9	0.14	0.02	7.7	1066	805.3	8.84	0.68
20	6.9	0.21	0.09	0.3	11	610.0	2.97	0.17
30	6.9	0.15	0.02	0.2	140	54.4	10.67	0.12

Table 6. Mean Imidacloprid concentration (μg/kg) in the soil at various depths and lateral distances from the emitter with three flow capacities.

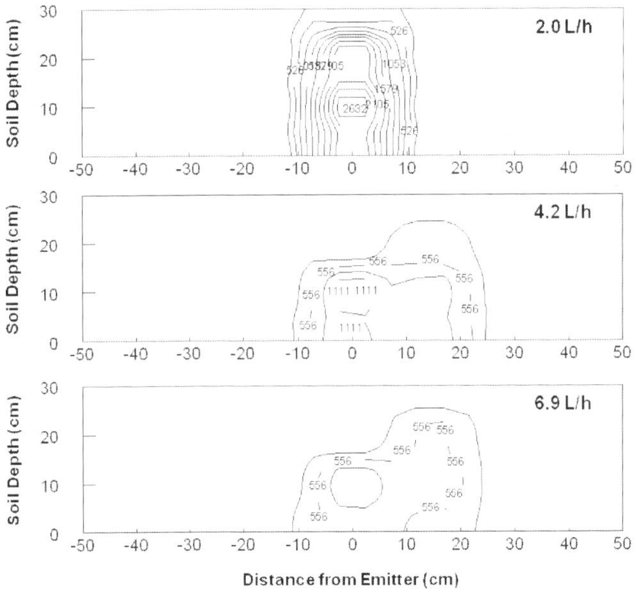

Fig. 11. Contour of Imidacloprid concentration in the soil within the 30 cm by 91.2 cm cross section area under the emitters at flow rates of 2.0, 4.2 and 6.9 L/h, respectively.

Soil Depth (cm)	Emitter flow (L/h)	Distance from emitter (cm)						
		-45.6	-30.4	-15.2	0	15.2	30.4	45.6
0	2.0	3700	8600	6333	40000	5533	7067	6867
10	2.0	6467	2733	5150	6000	3733	7800	3867
20	2.0	2533	2267	800	16067	15800	2200	1867
30	2.0	1333	933	600	1850	5000	900	467
0	4.2	18267	1667	6133	18000	4933	7067	5733
10	4.2	2450	13733	5067	4867	3067	2800	4800
20	4.2	6667	2267	3467	5200	9933	2800	2400
30	4.2	2800	14000	1600	1400	1600	1867	933
0	6.9	7600	5200	7133	6200	7067	7533	9333
10	6.9	5533	6533	4200	16867	3200	5933	5200
20	6.9	3733	2467	6267	5000	4733	4000	3067
30	6.9	467	533	5900	3067	1467	1800	1467

Table 7. Mean EPF concentration (CFU/g) in the soil at various depths and lateral distances from the emitter with three flow capacities. (new data)

The EPF had better distribution uniformity than Imidacloprid. The fungus spores of EPF were found at every sampling location for all three emitter flow-capacities (Table 7). Among the 28 sampling locations, the minimum number of spores was 467 CFU/g,which was located at 30 cm depth and 45.6 cm away from the emitter with both emitter flow-capacities of 2.0 and 6.9 L/h. Throughout the entire 91.2 cm wide and 30 cm deep section, the CV of EPF concentration was 130, 88, and 64% for the 2.0, 4.2 and 6.9 L/h emitters, respectively. Contours of EPF concentrations also showed that 6.9 L/h emitters had a better distribution uniformity of EPF than the 4.2 L/h emitters, while the 4.2 L/h emitters had a better distribution uniformity of EPF than the 2.0 L/h emitters (fig. 12). Therefore, higher flow could reduce variation in EPF distribution in soil.

Compared to their creamy-white color when healthy (Figure 13a), wax moth larvae turned brick red in color when infected with nematodes (Figure 13b). Detection of applied nematodes in soil samples was based on color changes in wax moth larvae. No moth larvae became infected when exposed to soil samples from untreated locations. Table 8 shows the presence of nematodes at different distances in the soil laterally and vertically away from the emitters with three different flow capacities. Among 28 locations, nematodes were found at 15 locations for the 2.0 L/h emitters, 15 locations for the 4.2 L/h emitters, and 20 locations for the 6.9 L/h emitters. Except for the absence of nematodes at one location 15.2 cm laterally and 30 cm vertically away from the 4.2 L/h emitters, nematodes were present at all 12 locations in the 30.4 cm wide and 30 cm depth area under all three emitter capacities. Nematodes moved further laterally in deeper soil locations when discharged from 6.9 L/h emitters as compared with 2.0 and 4.2 L/h emitters. Therefore, the higher presence of nematodes laterally at greater depths may be related to water application rates.

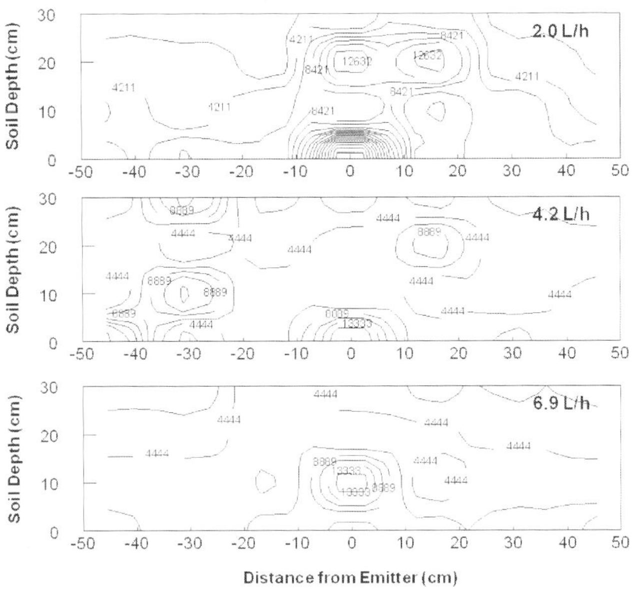

Fig. 12. Contour of EPF concentration in the soil within the 30 cm by 91.2 cm cross section area under the emitters at flow rates of 2.0, 4.2 and 6.9 L/h, respectively.

(a) No nematodes presented in soil (b) Nematodes presented in soil

Fig. 13. Comparison of wax moth larva color between (a) no nematodes and (b) nematodes found in soil samples. The soil sample shown in (a) was taken from the location 45.6 cm laterally and 20 cm vertically from the 4.2 L/h emitter. The soil sample shown in (b) was taken from the location 15.2 cm laterally and 20 cm vertically from the 4.2 L/h emitter.

Soil Depth (cm)	Emitter capacity (L/h)	Distance from emitter (cm)						
		-45.6	-30.4	-15.2	0	15.2	30.4	45.6
0	2.0	No	Yes	Yes	Yes	Yes	No	No
10	2.0	No	Yes	Yes	Yes	Yes	Yes	No
20	2.0	No	Yes	Yes	Yes	Yes	No	No
30	2.0	No	No	Yes	Yes	Yes	No	No
0	4.2	No	Yes	Yes	Yes	Yes	No	No
10	4.2	Yes	Yes	Yes	Yes	Yes	No	No
20	4.2	No	Yes	Yes	Yes	Yes	No	No
30	4.2	No	No	Yes	Yes	No	No	No
0	6.9	No	Yes	Yes	Yes	Yes	No	No
10	6.9	Yes	Yes	Yes	Yes	Yes	Yes	No
20	6.9	Yes	Yes	Yes	Yes	Yes	Yes	No
30	6.9	No	Yes	Yes	Yes	Yes	No	No

Table 8. Presence of nematodes in the soil at various depths and lateral distances from the emitter with three flow capacities.

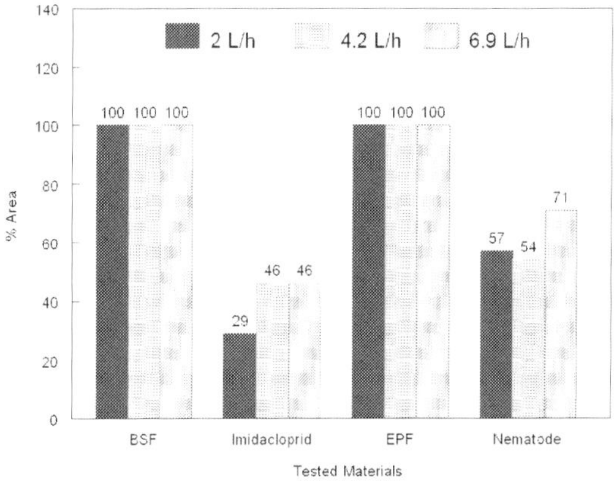

Fig. 14. Percent presence area of tested materials (BSF, Imidacloprid, ERF and nematodes) in the 30x91.2 cm^2 soil area under emitters with the flow capacities of 2, 4.2 and 6.9 L/h. (Change % Area to % Presence Area)

In the 30 cm by 91.2 cm sampling area under each emitter, percent area of the BSF and EPF presence was 100 for all three emitter flow capacities (fig. 14). Of the four materials tested in the soil, Imidacloprid had the lowest percent area for all three emitter flow capacities. The percent presence area of Imidacloprid was 29, 46 and 46 at 2.0, 4.2 and 6.9 L/h emitter flow rates, respectively. Presence area of Imidacloprid at 4.2 and 6.9 L/h was considerably higher than that at 2.0 L/h, while the flow capacities of 4.2 and 6.9 L/h did not significantly affect the percent presence of Imidacloprid in the sampling area. Flowable Imidacloprid did not extend laterally in the soil. More than half the tested area showed the presence of nematodes for all three flow rates. The nematode presence area was 71% at 6.9 L/h flow rate while it was 57 and 54% at 2.0 and 4.2 L/h flow rates, respectively. In general, Imidacloprid and nematodes could be distributed in a larger area in the soil by higher flow capacity emitters while distribution of BSF and EPF in the soil was not apparently affected by the emitter flow capacity. Nematodes, as living organisms, travel substantial distances in soil when target insects are present (Grewal et al., 2005). Thus, the potential area that nematodes presented in the soil should be much larger than the 30 cm by 91.2 cm.

Previous tests for the distribution throughout drip lines reported that the recovery rate of EPF and nematodes were very low for all three flow-capacity emitters. The recovery rate of EPF discharged from 2.0, 4.2 and 6.9 L/h flow capacity emitters was 9.0, 6.0 and 9.5%, respectively. For the same flow capacity emitters, recovery rate for nematodes was 41, 42 and 47%, respectively. Presence of EPF and nematodes in the entire area of 30 cm by 91.2 cm area under the emitters verified that drip irrigation systems could be an alternative method to apply suspendible microbial pesticides. Also, application of these microbial pesticides avoids the potential leaching damage normally rendered by the chemical pesticides; however, rates of EPF and nematode application that effectively control target pests must first be determined to compensate for their low recovery rates.

4. Conclusions and summary

Drip irrigation uniformly dispensed the water-soluble BSF, water-dispersible insecticide Imidacloprid and suspended nematodes throughout drip lines, but not the suspended powder formulation of the microbial insecticide EPF or the suspended granular formulation of the microbial fungicide SF. The uniformity of distribution of the various test agents throughout the drip line varied with their physical properties of the individual product formulation. The distribution uniformity of EPF discharged from emitters throughout the drip line was the lowest among the five materials tested, followed by SF, Imidacloprid, nematodes, and BSF.

Except for BSF and Imidacloprid, flow capacity of emitters affected the distribution uniformity of the other test agents throughout drip lines. Among the three emitters tested, EPF had the highest DU at 2.0 L/h flow rate. The uniformity of SF and nematode distribution throughout drip lines increased as the flow capacity of emitters increased. For the emitters with flow rates ranging from 2.0 to 6.9 L/h, DU averaged over 0.95 for BSF, over 0.80 for nematodes, over 0.75 for Imidacloprid, ranged from 0.44 to 0.68 for SF, and ranged from 0.33 to 0.56 for EPF.

Emitter size and flow capacity affected the recovery rates of Imidacloprid and SF discharged throughout the drip line, but not of BSF, EPF and nematodes. The recovery rates greatly varied with the physical properties of the individual product formulation. For the emitters with flow rates ranging from 2.0 to 6.9 L/h, the recovery rate was below 9.5% for EPF, 21% for SF, 50% for nematodes, and 78% for Imidacloprid.

Active agent distribution patterns in soil varied with the formulation. Water-insoluble microbial insecticides, EPF and nematodes, exhibited better distribution patterns than water-soluble systemic insecticide, Imidacloprid. EPF spores were found in the entire 91.2 cm by 30 cm cross-section under all three emitter flow-capacities. Imidacloprid showed a very narrow distribution pattern directly under the emitters.

Active-agent distribution patterns in the soil also varied with emitter flow capacity. EPF was present in the entire 30 cm by 91.2 cm area under the emitter for all three emitter flow capacities. Nematodes presented 57% of the area at 2.0 L/h flow rate, 54% at 4.2 L/h flow rate and 71% at 6.9 L/h flow rate, while Imidacloprid presented 29, 46 and 46% area at 2.0, 4.2 and 6.9 L/h emitter flow rate, respectively.

Uniformity of EPF and nematode distributions in the soil improved as emitter flow capacity increased. Presence area of Imidacloprid at 4.2 and 6.9 L/h was significantly higher than that at 2.0 L/h while there was no significant difference in the presence of Imidacloprid in the sampling area between flow capacities of 4.2 and 6.9 L.

These results demonstrated that drip irrigation could be a viable alternative method for water-soluble pesticide applications. However, the use of drip irrigation systems for the delivery of suspended powders and granular agents, e.g. EPF and SF, for pest control may be limited because of their poor uniformity and low recovery rates throughout drip lines. Any materials with sizes larger than the width or depth of emitter flow paths would clog emitters and should not be applied through drip irrigation systems.

5. References

Ajwa, H. A.; Trout, T.; Mueller, J.; Wilhelm, S.; Nelson, S. D.; Soppe, R. & Shatley, D. (2002). Application of alternative fumigants through drip irrigation systems. *Phytopathology*, 92(12):1349-1355.

ASAE Standards, EP405.1, APR1988 (R2008). (2008). Design and Installation of Microirrigation Systems. St. Joseph, MI:ASABE.

Barber, J.A.S. & Parkin, C. S. (2003). Fluorescent tracer technique for measuring the quantity of pesticide deposited to soil following spray applications. *Crop Protection*, 22(1):15-21, 42.

Becker, J. O.; Monson, M.; Van Gundy, S. D. & Schroth, M. N. (1989). Drip irrigation as a delivery system for infestation of Þeld plots with nematodes. *J. Nematol.*, 21: 524-529.

Bedding, R.A. & Akhurst, R. J. (1975). A simple technique for the detection of insect pathogenic nematodes in soil. *Nematologica*, 21:109-110.

Bryla, D. R.; Banuelos, G. S. & Mitchell, J. P. (2003). Water requirements of subsurface drip-irrigated faba bean in California. *Irrigation Science*, 22: 31-37.

Bracy, R.P; Parish, R.L. & Rosendale, R.M. (2003). Fertigation uniformity affected by injector type. *Hortechnology*, 13:103-105.

Byrne, F.J. & Toscano, N.C. (2006). Uptake and persistence of imidacloprid in grapevines treated by chemigation. *Crop Protection*, 25: 831–834.

Camp, C.R.; Sadler, E.J. & Busscher, W.J. (2003). A comparison of uniformity measures for drip irrigation systems. *Transactions of the ASAE*, 40(4):1013-1020.

Cassel, D.K. & Nielsen, D.R. (1986). Field capacity and available water capacity. In: *Methods of Soil Analysis. Part 1. Physical and Mineralogical Methods. 2nd ed. Agron. Monogr. no. 9*. A. Klute (ed.). p. 901–926. ASA and SSSA, Madison, WI.

Castle, S.J.; Byrne, F.J.; Bi J.L. & Toscano, N.C. (2005). Spatial and temporal distribution of Imidacloprid and thiamethoxam in citrus and impact on *Homalodisca coagulata* populations. *Pest Management Science*, 61:75-84.

Clark, G. A.; Lamm, F. R. & Rogers, D.H. (2005). Sensitivity of thin-walled drip tape emitter discharge to water temperature. *Applied Engineering in Agriculture*, 21(5):855-863.

Cox, L.; Koskinen, W. & Yen, P. (1997). Sorption-desorption of imidacloprid and its metabolites in soils. *J. Agric. Food Chem.*, 45(4): 1468–1472.

Dorner, J.W. & Lamb, M.C. (2006). Development and commercial use of afla-guard®, an aflatoxin biocontrol agent. *Mycotoxin Research*, 21:33-38.

Ellsbury, E. E.; Jackson, J. J.; Woodson, W. D.; Beck, D. L. & Strange, K. A. (1996). Efficacy, application distribution, and concentration by stemflow of *S. carpocapsae* (Rhabditida: Steinernematidae) suspension applied with a lateral-move irrigation system for corn rootworm (Coleoptera: Chrysomelidae) control in maize. *Journal of economic entomology*, 89:74-81.

Felsot, A.S.; Cone, W.; Yu, J. & Ruppert, J.R. (1998). Distribution of Imidacloprid in soil following subsurface drip chemigation. *Bull. Environ. Contam. Toxicol.*, 60: 363-370.

Felsot, A.; Evans, R. G. & Tallman, L. (2000). Soil distribution and plant uptake of imidacloprid under drip and furrow irrigation. *Proceedings of the 4th Decennial National Irrigation Symposium*, pp. 416-427. Phoenix, AZ. R. G. Evans, B. L. Benham, and T. P. Trooien, eds. American Society of Agricultural Engineers, St. Joseph, MI.

Felsot, A. S.; Evans, R. G. & Ruppert, J. R. (2003). Field studies of Imidacloprid distribution following application to soil through a drip irrigation system. Chapter 13 (pp. 189-205) in *Terrestrial Field Dissipation Studies: Purpose, Design and Interpretation*. E. L. Arthur, A. C. Barefoot, and V. E. Clay eds. ACS Symposium Series No. 842, American Chemical Society, Washington, DC. ISBN# 0-8412-3769-7. 368 pp.

Fleischer, S.J.; Orzolek, M.D.; De Mackiewicz, D. & Otjen, L. (1998). Imidacloprid effects of Acalymma vittata (Coleoptera: Chrysomelidae) and bacterial wilt in cantaloupe. *Journal of economic entomology*, 91(4): 940-949.

Fuxa, J.R. & Richter, A.R. (1999). Classical biological control in an ephemeral crop habitat with Anticarsia gemmatalis nucleopolyhedrovirus. *Biol. Control*, 44: 403–419.

Gopal, M.; Gupta, A.; Sathiamma, B. & Radhakrishnan Nair, C.P. (2001). Control of the coconut pest Oryctes rhinoceros L. using the Oryctes virus. *Insect Sci.*, 21(4): 93–101.

Grabow, G. L.; Huffman, R. L.; Evans, R.O.; Jordan, D. L. & Nuti, R.C. (2006). Water distribution from a subsurface drip irrigation system and dripline spacing effect on cotton yield and water use efficiency in a coastal plain soil. *Transactions of the ASABE*, 49(6): 1823-1835.

Grewal, P.S.; Power, K.T. & Shetler, D.J. (2001). Neonicotinoid insecticides alter diapause behavior and survival of overwintering white grubs (Coleoptera: Scarabaeidae). *Pest Management Science*, 57:852–857.

Grewal, P. S.; Ehlers, R. U. & Shapiro-Ilan, D. I. (2005). *Nematodes as Biocontrol Agents*, CABI Publishing, CAB International, Wallingford, Oxfordshire, UK, 505pp.

Hajek, A.E.; McManus, M.L. & Ju´nior, I.D. (2007). A review of introductions of pathogens and nematodes for classical biological control of insects and mites. *Biological Control*, 41: 1-13

Jaynes, D. B.; Rice, R. C. & Hunsaker, D. J. (1992). Solute transport during chemigation of a level basin. *Transactions of the ASAE*, 35(6): 1809-1815.

Keller, J. & Bliesner, R.D. (2000). *Sprinkler and Trickle Irrigation*. The Blackburn Press, Caldwell, NJ. 652 p.

Kramer, I. & Grunder, J. (1998). Efficacy of EPNs application by irrigation systems on black vine weevil larvae control. *International Organization for Biological Control (IOBC) Bulletin*, 21: 179-182.

Kruse, E.G. (1978). Describing irrigation efficiency and uniformity. *J. Irrig. Drain. Div. ASCE*, 104(IR):35-41.

Lamm, F. R.; Ayars, J. E. & Nakayama, F. S. (2007). *Microirrigation for Crop Production - Design, Operation and Management*. Elsevier Publications. 608 pp.

Lamm, F. R.; Stone, L. R., Manges, H. L. & O'Brien, D. M. (1997). Optimum lateral spacing for subsurface drip-irrigated corn. *Transactions of the ASAE*, 40(4):1021-1027.

Lamm, F. R. & Trooien, T. P. (2003). Subsurface drip irrigation for corn production: A review of 10 years of research in Kansas. *Irrigation Science,* 22(3-4):195-200.

Leib, B.G. & Jarrett, A.R. (2003). Comparing soil pesticide movement for a finite-element model and field measurements under drip chemigation. *Computers and Electronics in Agriculture,* 38: 55-69.

Leib, B.G.; Jarett, A.R.; Orzolek, M.D. & Mumma, R.O. (2000). Drip chemigation of Imidacloprid under plastic mulch increased yield and decreased leaching caused by rainfall. *Transactions of the ASAE,* 43(3):615-622.

Li, J.; Meng, Y. & Li, B. (2007). Field evaluation of fertigation uniformity as affected by injector type and manufacturing variability of emitters. *Irrigation Science,* 25:117-125.

Lumsden, R.D. & Locke, J.C. (1989). Biological control of damping-off caused by *Pythium ultimum* and *Rhizoctonia solani* with *Gliocladium virens* in soilless mix. *Phytopathology,* 79:361-366.

Plaut, Z.; Carmi, A. & Grava, A. (1996). Cotton root and shoot responses to subsurface drip irrigation and partial wetting of the upper soil profile. *Irrigation Science,* 16: 107-113.

Reding, M. E.; Zhu, H. & Derksen, R. C. (2008). Drip chemigation with Imidacloprid and nematodes for control of scarab larvae in nursery crops. *J. Environ. Hort.,* 26: 93-100.

Reed, D. K.; Reed, G. L. & Creighton, C. S. (1986). Introduction of entomogenous nematodes into trickle irrigation systems to control striped cucumber beetle (Coleoptera: Chrysomelidae). *Journal of economic entomology,* 79:1330-1333.

Schneider, S.M.; Ajwa, H.A. & Trout, T.J. (2006). Chemical alternatives to methyl bromide for vineyard replant conditions. *American Journal of Enology and Viticulture,* 57(2): 183-193.

Van Iersel, M. W.; Oetting, R. D. & Hall, D. B. (2000). Imidacloprid applications by subirrigation for control of silverleaf whiteßy (Homoptera: Aleyrodidae) on poinsettia. *Journal of economic entomology,* 93: 813-819.

Veen, K.H. & Ferron, P. (1966). A selective medium for the isolation of Beauveria tenella and Metarrihizium anisopliae. *J. Invert. Pathol.,* 8:268-269.

Wang, X.; Zhu, H.; Reding, M.E.; Locke, J.C.; Leland, J.E.; Derksen, R.C.; Spongberg, A.L. & Krause, C.R. (2009). Delivery of chemical and microbial pesticides from drip irrigation emitters. *Applied Engineering in Agriculture,* 25(6): 883-893.

Wennemann, L.; Cone, W.W.; Wright, L.C.; Perez, J. & Conant, M.M. (2003). Distribution patterns of entomopathogenic nematodes applied through drip irrigation systems. *Journal of economic entomology,* 96(2):287-291.

Woodring, J.L. & Kaya, H.K. (1988). *Steinernematid and Heterorhabditid Nematodes: A handbook of techniques, Southern cooperative Series Bulletin 331.* Arkansas Agricultural Experiment Station, Fayetteville, AR.

Wu, I.P.; Howell, T.A. & Hiler, E.A. (1979). *Hydraulic design of drip irrigation systems.* Hawaii Agricultural Experimental Station Technical Bulletin 105. Honolulu, Hawaii: University of Hawaii.

Zhu, H.; Lamb, M.C.; Butts, C. L. & Blankenship, P. D. (2004). Improving peanut yield and grade with surface drip irrigation in undulating fields. *Transactions of the ASAE,* 47(1): 99-106.

Zhu, H.; Derksen, R.C.; Krause, C.R.; Fox, R.D.; Brazee, R.D. & Ozkan, H.E. (2005). Effect of solution pH conditions on fluorescence of spray deposition tracers. *Applied Engineering in Agriculture,* 21(3):325-329.

Zhu, H.; Fox, R.D.; Ozkan, H.E.; Brazee, R.D. & Derksen, R.C. (1998). Time delay for Injection sprayers. *Transactions of the ASAE,* 41(3): 525-530.

Part 2

Pesticides Contamination and Exposure

10

International Food Safety Standards and the Use of Pesticides in Fresh Export Vegetable Production in Developing Countries: Implications for Farmer Health and the Environment

Julius J. Okello[1] and Scott M. Swinton[2]
[1]University of Nairobi
[2]Michigan State University
[1]Kenya
[2]USA

1. Introduction

Most developing country farmers producing for international markets rely on pesticides for agricultural production (Thrupp et al, 1995, Maumbe and Swinton, 2003). The warmth and humidity of tropical climates exacerbates the pest and disease problems (Okello, 2005). Due to standards for cosmetic quality in export markets for fresh fruits and vegetables, the use of pesticides has been especially pronounced in production of these products in the tropics.

Production and export of fresh produce from developing countires have witnessed major growth in many developing countries seeking to diversify their production from staples to high value commodities. Growth has especially been greatest in the fresh fruits and vegetables (FFV) and in the flower subsectors. In Africa, for instance, exports of FFV experienced a spurt in growth in the 1980s and 1990s as markets for major traditional exports (e.g., coffee, tea and cocoa) experienced a downturn. Most of these non-traditional exports were destined to Europe (with UK, Holland, Germany, and Italy being the leading importers) (Okello et al, 2008). Figure 1 presents the trends in exports of green beans, a major non-traditional export, by three of the leading exporters of fresh vegetables from Africa. It shows an increase in exports of green beans between 2000 and 2006 in all these countries.

Kenya is one of the leading exporters of fresh vegetables to Europe, and especially the United Kingdom (UK). Figure 2 shows the recent expansion of green bean exports, highlighting the growth in those destined for the UK.

The strong expansion in green bean exports is largely targeted at European consumers who demand aesthetic quality attributes such as spotlessness that generally encourage increased use of pesticides (Farina and Reardon, 2000). The demand for cosmetic quality attributes (color, shape, spotlessness) has been held responsible for increasing pesticide use in the production of fresh exports from developing countries. Thrupp et al (1995) and Ohayo-Mitoko (1997) document cases of widespread use of pesticides in Asia and Kenya respectively. Excessive use of pesticides in Kenyan horticultural industry has also been

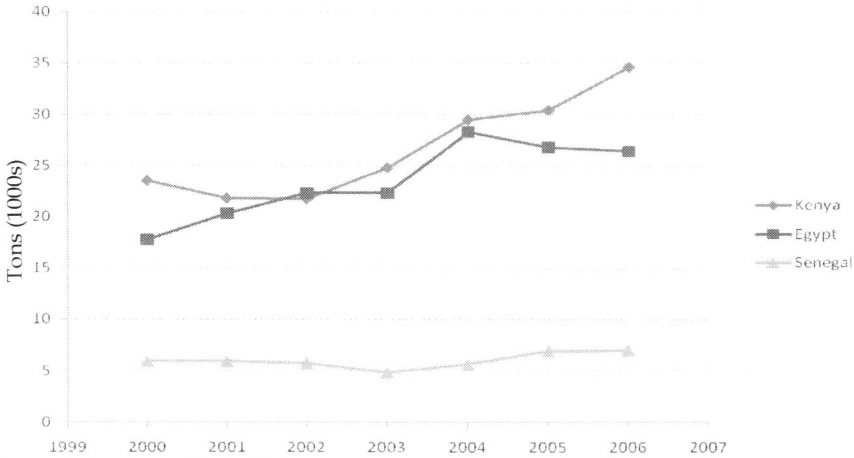

Source: Adapted from Okello (2010)

Fig. 1. Trends in green bean exports ('000 tons) to Europe from some leading African countries, 2000-2006

reported by Mwanthi and Kimani (1990), Okado (2001) and Jaffee (2003). These studies suggest that many Kenyan fresh export vegetable farmers used pesticides indiscriminately, in some cases, applying pesticides meant for other crops (such as coffee) on fresh vegetables. Concern with the health consequences of excessive use of pesticides on consumers' medical health and safety of farm workers and the environment in general led developing country governments to revise their pesticide residue standards. These revised international food safety standards (IFSS) have introduced a new order in the use of pesticides in production of fresh vegetables destined for sale in developed countries. They require that only pesticides that are safe to farmers and farm-workers, other non-target species and the consumers be used in production of vegetables for exports. However, the safer pesticides are often either more expensive or less efficacious (Jaffee, 2003). At the same time, farmers and pesticide applicantors are required, under IFSS, to handle, apply and discard leftover pesticides safely in order to reduce the hazards they pose to non-target animal and plant species. These requirements are reinforced by farmer training on safe use, storage and disposal of pesticides and enforced via close monitoring for compliance. African analysts have alleged that the expected benefits to European consumers would impose unacceptable costs on African producers, especially smallholders (Mungai, 2004). Hence, the welfare effects of African producers compliance with European IFSS have been a subject of intense debate.

Theoretically, IFSS are expected to induce some changes in pesticide usage and in the returns farmers receive from beans. Such changes can affect the profit margins earned but can also theoretically reduce the costs of health impairments as a result of reduced exposure to toxic pesticides. This chapter examines the effect of implementation and enforcement of IFSS in the green bean industry in Kenya. It specifically discusses these effects in the context of benefits to farm households and the environment and costs of complying with IFSS. The chapter focuses on health costs on exposure to pesticides, the use of environmentally-friendly pest and disease control strategies and the changes in consumer margins resulting from compliance with IFSS.

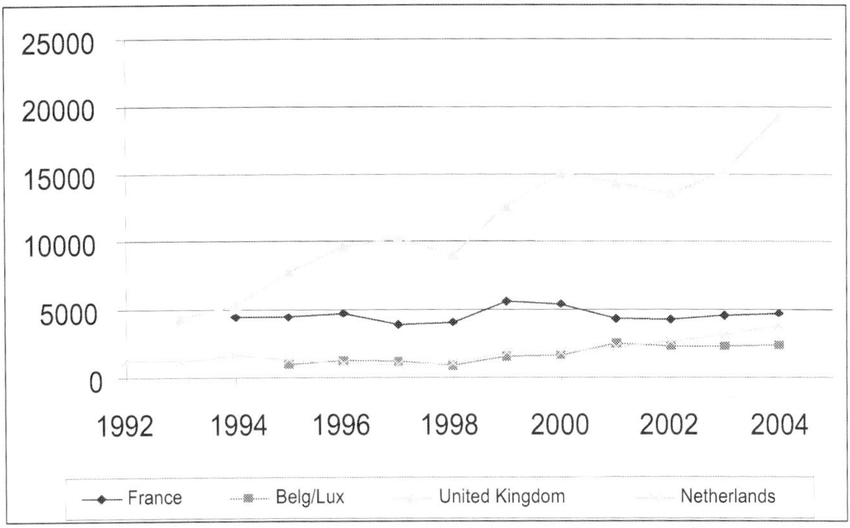

Source: Okello et al (2008)

Fig. 2. Major destination markets for Kenyan green beans (tons) 1992-2004

This chapter draws from a study conducted in Kenya in 2003-2004 involving smallholder farmers growing green beans for export to the European markets. Kenya is one of leading exporters of green beans to Europe, especially the United Kingdom (UK). Europe provides an interesting case study because major European retailers (e.g., Tesco, Waitrose, Mark & Spencer and Sainsbury's) have developed some of the most stringent IFSS. The rest of this paper is organized as follows. Section 2 provides historical overview of the Kenyan green bean industry and highlights the changes in export standards. Section 3 outlines the study methods while Section 4 discusses the results. Section 5 concludes.

2. Historical perspectives of the Kenyan green bean industry and the IFSS

Kenya's fresh export vegetable industry is one of the oldest in Africa, having started in the 1950s with off-season exports of fresh fruits and vegetables (FFV) to the UK (McCulloh and Ota, 2003, Okello, 2010). The shipments started with the launch of regular passenger flights between Kenya and Western Europe. The first consignment was flown to the UK in 1957. Subsequently, few hundred tons of FFV were annually shipped to a single wholesaler in London's Convent Garden Market from where they were eventually sold to high-class hotels, restaurants and department stores (Okello, 2010). The growth of the horticulture industry accelerated considerably during the 1970s. By 1975, annual exports of FFV had surpassed 10,000 tons. Trade in fresh vegetables (led by green beans) expanded most rapidly during the 1980s and 1990s. By the end of 1990s, Kenya was exporting close to 2 million tons of fresh vegetables annually. The expansion in trade was accompanied by the broadening of the destinations.

Kenya traditionally channels virtually all of its fresh vegetable exports to Western Europe, with very small quantities going to Australia/New Zealand, South Africa, and Dubai. The bulk of the exports go to the UK, Holland, France and Germany. The UK is still the biggest market for Kenyan vegetables absorbing more than 60% of Kenya's green beans per year. Within the UK, the leading retailers of Kenyan beans are Waitrose, Tesco, Marks and Spencer, and Sainsbury's. These major retailers control major share of fresh export business especially in the UK. Indeed, retailers/supermarkets control 70% Kenya green bean trade and 100% of the high-care pre-packed "ready to eat" fresh vegetable trade in general.

Majority of the leading European retailers developed very stringent standards relating to pesticide usage, among others, in response European food safety scandals of the 1990s. They have subsequently passed on these standards to sourcing agents or suppliers in developing countries. Developing-country suppliers have in turn developed their own code of practices relating to how pesticides may be handled, applied, and stored. Thus a developing country farmer is often subject to diverse standards ranging from international to domestic, with the latter induced by the former. Table 1 presents the kinds of standards a green bean farmer growing beans for a European retailer will typically be subject to. The domestic industry, private and public standards are usually drawn from the foreign standards especially those of the markets targeted by the exporter. Most green bean family farmers therefore comply with standards that encompass the requirements of UK industry standards (e.g., British Retail Consortium (BRC) and Global Good Agricultural Practices (GlobalGAP)), private retailer standards (e.g., Nature's Choice and Farm to Fork) and public sanitary and phytosanitary standards (SPS).

Foreign standards	*Domestic standards*
British Retail Consortium (BRC)	*i) Industry*
GlobalGAP	KenyaGAP
Ethical Trading Initiative Initiative	Horticultural Ethical Business
HACCP	*ii) Exporter code of practices*
Tesco's Nature's Choice	*iii) Public*
Marks & Spencer's Farm to Fork	Kenya Bureau of Standards
Sanitary and Phytosanitary Standards (SPS)	HCDA code of practices

Source: Adapted from Okello et al (2008);
HCCP = Hazard Analysis and Critical Control Points; HCDA=Horticultural Crop Development Authority

Table 1. Array of food safety standards in operation in Kenyan green bean industry

The diverse standards are primarily aimed at promoting practices that encourage farmers and pesticide applicators to adopt practices that protect them and the environment from hazards of pesticide exposure. These practices include i) wearing full pesticide protective gear, ii) handling pesticides in ways that ensure safety to farm family members and farm-workers, iii) bathing immediately after spraying or when pesticides accidentally come into contact with the skin, iv) storing pesticides away from foodstuffs in fully secured pesticide storage units with adequate ventilation, v) disposing of pesticide containers and leftover pesticides in ways that do not threaten the health of humans or animals, iv) discontinuing the use of unapproved (usually more toxic) pesticides, and v) using pesticides only when needed (especially when pest scouting reveals the need to apply them).

Farmers and farm-workers can get exposed to pesticides through four primary routes namely ingestion, inhalation, dermal absorption, and absorption through the eyes. Okello and Swinton (2010) highlight the various ways in which individuals in a farm situation can get exposed to pesticides. These include entry into freshly sprayed field, eating while spraying pesticides, and skin contact with liquid, powder or aerosol forms of pesticides. Exposure to toxic pesticides can result in health hazards in the form of acute or chronic illnesses (Maumbe and swinton, 2003). Common pesticide induced illnesses include skin irritation, eye irritation, gastrointestinal irritation, respiratory irritation, headaches, shortness of breath, dizziness, cancer, neurological problems, stillbirth and abortion.

The rationale behind enforcing IFSS was that they can help reduce the hazards posed to farmers' health and the environment by pesticides. Past studies have documented strategies that avert exposure to pesticides. Such strategies include wearing pesticide protective clothing during mixing and application, using properly secured pesticide storage units and disposing of pesticides in secured disposal pits (Antle and Capalbo, 1994; Maumbe and Swinton, 2003). Other exposure averting strategies include observing the interval between the application of pesticides and date of harvest, washing hands before eating, washing the protective clothing before next use, and combining pesticide application with other pest and disease control strategies. However, farmers can also reduce the health risks of pesticides after exposure has occurred by using mitigating strategies such as washing off pesticides from skin when there is accidental contact, removing clothes and taking a bath when there is accidental leakage of the spray pump, and taking medication. IFSS promote the use of these pesticide exposure averting and mitigating strategies. So how does compliance with this array of standards affect farmer health costs of exposure to pesticides, the use of environmentally-friendly practices and farmer profit margins? We turn to these questions after briefly presenting the methods used in this study.

3. Study methods

3.1 Theoretical framework

In order to examine the effect of IFSS on Kenyan fresh produce industry, we categorized green bean farmers into two groups namely, growers who supply exporters that monitor and enforce IFSS (i.e., monitored farmers) and those that supply non discerning exporters (non-monitored farmers). Pesticide usage may benefit farmers because it enhances the aesthetic quality of the produce, potentially enabling farmers to sell more quantity at higher prices. However, pesticide exposure may also be harmful to farmer health. The relationship between pesticide handling and usage by a farmer and health status ($h(.)$) can be expressed algebraically as a function of farmer specific variables (f), behavioral variables (b), exposure to pesticides ($e(.)$) which increases with pesticide inputs use (x), but decreases with pesticide exposure averting (a) and mitigating (m) variables. Health outcome is also assumed to be affected by doctor-prescribed and self administered treatment expenses (Φ) and institutional factors (z). Thus following previous authors (Cole et al, 1998; Strauss and Thomas, 1998; Hurley et al, 2000):

$$h=h[f, b, e(x, a, m), \Phi(e), z) \tag{1}$$

Equation 1) implies that the health outcome of a farmer depends upon the set of strategies employed during pesticide handling and application, among other factors. These strategies specifically include pesticide averting (a) (e.g., protecting clothing, secured pesticide storage units, fenced disposal pits, and the use alternative pest management strategies) and mitigating (m) strategies (treatment, washing pesticides off the skin when there is accidental contact, removing clothes and bathing when there is accidental leakage of the spray pump). The farmer uses pesticide and non-pesticide inputs to produce output (q) given by:

$$q=q[x,v,T,k,z] \tag{2}$$

where q is the output of beans, x is a vector of pesticide inputs, v are non-pesticide inputs (e.g., fertilizer); T is the total effective field labor requirement comprising effective family labor, ($l(h)$) and hired labor (r). Since exposure to toxic pesticides impairs health, we assume that effective family labor depends on the health outcomes. For instance, illness resulting from pesticide exposure is likely to cause loss in labor time as the victim recovers. We assume that the hired labor bears the cost of health impairments due to exposure to pesticide via inability to work when sick. Finally, k is a vector of capital factors, and z is as earlier defined.

Farmers who are monitored for compliance with IFSS produce beans under contracts that specify output volume, output price and other non-pesticide inputs; hence these are assumed to be predetermined. The farmer's optimization problem is to choose x, a, m and Φ to minimize the combined production and health costs subject to labor availability and contracted output level and quantity q^o. The optimization function can be expressed as:

$$\underset{x,a,m,\Phi}{Min}\ c(x,a) = w_x x + w_a a + w_m m + w_\Phi \Phi \tag{3}$$

$$\text{s.t.} \qquad q \geq q^o$$

$$T \geq l(h) + r$$

The farmer minimizes the combined costs subject to two constraints namely that, i) it produces no less than the contracted output ($q \geq q^0$) and ii) the total effective field labor is at least equal the sum of family and hired labor ($T \geq l(h)+ r$). This optimization problem can be expressed mathematically by a Lagrangean function (L) as:

$$L = w_x x + w_a a + w_m m + w_\Phi \Phi + \varphi\{q^0 - q(.)\} + \lambda(T - l(h) - r) \tag{4}$$

The Lagrange multiplier φ represents the marginal value of added output while λ is the marginal cost of labor. We assume that the cost and production functions are concave and that exposure to pesticides leads to poor health while using strategies that avert or mitigate exposure to pesticides improves the health status of the farm household. At the same time, doctor-prescribed or self administered treatment expenses improve health status. We further assume that improved health outcome increases the availability of effective family labor.

Solving the Equation 4) yields the medical health and input demand functions (Okello and Swinton, 2010; Okello and Okello, 2010). The input demand functions include the demand for pesticide exposure averting strategies (a). One such strategy is the use of alternative pest management strategies, which reduces overdependence on the chemical control of pests and diseases and is therefore friendly to the environment. Comparative static analysis shows that optimal use of pesticides requires consideration of farmer health whose costs can be reduced by using less toxic pesticides, employing more pesticide exposure averting and mitigating strategies and relying on medical treatment.

Green bean exporters supplying major EU supermarkets train their farmers on safe use, storage, and disposal of pesticides as well as the need to use alternative strategies of managing pests and diseases. As part the training, farmers are educated on health and environmental effects of pesticide exposure and on safe use of pesticides. *We therefore hypothesize that green bean farmers who comply with IFSS benefit by incurring lower health costs of exposure to pesticides. In addition, we hypothesize that monitoring compliance with IFSS increases the use of more environmentally friendly strategies of managing pests and diseases in green beans.*

The use of alternative pest management strategies and switching to safer approved pesticides may however present a challenge to green bean growers. First, the new (safer) pesticides may be less effective in controlling the target pests and diseases (i.e., less efficacious). Second, the new approved/safer pesticides tend to be more expensive than the unapproved ones (Jaffee, 2003; Okello, 2005). Consequently, the switch to new/approved pesticides may increase the cost of production. In sum, if the approved pesticides are less efficacious and more expensive, the overall effect of complying with the standards will be a reduction in margins earned from the sale of the beans. *We therefore hypothesize that farmers who comply with IFSS will receive lower margins than their counterparts.*

3.2 Study area and data

This paper draws from a study based on 180 green bean family farmers in Kirinyaga and Kerugoya-Kutus districts (located in Central province of Kenya) and Kangundo district (located in Eastern province of Kenya). The study was conducted between October 2003 and June 2004. A list of major green bean growing villages (primary sampling units) was drawn. From the list, 30 villages having both IFSS compliant and non-compliant farmers were randomly selected. Six farmers were then randomly sampled from each of the 30 villages, stratified by compliance with IFSS, giving a total of 180 farmers. Information was collected through personal interviews using questionnaires. Information on pesticides used was also collected.

Pesticide toxicity of the pesticides reported by farmers was looked up from the World Health Organization (WHO) toxicity classification and the pesticides categorized as class 1 (very toxic), class 2 (toxic), class 3 (slightly toxic) and class 4 (unharmful) (World Health Organization, 2005). The WHO class 4 pesticides were omitted from further analysis because they are not usually considered hazardous to users (Maumbe and Swinton, 2003).

Detailed cost and output data was gathered from six carefully selected green bean farmers in the study areas. Of the six farmers, three were producing beans for exporters who supplied UK supermarkets and hence routinely monitored compliance with IFSS under a contract while the other three sold their beans in the spot market, hence no IFSS monitoring. All the six farmers had 0.5 acres under green beans, which was the mean farm size for small family farms for the last crop of green beans grown in 2003.

3.3 Empirical methods

In order to address the three hypotheses above, this study used a combination of quantitative techniques. The first technique used was the cost of illness approach. Under this approach, health cost of exposure to pesticides is approximated by the direct and indirect costs incurred as a result of being sick from pesticide poisoning. Specifically, farmers were asked if they experienced pesticide-induced illness symptoms immediately following the application of pesticides in beans. When the answer was affirmative, the farmer was then asked what the illness was, the time taken to recover, time of travel to medical health facility and the cost of medicines. Time lost due to illness was then converted into monetary values using prevailing wage rate. Thus both the direct and indirect costs of pesticide exposure were collected. The indirect costs were approximated by the days lost (when farmer could not work in the field due to pesticide-induced illness). The direct costs were, on the other hand, measured by the medical doctor-prescribed and self-administered treatment (including consultation fee and cost of medicine) and the cost of travel to health facility. These indirect and direct costs were then summed to obtain the total health costs of exposure to pesticides which was then used as a dependent variable in a health cost regression model to test the benefit of monitoring farmers for compliance with IFSS on the costs of exposure to pesticides.

The health cost model also included several conditioning variables namely, farmer specific variables (e.g., *age*, *gender*, and *education*); farmer's behavioral characteristics (*alcohol intake*, *cigarette smoking*); institutional characteristics (e.g., *distance to health facility*); pesticide exposure enhancing variables (quantity of *class 1 pesticides* used, quantity of *class 2 pesticides* used, quantity of *class 3 pesticides* used, and dummies for *pesticide applicator and mixer*, *keeping unwashed gear in the house* and *drink spraying pesticides*); exposure averting and mitigating variables (e.g., *sprayer maintenance*, *number of gear items used*, and dummies for *pest scouting*, *change clothes*, and *washing gear* before next use.

The second technique, the Poisson regression model, was used to test the effect of monitoring farmers for IFSS on the use of alternative pest and disease management strategies. The dependent variable in this model was the number of alternative strategies used by the farmer to manage pests and diseases in green beans. The strategies considered included soil testing for pests, crop rotation, use of pest/disease resistant varieties, fallowing, mulching, uprooting/burning infected plants, pest scouting, using trap crops, use of biological/natural pesticides, use of beneficial insects/organism. The conditioning variables used in this model are similar to those used in the cost of illness model.

Both the health cost and the Poisson models used survey regression technique with *village* as the primary sampling unit to control for clustering effect of the variance at the village level. Survey regression also has the added advantage of generating estimates that are robust to heteroskedasticity (Vittinghoff et al., 2005, pp. 309-310). In estimating both models, we dropped the practices variables most emphasized under the IFSS (namely, pest scouting, sprayer maintenance, and use of protective gear) because including them alongside the monitoring variable resulted in "double counting". Other variables that added little information to the two models were also dropped when a Wald joint exclusion restriction tests showed that they had no significant effect on the models' explanatory power. Some of the observations contained zero values hence could not be directly transformed into natural logs. These were *health cost, class1 pesticides, class 2 pesticides* and *class 3 pesticides*. Less than six observations in the *health cost* and *class 2 pesticides* had zero observations, hence they were log-transformed after first adding 0.5 to all observations. However, *class 1 pesticides* and *class 3 pesticides* had several zero values and were therefore log-transformed after performing Battese (1997) dummy variable transformation method in order to identify zero-valued observations without bias to the analysis.

The third technique used was the simple gross margin analysis. It involved computing the revenues and costs of producing green beans among the farmers that were monitored for compliance with IFSS and their counterparts. The revenues and variable costs of producing beans were computed for each of the three selected farmers and then averaged over each category of famers to obtain the gross margin/acre. The revenues and costs were measured for the last crop of beans grown in 2003. Two sets of prices were used to compute green bean revenues namely, the contract price for the monitored farmers (i.e., growing beans under contract) and the season's average spot market price for the non-monitored farmers.

Table 2 summarizes the statistics for the key variables and also presents the results of paired t-tests of equality of means between monitored and non- monitored farmers.

	Monitored (N=92)		Unmonitored (N=83)		Test of Means	
Variable	Mean	Std. Dev	Mean	Std. Dev	t-stat	p-val
Dependent variables						
log health cost (Kshs*)	3.25	2.85	3.88	2.54	1.53	0.063
skin irritations (count)	1.58	2.79	2.67	4.73	1.89	0.030
eye irritations (count)	0.84	2.78	1.24	3.23	0.89	0.188
total irritations (count)+	2.72	5.44	4.34	7.61	1.63	0.052
apmp (count)	3.82	0.24	2.82	0. 20	-3.23	0.000
Farmer specific and institutional variables						
Female head of household	0.79	0.42	0.80	0.44	0.23	0.411
log age (years)	3.65	0.27	3.56	0.31	-1.81	0.964
log education (years)	1.95	0.09	2.14	0.39	-1.80	0.036
log income	11.37	0.10	11.08	0.10	-2.11	0.982
log distance to clinic (hours)	3.37	0.08	3.47	0.07	0.88	0.190
cigarette smoking (0,1)	0.32	0.04	0.30	0.05	0.30	0.616
alcohol intake (0,1)	0.38	0.05	0.31	0.05	-0.92	0.177
total income ('000 Kshs)	133.6	143.6	95.4	106.5	-1.98	0.025
marketable bean output (kg/acre)	1706	1258	795	740	-5.73	0.000
Exposure enhancing variables						
log class 1 pesticides (g/farm)	-0.47	0.13	-0.20	0.20	1.14	0.127
log class 2 pesticides (g/farm)	5.86	0.14	6.12	0.12	1.40	0.082
log class 3 pesticides (g/farm)	1.11	0.32	1.01	0.37	0.22	0.588
drinks spraying (0,1)	0.17	0.38	0.13	0.33	-0.89	0.813
unwashed gear in house (0,1)	0.26	0.50	0.30	0.50	0.55	0.292
primary applicator (0,1)	0.55	0.50	0.63	0.50	1.25	0.106
primary mixer (0,1)	0.74	0.46	0.75	0.48	0.12	0.553
Exposure averting and mitigating variables						
wash gear (0,1)	0.69	0.05	0.67	0.05	-0.30	0.619
change clothing & wash (0,1)	0.20	0.41	0.22	0.41	0.17	0.434
sprayer maintenance (count/year)	1.03	0.15	0.60	0.15	-2.29	0.011
pest scouting (0,1)	0.78	0.41	0.61	0.49	-2.46	0.008

Source: Adapted from Okello and Swinton (2010)
* Kenya Shillings (Exchange rate in 2003 was US$ 1 = Kshs 78).
+ Total includes gastrointestinal irritations, which were too infrequent to count separately.

Table 2. Summary statistics of key variables used in the analysis

As shown, monitored farmers experienced lower health costs of exposure to pesticide and used, on average, significantly higher number of alternative pest management practices than the unmonitored. Mean health costs associated with pesticide exposure were Kshs 186 for monitored farmers and Kshs 261 for the non-monitored. At the same time farmers who comply with IFSS received higher but more variable bean yield and income than their counterparts.

Notably, Table 2 shows that there are significant differences in the incidence of acute pesticide-induce illness between the monitored and non-monitored farmers. The former had lower mean number of skin irritations than the latter. Indeed, the average count of total[1] acute pesticide-induced irritations experienced by green bean farmers was much lower and strongly statistically significant among monitored farmers than their counterparts.
With regard to pesticide use, however, there are no significant differences between monitored and non-monitored farmers, except for class 2 pesticides. Indeed t-tests of difference in mean quantity of active ingredients per farm applied by monitored and non-monitored farmers are insignificant. This finding suggests that monitored farmers are still using toxic class 1 pesticides as much as their non-monitored counterparts. The continued use of class 1 pesticides by monitored farmers in controlling pests and diseases in beans is largely due to the dilemma monitored farmers face. The very market that demands the use of less toxic (but often less efficacious) pesticides also demands aesthetic quality attributes (e.g., spotless and straight beans) which are hard to meet without chemical control of pests and diseases. In the section below, we investigate these findings further using regression and gross margin analysis to specifically test if they are caused by monitoring farmers for compliance with IFSS.

4. Results

4.1 Effect of IFSS on pesticide-related cost of illness

As hypothesized, monitoring farmers for compliance with IFSS significantly reduces pesticide related health costs (Table 3). It reduces the log of health costs of exposure to pesticides by 0.80 units. Several other factors also condition the health costs of exposure to pesticides. Among the farmer specific variables, education reduces health costs while and income increases it. An additional year of education beyond the mean of two reduces health costs of pesticide exposure by 18 percent. The elasticity of health cost with respect to income is 0.53 implying that income is associated with increased health costs of pesticide induced illnesses. That is, farmers who earn greater income from beans (hence likely to have grown larger areas) get more exposed to pesticides and hence incur higher health costs.
As expected, results also show that the primary pesticide applicators incur higher health costs. Although not surprising, it corroborates previous findings by Harper and Zilberman (1992) from U.S. agriculture. Among the mitigating and averting variables, pesticide applicators who change clothing contaminated by pesticides and wash off the pesticides from their bodies (*change clothing*) following accidental leakage by the spray pump experience lower cost of pesticide illness than those who do not, presumably because they reduce duration of skin contact with pesticides.

4.2 Effect of IFSS on the use of alternative pest and disease management strategies

The finding that the use of pesticide averting and mitigating strategies reduces health costs of exposure to pesticides led us to investigate further whether monitoring farmers for compliance with IFSS increases the use of such strategies. In particular, we focused on strategies that are likely to benefit the environment namely, the alternative disease and pest

---

[1] Total irritations is the sum of the count of skin, eye, and gastrointestinal irritation experienced by a farmer soon after applying pesticides in green beans

management strategies. We therefore estimated survey Poisson regression model to test the effect of monitoring farmers for compliance with IFSS on number of such strategies used.

Independent variables	Coefficient	p-value
Farmer specific and institutional variables		
female head of the household	-0.055	0.923
log age	-1.524	0.040
education	-0.184	0.009
log income	0.539	0.026
log distance to clinic	-0.444	0.113
cigarette smoking	-0.708	0.125
alcohol intake	0.808	0.056
monitored farm	-0.799	0.055
Exposure enhancing variables		
log class 1 pesticides	-0.098	0.757
log class 2 pesticides	0.002	0.987
log class 3 pesticides	-0.535	0.120
primary applicator	1.184	0.017
unwashed gear in house	0.294	0.506
drinks spraying	0.021	0.965
Exposure averting and mitigating variables		
change clothing & wash	-0.988	0.037
Battese dummy variables		
class 1 pesticides	-1.433	0.513
class 3 pesticides	-3.524	0.106
constant	10.900	0.027
F statistic	2.44	
p-value	0.003	
R-squared	0.194	

Dependent variable: Natural log of farmer's direct and indirect health costs* of pesticide exposure in Kenya Shillings
* To health cost variable, we added 0.5 KSh to all observations before log transformation to accommodate 6 cases of zero health costs; likewise we added 0.5 g/farm to observations of Class 2 pesticide quantity due to 1 zero value. For Class 1 and 3 pesticide quantities, we created new dummy variables to signal zero values, replacing the pesticide quantity zeroes with 1's so that their log transformed values were zero (Battese, 1997) in order to remove bias from the estimation.
Source: Adapted from Okello and Swinton (2010)

Table 3. Determinants of pesticide-related health costs among Kenyan green bean growers, 2004 - Survey OLS regression

We selected a Poisson regression model because the dependent variable, namely the number of pest and disease management practices (*apmp*) used by a farmer, is a count variable. The results of this exercise are presented in Table 4.

As hypothesized, monitoring farmers for compliance with IFSS increases the expected number of alternative pest and disease management strategies used by the farmers. Other things being equal, the expected number of such alternative pest and disease management practices (*apmp*) used by monitored farmers is approximately 30% higher than for those that are not.

Independent variables	Coefficient	p-value
Farmer specific and institutional variables		
female head of household	0.280	0.079
log age	0.202	0.264
log education	0.102	0.081
Log plotsize	-0.134	0.225
Log income	0.022	0.548
cigarette smoking	-0.061	0.007
alcohol intake	0.016	0.597
log distance to clinic	0.197	0.008
EU-PS comply	0.302	0.005
Extension	0.005	0.008
Ownradio	0.206	0.019
Experience	0.004	0.033
Exposure enhancing variables		
log class 1 pesticides	-0.014	0.669
log class 2 pesticides	0.068	0.039
log class 3 pesticides	-0.004	0.798
F statistic	5.19	
p-value	0.001	
N	175	

Dependent variable: Number of alternative pest and disease management practices used by the farmer
Source: Adapted from Okello and Okello (2010)

Table 4. Effect of EU-PS on use of alternative pest management practices, 2004 - survey Poisson regression

A number of the conditioning variables also affect the number of *apmp* used by farmers. Among the farmer specific variables, *gender* and *education* increase the expected number of *apmp* used. The expected number of alternative pest management practices used by male farmers is approximately 28 percent higher than for female farmers. At the same time, increasing the mean years of education from 2 to 3 increases the number of *apmp* by 10 percent. Results also show that cigarette smoking reduces the expected number of *apmp* used by farmers.

A number of institutional factors also affect the number of alternative pest and disease management strategies in green beans. These include distance to clinic, extension, radio and experience. These factors increase the expected number of *apmp* used in controlling pests and diseases in beans. Other things being equal, increasing the log of distance to the clinic by one hour of travel time increases the number of *apmp* by about 20 percent while an additional contact with public extension personnel increases the expected number of *apmp* by 0.5 percent. The results further show that experience in growing beans increases the expected number of *apmp* used, probably because of familiarity with the hazards of exposure to pesticides. Each additional year of farming experience beyond the mean increases the expected number of *apmp* used by 0.4 percent. Among the exposure enhancing variables, results show that the use of the more toxic class 2 pesticides increases the expected number of *apmp* used.

4.3 Effect of IFSS on margins earned by green bean farmers

A major concern of many developing country exporters upon the onset of IFSS was that the standards would marginalize the poor family farmers (Mungai, 2004; Okello, 2005). Hence we investigated how compliance with IFSS affected the profit margins earned by farmers. Figure 2 presents the results of the analysis. As shown, monitored farmers received higher revenues from growing beans than their counterparts. The higher revenues obtained by monitored farmers resulted primarily from selling more beans due to better access to UK market than the unmonitored. Interestingly, monitored farmers did not receive premium price from their buyers despite having to comply with the IFSS (Okello, 2005). The average price of beans paid to monitored farmers under the contract remained stable at KShs 40/kg during the period.

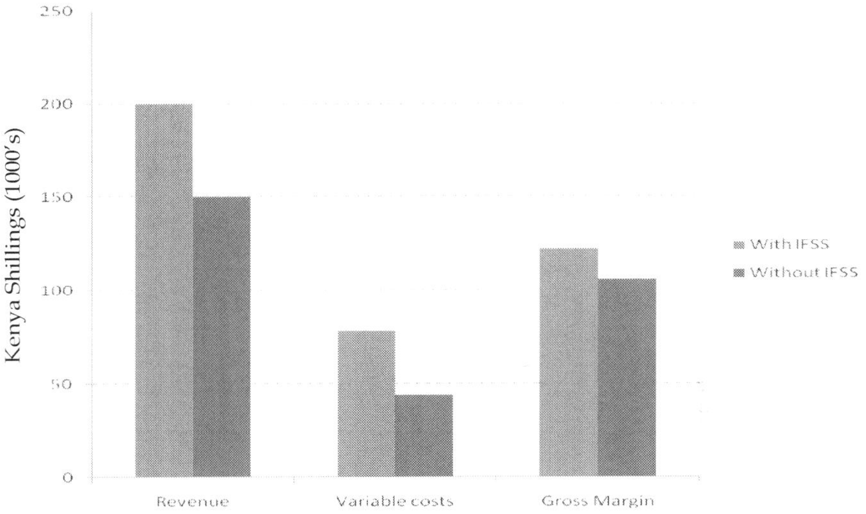

Source: Okello and Swinton (2005)

Fig. 2. Comparison of revenues, costs and gross margins of producing green beans with and with IFSS, Kenya, 2005 (in thousand Ksh)

Monitored farmers incurred higher variable costs of producing beans, as expected. The high variable costs likely resulted from new, more expensive pesticide products. For instance, at the time of this study, farmers were required to switch from the use of Dimethoate (a class 2 fungicide with long required interval between spraying and harvest) to Ortiva (a class 3 fungicide with a shorter interval). However, Ortiva (the new pesticide) cost 2.5 times as much as Dimethoate. Indeed, switching to approved pesticides contributed to 20% of the difference in costs of producing green beans between the monitored and non-monitored farmers.

Overall, contrary to our theoretical expectations, monitored farmers received higher margins from growing beans than their counterparts. The average gross margin for monitored farmers was 13% higher than for their counterparts indicating that IFSS does not really marginalize the poor family farmers who grow beans under strict monitoring for compliance with IFSS.

5. Conclusions

This research finds that compliance with international food safety standards (IFSS) reduces the incidence of pesticide-induced acute illness and the health costs associated with exposure to pesticides. Farmers who were monitored for IFSS compliance incurred much lower health costs and also experienced much fewer acute pesticide-induced illnesses than those who were not. The paper also finds that compliance with IFSS promotes the use of integrated pest and disease management strategies. The judicious use of pesticides has implications for sustainable production of non-traditional fresh exports in developing countries as it reverses the "circle of poison" pattern reported in Asia and instead promotes a circle of virtuous pesticide care in use. We further find that while compliance with IFSS increases the cost of green bean production, higher revenues result in the profit margins earned by IFSS compliant farmers that exceed their domestic counterparts who are not monitored for compliance. These higher margins emanate from greater access to the export market by farmers monitored under contracts.

This study therefore concludes that compliance with IFSS therefore brings health and environmental benefits in addition to the acknowledged access to high value overseas markets. Contrary to early concerns that IFSS compliance would marginalize poor family farmers, IFSS compliance has brought financial gains and many Kenyan smallholder farmers have found cooperative ways to gain access to these export markets (Okello and Swinton, 2007). Despite the health and financial gains, IFSS compliant farmers do not use less toxic pesticide active ingredients in fresh vegetable production than their non-monitored counterparts. The international market demand for clean, well-formed and spotless vegetables continues to require rigorous pest control that favors the use of efficacious pesticides.

6. References

Antle, J. M. & Capalbo, S.M. (1994). Pesticides, productivity and farmer health: Implications for regulatory policy and agricultural research. *American Journal of Agricultural Economics*, 76: 598-602.

Antle, J. M. & Pingali, P. (1994). Pesticides, productivity and farmer health: A Philippine case study. *American Journal of Agricultural Economics* 76: 418-430.

Battese, G.E. (1997). A note on the estimation of Cobb-Douglas production functions when some explanatory variables have zero values. *Journal of Agricultural Economics*. 48:250-252

Cole, D., Carpio, C. F., Julian, J.A. & Leon, N. (1998). "Health impacts of pesticide use in Carchi farm populations". In *Economic, Environmental and Health Tradeoffs in Agriculture: Pesticides and Sustainability of Andean Potato Production*. ed. C.C. Crissman, J.M. Antle, and S.M. Capalbo. Boston: Kluwer

Dolan, C. & Humphrey, J. (2000). Governance and trade in fresh vegetables: Impact of UK supermarkets on the African horticultural industry. *Journal of Development Studies* 37: 147-177.

Farina, E. M. M. Q. & Reardon, T. (2000). Agrifood grades and standards in extended Mercosur: their role in the changing agrifood systems. *American Journal of Agricultural Economics* 82: 1170-76

Harper, C. R. & Zilberman, D. (1992). Pesticides and Worker Safety. *American Journal of Agricultural Economics* 72: 68-78.

Jaffee, S. (2003). From Challenge to Opportunity: The Transformation of the Kenyan Fresh Vegetable Trade in the Context of Emerging Food Safety and Other Standards. PRMTR Division, World Bank. Washington, DC (May 2003).

Jaffee, S. & Morton, J. (1995). *Marketing Africa's High Value Foods: Comparative Experiences of Emergent Private Sector*. World Bank. Washington DC.

Jensen, M. F. (n.d). Food safety requirements and smallholders: A case study of Kenyan fresh produce exports. The Royal Veterinary and Agricultural University. Unpublished Draft.

Maumbe, B. M. & Swinton, S.M. (2003). Hidden costs of pesticide use in Zimbabwe's smallholder cotton growers. Social Science & Medicine, 57:1559-1571.

Mungai, N. (2004). EU rules could destroy horticulture: the protocol on good agricultural practices will have a profound impact on both large and small-scale farmers, although the biggest impact will be on the latter. *Daily Nation*. May 7, 2004, p11.

Mwanthi, M.A. & Kimani, V.N. (1990). Agrochemicals: a potential health hazard among Kenya's small-scale farmers. (In G. Forget, T. Goodman, and A. de Villiers (Eds.) Impact of Pesticides on Health in Developing Countries (pp 21-34). Ottawa: IDRC).

Okado, M.(2001). Background Paper on Kenya off-season and specialty fresh fruits and vegetables: Lessons of experience from the Kenya horticultural industry. Presented at United Nations Conference on Trade and Development Regional workshop for horticultural economies in Africa. Nairobi, Kenya. May 29-31, 2001 (URL): http//r0.unctad.org/infocomm/diversification/nairobi/Keny0901.doc.

Ohayo-Mitoko, G. J. A. (1997). Occupational pesticide exposure among Kenyan agricultural workers: an epidemiological and public health perspectives. Dissertation, Wagenigen Agricultural University.

Okello, J.J. 2010. Implications of extension of US trade preferences for Agricultural export market development in Africa - the case of horticulture in Kenya. A report submitted to Oxfam- America, March 29, 2010.

Okello, J.J. and R.M Okello. 2010. Do EU pesticide standards promote environmentally-friendly production of fresh export vegetables in developing countries? Evidence from Kenyan green bean industry. *Environment, Development and Sustainability*, 12 (3): 341-355

Okello, J.J. and S.M Swinton. 2010. From circle of poison to circle of virtue: pesticides, export standards and Kenya's green bean industry. *Journal of Agricultural Economics*, 61, 209-224

Okello, J.J., C. Narrod and D. Roy. 2008. "Compliance with international foods safety standards is not a fantasy: Evidence African green bean growers" *Fresh Perspectives*. No. 15, April 2008. Available for download at:

Okello, J. J. and S. M. Swinton. 2007. Compliance with international food safety standards in Kenya's green bean industry: comparison of a small and a large scale family farm producing for export. *Review of Agricultural Economics* 29: 269-285.

Okello, J.J. 2005. "Compliance with international food safety standards: The case of green bean production in Kenyan family farms." Ph.D. dissertation, Agricultural Economics, Michigan State University, East Lansing, Michigan, USA.

Okello, J.J. and S. M. Swinton. 2005. A race to the top, but at what cost? Kenyan green bean farmers' struggle to comply with international food safety standards. Selected poster paper presented at American Agricultural Economics Association meeting, Providence, Rhode Island, 24-27, 2005

Pingali, P.L. & Rogers, P.A. 1995. Impact of Pesticides on Farmer Health and the Rice Environment. (Boston: Kluwer Academic Publishers)

Strauss, J. & Thomas, D. (1998). Health, nutrition and economic development. *Journal of Economic Literature*, 82, 766-817 Thrupp, L. A., G, Bergeron, and W. F. Waters.1995. *Bittersweet Harvest for Global Supermarkets: Challenges in Latin America's Export Boom.* Washington, DC: Natural Resources Institute.

Vittinghoff, E., Glidden, D.V., Shiboski, S.C., & McCulloch, C.E. (2005). *Regression Methods in Biostatistics: Linear, Logistic, Survival and Repeated Measures Models.* New York: Springer.

World Health Organization. (2005). The WHO recommended classification of pesticides by hazard and guidelines to classification 2004. (Geneva: World Health Organization)

11

Pesticide Residues in Agricultural Products of Slovene Origin Found in 2001-2009

Helena Baša Česnik, Špela Velikonja Bolta and Ana Gregorčič
Agricultural Institute of Slovenia, Central Laboratories
Slovenia

1. Introduction

In the production of cereals, fruit and vegetables the appropriate protection from harmful organisms, which tend to appear in inappropriate places at inappropriate times, is needed. At present plant protection is based on the use of plant protection products (PPPs) which, when properly used, assure the most economical way of producing adequate quantities of high quality food. Incorrect and uncontrolled use of PPPs may cause great harm to people, animals and the environment. Agricultural experts have been constantly trying to develop new technologies of healthy food production including the development of PPPs which would be friendlier to people, animals and the environment.

The task of the government is to control the proper use of PPPs, which insures healthy food on the market (Akiyama et al., 2002; Andersen J. H. & Poulsen M. E., 2001; Dejonckheere et al., 1996; Dogheim et al. 2002; Fernandez et al., 2001; Ripley et al., 2000). This is why the Agricultural Institute of Slovenia was determining pesticide residues in agricultural products of Slovene producers prior to the market, i.e. after picking, digging or harvesting and in storage in accordance to Slovenian legislation (RS 1999a; 1999b; 2001; 2004a; 2004b; 2007a; 2007b; 2009). The samples were taken randomly in eight production areas in Slovenia: Celje, Koper, Kranj, Nova Gorica, Novo mesto, Murska Sobota, Maribor, and Ljubljana. Each year (except in 2009) analyses of pesticide residues were performed on potato, lettuce and apple samples due to the characteristic nutrition of Slovenes (the Slovene Food Basket has not yet been demarcated). Selection of other agricultural commodities and active substances followed the guidelines given in the European Union monitoring recommendations (EC 2001; 2002a, 2002b; 2004; 2005; 2006; 2007; 2008a; 2008b).

Control of pesticide residues in agricultural products prior to the market allows assessment of the conformity of production with good agricultural practice and the determination of sources and/or causes of residues found. Random choice of producers enables a statistical approach to the estimation of food safety on the Slovene market.

The results are intended to:

- Determine the conformity with the legally prescribed maximum residue levels (MRLs)
- Determine the conformity of the conventional, integrated and ecological production with good agricultural practice
- Determine the sources and/or causes of residues found

Legally prescribed MRLs are defined on the basis of field trials in accordance with good agricultural practice. Consideration of the prescribed way to use PPPs and the pre-harvest interval are therefore of key importance.
For monitoring purposes quick and reliable multiresidual methods are needed that enable simultaneous determination of a wide spectrum of active substances. The methods mainly use three types of solvents for extraction: ethylacetate (Berrada et al., 2006; Čajka & Hajšlová, 2004; Ferrer et al., 2005; Sharif et al., 2006), acetonitrile (method also known as QuEChERS method) (Lehotay, 2007; Maštovská et al., 2005) or acetone (Díez et al., 2006; Pizzutti et al., 2009; Stan & Linkerhägner, 1996). Our laboratory used acetone because of its low toxicity, high volatility and miscibility with water. For better extraction of active substances we added petroleumether and dichlorometane to the acetone (Baša Česnik & Gregorčič, 2003; Baša Česnik et al., 2006). For the determination of the extracted active substances laboratories mainly use gas chromatography coupled to various detections, i.e. flame ionisation detection (FID), electron capture detection (ECD), nitrogen phosphor detection (NPD) and flame photometric detection (FPD). We have used gas chromatography coupled to mass spectrometry (GC/MS) which enables simultaneous and unequivocal qualitative and quantitative determination of active substances. In the case of thermally labile compounds liquid chromatography coupled to UV detection or fluorescence detection is used. We have used liquid chromatography coupled to tandem mass spectrometry (LC/MS/MS) which enables simultaneous and unequivocal qualitative and quantitative determination of active substances.

2. Experimental

In nine years of monitoring we analysed 1504 samples: 102 cereal samples (9 barley samples, 72 wheat samples, 1 millet sample, 2 spelt samples, 7 corn samples, 1 triticale sample, 5 rye samples and 5 oat samples), 494 fruit samples and 908 vegetable samples. The sampling is presented in Table 1.
For the determination of PPP residues we used four different testing methods:

- method for the determination of benzimidazoles: tiabendazol and the sum of benomil and carbendazim (in 2001-2007) (van Zoonen, 1996),
- method for the determination of the maneb group: maneb, mankozeb, metiram, propineb, thiram and zineb, the sum is expressed as carbon disulfide (in the years 2001-2009) (Baša Česnik & Gregorčič, 2006)
- multiresidual GC/MS method (in 2001-2009) (Baša Česnik & Gregorčič, 2003; Baša Česnik et al., 2006)

  In 2001 and 2002 the scope of analyses was: acephate, aldrin, azinphos-methyl, captan, carbofuran, chlorpyriphos, chlorpyriphos-methyl, cyhalotrin-lambda, DDT, deltamethrin, diazinon, dimethoate, endosulfan, endrin, fenitrothion, fenthion, fludioxonil, folpet, HCH-alpha, heptachlor, heptenophos, imazalil, iprodione, lindane, malathion, mecarbam, metalaxyl, methamidophos, methidathion, parathion, permethrin, phosalone, pirimiphos-methyl, procymidone, pyridaphenthion, quinalphos, thiabendazole, triazophos and vinclozolin.

  In 2003 the scope was extended with the following active substances: azoxystrobin, bromopropylate, chlorothalonil, cypermethrin, dichlofluanid, omethoate, oxydemeton-methyl, phorate, propyzamide and tolylfluanid.

Commodity	2001	2002	2003	2004	2005	2006	2007	2008	2009	Sum
Apples	15	30	36	70	17	36	43	38	/	285
Beans	/	/	/	/	/	/	/	8	/	8
Carrot	/	/	/	/	15	/	/	17	/	32
Cauliflower	/	/	10	/	/	11	/	/	17	38
Cereals	31	/	15	/	/	26	10	/	20	102
Cherries	/	/	/	/	/	/	10	/	/	10
Cucumbers	/	/	/	/	17	/	/	20	/	37
Eggplant	/	/	/	/	/	/	/	/	9	9
Endive	/	/	/	/	/	/	/	/	28	28
Grapes	/	/	15	/	/	20	/	/	/	35
Head cabbage	/	/	/	15	/	/	21	/	/	36
Leek	/	/	/	/	/	/	9	/	/	9
Lettuce	15	30	24	28	17	16	25	24	23	202
Peaches	/	/	/	/	/	/	20	/	/	20
Pears	/	30	/	/	12	/	/	21	/	63
Peas	/	/	/	/	/	4	/	/	/	4
Pepper	/	/	15	/	/	16	/	/	21	52
Potatoes	30	30	35	61	16	33	36	32	52	325
Spinach	/	/	/	/	7	/	/	6	/	13
Strawberries	30	/	/	13	/	19	19	/	/	81
String beans	/	30	/	/	14	/	/	/	/	44
Tomatoes	30	/	/	24	/	/	17	/	/	71
Sum	151	150	150	211	115	181	210	166	170	1504

Table 1. Sampling of agricultural products in the years 2001 to 2009

In 2004 the scope was extended with the following active substances: cyprodinil, diphenylamine, kresoxim-methyl, myclobutanil, pyrimethanil and spiroxamine.

In 2005 the scope was extended with the following active substances: bifenthrin, bupirimate, carbaryl, chlorpropham, pirimicarb, propargite, tolclofos-methyl, triadimefon and triadimenol.

In 2006 the scope was extended with the following active substances: cyromazine, penconazole, trifloxystrobin.

In 2007 the scope was extended with the following active substances: boscalid, dichlorvos, fenamidone, quinoxyfen, tebuconazole. Cyromazine was removed.

In 2008 the scope was extended with the following active substances: carboxin, chloridazon, clomazone, cyproconazole, diniconazole, fenbuconazole, indoxacarb, metconazole, methacrifos, metribuzin.

In 2009 the scope was extended with the following active substances: acrinathrin, dazomet, desmethylpirimicarb, dimethachlor, esfenvalerate, fenvalerate, flonicamid, fluquinconazole, HCH-betha, HCH-deltha, hexachlorobenzene, metalaxyl-M, metrafenone, oxadixyl, parathion-methyl, profenofos, quinoclamine, tetraconazole, tetradifon.

- multiresidual LC/MS/MS method (in 2006-2009) (Bossi et al., 2002; Ortelli et all., 2004; Lehotay et al., 2005)

  In 2006 the scope of analyses was: aldicarb, bentazone, cymoxanil, difenoconazole, fenazaquin, fenhexamid, fluroxypyr, imidacloprid, methiocarb, methomyl, phoxim, pymetrozine, spirodiclofen, tebufenozide, thiacloprid, thiamethoxam and zoxamide.

  In 2007 the scope was extended with the following active substances: acetamiprid, amidosulfuron, benalaxyl, bitertanol, clofentezine, cyromazine, dimethomorph, epoxiconazole, ethofumesate, famoxadone, fenpropidin, fenpropimorph, fenpyroximate, flufenacet, fluquinconazole, hexythiazox, iprovalicarb, lufenuron, metosulam, pendimethalin, prochloraz, propamocarb, propiconazole, pyridate, spinosad, terbuthylazine, thiophanate-methyl and trichlorfon.

  In 2008 the scope was extended with the following active substances: aldicarb sulfon, aldicarb sulfoxid, buprofezin, carbendazim, clopyralid, clothianidin, cycloxydim, desmedipham, flutriafol, foramsulfuron, iodosulfuron-methyl-sodium, isoxaflutole, linuron, malaoxon, metamitron, metazachlor, methiocarb sulfon, methiocarb sulfoxid, methoxyfenozide, napropamide, phenmedipham, prosulfocarb, prosulfuron, pyraclostrobin, rimsulfuron, tetraconazole, thifensulfuron-methyl, thiodicarb, triasulfuron, trifluralin and triflusulfuron-methyl.

  In 2009 the scope was extended with the following active substances: 2,4-D, amitrole, azinphos-ethyl, beflubutamid, benalaxyl M, bromoxynil, carbosulfan, chlortoluron, cyazofamid, demeton-S-methyl sulphone, dichloprop-P, diflufenican, dimethenamid-P, fenarimol, fenoxaprop-P-ethyl, fenoxycarb, fenthion sulfone, fenthion sulfoxide, fipronil, florasulam, fluazifop-P-butyl, fluazinam, fluorochloridone, flusilazole, hexaconazole, isoproturon, mandipropamid, MCPA, monocrotophos, nicosulfuron, oxamyl, paraoxon-methyl, phorate sulfone, phorate sulfoxide, propaquizafop, pyrazophos, teflubenzuron, tribenuron-methyl and trinexapac-ethyl. Fluquinconazole and tetraconazole were removed.

The trueness of testing methods was verified by recoveries which had to be from 70% to 120%.

The trueness was additionally verified by participation in the French inter-laboratory proficiency testing scheme BIPEA (Bureau interprofessionnel d´etudes analytiques) and CRL European Proficiency Tests.

In January 2005 determination of pesticide residues was accredited by the French accreditation body COFRAC.

3. Results and discussion

During the period, from 2001 to 2009, 1504 samples were analysed. **Sample portions below reporting level (RL), sample portions below or equal to MRLs and sample portions above MRLs** are presented in Figure 1, Figure 2 and Table 2.

In 946 samples (62.9%) PPP residues were not found, in 493 samples (32.8%) PPP residues were lower or equal to MRLs and in 65 samples (4.3%) PPP residues were above MRLs.

The highest portion of PPP residues, 50% and more, was found in fruit. The farmers have to protect fruit against rot, mould and insects. The highest portion of determined but not exceeding PPP residues, (residues lower or equal to MRLs), was found in cherry samples

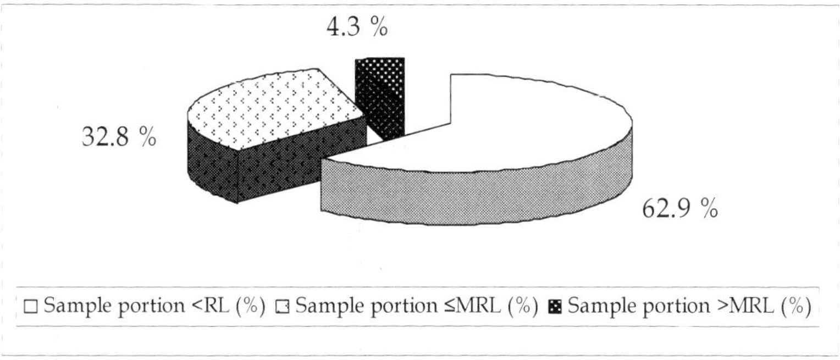

Fig. 1. Results of monitoring from 2001 to 2009

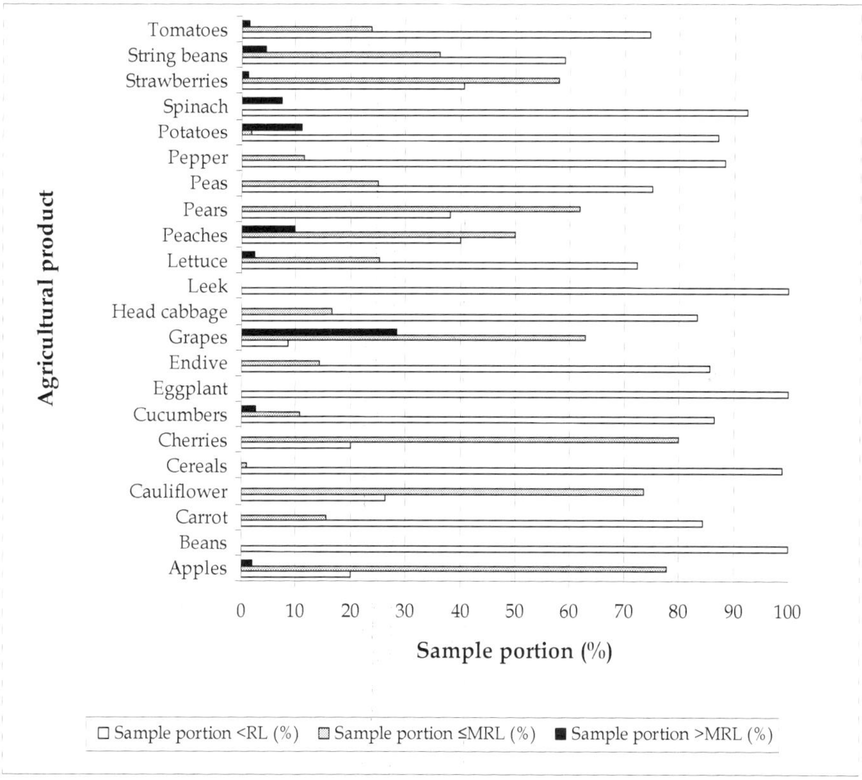

Fig. 2. Sample portions of PPP residues for each analysed matrix from 2001 to 2009

Commodity	Sample portion <RL (%)	Sample portion ≤MRL (%)	Sample portion >MRL (%)
Apples	20.0	77.9	2.1
Beans	100.0	0.0	0.0
Carrot	84.4	15.6	0.0
Cauliflower	26.3	73.7	0.0
Cereals	99.0	1.0	0.0
Cherries	20.0	80.0	0.0
Cucumbers	86.5	10.8	2.7
Eggplant	100.0	0.0	0.0
Endive	85.7	14.3	0.0
Grapes	8.6	62.9	28.6
Head cabbage	83.3	16.7	0.0
Leek	100.0	0.0	0.0
Lettuce	72.3	25.2	2.5
Peaches	40.0	50.0	10.0
Pears	38.1	61.9	0.0
Peas	75.0	25.0	0.0
Pepper	88.5	11.5	0.0
Potatoes	87.1	1.8	11.1
Spinach	92.3	0.0	7.7
Strawberries	40.7	58.0	1.2
String beans	59.1	36.4	4.5
Tomatoes	74.6	23.9	1.4

Table 2. Sample portions of PPP residues for each analysed matrix from 2001 to 2009

(80.0%), apple samples (77.9%), grape samples (62.9%), pear samples (61.9%), strawberry samples (58.0%) and peach samples (50.0%). Some fruit samples contained also exceeding PPP residues (residues above MRLs), i.e. grape samples (28.6%), peach samples (10.0%), apple samples (2.1%) and strawberry samples (1.2%).

The portion of PPP residues found in vegetables and cereals was less than 50%. Cauliflower was the only exception. In cauliflower the same active substance found in 2003, 2006 and 2009 was dithiocarbamates (maneb group). In cauliflower there are naturally present substances that give the same responses as dithiocarbamates. This is why we cannot say that cauliflower was really treated with dithiocarbamates. Besides dithiocarbamates, only one active substance was found in one sample (difenoconazole in 2006).

The highest portion of exceeding PPP residues (residues above MRLs) were found in grape samples (28.6%) and potato samples (11.1%).

MRL exceedances in grape samples were found for cyprodinil in 2006 and for fludioxonil in 2003 and 2006. This suggests that the farmers used PPP Switch authorised for grapes that contains both active substances. The national MRL for both compounds was 0.02mgkg^{-1} in

2003 and 2006. Today the European Community MRL for cyprodinil is 5mgkg^{-1} (Commission Regulation (EC) No. 459/2010) and for fludioxonil 2mgkg^{-1} (Commission Regulation (EC) No. 822/2009). The highest value obtained in grape samples was 0.40mgkg^{-1} for cyprodinil and 0.04mgkg^{-1} for fludioxonil. Taking into account today's MRLs none of the samples would be exceeding. The risk assessment performed with Pesticide Safety Directorate (PSD, York, UK), model for acute exposure for cyprodinil at concentration level 0.40mgkg^{-1} and Acceptable Daily Intake (ADI) 0.03mgkg^{-1} body weight^{-1} day^{-1} (Acute Reference Dose-ARfD for cyprodinil was not determined) showed that the National Estimate of Short Term Intake (NESTI) expressed in ADI percentage ranged from 2.5% for 7-10 years old children to 31.6% for adults. The risk assessment performed with the PSD model for acute exposure for fludixonil at concentration level 0.04mgkg^{-1} and ADI 0.37mgkg^{-1} body weight^{-1} day^{-1} (ARfD for fludixonil was not determined) showed that NESTI expressed in ADI percentage ranged from 0.0% for 7-10 years old children and residential elderly people to 0.3% for adults. ADIs were found on the internet (**http://ec.europa.eu/sanco_pesticides/public/index.cfm?event=activesubstance.detail** for cyprodinil and **http://ec.europa.eu/sanco_pesticides/public/index.cfm?event=activesubstance.detail** for fludioxonil), as well as the PSD model for acute exposure (**http://www.pesticides.gov.uk/approvals.asp?id=1687**). The risk assessment showed that the exceeding grape samples did not present any risk for health (NESTI in % of ADI was below 100%) and were therefore safe for consumers.

MRL exceedances in potato samples were found exclusively for dithiocarbamates in 2001-2004. Dithiocarbamates were the only active substances found in potato in these years. The reporting level for dithiocarbamates was 0.05mgkg^{-1}, which was at the same time the MRL for potato during these years (Official Gazette of the Republic of Slovenia No. 73/03). In January 2005 the MRL was raised to 0.1mgkg^{-1} (Commission Directive 2004/115/EC) and in 2008 it was raised to 0.3mgkg^{-1} (Commission Regulation (EC) No. 839/2008). If MRL 0.3mgkg^{-1} were valid from 2001 to 2004 only 8 samples instead of 36 would be exceeding. Among dithiocarbamates, ziram has the lowest ARfD (**http://ec.europa.eu/sanco_pesticides/public/index.cfm?event=activesubstance.selection**) which is 0.08mgkg^{-1} body weight^{-1} day^{-1}. The highest value obtained in potato samples for dithiocarbamates during 2001-2004 was 0.51mgkg^{-1}, which is equivalent to 1.02mgkg^{-1} of ziram. The risk assessment performed with the PSD model showed, that for that sample acute exposure for ziram was exceeding (NESTI in % of ARfD was above 100%) for infants (196.0% ARfD), toddlers (135.6% ARfD) and 4-6 year old children (102.1%ARfD). Among dithiocarbamates thiram has the highest ARfD (**http://ec.europa.eu/sanco_pesticides/public/index.cfm?event=activesubstance.selection**) which is 0.6mgkg^{-1} body weight^{-1} day^{-1}. The highest value 0.51mgkg^{-1}, obtained in potato samples for dithiocarbamates in the years 2001 to 2004, is equal to 0.81mgkg^{-1} of thiram. The risk assessment performed with the PSD model showed, that for that sample acute exposure for thiram was not exceeding (NESTI in % of ARfD was below 100%) for all groups. The highest acute exposure was 20.8% ARfD for infants and the lowest was 3.2% ARfD for adults. The active substances from the maneb group in potato were not determined separately and we cannot conclude if the samples presented any risk for the health of consumers. From 2005 to 2009 only two potato samples contained dithiocarbamates: one in 2006 and one in 2007 (dithiocarbamates content was 0.06 mgkg^{-1} for both samples). Results for 2005 to 2009 show that farmers had learned how to use PPPs containing active substances from the maneb group in accordance with good agricultural practice.

Annual results for the most frequently inspected commodities, i.e. apple, lettuce and potato, are presented in Tables 3-5 and Figures 3-5. Lettuce and potato were sampled each year while apples were not sampled in 2009. The highest percentage of apple samples with determined but not exceeding PPP residues was found in 2005 (88.2%) and the lowest in

Year	Sample portion <RL (%)	Sample portion ≤MRL (%)	Sample portion >MRL (%)
2001	26.7	73.3	0.0
2002	30.0	66.7	3.3
2003	16.7	83.3	0.0
2004	17.1	80.0	2.9
2005	5.9	88.2	5.9
2006	16.7	77.8	5.6
2007	18.6	81.4	0.0
2008	28.9	71.1	0.0

Table 3. PPP residues in apple samples for the period from 2001 to 2008

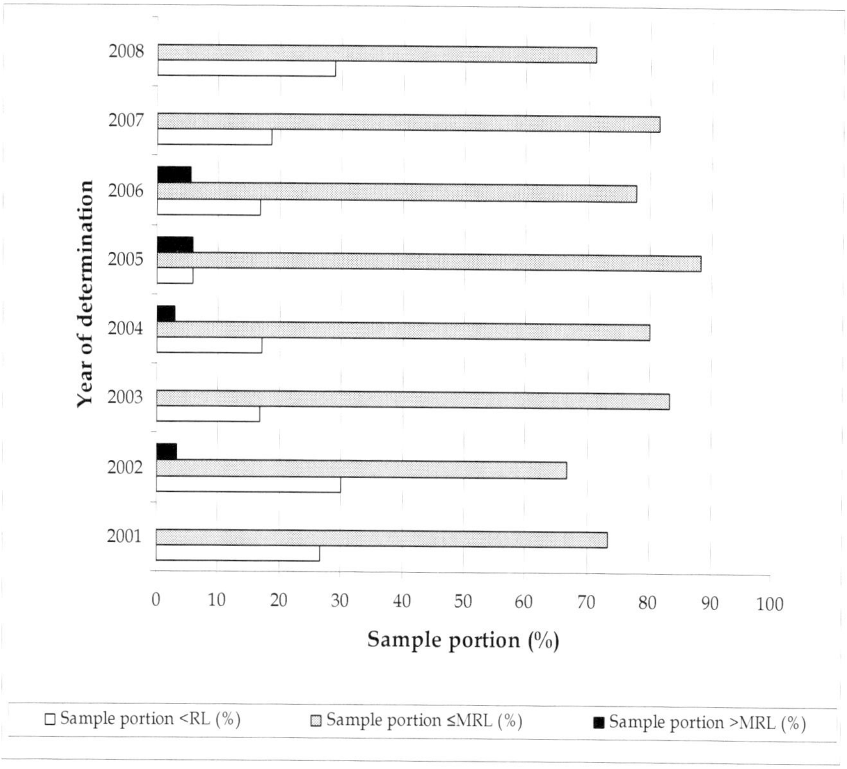

Fig. 3. Annual results for apple samples from 2001 to 2008

Year	Sample portion <RL (%)	Sample portion ≤MRL (%)	Sample portion >MRL (%)
2001	60.0	26.7	13.3
2002	63.3	33.3	3.3
2003	70.8	29.2	0.0
2004	42.9	57.1	0.0
2005	94.1	5.9	0.0
2006	93.8	6.3	0.0
2007	80.0	16.0	4.0
2008	83.3	12.5	4.2
2009	78.3	21.7	0.0

Table 4. PPP residues in lettuce samples for the period from 2001 to 2009

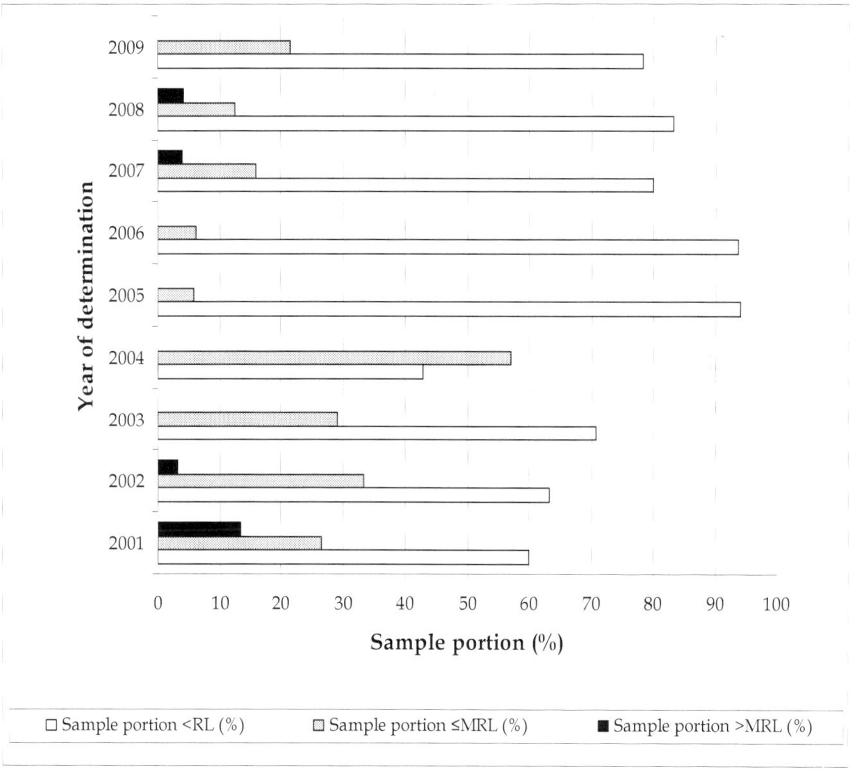

Fig. 4. Annual results for lettuce samples from 2001 to 2009

Year	Sample portion <RL (%)	Sample portion ≤MRL (%)	Sample portion >MRL (%)
2001	80.0	0.0	20.0
2002	56.7	3.3	40.0
2003	60.0	2.9	37.1
2004	91.8	0.0	8.2
2005	93.8	6.3	0.0
2006	93.9	6.1	0.0
2007	97.2	2.8	0.0
2008	100.0	0.0	0.0
2009	100.0	0.0	0.0

Table 5. PPP residues in potato samples for the period 2001 to 2009

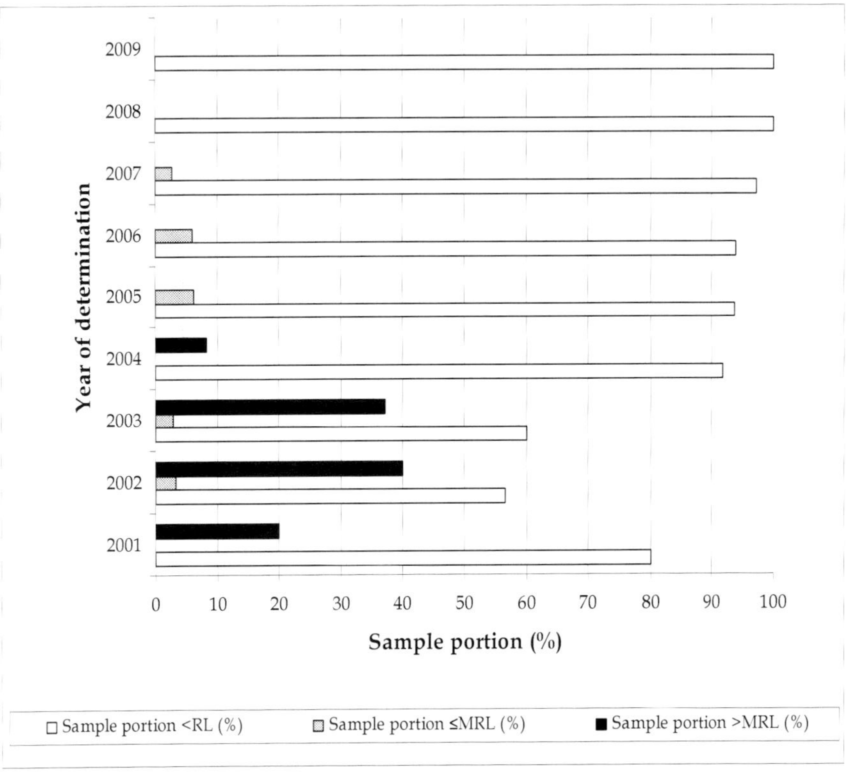

Fig. 5. Annual results for potato from 2001 to 2009

2002 (66.7%). The highest MRL exceedances in apple samples were found in 2005 (5.9%), exceedances were also found in 2002, 2004 and 2006. The highest percentage of lettuce samples with determined but not exceeding PPP residues was found in 2004 (57.1%) and the lowest in 2005 (5.9%). The highest MRL exceedances in lettuce samples were found in 2001 (13.3%), exceedances were also found in 2002, 2007 and 2008. MRL exceedances in potato samples were found in 2001 (20.0%), 2002 (40.0%), 2003 (37.1%) and 2004 (8.2%). From 2005 to 2009 MRL exceedances in potato samples were not found.

Multiple residues were found in 274 out of 1504 samples (18.2%). Residues of two active substances were determined in 125 samples (8.3%), residues of three active substances were determined in 86 samples (5.7%) and residues of four or more active substances were determined in 63 samples (4.2%). Multiple residues were mainly found in fruit samples, i.e. in grape samples (68.6%), in apple samples (55.1%), in pear samples (42.9%), in cherry samples (40.0), in strawberry samples (37.0%) and in peach samples (30.0%). According to our study the grapes were treated with different active substances, the sample in 2006 contained residues of nine active substances. In vegetable samples multiple residues were found in less than 10% of the matrix analysed. In some vegetable samples (beans, carrot, eggplant, head cabbage, leek, peas, potatoes, spinach and string beans) and in cereal samples multiple residues were not found. Results are presented in Figure 6 and in Table 6. Details about samples containing 5 or more active substances are presented in Table 7. Most of the pesticides, presented in Table 7, were insecticides or fungicides, some acaricides were also found.

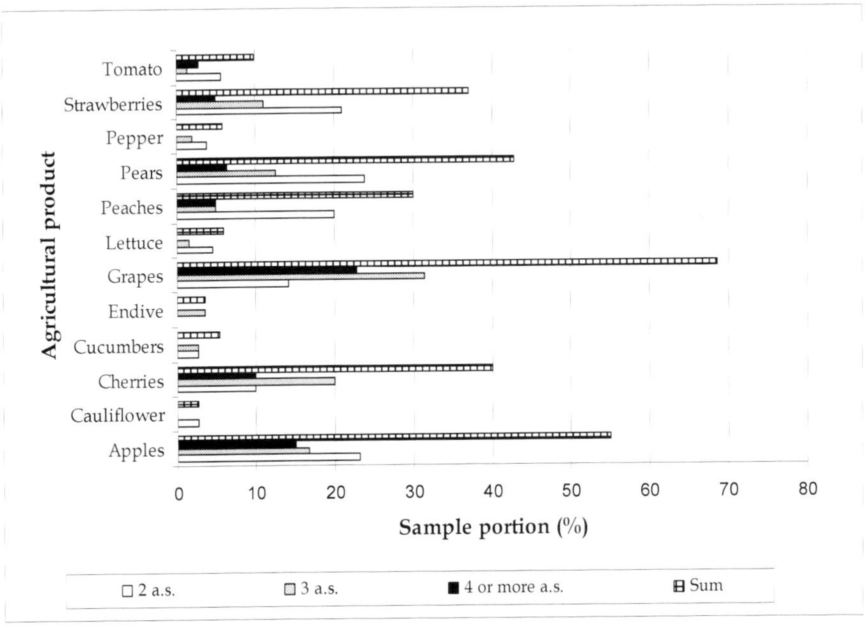

a.s. stands for active substances

Fig. 6. Distribution of samples with multiple residues from 2001 to 2009

Commodity	No. of samples 2 a.s.	Portion (%)	No. of samples 3 a.s.	Portion (%)	No. of samples 4 or more a.s.	Portion (%)	No. of samples sum multiple	Portion (%)
Apples	66	23.2	48	16.8	43	15.1	157	55.1
Beans	0	0.0	0	0.0	0	0.0	0	0.0
Carrot	0	0.0	0	0.0	0	0.0	0	0.0
Cauliflower	1	2.6	0	0.0	0	0.0	1	2.6
Cereals	0	0.0	0	0.0	0	0.0	0	0.0
Cherries	1	10.0	2	20.0	1	10.0	4	40.0
Cucumbers	1	2.7	1	2.7	0	0.0	2	5.4
Eggplant	0	0.0	0	0.0	0	0.0	0	0.0
Endive	0	0.0	1	3.6	0	0.0	1	3.6
Grapes	5	14.3	11	31.4	8	22.9	24	68.6
Head cabbage	0	0.0	0	0.0	0	0.0	0	0.0
Leek	0	0.0	0	0.0	0	0.0	0	0.0
Lettuce	9	4.5	3	1.5	0	0.0	12	5.9
Peaches	4	20.0	1	5.0	1	5.0	6	30.0
Pears	15	23.8	8	12.7	4	6.3	27	42.9
Peas	0	0.0	0	0.0	0	0.0	0	0.0
Pepper	2	3.8	1	1.9	0	0.0	3	5.8
Potatoes	0	0.0	0	0.0	0	0.0	0	0.0
Spinach	0	0.0	0	0.0	0	0.0	0	0.0
Strawberries	17	21.0	9	11.1	4	4.9	30	37.0
String beans	0	0.0	0	0.0	0	0.0	0	0.0
Tomatoes	4	5.6	1	1.4	2	2.8	7	9.9

a.s. stands for active substances

Table 6. Samples with multiple residues from 2001 to 2009

Active substances found from 2001 to 2009 are presented in Figure 7 and in Table 8. Dithiocarbamates were most frequently found (in 21.7% of all samples), followed by phosalone (in 7.9% of all samples) and diazinone (in 5.3% of all samples).

Dithiocarbamates (maneb group) were found each year and were expressed as carbon disulfide. Cauliflower and head cabbage naturally contain substances that during preparation liberate carbon disulfide and give the same responses as dithiocarbamates. Diazinon and phosalone were found each year from 2001 to 2007 but not in 2008 and 2009, captan was found each year except in 2003 and in 2009, fludioxonil each year except in 2005 and in 2008, folpet each year except in 2002 in 2008 and in 2009 and procymidone each year except in 2007 and in 2009.

Active substances exceeding MRLs in the period from 2001 to 2009 are presented in Figure 8 and in Table 9. Dithiocarbamates were most frequently exceeding (in 2.53% of all samples), followed by cyprodinil (in 0.47% of all samples) and tolylfluanid (in 0.40% of all samples). Dithiocarbamates (maneb group) were exceeding in 2001, 2002, 2003, 2004 and in 2008, tolylfluanid in 2004, 2005 and in 2006, chlorothalonil in 2005 and in 2007, dimethoate in 2001 and in 2002, fludioxonil in 2003 and in 2006 (Table 9).

The samples were taken randomly in **8 different production areas** in Slovenia. During the period from 2001 to 2009 the highest average percentage of active substances was found in the region of Nova Gorica (49.9%). In this region mainly fruit is produced. The lowest average percentage of active substances was found in the region of Kranj (22.3%). From 2001 to 2009 a reduced percentage of active substances was noticed in this region in spite of the increased number of active substances sought. Results are presented in Table 10.

Regional inspection showed that the highest average percentage of active substances exceeding MRLs, from 2001 to 2009, was in the regions of Maribor (6.7%) and Kranj (6.7%). In Kranj MRL exceedances occurred only in potato samples and did not occur after 2004. The lowest average percentage of active substances exceeding MRLs was in the region of Nova Gorica (0.5%) in spite of the fact that the same region had the highest average percentage of active substances found. Results are presented in Table 11.

Commodity	Test year	Multiple PPP residues from 1 sample (mgkg^{-1})		
Apple	2004	Captan 0.10 (F)	Diazinon 0.02 (I)	Maneb group 0.06 (F)
		Phosalone 0.14 (I)	Tolylfluanid 0.02 (F)	
Apple	2004	Captan 0.21 (F)	Chlorpyriphos-methyl 0.01 (I)	Diazinon 0.03 (I)
		Folpet 0.06 (F)	Tolylfluanid 0.09 (F)	
Apple	2004	Captan 0.17 (F)	Cyprodinil 0.02 (F)	Diazinon 0.02 (I)
		Maneb group 0.06 (F)	Phosalone 0.17 (I)	Tolylfluanid 0.18 (F)
Apple	2004	Captan 0.21 (F)	Cyprodinil 0.02 (F)	Diazinon 0.04 (I)
		Folpet 0.06 (F)	Maneb group 0.61 (F)	Phosalone 0.15 (I)
		Pyrimethanil 0.03 (F)	Tolylfluanid 0.17 (F)	
Apple	2005	Chlorpyriphos 0.04 (I)	Chlorpyriphos-methyl 0.03 (I)	Cyprodinil 0.02 (F)
		Diazinon 0.02 (I)	Maneb group 0.09 (F)	Pirimicarb 0.05 (I)
Apple	2005	Captan 0.36 (F)	Chlorpyriphos 0.09 (I)	Cyprodinil 0.01 (F)
		Diazinon 0.15 (I)	Maneb group 0.27 (F)	Tolylfluanid 0.73 (F)
Apple	2006	Captan 0.17 (F)	Maneb group 0.80 (F)	Pyrimethanil 0.02 (F)
		Spirodiclofen 0.02 (A)	Thiacloprid 0.01 (I)	
Apple	2006	Captan 0.34 (F)	Diazinon 0.06 (I)	Maneb group 0.23 (F)
		Tebufenozide 0.01 (I)	Tolylfluanid 0.24 (F)	
Apple	2006	Captan 0.16 (F)	Chlorpyriphos 0.17 (I)	Diazinon 0.14 (I)
		Maneb group 0.21 (F)	Phosalone 0.01 (I)	Spirodiclofen 0.06 (A)
		Tolylfluanid 0.05 (F)		
Apple	2006	Captan 0.26 (F)	Diazinon 0.01 (I)	Maneb group 0.15 (F)
		Phosalone 0.02 (I)	Spirodiclofen 0.02 (A)	

Apple	2006	Chlorpyriphos 0.07 (I)	Diazinon 0.02 (I)	Diphenylamine 0.02 (F)
		Maneb group 0.26 (F)	Phosalone 0.01 (I)	
Apple	2006	Captan 0.16 (F)	Chlorpyriphos 0.07 (I)	Cyprodinil 0.02 (F)
		Maneb group 0.12 (F)	Pyrimethanil 0.04 (F)	
Apple	2007	Captan 0.16 (F)	Cyprodinil 0.02 (F)	Diazinon 0.04 (I)
		Maneb group 0.10 (F)	Phosalone 0.57 (I)	
Apple	2007	Acetamiprid 0.02 (I)	Captan 0.51 (F)	Chlorpyriphos 0.07 (I)
		Pyrimethanil 0.03 (F)	Spirodiclofen 0.02 (A)	
Apple	2007	Diazinon 0.04 (I)	Maneb group 0.17 (F)	Phosalone 0.13 (I)
		Pyrimethanil 0.01 (F)	Trifloxystrobin 0.03 (F)	
Apple	2008	Boscalid 0.02 (F)	Maneb group 0.12 (F)	Methoxyfenozide 0.01 (I)
		Pyraclostrobin 0.01 (F)	Pyrimethanil 0.02 (F)	
Apple	2008	Acetamiprid 0.01 (I)	Boscalid 0.06 (F)	Captan 0.44 (F)
		Maneb group 0.96 (F)	Pyraclostrobin 0.02 (F)	Pyrimethanil 0.02 (F)
Grape	2006	Azoksystrobin 0.04 (F)	Chlorpyriphos 0.04 (I)	Cyprodinil 0.10 (F)
		Fenazaquin 0.03 (A)	Fenhexamid 0.33 (F)	Fludioxonil 0.03 (F)
		Folpet 0.09 (F)	Maneb group 0.10 (F)	Metalaxyl 0.05 (F)
Grape	2006	Cyprodinil 0.02 (F)	Folpet 0.42 (F)	Maneb group 0.12 (F)
		Phosalone 0.02 (I)	Pyrimethanil 0.53 (F)	Zoxamide 0.07 (F)
Grape	2006	Chlorothalonil 0.17 (F)	Cyprodinil 0.25 (F)	Fenhexamid 0.05 (F)
		Folpet 0.20 (F)	Tebufenozide 0.01 (I)	
Grape	2006	Chlorothalonil 0.39 (F)	Cyprodinil 0.01 (F)	Folpet 0.84 (F)
		Metalaxyl 0.18 (F)	Myclobutanil 0.02 (F)	
Pear	2005	Captan 0.09 (F)	Chlorpyriphos-methyl 0.03 (I)	Maneb group 0.07 (F)
		Phosalone 0.13 (I)	Tolylfluanid 0.28 (F)	
Pear	2005	Chlorpyriphos-methyl 0.01 (I)	Diazinon 0.04 (I)	Maneb group 0.43 (F)
		Procymidone 0.04 (F)	Tolylfluanid 0.17 (F)	
Pear	2008	Boscalid 0.36 (F)	Difenoconazole 0.01 (F)	Fluquinconazole 0.03 (F)
		Lufenuron 0.09 (I)	Maneb group 0.36 (F)	Pyraclostrobin 0.13 (F)
		Thiacloprid 0.09 (I)		
Strawberry	2004	Cyprodinil 0.03 (F)	Fludioxonil 0.01 (F)	Maneb group 0.37 (F)
		Metalaxyl 0.04 (F)	Pyrimethanil 0.27 (F)	
Strawberry	2004	Azoxystrobin 0.04 (F)	Bromopropylate 0.04 (A)	Cyprodinil 0.10 (F)
		Fludioxonil 0.11 (F)	Maneb group 0.14 (F)	Pyrimethanil 0.20 (F)
Strawberry	2006	Cyprodinil 0.20 (F)	Fludioxonil 0.13 (F)	Maneb group 0.25 (F)
		Metalaxyl 0.02 (F)	Tolylfluanid 0.01 (F)	

A-acaricide, F-fungicide, I-insecticide

Table 7. Agricultural products containing 5 or more active substances per sample from 2001 to 2009

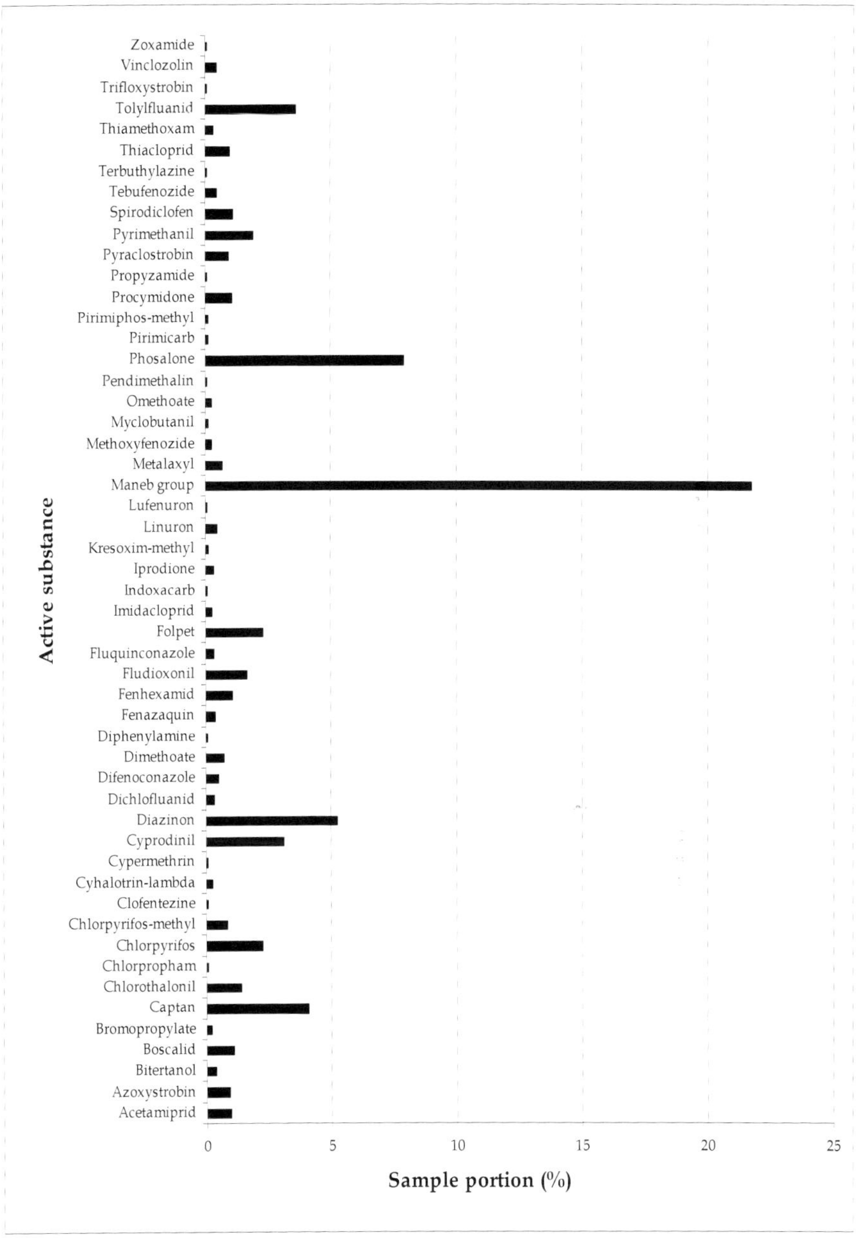

Fig. 7. Sample portion of active substances found in the period from 2001 to 2009

	Sample portion (%)								
Active substance	2001	2002	2003	2004	2005	2006	2007	2008	2009
Acetamiprid	n.a.	n.a.	n.a.	n.a.	n.a.	n.a.	1.9	6.6	n.d.
Azoxystrobin	n.a.	n.a.	0.3	0.8	n.d.	2.2	2.4	n.d.	0.6
Bitertanol	n.a.	n.a.	n.a.	n.a.	n.a.	n.a.	2.9	n.d.	n.d.
Boscalid	n.a.	n.a.	n.a.	n.a.	n.a.	n.a.	1.4	8.4	n.d.
Bromopropylate	n.a.	n.a.	0.6	0.3	n.d.	n.d.	n.d.	n.d.	n.d.
Captan	2.6	2.0	n.d.	5.8	7.0	4.4	5.7	3.6	n.d.
Chlorothalonil	n.a.	n.a.	n.d.	0.3	2.6	5.0	3.3	n.d.	0.6
Chlorpropham	n.a.	n.a.	n.a.	n.a.	0.9	n.d.	n.d.	n.d.	n.d.
Chlorpyrifos	n.d.	n.d.	n.d.	n.d.	3.5	4.4	7.6	3.6	n.d.
Chlorpyrifos-methyl	n.d.	1.3	0.6	1.7	2.6	n.d.	n.d.	n.d.	n.d.
Clofentezine	n.a.	n.a.	n.a.	n.a.	n.a.	n.a.	n.d.	0.6	n.d.
Cyhalotrin-lambda	0.7	n.d.	n.d.	0.3	n.d.	n.d.	1.0	n.d.	n.d.
Cypermethrin	n.a.	n.a.	n.d.	n.d.	n.d.	n.d.	n.d.	0.6	n.d.
Cyprodinil	n.a.	n.a.	n.a.	3.3	2.6	10.5	4.8	n.d.	1.8
Diazinon	4.0	4.7	3.3	6.4	10.4	5.0	4.8	n.d.	n.d.
Dichlofluanid	n.a.	n.a.	0.6	0.6	n.d.	0.6	n.d.	n.d.	n.d.
Difenoconazole	n.a.	n.a.	n.a.	n.a.	n.a.	1.1	1.9	1.2	n.d.
Dimethoate	1.3	3.3	0.3	n.d.	n.d.	n.d.	1.4	n.d.	n.d.
Diphenylamine	n.a.	n.a.	n.a.	n.d.	n.d.	0.6	n.d.	n.d.	n.d.
Fenazaquin	n.a.	n.a.	n.a.	n.a.	n.a.	2.8	n.d.	0.6	n.d.
Fenhexamid	n.a.	n.a.	n.a.	n.a.	n.a.	3.3	4.8	n.d.	n.d.
Fludioxonil	0.7	0.7	1.4	2.5	n.d.	2.8	0.5	n.d.	1.8
Fluquinconazole	n.a.	n.a.	n.a.	n.a.	n.a.	n.a.	n.d.	3.0	n.d.
Folpet	0.7	n.d.	2.2	1.4	0.9	10.5	0.5	n.d.	n.d.
Imidacloprid	n.a.	n.a.	n.a.	n.a.	n.a.	1.1	1.0	n.d.	n.d.
Indoxacarb	n.a.	n.a.	n.a.	n.a.	n.a.	n.a.	n.a.	n.d.	0.6
Iprodione	2.0	n.d.	n.d.	n.d.	n.d.	n.d.	n.d.	n.d.	1.2
Kresoxim-methyl	n.a.	n.a.	n.a.	n.d.	n.d.	0.6	n.d.	n.d.	n.d.
Linuron	n.a.	n.a.	n.a.	n.a.	n.a.	n.a.	n.a.	1.2	n.d.
Lufenuron	n.a.	n.a.	n.a.	n.a.	n.a.	n.a.	1.4	1.8	0.6
Maneb group	17.2	42.0	15.0	15.5	20.9	22.7	10.5	12.7	11.8
Metalaxyl	n.d.	1.3	n.d.	0.3	n.d.	2.2	0.5	1.2	n.d.
Methoxyfenozide	n.a.	n.a.	n.a.	n.a.	n.a.	n.a.	n.a.	2.4	n.d.
Myclobutanil	n.a.	n.a.	n.a.	n.d.	n.d.	1.1	n.d.	n.d.	n.d.
Omethoate	n.a.	n.a.	n.d.	n.d.	n.d.	n.d.	1.9	n.d.	n.d.

Pendimethalin	n.a.	n.a.	n.a.	n.a.	n.a.	n.a.	n.d.	**0.6**	n.d.
Phosalone	**4.0**	**17.3**	**4.4**	**5.3**	**7.8**	**10.5**	**11.4**	n.d.	n.d.
Pirimicarb	n.a.	n.a.	n.a.	n.a.	**0.9**	n.d.	n.d.	**0.6**	n.d.
Pirimiphos-methyl	**0.7**	**0.7**	n.d.	n.d.	n.d.	n.d.	n.d.	n.d.	n.d.
Procymidone	**2.0**	**2.7**	**0.6**	**0.6**	**0.9**	**1.7**	n.d.	**0.6**	n.d.
Propyzamide	n.a.	n.a.	n.d.	n.d.	n.d.	n.d.	**0.5**	n.d.	n.d.
Pyraclostrobin	n.a.	n.a.	n.a.	n.a.	n.a.	n.a.	n.a.	**8.4**	n.d.
Pyrimethanil	n.a.	n.a.	n.a.	**1.7**	n.d.	**6.6**	**2.9**	**3.0**	n.d.
Spirodiclofen	n.a.	n.a.	n.a.	n.a.	n.a.	**3.3**	**3.8**	**1.8**	n.d.
Tebufenozide	n.a.	n.a.	n.a.	n.a.	n.a.	**2.2**	n.d.	**1.8**	n.d.
Terbuthylazine	n.a.	n.a.	n.a.	n.a.	n.a.	n.a.	**0.5**	n.d.	n.d.
Thiacloprid	n.a.	n.a.	n.a.	n.a.	n.a.	**1.7**	**2.9**	**3.6**	n.d.
Thiamethoxam	n.a.	n.a.	n.a.	n.a.	n.a.	n.d.	**0.5**	**0.6**	**1.8**
Tolylfluanid	n.a.	n.a.	n.d.	**7.2**	**13.9**	**7.2**	n.d.	n.d.	n.d.
Trifloxystrobin	n.a.	n.a.	n.a.	n.a.	n.a.	n.a.	**0.5**	n.d.	n.d.
Vinclozolin	**4.6**	n.d.	n.d.	n.d.	n.d.	n.d.	n.d.	n.d.	n.d.
Zoxamide	n.a.	n.a.	n.a.	n.a.	n.a.	**0.6**	n.d.	n.d.	n.d.

n.a. means not analysed
n.d. means not detected

Table 8. Annual sample portions of active substances found in the years from 2001 to 2009

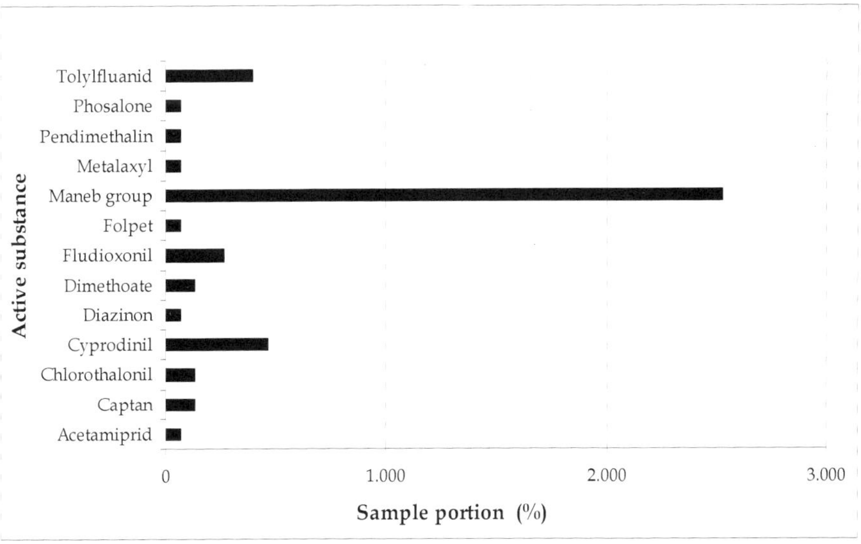

Fig. 8. Sample portion of active substances exceeding MRLs in the period from 2001 to 2009

	Sample portion (%)								
Active substance	**2001**	**2002**	**2003**	**2004**	**2005**	**2006**	**2007**	**2008**	**2009**
Acetamiprid	n.a.	n.a.	n.a.	n.a.	n.a.	n.a.	**0.5**	n.e.	n.d.
Captan	n.e.	n.e.	n.d.	n.e.	n.e.	n.e.	**1.0**	n.e.	n.d.
Chlorothalonil	n.a.	n.a.	n.d.	n.e.	**0.9**	n.e.	**0.5**	n.d.	n.e.
Cyprodinil	n.a.	n.a.	n.a.	n.e.	n.e.	**0.3**	n.e.	n.d.	n.e.
Diazinon	n.e.	n.e.	n.e.	n.e.	**0.9**	n.e.	n.e.	n.d.	n.d.
Dimethoate	**0.7**	**0.7**	n.e.	n.d.	n.d.	n.d.	n.e.	n.d.	n.d.
Fludioxonil	n.e.	n.e.	**2.0**	n.e.	n.d.	**0.6**	n.e.	n.d.	n.e.
Folpet	n.e.	n.d.	n.e.	n.e.	n.e.	n.e.	**0.5**	n.d.	n.d.
Maneb group	**4.6**	**8.0**	**8.7**	**2.4**	n.e.	n.e.	n.e.	**0.6**	n.e.
Metalaxyl	n.d.	**0.7**	n.d.	n.e.	n.d.	n.e.	n.e.	n.e.	n.d.
Pendimethalin	n.a.	n.a.	n.a.	n.a.	n.a.	n.a.	n.d.	**0.6**	n.d.
Phosalone	n.e.	**0.7**	n.e.	n.e.	n.e.	n.e.	n.e.	n.d.	n.d.
Tolylfluanid	n.a.	n.a.	n.d.	**1.4**	**0.9**	**1.1**	n.d.	n.d.	n.d.

n.a. means not analysed
n.d. means not detected
n.e. means not exceeding

Table 9. Annual sample portions of active substances exceeding MRLs in the years from 2001 to 2009

	Sample portion of active substances found (%)									
Region / Year	**2001**	**2002**	**2003**	**2004**	**2005**	**2006**	**2007**	**2008**	**2009**	**Average**
Celje	35.7	55.0	52.2	48.8	33.3	31.3	23.5	14.3	22.2	35.1
Koper	21.4	58.3	52.9	41.2	50.0	53.3	52.6	40.0	33.3	44.8
Kranj	47.6	41.2	22.2	36.4	20.0	10.5	8.7	10.5	4.0	22.3
Ljubljana	22.2	53.6	29.2	41.7	20.0	39.1	25.7	20.7	23.1	30.6
Maribor	20.8	59.3	46.7	55.9	31.8	56.8	41.5	41.0	18.5	41.4
Murska Sobota	15.0	40.0	23.5	55.0	40.0	37.0	30.4	0.0	0.0	26.8
Nova Gorica	22.2	75.0	50.0	63.6	50.0	78.6	57.1	42.9	9.5	49.9
Novo mesto	29.0	50.0	66.7	40.0	36.4	53.3	61.3	35.5	15.8	43.1

Table 10. Active substances found in the years from 2001 to 2009 for different regions

	Sample portion of active substances exceeding MRLs (%)									
Region / Year	**2001**	**2002**	**2003**	**2004**	**2005**	**2006**	**2007**	**2008**	**2009**	**Average**
Celje	0.0	10.0	13.0	0.0	0.0	6.3	0.0	4.8	0.0	3.8
Koper	0.0	16.7	0.0	0.0	0.0	0.0	5.3	0.0	0.0	2.4
Kranj	23.8	11.8	11.1	13.6	0.0	0.0	0.0	0.0	0.0	6.7
Ljubljana	0.0	10.7	12.5	2.8	10.0	4.3	0.0	0.0	0.0	4.5
Maribor	8.3	7.4	20.0	5.9	3.1	10.8	2.4	2.6	0.0	6.7
Murska Sobota	0.0	0.0	5.9	10.0	0.0	3.7	0.0	0.0	0.0	2.2
Nova Gorica	0.0	0.0	0.0	0.0	0.0	0.0	4.8	0.0	0.0	0.5
Novo mesto	3.2	16.7	11.1	0.0	0.0	6.7	3.2	0.0	0.0	4.5

Table 11. Active substances exceeding MRLs in the years from 2001 to 2009 for different regions

4. Comparison of Slovenia with the European Union, Norway, Iceland and Liechtenstein

In the European Union, Norway, Iceland and Liechtenstein 58003 of samples were tested in monitoring from 2001 to 2006, i.e. 4829 cereal samples (oat, rice, rye, wheat), 20978 fruit samples and 32196 vegetable samples **(http://ec.europa.eu/food/fvo/specialreports/pesticides_index_en.htm).** The sampling is presented in Table 12.

Sample portions below reporting level (RL), sample portions below or equal to MRLs and sample portions above MRLs are presented in Figure 9, Figure 10 and Table 13.

PPP residues were not found in 33734 samples (58.2%), in 22782 samples (39.3%) PPP residues were lower or equal to MRLs and PPP residues were above MRLs in 1487 samples (2.6%).

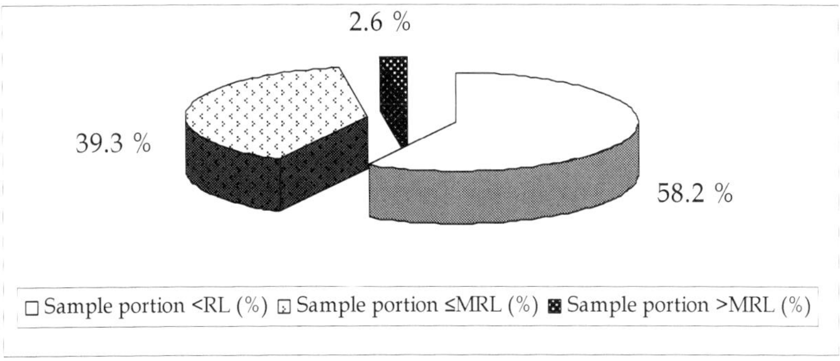

Fig. 9. Results of monitoring from 2001 to 2006 in the European Union, Norway, Iceland and Liechtenstein

Commodity	2001	2002	2003	2004	2005	2006	Sum
Apples	2641	/	/	3133	/	/	5774
Beans	/	896	/	/	1122	/	2018
Carrot	/	1457	/	/	1759	/	3216
Cauliflower	/	/	631	/	/	1014	1645
Cereals	/	/	1656	795	847	1531	4829
Cherries	/	/	/	/	/	/	/
Cucumbers	/	/	1150	/	1555	/	2705
Eggplant	/	/	706	/	/	960	1666
Endive	/	/	/	/	/	/	/
Grapes	1721	/	2163	/	/	2479	6363
Head cabbage	/	/	/	918	/	/	918
Leek	/	/	/	769	/	/	769
Lettuce	1838	/	/	2301	/	/	4139
Peaches/Nectarines	/	1190	/	/	/	/	1190
Pears	/	1330	/	/	2001	/	3331
Peas	/	/	519	/	/	853	1372
Pepper	/	/	1754	/	/	2248	4002
Potatoes	/	1502	/	/	1909	/	3411
Spinach	/	644	/	/	1010	/	1654
Strawberries	1652	/	/	2668	/	/	4320
String beans	/	/	/	/	/	/	/
Tomatoes	2016	/	/	2665	/	/	4681
Sum	9868	7019	8579	13249	10203	9085	58003

Table 12. Sampling of agricultural products in the years from 2001 to 2006 in the European Union, Norway, Iceland and Liechtenstein

The highest portion of determined but not exceeding PPP residues, (residues lower or equal to MRLs), was found in pear samples (68.8%), grape samples (62.2%), strawberry samples (58.6%), apple samples (53.5%) and peach/ nectarine samples (45.5%). Fruit samples contained also exceeding PPP residues (residues above MRLs), i.e. grape samples (3.5%), peach/nectarine samples (3.1%), strawberry samples (3.0%), apple samples (1.5%) and pear samples (1.1%). The highest portion of PPP residues was found in fruit, just like in Slovenia. The highest portion of PPP residues exceeding MRLs were found in spinach samples (8.9%) and bean samples (7.7%). The matrixes with exceeding MRLs were different from commodities in Slovenia.

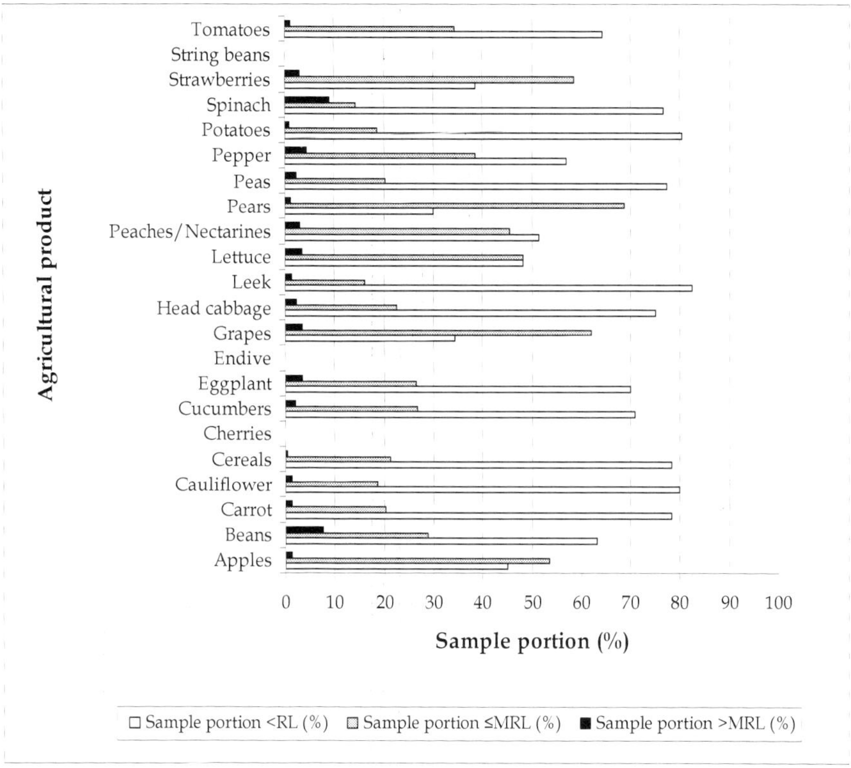

Fig. 10. Sample portions of PPP residues for each commodity from 2001 to 2006 in the European Union, Norway, Iceland and Liechtenstein

5. Conclusions

In Slovenia, during the monitoring from 2001 to 2009 the following samples were analysed:

- 102 cereal samples: PPP residues were not found in 101 samples (99.0%), 1 sample (1.0%) contained PPP residues lower or equal to MRLs, residues exceeding MRLs were not determined
- 494 fruit samples: 19 samples (3.8%) exceeded MRLs, 348 samples (70.4%) contained PPP residues lower or equal to MRLs, PPP residues were not found in 127 samples (25.7%)
- 908 vegetable samples: 46 samples (5.1%) exceeded MRLs, 144 samples (15.9%) contained PPP residues lower or equal to MRLs, PPP residues were not found in 718 samples (79.1%)

In the European Union, Norway, Iceland and Liechtenstein during the monitoring from 2001 to 2006 the following samples were analysed:

- 4829 cereal samples: 21 samples (0.4%) exceeded MRLs, 1024 samples (21.2%) contained PPP residues lower or equal to MRLs, PPP residues were not found in 3784 samples (78.4%)

Commodity	Sample portion <RL (%)	Sample portion ≤MRL (%)	Sample portion >MRL (%)
Apples	45.0	53.5	1.5
Beans	63.4	28.9	7.7
Carrot	78.4	20.2	1.4
Cauliflower	79.9	18.8	1.3
Cereals	78.4	21.2	0.4
Cherries	0.0	0.0	0.0
Cucumbers	70.9	26.9	2.2
Eggplant	69.9	26.5	3.5
Endive	0.0	0.0	0.0
Grapes	34.4	62.2	3.5
Head cabbage	75.1	22.7	2.3
Leek	82.4	16.3	1.3
Lettuce	48.2	48.2	3.6
Peaches/Nectarines	51.4	45.5	3.1
Pears	30.1	68.8	1.1
Peas	77.5	20.3	2.3
Pepper	57.0	38.6	4.4
Potatoes	80.4	18.7	0.9
Spinach	76.8	14.3	8.9
Strawberries	38.5	58.6	3.0
String beans	0.0	0.0	0.0
Tomatoes	64.4	34.4	1.2

Table 13. Sample portions of PPP residues for each commodity from 2001 to 2006 in the European Union, Norway, Iceland and Liechtenstein

- 20978 fruit samples: 506 samples (2.4%) exceeded MRLs, 12408 samples (59.1%) contained PPP residues lower or equal to MRLs, PPP residues were not found in 8064 samples (38.4%)
- 32196 vegetable samples: 960 samples (3.0%) exceeded MRLs, 9350 samples (29.0%) contained PPP residues lower or equal than MRLs, PPP residues were not found in 21886 samples (68.0%).

Levels of pesticide residues in agricultural products in Slovenia from 2001 to 2009 do not give any cause for alarm. In general the portion of exceedances is slightly higher (4.3%) than in the European Union, Norway, Iceland and Liechtenstein (2.6%) Exceedances in Slovenia have been reduced over the years. Otherwise in both monitoring results the highest portion of exceedances was found in vegetables and the lowest in cereals.

Healthy food in the Slovenian market can be achieved by frequent and accurate control of agricultural products. In future analytical work should be extended to other active substances, determination of possible metabolites and improvement of sensitivity of testing methods, i.e. lowering limit of quantification.

6. Acknowledgements

The authors thank those who contributed to the work: Ms. Mateja Fortuna, M. Sc. Zvonko Bregar and other co-workers at the Central Laboratories of Agricultural Institute of Slovenia. For financial support we express our thanks to the Phytosanitary Administration of the Republic of Slovenia (MAFF) and Inspectorate of the Republic of Slovenia for Agriculture, Forestry and Food (MAFF).

7. References

Akiyama, Y.; Yoshioka, N. & Tsuji, M. (2002). Pesticide residues in agricultural products monitored in Hyogo prefecture, Japan, FYs 1995-1999. *Journal of AOAC International*, Vol. 85, No. 3 (May 2002), pp. 692-213, ISSN1060-3271

Andersen, J. H. & Poulsen, M. E. (2001). Results from the monitoring of pesticide residues in fruit and vegetables on the Danish market, 1998-99. *Food Additives and Contaminants*, Vol. 18, No. 10 (October 2001), pp. 906-931, ISSN 0265-203X

Baša Česnik, H. & Gregorčič, A. (2003). Multirezidualna analizna metoda za določevanje ostankov pesticidov v sadju in zelenjavi. *Research Reports Biotechnical Faculty University of Ljubljana, Agriculture, Zootechny*, Vol. 82, No. 2 (December 2003), pp.167-180, ISSN 1408-3434

Baša Česnik, H. & Gregorčič, A. (2006): Validation of the method for the determination of dithiocarbamates and thiuram disulphide on apple, lettuce, potato, strawberry and tomato matrix. *Acta Chimica Slovenica*, Vol. 53, No. 1 (March 2006), pp. 100-104, ISSN 1318-0207

Baša Česnik, H.; Gregorčič, A.; Velikonja Bolta, Š. & Kmecl, V. (2006). *Food Additivs and Contaminants*,Vol. 23, No. 2 (February 2006), pp. 164-173, ISSN 0265-203X

Berrada, H.; Fernández, M.; Ruiz, M. J.; Moltó, J. C. & Mañes, J. (2006). Exposure assessment of fruits contaminated with pesticide residues from Valencia, 2001-03. *Food Additives and Contaminants*, Vol. 23, No. 7 (July 2006), pp. 674-682, ISSN 0265-203X

Bossi, R.; Vejrup, K. V.; Mogensen, B. B. & Asman, A. H. (2002). Analysis of polar pesticides in rainwater in Denmark by liquid chromatography–tandem mass spectrometry. *Journal of Chromatography A*, Vol. 957, No. 1 (May 2002), pp.27-36, ISSN 0021-9673

Čajka, T.& Hajšlová, J. (2004). Gas chromatography-high-resolution time-of-flight mass spectrometry in pesticide residue analysis: advantages and limitations. *Journal of Chromatography A*, Vol. 1058, No. 1-2 (November 2004), pp. 251-261, ISSN 0021-9673

Dejonckheere, W.; Steurbaut, W.; Drieghe, S.; Verstraeten, R. & Braeckman, H. (1996). Monitoring of pesticide residues in fresh vegetables, fruits, and other selected food items in Belgium, 1991-1993. *Journal of AOAC International*, Vol. 79, No. 1 (Jan-Feb 1996), pp. 97-110, ISSN 1060-3271

Díez, C.; Traag, W.A.; Zommer, P.; Marinero, P. & Atienza, J. (2006). Comparison of an acetonitrile extraction/partitioning and »dispersive solid-phase extraction« method with classical multi-residue methods for the extraction of herbicide residues in barley samples. *Journal of Chromatography A*, Vol. 1131, No. 1-2 (October 2006), pp. 11-23, ISSN 0021-9673

Dogheim, S. M.; El-Marsafy, A. M.; Salama, E. Y.; Gadalla, S. A. & Nabil, Y. M. (2002). Monitoring of pesticide residues in Egyptian fruits and vegetables during 1997. *Food Additives and Contaminants*, Vol. 19, No. 11 (November 2002), pp. 1015-1027, ISSN 0265-203X

EC (2001). Commission Recommendation 2001/42/EC of 22 December 2000 concerning a coordinated Community monitoring programme for 2001 to ensure compliance with maximum levels of pesticide residues in and on cereals and certain products of plant origin, including fruit and vegetables OJ L 11, 16.01.2001, pp. 40-45

EC (2002a). Commission Recommendation of 27 December 2001 concerning a coordinated Community monitoring programme for 2002 to ensure compliance with maximum levels of pesticide residues in and on cereals and certain other products of plant origin , OJ L 2, 04.01.2002, pp. 8-12

EC (2002b). Commission Recommendation 2002/663/EC of 19 August 2002 concerning a coordinated Community monitoring programme for 2003 to ensure compliance with maximum levels of pesticide residues in and on cereals and certain other products of plant origin, OJ L 225, 22.08.2002, pp. 29-33

EC (2004). Commission Recommendation of 9 January 2004 concerning a coordinated Community monitoring programme for 2004 to ensure compliance with maximum levels of pesticide residues in and on cereals and certain other products of plant origin, OJ L 16, 23.01.2004, pp. 60-64

EC (2005). Commission Recommendation of 1 March 2005 concerning a coordinated Community monitoring programme for 2005 to ensure compliance with maximum levels of pesticide residues in and on cereals and certain other products of plant origin and national monitoring programmes for 2006, OJ L 61, 08.03.2005, pp. 31-36

EC (2006). Commission Recommendation of 18 January 2006 concerning a coordinated Community monitoring programme for 2006 to ensure compliance with maximum levels of pesticide residues in and on cereals and certain other products of plant origin and national monitoring programmes for 2007, OJ L 19, 24.01.2006, pp. 23-29

EC (2007). Commission Recommendation of 3 April 2007 concerning a coordinated Community monitoring programme for 2007 to ensure compliance with maximum levels of pesticide residues in and on cereals and certain other products of plant origin and national monitoring programmes for 2008, OJ L 96, 11.04.2007, pp. 21-27

EC (2008a). Commission Recommendation of 4 February 2008 concerning a coordinated Community monitoring programme for 2008 to ensure compliance with maximum levels of pesticide residues in and on cereals and certain other products of plant origin and national monitoring programmes for 2009, L 36, 09.02.2008, pp. 7-15

EC (2008b). Commission Regulation of 5 December 2008 concerning a coordinated multiannual Community control programme for 2009, 2010 and 2011 to ensure compliance with maximum levels of and to assess the consumer exposure to pesticide residues in and on food of plant and animal origin, OJ L 328, 06.12.2008, pp. 9-17

Fernandez, M.; Picó, Y. & Mañes, J. (2001). Pesticide residues in oranges from Valencia (Spain), *Food Additives and Contaminants,* Vol. 18, No. 7 (July 2001), pp. 615-624, ISSN 1944-0049

Ferrer, I.; García-Reyes, J. F.; Mezcua, M.; Thurman, E. M. & Fernández-Alba, A. R. (2005). Multi-residue pesticide analysis in fruits and vegetables by liquid chromatography-time-of-flight mass spectrometry. *Journal of Chromatography A*, Vol. 1082, No. 1 (July 2005),pp. 81-90, ISSN 0021-9673

Lehotay, S. J.; de Kok, A.; Hiemstra, M. & Bodengraven, P. (2005). Validation of a Fast and Easy Method for the Determination of Residues from 229 Pesticides in Fruits and Vegetables Using Gas and Liquid Chromatography and Mass Spectrometric Detection. *Journal of AOAC International.*,Vol. 88, No. 2 (March 205), pp. 595-614, ISSN 1060-3271

Lehotay, S. J. (2007). Determination of pesticide residues in foods by acetonitrile extraction and partitioning with magnesium sulfate: collaborative study. *Journal of AOAC International*, Vol. 90, No. 2 (March/April 2007), pp. 485-520, ISSN 1060-3271

Maštovská, K.; Lehotay, S. J. & Anastassiades, M. (2005). Combination of analyte protectants to overcome matrix effects in routine GC analysis of pesticide residues in food matrixes. *Analytical Chemistry*, Vol. 77, No. 24 (November 2005), pp. 8129-8137, ISSN 0003-2700

Ortelli, D.; Edder, P. & Corvi, C. (2004). Multiresidue analysis of 74 pesticides in fruits and vegetables by liquid chromatography–electrospray–tandem mass spectrometry. *Analytica Chimica Acta*, Vol. *520*, No. 1-2 (August 2004), pp. 33-45, ISSN 0003-2670

Pizzutti, I. R.; Vreuls, R. J. J.; de Kok, A.; Roehrs, R.; Martel, S.; Friggi, C. A. & Zanella, R. (2009). Design of a compressed air modulator to be used in comprehensive multidimensional gas chromatography and its application in the determination of pesticide residues in grapes. *Journal of Chromatography A*, Vol. 1216, No. 15 (April 2009), pp. 3305-3311, ISSN 0021-9673

Ripley, B. D.; Lissemore, L. I.; Leishman, P. D. & Denomme, M. A. (2000). Pesticide residues on fruits and vegetables from Ontario, Canada, 1991-1995. *Journal of AOAC International*, Vol. 83, No. 1 (January 2000), pp. 196-213, ISSN 1060-3271

RS (1999a). Decree on Monitoring of Pesticide in Foodstuffs and Agricultural Products, Official Gazette of the Republic of Slovenia No. 13, 05.03.1999, pp. 1168 - 1189

RS (1999b). Regulation on Maximum Residue Values of Pesticides in or on Plants and Foodstuffs of Plant Origin, Official Gazette of the Republic of Slovenia No. 54, 08.07.1999, pp. 6771 - 6802

RS (2001). Law on Plant Protection Products, Official Gazette of the Republic of Slovenia No. 11, 16.02.2001, pp. 1163 - 1175

RS (2004a). Law on Plant Protection Products, Official Gazette of the Republic of Slovenia No. 98, 09.09.2004, pp. 11885-11899

RS (2004b). Regulation on Pesticide Residues in or on Foodstuffs in Agricultural Products, Official Gazette of the Republic of Slovenia No. 84, 30.07.2004, pp. 10210-10226

RS (2007a). Law on Plant Protection Products, Official Gazette of the Republic of Slovenia No. 35, 18.04.2007, pp. 5017-5031

RS (2007b). Regulation on changes and completion of Regulation on Pesticide Residues in or on Foodstuffs and Agricultural Products No. 108, 27.11.2007, pp. 14834

RS (2009). Regulation on the implementation of EC Parliament and EC Council Regulation on limit values of pesticide residues in or on the food or forage of plant and animal origin, Official Gazette of the Republic of Slovenia No. 16, 27.02.2009, pp. 2081-2082

Sharif, Z.; Che Man, Y. B.; Hamid, N. S. A. H. & Keat, C. C. (2006). Determination of organochlorine and pyrethroid pesticides in fruit and vegetables using solid phase extraction clean-up cartridges. *Journal of Chromatography A*, Vol. 1127, No. 3 (August 1992), pp. 254-261, ISSN 0021-9673

Stan, H.-J. & Linkerhägner, M. (1996). Pesticide residue analysis in foodstuffs applying capillary gas chromatography with atomic emission detection State-of-the-art use of modified multimethod S19 of the Deutsche Forschungsgemeinschaft and automated large-volume injection with programmed-temperature vaporization and solvent venting. *Journal of Chromatography A*, Vol. 750, No. 1-2 (October 1996), pp. 369-390, ISSN 0021-9673

van Zoonen, P. (1996). Benomyl/Carbendazim/Thiabendazole, *Analytical Methods for Pesticide residues in Foodstuffs*, van Zoonen, P. (Ed.), General Inspectorate for Health Protection Ministry of public Health, Welfare and Sport Netherlands, 2. Part, p. 1-4, Bilthoven: Rijksinstituut voor Volkgezondheid en Milieu (RIVM)

12

Pesticides in Agricultural Products: Analysis, Reduction, Prevention

Mohammad Shokrzadeh and Seyed Soheil Saeedi Saravi
Depatment of Toxicology-Pharmacology, Faculty of Pharmacy, Mazandaran University of Mediccal Sciences, Sari
Iran

1. Introduction

Food is the basic necessity of life and food contaminated with toxic pesticides is associated with severe effects on the human health. Hence it is pertinent to explore strategies that address this situation of food safety especially for the developing countries where pesticide contamination is widespread due to indiscriminate usage and a major part of population lives below poverty line (Kaushik et al., 2009).

It is therefore of significance to evaluate simple, cost effective strategies to enhance food safety from harmful pesticides for poor populace.

As FAO[1] defined, pesticide is any substance or mixture of substances intended for preventing, destroying, repelling or mitigating any pest, including vectors of human or animal disease, unwanted species of plants or animals causing harm during or otherwise interfering with the production, processing, storage, transport or marketing of food, agricultural commodities, wood and wood products or animal feedstuffs, or substances which may be administered to animals for the control of insects, arachnids or other pests in or on their bodies. The term includes substances intended for use as a plant growth regulator, defoliant, desiccant or agent for thinning fruit or preventing the premature fall of fruit, and substances applied to crops either before or after harvest to protect the commodity from deterioration during storage and transport (Food and Agriculture Organization of the United Nations, 2002). A pesticide may be a chemical substance, biological agent (such as a virus or bacterium), antimicrobial, disinfectant or device used against any pest. Pests include insects, plant pathogens, weeds, molluscs, birds, mammals, fish, nematodes (roundworms), and microbes that destroy property, spread disease or are a vector for disease or cause a nuisance. Although there are benefits to the use of pesticides, there are also drawbacks, such as potential toxicity to humans and other animals.

Pesticides can be classified by target organism, chemical structure, and physical state. Pesticides can also be classed as inorganic, synthetic, or biologicals (biopesticides) (Council on Scientific Affairs, American Medical Association, 1997), although the distinction can sometimes blur. Biopesticides include microbial pesticides and biochemical pesticides (EPA,

---

[1] Food and Drug Administration

2009). Plant-derived pesticides, or 'botanicals', have been developing quickly. These include the pyrethroids, rotenoids, nicotinoids, and a fourth group that includes strychnine and scilliroside (Kamrin, 1997). Many pesticides can be grouped into chemical families. Prominent insecticide families include organochlorines, organophosphates, and carbamates. Organochlorine hydrocarbons (e.g. DDT) could be separated into dichlorodiphenylethanes, cyclodiene compounds, and other related compounds. In addition, Pesticides can be classified based upon their biological mechanism function or application method.

Agricultural production has been accompanied by continuous growth in the number and quantity of agrochemicals applied to crops. Pesticide use is associated with environmental contamination and human health problems worldwide. Pesticides (insecticides, fungicides, etc.) are used globally for the protection of food, fibber, human health and comfort (Winteringham, 1971). Currently, more than 800 pesticide active ingredients in a wide range of commercial products are registered for use in agriculture to meet food supply demands. Pesticides are essential in modern agricultural practices but, due to their biocide activity and potential risk to the consumer, the control of pesticide residues in foods is a growing source of concern for the general population and environment. Over 98% of sprayed insecticides and 95% of herbicides reach a destination other than their target species, including non-target species, air, water and soil (Miller, 2004). Pesticide drift occurs when pesticides suspended in the air as particles are carried by wind to other areas, potentially contaminating them. Pesticides are one of the causes of water pollution, and some pesticides are persistent organic pollutants and contribute to soil contamination.

In addition, pesticide use reduces biodiversity, reduces nitrogen fixation, contributes to pollinator decline, destroys habitat (especially for birds), and threatens endangered species. It also happens that some of the pest adapt to the pesticide and don't die. What is called pesticide resistance, to eliminate the offspring of this pest, will be needed an new pesticide or an increase the dose of pesticide. This will cause a worsening of the ambient pollution problem (Rockets, 2007; Hackenberg, 2007; Wells, 2007; Haefeker, 2000; Zeissloff, 2001; Palmer et al., 2007; Winteringham, 1971).

Pesticides can be dangerous to consumers, workers and close bystanders during manufacture, transport, or during and after use (U.S. Environmental Protection Agency, 2007). Particular uncertainty exists regarding the long-term effects of low-dose pesticide exposures. Current surveillance systems are inadequate to characterize potential exposure problems related to pesticide usage or pesticide-related illnesses. The WHO[2] and the UN[3] Environment Program estimate that each year, 3 million workers in agriculture in the developing world experience severe poisoning from pesticides, about 18,000 of whom die (Winteringham, 1971). According to one study, as many as 25 million workers in developing countries may suffer mild pesticide poisoning yearly. There have been many studies of farmers intended to determine health effects of occupational pesticide exposure. Associations between non-Hodgkin lymphoma, leukemia, prostate cancer, multiple myeloma, and soft tissues sarcoma have been reported in studies, with less association found for other cancers (Jeyaratnam, 1990; McCauley, 2006).

[2] World Health Organization
[3] United Nation

We are exposed to pesticides in the food and water we consume and in the air we breathe. The three main entry routes for pesticides into the body are dermal, (exposure through the skin or eyes), respiratory (inhalation into the lungs), and oral (ingestion by mouth). The amounts of pesticides that remain as residues in food are miniscule-sometimes only 1 millionth of a kilogramme. Thus, contact with the concentrated product during mixing and loading presents the greatest risk for exposure. The level of absorption depends on the properties of the pesticide, its formulation, and parts of the body exposed. Following exposure, human health risks from pesticides and herbicides may be caused by acute (short term) or chronic (long term) exposures (Rell & Galvin, 2008).

However, the chapter tries to analysis of pesticide residues in agricultural products, determination of effective methods to reduce the residues, and prevention of food contamination, as vital works for worldwide health safety.

2. Analysis

In order to be able to set a maximum level, not only do the health risks have to be assessed but the residues must also be determinable analytically. Reliable residue analytical methods are necessary to measure the magnitude of residue in a commodity, and to enforce legal residue limits (tolerances). The Suitable analytical methods for the detection of pesticide residues in agricultural products includes the below stages:

2.1 Sampling, transport, processing and storage of samples

2.1.1 Sampling

Laboratory samples should be taken in uncontaminated status and collected in non-absorbable and non-leakage containers. Where it is impractical to take primary samples randomly within a lot, the method of sampling must be recorded.

2.1.2 Laboratory sample transportation

Samples must be transported under appropriate conditions to the laboratory in clean containers and robust packaging. Polythene bags, ventilated if appropriate, are acceptable for most samples but low permeability bags (e.g. nylon film) must be used for samples to be analyzed for residues of fumigants. Samples of commodities pre-packed for retail sale should not be removed from their packaging before transport. Very fragile or perishable products (e.g. ripe raspberries) may have to be frozen to avoid spoilage and then transported in 'dry ice' or similar, to avoid thawing in transit. Samples that are frozen at the time of collection must be transported without thawing. Samples that may be damaged by chilling (e.g. bananas) must be protected from both high and low temperatures.

Rapid transportation to the laboratory, preferably within one day, is essential for samples of most fresh products. The condition of samples delivered to the laboratory should approximate to that acceptable to a discerning purchaser, otherwise samples should normally be considered unfit for analysis.

Samples must be identified clearly, in a way that prevents inadvertent loss or confusion of labelling. The use of marker pens containing organic solvents should be avoided for labelling bags containing samples to be analyzed for fumigant residues, especially if an electron capture detector is to be used (Pihlström et al., 2009).

2.1.3 Sample preparation and processing prior to analysis

Sample preparation, sample processing and sub-sampling to obtain analytical portions should take place before visible deterioration occurs. This is particularly important when the analytical result is to be used to assess consumer intake. Canned, dried or similarly processed samples should be analyzed within the stated shelf life.

Sample processing and storage procedures should be demonstrated to have no significant effect on the residues present in the analytical sample. Where there is evidence that comminuting (cutting and homogenization) at ambient temperature has a significant influence on the degradation of certain pesticide residues, it is recommended that samples are homogenized at low temperature (e.g. frozen and/or in the presence of 'dry ice'). Where comminuting is known to affect residues (e.g. dithiocarbamates or fumigants) and practical alternative procedures are not available, the test portion should consist of whole units of the commodity, or segments removed from large units. For all other analyses, the whole laboratory sample (in most cases 1-2 kg) needs to be comminuted. All analyses should be undertaken within the shortest time practicable, to minimize sample storage. Analyses for residues of very labile or volatile pesticides should be started, and the procedures involved in potential loss of analyte completed, on the day of sample receipt. In any case, sample comminuting should ensure that the sample is homogeneous enough so that sub-sampling variability is acceptable. If this is not achievable, the use of larger test portions should be considered (Pihlström et al., 2009).

If a single analytical portion is unlikely to be representative of the analytical sample, replicate portions must be analyzed, to provide a better estimate of the true value.

2.2 Pesticide standards and calibration solutions

2.2.1 Identity, purity, and storage of standards

'Pure' standards of analytes should be of known purity and each must be uniquely identified and the date of receipt recorded. They should be stored at low temperature, preferably in a freezer, with light and moisture excluded, i.e. under conditions that minimize the rate of degradation. Under such conditions, the supplier's expiry date, which is often based on less stringent storage conditions, may be replaced, as appropriate for each standard, by a date allowing for storage up to 10 years. The pure standard may be retained if its purity is shown to remain acceptable. The purity should be checked by the allocated time after which a 'Pure' standard may be retained if its purity is shown to remain acceptable and a new expiry date is allocated. Ideally, the identity of freshly acquired 'Pure' standards should be checked if the analytes are new to the laboratory (Pihlström et al., 2009).

2.2.2 Preparation and storage of stock standards

When preparing stock standards (solutions, dispersions or gaseous dilutions) of 'Pure' standards of analytes and internal standards, the identity and mass (or volume, for highly volatile compounds) of the 'Pure' standard and the identity and amount of the solvent (or other diluents) must be recorded. The solvent(s) must be appropriate to the analyte (solubility, no reaction) and method of analysis. Moisture must be excluded during equilibration of the 'Pure' standard to room temperature before use and concentrations must be corrected for the purity of the 'Pure' standard.

Not less than 10 mg of the 'Pure' standard should be weighed using a 5 decimal place balance. The ambient temperature should be that at which the glassware is calibrated, otherwise preparation of the standard should be based on mass measurement. Volatile liquid analytes should be dispensed by weight or volume (if the density is known) directly into solvent. Gaseous (fumigant) analytes may be dispensed by bubbling into solvent and weighing the mass transferred, or by preparing gaseous dilutions (e.g. with a gas-tight syringe, avoiding contact with reactive metals).

Stock standards must be labelled indelibly, allocated an expiry date and stored at low temperature in the dark in containers that prevent any loss of solvent and entry of water. Currently available data show that stock standards of the large majority of pesticides in toluene and acetone are stable for at least 5 years in the freezer when stored in tightly closed glass containers.

For suspensions (e.g. dithiocarbamates) and solutions (or gaseous dilutions) of highly volatile fumigants that should be prepared freshly, the accuracy of the solution should be compared with a second solution made independently at the same time (Pihlström et al., 2009).

2.2.3 Preparation, use and storage of working standards:

When preparing working standards, a record must be kept of the identity and amount of all solutions and solvents employed. The solvent(s) must be appropriate to the analyte (solubility) and method of analysis. The standards must be labelled indelibly, allocated an expiry date and stored at low temperature in the dark in containers that prevent any loss of solvent and entry of water. Septum closures are particularly prone to evaporation losses (in addition to being a source of contamination) and should be replaced as soon as practicable after piercing, if solutions are to be retained. Following equilibration to room temperature, solutions must be re-mixed and a check made to ensure that no analyte remains non-dissolved, especially where solubility at low temperatures is limited.

At method development or validation, or for analytes new to the laboratory, the response detected should be shown to be due to the analyte, rather than to an impurity or artefact. If the techniques used can lead to degradation of the analyte during extraction, clean-up or separation, and they generate a product that is commonly found in samples but which is excluded from the residue definition, positive results must be confirmed using techniques that avoid this problem (Pihlström et al., 2009).

2.2.4 Testing and replacement of standards:

Whenever any standard is used beyond its expiry date its stability should be verified. Existing stock and working solutions may be tested against newly prepared solutions by comparing the detector responses obtained from appropriate dilutions of individual standards or mixtures of standards. The purity of an old 'Pure' standard may be checked by preparing a new stock standard and comparing the detector responses obtained from freshly prepared dilutions of old and new stock standards. Inexplicable differences in apparent concentration between old and new standards must be investigated.

The means from at least three replicate measurements for each of two solutions (old and new) should not normally differ by more than ±10%. The mean from the new solution is taken to be 100%. If the mean response of the old standard differs by more than ±10% from the new, storage time or conditions must be adjusted as necessary on the basis of the results and should be checked against a second solution independently prepared from the first one. The use of an internal standard may reduce the number of replicate injections required to achieve a ±10% difference (Pihlström et al., 2009).

2.3 Extraction and concentration

2.3.1 Extraction conditions and efficiency

Test portions should be disintegrated thoroughly during extraction to maximize extraction efficiency, except where this is known to be unnecessary or inappropriate (e.g. for determination of fumigants or surface residues). Temperature, pH, etc., must be controlled if these parameters affect extraction efficiency, analyte stability or solvent volume. To improve the extraction efficiency of low moisture containing commodities (cereals, dried fruits), it is recommended to add water to the samples before extraction is carried out. However, the time between addition of water and extraction should be controlled in order to avoid any significant losses of pesticides. For instance in a method, a non-fatty test portion is blended with acetone and filtered; pesticides are transferred from aqueous filtrate to organic phase by shaking with petroleum ether and dichloromethane (CH_2Cl_2). After drying, organic phase is concentrated in the presence of petroleum ether and then acetone to remove dichloromethane; an aliquot of concentrated organic phase is injected into various GC[4] systems for determination of a wide variety of pesticide residues (Dehghan et al., 2010).

2.3.2 Extract concentration and dilution to volume

Great care must be exercised when extracts are evaporated to dryness, as trace quantities of many analytes can be lost in this way. A small volume of high boiling point solvent may be used as a 'keeper' and the evaporation temperature should be as low as practicable. Frothing and vigorous boiling of extracts, or dispersion of droplets, must be avoided. A stream of dry nitrogen or vacuum centrifugal evaporation is generally preferable to the use of an air stream for small-scale evaporation, as air is more likely to lead to oxidation or to introduce water and other contaminants.

Where extracts are diluted to a fixed volume, accurately calibrated vessels of not less than 1 ml capacity should be used and further evaporation avoided.

Analyte stability in extracts should be investigated during method validation. Storage of extracts in a refrigerator or freezer will minimize degradation but potential losses at the higher temperatures of an auto-sampler rack should not be ignored (Pihlström et al., 2009).

2.4 Confirmation of results

Negative results (residues below the RL[5]) can be considered confirmed if the recovery and LCL[6] measurement for the batch are acceptable. Negative results for represented analytes are supported only indirectly by the recovery and LCL data for representative analytes and must be interpreted with caution.

---

[4] Gas Chromatography
[5] reporting level
[6] Lowest calibrated level

Confirmation of positive results (residues at or above the RL) for represented analytes (i.e. those with no concurrent calibration and recovery) should be supported by the appropriate concurrent calibration and recovery determinations. Confirmation is not mandatory for all positive results and must be decided by the laboratory on a case-by-case basis.
Suspected MRL7 exceeds or unusual residues must be identified. The use of a highly specific detection system, such as mass spectrometry, is recommended (Pihlström et al., 2009).

2.4.1 Identification

Selective detectors employed with GC or LC[8] such as ECD[9], FPD[10], NPD[11] and fluorescence, offer only limited specificity. Their use, even in combination with different polarity columns, does not provide unambiguous identification. These limitations may be acceptable for frequently found residues, especially if some results are also confirmed using a more specific detection technique. Such limitations in the degree of identification should be acknowledged when reporting the results.
The common identification and measurement systems are listed below:

i. Chromatographic separation system (GC or LC such as ECD, FPD, NPD and fluorescence):

Generally, chromatography is a method for physically separation of components of mixtures and substances, based on differences in distribution in mobile and solid phases. Separation is resulted from different tendency of components to solid phase. Gas chromatography and liquid chromatography are the most common methods for pesticide residues analysis. In Gas-Liquid Chromatography (GLC) method, identification and detection of pesticide residues via standard detectors (ECD, FPD, NPD and fluorescence)are confirmed followed by extraction with a suitable solvent, volatile derivation, injection of the volatile derivate to the GC instrument and gaseous mobile phase pass from the column. Hence, GC can be combined with mass spectrometry to achieve high quality chromatograms with no interferences caused by overlapping/unresolved peaks from co-extracted compounds (Shokrzadeh, 2007).

ii. Mass spectrometry (MS) coupled to chromatography (GC-MS and LC-MS):

Mass spectrometry technique in combination with chromatography is the most common and proper identification method, which has a high selectivity and specificity. Mass spectrometry coupled to GC can identify and measure the low residue level of insecticides and poisons (Shokrzadeh, 2007).
For GC-MS procedures, the chromatographic separation should be carried out using capillary columns. For LC-MS procedures, the chromatographic separation can be performed using any suitable LC column. In either case, the minimum acceptable retention time for the analyte(s) under examination should be at least twice the retention time corresponding to the void volume of the column. The retention time (or relative retention time) of the analyte in the sample extract must match that of the calibration standard (may need to be matrix matched) within a specified window after taking into consideration the

[7] maximum residue limits
[8] Liquid chromatography
[9] Electron Capture Detection
[10] Flame Photometric Detection
[11] Nitrogen-Phosphor Detection

resolving power of the chromatographic system. The ratio of the chromatographic retention time of the analyte to that of a suitable internal standard, i.e. the relative retention time of the analyte, should correspond to that of the calibration solution with a tolerance of ±0.5% for GC and ±2.5% for LC (Pihlström et al., 2009).

iii. UV-spectrometry:

The residues of insecticides and pesticides can be analyzed by UV spectrometry system. The limiting factor for use of this analyzing system is its unspecificity for some insecticides (Shokrzadeh, 2007).

2.5 Reporting of results

2.5.1 Expression of results

Results should normally be expressed as the chemical name defined by the MRL and in mg/kg. Residues below the Reporting Limit should be reported as <RL mg/kg.

2.5.2 Calculation of results

In general, residues data are not to be adjusted for recovery. If they are adjusted for recovery, then this must be stated. In this case they should be adjusted using the mean value from 3 recoveries performed in the same matrix, and analyzed in the same batch of samples. Where confirmed data are derived from a single test portion (i.e. the residue is not violative), the reported result should be that derived from the detection technique considered to be the most accurate. Where results are obtained by two or more equally accurate techniques, the mean value may be reported.

Where two or more test portions have been analyzed, the arithmetic mean of the most accurate results obtained from each portion should be reported. Where good comminuting and/or mixing of samples have been undertaken, the RSD of results between test portions should not exceed 30% for residues significantly above the LOQ[12]. Close to the LOQ, the variation may be higher and additional caution is required in deciding whether or not a limit has been exceeded (Pihlström et al., 2009).

2.5.3 Rounding of data

It is essential to maintain uniformity in reporting results. In general, results ≥0.01 and <10 mg/kg should be rounded to two significant figures; results ≥10 mg/kg may be rounded to three significant figures or to a whole number. Reporting limits should be rounded to 1 significant figure at <10 mg/kg and two significant figures at ≥10 mg/kg. These requirements do not necessarily reflect the uncertainty associated with the data. Additional significant figures may be recorded for the purpose of statistical analysis. In some cases the rounding may be specified by, or agreed with the customer/stakeholder of the monitoring (Pihlström et al., 2009).

2.5.4 Qualifying results with uncertainty data

To this end, laboratories should have available sufficient data derived from method validation/verification, inter-laboratory studies (e.g. proficiency tests) and in-house quality control tests, which are applied to estimate the uncertainties.

---

[12] Limit of quantization

Measurement uncertainty is a quantitative indicator of the confidence in the analytical data and describes the range around a reported or experimental result within which the true value can be expected to lie within a defined probability (confidence level). Uncertainty ranges must take into consideration all sources of error (Pihlström et al., 2009).

3. Reduction

Pesticide residues in or on plants may be unavoidable even when pesticides are used in accordance with Good Agricultural Practice (Uysal-Pala & Bilisli, 2006; Processing studies (Appendix E), 1997). Further while remarkable progress has been made in the development of effective pesticides, the fact remains that a very small fraction of all applied pesticides is directly involved in the pesticidal mechanism. This implies that most of the applied pesticides find their way as 'residue' in the environment into the terrestrial and aquatic food chains where they undergo concentration and exert potential, long term, adverse health effects (Winteringham, 1971).

Pesticide residues are influenced by processing or household preparation stages such as washing, peeling and cooking etc. (Dikshit et al., 2003; Petersen et al., 1996). Processing studies allow a better estimate of the consumer exposure to the residues (Uysal-Pala, & Bilisli, 2006). Studies into effects of storage and some commercial processing techniques on residues in food are a part of the registration requirements for pesticides in many countries. Reviewing the extensive literature showed that in most cases these steps lead to large reductions in residue levels in the prepared food, particularly through, washing, and peeling, fermentation, refrigeration and some other operations. The behaviour of residues in storage and processing can be rationalized in terms of the physico-chemical properties of the pesticide and the nature of the process. Recommendations are provided for the conduct of storage or processing studies on fate of pesticide residues in food so that data obtained is relevant, comparable and may be extrapolated to other situations (Holland et al., 1994).

Food processing techniques implies the set of methods and techniques used to transform raw ingredients into food or to transform food into other forms for consumption by humans or animals either in the home or by the food processing industry. The most common food processing techniques can aid in pesticide reduction. The techniques are described:

3.1 Baking

Baking is the technique of prolonged cooking of food by dry heat normally in an oven. It is primarily used for the preparation of bread, cakes, pastries and pies, tarts, and quiches. It is also used for the preparation of baked potatoes; baked apples; baked beans. For instance, commercially produced bread is an important component of every day diet in many countries. During bread making process, flour is subjected to biological (fermentation) and physical (baking) transformation (Sharma et al., 2005).

3.2 Dairy product manufacture

Milk and milk products form a main constituent of the daily diet. Butter, cheese and yoghurt are the popular dairy products (Abou-Arab, 1999). Hence it is important to study the effect of milk products manufacture on the concentration of pesticides and their metabolites.

3.3 Drying

Drying is the oldest method of preserving food. As compared with other methods, drying is quite simple. Food can be dried in several ways, for example, by the sun or in an oven or a food dryer can also be used. Drying has been found to reduce the pesticide residues considerably (Kaushik et al., 2009).

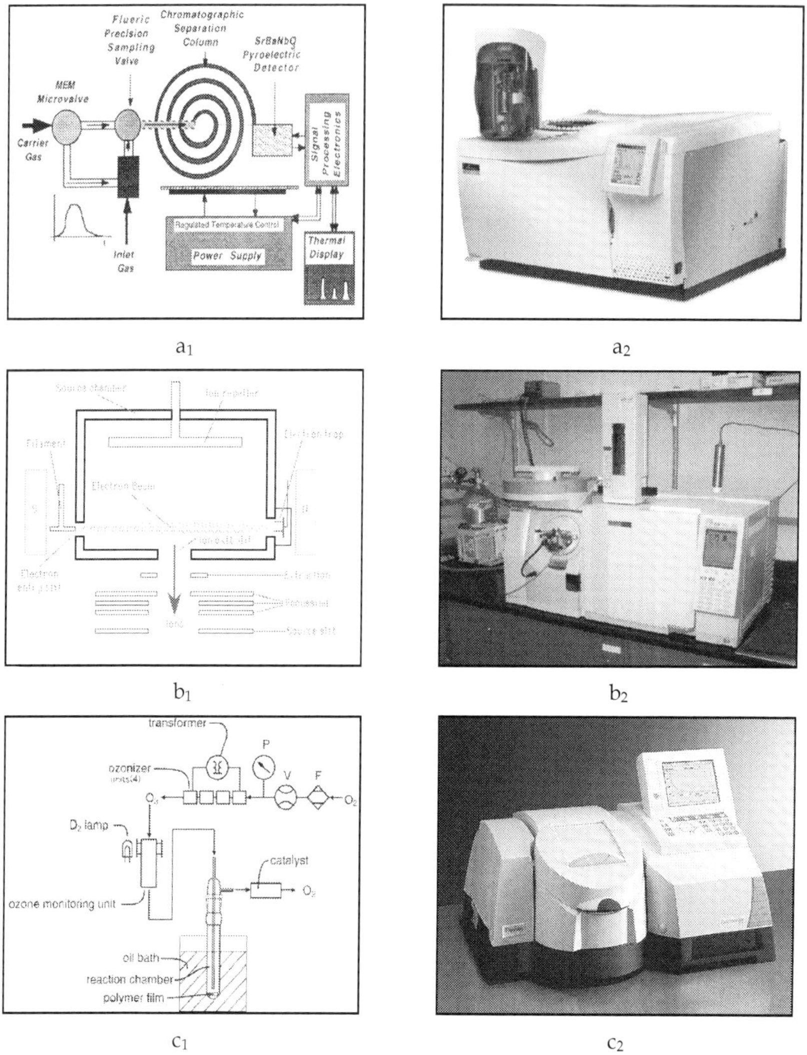

Fig. 1. The Gas chromatography (GC), Gas chromatography-Mass spectroscopy (GC-MS), UV-spectroscopy schematics and instruments: (a_1) GC components, (a_2) GC instrument, (b_1) GC-MS components, (b_2) GC-MS instrument, (c_1) UV-spectroscopy components, (c_2) UV-spectroscopy instrument.

3.4 Fermentation

Fermentation is a simple process during which the enzymes hydrolyze most of the proteins to amino acids and low molecular weight peptides; starch is partially converted to simple sugars which are fermented primarily to lactic acid, alcohol and carbon dioxide (Pardez-Lopez et al., 1991). Fermentation has been studied for reduction in pesticide residues.

In a fermentation method, the vegetables and fruits (e.g. cucumbers) are fermented in experimental, closed-top and dark fibreglass tanks. The surface halophilic and acid-tolerant bacteria (i.e. *Lactobacillus plantarum*) and fungi were as starter to establish lactic acid fermentation.

In a study, Fermentation of cucumbers lead to significantly lower amount of benomyl and mancozeb, compared with nonshrub cucumbers (Shokrzadeh & Saeedi Saravi, 2008).

3.5 Freezing and refrigeration

Freezing food is a common method of food preservation which slows both food decay and most chemical reactions (Kaushik et al., 2009). In refrigeration method, the samples which were subjected to cold storage procedure were kept at +4 °C in the refrigerator in polyethylene bags and darkly condition.

In a study it was found that when cucumbers contaminated at level of 2 L / 1000 L in water were refrigerated the reduction of residues were 5.3%, 22.4%, 43.7%, 52% and 68.2% after 2, 4, 6, 8 and 10 days with diazinon and 14.7%, 28.2%, 49.1%, 63% and 74.7% loss after these specific days with malathion, respectively (Dehghan et al., 2010).

3.6 Infusion

Tea and coffee are popular beverages throughout the world. A cup of tea that cheers can also be an important route of human exposure to pesticide residues. It is therefore important to evaluate the percent transfer of pesticide residue from dried (made) tea to tea infusion, as tea is subjected to an infusion process prior to human consumption (Jaggi et al., 2000).

In addition, this procedure can be occurred to reduce the heavy metal residues in agricultural products. For example, despite of a high level of heavy metals, such as lead and Cadmium in dried tea leaves, the possibility of the entry of these metals into the tea infusion was minimal. There was an excessive amount of lead and Cadmium in all dried leaf samples, but these levels were not found in the tea infusions, as the time of infusion was less than 1 hour, which was not sufficient to release metals from leaves (Shokrzadeh et al., 2008).

3.7 Juicing

Commercial juicing operations generally use whole fruit. The residue levels in juices from fruit will depend on the partitioning properties of the pesticide between the fruit skins/pulp and the juice. The pulp or pomace by-products, which often include the skin, retain a substantial proportion of lipophilic residues. Thus, moderately to highly lipophilic pesticides such as parathion, folpet, captan and synthetic pyrethroids are poorly transferred into juices and the residues are further reduced by clarification operation, such as centrifugation or filtering (Holland et al., 1994).

3.8 Malting

Malting is a process applied to cereal grains; it is a combination of two processes, notably germination and the kiln-drying process. The fate of pesticides was determined from barley to malt. The amount remaining after malting ranged from 13-51% for fenitrothion and nuarimol. The stages of malting steeping, germination and kilning contributed to loss of pesticide residues (Navarro et al., 2007).

3.9 Milling

The milling of grains substantially removes the pesticide residues. Most residues are present in the outer portions of the grain, and consequently levels in bran are consistently higher than in wheat. Even for the pesticides which can enter the grain by translocation, residues are higher in the bran than in the flour (Holland et al., 1994).

3.10 Peeling

Peeling is an important step in the processing of most fruits and vegetables. Chemical peeling (mostly lye peeling), mechanical peeling (mainly abrasion peeling), steam peeling and freeze peeling are conventional methods for peeling in the processing of fruits and vegetables (Toker & Bayindirli, 2003). A majority of the insecticides or fungicides applied directly to crops undergo very limited movement or penetration of the cuticle. It therefore follows that residues of these materials are confined to the outer surfaces where they are amenable to removal in peeling, operation. Peeling fresh fruits such avocado, bananas, citrus, cucumber, kiwifruit, mango and pineapple achieves virtually complete removal of residues from the fruit. For instance, removal of the skin, by peeling, leaves cucumber tissue below the waxy layer, therefore resulting in reduction of diazinon and malathion residues by more than 45.9% and 60.6% (Dehghan et al., 2010).

3.11 Storage

Grains are frequently stored long term (3-36 months) at ambient temperatures in bulk silos where insecticides may be applied post-harvest to reduce losses from storage pests (Holland et al., 1994; Joint Meeting on Pesticide Residues (JMPR), 1981). Grain based foods therefore have the potential to be a major source of residues in the diet for these insecticides. Studies on grain following post-harvest treatments with insecticides have generally shown that residues only decline rather slowly (Holland et al., 1994; Snelson, 1987). Residues of the more lipophilic materials tend to remain on the seed coat although a proportion can migrate through to the bran and germ which contain high levels of triglyceride (Anderegg & Madisen, 1983; Holland et al., 1994).

For example, a post-harvest interval (storage for 10 days) reduced diazinon in cucumbers by 94.7%. Overall disappearances of 97.8% of initial dose of malathion from cucumbers respectively were obtained (Dehghan et al., 2010).

3.12 Thermal processing (canning, cooking)

Application of heat to food commodities is commonly done through ordinary cooking, pressure cooking, microwave cooking, frying, sterilization and canning.

The commercial canning process in its various forms combines elements of washing, peeling, juicing, cooking and concentration.

Cooking is the act of preparing food for eating by the application of heat. It encompasses a vast range of methods depending on the customs and traditions, availability and the affordability of the resources. Literature is replete with work on effect of cooking on pesticide residues dissipation.

3.13 Washing

Washing is the most common form of processing which is a preliminary step in both household and commercial preparation. Loosely held residues of several pesticides are removed with reasonable efficiency by varied types of washing processes (Street, 1969). The effects of washing depend on location of the residue in cucumbers, water solubility of the pesticide, temperature and type of wash. Polar, water soluble pesticides are more readily removed than low polarity materials (Kaushik, et al., 2006; Elkins, 1989). The Samples were washed following two washing procedures, soaking and dipping in water (at temperature 25-30 °C for 10 minutes) and detergent-washing.

For example, washing of cucumbers by dipping in water for 10 minutes reduced diazinon and malathion residues to 81.8% and 65.6% of initial concentration. Also, 34.6% and 44.2% loss in residues obtained with detergent-washing. Thus, washing with detergent was more effective to reduce the pesticides residue than water-washing. The investigation indicated that washing with water and or detergent solutions were necessary to decrease the intake of pesticide residues (Dehghan et al., 2010).

3.14 Wine making

During the manufacture of wine in addition to the transfer of residues from the grapes into the must, stability of residues to the fermentation and fining processes are important factors. Fermentation on the skins as carried out in red wine production is likely to lead to higher residues in raw wine. Residues in must may be absorbed to the solids produced during fermentation and thus be lost in the fining processes. However, a range of pesticides with suitable solubility and stability can give rise to residues in wine (Holland et al., 1994). This process usually is less used in some Muslim societies and countries; because wine consumption is banned for Muslims, based on the religious principles and costumes.

4. Prevention

In assessing dietary exposure to the chronic toxicological effects of pesticides, most regulatory authorities consider some measure of typical food intake, such as mean or median food consumption values. But for compounds that might be acutely toxic, it is important to know if the dietary intake over a relatively short period of time (such as a day) is safe. By examining exposure at such an extreme, acute assessment protects the safety of people who ingest more pesticide residues than virtually anyone else in the population.

The risk from pesticide residues is evaluated on the basis of two toxicological limit values: the ADI and the ARfD. ADI stands for 'Acceptable Daily Intake'. It indicates the amount of a substance which can be ingested daily over a lifetime by consumers without any appreciable health risk. The ADI is used to assess the chronic risk and the Acute Reference Dose (ARfD) to assess the acute risk. The ARfD is a comparatively new risk assessment tool. It is defined as the amount of a substance which a consumer can ingest from one meal or several meals spread over the day without any appreciable health risk (Whitford et al., 2006).

The chapter emphasizes the fact that the advantages associated with the application of pesticides in enhancing the agricultural productivity must be weighed against the possible health hazard arising from the toxic pesticide residues in food. First and foremost the application of pesticides should be in compliance with good agricultural practices, using only the required amounts. Further the current shift in world opinion from 'chemical farming' towards 'organic farming' is a sustainable approach to minimize the damage posed by widespread contamination of environment by pesticides. However, the challenge lies in achieving food safety in developing countries where the indiscriminate application of pesticides results in the presence of residues in food commodities. However, due to several socio-cultural and technical reasons, diffusion and acceptance of this approach among the farming community in developing countries like India has been very slow. Hence, it becomes urgent and important in the transient phase that some pragmatic solution should be developed. So in the transitory phase it is important to address the concern of food safety through suitable processing techniques and appropriate storage period that enhance food safety even in developing countries especially for the poor populace which cannot afford the expensive organic food. In this background common and simple processing techniques acquire significance for reducing the harmful pesticide residues in food. The effects of processing pesticide residues in food is an area where available information should be consolidated and missing information needs to be obtained through further research.

The following suggestions are presented for safety improvement along with daily progress of industries and industrial pollution:

- Initial focus on dietary risks and food chain;
- Determination of hazard effects of pesticides on human and animal health;
- Determination of standard limit values, such as ADI and ARfD;
- Frequent measurement of pesticide residues in agricultural products;
- Attention shifts to occupational risks;
- Public scrutiny over residential risks;
- Early federal laws (landscape and grounds management policy, pesticide reduction strategies);
- Study requirements and scientific education (to farmers, industries, etc.);
- Determination of testing guidelines;
- Establishment of reference laboratory equipped with analytical apparatuses, such as GC, GC-MS, HPLC[13], etc;
- Optimization the food processing techniques (with regard to pesticide residue dissipation and nutrient content);
- Informing people about the reductive effects of house holding procedures (i.e. storage, washing, peeling, fermentation) on pesticide residues in agricultural products;
- Use of biological pest controls (such as pheromones and microbial pesticides), genetic engineering, and methods of interfering with insect breeding, as safer alternatives to chemical pesticides.

Overall, the chapter concerns to determine pesticide residue levels in agricultural products and compare it with standard limit values. The safety assessments can manifest the status of pesticide contamination in agricultural products, which is affected by water, soil and air pollution, and lead to apply prevention methods supported by governments (Shokrzadeh et al., 2009).

---

[13] high performance liquid chromatography

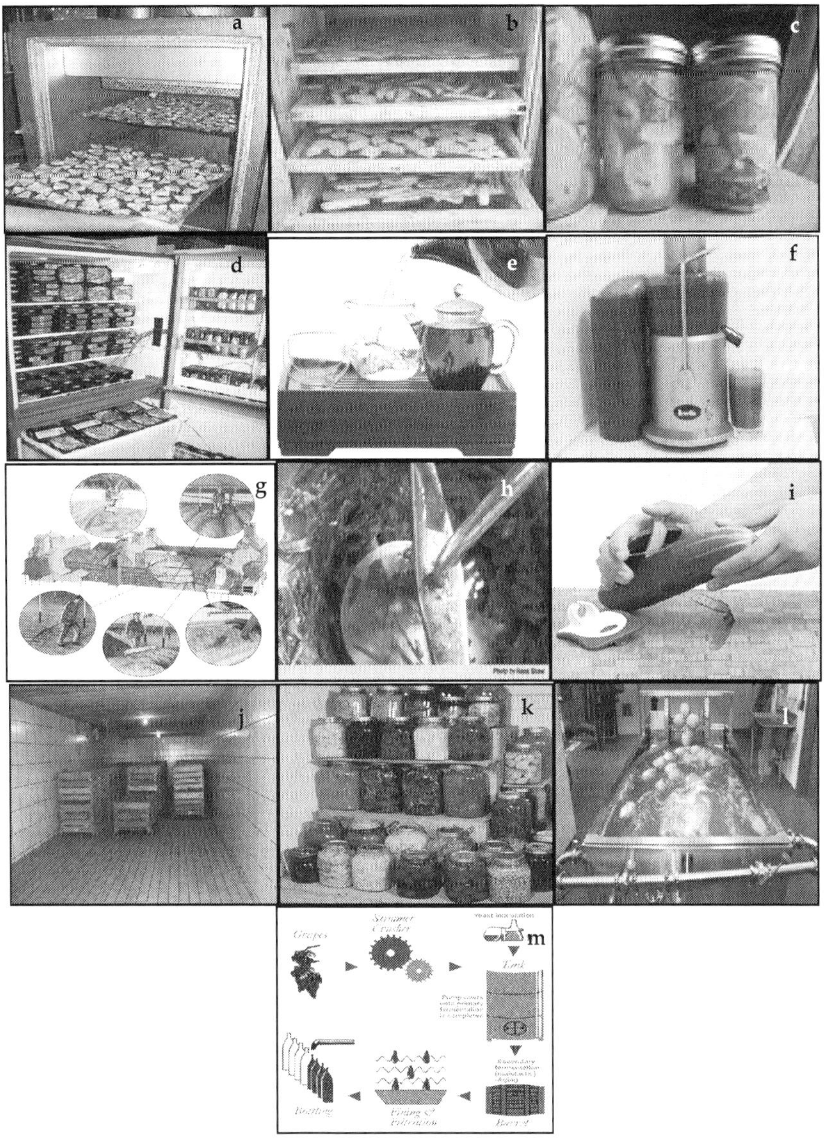

Fig. 2. Different processes for reduction of pesticide residues in agricultural products: (a) baking, (b) drying, (c) fermentation, (d) freezing, (e) infusion, (f) juicing, (g) malting, (h) milling, (i) peeling, (j) storage, (k) canning, (l) washing, (m) wine making.

5. References

Abou-Arab, A.A.K. (2002). Degradation of organochlorine pesticides by meat starter in liquid media and fermented sausage. *Food and Chemistry Toxicology* 40: 33-41.

Anderegg, B.M. & Madisen, L.G. (1983). *Journal of Economic Entomology* 76: 733–736.

Dehghan, A., Ghorbani, M., Maghsoudlou, Y., Saeedi Saravi, S.S., Babaee, Z.A., Shokrzadeh, M. (2010). Effects of Washing, Peeling, Storage and Fermentation on Residue Contents of Malathion and Diazinon in Cucumbers Grown in Greenhouses. *Food Chemistry*, submitted.

Dikshit, A.K., Pachaurý, D.C. & Jindal, T. (2003). Maximum residue limits and risk assessment of betacyfluthrin and imidacloprid on tomato. *Bulletin of Environment and Contamination Toxicology* 70: 1143-1150.

Elkins, E.R. (1989). Effect of commercial processing on pesticide residues in selected fruits and vegetables. *Journal of the Association of Official Agricultural Chemists* 72(3): 533–535.

Food and Agriculture Organization of the United Nations (2002). International Code of Conduct on the Distribution and Use of Pesticides, 2007.

Hackenberg, D. (2007). Letter from David Hackenberg to American growers from March 14, 2007. Plattform Imkerinnen-Austria.
URL: *WWW.imkerinnen.at.*

Haefeker, W. (2000). Betrayed and sold out-German bee monitoring.
URL: *WWW.beekeeping.com.*

Holland, P.T., Hamilton, D., Ohlin, B. & Skidmore, M.W. (1994). Effects of storage and processing on pesticide residues in plant products. IUPAC Reports on Pesticides (31). *Pure and Applied Chemistry* 66(2): 335-356.

Jaggi, S., Sood, C., Kumar, V., Ravindranath, S.D. & Shanker, A. (2000). Loss of quinalphos during tea processing. *Pestology* 24(12): 42-46.

Joint Meeting on Pesticide Residues (JMPR) (1981). Monographs of toxicological evaluations monograph 011. Delta.

Kamrin, M.A. (1997). *Pesticide Profiles: toxicity, environmental impact, and fate.* CRC Press.

Kaushik, G., Satya, S., Naik, S.N. (2009). Food processing a tool to pesticide residue dissipation-A review. *Food Research International* 42: 26-40.

Jeyaratnam, J. (1990). Acute pesticide poisoning: a major global health problem. *World Health Stat Q* 43 (3): 139-144.

McCauley, L.A., Anger, W.K., Keifer, M., Langley, R., Robson, M.G. & Rohlman, D. (2006). Studying health outcomes in farm worker populations exposed to pesticides. *Environmental Health Perspectives* 114 (3): 953-960.

Miller, G.T. (2004), *Sustaining the Earth*, 6th ed. Thompson Learning, Inc. Pacific Grove, California.

Navarro, S., Perez, G., Navarro, G. & Vela, N. (2007). Decline of pesticide residues from barley to malt. *Food Additives and Contaminants* 24(8): 851–859.

Palmer, W.E., Bromley, P.T. & Brandenburg, R.L. (2007). *Wildlife & pesticides-Peanuts*. North Carolina Cooperative Extension Service.

Pardez-Lopez, O., Gonzales-Casteneda, J. & Carabenz-Trejo, A. J. (1991). Influence of solid substrate fermentation on the chemical composition. *Journal of Fermentation and Bioengineering* 71: 58-62.

Petersen, B., Tomerlin, J.R. & Barraj, L. (1996). Pesticide degradation: Exceptions to the Rule. *Food Technology* 50(5): 221-223.

Pihlström, T., et al. (2009). Method validation and quality control procedures for pesticide residues analysis in food and feed.

Processing studies (Appendix E) (1997). Commission of the European Communities, Directorate General for Agriculture, 7035/VI/95rev.5.

Rell M.J., Galvin J.R. (2008). Safe Application Of Pesticides And Herbicides. Connecticut Department of Public Health.
URL: WWW.ct.gov/dph.

Rockets, R. (2007). *Down On The Farm? Yields, Nutrients And Soil Quality.*
URL: *WWW.Scienceagogo.com.*

Sharma, J., Satya, S., Kumar, V. & Tewary, D.K. (2005). Dissipation of pesticides during bread making. *Journal of Chemical Health and Food Safety*, 17-22.

Shokrzadeh, M., Saeedi Saravi, S.S. & Zehtab Yazdi, Y. (2009). Lindane residues in cultivated cucumber and in the most consumed fish in Caspian Sea (Iran). *Toxicology and Industrial Health* 25(8): 517-523.

Shokrzadeh, M. & Saeedi Saravi, S.S. (2008). Measurement of residues of benomyl and mancozeb pesticides in shrub and nonshrub cucumbers sampled from different regions of Mazandaran province (Iran). *Toxicological and Environmental Chemistry* 91(4): 627-630.

Shokrzadeh, M., Saberyan, M. & Saeedi Saravi, S.S. (2008). Assessment of lead (Pb) and cadmium (Cd) in 10 samples of Iranian and foreign consumed tea leaves and dissolved beverages. *Toxicological and Environmental Chemistry* 90(5): 879-883.

Shokrzadeh, M. (2007). *Veterinary toxicology*. Institute of Scientific and Practical University, Jahad Keshavarzi Publication.

Snelson, J.T. (1987). Grain protectants. ACIAR monograph series (Australia) (Vol. 3, pp. 448 p). Australian Centre for International Agricultural Research, Canberra, Australia.

Street, J.C. (1969). Methods of removal of pesticide residues. *Canadian Medical Association Journal* 100: 154–160.

Toker, I. & Bayindirli, A. (2003). Enzymatic peeling of apricots, nectarines and peaches. *Lebensmittel-Wissenschaftund-Technologie* 36(2): 215–221.

Uysal-Pala, C. & Bilisi, A. (2006). Fate of endosulfan and deltamethrin residues during tomato paste production. *Journal of Central European Agriculture* 7(2): 343-348.

Wells, M. (2007). Vanishing bees threaten U.S. crops.
URL: *WWW.bbc.co.uk.*

Whitford, F., Kronenberg, J., Lunchick, C., Driver, J., Tomerlin, R., Wolt, J., et al. (2006). Pesticides and Human Health Risk Assessment, Purdue Pesticide Program, Purdue University Cooperative Extension Service.

Winteringham, F.P.W. (1971). Some global aspects of pesticide residue problems, *Israel Journal of Entomology* 6(2): 171–181.

Zeissloff, E. (2001). Schadet imidacloprid den bienen (in German). URL: *WWW.beekeeping.com.*

13

Pesticide Residues in Fruits and Vegetables

B. M. Keikotlhaile, and P. Spanoghe
Ghent University
Belgium

1. Introduction

The aim of this chapter is to describe the presence of pesticide residues in fruits and vegetables, mainly how they are introduced, dissipated, degraded, affected by food processing techniques and their risk assessment.

Fruits and vegetables are important components of the human diet since they provide essential nutrients that are required for most of the reactions occurring in the body. A high intake of fruits and vegetables (five or more servings per day) has been encouraged not only to prevent consequences due to vitamin deficiency but also to reduce the incidence of major diseases such as cancer, cardiovascular diseases and obesity. Like other crops, fruits and vegetables are attacked by pests and diseases during production and storage leading to damages that reduce the quality and the yield. In order to reduce the loss and maintain the quality of fruits and vegetables harvest, pesticides are used together with other pest management techniques during cropping to destroy pests and prevent diseases. The use of pesticides have increased because they have rapid action, decrease toxins produced by food infecting organisms and are less labour intensive than other pest control methods. However, the use of pesticides during production often leads to the presence of pesticide residues in fruits and vegetables after harvest.

The presence of pesticide residues is a concern for consumers because pesticides are known to have potential harmful effects to other non-targeted organisms than pests and diseases. The major concerns are their toxic effects such as interfering with the reproductive systems and foetal development as well as their capacity to cause cancer and asthma (Gilden et al, 2010). Some of the pesticides are persistent and therefore remain in the body causing long term exposure. The concern has led to governments setting up monitoring systems in order to assess the safety situation and make informed decisions when passing legislation.

2. Pesticides fate after application to fruits and vegetables

Fate refers to the pattern of distribution of an agent, its derivatives or metabolites in an organism, system, compartment or (sub) population of concern as a result of transport, partitioning, transformation or degradation (OECD, 2003). After pesticides are applied to the crops, they may interact with the plant surfaces, be exposed to the environmental factors such as wind and sun and may be washed of during rainfall. The pesticide may be absorbed by the plant surface (waxy cuticle and root surfaces) and enter the plant transport system (systemic) or stay on the surface of the plant (contact). While still on the surface of the crop, the pesticide can undergo volatilization, photolysis chemical and microbial degradation. These processes are illustrated in Figure 1. All these processes can reduce the original pesticides concentration but can also introduce some metabolites in the crops.

Volatilisation of the pesticide usually occurs immediately after application in the field. The process depends on the vapour pressure of the pesticide. Pesticides with high vapour pressure tend to volatilize rapidly into the air while those with low vapour pressure remain longer on the surface. Volatilization rate also depends on the environmental factors such as wind speed and temperature. The faster the wind speed and the higher the temperature the more the pesticide will evaporate. Photolysis occurs when molecules absorb energy from the sunlight resulting in pesticide degradation. The indirect reaction can also be caused by some other chemicals being broken by the sunlight and their products reacting with pesticides in turn. Some pesticides may be degraded by microbial metabolism. Micro-organisms can use pesticides as nutrients thereby breaking them into carbon dioxide and other components (Holland and Sinclair, 2004). Because of difference between naturally occurring organic chemicals and pesticide structures, they cannot be assimilated by the microbes but they may be altered at reactive sites. The products formed may be less or more toxic than the parent chemical.

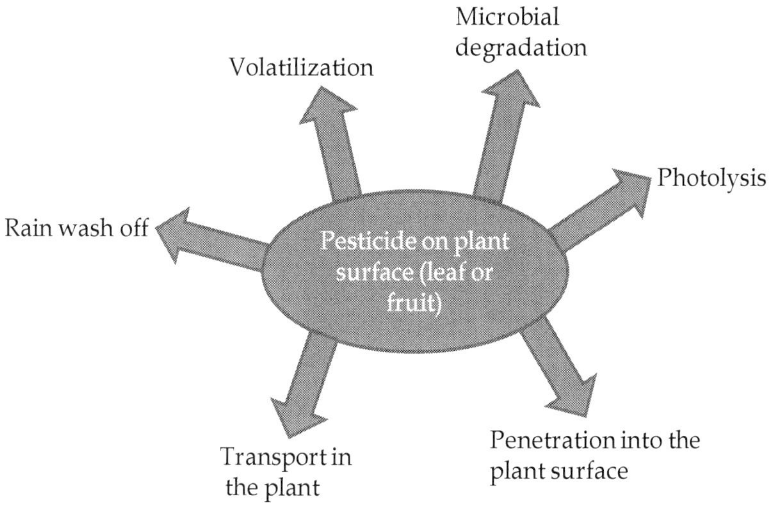

Fig. 1. Fate of pesticides in plant surfaces chemical

Although degradation of pesticides is influenced by different environmental processes, Celik et al, (1995) concluded that under natural field conditions volatilization is the main process that affects pesticides. These researchers applied six pesticides (azinphos-methyl, ethion, diazinon, methidathion, phosalone and pirimicarb) to apples and found that volatilization was the dominant process followed by solar irradiation. Bacterial degradation had the lowest influence except for phosalone. Pirimicarb was highly degraded by solar irradiation. Rain wash off can also be very important when it occurs shortly after application.

3. Monitoring

The purpose of pesticide monitoring programs is to ensure that in fruits and vegetables do not exceed maximum residues levels (MRLs) allowed by the government, no misuse of pesticides that could result in unexpected residues in food and that good agricultural practices (GAP) are maintained. Some programmes, mostly in developing countries, are carried out due to the demands by international trade. The results from these monitoring programmes are also used by regulatory bodies for future developments in setting MRLs and risk assessment exercises for public health. In most countries, the monitoring programs are organised by a single agency designated as the competent authority. The agency designs a monitoring plan based on the previous data available from dietary consumption and risk assessment exercises or pesticide usage in the available fruits and vegetables. In the European Union (EU) there is a coordinated programme for all the member countries to follow from the European commission and the member states national programs. The results are then yearly as a single report by the European Food Safety Authority (EFSA).

In the case of international trade, the monitoring plan is also influenced by the trading partners. For example, partners trading with the EU (normally referred to as third countries) have to incorporate EU standards to their food control programmes. In addition to monitoring, the agencies can engage in follow up sampling (enforcement actions) where some discrepancies had been observed. Laboratories carrying out pesticide residues analysis should be accredited or have started accreditation procedures to some quality standard. Pesticide legislation in developing countries is generally lacking or not implemented and this also affects pesticide monitoring since it relies on legislation to be effective (Ecobichon 2001). The other problem is lack of trained personnel to enforce laws and monitor the use of pesticides and residue levels in food and the environment. However, pesticide monitoring in some developing countries with high agricultural output is driven by international trade. Failure to adhere to trade standards can result in a loss of revenue for the population supported by the affected agricultural industry. This can be illustrated by using the Kenya's green bean farmers. These Kenyan bean farmers implemented developed country pesticide standards and are required by the UK retailers to show evidence of compliance with UK pesticide legislation (Okello 2001). In the same study, it was also noted that since the 1990s the arrangement saw Kenya increasingly becoming one of the leading countries in green bean production and supplier to developing countries. This also saw the benefits of reduced pesticide related cost of illnesses and incidences of acute symptoms of pesticide exposure in monitored farmers than compared to unmonitored farmers. This was attributed to the education the farmers received about the use and handling of pesticide as well as adhering to protective measures.

4. Maximum residue level

Maximum residue levels are the highest levels of residues expected to be in the food when the pesticide is used according to authorised agricultural practices (EFSA 2010). The MRLs are always set far below levels considered to be safe for humans. It should be understood that MRLs are not safety limits, a food residue can have higher level than MRL but can still be safe for consumption. Safety limits are assessed in comparison with acceptable daily intake (ADI) for short term exposure or acute reference dose (ARfD). MRLs are subject to legal requirements in most of the countries. In developed regions like Europe the

responsibility of the legislation is lead by the European Commission (EC) with input from the member states, EFSA and the standing committee on the Food Chain and Animal Health. In the US, the leading agency is Environmental Protection Agency (EPA) with input from the United States Department of Agriculture (USDA) and the Scientific Advisory Panel while in New Zealand the leading agency is the New Zealand Food Safety Authority (NZFSA) with input from the Environmental Risk Management Authority.
MRL setting can be the responsibility of one or more authorities in a country and normally involves the health, agriculture and environmental agencies. MRL enforcement can be a responsibility of one or more agencies and may also depend on different food types. MRL setting is based on the national registered good agriculture practice (GAP) data combined with the estimated likely residue from the supervised trials mean residue (STMR), ADI and ARfD. The information is then evaluated by the risk assessment agency like EFSA in EU or JMPR for CODEX Alimentarius. The JMPR procedure is shown in Figure 2. Where national or regional MRLs are not available, internationally recognised bodies such as the United Nations Codex Alimentarius Commission MRLs can be used as guidance. MRLs are generally published in open literature or websites of the regulatory bodies for public usage. MRLs may be exceeded because of pesticide misuse, false positives due to naturally occurring substances, differences in national MRLs, lack of registered pesticides and incorrect pesticide application (EFSA, 2010).

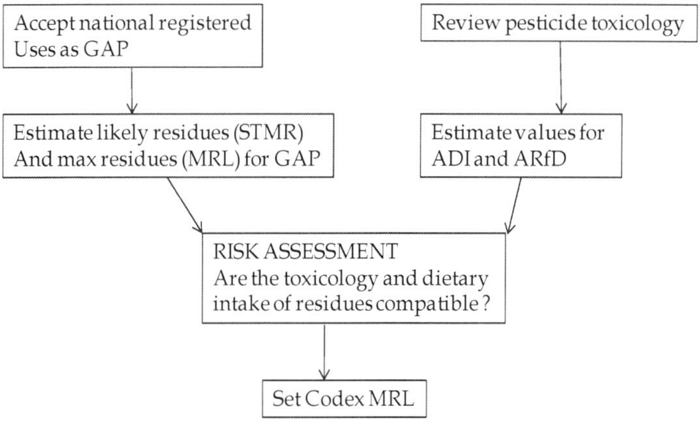

GAP - Good Agriculture Practice
STMR - Supervised Trials Mean Residue

Fig. 2. Procedure for setting JMPR MRLs

The emerging trend is to harmonize MRL in each region or globally and is highly supported by international organisations such as FAO, WHO, CCPR and OECD. In the EU the MRLs are already harmonised as from the beginning of September 2008 under the new regulation EC No. 396/2005 (OECD 2010). In developing regions like Africa, efforts were initiated under the Global MRL Harmonization Initiative - Africa Project that was supported by US Department of Agriculture - Foreign Service, IR-4 Project and USEPA. A summary of the questionnaire from the same project indicated that most of the African countries have adopted the CODEX MRLs with South Africa establishing some of its own in addition (Anonymous 2009).

5. Food processing

Fruits and vegetables like other foods pass through culinary and food processing treatments before they are consumed. The effects of these culinary and food processing techniques have been investigated by various researchers and they have been found to reduce the pesticide residue levels except in cases where there is concentration of the product like in juicing frying and oil production. Some toxic metabolites may be produced during processing treatments, especially thermal processing. One of the extensively studied metabolite is ETU that result from thermal processing of dithiocarbamates. However, the consumers can still be encouraged to employ those processing methods that reduce pesticide residues.

Food processing studies often results in transfer factors or food processing factors (PF) of the pesticide residue in the transition from raw agriculture commodity to the processed product. These processing factors are expressed as the concentration of pesticide after processing divided by the concentration before processing. Some processing factors are available in public literature while others are only available from the pesticide registering bodies. Processing studies have become a part of pesticide registration requirements. Effect of processing in fruits and vegetables are said to be influenced by the physico-chemical properties of the pesticide as well as the processing method (Holland et al., 1994).

Processing factors for a particular processing technique and a group of pesticides are not easily available in literature. These become important when researchers want to perform risk assessment for a group of pesticide in the population. An example can be illustrated by risk assessment of exposure of organophosphorus pesticides in the Dutch diet (Boon et. al, 2008). The authors could not find the general processing factor for a group of organophosphorus pesticide. However, they managed to derive the general processing factors for washing (0.76), peeling (0.44) and canning(0.74) for fruits and vegetables. The authors could not find the general processing factor for a group of organophosphorus

Processing	R*	95% CI	99.5% CI
Baking	1.38	0.91 -2.09	0.76 - 2.51
Blanching	0.21	0.10 - 0.44	0.07 - 0.61
Boiling	0.82	0.58 - 1.15	0.50 - 1.33
Canning	0.71	0.46 - 1.09	0.38 - 1.31
Frying	0.1	0.02 - 0.46	0.01 - 0.90
Juicing	0.59	0.32 - 1.09	0.24 - 1.42
Peeling	0.41	0.30 - 0.54	0.27 - 0.61
Washing	0.68	0.52 - 0.82	0.52 - 0.89

R* - processing factor
CI - confidence interval

Table 1. Average processing factors for different processing methods

pesticide their risk assessment, however, they managed to derive the general processing factors for washing (0.76), peeling (0.44) and canning(0.74) for fruits and vegetables. We attempted to summarize the processing factors for fruits and vegetables according to different processing methods using meta-analysis (Keikotlhaile et al, 2010). The results are shown in Table 1. However, the results were generalized and we recommended that the same procedure could be used for a group pesticides applied to similar vegetables for more refined processing factors.

6. Risk assessment

Risk assessment of chemicals is described as a process intended to calculate or estimate the risk to a given target organism, system or (sub) population, including identification of attendant uncertainties, following exposure to a particular agent taking into account the inherent characteristics of the agent of concern as well as the characteristics of the specific target system (OECD 2003). Risk assessment process includes four steps: hazard identification, hazard characterisation (dose-response assessment), exposure assessment and risk characterisation. In that context the risk assessment of pesticide residues in fruits and vegetables is tackled.

6.1 Hazard identification

Hazard identification is the first step in risk assessment and it involves the identification of the type and the nature of adverse effects that an agent has as inherent capacity to cause in an organism, system or (sub) population (OECD 2003). Recent regulations require that hazard identification be performed before a pesticide can be approved for usage in agriculture or other areas. Therefore the information on hazards posed by pesticides is readily available from the pesticide registering bodies and on their websites for public usage. Most of the information is also available from international organisations such as JMPR, OECD and EC. The hazards that have been identified concerning pesticide include reproductive and endocrine disruption, neurodevelopmental delays, immune system, cancer and respiratory distress (Gilden 2010). Studies are carried out in test organisms (microbial, cells or animals) and the exposure level is increased until an adverse effect is produced. The highest dose of the pesticide that does not cause detectable toxic effects on the test organisms is called the no-observed-adverse-effect-level (NOAEL) and is expressed in milligrams per kilogram of body weight per day (WHO 1997)). This is important because it is used in calculation of the ADI or the ARfD.

6.2 Hazard characterisation

Hazard characterisation is the qualitative and, wherever possible, quantitative description of the inherent properties of an agent or situation having the potential to cause adverse effects. This should, wherever possible, include a dose response assessment and its attendant uncertainties (OECD 2003). Hazard characterisation involves comparing the pesticide exposure concentration with the ADI or the ARfD. The ADI is the estimate of the amount of a substance in food (mg/kg body weight/day) that can be ingested daily over a lifetime without appreciable health risk to the consumer (WHO 1997). ADI is calculated by dividing the NOAEL for animal studies with an uncertainty factor of 100 to convert to a safe level for humans. A factor 100 (10 x 10) mostly used to account for species differences and individual variability in sensitivity to the chemicals (Renwick 2002). ARfD is the estimate of the amount of a substance in food that can be ingested over a short period of time, usually during one meal or one day, without appreciable health risk to the consumer (WHO 1997).

6.3 Exposure assessment

Evaluation of the concentration or the amount of a particular agent that reaches a target organism, system or (sub) population in a specific frequency for defined duration (OECD 2003). The potential intake or consumption of pesticide residues is divided by the body weight and compared to ADI or ARfD in exposure assessment.

Exposure = (Concentration of pesticide residue x Food consumed)/ body weight

The input data used in exposure assessment comes from supervised field residue trials, national pesticide monitoring programs and food consumption surveys. The residue levels from pesticide monitoring programs mighty not cover the whole food supply but they are always available in most countries and they reflect samples available for consumers. However, targeted sampling data may over-estimate exposure because it is biased against suspect samples.

6.3.1 Consumption data

Food consumption data are essential component of dietary risk assessment. The data used depend upon the type of population being assessed: children, special ethnic groups, geographical regions and estimation of the quantity of food eaten. Food consumption data may be obtained during food supply surveys (food balance sheets), household inventories, household food use and individual food intake surveys (Hamilton, 2004). According to EFSA guidance document on collection of food consumption data (EFSA, 2009), there are four types of dietary assessment methods, namely: diet history, food frequency, dietary records and dietary recall. In diet history, the history of the whole daily food intake of an individual and the usual meal pattern is assessed over a period of days, months and up to one year. Food frequency involves asking the consumers to estimate the usual frequency of consumption during a specified time for the foods that are listed on the questionnaire. In dietary records, the consumers weigh and record all the food including beverages before eating and also the leftovers after eating. Dietary recall involves asking the consumers to recall the actual food intake for the past 24 or 48 hours or previous days. The quantities are described using household measures, food models or photographs. The most common dietary recall method is the 24-hour recall. The methods that are suitable for both acute and chronic risk assessment are dietary records and dietary recall.

The most appropriate source is the one that measures actual consumption instead of available food supply. Average daily consumption is the most used in exposure assessment calculations, however there are others such as percentile consumption values, average consumption (weekly, monthly, etc) and long term consumption habits. The latter is mostly important in calculation of chronic exposure. In cases where national food consumption data are not available, food balance sheets from FAO can be used even though they might be too generalised.

6.3.2 Dietary exposure models

Dietary intake exposure models are mainly conducted in deterministic and probabilistic assessment. Deterministic exposure assessment is based on single point estimate, usually the mean or worst case scenario (97.5 percentile). Probabilistic exposure assessment is based on the probability of occurrence of the risk and results in a distribution of risk values. Deterministic exposure is generally used as a low tier approach to determine whether there is a course for concern for the defined exposure. It is easy to perform and requires less time to complete. The disadvantage is that it gives single estimate of the risk and does not give an insight of other possible risks for lower levels. Therefore it does not contain information about variability in potential exposure to the exposed population. Probabilistic assessment is based on simulations of potential exposures using computer software and allows more inputs to come up with the final exposure. Most of these distributional models are based on Monte Carlo simulations and are referred to as Monte Carlo models (Hamilton, 2004). These distributional models provide a range of risks throughout the population distribution and provide quantitative information about variability and uncertainty. The disadvantage is that they require time and resources for additional data generation. A brief overview is outlined by Hamilton et al., (2004). Since deterministic models gives an over-estimated exposure assessment by assuming all time consumption of higher concentration the pesticide a more realistic approach of probabilistic assessment is preferred when resources allow.

6.4 Risk characterisation

The qualitative and, wherever possible, quantitative determination, including attendant uncertainties, of the probability of occurrence of known and potential adverse effects of an agent in a given organism, system or (sub) population, under defined exposure conditions (OECD 2003). The international estimate daily intake (IEDI) has been used to characterise the risk of pesticides. It is expressed as:

$$IEDI = \sum STMR \times E \times P \times F$$

Where
STMR = supervised trial median residue level
E = Edible portion
P = processing factor
F = consumption of the food commodity
When the IEDI is more than the ADI the food involved is considered a risk to the concerned consumers. For the national estimated short term intake (NESTI), the risk characterisation is compared with the ARfD.

7. Future work

In pesticide residues research, future work involves mainly the improvement of risk assessment of dietary exposure methods and harmonisation of data collection in as many countries as possible. The methods are also aimed at incorporating all the factors that contribute to exposure assessment in the final model predictions so that it can be realistic. Common methods of dietary exposure assessment were based on deterministic calculations and those have been found to have short comings of only providing exposure for average consumers while excluding higher consumers. The most preferred method is the probabilistic risk assessment since it considers all exposure throughout the entire consumer distribution. Recently risk assessment studies have focused on simultaneous exposure to multiple pesticides instead of only on a single pesticide (Van Klaveren and Boon, 2009). In their paper, the authors discuss the risk trade-offs, risk benefits and the use of integrated probabilistic risk assessment model (IPRA). The model integrates exposure and health effect modelling while incorporating variability and uncertainty.

8. References

Anonymous (2009). Questionnaire Summary: Global MRL Initiative - Africa. Alexandria, Egypt.09.09.2010, available from Http: ir4.rutgers.edu/GMUS/MRLworkshop.htm

Boon, P.E.; Van der Voet, H.; Van Raaij, M.T.M. & Van Klaveren, J.D. (2008). Cumulative risk assessment of the exposure to organophosphorus and carbamates insectidies in the Dutch diet. *Food and Chemical Toxicology*, 46(9), 3090 - 3098

Celik, S.; Kunc, S. & Asan T. (1995). Degradation of some pesticides in the field and effect of processing. *Analyst*, 120, 1739 - 1743

Ecobichon, D.J. (2001). Pesticide use in developing countries. *Toxicology*, 160, 27 - 33

EFSA (2009). General principles for the collection of national food consumption data in the view of a pan-Europe dietary survey. *EFSA Journal 2009*, 7(12), 1435

EFSA (2010). 2008 Annual report on pesticide residues according to article 32 of regulation (EC) No 396/2005. *EFSA Journal 2010*, 8(6), 1646

Gilden, R. C.; Huffling, K. & Sattler B. (2010). Pesticides and Health Risks. *JOGNN*, 39, 103 - 110

Hamilton, D.; Ambrus, A.; Dieterle R.; Felsot, A.; Harris, C.; Petersen B.; Racke, K.; Wong, S.; Gonzalez R.; Tanaka, K.; Earl, M.; Roberts G. & Bhula, R. (2004). Pesticide residues in food - acute dietary exposure. *Pest Management Science*, 60, 311 - 339

Holland, J. & Sinclair, P. (2004). Environmental fate of pesticides and the consequences for residues in food and drinking water, in: *Pesticide Residues in Food and Drinking Water: Human Exposure and Risks*, Hamilton D. & Crossley S. (Ed.), 27 - 62, John Willey &Sons LTD, 0-471-48991-3, West Sussex, England

Holland, P.T. (1994). Effects of storage and processing on pesticide residues in plant products. *Pure and Applied Chemistry*, 66, (2), 335 - 356

Keikotlhaile, B.M.; Spanoghe P. & Steurbaut W. (2010). Effects of food processing on pesticide residues in fruits and vegetables: a meta-analysis approach. *Food and Chemical Toxicology*, 48, 1 - 6

OECD (2003). Descriptions of selected key generic terms used in chemical hazard/risk assessment. *OECD Series on Testing and Assessment*, No. 44

OECD (2010). OECD survey on pesticide maximum residue limit (MRL) policies: survey results. *OECD Series on Pesticides*, No. 51

Okello, J.J. & Swinton, S.M. (2010). From circle of poison to cicle of virtue: pesticides, export standards and kenya's green bean farmers. *Journal of Agriculture Economics*, 61, 209 - 224

Renwick, A.G. (2002). Pesticide residue analysis and its relationship to hazard characterisation (ADI/ARfD) and intake estimations (NEDI/NESTI). *Pest Management Science*, 58, 1073 - 1082

Van Klaveren, J.D. & Boon, P.E. (2009). Probabilistic risk assessment of dietary exposure to single and multiple pesticide residues or contaminants: summary of the work performed within the SAFE FOODS project. *Food and Chemical Toxicology*, 47, 2879 - 2882

WHO (1997). *Guidelines for predicting dietary intake of pesticide residues*. WHO, Switzerland.

14

A Risk Assessment Study of Greek Population Dietary Chronic Exposure to Pesticide Residues in Fruits, Vegetables and Olive Oil

Ioannis N. Tsakiris[3], Maria Toutoudaki[1], Manos Kokkinakis[2], Mitlianga Paraskevi[3], and Aristides M. Tsatsakis[1],
[1]*Laboratory of Toxicology, Department of Medicine, University of Crete, Voutes 71409, Heraklion*
[2]*Technological Education Institute (T.E.I) of Crete, Food Quality Laboratory, K. Palama and I. Kakridi Str., 722 00 Ierapetra, Crete*
[3]*TEI of Western Macedonia, Florina Branch, Department of Agricultural Products Management and Quality Control, Terma Kontopoulou 53100 Florina*
Greece

1. Introduction

The term Mediterranean diet refers to the diet regime adopted by people living around the Mediterranean Sea, and especially Greeks and Cretans. This diet regime allows the daily consumption of bread, cereals, pasta, rice, potatoes, pulses, nuts, vegetables, milk, cheese, yogurt, olives and olive oil. Consumption of eggs, chicken, fish and sweets is allowed a few times weekly, while consumption of red meat is allowed only a few times monthly. Moderate consumption of wine is allowed daily with the meals. Overall, the Mediterranean diet is characterized by low consumption of fats, especially saturated fats and a high consumption of carbohydrates that come from cereals, bread, rice, pasta, fruits and vegetables. Milk products are consumed regularly in small quantities.

One of the major diet components, especially in the summer, is grapes. They are characterized by high nutritional value as they supply several antioxidants and polyphenolic compounds to human organism. Many published papers report high concentrations of phenolic compounds such as flavonoids and resveratrol, especially in the red varieties (Dani et al., 2007).

The main source of fat is olive oil, which is rich in monounsaturated fatty acids. As already mentioned fruits and vegetables are major constituents of the Greek diet, and although high consumption of these components is thought to be beneficial for the human organism, it is also a source of exposure to the various pesticides used for the protection of the crops (Coxam & Puel, 2010) (Roche et al., 2000).

Many insects and pathogens (i.e. *Botrytis cinerea, Uncicula necator* and *Plasmopara viticola, Dacus olea*) affect the aforementioned cultivations and their occurrence demands the use of pesticides in order to eliminate economic losses. The key issues during crop protection planning are to use the pesticides at the right stage of crop (i.e flowering) and to keep the levels of pesticides residues below the Maximum Residues Limits (MRLs) at harvest stage. Pesticide residues in fruits, vegetables and their processed products may also appear as a result of contaminated agro-inputs, drift from adjoining fields, as well as inappropriate or abusive treatments. Residual pesticides present in agricultural products may result to health hazards and affect the sensory quality of processed products. In the case of wine for example, they interfere with fermentative micro flora (Patil et al., 2009). It is therefore evident that it is necessary to monitor the levels of pesticide residues in specific agricultural products, in order to avoid risks for consumers.

Currently used pesticides differ from those used in previous decades. New groups of pesticides are developed and tested for plant protection, such as strobilurins (i.e azoxystrobin and pyraclostrobin) and imidazolinones (i.e fenamidone). Developed countries have established new laws in order to: 1) scale down the use of pesticides 2) protect the consumers (lower concentration limits in food) 3) protect farmers during application (reduced risk to humans) and 4) protect the environment (non target organisms and environmental resources). According to current tendency pesticides characterized by broad spectrum of action (fungicides), are more focused to target organisms (insecticides), and characterized by special penetration and redistribution properties. Additionally due to the withdrawal of many active ingredients as a result of the review process based on Council Directive 91/414/EEC, the use of remaining pesticides is more intensive compared to previous years.

Due to globalization, agricultural products with pesticides residues may travel all over the world. Thus consumers are increasingly conscious of the potential environmental and health problems associated with the build-up of toxic chemicals, particularly in food products. Exposure to pesticides through consumption of fruits, vegetables and olive oil is almost continuous, especially for Mediterranean people. In the recent years conventional cultivation of fruits is replaced by more controlled systems such as Integrated Crop Management (ICM) and biological (organic) cultivation in order to eliminate the exposure (Tsakiris et al., 2004) (Tsatsakis & Tsakiris, 2010).

Hereby, the status of pesticides residues in grapes, olive oil and vegetables (peppers, cucumbers, tomatoes and eggplants) is presented, focusing in organic (grapes, olive oil) and ICM cultivation (vegetables). Moreover monitoring results are used in order to estimate the dietary intake and cumulative risk from the detected pesticides in Greek people. Our data are based: 1) on the results reported by certification organizations in Greece and 2) on the unpublished results from monitoring programs of Toxicology Laboratory of the University of Crete.

2. Critical points for the determination of pesticides in agricultural products

Pesticide monitoring is performed by National and private laboratories, operating in each European country. National laboratories monitor pesticides and inform the European Union in order to publish specific pesticide reports. Private laboratories in each country perform routine analysis for major pesticide categories, with agricultural samples originating from independent farmers, farmer groups, super-markets or factories that export vegetables to other European countries. The majority of samples are analyzed as a prerequisite for certification purposes. Hence, when all agricultural products exported to other European countries are certified and the requirements of each certification scheme are strictly followed, public health is not jeopardized.

A great number of analytical methods for the determination of pesticide residues in agricultural products, especially in those that are extensively used by the Mediterranean diet, have been published up to now.

The analysed agricultural products may be divided into the three following categories: 1) Fresh fruits and vegetables 2) Alteration products of fruits and vegetables i.e. wine 3) Olive oil. Each category requires a different sample treatment process and presents a variable degree of difficulty in the analysis of residues.

Analytical methods for determining pesticide residues in two first categories involve several extraction and purification steps in order to remove the huge amount of potentially interfering compounds that are present at higher concentrations than the pesticides residues themselves. Many of them are characterized by laborious and time consuming sample preparation and also require non-environmentally friendly solvents. What's more, some of them demand expensive analytical instruments (i.e. LC-MS/MS) and this reduces their use for screening purposes.

For the detection of volatile and thermally stable pesticides gas chromatography (GC) in combination with appropriate detectors such as electron-capture (ECD), nitrogen-phosphorus (NPD) and mass spectrometry (MS) is most commonly used. On the other hand, for the detection of pesticides with low volatility and high polarity, methods based on liquid chromatography (LC) coupled to mass spectrometry are more common (Venkateswarlu et al., 2007).

The most widely used method for multi-class residues analysis is the QUECHERS method (Paula et al., 2007). The extraction procedure is based on liquid–liquid partitioning with acetonitrile followed by a cleanup step with dispersive-SPE. The great advantages of this method are the simplicity, the low cost of implementation and the short analysis time. A new method based on QUECHERS was developed for the evaluation of multiple pesticide residues in grapes, must and wine using the similar extraction procedures and low-pressure gas chromatography coupled to mass spectrometry (Cunhaa et al., 2009). Olive oil is a more difficult sample compared to the other products of this study, not only because of the relatively high amount of lipids that elute from the clean up system, but also because of the potential lipid interference at the GC determinate step (Lentza-Rizos, 1994). Another problem faced during olive oil analysis for pesticide residues is the fact that multiresidues analytical methods are highly desirable, although the different natures, classes and physicochemical properties of pesticides used in olive groves eventually hamper the development of such methodologies (Garcia-Reyes et al., 2007).

3. Pesticides residues in fruits, vegetables and olive oil

3.1 Monitoring data

Grapes peppers, tomatoes, cucumbers eggplants and olive oil were selected for pesticide monitoring and risk assessment since large amounts of these products are consumed daily in Mediterranean countries, especially during the summer period. The target of the study was not only to confirm that consumption of these products was safe for adults, but also for more sensitive population groups such as children. The first step of this study was to collect the raw data from pesticides residues analysis. Part of the results (olive oil and grapes) was obtained from Bio-Hellas monitoring program. Bio-Hellas is the biggest Control and Certification organism in Greece which supervises a great variety of agricultural products such grapes, olives, olive oil, fresh fruits, vegetables etc. Monitoring for grapes was performed for 381 pesticides by GC/MS-MS and HPLC/MS-MS. During 2005-2009, some 234 samples were analyzed. Monitoring in olive oil from biological cultivation of olives was performed during 2008 - 2009 for the following, most frequently used pesticides in conventional cultivation: Chlorpyriphos, Cyfluthrin, a-Cypermethrin, Cyhalothrin, Deltamethrin, Diazinon, Dimethoate, Endosulfan (Endosulfan-a, Endosulfan-b, Endosulfan sulfate), Fenthion (Fenthion oxon, Fenthion sulfone, Fenthion sulfoxide, Fenthion o sulfone, Fenthion o sulfoxide), Malathion/Malaoxon, Methidathion, Methomyl, Parathion. The results for peppers (60 samples), tomatoes (45 samples), cucumbers (50 samples) and eggplants (45 samples) came from the monitoring program of Laboratory of Toxicology. Monitoring was performed for 175 pesticides by HPLC/MS-MS. The monitoring period for vegetables was 2008 - 2009 and the results are unpublished. All samples were analyzed in line with National Accreditation System SA ESYD laboratories (EN ISO/IEC 17025:2005) in accordance to standard official analytical methods.

As stated at 834/2007 EC biological products should be free of pesticides residues. Moreover according to 396/2005 EC the levels of detected pesticides residues should not exceed the MRLs. Thus the monitoring for pesticide residues is a crucial part of the inspection procedure in every cultivation system (biological cultivation, ICM and conventional cultivation). A very important stage of this procedure is the selection of the appropriate pesticides in the monitoring program. The available new analytical techniques allow the supervising agronomist to select multiresidue methods enabling simultaneous analysis of pesticides from different groups. This is very important because certification authorities are able to detect apart from the pesticides used in specific cultivations other pesticides used in wide spectrum of other crops. In this way contaminated samples (bad agricultural and industrial practice, drift etc) are more easily detected. Thus the reliability of the system is increased and the risk assessment studies are enhanced with a great amount of crucial data.

3.2 Monitoring results

A total of 815 samples from the selected agricultural products were analyzed during the monitoring period. The prevalence of positive samples of pesticide is shown in table 1. The percentage of residual pesticide detection was very low (4.62% was the maximum value) in samples from biological cultivations (grapes and olive oil) compared to vegetables (55.56% was the maximum value).

The levels of detected pesticides residues exceeded the MRLs only in the cases of dimethoate (1 grape sample), endosulfan (3 samples in olive oil), a-cypermethrin (2 samples in olive oil), chlorpyrifos (1 sample in olive oil), diazinon (1 sample in olive oil), in oxamyl (2 samples in peppers, 4 samples in tomatoes and 7 samples in cucumbers) and methamidophos (5 samples in cucumbers). According to these data only 3.06% of samples, from all food commodities, exceeded the MRLs.

In grapes dithiocarbamates and benzimidazoles (carbendazim and benomyl) were the most frequently detected pesticide categories. Fenthion and endosulfan were the most frequently detected pesticides in olive oil. Thiacloprid, spinosad, oxamyl and pyrimiphos methyl were detected in peppers, oxamyl, diethofencarb and carbendazims were detected in tomatoes, endosulfan, spinosad, azoxystrobin, metalaxyl and pyrimethanil in cucumbers and carbendazims and thiamethoxam in eggplants were the most frequently detected pesticides in the vegetable group of commodities.

Food commodity	Pesticide	Pos	%	Average concentration (mg/kg)	Max value (mg/kg)	MRLs (mg/kg)
Grapes (n=234)	Diothiocarbamates	8	3.42	0.274	0.84	5
	Carbendazim + Benomyl	4	1.71	0.055	0.1	0.3
	Iprovalicarb	2	0.85	0.032	0.04	2
	Myclobutanil	2	0.85	0.035	0.04	1
	Fenbutatin oxide	2	0.85	0.095	0.14	2
	Fenarimol	1	0.43	0.010	0.01	0.3
	Penconazole	1	0.43	0.020	0.02	0.2
	Pirimiphos methyl	1	0.43	0.010	0.01	0.05
	Deltamethrin	1	0.43	0.020	0.02	0.2
	Dimethomorph	1	0.43	0.050	0.05	3
	Dimethoate-omethoate	1	0.43	0.060	0.06	0.02
	Tetraconazole	1	0.43	0.010	0.01	0.1
	Boscalid	1	0.43	0.260	0.26	5
	Kresoxim methyl	1	0.43	0.010	0.01	1
	Tebuconazole	1	0.43	0.010	0.01	2
	Endosulfan sulfate	1	0.43	0.010	0.01	0.5
	Endosulfan a and b	1	0.43	0.020	0.02	0.5
Olive oil (n=381)	Fenthion	13	4.62	0.015	0.038	1
	Endosulfan sulfate	11	2.88	0.034	0.071	0.05
	Endosulfan-total	10	2.62	0.034	0.079	0.05
	Fenthion total	10	2.62	0.022	0.035	1
	a-Cypermethrin	7	1.83	0.055	0.200	0.05
	Chlorpyrifos	3	0.78	0.027	0.055	0.05
	Fenthion sulphoxide	2	0.52	0.022	0.022	1
	Fenthion o sulphoxide	2	0.52	0.022	0.022	1
	λ-cyhalothrin	1	0.26	0.050	0.050	0.5
	Diazinon	1	0.26	0.024	0.024	0.01
	Endosulfan-a	1	0.26	0.006	0.006	0.05
	Endosulfan-b	1	0.26	0.002	0.002	0.05
	Methidathion	1	0.26	0.021	0.021	1
Peppers (n=60)	Thiachloprid	31	51.67	0.026	0.04	1
	Spinosad A+D	16	26.67	0.310	0.61	2
	Oxamyl+Oxamyl-Oxine	10	16.67	0.018	0.09	0.02
	Pyrimiphos	10	16.67	0.020	0.020	1

Food commodity	Pesticide	Pos	%	Average concentration (mg/kg)	Max value (mg/kg)	MRLs (mg/kg)
	methyl					
	Pyriproxyfen	7	11.67	0.020	0.020	1
	Indoxacarb	5	8.33	0.020	0.020	0.3
	Pymetrozine	5	8.33	0.050	0.050	1
	Thiamethoxam	5	8.33	0.010	0.010	0.5
	Fenarimol	4	6.67	0.020	0.020	0.5
	Malathion	3	5.00	0.010	0.010	0.1
	Endosulfan a	2	3.33	0.010	0.010	1
	Endosulfan b	2	3.33	0.020	0.020	1
	Endosulfan sulfate	2	3.33	0.090	0.090	1
Tomatoes (n=45)	Oxamyl+Oxamyl-Oxine	25	55.56	0.009	0.12	0.02
	Diethofencarb	16	35.56	0.015	0.02	1
	Carbendazim + Benomyl	10	22.22	0.030	0.04	0.5
	Famoxadone	5	11.11	0.030	0.030	1
	Mepanipyrim	5	11.11	0.030	0.030	1
	Pyrimethanil	5	11.11	0.050	0.050	1
	Thiacloprid	3	6.67	0.020	0.020	0.5
Cucumbers (n=50)	Endosulfan sulfate	15	30.00	0.030	0.040	0.05
	Spinosad A+D	12	24.00	0.060	0.060	1
	Endosulfan b	11	22.00	0.010	0.010	0.05
	Azoxystrobin	10	20.00	0.010	0.010	1
	Diothiocarbamates	10	20.00	0.330	0.330	2
	Endosulfan a	10	20.00	0.010	0.010	0.05
	Metalaxyl	10	20.00	0.026	0.030	0.5
	Pyrimethanil	10	20.00	0.245	0.43	1
	Iprodione	8	16.00	0.050	0.050	2
	Carbendazim + Benomyl	7	14.00	0.020	0.020	0.1
	Oxamyl+Oxamyl-Oxine	7	14.00	0.04	0.040	0.02
	Bifenthrin	5	10.00	0.020	0.020	0.1
	Dimethomorph	5	10.00	0.020	0.020	1
	Methamidophos	5	10.00	0.020	0.020	0.01
	Tolyfluanid	3	6.00	0.001	0.010	2
Eggplants (n=45)	Carbendazim + Benomyl	11	24.44	0.030	0.030	0.5
	Thiamethoxam	10	22.22	0.010	0.010	0.2
	Thiacloprid	9	20.00	0.032	0.06	0.5
	Fludioxonil	6	13.33	0.030	0.030	1

Food commodity	Pesticide	Pos	%	Average concentration (mg/kg)	Max value (mg/kg)	MRLs (mg/kg)
	Cypermethrin	5	11.11	0.020	0.020	0.5
	Iprodione	5	11.11	0.130	0.130	5
	Myclobutamil	5	11.11	0.020	0.020	0.3
	Cyprodinil	4	8.89	0.070	0.070	0.5

Table 1. Positive (pos) and %positive (%) detections of pesticide residues, average concentrations, max values and MRLs in grape samples from organic cultivations during 2005-2009, in olive oil samples from organic cultivations of olives and in vegetables (pepper, tomatoes, cucumbers and eggplants) from ICM cultivations during 2008-2009 (all samples are from Greece)

Table 2 presents the pesticides detected in two or more specific food commodities. Endosulfan was detected in all food items except tomatoes and eggplants. In biological cultivations it was detected only during 2008. Benzimidazoles represented by carbendazim and benomyl were detected in all food commodities except olive oil and peppers. The most frequently detected chemical categories, among all monitoring products, were organophosphates, organochorides, carbamates and pyrethrins.

4. Risk assessment

4.1 Dietary intake and cumulative risk assessment by Greek consumers

The results from monitoring program were used for the assessment of Greek consumers' risk. The risk assessment was performed for the four most frequently detected chemical categories. The assessment of consumers' exposure was based on Estimated Daily Intake (EDI) which was compared to Acceptable Daily Intake (ADI) and was expressed as a percentage of it (chronic dietary exposure). The calculation of EDI, expressed in mg/kg body weight/day, was based on the following equation:

EDI = R(mean concentration of the residue in the food commodity in mg/kg) *C(daily consumption kg/person/day)*EP(edible portion, 0 to 1 values) *PF(Processing factor for the specific food commodity, 0 to 1 values)/BW(body weight, kg)

All calculations for the determination of EDI were according to international guidelines (Iñigo-Nuñez et al., 2010). Residue levels used were those derived from the mean of detected samples for each food commodity. The average daily intake used for grapes was 82.19 g, for olive oil was 43.56 g, for tomatoes 258.35, and for the cucumbers, peppers and eggplants 360.27g. These intakes were based on FAO foodbalance sheet for 2007 (http://faostat.fao.org/site/368/DesktopDefault.aspx?PageID=368#ancor, Accessed in 23/08/2010). The value of EP for all food commodities was 1 in order to represent the local practice in food consumption.

Moreover the effects of processing factors were not taken into account in any case (PF=1). The body weight used for all calculations was 60 kg. The ADI values for pesticides were taken from official EU Pesticides Database.

The Hazard Index (HI) method was used for risk assessment of the mixtures of the detected pesticides belonging to the same chemical group (organophosphates, organochoride, carbamate and pyrethrins). HI was calculated according to the following equation:

$$HI = EDI_1/ADI_1 + EDI_2/ADI_2 + \ ..... \ + EDI_n/ADI_n = \sum_{i=1}^{n} \frac{EDI_i}{ADI_i}$$

EDI_i is the EDI of each active ingredient of each chemical group and ADI_i is the corresponding acceptable daily intake (Amvrazi & Albanis, 2009). If the hazard index exceeds 1, the mixture has exceeded the maximum acceptable level (e.g. ADI or ARfD) and there might thus be a risk. The fractions (EDI_1/ADI_1 etc.) are sometimes called the hazard quotients, HQ.

Based on the reported calculations, table 3 presents the cumulative risk assessment of the intake for the monitored food commodities. In the case of endosulfan where the MRLs were expressed as the sum of alpha- and beta-isomers and endosulfan-sulphate, the EDI was based on the average values of the reported values in table 1. The same practice was followed for fenthion and its metabolites (oxygen analogue, their sulfoxides and sulfone).

As can been seen in table 3 consumers exposure to pesticides do not exceeded the ADI in any of the reported cases. The hazard index of carbamates (cumulative risk) estimated for the adults (60kg) do not exceeded the value of 1 (0.38993) but because of the high value may constitute a risk for the consumers. The hazard index of organophosphorus pesticides estimated for the Greek adult population was 0,34361. The EDI/ADI values for organophosphorus pesticides ranged from 0.001 (fenthion) to 0,12 (methamidophos). Hazard indexes of organochlorine pesticides and pyrethrins were 0,08840 and 0,05288 respectively. Results indicate that there is a negligible risk associated with the exposure via the consumption of selected agricultural products for orgaclorines and pyrethrins. Special attention should be given to organophosphates and carbamates.

An acute dietary exposure assessment was also performed based on the monitoring results.

Food Commodity / Pesticide	Grapes	Olive Oil	Peppers	Tomatoes	Cucumbers	Eggplants
Carbendazim + Benomyl	+	nd	nd	+	+	+
Endosulfan	+	+	+	nd	+	nd
Oxamyl	nd	nd	+	+	+	nd
Thiachloprid	nd	nd	+	+	nd	+
Dithiocarbamates	+	nd	nd	nd	+	nd
Iprodione	nd	nd	nd	nd	+	+
Pyrimethanil	nd	nd	nd	+	+	nd
Spinosad	nd	nd	+	nd	+	nd
Thiamethoxam	nd	nd	+	nd	nd	+

Table 2. Pesticides detected in more than two food commodities.
[n.d: not detected, +:detection]

In this case intake for each pesticide was compared to an acute reference dose (ARfD). The estimated short-term intake (ESTI) was used to estimate acute dietary exposure. For the calculation of intake the maximum reported values of residues for each pesticide (in mg/kg) were multiplied by previously reported food consumption for each food commodity and was divided by body weight (60 kg) as described by following equation (Tsoutsi et al., 2008):

ESTI = HR(Highest residue value in mg/kg) * C(daily consumption kg/person/day) / BW(body weight, kg)

The risk of exposure was considered as insignificant in the cases where the estimated exposure was equal or lower than to ARfD. Table 4 presents the estimated short term intake of pesticides residues by adults in Greece. In this table values for dithiocarbamates and benzimidazoles were not estimated due to the fact that the parent compound was not possible to be determined. In the case of metabolites the ARfD of mother compound was used. The ARfD values for pesticides were taken from official EU Pesticides Database.

In samples from biological cultivations (grapes and olive oil) the values of ESTI as a percentage of ARfD ranged from 0.01 up to 0.82 indicating a minimum acute risk from the detected pesticides. Despite the fact that the same values were estimated for vegetables, the reported values of ESTI as a percentage of ARfD for methamidophos in cucumbers and of oxamyl in peppers, tomatoes and cucumbers point out an acute risk for the consumers.

4.2 Case study: dietary exposure of Greek consumers to dithiocarbamates by consumption of grapes

The levels of dithiocarbamate residues in grapes, which are the most frequently detected class of pesticides, were used for further research and preliminary assessment of consumers risk resulting from exposure via grapes consumption. According to previously published papers, in order to evaluate the potential risk of dietary exposure to dithiocarbamates, the estimated intakes need to be compared to a toxicological reference value of one compound within the dithiocarbamate class or to a group of compounds with the same mechanism of toxicity, that are present in the analyzed food (Caldas et al., 2004) (Caldas et al., 2006).

The registered dithiocarbamates in Greece are mancozeb, maneb, metiram, propirined, thiram, ziram, and metam sodium. Ziram and metam sodium were not register for grapes.

Thiram is only registered for grapes used for the production of wine. Mancozeb, maned and metiram belong to Ethylen bis-dithiocarbamate (EBDC) group, propineb in Propylene bis-dithiocarbamate and thiram in dimethyl dithiocarbamate.

According to local grape protection practice, the chemical groups of compounds and the toxicological profiles, the following exposure scenarios were visualized: The CS_2 detected in the samples was derived from the use of a) thiram or propineb alone, b) of at least one compound of mancozeb, maned and metiram.

Residue levels used for the risk assessment were those derived from mean, median and 90th percentile in mg/kg. According to EFSA statement for the table grapes, the most critical European consumer was identified as a German child with body weight 16.15 Kg, eating in one occasion 211.5 g of table grapes (13.1 g table grapes per kg body weight) (http://www.efsa.europa.eu/en/scdocs/doc/1590.pdf, Accessed in 23/08/2010). In the same report it was also stated that for adult population the intake was less critical (maximum food intake for adults 6.35 g /Kg body weight, for a consumer with 63 kg body weight). Thus our exposure calculations were based on previously reported body weight and consumption data.

The risk assessment again was performed by the HI method. As can been seen in table 5, chronic dietary intake of dithiocarbamates by children and adult population did not overcome the ADI for any one of the suggested scenarios, except in the case of propineb in children population, with the EDI estimated as the 90th percentile value of pesticides residues levels.

Chemical Group	Pesticide	ADI in mg/kg / bw/day	EDI in mg/kg / bw/day	EDI/ADI	Food Commodity
Carbamate	Oxamyl	0.001	0.00004	0.03875	Tomatoes
	Oxamyl	0.001	0.00111	0.10808	Peppers
	Oxamyl	0.001	0.00024	0.24018	Cucumbers
	Iprovalicarb	0.015	0.00004	0.00292	Grapes
$\sum \frac{EDI}{ADI} = HI$				0.38993	
Organophosphate	Methamidophos	0.001	0.00012	0.12009	Cucumbers
	Diazinon	0.0002	0.00002	0.08712	Olive Oil
	Dimethoate-omethoate	0.001	0.00008	0.08219	Grapes
	Pirimiphos methyl	0.004	0.00012	0.03002	Peppers
	Methidathion	0.001	0.00002	0.01525	Olive Oil
	Pirimiphos methyl	0.004	0.00001	0.00342	Grapes
	Malathion	0.03	0.00006	0.00200	Peppers
	Chlorpyrifos	0.01	0.00002	0.00196	Olive Oil
	Fenthion	0.007	0.00001	0.00156	Olive Oil
$\sum \frac{EDI}{ADI} = HI$				0.34361	
Organochlorine	Endosulfan	0.006	0.00026	0.04253	Cucumbers
	Endosulfan	0.006	0.00024	0.04003	Peppers
	Endosulfan	0.006	0.00002	0.00342	Grapes
	Endosulfan	0.006	0.02000	0.00242	Olive oil
$\sum \frac{EDI}{ADI} = HI$				0.08840	
Pyrethrins	a-Cypermethrin	0.015	0.00040	0.02686	Olive oil
	Bifenthrin	0.015	0.00012	0.00801	Cucumber
	a-Cypermethrin	0.015	0.00012	0.00801	Eggplants
	λ-cyhalothrin	0.005	0.00004	0.00726	Olive oil
	Deltamethrin	0.010	0.00003	0.00274	Grapes
$\sum \frac{EDI}{ADI} = HI$				0.05288	

Table 3. Cumulative intake of carbamate, organophosphate, organochlorine and pyrethrins pesticides detected in all samples during monitoring period based on HI method.[n.a: not available, n.e: not estimated]

Food Commodity	Pesticide	ARfD in mg/kg/ bw/day	ESTI in mg/kg/ bw/day	ESTI as % of ADI
Grapes (n=234)	Dimethoate-omethoate	0.01	0.00008	0.82
	Deltamethrin	0.01	0.00003	0.27
	Endosulfan a and b	0.02	0.00003	0.14
	Fenarimol	0.02	0.00001	0.07
	Endosulfan sulfate	0.02	0.00001	0.07
	Tebuconazole	0.03	0.00001	0.05
	Tetraconazole	0.05	0.00001	0.03
	Penconazole	0.5	0.00003	0.01
	Pirimiphos methyl	0.15	0.00001	0.01
	Dimethomorph	0.6	0.00007	0.01
	Diothiocarbamates	n.a	0.00115	n.e
	Carbendazim + Benomyl	n.a	0.00014	n.e
	Iprovalicarb	n.a	0.00005	n.e
	Myclobutanyl	n.a	0.00005	n.e
	Fenbutatin oxide	n.a	0.00019	n.e
	Boscalid	n.a	0.00036	n.e
	Kresoxim methyl	n.a	0.00001	n.e
Olive oil (n=381)	λ-cyhalothrin	0.0075	0.00004	0.48
	a-Cypermethrin	0.04	0.00015	0.36
	Endosulfan-total	0.02	0.00006	0.29
	Fenthion	0.01	0.00003	0.28
	Endosulfan sulfate	0.02	0.00005	0.26
	Fenthion total	0.01	0.00003	0.25
	Fenthion sulphoxide	0.01	0.00002	0.16
	Fenthion o sulphoxide	0.01	0.00002	0.16
	Methidathion	0.01	0.00002	0.15
	Diazinon	0.025	0.00002	0.07
	Chlorpyrifos	0.1	0.00004	0.04
	Endosulfan-a	0.02	0.00000	0.02

Food Commodity	Pesticide	ARfD in mg/kg/ bw/day	ESTI in mg/kg/ bw/day	ESTI as % of ADI
	Endosulfan-b	0.02	0.00000	0.01
Peppers (n=60)	Oxamyl+Oxamyl-Oxine	0.001	0.00054	54.04
	Endosulfan sulfate	0.02	0.00054	2.70
	Thiachloprid	0.03	0.00024	0.80
	Fenarimol	0.02	0.00012	0.60
	Endosulfan b	0.02	0.00012	0.60
	Pymetrozine	0.1	0.00030	0.30
	Endosulfan a	0.02	0.00006	0.30
	Indoxacarb	0.125	0.00012	0.10
	Pyrimiphos methyl	0.15	0.00012	0.08
	Malathion	0.3	0.00006	0.02
	Thiamethoxam	0.5	0.00006	0.01
	Pyriproxyfen	10	0.00012	0.00
	Spinosad A+D	n.a	0.00366	n.e
Tomatoes (n=45)	Oxamyl+Oxamyl-Oxine	0.001	0.00052	51.67
	Carbendazim + Benomyl	0.02	0.00017	0.86
	Thiacloprid	0.03	0.00009	0.29
	Famoxadone	0.2	0.00013	0.06
	Diethofencarb	n.a	0.00009	n.e
	Mepanipyrim	n.a	0.00013	n.e
	Pyrimethanil	n.a	0.00022	n.e
Cucumbers (n=50)	Oxamyl+Oxamyl-Oxine	0.001	0.00024	24.02
	Methamidophos	0.003	0.00012	4.00
	Endosulfan sulfate	0.02	0.00024	1.20
	Carbendazim + Benomyl	0.02	0.00012	0.60
	Bifenthrin	0.03	0.00012	0.40
	Endosulfan b	0.02	0.00006	0.30
	Endosulfan a	0.02	0.00006	0.30

Food Commodity	Pesticide	ARfD in mg/kg/ bw/day	ESTI in mg/kg/ bw/day	ESTI as % of ADI
	Metalaxyl	0.5	0.00018	0.04
	Dimethomorph	0.6	0.00012	0.02
	Tolyfluanid	0.25	0.00006	0.02
	Spinosad A+D	n.a	0.00036	n.e
	Azoxystrobin	n.a	0.00006	n.e
	CS2	n.a	0.00000	n.e
	Pyrimethanil	n.a	0.00258	n.e
	Iprodione	n.a	0.00030	n.e
Eggplants (n=45)	Thiacloprid	0.03	0.00036	1.20
	Carbendazim + Benomyl	0.02	0.00018	0.90
	Cypermethrin	0.04	0.00012	0.30
	Thiamethoxam	0.5	0.00006	0.01
	Fludioxonil	n.a	0.00018	n.e
	Iprodione	n.a	0.00078	n.e
	Myclobutamil	n.a	0.00012	n.e
	Cyprodinil	n.a	0.00042	n.e

Table 4. Estimated short term intake of residual pesticides by Greek adults (60 kg bw) expressed as percentage of ARfD

The lowest value of EDI% of ADI was obtained when median value of pesticides residues levels was used in the calculation and increased when the mean value was used. The highest value was observed when 90th percentile was used. The exposure of children population was always higher than that of an adult.

5. Conclusions

- Only 3.06% of total samples (815) exceeded the MRLs. The problem occurred in the case of dimethoate (1 grape sample), endosulfan (3 samples in olive oil), a-cypermethrin (2 samples in olive oil), chlorpyrifos (1 sample in olive oil), diazinon (1 sample in olive oil), in oxamyl (2 samples in peppers, 4 samples in tomatoes and 7 samples in cucumbers) and methamidophos (5 samples in cucumbers).
- The most frequently detected pesticides for each food commodity were dithiocarbamates for grapes (3.42%), fenthion in olive oil (4.62%), thiacloprid in peppers (51.67%), oxamyl in tomatoes (55.56%), endosulfan sulfate in cucumbers (30%), and carbendazim group in eggplants (24.44%).
- The most frequently detected chemical categories, among all monitored products, were organophosphates, organochlorine, carbamate and pyrethrins.

Risk assessment parameters		Dithiocarbamate		
		EBDC	Propineb	Thiram
ADI (in CS_2 mg/kg / bw/day)		0.0169	0.0036	0.0063
Child	EDI^{median}	0.0025		
	EDI^{median} % of ADI	15.07	69.32	40.35
Adult	EDI^{median}	0.0012		
	EDI^{median} % of ADI	7.30	33.60	19.56
Child	EDI^{mean}	0.0035		
	EDI^{mean} % of ADI	21.09	97.04	56.49
Adult	EDI^{mean}	0.0017		
	EDI^{mean} % of ADI	10.22	47.04	27.38
Child	EDI^{90th}	0.0061		
	EDI^{90th} % of ADI	36.24	166.71	80.82
Adult	EDI^{90th}	0.0029		
	EDI^{90th} % of ADI	17.56	80.82	47.04

Table 5. Risk assessment of exposure of child (16.15 Kg body weight) and adult (63 Kg body weight) population to the dithiocarbamate pesticides (mean, median and 90th percentile value of pesticides residues values) in EDI% of ADI based on 211.5 g and 400g consumption respectively, of table grapes originate from biological cultivations. [The transform of ADI of each dithiocarbamate to CS_s was based on molecular weight factor reported in E.D Caldas et. al 2004., CS_2: Carbon disulfide, $EDI^{mean,mediam,90th}$:Estimated Daily Intake in CS_s (mg/ kg / bw / day) based on mean, median and 90th percentile value of pesticides residues levels respectively.]

- According to selected parameters for the risk assessment (food commodities, consumption, body weight,) pyrethrins (HI value 0.05288) and organochlorine (HI value 0.08740) present a negligible hazard for the consumers. Organophoshates with estimated HI value of 0.34361 and carbamates with estimated HI value 0.38993, which were also far below 1, do not expect to constitute a risk but further attention should be given. A long term monitoring program from different areas and different times of sampling is necessary in order to reach more accurate conclusions.
- All samples from biological cultivations (grapes and olive oil) presented a minimum acute risk from detected pesticides.
- The acute risk from methamidophos in cucumbers (ESTI was 4% of ARfD) and especially from oxamyl in peppers(ESTI was 54.04% of ARfD)and in tomatoes (ESTI 51.67% of ARfD) is high.
- The chronic dietary intakes of dithiocarbamates by adult and children through the consumption of grapes originating from organic cultivations did not overcome the toxicologicaly acceptable levels.

- Grapes originating from organic cultivation may consist a reliable choice for the consumers and especially for the most sensitive groups such as children.
- A wide spectrum of pesticides should be included in monitoring programs in order to eliminate the possibility of contaminated samples to reach the markets store. The usage of multiresidue methods characterized by simultaneous analysis of pesticides from different groups could enchase the risk assessment studies with a great amount of crucial data.

6. Acknowledgment

The authors acknowledge BIO-HELLAS Control Certification Body for providing the data regarding the monitoring program biological olive oil and grape residues. The study was sponsored by the research funds of the Center of Toxicological Sciences and Reaserch of the University of Crete.

7. References

Amvrazi, E.G & Albanis T.A. (2009). Pesticide residue assessment in different types of olive oil and preliminary exposure assessment of Greek consumers to the pesticide residues detected. *Food Chemistry*. 113, 253-261

Caldas, E.D.; Boon P.E. & Tressou, J. (2006). Probabilistic assessment of the cumulative acute exposure to organophosphorus and carbamate insecticides in the Brazilian diet. *Toxicology*. 222, 132-142

Caldas, E.D.; Miranda, M.C.C.; Conceicao, M.H. & de Souza L.C.K.R. (2004) Dithiocarbamates residues in Brazilian food and the potential risk for consumers. *Food and Chemical Toxicology*. 42, 1877-1883

Coxam, V.; Puel, C. & Davicco, M.J. (2010). *Olives and Olive Oil in the Prevention of Osteoporosis*, In: *Olives and Olive Oil in Health and Disease Prevention*. Eds. V. R. Preedy, R.R. Watson, 1195-1203, Oxford Academic Press

Cunhaa, S.C.; Fernandesa, J.O.; Alvesb A.P. & Oliveiraa, M.B.P.P. (2009). Fast low-pressure gas chromatography–mass spectrometry method for the determination of multiple pesticides in grapes. musts and wines. *Journal of Chromatography A*. 1216, 119-126

Dani, C.; Oliboni, L.S.; Vanderlinde, R.; Bonatto, D.; Salvador, M. & Henriques, J.A.P. (2007). Phenolic content and antioxidant activities of white and purple juices manufactured with organically or conventionally produced grapes. *Food and Chemical Toxicology*, 45, 2574–2580

Garcia-Reyes, J.F.; Ferres, C.; Gomez-Ramos, M.J.; Molina-Diaz, A. & Fernandez-Alba, A.R. (2007). Determination of pesticide residues in olive oil and olives. *Trends in Analytical Chemistry*. 26 (3), 239–251

Iñigo-Nuñez, S.; Herreros, M.A.; Encinas, T.& Gonzales-Bulnes, A. (2010) Estimated daily intake of pesticides and xenoestrogenic exposure by fruit consumption in the female population from a Mediterranean country (Spain). *Food Control*. 21, 417-477

Lentza-Rizos, Ch. (1994). Monitoring of pesticide residues in olive products: Organophosphorus insecticides in olives and oil. *Journal of Association of Official Analytical Chemists International*. 77, 5

Patil, S.H.; Banerjee, K.; Dasgupta, S.; Oulkar, D.P.; Patil, S.B.; Jadhav, M.R.; Savant, R.H.; Adsule, P.G. & Deshmukh, M.B. (2009). Multiresidue analysis of 83 pesticides and

12 dioxin-like polychlorinated biphenyls in wine by gas chromatography–time-of-flight mass spectrometry. *Journal of Chromatography A*, 1216, 2307-2319

Payá, P.; Anastassiades, M.; Mack, D.; Sigalova, I.; Tasdelen, B.; Oliva, J. & Barda, A. (2007). Analysis of pesticide residues using the Quick Easy Cheap Effective Rugged and Safe (QuEChERS) pesticide multiresidue method in combination with gas and liquid chromatography and tandem mass spectrometric detection. *Analytical and Bioanalytical Chemistry*. 389, 1697-1714

Roche, M.H.; Gibney, M.J.; Kafatos, A.; Zampelas, A. & Williams C. M. (2000). Beneficial properties of olive oil. *Food Research International*, Volume 33, Issues 3-4, Pages 227-231

Tsakiris, I.N.; Danis, T.G.; Stratis, I.A.; Nikitovic D.; Dialyna I.A.; Alegakis A.K. & Tsatsakis A.M. (2004). Monitoring of pesticide residues in fresh peaches produced under conventional and integrated crop management cultivation. *Food Additives & Contaminants*. 21(7), Jul, 670-7

Tsatsakis A.M. & Tsakiris I.N. *Fenthion Dimethoate and Other Pesticides in Olive Oils of Organic and Conventional Cultivation*. In: *Olives and Olive Oil in Health and Disease Prevention*. Eds V. R. Preedy, R. R. Watson, 415-424 Oxford Academic Press

Tsoutsi, C.S.; Konstantinou, I.K. & Hela, D.G. (2008). Organophosphorus pesticide residues in Greek virgin olive oil: levels, dietary intake and risk assessment. *Food Additives and Contaminants*. Vol.25, No. 10, October, 1225-1236

Venkateswarlu P.; Mohan K.R.; Kumar C.R. & Seshaiah, K. (2007). Monitoring of multi-class pesticide residues in fresh grape samples using liquid chromatography with electrospray tandem mass spectrometry. *Food Chemistry*, 105, 1760-1766

15

Pesticides Surveillance on Surface Waters: Developing a Method for Watersheds Prioritization

Carolina Menezes and Léo Heller
Universidade Federal de Minas Gerais
Brazil

1. Introduction

The occurrence of pesticides in drinking water is a matter of growing concern in several parts of the world, mainly in developing countries, due to the possible adverse effects on human health and the environment. Pesticides applied in the agriculture, important source of contamination, can persist for several years in soil, be retained in agricultural products, be dispersed by the wind or reach the surface and ground waters (Tomita & Beyruth, 2003). Traces of pesticides can be found, both in soil and water sources, depending on their physical and chemical properties, their application and dose patterns, as well as the local environmental characteristics.

In Brazil, pesticide monitoring in surface water, both by water supply operators and health authorities, is still rare and insufficient for assessing health risks, often not accomplishing the legal issues regarding the quality control of raw waters (Brazil, 2004b). Inclusion of the whole set of these parameters in monitoring plans is unusual, mainly due to high costs of the analyses and to necessity of specialized manpower. There is also a lack of data on pesticides use in agriculture, in the several Brazilian regions, as well as of studies that associate their use with presence in water and with health risks. Additionally, low potential of pesticides removal in most of the Brazilian water treatment plants completes the picture of lack of reliable information on health risks associated with pesticides exposure by water ingestion.

The lack of data and of financial and material resources to comprehensive pesticide surveillance suggests tracing other alternatives to the evaluation of potentiality of pesticides occurrence in surface water. Thus, the aim of this chapter is to discuss a method for prioritization of surveillance actions of pesticides in surface waters, through multicriteria analysis. In the chapter, firstly we discuss concepts related to the subject. Secondly, the structure of the method is presented and, thirdly, we describe an application in Rio Grande basin, Minas Gerais, Brazil. Finally, we draw some conclusions on the study.

2. Pesticides in water and health

Health risks from exposure to pesticides, especially on child health, is a well known issue, with a number of epidemiological studies. These studies are consistently reporting increased risks between pesticide exposures and childhood leukemia, brain cancer, neuroblastoma, non-Hodgkin's lymphoma, Wilms' tumor, and Ewing's sarcoma (Infante-Rivard & Weichenthal, 2001).

Studies have been developed focusing on the effects of pesticides in water on human health, particularly workers´ health, however, with many gaps in knowledge of the relationship between pesticides and risks to human health, especially regarding presence of pesticides in water.

Although developed countries concentrate the bulk of pesticides consumption, most of poisonings and deaths caused by pesticides occurs in developing countries, and countries like Brazil shows an unsafe situation concerning human and environmental risks due to pesticides in water. This reality can have association with a combination of factors: excessive use of these substances; inadequate and unsafe occupational practices including inefficient use of personal protective equipment; low educational qualification of rural workers; improper labelling; inadequate or nonexistent infrastructure for washing utensils; improper handling of wastes and empty containers; use of containers for storing food and water; high productionist pressures from distributors and producers. A weak oversight of law enforcement, poor technical assistance to rural production and poor health care facilities complements the framework of health risks from pesticides in developing countries (FAO, 1999 apud Campanhola & Bettiol, 2002; Moreira et al., 2002).

Water contamination by pesticides can result from numerous non-point sources and agriculture is identified as the largest contributor (Tomita & Beyruth, 2003). Pesticides applied on agricultural crops can persist for several years in soil and can reach surface waters through superficial fluxes and leaching, contaminate groundwater by percolation, disperse in atmosphere or accumulate as residues in food. Regarding water contamination, when applied in agriculture, pesticides may reach surface water through transport dissolved in water or by transport associated with suspended sediment.

Understanding the behaviour and persistence of pesticides in the environment depends on knowledge of their specific chemical and physical properties; characteristics of their application vis-a-vis types of agricultural crops; environmental and climatic conditions (Brazil, 2002b; Luchini & André, 2002; Martins et al., 2004). Depending on the characteristics of a particular pesticide, alone or associated with other substances, high degree of persistence, bioaccumulation and toxicity can be found (Martins et al., 2004).

Once reached the water source, persistence of pesticides in water and water characteristics, such as pH, temperature, turbidity, suspended solids, flow and depth will determine the potential for ingestion of these micro contaminants (Kamrin, 1997 apud Martins et al., 2004).

It is relevant to point out that some monitoring data may be poor indicators of water pollution by pesticides, mainly when substances are adsorbed to suspended solids. In this line, results reported as "undetectable" may be due to inadequate procedures, analysis or sampling. Values associated with sediments can be generally much higher than those recorded in water samples. Thus, difficulties in the evaluation and quantification of pesticides should not be neglected. A possible procedure, adopted by some water control agencies, is the use of various types of samples (water + sediment + biota), that could ensure obtaining more consistent results (D'Amato et al., 2002).

In some cases, there are difficulties for assessing the source and the when the pollution occurred, due to the persistence of pesticides. Their presence in water either may be the result of a recent discharge, of air transport over long distances after crops application, or a persistent residue remaining from a very old application. This is the case of DDT, for instance, which still can be found in many countries, although its use was officially prohibited several years ago (Rissato et al., 2006).

Regarding legislation for drinking water surveillance, there is not necessarily a standard procedure among countries, although international directives related to control of water quality, like the WHO guidelines and EU directives, have long been considered by several countries in their national policies for water and health promotion. However, translation of these international recommendations is not easily translated to national realities, for the establishment of drinking water standards or of surveillance procedures, given the complex nature of the concepts involved, the need of reliable research involved in the proposal and review them.

World Health Organization - WHO, through the 3rd edition of the Guidelines for Drinking-Water Quality, establishes guidelines on how to provide access to safe water and with acceptable risk to human health. Regarding pesticides, the document presents recommended limits for concentration of those most relevant and with evidenced health risks (WHO, 2004). In the EU, the European Directive No. 98/83 (European Council, 1998) establishes a limit of 0.1 μg/L as the maximum concentration level for any pesticide individually, 0.5 μg/L for total pesticides in drinking water and 1-3 μg/L in surface waters.

In Brazil, Decree 518/2004 of the Ministry of Health (Brazil, 2004b), in establishing procedures and responsibilities related to control and surveillance of drinking water and quality standards, presents concentration limits of 22 pesticides, considered potential health hazards. Most of them consist of organochlorines, pyrethroids and organophosphates and only one carbamate. Moreover, this ordinance recommends identification of activity of the enzyme acetylcholinesterase, in order to assess the presence of organophosphate and carbamate insecticides in water.

Regarding monitoring, the referred Brazilian legislation establishes the minimum number and minimum frequency of samples for water quality control, depending on the sampling point, the population served and the type of source. For pesticides control, at least one sample every six months should be collected, in the effluent of the treatment plant, both for surface and groundwater sources. Systems supplied by surface water sources are due to raw water samples, each six months, for analysis according to the environmental legislation, assessing compatibility between raw water characteristics and the type of water treatment (Brazil, 2004b, Art. 29). Brazilian environmental legislation classifies surface water bodies, according to their uses and water quality requirements, and establishes standards for effluent discharge, setting limits for some pesticides, with close association with the Decree MS 518/2004.

Despite the potential risks associated to presence of pesticides in water and the scope of legislation on this subject, there is great deficiency in Brazil related to adequate control by water providers, and mainly surveillance by health authorities. Besides that, there is no formalised methodology for visioning priority basins, seeking at rationalising efforts in surveillance, based on land use and occupation; application practices; crop production or raw water quality.

3. The method

Decision-making on the best practices for water quality surveillance on pesticides presence is typically a complex situation, involving multiple stakeholders and factors. In cases like this, it is important that decision-making, involving multiple actors and multiple uses of water, consider the political, social, economic, financial, hydrological, environmental and engineering factors, among others, leading solutions that best reconcile interests and assumptions (Braga & Gobetti, 2002).

Decision-making can be defined as an effort to solve the dilemma of conflicting objectives, whose presence prevents the existence of "optimal solution", leading in search of the "best solution agreement." The complexity of decision-making requires a qualified approach and justifies the use of methods for decision support. The multicriteria methods encompass tools for subsidizing the decision process, taking into account a number of different factors, from different analytical dimensions, through using of qualitative and/or quantitative approaches. They provide a basis for discussion, especially in cases of conflicts between decision-makers, contributing to an integrated analysis of a large number of data, interactions, and goals. On the other hand, as an inconvenience, there is a lack of an overall methodology, which overcomes all the limitations of each method (Vilas Boas, 2005).

As detection of several critical factors is crucial for selection of water systems with potential health risks from pesticides presence, we developed a sequence of steps looking for application of multicriteria method, based on a theoretical risk of contamination of surface waters by pesticides applied in agricultural areas, as follows:

Firstly, information on pesticides (toxicological and environmental classifications, physical and chemical properties, effects on human health and on the environment, etc) was gathered.

Then, a theoretical model was developed, aiming at explaining influences of the characteristics of the environment, as soil type, rain, hydrography, topography and particularities of agriculture practices, on the potential of pesticides dispersion in the environment and occurrence in surface waters. The key intervening factors were identified in the model and five of them were selected to feed the multicriteria analysis.

The multicriteria method adopted was the TOPSIS (Technique for Order Preference by Similarity to Ideal Solution), developed by Hwang and Yoon (1981 *apud* Pomerol & Barba-Romero, 1993), which evaluates the distance in relation to an ideal and to an anti-ideal pattern, through a geometric notion of the best one. The solutions recommended are that closest to the ideal solution, by a proximity measure (Pomerol & Barba-Romero, 1993; Braga & Gobetti, 2002), according to the following equations:

- Distance to the ideal

$$d_p^M(a_i) = \left[\sum_j w_j^p * \left| a_j^M - a_{ij} \right|^p \right]^{1/p} \quad \text{for } p \geq 1 \tag{1}$$

- Distance to the anti-ideal

$$d_p^m(a_i) = \left[\sum_j w_j^p * \left| a_j^m - a_{ij} \right|^p \right]^{1/p} \quad \text{for } p \geq 1 \tag{2}$$

Based on the equations (1) and (2), the similarity rate is calculated by:

$$D_p(a_i) = \frac{d_p^m(a_i)}{d_p^M(a_i) + d_p^m(a_i)}, \quad \text{for } p \geq 1 \tag{3}$$

$d^M{}_p(a_i)$: distance of Minkovski between $a^M{}_i$ and a_{ij};

$d^m{}_p\ (a_i)$: distance of Minkovski between $a^m{}_i$ and a_{ij};
j: analyzed criterion;
w_j: weight of j criterion;
$a^M{}_j$: point of ideal of j criterion (maximum value);
$a^m{}_i$: point of anti-ideal of j criterion (minimum value);
a_{ij}: point of i alternative and j criterion;
p: value that defines the distance type;
$D_p\ (a_i)$: similarity rate;
$d^M{}_p\ (a_i)$: distance of Minkovski to ideal;
$d^m{}_p\ (a_i)$: distance of Minkovski to anti-ideal.
The value of D_p varies from 0, for the anti-ideal point, to 1, for the ideal point. In this research two values of p ($p = 1$ and $p = 2$) were considered.
By the similarity rate calculated for each system, the proposed solutions are ordered in a ascending list, in which the value closest to one, obtained in the similarity rate measure - $D_p(a_i)$, corresponds to the best solution. For each of the five selected criteria a weight range was attributed, in a scale from 0 to 10.
Next, the validation of the method was carried out by an application in five sub-basins of Grande River Basin, in Minas Gerais, Brazil: GD 3 (around Furnas's Dam), GD 4 (Verde River Basin), GD 5 (Sapucaí River Basin), GD 6 (Pardo River and Mogí-Guaçu River Basins), GD 7 (around Peixoto's Dam and Sapucaí Stream). The values of each criterion for each sub-basin were obtained and, in order to develop a sensitive analysis, the weights were varied.

4. Defining criteria and weights

The method considered intrinsic factors related to the pesticides and environmental aspects, regarding the assessment and comparison of different potential of pesticide dispersion. Five factors were considered for the application of multicriteria analysis.
Selection and prioritization of sub-basins with higher potential risk considered weights for each alternative. Weighting for pesticides risk was based on a literature review, highlighting the major factors involved in persistence and mobility of pesticides in the environment.
The following sections present the criteria and assumptions adopted:

4.1 Proportional cultivation area of the main agricultural cultures of the sub-basin.

It was assumed that the larger the agricultural occupation in the sub-basin is, the bigger is the likelihood of pesticides use on agricultural crops.
The calculation of this factor considered the significant cultures existent in the sub-basin, both in terms of cultivated area, for the cultures that have the characteristic of demanding great areas, or in terms of cultures that demand smaller areas with great productivity or with possibility of high use of pesticides. The main crops in each selected sub-basin were identified, based on the extent of area per crop and their average productivity (data obtained in FAEMG, 2005). Monitoring Report of Surface Water in the Rio Grande Basin in 2004 (IGAM, 2005) provided information on total area of each sub-basins.
In order to establish a grade of importance for the different crops, we established an association between proportions of area covered by the crop and sub-basin area and the crop relevance for the specific sub-basin (Table 1).
Criterion for inclusion of the crop in the analyses considered those with classifications "main crop" and "less relevant crop". A range of weight from 8 to 9 was attributed to this criterion.

Crops	Main crop (%)	Less relevant crop (%)	Crop not relevant (%)
Orange and tomato	> 0.1	0.02 - 0.1	< 0.02
Rice, banana, potato and sugar cane	> 0.5	0.2 - 0.5	< 0.2
Bean	> 1.0	0.5 - 1.0	< 0.5
Coffee and corn	> 4.0	2.0 - 4.0	< 2.0

Table 1. Criteria for selection of the relevant cultures, in percentage of area by crop culture in sub-basins of Rio Grande.

4.2 Proportion of group 1 pesticides.

Group 1 pesticides reflects high potential of surface waters contamination through transport associated with sediment or dissolved in water. The probability of use of this group of pesticides in a certain watershed suggests high risk of deterioration of surface water quality regarding human consumption, according to the method of Goss (1992 *apud* Dores & De-Lamonica-Freire, 2001).

By the identification of the significant crops in each basin, and based on papers (Kammerbauer and Moncada, 1998; Larini, 1999; Laabs *et al.*, 2000; Dores and De-Lamonica-Freire, 2001; Cerejeira *et al.*, 2003; Brazil, 2004a), it was possible to estimate, the pesticides of groups 1, 2 and 3 (high, medium and low potential of contamination of surface waters, respectively) likely to be applied and to calculate the proportion of those pesticides with potential use in each crop, for each potential group of contamination of surface waters. Thus, the representative value of the group 1 for each sub-basin was estimated by the average of the percentage of group 1 use, referring to each culture considered significant for the sub-basin. A range of weight from 7 to 9 was adopted for this factor, depending on the level of information available about which pesticides used on agricultural crops in each sub-basin.

The lack of consistent information in governmental agencies about pesticides uses on agricultural crops had required consultation to other databases such the Pesticide Information System (Sistema de Informações sobre Agrotóxicos - ANVISA), which identifies the cultures allowed to apply of specific active ingredients and guidelines for pesticide application in agricultural crops.

4.3 Proportion of municipal districts that have water treatment plants with techniques that allow at least partial removal of pesticides.

The water treatment method influences considerably the drinking-water quality and the health risks, regarding the exposure to pesticides.

Conventional treatment (coagulation, flocculation, sedimentation and rapid filtration) does not ensure satisfactory performance in removal of most of pesticides. So, use of advanced treatment, such as adsorption by activated carbon or membrane filtration is recommended. These last types of water treatment techniques, however, are very seldom used in Brazil.

The criterion consists of the identification of proportions of water treatment techniques within the sub-basin that have at least some potential of pesticides removing, in order to qualitatively assessing risks to human health. The proportion was calculated in relation of the total number of municipalities in the basin.

The likely lack of use of advanced techniques in water treatment plants in the municipalities of Rio Grande basin required adoption of an alternative approach, by consideration of plants with conventional treatment or direct filtration and plants with treatment level "less" than these techniques, due to the ability of the former to better removing suspended solids and, consequently, some pesticides associated. So, municipalities with simplified or conventional treatment would be in a safer situation than those without treatment or with single disinfection, regarding pesticides presence in drinking-water, even with differences in degree of raw water contamination.
The current techniques of treatment available in municipality of each sub-basin had been listed initially, based on information from the Water and Sanitation Company of Minas Gerais - COPASA, which attends most cities in the region, and from IBGE (2000), for those municipalities not provided by COPASA. The IBGE database provides the following categories for treatment process: "conventional treatment or non-conventional", "treatment with simple disinfection" and "no treatment". On the other side, COPASA database classifies the types of treatment in "conventional", "direct filtration" and "well with chlorination and fluoridation". Due to these divergent classifications, we considered the first category of IBGE and the two former categories of COPASA as "system with partial removal potential of pesticides" and the others as "system without conditions for pesticides´ removal", as Table 2.

Database	Systems with partial removal potential of pesticides	Systems without conditions for removal of pesticides
COPASA	"conventional"; "direct filtration".	"well with chlorination and fluoridation".
IBGE	"conventional or non-conventional treatment ".	"treatment with simple disinfection"; "no treatment".

Table 2. Grouping of the conditions of drinking water treatment.

As that factor influences positively the health risk, its value was inverted, in order to feed the analysis. A range weight from 4 to 5 was adopted.

4.4 Medium slope of the sub-basin

The topography of the basin, together with the soil type and vegetation, can interfere in the contaminants flow in the basin. The steeper the topography, the larger the run-off potential and the resulting transport of sediments to water sources.
The areas of each sub-basin can be analysed according to a range of slopes, varying from flat terrain to hilly. Table 3 presents the punctuation of the sub-basin according to the slope range value of the, in a scale from 1 to 5.

Relief class	Slope range	Score
Flat	< 3 %	1
Mild wavy	3 a 12 %	2
Wavy	12 a 24 %	3
Highly wavy to hilly	24 a 45 %	4
Hilly	> 45 %	5

Table 3. Medium slope of the sub-basin punctuation according to the range of slopes

Obviously, the greater the slope of a watershed, the greater the possibility of run-off and pesticides discharge in watercourses. In the study, the use of this criterion was based on information provided by the Water Management Institute of Minas Gerais - IGAM.
The range from 6 to 8 was considered for the weight.

4.5 Annual maximum intensity of rain in the sub-basin

When the period of application coincides with intense rains, the probability of carrying products in the watershed increases, requiring eventual reapplication, due to soil washing.
Regarding this criterion, the annual maximum rain intensity measured in rain gauge stations was calculated in each sub-basin, through the software Hidro, available in the Brazilian Water National Agency (Agência Nacional das Águas - ANA) website.
In one hand, heavy rains cause carrying of larger amount of substances into groundwater or surface water, in the other, total rainfall contributes to the dissolution of these substances, reducing its concentration in the watercourse. Due to these considerations, the maximum annual precipitation in each sub-basin was adopted. A range of weights from 6 to 8 was attributed.
Table 4 summarises the selected criteria list and their range of weights.

	Criteria	Weight (w)
I	Proportional cultivation area of the main agricultural cultures of the sub-basin.	8 to 9
II	Proportion of group 1 pesticides.	7 to 9
III	Proportion of municipal districts that have water treatment plants with techniques that allow at least partial removal of pesticides.	4 to 5
IV	Medium slope of the sub-basin.	6 to 8
V	Annual maximum intensity of rain in the sub-basin.	6 to 8

Table 4. Selected criteria and range of weight

All the criteria, except criterion III, can increase the potentiality of pesticides contamination in the sub-basin, as the respective weight increases.

5. Application in Rio Grande Basin, Brazil

Rio Grande basin, located in the southern region of Minas Gerais, state at the Southeast Region of Brazil, presents a profile of important agricultural producer and has a cluster of agricultural crops, among the selected in this study, in nearby areas. Comparing the region with other similar in the state of Minas Gerais, based on agricultural practices, the South Region seems to produce crops with highest potential for health risks in water, and for this reason was chosen for the study.
Through research of the municipalities in southern Minas Gerais with highest crop production and productivity, we identified five sub-basins for study. These sub-basins are part of the Rio Grande Basin, sub-basin of the Paraná River Basin. The Rio Grande has a length of 1,360 km and the basis has 143,000 km^2.
The application of the method for the five sub-basins of the Grande River Basin, selected for study, resulted in a priority order for surveillance of pesticides presence in surface waters.

Table 5 lists these sub-basins, indicating the most cultivated crops within its coverage area in order to facilitate evaluation of potential risk from pesticide use in their production.

Sub-basins		Main crops	Other relevant crops	Other considered crops
GD 3	Around Furnas Reservoir	Potato, coffee, sugar cane, bean and corn.	Rice, orange and tomato.	Banana.
GD 4	Verde River Basin	Potato, coffee, bean and corn.	Banana.	Sugar cane and tomato.
GD 5	Sapucaí River Basin	Banana, potato, coffee, bean and corn.	Rice, orange and tomato.	Sugar cane.
GD 6	Pardo River and Mogi-Guaçu River Basins	Potato, coffee, sugar cane and corn.	Bean.	Banana e tomato.
GD 7	Around Peixoto's Reservoir and Sapucaí Stream	Coffee, sugar cane and corn.	Rice, bean e orange.	Potato and tomato.

Table 5. Pre-selection of sub-basins of the Rio Grande Basin - MG.
GD = Term used by IGAM for sub-basins of the Rio Grande Basin.
Sources: FAEMG, 2005; IGAM, 2005.

Sub-basins GD 3, GD and GD 4 5 converge to Furnas Reservoir and sub-basin GD 7 to Peixoto´s Reservoir, both built for energy generation. Regarding the sub-basin GD 6, it is part of the basin of Mogi Guaçu, in Sao Paulo State.
Classification of pesticides potentially used in the basin resulted in grouping in high, medium and low potential for contamination of surface waters, according to the method of Goss (1992), as shown in Table 6.

Potential of surface water contamination	High	Medium	Low
Associated with the sediment transported in suspension	Chlorpyrifos, endosulfan, glyphosate, lindane, pendimethalin, trifluralin	Aldrin, atrazine, DDT, heptachlor, metolachlor, parathion-methyl, permethrin, simazine	2,4-D, alachlor, bentazone, cyanazine, dieldrin, hexachlorobenzene, malathion, methoxychlor, molinate, propanil
Dissolved in water	Aldicarb, atrazine, carbofuran, lindane, simazine	2,4-D, alachlor, cyanazine, chlorpyrifos, glyphosate, malathion, metolachlor, molinate, parathion-methyl, trifluralin	Chlordane, endrin, endosulfan, permethrin

Table 6. Grouping of pesticides based on the method of Goss.
Sources: WHO, 2004; BRASIL, 2004a; GOSS, 1992 (apud DORES and DE-LAMONICA-FREIRE, 2001)

According to Goss criteria, pesticides applied in agriculture, regarding their potential to reach surface water, can be classified in those that can be transported dissolved in water and those associated with the sediment transported in suspension.
Figure 1 illustrates the proportion of group 1 pesticides (high potential for contamination of surface waters) used in the five sub-basins of the Rio Grande - MG, stratified with those possibly spread dissolved in water and those that may be associated with sediment transport. This information was used to application of criterion II.

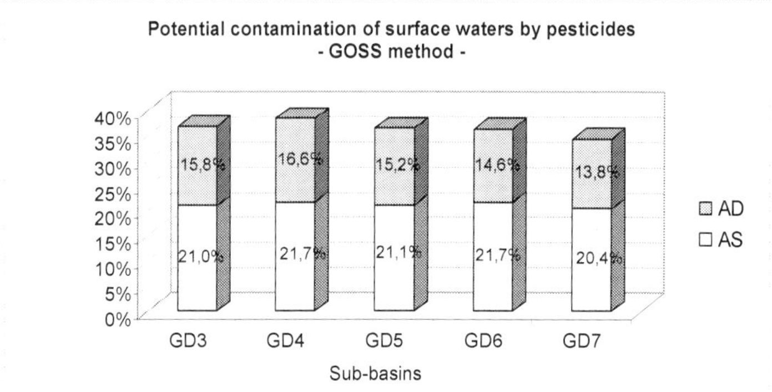

AD = high potential for transport dissolved in water, AS = high potential for transport associated with sediment.

Fig. 1. Proportion of pesticide use group 1 (high potential for contamination of surface waters) in sub-basins of the Rio Grande - MG.

Sources: KAMMERBAUER & MONCADA, 1998; LARINI, 1999; LAABS et al., 2000; DORES & DE-LAMONICA-FREIRE, 2001; BRASIL, 2002; CEREJEIRA et al., 2003; BRASIL, 2004a; MARTINS e tal., 2004; MINAS GERAIS, 2004; IGAM, 2005; FAEMG, 2005.

According to the crops identified as the main crops in each sub-basin, we can identify a proportion of 34% to 39% of pesticides group 1 - high contamination potential of surface water. The largest part of these pesticides relates to the potential of association with sediments. There is no high variation of this potential contamination between sub-basins, except for GD 4.
Table 7 shows the proportion of municipalities that have water treatment plants that ensure at least partial removal of pesticides and those without potential for removal of pesticides, as defined in Criterion III.
Out of 138 municipalities surveyed, near 85% has a potential for removal of pesticides at least partially. It can be highlighted that sub-basin GD 6, in the region of rivers Pardo and Mogi-Guaçu, stands out with a more favourable condition in terms of water treatment, while sub-basin GD 4 shows the lowest percentage of cities with potential to at least partial removal of pesticides.
Regarding the criterion IV, Table 8 shows the scores adopted in each sub-basin of the Rio Grande - MG, based on ranges of average slopes of the land.
Sub-basins GD 3 and GD 7 have low average slope, which is coherent with the fact of including areas near or inside reservoirs. Sub-basins GD 5 and GD 6, on the other hand, have an average slope steeper compared to the other sub-basins, with classes of relief from

Sub-basin	Pesticides potential removal			
	Partial potential		No potential	
	municipalities	%	municipalities	%
GD3	30	83.3%	6	16.7%
GD4	18	78.3%	5	21.7%
GD5	35	85.4%	6	14.6%
GD6	19	95.0%	1	5.0%
GD7	15	83.3%	3	16.7%
Total	117	84.8%	21	15.2%

Table 7. Proportion of municipalities of Minas Gerais with the potential removal of pesticides, by sub-basin of the Rio Grande - MG
Sources: IGAM, 2005, COPASA, 2005; IBGE, 2000.

Average slope of the sub-basin	GD3	GD4	GD5	GD6	GD7
Range slope	12 to 24%	12 to 24%	24 to 45%	24 to 45%	3 to 12%
Score	3	3	4	4	2

Table 8. Average slope of the sub-basins of the Rio Grande - MG.
Source: IGAM, 2006.

Criteria	Un.	Weight	Sub-basins				
			3	4	5	6	7
I	%	8	27.71	15.11	13.88	18.66	14.65
II	%	7	38.08	39.74	37.45	36.74	34.94
III	%	4	75.00	69.57	65.85	85.00	83.33
IV	-	6	3	3	4	4	2
V	mm	7	80.15	76.13	75.88	76.74	73.58

Table 9. Scoring and weighting of criteria for each sub-basin.
Source: IBGE, 2000; ANA, 2005; BRASIL, 2002; BRASIL, 2004a; MARTINS et al., 2004; COPASA, 2005; FAEMG, 2005; IGAM, 2005; IGAM, 2006.

strongly wavy to hilly. Analyzing criterion IV, sub-basins GD5 and GD6 have higher potential for dispersal of pesticides within their basins, with possibility of contamination of surface waters.

Criterion V, in turn, considers the influence of rainfall on the potential dispersal of pollutants in the environment. Sub-basin GD 3 shows the highest maximum annual rainfall among the selected areas.

Table 9 summarises values obtained in each sub-basin for each of the five criteria, and the weight given to the criteria considered in multicriteria analysis.

Table 10 shows the final values for each criterion in each of the five sub-basins, together with their respective weights parameterized, and the values for the "ideal" and "anti-ideal."

Criteria	Weight	Sub-basins					Ideal	Anti-ideal
		3	4	5	6	7		
I	0.250	0.319	0.165	0.152	0.204	0.160	**0.319**	**0.152**
II	0.219	0.202	0.211	0.200	0.200	0.188	**0.211**	**0.188**
III	0.125	0.206	0.214	0.226	0.175	0.179	**0.226**	**0.175**
IV	0.188	0.188	0.188	0.250	0.250	0.125	**0.250**	**0.125**
V	0.219	0.210	0.199	0.198	0.201	0.192	**0.210**	**0.192**

Table 10. Identification of Ideal and Anti-ideal for each criterion and each sub-basin.

Finally, Table 11 shows sub-basins ordering for pesticides surveillance in surface waters, based on the application of the multicriteria method.

Sub-basin	$p = 1$				$p = 2$			
	$d^{M}_{1}(a_i)$	$d^{m}_{1}(a_i)$	$D_1(a_i)$	TOPSIS	$d^{M}_{1}(a_i)$	$d^{m}_{1}(a_i)$	$D_1(a_i)$	TOPSIS
GD 3	1,608	6,411	0.800	1	1,213	4,375	0.783	1
GD 4	5,397	2,621	0.327	4	4,029	1,407	0.259	4
GD 5	4,649	3,370	0.420	3	4,183	2,446	0.369	3
GD 6	3,947	4,071	0.508	2	2,961	2,695	0.476	2
GD 7	7,776	243	0.030	5	4,689	204	0,042	5

Table 11. Hierarchy of systems by TOPSIS method (p = 1 and p = 2).

Figure 2 presents the descriptive statistics of the values of the similarity rate - Dp, considering 108 possibilities, for each value of p ($p = 1$ and $p = 2$). It can be observed that, for $p = 1$, the values of similarity rate of the sub-basin GD 3 are closer to the unit (average of 0.789) and that this value disagrees with the other sub-basins. The sub-basins GD 5 and GD 6 present a common range of values, between 0.45 and 0.50. Regarding sub-basin GD 7, the Dp values are significantly below the others and present smaller standard deviation. The sub-basin GD 5 presents the highest standard deviation, of 0.0247.

Comparing the diagrams presented in Figure 2, it can also be observed that the values of the similarity rate for $p = 1$ present smaller standard deviation than when $p = 2$. Dp values for the sub-basins GD 4 (Verde River Basin), GD 5 (around Furnas's Dam) and GD 6 (Pardo River and Mogí-Guaçu River Basins) are closer among themselves in $p = 1$ than in $p = 2$.

Despite the simulation with 108 possibilities, significant differences between the results were not observed, demonstrating the stability of the method.

At the end, multicriteria analysis by the method TOPSIS allowed the hierarchy of sub-basins studied in terms of priority for surveillance the presence of pesticides in water, as shown in Table 12.

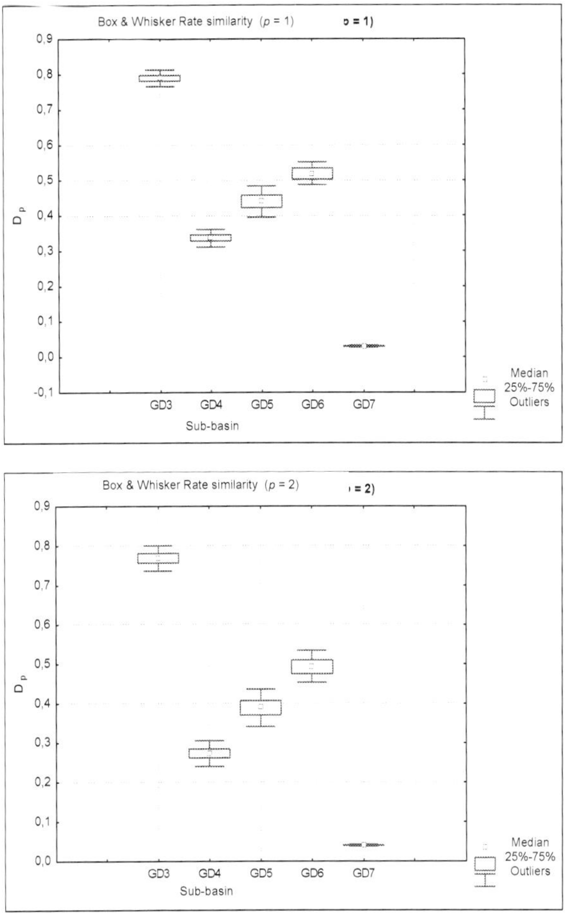

Fig. 2. Box & Whisker Rate similarity - Dp, for p = 1 and p = 2.

Priority	Sub-basin
1	GD 3
2	GD 6
3	GD 5
4	GD4
5	GD 7

Table 12. Prioritization of sub-basins of the Rio Grande - MG to monitoring the presence of pesticides in water.

6. Conclusion

The method for prioritization of pesticides surveillance in surface waters looks at intrinsic factors of the active ingredients and environmental factors that could influence in the dynamics of the pesticides in the environment and human health risks.

The application of the method in the south area of Minas Gerais allowed the ordering of priority sub-basins for pesticides surveillance, suggesting sub-basins GD 3 (around Furnas's Dam) and GD 6 (Pardo River and Mogi-Guaçu River Basins) as priorities. The validation performed enabled the evaluation and adjustment of the method, mainly regarding the availability of information. Another issue revealed by this application is the need of generating information to best feed the model and improve its outcomes.

The method showed as a practical alternative for the environmental surveillance, targeting priority areas, which is important in realities with limitations of technical, material or personnel resources. Moreover, its structure allows the application in other different areas and for other pollutants.

Difficulty in data gathering was observed, associated with dispersion of information in different public and private organizations. This situation is critical to fill the failures in planning environmental surveillance, particularly on drinking-water quality. Another crucial issue the research points out is the lack of effective integration among the public sectors with interface with the problem - health, water supply, water resources, agriculture - in order to ensure an adequate epidemiological and environmental surveillance of pesticides in water.

7. References

Agência Nacional de Águas - ANA. (2005). *Hidroweb* – Sistema de Informações Hidrológicas. http://hidroweb.ana.gov.br/ (accessed 25 feb. 2006).

Braga, B. & Gobetti, L. (2002). Análise multiobjetivo. In: PORTO, RL. (org.) *Técnicas quantitativas para o gerenciamento de recursos hídricos*. UFRGS (ed.), 2nd edn, Porto Alegre, Brazil. cap. 7, pp. 361-420.

Brazil (2002). Ministério da Agricultura, Pecuária e Abastecimento. Empresa Brasileira de Pesquisa Agropecuária - EMBRAPA Cerrados. *Potencial de impacto da agricultura sobre os recursos hídricos na região do Cerrado*. Documento 56. Dec., 2002.

Brazil (2004a). Ministério da Saúde. Agência Nacional de Vigilância Sanitária. *Sistema de informações sobre agrotóxicos*. http://www4.anvisa.gov.br/agrosia/asp/default.asp.

Brazil (2004b). Ministério da Saúde. *Portaria n. 518*, de 23 de março de 2004. Estabelece os procedimentos e responsabilidades relativas ao controle e vigilância da qualidade da água para consumo humano e seu padrão de potabilidade, e dá outras providências. http://www.funasa.gov.br/sitefunasa/legis/pdfs/portarias_m/pm1518_2004.pdf (accessed 25 oct. 2004).

Campanhola, C.; Bettiol, W. (2002). Panorama sobre o uso de agrotóxicos no Brasil. In: Ministério do Meio Ambiente (org.). Fórum Nacional de Secretários de Agricultura. *Programa de Defesa Ambiental Rural – textos orientadores*.

Cerejeira, M.J.; Viana, P.; Batista, S.; Pereira, T.; Silva, E.; Valério, M. J.; Silva, A.; Ferreira, M. & Silva-Fernandes, A.M. (2003). Pesticides in Portuguese surface and ground waters. *Water Research*, (37), 1055-1063.

Companhia de Saneamento de Minas Gerais - COPASA (2005). *Tipos de técnicas de tratamento de água por município operado pela COPASA.xls*. Belo Horizonte.

D'amato, C.; Torres, J. P. M.; Malm, O. (2002). DDT (Dicloro Difenil Tricloroetano): toxicidade e contaminação ambiental - uma revisão. *Química Nova*, v. 25, n. 6, p. 995-1002.

Dores, E.F.G. & De-Lamonica-Freire, E.M. (2001). Contaminação do ambiente aquático por pesticidas. Estudo de caso: Águas usadas para consumo humano em Primavera do Leste, Mato Grosso - Análise preliminar. *Química Nova*, (24)2, 27-36.

European Council (1998). *Council Directive 98/83/EC* of 3 November 1998 on the quality intended for human consumption. Official Journal of the European Comunities. L 330, p. 32-54.

Federação da Agricultura e Pecuária do Estado de Minas Gerais - FAEMG (2005). *Informações sobre o setor*. Dados de 2000 a 2004. http://www.faemg.org.br/ (accessed 15 aug. 2005).

Food and Agriculture Organization of the United Nations - FAO (1999). *Guidelines for the management of small quantities of unwanted and absolete pesticides*. Roma: FAO, 1999. 25p. (Boletim de Pesquisa, 2) *apud* Campanhola, C.; Bettiol, W. Panorama sobre o uso de agrotóxicos no Brasil. In: Ministério do Meio Ambiente (org.). Fórum Nacional de Secretários de Agricultura (2002). *Programa de Defesa Ambiental Rural - textos orientadores.*

Goss, D.W. (1992). Screening procedure for soils and pesticides for potential water quality impacts. *Weed Technology*, (6)3, 701-708 *apud* Dores, E.F.G. & De-Lamonica-Freire, E.M. (2001). Contaminação do ambiente aquático por pesticidas. Estudo de caso: Águas usadas para consumo humano em Primavera do Leste, Mato Grosso - Análise preliminar. *Química Nova*. (24)2, 27-36.

Hwang, C. & Yoon, K. (1981). *Multiple attribute decision making*. Methods and applications survey. Springer *apud* Pomerol, J-C. & Barba-Romero, S. (1993). *Choix multicritère dans l'entreprise* - principes et pratique. Hermès, Paris.

Infante-Rivard, C. & Weichenthal, S. (2007). Pesticides and childhood cancer: an update of Zahm and Ward's 1998 review. *Journal of Toxicology and Environmental Health, Part B*, 10, 81-99, 2007, ISSN: 1093-7404 print / 1521-6950 online

Instituto Brasileiro de Geografia e Estatística - IBGE (2000). *Programa Nacional De Saneamento Básico - PNSB*. Controle de qualidade na captação de água procedente de adutora de água bruta, segundo os distritos-sede dos municípios - Minas Gerais 2000. http://www.fjp.gov.br/produtos/cei/saneamento/Agua/AGUA-Tabela_5.1.xls (accessed 12 mar. 2006).

Instituto Mineiro de Gestão das Águas - IGAM (2000). *Unidades de planejamento e de gestão dos recursos hídricos do estado de minas gerais*. www.ana.gov.br/.../MG/ Outros/unidades%20planejamento%20tabela% 20final.doc (accessed 5 june 2006).

Instituto Mineiro de Gestão das Águas - IGAM (2005). *Relatório de monitoramento das águas superficiais na bacia do Rio Grande em 2004* / Instituto Mineiro de Gestão das Águas. IGAM, Belo Horizonte.

Kammerbauer, J. & Moncada, J. (1998). Pesticide residue assessment in three selected agricultural production systems in the Choluteca River Basin of Honduras. *Environmental Pollution*, (103), 171-181.

Kanrim, M. A. (1997). Pesticides profiles: Toxicity Environmental Impact and Fate. Florida, USA: Lewis Publishers *apud* Martins, M. D., Fernandes, C. S., Valente, J. T. Water contamination by pesticides. Case study: pesticides research in the Lower Cávado River Basin. In: World Water Congress, 4, 2004, Marrakesh. [*Anais eletrônicos...*] Marrakesh: IWA, 2004. 1 CD-ROM.

Laabs, V.; Amelung, W.; Pinto, A.; Altstaedt, A. & Zech W. (2000). Leaching and degradation of corn and soybean pesticides in an Oxisol of the Brasilian Cerrados. *Chemosphere*, (41), 1441-1449.

Larini, L. (1999). *Toxicologia dos Praguicidas*. Manole, São Paulo.

Luchini, L. C.; Andréa, M. M. de (2002). Dinâmica de agrotóxicos no ambiente. In: Brasil. Ministério do Meio Ambiente (org.). Fórum Nacional de Secretários de Agricultura. *Programa de Defesa Ambiental Rural - textos orientadores*.

Martins, M.D.; Fernandes, C.S. & Valente, J.T. (2004). Water contamination by pesticides. Case study: pesticides research in the Lower Cávado River Basin. In: World Water Congress, 4, 2004, Marrakesh. [*Proceedings...*] IWA, Marrakesh. 1 CD-ROM.

Minas Gerais (2004). Secretaria de Estado da Agricultura, Pecuária e Abastecimento. Empresa de Assistência Técnica e Extensão Rural do Estado de Minas Gerais - EMATER. *Principais agrotóxicos aplicados por cultura*. Mimeografado.

Moreira, J. C.; Jacob, S. C.; Peres, F.; Lima, J. S.; Meyer, A.; Oliveira-Silva, J. J.; Sarcinelli, P. N.; Batista, D. F.; Egler, M.; Faria, M. V. C.; Araújo, A. J. de; Kubota, A. H.; Soares, M. De O.; Alves, S. R.; Moura, C. M.; Curi, R. (2002). Avaliação integrada do impacto do uso de agrotóxicos sobre a saúde humana em comunidade agrícola de Nova Friburgo, RJ. *Ciência & Saúde Coletiva*, v. 7, n. 2, p. 299-311.

Pomerol, J-C & Barba-Romero, S. (1993). *Choix multicritère dans l'entreprise* - principes et pratique. Hermès, Paris.

Rissato, S.R.; Galhiane, M.S.; Ximenes, V.F.; Andrade, R.M.B.; Talamoni, J.L.B; Almeida, M.V. & Libânio, M. (2006). Organochlorine pesticides and polychlorinated biphenyls in soil and water samples in the Northeastern part of São Paulo State, Brazil. *Chemosphere* (Oxford), Vol. 65, No. 11, p. 1949-1958.

Tomita, R.Y. & Beyruth, Z. (2003). Toxicologia de agrotóxicos em ambiente aquático. *O Biológico*. São Paulo, (64)2, 135-142. http://www.geocities.com/~esabio /tomita.htm (accessed 25 may 2004).

Vilas Boas, C. L. (2005). Análise da Aplicação de métodos multicritérios de apoio à decisão (MMAD) na gestão de recursos hídricos. In: Simpósio Brasileiro De Recursos Hídricos, 16, João Pessoa. [*Proceedings...*] João Pessoa: ABRH, 2005. 1 CD-ROM

World Health Organization - WHO. (2004). *Guidelines for Drinking-water Quality*. 3ed., v. 1.

16

Herbicide Contamination of Freshwater Ecosystems: Impact on Microbial Communities

Villeneuve A.[1], Larroudé S.[2,3] and Humbert J.F.[2]
[1]*Institut Pasteur, Collection des cyanobactéries, Paris*
[2]*INRA, UMR Bioemco, Site de l'ENS*
[3]*ASCONIT Consultants, Lyon*
France

1. Introduction

Human society relies on surface freshwater ecosystems for many goods and services (drinking water, recreational facilities...), which places these systems at the center of a web of ecological, economic and political interests (Wilson & Carpenter, 1999). They are also being subjected to increasing pressure resulting from anthropogenic activities, including contamination by a variety of mineral and organic pollutants. Most of the organic pollutants are herbicides (e.g. Kreuger, 1998; Dorigo et al., 2007), which are used not only in agriculture but also for many other purposes (ranging from domestic use in gardens to maintaining railway tracks weedfree, for example). These herbicides can enter aquatic ecosystems as a result of terrestrial runoff, and to a lesser extent, of direct application and aerial spraying (e.g. Carter, 2000). Microbial communities in freshwater ecosystems are not directly targeted, but these communities are exposed to herbicides and can be directly or indirectly affected by these compounds. For instance, many commercial herbicides act by binding to Photosystem II (PSII), which is a pigment-protein membrane complex (see e.g. Schuler & Rand, 2008). PSII inhibitors have a direct impact on photosynthetic aquatic microorganisms that contain the same PSII apparatus as the terrestrial weeds targeted by these herbicides. In addition to this direct impact on photosynthetic microorganisms, as a result of the strong interactions that occur between all these aquatic microorganisms, herbicides can also have an indirect impact on non-photosynthetic species that are not susceptible to PSII inhibitors.

These effects on microbial communities can have a critical impact on the overall functioning of freshwater ecosystems, because prokaryotic and eukaryotic microorganisms are major players in these systems. Indeed, microorganisms contribute to most of the primary production in these systems, which is closely related to the total productivity (Aoki, 2003). They are also involved in nutrient cycling and decomposition, via the microbial loop (see, for example, the review of Fenchel, 2008). In lentic ecosystems (lakes, reservoirs, ponds...), most of these processes (primary production and nutrient cycling/decomposition) occur in the water column, and involve planktonic communities. On the other hand, in lotic ecosystems (rivers, streams...), these processes occur mainly in the biofilms located on the surface of sediments and plants, which involve benthic microbial communities. Depending on the kind of freshwater ecosystem concerned, both these communities must be considered when attempting to evaluate the impact of herbicides on microbial communities.

In this paper, we will start by brief overview of herbicide contamination of freshwater ecosystems, focusing particularly on a few specific countries in different geographical zones. We will then go on to present the main methods used to assess the impact of herbicides on microbial communities, ranging from single species tests to field studies. The third part of this review will focus on molecular methods that have been used in the last few years to evaluate the impact of herbicides on freshwater microbial communities. The fourth section will provide a brief summary of what is known about the impact of herbicides on freshwater microbial communities, and also on how these communities respond to this selective pressure. We will then look at the data available on glyphosate, which is one of the herbicides most often found in freshwater ecosystems. Finally, we will try to relate herbicide contamination and its impact on freshwater microbial communities to global changes, which are also impacting on these communities.

2. A rapid review of the contamination of aquatic ecosystems

2.1 General aspects

The information available about herbicide contamination of surface freshwater ecosystems varies considerably from country to country. A lot of information is available for countries in Europe and North America, but the situation is more patchy for Asia and South America, and there is almost no general or local data for Africa. This can be explained by the fact that so far the heavy use of herbicides has largely been confined to North and Latin America, Europe, Japan, and Australia, although their use is rapidly increasing in many developing countries (Ware & Whitacre, 2004). From all these data, it can be seen that even though the amount of herbicides entering the surface water varies considerably between regions, all surface freshwater ecosystems are now contaminated by herbicides. However, the concentrations of these herbicides in water vary considerably depending on the size and land use of the watersheds (e.g. Neumann et al., 2003), and also on season and climatic events. For example, peak contamination of rivers occurs just after discharge events, with relatively low levels between these events (e.g. Spalding & Snow, 1989; Botta et al., 2009). In agricultural landscapes, the amount of pesticides contaminating surface water also depends on the methods and levels of application, and more generally on agricultural practice (Huber et al., 2000). Finally, in a recent paper of Wittmer et al. (2010), it has been shown that five types of pesticide concentration patterns in surface water samples can be distinguished, and that these patterns reveal the relative contributions made by urban and agricultural land to the contamination of water by herbicides. This paper also showed that in mixed land use catchment areas, the contamination of surface water by biocides from urban areas was at least as great as that from agricultural areas. Thus, both urban and agricultural contaminations contribute to the contamination of surface water, although the most dramatic levels occur in areas characterized by intensive agriculture (see below).

2.2 Water contamination in some specific countries

In France, the "Institut Français de l'Environnement" (French Institute for Environment; now known as the Service de l'Observation et des Statistiques, SOeS) has published data on the contamination of water by pesticides for every year since 1998. Their annual report is based on monitoring 453 pesticides at 2023 sampling points (groundwater and rivers). In 2007, pesticides were detected at almost 91% of the sampling points, but usually at mean annual concentrations of <0.5 μg/L. The highest concentrations were found in regions with intensive agriculture (South-West, Center-North and North of France) and the lowest in regions (South-East and South of Massif Central) characterized by less intensive agriculture or by the presence of large areas of natural environments. The pesticide most often detected in French streams was AMPA (aminomethyl phosphonic acid), which is the primary degradation product of glyphosate. The second and the third most frequently identified pesticides in streams were diuron and glyphosate, which are both herbicides. More generally, the 15 most frequently occurring pesticides were all herbicides or degradation products of herbicides. Finally, changes in herbicide concentrations in French streams from 1997 to 2007, indicate that there has been no significant decrease in the herbicide pollution of streams, but that there have been changes in the dominant compounds found in these ecosystems. For example, atrazine, which was the most frequently-found herbicide in 1997, have been less often detected since it has been banned and replaced by glyphosate (Dubois et al., 2010). Data of this type is not available for lentic ecosystems, for which only occasional analyses have been performed. To compare the situation in France with that in another European country, a recent review of Greek freshwater ecosystems has shown that atrazine, simazine, metolachlor and alachlor were the most frequently detected herbicides (Konstantinou et al., 2006). Glyphosate was not mentioned in this review, but we do not know if this was because most of the data were obtained before massive use of this herbicide began, or because Greece does not have large areas of corn production. However these studies have shown that the levels of contamination of major European rivers by herbicides such as atrazine or simazine, are similar in all countries.

We do yet have overall data for the level of contamination of surface freshwater ecosystems by herbicides in China, but three papers have provided some indications of the level of herbicide contamination in Chinese surface freshwater. Gfrerer et al. (2002a,b) have shown that triazines were detected throughout the year in three rivers from China, with a concentration peak in June due to their use in agriculture. But, apart from the June values, the concentrations of triazine were quite low compared to those in European rivers. On the other hand, in an analysis of 14 different organic pesticides and their metabolites in Lake Taihu, Na et al. (2006) reported very high concentrations of atrazine. This herbicide has been widely used in China since the 1980s, with an annual increase of 20% (*in* Na et al., 2006). More data are available for another Asian country, Japan. For example, in a study on the contamination of the lake Biwa, Sudo et al. (2002 & 2004) have shown that three of the pesticides most frequently found in this lake and in the rivers flowing into it were herbicides (simetryn, molinate and bromobutide), which are all used in paddy fields, and so found in numerous rice producing areas.

It appears that triazines and in particular atrazine and simazine, are very frequently found in surface freshwater ecosystems worldwide, as are glyphosate and its degradation product, AMPA. But, local differences do occur, and are related to the kind of agriculture or to the banning of specific compounds.

3. The main methods used to estimate the impact of herbicides on microbial communities

Ecotoxicology deals with the evaluation of the processes and mechanisms by which toxicants are dispersed, and with their effects on populations, communities and ecosystems. This implies the need for studies at different geographical and time scales, which has resulted in the development of a wide variety of methods (Fig. 1)

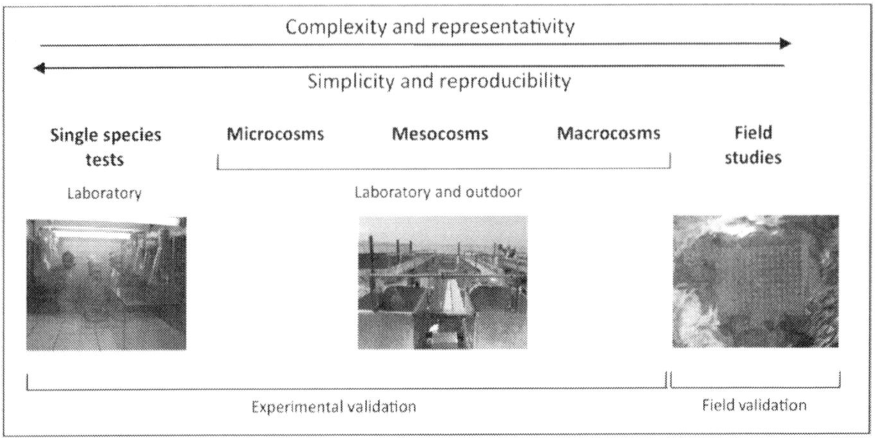

Fig. 1. Various levels of biological complexity in the experimental methods used in ecotoxicology (from Caquet et al., 1996)

Many studies are based on single-species tests, and these are usually considered to concern environmental toxicology rather than ecotoxicology (Ramade, 2007). These studies are fundamentally reductionist, and they give priority to the reproducibility of the data acquired (single-species tests and microcosm experiments). In such studies, the number of experimental parameters is limited, which makes them easier to control. In contrast, studies conducted in mesocosms or in macrocosms are closer to those performed in ecosystems, which makes them more realistic. They are designed to provide an appropriate way to assess the real impact of a pollutant in a natural ecosystem. However, these studies are more difficult to conduct, and often less reproducible, because of the large number of factors and processes involved, many of which are not easy to control.

3.1 Single species tests

Ecotoxicity tests are a useful way of determining the effect of one or more products on selected organisms, under defined conditions (Keddy et al., 1994). These tests can be classified according to their duration, compared to the life cycle of the organism, and to the complexity of the biological community studied. Acute toxicity tests cover a relatively short part of the life cycle of the organism. Bacteria, protists and algae have life cycles lasting 24-48 hours. In contrast, chronic toxicity tests involve the repeated exposure of organisms to low doses of a contaminant over a long period.

In order to characterize the toxicity of a substance in a regulatory context, it is necessary to evaluate both its acute and chronic toxicity at several trophic levels. Various estimators have been defined in order to do this:

- EC_{50}: The "half maximal effective concentration" is the concentration of a toxicant that induces a response halfway between the baseline and maximum response after some specified exposure time.
- LC_{50}: The Lethal Concentration 50 is the concentration of a toxicant that causes 50% mortality

- NOEC: No Observed Effect Concentration.
- LOEC: Lowest Observed Effect Concentration.

In addition to these indicators, water resource managers also use the concepts of:

- PNEC: Predicted No Effect Concentration
- PEC: Predicted Effect concentration

The EC_{50} value is estimated using dose-response relationships. Generally, increasing the dose of a toxicant causes a proportional increase in its toxic effect on the biological parameters investigated. The dose-response curve is based on the following assumptions:

- The response increases with dose,
- There is a threshold dose below which there is no measurable effect.

In aquatic ecosystems, algae are known to be sensitive to many chemicals, and using these organisms in test batteries has been shown to improve the capacity to predict the most sensitive ecosystem responses. The importance of these organisms in the primary production of most aquatic ecosystems justifies their choice for risk assessment. The usefulness of single or multi-species tests to characterize the ecotoxicological properties (EC50, NOEC, LOEC...) of potentially polluting molecules has been proven. Performed in the laboratory under controlled conditions, they meet repeatability, reproducibility, reliability and robustness criteria. However, their use for assessing and predicting effects on natural ecosystems is limited because of their lack of ecological realism. To address this aspect, it is necessary to work at a higher level of biological organization, i.e. with more complex experimental systems.

3.2 Microcosms, mesocosms and macrocosms

All these experimental systems, which differ by their volume (ranging from several mL to several m^3) and their biological complexity (from a few strains to complete trophic networks), are used to study the impact of a toxicant on communities. These systems take into account both positive and negative interactions between organisms and also those with various abiotic factors. They can be used to establish an "exposure-response" relationship, rather than just a "concentration-effect". Because of the relatively large size of these systems, the tests can be conducted with long exposure times, which makes them more suitable for studying the effects of contaminants on the dynamics of species succession, and on ecological processes, such as the cycling of matter and energy flows (Guckert, 1993). These systems can also be used to evaluate the impact on diversity and functioning of the intensity of the contamination (concentration and form of contaminants) and the exposure dynamics (duration, frequency), whilst simultaneously investigating the fate and the effects of pollutants (Belanger et al., 2000). These methods can also be used to identify the temporary, progressive or persistent effects of contaminants on communities (Rand et al., 2000; Belanger et al., 2002), by measuring their resilience, and also to distinguish between the direct and indirect ecological effects of disturbances caused by contaminants (Belanger et al., 2000; Culp et al., 2000; Hense et al., 2003).

Just to provide recent examples of the variety of systems used, 3L liter Pyrex Erlenmeyer flasks have been used to study the response of microbial communities following exposure to glyphosate (Pesce et al., 2009a), indoor experimental channels have been used to study the combined effect of physical factors and exposure to diuron on benthic microbial communities (Villeneuve, 2008), and artificial outdoor mesocosms (surface area 25 m^2) have been used by Vera et al. (2010) to evaluate the impact of Roundup on periphyton communities.

To conclude, all these systems can be defined as essential intermediates between single-species tests and *in situ* studies (Caquet et al., 2000) in exploring the effects of toxicants and their potential interaction with environmental factors.

3.3 Field studies to evaluate the impact of contaminants on microbial communities

Numerous studies have attempted to evaluate the impact of herbicide contamination on microbial communities in freshwater ecosystems (see, for example, the reviews of DeLorenzo et al., 2001; Ricciardi et al., 2009). From these studies, it appears that one of the difficulties that arises is that microbial communities are subjected simultaneously to herbicides and to many other environmental factors and processes (see §5.3). In order to evaluate the direct impact of herbicides on microbial communities, Blank et al. (1988) have developed a very interesting concept, known as PICT (Pollution Induced Community Tolerance). This concept is based on the fact that communities contain species with differing sensitivities to various chemicals (see for example for diatoms, Debenest et al., 2009). After exposure to these pollutants for long enough, sensitive species or strains will be eliminated and tolerant ones will be selected, and the resulting restructured community will become more tolerant. Increased tolerance in a community can therefore be regarded as indicative of a direct impact of a pollutant on this community (for example for microalgal communities, see Bérard et al., 2003; Schmitt-Jansen & Altenburger, 2005). This pollution-induced community tolerance can be estimated by means of various different metabolic tests.

4. Molecular methods used to evaluate the impact of herbicides on microbial communities

Over the past 15 years, molecular tools have been increasingly used in ecotoxicological studies, mainly to assess the impact of pesticides on the structure and composition of microbial communities. Two approaches can be distinguished (Dorigo et al., 2005). The first group of tools uses an initial PCR to target one or several specific genes in the microbial communities, whereas the second group consists of non-PCR-approaches.

4.1 Methods based on an initial PCR amplification

The Polymerase Chain Reaction (PCR) is used to amplify target sequences in microbial communities. Depending on the choice of the primers, these sequences may be more or less specific, i.e. they target different taxonomic levels, ranging from a single species to the entire prokaryotic community. In most cases, rRNA genes, in particular 16S (prokaryotes) and 18S (eukaryotes) rRNA, are targeted, because they contain highly-conserved domains interspersed with more variable regions (Gutell et al., 1994). However other genes can also be used, for example those linked to a particular function, such as a herbicide degradation gene (Lee et al., 2005; de Lipthay et al., 2002).
After an initial PCR amplification, two main kinds of method are used to characterize the amplified sequences. The first is based on sequencing these amplicons and identifying the sequences obtained, for example by comparison with sequences available in databases (GenBank™). New sequencing technologies (454 sequencing) now make direct sequencing of the amplicons after PCR possible, but so far, sequencing is always preceded by a cloning step. Amplicon sequencing is used to compare the composition of different communities, for example with more or less exposure to a herbicide, and it provides a species identification

when rRNA genes are used, as has been successfully done in several ecotoxicological studies of herbicides (e.g. Dorigo et al., 2002).

The second kind of methods used after PCR amplification are known as fingerprinting methods, because they provide an overview of a microbial community (like a bar code), making it easy to compare several communities. These methods include SSCP (Single Strand Conformation Polymorphism), ARISA (Automated Ribosomal Intergenic Spacer Analysis), RAPD (Random Amplified Polymorphic DNA), T-RFLP (Terminal Fragment Length Polymorphism), and AFLP (Amplified fragment-Length Polymorphism), but DGGE (Denaturing Gradient Gel Electrophoresis) and TGGE (Temperature Gradient Gel Electrophoresis) are the ones that have been most frequently used in ecotoxicological studies involving microbial communities. These two methods were first introduced 15 years ago (Muyzer, 1999). They are based on the separation of the amplified fragments of DNA on a linear gradient of increasing chemical denaturants of urea and formamide (DGGE), or on a linear temperature gradient (TGGE). After the migration, a band pattern will characterize each amplified sample, and is used to compare them. Moreover, individual bands can be excised, reamplified and sequenced, and thus identified (Riemann and Winding, 2001). DGGE/TGGE approaches have been used in a huge number of studies to characterize the impact of herbicides on microbial communities (e.g. Lyautey et al., 2003; Schauer et al., 2003; Pesce et al., 2009a; Tadonléké et al., 2009). These methods provide a quick way of comparing several samples, but they have their limitations. In particular, it is very difficult to compare large numbers of complex band patterns across several migration gels (Dorigo et al. 2005). Other limitations are due to the fact that different DNA fragments often co-migrate, which limits the sensitivity of these methods. Finally, only dominant species are detected well, which makes it difficult to highlight changes in the presence/absence of rare species.

4.2 Non-PCR based molecular methods

These methods have so far been used less often than those based on the PCR, but the rapid development of "omic" approaches in microbial ecology suggests that they will probably soon be used in ecotoxicology.

In recent years, a few papers have been published reporting the use of the FISH (Fluorescence in situ hybridization) technique, which is based on the use of fluorescent probes to quantify different phylogenetic groups by fluorescence microscopy. This method was used, for example, by Pesce et al. (2006), in a study of the effect of the phenylurea herbicide diuron on river microbial communities. However, this method can also be used for the *in-situ* detection of bacterial strains that are able to degrade herbicides, as for example by Grenni et al. (2009).

The DNA reassociation kinetics method has been also used to compare the genetic diversity of bacterial communities in more or less polluted soils (Gans et al., 2005), but not so far applied to freshwater ecosystems. This method is unlikely to be used in the near future in ecotoxicological studies in natural environments, due to the technical difficulties of applying this method in this context, and also due to the difficulty of analyzing the results.

The use of microarrays is also a very promising method developed in the past. This method is based on the use of slides (microarrays) containing a very large number of microscopic spots of DNA oligonucleotides probes (from several hundred to several thousand). These microarrays can be used to detect and quantify the species in a community after extracting the DNA and hybridizing it on the microarray. However, they are usually used to study the gene expression of a strain under various stressful conditions and in mixed microbial communities (Dennis et al., 2003). This method has also been used by Oldenburg et al. (2003) to monitor the diversity of the genes involved in the nitrogen cycle. However, the development of high throughput sequencing methods and the concomitant decrease in their cost will make probably the use of microarrays obsolete within a few years

4.3 Promising new molecular tools for ecotoxicological studies

All the methods described above have been in use for more than a decade, but there are many more recently developed methods, which look very promising for studying the impact of herbicides on freshwater microbial communities, and the responses of these communities.

Real Time PCR (RT PCR or qPCR) is not really a recent method, but its routine use in the laboratory has dramatically increased in the last year. This method can be used to determine the number of copies of target genes or the expression of target genes in a sample. For example, qPCR has been recently used by Bopp & Lettieri (2007) to study gene regulation in a marine diatom exposed to polycyclic aromatic hydrocarbons. Another example is the work of Gonod et al. (2006), who used qPCR to study the ability of bacterial communities in soil to degrade the 2,4-D (2,4-dichlorophenoxyacetic acid) herbicide. Although so far there have been no publications involving the use of this method in freshwater environments, it will probably soon be widely used in ecotoxicological studies of microbial communities.

So far, there has also been no paper reporting the use of isotope-labeling experiments (DNA-SIP, RNA-SIP...) to study, for example, the bacterial degradation of herbicides in aquatic environments. However in recent review, Neufeld et al. (2007) disclaimed that isotope-labeling experiments (DNA-SIP, RNA-SIP...) are coming of age, and are likely to be very useful for studying the degradation of pollutants by microbial communities. A paper by Park et al. (2006) also highlighted the potential of Stable Isotope Probing for enhancing the resolution of ecotoxicological assessment.

Finally, the introduction of "omic" methods opens a new area in the assessment of the impact of pesticides on microbial communities. The use of metagenomic, metatranscriptomic and metaproteomic methods in recent years has shown that environmental microbial ecology is already entering the "Omics" era (see for example Nelson et al., 2006; Vandenkoornhuyse et al., 2010). Despite some restrictions, there is no doubt that all these methods will soon be major tools for ecotoxicological studies.

5. Impact of herbicides on freshwater microbial communities

The impact of herbicide on freshwater microbial communities depends on a large variety of factors linked to the pollutant chemical, its mechanism of action, and the time and the level of exposure... However, the impact is also modulated by interactions with physical (flow, temperature...), chemical (nutrients, organic matter, salinity...), and biological factors and processes (the presence of grazers...).

The impact of herbicides on microbial communities can be estimated using structural and functional descriptors, which provide information about the short- and long-term effects of these compounds (Fig. 2).

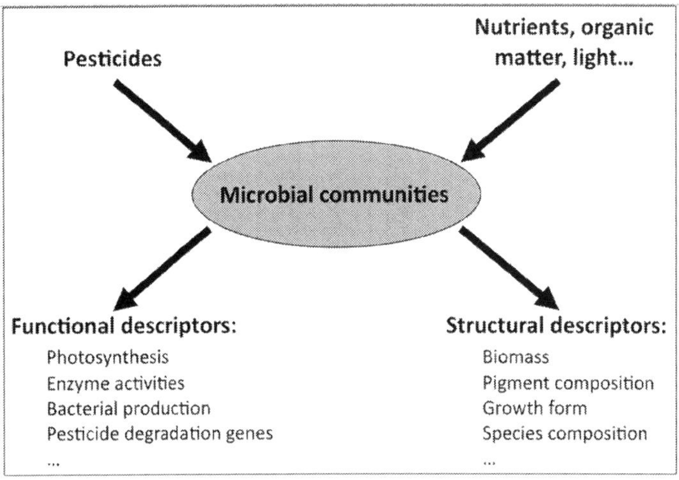

Fig. 2. Flow chart of the causes and effects of chemical and physical disturbances on microbial communities (adapted from Sabater et al., 2007)

5.1 Impact of herbicides on the composition and structure of freshwater microbial communities

Many authors have highlighted the fact that herbicides have significant effects on the composition and structure of freshwater microbial communities, by estimating the species richness and the diversity of autotrophic and heterotrophic microorganisms, and also by estimating their biomass and abundance.

Herbicides can have a direct impact on the composition of freshwater microbial communities by selecting those with greater herbicide resistance (Dahl & Blanck, 1996; Pérès et al., 1996; DeLorenzo et al., 1999). For example, Dorigo et al. (2007) found that there were changes in the composition of microbial communities in a small river displaying an upstream-downstream pollution gradient, and that species located in the polluted downstream area were more tolerant towards herbicides than those located in the upstream-unpolluted area. Recently, Vera et al. (2010) have shown that diatoms were more susceptible than cyanobacteria to glyphosate in periphyton communities, and consequently that over the long term a shift occurred in the composition of these communities. In the same way, Lürling & Roessink (2006) showed, by an experimental approach, that *Scenedesmus* (green algae) out competed *Microcystis* (Cyanobacteria) in the absence of metribuzin, a photosynthesis-inhibiting herbicide, whereas the reverse was true in the presence of this herbicide. However, herbicides can also have an indirect impact on the species composition of these communities by modifying the equilibria between species and also the interactions between them (as a result of effects on potential competitors or predators).

In addition to photosynthetic microorganisms, which are directly impacted by herbicides because they share some physiological metabolisms with terrestrial plants, these compounds can also have direct impacts on bacteria. For example, in a study of Foley et al. (2008), DGGE fingerprinting showed that freshwater bacterial communities exposed to acetochlor, an inhibitor of cell growth and division, displayed reduced diversity. However, contradictory results have been obtained, and, for example Tadonleke et al. (2009) did not find any significant impact of diuron, a phenylurea herbicide, on bacteria, whereas previous studies had seemed to do so.

In addition to their effects on the composition of microbial communities, herbicides can also affect the viability and abundance of microorganisms and some other cellular parameters. For example, Ricart et al. (2009) found that there was a reduction of the biovolume of diatoms at low concentrations of diuron.

Herbicide exposure usually has a harmful effect, but some microalgae bioassays of various compounds have also revealed beneficial effects (Haglund, 1997; Franqueira et al., 1999; Wong, 2000; Rioboo et al., 2002; Yoshida et al., 2003). Stimulation of microalgal growth was, for example, found at the lowest paraquat concentrations tested in the study of Prado et al. (2009). Some papers have suggested that the growth stimulation observed in the presence of herbicides, such as glyphosate, may result from the use of degradation products of this compounds as sources of carbon, phosphorus or nitrogen (Wong, 2000). In the same way, bacterial species that are able to degrade herbicides are also stimulated by their presence, thus enabling them to become dominant in bacterial communities (Macur et al., 2007).

5.2 Impact of herbicides on the functional aspects of freshwater microbial communities

In addition to changes occurring in the composition and structure of microbial communities, their metabolic activities (respiration, photosynthesis, enzyme activities) can also be modified by herbicide application. These modifications may or may not be linked to changes in the composition of the community.

Ricart et al. (2009) observed a marked decrease in the photosynthetic efficiency of biofilms after a long-term exposure to low concentrations of diuron. There was also an increase in the proportion of dead bacteria and in the leucine-aminopeptidase bacterial activity with increasing concentrations of diuron. This increase in peptidase activity may be explained by the release of algal material as a result of cell lysis, which might provide bacteria with organic proteinaceous compounds. Similar results have been observed for other photosystem II inhibitor herbicides in freshwater microbial communities (Guash et al., 1997, Pesce et al., 2009; Villeneuve, 2008).

The microbial community could also acclimate to the new conditions by the induction of enzyme activity, i.e. gene expression. For example by Lipthay et al. (2002) observed the induction of *tfdA* transcription in aquatic microbial communities during acclimation to 2,4-D herbicide acid (a phenoxy herbicide, which affects protein synthesis and cell division). The *tfdA* gene is known to encode a 2,4-D/2-oxoglutarate dioxygenase involved in the degradation of 2,4-D to dichlorophenol.

Method dealing with changes in the composition and structure of microbial communities, and in some of their physiological capacities when they are exposed to herbicide pollution, provide complementary information. Indeed, an impact on some specific functions may not necessarily affect the structure of the community and *vice versa*. For example, Widenfalk et al. (2008) showed that no effect of glyphosate could be detected in microbial communities from sediments exposed to this herbicide using descriptors such as bacterial production or microbial biomass, but that changes in the composition of the bacterial community could be detected using T-RFLP.

5.3 Interactions between environmental selective pressures and herbicide contamination

Many studies have reported the effects of contaminants on natural freshwater microbial communities, but in all these studies, the main difficulty was distinguishing between the impact of contaminants and that of other environmental factors and processes. It is well known, for example, that the response of microbial communities to herbicides depends on the composition of the microbial community, which is itself influenced by (i) physico-chemical and other factors such as nutrient resource availability (DeLorenzo et al., 2001; Shabana et al., 2001), temperature (Bérard et al., 1998), light history (Guasch & Sabater, 1998), current velocities for periphyton communities (Villeneuve, 2008) and previous pollutant exposures (Dorigo et al., 2004), and (ii) biological interactions between microorganisms and biotic factors, such as predation by grazers (Muñoz et al., 2001). All these factors introduce seasonal variations in the structure and composition of the microbial communities of freshwater ecosystems, which influence their response to exposure to a contaminant. For example, Guash and Sabater (1998) showed that the differences in algal diversity linked to the light history of a biofilm can lead to differing responses to the presence of herbicides. Seasonal factors or spatial variability can also result in communities displaying differing diversity and differing susceptibilities to contaminants (Dorigo et al., 2009), including their ability to degrade pesticides (Pesce et al., 2009b). In a long-term laboratory experiment to test the effect of atrazine after grazing Muñoz et al. (2001) demonstrated that grazers increased the toxic effect of atrazine on river biofilms. In fact, in a context of heavy grazing, atrazine at 14 µg/L caused a greater decrease in periphyton carbon incorporation than either of these two factors individually. They also showed that grazing hindered adaptation to atrazine exposure. On the other hand, no significant interactions were found between diuron and grazers in a river biofilm community (López-Doval et al., 2010).

6. Special focus on the impact of glyphosate on freshwater microbial communities

Glyphosate (N-(phosphomethyl)glycine), the active ingredient in the commercial product Roundup, is a broad spectrum, non-selective and post-emergent herbicide, developed in 1971 by Monsanto (Franz et al., 1997). Since 1997, its agricultural use has increased considerably as a result of the introduction of genetically-engineered "Roundup Ready" glyphosate tolerant varieties of soybean, cotton and maize (Giesy et al., 2000; Woodburn, 2000). It has become one of the most commonly-used herbicides for agricultural weed control worldwide (Giesy et al., 2000; Kolpin et al., 2006), and also for domestic and industrial weed control in gardens or along rail tracks (Woodburn, 2000; Kolpin et al., 2006; Ghanem et al., 2007; Botta et al., 2009).

6.1 The mechanism of action and different formulations of glyphosate

Glyphosate acts by inhibiting 5-enolpyruvyl-shikimate-3-phosphate synthase (EPSPS), which is involved in a pathway common to all higher plants as well as to some microorganisms, the shikimate pathway (Amrhein et al., 1980; Carlisle & Trevors, 1988). As a result, it disrupts the biosynthesis of aromatic amino acids required by those organisms, for instance, to synthesize peptides, secondary metabolites or vitamins.

The initial glyphosate molecule is a free acid, and it is converted into a salt form, such as isopropylamine (IPA), monoammonium, diammonium, or potassium salts (in newer formulations), to make it water soluble and thus to able to move into the plant (Duke, 1988; WHO, 1994). There are numerous commercial formulations, most of which are based on the IPA salt (Table 1).

Product	Company	Formulation	
		g AE/L	Salt
Touchdown HiTech	Syngenta	500	Potassium
Roundup WeatherMax	Monsanto	540	Potassium
Roundup PowerMax	Monsanto	540	Potassium
Durango DMA/Duramax	Dow AgroSciences	/	Dimethylamine
Touchdown Total	Syngenta	500	Potassium
Roundup Ultra Max	Monsanto	450	IPA
Cornerstone	Agriliance	356	IPA
Credit	Nufarm	356	IPA
Credit Extra	Nufarm	360	IPA
Glyfos	Cheminova	360	IPA
Glyfos X-tra	Cheminova	356	IPA
Gly Star Original	Albaugh / AgriStar	356	IPA
Glyphomax	Dow AgroSciences	356	IPA
Glyphomax Plus	Dow AgroSciences	360	IPA
Honcho	Monsanto	360	IPA
Makaze	UAP	360	IPA
Roundup Original	Monsanto	356	IPA
Touchdown IQ	Syngenta	360	Diammonium

Table 1. Some of the commercial formulations of glyphosate. The acid equivalent (AE) is the corresponding weight of glyphosate acid, which is the active ingredient.

Those commercial formulations generally include a surfactant, which enhances the ability of glyphosate to penetrate through the cuticular waxes on target plants. With the exception of Roundup, in which the surfactant is known to be polyethyloxylated tallowamine (POEA), information about the surfactants involved are generally not clearly stated by the manufacturers. Furthermore, several studies have shown that the surfactant molecules also make the commercial product (in particular, Roundup) more toxic than the technical grade compound (salt or free acid forms) (Powell et al., 1991; Diamond & Durkin, 1997; Giesy et al., 2000; Tsui & Chu, 2003).

6.2 Glyphosate in the environment: degradation and occurrence in freshwaters

Glyphosate has long been assumed to be safe for the environment due to its supposed rapid biodegradation by soil microorganisms and/or the fact that it is tightly adsorbed by soil particles (Sprankle et al., 1975; Rueppel et al., 1977; Carlisle & Trevors, 1988; Pipke &

Amrhein, 1988; Kolpin et al., 2006). The main degradation pathway appears to involve splitting the C-N bond to produce aminomethyl phosphonic acid (AMPA) (Rueppel et al., 1977). However, despite being biodegraded in the soil, glyphosate is frequently found in freshwater ecosystems. In these ecosystems, the herbicide is dissipated by the same microbial biodegradation as in the soil, and additionally, through its adsorption onto sediments with the subsequent microbial breakdown of bound residues under anaerobic conditions (Tooby, 1985; Mallat & Barceló, 1998). This degradation generally occurs more slowly in water, because it contains fewer microorganisms than soil (Ghassemi et al., 1981). Glyphosate and its main degradation product AMPA are among the most common pesticides detected in water-pollution monitoring (Scribner et al., 2007). However, although glyphosate is the main source of AMPA, it is not its only source (Trass & Smit, 2003), since phosphonate compounds can also be degraded to form AMPA (Novack, 1997; Skark et al., 1998; Kolpin et al., 2006; Botta et al., 2009).

Most studies report fairly low concentrations in surface freshwater ecosystems in regions where glyphosate use is very high (Table 2). However, a few studies have reported very high glyphosate concentrations, such as 700 or 1700 μg/L in some ecosystems (Table 2). These very large differences in concentrations may arise from differences in (i) the proximity of the site of application to water-bodies, (ii) terrain and soil types and also (iii) the contribution of overspray (WHO, 1994). In general, these differences suggest the possibility of short-term pulse phenomena, which are likely to occur in water contamination as previously found for other herbicides (e.g. Wauchope, 1978; Shipitalo & Owens, 2006; Gilliom, 2007).

6.3 Impact of glyphosate on aquatic microorganisms

There have been numerous studies of the toxicity of glyphosate on freshwater prokaryotic and eukaryotic photosynthetic microorganisms (Table 3), but most of them have been based on the use of isolated strains in laboratory experiments.

As Table 3 shows, a lot of data is available about the toxicity of glyphosate towards the *Chlorophyceae* and *Cyanobacteria*, and to a lesser extent towards diatoms. These data show that widely differing EC_{50} values have been found, depending on the species and formulations involved. For example, the *Chlorophyceae* exhibited EC_{50} values ranging between 2.1 and 1082 mg/L and *Chlorella pyrenoidosa*, EC_{50} values of between 3.5 and 590 mg/L. These wide differences could arise as a result of differences in study design, such as the duration of the experiments, the glyphosate formulation used (e.g. Sáenz et al., 1997), which is not always clearly identified, and the EC_{50} measurements, which may be based on the biomass, the chlorophyll concentration, or on the carbon uptake. In the same way, the *Cyanobacteria*, which are photosynthetic prokaryotes, exhibited a wide range of EC_{50} values varying from 2 to 1183 mg/L. However, despite this variability in EC_{50} values, *Cyanobacteria* in general appear to be less sensitive to glyphosate than eukaryotic photosynthetic microorganisms (Powell et al., 1991). This tolerance of glyphosate could be explained by various mechanisms, notably the ability of *Cyanobacteria* to metabolize phosphonate (Forlani et al., 2008). Little data is available about the toxicity of the main degradation product AMPA. As far as we are aware, the only values reported for its EC_{50} (72 h) are 72.9 mg/L and 79.7 mg/L for the *Chlorophycea* species *Scenedesmus subpicatus* (Trass & Smit, 2003; AFSSA, 2006). These values are consistent with the hypothesis that the toxicity of AMPA is equal to or less than that of glyphosate (Carlisle & Trevors, 1988; Giesy et al., (2000).

	Time period	Glyphosate (µg/L)		AMPA (µg/L)		Reference
		Min	Max	Min	Max	
Canada						
Rivers, streams & wetlands	2004-05	1	41	/	66	Struger et al., 2008
Urban & rural watersheds	2007	0.02	12	0.04	1.3	Byer et al., 2008
USA						
Stream	1987	35	1237	<1.0	10	Monsanto, 1990
Pond	1987	90	1700	2	35	Monsanto, 1990
Streams, rivers, lakes, wetlands & vernal pools	2001-06	0.02	427	0.02	41	Scribner et al., 2007
Stream	2002	0.1	8.7	0.4	3.6	Battaglin et al., 2005
Streams & wastewater effluents	2002	<0.1	2.2	<0.1	3.9	Kolpin et al., 2006
Vernal pools	2005-06	<0.02	328	<0.02	41	Battaglin et al., 2009
Argentina						
Streams	2003-04	100	700	/	/	Peruzzo et al., 2008
France						
NR	1999	/	6.06	/	5.05	Horth & Blackmore, 2009
NR	2000	/	35	/	2.99	Horth & Blackmore, 2009
Streams & rivers	2003-04	2	165	2.1	48.1	IFEN, 2006
River	2003	0.23	0.74	0.17	3.76	Pesce et al., 2008
NR	2004	/	50	/	17	Horth & Blackmore, 2009
Streams & rivers	2005	2.1	17	2.1	18.8	IFEN, 2007
NR	2005	/	9.6	/	30	Horth & Blackmore, 2009
Streams & rivers	2006	2.0	34.0	2.2	27.5	Horth & Blackmore, 2009; IFEN, 2009
Sweden						
NR	2000-08	0.06	13	0.07	4	Horth & Blackmore, 2009

NR = type of water not reported; / = no data.

Table 2. Data available on glyphosate and AMPA concentrations in surface water

	Glyphosate formulation	EC_{50} (mg/L)	European toxicity classification	Reference
Bacillariophyceae				
Nitzschia sp.	IPA salt	<2.8	Toxic to very toxic	Peterson et al., 1994
Cyclotella meneghiana	IPA salt	<2.8	Toxic to very toxic	
Navicula pelliculosa	96.6%	39.9	Harmful	Hughes, 1987b [M]
	Acid	17	Harmful	Office of pesticides programs, 2000; Smyth et al., 1996[M]
		42	Harmful	Smyth et al., 1996[M]
	IPA salt	38.6	Harmful	Office of pesticides programs, 2000
Chlorophyceae				
Scenedesmus obliquus	95%	55.9	Harmful	Ma, 2002
	Knockdown	80	Harmful	Ermis & Demir, 2009
Scenedesmus quadricauda	95%	70.5	Harmful	Ma et al., 2003
	IPA salt	>>2.8		Peterson et al., 1994
		7.2	Toxic	Sáenz & Di Marzio, 2009
		9.02	Toxic	Sáenz et al., 1997
	Rondo	7.2	Toxic	Sáenz et al.1997
	Roundup	120	NC	Sáenz & Di Marzio, 2009
Scenedesmus acutus	97.5%	24.5	Harmful	Vendrell, et al., 2009
	IPA salt	10.2	Harmful	Sáenz, et al., 1997
		10.2	Harmful	Sáenz & Di Marzio, 2009
	Rondo	9.08	Toxic	Sáenz, et al., 1997
Scenedesmus subscapitatus	97.5%	26	Harmful	Vendrell et al., 2009
	IPA salt	72.9	Harmful	Tomlin, 1997
Chlorella saccharophila	97.5%	40.6	Harmful	Vendrell et al., 2009
Chlorella vulgaris	97.5%	41.7	Harmful	Vendrell et al., 2009
	95%	4.7	Toxic	Ma et al., 2002
	IPA salt	13.1	Harmful	Sáenz & Di Marzio, 2009
Chlorella pyrenoidosa	96.7%	590	NC	Maule & Wright, 1984

	Glyphosate formulation	EC_{50} (mg/L)	European toxicity classification	Reference
	95%	3.5	Toxic	Ma & Wang, 2002
		3.5	Toxic	Ma, 2002
	Roundup	189	NC	Hernando et al., 1989
	NR	380	NC	Anton et al., 1993
	NR	1082	NC	Anton et al., 1993
Chlorella fusca	NR	377	NC	Faust et al., 1993
Chlorococcum hypnosporum	96.7%	68	Harmful	Maule & Wright, 1984
Pseudokirchneriella subcapitata [1]	96.6%	12.5	Harmful	Office of pesticide programs, 2000
		13.8	Harmful	Hughes, 1987c [M]
		460	NC	Smyth et al., 1995 [M]
		485	NC	Smyth et al., 1995 [M]
		270	NC	Cedergreen & Streibig, 2005
	95%	21.8	Harmful	Bozeman & Koopman , 1989
		129	NC	Pereira et al., 2009
	IPA salt	24.7	Harmful	Tsui & Chu, 2003
		41	Harmful	Tsui & Chu, 2003
	Roundup	5.8	Toxic	Tsui & Chu, 2003
		2.1	Toxic	LISEC, 1989a [M]
		8	Toxic	LISEC, 1989a [M]
		64.7	Harmful	Cedergreen & Streibig, 2005
	Spasor	71	Harmful	Pereira et al., 2009
	Sting	2.5	Toxic	LISEC, 1989b [M]
Cyanobacteria				
Microcystis aeruginosa	Acid	110	NC	López-Rodas et al., 2007
	NR	169**	NC	Forlani, 2008
Anabaena sp.	NR	338**	NC	Forlani, 2008
Anabaena flos aquae	96.7%	304	NC	Maule & Wright, 1984
	96.6%	11.7	Harmful	Hughes, 1987a[M]
	Acid	15	Harmful	Smyth et al., 1996[M]
Anabaena varialis	Acid	2	Toxic	Hutber et al., 1979

	Glyphosate formulation	EC_{50} (mg/L)	European toxicity classification	Reference
Aphanocapsa 6308	Acid	2	Toxic	Hutber et al., 1979
Aphanocapsa 6714	Acid	100	Harmful	Hutber et al., 1979
Nostoc sp.	Acid	2	Toxic	Hutber et al., 1979
Nostoc punctiform	NR	1183*	NC	Forlani, 2008
Aphanizomenon flos-aquae	IPA salt	<2.85	Toxic to very toxic	Peterson et al., 1994

[1] formerly, *Selenastrum capricornutum;* NR= not reported; >>= considerably more than; NC= no classification for this range; * estimation of EC_{50} value based on graph; [M] = unpublished studies performed by the Manufacturers (Monsanto, Zeneca, Malcolm Pirnie)

Table 3. Overview of aquatic microorganism toxicity of various formulations of glyphosate

All these toxicity tests are simple ways of assessing the direct impacts of glyphosate on freshwater microorganisms, but they cannot be used to assess the real impact of this compound on natural populations and communities in freshwater ecosystems. For this reason, many experimental studies have attempted to use complex communities. Two papers by Goldsborough & Brown (1988) and Austin et al. (1991) found that glyphosate had no major impact on periphyton communities. However, Austin et al. (1991) suggested that the degradation of glyphosate could increase the concentration of soluble phosphorus, and consequently increase the biomasses of the communities. Subsequently, Vera et al. (2010) showed that the periphytic colonization of substrata was delayed in Roundup-contaminated large mesocosms, and that this phenomenon could be attributed to a direct effect of the contaminant. In addition they showed that in periphyton communities Roundup produced a long-term shift in the typology and functioning of contaminated mesocosms, which was consistent with data available from natural lakes in the Argentine Pampas. Another recent paper by Pérez et al. (2007) has also demonstrated an impact of Roundup on the structure of the phytoplankton and periphyton communities, and more particularly a decrease in the total micro- and nanophytoplankton and an increase in picocyanobacteria. Moreover, they have shown that these changes are more likely to be due to a direct toxic effect of glyphosate rather than to an indirect effect *via* a phosphorus enrichment of the streams linked to the degradation of this compound.

To conclude, herbicides containing glyphosate seem to have direct toxicological effects on non-target periphyton and phytoplankton communities in freshwater ecosystems, as well as indirect effects via the eutrophication potential of glyphosate degradation. These communities constitute the basis of food webs in these ecosystems, and so it would appear that glyphosate can potentially impact the overall functioning of freshwater ecosystems. This potential impact is reinforced by the fact that this compound is also known to be toxic for fish (e.g. Langiano & Martinez, 2008; Cavalcante et al., 2008), and could be involved in a trophic cascading process (Bengtsson et al., 2004).

7. Conclusions

As we have seen, a lot of data is available about the contamination of freshwater ecosystems by herbicides and also on the direct or indirect impact of these compounds on microbial communities living in these ecosystems. These compounds appear to affect the structure and composition of these communities, and also the metabolism of the microorganisms involved. It is very difficult to evaluate the consequences of such changes on the whole functioning of freshwater ecosystems, but there can be no doubt that it is significantly affected by herbicide contamination, because microbial communities play a key role in these ecosystems. This impact is probably reinforced by the fact that freshwater ecosystems are simultaneously subjected to other selective pressures. For example, herbicide pollution is generally concomitant with pollution by mineral nutrients (phosphorus and nitrogen), which also influence the structure and the functioning of microbial communities. In the same way, even though only a small number of papers have been published on the topic, we know that climate change is already affecting aquatic microbial communities (e.g. Falkowski & Oliver, 2007; Jöhnk et al., 2008).

The situation concerning herbicide contamination of freshwater ecosystems will change in the next few years, as a result of changes in agricultural practice. In a developed country like France, a drastic reduction of herbicide (50%) use is planned within a short time scale (<10 years). The European community as a whole is also engaged in reducing the use of pesticides in agriculture. In contrast, demographic pressure and Biofuel production in some areas of the world, such as Asia and South America, are leading to increasingly intensive agricultural practices, and consequently to growing herbicide pollution of freshwater ecosystems. These divergent situations will probably result in sharply contrasting situations with regard to the herbicide pollution of aquatic ecosystems in these different regions.

8. References

AFSSA (2006). Database of the INRA: AGRITOX. http://www.dive.afssa.fr/agritox/index.php (accessed in July 2010).

Amrhein, N.; Schab, J. & Steinrücken, H.C. (1980). The mode of action of the herbicide glyphosate. *Naturwissenschaften*, 67, 356-357.

Anton, F.A.; Ariz, M. & Alia, M. (1993). Ecotoxic effects of four herbicides (glyphosate, alachlor, chlortoluron and isoproturon) on the algae *Chlorella pyrenoidosa* chick. *Science of the Total Environment*, Supplement, 845-851.

Aoki, I. (2003). Diversity-productivity-stability relationship in freshwater ecosystems: Whole-systemic view of all trophic levels. *Ecological Research*, 18, 397-404.

Austin, A.P.; Harris, G.E. & Lucey, W.P. (1991). Impact of an organophosphate herbicide (Glyphosate®) on periphyton communities developed in experimental streams. *Bulletin of Environmental Contamination and Toxicology*, 47, 29-35.

Battaglin, W.A.; Kolpin, D.W.; Scribner; E.A.; Kuivila, K.M. & Sandstrom, M.W. (2005). Glyphosate, other herbicides and transformation products in midwestern streams, 2002. *Journal Of The American Water Resources Association*, 41, 323-332.

Battaglin, W.A.; Rice, K.C.; Focazio, M.J.; Salmons, S. & Barry, R.X. (2009). The occurrence of glyphosate, atrazine, and other pesticides in vernal pools and adjacent streams in Washington , DC, Maryland, Iowa, and Wyoming, 2005–2006. *Environmental Monitoring and Assessment*, 155, 281-307.

Belanger, S.E.; Bowling, J.W.; Lee, D.M.; LeBlanc, E.M.; Kerr, K.M.; McAvoy, D.C.; Christman, S.C. & Davidson, D.H. (2002). Integration of aquatic fate and ecological responses to linear alkyl benzene sulfonate (LAS) in model stream ecosystems. *Ecotoxicology and Environmental Safety*, 52, 150-171.

Belanger, S.E.; Guckert, J.B.; Bowling, J.W.; Begley, W.M.; Davidson, D.H.; LeBlanc, E.M. & Lee, D.M. (2000). Responses of aquatic communities to 25-6 alcohol ethoxylate in model stream ecosystems. *Aquatic Toxicology*, 48, 135-150.

Bengtsson, G.; Hansson, L.A. & Montenegro, K. (2004). Reduced grazing rates in *Daphnia pulex* caused by contaminants: Implications for trophic cascades. *Environmental Toxicology and Chemistry*, 11, 2641-2648.

Bérard, A.; Dorigo, U.; Mercier, I.; Becker-van-Slooten, K.; Grandjean, D. & Leboulanger, C. (2003). Comparison of the ecotoxicological impact of the triazines Irgarol 1051 and atrazine on microalgal cultures and natural microalgal communities in Lake Geneva. *Chemosphere*, 53, 935-944.

Bérard, A.; Pelte, T.; Menthon, E.; Druart, J.C. & Bourrain, X. (1998). Characterisation of phytoplankton from two limnic systems contaminated by a herbicidal photosynthetic inhibitor. The PICT method (Pollution-Induced Community Tolerance): application and significance. *Annales de Limnologie-International Journal of Limnology*, 34, 269-282.

Blank, H.; Wängberg, S.A. & Molander, S. (1988). Pollution-induced community tolerance-a new ecotoxicological tool. In: *Functional testing of aquatic biota for estimating hazards of chemicals*J. Cairns, Jr. and J.R. Pratt, Editors, *ASTM STP 988*, American Society for Testing and Materials, Philadelphia, pp. 219–230.

Bopp, S.K. & Lettieri, T. (2007). Gene regulation in the marine diatom *Thalassiosira pseudonana* upon exposure to polycyclic aromatic hydrocarbons (PAHs). *Gene*, 396, 293-302.

Botta, F.; Lavison, G.; Couturier, G.; Alliot, F.; Moreau-Guigon, E.; Fauchon, N.; Guery, B.; Chevreuil, M. & Blanchoud, H. (2009). Transfer of glyphosate and its degradate AMPA to surface waters through urban sewerage systems. *Chemosphere*, 77, 133-139.

Bozeman, J.; Koopman, B. & Bitton, G. (1989). Toxicity testing using immobilized algae. *Aquatic Toxicology*, 14, 345-352.

Byer, J.D.; Struger, J.; Klawunn, P.; Todd, A. & Sverko, E. (2008). Low cost monitoring of glyphosate in surface waters using the ELISA method: An evaluation. *Environmental Science and Technology*, 42, 6052-6057.

Caquet, T.; Lagadic, L. & Sheffield, S. R. (2000). Mesocosms in ecotoxicology (1): Outdoor aquatic systems. *In: Reviews of Environmental Contamination and Toxicology*. 165, 1-38.

Caquet, T.; Lagadic, L.; Jonot, O.; Baturo, W.; Kilanda, M.; Simon, P.; LeBras, S.; Echaubard, M. & Ramade, F. (1996). Outdoor experimental ponds (mesocosms) designed for long-term ecotoxicological studies in aquatic environment. *Ecotoxicology and Environmental Safety*, 34, 125-133.

Carlisle, S.M. & Trevors, J.T. (1988). Glyphosate in the environment. *Water, Air and Soil Pollution*, 39, 409-420.

Carter, A. (2000). How pesticides get into water - and proposed reduction measures. *Pesticide Outlook*, 2000, 149-156.

Cavalcante, D.G.S.M.; Martinez, C.B.R. & Sofia, S.H. (2008). Genotoxic effects of Roundup® on the fish *Prochilodus lineatus*. *Mutation Research/Genetic Toxicology and Environmental Mutagenesis*, 655, 41-46.

Cedergreen, N. & Streibig, J.C. (2005). The toxicity of herbicides to non-target aquatic plants and algae: assessment of predictive factors and hazard. *Pesticide Management Science*, 61, 1152-1160.

Culp, J. M.; Podemski, C. L.; Cash, K. J. & Lowell, R. B. (2000). A research strategy for using stream microcosms in ecotoxicology : integrating experiments at different levels of biological organization with field data. *Journal of Aquatic Ecosystem Stress and Recovery*, 7, 167-176.

Dahl, B. & Blanck, H. (1996). Toxic effects of the antifouling agent Irgarol 1051 on periphyton communities in coastal water microcosms. *Marine Pollution Bulletin*. 32, 342-350.

de Lipthay, J.R. ; Aamand, J. & Barkay, T. (2002). Expression of *tfdA* genes in aquatic microbial communities during acclimation to 2,4-dichlorophenoxyacetic acid. *FEMS Microbiology Ecology*, 40, 205-214.

Debenest, T.; Pinelli, E.; Coste, M.; Silvestre, J.; Mazzella, N.; Madigou, C. & Delmas, F. (2009). Sensitivity of freshwater periphytic diatoms to agricultural herbicides. *Aquatic Toxicology*, 93, 11-17.

DeLorenzo, M.E.; Scott, G. I. & Ross, P.E. (1999). Effects of the agricultural pesticides atrazine, deethylatrazine, endosulfan, and chlorpyrifos on an estuarine microbial food web. *Environmental Toxicology and Chemistry*, 18, 2824-2835.

DeLorenzo, M.E.; Scott, G.I. & Ross P.E. (2001). Toxicity of pesticides to aquatic microorganisms: a review. *Environmental Toxicology and Chemistry*. 20, 84-98.

Dennis, P.; Edwards, E.A.; Liss, S.N. & Fulthorpe R. (2003). Monitoring gene expression in mixed microbial communities by using DNA microarrays. *Applied and Environmental Microbiology*, 69, 769-778.

Diamond, G.L. & Durkin, P.R. (1997). Effects of surfactants on the toxicitiy of glyphosate, with specific reference to Rodeo. Prepared under USDA FS Contract No. 53- 3187-5-12. Final Report.

Dorigo, U.; Bérard, A. & Humbert, J.F. (2002). Comparaison of eukaryotic phytobenthic community composition in a polluted river by 18S rRNA gene cloning and sequencing. *Microbial Ecology*, 44,372-380.

Dorigo, U.; Bourrain, X.; Bérard, A. & Leboulanger, C., (2004). Seasonal changes in the sensitivity of river microalgae to atrazine and isoproturon along a contamination gradient. *Science of the Total Environment*. 318, 101–114.

Dorigo, U.; Leboulanger, C.; Bérard, A.; Bouchez, A.; Humbert, J.F. & Montuelle, B. (2007). Lotic biofilm community structure and pesticide tolerance along a contamination gradient in a vineyard area. *Aquatic Microbial Ecology*, 50, 91-102.

Dorigo, U.; Lefranc, M.; Bouchez, A.; Montuelle, B. & Humbert, J.F. (2009). Spatial heterogeneity of periphytic microbial communities in a small pesticide-polluted river. *FEMS Microbial Ecology*, 67, 491-501

Dorigo, U.; Volatier, L. & Humbert, J.F. (2005). Molecular approaches to the assessment of biodiversity in aquatic microbial communities. *Water Research*, 39, 2207-2218.

Dubois, A.; Lacouture, L. & Feuillet, C. (2010). Les pesticides dans les milieux aquatiques - données 2007. Etude et documents. Commissariat Général au développement durable-Service de l'observation et des statistiques, 49 pp. (*in French*)

Duke, S.O. (1988). Glyphosate, In: *Herbicides: chemistry, degradation and mode of action, Vol. 3*, Kearney P.C. & Kaufman D.D. (Ed), 1-70, Marcel Dekker, New York, NY.

Ermis, U.B. & Demir, N. (2009). Toxicity of glyphosate and ethoxysulfuron to the green microalgae (*Scenedesmus obliquus*). *Asian Journal of Chemistry*, 21, 2163-2169.

Falkowski, P.G. & Oliver, M.J. (2007). Mix and match: how climate selects phytoplankton. *Nature Reviews Microbiology*, 5, 813-820.

Faust, M.; Altenburger, R.; Boedeker, W. & Grimme, L. (1993). Additive effects of herbicide combinations on aquatic non-target organisms. *The Science of the Total Environment*, Supplement, 941-952.

Fenchel, T. (2008). The microbial loop - 25 years later. *Journal of Experimental Marine Biology and Ecology*, 366, 99-103.

Foley, M.E.; Sigler, V. & Gruden, C.L. (2008). A multiphasic characterization of the impact of the herbicide acetochlor on freshwater bacterial communities. *The ISME Journal*. 2, 56-66.

Forlani, G.; Pavan, M.; Gramek, M.; Kafarski, P. & Lipok, J. (2008). Biochemical bases for a widespread tolerance of Cyanobacteria to the phosphonate herbicide Glyphosate. *Plant & Cell Physiology*, 49, 443-456.

Franqueira, D.; Cid, A.; Torres, E.; Orosa, M. & Herrero, C. (1999). A comparison of the relative sensitivity of structural and functional cellular responses in the alga Chlamydomonas eugametos exposed to the herbicide paraquat. *Archive of Environmental Contamination and Toxicology*, 36, 264–269.

Franz, J.E.; Mao, M.K. & Sikorski, J.A. (1997). Glyphosate: A unique global herbicide. *American Chemistry Society, ASC Monographs*, 189, 1-600.

Gans, J.; Wolinsky, M. & Dunbar, J. (2005). Computational improvements reveal great bacterial diversity and high metal toxicity in soil. *Science*, 309, 1387–1390

Gfrerer, M.; Martens, D.; Gawlik, B.M.; Wenzl, T.; Zhang, A.; Quan, X.; Sun, C.; Chen, J.; Platzer, B.; Lankmayr, E. & Kettrup, A. (2002a). Triazines in the aquatic systems of the Eastern Chinese Rivers Liao-He and Yangtse. *Chemosphere*, *47*, 455-66.

Gfrerer, M.; Wenzl, T.; Quan, X.; Platzer, B. & Lankmayr, E. (2002b). Occurrence of triazines in surface and drinking water of Liaoning Province in Eastern China. *Journal of Biochemical and Biophysical methods*, 53, 217-28.

Ghanem, A.; Bados, P.; Estaun, A.R.; de Alencastro, L.F.; Taibi, S.; Einhorn, J. et al. (2007). Concentrations and specific loads of glyphosate, diuron, atrazine, nonylphenol and metabolites thereof in French urban sewage sludge. *Chemosphere*, 69, 1368-1373.

Ghassemi, M.; Fargo, L.; Painter, P.; Quinlivan, S.; Scofield, R.; Takata., A. et al. (1981). Environmental fates and impacts of major forest use pesticides. EPA contract No. 68-02-3174. Environmental Protection Agency, Washington, DC. Washington, DC.

Goldsborough, L.G. & Brown D.J. (1988). Effect of Glyphosate (Roundup® formulation) on periphytic algal phosotosynthesis. *Bulletin of Environmental Contamination and Toxicology*, 41, 253-260.

Giesy, J.P.; Dobson, S. & Solomon, K.R. (2000). Ecotoxicological risk assessment for Roundup herbicide. *Review of Environmental Contamination and Toxicology*, 167, 35-120.

Gilliom, R.J. (2007). Pesticides in US streams and groundwater. *American Chemistry Society, Environmental Science and Technology*, 15, 3408-3414.

Gonod, L.; Martin-Laurent, F. & Chenu, C. (2006). 2,4D impact on bacterial communities, and the activity and genetic potential of 2,4D degrading communities in soil. *FEMS Microbiology Ecology*, 58, 529-537

Grenni, P.; Gibello, A.; Barra Caracciolo, A.; Fajardo, C.; Nande, M.; Vargas, R.; Sacca, M.L.; Martinez-Inigo, M.J.; Ciccoli, R. & Martin M. (2009). A new fluorescent oligonucleotide probe for in situ detection of s-triazine-degrading *Rhodococcus wratislaviensis* in contaminated groundwater and soil samples. *Water Research*, 43, 2999-3008

Guasch, H. & Sabater, S. (1998). Light history influences the sensitivity to atrazine in periphytic algae. *Journal of Phycology*, 34, 233-241

Guasch, H.; Munoz, I.; Roses, N. & Sabater, S. (1997). Changes in atrazine toxicity throughout succession of stream periphyton communities. *Journal of Applied Phycology*, 9, 137-146.

Guckert, J.B. (1993). Artificial streams in ecotoxicology. In: Lamberti G.A. et Steinman A.D. Research in artificial streams : applications, uses and abuses. *Journal of National American Benthology Society*, 12, 350-356.

Gutell, R.R. (1994). Collection of small subunit (16S- and 16S-like) ribosomal RNA structures. *Nucleic Acids Research*, 22, 3502-3507

Haglund, K. (1997). The use of algae in aquatic toxicity assessment. In: Round, F.E., Chapman, D.J. (Eds.), Progress in Phycological Research. Biopress Ltd., Bristol (England), 181–212.

Hense, B. A.; Juttner, I.; Welzl, G.; Severin, G. F.; Pfister, G., Behechti, A. & Schramm, K. W. (2003). Effects of 4-nonylphenol on phytoplankton and periphyton in aquatic microcosms. *Environmental Toxicology and Chemistry*, 22, 2727-2732.

Horth, H. & Blackmore, K. (2009). Survey of glyphosate and AMPA in Europe, WRc Report UC 8073.02, WRc plc, Frankland Road, Blagrove, Swindon, Wiltshire, SN5 8YF, England (Confidential report to Monsanto Europe).

Huber, A.; Bach, M. & Frede, H.G. (2000). Pollution of surface waters with pesticides in Germany: modeling non-point source inputs. *Agriculture, Ecosystems and Environment*, 80, 191-204.

Hughes, J. (1987a). The toxicity of glyphosate technical to *Anabaena flos-aquae*: Lab Project ID: 1092-02-1100-4. Unpublished study prepared by Malcolm Pirnie, 23, In: US EPA. (1993). Reregistration Eligibility Decision (RED) for Glyphosate. EPA 738-R-93-014 Office of Preventions, Pesticides and Toxic Substances. September 1993.

Hughes, J. (1987b). The toxicity of glyphosate technical to *Navicula pelliculosa*: Lab Project ID: 1092-02-1100-2. Unpublished study prepared by Malcolm Pirnie, 23, In: US EPA. (1993). Reregistration Eligibility Decision (RED) for Glyphosate. EPA 738-R-93-014 Office of Preventions, Pesticides and Toxic Substances. September 1993.

Hughes, J. (1987c). The toxicity of glyphosate technical to *Selenastrum capricornutum*: Lab Project ID: 1092-02-1100-1. Unpublished study prepared by Malcolm Pirnie, 23, In: US EPA. (1993). Reregistration Eligibility Decision (RED) for Glyphosate. EPA 738-R-93-014 Office of Preventions, Pesticides and Toxic Substances. September 1993.

Hutber, G.N.; Rogers, L.J. & Smith, A.J. (1979). Influence of pesticides on the growth of cyanobacteria. *Zeitschrift für allgemeine Mikrobiologie*, 19, 397-402.

IFEN (2006). Les pesticides dans les eaux- Données 2003 et 2004 (Pesticides in water, Data from 2003 and 2004), Institut Français de l'Environnement, Orléans, France (*in French*).

IFEN (2007). Les pesticides dans les eaux- Données 2005 (Pesticides in water, Data from 2005), Institut Français de l'Environnement, Orléans, France (*in French*).

IFEN (2009). Les pesticides dans les eaux (Pesticides in waters), http://www.stats.environnement.developpement-durable.gouv.fr/donnees-essentielles/eau/les-pesticides-dans-les-eaux.html (accessed in July 2010).

Jöhnk, K.D.; Huisman, J.; Sharples, J.; Sommeijer, B.; Visser, P. & Strooms, M. (2008). Summer heatwaves promote blooms of harmful cyanobacteria. *Global Change Biology*, 14, 495-512.

Keddy C.; Greene J.C., & Bonnell M.A. (1994). Examen des biotests effectués sur des organismes entiers pour l'évaluation de la qualité des sols, des sédiments et des eaux douces au Canada. Environnement Canada, Direction générale de la Conservation des écosystèmes, Ottawa, Ontario, Etude N°198.

Kolpin, D.W.; Thurman, E.M.; Lee, E.A.; Meyer, M.T.; Furlong, E.T.; Glassmeyer, S.T., et al. (2006). Urban contributions of glyphosate and its degradate AMPA to streams in the United States. *Environmental Science and Technology*, 354, 191-197.

Konstantinou, I.K.; Hela, D.G. & Albanis T.A. (2006). The status of pesticide pollution in surface waters (rivers and lakes) of Greece. Part I. Review on occurrence and levels. *Environmental Pollution*, 141, 555-570.

Kreuger, J. (1998). Pesticides in stream water within an agricultural catchment in southern Sweden. *The Science of the Total Environment*, 216, 227-251.

Langiano, V.C. & Martinez, C.B.R. (2008). Toxicity and effects of a glyphosate-based herbicide on the neotropical fish *Prochilodus lineatus*. *Comparative Biochemistry and Physiology Part C: Toxicology & Pharmacology*, 147, 222-231.

LISEC (1989a). Alga, growth inhibition test. Effect of MON 2139 on the growth of *Selenastrum capricornutum*. Unpublished report submitted by Monsanto, In: WHO (1994). Glyphosate - Environmental Health Criteria No 159. International Programme on Chemical Safety (IPCS), Geneva

LISEC (1989b). Alga, growth inhibition test. Effect of MON 8755 on the growth of *Selenastrum capricornutum*. In Unpublished report submitted by Monsanto, In: WHO (1994). Glyphosate - Environmental Health Criteria No 159. International Programme on Chemical Safety (IPCS), Geneva.

López-Rodas, V.; Flores-Moya, A.; Maneiro, E.; Perdigones, N.; Marva, F.; Marta, G.E., et al. (2007). Resistance to glyphosate in the cyanobacterium *Microcystis aeruginosa* as result of pre-selective mutations. *Evolutionary Ecology*, 21, 535-547.

Lee, T.H.; Kurata, S.; Nakatsu, C.H. & Kamagata, Y. (2005). Molecular analysis of bacterial community based on 16S rDNA and functional genes in activated sludge enriched with 2,4-dichlorophenoxyacetic acid (2,4-D) under different cultural conditions. *Microbial Ecology*, 49, 151-162.

López-Doval, J.C. ; Ricart, M. ; Guasch, H. ; Romani, A.M. ; Sabater, S. & Muñoz, I. (2010). Does grazing pressure modify diuron toxicity in a biofilm community. *Archives of Environmental Contamination and Toxicology*, 58, 955-962.

Lürling, M. & Roessink, I. (2006). On the way to cyanobacterial blooms: Impact of the herbicide metribuzin on the competition between a green alga (*Scenedesmus*) and a cyanobacterium (*Microcystis*). *Chemosphere*, 65, 618-626.

Lyautey, E.; Teissier, S.; Charcosset, J. Y.; Rols, J.L. & Garabetian, F. (2003). Bacterial diversity of epilithic biofilm assemblages of an anthropised river section, assessed by DGGE analysis of a 16S rDNA fragment. *Aquatic Microbial Ecology*, 33, 217-224.

Ma, J. (2002). Differential sensitivity to 30 herbicides among populations of two green algae *Scenedesmus obliquus* and *Chlorella pyrenoidosa. Bulletin of Environmental Contamination and Toxicology*, 68, 275-281.

Ma, J. & Wang, S. (2002). A quick, simple, and accurate method of screening herbicide activity using green algae cell suspension cultures. *Weed Science Society of America*, 50, 555-559.

Ma, J.; Lin, F.; Wang, S. & Xu, L. (2003). Toxicity of 21 herbicides to the green alga *Scenedesmus quadricauda. Bulletin of Environmental Contamination and Toxicology* 71, 594–601.

Ma, J.; Xu, L.; Wang, S.; Zheng, R.; Jin, S.; Huang, S. et al. (2002). Toxicity of 40 herbicides to the green alga *Chlorella vulgaris. Ecotoxicology and Environmental Safety*, 132, 128-132.

Mallat, E. & Barceló, D. (1998). Analysis and degradation study of glyphosate and of aminomethylphosphonic acid in natural waters by means of polymeric and ion-exchange solid-phase extraction columns followed by ion chromatography-post column derivatization with fluorescence detection. *Journal of Chromatography A*, 823, 129-136.

Maule, A. & Wright, S.J. (1984). Herbicide effects on the population growth of some green algae and cyanobacteria. *Journal of Applied Bacteriology*, 57, 369-379.

Macur, R.E.; Wheeler, J.T.; Burr, M.D. & Inskeep W.P. (2007). Impacts of 2,4-D application on soil microbial community structure and on populations associated with 2.,4-D degradation. *Microbiological Research*, 162, 37-45.

Monsanto (1990). Dissipation of glyphosate and aminomethylphosphonic acid in forestry sites (Unpublished report No. MSL-9940), In: WHO (1994). Glyphosate - Environmental Health Criteria No 159. International Programme on Chemical Safety (IPCS), Geneva.

Muñoz, I.; Real, M.; Guasch, H.; Navarro, E. & Sabater, S. (2001). Effects of atrazine on periphyton under grazing pressure. *Aquatic Toxicology*, 55, 239-249.

Muyzer, G. (1999). DGGE/TGGE a method to identifying genes from natural ecosystems. *Current Opinion in Microbiology*, 2, 317 - 322

Na, T.; Fang, Z.; Zhanqi, G.; Ming, Z. & Cheng, S. (2006). The status of pesticide residues in the drinkingwater sources in Mailiangwan bay, Taihu lake of China. *Environmental Monitoring and Assessment*, 123, 351-370.

Nelson, K. ; Methé, B.A. & Kowalchuk, G.A. (2006). Environmental Microbial Ecology in an « Omics » Era. Summary report of MicroEnGen-II (Shanghai, June 12–15, 2006). *Microbial Ecology*, 53, 369-370.

Neufeld, J.D. ; Wagner, M. & Murrell, J.C. (2007). Who eats, what, where and when? Isotope-labelling experiments are coming of age. *The ISME Journal*, 1, 103-110.

Neumann, M. ; Liess, M. & Schultz, R. (2003). A qualitative method for monitoring water quality in temporary channels or point sources and its application to pesticide contaminbation. *Chemosphere*, 51, 509-513.

Novack, B. (1997). Determination of phosphonates in natural waters by ion-pair high performance liquid chromatography. *Journal of Chromatography A*, 773, 139-146.

Office of Pesticide Programs. (2000). Pesticide Ecotoxicity Database. Environmental Fate and Effects Division, U.S.EPA, Washington, DC.

Oldenburg, G.; Griner, E.M.; Francis, C.A. & Ward, B.B. (2003). Oligonucleotide microarray for the study of functional gene diversity in the nitrogen cycle in the environment. *Applied of Environmental Microbiology*. 69, 1159-1171.

Park, J. ; Congeevaram, S. ; Ki, D.W & Tiedje, J.M. (2006). Use of stable isotope probing in selective isolating target microbial community genomes from environmental samples for enhancing resolution in ecotoxicological assessment. *Molecular & Cellular Toxicology*, 2, 11-14.

Pereira, J. L.; Antunes, S.C.; Castro, B.B.; Marques, C.R.; Gonçalves, A.M.; Pereira, R. et al. (2009). Toxicity evaluation of three pesticides on non-target aquatic and soil organisms : commercial formulation versus active ingredient. *Ecotoxicology*, 18, 455-463.

Pérès, F.; Florin, D.; Grollier, T.; Feurtet-Mazel, A.; Coste, M.; Ribeyre, F.; Ricard, M. & Boudou, A. (1996). Effects of the phenylurea herbicide isoproturon on periphytic communities in freshwater indoor microcosms. *Environmental Pollution*. 94, 141-152.

Pérez, G.L. ; Torremorell, A. ; Mugni, H. ; Rodriguez, P. ; Vera, M.S. ; do Nascimento, M. ; Allende, L. ; Bustingorry, J. ; Escaray, R. ; Ferraro, M. ; Izaguirre, I. ; Pizarro, H. ; Bonetto, C. ; Morris, D.P. & Zagarese, H. (2007). Effects of the herbicide Roundup on freshwater microbial communities : A mesocosm study. *Ecological Applications*, 17, 2310-2322.

Peruzzo, P.J.; Porta, A.A. & Ronco, A.E. (2008). Levels of glyphosate in surface waters, sediments and soils associated with direct sowing soybean cultivation in north pampasic region of Argentina. *Environmental Pollution*, 156, 61-66.

Pesce, S.; Batisson, I.; Bardot, C.; Fajon, C.; Portelli, C.; Montuelle, B. & Bohatier, J. (2009a). Response of spring and summer riverine microbial communities following glyphosate exposure. *Ecotoxicology and Environmental Safety*, 72, 1905-1912

Pesce, S.; Martin-Laurent F.; Rouard N.; & Montuelle, B. (2009b). Potential for microbial diuron mineralization in a small wine-growing watershed: from treated plots to lotic receiver hydrosystem. *Pesticide Management*, 65, 651-657.

Pesce, S.; Fajon, C.; Bardot, C.; Bonnemoy, F.; Portelli, C. & Bohatier, J. (2006). Effects of the phenylurea herbicide diuron on natural riverine microbial communities in an experimental study. *Aquatic Toxicology*, 78, 303-314.

Pesce, S.; Fajon, C.; Bardot, C.; Portelli, C. & Bohatier, J. (2008). Longitudinal changes in microbial planktonic communities of a French river in relation to pesticide and nutrient inputs. *Aquatic Toxicology*, 86, 352-360.

Peterson, H.G.; Boutin, C.; Martin, P.A.; Freemark, K.E.; Ruecker, N.J.; Moody, M.J. et al. (1994). Aquatic phyto-toxicity of 23 pesticides applied at expected environmental concentrations. *Aquatic Toxicology*, 28, 275-292.

Pipke, R. & Amrhein, N. (1988). Degradation of the phosphonate herbicide glyphosate by *Arthrobacter atrocyaneus* ATCC 13752. *Applied and Environmental Microbiology*, 54, 1293-1296.

Powell, H.A. ; Kerbby, N.W. & Rowell, P. (1991). Natural tolerance of cyanobacteria to the herbicide glyphosate. *New Phytologist*, 119, 421-426.

Prado, R.; Rioboo, C.; Herrero, C. & Cid, A. (2009). The herbicide paraquat induces alterations in the elemental and biochemical composition of non-target microalgal species. *Chemosphere*, 76, 1440-1444

Ramade, F. (2007). Introduction à l'écotoxicologie. Ed : Dunod, Paris. 550pp.

Rand, G. M.; Clark, J. R. & Holmes, C. M. (2000). Use of outdoor freshwater pond microcosms: I. Microcosm design and fate of pyridaben. *Environmental Toxicology and Chemistry*, 19, 387-395.

Ricart, M.; Barcelo, D.; Geiszinger, A.; Guasch, H.; Lopez de Alda, M.; Romani, A.M.; Vidal, G., Villagrasa, M. & Sabater S. (2009). Effects of low concentration of the phenylurea herbicide diuron on biofilm algae and bacteria. *Chemosphere*, 76: 1392-1401.

Ricciardi, F.; Bonnineau, C.; Faggiano, L.; Geizinger, A.; Guasch, H.; Lopez-Doval, J.; Muñoz, I.; Ricart, M.; Romani, A. & Sabater, S. (2009). Is chemical contamination linked to the diversity of biological communities in rivers? *Trends in Analytical Chemistry*, 28, 592-602.

Riemann, L. & Winding, A. (2001). Community dynamics of free-living and particle-associated bacterial assemblages during a freshwater phytoplankton bloom. *Microbial Ecology*, 42, 274-285.

Rioboo, C.; González, O.; Herrero, C. & Cid, A. (2002). Physiological response of freshwater microalga (*Chlorella vulgaris*) to triazine and phenylurea herbicides. *Aquatic Toxicology*, 59, 225–235.

Rueppel, M.L.; Brightwell, B.B.; Schaefer, J. & Marvel, J.T. (1977). Metabolism and degradation of glyphosate in soil and water. *Journal of Agricultural and Food and Chemistry*, 25(3), 517-528.

Sabater, S.; Guasch, H.; Ricart, M.; Romani, A.; Vidal, G.; Klunder, C. & Schmitt-Jansen, M. (2007). Monitoring the effect of chemicals on biological communities. The biofilm as an interface. *Analytical and Bioanalytical Chemistry*, 387, 1425-1434.

Sáenz, M.E. & Di Marzio, W.D. (2009). Ecotoxicidad del herbicida Glifosato sobre cuatro algas clorófitas dulceacuícolas, *Limnetica*, 28, 149-158.

Sáenz, M.E. ; Di Marzio, W.D. ; Alberdi, J.L. & del Carmen Tortorelli, M. (1997). Effects of technical grade and a commercial formulation of glycphosate on algal population growth. *Bulletin of Environmental Contamination and Toxicology*, 59, 638-644.

Schauer, M.; Balague, V.; Pedros-Alio, C. &. Massana, R. (2003). Seasonal changes in the taxonomic composition of bacterioplankton in a coastal oligotrophic system. *Aquatic Microbial Ecology*, 31, 163-174.

Schmitt-Jansen, M. & Altenburger, R. (2005). Predicting and observing responses of algal communities to photosystem II - Herbicide exposure using pollution-induced community tolerance and species-sensitivity distributions. *Environmental Toxicology and Chemistry*, 24, 304-312.

Schuler, L.J. & Rand, G.M. (2008). Aquatic risk assessment of herbicides in freshwater ecosystems of south Florida. *Archives of Environmental Contamination and Toxicology*, 54, 571-583.

Scribner, E.A.; Battaglin, W.A.; Gilliom, R.J. & Meyer, M.T. (2007). Concentrations of glyphosate, its degradation product, aminomethylphosphonic acid, and glufosinate in ground- and surface-water, rainfall, and soil samples collected in the United States, 2001–06- Scientific Investigations Report 2007–5122, U.S. Geological Survey, Reston, Virginia.

Shabana, Y.M.; Elwakil, M.A. & Charudattan, R. (2001). Effect of nutrittion and physical factors on mycelial growth and production of pigments and non-chromatic UV-absorbing compounds of *Alternaria eichhorniae*. *Journal of Phytopathology*, 149, 21-27

Shipitalo, M.J. & Owens, L.B. (2006). Tillage system, application rate, and extreme event effects on herbicide losses in surface runoff. *Journal of Environmental Quality*, 35, 2186-2194.

Skark, C.; Zullei-Seibert, N.; Schottler, U. & Schlett, C. (1998). The occurrence of glyphosate in surface water. *International Journal of Environmental Analytical Chemistry*, 70, 93-104.

Smyth, D.; Kent, S.; Morris, D. et al. (1996). Glyphosate acid: toxicity to the freshwater diatom (*Navicula pelliculosa*): Lab Project Number: BL5673/B: AB0503/K. Unpublished study prepared by Zeneca Limited, 25, In: Durkin, P.R. (2003) Glyphosate, Human health and ecological risk assessment. USDA forest service, Arlington.

Smyth, D.; Kent, S.; Morris, D. et al. (1995). Glyphosate acid: acute toxicity to the green alga (*Selenastrum capricornutum*): Lab Project Number: BL5550/B: AB0503/B. Unpublished study prepared by Zeneca Limited, 24, In: Durkin, P.R. (2003) Glyphosate, Human health and ecological risk assessment. USDA forest service, Arlington.

Smyth, D.; Shillabeer, N.; Morris, D. et al (1996). Glyphosate acid: acute toxicity to the blue-green Alga (*Anabaena flos-aquae*): Lab Project Number: BL5698/B: AB0503/J. Unpublished study prepared by Zeneca Limited., 25, In: Durkin, P.R. (2003) Glyphosate, Human health and ecological risk assessment. USDA forest service, Arlington.

Spalding, R.F. & Snow, D.D. (1989). Stream levels of agrichemicals during a spring discharge event. *Chemosphere*, 19, 1129-1140.

Sprankle, P.; Meggitt, W.F.; Penner, D.; Science, S.W. & May, N. (1975). Adsorption, mobility, and microbial degradation of glyphosate in the soil. *Weed Science Society of America*, 23, 229-234.

Struger, J.; Thompson, D.; Bozena, S.; Martin, P.; Mcdaniel, T.; Marvin, C. et al. (2008). Occurrence of glyphosate in surface waters of southern Ontario. *Bulletin of Environmental Contamination and Toxicology*, 80, 378-384.

Sudo, M. ; Kawachi, T. ; Hida, Y. & Kunimatsu T. (2004). Spatial distribution and seasonal changes of pesticides in lake Biwa, Japan. *Limnology*, 5, 77-86.

Sudo, M. ; Kunimatsu, T. & Okubo, T. (2002). Concentration and loading of pesticides resisdues in lake Biwa basin (Japan). *Water Research*, 36, 315-329.

Tadonléké, R.; Le Berre, B.; Perreau, F. & Humbert, J.F. (2009). Responses of lake bacterioplankton activities and composition to the herbicide diuron. *Aquatic Toxicology*, 94, 103-113.

Tomlin, C.D. (1997). *The pesticide manual 11th*, British Crop Protection Council, Farnham Surrey, UK.

Tooby, T.E. (1985). Fate and biological consequences of glyphosate in the aquatic environment, In: *The herbicide glyphosate*, Grossbard, E. & Atkinson, D. (Ed.), Butterworth and Co. London.

Trass, T.P. & Smit, C.E. (2003). Environmental risk limits for aminomethylphosphonic acid (AMPA) RIVM report 601501018/2003. National Institute of Public Health and the Environment. Bilthoven, The Netherlands.

Tsui, M.T. & Chu, L.M. (2003). Aquatic toxicity of glyphosate-based formulations: comparison between different organisms and the effects of environmental factors. *Chemosphere*, 52, 1189-1197.

Vandenkoornhuyse, P.; Dufresne, A.; Quaiser, A.; Gouesbet, G.; Binet, F.; Francez, A.J.; Mahé, S.; Bormans, M.; Lagadeuc, Y. & Coué, I. (2010). Integration of molecular functions at the ecosystemic level: breakthroughs and future goals of environmental genomics and post-genomics. *Ecology Letters*, 13, 776-791.

Vendrell, E.; Ferraz, D.G.; Sabater, C. & Carrasco, J.M. (2009). Effect of glyphosate on growth of four freshwater species of phytoplankton : A microplate bioassay. *Bulletin of Environmental Contamination and Toxicology*, 82, 538-542.

Vera, M.S.; Lagomarsino, L.; Sylvester, M.; Pérez, G.L.; Rodriguez, P.; Mugni, H.; Sinistro, R.; Ferraro, M.; Bonetto, C.; Zagares, H. & Pizarro, H. (2010). New evidence of Roundup (glyphosate formulation) impact on periphyton community and the water quality of freshwater ecosystems. *Ecotoxicology*, 19, 710-721

Villeneuve, A. (2008). Effets conjoints de facteurs physiques et chimiques sur la structure et la composition du periphyton : une approche multi-échelle. These de L'université de Savoie.

Ware, G. W., & Whitacre, D. M. (2004). *The Pesticide Book* (6th ed). Willoughby, Ohio: Meister Media Worldwide, 496 pp.

Wauchope, R.D. (1978). The pesticide content of surface water draining from agricultural fields: A review. *Journal of Environmental Quality*, 7, 459-472.

WHO (1994). Glyphosate - Environmental Health Criteria No 159. International Programme on Chemical Safety (IPCS), Geneva.

Widenfalk, A.; Bertilsson, S.; Sundh, I. & Goedkoop, W. (2008). Effects of pesticides on community composition and activity of sediment microbes – Responses at various levels of microbial community organization. *Environmental Pollution*, 152, 576-584.

Wilson, M.A. & Carpenter S.R. (1999). Economic valuation of freshwater ecosystem services in the United States: 1971-1997. *Ecological Applications*, 9, 772-783.

Wittmer, I.K.; Bader, H.-P.; Scheidegger, R.; Singer, H.; Lück, A.; Hanke, I.; Carlsson, C. & Stamm, C. (2010). Significance of urban and agricultural land use of biocide and pesticide dynamics in surface waters. *Water Research*, 44, 2850-2862.

Wong, P.K. (2000). Effects of 2, 4-D, glyphosate and paraquat on growth, photosynthesis and chlorophyll-a synthesis of *Scenedesmus quadricauda* Berb 614. *Chemosphere*, 41, 177-182.

Woodburn, A.T. (2000). Glyphosate : production , pricing and use. *Pest Management Science*, 56, 309-312.

Yoshida, K.; Igarashi, E.; Mukai, M.; Hirata, K. & Miyamoto, K. (2003). Induction of tolerance to oxidative stress in the green alga, *Chlamydomonas reinhardtii*, by abscisic acid. *Plant Cell Environment*, 26, 451–457.

17

Pesticides Reaching the Environment as a Consequence of Inappropriate Agricultural Practices in Argentina

Andrés H. Arias[1,2], Natalia S. Buzzi[1,2], Marcelo T. Pereira[1] and Jorge E. Marcovecchio[1,2]
[1]*Universidad Nacional del Sur,*
[2] *Instituto Argentino de Oceanografía, CCT-CONICET, Bahía Blanca*
Argentina

1. Introduction

Recent water quality inventories show that agricultural non-point source pollution (NPS) is the leading source of water quality impacts to surveyed rivers and lakes, and also a major contributor to groundwater contamination and wetlands degradation (Blankenberg et al. 2006; Haarstad and Braskerud 2005; Bergstrom 2004; Thiere and Schulz 2004; Schreiber et al. 2001; Huber et al. 2000; Luo and Zhang 2010).

Pest management is one of the main scopes of pesticides usage, because more than 45% of annual food production is lost due to pest infestation. In particular, at tropical climates this is enhanced due to prevailing high temperature and humidity. However, the sporadic use has been leading to significant consequences not only to public health but also to food quality resulting in an impact load on the environment and hence the development of pest resistance (Giupponi and Rosato 1999; Luo and Zhang 2010)

Through overuse, misuse and losses due to the inappropriate application of pesticides there is considerable waste contributing to the environmental burden (Li et al. 2006; Giupponi and Rosato 1999; Hu et al. 2010; Marco and Kishimba 2007; Luo and Zhang 2010). It is well known that most of the applied pesticides are subject to many transport and conversion processes. Thus, they do not remain at their target site but often enter aquatic environments via soil percolation, air drift or surface run-off, affecting abundance and diversity of non-target species, producing complex effects on the ecosystems and, altering trophic interactions (Islam and Tanaka, 2004). Pesticides overuse also destroys the healthy pool of bio-control agents that normally co-exist with the vegetation. Simultaneously, some soil biological functions such as the bioavailability of nutrients and organic matter decomposition could also be altered (Hendrix 1996; Guo et al. 2009). For instance, the herbicides can influence soil microbial biomass carbon (MBC) and metabolic quotient (qCO), variables directly related to soil quality (Reis et al. 2009). In addition, agrochemical application on soybean shoots affects the activity of soil microorganisms in the plant rhizosphere (Reis et al. 2009).

The bulk of pesticides worldwide used is herbicides and there is almost no knowledge of their impact on potential non-target plant species, especially rare or endemic species

(McLaughlin and Mineau 1995). For example, granular insecticides such as carbofuran are very efficient at killing a large proportion of the songbird population breeding on the edge of fields where they are applied (De Silva et al. 2010; Tataruch et al. 1998).

These compounds affect the whole ecosystem by entering into the food chain and polluting the soil, air, ground and surface water. As an immediate consequence, humans are exposed to pesticides found in the media. These come to human contact by different routes of exposure such as inhalation, ingestion and dermal contact. As is widely stated, recent and chronic pesticides residues are being detected in child blood, human adipose tissue and breast milk (Perez-Maldonado et al. 2010; Waliszewski et al. 2010; Waliszewski et al. 2009; Wang et al. 2009; Lopez-Espinosa et al. 2008; Ntow et al. 2008). These exposures could ultimately result in acute and chronic health problems, increasing the incidence of cancer, chronic kidney diseases, suppression of the immune system, sterility among males and females, endocrine disorders, neurological and behavioral disorders, especially among children, have been attributed to chronic pesticide poisoning (Gbaruko et al. 2009; Lin et al. 2009; Casabe et al. 2007; Bila and Dezotti 2007; Aronson et al. 2010; Panegyres et al. 2010; Rull et al. 2009; Silva and Gammon 2009).

As an agricultural country, Argentina was estimated from 26 to 30 million hm2 of farmland using pesticides in the period 2005-2008. It is stated that during pesticides application, up to 30-50% of the applied amount can be lost to the air. Assuming an average dose of a 4.3 kg.hm-1, the amount of pesticides used range between 112 to 129 million Kg with losses to the environment ranging from 33.6 to 64.5 million Kg per year (this means a minimum of about 1 Kg of pesticide in the environment per habitant). In addition, this estimate is not considering that the number of treatments has increased in the last decade, with 66% of the cropped area using two or more herbicide types, and 80% using two or more insecticides during treatment.

To sum up, the existing evidence arguments in favour of revising and restructuring the national agriculture policy and monitoring programs. As a first step, this chapter will discuss some elements of the Argentinean agricultural practices and policies, and then it will focus in reviewing pesticides' occurrence in the environment.

2. Country and agricultural practices

The Argentinean Republic is a country with a total area of 3.761.274 Km^2, including 964.000 Km^2 of islands and Antarctic lands. Its population was of 36.260.130 inhabitants in 2001 (the year of the last census), with an average density of 13 inhab/km^2. Despite this, 46% of the population is located around the country's capital city, Buenos Aires. Argentina reaches up to 89.9% of urban population. 77% of housing has access to drinking water through the national network and 59% of people have social medical insurance. The % of literacy is up to 97.8% (older than ten years) and about 34% complete a secondary education (12 years). The last census also showed 14.3% of homes below the line of poverty (Health Ministry, Argentina, 2003).

The combination of adequate temperatures, soil richness and rainfalls provides, as a consequence, top quality lands for agricultural use. The permanent population growth causes an increase on natural resources pressure that conduces to an overexploitation of them. By 2002, the country counted with 332.057 agricultural exploitations (AE), including an area of about 172.105.798 Ha. The major area was sowed with cereal and oilseeds. AE`s population reached up to 1.447.365, mainly constituted of workers.

About the pesticides regulation in Argentina, the government implemented the Rotterdam Convention by the year 2000. The Convention regulates the import and export of certain hazardous chemicals and pesticides. It is based on the fundamental principle of Prior Informed Consent (PIC), meaning that under the Convention, a chemical listed in the Convention may only be exported with the importer's prior consent. The Convention establishes a procedure to disseminate the decisions taken by the importing countries, thus implementing the PIC principle in the international trade in chemicals. It contains provisions requesting detailed information on the chemicals so that these decisions may be taken once data are available on the properties and the incidence of these products in particular on human health and the environment.

By the date of this convention, Argentina already had several pesticide regulations' laws. As a result, the Rotterdam Convention did not prohibit additional pesticides. The present existing regulation for pesticides in Argentina is summarized in Table 1.

Statistics reveals that the pesticide's usage is well regulated by different government offices, upon the destination by which they are registered in the National Phytosanitary Pesticides' Registry. In general, the province legislations agree with the national directives, however there are a few provinces and cities which run their own and independent registry. This reflects a kind of incoherence in the current legislation.

About pesticide's toxicological information, there are 21 Toxicological Assistance Centers in Argentina, offering several information services pointed to prevention, diagnosis and poisoning treatments. Since 1999, it was established a unique Registry of Toxicological Statistics, established in agreement with the INTOX Project of the International Programme Chemical Safety of the United Nations.

3. Country economy and probable agriculture evolution

The agriculture participation in the Gross Domestic Product (GDP) for 2002 was about 6 %. Inside this contribution area, the agriculture was the highest contributor (63%), followed by stockbreeding (31%). The sector's evolution estimation is of continuous growth (in magnitude of sowed surface and crops), about 16% by the end of 2010 and 9% by 2016. The estimated production for 2010 reaches 100 millions of Tons, and is expected to grow up to 150 millions in five years. These projections define a strong agriculture sector, which represents country wealth but also an environment threat.

3.1 Evolution of the pesticides' market

Pesticide use has increased dramatically over the last four decades to an estimated 2.59 x 109 kg of active ingredient used globally during1995 (Golfinopoulos et al. 2003). The gradual change in Argentina towards modern and intensive agricultural activities has led to an increase in the use of pesticides (Figure 1). Between the years 1994 and 2002 there was a 130% increase in the consumption of phytosanitary products, from 80 to almost 180 million liters (Miglioranza et al. 2003a). The physical volume commercialized during the present decade was about three times from the 90's, with a drastic increase in consumables for Glyphosate (N-(phosphonomethyl) glycine), a broad-spectrum systemic herbicide used to kill weeds, especially perennials, in particular for soybean cultivation by direct sowing.

In Argentina, the major pesticide class is Herbicides (commonly known as a weedkillers; substances used to kill unwanted plants). During 2006, 71% of the pesticides' market was

Compound	Restriction Level	Details of the restriction	Year
2,4,5-T	P	Decree-law 2.121/90	1990
ALDICARB	SR	Restrictions in soil usage: Decree-law 2.121/90	1990
ALDRIN	SR	Prohibited for bovine and pigs: Decree-law 2.143/68	1968
ALDRIN	P	Decree-law 2.121/90	1990
ALFANAFTIL-TIOUREA (ANTU)	P	Prohibited as rat poison. Law 7.292/98	1998
AMINOTRIAZOL	SR	Prohibited in Tobacco: Law 80/71	1971
ARSENIC	P	Decree-law 2.121/90	1990
ARSENIC and arsenic compounds	P	Prohibited as rat poison. Law 7.292/98	1998
LEAD ARSENATE	P	Decree-law 2.121/90	1990
AZINFOS, METIL	SR	Prohibited in horticulture and fruit trees. Resolution 10/91	1991
BARIUM compounds	P	Prohibited as rat poison. Law 7.292/98	1998
MERCURY dichloride	SR	Prohibited in Tobacco: Law 80/71	1971
METHYL BROMIDE	SR	Prohibited in public (urban or domestic) pest control. Res. 280/98	1998
CANFECLOR	SR	Prohibited for bovine and pigs: Decree-law 2.143/68	1968
CANFECLOR	SR	Prohibited as weevil insecticide. Law 47/72 and in the entire life cycle of cereals and leguminous plants. Law 79/72.	1972
CAPTAFOL	P	Decree-law 2.121/90	1990
CARBOFURAN	SR	Prohibited for pear tree and apple tree . Res. 10/91	1991
CLORDANE	P	Decree-Law 2.143/68. Law 18.073/69. Decree-Law 2.678/69	1969
CLORDANE	SR	Prohibited in Tobacco: Law 80/71	1971
CLORDANE	SR	Prohibited as weevil insecticide. Law 47/72	1972
CLORDANE	SR	Prohibited in grassland and fodder. Law 18.073/69. Decree-Law 2.678/69	1969
CLORDANE	SR	Prohibited in the entire life cycle of cereals and leguminous plants. Law 79/72	
CLORDANE	SR	Legal usage: ants insecticide and soil treatment.	
CLORDANE	P	Prohibited for domestic insecticides. Disposition 7.292/98	1998
CLOROBENCILATHE	P	Decree-law 2.121/90	1990
DAMINOZIDE	S	Decree-law 2.121/90	1990
		Allowed for chrysanthemum controlled production. Resolution:175/91.	1991
D.D.T.	SR	Prohibited for bovine and pigs: Decree-law 2.143/68	1968
D.D.T.	P	Decree-law 2.121/90	1990
D.D.T.	P	Resolution 133/91	1991
D.D.T.	P	Prohibited for domestic insecticides. Disposition 7.292/98	1998
ETHYLENE DIBROMIDE	P	Decree-law 2.121/90	1990
DICLORVOS	R	Prohibited for domestic insecticides. Disposition 7.292/98	
DIELDRIN	P	Law 22.289/80	1980
DINOCAP	S	Decree-law 2.121/90	1990
DISULFOTON	SR	Prohibited for apple tree and peach tree. Resolution:10/91	1991
ENDRIN	SR	Prohibited for bovine and pigs: Decree-law 2.143/68	1968
ENDRIN	P	Decree-law 2.121/90	1990
STRYCHNINE SULFATE	P	Decree-law 2.121/90	1990
STRYCHNINE	P	Prohibited as rat poison. Law 7.292/98	1998
ETHYL AZINFOS	SR	Prohibited in horticulture and fruit trees. Resolution 10/91	1991
ETION	SR	Prohibited for apple tree and pear tree. Resolution:10/91	1991
PHENYLMERCURY ACETATE	SR	Prohibited in Tobacco: Law 80/71	1971
METALLIC PHOSPHYTES	P	Prohibited as rat poison. Law 7.292/98	1998
WHITE PHOSPHOROUS	P	Prohibited as rat poison. Law 7.292/98	1998
H.C.B.	SR	Prohibited for bovine and pigs: Decree-law 2.143/68	1968
H.C.B.	SR	Prohibited as weevil insecticide. Law 47/72	1972
		Allowed for seed treatment. Res. 10/91.	1991
H.C.H.	P	Law 22.289/80.	1980
H.C.H.	P	Prohibited for domestic insecticides. Disposition 7.292/98	1998
HEPTACHLORO	P	Decree-Law 647/68. Law 18.073/69. Decre-Law 2.678/69.	1969
HEPTACHLORO	P	All usages prohibited. Res. IASCAV 27/93 -	1993
HEPTACHLORO	P	Prohibited for domestic insecticides. Disposition 7.292/98	1998
LINDANE	SR	Prohibited for bovine and pigs: Decree-law 2.143/68	1968
LINDANE	SR	Prohibited in Tobacco: Law 80/71. Prohibited as weevil insecticide. Law 47/72. Allowed for: ants and grasshooper insectice/soil and seeds treatment	1971 1972
LINDANE	SR	Allowed for louse treatment. Res. 133/91	1991
LINDANE	P	Prohibited for domestic insecticides. Disposition 7.292/98	1998
METOXICHLOR	P	Prohibited for domestic insecticides. Disposition 7.292/98	1998
METOXICHLOR	SR	Prohibited as weevil insecticide. Law 47/72 and in the entire life cycle of cereals and leguminous plants. Law 79/72. Prohibited for bovine and pigs: Decree-law 2.143/68.	1968 1972
MIREX	P	Prohibited for all usages. Res. 627/99	1999
MONOCROTOFOS	SR	Prohibited for alfalfa crops. Res. 396/96 . Prohibited in horticulture and fruit trees. Resolution 10/91	1991 1996
MONOCROTOFOS	P	Prohibited for all usages. Res. 627/99	1999
MONOFLUORO-ACETHAMIDE	P	Prohibited as rat poison. Law 7.292/98	1998
SODIUM MONOFLUORO ACETATE	P	Prohibited as rat poison. Law 7.292/98	1998
PARATHION	P	Res. 7/96	1996
PARATHION (ETHYL)	P	Res. SAGYP 606/93	1993
PARATION (METHYL)	P	Res. SAGYP 606/93	1993
PCP	P	Res. 356/94	1994
THALLIUM compounds	P	Prohibited as rat poison. Law 7.292/98	1998

P: Prohibited; SR: Severely Restricted; S: Suspended

Table1. Restricted Pesticides in Argentina (adapted from information provided by the Health Ministry of Argentina, García et al., 2003).

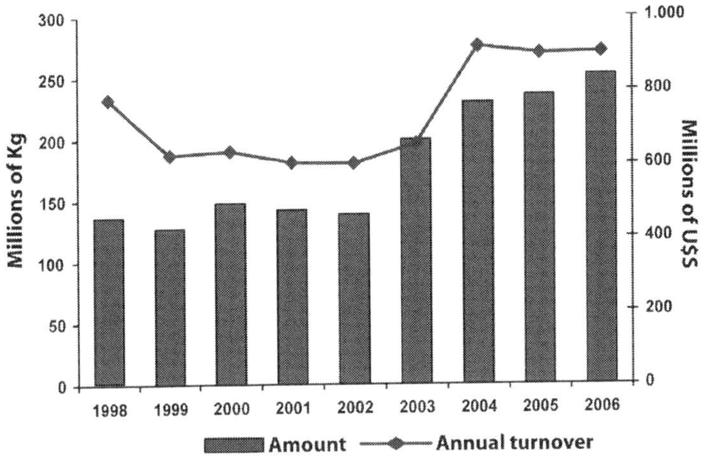

Fig. 1. Evolution of the Argentinean pesticides market (data from the Argentinean House of Agricultural and Livestock Health and Fertilizers)

dominated by herbicides, in particular by Glyphosate (46%), the principal herbicide used in the soybean cultivation (Figure 2).

Apart from these, there is another item hard to document: the illegal commercialization of pesticides. One of the documents delivered by the National Program of Chemical risks (Health Ministry, Argentina) identified the following compounds as the main products of this commerce: Pentachlorophenol, Parathión, DDTs, HCHs and Daminozide (Alar, Kylar, B-NINE, DMASA, SADH, B 995), a plant growth regulator usually sprayed on fruits (in Argentina for apples and pears) to regulate their growth, make their harvest easier, and enhance their color.

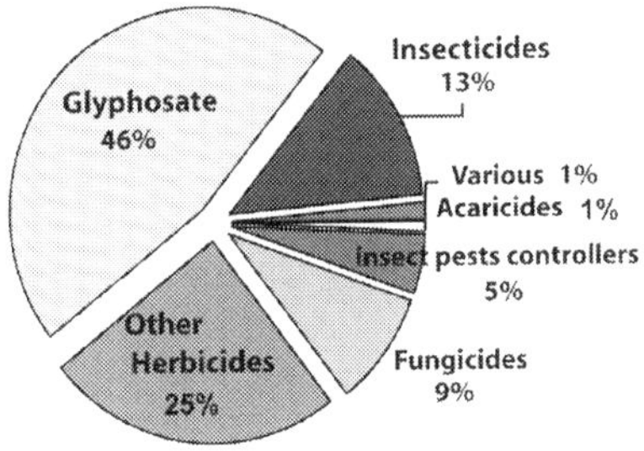

Fig. 2. Distribution of the Argentinean pesticides market for the year 2006 (data from the Argentinean House of Agricultural and Livestock Health and Fertilizers)

4. Factors affecting the application of pesticides

The technology used for their application is a critical factor when considering waste of pesticides and unnecessary environmental contamination. The design of nozzles is one of the major variables that could save up to 70% of pesticides compared to usual farmer's practice (Mathews, 1998). Although the Argentinean pesticide spraying systems do not use droplet size < 200 um -avoiding much of the airborne drifting-, pesticide application management usually relies on farmers personal decisions, with scarce regulatory control. Farmers and sprayers equipment operators still have wrong notion that high volumes, high pressure and high doses being perceived as the most appropriate ways for pesticide application (Abilash & Singh, 2009). In the pest's chemical control, the traditional method use nozzles of high volume rates, which require great volume of water to reach to the dripping point. Nozzles are normally not replaced and are even enlarged on purpose to achieve higher flow rates. The distribution patterns under these conditions are uneven; leaving sections with no pesticide coverage and others receiving overdoses (Mathews, 1998).

The common factors which alter pesticides efficiency, in particular in non industrialized producers are:

- Nozzles' poor calibration. Usually the equipment is inappropriate or obsolete.
- Inappropriate tractor speed which is reflected in unsatisfactory turbine's air volumes and pulverized water.
- Late or inadequate fruit pruning which facilitates infestation (in fruit production).
- Inadequate choice of the treatment opportunity time. This reflects ignorance of the pest's biology.
- Poor weather conditions: weather plays a major factor in affecting timing decisions; these conditions also play a significant role in the occurrence of spray drift. This is a major concern because it diverts the pesticide from the intended target, reduces efficacy, and deposits pesticide where it is not needed or wanted. The factors influencing spray drift contamination are: operating pressure, nozzle, type, orientation, orifice size, wind speed, wind direction, temperature, relative humidity and atmospheric stability. When a pesticide drifts, it may cause both environmental and economic damage through injury to susceptible vegetation, harm to wildlife, deposition of illegal residues on crops, and contamination of water supplies.
- Inadequate choice of the pesticide product in order to the target pest or absence of new pesticides application after precipitations events

Advanced producers are less affected by the abovementioned factors. Many of them use control techniques based on sanitary, environmental and economic criteria; in e.g., the sexual confusion of the *Carpocapsa* pest technique. These producers have integrative corporations that include many steps of the production process or are close-related to export agents. These conditions put them in the center of the commercial scenario, in a way that pesticides control and regulations must be thoroughly observed.

However, recent projections of several agriculture and livestock trend researchers for one of the most important agriculture regions (Alto Valle, Argentina, Huerga & San Juan, 2004) point that about 5% of crops are performed under "organic" practices (no pesticides usage), 10% under EUREGAP25 practices (=EURE Good Agricultural Practics), 45% under direct responsibility of the exporting corporation -with own protocols- and 40% with no particular system.

5. Potential risks of inappropriate pesticides' usage

There are at least three potentially dangerous items to consider about pest control:

1. Farm worker and rural population intoxications

2. Food contamination
3. Environment and natural resources contamination.

At present, is feasible to avoid these negative effects of pest controlling by regulation, control, education and technological assistance of the producer, executed by both, particular producers and government agents. However, in Argentina, this is hardly achieved and serious risk situations for producers, farms and food consumers are widely observed. Problems are well identified, so as well solutions, then, key questions arise: Why solutions are not applied? Who should be the main funding agent for these secure practices?

The last Agriculture and livestock census in 2002 showed that only 40% of farms were doing pest monitoring and only 10% accurately protecting their workers (Table 2). A key factor for the inclusion of good agricultural practices is the socioeconomic condition of the farm. For example, in the Alto Valle is clearly observed the two scenarios: worked land and abandoned scrubland. The last one usually belongs to low socioeconomic producers stroked by the 2002 crisis. Scrublands are then normal pest reservoirs, which diminish the results of the pest control efforts. A second commonly found scenario is traditionally hand-worked farms, which result inappropriate for pesticides' modern application techniques (Huerga & San Juan, 2004). As a result, the low socioeconomic condition of many farms affects the whole group of producers, including the ones who adopted good agricultural practices (GAPs).

Good Agricultural Practices (GAPs) adoption for pest control

Provinces and cities	Farms (N°)	Area (ha)		Use agrochemicals according to pre-harvest intervals		Assessment of the balance status between pests and diseases		Adopt organic control practices		Promote integrated pest management (IPM)		Application by specially trained and knowledgeable persons		Empty containers management	
		Total	Sowed	N°	%	N°	%	N°	%	N°	%	N°	%	N°	%
Buenos Aires															
Colón	402	79.560	63.305	5	1,2	129	32,1	8	2,0	18	4,5	18	4,5	113	28,1
Pergamino	1.117	285.991	231.580	249	22,3	395	35,4	21	1,9	172	15,4	124	11,1	121	10,8
Rojas	584	183.273	149.897	129	22,1	277	47,4	13	2.2	94	16,1	61	10,4	104	17.8
Salto	516	139.423	121.261	83	16,1	149	28.9	33	6,4	31	6,0	58	11,2	76	14,7
San Nicolás	293	53.104	43.475	4	1,4	79	27.0	4	1,4	8	2,7	37	12,6	61	20,8
Córdoba															
Marcos Juárez	2.077	833.774	708.470	588	28,3	553	26,6	44	2,1	279	13,4	498	24,0	511	24,6
Gral. San Martín	786	343.195	297.971	43	5,5	54	6,9	7	0,9	165	21,0	137	17,4	117	14,9
Río Segundo	1.420	495.255	437.263	362	25,5	372	26.2	8	0,6	205	14,4	606	42,7	417	29,4
Mendoza															
Junín	1.589	20.322	13.932	424	26,7	53	3,3	4	0,3	7	0,4	303	19,1	172	10,8
Santa Fe															
Constitución	1.641	869.216	224.498	54	3,3	458	27,9	32	2,0	263	16,0	200	12,2	269	16,4
Iriondo	1.435	297.495	261.657	341	23,8	765	53,3	43	3,0	258	18,0	424	29,5	387	27,0
Tucuman															
La Cocha	660	47.325	33.758	164	24,8	90	13,6	9	1,4	57	8,6	283	42,9	400	60,6
Lules	159	12.230	9.230	31	19,5	17	10,7	2	1,3	3	1,9	29	18,2	31	19,5

Table 2. Good Agricultural Practices for Pest Control. Data from the Agriculture and Livestock breeding National Census (2002), adapted from the Agriculture Zonal Study, Huerga & San Juan, 2004.

Farm worker, rural population intoxications and pesticide food contamination are away of the scope of this chapter. However these risks are close linked to the environment pollution. There are two major pathways for pesticides reaching the environment:

1. Accidental spills, as punctual sources, generally registered in farm lands but can also occur away from rural environments (acute incidents, generally characterized as point source pollutions).

2. Continuous spills in lower concentrations due to pesticides applications to sowed land etc (chronic releases to the environment, generally characterized as non-point source pollution).

The following section deals with scientifically documented occurrence of pesticides in environmental compartments of Argentina, as a result of the abovementioned sources.

6. Pesticides reaching Argentinean environments

6.1 Water

Most of the Earth's living resources are found in specific geographical locations such as the global coastal environment and the catchment basins of large river systems. Furthermore, more than 3 billion people live in close proximity to these regions and are dependent upon it for either part or much of their food supply and industrial raw materials (Moore et al. 2004). The consequence of this situation is that much of the industrial, domestic and agricultural wastes are ultimately transported into aquatic environments, generating ecosystem changes and habitat destruction. These environments are of the greatest biological and economic significance and are quite likely to impact on human life's quality. In Argentina, the POPs environmental monitoring data is usually related to densely populated areas along the major rivers such as the Río de la Plata, Paraná and Reconquista. The bulk of this information relates to freshwater environments, in detriment of coastal marine areas, which have historically received less attention. There is a general lack of large-scale regional water monitoring programs in the country. Further, only potentially contaminated ecosystems are usually monitored. Scarce multidisciplinary approaches considering simultaneous evaluation of a number of factors and processes are found in the country; in fact, the historic freshwater quality monitoring has been developed on water chemistry and bacteriology, with measurements of only the main variables required for the determination of quality indexes (Salibián, 2006).

6.1.1 De la Plata River

De la Plata river is an international river (Buenos Aires province), which is part of the second largest hydro geographic system of South America, after the Amazon, and the fifth largest in the world. It covers an area of about 38,800 km^2 and drain a 3,170,000 km^2 basin. It is situated on the East coast of the country, delimited by Argentina and Uruguay. The tidal river is used for commercial fishing, angling, shipping to and from several mayor ports, recreation and tourism, especially along the sandy north shore. The waters of the river are a depository for many raw wastes and effluents from industries, cites and towns, and from the disposal of dredging spoils. Extensive chemical and physical pressures occur on this river, the same as other estuarial systems surrounded by cities and industries (Kurucz et al. 1998). In 1991, Janiot et al. observed the presence of six chlorinated pesticides (alpha, gamma and delta hexachlorociclohexane (HCH), heptachlor-epoxide, p,p'DDE and p,p'DDD) in sixteen water samples (water and suspended material) from the South shore and along a transect in the outer boundary of the river. Levels were higher near the coast, indicating the HCHs as the most frequent isomers. As levels decreased with increasing shoreline's distance, the observed distribution supported the hypothesis of a dilution effect in the river as a whole. More recent studies at the South coastal shoreline, showed that high concentrations of chlorinated pesticides (Mirex, Lindane (γHCH), Dieldrin, Aldrin, etc.; 6,8-80 ng/L) in the water column were limited to the mix areas of discharge and tributaries

(FREPLATA. 2004). A similar behavior was previously observed in the discharge area of *Río Santiago* River (a highly industrialized -8 km long- tributary of De *La Plata* River); however, this spot was influenced by the discharge of other tributary, which receives chronic inputs from one of the major petrochemical complexes of the area. Relatively high organochlorine levels were also detected at coastal sampling stations from *La Plata* Harbor. The release of hydrophobic organochlorines from sediments disturbed by dredging and ship transit could have contributed to the high values of this place. As well as Janiot et al., this study pointed γHCH as dominant in the dissolved phase which was widely detected, in concentrations ranging between 0.9 and 61 ng/L (Colombo et al., 1990). In the same work, an hexachloro component of technical Chlordane was also detected at several stations in relatively high levels (0.4-28 ng/L). On the contrary, other chlorinated pesticides showed lower levels or were under the detection limit at several stations.

Geographically moving to the south, at the outer zone of *de la Plata* River's Estuary is located the *Samborombón* Bay and the maritime front (Atlantic Ocean). The available information at these sites showed that, unless some punctual exceptions, pesticides in the water column were moderate to low (1.8±2.7 ng/L), away from the maximum limits suggested for the protection of the biota. Again, γHCH was the most frequent compound; however it was under the guide value in all cases (FREPLATA. 2004).

More than the 97% of the total water in *De La Plata River* comes from *Paraná* and *Uruguay* rivers. 2004 monitoring surveys showed that the contribution of organochlorinated and organophosphorous pesticides, PCBs and PAHs from these rivers were of little significance, as their concentrations in water were under the detection limits of the used techniques. Only PCBs were detected in a station near *Paraná de las Palmas* (46 ng/L, FREPLATA. 2004). In general, chlorinated pesticides occurred at very low concentration in water samples collected along the *Uruguay* River. As in *De La Plata* River, HCH isomers were the most commonly detected compounds, with concentrations ranging from the detection limits to 10 ng/L. Heptachlor, Heptachlor-epoxide, Aldrin, Dieldrin, p-p´DDE and p-p´DDT were occasionally encountered while o-p´DDE, p-p´DDD, o-p´DDD and o-p´DDT were never detected (Janiot et al. 1994).

In the lower *Paraná* River's Delta the concentrations of chlorinated pesticides in water were low and similar between sites (3.9±5.1 ng/L), as informed by Cataldo el al. in 2001. In this survey, γHCH accounted for 45-63% of total chlorinated pesticides, reflecting the relatively high water solubility of this chemical (7 ppm). Pollutant concentrations followed a clear geographic pattern with the highest values in the densely populated area of the *Reconquista* and *Luján* Rivers, lower levels in the *San Antonio*, and lowest loadings in the remote *Paraná de las Palmas* River. This gradient adequately matched the pattern of mortality rates of the mollusk *Corbícula flumínea*, which were highest in the *Reconquista-Luján* Rivers (40 ± 93%) and lowest (and not significantly different from the control) in the *Paraná* River (3.3 ± 23%).

6.1.2 *Reconquista* River

The Reconquista River is a typical lowland watercourse located at the Buenos Aires Province. More than 3 million people (10% of the population of the country) are settled on its basin. It is located in a temperate subtropical region, crossing what is known as the Pampa Region, sedimentary, flat in topography, with a total surface of 167,000 Ha and 50 km in length. It flows into the *Luján* River. For over a century, complex mixtures of domestic, agricultural, industrial solids and liquid wastes (mostly untreated) have been dumped in the river, which has thus become a typical example of the adverse impact of human activities on the health of aquatic environments (Salibián, 2006). For instance, Rovedatti et al. indicated that over 60 samples analyzed 35% presented organochlorine pesticides in a concentration greater than 0.1μg/L (2001). On the opposite, in the same study organophosphates were found in no case, possibly due to their low persistence because of their short half-lives in aquatic environments. Some of the detected compounds were DDT and its metabolite DDE, γ-Chlordane, Heptachlor and HCH isomers. They did not find temporal or spatial trends and there was not a relationship between the time of samplings and the fumigation season for farming purposes. At all locations, pesticides levels were found to be between 40 and 400 times higher than the legal limits established for protection of aquatic life. Recently, new studies demonstrated the effects of poor water quality and environmental deterioration on biomarkers of native fishes species, however in that paper, organochlorine compounds were not found in the water samples (de la Torre, 2007). This fact could be explained due to the low water solubility of organo-halogenated compounds, however, it was clearly documented the occurrence of the "*hit and run*" effect: although the xenobiotic is not present, it is possible to measure its effects in biota.

6.1.3 *Pergamino-Arrecifes* system

Soybean production in Argentina has increased over the last decade, currently with 10,000,000 hectares of sowed land. A total of 95% of this area corresponds to a transgenic variety of glyphosate tolerant soybean, which is cultivated by direct sowing (Pengue, 2005). Recent monitoring surveys showed that the levels of glyphosate in water, from a transgenic soybean cultivation area located near to tributaries streams of the *Pergamino-Arrecifes* system in the north of the province of Buenos Aires, ranged from 0.10 to 0.70 mg/L (Peruzzo et al. 2008). The authors concluded that temporal variation of glyphosate levels depended directly on the time of application and the rain events. This emerges as a documented example of pesticides reaching the environment as a consequence of inappropriate agriculture practices.
There is scarce available information about pesticides pollution for other regions of Argentina. In the Central and Midwest region of the country (i.e.; San Juan and San Luis provinces), on account of the intense agricultural activity near the rivers that drain the region, the presence of high pesticides' concentrations is likely to occur.
In 2003, Baudino et al. performed a survey at San Juan and after analyzing the overall mean value for the concentration of OCs in water sources, only βHCH (6.556 μg/L) and Dieldrin (5.354 μg/L) were above the maximum permissible value recommended by international organizations (E. C. Council Directive 1980). The authors concluded that the San Juan River basin was the most contaminated area, possibly due to the higher population density, larger cultivated area and industrial complexes.
In San Luis province, water samples coming from agricultural-livestock areas principally indicated an homogeneous distribution of the pesticides found in the area with a clear predominance of HCH isomers and DDT analogs over chlorodines. A prevalence of 4, 4-DDE was observed, suggesting old DDT inputs. Further, this was corroborated by local farmers (Luco et al. 1992), as DDT was a common pesticide in the past.
At the South region of the country (i.e.; *Río Negro* province), organochlorinated and organophosphorous pesticides have also been detected. Monitoring stations located at Negro River basin's origin, confluence of *Limay* and *Neuquén* Rivers, detected the presence of α-HCH, γ-HCH and Parathion, but in non toxic concentrations for aquatic biota. DDT and metabolites have also been detected but in much less quantity (Natale et al., 1995). Added to this, near this region, ground water samples from fruit production farms belonging to the

Valley of *Neuquen* River frequently showed organophosphorus pesticides levels that exceeded the acute toxicity risk ratios for aquatic life protection (Loewy et al. 2003). It was found that some pesticides, as Azinphos methyl, had a high detection frequency - 66% of the samples- with concentrations varying from non detectable to 48.9 ppb. Dimethoate, Metidathion and Phosmet were also detected but in less frequency and values. Finally, these authors found that pesticides in ground water samples followed seasonal variations and temporal trends.

6.2 Sediments

Regional POP information for sediments is also dominated by chlorinated pesticides. Overall, as observed for waters, sediment data indicates a complex situation in densely populated areas affected by urban-industrial inputs. The applied pesticides can be transported through surface runoff, leaching, and vapor phase and generally, estuarine and marine sediments are the temporary or long-term ultimate sinks for most of OCs. Consequently, these sediments act later as secondary sources of these substances reaching the ocean and biota. The most frequently reported POPs are DDTs, HCHs, PCBs, and heptachlors, however, concentrations show a large variability.

The FREPLATA 2004 program showed the occurrence of pesticides in sediments from the discharge of *De la Plata* River basin. The organochlorines' levels were around 1.9±3.84 ng/g. In the middle and external area of the river, the concentration was under the detection limit of the method. Contrarily, on the South coast of this river, the values of toxic compounds were higher than the levels suggested for the protection of aquatic biota. It was found that pesticide levels diminished by distance from coast line and from tributaries discharge's sites. Similarly, littoral affluents between *De La Plata* River and Necochea City showed concentratios of 3.1±6.5 ng/g, with maximum values of 12-31 ng/g in *Atalaya* and *Mar del Plata* harbor. In general Chlordane, DDT and its derivates predominated among OCs.

In 2001, pesticides monitoring in the *Reconquista* and *Luján* Rivers showed levels around 2.8±3.9 ng/g, being the trans-Chlordane the most abundant (Cataldo el a., 2001). As previously shown in sediments sampled from *De la Plata* River (Colombo et al., 1990), the abundance of DDD and DDE, relative to the parent compound, indicated that DDT was being readily metabolized in the sediments.

As expected by anthropogenic pressure, the highest loadings of pollutants occurred in areas located closer to the urbanized area, decreasing toward the more remote sites. The same behavior -as abovementioned for water- was observed in *Corbicula fluminea* mortality. As expected for hydrophobic substances, all sediment samples were enriched in organic contaminants relative to water, then, in agreement with their highest pollution levels the strongest toxicity responses were obtained with them (Cataldo et al., 2001).

In relation with organophosphorus pesticides, levels of Glyphosate in sediment and soils from a transgenic soybean cultivation area located near to tributaries streams of the Pergamino-Arrecifes system in the north of the Buenos Aires' province were between 0.5 and 5.0 mg/Kg (Peruzzo et al. 2008).

Results from two creeks of the Southeast Argentina region showed similar total OCs concentrations in sediments, in the range from 6 to 25 ng/g (dry wt.), being below the sediment quality criteria demanded for wildlife protection. ΣEndosulfans, ΣDDTs and ΣChlordanes were the main OCs' group, with Endosulfan sulfate being the most frequent and abundant compound. The predominance of metabolites with respect to parent

compounds suggested a contamination mainly by runoff from aged and weathered agricultural soils (Miglioranza et al. 2004). The latter shows another clue about pesticides inputs to the environment due to inappropriate application practices.
Moving toward the south, the Southwest coastal area of Buenos Aires province presented total OCs levels from non-detectable to 166.5 ng/g (d.w.) (Menone et al. 2001; Arias et al. 2010). In example, in terms of average concentration, the major pesticides detected in sediments from *Bahía Blanca* Estuary were Mirex > Heptachlor-epoxide > Metoxychlor > δ-HCH > Endosulfan I > α-HCH > Heptachlor > DDE > DDD (Arias et al. 2010).
Macroinvertebrate bioturbation affects the fate and partitioning of sediment-bound contaminants in sediment profiles, pore water and the water column. These factors increase the rate of important physicochemical processes that occur at the sediment-water interface such as diffusion, desorption, degradation, and resuspension of organic and inorganic compounds (Ciarelli et al. 2000). The burrowing crab, *Neohelice granulata*, is a bioturbator widely distributed in SW Atlantic estuaries. These crabs inhabit almost all the zones of the intertidal, the soft bare sediment flats and the lower salt marsh zones (Iribarne et al. 1997). Crab beds act as sinks for OC pesticides in SW Atlantic coastal environments. Sediments from these sites from the northeastern of the country exhibited total OC pesticide concentrations in the order of other Argentinean coastal environments (Menone et al. 2000; Menone et al. 2004). β- and γ-isomers of Hexachlorocyclohexane, γ-Chlordane, Dieldrin and p,p-DDE were the dominant OC pesticides detected in all sediment samples, while Aldrin, Metoxichlor and p,p´-DDD were below the detection limits (Menone et al. 2006).

6.3 Soils

Extensive agricultural practices can cause soils' degradation by means of hydric and wind erosion, structure deterioration, salinization, fertility diminution and desertification. Moreover, soils are natural sinks for persistent and lipophilic compounds that strongly adsorb to organic carbon and remain relatively immobile in this reservoir. Pollutants enter the soil either by deposition from air, drift or by washing-off from plant surface during rainfall or irrigation. The proportion of applied pesticide reaching the target pest has been found to be less than 0.3%, thus leaving over 99% elsewhere in the environment (Pimentel 1995). POP monitoring in soils is also limited within the region. There are no regional monitoring programs, and most data refer to agricultural areas.
Agricultural soils in the Southeastern region of Argentina could be an important source (if not the major) for OC pesticides. In 2003, the highest values for OCs (656.1 ng/g d.w.) were found in the most superficial layers of soil, even though at sites which have never received direct OCs application (Miglioranza et al., 2003). At sowing lands, OCs levels were of 30.19 ng/g dry wt in the surface horizon. The pattern of OCs distribution was similar at all sampled soils, with DDT and metabolites > HCHs > Heptachlor > Chlordanes. In this work the authors concluded that volatilization could have been one of the major causes of pesticide loss from the sowing target area, due to inappropriate management practices (Miglioranza et al. 2003b).
In the same geographical area, other researchers found that total OCs levels in soil from conventional farm were greater than those from organic farm, but with the same distribution pattern - DDT and derivatives as the major compounds- (Gonzalez et al. 2005).
In 2005 Andrade et al. studied the concentrations of organochlorine and organophosphorus pesticides in soils from the South of Buenos Aires province. Results showed that the

horticulture dedicated soils contained higher pesticide levels than the wheat, soybean and sunflower dedicated ones. This fact probably reflected the differences in soil management according to the different crops. The studied soils contained DDT and their metabolites, Heptachlor-epoxide, Dieldrin, Endrin, Lindane, Malathion and Parathion. In the same study the authors concluded that the soil pesticide accumulation was principally due to the age and persistence in the application of these products (Andrade et al. 2005).
Concentrations of OCs in the Patagonia region are of importance on account of their massive past and/or present use in fruits and vegetable production. Recent studies showed that levels in *Río Negro* Valley reached up to 492 ng/g and 3.43 ng/g for DDTs and Endosulphanes, respectively (Mitton et al., 2010). The authors found that in different vegetables and fruit plants of this region, OCs were more abundant in the roots than in the aerial parts. In this way, they demonstrated the translocation of toxic compounds from the soils and the capacity of plants for bioaccumulating highly hydrophobic compounds.

6.4 Air

POPs monitoring in air as well as studies about volatilization of persistent pollutant from the sowed land are scarce in Argentina. Stronger efforts have to be made in this research discipline in order to test the volatilization/air drift and atmospheric transport from other matrices.

6.5 Biota

Available data on POPs in Argentina for both aquatic and terrestrial animals is poor when comparing to other regions of the world. Among this information, aquatic organisms are by far the most studied organisms; in special bivalves and fish are the preferred ones. In Argentina, the most comprehensive program of POP monitoring in coastal organisms was the global Mussel Watch (Farrington and Tripp 1995). In this monitoring, the highest levels of POPs were obtained in mussels from the coastal Hudson City (southern Buenos Aires province). On the one hand, among the reported POPs, PCBs dominated, followed by Chlordanes and DDTs. On the other hand, in 2006, Endosulfan sulfate, Chlordanes, HCH isomers and DDT compounds dominated in tissues and ingested food of fish (*Cynoscion guatucupa*), from *Bahía Blanca* Estuary (Lanfranchi et al., 2006). These authors also identified α-Chlordane, Heptachlor and p,p´-DDE as the major bioaccumulated and biomagnified OCs.
Recent studies in birds from the arid-semiarid Midwest region of Argentina showed the presence of several OCs in fat tissue [ΣHCH range: ND to 3168.41 ng/g fat, ΣCHL range: ND to 4961.66 ng/g fat, ΣALD range: 287.07 to 9161.70 ng/g fat, ΣDDT range: 1068.98 to 6479.84 ng/g fat] with the exception of p,p'-DDT. Total OCs concentration in all bird species ranged from 2684.91 to 19231.91 ng/g fat (Cid et al. 2007). This point to an immediate threat, since in a previous study, Gil et al. indicated concentration of pesticides under the detection limit of the method at Patagonia coasts, corresponding the higher value for pp´-DDE. In the same study, the concentration in mammals did not exceed 0.1 mg/L (Gil et al., 1997).

7. Conclusions

About the regulatory point of view, Argentina's government implemented the Rotterdam Convention by the year 2000. Despite this, by the date of this convention, Argentina already had several pesticide regulations' laws. This reflects a strong national regulatory framework, however, there are a few provinces and cities which run their own and independent registry and regulations. This reflects a kind of incoherence in the current legislation. Added to this, some of the more severely regulated pesticides have been identified as main compounds in the illegal trade-market (Pentachlorophenol, Parathión, DDTs, HCHs and Daminozide) and in environmental samples.

About *Good Agricultural Practices*'s adoption, although many steps have been taken, Argentina still has a long way to cover. As a matter of fact, recent projections of several agriculture and livestock trend researchers for one of the most industrialized agriculture regions showed up to 40% of the producers with no particular system of Agricultural practices.

According to the environmental point of view, the most relevant problems with pesticides arise from the improper use, disposal, and maintenance of the available stock. The discharge of untreated effluents throughout the whole region is recognized as a major input pathway of POPs into the environment. There is enough scientific information to conclude that pesticides are extensively reaching the environment, with levels in some compartments exceeding the permitted values to protect wildlife. These levels include several already banned pesticides.

It came clear that POP's survey, inventory and monitoring are still poorly developed in the country. Argentina lacks of routine monitoring programs and most of the available data were generated by punctual studies in urbanized areas. The bulk of the information corresponded to aquatic animals, waters and sediments, with scarce information regarding soils and atmosphere. Finally, it is recommended the consideration of the actual relevance of atmospheric transport of pesticides, since they were usually identified in pristine regions or geographically distant areas from the pesticides application points.

8. References

Abhilash, P.C. & Singh, N. (2009). Pesticide use and application: An Indian scenario. *Journal of Hazardous Materials*. 165, 1-12, 0304-3894.

Andrade, M.; Covelo, E. & Alonso Vega, M. (2005). Influencia del manejo agrícola intensivo en la contaminación del suelo, In *Revista Pilquen*. Sección Agronomía, Pozzo Ardizzi, M. Cristina (Ed.), pp17, Año VII, N° 7, Centro Universitario Regional Zona Atlántica (CURZA) de la Universidad Nacional del Comahue, ISSN 1851-2852.

Arias, A.; Pereyra, M. & ·Marcovecchio J. (2010). Environ Multi-year monitoring of estuarine sediments as ultimate sink for DDT, HCH, and other organochlorinated pesticides in Argentina. *Environmental Monitoring and Assessment*, DOI 10.1007/s10661-010-1315-9.

Aronson, K J; Wilson, J W L; Hamel, M; Diarsvitri, W; Fan, W L; Woolcott, C; Heaton, J P W; Nickel, J C; Macneily, A, & Morales, A. (2010). Plasma organochlorine levels and prostate cancer risk. *Journal of Exposure Science and Environmental Epidemiology*, 20, 5, 434-445, 1559-0631.

Baudino, O.; Suero, E.; Augusto, M.; Gimenez, M. & Flores, N. (2003). Monitoring organochlorine pesticides in surface and ground water in San Juan Argentina. *Journal of the Chilean Chemical Society*, 48, 2, 7-12, ISSN 0167-6369 (print version); ISSN 1573-2959 (electronic version)

Bergstrom, L F. (2004). Symposium 2 part 2: Food production for a growing world population - The impact of food production on soils and groundwater resources.*Journal of Food Science*, 69, 9, R181-R184, 0022-1147.

Bila, D M & Dezotti, M. (2007). Endocrine disrupters in the environment: Part 1 - Effects and consequences.*Quimica Nova*, 30, 3, 651-666, 0100-4042.

Blankenberg, A G B; Braskerud, B, & Haarstad, K. (2006). Pesticide retention in two small constructed wetlands: treating non-point source pollution from agriculture runoff.International *Journal of Environmental Analytical Chemistry*, 86, 3-4, 225-231, 0306-7319.

Casabe, N; Piola, L; Fuchs, J; Oneto, M L; Pamparato, L; Basack, S; Gimenez, R; Massaro, R; Papa, J C, & Kesten, E. (2007). Ecotoxicological assessment of the effects of glyphosate and chlorpyrifos in an Argentine soya field.*Journal of Soils and Sediments*, 7, 4, 232-239, 1439-0108.

Cataldo, D.; Colombo, J.; Boltovskoy, D.; Bilos, C. & Landoni, P. (2001). Environmental toxicity assessment in the Paraná river delta (Argentina): simultaneous evaluation of selected pollutants and mortality rates of *Corbícula fluminea* (Bivalvia) early juveniles. *Environmental Pollution*, 112, 3, 379-389, ISSN 0269-7491.

Ciarelli, S.; Kater, B. & van Straalen, N. (2000). Influence of bioturbation by the anphipod *Corophium volutator* on fluoranthene uptake in the marine polychaete *Nereis virens*. *Environmental Toxicology and Chemistry*, 19, 2, 1575–1581, ISSN 0730-7268.

Cid, F.; Antón, R.; Caviedes-Vidal, E. (2007). Organochlorine pesticide contamination in three bird species of the Embalse La Florida water reservoir in the semiarid midwest of Argentina. *Science of the Total Environment*, 385, 1-3, 86–96, ISSN 0048-9697.

Colombo, J.; Khall, M.; Arnac, M.; Horth, A. & Catoggio, J. (1990). Distribution of Chlorinated Pesticides and Individual Polychlorinated Biphenyls in Biotic and Abiotic Compartments of the Rio de La Plata, Argentina. *Environmental Science & Technology*, 24, 4, 498-505, ISSN 0013-936X.

de la Torre, F.; Salibián, A. & Ferrari, L. (2007). Assessment of the pollution impact on biomarkers of effect of a freshwater fish. *Chemosphere*, 68, 8, 1582–1590, ISSN 0045-6535.

De Silva, P M C S; Pathiratne, A, & van Gestel, C A M. (2010). Toxicity of chlorpyrifos, carbofuran, mancozeb and their formulations to the tropical earthworm Perionyx excavatus.*Applied Soil Ecology*, 44, 1, 56-60, 0929-1393.

Farrington, F. & Tripp, B. (1995). International Mussel Watch Project. Initial Implementation Phase. Final Report. NOAA Technical Memorandun NOS ORCA 95, 63 p and 5 append..

FREPLATA. (2004). "Análisis Diagnóstico Transfronterizo del Río de la Plata y su Frente Marítimo". Documento Técnico. Proyecto Protección Ambiental del Río de la Plata y su Frente Marítimo, p301, Proyecto PNUD/GEF/RLA/99/G31.

García, S.I.; Bovi-Mitre, G., Moreno I., Eiman Grossi, M., Digón, A., de Titto, E. (2003). Regional Workshop on Pesticides Toxicology and Data Collection Protocols.

Argentina Health Ministry, Sanitary Program Secretariat, Buenos Aires, Argentina.

Gbaruko, B C; Ogwo, E I; Igwe, J C, & Yu, H T. (2009). Organophosphate induced chronic neurotoxicity: Health, environmental and risk exposure issues in developing nations of the world.African *Journal of Biotechnology*, 8, 20, 5137-5141, 1684-5315.

Gil, M.; Harvey, M.; Beldoménico, H.; García, S.; Commendatore, M.; Gandini, P.; Frere, E.; Yorio, P.; Crespo, E. & Esteves, J. (1997). Contaminación por metales y plaguicidas organoclorados en organismos marinos de la zona costera patagónica. *Informes Técnicos del Plan de Manejo Integrado de la Zona Costera Patagónica (Puerto Madryn, Argentina)*, 32, 1-28.

Giupponi, C & Rosato, P. (1999). Agricultural land use changes and water quality: A case study in the watershed of the Lagoon of Venice.*Water Science and Technology*, 39, 3, 135-148, 0273-1223.

Golfinopoulos, S.; Nikolaou, A.; Kostopoulou, M.; Xilourgidis, N.; Vagi, M. & Lekkas, D. (2003). Organochlorine pesticides in the surface waters of Northern Greece. *Chemosphere*, 50, 4, 507–516, ISSN 0045-6535.

Gonzalez, M.; Miglioranza, K.; Aizpún de Moreno, J. & Moreno, V. (2005). Evaluation of conventionally and organically produced vegetables for high lipophilic organochlorine pesticide (OCP) residues. *Food and Chemical Toxicology*, 43, 2, 261–269, ISSN 0278-6915.

Guo, H; Chen, G F; Lv, Z P; Zhao, H, & Yang, H. (2009). Alteration of microbial properties and community structure in soils exposed to napropamide.*Journal of Environmental Sciences-China*, 21, 4, 494-502, 1001-0742.

Haarstad, K & Braskerud, B C. (2005). Pesticide retention in the watershed and in a small constructed wetland treating diffuse pollution.*Water Science and Technology*, 51, 3-4, 143-150, 0273-1223.

Hendrix, P F. (1996). Nearctic earthworm fauna in the southern USA: Biodiversity and effects on ecosystem processes.*Biodiversity and Conservation*, 5, 2, 223-234, 0960-3115.

Hu, G C; Luo, X J; Li, F C; Dai, J Y; Guo, J Y; Chen, S J; Hong, C; Mai, B X, & Xu, M Q. (2010). Organochlorine compounds and polycyclic aromatic hydrocarbons in surface sediment from Baiyangdian Lake, North China: Concentrations, sources profiles and potential risk.*Journal of Environmental Sciences-China*, 22, 2, 176-183, 1001-0742.

Huber, A; Bach, M, & Frede, H G. (2000). Pollution of surface waters with pesticides in Germany: modeling non-point source inputs.*Agriculture Ecosystems & Environment*, 80, 3, 191-204, 0167-8809.

Huerga, M. & San Juan, S.(2004). Pests Control in Agriculture, Argentina. World Bank/Investment Center FAO, Buenos Aires, Argentina.

Iribarne, O.; Bortolus, A. & Botto, F. (1997). Between-habitats differences in burrow characteristics and trophic modes in the southwestern Atlantic burrowing crab *Chasmagnathus granulata*. *Marine Ecology Progress Series*, 155, 132–145, ISSN 0171-8630.

Islam, M. & Tanaka, M. (2004). Impacts of pollution on coastal and marine ecosystems including coastal and marine fisheries and approach for management: a review and synthesis. *Marine Pollution Bulletin*, 48, 7-8, 624–649, ISSN 0025-326X.

Janiot, L.; Orlando, A. & Roses, O. (1991). Niveles de plaguicidas clorados en el Río de la Plata. *Acta Farmacéutica Bonaerense*, 10, 1, 15-23, ISSN 0326-2383.

Janiot, L.; Sericano J. & Roses, O. (1994). Chlorinated pesticide occurrence in the Uruguay river (Argentin-Uruguay). *Water, Air and Soil Pollution*, 76, 3-4, 323-331, ISSN 0049-6979.

Kurucz, A.; Masello, A.; Méndez, S.; Cranston, R. & Wells, P. (1998). Calidad ambiental del Río de La Plata, Capítulo 3, 71-86, In *El Río de la Plata, Una revisión ambiental. Un informe de Antecedentes del Proyecto EcoPlata*, Wells, P. & Daborn, G. (Eds.), 256p., Dalhousie University, Halifax, Nova Scotia, Canada.

Lanfranchi, A.; Menone, M.; Miglioranza, K.; Janiot, L.; Aizpún, J. & Moreno, V. (2006). Striped weakfish (*Cynoscion guatucupa*): A biomonitor of organochlorine pesticides in estuarine and near-coastal zones. *Marine Pollution Bulletin*, 52, 1, 74–80, ISSN 0025-326X.

Li, R Q; Dong, M; Peng, H; Zhang, L L; Cui, Q G, & He, W M. (2006). Agricultural expansion in Yunnan Province and its environmental consequences.*Chinese Science Bulletin*, 51, 136-142, 1001-6538.

Lin, A Y C; Huang, S T Y, & Wahlqvist, M L. (2009). Waste management to improve food safety and security for health advancement.*Asia Pacific Journal of Clinical Nutrition*, 18, 4, 538-545, 0964-7058.

Loewy, R.; Carvajal, L.; Novelli, M. & Pechen de D'Angelo, A. (2003). Effect of Pesticide Use in Fruit Production Orchards on Shallow Ground Water. *Journal of Environmental Science and Health*, Part B, 38, 3, 317–325, ISSN (printed version) 0360-1234, ISSN (electronic version) 1532-4109.

Lopez-Espinosa, M J; Lopez-Navarrete, E; Rivas, A; Fernandez, M F; Nogueras, M; Campoy, C; Olea-Serrano, F; Lardelli, P, & Olea, N. (2008). Organochlorine pesticide exposure in children living in southern Spain.*Environmental Research*, 106, 1, 1-6, 0013-9351.

Luco, J.; Aguilar, E.; Silva, P.; Baudino, O. & Gonzalez, D. (1992). Niveles de Pesticidas Organoclorados en Aguas de la Provincia de San Luis (República Argentina). *Acta Farmacéutica. Bonaerense*, 11, 3, 121-8, ISSN 0326-2383.

Luo, Y Z & Zhang, M H. (2010a). Spatially distributed pesticide exposure assessment in the Central Valley, California, USA.*Environmental Pollution*, 158, 5, 1629-1637, 0269-7491.

Luo, Y Z & Zhang, M H. (2010b). Spatially distributed pesticide exposure assessment in the Central Valley, California, USA.*Environmental Pollution*, 158, 5, 1629-1637, 0269-7491.

Marco, J.A.M. & Kishimba, M.A. (2007). Organochlorine pesticides and metabolites in young leaves of Mangifera indica from sites near a point source in Coast region, Tanzania.*Chemosphere*, 68, 5, 832-837, 0045-6535.

Mathews, G.A. (1998). Application Equipment with Particular Reference to Smallscale Operations, UNIDO Report.

McLaughlin, A & Mineau, P. (1995). The impact of agricultural practices on biodiversity.*Agriculture Ecosystems & Environment*, 55, 3, 201-212, 0167-8809.

Menone, M.; Aizpun de Moreno, J.; Moreno, V.; Lanfranchi, A.; Metcalfe, T. & Metcalfe, C. (2001). Organochlorine pesticides and PCBs in a southern Atlantic coastal lagoon watershed, Argentina. *Archives of Environmental Contamination and Toxicology*, 40, 3, 355-362, ISSN 0090-4341 (print version), ISSN 1432-0703 (electronic version).

Menone, M.; Bortolus, A.; Botto, F.; Aizpún de Moreno, J.; Moreno, V.; Iribarne, O.; Metcalfe, T. & Metcalfe, C. (2000). Organochlorine contaminants in a coastal lagoon in Argentina: analysis of sediment, crabs and cordgrass from two different habitats. *Estuaries*, 23, 4, 583-592, ISSN 0160-8347.

Menone, M.; Miglioranza, K.; Botto, F.; Iribarne O.; Aizpún de Moreno, J. & Moreno, V. (2006). Field accumulative behavior of organochlorine pesticides.The role of crabs and sediment characteristics in coastal environments. *Marine Pollution Bulletin*, 52, 12, 1717-1724, ISSN 0025-326X.

Menone, M.; Miglioranza, K.; Iribarne, O.; Aizpún de Moreno, J. & Moreno, V. (2004). The role of burrowing beds and burrows of the SW Atlantic intertidal crab Chasmagnathus granulata in trapping organochlorine pesticides. *Marine Pollution Bulletin*, 48, 3-4, 240-247, ISSN 0025-326X.

Miglioranza, K.; Aizpón de Moreno, J. & Moreno, V. (2003a). Trends in Soil Science: Organochlorine Pesticides in Argentinean Soils. *Journal of Soils & Sediments*, 4, 3, 264-265, ISSN 1439-0108.

Miglioranza, K.; Aizpún de Moreno, J. & Moreno, V. (2003b). Dynamics of organochlorine pesticides in soils from a southeastern region of Argentina. *Environmental Toxicology and Chemistry*, 22, 4, 712-717, ISSN 0730-7268.

Miglioranza, K.; Aizpun de Moreno, J. & Moreno, V. (2004). Land-based sources of marine pollution: organochlorine pesticides in stream systems. *Environmental Science and Pollution Research International*, 11, 4, 227-32, ISSN 0944-1344 (print version) ISSN: 1614-7499 (electronic version).

Mitton, F.; Gonzalez, M. & Miglioranza K. (2010). Dinámica de DDTs y endosulfanes en plantas comestibles cultivadas en suelos de Río Negro, Patagonia. *III Congreso Argentino de la Sociedad de Toxicología y Química Ambiental*, SETAC 2010, p.113, Santa Fe, Argentina.

Moore, M.; Depledge, M.; Readman J. & Leonard D. (2004). An integrated biomarker-based strategy for ecotoxicological evaluation of risk in environmental management. *Mutation Research/Fundamental and Molecular Mechanisms of Mutagenesis*, 552, 1-2, 247-268, ISSN 0027-5107.

Natale, O.; Gómez, C.; Vermeulen, J.; Casamiquela, C.; D´Angelo, A.; Loewy, R.; Alcalde, R.; Vernier, B.; Schultz, F. & Cardot, L. (1995). Anexo 2, Estudio de caso: Plaguicidas en el Río Negro, Argentina. In: Manual de Evaluación y Manejo de sustancias tóxicas en aguas superficiales, 5409/C44/40272; CEPIS-OPS-INCYTH.

Ntow, W J; Tagoe, L M; Drechsel, P; Kelderman, P; Gijzen, F J, & Nyarko, E. (2008). Accumulation of persistent organochlorine contaminants in milk and serum of farmers from Ghana.*Environmental Research*, 106, 1, 17-26, 0013-9351.

Panegyres, P.K.; Gray, V.; Barrett, L., & Perceval, S. (2010). Neurological disorders in a rural Western Australian population.*Internal Medicine Journal*, 40, 3, 209-213, 1444-0903.

Pengue, W. (2005). Transgenic crops in Argentina: the ecological and social debt. *Bulletin of Science, Technology & Society*, 25, 4, 314-32, ISSN 0270-4676 (print version), ISSN 1552-4183 (electronic version).

Perez-Maldonado I.N; Trejo, A; Ruepert, C; Jovel, R D; Mendez, M P; Ferrari, M; Saballos-Sobalvarro, E; Alexander, C; Yanez-Estrada, L; Lopez, D; Henao, S; Pinto, E R, & Diaz-Barriga, F. (2010). Assessment of DDT levels in selected environmental media and biological samples from Mexico and Central America.*Chemosphere*, 78, 10, 1244-1249, 0045-6535.

Peruzzo, P.; Porta, A. & Ronco, A. (2008). Levels of glyphosate in surface waters, sediments and soils associated with direct sowing soybean cultivation in north pampasic region of Argentina. *Environmental Pollution*, 156, 1, 61-66, ISSN 0269-7491.

Pimentel, D. (1995). Amounts of Pesticides reaching target pests: environmental impacts and ethics. *Journal of Agricultural and Environmental Ethics*, 8, 1, 17-29, ISSN 1187-7863 (print version), ISSN 1573-322X (electronic version).

Reis, M R; Silva, A A; Freitas, M A M; Pereira, J L; Costa, M D; Picanco, M C; Ferreira, E A; Belo, A F; Coelho, A T C P, & Silva, G R. (2009). Impact of Glyphosate Associated with Insecticide and Fungicide Application on the Microbial Activity and Potential Phosphate Solubilization in Soil Cultivated with Roundup Ready (R) Soybean.*Planta Daninha*, 27, 4, 729-737, 0100-8358.

Rovedatti, M.; Castañé, P.; Topalián, M. & Salibián, A. (2001). Monitoring of organochlorine and organophosphorus pesticides in the water of the Reconquista river (Buenos Aires, Argentina). *Water Research*, 35, 14, 3457–3461, ISSN 0043-1354.

Rull, R P; Gunier, R; Von Behren, J; Hertz, A; Crouse, V; Buffler, P A, & Reynolds, P. (2009). Residential proximity to agricultural pesticide applications and childhood acute lymphoblastic leukemia.*Environmental Research*, 109, 7, 891-899, 0013-9351.

Salibián, A. (2006). Ecotoxicological Assessment of the Highly Polluted Reconquista River of Argentina. *Reviews of Environmental Contamination & Toxicology*, 185, 35-65, ISSN 0179-5953.

Schreiber, J D; Rebich, R A, & Cooper, C M. (2001). Dynamics of diffuse pollution from US southern watersheds.*Water Research*, 35, 10, 2534-2542, 0043-1354.

Silva, M H & Gammon, D. (2009). An Assessment of the Developmental, Reproductive, and Neurotoxicity of Endosulfan.*Birth Defects Research Part B-Developmental and Reproductive Toxicology*, 86, 1, 1-28, 1542-9733.

Tataruch, F; Steineck, T, & Frey, H. (1998). Carbofuran poisoning in wildlife (birds of prey, songbirds and carnivores) in Austria.*Wiener Tierarztliche Monatsschrift*, 85, 1, 12-17, 0043-535X.

Thiere, G & Schulz, R. (2004). Runoff-related agricultural impact in relation to macroinvertebrate communities of the Lourens River, South Africa.*Water Research*, 38, 13, 3092-3102, 0043-1354.

Waliszewski, S M; Melo-Santiesteban, G; Villalobos-Pietrini, R; Gomez-Arroyo, S; Amador-Munoz, O; Herrero-Mercado, M, & Carvajal, O. (2009). Breast Milk Excretion

Kinetic of beta-HCH, pp'DDE and pp'DDT.*Bulletin of Environmental Contamination and Toxicology*, 83, 6, 869-873, 0007-4861.

Waliszewski, S M; Quintana, R V; Corona, C A; Herrero, M; Sanchez, K; Aguirre, H; Aldave, I A; Arroyo, S G, & Pietrini, R V. (2010). Comparison of Organochlorine Pesticide Levels in Human Adipose Tissue of Inhabitants from Veracruz and Puebla, Mexico.*Archives of Environmental Contamination and Toxicology*, 58, 1, 230-236, 0090-4341.

Wang, R Y; Jain, R B; Wolkin, A F; Rubin, C H, & Needham, L L. (2009). Serum Concentrations of Selected Persistent Organic Pollutants in a Sample of Pregnant Females and Changes in Their Concentrations during Gestation.*Environmental Health Perspectives*, 117, 8, 1244-1249, 0091-6765.

18

Bioconcentration of Pesticides in Fish from Rivers and Lakes

Taizo Tsuda
Lake Biwa Environmental Research Institute
Japan

1. Introduction

Pesticide contamination of river and lake waters from agriculture use is a problem of worldwide importance. Many field data on the pesticide contamination of surface waters and aquatic organisms in rivers and lakes (Amaraneri & Pillala, 2001, Abdel-Halim et al., 2006; Agradi et al., 2000; California Environmental Protection Agency, 2002; California Regional Water Quality Control Board, 2002; California State Water Resources Control Board, 2002; Domagalski, 1996; 1997; Environment Canada, 2002; Ganapathy et al., 1997; Gfrerer et al., 2002a; 2002b; Harman-Fetcho et al., 1999; Hall, 2003; Laabs et al., 2002; Lekkas et al., 2004; Leong et al., 2007; Mansour et al., 2001; Mansour & Sidky, 2003; McConnell et al., 2004; Oros et al., 2003; Rovedatti et al., 2001; Struger et al., 2004; Sudo et al., 2002a; 2002b; 2004; Tanabe et al., 2001; Tsuda et al., 1996a; 1997a; 1998; 1999; Vitanov et al., 2003; Washington State Department of Ecology, 1999; 2000) have been reported in the world.
This chapter consisted of (1) Field surveys on pesticide contaminations in rivers and lakes, (2) Bioconcentration of pesticides in the field fish (3) Bioconcentration of pesticides in fish by laboratory experiments (4) Evaluation of the pesticide contamination in the field fish by their laboratory bioconcentration potential data

1. Diazinon, fenitrothion, malathion and fenthion were selected as insecticides and atrazine, simazine, simetryn, molinate and benthiocarb, mefenacet and pretilachlor as herbicides. Surveys on the contamination of the 4 insecticides (Abdel-Halim et al., 2006; Ministry of the Environment, Japan, 2001; Mansour & Sidky, 2003; Ohtsuki, 1994; Tsuda et al., 1992a; 1994; 1998; 2009) and the 7 herbicides (Chiba Prefecture, 2002; 2003; Kanagawa Prefecture, 2000; 2001; Ministry of the Environment, Japan, 1993; 1999; Takino et al., 1998; Tsuda et al., 1996a; 1997a; 2009; Watanugi et al., 1993) in water and fish from rivers and lakes in the world were reviewed from literatures in the past.
2. Bioconcentration factor (BCF) of each 11 pesticide in the field fish was calculated as its bioconcentration potential from the data on the pesticide concentration in the water and fish from the rivers and lakes in the world.
3. Laboratory BCF data of the 11 pesticides in fresh-water fish were reviewed from literatures in the past for the 4 insecticides (Allison & Hermanutz, 1977; De Bruijn & Hermens, 1991; Escartin & Porte, 1996; Fisher, 1985;Goodman et al., 1979; Kanazawa, 1975; 1978; 1981; 1983; 1987; Keizer et al., 1991; 1993; Lockhart et al., 1983; Miyamoto et al., 1979; Nihon Kagaku-busshitsu Anzen-Jyohou Center, 1992; Sancho et al., 1992; 1994; Seguchi & Asaka, 1981; Takimoto et al., 1984; 1987; Tsuda et

Laboratory BCF data of the 4 insecticides in fresh-water fish are shown in Fig. 3 for diazinon (Allison & Hermanutz, 1977; Goodman et al., 1979; Kanazawa, 1975; 1978; 1981; 1983; Keizer et al., 1991; 1993; Nihon Kagaku-busshitsu Anzen-Jyohou Center, 1992; Seguchi & Asaka, 1981; Sancho et al., 1992; Tsuda et al., 1989a; 1990; 1995; 1997b; 1997c), Fig. 4 for fenitrothion (De Bruijn, & Hermens, 1991; Escartin & Porte, 1996; Fisher, 1985; Kanazawa, 1975; 1981; 1983; 1987; Lockhart et al., 1983; Miyamoto et al., 1979; Nihon Kagaku-busshitsu Anzen-Jyohou Center, 1992; Sancho et al., 1994; Takimoto et al., 1984; 1987; Tsuda et al., 1989a; 1990; 1995; 1997b; 1997c), Fig. 5 for malathion (Tsuda et al., 1989a; 1990; 1997b) and fenthion (De Bruijn & Hermens, 1991; Tsuda et al., 1992b; 1993; 1995; 1996b; 1997c). The average BCF value of each insecticide was 100 (n=12) for diazinon, 170 (n=10) for fenitrothion, 20 (n=2) for malathion and 340 (n=6) for fenthion. The order of the 4 insecticides in the BCF values was fenthion > fenitrothion > diazinon > malathion.

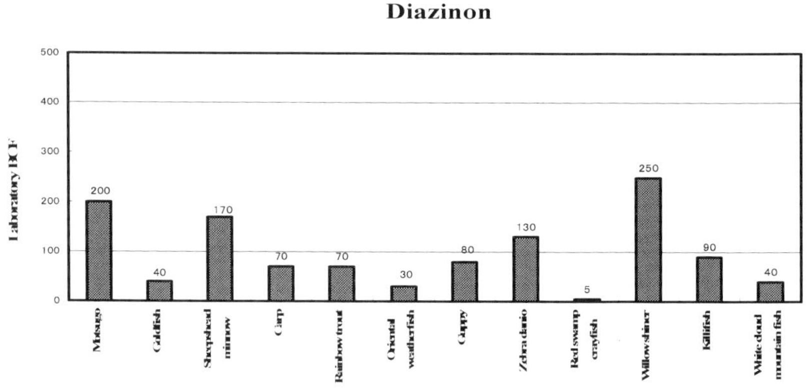

Fig. 3. Bioconcentration of diazinon in fresh-water fish

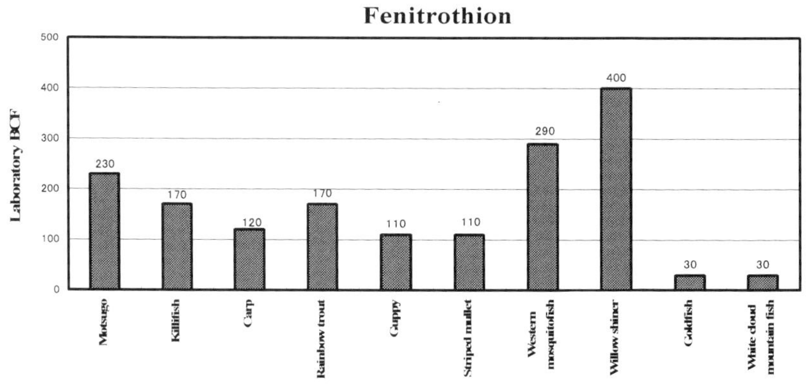

Fig. 4. Bioconcentration of fenitrothion in fresh-water fish

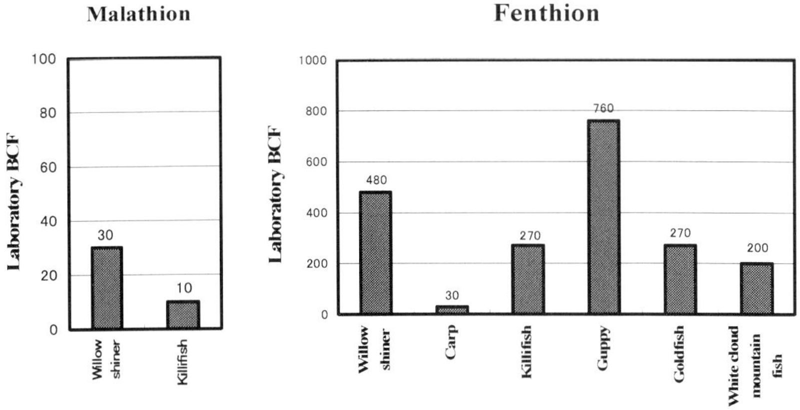

Fig. 5. Bioconcentration of malathion and fenthion in fresh-water fish

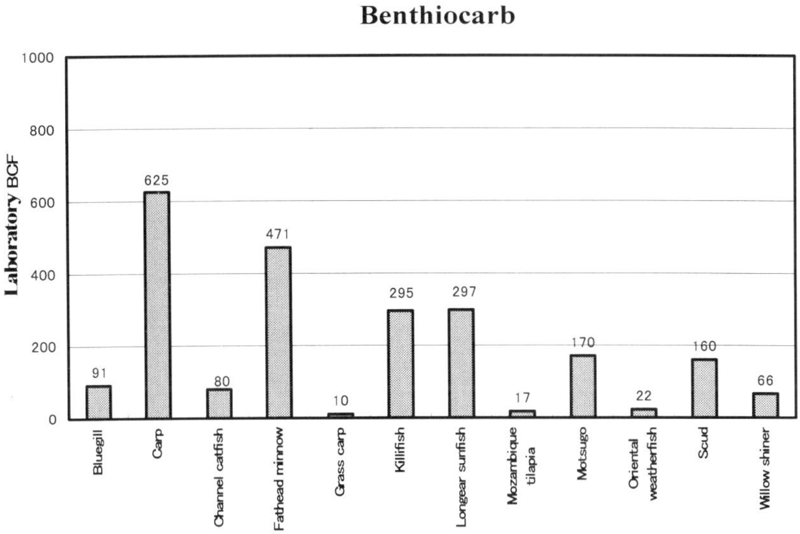

Fig. 6. Bioconcentration of benthiocarb in fresh-water fish

Those of the 7 herbicides in fresh-water fish are shown in Fig. 6 for benthiocarb (Kanazawa, 1981; 1983; Sanders & Hunn, 1982; Tsuda et al., 1988; 1989b; 1997d; Wang et al., 1992) and Fig. 7 for simetryn (Tsuda et al., 1988; 1989b; Xu & Zhang, 1989), molinate (Kanazawa, 1981; 1983; Martin et al., 1992; Tsuda et al., 1999), mefenacet (Tsuda et al., unpublished data), pretilachlor (Tsuda et al., unpublished data), simazine (Tsuda et al., 1992c) and atrazine (Isensee, 1976; Kearney et al., 1977; Gunkel & Streit, 1980; Gorge & Nagel, 1990; Du Preez & Van Vuren, 1992). The average BCF value of each herbicide was 192 (n=12) for benthiocarb,

35 (n=3) for simetryn, 19 (n=3) for molinate, 21 (n=2) for mefenacet, 33 (n=2) for pretilachlor, 3.9 (n=1) for simazine and 8.3 (n=3) for atrazine. The order of the 7 herbicides in the BCF values was benthiocarb > simetryn, pretilachlor ≧ mefenacet, molinate > atrazine, simazine.

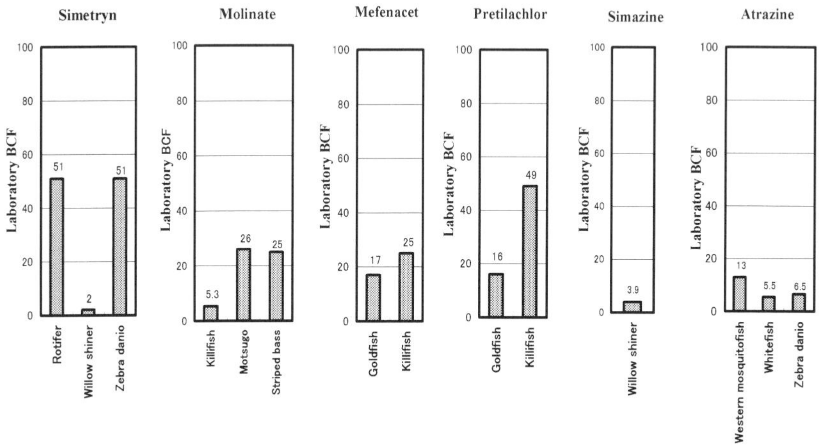

Fig. 7. Bioconcentration of simetryn, molinate, mefenacet, pretilachlor, simazine and atrazine in fresh-water fish

For benthiocarb, simetryn and atrazine, their bioconcentration in muscle and viscera (liver, kidney and gallbladder) of two species of fish (carp and bream) (Du Preez & Van Vuren, 1992; Tsuda et al., 1989b) is shown in Fig. 8. BCF values of benthiocarb were 26 in muscle, 63 in liver, 73 in kidney and 63 in gallbladder. Similarly, those of simetryn were 2.4 in muscle, 14 in liver, 8.1 in kidney and 11 in gallbladder. The order of the BCF values in the 4 parts of the carp for benthiocarb was slightly different from that of simetryn. But for both herbicides, the values of BCF in the viscera were higher than those in the muscle. Further in the bream, the BCF value in the liver was higher than that in the muscle.

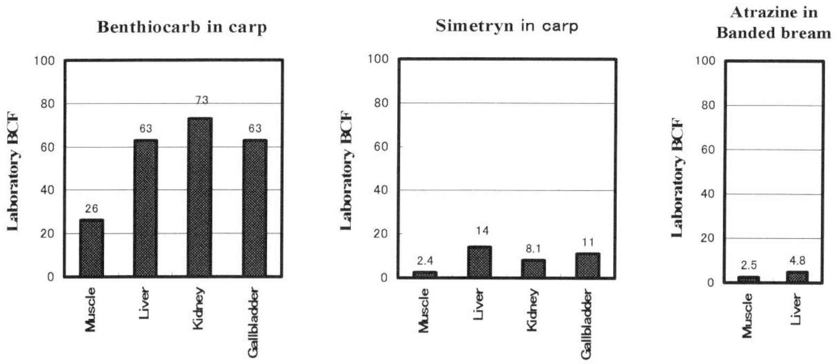

Fig. 8. Bioconcentration of benthiocarb, simetryn and atrazine in muscle and viscera of fresh-water fish

4. Evaluation of the pesticide contaminations in the field fish by their laboratory BCF data

The contaminations of the 10 pesticides in the field fish were evaluated by comparing the field BCF data with the laboratory BCF data.

The field BCF data of the 4 insecticides in the field fish (Table 5) and the laboratory BCF data (Figs. 3 - 5) are summarized and compared in Fig. 9. The field BCF data of the 4 insecticides were nearly equal to the laboratory BCF data. Similarly, the field BCF data (Table 6) and the laboratory BCF data (Figs. 6 - 7) of the 6 herbicides except atrazine are summarized and compared in Fig. 10. The field BCF data of the 4 insecticides and the 5 herbicides except simazine were nearly equal to the laboratory BCF data. It was revealed that the contamination of 9 pecticides except simazine in fish from the rivers and the lakes was approximately predicted by the laboratory BCF data.

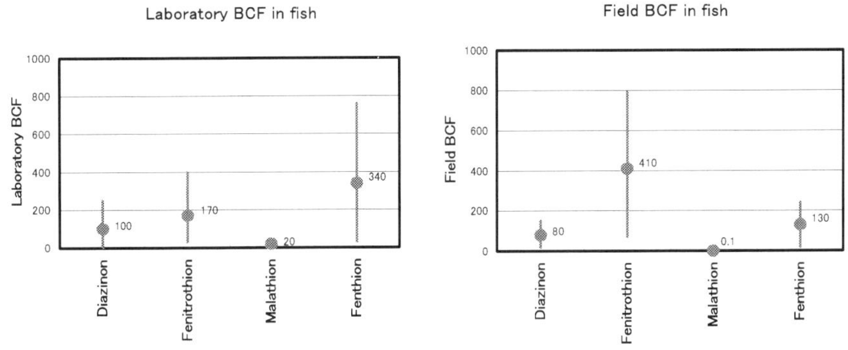

Fig. 9. Comparison of laboratory BCF data and field BCF data for 4 insecticides

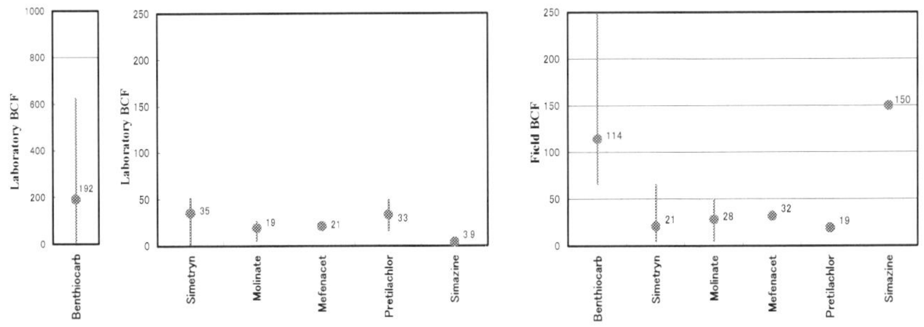

Fig. 10. Comparison of laboratory BCF data and field BCF data for 6 herbicides

Field BCF data of the 5 herbicides (molinate, bromobutide, simetryn, pretilachlor and mefenacet) in the fish from Lake Biwa (Fig. 2) and the laboratory BCF data are shown in Fig. 11. The average field BCF values were nearly equal to the average laboratory BCF values for molinate, bromobutide and pretilachlor but slightly lower for simetryn and slightly higher for mefenacet. The differences in the field and laboratory BCF values of simetryn and mefenacet are not wide, so both of the field and laboratory BCF data are considered to be the same levels for all of the 5 herbicides. From the comparison shown in Fig. 11, it was clarified that the contamination of the 5 herbicides in the fish from Lake Biwa could be approximately estimated by the laboratory BCF.

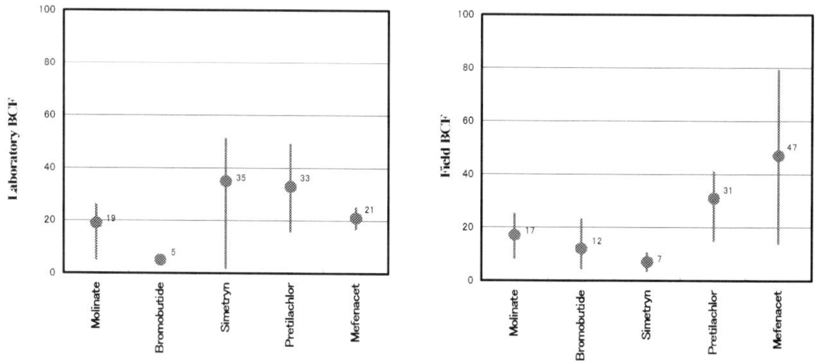

Fig. 11. Comparison of laboratory BCF data and field BCF data for 5 herbicides

5. References

Agradi, E., Baga, R., Cillo, F., Ceradini, S. & Heltai, D. (2000). Environmental contaminants and biochemical response in eel exposed to Po river water, Chemosphere Vol. 41: 1555-1562.

Amaraneri, S.R. & Pillala, R.R. (2001). Concentrations of pesticides in tissues of fish from Kolleru Lake in India, *Environ. Toxicol.* Vol. 16: 550-556.

Allison, D.T. & Hermanutz, R.O. (1977). Toxicity of diazinon to brook trout and fathead minnows, *Ecol. Res. Ser., EPA-600/3-77-060,* Environ. Res. Lab., U.S. EPA.

Abdel-Halim, K.Y., Salama, A.K., El-khateeb, E.N. & Bakry, N.M. (2006). Organophosphorus pollutants (OPP) in aquatic environment at Damietta Governorate, Egypt: Implications for monitoring and biomarker responses, *Chemosphere,* Vol. 63: 1491-1498.

California Environmental Protection Agency (2002). *Rice pesticides use and surface water monitoring 2002,* California Environmental Protection Agency.

California Regional Water Quality Control Board (2002). *Central Valley Region Draft Staff Report on Recommended Changes to California's Clean Water Act Section 303(d) List Appendix B.,* California Environmental Protection Agency.

California State Water Resources Control Board (2002). *Dissolved pesticides in the Alamo River and the Salton Sea, California, 1996-97, Open-File Report 02-232,* California Environmental Protection Agency.

Chiba Prefecture (2002). *Survey results on endocrine disrupter etc. in fiscal year 2001,* Kanagawa Prefecture (in Japanese).

Chiba Prefecture (2003). *Survey results on endocrine disrupter etc. in fiscal year 2002,* Kanagawa Prefecture (in Japanese).

De Bruijn, J & Hermens, J.: Uptake and elimination kinetics of organophosphorous pesticides in the guppy (*Poecilia reticulata*), (1991). Correlations with the octanol/water partition, *Environ. Toxicol. Chem.* Vol. 10: 791-804.

Domagalski, J. (1996). Pesticides and pesticide degradation products in stormwater runoff: Sacramento River Basin, California, *Water Resources Bulletin* Vol. 32: 953-964.

Domagalski, J. (1997). Results of a prototype surface water network design for pesticides developed for the San Joaquin River Basin, California, *Journal of Hydrology* Vol. 192: 33-50.

Du Preez, H.H. & Van Vuren, J.H.J. (1992). Bioconcentration of atrazine in the Banded tilapia, *Tilapia sparrmanii. Comp. Biochem. Physiol.,* Vol. 101C, 651-655.

Environment Canada (2002). *Pesticide concentrations in the Don and Humber River Watersheds (1998-2000),* Environment Canada.

Escartin, E. & Porte, C. (1996). Bioaccumulation, metabolism and biochemical effects of the organophosphorus pesticide fenitrothion in Procambarus clarkii. *Environ, Toxicol. Chem.* Vol. 15: 915-920.

Fisher, S.W. (1985). Effects of pH upon the environmental fate of [14C] fenitrothion in an Aquatic Microcosm, *Ecotoxicol. Environ. Saf.* Vol. 10: 53-62.

Ganapathy, C., Nordmark, C., Bennett, K., Bradley, A., Feng, H., Hernandez, J. & White, J. (1997). *Temporal distribution of insecticide residues in four California rivers. Environment Hazard Assessment Program,* Environmental Monitoring and Pest Management Branch, California Department of Pesticide Regulation.

Gfrerer, M., Martens, D., Grawlik, B.M., Wenzl, T., Zhang, A., Quan, X., Sun, C., Chen, J., Platzer, B., Lankmayr, E. & Kettrup, A. (2002a). Triazines in the aquatic systems of the Eastern Chinese Rivers Liao-He and Yangtse, *Chemosphere* Vol. 47: 455-466.

Gfrerer, M., Wenzl, T., Quan, X., Platzer, B. & Lankmayr, E. (2002b). Occurrence of triazines in surface and drinking water of Liaoning Province in Eastern China, *J. Biochem. Biophys. Methods* Vol. 53: 217-228.

Goodman, L.R., Hansen, D.J., Coppage, D.L., Moore, J.C. & Matthews, E. (1979) Diazinon: Chronic toxicity to, and brain acetylcholinesterase inhibition in the sheepshead minnow, *cyprinodon variegates, Trans. Am. Fish. Soc.* Vol. 108: 479-488.

Gorge, G. & Nagel, R. (1990). Kinetics and metabolism of 14C-lindane and 14C-atrazine in early life stages of zebrafish (*Brachydanio rerio*), *Chemosphere* Vol. 21, 1125-1137.

Gunkel, G. & Streit, B. (1980). Mechanism of bioconcentration of a herbicide (Atrazine, S-triazine) in a freshwater mollusc (*Ancylus fluviatilis Mull.*) and a fish (*Coregonus fera Ju*), *Wat. Res.* Vol. 14, 1573-1584.

Harman-Fetcho, J.A., McConnell, L.L. & Baker, J.E. (1999). Agricultural pesticides in the Patuxent River, a tributary of the Chesapeake Bay. *J. Environ. Qual.* Vol. 28: 928-938.

Hall, L.W.Jr.(2003). Analysis of diazinon monitoring data from the Sacramento and Feather River watersheds: 1991-2001, *Environmental Monitoring and Assessment*, Vol. 86: 233-253.

Isensee, A.R. (1976). Variability of aquatic model ecosystem-derived data, *Int. J. Environ. Stud.* Vol. 10, 35-41.

Kanazawa, J. (1975). Uptake and excretion of organophosphorus and carbamate insecticides by freshwater fish, Motsugo, *Pseudorasbora pava*, *Bull. Environ. Contam. Toxicol.* Vol. 14: 346-352.

Kanazawa, J. (1978). Bioconcentration ratio of diazinon by freshwater fish and snail, *Bull. Environ. Contam. Toxicol.* Vol. 20: 613-617.

Kanazawa, J. (1981). Measurement of the bioconcentration factors of pesticides by freshwater fish and correlation with physicochemical properties or acute toxicities, *Pestic. Sci.* Vol. 12: 417-424.

Kanazawa, J. (1983). A metod of predicting the bioconcentration potential of pesticides by using fish, *Jpn. Agric. Res. Q.* Vol. 17: 173-179.

Kanagawa Prefecture (2000). *Survey results on endocrine disrupter etc. in fiscal year 1999*, Kanagawa Prefecture (in Japanese).

Kanagawa Prefecture (2001). *Survey results on endocrine disrupter etc. in fiscal year 2000*, Kanagawa Prefecture (in Japanese).

Kearney, P.C., Oliver, J.E., Helling, C.S., Isensee, A.R. & Kontson, A. (1977). Distribution, movement, persistence and metabolism of N-nitrosoatrazine in soils and a model aquatic ecosystem, *J. Agric. Food. Chem.* Vol. 25, 1177-1181.

Keizer, J., D'Agostino, G. & Vittozzi, L. (1991). The importamce of biotransformation in the toxicity of xenobiotics to fish. I. Toxicity and bioaccumulation of diazinon in guppy (*Poecilia reticulata*), *Aquat. Toxicol.* Vol. 21: 239-254.

Keizer, J., D'Agostino, G., Nagel, R., Gramenzi, F. & Vittozzi, L. (1993). Comparative diazinon toxicity in guppy and zebra fish: Different role of oxidative metabolism, *Environ. Toxicol. Chem.* Vol. 12: 1243-1250.

Laabs, A., Amelung, W., Pinto, A.A., Wantzen, M., Silva, C.J. & Zech, W. (2002). Pesticides in surface water, sediment and rainfall of the northeastern Pantanal basin, Brazil, *J. Environ. Qual.*, Vol. 31: 1636-1648.

Lekkas, T., Kolokythas, G., Nikolaou, A., Kostopoulou, M., Kotrikla, A., Gatidou, G., Thomaidis, N.S., Golfinopoulos, S., Makri, C., Babos, D., Vagi, M., Stasinakis, A., Petsas, A. & Lekkas, D.F. (2004). Evaluation of the pollution of the surface waters of Greece from the priority compounds of List II, 76/464/EEC Directive, and other toxic compounds, *Environ. Int.* Vol. 30: 995-1007.

Leong, K.H., Tan, L.L.B., Mustafa, A.M. (2007). Contamination levels of selected organochlorine and organophosphate pesticides in the Selangor River, Malaysia between 2002 and 2003, *Chemosphere* Vol. 66: 1153-1159.

Lockhart, W.L., Metner, D.A.,Billeck, B.N., Rawn, G.P. & Muir, D.C.G. (1983). Bioaccumulation of some forestry pesticides in fish and aquatic plants, in Garner, W.Y., Billeck, B.N. & Muir, D.C.G.: (ed.), *Proc. Symp. Implications of Chemical and Biol. Agents in Forestry Applic., Chem. Soc.* No. 238: 297-318.

Mansour, S.A., Mahram, M.R. & Sidky, M.S. (2001). Ecotoxicological studies. 4. Monitoring of pesticides residues in the major components lake Qarun, *Egypt. J. Acad. Soc Environ. Dev.* Vol. 2: 83-116.

Mansour, S.A. & Sidky, M.M. (2003). Ecotoxicological studies. 6. The first comparative study between Lake Qarun and Wadi El-Rayan wetland (Egypt), with respect to contamination of their major components, *Food Chemistry* Vol. 82: 181-189.

McConnell, L.L., Harman-Fetcho, J.A. & Hagy III, J.D. (2004). Measured concentraions of herbicides and model predictions of atrazine fate in the Patuxent River Estuary, *J. Environ. Qual.* Vol. 33: 594-604.

Martin, J.F., Bennett, L.W. & Anderson, W. (1992). Off-flavor in commerical catfish ponds resulting from molinate contamination, *Sci. Total Environ.* Vol. 119: 281-287.

Miyamoto, J., Takimoto, Y. & Mihara, K. (1979). Metabolism of organophosphorus insecticides in aquatic organisms, with special emphasis on fenitrothion, in Khan, M.A.Q.& Menn, J.J. (ed.), *Pesticide and Xenobiotic Metabolism in Aquatic Organisms,* ACS (Am. Chem. Soc.), Symp. Ser. 99: 3-20.

Ministry of the Environment, Japan (1993). *Chemicals in the environment (Fiscal year 1993),* Ministry of the Environment, Japan (in Japanses).

Ministry of the Environment, Japan (1999). *Survey results on pesticide etc. residues in the environment,* Ministry of the Environment, Japan (in Japanese).

Ministry of the Environment, Japan (2001). *Survey results on environmental movements of pesticides (Fiscal year 2000),* Ministry of the Environment, Japan (in Japanese).

Nihon Kagaku-busshitsu Anzen-Jyohou Center (1992). *Kashinhou no Kizon-Kagakubusshitsu Anzensei-Tenken Data-Shuu,* Tsuusanshou kagakuhin-Anzen Ka (supervision), Kagakuhin-Kensa-Kyokai (edition), Nihon Kagaku-busshitsu Anzen-Jyohou Center (in Japanese).

Ohtsuki, A. (1994). *Tamagawa Chu-Karyu-ryuiki niokeru Nouyaku-Kigen no Iou-ganyu-Yukibutsu no Noudo-Bunpu to Sono Seibutsu-Noushuku-Kikou nikansuru Kenkyu Seika No. 161,* Zaidan-Houjin Tokyu-Kankyo-Jyoka-Zaidan (in Japanese).

Oros, D.R., Jarman, W.M., Lowe, T., David, N., Lowe, S. & Davis, J.A. (2003). Surveillance for previously unmonitored organic contaminants in the San Francisco Estuaruy, *Marine Pollution Bulletin* Vol. 46: 1102-1110.

Rovedatti, M.G., Castañé, P.M., Topalián, M.L. & Salibián, A. (2001). Monitoring of organochlorine and organophosphorous pesticides in the water of the Reconquista River (Buenos Aires, Argentina), *Wat. Res.* Vol. 35: 3457-3461.

Sanders, H.O. & Hunn, J.B. (1982). Toxicity, bioconcentration and depuration of the herbicide Bolero 8EC in freshwater invertebrates and fish, *Bull. Jpn. Soc. Fish./ Nippon Suisan Gakkaishi* Vol. 48, 1139-1143.

Sancho, E., Ferrando, M., Andreu, E. & Gamon, M. (1992). Acute toxicity, uptake and clearance of diazinon by the European eel, *Anguilla Anguilla, J. Environ. Sci. Health* Vol. B27: 209-221.

Sancho, E., Ferrando, M.D., Andreu-Moliner, E. & Gamon, M. (1994). Fenitrothion Upake and bioconcentration in the European eel (*Anguilla anguilla*), *Toxicol. Environ. Chem.* Vol. 44: 81-87.

Seguchi, K. & Asaka, S. (1981). Intake and excretion of diazinon in freshwater fishes, *Bull. Environ. Contam. Toxicol.* Vol. 27: 244-249.

Struger, J., L'Italien, S. & Sverko, E. (2004). In-use pesticide concentrations in surface waters of the Laurentian Great Lakes, 1994-2000, *J. Great Lakes Res.* Vol. 30: 435-450.

Sudo, M., Kunimatsu, T. & Okubo, T. (2002a). Concentration and loading of pesticide residuues in Lake Biwa basin (Japan), *Wat. Res.* Vol. 36: 315-329.

Sudo, M., Okubo, T., Kunimatsu, T., Ebise, S., Nakamura, M. & Kaneki, R. (2002b). Inflow and outflow of agricultural chemicals in Lake Biwa, *Lakes & Reservoirs: Research and Management* Vol. 7: 301-308.

Sudo, M., Kawachi, T., Hida, Y. & Kunimatsu, T. (2004). Spatial distribution and seasonal changes of pesticides in Lake Biwa, Japan, *Limnology* Vol. 5: 77-86.

Takimoto, Y., Ohshima, M., Yamada, H. & Miyamoto, J. (1984). Fate of fenitrothion in several developmental stages of the killifish (*Oryzias latipes*), *Arch. Environ. Contam. Toxicol.* Vol. 13: 579-587.

Takimoto, Y., Ohshima, M. & Miyamoto, J. (1987). Comparative metabolism of fenitrothion in aquatic organisms. I. Metabolism in the Euryhaline fish, *Oryzias latipes* and *Mugil cephalus*, *Ecotoxicol. Environ. Saf.* Vol. 13: 104-117.

Takino, A., Tsuda, T., Kojima, M., Harada, H. & Nakajima, A. (1998). Survey on herbicides in water and fish from rivers in Shiga Prefecture, *Rep. Shiga Pref. Inst. Pub. Hlth. & Environ. Sci.* Vol. 33: 96-98.

Tanabe, A., Mitobe, H., Kawata, K., Yasuhara, A. & Shibamoto, T. (2001). Seasonal and spatial studies on pesticides residues in surface waters of the Shinano River in Japan, *J. Agric. Food Chem.* Vol. 49: 3847-3852.

Tsuda, T., Aoki, S., Kojima, M. & Harada, H. (1988). Bioconcentration and excretion of benthiocarb and simetryne by willow shiner, *Toxicol. Environ. Chem.* Vol. 18: 31-36.

Tsuda, T., Aoki, S., Kojima, M. & Harada, H. (1989a). Bioconcentration and excretion of diazinon, IBP, malathion and fenitrothion by willow shiner, *Toxicol. Environ. Chem.* Vol. 24: 185-190.

Tsuda, T., Aoki, S., Kojima, M. & Harada, H. (1989b). Bioconcentration and excretion of benthiocarb and simetryne by carp, *Wat. Res.* Vol. 23, 529-531.

Tsuda, T., Aoki, S., Kojima, M. & Harada, H. (1990). Bioconcentration and excretion of diazinon, IBP, malathion and fenitrothion by carp, *Comp. Biochem. Physiol.* Vol. 96C: 23-26.

Tsuda, T., Aoki, S., Kojima, M. & Fujita, T. (1992a). Pesticides in water and fish from rivers flowing into Lake Biwa (II), *Chemosphere* Vol. 24: 1523-1531.

Tsuda, T., Aoki, S., Kojima, M. & Fujita, T. (1992b). Accumulation and excretion of organophosphorous pesticides by willow shiner, *Chemosphere* Vol. 25: 1945-1951.

Tsuda, T., Aoki, S., Kojima, M. & Fujita, T. (1992c). Accumulation and excretion of pesticides used in golf courses by carp (*Cyprinus carpio*), *Comp. Biochem. Physiol.* Vol. 101C: 63-66.

Tsuda, T., Aoki, S., Kojima, M. & Fujita, T. (1993). Accumulation and excretion of organophosphorous pesticides by carp *Cyprinus carpio*, *Comp. Biochem. Physiol.* Vol. 104C: 275-278.

Tsuda, T., Aoki, S., Kojima, M. & Fujita, T. (1994). Pesticides in water and fish from rivers flowing into Lake Biwa (III), *Toxicol. Environ. Chem.* Vol. 41: 85-90.

Tsuda, T., Aoki, S., Inoue, T. & Kojima, M. (1995). Accumulation and excretion of diazinon, fenthion and fenitrothion by killifish: comparison of individual and mixed pesticides. *Wat. Res.* Vol. 29: 455-458.

Tsuda, T., Inoue, T., Kojima, M. & Aoki, S. (1996a). Pesticides in water and fish from rivers flowing into Lake Biwa, *Bull. Environ. Contam. Toxicol.* Vol. 57: 442-449.

Tsuda, T., Kojima, M., Harada, H., Nakajima, A. & Aoki, S. (1996b). Accumulation and excretion of fenthion, fenthion sulfoxide and fenthion sulfone by killifish. *Comp. Biochem. Physiol.* Vol. 113C: 45-49.

Tsuda, T., Kojima, M., Harada, H., Nakajima, A. & Aoki, S. (1997a). Herbicides in water and fish from rivers flowing into Lake Biwa. *Toxicol. Environ. Chem.* Vol. 61: 243-249.

Tsuda, T., Kojima, M., Harada, H., Nakajima, A. & Aoki, S. (1997b). Acute toxicity, accumulation and excretion of organophosphorous insecticides and their oxidation products in killifish. *Chemosphere* Vol. 35: 939-949.

Tsuda, T., Kojima, M., Harada, H., Nakajima, A. & Aoki, S. (1997c). Relationships of bioconcentration factors of organophosphate pesticides among species of fish, *Comp. Biochem. Physiol.* Vol. 116C: 213-218.

Tsuda, T., Kojima, M., Harada, H., Nakajima, A. & Aoki, S. (1997d). Acute toxicity, accumulation and excretion of benthiocarb and its degradation products in killifish, *Bull. Environ. Contam. Toxicol.* Vol. 58, 603-610.

Tsuda, T., Kojima, M., Harada, H., Nakajima, A. & Aoki, S. (1998). Pesticides and their oxidation products in water and fish from rivers flowing into Lake Biwa, *Bull. Environ. Contam. Toxicol.* Vol. .60: 151-158.

Tsuda, T., Takino, A., Kojima, M. & Harada, H. (1999). Accumulation and excretion of molinate, bromobutide and their degradation products in fish, *Toxicol. Environ. Chem.* Vol. 73: 237-246.

Tsuda, T., Nakamura, T., Inoue, A. & Tanaka, K. (2009). Pesticides in water, fish and shellfish from littoral area of Lake Biwa, *Bull. Environ. Contam. Toxicol.* Vol. 82: 716-721.

Tsuda, T., Takino, A., Kojima, M., Harada, H. & Nakajima, A., unpublished data.

Vitanov, N.K., Lekova, K.I. & Dobreva, N.I. (2003). Monitoring river water in the lower Danube for atrazine contamination, *Acta Chromatographica* Vol. 13, 230-242.

Wang, Y.S., Jaw, C.G, Tang, H.C., Lin, T.S. & Chen, Y.L. (1992). Accumulation and release of herbicides butachlor, thiobencarb and chlomethoxyfen by fish, clam and shrimp, *Bull. Environ. Contam. Toxicol.* Vol. 48: 474-480.

Watanugi, F., Namiki, T. & Tsukabayashi, H. (1993) Fates of paddy herbicides in water course and residues in fishes (2), *Rep. Ishikawa Pref. Inst. Pub. Hlth. & Environ. Sci.* Vol. 30: 121-130 (in Japanese).

Washington State Department of Ecology (1999). *Lake Whatcom watershed cooperative drinking water protection project - Results of 1998 water, sediment and fish tissue sampling -, Publication No.99-337,* Washington State Department of Ecology.

Washington State Department of Ecology (2000). *Washington State Pesticide Monitoring Program - 1997 Surface Water Sampling Report*, Washington State Department of Ecology.

Xu, Y., Lay, J.P. & Zhang, Y. (1989). Uptake and transfer of 14C-simetryne through the laboratory freshwater food chain, *Chin. J. Oceanol. Limnol.* Vol. 7: 10-16.

19

Organochlorine Pesticides Residues for Some Aquatic Systems in Albania

Aurel NURO and Elda MARKU
Tirana University, Faculty of Natural Sciences, Chemistry Department
Albania

1. Introduction

The Republic of Albania is situated in Southeastern Europe, in the western part of Balkan Peninsula facing the Adriatic Sea and the Ionian Sea. The coastal area has a surface area of 7000 km^2 or 25% of the territory. The coast line is 476 km. Marine and coastal environment constitutes resources of high economic and ecological values for the country. Due to the mismanagement of these resources in years, considerable amounts of wastes have been discharged directly or through river flows and atmospheric deposits, into the sea. Adriatic and Ionic Sea as part of the Mediterranean Sea, which does not have an open linkage with the oceanic waters, are inclined to be considerably affected by pollution, due to the absence of its dilution. The highest impacts on marine waters in Albania are related to eutrophication, pollution from heavy metals and POPs, illegal fishing, overexploitation of aquatic fauna and degradation of coastal zones, mostly due to the uncontrolled development.

Shkodra Lake and Ohrid Lake are the most important and multi-dimensional freshwater resources of Albania. These lakes are also considered as the most important transboundary areas in Albania. Shkodra Lake is one of the most important multi-dimensional resources of this region as fishing, hunting, tourism, recreation and aquatic sports resource. Shkodra Lake is considered as one of the most important transboundary lakes in Albania. It is the biggest lake in Balkan region. The lake area varies between 354 km^2 in low water periods and 506 km^2 at high water periods. In the maximum, in the Montenegro territory there the lake area is 340 km^2 and 166 km^2 in Albania. The most important tributaries, that are sources of water for the lake, are: Moraca, Crnojevica, Orahovstica, Crmnicka, Karatuna, Baragurska rivers (Montenegro side) and Virstica, Proni, Tata, Riola, Vraka, Bunusi and Kiri rivers (from the Albanian side). Occasionally, Shkodra Lake receives water from river Drini too (Albania). This happens when the water level of river is higher than the water level of the lake. During the maximum of water level, the lake depth is over 12 m, and during the minimum is around 8 m. Water level oscillates between 4.7 and 10 m above the sea level. Buna River is the only emissary of Shkodra Lake with medium flow over 300m^3/s. The water catchments basin of Shkodra Lake is 5179 km^2, 1025 km^2 of which belongs to the Albanian territory. 45 kinds of fish are leaving in the Lake and its effluents. Half of them are permanent and the others migrate through Buna River to the Adriatic Sea. The most characteristic fish is *Cuprinos carpio,* besides *Anguilla anguilla, Mugil cephalus, Mugil ramada,*

Marone labrax, Platichthys flessus italicus, Pachycilon pictum, Alosa fallax nilotica, Acipenser sturio, etc. (Dhora 1995).

Buna River is one of the most important Mediterranean rivers. Also, thanks to the waters from the Drini River, the Buna ranks second place among all tributaries to the Adriatic, measured by the annual discharge, after the Po in Italy (with 352 m^3/s). Both rivers together are determinative on Adriatic Sea water balance. Out flowing from Shkodra Lake, Buna immediately joins Drini River water and both rivers discharge into southeast Adriatic Sea. Shkodra Lake - Drini River - Buna River hydrographical complex is very complicated and unique for its hydraulic regime in the world hydrography. Despite being short, the river has quite a large watershed, covering 5,187 km^2, because the whole drainage area of Shkodra Lake, the largest lake in southeastern Europe, is also part of it (Pano & Abdyli, 1984).

The Ohrid Lake is one of the oldest lakes in the world. The basin of the Ohrid Lake together with the lake belong to the geo-tectonic zone of the South-Eastern holes of Albania. The Ohrid Lake is part of the so-called Desaret Lakes. The flora and fauna of the Ohrid Lake is unique in the Balkans and wider. The lake's biota is constrained by several abiotic factors, among which the geographic location, the water basin play an important role. The low temperature lake's water makes the biologic processes be developed slowly, while the plankton (the phytoplankton and zooplankton) are characteristic and interesting (Topi et al. 2006). The number of the endemic kinds is high and makes up about 60% of the organisms that live in the lake. About 60% of the fishes in the lake are endemic, particularly the various kinds of fishes such as *Salmo letnica, Salmo truta, Salmothynus ohridanus, Cyprinuscarpio, Leuciscus cephalus albus, Alburnus alburnus alborella, Rutilus rubillioohridanus, Cobitis taenia ohridana, Anguilla anguilla*, etc,. Water basin of Ohrid Lake is huge. Prespa Lake is therefore considered to be a part of the Lake Ohrid watershed.

Complex Lagoonary of Butrint is located 4 km south of Saranda, it is 140,000 m^2, opposite to the island of Corfu and near the Greek-Albanian border. The area is mainly composed of wetland and lakes and crossed by two rivers: Pavllo Bistrica in the north and south. In the east, the Mountain Mile Complex (824 m) separates the interior of the complex Lagoonary Albania. Less than ten kilometers from the island of Corfu, Butrinti was linked to the Mediterranean by the Vivari canal, which ran from the Butrinti Lake to the Ionian Sea.The area has a great value for its scenery and contains a variety of natural habitats, semi-natural and artificial. These include areas covered with oak, typical Mediterranean scrub, the salt lake Bufi (ray), salt water lagoon of Butrint, the salty marshland Alinures, spills and Pavllos Bistrica River, halophytic lands with plants (which live in the salty earth) and agricultural land. Beside it is the Butrinti (Bouthroton) site which is full of ancient history and is one of Albania's most popular tourist attractions. Mussels *Mytulus Galloprovinciulis* from the lake are famous all over the world, as fresh and wholesome, and include a variety of sizes and natural flavored.

Aquatic areas showed above are some of most important aquatic areas in Albania. For these areas are studied different type of samples for evaluating levels and distribution of organochlorinated pesticides.

What are organochlorinated pesticides in short terms?

The modern history of pesticides dates back to World War II when for the first time the insecticidal properties of DDT were recognized. DDT was first introduced on a large scale to fight fleas, lice, flies and mosquitoes and reduce the spread of insect borne diseases such as malaria and yellow fever. Many public health benefits have been realized by the use of pesticides, but their potential impact on the environment is substantial too (Di Muccio, 1996). Over 800 pesticide active ingredients are formulated in about 21000 different commercial products. Most of organochlorinated pesticides have been progressively restricted and then banned in the 1970s in most industrialized countries, a widespread environmental pollution has resulted from their before use. Organochlorinated pesticides are usually big molecules with a big number of chlorine atoms attached their molecules; they can persist without changes for many years in application areas. Water irrigations can take away pesticides from these application areas mainly to the surface waters. Organochlorinated pesticides solubility in water it's limited (depend on their Kow), they are mostly lipophilic compounds so they attend to connect in the suspend matter, to precipitate in sediments, to accumulated and concentrated in biota of aquatic systems.

Albania is a country rich with lakes, rivers, and many effluents. Shkodra Lake, Ohrid Lake and Prespa Lakes are the large lakes in Albania and Balkan Peninsula. Water basin of these lakes is huge with many effluents and close with them has many fields used for agricultural purposes. Shkodra Lake and Adriatic Sea are connected between each other with Buna River. The most important rivers are Drini, Buna, Mati, Ishmi, Erzeni, Shkumbini, Semani, Vjosa and Bistrica. All rivers start in the east of Albania, in mountain areas, and finish to the Adriatic Sea. They are very fast (except Buna River) so the pesticide residues or other materials deposited in the sea. For many industries in Albania (also Lindane Plant), the waste of them were discharged for many years directly to the sea. These areas are located especially close to Durres and Vlora bay. Before 90′ organochlorinated pesticides were used widely in Albania for agricultural purposes. The main agricultural areas were in the western of the country, but almost every were had been developed different directions of agricultural (fruits, corns, vegetables, etc.). The districts where the larger quantities have been used are Fieri, Lushnja, Tirana, Vlora. The most used chlorinated pesticides were DDT, Lindane, HCB, Aldrins and Heptachlors. Lindane was produced in the country at Lindane Plant, in Porto-Romano, near Durresi City. Other chlorinated pesticides were imported mostly from eastern countries. The scale of their use after 90′ in agriculture has decreased, due the change of soil structure. Emigration of many peoples in western country and also free movement inside the country were two main factors that impact directly in agriculture areas and it's developing. Use of pesticides generally has decreased, because of large areas were not using for agricultural purposes and other areas were used for building new houses and industries. Except this, many chemical industries, include Lindane Plant, were stopped or destroyed. The former has generated the expired pesticides, which due to the inappropriate conditions of conservation and storage have been damaged. The other part of expired or out of use pesticides, to be disposed of, has been distributed in various districts of the country. In the country after 90′ had a large amount of stocks of pesticides (1000 ton). About 45% of all pesticides in the country have been evaluated as "hazardous class". Different projects were done last years for isolate and eliminate pesticides and other pollutants. These wastes are not pretreated; it is very difficult to find precise data on the amount of industrial solid wastes stores in various parts of the country. Mismanagement of oddments pesticides, for some years after 90′ was another source of pesticides contamination. Before use of pesticides, their persistence, water irrigation, their ability to concentrate in sediments and biota adding other factors suggest us to study levels of organochlorinated pesticides in above aquatic systems.

For studding organochlorinated pesticides in environmental samples are developed many methods and techniques. Pesticide residues in environmental samples are found in very lower levels. Qualitative and quantitative analyze of organochlorinated compounds performed usually by modern chromatographic techniques. Gas chromatography (GC) technique equipped with electron-capture detector (ECD) or mas-spectrometry detection (MS) are most favorable technique for pesticide residue analyze. Many columns and other part of GC are suggested for pesticide analyze. For gas chromatographic technique in mostly of samples needed different steps before injecting in GC. Sample preparation prior to the determination of many environmental pollutants including organic pesticides in water, sediment, and biota samples usually consists of many steps because of the complexity of the matrix. Extraction of the target compounds and clean-up for the extract are the most critical steps in the analytical procedure when it comes to complete recovery of the target substances. For extraction of organochlorinated pollutants different methods are used: from classical Soxhlet extraction to automated extractions method as Accelerated Solvent Extraction (ASE) and Supercritical Fluid Extraction (SFE). For solid samples ultrasonic extraction is widely used, as well. The main goal of the clean-up step(s) is to remove substances that could interfere with the final determination and quantization of target compounds. Removal of interfering substances can be accomplished in many different ways. For example, copper is often used as sorbent for retaining elementary sulphur from sediment samples. Solid-phase extraction (SPE) is still the dominant method for extracts purification. A large number of sorbents are used for the isolation of organic compounds from the extract solutions include alumina, Florisil, ion-exchange resins, silica gel and many silica-based sorbents.

2. Material and methods

2.1 Chemicals

Organic solvents and reagents. n-Hexane, dichlorometane of special grade for pesticide residue analysis were purchased from Merck, Germany. Organic solvents particularly dichlorometane which is toxic, were handled with care observing safety precautions, using efficient fume hoods and wearing protective gloves. Anhydrous sodium sulphate (Na_2SO_4) from Merck, Germany. Florisil (≥ 400 mesh ASTM) and silicagel (60-100 mesh ASTM) were purchased from Merck, Germany. H_2SO_4 with 95-97% purity for GC analyses was purchased by Merck. The sodium sulfate, florisil and silica gel were pre-extracted with hexane/dichloromethane (4/1) in a Soxhlet extractor, dried and were rinsed with hexane/dichloromethane (4/1) just before utilization. All glassware was rigorously cleaned with detergent followed by pyrolysis at 300°C.

Pesticides solution and standarts. Organochlorinated pesticides stock solutions from 5 mM concentration were donated by IAEA/MEL-Monaco. Standard solutions of pesticides and PCB-s were prepared by dissolving their stock solutions in n-hexane in concentartion 50 ng/μl and storing them in refrigerator in glass bottles with PTFE-faced screw caps. The organchlorine pesticides detected were HCHs (a-, b-, g- and d-isomers) and the DDT-related chemicals (o,p-DDE, p,p-DDE, p,p-DDD, p,p-DDT), Hexachlorobenzene (HCB), Heptachlor, Heptachlor epoxide, Methoxychlor and Mirex.

IAEA-383 homogenized sediment sample donated by IAEA/MEL-Monaco, France (International Atomic Energy Agency/ Marine Environmental Laboratory) was used as CRM for sediment samples. IAEA-435 tuna fish was used as biota CRM (provided by IAEA/MEL-Monaco, France International Atomic Energy Agency/Marine Environmental Laboratory).

2.2 Study areas

Study areas were: Adriatic Sea (three sampling stations; Vlora Bay, Porto-Romano and Velipoja), Ohrid Lake, Shkodra Lake, Butrinti Lake and Buna River. These areas were selected in different locations of Albania as representative of huge hydrological systems found in the country. Samples were collected for a period of five years (2005-2010). Data presented here shown levels and distribution of organochlorinated pesticides in water, sediments and biota for selected stations. Sampling was done based in Reference Method No 6, UNEP, 1993. Fig. 1. Geographic position of studied aquatic locations in Albania.

2.3 Water

Water sampling. Water samples were sampling at Vlora Bay (Adriatic Sea), Butrinti Lake, Shkodra Lake and Buna River. Water samples were taken on surface of water in PTFE bottle. For organochlorinated pesticide analyze were taken 1 L of water from each station. Water samples were transported to the laboratory and stored in a refrigerator in +4 °C.

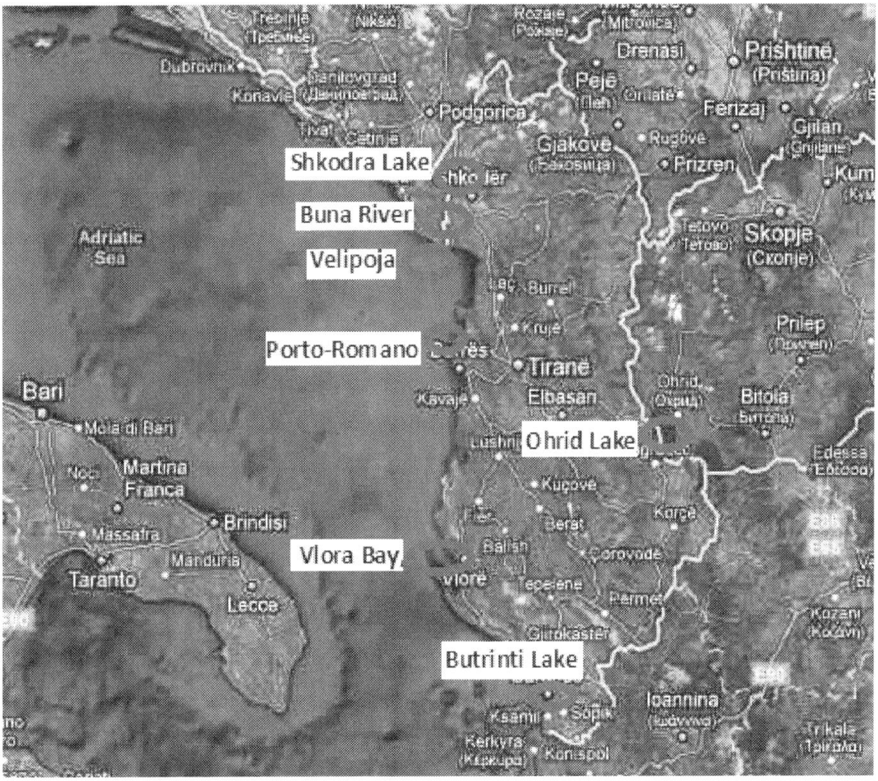

Fig. 1. Geographic position of studied aquatic locations in Albania.

Preparation of water samples for GC analysis. Liquid-liquid extraction was used for the extraction of organochlorine pesticide residues from water samples. 1 L of water, 10 µl PCB-

29 as internal standard and 20 mL n-hexane (extracting solvent) were added in a separatory funnel. After separation the organic phase was dried with 5 g Na_2SO_4 anhidrous, for water removing. A Florisil column was used for the sample clean-up. The extract was concentrated at 1 ml for analyzing by GC-ECD.

2.4 Sediments

Sediment sampling. Sediment samples collected at different aquatic locations of Albania; Adriatic Sea (Vlora Bay, Velipoja), Shkodra Lake and Buna River. Standard sediment Van Veen grab with 440 cm^3 of volume was used for sediment sampling. Twenty two stations among Vlora Bay were choosing for this study and three stations in different depth of Velipoja sea shore (Adriatic Sea stations). Six stations were chosen to represent different conditions of the Shkodra Lake and seven stations in Buna River. For all sediment stations are analysed surface corer with a layer of 5 cm. Samples were transported to the laboratory and stored in a refrigerator in +4°C. The samples were air-dried, grinded and sieved. Particles of 0.063 mm were taken for analysis.

Sediment treatment. The sediment samples (1 gram dry weight) were transferred into glass ampoules for extracting with dichloromethane-hexane (1/3), after a standard solution of PCB-29 was added as internal standard. Extraction was carried in ultrasonic bath in 30^0C for 30 minutes. The preliminary clean-up of sample extract was carried out with elemental Hg to remove sulphur compounds from the sediment samples. Further clean up was performed using a column made of two layers. The upper layer was silica gel activated in 300^0C and impregnated with 45% sulphuric acid. The down layer of the column was made of florisil activated in 250 ^{0} C and deactivated with 5% distilled water. The elution was carried on with 12 ml dichloromethane-hexane (1/4). The final volume of samples was 1ml using Kuderna-Danish for evaporate. IAEA 383 certified sediment sample for chlorinated pollutants was used for method validation.

2.5 Biota

Biota samples. Fish specimens were taken in random mode from the catch of local fishermen in Vlora Bay, Velipoja, Shkodra Lake and Ohrid Lake. A total of 32 fish species and 102 specimens has been analyzed, respectively 9 species with 26 specimens from Vlora Bay, 10 species with 33 specimens from Velipoja, 9 species with 24 specimens from Shkodra Lake and 7 species with 19 specimens from Ohrid Lake. Fish species for all stations were: Vlora Bay (*Engraulis encrasicholus, Lithognathus mormyrus, Solea vulgaris, Mullus barbatus, Mugil cephalus, Trachurus trachurus, Belone belone gracilis, Dicentrarchus labrax, Lichia amia, Mytulus galloprovincialius);* Porto-Romano (*Mytulus galloprovincialius);* Velipoja *(Mullus barbatus, Sphyraena sphyraena, Pargus coeruleosticus, Uranoscopus scaber, Merluccius merluccius, Scomber scombrus, Dicentrarchus labrax, Boops boops, Trigla lyra);* Shkodra Lake *(Cyprinus carpio, Anguilla Anguilla, Carassius gibelio, Pseudorasbora parva, Scardinius knezevici, Mugil cephalus, Alburnus scoranza, Allosa agone);* Ohrid Lake *(Salmo letnica, Salmo ohridanus, Squalius cephalus, Salmo truta, Cyprinus carpio, Alburnus scoranza, Rutilus ohridanus) and* Butrinti Lake *(Mytulus galloprovincialius).* The mussels, *Mytilus Galloprovincialus* were collected in Vlora Bay, Porto-Romano, and Butrinti Lake with a number of 75 species for each station. Before treatment mussels were divided in five groups with 15 members' based in their length. The biota (fish and mussels) samples were transferred to an aluminum container and stored at -10°C.

Preparation of biota samples for GC analysis. For all biota samples were analyzed tissues. The method was based on EN 1528/1/2/3/4 (2000) for determination of organochlorinated pesticides and PCBs in biota samples. Prior analyzing biota samples were homogenized with anhydrous sodium sulphate. 1 g fresh weight of organism tissue was extracted by ultrasonic bath assisted extraction with 20 ml mixture of hexane/dichloromethane (3/1). The extract was purified firstly by shaking with 15g silica gel treated with 45% sulfuric acid for lipid hydrolysis. After filtration, the extract was concentrated in Kuderna Danish to 5 ml volume and a second purification on a column of Florisil with 5% water was performed. (Petrick et al, 1988; Koci, 1997) Organochlorinated pesticides were eluted with 10 ml of a mixture of hexane/dichloromethane (5/1). The extract was concentrated at 1 ml for analyzing by GC-ECD. Homogenized tuna sample IAEA 435 was used for method validation.

2.6 Apparatus and chromatography

Gas chromatographic analyses were performed with an HP 6890 Series II gas chromatograph equipped with a ^{63}Ni electron-capture detector and a split/splitless injector. The column used was a HP-5 capillary column [low/mid polarity, 5% (phenyl methyl siloxane)] (25 m x 33 mm I.D.x 25 mm film thick). The split/splitless injector and detector temperatures were set at 280^{0}C and 320^{0}C, respectively. Carrier gas was He at 1.8 ml/min and make-up gas was nitrogen 28,2 ml/min. The initial oven temperature was kept at 60^{0}C for 4 min, which was increased, to 200^{0}C at 20^{0}C/min, held for 7 min, and then increased to 280^{0}C at 4^{0}C/min for 20 min. The temperature was finally increased to 300^{0}C, at 10 ^{0}C/min, held for 7 min. The total run time was 38 min. Injection volume was 2 µl, when splitless injections were made. Organochlorine pesticide quantification was performed by internal standard method. The relative response factors of the different compounds were determined by injecting the standard solutions of organochlorinated pesticides spiked with the same solution of internal standards PCB 29 as that used for spiking the samples. Procedural of blanks were regularly performed and all results presented are corrected for blank levels. Quality assurance procedures included the analyses of marine sediment reference materials IAEA 385 and tuna fish homogenized certified sample IAEA 435 to determine the precision and the accuracy of the method (Schantz et al, 1993; Marku and Nuro, 2005).

3. Results and discussion

Data presented below are interpreted in order of aquatic locations selected. These locations have own characteristics, but except this, are way to see status of Albania aquatic body in order of organochlorinated pesticides. All data represent mean value for a group of experimental for a selected station or species.

Vlora Bay. In Vlora Bay were analyzed water, sediment, fish and mussel samples. Organochlorinated pesticides level in water samples of Vlora Bay (Figure 2) were with a minimum value of 17,1 ng/L to the maximum value of 53,4 ng/L. Mean value for analyzed water samples was 28,9 ng/L. Higher values of pesticide residues were found in water samples in the center of bay compared with water samples near seashore. Distribution for these found levels in different stations it's connected from water current inside and outside bay. Total of organochlorinated pesticides was higher also for sediment station number 43, 44 and 45 (Figure 3) sampling in the center of bay. Minimum value for organochlorine pesticides residues in sediment sampling in Vlora Bay were 6,7 ng/g to 138,7 ng/g, mean value was 34,42 ng/g dry sediment. Two are the most favorable factors for this distribution of pesticides in sediments of bay: Before uses of pesticides for agricultural issues in Vlora region (mostly as insecticides). Applied pesticides in these areas after irrigation go throughout the sea. The current of water into the bay is the second factor. Pesticide stability, their chemistry and sediment conditions are factors affect levels of pesticide residues in water and sediment samples. Figure 4 shown data about the total of organochlorine pesticides in fish samples of Vlora Bay. The sample of *Mugil cephalus* with 107.4 ng/g f.w. (fresh weight) was the most polluted one among the fish species analyzed for organochlorined pesticides in Vlora Bay. The sample of *Solea vulgaris* from the same area was the "cleanest" one with 37,5 ng/g f.w. Mean value for pesticide residues in fish samples was 67,1 ng/g f.w. Found levels in fish samples were almost 2 - 5 time higher than levels found in water and sediments of Vlora Bay. About levels of pesticide residues found in fish samples it's evident higher levels than water and sediments but we must evaluate also fish species, their position in food chain and their age, sex, ect,. Very interests were found levels of pesticide residues in mussels *Mytulus galloprovincialius* sampling in Vlora Bay. Minimum value was 152,8 ng/g f.w; maximum value was 379,1 ng/g f.w. and mean value 259,7 ng/g f.w. Mussels which are water altering bivalves have been shown to concentrate many organic contaminants, providing a direct representation of pollutant bioavailability are used mostly as bio-indicators of marine areas. Levels of orgnochlorinated pesticides found in these organisms are related with levels of these pollutants in water by a factor of 10 times. This connection is shown also for the fact that higher polluted were mussels of group five that are higher in age and weight, and minimum of pollution for mussels of group two that were smaller one.

Profile of organochlorine pesticides is the same for all type of samples analysed in Vlora Bay (Figure 6). *p,p'*-DDE was found in greatest concentration among pesticides for all samples. The second pollutants were DDT metabolites, *o,p'*-DDE and *p,p'*-DDD. DDT levels were comparable with DDE for sediment samples. This is not linked with uses of DDT at last for agricultural purpose because after 90' in our country it is banded. Degradation of DDT in sediment is slow. Lindane was not the first isomer in total HCH contribution. *d*-HCH and *b*-HCH concentrations were higher than the other isomers, because of their tendencies to accumulate in fatty tissue. It's evident the level of *b*-HCH in mussel samples. HCB concentration was higher mostly in fish and mussel samples, because of its Kow and bioavability for concentration in upper steps of food chain.

Porto-Romano, Durres, Adriatic Sea. *Mytulus galloprovincialius* mussel samples were analyzed in Porto-Romano, Adriatic Sea. Mean level of organochlorinated pesticides in mussel samples was 220,6 ng/g f.w.; minimum was for second group with 107,4 ng/g f.w. and maximum for fifth group with 397,5 ng/g f.w. (Figure 7). Levels of organochlorinated pesticides in Porto-Romano, Durres were comparable with levels found for mussels analyzed in Vlora Bay. For this station the old industries located in this area such as Lindane Chemical Plant of Porto-Romano, Durres represent the major polluting factor, followed by agricultural contribution and other possible sources of contamination such as the water-based contribution, atmospheric factor, etc. There is comparability between pesticide residues data in Vlora and Porto-Romano obtain for other areas of Adriatic Sea.

The most frequently detected pesticides for Porto-Romano were *p,p*-DDE (Figure 8). *p,p*-DDT was in higher levels than o,p-DDE and p,p-DDD, also in higher levels than p,p-DDT levels found in mussel samples in Vlora Bay. This would have principally agricultural component although there are also other possible sources of contamination such as

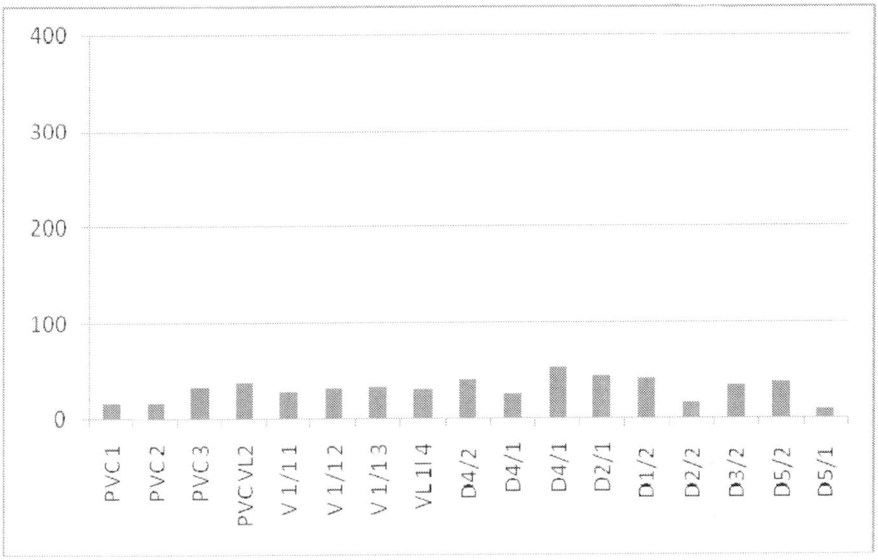

Fig. 2. Total of organochlorinated pesticides in water samples of Vlora Bay

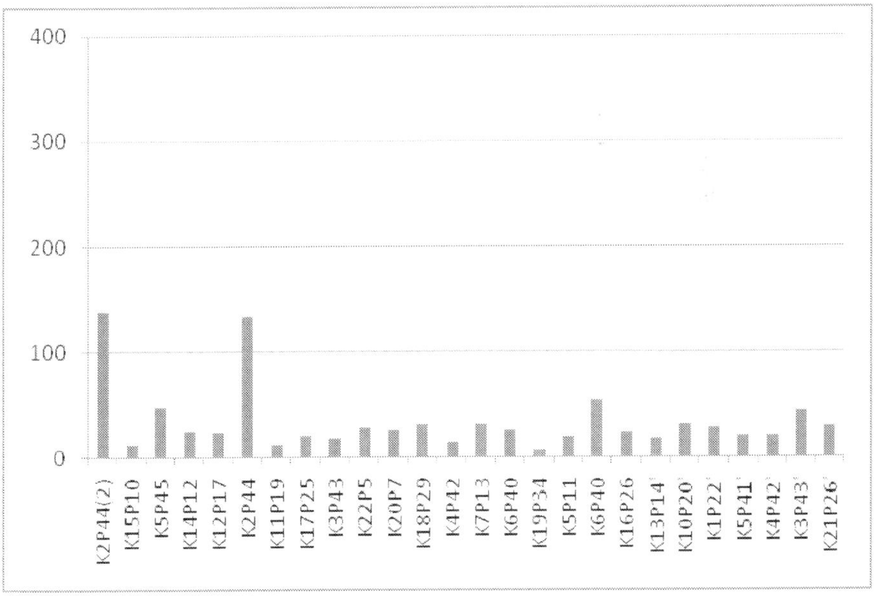

Fig. 3. Total of organochlorinated pesticides in sediment samples of Vlora Bay

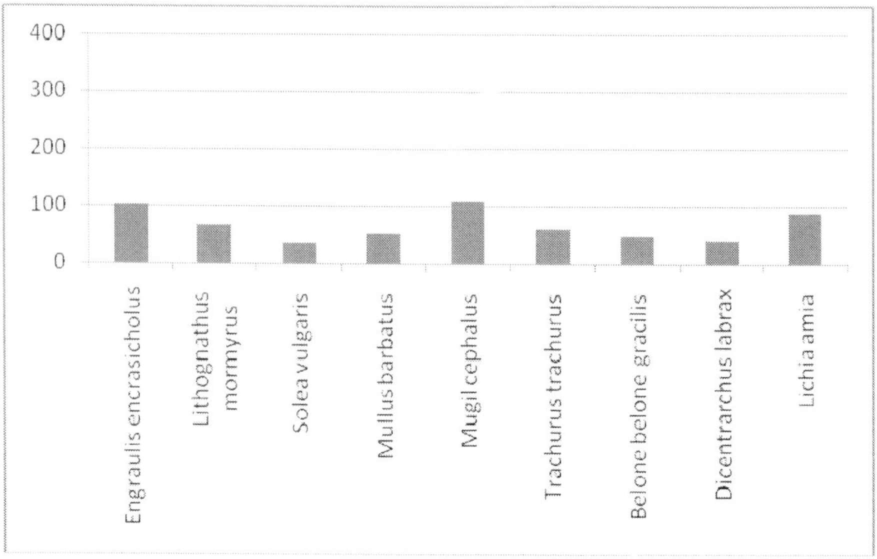

Fig. 4. Total of organochlorinated pesticides in fish samples of Vlora Bay

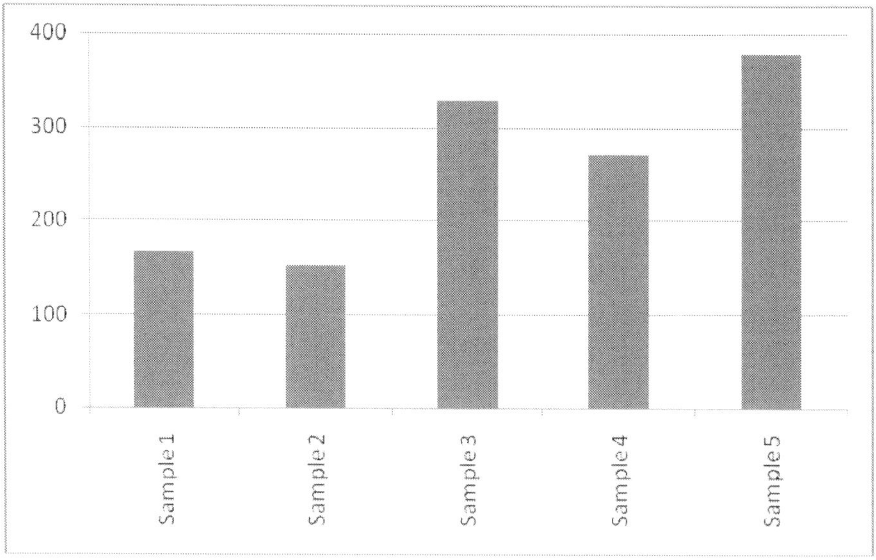

Fig. 5. Total of organochlorinated pesticides in mussels *Mytulus galloprovincialius* of Vlora Bay

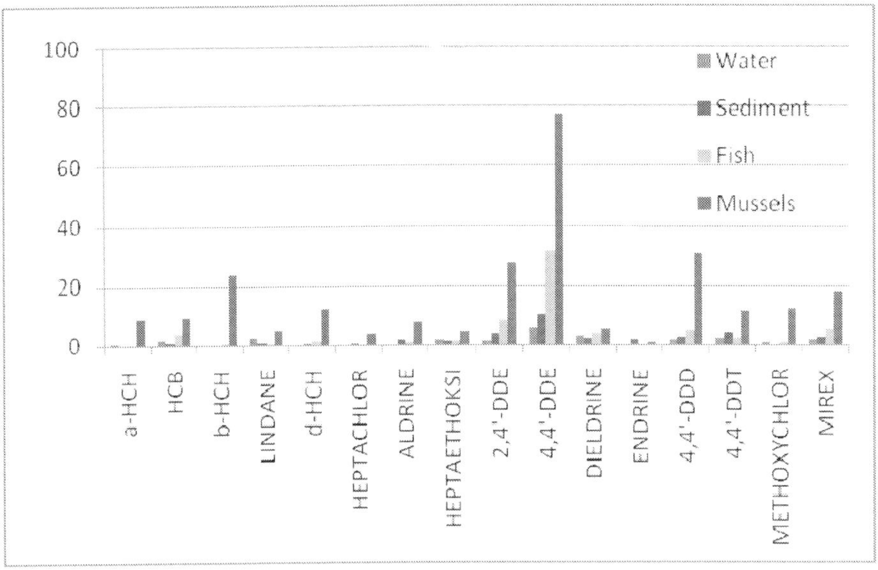

Fig. 6. Average of organochlorinated pesticides in water, sediment, fish and mussel samples of Vlora Bay

mismanagement of oddments pesticides, irrigation of surfaces where these pesticides were used by rainwater, drift, water-based contribution and atmospheric component. HCHs were expected to be found for analyzed samples because this station it's near old Lindane Chemical Plant in Porto-Romano. Was interest the fact that Lindane was in lower levels than *b*-HCH and *d*-HCH isomer. *b*-HCH concentration was higher than the other isomers' concentrations because of its higher tendencies to accumulate in fatty tissue. The *a*/*g* HCH ratio, an indicator of current technical HCH application, was lower. Again this likely reflects recent use of HCH formulations, or possibly ongoing releases of HCH isomers from waste repositories.

Velipoja, Adriatic Sea. Sediment and fish samples were sampling in Velipoja, third location of Adriatic Sea, in North Albania. In Figure 9 are shown data about organochlorinated pesticides in sediment samples of Velipoja. Sediment samples were choosing in different depth of Adriatic Sea in front of Buna river mouth. It was evident that Buna rivermouth (in Adriatic Sea) was the most polluted station with 173,8 ng/g dry sediment. Lower concentration of pesticide residues were found going to the higher depth of Adriatic Sea (far away from Buna river mouth). Sediment sampled in 100 m depth was lower level for total of organochlorinated pesticides with 25,2 ng/g dry sediment. Influence of Buna River is main factor for this distribution of pesticide residues in sediments of Adriatic Sea for this station.

Buna River and its water basin are located in a huge agricultural region. Before use of pesticides, rain fall of these areas, sedimentation process and sedimentation transport, pesticides stability and lower velocity of pesticides degradability are factors that affect in the profile and in the total of pesticide residue in Vleipoja station. Total concentrations of pesticides for fish samples in Velipoja were in range from 43.9 ng/g f.w. to 91.2 ng/g f.w. respectively for *Raja clavata* and *Uranoscopus scaber* (Figure 10). Mean value were 58,1 ng/g

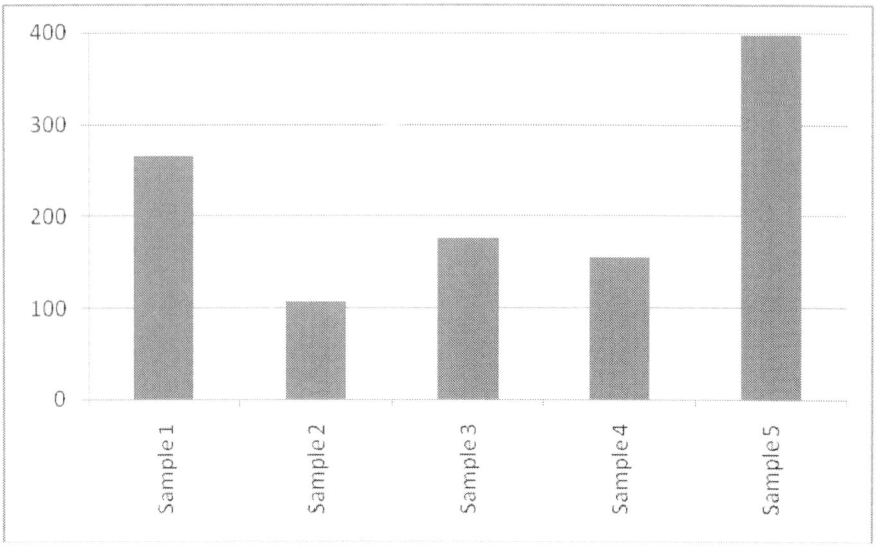

Fig. 7. Total of organochlorinated pesticides in mussels *Mytulus galloprovincialius* of Porto-Romano, Durres, Adriatic Sea

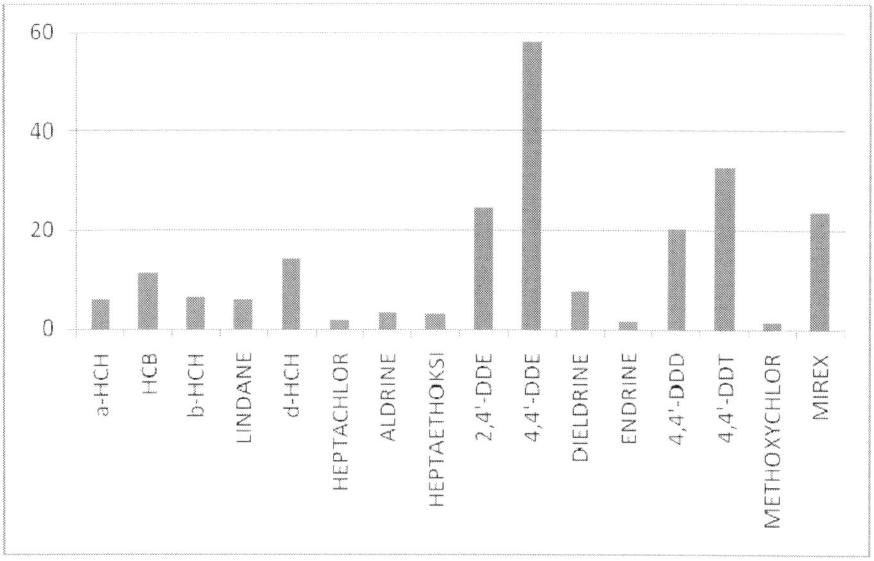

Fig. 8. Profile of organochlorinated pesticides in mussels *Mytulus galloprovincialius* of Porto-Romano, Durres, Adriatic Sea

f.w. Found value were comparable with levels of pesticide residue in fish samples of Vlora Bay. Profile of organochlorine pesticide dedected for sediment and fish samples of Velipoja are shown in Figure 11. For fish sampes could saw that is the same profile found in samples of Vlora Bay or Porto-Romano. The higher pesticide concentrations were for DDT and its metabolite and HCH isomers. p,p-DDE concentration was higher than other pesticides in fish samples of Velipoja followed from *b*-HCH concentration. Pesticide profile for sediments of Velipoja it's quite different from profile of fish samples and also different from profile found for sediments of Vlora Bay. Were evident higher concentration of HCHs and Heptachlors followed from DDTs and Aldrine. Apart from above factors this station could be affected from other factors. Another important factor could be water current in this part of Adriatic Sea.

Shkodra Lake. In Shkodra Lake were sampling water, sediment and fish samples. In Figure 12 are shown total of organochlorinated pesticides in water samples of Shkodra Lake. Pesticide residues in water samples ranged from 17,6 to 83,4 ng/L. Shkodra Lake has a huge water basin. The concentrations of the organochlorinated pesticides (except DDTs and Lindane) suggest little use of these pesticides in adjacent agricultural areas. Some residues detected in the lake sediments may be also the result of agriculture runoff. Water cycle and water current in Shkodra Lake could be affect in levels of pesticide residues and profile of them in different stations of water samples in Shkodra Lake. The total concentrations of organochlorinated pesticides for sediment stations are shown in the Figure 13. The highest values correspond to Station 5 with 98.6 ng/g dry sediment. Station 5 is situated close to the main bridge of the lake, where Buna River begins to flow. The lower concentration was for Station 6 with 29,6 ng/g dry sediment. Station 6 was close to Vraka village (500 m far away). Note that sediments are rockier in this side of lake. The total concentrations of chlorinated pesticides in the different fish samples of Shkodra Lake are shown in Figure 14. The samples

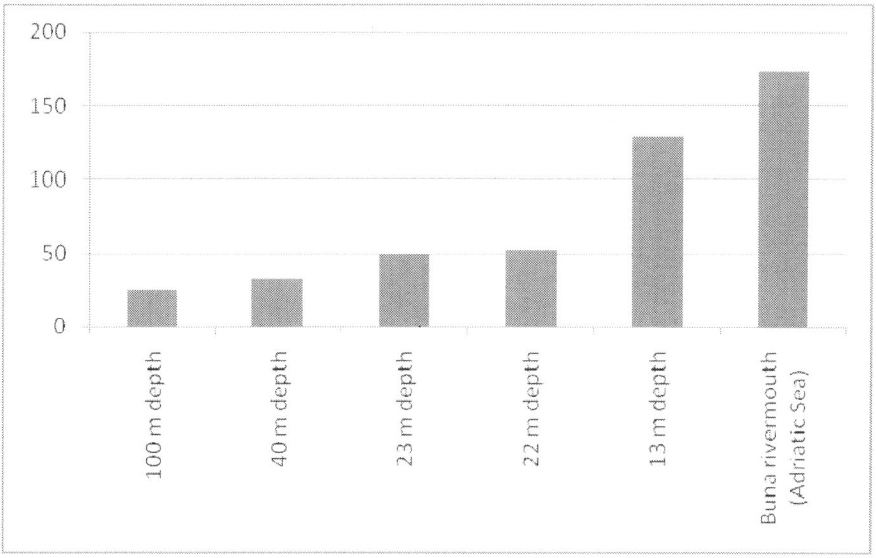

Fig. 9. Total of organochlorinated pesticides in sediments of Velipoja, Adriatic Sea

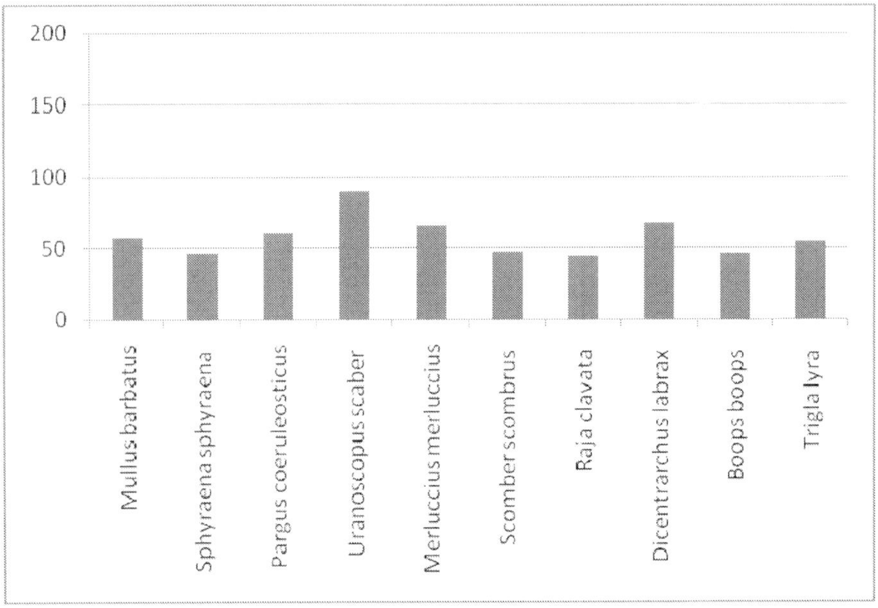

Fig. 10. Total of organochlorinated pesticides in fish samples of Velipoja, Adriatic Sea

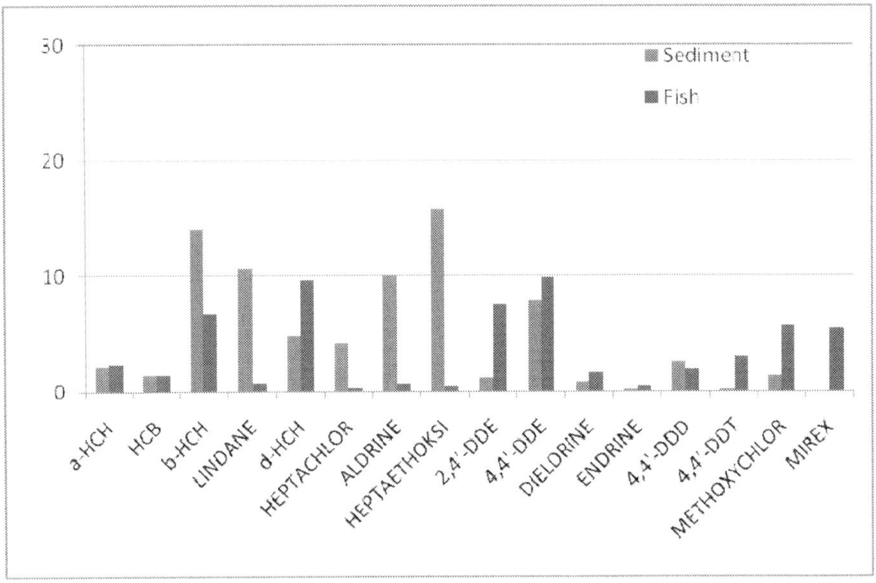

Fig. 11. Average of organochlorinated pesticides in sediment and fish samples of Velipoja, Adriatic Sea

of *Allosa agone* fom Shkodra Lake had the maximum concentration with 80,5 ng/g f.w. and *Rutilus ohridanus* the minimum with 31,2 ng/g f.w. These levels suggest correlations with water and sediment levels found for Shkodra Lake. Found pesticide residue levels for fish samples of Shkodra Lake were comparable with levels of fish samples of Adriatic Sea. Most of fish species are migratory and stay in the lake not all over the year. They migrate to Adriatic Sea, through Buna River, which is a natural communicating channel to the sea. Before use of pesticide for agricultural reasons, individual pesticide properties and pesticide residues from rain water flow and Shkodra Lake effluents are factors for levels ofound in samples of lake.

The mean concentrations of analyzed individual organochlorinated pesticides in water, sediment and fish samples of Shkodra Lake are shown in Figure 15. The concentrations of DDT & metabolites and HCH isomers were higher than other pesticides for both sediment sampling. The highest concentrations found in sediments were for 4, 4'-DDT at with 21.7 ng/g dry weight because of lower degradation process in sediments. Aldrine, Heptachlor, Mirex and Heptachlorepoxide were present in very low concentrations, sometimes even non detectable. DDT and metabolites concentrations in fish species show a different profile from the sediments. The concentrations of p,p'-DDT are lower than the concentrations of its metabolites. The higher value of ΣDDTs' is found for both sampling in Carp (*Cyprinos carpio*), which is the famous characteristic fish (non migratory) of Shkodra Lake. Lindane concentrations in fish are in the same range with concentrations reported for Adriatic fish (Albanian coast). Differently to the sediments, the profile of HCH isomers in some migratory fish shows relatively high values of Lindane, compared with other HCH isomers.

Buna River. In Buna River were analyzed water and sediment samples. Organochlorinated pesticide levels in water samples are shown in Figure 16. Maximum level was for Middle of

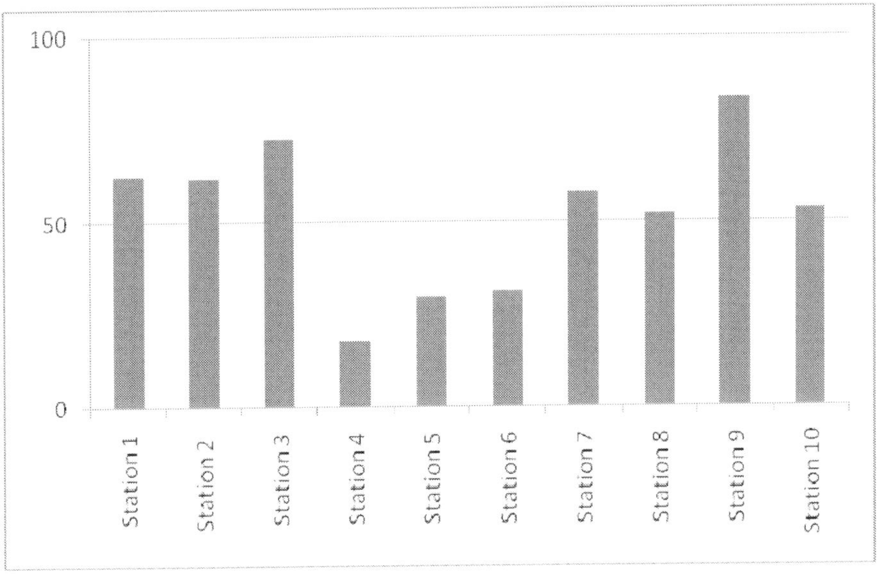

Fig. 12. Total of organochlorinated pesticides in water samples of Shkodra Lake

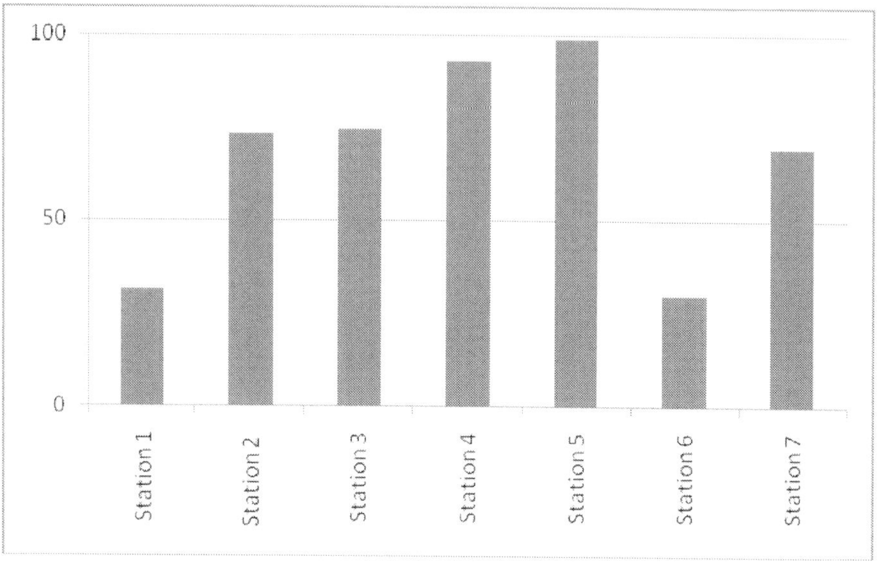

Fig. 13. Total of organochlorinated pesticides in sediments of Shkodra Lake

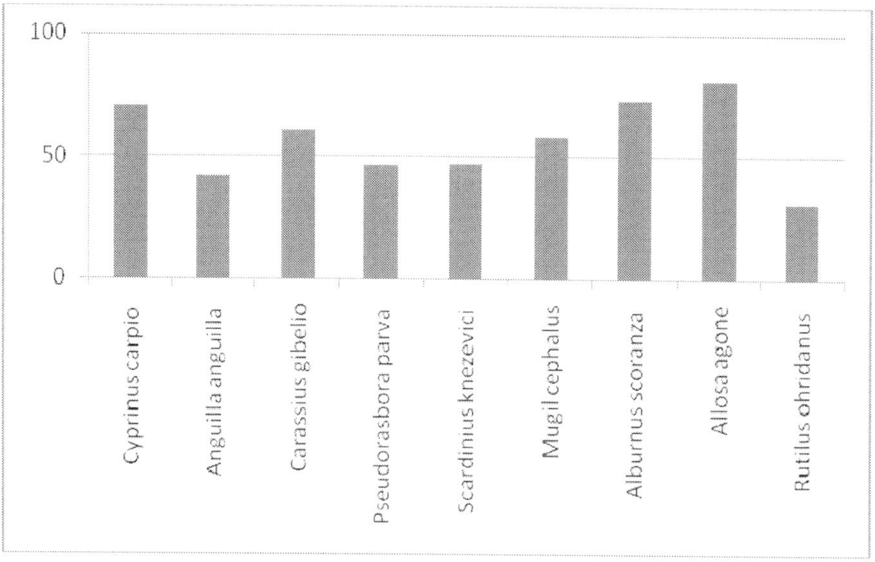

Fig. 14. Total of organochlorinated pesticidesin fish samples of Shkodra Lake

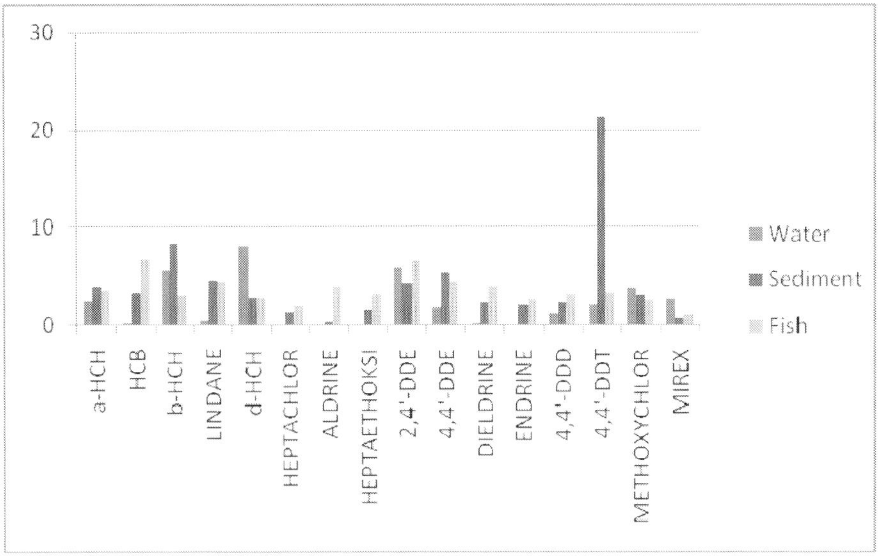

Fig. 15. Average of organochlorinated pesticides in water, sediments and fish samples of Shkodra Lake

Buna sample with 101,2 ng/L, minimum value was for Buna Start station with 38,8 ng/L and mean value was 57,9 ng/L. It's evident the fact that levels of water samples are in almost in the same levels with sediment sample in Buna River. Buna River had high sedimentation because its hydrogeology, water volume and water rate, adding anthropogenic factor. Sediment samples were taken in the surface of river bed. Their depths were no more than 5 cm, (data found are evidence for pollution in last year's). Organochlorinated pesticides were found in all sediment samples studied. Run off of agricultural land near the river it's not the only factor. Other sources could be water and water matter of Drini River and Shkodra Lake. Total concentration of organochlorinated pesticides was shown in Figure 17. Mean value for pesticide residue levels in sediment of Buna River was 65,0 ng/g dry sediment. The highest level was found for the sample 800 m from the river mouth with 116,6 ng/g dry weight. The lowest levels was found for station 200 m from river mouth with 16,7 ng/g dry weight. "Buna start" station (point of connection between Buna River and Shkodra Lake) was 75.5 ng/g dry sediment. Levels of pesticide residues in Buna rivermouth was 70,0 ng/g dry sediment, lower than concentration found in Buna rivermouth in Adriatic Sea. Water current and different sedimentation rate are the main factors for this fact. Concentration levels were the same with levels of organochlorinated pesticides found in sediment samples of Shkodra Lake (Marku & Nuro, 2005). Organchlorinated pesticides profile was shown in Figure 18. It's evident the fact that DDE has the higher concentration than other pesticides in water and sediment samples DDT use was banded in our country after 90'. Degradation of DDT, DDE stability and their chemical and physical properties can affect the distribution of DDT and their metabolites. DDT was the second of this group not because of recent use but because of the lower degradation rate for DDT in sediments. Levels of DDTs in sediment sample in Buna River were lower than DDT levels found in sediment samples of

Shkodra Lake. HCB is the second contributor because of their before use of this insecticide in fruit trees. HCHs are found in all sediment samples. In addition the higher concentrations were for b-HCH in sediment samples. This is connected with other factors such as chemistry of HCHs, and atmospheric depositions, rather than agriculture runoff. HCHs data correlate with levels and distribution of these pollutants in sediment samples of Shkodra Lake. Lindane was used as insecticide in adjacent agricultural but their level was lower.

Ohrid Lake. In Ohrid Lake were analyzed fish samples. Sum of organochlorinated pesticides in fish samples was shown in Figure 19. Mean value was 69,2 ng/g f.w. *Rutilus ohridanus* had the higher value with 87.8 ng/g f.w. Minimum concentration for Ohrid Lake fishes was found for the sample of *Cyprinus carpio* with 40.3 ng/g f.w. Pesticide residue levels for fish samples of Ohrid Lake were in the comparable levels with fish samples of Shkodra Lake. Distribution of pesticides in fish samples of Ohrid Lake were also the same with profiles of pesticide residues in Shkodra Lake. The same origins of organochlorinated pesticides suggest this fact. Run off of agricultural land near the lakes and in its water basin is main factor for levels and distribution found in Ohrid Lake. Water cycle in Ohrid Lake and analyzed fish species are also important factors for data collected.

Butrinti Lake. In Butrinti Lake were analyzed sample of water and *Mytulus Galloprovincialius* mussels. In water samples of Butrinti Lake mean value of organochlorinate pesticides total was 14,0 ng/L, with a minimum of 7,3 ng/L and maximum of 30,7 ng/L (Figure 21). Levels found for water samples in Butrinti Lake were lower from levels found in Vlora Bay, Adriatic Sea or Shkodra Lake. Levels of total for organochlorinated pesticides for mussel *Mytulus Galloprovincialius* of Butrinti Lake were also lower than levels found in mussel samples in Vlora and Porto-Romano (Figure 22). Except this was interest the high difference in profile of pesticide residues for water and mussel samples of

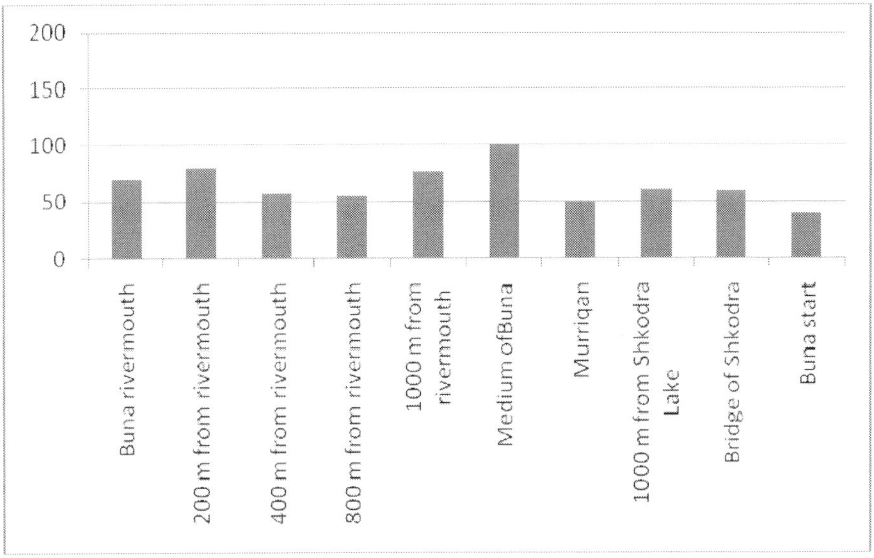

Fig. 16. Total of organochlorinated pesticides in water samples of Buna River

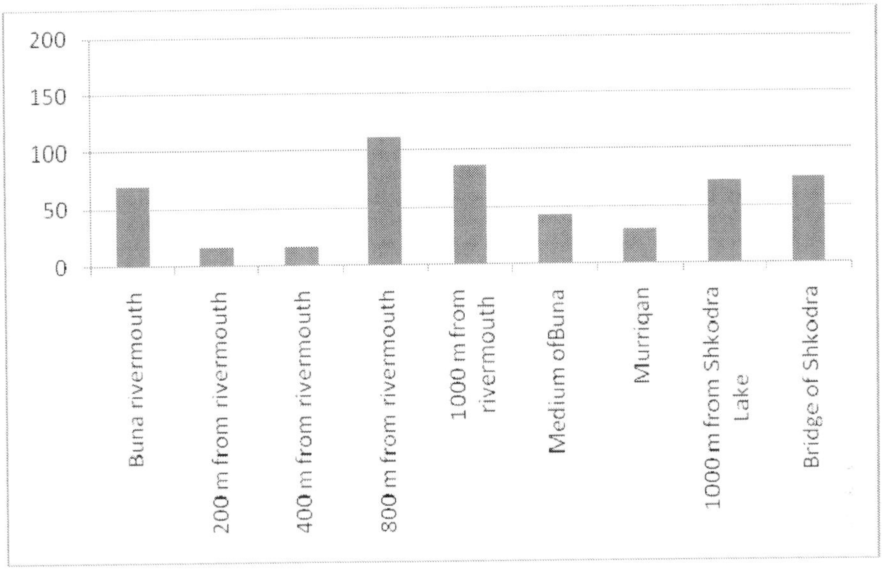

Fig. 17. Total of organochlorinated pesticides in sediments of Buna River

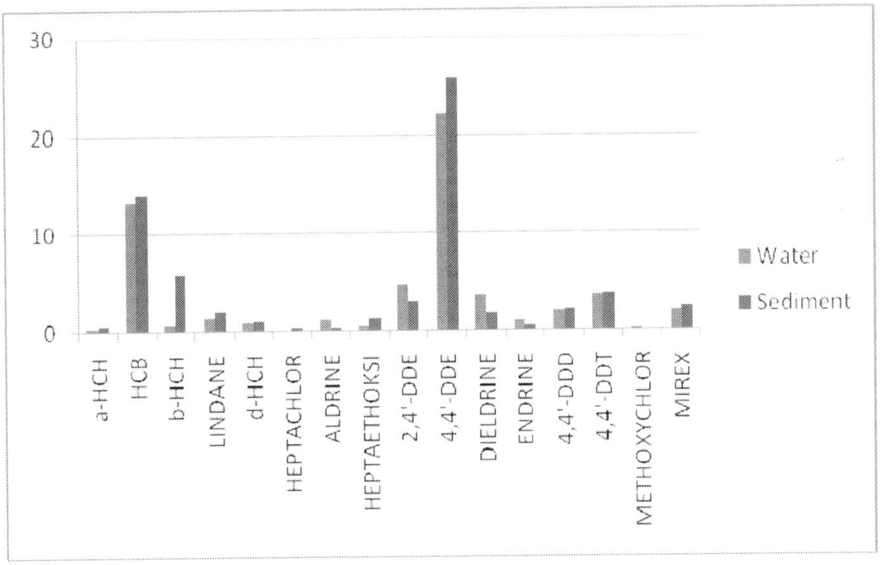

Fig. 18. Average of organochlorinated pesticides in water and sediment samples of Buna River

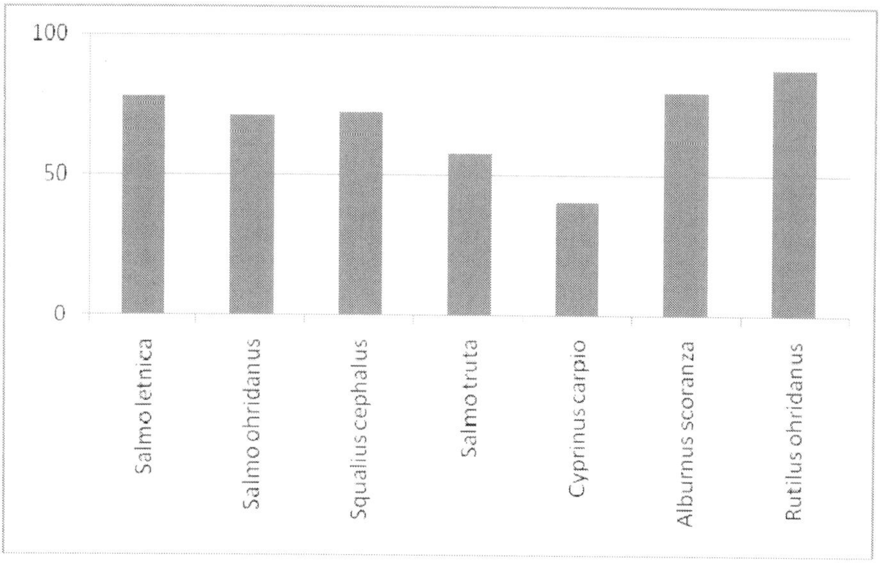

Fig. 19. Total of organochlorinated pesticides in fish samples of Ohrid Lake

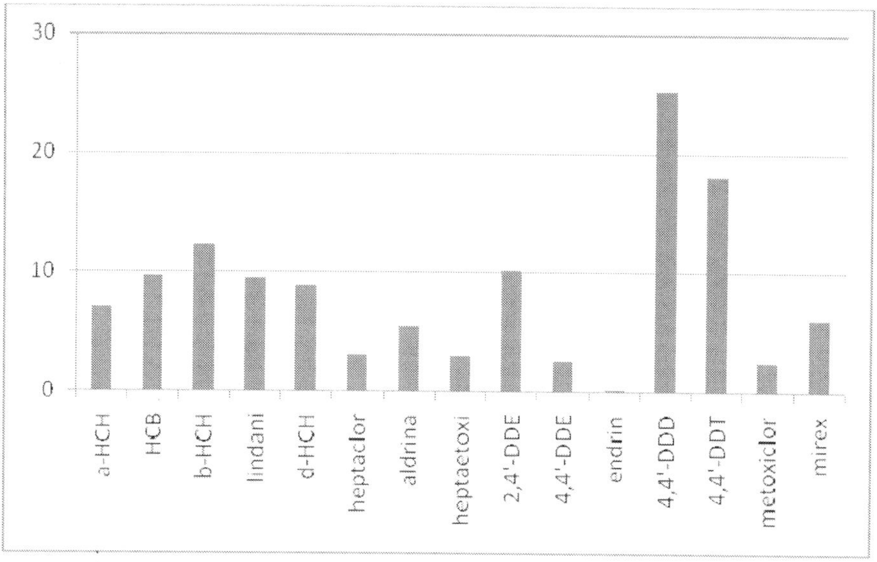

Fig. 20. Average of organochlorinated pesticides in fish samples of Ohrid Lake

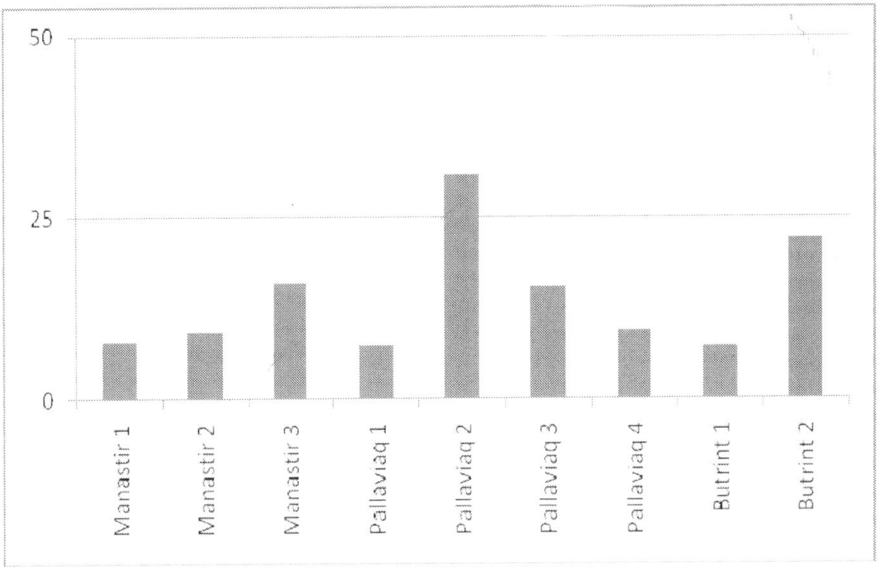

Fig. 21. Total of organochlorinated pesticides in water samples of Butrinti Lake

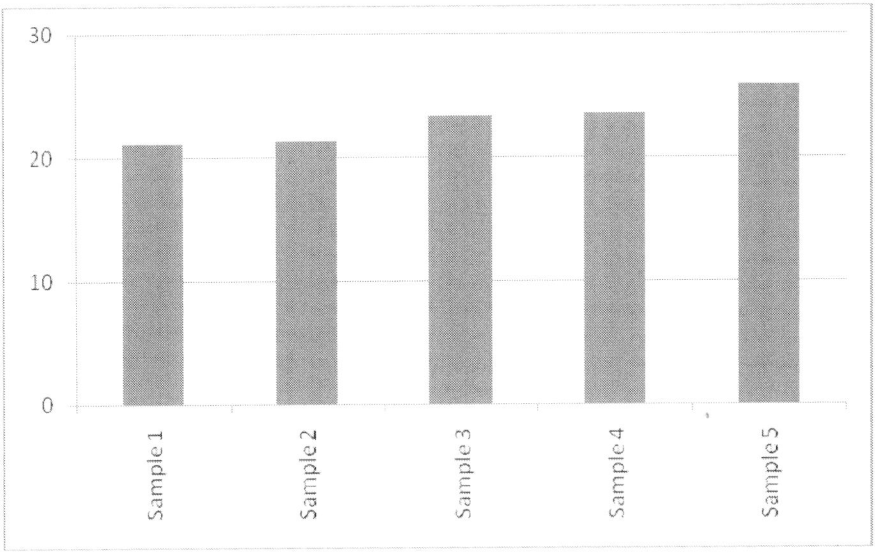

Fig. 22. Total of organochlorinated pesticides in mussel *Mytulus galloprovincialius* of Butrinti Lake

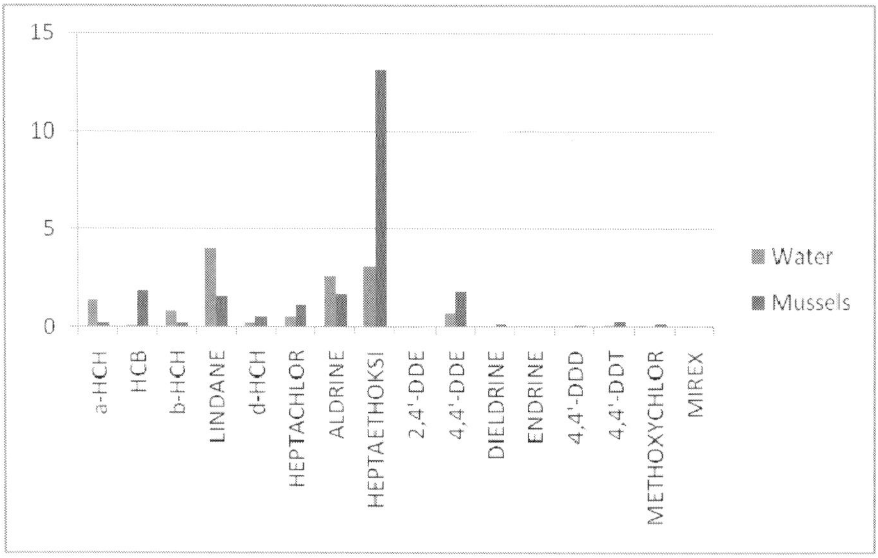

Fig. 23. Profile of organochlorinated pesticides in water and mussel samples of Butrinti Lake

Butrinti Lake compared with other same samples. Profile of pesticide residues in water and mussel samples of Butrinti Lake were almost the same (Figure 23). HCHs were major pollutants in water samples followed from Heptachlors and Aldrine. Heptachlors were the first pollutants in fish samples followed from HCHs. Lindane for both samples type was in higher concentration than other HCH isomers. HCB was found in higher levels in mussel samples than in water samples. DDE was found in lower levels. Pesticides were used for agricultural purposes also in agricultural areas in fields near Butrinti Lake before 90′. Butrinti Lake has a communication channel with Ionian Sea that can affect in found levels for pesticide residues in water samples of lake.

4. Conclusions

In this study were included different aquatic locations of Albania with their own characteristics. Difference in their position and the other characteristics followed from selected samples for each aquatic location can shown a correct thought about levels and distribution of organochlorined pesticides in Albania. In all studied aquatic locations were shown presence of organochlorinated pesticides. These compounds were used insensitive mostly as insecticide for agricultural purposes before 90′. Note the organochlorinated pesticides are not in use after 90′ in Albania, but their before use and their stability are main factors for founding them in all systems. Aquatic systems are affected from pesticide residues mainly as result of agriculture runoff. Discharging of many industrial and agricultural wastes directly in to the sea, rivers and lakes is another factor that affect in found levels of pesticide residues. Organochlorinated pesticides are not hydrophilic compounds for this reason levels of pesticides found in water samples were always lower the sediments of biota samples for the same locations. Levels of pesticides in water samples for Adriatic Sea, Buna River and Shkodra Lake were in the same levels. Butrinti Lake were in low concentration. Their geographic position include specifics for studied locations (water current, nearest from agricultural field, ect,.)

Sediment of aquatic systems considered as archive for organic pollutants, so interpreted of data found for pesticide residue were very interest. Concentration of pesticides in sediments of Buna river mouth (Adriatic Sea), station was higher than in other sediment samples because of Buna River influence. Going to the depth of Adriatic Sea concentrations were lower. Sediment of Shkodra Lake and Buna River were almost in the same range and distribution between them because of their geographic connection. It's shown a difference in profile of pesticides between Velipoja (Adriatic Sea) and Buna River because of water currents of Adriatic Sea influence. Water current influence was also shown for sediment samples of Vlora Bay.

Concentrations of organochlorine pesticides due to their use in Albania before 90' and also for their stability and affinity for their accumulation in fish tissue were main factors for found levels in fish samples. The highest values of total pesticides were found in some fish species, to concentrate many organic contaminants, because of pollutant bioavailability, but also species characteristics (specimen, age, sex, ect,.). Levels of pesticide residues found in fish samples of Vlora Bay was comparable to fish samples of Velipoja. Fish samples of Adriatic Sea were in the same levels of pesticide residues with levels of fish samples of Shkodra and Ohrid Lake, because it's known a migratory communications between both aquatic systems. Concentrations of organochlorine pesticides found in Shkodra Lake and Ohrid Lake are due to their use in Albania before 90', their stability and affinity for accumulation in animal tissue, as well as relatively limited rate of water circulation in these basins. Pesticides profile shows differences between the two lakes. DDT and their metabolite, as well as HCHs levels were in higher levels in samples of Ohrid Lake compare to the samples of Shkodra Lake. Levels of each chlorinated compound are connected mostly to the geographic position, water basin and water currents in these basins. Levels of organochlorinated pesticides were in the same range with other studies in Adriatic Sea, Shkodra Lake and Ohrid Lake. Mussels *Mytulus Galloprovincialius* were studied in two stations in Adriatic Sea (Vlora Bay and Porto-Romano) and Butrinti Lake. Mussel samples were most polluted compared with other type of samples because of they are water filtering and tend to concentrate in their tissue especially liophilic pollutants such are organochlorinated pesticides. Levels of pesticide residues found in mussels of Butrinti Lake were lower than levels found for mussel samples in Adriatic Sea.

For all samples DDTs were in higher concentrations compared with the other organochlorinated pesticide residues analyzed. DDE, metabolite of DDT was found in the higher level for all samples. Before use of DDT, degradation of DDT, DDE stability and their chemical and physical properties can affect in this fact. The concentrations HCH isomers were second compared with other pesticides for analyzed samples. Profiles of HCHs were not the same for all samples. In sediment samples the higher concentration was found for Lindane. Lindane was used as insecticide in adjacent agricultural but for a great number of biota samples and water samples was found in higher concentration beta isomer of HCH. This is connected with other factors such as chemistry of HCHs, and atmospheric depositions, rather than agriculture runoff. Beta isomer of HCH has a great potential of concentrations in lipids than other isomers. Hexachlorbenzene (HCB) was the third pollutant found almost for all samples. Before often use of this insecticide in fruit trees was the main factor. Aldrine, Heptachlor, and Methoxychlors were present in very low concentrations, sometimes even no detectable. Different concentrations and profile found for pesticides between different sampling stations and between different types of samples are connected with before use of pesticides and with their chemical-physical properties.

5. References

Di Muccio, (1996). Organoclorine, Pyrethrin and Pyrethroid Insecticides: Single Class, Multiresidue Analysis of. Pesticides. *Pesticides*. 6384-6411

Dhimitër Dhora (1995). *Shkodra Lake*

Dritan Topi, Pandeli Troja, Koste Koci, Elda Marku, Aurel Nuro (2006) "Same experimental dates about the levels of chlorinated pesticides and PCBs in the biota of Ohrid Lake". www.BALWOIS.net/Conference/Balwois 2006/articles

E.Marku, A. Nuro, (2005) Chlorinated pesticides in the sediments and fish species of Shkodra Lake. *Journal of Environmental Protection and Ecology*; Vol.6, No.3:539-549

EN 1528-1. (2000) Part 1: Fatty Food-Determination of Pesticides and polychlorinated biphenyls [PCBs]

EN 1528-2. (2000) Part 2: Extraction of fat, Pesticides and polychlorinated biphenyls [PCBs and -Determination of Fat Content

EN 1528-3. (2000). Part 3: Clean-up methods

EN 1528-4. (2000) Part 4: Determination, Confirmatory tests, miscellaneous

G.Rene van der Hoff, P. van Zoonen, (1999). Trace analysis of pesticides by gas chromatogaphy. *Journal of Chromatogaphy A*, 843 301–322

Koci, K. (1997) The trend of POP pollution in the Albanian Adriatical coast: Case study: PCBs (1992-1996) Proceedings of subregional awareness raising workshop on POPs. Slovenia

Pano N. Abdyli B. (1984) - Hydrology of Albania. Monography. Institute of Hydrometeorology. Academy of Science, Tirana, pp 94-146, 360-397.

Petrick, G., Schulz, D.E. and Duinker, J.C. (1988). Clean-up of environmental samles for analysis of Organochlorine compounds by gas chromatography with electron-capture detection. *Journal of Chromatogrpahy*, 435, 241-248.

Reference Method No 6, UNEP/FAO/IAEA/IOC: UNEP, (1993) Sampling of selected marine organisms and sample preparation for the analysis of chlorinated hydrocarbons.

Schantz, M. M., Parris, R. M., Kurz, J., Ballschmiter, K. and Wise, S.A., (1993). Comparison of methods for the gas-chromatographic determination of PCB congeners and chlorinated pesticides in marine reference materials. Fresenius *Journal of Analytical Chemistry*, 346. 766-778.

20

Distribution Characteristics of Organochlorine Pesticides in Soil and Groundwater of Different Irrigation Farmlands in Southeast Suburb of Beijing

HE Jiang-tao, MA Wen-jie, JIN Ai-fang and WEI Yong-xia
China University of Geosciences, Beijing
China

1. Introduction

Wastewater has been widely used in agriculture irrigation for nearly a century in the world, due to the scarcity and quality declination of water resources. The wastewater irrigation technologies in USA, Japan, Israel and some developed countries are more mature than other developing countries (Angelakis A N et al., 1999; U Pinto et al., 2010; US EPA, 2004). Although wastewater irrigation can supply a lot of nutrients to crop, such as Nitrogen and Phosphor sources, it can cause soil pollution, like heavy metals, organic pollution and so on. Therefore, reclaimed water, instead of wastewater, as agriculture irrigation water resources became more popular now, because not only can reclaimed water relieve the tensions of agriculture water resource, but also relieve the pressure of wastewater on environment (T.Asano et al., 1996; D.Levine et al., 2004; I.K.Kalavirouziotis et al., 2008; V.Reboll et al., 2000). However, the nature of reclaimed water is still a poor-quality water, the amounts of contaminants in reclaimed water are changed with treatment, still a lot of pollutants can not be effectively removed. Some matters in reclaimed water may benefit crops, such as N, P, K, some else may be harmful to the surrounding environment, crop, operator and consumer of agriculture products, like heavy metals, salts, and various carcinogenic, teratogenic, or mutagenic organic compounds, the long-term irrigation will bring harmful to soil and groundwater, causing a new source pollution (Francisco P et al., 2009; Friedler E et al., 2004; Gross A et al., 2007).

As a large agricultural country, the agricultural irrigation water consumption accounts for more than 70% of the total water consumption in China. Due to water shortage, wastewater irrigated area expanded rapidly from 1970s to the 1990s. By 1998, wastewater irrigated area has reached 361.8×10^4 hm^2, which accounting for 7.3% of the total irrigated area. The city of Beijing in China began to use wastewater irrigation since the early 50s of the 20th century, so far, the wastewater irrigation area in Beijing are nearly 80,000 hm^2 and the yearly wastewater irrigation is about 220 million m^3 which accounting for 20% of the city's wastewater emissions, 87% of the wastewater irrigation area are mainly in Tongzhou District, Daxing District and Chaoyang District (Zhang H Y et al., 2006; Hua X M et al., 1996). The quality and fertility of domestic sewage are better and beneficial to rice, but the industrial effluents are not as well, it contains some heavy metal salts, such as lead, chromium, arsenic, mercury, and some harmful ingredients, take chlorine, sulfur, phenol, cyanide for example, all of them are bad for rice growing, therefore, long-term use of wastewater irrigation is bound to lead a series of pollution problems.

In recent years, along with urban sewage treatment rate rising, reclaimed water reuse has been strengthened. Various studies and experimental demonstration bases have been performed, which greatly promoted the reclaimed water reuse in agricultural and associated research. As one of the most serious water shortage city in China, Beijing is constantly increasing the intensity use of reclaimed water irrigation. In 2006, about 199 million m^3 reclaimed water were used for agriculture irrigation in Beijing. By 2010, the Beijing center city reclaimed water production capacity will reach 1.4 million m^3 everyday, annual renewable water consumption will reach 600 million m^3, recycled water will become the major water sources for irrigation, industrial cooling, and ecological water use et al (LIU P B, 2007). However, even the reclaimed water quality has been greatly improved, it still has potential effects on crop growth, soil quality, and groundwater quality during the process of irrigation, especially some refractory organics will persist in the environment. These problems have received widely concern of the publics. The use of reclaimed water in China was only at the initial stage. Therefore, a lot of works need to be done to confirm the safety use of reclaimed water.

Organochlorine Pesticides (OCPs) are a kind of synthetic chemical pesticides, which can be divided into two branches, one is take benzene as raw materials, the other is cyclopentadiene, their physical and chemical properties are stability, both the persistence and bioaccumulation are strong, they are also difficult to degrade naturally in environment, but they can be a threat to ecosystems and human healthy through evaporation, migration, food chain transfer and other paths (Tan Z et al., 2008; Katsoyiannis A et al., 2005). Among 12 kinds of Persistent Organic Pollutants (POPs) listed in the Sidegemo Convention, 9 of them are OCPs (Liu C et al., 2008). Once these pollutants were released to the environment, they will exist in environment for many years. Even in some areas OCPs never been applied before, we can also find the existence of them, this indicates that the OCPs have stronger migration ability in space. Although some of the OCPs had been banned in China since 1983, the harm caused to the environment still exists due to their refractory properties.

OCPs are widely distributed in various environment media, such as soil sediments, plant tissues, animal organs and body. In which the soils are usually the main destination and accumulation sites for these pollutants, it often acts as pollutants carrier and offer the natural purification place. Therefore, it is important to study the pollution characteristics of OCPs in soils and assess the potential of groundwater contamination. Based on the above, 3 typical farmlands, i.e. wastewater irrigated farmland, reclaimed water irrigated farmland and groundwater irrigated farmland of Beijing suburb were selected for study. Both soil samples of vertical profiles, irrigation water and groundwater samples of each farmland were collected and analyzed. Through detail comparison of OCPs contents in each soil vertical profile and OCPs concentration in irrigation water and groundwater, the distribution characters of OCPs in soil profiles were summarized, the impact of irrigation water and controlling factors of OCPs migration were also discussed.

2. Materials and methods

2.1 Research area description

Both 3 farmlands selected in this study are all located in the south-east suburb of Beijing. The wastewater irrigated farmland (WIF) makes use of Liangshui River's water for irrigation. Liangshui River is the second-largest drainage river in Beijing. It collects domestic water and industrial wastewater from Beijing centre city. This WIF has a long history of wastewater irrigation, which began from the 20th century 60s and with an area of about 1.24×10^4 hm^2 so far. Irrigation periods of each year are from March to June and the middle September to December, respectively. The average water use is about 3×10^3 $m^3\cdot hm^{-2}\cdot a^{-1}$.

The reclaimed water irrigated farmland (RIF) selected in this study is located in Duozi village, Ciqu town, which is about 4 kilometers northwest to the WIF. Irrigation water used in this farmland was came from Tonghui irrigation canal which got through the farmland from north to south. Many branches of Tonghui irrigation canal provide sufficient reclaimed water for agriculture irrigation. This canal was built in 1958, which started from the Gaobeidian Lake, and ended in Liangshui River. The full length is about 17 kilometers. It was sewage canal historically which accepted sewage water from Tonghui River. In 1993, the largest sewage treatment plant in Beijing named Gaobeidian was completed and put into operation. From then, Tonghui irrigation canal began to accept reclaimed water. Therefore, this region was also wastewater irrigated area before 1993. The history of reclaimed water irrigation is no more than 20 years.

Groundwater irrigated farmland (GIF) is located in Yongle town, Tongzhou District of Beijing, which is about 18 kilometers southeast to the WRF. The irrigation water for this farmland is groundwater, extracting from 60m underground through the motor-pumped wells.

Fig. 1. Location of the study areas

2.2 Sample collection

Soil profiles with depth of 6.0 m were drilled using Eijkelkamp soil sampler in farmlands. Each farmland has 3 profiles, and they were distributed at the vertices of equilateral triangular with a side length of 1.0 m. Therefore, total 9 boreholes were drilled. A, B and C profiles were drilled in the WIF. D, E, and F profiles were drilled in the RIF. G, H, and I profiles were drilled in the GIF. Soil samples were collected from surface to 5.5 m underground every 0.5m. So, there were 12 samples taken from each borehole, and 108 soil samples named from A00 to I55 were collected from three farmlands. Soil samples were stored in brown glass bottles, sealed with film, and kept store no more than 20 days. Physical-chemical properties of soil samples such as Clay content, total amount of clay minerals, Cation exchange capacity (CEC), Total organic compounds (TOC), et al were analyzed. 14 kinds of OCPs including α-HCH, β-HCH, γ-HCH, δ-HCH, 4,4'-DDT, 2,4'-DDT, 4,4'-DDE, 4,4'-DDD, Hexachlorobenzene, Heptachlor, Aldrin, Heptachlor epoxide, Dieldrin and Endrin were also analyzed using Gas chromatography.

The groundwater and surface water samples near the profiles were also taken for analyses in each farmland. Water samples were placed in incubator to refrigeration and avoid light, sent to the laboratory at the same day and then stored in refrigerator at 4 °C.

2.3 Sample preparation

Soil sample preparation: According to the EPA 3550, 15g soil sample, 3g sodium sulfate anhydrous and 20 mL extraction solvent (1:1 acetone/ hexane mixed solvent) were added into a 40 mL vial. The soil was ultrasonically extracted at 50 °C in 400 W for 20 min to promote the diffusion of OCPs from the soil into the extractant, and then the vials were centrifuged at 3000 rpm for 3 min. After that, the organic phase was transfered to 100 mL glass bottle. Repeated this extraction twice, combined the extracts. The extracts then were dehydrated with sodium sulfate anhydrous, and concentrated to 1.0 mL using a rotary evaporator in water bath at 35-40 °C. OCPs in extracts were purified by prepared SPE florisil columns, and concentrated to 1.0 mL through nitrogen blow for further chromatographic analysis.

Water sample preparation: According to the EPA 3510, water samples were filtered with APFF glass fiber membrane at first. Then 1 L filtered water was put into separator funnel and added 20 mL cyclohexane. Then Liquid-liquid extracted for 10 min, held for 10 min, and transfered the organic phase. This extraction should be repeated twice, and combined the organic phase. The extracts then were dehydrated with sodium sulfate anhydrous, and concentrated to 1.0 mL using a rotary evaporator in water bath at 35-40 °C. OCPs in extracts were purified by prepared SPE florisil columns, and concentrated to 1.0 mL through nitrogen blow for further chromatographic analysis.

2.4 Analysis methods

The purified sample extracts for OCPs analysis were quantitatively analyzed using a HP-6890 gas chromatograph, equipped with electron capture detector (ECD) and a HP5 capillary column (30 m×0.32 mm×0.25 μm). The oven temperature was programmed begin from 80°C and kept for 2 min, then increased from 80°C to 185°C at the rate of 30 °C/min, and to 215 °C at the rate of 3 °C/min, held for 4 min and then to 225 °C at the rate of 1 °C/min and held for 2 min, finally increased to 290 °C at the rate of 20 °C/min and held for 2 min. Helium was the carrier gas at a flow rate of 1.0 mL/min. The temperatures of the injector and detector were set up at 250 °C and 320 °C, respectively. 1.0 μL of sample was injected under the splitless injection mode.

Instrument detection limits and method detection limits were list in table 1. Separation effect of mixed standard sample which contains 14 kinds of OCPs was shown in figure 2. Table 2 list the testing methods of soil physical-chemical properties.

3. Physical-chemical properties of soil profile in farmlands

3.1 Physical-chemical indexes

Soil Physical-chemical properties are important factors affecting the vertical migration of organic pollutants in vadose zone. 8 indexes of each sample including pH, Eh, soil moisture (%), EC, clay content (%), total clay minerals (%) total organic carbon content (%), and cation exchange capacity (CEC) were tested in this study. The vertical variations of 8 indexes were shown in Figure 3.

ITEM	Method limits (μg/kg)	Instrument limits (μg/L)
α-HCH	0.06	0.60
Hexachlorobenzene	0.07	0.65
β-HCH	0.04	0.40
γ-HCH	0.07	0.65
δ-HCH	0.60	5.75
4,4'-DDE	0.01	0.05
4,4'-DDD	0.04	0.35
2,4'-DDT	0.25	2.45
4,4'-DDT	2.00	19.5
Heptachlor	0.07	0.65
Aldrin	0.04	0.40
Heptachlor expoxide	0.04	0.35
Dieldrin	0.05	0.50
Endrin	1.00	9.65

Table 1. The detection limits of OCPs

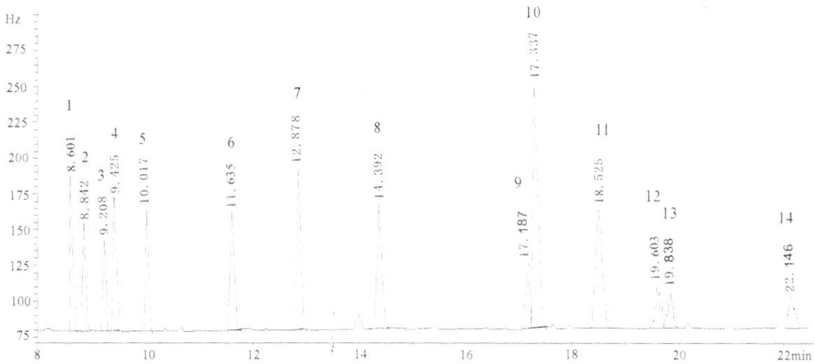

Fig. 2. 1. α-HCH, 2. Hexachlorobenzene, 3. β-HCH, 4. γ-HCH, 5. δ-HCH, 6. Heptachlor, 7. Aldrin, 8. Heptachlor expoxide, 9. 4,4'-DDE, 10. Dieldrin, 11. Endrin, 12. 4,4'-DDD, 13. 2,4'-DDT, 14. 4,4'-DDT.

Testing project	Testing method
Soil moisture	Velocity and mass method
pH	Potentiometric method
CEC	Ammonium chloride-ammonium acetate exchange and ammonium acetate exchange method
TOC	External heating potassium dichromate oxidation method
EC	Conductivity method
Eh	Potentiometric method
Grain composition	Hydrometer method
Clay mineral	X−ray diffraction method

Table 2. Testing projects and methods of physical-chemical properties

As Figure 3 shows, the soil physical-chemical properties of these 3 farmlands had no obvious differences in general. The soil moisture of WIF profiles varied between 18% - 39%, maximum soil moisture all appeared at the 1.5 m in 3 boreholes. The profiles of RIF and GIF basically had the same variation of soil moisture, mostly changed between 10.91% -36.98%, and most lower soil moisture appeared at 0-0 .5 m. The pH values of the three farmlands were basically around 7.7, rendered as a neutral environment, and slightly increased along the profiles. The soil redox potential variations with depth were consistent in 3 farmlands profiles, and oxidation environment played a leading role in the whole profiles. In contrast, RIF has higher soil redox potential values, GIF in the middle, WIF in the last. The CEC variations with depth were consistent in 3 farmlands profiles too. The CEC values of WIF and RIF were 2.13-18.00 cmol/kg and 4.90-16.44 cmol/kg respectively, and maximum value appeared in 1.0-1.5 m both for WIF and RIF. The CEC values of GIF were 3.37-9.74 cmol/kg, and maximum value appeared in 3.0 m. Compared with WIF and RIF, GIF has higher EC, the high values was in 1.5-3.0 m. Clay content and clay minerals contents were almost same, and their changes alone the profiles in 3 farmlands had no obvious differences too. However, we found that clay contents with slightly higher values were in 1.0-1.5m in WIF and RIF, and 2.0-3.0m in GIF. The changes of TOC content alone the profiles in 3 farmlands were more obvious than other indexes, significant reductions were found in almost all soil profiles, the average maximum TOC content occurred at top soils. Under the top layer, higher TOC content layers can be outlined in 1.5 m in WIF profiles and 3-4.5 m in RIF profiles. However, there were no obvious high TOC content layers in GIF profiles.

3.2 Profile textures

The profile soil textures of three farmlands based on soil grain size were shown in figure 4. Accumulated particles percentages were used to help us found out main particle composition. Particle sizes were divided into three classes, i.e. sand (>0.075 mm), silt (0.075-0.005 mm) and clay (<0.005 mm) respectively according soil classification standards.

As shown in figure 4, the main soil particles in three farmlands were silt, followed by clay. Only small percentages were sand. However, profiles in RIF had more sandy particles than GIF and WIF statistically. That to say, pollutants were more easier to migrate in RIF than in GIF and WIF, only from the view of soil textures.

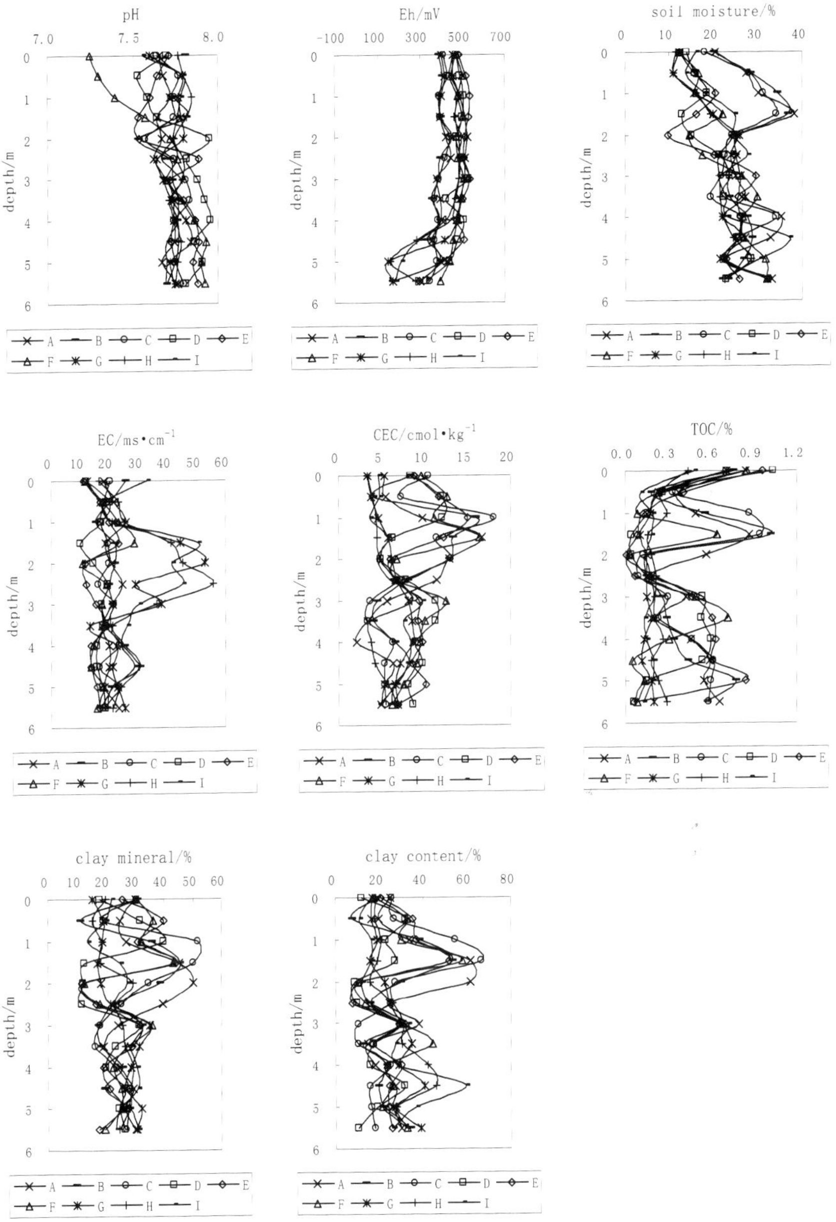

Fig. 3. Variation of soil Physical-chemical indexes with depth in three farmlands

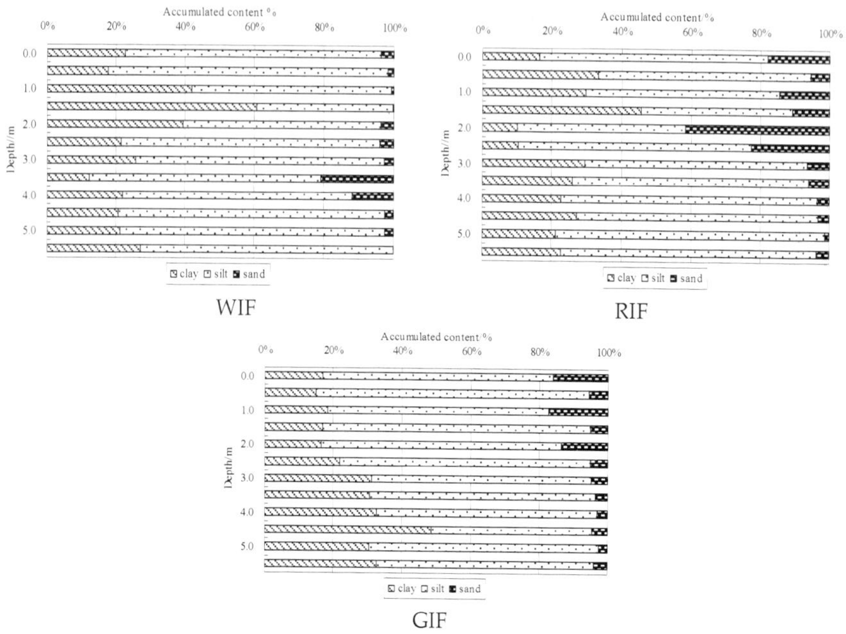

Fig. 4. Soil textures classification of three farmlands

4. The vertical distribution of OCPs in three farmlands

4.1 The distribution of OCPs in soil profiles of WIF

4.1.1 Vertical change of total OCPs

There were 14 kinds of OCPs detected in three boreholes, including α-HCH, β-HCH, γ-HCH, δ-HCH, 4,4'-DDT, 2,4'-DDT, 4,4'-DDE, 4,4'-DDD, Hexachlorobenzene (HCB), Heptachlor, Aldrin, Heptachlor epoxide, Dieldrin and Endrin. The changes of total OCPs along the profiles and the detected OCPs species numbers in 3 profiles were shown in figure 5.

Clearly we can see the highest ∑OCPs content was in top soil layers, and immediate drop in the whole profiles under the top layer. The maximum content was 30.93 μg/kg, which was at least 13 times of other layers. That means most OCPs were not easy to migrate vertically in these profiles. The detected OCPs numbers in surface soil were the most in 3 profiles, and decreased with depth. Under the surface, the greatest detected OCPs number was only 3. That's to say, 7-8 kinds of OCPs were detected in the surface soil, but only 3 kinds of OCPs detected under the surface. In additions, Heptachlor epoxide, Dieldrin and Endrin were not detected in three profiles.

Figure 6 described OCPs' compositions in surface soils. DDTs (mostly 4,4'-DDE) and HCHs (mostly δ-HCH) were the highest contents OCPs, which highest content were 20.03 μg/kg and 10.46 μg/kg, respectively, lower than the first class soil standard (<50 μg/kg) of Chinese soil environmental quality standard (GB15618-1995). Aldrin and HCB were also found in surface soils with lower contents. That's to say, DDTs and HCHs were the main pollutants in surface soils.

However, from the detection frequency of the 3 profiles (figure 7), we can see that Heptachlor had the highest detection frequency (66.7%), followed by DDTs (50.0%) and Aldrin (41.7%) in profile A. In profile B, DDTs had the highest detection frequency (66.7%), followed by Heptachlor (41.7%). Same to profile B, the OCPs which had the highest detection frequency in profile C was DDTs (58.3%), followed by Aldrin (50.0%) and Heptachlor (33.3%). That means except DDTs and HCHs, Heptachlor and Aldrin were the main pollutants in the profiles too.

4.1.2 Vertical variation of single OCP

DDTs, HCHs, Heptachlor, Aldrin were chosen as the main objects to discuss the vertical distribution of OCP, the variation were shown in figure 8.

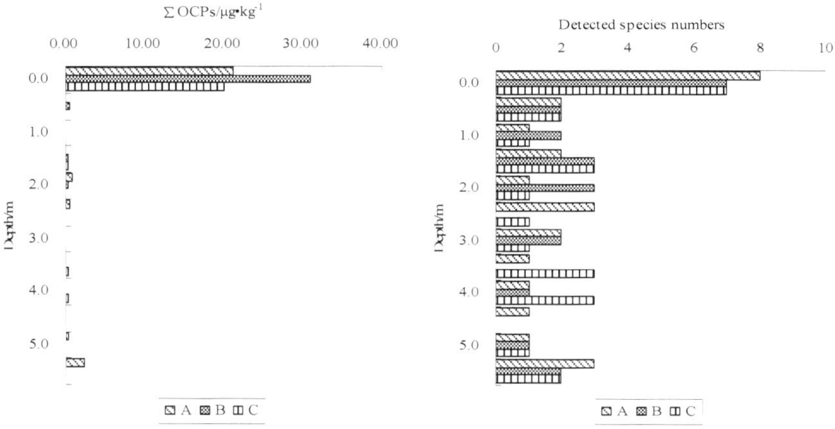

Fig. 5. Vertical distribution of ∑OCPs and detected OCPs species number in WIF

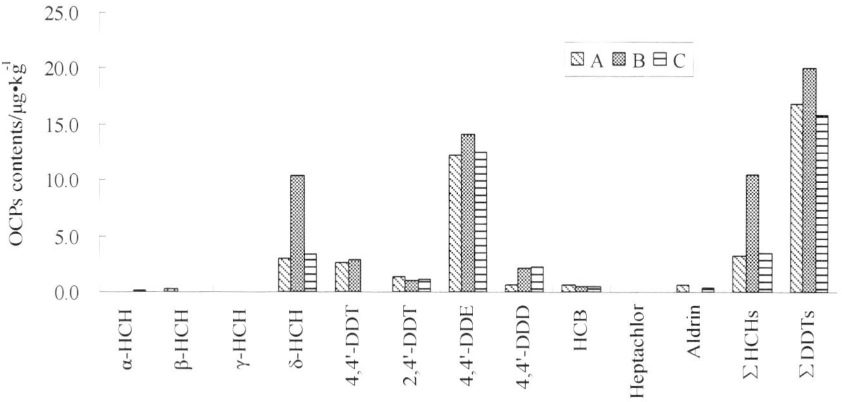

Fig. 6. Distribution of OCPs in surface soils of WIF

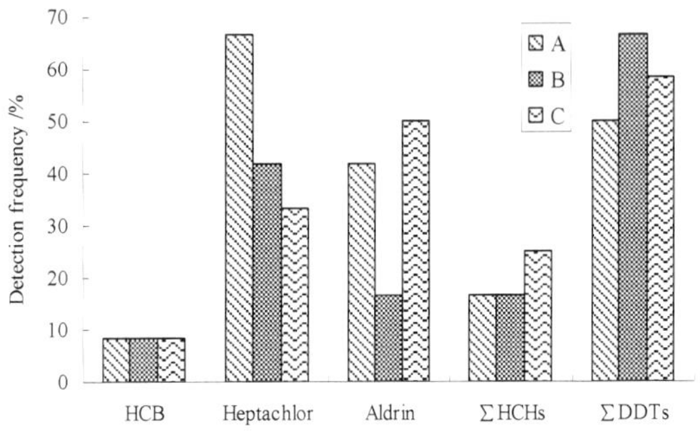

Fig. 7. Detection frequency of OCPs in WIF profiles

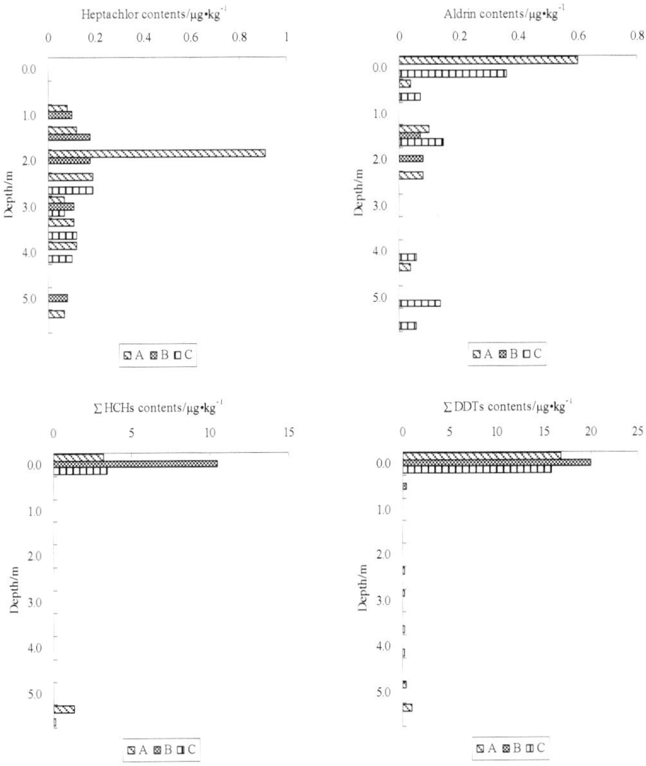

Fig. 8. Vertical distribution of single OCP in profiles of WIF

DDTs and HCHs were the main OCPs in WIF, their vertical distribution in generally took on sharp dropped look with depth and then kept very low content under the surface, which was same to the ∑OCPs. Aldrin contents in the profiles also dropped very fast, but didn't like DDTs and HCHs drop so much.

The changes of Heptachlor in the profiles were quite different from former mentioned OCPs, there was no Heptachlor detected in surface soils, but under the surface, Heptachlor were detected from 1.0 m to 5.5 m. Considered the physical-chemical properties of these OCPs, we can find out that DDTs have higher logK_{OW} and lower Henry's Law Constant, however, except higher logK_{OW}, Heptachlor also have higher Henry's Law Constant (table 3). That means both DDTs and Heptachlor were easy to be absorbed by soil and hard to migrate, but Heptachlor much easier to volatilize and lost in surface soil than DDTs. This can be a reasonable explanation for the difference.

OCPs	Molecular Weight	Water Solubility 25°C mg·L^{-1}	logK_{OW} (octanol-water)	Henry's Law Constant atm-m^3/mole	Vapor Pressure mm Hg
α-HCH	291	2	3.8	1.22E-05	-
β-HCH	291	0.24	3.78	4.40E-07	-
γ-HCH	291	7.3	3.72	5.14E-06	4.20E-05
δ-HCH	291	10	4.14	4.29E-07	3.52E-05
4,4'-DDT	354.5	0.0055	6.91	8.32E-06	1.60E-07
2,4'-DDT	354.5	0.085	6.79	7.41E-06	3.44E-12
4,4'-DDE	318	0.04	6.51	4.16E-05	7.43E-12
4,4'-DDD	320	0.09	6.02	6.60E-06	1.35E-06
Hexachlorobenzene	285	0.0062	5.73	0.0017	1.80E-05
Heptachlor	373.5	0.18	6.1	2.94E-04	4.00E-04
Aldrin	365	0.017	6.5	4.40E-05	1.20E-04
Heptachlor epoxide	389.2	0.2	4.98	2.10E-05	1.95E-05
Dieldrin	381	0.195	5.4	1.00E-05	5.89E-06
Endrin	381	0.25	5.2	6.36E-06	3.00E-06

Table 3. Physical-chemical properties of 14 kinds OCPs

4.2 The distribution of OCPs in soil profiles of RIF

4.2.1 Vertical change of total OCPs

The changes of total OCPs along the profiles in RIF and the detected OCPs species numbers in 3 profiles were shown in figure 9.

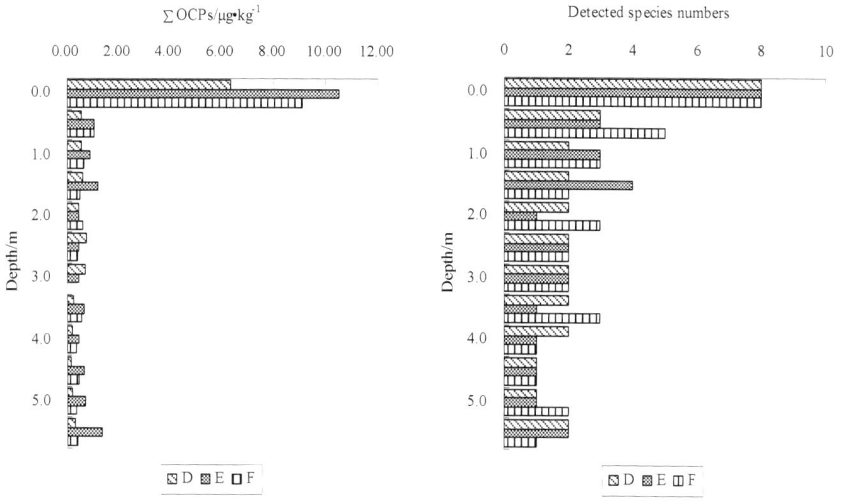

Fig. 9. Vertical distribution of ∑OCPs and the detected OCPs species numbers in RIF

From figure 9, we can see that the vertical variation of OCPs in three profiles were consistent, the contents decreased rapidly with depth, and then changed slowly, ended with consistent. The maximum contents all occurred in surface soils, of which the highest content was in profile E, reached 10.50 μg/kg. Compared the ∑OCPs in surface soil with WIF, the total OCPs in RIF was about 2 times lower. There were no apparent differences under the surface of two farmlands, but the total OCPs in RIF was slightly higher than WIF. The detected OCPs numbers in surface soil were the most in 3 profiles, and the amount decreased with depth. Under the surface, the greatest detected OCPs number were 5. That's to say, 8 kinds of OCPs were detected in the surface soil, but only 1-5 kinds of OCPs detected under the surface. In addition, Heptachlor epoxide, Dieldrin and Endrin were not detected in three profiles.

Figure 10 shows that, HCHs, DDTs, Heptachlor and Hexachlorobenzene were detected in the surface soils, the main pollutants were HCHs and DDTs, their highest contents were 6.82 μg/kg and 1.63 μg/kg, both of them were lower than the WIF, even in an order of magnitude. The lowest detected content was Hexachlorobenzene, and the value was 0.69 μg/kg. 4,4'-DDE was the main composition of detected DDTs and δ-HCH was the main composition of detected HCHs.

Figure 11 shows that, Heptachlor was detected in all three profiles and the detection frequency in profile D and E was 100%, followed by Aldrin and DDTs. The detection frequency of HCB was the lowest. Therefore the main OCPs in RIF were Heptachlor, Aldrin and DDTs.

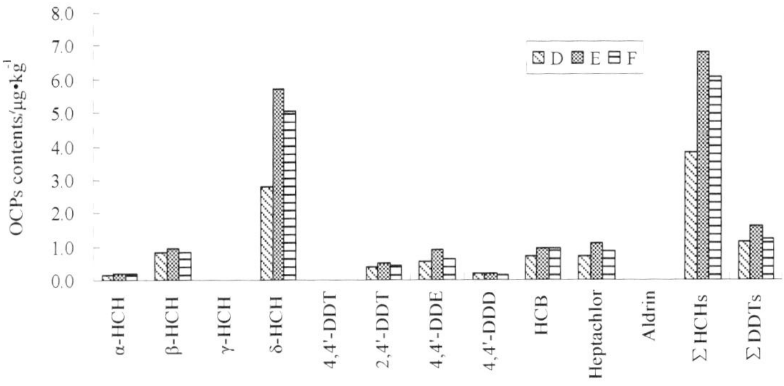

Fig. 10. Distribution of OCPs in surface soil of RIF

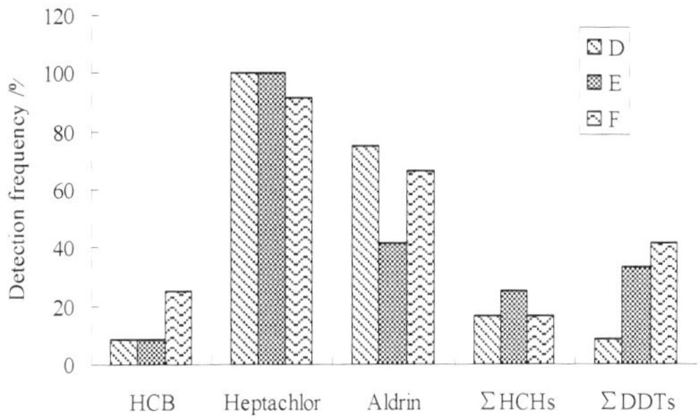

Fig. 11. Detection frequency of OCPs in RIF profiles

4.2.2 Vertical variation of single OCP

Heptachlor's, Aldrin, DDTs and HCHs were chosen as the main objects to discuss the vertical distribution of OCP, the variation were shown in figure 12.

DDTs and HCHs were the main OCPs in RIF, its vertical distribution in generally took on sharp drop look with depth and then kept very low content under the surface, which was same to the ∑OCPs changes, and also to WIF. Although Aldrin was not detected in RIF's surface soil, it was found in most sub-layers of the 3 profiles.

The changes of Heptachlor contents in the profiles were different from other OCPs. There was no apparent difference between top layer and sub layers. As we discussed before, both DDTs and Heptachlor were hard to migrate in vertical profiles. But Heptachlor was easier to volatilize in surface soil layer than DDTs. This supposed that surface soil layers of RIF maybe suffered more serious pollution relatively. That's why Heptachlor still can be detected in the surface soils. As we know, Octanol-water partition coefficient and Henry's Law Constant of Aldrin were close to Heptachlor. There was no Aldrin detected in surface soil may be due to volatilization.

In addition, heptachlor epoxide in all layers of three profiles was not detected which meant that there was no degradation reaction of Heptachlor. Dieldrin was not detected in three profiles which meant that Aldrin had not changed.

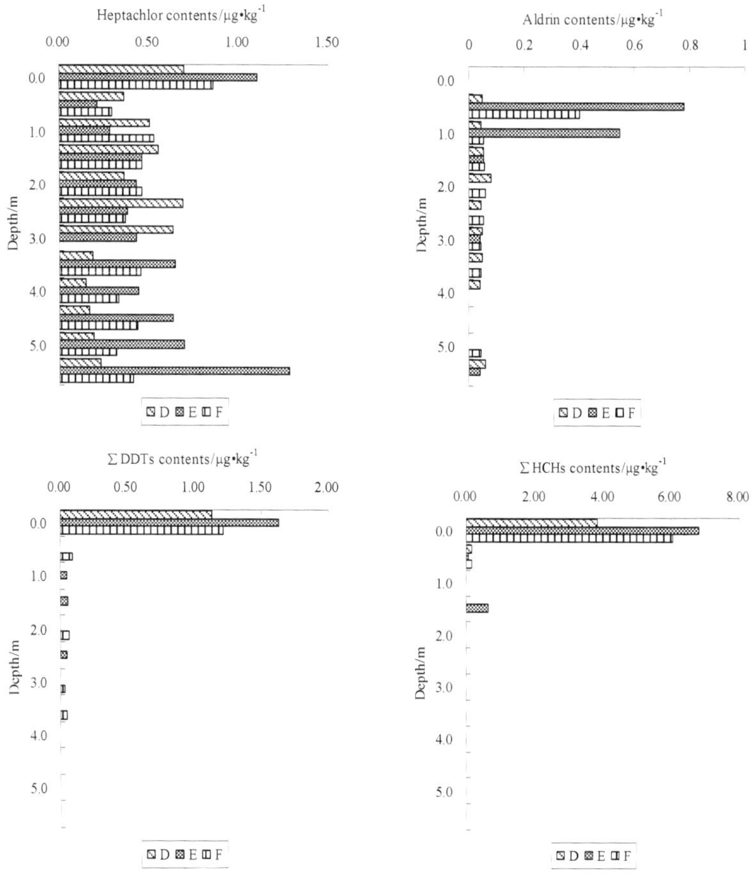

Fig. 12. Vertical distribution of single OCP in profiles of RIF

4.3 The distribution of OCPs in soil profiles of GIF

Groundwater is an ideal irrigation water resource. Using groundwater for irrigation can not only avoid new pollution sources to soil, also can dilute the pollutants in soil. Because the groundwater irrigated farmland has no history of wastewater irrigation, it can be regarded as background.

4.3.1 Vertical change of total OCPs

The distribution of total OCPs in soil profiles of GIF and detected OCPs species number of 3 profiles were shown in figure 13.

The vertical variations of ∑OCPs in boreholes of GIF were different from WIF and RIF. We can not see the typical drop of ∑OCPs contents from surface layer to sub layers. On the contrary, the maximum content was detected in 5.5 m, and the value was 6.77 μg/kg. But

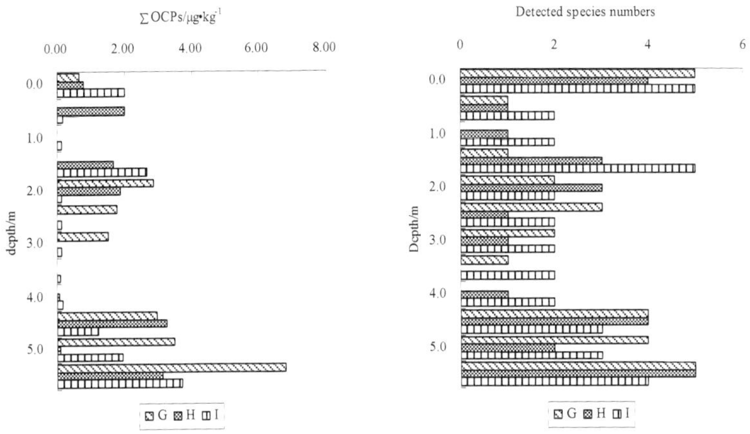

Fig. 13. Vertical distribution of ∑OCPs and detected OCPs species number in GIF

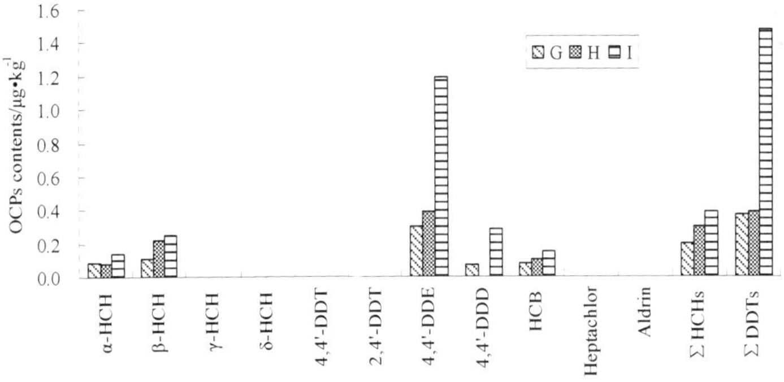

Fig. 14. Distribution of OCPs in surface soil of GIF

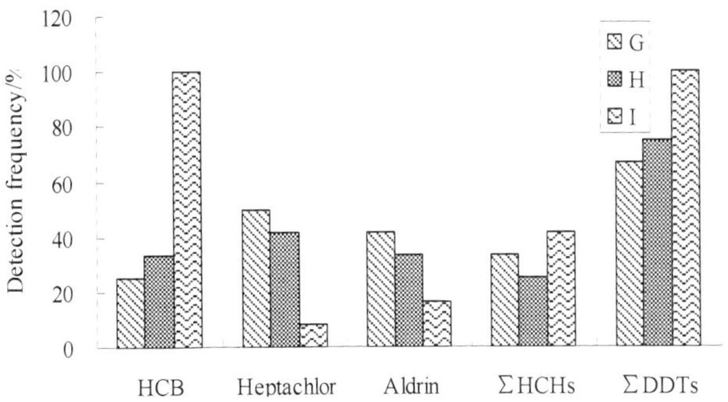

Fig. 15. Detection of frequency of OCPs in GIF profiles

this value was much lower than that of WIF and RIF. The detected OCPs species number's distributions were also different from WIF and RIF. No sharp drop occurred from top layer to sublayers. May be this just was the characters of background.
In figure 14, we can see that DDTs, HCHs and HCB were detected in the surface soil, and 4,4'-DDE was the main composition of DDTs with the maximum value of 1.19 μg/kg and β-HCH was the main composition HCHs with the maximum value of 0.25 μg/kg. However, figure 15 shows that except DDTs, HCHs and HCB, both Heptachlor and Aldrin had been detected in 3 profiles. The sequence of average detection frequency of OCPs was HCB> DDTs> Heptachlor, HCHs> Aldrin.

4.3.2 Vertical variation of single OCP

DDTs, HCHs, HCB, Heptachlor, Aldrin were chosen as the main objects to discuss the vertical distribution of OCP, the variation were shown in figure 16.
As we discussed before, DDTs, Heptachlor and Aldrin had the same distribution characters with WIF and RIF. Compared with DDTs, HCHs have lower Octanol-water partition coefficients, means that HCHs were lower hydrophobic and easier to migrate with infiltration water than DDTs. So, we found more HCHs detected in deep layers. The $\log K_{OW}$ of HCB is 5.73. This value is lower than DDTs' and higher than HCHs'. So HCB was no easy to migrate too. But HCH has highest Henry's Law Constant in 14 kinds OCPs detected in this study, which is 0.0017 atm-m^3/mole. We did not discuss HCB in WIF and RIF, because it was not main OCPs pollutant in that two farmlands. Actually, it was detected in 3 farmlands. The average contents in surface layer of WIF, RIF GIF were 0.49 μg/kg, 0.86 μg/kg, and 0.11 μg/kg respectively. Only because HCB's relative content was higher in GIF, so we discussed here.

4.4 Comparison of OCPs in soil profiles

Firstly, detected OCPs average contents in surface soil of 3 farmlands were compared in figure 17. We can see that there was slightly difference between detected species of OCPs in 3 farmlands. α-HCH, β-HCH, δ-HCH, 4,4'-DDT, 2,4'-DDT, 4,4'-DDE, 4,4'-DDD, HCB and Aldrin were detected in WIF, totally 9 species. α-HCH, β-HCH, δ-HCH, 2,4'-DDT, 4,4'-DDE, 4,4'-DDD, HCB and Heptachlor were detected in RIF, totally 8 species. α-HCH, β-HCH, 4,4'-DDE, 4,4'-DDD and HCB were detected in GIF, only 5 species. We can also see that the ∑OCPs of WIR was much higher than RIF and GIF, the average contents of 3 farmlands were 24.12 μg/kg, 8.64 μg/kg, and 1.15 μg/kg respectively. This indicated the irrigation water sources have great affects on the OCPs pollution in soil, despite that most OCPs came from agricultural pesticides application. Secondly, the average contents of OCPs in sub-layers of 3 farmlands were compared in figure 18. The results of detected species below surface soil of three farmlands showed that Heptachlor and Aldrin were the common detected OCPs, the detection frequency and contents of Heptachlor were much higher than other OCPs, and its downward movement in profiles was the most obvious one of all. The contents of OCPs under surface soil appeared as: GIF >RIF >WIF. That was quite different from surface layer. Since most OCPs were easy to be absorbed in surface layer, though the surface layer suffered different level of pollution, the sub-layers had no distinct difference. To explain this phenomenon, soil texture should be considered too. As we mentioned above, soil profiles textures of these 3 farmlands had no obviously difference. The percentage of clay and silt were more than 24.6% and 63.9% for 3 farmlands. Therefore these kinds of soil textures

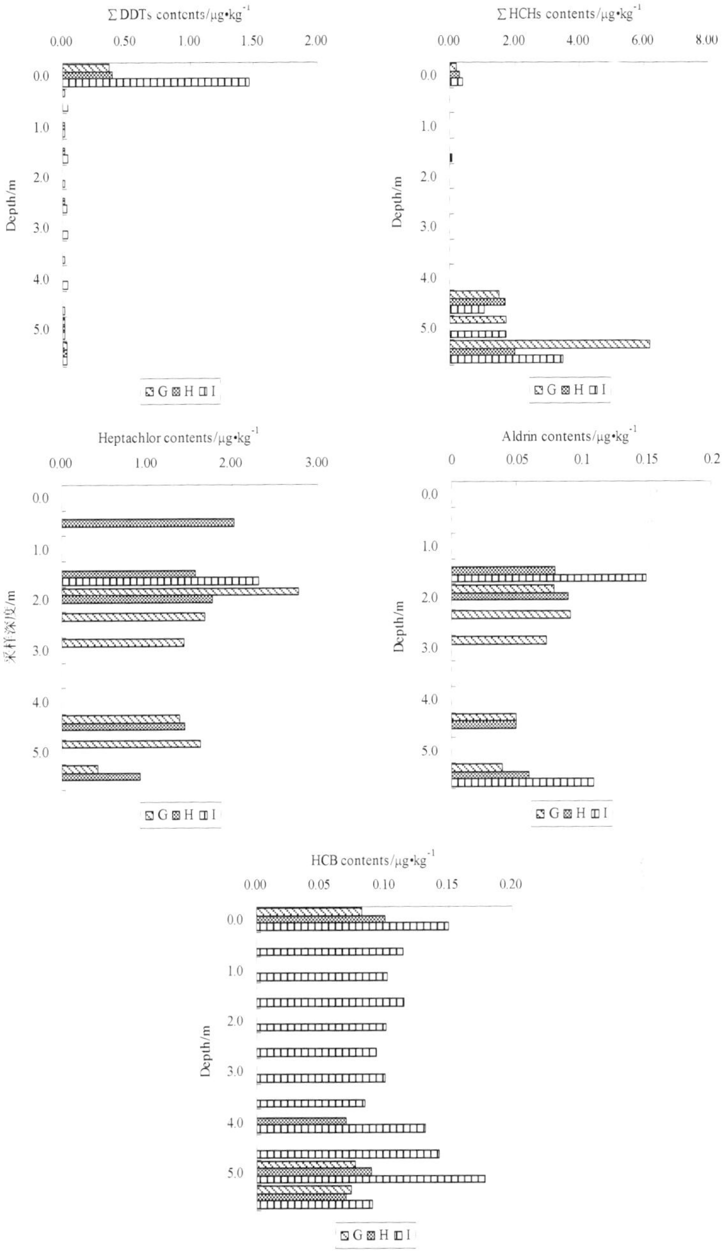

Fig. 16. Vertical distribution of single OCP in GIF

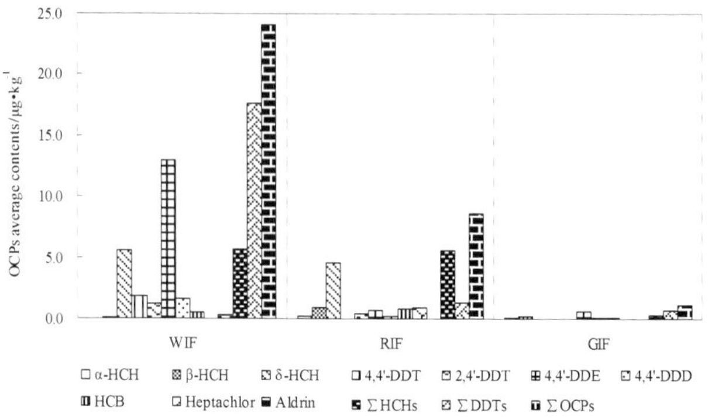

Fig. 17. Comparison of average OCPs contents in surface soil in 3 farmlands

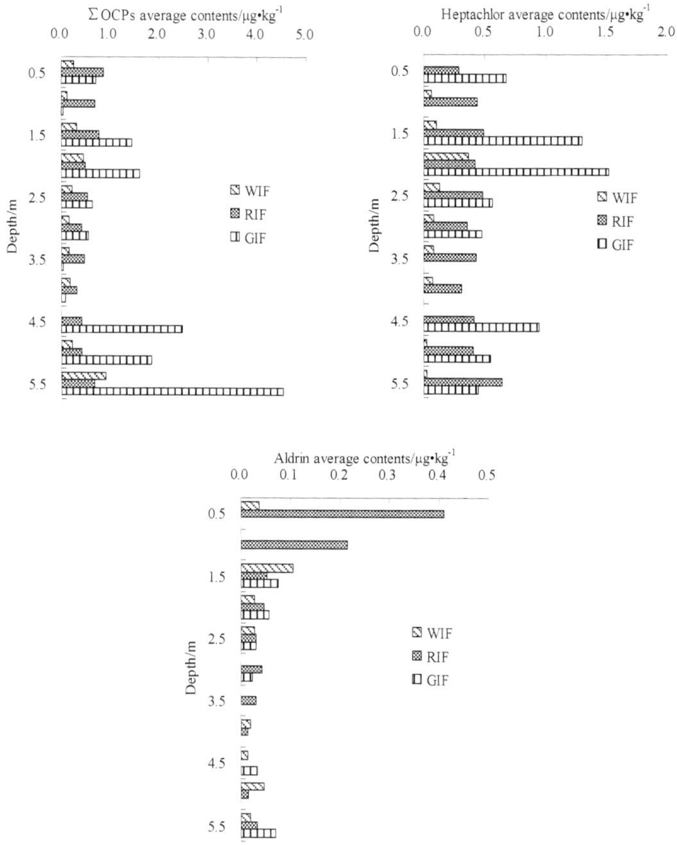

Fig. 18. Comparison of average OCPs contents in sub-layers in 3 farmlands

have strong ability to absorb the OCPs and prevent them from vertical migration. If we must give out the different of the 3 farmlands, we can say, pollutants were a little easier to migrate in RIF and GIF than in WIF, and the vertical changes of soil texture in WIF and RIF were larger than GIF.

On the whole, although none of the detected OCPs' contents in 3 farmlands was over the first class soil standard (<50 μg/kg) of Chinese soil environmental quality standard (GB15618-1995), OCPs had the ability of downward migration in profiles under the long-term irrigation both for wastewater and reclaimed water, and in most conditions, the main accumulation layer of OCPs was the surface soils.

5. OCPs in water samples of three farmlands

5.1 OCPs in water samples of WIF

5.1.1 OCPs in surface water

Table 4 was the concentration of OCPs detected in surface water of WIF. Only 5 species of OCPs including α-HCH, β-HCH, γ-HCH, Heptachlor, and Aldrin were detected in the irrigation water. The detected OCPs also were found in profiles. It's confirmed that irrigation water was a kind of OCPs pollution source, but the contribution degree still under discussion. In addition, DDTs was detected in the soil profiles, but not detected in surface water, indicating there were other sources of pollution, may be related to pesticide application. Other reason for that was pretreatment procedure of water samples before analysis. Since the water samples must be filtered by APFF glass fiber membrane to remove the suspend particles, the OCPs absorbed in the particles were also removed. This part OCPs were not count in the water analysis results. They would go into farmlands accompany with irrigation.

	α-HCH	β-HCH	γ-HCH	Heptachlor	Aldrin
WS-1	<0.60	1.47	<0.65	2.50	2.09
WS-2	0.85	3.17	0.82	1.90	1.81
WS-3	0.85	2.76	<0.65	2.06	2.79

Ps: WS means surface sample of wastewater irrigation farmland

Table 4. Concentration of OCPs in surface water of WIF (ng/L)

5.1.2 OCPs in groundwater

Table 5 shows the results of groundwater detection. β-HCH, Heptachlor, Aldrin and Dieldrin were detected in groundwater. These OCPs also detected in irrigation water and soil profiles except Dieldrin. Since Dieldrin was detected only once in one well, the detection had contingency. The consistency of detected species of OCPs supposed that wastewater irrigation was one of OCPs' pollution sources. These OCPs detected in groundwater were supposed to have relatively higher mobility. In contrast, other OCPs such as DDTs, which had high detection frequency in profiles but undetected in groundwater, were poor in vertical migration and easy to be trapped by soils. Because DDTs were a group of compounds with large molecular, and had strong fat-soluble and poor water-soluble chemical properties.

	β-HCH	Heptachlor	Dieldrin	Aldrin
WG-1	6.08	49.44	< 0.50	1.94
WG-2	2.57	< 0.65	< 0.50	0.96
WG-3	1.27	< 0.65	1.16	2.48
WG-4	1.97	< 0.65	< 0.50	< 0.40
WG-5	2.82	38.96	< 0.50	1.55
WG-6	2.22	< 0.65	< 0.50	0.91
WG-7	2.27	< 0.65	< 0.50	< 0.40

Ps: WG means groundwater of wastewater irrigation farmland

Table 5. Concentrations of OCPs in groundwater of WIF (ng/L)

5.2 OCPs in water samples of RIF

5.2.1 OCPs in surface water

The results of OCPs detected in irrigation water of RIF were listed in table 6. β-HCH, Heptachlor and Aldrin were detected in irrigation water. These three OCPs were also detected in soil profiles. In addition, DDTs, HCB which detected in soil profiles were not detected in surface water. This phenomenon was the same to WIF.

	β-HCH	Heptachlor	Aldrin
RS-1	0.54	1.87	3.40
RS-2	1.41	1.78	5.87
RS-3	1.26	1.80	2.42
RS-4	1.25	1.94	1.96

Ps: RS means surface sample of RIF

Table 6. Concentrations of OCPs in surface water of RIF (ng/L)

5.2.2 OCPs in sewage treatment plants effluent

Reclaimed water of 4 sewage treatment plants were detected several times in this study.
As shown in figure 19, ∑DDTs were detected in all plants with relative high concentration. It was the typical OCPs in wastewater and reclaimed water. The highest concentration of DDTs was appeared in the June from Gaobeidian secondary effluent, which was 97.8 ng/L. The main composition of DDTs was 4,4'-DDE, which accounted for at least 84% of DDTs. 2,4'-DDT was also detected occasionally, but the concentration was very low. Though Heptachlor was only detected in May, its concentration were much higher than other OCPs except in Gaobeidian secondary effluent. ∑HCHs were detected in all plants too, but the concentration were not as high as ∑DDTs, the highest concentration of HCHs was 20.2 ng/L which appeared at Xiaohongmen secondary effluent in May, and 4 kinds of isomers, α-HCH, β-HCH, γ-HCH, δ-HCH were detected occasionally. Despite of the low concentration, HCHs were considered as the typical OCPs of wastewater and reclaimed water for their high detection frequency. Like HCHs, HCB and Aldrin were detected in all plants too, and concentrations were not high. Since they were found in 3 farmlands, the irrigation water should not be excluded from the OCPs sources. Heptachlor epoxide and Dieldrin were only

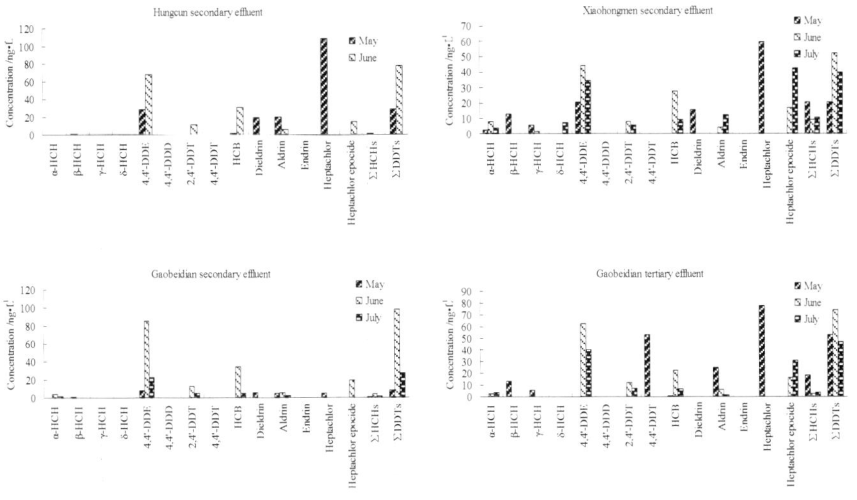

Fig. 19. Concentrations of OCPs in sewage treatment plants

detected occasionally, and were not detected in the 3 farmlands. Endrin was not found both in sewage treatment plants and farmlands.

Through comparing the irrigation water and effluent of sewage treatment plants, we could find that the OCPs detected in irrigation water were all detected in effluent of four plants, but the concentrations were lower than in sewage treatment plants' effluent. Pollutants could be removed effectively through sedimentation and degradation, during the long distance transportation in river channels. And these confirmed that wastewater and reclaimed water were the OCPs pollution sources for farmlands in some degree.

5.2.3 OCPs in groundwater

The concentrations of OCPs in groundwater of RIF were listed in table 7.

	α-HCH	β-HCH	Heptachlor	Aldrin
RG-1	<0.60	2.71	157.45	2.25
RG-2	2.60	9.63	24.79	1.63
RG-3	<0.60	<0.40	<0.65	<0.40
RG-4	1.72	12.90	<0.65	<0.40
RG-5	<0.60	<0.40	22.20	<0.40
RG-6	2.09	11.32	22.80	2.02

Ps: RG means groundwater of RIF

Table 7. Concentrations of OCPs in groundwater of RIF (ng/L)

α-HCH, β-HCH, Heptachlor and Aldrin were detected in the groundwater, the other substances were not detected. The difference between soil profiles and groundwater was that DDTs were not detected in groundwater samples, while a higher detection frequency in

HCHs, this meant that HCHs had stronger vertical migration than DDTs. Compared the test results of irrigation water, groundwater and effluent from sewage treatment, common detected pollutants were β-HCH, Heptachlor, and Aldrin.

5.3 OCPs in water samples of GIF

6 wells near the boreholes were selected to gather groundwater samples, but none of 14 kinds of OCPs were detected in them in the GIF. This supported that the OCPs pollution of soil in this farmlands had no relationship with groundwater irrigation. The detection in soil mainly came from pesticide application.

6. The relationship between soil properties and OCPs

The physical and chemical properties of OCPs in soil are mainly controlled by soil characteristics (Cheng H H et al., 1990), so it is important to study the vertical variety of soil properties in profiles. The solubility of OCPs in water is typically under 1 $mg \cdot L^{-1}$, while the distribution coefficient $\log K_{OC}$ is usually over 3, which shows that OCPs have a high fat-soluble and low water solubility, so the TOC will be an important factor in controlling the migration of OCPs. Furthermore, OCPs are mostly weak polar material, according to the rule of similarity, water with strong polar can play a significant inhibitory effect, thus, soil moisture is also the main factor in impacting the migration of OCPs. Montmorillonite is the most content material in soil of various components of clay mineral, it has the feature of double-crystal structure and large CEC, which can enhance the capacity to adsorb organic or inorganic ions. In order to quantitative descriptive the effects of soil properties on the concentration of OCPs, multiple regression analysis was taken in three farmlands, total content of OCPs was the dependent variable and soil properties were the independent variables in stepwise method. The results were in the table 8, 9, 10.

Model Summary

Model	R	R Square	Adjusted R Square	Std. Error of the Estimate
1[a]	0.423	0.179	0.155	6.28979
2[b]	0.693	0.480	0.448	5.08123
3[c]	0.779	0.606	0.569	4.49014

a. Predictors: (Constant), Soil moisture
b. Predictors: (Constant), Soil moisture, TOC
c. Predictors: (Constant), Soil moisture, TOC, EC

Table 8. Multiple regression analysis in WIF

The results in table 8 were used the stepwise removing, we can see the coefficient of determination in model 3 was 0.606 which was relatively high, while the TOC, soil moisture and EC were the independent variables. The regression equation was $y=30.541+19.428\,x1-0.665x2-0.954x3$, where y was the total OCPs, x1 was the TOC, x2 was the soil moisture content, x3 was the EC. These showed that TOC, soil moisture and soluble salt of soil were the impact factors of OCPs in WIF.

In table 9, we can see the coefficient of determination in model 3 was 0.575. The TOC, soil moisture and Eh were the independent variables. The regression equation was y=9.334+4.812x1-0.196x2-0.012x3, where y was the total OCPs, x1 was the TOC, x2 was the soil moisture content, x3 was the Eh. Showed that TOC and soil moisture of soil were the common impact factors in wastewater and RIF.

Model Summary

Model	R	R Square	Adjusted R Square	Std. Error of the Estimate
1[a]	0.540	0.292	0.271	2.00184
2[b]	0.717	0.514	0.484	1.68367
3[C]	0.759	0.575	0.536	1.59737

a. Predictors: (Constant), TOC
b. Predictors: (Constant), TOC, Soil moisture
c .Predictors: (Constant), TOC, Soil moisture, Eh

Table 9. Multiple regression analysis in RIF

Model Summary

Model	R	R Square	Adjusted R Square	Std. Error of the Estimate
1	0.583	0.340	0.321	1.27109

Predictors: (Constant), Eh

Table 10. Multiple regression analysis in GIF

In table 10, the coefficient of determination in model 1 was 0.340, while the Eh was the independent variable and the regression equation was y=4.521-0.008x, where y was the total OCPs, x was the Eh. These showed that Eh was the common factor in reclaimed water and GIF.

7. Source analyses of OCPs in three farmlands

Sources of pesticide in soil are mainly the following aspects: using for preventing pest; the input of irrigation water and the deposition of atmospheric particles. In recent years, due to rapid growth of food demand, pesticide application in most cases are excessive. Besides, solid wastes are piled up and dumped to the soil surface continually, thus, hazardous wastewater is continued to infiltrate into the soil, more and more hazardous gases and particulates landed into the soil with rain. When the content of harmful substances in soil exceed the soil's self-purification ability, the composition, structure and function of soil will be changed, microbial activity will be inhibited, and harmful substances or its decomposition products will be accumulated in soil gradually, finally absorbed by human body, when the extent great enough to threaten human health, the soil pollution is formed.

Commercial HCHs is a mixture of several closely-related compounds. The common components were 55%-80% of α-HCH, 5%-14% of β-HCH, 12%-14% of γ-HCH, 2%-10% of δ-HCH and 3%-5% other organochlorine pesticides. The dechlorination rate sequence in HCHs is: α-HCH> γ-HCH>δ-HCH> β-HCH, in which the α-HCH is the most unstable and has the fastest degradation rate; γ-HCH in agricultural soils can be easily changed into other HCHs by decomposition or biotransformation; β-HCH is a stable HCHs with the lowest solubility and difficult to evaporate(Chen L G et al., 2005; Rekha P N et al., 2004). In the three farmlands, δ-HCH had the highest detected concentration in all soil profiles and the γ-HCH had the lowest, one reason was the migration from surface soil where had the high concentration of δ-HCH, the other reason was the high solubility of δ-HCH(Prakash O et al., 2004), the undetected γ-HCH in profiles showed that there was no new input in recently, so the HCHs residues in soil was a legacy problem which caused by history wastewater irrigation and pesticide application.

Commercial DDTs is a mixture too. The major components include 4,4'-DDT (75%), 2,4'-DDT (15%), 4,4'-DDE (5%), 4,4'-DDD (<5%) and other substances, DDT in anaerobic condition can be degraded to DDD compounds, while degraded to DDE under aerobic conditions, the property of DDE is more stable and its degradation rate is proportional with the increasing of soil moisture, temperature and microbial activity(Hitch R K et al., 1992). In three farmlands, the content of 4,4'-DDE was the highest in DDTs in soil profiles, 4,4'-DDD was the lowest, because the 4,4'-DDE was harder to decompose than 4,4 ' -DDT and 4,4'-DDD, 4,4'-DDE had the highest percentage in DDTs indicating that most residual DDT in soil was the degradation products, which consistent with the degradation in environment after DDTs been banned, also consistent with the Eh test results (soil in three farmlands were the oxidation state). DDT in soil environment undergoes a long period of physical, chemical and biological change, the (DDE + DDD) / DDT ratio should be greater than 1, if the ratio is less than 1, indicating that new sources may input (Qiu X et al., 2004; Jaga K et al., 2003). The ratio in WIF was 5.33 and 2.25 in RIF, DDT was not detected in GIF, all these values were greater than 1, indicating that no new sources input recently.

8. Conclusions

From the analyses and studies of irrigation water, pollutants contents in soil profiles and the physical-chemical properties of soils, combined with statistical analysis software to analyze the relationship between contents of OCPs and the physical-chemical parameters of soil, conclusions are as follows:

The variations of soil physical-chemical indexes with depth in three farmlands were nearly consistent. The main soil particles were silt, followed by clay, and sand. Lithology changes in WIF and RIF were larger than in GIF. The proportion of silt and clay in RIF were less than in WIF, while the sand was greater than in WIF.

Surface soil was the accumulated layer of OCPs in WIF and RIF, both ∑OCPs and single OCP overall decreased with depth, but there was no this phenomenon in GIF. The average contents of ∑OCPs in surface soils of 3 farmlands were 24.11 μg/kg, 8.64 μg/kg, and 1.15μg/kg respectively. Clearly the sequence of total OCPs in soils was: WIF> RIF> GIF, which supposed that the content of OCPs had a great relationship with the quality of irrigation water.

OCPs detected in three farmlands were basically the same, and surface water and groundwater were also in good agreement. DDTs and HCHs were the main OCPs in surface soils which accounted for at least 80% of ∑OCPs. However, Heptachlor and Aldrin were common OCPs under the surface in profiles. This phenomenon supposed that Heptachlor and Aldrin were easier to migrate than DDTs and HCHs in soils profiles, and these two substances were a major threat to quality of groundwater.

DDTs and HCHs were not detected in irrigation water and groundwater, maybe caused by water sample pretreatment of filtration. OCPs detected in effluent of sewage plants confirmed that DDTs and HCHs were the main OCPs in wastewater and reclaimed water. But most OCPs could be removed effectively through sedimentation and degradation, during the long distance transportation in river channels.

Heptachlor and Aldrin were detected both in irrigation water and groundwater except groundwater of GIF. That means except for pesticides application, wastewater and reclaimed water were other important sources for OCPs in farmlands.

Using SPSS to analyze the soil properties and total OCPs, the results were that, TOC and soil moisture were the main impact factors of vertical distributions of OCPs in WIF and RIF, while Eh was the important factor in GIF.

Source analyses supposed that, there were no new application of DDTs and HCHs in 3 farmlands. The accumulations of DDTs and HCHs in surface soils were due to historical pesticides application and wastewater irrigation.

9. References

Angelakis AN, Marecos Do Monte MHF, Bontoux L,et al, J. (1993). The status of wastewater reuse practice in the Mediterranean basin: need for guidelines. *Water Resource*, 33., 10., 2201–2217, 0097-8078

CampDresser & McKee Inc. for the US Environmental Protection Agency, Washington, DC.(2004). Guidelines for Water Reuse. EPA/625/R-04/108.US EPA.

Chen L G, Ran Y, Xing B S, J. (2005). Contents and sources of polycyclic aromatic hydrocarbons and organochlorine pesticides in vegetable soils of Guangzhou, China. *Chemosphere*, 60., 7., 879-890, 0045-6535

Chen J, Wang X J, Tao S, J. (2005). The Influences of Soil Total Organic Carbon and Clay Contents on PAHs Vertical Distributions in Soils in Tianjin Area. *Research of Environmental Sciences*, 18., 4., 79-83, 1001-6929

Cheng H H, M. (1990). *Pesticides in the soil environment: processes, impacts, and modeling*. Soil Science Society of America, Inc, ISBN 0-89118-791-X, Madison, Wisconsin, USA

D.Levine and T.Asano, J. (2004). Recovering sustainable water from waste water. *Environ.Sci.Technol*, 38., 11., 201A-208A, 0013-936X

Francisco Pedrero, Juan Jose Alarcon, J.(2009). Effects of treated wastewater irrigation on lemon trees. *Desalination*, 246., 1-3., 631-639, 0011-9164

Friedler, J.(2004). Equality of individual domestic greywater streams and its implication for no-site treatment and reuse possibilities. *Environmental Technology*, 25., 9., 997-1008, 0959-3330

Gross A,Kaplan D,Baker K, J.(2007). Removal of chemical and microbiological contaminants from domestic greywater using a recycled vertical flow bioreactor(RVFB). Ecological Engineering, 31., 2., 107-114, 0925-8574

He Y W, Qu Z Q, Sun T H, J.(1995). Adsorption of polycyclic aromatic hydrocarbons on soils. *Chinese Journal of Applied Ecology*, 6., 4., 423-427, 1001-9332

Hitch R K, Day H R, J.(1992). Unusual persistence of DDT in some western USA soils. *Bulletin of Environmental Contamination and Toxicology*, 48., 2., 259-264, 0007-4861

Hua X M, Shan Z J, J.(1996). The production and application of pesticides and factor analysis of their pollution in environment in china., *Techniques and Equipment for Environmental Pollution Control*, 4., 2., 33-34, 1008-9241

I.K.Kalavrouziotis, P.Robolas, P.H. Koukoulakis et al, J.(2008). Effects of municipal reclaimed wastewater on the macro-and micro-elements status of soil and of Brassica oleracea var.Italica,and B.oleracea var.Gemmifera. *Agricultural Water Management*, 95., 4., 419-426, 0378-3774

Jaga K,Dharmani C, J.(2003). Global surveillance of DDT and DDE levels in human tissues. *International Journal of Occupational Medicine and Environmental Health*, 16., 1., 7-20, 1232-1087

Katsoyiannis A.,Salnara C, J.(2005). Persistent organic Pollutants(POPs) in the conventional activated sludge treatment Process: fate and mass balance. *Environmental Research*, 97., 3., 245-257, 0013-9351

Li X H, Ni J Z, Luo Y M, J.(2006). Sorption and Desorption of Phenanthrene by Soil Different Particle-Size Separates. *Soils*, 38., 5., 584-590, 0253-9829

Liu C, Chen J W, Yang Z F, et al, J.(2008). The investigation and evaluation of HCH residues in soils of Beijing suburbs. *Geophysical and Geochemical Exploration*, 32., 5., 567-570, 1000-8918

Ping L F, Lu Y M, J.(2005). Effects of organic matter on environmental behaviors of polycyclic aromatic hydrocarbons. Soils, 37., 4., 362-369, 0253-9829

Liu P B, J.(2007). Development of recycled water and measures in the city of Beijing. *China Water Resources*, 6., 37-39, 1000-1123

Prakash O, Suar M, Raina V,et al, J.(2004). Residues of hexachlorocyclohexane isomers in soil and water samples from Delhi and adjoining areas. *Current Science*, 87., 1., 73-77, 0011-3891

Qiu X,Zhu T,Li J,et al, J.(2004). Organochlorine pesticides in the air around the Taihu Lake, China. *Environ.Sci.Technol*, 38., 5., 1368-1374, 0013-936X

Rekha P N, Ambujam N K, Krishnani K K, et al, J.(2004). Groundwater quality in paper mill effluent irrigated area with special reference to organochlorine residues and heavy metals. *Bulletin of Environmental Contamination and Toxicology*, 72., 2., 312-318, 0007-4861

T.Asano and D.Levine, J.(1996). Wastewater reclamation, recycling and reuse: past, present and future. *Water Science and Technology*, 33., 10-11., 1-14, 0273-1223

Tan Z, Li C H, Yu L W, et al, J.(2008). Vertical distribution of organochlorine pesticide residue along soil profile in agricultural soils of Huizhou. *Chinese Journal of Eco-Agriculture*, 16., 5., 1134-1137, 1671-3990

U. Pinto, B.L. Maheshwari, H.S. Grewal, J.(2010). Effects of greywater irrigation on plant growth, water use and soil properties. *Resources, Conservation and Recycling*, 54., 7., 429–435, 0921-3449

V.Reboll,M.Cerezo,A.Roig ,et al, J.(2000). Influence of wastewater vs. groundwater on young citrus trees. *Science of Food and Agriculture*, 80., 10., 1441-1446, 0022-5142

Zhang H Y,Gao R T, Jiang S R, et al, J.(2006). Spatial Variability of Organochlorine Pesticides (DDTs and HCHs) in Surface Soils of Farmland in Beijing, China. *Scientia Agricultura Sinica*, 39., 7., 1403-1410, 0578-1752

Efficacy of Management Practices to Mitigate the Off-Site Movement and Ecological Risk of Pesticides Transported with Runoff from Agricultural and Turf Systems

Pamela J. Rice[1], Brian P. Horgan[2],
Cathleen J. Hapeman[1] and Laura L. McConnell[1]
[1]*United States Department of Agriculture, Agricultural Research Service*
[2]*University of Minnesota, Department of Horticulture*
United States of America

1. Introduction

Pest management in both agricultural and non-agricultural settings uses established practices and new technologies to control harmful and nuisance pests. Pesticides are an important tool in integrated pest management. Without pesticides a significant percentage of food and fiber crops would be lost, infectious diseases would increase, and valuable native habitats would be devastated by invasive species. In 2000 and 2001, annual world pesticide usage exceeded 2.3 billion kilograms of active ingredient (Kiely et al., 2004). The application of pesticides to targeted areas inevitably results in the transport of a portion of these chemicals and their degradation products to surrounding non-target areas. Pesticides are biologically active compounds designed to interfere with metabolic processes (Matsumura, 1985; Manahan, 1994). The detection of pesticide in soil, water and air (Hoffman et al., 2000; Goel et al., 2005; Harman-Fetcho et al., 2005; Loague et al, 2006; Lv et al., 2010; Riederer et al., 2010; Weber et al., 2010; Hayward et al., 2010), and reported adverse effects of pesticides to non-target organisms at environmentally relevant levels (Chandler et al., 1991; Clark et al., 1993; Margni et al., 2002; Schulz, 2004) has invoked public concern.

The off-site transport of pesticides and soil with runoff from agriculture is believed to be a large contributor to water quality degradation. It is estimated that 1 to 6% of soil-applied pesticides may be lost to aquatic environments in runoff and drainage from agricultural fields (Wauchope, 1978; Bengston et al., 1990). A number of pesticides have been detected in surface waters of agricultural watersheds (Baker and Richards, 1990; Thurman et al., 1992; Goolsby & Battaglin, 1993; Johnson et al., 1994) and research has demonstrated significant negative effects of pesticides on aquatic organisms and ecosystems resulting from nonpoint source agricultural runoff (Scott et al., 1990; Chandler and Scott, 1991; Savitz et al., 1994).

Tomatoes are one of the most economically important vegetables grown in the United States with average annual yields of 1.6 billion kilograms and 8.9 billion kilograms for fresh market and processing tomatoes, respectively, at an estimated value of nearly $2 billion (Davis et al., 1998). A cultivation method widely accepted by vegetable growers is the use of polyethylene mulch where a thin sheet of plastic is placed over a raised bed with bare-soil furrows between the beds. In 1999, polyethylene mulch was used on over 12 million hectares of land throughout the world; representing 450 thousand hectares in Europe, 9.7 million hectares in Asia, 200 thousand hectares in the Americas and 80 thousand hectares in Africa and the Middle East (Takakura & Fang, 2001). Polyethylene mulch is a preferred practice because it can warm the soil and control weeds; however, nearly 50 to 75% of the field is covered with an impervious surface which enhances runoff due to reduced water infiltration. Research has shown greater runoff volumes and soil erosion associated with polyethylene mulch relative to bare soil (McCall et al., 1988; Wan & El-Swaify, 1999). In addition, compared with conventional production agriculture, polyethylene mulch systems have an additional surface for pesticide adsorption/desorption which may enhance or impede chemical runoff and degradation rates (Vuik et al., 1990; Topp & Smith, 1992; Nerin et al., 1996).

Hairy vetch (*Vicia villosa* Roth) mulch, a vegetative cover crop, has been shown to be a profitable management practice for fresh market tomato production resulting in greater tomato yields and lower production costs than polyethylene mulch or bare soil (Kelly et al., 1995). Cover crops including hairy vetch have been shown to suppress weeds (Teasdale, 1996), act as a slow-release fertilizer (Ranells and Wagger, 1996), reduce soil erosion, increase soil organic matter, improve soil tilth (Smith et al., 1987), and increase soil infiltration and soil moisture (McVay et al., 1989). Field and laboratory studies have demonstrated that crop residues and vegetative mulches can reduce runoff, soil erosion, and the off-site transport of pesticides and nutrients from agricultural fields (Ghadiri et al., 1984; Dao, 1991; Sur et al., 1992; Zuzel and Pikul, 1993).

Approximately 32% of pesticide use in the United States results from non-agricultural pest control; including applications to protect structures, maintain lawns and landscapes, control weeds at roadsides and right-of-ways, maintain recreational areas and gardens, and repel and control nuisance and disease carrying pests. In 2000 and 2001, 19% of pesticide usage in the United States occurred in the home and garden sectors while 13% was accounted for by industrial, commercial and governmental sectors (Kiely et al., 2004).

More than 16 million hectares of land in the United States is estimated to be covered by tended lawn (Milesi et al., 2005). Managed turf is found in both private and public settings; as residential, commercial and public lawns, on golf courses and athletic fields, as sod farms, and in parks and cemeteries. In the late 1990s the urban/suburban pesticide market was estimated to be a billion dollars annually with pest control of turfgrass (e.g. professional lawn care, landscapers, nurseries, golf courses, sod farms and institutions) representing a significant portion of that market; approximately $500-$700 million (Joyce, 1998; Racke, 2000; Clark and Kenna, 2001). Golf courses contain some of the most intensely managed turf, which often requires multiple applications of pesticides at rates that may exceed those typically found in agricultural or home environments (Barbash and Resek, 1996; Gianessi and Anderson, 1996). Pesticides associated with the turfgrass industry have been detected in surface waters of urban watersheds (Cohen et al., 1999; Gilliom et al., 2006). Examples include spring and summer detections of carbaryl and diazinon at levels that exceeded criteria for protection of aquatic life (Hoffman et al., 2000), reports of dimethylamine salt of 2,4-dichlorophenoxyacetic acid (2,4-D), dicamba, and mecoprop in 85% of evaluated storm runoff events (Wotzka et al., 1994), and evidence of chlorpyrifos, diazinon and 2,4-D in surface waters throughout the year (Frick et al., 1998).

Fairways comprise approximately one-third of the managed turf of a typcial golf course (Watson et al., 1992; Lyman et al., 2007), which may be adjacent to surface waters such as streams, ponds and lakes. Golf course fairways and greens are often managed with core cultivation during the spring or fall to control thatch, stimulate root and shoot growth, alleviate surface compaction, and enhance water infiltration (Beard, 1973; White & Dickens, 1984; Turgeon, 1985; Carrow et al., 1987; Dunn et al., 1995, Callahan et al., 1998). Solid tine core cultivation requires a reduced amount of labor and is less disruptive to the surface of the turf but is believed to cause localized compaction. Cultivation with hollow tines typically involves removing cores from the turf, which are air-dried and brushed back into the open holes (Murphy et al., 1992). Despite the widespread use of core cultivation, little is known about the quantity of pesticides transported in runoff from turf managed with either solid tine or hollow tine core cultivation practices.

Evaluation of established and emerging management practices is important in order to understand their efficacy and sustainability. As benefits and improvements in management strategies are discovered they can be implemented; while practices with unexpected adverse consequences can be modified or replaced. Here we present research measuring the quantity of pesticides transported with runoff from agricultural (fresh market tomato production) and non-agricultural (golf course fairway turf) systems; to evaluate the influence of management practices to reduce the off-site transport of pesticides with runoff. Following quantification of pesticide mass transport, real-world runoff-to-surface water scenarios were used to extrapolate edge-of-plot runoff data to estimated environmental concentrations of pesticides in surface waters receiving the runoff. A comparison of surface water pesticide concentrations with published toxicity data assessed the ability of management practices to reduce ecological risk of pesticides from the evaluated agricultural and non-agricultural systems.

2. Materials and methods

2.1 Site description

Agricultural system: Runoff water was collected from tomato (*Lycopersicon esculentum* Mill) plots located at the Henry A. Wallace Beltsville Agricultural Research Center, Beltsville, Maryland, USA. The 2500-m^2 field is comprised of Mattapex silt loam (fine-silty, mixed, mesic Aquic Hapludults with 1.3 to 1.6% organic carbon content) with a 5 to 7% slope. The site was divided into sixteen plots, each with four raised beds (15 cm high, 27 m long, 0.9 m wide, and 1.5 m center-to-center), prepared in a north-south direction. A randomized complete block design was used to assign eight plots to tomato production and the remaining plots were planted with sweet corn (*Zea mays*). Tomato and corn plots were rotated annually to reduce pest pressure. Earthen berms were constructed around each plot to prevent water movement between the plots and to capture runoff only from the three central rows within each four-bed tomato plot.

Turf system: Experiments were conducted on 976 m^2 site located at the University of Minnesota Turf Research, Outreach and Education Center, Saint Paul, MN, USA. The soil was characterizes as Waukegan silt loam (sandy-skeletal, mixed superactive, mesic Typic Hapludolls) containing 55% silt, 29% sand, 16% clay and 3% organic carbon with a natural slope running east to west that was graded to 4% with less than 1% slope from north to south. The study area was sodded with L-93 creeping bentgrass (*Agrostis palustris* Huds.)

and divided into plots (24.4 m x 6.1 m, length x width), prepared in an east to west direction, 14 months prior to initiation of the reported studies. Prior to simulated precipitation events, plots were hydrologically isolated with removable berms, constructed from horizontally-split 10.2-cm schedule 40 PVC pipe, inverted to rest on the cut edges. Observations during runoff events showed no water movement under the PVC berms.

2.2 Management practices (treatments)

Agricultural system: During the month of September raised vegetable beds were constructed in plots assigned to the vegetative mulch treatment and planted with hairy vetch (*V. villosa*) seed on the beds and in the furrows between the raised beds. In early May the hairy vetch was finely chopped using a flail-mower to provide a vegetative mulch residue over the soil surface. By mid May, beds for the polyethylene treatments were constructed and drip irrigation lines were installed 8 to 10 cm from the plant row prior to the installation of the polyethylene. Furrows between raised beds covered with polyethylene remained as bare soil. During the last week of May, 'Sunbeam' tomato plant seedlings were transplanted in the center of each bed with a no-tillage planter to minimize the disruption of the mulches (polyethylene or hairy vetch residues). Immediately after transplanting, drip lines were added to the surface of the hairy vetch mulch beds approximately 8 to 10 cm from the plants. Tomato plants grown in the polyethylene and hairy vetch mulch received equal quantities of water through the irrigation system to maintain the plants during dry conditions. The quantity of water applied with the trickle irrigation system was not enough to produce surface runoff. Additional information on cultivation of hairy vetch cover crops and installation of polyethylene or hairy vetch mulch is given elsewhere (Abdul-Baki et al., 1996; Abdul-Baki et al., 1997; Rice et al., 2001).

Turf system: Creeping bentgrass turf was managed as a fairway with 1.25 cm height of cut (3 times weekly, clippings removed), topdressed with sand (weekly, 1.6 mm depth) and irrigated to prevent drought stress. The quantity of water applied with the maintenance irrigation was not enough to produce surface runoff. Core cultivation with either solid tines (0.95 cm diameter x 11.43 cm length) or hollow tines (0.95 cm internal diameter x 11.43 cm length) was performed twice (June 21st, Sept 28th) (Rice et al., 2010b). Cores removed with the HT were allowed to dry, broken into smaller pieces, and worked back into the turf. A back-pack blower and leaf rake removed the turf and thatch from the plot surface. Sand topdressing was not performed immediately after core cultivation or within a week of simulated precipitation and generation of runoff.

2.3 Pesticides

Agricultural system: Pesticides monitored in the tomato production experiment and reported in the present publication were as follows: Bravo® 720 fungicide (ISK Biosciences, Mentor, OH) containing 40.4% chlorothalonil (tetrachloroisophthalonitrile); Thiodan® 50 WP insecticide (FMC, Philadelphia, PA) containing 50% endosulfan (hexachlorohexahydromethano-2,4,3-benzodioxathiepin 3-oxide) and Asana® XL insecticide (Du Pont, Wilmington, DE) containing 8.4% esfenvalerate ((s)-cyano(3-phenoxyphenol)methyl(s)-4-chloro-alpha-(1-methylethyl) benzeneacetate). Details on the applied herbicide (metribuzin) and inorganic fungicide (copper hydroxide), can be found in previous publications (Rice et al., 2001; Rice et al., 2002). Properties of the active ingredients for pesticides reported in this manuscript are provided in Table 1.

	Water Solubility (20°C)	K_{OC}[b]	K_{OW}[c] (pH 7, 20°C)	Half Life (d)		
Pesticide[a]	(mg/L)	(ml/g)	(Log P)	Soil	Water-Sediment	Water
Chlorothalonil	0.81	850	2.94	22	0.1	0.1
Chlorpyrifos	1.05	8151	4.70	50	36.5	5
Endosulfan	0.32	11,500	4.75	50	---[d]	---
Esfenvalerate	0.001	5300	6.24	44	71	30
Flutolanil	8.01	735	3.17	233	320	90.5

[a]http://sitem.herts.ac.uk/aeru/footprint/en/index.htm.

[b]Soil organic carbon partition coefficient.

[c]Octanol-water partition coefficient.

[d]--- = no data.

Table. 1. Pesticide Properties.

Turf system: Commercially available pesticide products were tank mixed and applied at label rates to all plots perpendicular to runoff flow. The insecticide Dursban® 50W (Dow AgroSciences LLC, Indianapolis, IN) containing 50% chlorpyrifos (*O,O*-diethyl *O*-(3,5,6-trichloro-2-pyridinyl) phosphorothioate) and fungicide ProStar® 70WP (Chipco® Professional Products, Aventis CropScience) containing 70% flutolanil (*N*-[3-(1-methylethoxy) phenyl]-2-(trifluoromethyl) benzamide) are reported in the present publication. Properties of their active ingredient are provided in Table 1. Details on the tank mixed herbicides (2,4-D, dicamba, mecoprop-p), application equipment and spray characteristics are reported elsewhere (Rice et al., 2010a,b).

2.4 Precipitation

Agricultural system: A tipping-bucket rain gauge was used to measure the time and intensity of each precipitation event. Runoff resulting from natural precipitation was collected during the growing season, May to September. On occasions when rainfall runoff had not been produced within a week of pesticide applications, an overhead sprinkler system was used to create precipitation. The sprinkler system was erected around the field and between the plots with sprinkler nozzles that were 1.8 m above ground and spaced every 12 m, in order to give an even application of water to all plots simultaneously. This system delivered from 7.5 to 14.3 mm/hr rain intensities to the plots, which is similar to the average natural rainfall intensity (~ 11 mm/hr) during the growing season. Artificial rain events represented only 9 of the 41 precipitation events evaluated.

Turf system: A rainfall simulator was constructed following the design of Coody and Lawrence (1994) (United States patent 5,279,151), which delivered precipitation with a droplet size spectrum, impact velocity, and spatial uniformity characteristic of natural rainfall. Risers were spaced 3.7 m apart with nozzles and spinners suspended 2.7 m above the turf. Simulated precipitation was initiated 26 ± 13 h after pesticide application when the wind speeds averaged 0.8 ± 0.7 mps (1.8 ± 1.6 mph). Rain gauges (Taylor Precision Products) were distributed throughout each plot to quantify simulated precipitation. Measured rainfall rates were 29 ±6 mm/h; similar to storm intensities recorded in Minnesota, USA, during July through October. The duration of the simulated precipitation was 2.0 ± 0.5 h, which was chosen to assure 90 min of runoff had been generated from each plot. Details on the materials and dimensions of the simulator are provided elsewhere (Rice et al. 2010a,b).

2.5 Runoff collection

Runoff water samples and flow data were collected from either fiber-glass H flumes (agricultural study) or stainless-steel trapezoidal flumes (turf study) outfitted with bubble-tube and sample-collection ports using automated runoff samplers (ISCO model 6700) equipped with flow meters (Isco model 730, Lincoln) (Rice et al., 2001; Rice et al., 2010a). Water samples were deposited into glass bottles, removed from the samplers and stored at -20 °C until laboratory processing and analysis. Up to 24 300-ml samples were collected from each runoff event.

2.6 Runoff processing and pesticide analysis

Water samples resulting from overland flow in the agricultural study contained soil particulates while those from the turf study were relatively clear. As a result, runoff from the agricultural study was characterized for both dissolved- and particulate-phase pesticides while runoff from the turf study was characterized for dissolved-phase pesticides.

Agricultural system: Due to the small sample volume and the large numbers of water samples collected in this study, composite samples from each plot for each runoff event were analyzed for dissolved-phase and particulate-phase pesticide concentrations. A detailed description of the processing and analytical methods are given elsewhere (Rice et al., 2001). Briefly, filtered water (0.7 µm) was extracted using a Varian SPME III autosampler (Varian, Palo Alto, CA) equipped with either a 100 µm polydimethylsiloxane or 85 µm polyacrylate fiber (Supelco, Bellefonte, PA). Analyses were carried out using a Hewlett Packard 5890 Series II gas chromatograph (Hewlett Packard, Santa Clarita, CA) with an electron capture detector. Chromatographic conditions for chlorothalonil and endosulfan analyses were as follows: J&W DB-5 column, 30 m x 0.25 mm i.d., film thickness of 25 µm (J&W Scientific, Folsom, CA), temperature program, injector temperature 270 °C, 150 °C, initial temperature, 2 min hold, 3 °C/min to 200 °C, 5 min hold, 10 °C/min to 260 °C, 1 min hold, detector temperature 270 °C. The limits of detection were 3500 ng/L for chlorothalonil, 9.5 ng/L for α-endosulfan and 13 ng /L for β-endosulfan. Esfenvalerate was extracted from 5-ml of runoff water using liquid-liquid extraction with 3, 5-ml aliquots of ethyl acetate. Organic extracts were combined and residual water was removed using a 2-g column of anhydrous $MgSO_4$ and concentrated to 1 ml with high purity N_2 gas and spiked with PCB #204 (60 ng) as an internal standard. Calibration standards were prepared in de-ionized water and extracted using the same method as the samples. De-ionized water control samples and blank water (5 ml) spiked with 400 ng Asana were extracted along with runoff samples. The limits of detection and extraction efficiencies for esfenvalerate were 540 ng/L, 96.6±13.5%. Analyses were carried out using the same chromatographic conditions described for chlorothalonil and endosulfan, with the exception of the temperature program which was as follows: 150°C initial temperature, 2 min hold, 10°C/min to 200°C, 3.5°C/min to 270°C, 0.40°C min to 275°C, 3.5 min hold.

In order to determine the particle-phase pesticide concentration, 50 ml from each sample was combined into an integrated sample from each plot for a total of 8 integrated samples from each rain event. Integrated samples were filtered through a glass fiber filter (Whatman GF/F, 0.7 µm nominal pore size) using a stainless steel filter holder. A quarter of each filter was extracted with 3:1 dichloromethane (DCM): acetone (chromatographic grade) for 6 h using a Soxhlet apparatus. Extracts were cleaned up using an LC-Alumina-N 2 g (Supelco, Bellefonte, PA) cartridge topped with 1 g anhydrous $MgSO_4$ to remove color, particle material, and water. An additional 15-ml of 1:1 DCM:acetone was passed through the clean up column and combined with the extract. Extracts were reduced using N_2 gas and exchanged into isooctane. Extraction efficiency of the method was evaluated by spiking

filter papers used to filter runoff water from untreated soil with the target analytes and dibutyl chlorendate as a sample specific extraction efficiency determination. Recoveries ranged from 85.3 ± 4.9% to 91.2 ± 10.1%. Blank filter papers were also extracted and analyzed with samples and no interfering peaks were found. Extracts were analyzed using the chromatographic conditions described for dissolved-phase esfenvalerate, with the exception of the temperature program which was as follows: 150°C initial temperature, 2 min hold, 10°C/min to 200°C, 1 min hold, 3.0°C/min to 260°C, 10 min hold, 7.0°C/min to 280°C, 10 min hold. The method detection limits for the target analytes were: chlorothalonil 1.0 μg/L, α-endosulfan 0.63 μg/L, β-endosulfan 2.2 μg/L and esfenvalerate 49 μg /L.
Turf system: Water samples were processed by filtering 3-ml through a 0.45 μm nylon syringe filter (Whatman) followed by methanol (0.5 ml) to rinse the filter. Irrigation source water, background runoff water, and background runoff spiked with known quantities of pesticides served as blank and positive control samples. Each runoff sample was analyzed for pesticides. No samples were combined. Methanol rinsates of Petri dishes, containing pesticide residues for determination of actual application rates, were diluted with laboratory-grade organic-free water to 14% methanol to mimic the methanol and water content of the filtered runoff samples. Concentrations of each pesticide were measured by direct injection (500 μl) onto a high performance liquid chromatograph (Waters model 717plus autosampler and model 1525 binary pump) with a photodiode array detector (Waters model 2996: Waters) set at 230nm. Analytes were eluted from an Agilent C-18 column (150 mm long, 4.6 mm diameter, 5 μm packing) using two solvents [solvent A: laboratory-grade organic-free water (0.17% trifluoroacetic acid); solvent B: 82:18 methanol:acetonitrile] at a rate of 1 ml/min. Initial conditions, 60% B, were held for 2 min followed by a gradient ramped from 60 to 95% B in 23 min, a 3 min hold, then back to 60% B in 10 min with a 5 min hold. Recoveries were: 74 ± 23% for chlorpyrifos and 91 ± 8% for flutolanil. Method detection limits ranged from 2.5 to 3.7 μg/L. Limits of quantification for the target analytes were: chlorpyrifos 5.3 ± 0.9 μg/L and flutolanil 4.5 ± 0.8 μg/L.

2.7 Pesticide loads with runoff

Agricultural system: Seasonal pesticide loads represent the sum of dissolved- and particulate-phases for all runoff events collected from May to September. Loads from individual runoff events were based on a composite sample made up of combined individual flow-weighted samples that were collected throughout the runoff event for each plot. Dissolved-phase pesticide loads (mg/m^2) were calculated from measured runoff volumes from each plot (L/m^2) and the concentration of pesticides in the filtered runoff water (mg/L). Particulate-phase pesticide loads (mg/m^2) were quantified using measured runoff volumes for each plot (L/m^2), mass of total suspended solids per unit volume of runoff (g/L) and the concentration of the pesticide extracted from the particulates (mg/g).
Turf system: Pesticide loads (mg/m2) for individual runoff events were calculated from the sum of time-weighted samples collected throughout the runoff event. In summary, the mass of a pesticide transported with runoff for each time point was calculated from the measured pesticide concentration (mg/L) in the filtered runoff water, the flow rate at the time of sampling (L/min) and the time between samples (min) for the area of the turf plot (m^2). Graphical representation of runoff volumes and pesticide loads for individual samples throughout a runoff event are presented as hydrographs and chemographs elsewhere (Rice et al., 2010a,b).

2.8 Pesticide concentrations in surface water receiving runoff

Measured pesticide loads in edge-of-plot runoff were extrapolated to surface water concentrations using real-world runoff-to-surface water scenarios reported by others. Estimated environmental concentrations in the receiving surface water do not account for sorption or degradation of the pesticides and therefore were used as a relative comparison to evaluate the effectiveness of management practices to mitigate pesticide transport with runoff rather than represent a definitive available concentration.

Agricultural system: Dietrich & Gallagher (2002) measured dissolved copper concentrations in runoff from tomato fields managed with polyethylene mulch at the edge-of-field and in an adjacent creek receiving the runoff. Concentrations ranged from 20-236 μg/L in the edge-of-field runoff and were as high as 20 μg/L in a nearby creek. Using the lowest, highest and average chemical concentrations for the edge-of-field runoff and receiving surface water from their study, we determined nine dilution ratios (pesticide concentration in the runoff water to pesticide concentration in the creek) ranging from a 1:1 dilution to a 1:236 dilution, with a median dilution ratio of 1:15. Thus, in the present study, a 1:15 dilution was used to calculate environmentally realistic pesticide concentrations in surface water receiving edge-of-plot runoff from the polyethylene mulch plots.

A literature search afforded no similar dilution factor for hairy vetch mulch. Although edge-of-plot concentrations for hairy vetch mulch were also measured in our field experiments, using the 1:15 dilution to calculate surface water concentrations would not be appropriate; as the total pesticide loads in runoff from hairy vetch mulch were significantly less than the total pesticide loads in the runoff from polyethylene mulch. This decrease in load was primarily the result of smaller runoff volumes and less total soil lost as opposed to smaller pesticide concentrations in the runoff (Rice et al., 2001; Rice et al., 2002). Unlike concentration, load accounts for overall mass differences resulting from changes in both runoff volumes and quantity of soil lost with runoff. Therefore, pesticide concentrations in surface water associated with runoff from hairy vetch mulch were based on calculations using the percent reduction in pesticide load for edge-of-plot runoff from hairy vetch mulch relative to that from polyethylene mulch. For example, average seasonal loads of chlorothalonil in edge-of-plot runoff during the third season for polyethylene plots (70.1 mg/m^2/season) and hairy vetch plots (4.4 mg/m^2/season) represents a 94% reduction in mass of chlorothalonil entering the receiving surface water from hairy vetch plots. Therefore the creek concentration associated with hairy vetch mulch was calculated by reducing the polyethylene mulch creek concentration (192.4 μg/L, a 1:15 dilution of the edge-of-plot concentration 2885.4 μg/L) by 94%, resulting in a hairy vetch creek concentration of 12.1 μg/L, which represents the difference in mass of chlorothalonil entering the surface water.

Turf system: Pesticide loads in runoff from the turf were extrapolated to pesticide concentrations in surface water receiving the runoff based on sub-watershed characteristics and receiving surface water dimensions reported from a golf course located less than 20 miles from the study site (http://www.pca.state.mn.us/publications/stormwaterresearch-eaglelake.pdf, pond 4 sub-watershed). Using this real-world scenario, pesticide loads from the fairway turf plots (mg/m^2) were multiplied by the area (5641 m^2) of the golf course contributing runoff to the receiving surface water, and the estimated percentage of the golf course represented by fairway turf (0.33 or 33%) (Watson et al., 1992; Lyman et al., 2007), providing the overall mass of pesticide transported with runoff to the receiving surface water (440,000 L). Estimated pesticide concentrations of the surface water receiving runoff from fairway turf managed with solid tine or hollow tine core cultivation were compared to toxicological endpoints to evaluate which core cultivation practice would be the most efficient at mitigating ecological risk.

2.9 Statistical analysis

For both the agricultural and turf studies management practices were assigned to the plots using a randomized complete block design. Experiments were repeated for three consecutive field seasons with four replications of each treatment (type of mulch) for the agricultural study and for two simulated precipitation events with three replications of each treatment (type of core cultivation) for the turf study. Analyses of variance were performed to evaluate runoff volumes, soil loss and chemical loads, with the management practice as the single criteria of classification for the data (Steel & Torrie, 1997).

3. Results and discussion

3.1 Agricultural crop: fresh market tomato production

Precipitation and runoff volume. Runoff volumes were measured from each plot throughout three growing seasons (May to August). With the exception of a few runoff events, the volume of runoff collected from polyethylene mulch plots was 2 to 100 times more than runoff from hairy vetch plots for the 41 recorded runoff events. Precipitation and volumes of individual runoff events are provided in detail elsewhere (Rice et al., 2001). The seasonal water losses for the three growing seasons were 90.6, 55.4 and 146.0 mm per growing season for polyethylene plots and 36.8, 13.7 and 75.7 mm per growing season for hairy vetch plots; resulting in a 59, 75, and 48% reduction in runoff from tomatoes grown with the vegetative mulch (Figure 1A).

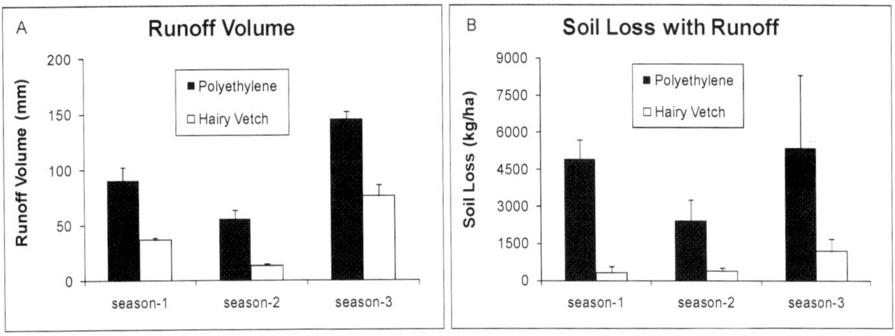

Fig. 1. Volume of runoff (A) and quantity of soil (B) transported in runoff from fresh market tomato production with polyethylene (plastic) and hairy vetch (vegetative) mulch.

Precipitation rate and infiltration rate determines the overall quantity of surface runoff. Vegetable production with polyethylene mulch may result in up to 75% of the soil surface covered with an impermeable film, depending on the width of the plastic covered tomato-beds and bare-soil furrows. McCall et al. (1988) and Wan & El-Swaify (1999) reported greater runoff volumes associated with the use of impermeable plastic mulch relative to bare soil. We have also observed that the addition of vegetative-mulch between polyethylene-covered tomato-beds reduces runoff volume compared to vegetable production with polyethylene-covered beds and bare-soil furrows (Rice et al. 2004, 2007). Plant residue of the vegetative mulch dissipates raindrop energy and increases surface roughness, both of which reduce the velocity of surface runoff allowing for greater infiltration within the furrows (Mannering & Meyer, 1963; Foster & Meyer, 1972). Elimination of the polyethylene mulch and use of vegetative-residue mulch on both the tomato-beds and in the surrounding furrows reduces runoff volumes do to greater infiltration as well as an increase in infiltration area since the impervious surface has been removed.

Soil erosion. The quantity of soil transported in the runoff ranged from 2.7 to 2000 kg/ha per event for polyethylene compared to 0.07 to 338 kg/ha per event for hairy vetch. Ten percent of the runoff events delivered >250 times more soil from the polyethylene plots (Rice et al., 2001). Average soil loss for the three growing seasons was 4921, 2418 and 5353 kg/ha per season for polyethylene plots and 328, 387 and 1179 kg/ha per season for hairy vetch mulch plots; representing a 93, 84 and 78% reduction in soil loss with runoff from the vegetative mulch (Figure 1B).

The significant reduction in soil erosion with runoff from tomatoes grown with hairy vetch mulch is likely the result of the hairy vetch plant residues dissipating raindrop energy and the anchoring characteristics of the roots providing increased structural stability in the vegetated soil (Mannering & Meyer, 1963; Foster & Meyer, 1972; Sur et al, 1992; Zuzel & Pikul, 1993). These factors contribute to greater infiltration of precipitation, reduced runoff volume and runoff velocity, and the significant reduction of suspended particulates transported with runoff. Measured flow rates of runoff from plots with polyethylene mulch were 1.2 to 7.5 times greater than from plots with hairy vetch mulch. Suspended sediment concentrations were 4 times greater in polyethylene runoff (polyethylene: 3,334 ± 0.87 mg/L; hairy vetch: 692 ± 1.5 mg/L).

Pesticides in runoff. Concentrations of pesticides associated with both the dissolved and particulate phases of the runoff were measured for each runoff event and combined with runoff and soil loss data to calculate total pesticide loads transported off-plot with runoff during each growing season. Chlorothalonil and endosulfan were measured in both the dissolved and particulate phases while esfenvalerate was absent from the dissolved phase (Figure 2). Greater loads of all three pesticides were associated with runoff from polyethylene mulch than hairy vetch mulch regardless of the phase (dissolved or particulate) evaluated. When total pesticide loads were considered, hairy vetch mulch reduced pesticide loads with runoff by 70 to 97% compared to polyethylene mulch. Details on the dissolved and particulate phase pesticide concentrations and loads for individual runoff events can be obtained elsewhere (Rice et al., 2001).

Although the fungicide (chlorothalonil) and insecticides (endosulfan, esfenvalerate) were applied to the tomato plants, inevitably, a percent of the applied pesticides are either washed off the foliage onto the mulch or are directly applied to the mulch during foliar application. Rainfall may interact with chemical residues in the top centimeter of soil, termed the extraction- or mixing-zone (ISU, 1992; Baker and Mickelson, 1994; Wauchope, 1996). The compromised infiltration capacity of the polyethylene-covered vegetable-beds and reduced leaching of pesticides below the extraction zone explains, in part, the increased availability of pesticides for transport with runoff from the plastic mulch system. The

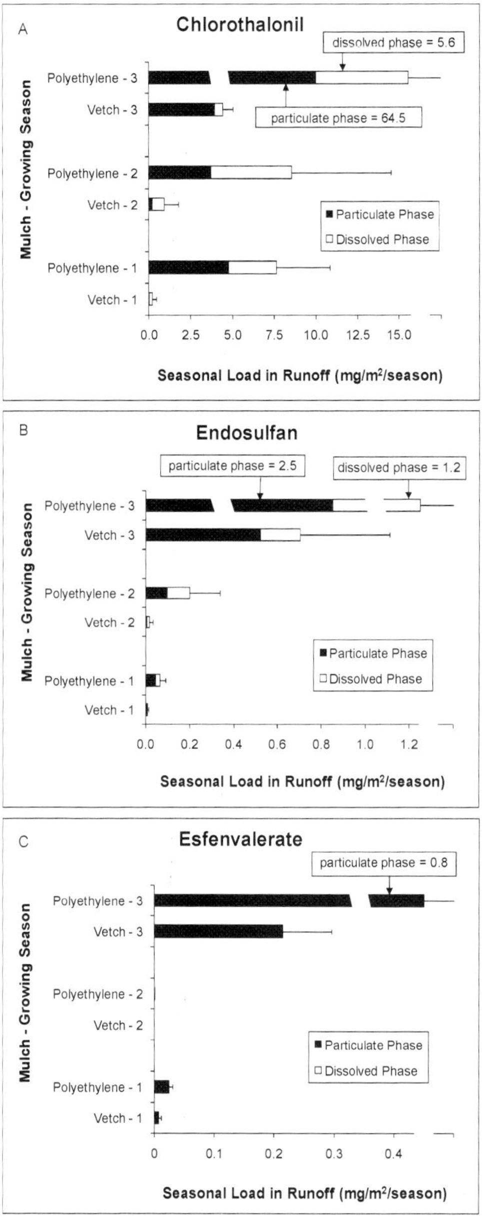

Fig. 2. Quantity of chlorothalonil (A), endosulfan (B) and esfenvalerate (C) transported in runoff from fresh market tomato production using polyethylene (plastic) or hairy vetch (vegetative) mulch.

adsorption and desorption of pesticides to the mulch will also influence the pesticides' availability to be transported with surface runoff. The adsorption of pesticides to polyethylene mulch, which has been shown to be influenced by the chemical properties of the pesticide as well as the density of the polyethylene (Vuik et al., 1990; Topp & Smith, 1992), may be playing an important role in the pesticide loading profile of the runoff. When a chemical is weakly adsorbed to the polyethylene mulch the loading profile would show a large pulse at the beginning of the runoff event followed by a sharp decline. Pesticides that are more strongly adsorbed to the polyethylene material would be illustrated by a slow bleed of chemicals off the plastic that may occur over the entire runoff event. McCall et al. (1988) reported the peak concentration of endosulfan in runoff from polyethylene mulch occurred at the beginning of the rainfall event compared to the peak concentration in the bare soil runoff that occurred in tandem with the peak flow. This illustrates endosulfan was more weakly adsorbed to polyethylene than soil. We found a great difference in the loading profile chemographs of endosulfan and chlorothalonil in the dissolved phase of runoff from polyethylene mulch compared to hairy vetch mulch (Rice et al., 2001). Statistical analysis revealed chlorothalonil loads in the dissolved phase of runoff were more associated with runoff volume than pesticide concentration in the runoff for both polyethylene and hairy vetch mulch (chlorothalonil: volume r^2 = 0.57 to 0.83, concentration r^2 = 0.00 to 0.06). Runoff volume and pesticide concentration in the runoff water were equally important for endosulfan (endosulfan: volume r^2 = 0.48 to 0.92, concentration r^2 = 0.71 to 0.99). Esfenvalerate was only measured in the suspended particulates, which corresponds to its very low water solubility (Table 1). Management practices that enhance or reduce either runoff volume or soil loss with runoff will influence the mass of pesticides transported with runoff. Additional studies on the sorption interactions of pesticides with polyethylene and plant residues of the hairy vetch will be needed to fully understand pesticide transport with runoff from these two mulch systems.

3.2 Turf: golf course fairway

Precipitation and runoff volume. Simulated precipitation and evaluation of resulting runoff occurred during the months of August and September while the turf was actively growing (mean air temperatures approximately 21 °C (70 °F)). Core cultivation with solid or hollow tines occurred 63 d prior to initiation of the first simulated rainfall event, while the period of time between the second core cultivation and simulated precipitation was only 2 d. The greater time lag between core cultivation and the runoff study for the first event was a result of delays in the construction of the rainfall simulator. Although differences were noted in the results of the 63 d (event - 1) and 2 d (event - 2) data the overall trends observed between solid tine and hollow tine core cultivation remained the same (Rice et al., 2010b). It is important to note the time between pesticide application and runoff was similar for both runoff events (26 ± 13 h).

Overall, runoff volume was reduced in fairway turf plots managed with hollow tine core cultivation compared to solid tine core cultivation. The total volume of runoff measured from turf managed with hollow tines was 10% (63 d) and 55% (2 d) less than the total volume quantified from turf managed with solid tines (63 d: hollow tine = 21.1 ± 6.2 mm, solid tine = 23.4 ± 7.4 mm; 2 d: hollow tine = 12.5 ± 0.9 mm, solid tine = 28.0 ± 16.1 mm) (Figure 3A). Hydrographs revealed reduce runoff volumes associated with hollow tine core cultivation in 81% (63 d) and 87% (2 d) of the samples (Figure 3B).

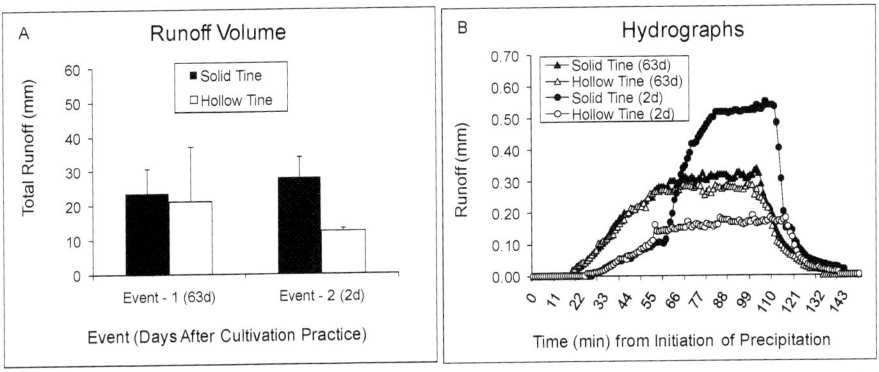

Fig. 3. Total volume of runoff (A) and runoff hydrographs (B) from golf course fairway turf managed with solid tine core cultivation or hollow tine core cultivation 63 d and 2 d prior to precipitation.

Solid tine core cultivation has been shown to result in localized compaction with the most sever compaction at the base of the zone of cultivation (Murphy et al., 1992). Hollow tine core cultivation has also been shown to result in compaction along the sidewalls and base of the coring; however, sidewall compaction dissipated while bottom compaction remained after 95 d (Petrovic, 1979). Our field observations and measurements showed greater infiltration with hollow tine core cultivation compared to solid tine core cultivation with the greatest difference between cultivation practices occurring shortly after treatment (2d) and decreasing with time (63 d) (Rice et al., 2010b). We speculate the greatest difference in soil physical properties was most prominent shortly after hollow tine or solid tine core cultivation; which diminished as roots grew, compaction dissipated, and core channels were covered or filled. Other researchers have reported greater air porosity and saturated water conductivity in turf managed with hollow tines compared to solid tines, and enhanced water infiltration in turf managed with hollow tine core cultivation compare to untreated turf (Murphy et al., 1992; Baldwin et al., 2006; McCarty et al., 2007).

The quantity of applied precipitation measured as surface runoff represented 28 to 62% of the applied rainfall depending on the type of core cultivation and time duration between core cultivation and precipitation (2d: hollow tine = 28 ± 2%, solid tine = 62 ± 25%; 63 d: hollow tine = 36 ± 11%, solid tine = 41 ± 13% (Rice et al., 2010b). Shuman (2002) observed 37 to 44 % of applied water resulted as runoff from fairways of Tifway bermudagrass (*Cynodon dactylon* (L.) Pers.), which received 50 mm of simulated precipitation, 2 d following irrigation to field capacity. Kauffman and Watschke (2007) observed three to 21% of applied simulated precipitation was measured as runoff from bentgrass and perennial ryegrass turf managed with hollow tine core cultivation. They attributed variations in runoff volumes to differences in antecedent soil moisture, slope and environmental conditions. For our experiments, soil moisture measurements for both cultivation practices were 46 ± 7% (n = 54) water holding capacity 3 h prior to initiation of the simulated precipitation and 67 ± 6% (n = 54) water holding capacity 2 h following simulated precipitation. The slope of each plot was 4% and simulated precipitation and collection of runoff were performed in side-by-side paired comparisons (hollow tine versus solid tine for each replication) with runoff data normalized to the measured quantity of precipitation applied to each plot (n = 11 per plot).

Therefore the reduction in percentage of applied precipitation resulting as surface runoff from the hollow tine plots versus solid tine plots is the effect of the core cultivation practices.

Pesticides in runoff. Plots receiving hollow tine core cultivation to manage thatch 63 d prior to runoff showed a 15% reduction in flutolanil loads (solid tine = 29.4 ± 11.2 mg/m^2, hollow tine = 24.9 ± 10.6 mg/m^2) while chlorpyrifos loads were similar (solid tine = 0.24 ± 0.01 mg/m^2, hollow tine = 0.28 ± 0.17 mg/m^2) (Figure 4). Following the second core cultivation, 2 d prior to initiation of simulated precipitation and runoff, hollow tine plots displayed a reduction in both flutolanil and chlorpyrifos loads relative to the solid tine plots with a 55% decline in total loads of flultolanil (solid tine = 29.0 ± 17.2 mg/m^2, hollow tine = 13.2 ± 0.2 mg/m^2) and a 57% reduction in total loads of chlorpyrifos (solid tine = 0.71 ± 0.34 mg/m^2, hollow tine = 0.31 ± 0.02 mg/m^2). Solid tine core cultivation pushes the soil aside to create channels while hollow tine core cultivation removes cores and returns the soil back to the turf. As a result one would anticipate increased soil compaction with the solid tine cultivation and greater accessibility of soil adsorptive sites with the hollow tine cultivation. This would influence infiltration and hydraulic conductivity (Murphy et al., 1992; Baldwin et al., 2006; McCarty et al., 2007) as well as pesticide availability for transport (Liu et al., 1995; Gardner et al., 2000; Raturi et al., 2005). Analysis of pesticide loads with runoff volumes and pesticide concentrations in the runoff showed loads were attributed to runoff volume more than chemical concentrations for both management practices (volume r^2 = 0.78 to 0.90, concentration r^2 = 0.05 to 0.22). The greater association of pesticide load with runoff volume explains in part the increased pesticide transport associated with the solid tine plots compared to hollow tine plots and the greater difference in pesticide loads between cultivation practices at 2 d compared to 63 d.

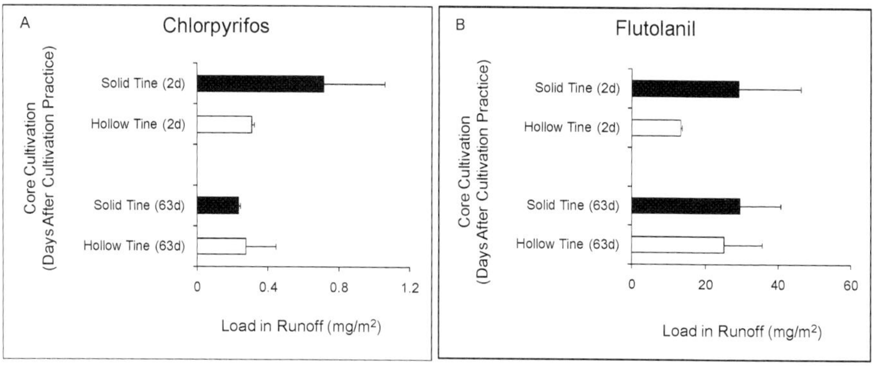

Fig. 4. Quantity of chlorpyrifos (A) and flutolanil (B) transported in runoff from golf course fairway turf managed with solid tine core cultivation or hollow tine core cultivation 63 d and 2 d prior to precipitation.

In addition to numerous environmental and management factors that contribute to the availability of pesticides for movement with overland flow, the physical and chemical properties of the pesticide will also influence the quantity of applied active ingredient observed in the runoff. A greater percentage of the applied flutolanil (less than 10%) was measured in the runoff compared to chlorpyrifos (less than 2%), which corresponds to the greater water solubility and smaller soil organic partition coefficient of flutolanil (Table 1). The quantity of chlorpyrifos and flutolanil quantified in the runoff are in range of the values reported by others who have observed less than 2 to 15% of applied pesticides measured in runoff from turf (Wauchope et al., 1990; Cole et al., 1997; Ma et al., 1999; Armbrust & Peeler, 2002). Chemical degradation was not influential in the present study as the time from chemical application to runoff (30 ± 8 h) was much less than the reported half lives of the compounds of interest (5 to 320 d) (Table 1). Chemographs of chlorpyrifos and flutolanil and additional information on the transport of applied nutrients and herbicides from managed turf are published elsewhere (Rice et al., 2009; Rice et al., 2010a,b). In general, chemographs of flutolanil and chlorpyrifos quickly diverged from the hydrograph compared to more water soluble herbicides, which had chemographs that closely resembled the runoff hydrograph during the first 50 minutes of runoff (Rice et al., 2010a).

3.3 Mitigation of ecological risk with management practices

The effectiveness of management practices to reduce ecological risk of pesticides transported in runoff from agricultural crops or managed turf was evaluated. Edge-of-plot runoff data was extrapolated to surface water concentrations using reported real-world scenarios of runoff from tomato production with polyethylene mulch into an adjacent creek (Dietrich & Gallagher, 2002.) and runoff from a golf course fairway into an adjacent pond (http://www.pca.state.mn.us/publications/stormwaterresearch-eaglelake.pdf, pond 4 sub-watershed). These estimated environmental concentrations of pesticides in the receiving surface waters were compared to published toxicity data. The toxicity of compounds to organisms can be evaluated using endpoints such as behavioral, reproductive and developmental effects as well as lethality. Although changes in behavior or effects on reproduction and development are more sensitive endpoints, a common and more readily quantifiable endpoint used in risk assessments is the median lethal concentrations (LC50) or the concentration of compound that results in mortality of fifty percent of the exposed organisms during a measured exposure period. A summary of median lethal concentrations used in the assessment of the agricultural and turf management practices are presented in Table 2.

Agricultural crop: tomato production. Concentrations of the fungicide chlorothalonil and the insecticides endosulfan and esfenvalerate in a surface water receiving runoff from tomatoes grown in polyethylene mulch or hairy vetch mulch are presented in Figure 5. Although the surface water concentrations represent a 1:15 dilution of the runoff, quantities of these pesticides in surface waters receiving runoff from polyethylene mulch exceeded the median lethal concentration for five fish (common carp, bluegill, striped bass, rainbow trout and fathead minnow), three crustaceans (freshwater prawn, giant river prawn and shrimp) and a mollusk (fingernail clam), during two or more of the growing seasons evaluated (Figure 5, Table 2). With a few exceptions, replacing polyethylene mulch with hairy vetch mulch reduced surface water concentrations of the pesticides to levels below the median lethal concentrations (Fig. 5). Exceptions were noted during the third season when pesticide loads were greatest and surface water concentrations of endosulfan exceeded the median leathal concentration for two fish (fathead minnow and striped bass) and two crusteaceans (shrimp and giant river prawn) and esfenvalerate exceeded the median lethal concentration for two fish (bluegill and rainbow trout) regardless of the type of mulch (polyethylene or hairy vetch). Concentrations of chlorothalonil, endosulfan and esfenvalerate in the surface water were below median lethal concentrations for three amphibians (green frog, bog frog and leopard frog) and two mollusks (bivalve and mussel).

Managed turf: golf course fairway. Estimated environmental concentrations of pesticides in a surface water receiving runoff from turf managed as a golf course fairway resulted in concentrations of chlorpyrifos from 1.0 to 3.0 µg/L and flutolanil from 55 to 123 µg/L. Specific concentrations associated with runoff from turf managed with solid tines compared to hollow tines are provided in Figure 6. With the exception of chlorpyrifos in the first runoff event occurring 63 days after core cultivation, replacing solid tine core cultivation with hollow tine core cultivation reduced concentrations of pesticides in the surface water receiving runoff. This included three herbicides that were co-applied with chlorpyrifos and flutolanil (data not shown) (Rice et al., 2010b). For the second runoff event, occurring 2 days after core cultivation, chlorpyrifos concentrations in the surface water associated with runoff from the plots managed with solid tines exceeded median lethal concentrations of three fish (common carp, bluegill and striped bass) and three crustaceans (copepod, shrimp and white river crayfish) (Table 2, Figure 6). Managing thatch with hollow tine core cultivation compared to solid tine core cultivation reduced surface water concentrations of chlorpyrifos to levels below the median lethal concentration of the copepod, white river crayfish, common carp and bluegill. However, the sensitivity of shrimp and striped bass to

Figure letter[a]	Scientific name	Common name	LC50 (µg/L) [exposure duration (d)][b] Chlorothalonil	Chlorpyrifos	Endosulfan	Esfenvalerate	Flutolanil
	Amphibians						
A	*Rana clamitans*	Green Frog	---[c]	235.9 [2][d,e]	15 [13][f]	---	----
B	*Rana limnocharis*	Bog Frog	245 [2][g]	2401 [2][g]	12 [2][g]	28 [2][g]	----
C	*Rana pipiens*	Leopard Frog	----	----	----	7.29 [4][h]	----
	Crustaceans						
D	*Copepoda*	Copepod Subclass	----	2.13 [2][i]	----	---	----
E	*Macrobrachium dayanum*	Freshwater Prawn	----	----	6.2 [1][j]	---	----
F	*Macrobrachium rosenbergii*	Giant River Prawn	----	----	0.2 - 0.93 [4][k]	---	----
G	*Paratya australiensis*	Shrimp	16 [4][l]	0.1 [3][m]	0.51 - 0.96 [2][n]	---	----
H	*Procambarus acutus acutus*	White River Crayfish	----	2 [4][o]	----	---	----
I	*Streptocephalus sudanicus*	Fairy Shrimp	----	3.48 [2][p]	----	---	----
	Fish						
J	*Cyprinus carpio*	Common carp	110 [2][q]	1.8 [1][r]	9.5 [4][s]	---	≥ 2900 [2][t]
K	*Lepomis macrochirus*	Bluegill	26 - 62 [4][u]	1.7 - 2.5 [4][v]	3.3 [1][v]	0.31 [4][w]	5400 [4][u]
L	*Morone saxatilis*	Striped Bass	----	0.58 [4][x]	0.22 - 0.43 [4][y]	2.17 [1][z]	----
M	*Oncorhynchus mykiss*	Rainbow trout	40.2 [1][aa]	15 [1][v]	8.89 [2][bb]	0.07 [4][u]	5400 [4][u]
N	*Pimephales promelas*	Fathead Minnow	----	120 - 170 [4][cc]	1.84 [1][dd]	0.616 [2][ee]	4800 [4][u]
	Mollusks						
O	*Lamellidens corrianus*	Bivalve	----	----	17 - 44 [4][ff]	---	----
P	*Lamellidens marginalis*	Mussel	----	----	6 - 40 [4][ff]	---	----
Q	*Sphaerium sp.*	Fingernail clam	----	----	----	1.6 [2][gg]	----

[a]Letters referenced in Figures 5 & 6. [b]Data and references available at http://cfpub.epa.gov/ecotox/ecotox_home.cfm. [c]--- = no data. [d]Effect measured for EC50 = stimulus avoidance. [e]Wacksman et al., 2006. [f]Harris et al., 1998. [g]Pan & Liang, 1993. [h]Materna et al., 1995. [i]Siefert, 1987. [j]Omkar & Murti, 1985. [k]Lombardi et al., 2001. [l]Davies et al., 1994. [m]Olima et al., 1997. [n]Hose & Wilson, 2005. [o]Carter & Graves, 1972. [p]Lahr et al. 2001. [q]Hashimoto & Nishiuchi, 1981. [r]Dutt & Guha, 1988. [s]Shivakumar & David, 2004. [t]Nishiuchi et al., 1985. [u]Office of Pesticide Programs, 2000. [v]Mayer & Ellersieck, 1986. [w]Fairchild et al., 1992. [x]Korn & Earnest, 1974. [y]Fujimura et al., 1991. [z]Geist et al., 2007. [aa]Davies & White, 1985. [bb]Capkin et al., 2006. [cc]Jarvinen & Tanner, 1982. [dd]Kleiner et al., 1984. [ee]Bouldin et al., 2004. [ff]Mane and Muley, 1984. [gg]Lozane et al., 1989.

Table 2. Median lethal concentrations (LC50) of the selected pesticides.

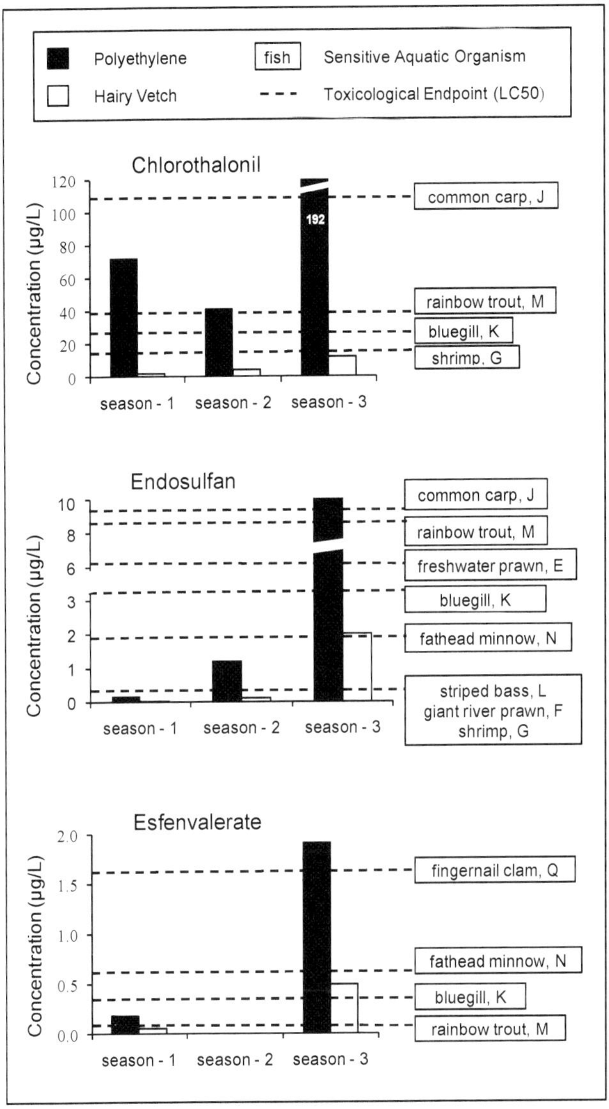

Fig. 5. Estimated environmental concentrations of chlorothalonil, endosulfan, and esfenvalerate in a surface water receiving runoff receiving runoff from fresh market tomato production with polyethylene (plastic) mulch or hairy vetch (vegetative) mulch. The broken lines represent the median lethal concentrations of sensitive aquatic organisms named in the attached boxes. Capital letters following the name of the aquatic organisms correspond to the letters given in the first column of Table 2, which provides the toxicological data in greater detail.

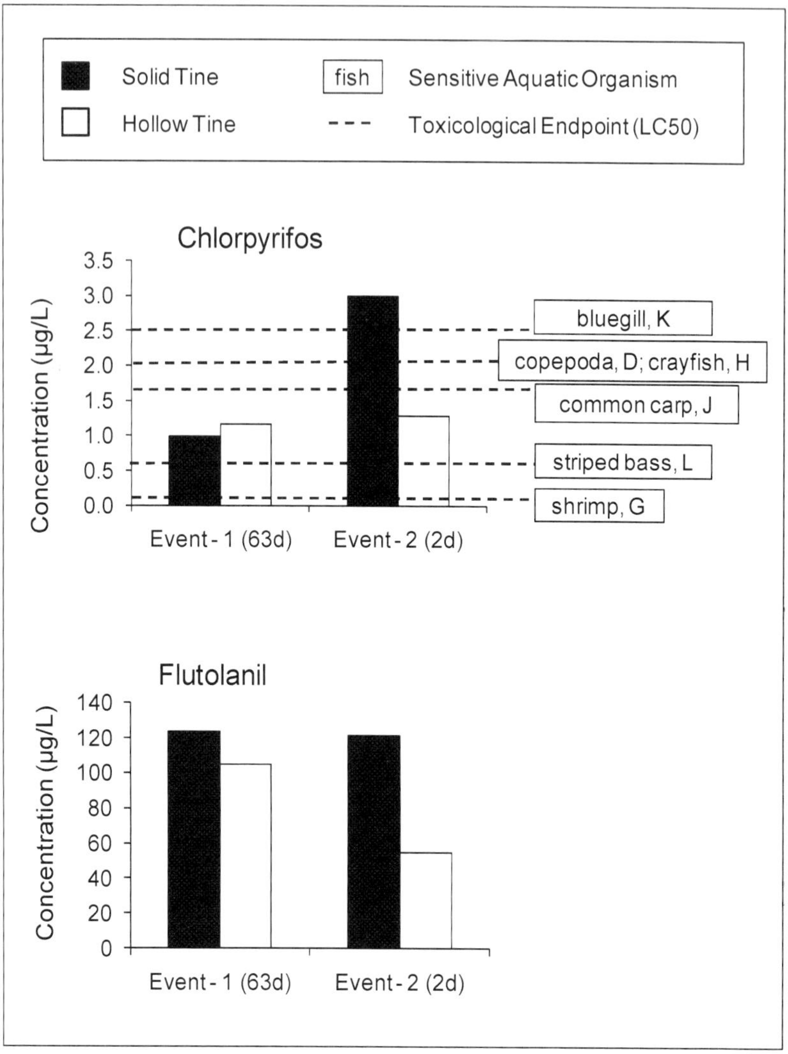

Fig. 6. Estimated environmental concentrations of chlorpyrifos and flutolanil in a surface water receiving runoff from creeping bentgrass turf managed as a golf course fairway with solid tine core cultivation or hollow tine core cultivation 63 days and 2 days prior to runoff. The broken lines represent the median lethal concentrations of sensitive aquatic organisms named in the attached boxes. Capital letters following the name of the aquatic organisms correspond to the letters given in the first column of Table 2, which provides the toxicological data in greater detail. Surface water concentrations of flutolanil were 24 to 98 times below the median lethal concentrations of the aquatic organisms evaluated, therefore no sensitive aquatic organisms are shown on the graph for flutolanil.

chlorpyrifos was great enough that estimated surface water concentrations exceeded the median lethal concentration regardless of the turf cultivation practice (solid tines or hollow tines). In contrast, surface water concentrations of flutolanil were 24 to 98 times below the median lethal concentrations of the four fish evaluated (common carp, bluegill, rainbow trout, and fathead minnow).

In summary, median lethal concentrations of chlorothalonil, chlorpyrifos, endosulfan, esfenvalerate and flutolanil for 17 organisms were compared with estimated environmental concentrations of surface waters that received runoff from either vegetable production with polyethylene mulch compared to hairy vetch mulch or golf course fairway turf managed with solid tine compared to hollow tine core cultivation. These organisms were chosen to represent amphibians, crustaceans, fish and mollusks that spend part or all of their life in freshwater creeks, streams, ponds or lakes. In 18 of the 39 exposure scenarios (Table 2) changing the management practice reduced the ecological risk of the pesticide in at least one of the seasons or events, bringing the surface water concentration below the reported median lethal concentration for the evaluated organism (Figures 5&6, Table 2, chlorothalonil: organisms G, J, K, M; chlorpyrifos: organisms D, H, J, K for event-2; endosulfan: organisms E, J, K, M for season-3 and F, G, L for season-2; esfenvalerate: organisms M for season-1 and N, Q for season-3). In some circumstances the sensitivity of the organism to the pesticide was great enough that estimated surface water concentrations exceeded the median lethal concentration regardless of the management practice (Figures 5&6, Table 2, chlorpyrifos: organisms G, L; endosulfan: organisms F, G, L, N for season-3; esfenvalerate: organisms K, M for season-3). Similarly, changes in management practice did not significantly influence the risk of pesticides to organisms with median lethal concentration above the estimated environmental concentration in the diluted surface water (Figures 5&6, Table 2, chlorothalonil: organism J for seasons-1&2, B (not shown on Figure 5); chlorpyrifos: organisms D, H, J, K for event-1 and A, B, I, M, N (not shown on Figure 6); endosulfan: organisms E, J, K, M for seasons-1&2, F, G, L for season-1 and A, B, O, P (not shown on Figure 5); esfenvalerate: organisms K, N, Q for season-1 and B, C, L (not shown on Figure 5); flutolanil: organisms J, K, M, N (not shown on Figure 6)). The toxicity of compounds to organisms can be evaluated using sublethal effects such as induction of enzyme systems, behavioral traits, or reproductive and developmental effects; which are often more sensitive than the end point of lethality (Klaassen, 1996). The impact of replacing impermeable polyethylene mulch with the vegetative hairy vetch mulch in vegetable production and replacing solid tine core cultivation with hollow tine core cultivation when managing turf will be further evident when more sensitive toxicological endpoints are evaluated.

4. Conclusions

The research described in this chapter measured the quantity of pesticides transported with runoff from agricultural systems (fresh market tomato production with polyethylene mulch or hairy vetch mulch) and turfgrass systems (golf course fairway turf managed with solid tine or hollow tine core cultivation) in order to evaluate the capacity of management practices to reduce the off-site transport of pesticides. Reported real-world runoff-to-surface water scenarios were used to extrapolate pesticide loads in runoff to estimated environmental concentrations of pesticides in surface waters receiving the runoff. Surface water concentrations of the pesticides were compared with published toxicity data to assess reductions in the ecological risk associated with implementation of the management practices. The reduced runoff volume and pesticide loads measured in runoff from the hairy vetch mulch and hollow tine core cultivation suggests these management practices are more sustainable for the agricultural and turf systems, respectively. This was further illustrated by reduced ecological risk in 18 of 39 pesticide exposure scenarios in which changing the management practice resulted in surface water concentration of pesticides below the reported median lethal concentrations for the evaluated aquatic organisms. The scenarios presented in this study do not represent absolute risk as they do not consider degradation and bioavailability of the pesticide, potential synergistic interactions between pesticides, effects of pesticides on other ecosystem components, or the life cycle of the species (Matthews et al., 2002). However, these assessments are informative for identifying management practices that reduce ecological risk by maintaining pesticides at targeted locations. According to the United States Environmental Protection Agency, integrated pest management relies on a combination of evaluations, practices, and options to manage pests with an effective and environmentally sensitive approach that is economical and presents the smallest potential hazard to people, property, and the environment (http://www.epa.gov/pesticides/factsheet/ipm.htm). Our results demonstrate that management practices can enhance the sustainability of intensely managed biotic systems; improving efficacy of pesticides at targeted locations while reducing adverse impacts to non-target organism. In addition, management practices can be instrumental in maintaining the use of pesticides as a tool in integrated pest management, by providing the least possible hazard of pesticides to the environment.

Acknowledgement - Data presented in this chapter were collected from two locations over a period of years. We gratefully acknowledge the contributions of numerous students, support staff and colleagues for their assistance: Aref Abdul-Baki, Christina Borgen, Troy Carson, Mike Dolan, Peter Downey, Shari Greseth, Jennifer Harman-Fetcho, Lynne Heighton, Rachel Herbert, Andrew Hollman, Allan Isensee, Craig Krueger, Jeff Lanners, Matthew McNearney, Jennifer Rittenhouse, Carl Rosen, Jon Sass, Ali Sadeghi, Alex Seeley, Kerry Sefton, Karli Swenson, John Teasdale and Robert Wevadu. We also thank the United States Golf Association Green Section Research for funding the turf research and Nelson Irrigation for donating supplies for construction of the turf rainfall simulator.

5. References

Abdul-Baki, A.; Teasdale, J. & Korcak, R. (1997). Nitrogen requirements of fresh-market tomatoes on hairy vetch and black polyethylene mulch. *HortScience*, 32, 217-221

Abdul-Baki, A.; Teasdale, J.; Korcak, R.; Chitwood, D. & Huettel, R. (1996). Fresh-market tomato production in a low-input alternative system using cover-crop mulch. *HortScience*, 31, 65-69

Armbrust, K. & Peeler, H. (2002). Effects of formulation on the run-off of imidacloprid from turf. *Pest Manag. Sci.* 58, 702-706

Baker, D. & Richards, R. (1990). Transport of soluble pesticides through drainage networks in large agricultural river basins, In: *Long Range Transport of Pesticides*, D.A. Kurtz (Ed.), pp. 241-270, Lewis Publ., Chelsa, MI, USA

Baker, J. & Mickelson, S. (1994). Application technology and best management practices for minimizing herbicide runoff. *Weed Technol.*, 8, 862-869

Baldwin, C.; Liu, H. & Brown, P. (2006). Effects of core cultivation tine entry angle on golf putting greens. *HortTechnology*, 16, 265-269

Barbash, J. & Resek, E. (1996). *Pesticides in Groundwater: Distribution, Trends, and Governing Factors*, Lewis, Boca Raton, FL, USA

Beard, J. (1973). *Turfgrass: Science and Culture*, Prentice-Hall, Engelwood Cliffs, NJ, USA

Bengtson, R.; Southwick L.; Willis G. & Carter C. (1990). The influence of subsurface drainage practices on herbicide losses. *Transactions of the ASAE*, 33, 415-418

Bouldin, J.; Milam, C.; Farris, J.; Moore, M.; Smith, S. & Cooper, C. (2004). Evaluating toxicity of Asana XL (esfenvalerate) amendments in agricultural ditch mesocosms. *Chemosphere*, 56, 677-683

Callahan, L.; Sanders, W.; Parham J.; Harper, C.; Lester, L. & McDonald, E. (1998). Cultural and chemical controls of thatch and their influence on rootzone nutrients in a bentgrass green. *Crop Sci.*, 38, 181-187

Capkin, E.; I. Altinok, I. & Karahan, S. (2006). Water quality and fish size affect toxicity of endosulfan, an organochlorine pesticide, to rainbow trout. *Chemosphere*, 64, 10, 1793-1800

Carrow, R.; Johnson, B. & Burns, R. (1987). Thatch and quality of Tifway bermudagrass turf in relation to fertility and cultivation. *Agron. J.*, 79, 524-530

Carter, F. & Graves, J. (1972). Measuring effects of insecticides on aquatic animals. *La.Agric.*, 16, 2, 14-15

Chandler, G. & Scott, G. (1991). Effects of sediment-bound endosulfan on survival, reproduction and larval settlement of meiobenthic polychaetes and copepods. *Environ. Toxicol. Chem.* 10, 375-382

Clark, J.; Lewis, M. & Pait, A.. (1993). Pesticide inputs and risks in coastal wetlands. *Environ. Toxicol. Chem.* 12, 2225-2233

Clark, J. & Kenna, M. (2001). Lawn and turf: Management and environmental issues of turfgrass pesticides. In: *Handbook of Pesticide Toxicology, 2nd ed.*, Krieger, R. & Krieger, W., (Eds.), pp. 203-241, Academic Press, San Diego, CA, USA

Cohen, S.; Svrjcek, A.; Durborow, T. & Barnes, N. (1999). Water quality impacts by golf courses. *J. Environ. Qual.*, 28, 798-809

Cole, J.; Baird, J.; Basta, N.; Huhnke, R.; Storm, D.; Johnson, G.; Payton, M.; Smolen, M.; Martin, D. & Cole, J. (1997). Influence of buffers on pesticide and nutrient runoff from bermudagrass turf. *J. Environ. Qual.*, 26, 1589-1598

Coody, P. & Lawrence, L. (1994). Method and system for conducting meso-scale rainfall simulations and collecting runoff. *U.S. Patent 5,279,151.* U.S. Patent Office, Washington, DC

Dao, T. (1991). Field decay of wheat straw and its effects on metribuzin and S-ethyl metribuzin sorption and elution from crop residues. *J. Environ. Qual.*, 20, 203-208

Davies, P.; Cook, L. & Goenarso, D. (1994). Sublethal responses to pesticides of several species of australian freshwater fish and crustaceans and rainbow trout. *Environ.Toxicol.Chem.*, 13, 8, 1341-1354

Davies, P. & White, R. (1985). The toxicology and metabolism of chlorothalonil in fish. I. Lethal levels for *Salmo gairdneri, Galaxias maculatus, G. truttaceus* and *G. auratus* and the fate of 14C-TCIN in *S. gairdneri*. *Aquat.Toxicol.*, 7, 1/2,93-105

Davis, R.; Hamilton, G.; Lanini, W.; Screen, T. & Osteen, C. (1998). The importance of pesticides and other pest management practices in U.S. tomato production. *USDA National Agricultural Pesticide Impact Assessment Program. Document No. I-CA-98.* U.S. Gov. Print Office, Washington, DC

Dietrich, A. & Gallagher, D. (2002). Fate and environmental impact of pesticides in plastic mulch production runoff: Field and laboratory studies. *J. Agric. Food Chem.,* 50, 4409-4416

Dunn, J.; Minner, D.; Fresenburg, B.; Bughrara, S.& Hohnstrater, C. (1995). Influence of core aerification, topdressing, and nitrogen on mat, roots and quality of 'Meyer'. Zoysiagrass. *Agron. J.,* 87, 891-894

Dutt, N. & Guha, R. (1988). Toxicity of few organophosphorus insecticides to fingerlings of bound water fishes, *Cyprinus carpio* (Linn.) and *Tilapia mossambica* Peters. *Indian J.Entomol.,* 50, 4, 403-421

Fairchild, J.; La Point, T.; Zajicek, J.; Nelson, M.; Dwyer, F. & Lovely, P. (1992). Population-, community- and ecosystem-level responses of aquatic mesocosms to pulsed doses of a pyrethroid insecticide. *Environ.Toxicol.Chem.,* 11, 1, 115-129

Foster, G. & Meyer, L. (1972). Erosion mechanics of mulches. *A.S.A.E. Paper no. 72-754.* American Society of Agricultural Engineers, St. Joseph, MI, USA

Frick, E.; Hippe, D.; Buell, G.; Couch, C.; Hopkins, E.; Wangsness, D.& Garrett, J. (1998). *Water quality in the Apalachicola-Chattahoochee-Flint River basins, Georgia, Alabama, and Florida, 1992-95.* U.S. Geological Survey Circular 1164, Reston, VA

Fujimura, R.; Finlayson, B. & Chapman, G. (1991). Evaluation of acute and chronic toxicity tests with larval striped bass, In: *Aquatic Toxicology and Risk Assessment,* Vol.14, Mayes, M. & Barron, M., (Eds.), pp. 193-211, ASTM STP 1124, Philadelphia, PA , USA

Gardner, D.; Branham, B. & Lickfeldt, D. (2000). Effect of turfgrass on soil mobility and dissipation of cyproconazole. *Crop Sci.,* 40, 1333-1339

Geist, J.; Werner, I.; Eder, K. & Leutenegger, C. (2007). Comparisons of tissue-specific transcription of stress response genes with whole animal endpoints of adverse effect in striped bass (*Morone saxatilis*) following treatment with copper and esfenvalerate. *Aquat.Toxicol.* 85, 1, 28-39

Ghadiri, H.; Shea, P. & Wicks, G. (1984). Interception and retention of atrazine by wheat (*Triticum aestivum*) stubble. *Weed Sci.,* 32, 24-27

Gianessi, L. & Anderson, J. (1996). *Pesticide use in U.S. crop production: National data report,* Washington, DC, National Center for Food and Agricultural Policy, unpaged

Gilliom, R.; Barbash, J.; Crawford, C.; Hamilton, P.; Martin, J.; Nakagaki, N.; Nowell, L.; Scott, J.; Stackelberg, E.; Thelin, G. & Wolock, D. (2006). *Pesticides in the nation's streams and ground water, 1992-2001.* Ci no. 1291. Revised Feb 2007. U.S. Geological Survey National Water-Quality Assessment Program, Reston, VA, USA (http://pubs.usgs.gov/circ/2005/1291)

Goel, A.; McConnell, L. & Torrents, A. (2005). Wet deposition of current use pesticides at a rural location on the Delmarva Peninsula: impact of rainfall patterns and agricultural activity. *J. Agric. Food Chem.,* 53, 7915-7924

Goolsby, D. & Battaglin, W. (1993). Occurrence, distribution and transport of agricultural chemicals in surface waters of the Midwestern United States. In Goolsby DA, Boyer LL, Mallard GE, eds, *Selected Papers on Agricultural Chemicals in Water Resources of the Midcontinental United States*. U.S. Geological Survey Open-File Report 93-418, pp 1-25

Harman-Fetcho, J.; Hapeman, C.; McConnell L.; Potter, T.; Rice, C.; Sadeghi, A.; Smith, R.; Bialek, K.; Sefton, K. & Schaffer, B. (2005). Pesticide occurrence in selected South Florida canals and Biscayne Bay during high agricultural activity. *J. Agric. Food Chem.*, 53, 6040-6048

Harris, M.; Bishop, C.; Struger, J.; Ripley, B. & Bogart, J. (1998). The functional integrity of northern leopard frog (*Rana pipiens*) and green frog (*Rana clamitans*) populations in orchard wetlands. II. Effects of pesticides and eutrophic conditions on early life stage. *Environ.Toxicol.Chem.*, 17, 7, 1351-1363

Hayward, S.; Gouin, T. & Wania, F. (2010). Levels and seasonal variability of pesticides in the rural atmosphere of southern Ontario. *J. Agric. Food Chem.* 58, 2, 1077-1084, 0021-8561

Hashimoto, Y. & Nishiuchi, Y. (1981). Establishment of bioassay methods for the evaluation of acute toxicity of pesticides to aquatic organisms. *J.Pestic.Sci.*, 6, 2, 257-264

Hoffman, R.; Capel, P. & Larson, S. (2000). Comparison of pesticides in eight U.S. urban streams. *Environ. Toxicol. Chem.*, 19, 2249-2258

Hose, G. & Wilson, S. (2005). Toxicity of endosulfan to *Paratya australiensis* Kemp (Decapoda: Atyidae) and *Jappa kutera* Harker (Ephemeroptera: Leptophlebiidae) in Field-Based Tests. *Bull.Environ.Contam.Toxicol.*, 75, 5, 882-889

Iowa State University. (1992). Understanding and reducing pesticide losses. *Iowa State Univ. Extn. Pm-1495*, Ames

Jarvinen, A.W., and D.K. Tanner. (1982). Toxicity of selected controlled release and corresponding unformulated technical grade pesticides to the fathead minnow *Pimephales promelas*. *Environ.Pollut.Ser.A*, 27, 3, 179-195

Johnson, W.; Kroll, R.; Plimmer, J. & Pait, A. (1994). The occurrence and distribution of pesticides in Chesapeake Bay. In: *Perspectives on Chesapeake Bay, 1994: Advances in estuarine sciences*, Nelson, S. & Elliott, P., (Eds.), pp. 105-145, CRC Publication No. 147, Chesapeake Research Consortium Inc., Edgewater, MD

Joyce, S. (1998). Why the grass isn't always greener. *Environ. Health Perspect.*, 106A, 378-383

Kauffman, G. & Watschke, T. (2007). Phosphorus and sediment in runoff after core cultivation of creeping bentgrass and perennial ryegrass turfs. *Agron. J.*, 99, 141-147

Kelly, T.; Lu, Y.-C.; Abdul-Baki, A. & Teasdale, J. (1995). Economics of a hairy vetch mulch system for producing fresh-market tomatoes in the Mid-Atlantic region. *J. Amer. Soc. Hort. Sci.*, 120, 854-860

Kiely, R.; Donaldson, D. & Grube, A. (2004). *Pesticide industry sales and usage, 2000 and 2001 market estimates*. Office of Pesticide Programs, U.S. Environmental Protection Agency, Washington, DC, USA, (http://www.epa.gov/opp00001/pestsales/01pestsales/usage2001.htm#3_1)

Klaassen, C. (1996). *Casarett & Doull's Toxicology: The Basic Science of Poisons*, 5th ed. McGraw-Hill, New York, NY, USA

Kleiner, C.; Anderson, R. & Tanner, D. (1984). Toxicity of fenitrothion to fathead minnows (*Pimephales promelas*) and alternative exposure duration studies with fenitrothion and endosulfan. *Arch.Environ.Contam.Toxicol.*, 13, 573-578

Korn, S. & Earnest, R. (1974). Acute toxicity of twenty insecticides to striped bass, *Morone saxatilis*. *Calif.Fish Game*, 60, 3, 128-131

Lahr, J.; Badji, A.; Marquenie, S.; Schuiling, E.; Ndour, K.; Diallo, A. & Everts, J. (2001). Acute toxicity of locust insecticides to two indigenous invertebrates from sahelian temporary ponds. *Ecotoxicol.Environ.Saf.*, 48, 1, 66-75

Liu, L.; Hsiang, T. & Eggens, J. (1995). Core cultivation and efficacy of benomyl applied to creeping bentgrass. *Agron. J.*, 87, 272-275

Loague, K. & Soutter, L. (2006). Desperately Seeking a Cause for Hotspots in Regional-Scale Groundwater Plumes Resulting from Non-Point Source Pesticide Applications. *Vadose Zone J.*, 5, 204-221

Lombardi, J.; Machado-Neto, J.; Brossi-Garcia, A.; Marques, H. & Kubo, E. (2001). Acute toxicity of the pesticides endosulfan and ametryne to the freshwater prawn *Macrobrachium rosenbergii* De Man. *Bull.Environ.Contam.Toxicol.*, 67, 5, 665-671

Lozano, S.; Brazner, J.; Knuth, M.; Heinis, L.; Sargent, K.; Tanner, D.; Anderson, L.; O'Halloran, S. & Ber, S. (1989). Effects, persistence and distribution of esfenvalerate in littoral enclosures. Rep.No.DU E104/PPA 06/7592A, U.S.EPA, Duluth & Univ.of Wisconsin-Superior, Superior, WI, USA, 5 p

Lv, J.; Shi, R.; Cai, Y.; Liu, Y.; Wang, Z.; Feng, J. & Zhao, M. (2010). Assessment of 20 organochlorine pesticides (OCPs) pollution in suburban soil in Tianjin, China. *Bulletin of Environ. Contamin. Toxicol.*, 85,2, 137-141, 0007-4861

Lyman, G.; Throssell, C.; Johnson, M.; Stacey, G.; Brown, C. (2007). Golf course profile describes turfgrass, landscape and environmental stewardship features. Online. *Applied Turfgrass Science,* doi:10.1094/ATS-2007-1107-01-RS

Ma, Q.; Smith, A.; Hook, J.; Smith, R. & Bridges, D. (1999). Water runoff and pesticide transport from a golf course fairway: observations vs. Opus model simulations. *J. Environ. Qual.*, 28, 1463-1473

Manahan, S. (1994). *Environmental Chemistry, 6th ed.* Lewis, Boca Raton, FL, USA

Mane, U. & Muley, D. (1984). Acute toxicity of endosulfan 35EC to two freshwater bivalve molluscs from Godavari River at Maharashtra State, India. *Toxicol.Lett.*, 23, 2, 147-155

Mannering, J. & Meyer, L. (1963). The effect of various rates of surface mulch on infiltration and erosion. *Soil Sci. Soc. Amer. Proc.*, 27, 84-86

Margni, M.; Rossier, D.; Crettaz, P. & Jolliet, O. (2002). Life cycle impact assessment of pesticides on human health and ecosystems. *Agric. Ecosyst. Environ.*, 93, 379-392

Materna, E.; Rabeni, C. & LaPoint, T. (1995). Effects of the synthetic pyrethroid insecticide, esfenvalerate, on larval leopard frogs (Rana spp.). *Environ.Toxicol.Chem.*, 14, 4, 613-622

Matsumura, F. (1985). *Toxicology of Insecticides*, Plenum Press, New York, USA

Matthews, H.; Lave, L. & MacLean, H. (2002). Life cycle impact assessment: A challenge for risk analysis. *Risk Anal.* 22, 852-860

Mayer, F. & Ellersieck, M. (1986). Manual of acute toxicity: Interpretation and data base for 410 chemicals and 66 species of freshwater animals. Resour.Publ.No.160, U.S.Dep.Interior, Fish Wildl.Serv., Washington, DC, USA, 505 p

McCall, E.; Scott, G. & Hurley, J. (1988). *A comparison of agricultural non-point source runoff from black plastic and conventional-till tomato test plots,* Project Agreement No. 14-08-0001-G1251, Technical completion report G1251-07, South Carolina Water Resources Research Institute, Columbia, SC, USA

McCarty, L.; Gregg, M. & Toler, J. (2007). Thatch and mat management in an established creeping bentgrass golf green. *Agron. J.*, 99, 1530-1537

McVay, K.; Radcliffe, D. & Hargrove, W. (1989). Winter legume effects on soil properties and nitrogen fertilizer requirements. *Soil Sci. Soc. Am. J.*, 53, 1856-1862

Milesi, C.; Running, S.; Elvidge, C.; Dietz, J. & Tuttle, B.; Nemani, R. (2005). Mapping and modeling the biogeochemical cycling of turf grasses in the United States. Environ. Manage., 36, 3, 426-438

Murphy, J.; Rieke, P. & Erickson, A. (1992). Core cultivation of a putting green with hollow and solid tines. *Agron. J.*, 85, 1-9

Nerin, C.; Tornes, A.; Domeno, C. & Cacho, J. (1996). Adsorption of pesticides on plastic films used as agricultural soil covers. *J. Agric. Food Chem.*, 44, 4009-4014

Nishiuchi, Y., H. Iwamura, and K. Asano. (1985). Toxicity of pesticides to some aquatic animals. VI. Acute toxicity of latest registered pesticides to some aquatic animals. *Aquat.Ecol.Chem./Seitai Kagaku*, 8, 1, 13-15

Office of Pesticide Programs. (2000). Pesticide Ecotoxicity Database (Formerly: Environmental Effects Database (EEDB)). Environmental Fate and Effects Division, U.S.EPA, Washington, DC, USA

Olima, C.; Pablo, F. & Lim, R. (1997). Comparative tolerance of three populations of the freshwater shrimp (*Paratya australiensis*) to the organophosphate pesticide, chlorpyrifos. *Bull.Environ.Contam.Toxicol.*, 59, 321-328

Omkar & Murti, R. (1985). Toxicity of some pesticides to the freshwater prawn, *Macrobrachium dayanum* (Henderson) (Decapoda, Caridea). *Crustaceana,* 49, 1, 1-6

Pan, D. & Liang, X. (1993). Safety study of pesticides on bog frog, a predatory natural enemy of pest in paddy field. *J.Hunan Agricult.Coll.*, 19, 1, 47-54

Petrovic, A. (1979). Dissertation Abstract ADG80-06178, Michigan State University; http://guides.lib.msu.edu/page.phtml?page_id=1312)

Racke, K. (2000). Pesticides for turfgrass pest management: Uses and environmental issues, In: *Fate and Management of Turfgrass Chemicals*, Clark, J. & Kenna, M., (Eds.), pp. 45-64, ACS Symposium Series, Vol. 743, American Chemical Society, Washington, DC, USA

Ranells, N. & Wagger, M. (1996). Nitrogen release from grass and legume cover crop monocultures and bicultures. *Agron. J.*, 88, 777-782

Raturi, S.; Islam, K.; Caroll, M. & Hill, R. (2005). Carbaryl, 2,4-D, and triclopyr adsorption in thatch-soil ecosystems. *Environ. Sci. Health, Part B,* 40, 697-710

Rice, P.; McConnell, L.; Heighton, L.; Sadeghi, A.; Isensee, A.; Teasdale, J.; Abdul-Baki, A.; Harman-Fetcho, J. & Hapeman, C. (2001). Runoff loss of pesticides and soil: A comparison between vegetative mulch and plastic mulch in vegetable production systems. *J. Environ. Qual.*, 30, 1808-1821

Rice, P.; Harman-Fetcho, J.; Teasdale, J.; Sadeghi, A.; McConnell, L.; Coffman, C.; Herbert, R.; Heighton, L. & Hapeman, C. (2004). Use of vegetative furrows to mitigate copper loads and soil loss in runoff from polyethylene (plastic) mulch vegetable production sytems. *Environ. Toxicol. Chem.*, 23, 3, 719-725

Rice, P.; Harman-Fetcho, J.; Sadeghi, A.; McConnell, L.; Coffman, C.; Teasdale, J.; Abdul-Baki, A.; Starr, J.; McCarty, G.; Herbert, R. & Hapeman, C. (2007). Reducing insecticide and fungicide loads in runoff from plastic mulch with vegetative-covered furrows. *J. Agric. Food Chem.*, 55, 1377-1384

Rice, P.; McConnell, L.; Heighton, L.; Sadeghi, A.; Isensee, A.; Teasdale, J.; Abdul-Baki, A.; Harman-Fetcho, J. & Hapeman, C. (2002). Comparison of copper levels in runoff from fresh-market vegetable production using polyethylene mulch or a vegetative mulch. *Environ. Toxicol. Chem.*, 21, 24-30

Rice, P. & Horgan, B. (2009). Fungicide and nutrient transport with runoff from creeping bentgrass turf. *International Turfgrass Soc. Res. J.*, 11, 61-75

Rice, P.; Horgan, B. & Rittenhouse, J. (2010a). Pesticide transport with runoff from creeping bentgrass turf: Relationship of pesticide properties to mass transport. *Environ. Toxicol. Chem.*, 29, 1209-1214

Rice, P.; Horgan, B. & Rittenhouse, J. (2010b). Evaluation of core cultivation practices to reduce ecological risk of pesticides in runoff from *Agrostis palustris*. *Environ. Toxicol. Chem.*, 29, 1215-1223

Riederer, A.; Smith, K.; Barr, D.; Hayden, S.; Hunter, R.; Ryan, B. (2010). Current and historical used pesticides in residential soil from 11 homes in Atlanta, Georgia, USA. *Arch. Environ. Contam. Toxicol.* 58, 4, 908-917, 0090-4341

Savitz, J.; Wright, D. & Smucker, R. (1994). Toxic effects of the insecticide Diflubenzuron (Dimilin) on survival and development of nauplii of the estuarine copepod, *Eurytemora affinis*. *Mar. Environ. Res.*, 37, 297-312

Schulz, R. (2004). Field studies on exposure, effects, and risk mitigation of aquatic nonpoint-source insecticide pollution: a review. *J. Environ. Qual.*, 33, 419-448

Scott, G.; Fulton, M.; Moore, D.; Chandler, G.; Bidleman, T.; Key, P.; Hampton, T.; Marcus, J.; Jackson, K.; Baughman, D.; Trim, A.; Williams, L.; Louden, C. & Patterson, E. (1990). *Agricultural insecticide runoff effects on estuarine organisms: Correlating laboratory and field toxicity testing with ecotoxicological biomonitoring*. USEPA Rep. CR-813138-02-1, Gulf Breeze, FL. USA

Shivakumar, R., and M. David. (2004). Endosulfan Induced Behavioural Changes in the Freshwater Fish, Cyprinus carpio. *J.Ecotoxicol.Environ.Monit.*, 14, 1, 65-69

Shuman, L. (2002). Phosphorus and nitrate nitrogen in runoff following fertilizer application to turfgrass. *J. Environ. Qual.*, 31, 1710-1715

Siefert, R. (1987). Effects of Dursban (chlorpyrifos) on aquatic organisms in enclosures in a natural pond. Final Report, U.S.EPA, Duluth, MN, USA, 214 p

Smith, M.; Frye, W. & Varco, J. (1987). Legume winter cover crops. *Adv. Soil Sci.*, 7, 95-139

Steel, R.; Torrie, J. & Dickey, D. (1997). *Principles and Procedures of Statistics: A Biometrical Approach*, 3rd ed. McGraw-Hill, New York, NY, USA

Sur, H.; Mastana, P. & Hadda, M. (1992). Effect of rates and modes of mulch application on runoff, sediment and nitrogen loss on cropped and uncropped fields. *Trop. Agric.*, 69, 319-322

Takakura, T. & Fang, W. (2001). Climate under cover, 2nd ed., . Kluwer Academic Publishers, Boston, MA, USA, http://ecaaser3.ecaa.ntu.edu.tw/weifang/Bio-ctrl/cuc-chap1.pdf

Teasdale, J. (1996). Contribution of cover crops to weed management in sustainable agricultural systems. *J. Prod. Agric.*, 9, 475-479

Thurman, E.; Goolsby, D.; Meyer, M.; Mills, M.; Pones, M. & Kolpin, D. (1992). A reconnaissance study of herbicides and their metabolites is surface water of the Midwestern United States using immunoassay and gas chromatography/mass spectrometry. *Environ. Sci. Technol.*, 26, 2440-2447

Topp, E. & Smith, W. (1992). Sorption of the herbicides atrazine and metolachlor to selected plastics and silicone rubber. *J. Environ. Qual.*, 21, 316-317

Turgeon, A. (1985). *Turfgrass Management, 2nd ed.*, Reston Publishing, Reston, VA, USA

Vuik, J.; van der Poll, J.; Vink, R. & de Vos, R. (1990). Adsorption experiments of etridiazole and oxamyl on polyethylene sheets and poly(vinyl chloride) tubing used in horticulture. *J. Agric. Food Chem.*, 38, 328-330

Wacksman, M.; Maul, J. & Lydy, M. (2006). Impact of atrazine on chlorpyrifos toxicity in four aquatic vertebrates. *Arch.Environ.Contam.Toxicol.*, 51, 4, 681-689

Wan, Y. & El-Swaify, S. (1999). Runoff and soil erosion as affected by plastic mulch in a Hawaiian pineapple field. *Soil Tillage Res.*, 52, 29-35

Watson, J.; Kaerwer, H. & Martin, D. (1992). The turfgrass industry. In *Turfgrass*, Waddington D., (Ed.), Agronomy Monograph 32. pp. 29-88, ASA, Madison, WI, USA

Wauchope, R. (1978). The pesticide content of surface water draining from agricultural fields - a review. *J. Environ. Qual.*, 7, 459-472

Wauchope, R. (1996). Pesticides in runoff: measurement, modeling and mitigation. *J. Environ. Sci. Health*, B31, 337-344

Wauchope, R.; Williams, R. & Marti, L. (1990). Runoff of sulfometuron-methyl and cyanazine from small plots: Effects of formulation and grass cover. *J. Environ. Qual.*, 19, 119-125

Weber, J.; Halsall, C.; Muir, D.; Teixeira, C.; Small, J.; Solomon, K.; Hermanson, M.; Hyung, H.; & Bidleman, T. (2010). Endosulfan, a global pesticide: A review of its fate in the environment and occurrence in the Artic. *Sci. Total Environ.* 408, 15, 2966-2984, 0048-9697

White, R. & Dickens, R. (1984). Thatch accumulation in bermudagrass as influenced by cultural practices. *Agron J.*, 76, 19-22

Wotzka, P.; Lee, J.; Capel, P.& Lin, M. (1994). Pesticide concentrations and fluxes in an urban watershed. *AWRA Technical Publication Series TPS-94-4*. American Water Resources Association, Herndon, VA, USA, pp 135-145

Zuzel, J. & Pikul, J. (1993). Effects of straw mulch on runoff and erosion from small agricultural plots in northeastern Oregon. *Soil Sci.*, 156, 111-117

22

Endosulfan in China: Usage, Emissions, and Residues

Hongliang Jia[1], Liyan Liu[2], Yeqing Sun[3] and Yi-Fan Li[4]
[1,3,4]*International Joint Research Center for Persistent Toxic Substances (IJRC-PTS), Dalian Maritime University*
[2,4]*IJRC-PTS, State Key Laboratory for Urban Resources and Environment, Harbin Institute of Technology*
[3]*Environmental System Biology Institute, Dalian Maritime University*
[4]*Science and Technology Branch, Environment Canada*
[1,2,3]*P. R. China*
[4]*Canada*

1. Introduction

Endosulfan (6,7,8,9,10,10-hexachloro-1,5,5a,6,9a-hexahy-dro-6,9-methano-2,3,4-benzodioxyanthiepin-3-oxide), an organochlorine pesticide, was first introduced in the 1950s by Hoechst AG and FMC Corporation. As a non-systemic and ingested insecticide and miticide, endosulfan was extensively used for over 30 years on a wide variety of food crops, such as grains, tea, fruits, and vegetables, on non-food crops, such as tobacco and cotton, and also used as a wood preservative (Sutherland et al, 2000). The commercial technical endosulfan consists of 70% α-endosulfan and 30% β-endosulfan (Maier-Bode, 1968; Rice et al., 1997), which are with similar insecticidal properties but different physicochemical properties (Peterson & Batley, 1993; Schmidt et al., 1997). Main trade names of endosulfan are Beosit, Chlortiepin, Cyclodan, Devisulphan, Endocel, Endosol, Hildan, Insectophene, Malix, Rasayansulfan, Thifor, Thimul, Thiodan, Thionex, Thiosulfan. In the environment, the cyclic sulfite group of endosulfan can be hydrolyzed to form a less toxic endosulfan diol (Peterson & Batley, 1993) or oxidized to the corresponding endosulfan sulfate (Chandler & Scott, 1991; Miles & Pfeuffer, 1997). It has been reported that endosulfan is genotoxic in mammalian cells (Chaudhuri et al., 1999; ASTDR, 2000). Endosulfan is extremely toxic to fish and aquatic invertebrate (Verschureren, 1983; Sunderam et al., 1992) and it has been implicated increasingly in mammalian gonadal toxicity (Sinha et al., 1995), genotoxicity (Sutherland et al., 2002), and neuro toxicity (Southan & Kennedy, 1995). In India, the National Institute of Occupational Health conclusively proved that endosulfan was a causative factor in the incidence of all crippling illness in the Kasargode area of Kerela where this insecticide was sprayed aerially in the cashew plantations (Devakumar, 2002). α- and β-endosulfan, and their metabolize endosulfan sulfate are generally considered to be equally toxic and are classified by the US Environmental Protection Agency as priority pollutants (Keith & Telliard, 1979), and endosulfan was recently banned for use in the USA (Lubick, 2010).

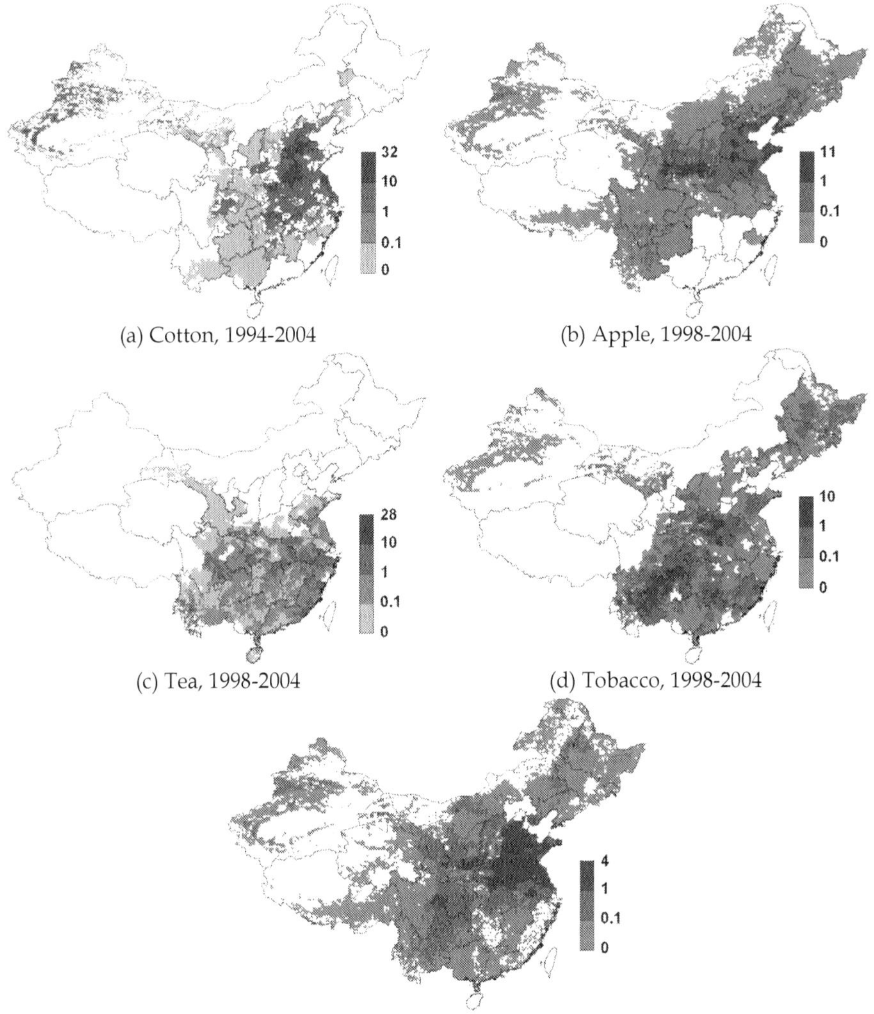

(a) Cotton, 1994-2004

(b) Apple, 1998-2004

(c) Tea, 1998-2004

(d) Tobacco, 1998-2004

(e) Wheat, 1998-2004

Fig. 2. Distribution of endosulfan usage in Chinese crops from 1994 to 2004 with 1/4° longitude by 1/6° latitude resolution (unit: t/cell).

3.2 Spatial endosulfan use distribution

In order to obtain a high resolution of spatial distribution of Chinese endosulfan usage, gridded lands with a 1/4° longitude by 1/6° latitude resolution for different crops (Li et al., 2001) was used to allocate the use of this insecticide, and the gridded Chinese endosulfan usage inventories with the same resolution were compiled (Jia et al., 2009a). These usage inventories were included endosulfan usage on cotton from 1994 to 2004 (Fig, 2 (a)),

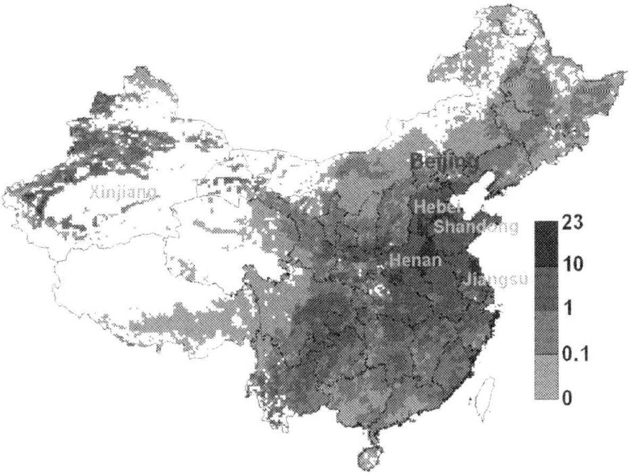

(a)

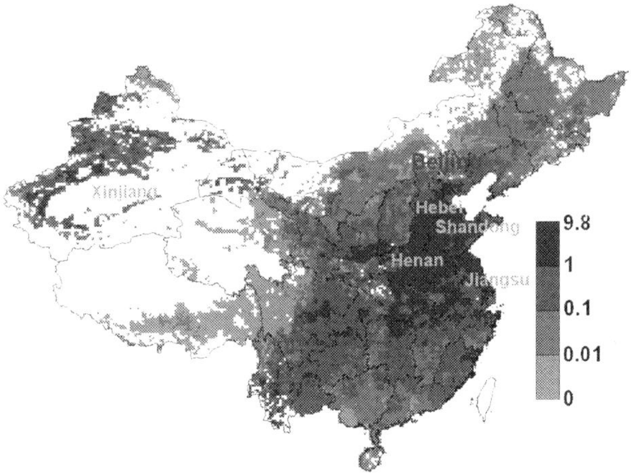

(b)

Fig. 3. Distribution of total α-endosulfan (a) and β-endosulfan (b) usage in China from 1994 to 2004 with 1/4° longitude by 1/6° latitude resolution (unit: t/cell).

endosulfan usage on apple, tea, tobacco, and wheat from 1998 to 2004 (Fig, 2 (b) - (e)). These maps show that the dense use of endosulfan on cotton in China was in the central east of China, including Provinces of Hebei, Henan, Shandong, and Jiangsu. The use of endosulfan on apple in China was also mainly in the central east of China, including provinces of Liaoning, Hebei, Henan, Shandong, Shaanxi, and Shanxi. The dense use of endosulfan on tea was in the south of China, and the dense use of endosulfan on tobacco in China was in Provinces of Yunnan, Guizhou, and Henan, where tobacco widely grows. The area of major use of endosulfan on wheat in China was in the central east of China, including Provinces of Hebei, Henan, Shandong, Jiangsu, and Anhui.

3.3 Usage data for α- and β-endosulfan

Chinese technical endosulfan usage from 1994 to 2004 was estimated to be around 25700 t, as the proportion of α-endosulfan and β-endosulfan is 7: 3, the usage was 17990 t for α-endosulfan and 7710 t for β-endosulfan in this period. Gridded usage inventories for α-endosulfan and β-endosulfan on a 1/4° longitude by 1/6° latitude grid system are given in Fig. 3, which indicates that the most intensive use of endosulfan in China was in the south of Hebei Province, west of Shangdong Province, east of Henan Province, north of Anhui Province, east of Jiangsu Province, and some areas in Yunnan Province and Xinjiang Autonomous Region.

4. Modelling endosulfan in Chinese soil and air

4.1 Method

The simplified gridded pesticide emission and residue model (SGPERM) (Li et al., 2004b) has been used to estimate the emission and the residues for different pesticides in different areas, such as for toxaphene in the United States (Li et al., 2001), for α-HCH (Li et al., 2000) and β-HCH (Li et al., 2003) on a global scale. Details of the SGPERM can be found elsewhere (Li et al. 2004b; Jia et al., 2009b), and only a brief description is given here.

The model considers a single plant/harvest cycle in each year and two different emission scenarios for simplicity. The first one is a spraying event and the second is a tilling event. Both events were assumed to happen at the same time.

Considering two consecutive years, Year 1 and Year 2, we suppose that $U^{(1)}$ t of endosulfan is used in Year 1, and the emission in Year 1 will be (in the unit of t)

$$E_{sp}{}^{(1)} = U^{(1)} F_{sp} \tag{1}$$

where F_{sp} is annual emission factor for spraying event. Soil is free of the pesticides considered before its use. After spray, the residue in soil becomes (in the unit of t)

$$R_1{}^{(1)} = U^{(1)} - E_{sp}{}^{(1)} \tag{2}$$

The residues of all endosulfan applied are assumed to be mixed and stay in the soil until the next tilling event in the next year. In Year 2, the residue due to the previous year is given by

$$R_1{}^{(2)} = R_1{}^{(1)} e^{-\mu t} \tag{3}$$

where μ is the degradation rate constant of endosulfan in soil (in the unit of year−1), and equal to $\ln 2/t_{1/2}$ with $t_{1/2}$ being the half-life of degradation (in year) for the pesticide in soil. t is the time period and equal to 1 year in our study.

Suppose $U^{(2)}$ of endosulfan are used in Year 2, and the emission due to spray in Year 2 will be

$$E_{sp}{}^{(2)} = U^{(2)} F_{sp} \tag{4}$$

Emission due to the tilling is given by

$$E_{tl}^{(2)} = R_1^{(2)} F_{tl} \tag{5}$$

where F_{tl} is annual emission factor for tilling event. After both the events in Year 2, the emission is given by $E^{(2)} = E_{sp}^{(2)} + E_{tl}^{(2)}$ and the residue by

$$R_2^{(2)} = (U^{(2)} - E_{sp}^{(2)}) + (R_1^{(2)} - E_{tl}^{(2)}) \tag{6}$$

Therefore, in Year 2, there are two values for residues, $R_1^{(2)}$ and $R_2^{(2)}$. The former is the residues of endosulfan remaining in the soil from the previous year and the latter is residues of endosulfan remaining in the soil from the previous year plus the residues due to the current use of the pesticide. Obviously, $R_1^{(2)} - E_{tl}^{(2)}$ and $R_2^{(2)}$ are the minimum and maximum values of residue in the same year, respectively. The unit of all U, E, and R is in t, given total usage, emission, and residue in 1 year, respectively. The residue in soil decreases over time by degradation until the sudden increase in the next year due to application of the pesticide. Equations for Year 2 were used for any year after Year 2.

It appears that the calculation of emissions on a grid system covering the whole China requires considerable human and computer resources. The difficulty in undergoing such a large number of calculations is that some of the input data required for the calculations, such as the timing and mode of application and the degradation rates in the soil. Within the scope of the present study, a number of assumptions have been made for simplicity. In present study, only one plant/harvest cycle is assumed in a given year, and emission of endosulfan residues in the agricultural soil are assumed to start at the time of tilling in each year. The errors created by these assumptions are reduced since only annual emissions are considered here. But the re-emission by endosulfan deposition on soil and water transportation in the atmosphere has not been considered here, so in this case, only first-time emissions are estimated. Besides emission to the air, the residues of the pesticides in soil are lost over time by degradation.

To calculate the emission and residue of endosulfan, emission factors are crucial parameters in the SGPERM. Emission factors of the pesticides represent the emission potential of pesticides in a geographical area, and are used to calculate the emission. In present study, a simplified approach is taken for estimating the annual emission factor by considering two different emission factor scenarios: one is spraying event (F_{sp}) and the other is tilling event (F_{tl}). The definition of the emission factor is given as (Li et al., 2000)

$$F_k = \frac{E_k}{V_k} \tag{7}$$

where the subscript k = s corresponds to a spraying event and k = t corresponds to a tilling event. In the case of a spraying event, Vs is the current usage of pesticide as a spray, E_{sp} is the amount of pesticide emitted to the atmosphere due to the spraying event, and the emission factor (F_{sp}) is defined as the ratio of the emission (E_{sp}) to the usage (U_{sp}). In the case of a tilling event, U_t is the current usage of pesticide applied using a soil incorporation mode (if any) plus any pesticide residues remaining in the soil due to applications in previous years from all application modes. E_{tl} is the amount of pesticide emitted to the atmosphere, and the emission factor (F_{tl}) is defined as the ratio of the emission (E_{tl}) to the amount of current usage of pesticide applied by a soil treatment mode plus the residues in the soil. This definition is convenient for considering the emissions due to several consecutive years of pesticide application (Li et al., 2000).

The emission factors of organochlorine pesticides are a very complicated function of a large number of variables. They depend on a number of factors such as the physical and chemical properties of pesticides, application modes, the local meteorological conditions when they applied, the soil type, and the temperature (Li et al., 2000; Li et al., 2001).

Some prior researchers performed on emission of pesticides on a local scale (Baas & Huygen, 1992; Baas, 1994; Baart et al., 1995; Dörfler et al., 1991; Rüdel, 1997), and Rüdel (1997) concluded the volatilisaion rates of endosulfan were 50-70% from soil surfaces after this pesticide application. Further, Scholtz et al. tailed estimated emission factors for 20 pesticides for North America (1997) and on a global scale (1998). It had been realized that most pesticides emissions from soil residues happened due to high temperature (Harner et al., 2001). In this case, the emission factor (F_{tl}) defined as the ratio of annual emission (E_{tl}) to the amount of residues can be treated as the annual emission factor of endosulfan due to tilling or soil-air exchange. This will not create significant errors since only annual emissions are considered in this study.

Emission factors of endosulfan for China area are obtained from Scholtz et al. (1998), and were organized into a 1/6° x 1/4° grid system. The emission factors range from 0.41 to 0.55 for the spray mode are given in Fig. 4 (a) and at a tilling mode, the range is from 0.068 to 0.092 (Fig. 4 (b)). The areas with the highest emission factors are Xinjiang autonomous region, northwest of China, and Hainan province, the south of China.

It appears that the calculation of emissions on a grid system covering the whole China requires considerable human and computer resources. The difficulty in undergoing such a large number of calculations is that some of the input data required for the calculations, such as the timing and mode of application and the degradation rates in the soil. Within the scope of the present study, a number of assumptions have been made for simplicity. Emission and residue of endosulfan can be calculated by using the simplified gridded pesticide emission and residue model (SGPERM) (Li et al., 2004b), and in order to calculated the air/soil concentrations, a module were added to SGPERM. The detailed descriptions can be found elsewhere (Jia et al., 2009b). In present study, only one plant/harvest cycle is assumed in a given year, and emission of endosulfan residues in the agricultural soil are assumed to start at the time of tilling in each year. The errors created by these assumptions are reduced since only annual emissions are considered here. But the re-emission by

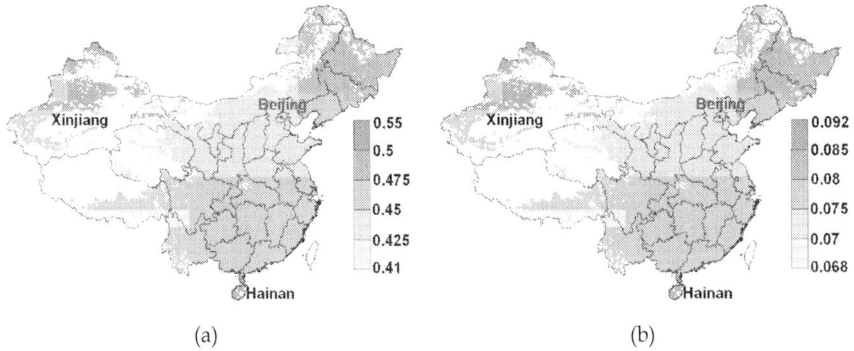

Fig. 4. Emission factors of endosulfan for (a) spray event, and (b) tilling event in China with 1/4° longitude by 1/6° latitude resolution.

endosulfan deposition on soil and water transportation in the atmosphere has not been considered here, so in this case, only first-time emissions are estimated. Besides emission to the air, the residues of the pesticides in soil are lost over time by degradation.

4.2 Temporal trends

Temporal trends of emissions to air and the highest and the lowest residues in Chinese agricultural soil for α- + β-endosulfan from 1994 to 2004 are given in Fig. 5. The annual emission were approximately 600 t from 1994 to 1997, 1,200 t for each year from 1998 to 2004, and the annual highest residue was approximately 629 t in 2004 with the lowest residue in the same year being approximately 120 t, around one fifth of the highest value. The highest residues of these 2 isomers appeared after the application of endosulfan, and the lowest residues occurred before the first application of this pesticide in the year.

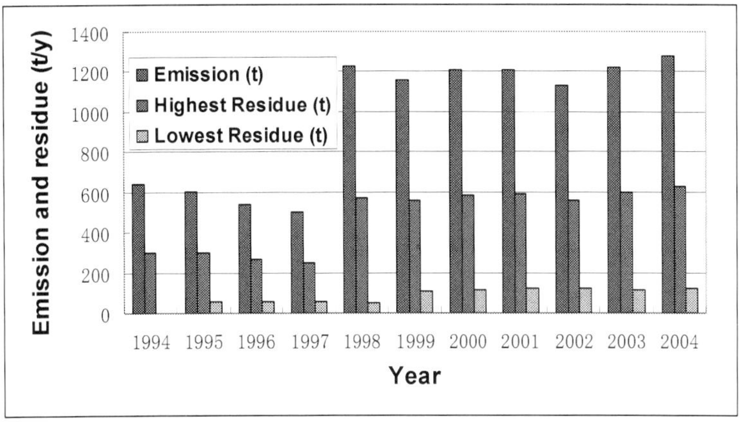

Fig. 5. Annual endosulfan emission and residue in China from 1994 to 2004 (Jia et al., 2009b)

4.3 Residues in soil and soil concentrations

The annual average soil concentrations of endosulfan in agricultural soil of each cell can be determined by using its residue inventories, given by

$$C_{soil} = R / (ALD) \tag{8}$$

where A is the area of the agricultural land in a grid cell, L is the depth of the soil (assuming that endosulfan residues in each grid cell are contained in the top 15-cm layer of cropland soil), and D is the density of the soil. The gridded soil density with 1/4° longitude by 1/6° latitude resolution was interpolated from a soil density dataset with 1° longitude by 1° latitude resolution, available at http://islscp2.sesda.com/ISLSCP2_1/data/hydrology_soils/islscp2soils-1deg. The soil densities in the model domain are from 0.53 to 1.79 g/cm³. R is either $R_1^{(2)}$ or $R_2^{(2)}$, but with a unit of ng in this equation. The unit for C soil is usually given in ng/g dry weight (dw). There are also two values for C soil, the lowest value when $R = R_1^{(2)}$ and the highest when $R = R_2^{(2)}$.

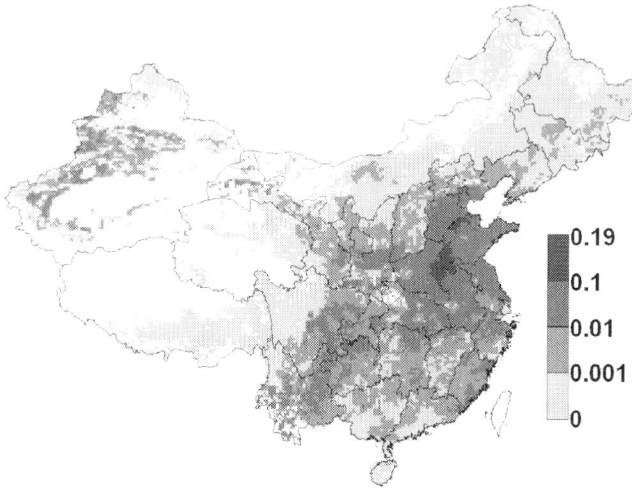

(a)

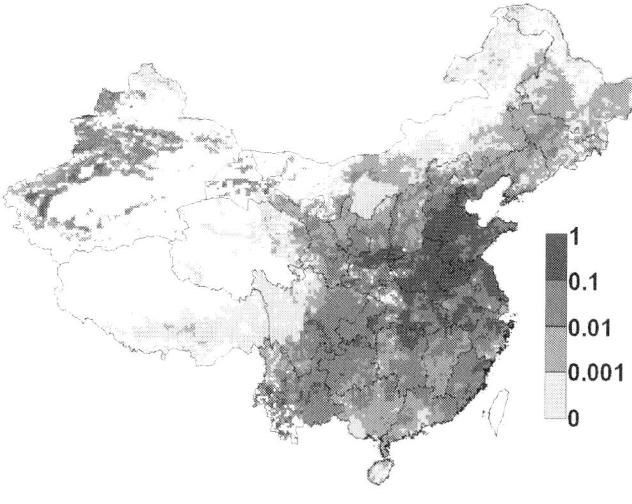

(b)

Fig. 6. Distribution of (a) the lowest and (b) the highest residuest of endosulfan in Chinese agricultural soil in 2004 with 1/4° longitude by 1/6° latitude resolution. (unit: t/cell)

Distribution of the lowest and the highest residues of endosulfan (α- + β-isomer) in Chinese agricultural soil in 2004 with 1/4° x 1/6° longitude and latitude resolution are shown in Fig. 6, which indicating heavier residues of these 2 isomers located in the central east China, including Henan, Shandong, Anhui provinces, and the south of Hebei, north and east Jiangsu, parts of Hubei, Hunan provinces. In 2004 the lowest and the highest residues endosulfan were 120 and 629 t, respectively in Chinese cropland where endosulfan had been applied.

Fig. 7 and Fig. 8 depictes the gridded lowest and highest concentration of α-endosulfan and β-endosulfan in Chinese agricultural soil in 2004 with 1/4° x 1/6° longitude and latitude resolution, respectively. The densest concentrations of endosulfan in Chinese agricultural soil were in Yunnan Province, where tobacco planning is a common agricultural practice. Other regions with dense soil concentrations included northern Gansu, southern Anhui, north of Fujian, and parts of Xinjiang Autonomous Region. As shown in Fig. 7 and Fig. 8, while the highest concentration is only several times higher than the lowest concentration for β-endosulfan, the difference between the lowest and the highest concentration for α-endosulfan in agricultural soil is very large, reaching as high as 2-3 orders of magnitude. The much longer (135 days) half-life for β-endosulfan than α-endosulfan (35 days) leads to a more stable soil concentration in soil for β-endosulfan.

4.4 Emissions to air and air concentrations

The annual average concentration of endosulfan in air can be calculated by using a simplified Gaussian formula (Jiang & Wu, 1990).

$$C_{air} = E / (ut\Delta l\Delta h) \tag{9}$$

where C air is air concentration in each grid cell (in unit pg/m³), E is annual emission in the grid cell (=E_{sp} + E_{tl}) (t before, but needs to be transferred to pg here), t is the time period, and equal to 1 year here, Δl is the dimension of the grid cell (24 km used here), Δh the height of the boundary layer, which is assumed to be 1,000 m in this study, and u is the annual mean wind speed (in m/year), which was obtained from the National Centers for Environmental Prediction (NCEP) reanalysis at standard atmospheric pressure levels (Kalnay et al. 1996). The annual wind speeds (in m/s) in the model domain were from 2.7 to 8.0. It is worthwhile to point out first that the air concentration estimated using our model is an annual average, and actual air concentration will vary from season to season. Secondly, our model does not consider the transport of the insecticide in the atmosphere, which could be very important in some areas.

Emissions of α- and β-endosulfan on a 1/4° longitude by 1/6° latitude resolution in 2004 are shown in Fig. 9. Total emissions of α- and β-endosulfan in 2004 were estimated to be 890 t and 390 t, respectively, and the highest emissions per each grid cell were 1.45 t for α-endosulfan and 0.64 t for β-endosulfan.

Distribution of air concentrations for these 2 isomers in 2004 with 1/4° x 1/6° longitude and latitude resolution are shown in Fig. 10. The highest α-endosulfan concentrations in air were in the Shandong, Henan, Shanxi, and Shaanxi provinces, which are the wheat and cotton growing area.

(a)

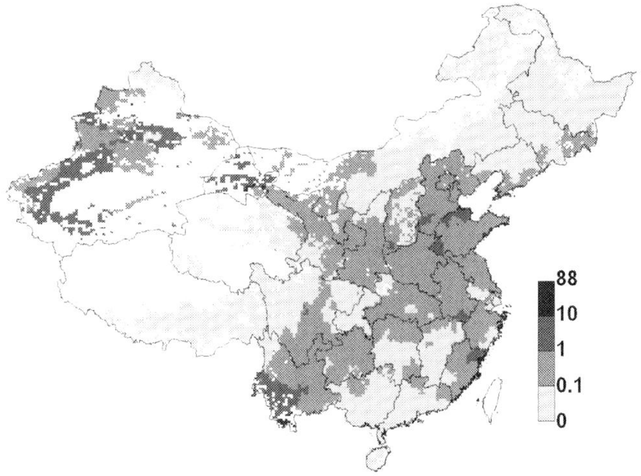

(b)

Fig. 7. Distribution of of (a) the lowest and (b) the highest concentration (ng/g dw) of α-endosulfan in Chinese agricultural soil in 2004 with 1/4° longitude by 1/6° latitude resolution.

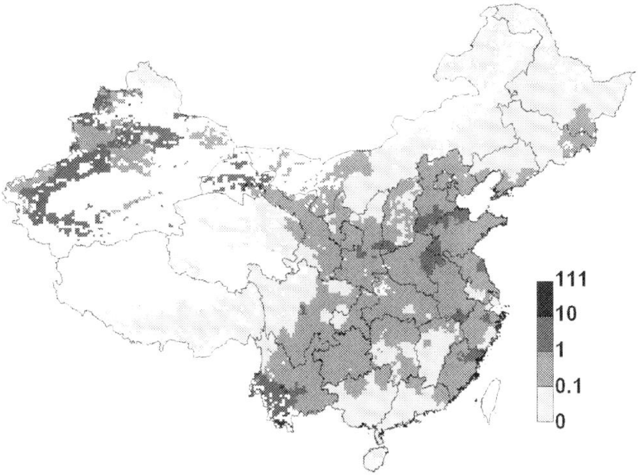

(a)

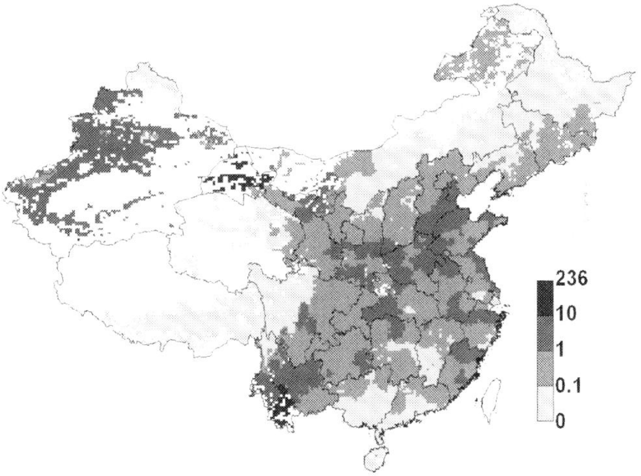

(b)

Fig. 8. Distribution of of (a) the lowest and (b) the highest concentration (ng/g dw) of β-endosulfan in Chinese agricultural soil in 2004 with 1/4° longitude by 1/6° latitude resolution.

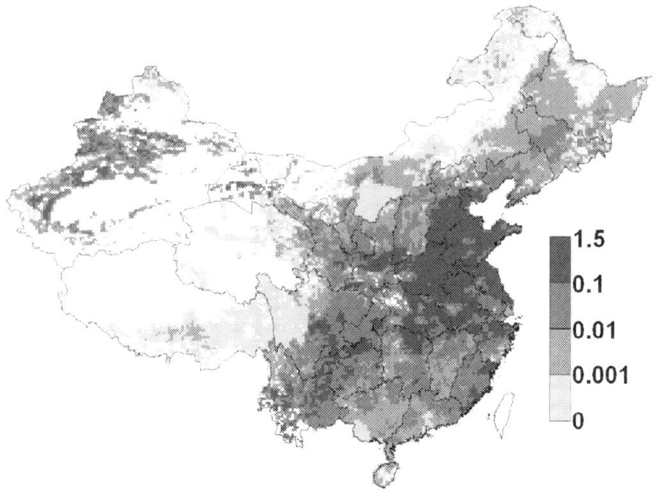

(a)

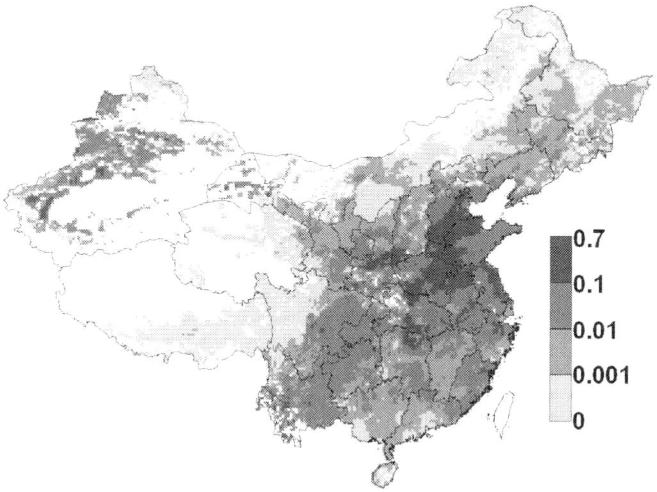

(b)

Fig. 9. Distribution of endosulfan emissions in China in 2004 with 1/4° longitude by 1/6° latitude resolution. (unit: t/cell)

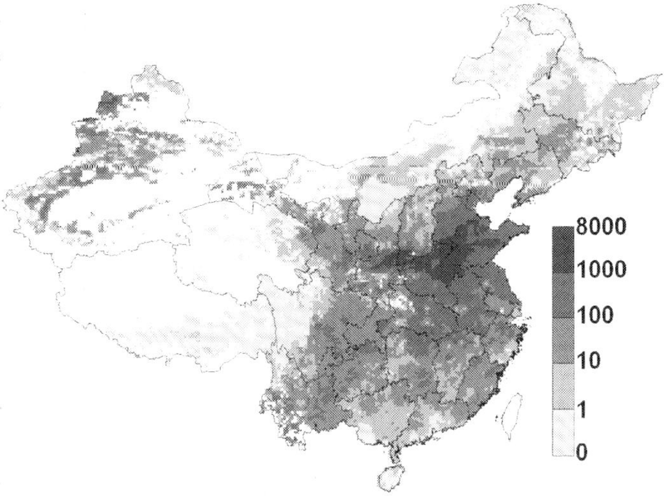

(a)

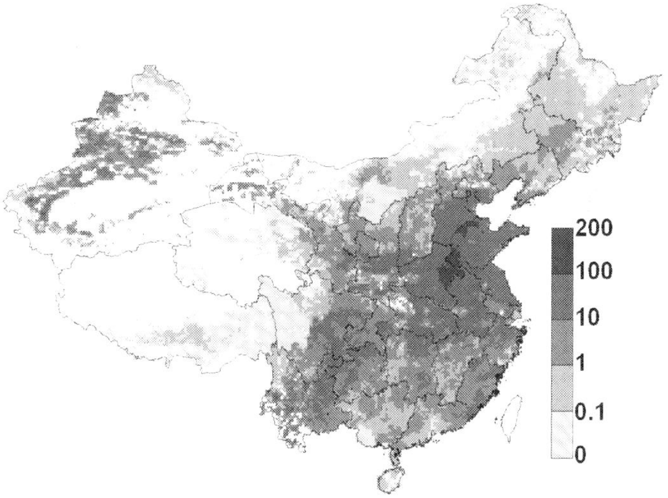

(b)

Fig. 10. Distribution of annual mean concentration of endosulfan (pg/m^3) in Chinese air in 2004 with 1/4° longitude by 1/6° latitude resolution.

endosulfan indicated their longer half-life in soil for β-endosulfan than α-endosulfan in soil. Besides, Paired-Sample T test was made to address the possible difference between monitoring and modeling data for both α- and β-endosulfan separately, and the results show that, at the 0.05 level, no significantly differences were found between modeled and monitoring results.

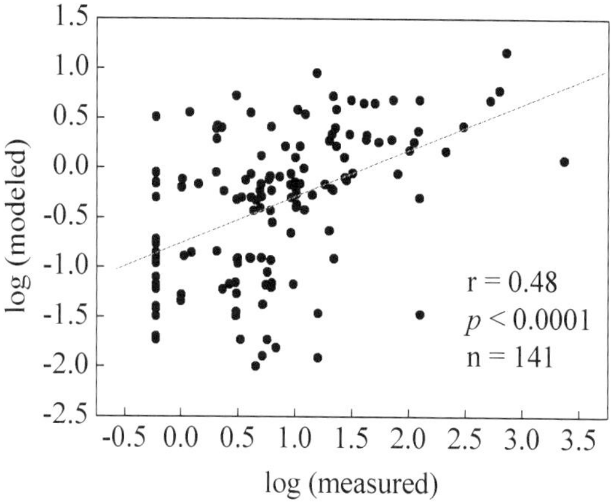

(a)

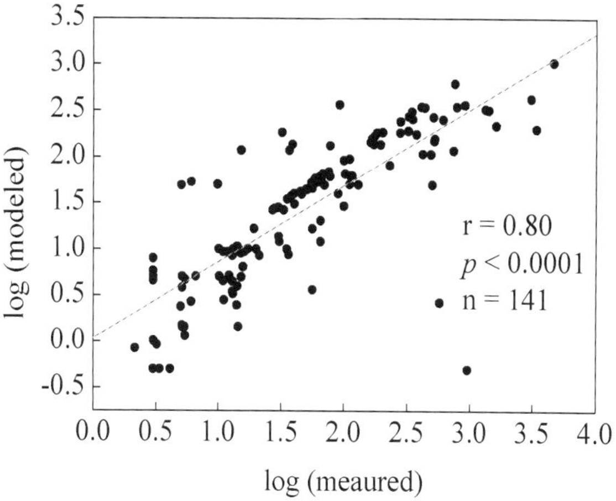

(b)

Fig. 13. Comparison between monitoring soil concentrations in 2005 and modeling data in 2004 for (a) α-endosulfan, (b) β-endosulfan. The monitoring data were from Liu et al (2009).

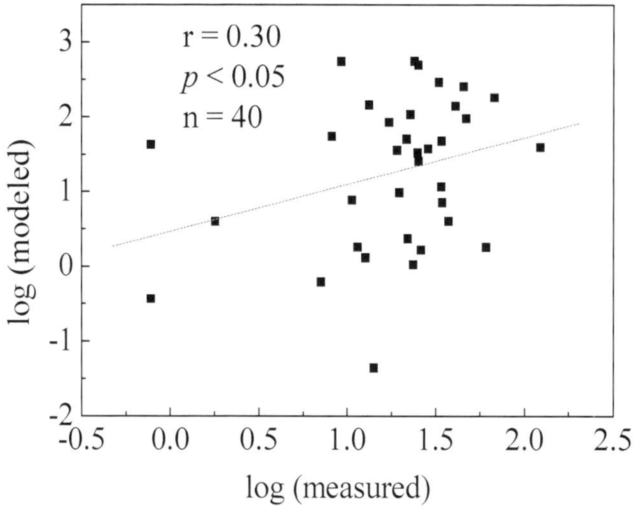

(a)

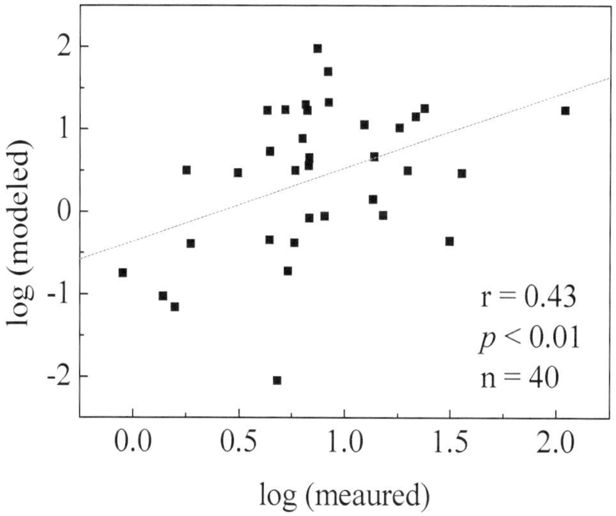

(b)

Fig. 14. Comparison between monitoring air concentrations in 2005 (Liu et al. 2009) and modeling data in 2004 for (a) α-endosulfan, (b) β-endosulfan.

6.2 Comparison between concentrations in air

Fig. 14. depicts correlation between modeling and monitoring concentrations (log-scal) in air across China. The measured soil concentration data for α- and β-endosulfan at the 40 monitoring sites in 2005 (Liu et al. 2009). were compared to their corresponding modeled concentration data in 2004 and a good consistence was found. Firstly, Paired-Sample T test was made to address the possible difference between monitoring and modeling data for both α- and β-endosulfan separately, and the results show that, at the 0.05 level, no significantly differences were found between modeled and monitoring results. Secondly, regression analysis between the monitoring and modeled data indicated a weak correlations with R = 0.30 for α-endosulfan and R = 0.43 for β-endosulfan. Besides, GM of measured air concentration of α-endosulfan from 40 sampling site was 18 pg/m^3, which is the same as our modeling results (18 pg/m^3), and for β-endosulfan, GM of measured concentration was 7 pg/m^3, which is also close to the modeled data (3 pg/m^3).

7. Conclusion and future work

The following conclusions were derived from this chapter:

- Chinese endosulfan usage inventories were developed with a high resolution of 1/4° longitude by 1/6° latitude, which localized the source of this pesticide in China.
- Chinese endosulfan inventories for emissions to air and residues in soil were calculated for both α- and β-endosulfan by using the SGPERM, and the inventories for air and soil concentrations were also compiled for these two isomers.
- Soil and air samples were collected from 141 and 40 sites over China, respectively, for α- and β-endosulfan and their concentrations in these two media were measured. The results gave general spatial trends of the occurrence of these two isomers in Chinese air and soil.
- Comparison between the measured and modeled results for both α- and β-endosulfan led to a good consistence between them, indicating the accuracy of the developed inventories.

This research has paved the way for further research for endosulfan in future, which includes development of inventories of emissions to air, residues in soil, and air and soil concentrations for endosulfan sulfate, the metabolize of both α- and β-endosulfan, the transport and the fate of endosulfan in Chinese environment, and risk assessment of this pesticide to the health of humans and wild lives. This work is also beneficial to the study of endosulfan on a global scale, including development of global gridded emission/residue inventories for this pesticide.

8. References

ASTDR. (2000). Toxicological Profile for Endosulfan. US Department of Health and Human Services, Agency for Toxic Substances and Disease Registry, Atlanta, Ga., USA.

Baart, A.; Berdowski, J. & van Jaarsveld, J. (1995). The calculation of atmospheric deposition of contaminants on the North Sea (in Dutch), Rep. TNO-MW-R95-138, TNO Inst. of Environ. Sci., Delft, The Netherlands.

Baas, J. (1994). Emission of pesticides from orchards to the atmosphere (in Dutch), Rep. IMWTNO R 94-040a, TNO Lnst. of Environ. Sci., Delft, The Netherlands.

Bass, J. & Huygen, C. (1992) Emission of pesticides from greenhouses to the atmosphere (in Dutch), Rep. IMW-TNO R 920304, TNO Inst. of Environ. Sci., Delft, The Netherlands.

Chandler, G. & Scott, G. (1991). Effects of sediment-bound endosulfan on survival reproduction and larval settlement of meiobenthic polychaetes and copepods, *Environ Toxicol Chem* 10: 375-382.

Chaudhuri, K.; Selvaraj, S. & Pal, A. (1999). Studies on the genotoxicity of endosulfan in bacterial systems, *Mutat Res* 439: 63–67.

Chinese Network for Public Science and Technology. http://database.cpst.net.cn/

Devakumar, C. (2002). Endosulfan aerial spray controversy in Kerela. *Pestic Res* J 14: 343–344.

Dörfler, U.; Adler-Körler, R.; Schneider, P.; Scheunert, I. & Korte, F. (1991). A laboratory model system for determing the volatility of pesticides from soil and plant surfaces, *Chemosphere* 23: 484-496.

Halsall, C.; Bailey, R.; Stern, G.; Barrie, L.; Fellin, P. & Muir, D. (1998). Multi-year observations of organohalogenated pesticides in the Arctic atmosphere, *Environ Pollut* 102: 51–62.

Harner, T.; Bidleman, T.; Jantunen, L. & Mackay, D. (2001). Soil- air exchange models of persistent pesticides in the U. S. Cotton Belt, *Environ Toxicol Chem* 20: 1612-1621.

Hung, H.; Halsall, C.; Blanchard, P.; Li, H.; Fellin, P. & Stern, G. (2002). Temporal trends of organochlorine pesticides in the Canadian Arctic atmosphere, *Environ. Sci. Technol* 36: 862–868.

Jia, H.; Li, Y.; Wang, D.; Cai, D.; Yang, M.; Ma, J. & Hu, J. (2009a). Endosulfan in China 1-gridded usage inventories, *Environ Sci Pollut Res* 16: 295-301.

Jia, H.; Liu, L.; Sun, Y.; Sun, B.; Wang, D.; Su, Y.; Kannan, K. & Li, Y. (2010). Monitoring and modeling endosulfan in Chinese surface soil, *Environ Sci Technol* 44: 9279-9284.

Jia, H.; Sun, Y.; Li, Y.; Tian, C.; Wang, D.; Yang, M.; Ding, Y. & Ma, J. (2009b). Endosulfan in China 2-emissions and residues, *Environ Sci Pollut Res* 16: 302-311.

Jiang, W. & Wu, X. (1990). A linked three-dimensional PBL and dispersion model in coastal regions, *Bound-lay Meteorol* 53: 43-62.

Kalnay, E.; Kanamitsu, M.; Kistler, R.; Collins, W.; Deaven, D.; Gandin, L.; Iredell, M.; Saha, S.; White, G.; Woollen, J.; Zhu, Y.; Leetmaa, A.; Reynolds, B.; Chellia, M.; Ebisuzaki, W.; Higgins, W.; Janowiak, J.; Mo, K.; Ropelewski, C.; Wang, J.; Jenne, R.; & Joseph, D. (1996). The NCEP/NCAR reanalysis project, *Bull Am Meteorol Soc* 77: 437-471.

Keith, L.; & Telliard, W. (1979). Priority pollutants: I-a perspective view, *Environ. Sci. Technol* *13*: 416-423.

Li, Y. & Bidleman, T. (2003). Correlation between global emissions of α-hexachlorocyclohexane and its concentrations in the arctic air, *J Environ Informatics* 1: 52-57.

Li, Y.; & MacDonald, R. (2005). Sources and pathways of selected organochlorine pesticides to the Arctic and the effect of pathway divergence on HCH trends in biota: A review, *Sci Total Environ* 342: 87-106.

Li, Y.; Bidleman, T. & Barrie, L. (2001). Toxaphene in the United States: (2) Emissions and Residues, *J Geophys Res* 106(D16): 17,929-17,938.

Li, Y.; Macdonald, R.; Ma, J.; Hung, H. & Venkatesh, S. (2004a). α-HCH Budget in the Arctic Ocean: The Arctic Mass Balance Box Model (AMBBM), *Sci Total Environ* 324: 115-139.

Li, Y.; Scholtz, M. & Heyst, B. (2003). Global gridded emission inventory of β-hexachlorocyclohexane, *Environ Sci Technol* 37: 3493-3498.

Li, Y.; Scholtz, M. & Van-Heyst, B. (2000). Global gridded emission inventory of α-hexachlorocyclohexane, *J Geophys Res* 105: 6621-6632.

Li, Y.; Venkatesh, S. & Li D. (2004b). Modeling global emissions and residues of pesticides, *Environ Model Assess* 9: 237-243.

Liu, X.; Zhang, G.; Li, J.; Yu, L.; Xu, Y.; Li, X.; Kobara, Y. & Jones, K. (2009). Seasonal patterns and current source of DDTs, chlordanes, hexachlorobenzene, and endosulfan in the atmosphere of 37 Chinese cities, *Environ Sci Technol* 43: 1316-1321.

Lubick, N. (2010). Endosulfan's exit: U. S. EPA pesticide review leads to ban, *Science* 328: 1466.

Maier-Bode, H. (1968). Properties, effect, residues and analytics of the insecticide endosulfan, *Residue Rev* 22: 1-44.

Miles, C. & Pfeuffer, R. (1997). Pesticides in Canals of south Florida, *Arch Environ Contam Toxicol* 32: 337-345.

Patton, G.; Walla, M.; Bidleman, T. & Barrie, L. (1991). Polycyclic aromatic and organochlorine compounds in the atmosphere of northern Ellesmere Island, Canada, *J Geophys Res* 96: 10867–10877.

Peasant Daily. (2001). June 8th, 2001, in Chinese.

Pesticide Electronic Handbook. (2006). http://www.chinapesticide.gov.cn/

Peterson, S. & Batley, G. (1993). The fate of endosulfan in aquatic ecosystems, *Environ Pollut* 82: 143-152.

Pozo, K.; Harner, T.; Lee, S.; Wania, F.; Muir, D. & Jones, K. (2009). Seasonally resolved concentrations of persistent organic pollutants in the global atmosphere from the first year of the GAPS study, *Environ Sci Technol* 43: 796-803.

Pozo, K.; Harner, T.; Wania, F.; Muir, D.; Jones, K. & Barrie, L. (2006). Towards a global network for persistent organic pollutants in air: Results from the GAPS Study, *Environ Sci Technol* 40: 4867-4873.

Rice, C.; Chernyak, S. & McConnell, L. (1997). Henry's Law Constants for Pesticides Measured as a Function of Temperature and Salinity, *J Agric Food Chem* 45: 2291-2298.

Rüdel, H. (1997). Volatilisation of pesticides from soil and plant surfaces, *Chemosphere* 35: 143-152.

Saiyed, H.; Dewan, A. & Bhatnagar, V. (2003). Effect of endosulfan on male reproductive development, *Environ Health Perspect* 111: 1958-1962.

Schmidt, W.; Hapeman, C.; Fettinger, J.; Rice, C. & Bilboulian, S. (1997): Structure and asymmetry in the isomeric conversion of β- to α-endosulfan, *J Agric Food Chem* 26: 1101-1106.

Scholtz, M.; McMillan, A.; Li, Y.; Ting, N.; Taylor, A. & Van Heyst, B. (1998). Global pesticide emissions: Development of pesticide emission factors on a global 1°x1° longitude-latitude grid and global emissions of Hexachlorocyclohexane, Final Rep. CGEIC-1998-2, pp. 1-89, Can. Global Interpretation Cent, Toronto, Ont.

Scholtz, M.; McMillan, A.; Slama, C.; Li, Y.; Ting, N. & Davidson, K. (1997). Pesticide emissions modeling: development of a north American pesticide emissions inventory, Rep. CGEIC-1997-1, pp. 1-242, Can. Global Interpretation Cent., Toronto, Ont.

Sinha, N.; Narayan, R.; Shanker, R. & Saxena, D. (1995). Endosulfan induced biochemical changes in the testis of rats, *Vet Hum Toxicol* 37: 547-551.

Smimonich, S. & Hites, R. (1995). Global distribution of persistent organochlorine compounds, *Science* 269: 1851-1854.

Southan, S.; & Kennedy, I. (1995). Dissipation of the pesticide endosulfan from cotton field run-off water using an improved apparatus for the simulation of field conditions. In *Proceedings of the 4th RACI Environmental Chemistry Conference* (pp. 104-109). Darwin, Australia.

Su, Y.; Hung, H.; Blanchard, P.; Patton, G.; Kallenborn, R.; Konoplev, A.; Fellin, P.; Li, H.; Geen, G.; Stern, G.; Rosenberg, B. & Barrie, L. (2008). A circumpolar perspective of atmospheric organochlorine pesticides (OCPs): Results from six Arctic monitoring stations in 2000-2003, *Atmos Environ* 42: 4682-4698.

Sunderam, R.; Cheng, D. & Thomson, G. (1992). Toxicity of endosulfan to native and introducted fish in Australia, *Environ Toxicol Chem* 11: 1469-1476.

Sutherland, T.; Horne, I.; Lacey, M.; Harcourt, R.; Russel, R. & Oakeshott, J. (2000). Enrichment of an endosulfan-degrading mixed bacterial culture, *Appl Environ Microbiol* 66: 2822–2828.

Sutherland, T.; Weir, K.; Lacey, M.; Horne, I. & Russel, R. (2002). Gene cloning and molecular characterization of a two enzyme system catalyzing the oxidative detoxification of β-endosulfan. *J Appl Microbiol* 92: 541-548.

UNEP (United Nations Environment Programme). (2007). Guidance on the Global Monitoring Plan for Persistent Organic Pollutants, Preliminary Version February 2007. Amended in May 2007. Stockholm Convention Secretariat, Geneva, pp 147.

Verschureren, K. (1983). Endosulfan. In *Handbook of environmental data on organic chemicals* (pp. 604-607). New York : Van Nostrand Reinhold.

WHO. (1984). World Health Organization, Environmental Health Criteria 40.

Zhang, G.; Chakraborty, P.; Li, J.; Sampathkumar, P.; Balasubramanian, T.; Kathiresan, K.; Takahashi, S.; Subramanian, A.; Tanabe, S. & Jones, K. (2008). Passive atemospheric sampling of organochlorine pesticides, polychlorinated biphenyls, and polybrominated diphenyl ethers in urban, rural, and wetland sites along the coastal length of India, *Environ Sci Technol* 42: 8218-8223.

23

Hexachlorocyclohexanes in Arctic and Antarctic Marine Ecosystems

Alessandra Cincinelli[1], Tania Martellini[1] and Simonetta Corsolini[2]
[1]*Dipartimento di Chimica, Ugo Schiff, University of Florence*
[2]*Dipartimento Scienze Ambientali G. Sarfatti, University of Siena*
Italy

1. Introduction

Pesticides are a group of chemicals made for the purpose of killing or otherwise deterring "pest" species. The word pesticides may refer to insecticides, fungicides, herbicides or other pest control formulations. Introduced in 1940s, organochlorine pesticides (OCPs) were widely used as insecticides in agriculture and pest control until research and public concern regarding the hazards of their use and adverse effects in the environment led to government restrictions and bans. Two International legally binding instruments have been negotiated and concluded: the Protocol to the regional United Nations Economic Commission for Europe Convention on Long Range Transboundary Air Pollution (CLRTAP) on Persistent Organic Pollutants (POPs), opened for signature in June 1998 and entered into force on 23 October 2003 and the Stockholm Convention on POPs, opened for signature in May 2001 and entered into force on 17 May 2004. Both these agreements identify POPs that should be banned and/or phased out or whose use or emissions should be restricted, they include industrial chemicals and by-products such as PCBs, hexachlorobenzene, dioxins and furans, and a number of OCPs such as aldrin, chlordane, DDT, dieldrin, endrin, heptachlor, hexachlorobenzene, mirex and toxaphene. All together are often called the "dirty dozen" (Stockholm Convention, 2004). The hexachlorocyclohexanes (HCHs) are covered by the UNECE Protocol but not the Stockholm Convention. For several listed substances, some limited use is allowed, for example DDT for fighting malaria. Despite the actions of these two Conventions, POPs are still present at high levels in the polar regions and will require vigilant action in the continuing implementation of the Conventions to prevent further contamination of these rich and productive ecosystems.

OCPs are organic chemical substances which possess a particular combination of physical and chemical properties and once released into the environment they remain intact for long periods of time (from weeks to decades). They are persistent that is they resist to environmental degradation through chemical, biological and photolytic processes. They are toxic to both humans and wildlife and accumulate in the fatty tissue of living organisms (bioaccumulation), and are found at higher concentrations at higher levels of the food webs (biomagnification). They are subjected to long range transport (LRT) and can be found in remote regions, including Arctic and Antarctica, where they have never been used or produced (i.e.: Su et al., 2006; Bargagli, 2008; Corsolini, 2008; Rigét et al., 2010; Donaldson et al., 2010)

The transport of POPs in the Northern and Southern hemispheres is a well documented phenomenon. POPs are thought to be transported such long distances by a variety of processes including:

- a process known as global distillation (Wania and Mackay, 1993; 1996), according to which POPs with sufficiently high vapour pressure evaporate from the warmer regions, (where they are used or released), and then move through the atmosphere, condense at colder and high latitudes, finally concentrating in Arctic and Antarctic. According to this process, POPs of higher volatility like HCHs may migrate faster towards the poles than those of lower volatility such as DDTs;
- migration through ocean circulation;
- deposition by means of wet (snow, rain, mist) or dry depositions (i.e. atmospheric processes (particle settling)) onto terrestrial and aquatic surfaces;
- transport through migratory animals which are thought to offload their body burdens into polar ecosystems through their excretion and during body decomposition (Wania, 1998).

The fate and different transport routes of a POP are strongly influenced by its specific physical and chemical properties such as water solubility, vapour pressure (VP), Henry's law constant (H), octanol-water partition coefficient (K_{ow}) and the carbon-water partition coefficient (K_{oc}).

2. POPs in polar regions

The polar regions are of great intrinsic value and vital importance for the conservation of biological diversity.

Even though the Antarctic is still the region of the earth that is the least influenced by human activity, strict regulation is needed to maintain its untouched and pristine condition today. Most of the Antarctic region is situated south of 60°S latitude and is governed in accordance with the International legal regime of the Antarctic Treaty System. The Protocol on Environmental Protection to the Antarctic Treaty, which came into effect in 1998, designates Antarctica as an internationally important natural reserve devoted to peace and science, and provides a comprehensive environmental management regime. The Treaty area cover the continent itself as well as the archipelagos of the South Orkney Islands, South Shetland Islands, Peter Isalan, Scott Island and Balleny Island. Antarctica and the Southern Ocean are a remote region with no indigenous human population and no industrial and agriculture activities. Human impact is concerned largely with scientific investigations and the logistic operations in support of these. As a result of Antarctica's designation as a Special Conservation Area, many countries that maintain research stations in Antarctica have improved management practises and developed strategies to reduce environmental disturbances. The protection of the Antarctic Environment is the primary responsibility of the Antarctic Treaty Parties and the release of POPs into the Antarctic environment is incompatible with the comprehensive approach of the Protocol. Therefore, the obligation to monitor the introduction of these substances is vested upon the Parties to the Protocol and practices have been improved and importation of specific POPs has been prohibited. However, LRT remains an important process by which POPs contaminate the Antarctic environment.

In the last decades, extensive international cooperation has also been developed in several fields in the Arctic region as well. The Arctic Monitoring and Assessment Programme

(AMAP) is the first international programme to design and develop monitoring programmes to study the sources, transport mechanisms and pathways, levels, fate and behaviour of most of the groups of contaminants, including POPs, in the Arctic environment and its ecosystems (atmospheric, marine, terrestrial, humans). Further aims of the AMAP are to prevent releases of radioactive substances and emissions of other hazardous chemicals and provide scientific advice on actions to be taken in order to support Arctic governments in their efforts to take remedial and preventive actions relating to contaminants. AMAP is one of the six scientific Working Groups of the Arctic Council (Arctic Contaminants Action Program - ACAP, AMAP, Conservation of Arctic Flora and Fauna - CAFF), Emergency, Prevention, Preparedness and Response - EPPR), Protection of the Arctic Marine Environment - PAME and SDWG) a forum of cooperation between the eight Arctic countries (Canada, Greenland (Denmark), Finland, Iceland, Norway, Russia, Sweden and United States), and also between national governments and indigenous peoples.

POPs are now distributed in the global environment and their accumulation of organochlorine pesticides in Northern and Southern latitudes has been extensively documented. In particular, HCHs have been extensively used throughout the world and this factor in combination with a relatively high vapour pressure, a low octanol-air partition coefficient, a low Hanry's law constant and the highest water solubility of all organochlorines (Harner *et al.*, 1999, Brubaker and Hites, 1998, Gregor and Gummer, 1989, Kucklick et al., 1991) is reflected in the relative abundance of these compounds in Arctic and Antarctic ecosystem compartments.

Most of the associated literature focuses on the occurrence and levels of HCHs in air and on the study of air/water exchanges fluxes (UNECE, 1998; Hargrave et al.,1997), while only a little number of papers can be found dealing with water samples probably due to the difficulty involved in determination of very low concentrations of these contaminants in the Arctic and Antarctic seawaters. However seawater is an almost unique passage where pollutants transfer from the atmosphere or rivers to the shallow water. It is also a significant path way for OCPs accumulated in the plankton and therefore enter the terrestrial food webs of the polar regions (Cai et al., 2010).

This review reports the levels, trends and distribution of HCHs in water and biota of both polar regions. The accumulation in marine organisms will be described; in particular, krill, fish, seabirds and seals will be considered. Moreover, the α-HCH enantiomeric composition in the Arctic and Southern oceans will be reviewed in order to evidence the presence of environmental biochemical processes.

2.1 Hexachlorocyclohexanes

Hexachlorocyclohexane (HCH), also known as benzene hexachloride (BHC), is one of the most widely studied pesticides with respect to its environmental fate and effects (Breivick et al., 1999). It is an organochlorine insecticide that is available in two commercial formulations: technical grade and lindane.

Technical HCH was heavily used and it is an ubiquitous pesticide introduced in world war II and consists of a mixture of different isomers α-HCH (60-70%), β-HCH (5-12%), γ-HCH (10-15%), δ-HCH (6-10%) and ε-HCH (3-4%) (Kutz et al., 1991). Because of its low cost and high effectiveness, HCH was one of the most widely used insecticides in the world (Li, 1998). The insecticidal properties of HCH were first discovered in Europe in 1941-1942, however, in 1944 it was found that the γ-isomer is the only HCH isomer responsible for these properties (Hardie, 1964). Lindane is the γ -HCH (>99% pure) (UNEP, 2006). Lindane has been used as a broad-spectrum insecticide, which acts by contact, against a wide range of insects. Its main uses include treatment of seeds, on crops, in warehouses, in forestry, on domestic and agricultural animals, and for pest control of scabies and lice on humans (WHO, 1991).

	α-HCH	β-HCH	γ-HCH
Molecular weight	290.85	290.85	290.85
Structural code Axial/Equatorial	AAEEEE	EEEEEE	AAAEEE
Melting point (°C)	159.5	309.5	112.5
Boiling point (°C)	288 at 760 mmHg[1]	60 at 0.5 mmHg[2]	323.4 at 760 mmHg[1]
Vapour Pressure[1]	4.5×10^{-5} mmHg at 25°C	3.6×10^{-7} mmHg at 20°C	4.2×10^{-5} mmHg at 20°C
Henry's law constant[3]	$LogH_{\alpha}$= 10.13(±0.29)-3098(±84)/T	$LogH_{\beta}$= 9.96(±0.23)-3400(±68)/T	$LogH_{\gamma}$= 10.14(±0.59)-3208(±161)/T
Log K_{ow}[4]	3.94	3.9	3.9
Log K_{oc}[5]	3.57	3.57	3.57

Table 1. Physical and chemical properties of key HCHs (References: 1.HSDB, 2003; 2. Lide, 1991; 3. Sahsuvar et al., 2003; 4. Shen et al., 2004; Willett et al., 1998; 5. Weiss, 1986)

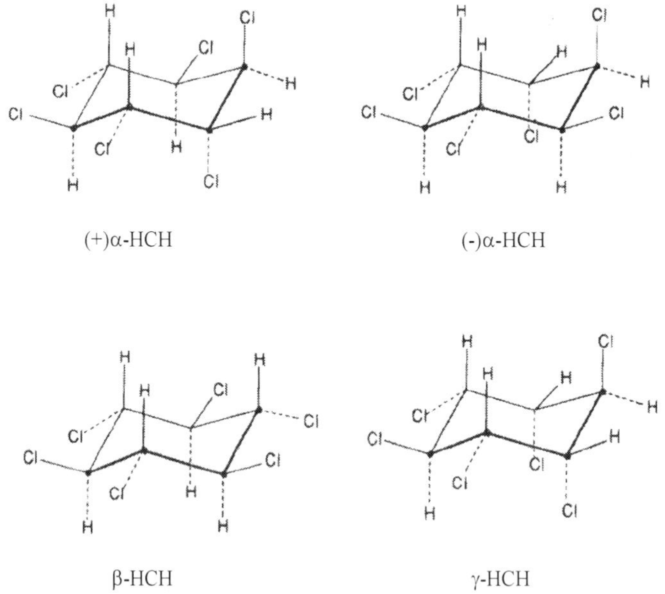

Fig. 1. Structure of the principal HCH isomers (α, β and γ)

Because of the environmental and biological persistence of HCHs, many developed countries banned or restricted technical HCH usage since 1970s, a ban followed by many developing countries in 1980s (Voldner and Li, 1993). In 2009, an International ban on the use of lindane in agriculture was implemented under the Stockholm Convention on POPs. A specific exemption allows for it to continue to be used in second-line treatments for the head lice and scabies for 5 more years.

Distribution, transportation and fate of HCHs in the environment is strongly influenced by the different physical and chemical properties of the various isomers and these are, to a great extent, responsible for the vast differences in the fates of the α, β and γ-isomers.

The chemical structures and chemical-physical properties of the key HCHs are reported in Table 1 and Figure 1.

2.2 Antarctica

2.2.1 The marine environment

The Southern Ocean occupies a belt between 55-70°S. The more southerly part of this zone is covered by sea ice for all or part of the year. Water temperatures vary very slightly in both time and space. Summer temperatures may be as high as 3-4°C in the northern part of the Southern Ocean, decreasing to slightly below zero close to the continent. The annual range of water temperature is typically only a few degrees. Despite this apparently hostile environment, the Southern Ocean supports a diverse biota, in strong contrast to Antarctic terrestrial ecosystems that is a cold desert. The region supports a number of fisheries, some of which are of global importance.

2.2.2 HCHs in Antarctic seawater

Relatively few investigations have been carried out to determine HCHs content in Antarctic seawater. The first HCHs data in seawater samples were reported by Tanabe et al. (1982a, b) who found the sum of HCHs ranging from 260 to 920 pg/L during an Antarctic supply voyage between Japan and the Syowa Research Station in 1980-1981. Similar total HCH concentrations (210-930 pg/L) were determined in seawater samples collected close to Syowa Station, not showing strong variations between samples collected under fast ice and at the outer margins of pack ice (Tanabe et al., 1983). Authors found mean α-HCH concentrations of 3.2 pg/L in 2002 and 1.41 pg/L in 2003-2004, showing little spatial and temporal variability during each sampling period. These values were lower than those measured in 1989-1990 (18-43 pg/L) (Iwata et al., 1993), 1997 (3.6-15 pg/L), 1997-1998 (5.1-28 pg /L) (Jantunen et al., 2004), and 1999 (2-9.6 pg/L) (Lakaschus et al., 2002).

Desideri et al. (1991) and Cincinelli et al (2009) measured the concentrations of HCHs in surface seawater samples collected at Terra Nova Bay in the spring-summer periods 1988-1989 and 2003-2004, respectively. They found mean α-HCH concentrations of 129 (±42) and 3.13(±0.89) pg/L, and mean γ-HCH concentrations of 562 (±335) and 7.11 (±1.22) pg/L, respectively. Desideri et al. (1991) found a mean of 147±25 pg/L of a-HCH and 470±229 pg/L of g-HCH in sea-ice samples and these values were higher than those measured in seawater samples; they hypothesized that POPs are accumulated during winter in the sea ice and then released into the seawaters during the seasonal ice melting, wich contribute to the Summer POP increasing concentration in the seawater. This hypothesis was also confirmed by Cincinelli et al. (2009), who found higher HCH concentrations in the seawater samples collected at Terra Nova Bay than those measured later in the Ross Sea (α-HCH 1.08±0.40 and γ-HCH 2.12±1.08), as well by Dickhut et al. (2005), who detected higher levels of γ-HCH in surface water collected during the early part of the summer in the vicinity of Palmer Station on the western Antarctic Peninsula in 2002. These authors found concentrations of α-HCH ranging from 1.65 to 4.54 (mean value 3.20±0.82) pg/L, and concentrations of γ–HCH ranging from 0.90 to 10.6 (mean value 4.09±4.08) pg/L in seawater samples.

2.2.3 HCHs in Antarctic biota

The scientific literature on the presence of HCH isomers in Antarctic organisms is very scarce, owing to the difficulty of collecting biotic samples in such an extreme environment, the distance from any part of the world, the very high cost of scientific expeditions and the need to be part of one of those to be allowed to reach the continent and collect organism samples, whose collection need special internationally valid permits released by a commission of the Antarctic Treaty. Moreover, data are often reported in different ways and thus no comparisons are allowed owing to different tissues, unit of measures (on a lipid, wet, or dry wt), chemicals analyzed, and species. The presence of HCHs in Antarctic marine organisms has been reported in few articles since 1960s.

The presence of HCHs in krill (*Euphausia superba*) and in Emperor penguin (*Aptenodytes forsteri*) feathers was reported first by Sen Gupta et al (1996); they detected 141.3±9.8 - 164±16.6 pg/g dry wt of α+γ-HCHs in krill, and 108.7±7.6 - 112.5±8.6 pg/g dry wt in penguin feathers collected in 1987 near the Indian Station Daksin Gangotri (70°05'S, 12°00'E). They reported that the γ-isomer was the most abundant in both species. Twenty years later, Bengston-Nash et al. (2008) found 0.03 ng/g wet wt of HCHs; the most abundant isomers were α-HCH > β-HCH > δ-HCH (14.2, 9.3, and 6.9 pg/g wet wt, respectively). HCH concentrations in the Antarctic Peninsula were 0.009 ng/g lipid wt in samples collected in 2001 (Chiuchiolo et al. 2004) and 0.25 ng/g wet wt (0.14-0.35 ng/g wet wt) in samples collected in 2005 at King George Is. (South Shetlands). Concentrations in krill samples collected in the Ross Sea were reported by Corsolini et al. (2006) and Cincinelli et al., (2009) and they were 0.28±0.04 ng/g wet wt and 0.11±0.07 ng/g wet wt, respectively. Notwithstanding the difference in the unit of measure, samples collected in 1980s seems to be more contaminated than those collected in the 2000s; moreover, the Indian Sector of the Southern Ocean showed lower levels than the Ross Sea in 2000s.

Data in fish are very scarce. A paper reported 1.35±0.72 ng/g wet wt of ΣHCHs in the emerald rockcod (*Trematomus bernacchi*) muscle, where γ-HCH was the principal contributor (1.23±0.67 ng/g) (Corsolini et al., 2006). Usually β-HCH is stable in animals, but it is less volatile than α- and γ-HCH, thus it can reach the Polar Regions less easily than other HCH isomers, due to the global fractionation (Wania and Mackay, 1993). β-HCH concentration in Arctic atmosphere (Li et al., 2003) and in some species of seal and whale (Willett et al., 1998) is very low in comparison with the more volatile α- and γ-HCHs. Several industrial countries such as Canada, European Countries and the U.S. have banned HCHs since the 1970s. However, a few developing countries from tropical belt continued to use Lindane (pure γ-HCH) until the 1990s (Li et al., 1996, 2003; Senthil Kumar et al., 2002); this would have influenced γ-HCH occurrence in Antarctic food webs.

The HCH contamination in seabirds will be discussed in reference to penguins; in fact, among seabirds, penguins spend all their life cycle in the Antarctic Ocean; other seabird species nest in Antarctica during Summer, but they overwinter northward, often in non-Antarctic seawaters. Thus they can accumulate higher burdens of contaminants and they do not reflect exactly the contamination status of the Antarctic ecosystems (Corsolini, 2009). Penguins feed mainly on krill and they are a good indicator of HCH contamination; many different chemical families have been studies in various species since 1960s (for a review see Corsolini, 2009). Data of HCH presence in their tissues were first reported in 1985: Schneider et al. (1985) detected γ-HCH in Adélie penguin fat (*Pygoscelis adéliae*) (73 ng/g lipid wt), and Emperor penguin fat (26 and 118 ng/g lipid wt). The figure 2 shows the ΣHCH concentrations in Adélie, Chinstrap (*Pygoscelis antarctica*) and Gentoo (*Pygoscelis papua*) penguins from different regions of Antarctica and collected from 1988 to 2005. Levels ranged between below the detection limit in Adélie penguin muscle from King George Island to 5 ng/g wet wt in blood of Gentoo penguin from the same area (Inomata et al., 1996). The minimum and maximum concentrations found ranged from 0.25 to 1.32 ng/g wet wt in Adélie penguin, from 0.17 to 2.28 ng/g wet wt in Chinstrap penguin, and from 0.1 to 5 ng/g wet wt in Gentoo penguin, showing a weak increase of concentration in Gentoo > Chinstrap > Adélie penguins. These penguin species nesting at King George Island share the same environment and have adopted a fine ecological segregation to reduce niche overlap and food competition (Trivelpiece et al., 1987). In fact, all the Pygoscelid species are usually very synchronous in nesting, while the breeding chronologies of these populations are asynchronous (Trivelpiece et al., 1987). Chicks hatched at approximately 2 week intervals, with Adélie penguin being the earliest and Chinstrap penguin the latest to hatch annually. Asynchronous breeding chronologies greatly reduce competition for food between species during chick rearing. At the same time, this asynchrony may affect the HCH accumulation;

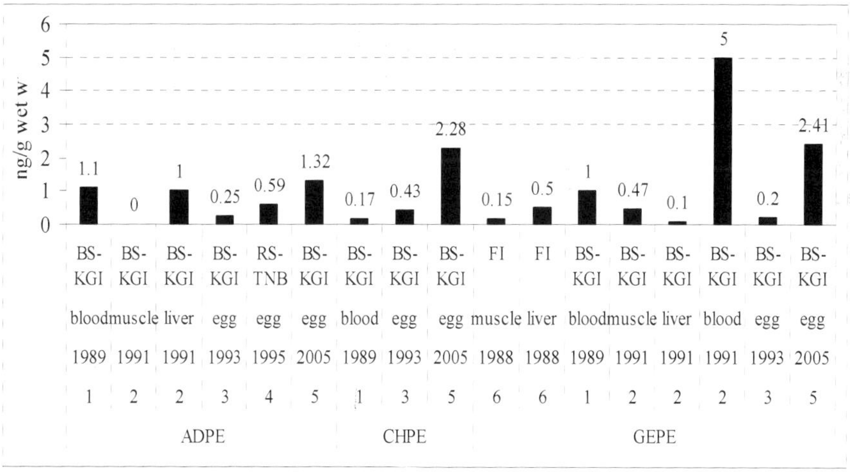

Fig. 2. ΣHCH concentrations (ng/g wet wt) in tissues of Adélie (ADPE), Chinstrap (CHPE), and Gentoo (GEPE) penguins from King George Island (BS-KGI) and Falkland Islands (FI) (YoS = year of sampling; Refs = references: 1. Lara et al., 1990; 2. Inomata et al, 1996; 3. Wanwimolruk et al., 1999; 4. Corsolini et al., 2006; 5. Cipro et al., 2010; 6. de Boer et al., 1991).

in fact, large amounts of HCH pesticides are used in tropical and temperate regions during Summer, with resulting HCH increasing levels in the Antarctic environment in Summer and Autumn (Larsson et al., 1992). The seasonality of HCH transport to Polar Regions has already been reported, and it affects also the accumulation in Antarctic organisms (Sen Gupta et al., 1996; Corsolini et al., 2000). Interestingly, the HCH concentrations in Adélie, Chinstrap and Gentoo penguin eggs increased from 1993 (Wanwimolruk et al., 1999) to 2005 (Cipro et al., 2010) (figure 3). This pattern may be due to the historical use of HCH-based pesticides worldwide and in particular in American, European and some Asiatic countries: after a massive use until the 1970s and 1980s, governments started to ban their use and production. First, levels continued to increase as a result of their global transport and dispersion, but then concentrations in biota showed a light decrease followed by the reduced use. The new increase observed at the end of 1990s can be as a result of a couple of reasons. The first reason can be the slow release of these chemicals from legal or illegal stocks. Secondly, the releases from the final sink as deep oceanic sediments and waters that may follow natural cycles in the marine ecosystems. Increasing concentration trends have been observed in other Antarctic species and also for other chemicals (Aono et al., 1997; Corsolini, 2009; Corsolini, 2011).

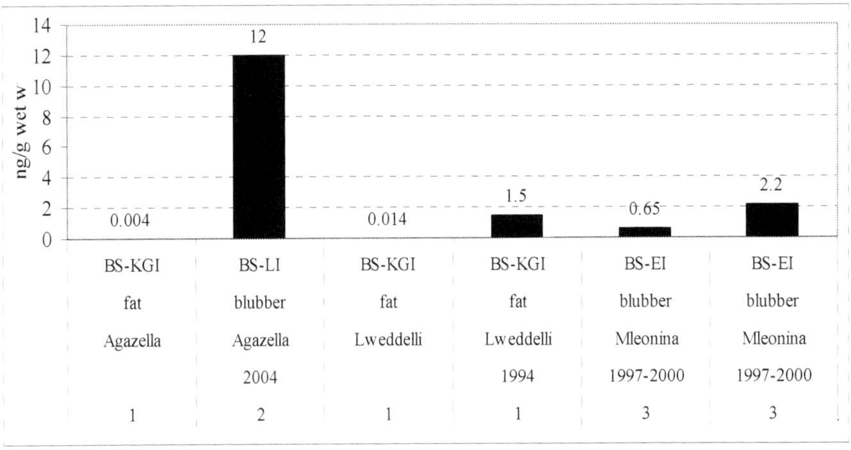

Fig. 3. ΣHCH concentrations (ng/g wet wt) in tissues of fur seal (Agazella), Weddell seal (Lweddelli), and elephant seal (Mleonina) from King George Island (BS-KGI), Livingston Island (BS-LI), and Elephant Island (BS-EI) (YoS = year of sampling; Refs = references: 1. Vetter et al., 2001; 2. Schiavone et al., 2009; 3. Miranda-Filho et al., 2007).

There are few articles that report HCH levels in Antarctic seals. HCH were very low in the fur seal (*Arctocephalus gazella*) from King George Island, 0.04 ng/g wet wt, and in the Weddell seal (*Leptonychotes weddelli*), 0.014 ng/g wet wt (Vetter et al., 2001). HCH higher concentrations were reported in specimens of Weddell seal and elephant seal (*Mirounga leonina*). Schneider et al. (1985) detected 13 and 20 ng/g lipid wt in the Weddell seal blubber, and 39 and 103 ng/g lipid wt in the Crabeater seal blubber (*Lobodon carcinophagus*), collected in 1981. The β-HCH prevailed in samples of fur seal from Livingstone island, where the isomer pattern abundance was β-HCH > α-HCH > γ-HCH > δ-HCH (Schiavone et al., 2009). The pattern was γ-HCH> α-HCH > β-HCH in the elephant seal (Miranda-Filho et al., 2007). A temporal trend of HCH concentration in Antarctic organisms is not evident mainly because of the paucity of data available. Extreme environments promote the selection of peculiar adaptations like the use of lipids to store energy, to protect against cold, to aid buoyancy in fish; the presence of this lipid component may affect the concentrations of lipophylic contaminants such as HCHs in relation to the season and period of the biological life cycle during which sampling is carried out. Therefore, HCH concentrations may vary depending on all such factors. Similar patterns have been reported also for other families of POPs in most of the species analyzed and collected in Antarctic seawaters (for a review see Corsolini et al., 2009).

2.3 Arctic

2.3.1 The marine environment

The Arctic region can be defined as the area north of the Arctic circle (63°33'N). It covers an area of approximately 13.4 million km^2 and large tracks of land are covered by glacial ice. The Arctic includes the Arctic Ocean and parts of Canada, Greenland, Russia, the United States (Alaska), Iceland, Norway, Sweden and Finland. The Arctic marine area includes the Arctic Ocean, the adjacent shelf seas (Beaufort, Chukchi, East Siberian, Laptev, Kara and the Barents Sea), the Northern Seas (Greenland, Norwegian and Iceland seas), the Labrador Sea, Baffin Bay, Hudson Bay, the Canadian Arctic Archipelago and the Bering Sea. The connection with the shallow Bering Sea occurs through the narrow Bering Strait, while the main connection with the Atlantic Ocean is via the deep Fram Strait and the Nordic Seas. The Arctic Ocean is divided into two deep basins, the Eurasian and the Canadian by the transpolar Lomonosov Ridge (AMAP 1998), extending as a submarine bridge about 1060 miles from Siberia to the northwestern tip of Greenland. Parallel to it there are two shorter ridges: the Alpha Ridge on the North American side, defining the Canada and Makarov basins, and the mid-ocean ridge on the Eurasian side, defining the Nansen and Fram basins. Due to ice coverage the temperature of the Arctic Ocean is close to freezing point year round. Surface water salinities vary between 30-33 in the Arctic Ocean. The salinity is lower in the summer due to the input of freshwater from rivers and terrestrial ice melting (AMAP, 1998).

The Arctic ocean is considered a sink for global pollution because of the flow of oceanic and atmospheric currents. It is a fragile ecosystem threatened by land-based sources of pollution particularly POPs and heavy metals (Lystsov, 2006). Principal loadings of HCHs to the Arctic Ocean during the last decades occurred by atmospheric transport and air-water exchange, precipitation and riverine input, and migration through north flowing ocean currents (Li et al., 2004). However the relative inputs by these pathways varied over time and differed for the eastern and western sides of the Arctic Ocean which have been termed the North American Arctic Ocean (NAAO) and Eurasian Arctic Ocean (EAO) (Bidleman et al., 2007).

2.3.2 HCHs in Arctic sea-water

Seawater samples were collected from a few limited number of cruises under taken throughout the 1980s to 2000s in different regions of the Arctic Ocean. The most extensive database on HCH concentrations in Arctic seawater samples belongs to the Bering and

Chukchi Seas where several cruise expeditions occurred, mainly organised by Japan, USA, Canada and former USSR.

An USSR/US investigation was begun in 1984 in the Bering and Chukchi Seas to study the transport of agricultural chemicals such as pesticides and other persistent pollutants (Chernyak et al., 1995). Results showed that α-HCH concentrations ranged from 810 to 1220 pg/L on the transect from the Sea of Japan to the Bering Sea, and a trend of increasing concentration with increasing latitude was observed (r^2= 0.88). This correlation is proved to be even stronger when average concentrations and latitudes from the Bering and Chukchi Seas and the Chirikov Basin were included in the regression (r^2= 0.99) (Chernyak et al., 1992), an increasing of with α-HCH concentrations from 800 pg/L in the South China Sea to 2500 pg/L in the Chukchi Sea. This trend for α-HCH concentration may reflect the effect of much colder surface water temperatures in the polar seas, because the Henry's law constant of α-HCH decreases with decreasing water temperature, thus favouring deposition to the water phase. However the same trend was not observed for the γ- and β- isomers which ranged from 770 to 1150 pg/L and from 80 to 740 pg/L, respectively.

Mean concentrations of α-HCH and γ-HCH of 7.1 and 0.8 ng/L, respectively, were reported by Patton et al. (1989) for four seawater samples collected over a depth of 1-10 m from the Ice island in the Beaufort Sea in June 1987. In the same area Hargrave et al. (1988) measured α-HCH levels in seawater samples at different depths and found average values of 4249 pg/L at 0-60 m, 2030 pg/L at 75-200 m and 320 pg/L at a depth >200 m in May 1986, average values of 5430 pg/L at 0-60 m, 2230 pg/L at 75-175 m in August 1986 and average values of 2820 pg/L at 10 m and 1440 pg/l at 110 m in June 1987. In general, concentrations of HCHs were maximum in the upper 60 m layer with decreasing values towards greater depths. The observed vertical distribution of α-HCH and other OCPs observed in this study indicated a source in the upper low salinity surface layer, probably a direct exchange between atmosphere and ocean (sea ice). In fact, most relevant inputs of pesticides into the Arctic marine environment may be atmospheric, riverine and oceanic transport even if the relative importance of each of these sources is difficult to assess.

Similar concentrations of ΣHCH (5-7 ng/L) in the Beaufort Sea were also reported by McDonald and McLaughlin (1993, 1994) for surface water samples collected between 1992 and 1993. Graphical presentation of data in Mc Donald et al. (2000) indicated levels for both isomers of 0.4-0.8 ng/L.

On the third Soviet-American Joint Ecological Expedition to the Bering and Chukchi Seas (August 1988), seawater samples were collected and analysed for OCPs. Average α-HCH concentrations in surface water samples were 2.4 ng/L, and average γ-HCH concentrations were 0.6 ng/L (Hinckley et al., 1991). In the same seas, Iwata et al (1993) collected seawater samples during the period between April 1989 and August 1990 reporting no differences between the mean concentration values of α-HCH (1.5 ng/L(range 1.2-1.9 ng/L) and 1.4 ng/L (range 1.3-1.6 ng/L)) and γ-HCH (0.190 ng/L (range 0.160-0.230 ng/L) and 0.180 ng/L (range 0.150-0.220 ng/L)), in the Bering and Chukchi Seas, respectively.

Paired air and water samples were collected at Resolute Bay (74°N, 95°W) in summer 1992 to estimate the direction of gas exchange of HCHs and investigate possible loss processes in the water column (Bidleman et al. ,1995; Falconer et al., 1995). Average concentrations of α-HCH and γ-HCH in ocean surface water were 4.75±0.9 and 0.44±0.11 ng/l, respectively. Water/air fugacity ratios were 1.03 for γ-HCH and 1.57 for α-HCH, indicating a slight potential for volatilization of α-HCH. The average concentration of the sum of α-HCH and γ-HCH (5100 pg/L) (Bidleman et al., 1995; Falconer et al., 1995) was higher than values from Bering Sea and Chukchi Sea (1600 -3400 pg/L) reported by Iwata et al.(1993) and Kawano et al.(1988) (mean α-HCH 2800 pg/L, γ-HCH 610 pg/L).

During August-September 1993 a joint Russian -US expedition to the Bering Sea and Chukchi Sea took place and surface water samples (from 1 to 2 m) were collected from 21 sites. Highest water concentrations were observed for HCH in open waters north and south of the Bering strait, both regions being similar (α-HCH 2.2 ng/L and lindane 0.35 ng/L) (Strachan et al., 2001). The water-air fugacity for the HCHs indicated that α-HCH is degassing in both Bering and Chukchi Seas while γ-isomer is degassing in Bering Sea but it is close to equilibrium in the Chukchi Sea (Strachan et al., 2001).

In the summers of 1993 and 1994, seawater samples from the surface layer (40–60 m) were collected to determine the spatial distribution of organochlorine pesticides on expeditions that crossed the Arctic Ocean from the Bering and Chukchi Seas to the North Pole, to a station north of Spitsbergen, and then south into the Greenland Sea (Jantunen and Bidleman, 1998). In the upper 40 m of the northern Chukchi Sea, α-HCH and γ-HCH averaged 2.06±0.48 ng/L and 0.43±0.09 ng/L. In the polar mixed layer (60 m) of the western Arctic Ocean α-HCH and γ-HCH averaged 2.42±0.23 ng/L and 0.47±0.11 ng/L. Concentrations were 2–3 times lower than these means at two stations near Spitsbergen and one station in Greenland Sea, averaging 0.87±0.22 ng/L for α-HCH and 0.20±0.03 ng/L for γ-HCH. Thus, HCHs in the upper 40–60 m increased from the Chukchi Sea to the pole, and then decreased toward Spitsbergen and Greenland Sea.

Similar results were found by Harner et al. (1999) who measured HCH concentrations in seawater samples collected during a cruise abroad the Swedish icebreaker "Oden" in July-September 1996 and found mean concentrations in surface water of 910 pg/L (range 350-1630 pg/L) for α-HCH and 270 pg/L (range 120-400 pg/L) for γ-HCH. Both HCHs increased with latitude between 74°-88°N (r^2= 0.58 and 0.69 for α-HCH and γ-HCH, respectively). Same authors also observed that mean surface concentrations of HCHs in the eastern Arctic ocean were lower than those in the western Arctic.

Further water samples from the Bering and Chukchi Seas were collected and analysed by Yao et al. (2002) during the first Chinese Arctic Research Expedition from July to September 1999. They investigated the distribution and composition of organochlorine pesticides (including HCHs, heptachlor, heptachlor epoxide, aldrin, endosulfan I, p,p'-DDE, dieldrin, endrin, p,p'-DDD, endosulfan II, p,p'-DDT, endrin aldehyde, and endosulfan sulphate) and found that the most abundant pesticide in the Arctic seawater was α-HCH, whose concentration was usually one or two magnitude grade greater than other contaminants. α-HCH concentrations ranged between 156 and 683 pg/L and between 157 and 662 pg/L in the Bering Sea and Chukchi Sea, respectively. The average of ΣHCHs (α, β, γ, δ) was nearly equal in the Bering Sea (mean concentration 412.7 pg/L) and in the Chukchi Sea (mean concentration 445.8 pg/L), showing no latitudinal difference of these two regions. Compared with previously reported studies, concentrations of OCPs in these regions were much lower than the levels in the last decades. The α/γ-HCH ratio was 5.0±1.8 and 3.4±1.0 for the Bering and Chukchi Seas, respectively, which indicated the different pesticide composition in these regions. The ratio of α/γ in Bering Sea suggested a technical HCH mixture indicating that OCPs were mostly transported from the low latitude. However the observed low α/γ ratio observed in the Chukchi Sea might imply the presence of a possible emission source of lindane nearby this region.

Hexachlorocyclohexanes (HCHs) in the surface water of the Canadian Archipelago and south Beaufort Sea were measured in summer 1999 (Bidleman et al., 2007). Overall concentrations of HCH isomers were in order of abundance: α-HCH (ranging between 1.1 and 5.4, mean value 3.5±1.2 ng/L) > γ-HCH (ranging between 0.19 and 0.45, mean value 0.31±0.07 ng/L) > β-HCH (ranging between 0.056 and 0.16, mean value 0.10±0.03 ng/L). Concentrations also varied latitudinally for α-HCH and γ-HCH ($p < 0.002$) but not for β-HCH.

A recent study reports HCHs data collected on the FS Polarsten during the cruise ARKXX in the North Atlantic and Arctic Ocean in 2004 (Lohmann et al., 2009) and shows the α-HCH concentrations near 1 pg/L in many samples < 80°N to mostly >20 pg/L above 80°N. The concentrations of γ-HCH were generally lower than α-HCH ranging from <1 to 20 pg/L. These authors confirmed that concentrations of HCHs have continued to decline in the last few years.

The most recent data on HCHs concentrations in Arctic ocean are presented by Cai et al. (2010), who successfully applied the new analytical method developed by Qiu and Cai (2010) based on the combination of solid phase extraction and headspace solid phase microextraction (HS-SPME), for the determination of 17 ultra trace OCPs. Surface seawater samples were collected during the third Chinese Arctic expedition cruise from July to September 2008 on board the R/V "Xuelong" The track covered the Japan sea, Okhotsk sea, Bering sea and the zone to the North of the Bering Strait including the Chukchi sea, Canadian Basin and Arctic ocean. Cai et al. (2010) found that among the organochlorine pesticides, HCHs, especially α-HCH and γ-HCH, were the most predominant in the Arctic surface water body as found by other authors (Iwata et al., 1993; Chernyak et al., 1995; Jantunen et al., 1995; Yao et al., 2002; Weber et al., 2006). This trend might be attributable to the higher historic usage of HCHs compared to the other investigated compounds (Li and McDonald, 2005). In the Bering Sea, the surface water concentrations were found to be 0.065-0.2671 ng/L for α-HCH, 0.0775-0.8075 ng/L for β– HCH and 0.0725-0.7175 ng/L for γ-HCH. Respect to concentrations values reported earlier for this area, γ-HCHs values reported by Cai et al (2010) showed a level comparable to that reported in 1999, but a slightly decreasing trend was observed for α-HCH. Water samples collected in the western Arctic Ocean presented concentrations of 2.07-2.63 ng/L and 0.33-070 ng/L for α-HCH and γ-HCH, respectively. Results for α-HCH in the Chukchi Sea (0.0583-0.3926 ng/L) were slightly lower than those in 1999 (Yao et al., 2002) while γ-HCH concentration level were fairly comparable indicating that a state of equilibrium of γ-HCH was achieved recently. β-HCH showed higher concentrations respect to previous data in the Chukchi Sea and authors (Cai et al., 2010) attributed this result to the extremely low sensitivity of detection which could lead to higher error in the integration of the peak area.

2.3.3 HCHs in Arctic biota

The POP presence in Arctic organisms has been investigated during the last decades and many articles have been published in international scientific journals. The easier access to the area with respect to the Antarctic region is responsible of the high number of studies on Arctic biotic ecosystems. The interest in this polar region increased when the presence of very high concentrations of POPs was detected in those human populations that live the further north lands of Europe, Asia and America (for a comprehensive study see the AMAP Report 2009, AMAP, 2009). In the framework of the AMAP, many researches have been carried out. Recently, reviews of POP presence and trends in the Arctic have been published (Muir et al., 2010). The occurrence of HCHs in the Arctic organisms will be examined in the same classes reported for the Antarctic organisms, that is pelagic crustaceans (krill), fish, seabirds, and seals, in order to allow a discussion.

Borgå et al. (2005) reported 26.1±2.3 ng/g lipid wt of ΣHCHs in the Arctic krill *Thysanoessa inermis*. The isomer abundance followed the pattern α-HCH > δ-HCH > β-HCH (α-HCH made up more than 60% of the ΣHCHs); samples were collected in 1999 in the Greenland Sea and NW and NE of Svalbard Islands. The extracted organic matter in these samples was 10.7±0.4. The comparison with concentrations in Antarctic krill showed a higher concentration in Arctic krill. These two species of Euphausiacea are important species in their trophic webs being food of many species of fish, seabirds, seals, and whales.

Fish species collected in the Arctic region and analyzed for HCHs were mostly the Arctic cod (*Arctogadus glacialis*), the Polar cod (*Boreogadus saida*), the Greenland halibut (*Reinhardtius hippoglossoides*), and others. ΣHCH concentrations in Polar cod muscle from the Canadian Arctic were 90.2±13.7 ng/g lipid wt (Moisey et al., 2001), 40±3.2 ng/g lipid wt (Hoekstra et al., 2003), and 10 ng/g lipid wt (Kelly et al , 2008); concentrations in the Greenland halibut muscle were as 81 ng/g lipid wt in (Fisk et al., 2002), and 53 ng/g lipid wt in Greenland shark (*Somniosus microcephalus*) liver (Fisk et al., 2002). Levels in Polar cod were of the same order of magnitude, but a decreasing trend can be observed during 2000s. Sinkkonen et al. (2000) analyzed Polar cod liver in specimens collected from 1987 to 1998 in the Norwegian Arctic and they detected 4-23.3 ng/g lipid wt of α-HCH and 2.9-8.1 ng/g lipid wt of δ-HCH. Their decreasing trend were very significant ($p = 0.001$) during this time span; these Authors reported that α-HCH concentrations declined faster than those of δ-HCH, in agreement with observations for Arctic air and water (Li et al., 1998; Bidleman et al., 1995; Jantunen and Bidleman, 1995). The HCH isomer abundance were α-HCH > β-HCH > δ-HCH in Polar cod (Sinkkonen et al., 2000; Moisey et al., 2001) and δ-HCH > α-HCH > β-HCH in the Antarctic *T. bernacchi* (Corsolini et al., 2006). These patterns could be interpreted as an indication that the use of technical HCH (containing 60-70% of α-HCH) has decreased faster than that of pure lindane (Li et al., 1998). Li et al. (1998) reported that there were two significant drops of HCH concentrations in the Arctic air: in 1982-1983 and in 1990-1992 and they followed the ban of technical HCH use in China (1983), India and ex-Soviet Union (1990). This decrease was not followed by a decrease of α-HCH concentrations in Arctic seawater. These speculations might help to interpret the different patterns found in Arctic and Antarctic fish. In fact, chemical concentrations and patterns in the two polar regions might be influenced by the air mass movement and the different use of HCHs in the two hemispheres, in agreement with the model proposed by Wania et al. (1999), suggesting that levels in the Southern Ocean are higher than those in tropical seas. Anyway, α- and δ-HCH were reported to be decreasing in the Arctic (Rigét et al., 2010).

A study on the occurrence of HCH in several species of seabirds from the Northwater Polynia (NOW) reported ΣHCH concentrations in various species: 222±19.9 ng/g lipid wt in dovekie, 84.5±9.6 ng/g lipid wt in thick-billed murre, 285±46.7 ng/g lipid wt in black guillemot, 47.3±6.3 ng/g lipid wt in black-legged kittiwake, 442.7±51.9 ng/g lipid wt in glacous gull, 143.0±32.7 ng/g lipid wt in ivory gull, 65.1±5.8 ng/g lipid wt in northern fulmar (Moisey et al., 2001). These Authors reported the α, β and γ-isomer patterns of abundance was β-HCH > α-HCH > δ-HCH in all these species.

Concentrations of ΣHCHs collected from 1975 to 2003 were 2.0±0.4 - 5.7±0.7 ng/g wet wt in northern fulmars, 4.0±0.4 - 7.7±0.6 ng/g wet wt in black-legged kittiwake, and 9.2±1.3 - 18.6±2.0 ng/g wet wt in thick-billed murre (Braune, 2007 a, b). The isomer abundance patterns were α-HCH > β-HCH in northern fulmar and thick-billed murre collected from 1975 to 1993, and β-HCH > α-HCH in northern fulmar and thick-billed murre collected in 1998 and 2003, and in black-legged kittiwake. Braune (2007 a, b) noted significantly increasing concentrations of β-HCH in northern fulmars and thick-billed murres, and ΣHCH in black-legged kittiwakes and northern fulmars; the increasing ΣHCH concentrations in the northern fulmars and black-legged kittiwakes were owed to β-HCH.

Other articles reported ΣHCH levels ranging from <1-170 ng/g lipid wt (white-winged scoter and common eider: Kelly et al., 2008; herring gull, common guillemot, Atlantic puffin, and black-legged kittiwake: Helgason et al., 2008; black guillemot: Vorkamp et al. 2004; northern fulmar: Knudsen et al., 2007; glacous gull: Verrault et al., 2005; peregrine falcon: Vorkamp et al. 2009), and 24-80 ng/g wet wt (northern fulmar, glacous-winged gull, and tufted puffin: Ricca et al., 2008). Concentrations exceeding 100 ng/g lipid were detected in black guillemot eggs from east Greenland (Vorkamp et al., 2004), glacous gull plasma from Svalbard Islands (Verrault et al., 2005), and ivory gull eggs from Canada (Braune et al., 2007a, b). A decreasing or stable temporal trend of ΣHCHs, α-, β-, and δ-HCH concentrations during the last decade was reported in all these species except fulmar and thick-billed murre eggs, and the rate of decrease varied among species and geographical areas (Rigét et al., 2010).

Pinnipeds may accumulate high amount of chemicals and different concentrations in their tissues are often evident and depending on their diet. ΣHCH concentrations in seal blubber were 150.5±13.1 ng/g lipid wt in ringed seal collected in the NOW in 1998 (Moisey et al., 2001), 145 ng/g lipid wt in specimens from the Canadian Arctic (Kelly et al., 2008), and 190±50 ng/g lipid wt in the same species collected in 1999-2000 (Hoekstra et al., 2003). Lower concentrations were detected in ringed seal from Greenland: 67 ng/g lipid wt (Vorkamp et al., 2004) and 40 ng/g lipid wt (Vorkamp et al., 2008). Vetter et al. (2001) reported ΣHCH concentrations in three species of Arctic seals and in three species of Antarctic seals: they were 659 ng/g wet wt in grey seal from the Baltic Sea, 5-11 ng/g wet wt in grey seal from Iceland, and 181 ng/g wet wt in harp seal from the North Sea. Values found in samples from Iceland were of the same magnitude than those found in Antarctic species (Weddell seal = 14 ng/g lipid wt, fur seal = 4 ng/g wet wt, elephant seal = not detected). The concentrations of ΣHCHs and HCH α, β and γ-isomers were reported to be decreasing in ringed seals (Rigét et al., 2010). Riget et al. (2008) reported annual decreases in ringed seals from East and West Greenland from 1986 to 2006 that were 9.1–11.7%, 1.4–3.9% and 6.0–6.4% for α, β and γ-HCH, respectively. δ-HCH was the less abundant isomer both in West Greenland juvenile seals (4.95 ng/g lipid wt in 1995 and 1.91 ng/g lipid wt in 2002) and in East Greenland juvenile seals (7.89 ng/g lipid wt in 1986 and 2.57 ng/g lipid wt in 2006); adult seals showed larger temporal variation (9.3 ng/g lipid wt in 1994 and 2.7 ng/g lipid wt in 2002).

2.4 Enantiomer fraction of α-HCH in Antarctic and Arctic marine environment

α-HCH is a chiral compound and thus exists in two enantiomers forms. Enantiomers are stereoisomers in which the atoms are arranged such that the molecules are mirror images of each other. The two enantiomers can rotate polarized light in different directions. Two

principal metrics are used for reporting enantiomer composition of chiral POPs. The enantiomer ratio (ER) and the enantiomer fraction (EF). Most earlier studies used ER, consisting in the separation of these enantiomers by GC and subsequent calculation of the peak area or height ratio of the (+) ad (-) enantiomers. The ER ranges from zero to infinity, with a racemate having an ER of 1. The EF, now more commonly used, is defined as the ratio of the (+)enantiomer to the sum total enantiomer concentrations. EF ranges between zero to infinity with a racemic value of 0.5. EF are preferred to ER as the EF is bounded and a deviation from the racemic value in one direction is the same as in the other (Harrad, 2009).
Chirality has been used to detect, characterize and differentiate biotic and abiotic transformation processes. Biotic processes such as microbial degradation, enzymatic processes or biological uptake may be enantioselective, causing the observed ER or EF to vary from the racemic value of 1.0 or 0.5, respectively; while abiotic processes such as hydrolysis and photolysis are not enantioselective (Helm, 2000) and affect both enantiomers of achiral compound in the same way.
The enantiomer composition of α-HCH in Antarctic seawater samples is not well documented; in fact only Jantunen et al. (2004) studied the influence of latitude on EF ratio on seawater samples collected during the South African National Antarctic Expedition (December 1997 - February 1998). Same authors reported EF values ranging from 0.477 to 0.515 and evidenced a significant regression of EF versus latitude (R^2=0.28, p≤0.005) with a slight preferential tendency to degradate of (-) enantiomer at the lower latitudes (EF ≥0.500), versus racemic or depletion of the (+) enantiomer (EF≤0.500) at the higher latitudes.
Enantiomeric fractions of chiral α-HCH in Antarctic biota were studied by Corsolini et al. (2006). They reported average EF values of 0.44±0.01, 0.49±0.01 and 0.58±0.04 in krill, emerald rockcod and Adélie penguin eggs, respectively. According to these results, these authors suggested enantioselective biotrasformation at lower trophic animal, with a decrease in the (+)α-HCH enantiomer compared to the (-)α-HCH. The (+)α-HCH contribution increased by 14% from lower to the higher trophic level (from krill to penguin): the proportion of the (+) enantiomer increased from 44% to 58%, suggesting an enantioselective biotransformation up the food web. Accumulation of (+)α-HCH in the higher trophic levels was already reported for marine mammals and polar bear (Iwata et al., 1998; Wiberg et al., 2000; Kallenborn and Huhnerfuss, 2001).
There has been an extensive analysis of α-HCH enantiomer composition in the Arctic Ocean. Falconer et al. (1995) found ER of 0.93 ± 0.06 in Resolute Bay in August -September 1992, showing that the (+) enantiomer was depleted in seawater and suggesting a microbial degradation of HCHs. A microbial degradation was also observed in seawater samples collected on a cruise along the Eastern Arctic Ocean in July-September 1996 (Harner et al., 1999), where the ER values ranged between 0.72 to 0.94 (mean value 0.87± 0.06) in surface water and decreased with depth.
The ER of α-HCH ((+)-α -HCH / (-)- α -HCH) in Arctic seawater was reported by Jantunen and Bidleman (1996, 1997), who found ER generally > 1.00 in the Bering-Chukchi Seas, indicating preferential degradation of (-)α -HCH, whereas depletion of the (+)α -HCH in the Arctic Ocean and Greenland Sea, with ERs < 1.00. One hypothesis to explain this different enantiomer depletions could be due to the different microbial populations in these regions. Although generally HCHs are measured in the dissolved fraction, levels of HCHs in water were high enough to allow ER values to be measured on the filters of the large volume samples. The results showed different enantioselective degradation in the dissolved and

particulate phase suggesting that different microbial populations are involved in these two phases. Authors also noted that ERs for α-HCH decreased with depth (Jantunen and Bidleman 1996, 1997) as observed by Harner et al. (1999) in the Eastern Arctic ocean, where the ER ranged from 0.72 to 0.94 (mean 0.87 ±0.06) in 21 surface water samples indicating selective degradation of (+)α-HCH and microbial degradation was suggested as the major removal process of HCHs from the water column. The reversal of enantiomer preference observed respect to the early 1990s could be due to changes in microbial degradation in the water column with subsequent changes over time leading to consistent enrichment of the (-)enantiomer (Harrad, 2009).

Tracer studies have determined that the "ventilation age" of the 250-1000 m water in the Nansen Basin of the eastern Arctic Ocean is in the range of 12-20 years (Wallace et al., 1992). The ventilation age is the time since the water was at the surface and able to exchange gases with the atmosphere. This information, combined with measurements of HCH concentrations in 1996 (Harner et al. 1999) and 1979 (Gaul et al., 1992), allows us to estimate the removal rates from the water column, which are due to microbial degradation, hydrolysis and sedimentation. The latter is negligible, and microbial degradation is 3-10 times faster than hydrolysis (Harner et al. 1999).

In a recent paper, Lohmann et al. (2009) found depletion of the (+) enantiomer, with EFs ranging from 0.432-0.463, and increased from west to east in the Archipelago.

3. Conclusions

The concentration levels of α- and γ-HCH in surface waters of Antarctica are much lower (by 1-2 orders of magnitude) than those in the Arctic, due to the remoteness of this continent from populated and industrial regions relative to the Arctic. A decreasing trend was observed for α-HCH and γ-HCH concentration in both Arctic and Antarctic oceans, corresponding to the global bans on HCHs. However the recent declining usage and atmospheric inputs of α-HCH isomer have caused the exchange to reverse, and made the Arctic ocean a source of α-HCH to the atmosphere rather than the major sink for LRT of it.

A comparison between levels and trends in Arctic and Antarctic biota indicates higher contamination levels in the Arctic organisms in relation to the geographical isolation of the Antarctic continent and Southern Ocean, which make difficult for chemicals to reach this region. It might be that the global transport to Antarctica and equilibrium between phases of HCHs follow different mechanisms of time scale in the two Polar Regions; Antarctica could show a delay in the chronological steps that characterize the HCH distribution globally and in the marine ecosystems. The Arctic is showing a decreasing temporal trend of contamination followed the peak occurred in the past years and in relation to the great HCH use/production and following ban. Thus, a decreasing or stable temporal trend of ΣHCHs, α–, β–, and δ-HCH concentrations during the last decade was reported in seawater and most of the studied species, and the rate of decline varied among species and geographical areas.

By comparing available data, it emerges a different pattern in the Antarctica where a temporal trend is not clearly recognizable, owing to the paucity of data and unevenness of report features and style. In this regard, compared to the Arctic, it could be hypothesized a slight delay in the transport, accumulation, and decrease of HCH in the southern polar region, that can be an effect of the geography and of the chemical transport pathways, that affect the distribution on the marine ecosystems.

4. References

AMAP, 1998. AMAP Assessment Report: Arctic Pollution Issues. Arctic Monitoring and Assessment Programme, Oslo, Norway, pp 56

AMAP. 2009. AMAP assessment 2009: Human health in the Arctic., ed. (AMAP), AMAP, 256. Oslo, Norway

Aono, S.; Tanabe, S.; Fujise, Y.; Kato, H. & Tatsukawa, R. (1997). Persistent organochlorine in minke whale (Balemoptera acutorostrata) and their prey species from the Antarctic and the North Pacific. *Environmental Pollution*, 98, 1: 81-89.

Bargagli, R. (2008) Environmental contamination in Antarctic ecosystem. *Science of the Total Environment*, 400, 212-226

Bengtson Nash, S.M.; Poulsen, A.H.; Kawaguchi, S.; Vetter, W.& Schlabach, M. (2008) Persistent organohalogen contaminant burdens in Antarctic krill (Euphausia superba) from the eastern Antarctic sector: A baseline study. *Science of The Total Environment*, 407, 1: 304-314.

Bidleman, T.F.; Falconer, R.L.; Walla, M.D. (1995a) Toxaphene and other organochlorine compounds in air and water at resolute Bay, N.W.T., Canada. *Science of Total Environment*, 160-161, 55-63.

Bidleman, T.F.; Jantunen, L.M.; Falconer, R.L.; Barrie, L.A. & Fellin, P. (1995b). Decline of hexachlorocyclohexane in the Arctic atmosphere and reversal of air-sea gas exchange. Geophys. Res. Lett. 22, 219-222.

Bidleman, T.F.; Kylin, H.; Jantunen, L.M.; Helm, P.A. & Macdonald, R.W. (2007) Hexachlorocyclohexanes in the Canadia Archipelago a. Spatial distribution of α, β and γ-HCHs in surface water. *Environmental Science& Technology*, 41, 2688-2695.

Borga, K. & T.F. Bidleman. (2005) Enantiomer fractions of organic chlorinated pesticides in Arctic marine ice fauna, zooplankton, and benthos. *Environmental Science& Technology*, 39, no. 10, 3464-3473.

Braune B.M. (2007a) Temporal trends of organochlorines and mercury in seabird eggs from the Canadian Arctic, 1975–2003. *Environment Pollution*, 148, 599–613.

Braune, B.M.; Mallory, M.L.; Gilchrist, H.G.; Letcher, R.J & Drouillard, K.G. (2007b). Levels and trends of organochlorines and brominated flame retardants in ivory gull eggs from the canadian arctic, 1976 to 2004. *Science of The Total Environment*, 378, 403-417.

Breivik, K.; Pacyna, J.M. & Munch, J. (1999) Use of α, β and γ--hexachlorocyclohexane in Europe 1970-1996. *Science of the Total Evironment*, 239, 151-163.

Brubaker, W.W. and Hites, R.A. (1998). OH reaction kinetics of gas-phase α- and γ–hexachlorocyclohexane and hexachlorobenzene. *Environmental Science & Technology*, 32,766-769.

Cai, M.G.; Qiu, C.R.; Shen, Y.; Cai, M.H.; Huang, S.Y.; Qian, B.H.; Sun, J.H. & Liu, X.Y. (2010).Concentration and distribution of 17 organochlorine pesticides (OCPs) in seawater from the Japan sea northward to the Arctic Ocean. *Science China, Chemistry*, 5, 1033-1047.

Chernyak, S.M.; McConnell, L.L. & Rice, C.P. (1995). Fate of some chlorinated hydrocarbons in Arctic and far Eastern ecosystems in the Russian Federation. *Science of Total Environment 160-161, 75-85.*

Chernyak, S.M.; Vronskaya, V.M. & Kolobiva, T.P. (1992) Migratory and bioaccumulative peculiarities in the biogeochemical cycling of chlorinated hydrocarbons. In P.N.Nagel (ED), Results of the third joint US Bering and Chukchi Seas Expeditions

(BERPAC)-Summer 1988. US Dept of Interior, Fish and wildlife service, Washingto DC, pp 279-284.

Chiuchiolo, A.M.; Dickhut, R.M.; Cochran, M.A.; & Ducklow, H.W., (2004). Persistent Organic Pollutants at the Base of the Antarctic Marine Food Web. *Environmental Science & Technology* 38 (13), 3551–3557

Cincinelli, A.; Martellini, T.; Del Bubba, M.; Lepri, L.; Corsolini, S.; Borghesi, N.; King, M.D. & Dickhut, R.M. (2009) Organochlorine pesticide air-water exchange and bioconcentration in krill in the Ross Sea. *Environmental Pollution*, 157, 2153-2158.

Cipro, C.V.Z., Taniguchi, S. & Montone, R.C. (2010). Occurrence of organochlorine compounds in euphausia superba and unhatched eggs of pygoscelis genus penguins from admiralty bay (King George island, Aantarctica) and estimation of biomagnification factors. *Chemosphere* 78, no. 6, 767-771.

Corsolini, S. & Focardi, S. (2000). Bioconcentration of polychlorinated biphenyls in the pelagic food chain of the Ross sea. Berlin: Springer-Verlag.

Corsolini, S.; Covaci, A.; Ademollo, N.; Focardi, S. & Schepens, P. (2006) Occurrence of organchlorine pesticides (OCPs) and their enantiomeric signatures, and concentrations of polybrominated diphenyl ethers (PBDEs) in the Adélie penguin food web, Antarctica. *Environmental Pollution*, 140, 371-382.

Corsolini, S. (2009) Industrial contaminants in antarctic biota. *Journal of Chromatography A* 1216, 3, 598-612.

Corsolini S. (2011). Contamination Profile and Temporal Trend of POPs in Antarctic Biota. In: B.G. Loganathan & P.K-S. Lam eds, Global Contamination Trends of Persistent Organic Chemicals, Taylor and Francis Group LLC., 6000 Broken Sound Pkwy, Boca Raton, FL 33487, in press.

De Boer, J. & Wester, P. (1991). Chlorobiphenyls and organochlorine pesticides in various sub-antarctic organisms. *Marine Pollution Bulletin*, 22, no. 9: 441-447.

Desideri, P.G.; Lepri, L.; Santianni, D.; Checchini, L. (1991) Chlorinated pesticides in seawater and pack ice in Terra Nova Bay (Antarctica). *Annali di Chimica*, 81, 533-540.

Dickhut, R.M.; Cincinelli, A.; Cochran, M. & Ducklow, H.W. (2005) Atmospheric concentrations and air-water flux of organochlorine pesticides along the Western Antarctic Peninsula. *Environmental Science & Technology*, 39, 465-470.

Donaldson, S:G.; Van Oostdam, J.; Tikhonov, C.; Feeley, M.; Armstrong, B.; Ayotte, P., Boucher, O.; Bowers, W.; Chan, L.; Dallaire, F.; Dallaire, R.; Dewailly, E.; Edwards, J.; Egeland, G.M.; Fontaine, J.; Furgal, C.; Leech, T.; Loring, E.; Muckle, G.; Nancarrow, T.; Pereg, D.; Plusquellec,P.; Potyrala, M.; Receveur, O.& Shearer, R.G. (2010) Environmental contaminants and human health in the Canadian Arctic. *Science of Total Environment*, 408, 5165-5234.

Falconer, R.L.; Bidleman, T.F. & Gregor, D.J. (1995) Air-water gas exchange and evidence for metabolism of hexachlorocyclohexanes in Resolute Bay, N.W.T. *Science of Total Environment*, 160-161, 65-74.

Fisk, A.T.; Tittlemier, S.A.; Pranschke, J.L. & Norstrom, R.J (2002). Using anthropogenic contaminants and stable isotopes to assess the feeding ecology of greenland sharks. *Ecology*, 83, no. 8, 2162-2172.

Gaul, H. (1992) Temporal and spatial trends of organic micropollutants in the seawater of the Baltic Sea, the North Sea and the Northeast Atlantic. *ICES Marine Science Symposium* 195, 110-126

Gregor, D.J. & and Gummer, W.D. (1989) Evidence of atmospheric transport and deposition of organochlorine pesticides and polychlorinated biphenyls in Canadian Arctic snow. *Environmental Science & Technology*, 23, 56 1-565.

Hardie, D.W.F. (1964) Benzene hexachloride. In: Kirk, R.E. and Othmer, D.F. (eds.) Encyclopedia of Chemical Technology, 2nd edition. New York: John Wiley & Sons, vol. 5, pp. 267-281.

Hargrave, B. T.; Vass, W. P.; Erickson, P. E. & Fowler, B. R. (1988) Atmospheric transport of organochlorines to the Arctic Ocean. *Tellus*, 40B, 480-493.

Hargrave, B.T., Barrie, L.A., Bidleman, T.F. & Welchs, H.E. (1997) Seasonality in exchange of organochlorines between Arctic air and seawater. *Environmental Science & Technology*, 31, 3258-3266.

Harner, T.; Kylin, H.; Bidleman, T. F. & Strachan, W. M. J. (1999) Removal of α- and γ-hexachlorocyclohexane and enantiomers of α-hexachlorocyclohexane in the Eastern Arctic Ocean. *Environmental Science & Technology, 33*, 1157-1164.

Harrad, S. Persistent Organic Pollutants, Editor Stuart Harrad, 2009, 288pages. John Wiley &sons ISBN-10: 1-4051-6930-3.

Helgason, L.B.; Barrett, R.; Lie, E.; Polder, A.; Skaare, J.U. & Gabrielsen, G.W. (2008). Levels and temporal trends (1983–2003) of persistent organic pollutants (POPs) and mercury (Hg) in seabird eggs from northern Norway. *Environmental Pollution*, 155, 190–198.

Helm, P.A.; Diamond, M.L.; Sernkin, R. & Bidleman, T.F. (2000). Degradation as a loss mechanism in the fate of α-hexachlorocyclohexane in arctic watershelds. Environmental Science & Technology, 34: 812-818.

Hinckley, D.A.; Bidleman, T.F. & Rice, CP. (1991) Atmospheric organochlorine pollutants and air-sea exchange of hexachlorocyclohexane in the Bering and Chukchi seas. *Journal of Geophysical Research*, 96, 7201-7213.

Hoekstra, P.F.; O'hara, T.M.; Fisk, A.T.; Borga, K.; Solomon, K.R. & Muir, D.C.G. (2003). Trophic transfer of persistent organochlorine contaminants (OCs) within an Arctic marine food web from the southern Beaufort-Chukchi seas. *Environmental Pollution*, 124, 509-522.

HSDB. 2003. Hexachlorocyclohexanes. Environmental standards and regulations. Bethesda, MD: Hazardous Substances Data Bank.

Inomata, O.N.K.; Montone, R.C.; Lara, W.H.; Weber, R.R. & Toledo, H.H.B. (1996). Tissues distribution of organochlorine residues - PCBs and pesticides - antarctic penguins. *Antarctic Science* 8, no. 3: 253-255.

Iwata, H.; Tanabe, S.; Sakai, N. & Tatsukawa, R. (1993) Distribution of persistent organochlorines in the oceanic air and surface seawater and the role of ocean on their global transport. *Environmental Science & Technology*, 27(6) 1080-1098.

Iwata, H.; Tanabe, S.; Iida, T.; Baba, N.; Ludwig, J.P. & Tatsukawa, R., (1998). Enantioselective bioaccumulation of α-hexachlorocyclohexane in Northern fur seals and double-crested cormorants: effects of biological and ecological factors in the higher trophic levels. *Environmental Science & Technology* 32, 2244-2249

Jantunen, L.M. & Bidleman, T.F. (1995) Reversal of the air-water gas exchange direction of hexachlorocyclohexanes in the Bering Chukchi Sea: 1993 versus 1988. *Environmental Science & Technology* , 29, 1081-1089.

Jantunen, L. M. & Bidleman, T. F. (1996) Air-water gas exchange of hexachlorocyclohexanes (HCHs) and enantiomers of α-HCH in Arctic regions. *J. Geophys. Res.*, 101, 28837-28846. Corrections: *J.Geophys. Res.* 1997, 102, 19279-19282.

Jantunen, L.M. & Bidleman. T. F. (1998) Organochlorine pesticides and enantiomers of chiral pesticides in Arctic Ocean water. *Arch. Environ. Contam. Toxicol., 35*, 218-228.

Jantunen, L.M.; Kylin, H. & Bidleman, T.F. (2004) Air–water gas exchange of α–hexachlorocyclohexane enantiomers in the South Atlantic Ocean and Antarctica. *Deep Sea Research Part II: Topical Studies in Oceanography*, 51 (22-24), 2661-2672.

Kallenborn, R. & Hurhnerfuss, H. (2001) Chiral Environmental Pollutants: Trace Analysis and Ecotoxicology. Springer-Verlag, Heidelberg, Germany.

Kawano, M.; Inoue, T.; Wada, T.; Hidaka, H. & Tatsukawa, R. (1988) Bioconcentration and residue patterns of chlordane compounds in marine animals: invertebrates, fish, mammals and seabirds. *Environmental Science & Technology*, 22, 792-797.

Kelly, B.C.; Ikonomou, M.G.; Blair, J.D. & Gobas, F.A.P.C. (2008). Hydroxylated and methoxylated polybrominated diphenyl ethers in a Canadian Arctic marine food web. *Environmental Science & Technology*, 42, no. 19, 7069-7077.

Knudsen, L.B.; Borga, K.; Jørgensen, E.H.; van Bavel, B., Schlabach, M.; Verreault, J.; et al. (2007) Halogenated organic contaminants and mercury in northern fulmars (Fulmarus glacialis): levels, relationships to dietary descriptors and blood to liver comparison. *Environmental Pollution*, 146, 25–33.

Kucklick, J.I.D.; Hinckley, D.A. & Bidleman, T.F. (1991) Detemination of Henry's law constants for hexachlorocyclohexanes in distilled water and artifcial seawater as a function of temperature. *Marine Chemistry*, 34: 197-209.

Kutz, F.W.; Wood, P.H. & Bottimore, D.P. (1991) Organochlorine pesticides and polychlorinated biphenyls in human adipose tissue. *Reviews of Environmental Contamination & Toxicology*, 120, 1-82.

Lakaschus, S.; Weber, K.; Wania, F.; Bruhn, R. & Schrems, O. (2002) The air-sea equilibrium and time trend of hexachlorocyclohexanes in the Atlantic Ocean between the Arctic and Antarctica. *Environmental Science & Technology*, 36, 138-145

Lara, W.H.; Barretto, H.H.C.; Inomata, O.N.K.; Montone, R. & Weber, R.R. (1990). Organochlorine residue in antarctic penguins. *Pesq. antart. bras. 2*, no. 1, 1-6.

Larsson, P.; Jarnmark, C.; Sodergren, A. (1992) PCBs and chlorinated pesticides in the atmosphere and aquatic organisms of Ross island, Antarctica. *Marine Pollution Bulletin*, 25, 9-12: 281-287.

Li, Y.F., McMillan, A., & Scholtz, M.T. 1996. Global HCH usage with 1°x1° longitude/latitude resolution. *Environmental Science & Technology, 30*, 3525-3533.

Li, Y.F.; Bidleman, T.F.; Barrie, L.A. & McConnell, L.L. (1998) Global hexachlorocyclohexane use trends and their impact on the Arctic atmospheric environment. *Geophysical Research Letters*, 25(1) 39-41.

Li, Y.F.; Trevor Scholtz, M. & Van Heyst, B.J. (2003) Global gridded inventories of β-hexachlorocyclohexane. *Environmental Science & Technology*, 37, 3493-3498.

Li, Y.F.; Macdonald, R. W.; Ma, J. M.; Hung, H. & Venkatesh, S. (2004) Historical α-HCH budget in the Arctic Ocean: the Arctic mass balance model (AMBBM). *Science of Total Environment*, 324, 115-139.

Lide DR. 1991. CRC handbook of chemistry and physics: A ready-reference book of chemical and physical data. Boston, MA: CRC Press, 35-195.

Lohmann, R.; Gioia, R.; Jones, K. C., Nizzetto, L., Temme, C., Xie, Z.; Schulz-Bull, D.; Hand, I.; Morgan, E. & Jantunen, L. (2009) Organochlorine pesticides and PAHs in the surface water and atmosphere of the north Atlantic and Arctic Ocean. *Environmental Science & Technology*, 43 (15). 5633-5639.

Lystsov, V.N. (2006). Arctic ocean p.193-211 in: UNEP/GPA (2006), The State of the marine evironments:regional assessments. UNEP/GPA, The Hague.

Macdonald, R.W.& McLaughlin, F.A. (1993,1994) Long range transport of contaminants to the Canadian basin. In environmental studies n.70, 72.Synopsis of research conducted under the 1992-1993. anad 1993-1994 Northern Contaminants Program, eds. Murray, JL, Shearer, RG, pp 15-20, Indian and Nortehrn Affairs Canada Publ, Ottawa.

Miranda-Filho, K.C.; Metcalfe, T.L.; Metcalfe, C.D.; Robaldo, R.B.; Muelbert, M.N.M.C.; Colares, E.P.; Martinez, P.E. & Bianchini, A.(2007) Residues of persistent organochlorine contaminants in southern elephant seals (mirounga leonina) from elephant island, Antarctica. *Environmental Science & Technology*, 41, no. 11: 3829-3835.

Moisey, J., Fisk, A.T.; Hobson, K.A. & Norstrom, R.J. (2001). Hexachlorocyclohexane (HCH) isomers and chiral signatures of α-HCH in the Arctic marine food web of the northwater polynya. *Environmental Science & Technology* , 35, no. 10, 1920-1927.

Muir, D., de Wit C., et al. 2010. Levels, trends and effects of legacy and new persistent organic pollutants in the Arctic: An AMAP Assessment. *Science of The Total Environment*, 408(15), 2851-3052.

Patton, G. W.; Hinckley, D. A.; Walla, M. D.; Bidleman, T. F. & Hargrave, B. T. (1989).Airborne organochlorines in the Canadian high Arctic. *Tellus*, 41B, 243-255.

Qiu, C. & Cai, M. (2010) Ultra trace analysis of 17 organochlorine pesticides in water samples from the Arctic based on the combination of solidphase extraction and headspace solid-phase micro-extraction-gas chromatography-electron capture detector. *Journal of chromatography A*, 1217, 1191-1202.

Ricca, M.A.; Miles, A.K. & Anthony, R.G. (2008) Sources of organochlorine contaminants and mercury in seabirds from the Aleutian archipelago of Alaska: inferences from spatial and trophic variation. *Science of the Total Environment*, 406, 308–23.

Rigét, F.; Vorkamp, K.; Dietz, R. & Muir, D.C.G. (2008) Levels and temporal trends of HCH isomers in ringed seals from West and East Greenland, *Journal Environmental Monitoring* 10, 935–940

Rigét, F.; Bignert, A.; Braune, B.; Stow. J. & Wilson, S. (2010). Temporal trends of legacy POPs in arctic biota, an update. *Science of the Total Environment*, 408, 15: 2874-2884.

Sahsuvar, L.; Helm, P.A.; Jantunen, L.M. & Bidleman, T.F. (2003) Henry's law constants for α, β, γ-hexachlorocyclohexanes (HCHs) as a function of temperature and revised estimates of gas exchange in Arctic regions. *Atmospheric Evironment*, 37, 983-992.

contaminants in the polar bear food chain. *Environmental Science & Technology* 34, 2668-2674.

Willet, K.; Ulrich, E. & Hites, R. (1998) Differential toxicity and environmental fates of hexachlorocyclohexane isomers. *Environmental Science & Technology*, 32, 15: 2197-2207.

Yao, Z.W.; Jiang G.B. & Xu, H.Z. (2002) Distribution of organochlorine pesticides in seawater of the Bering and Chukchi sea. *Environmental Pollution*, 116, 49-

24

Sublethal Effects of Pyrethroids on Insect Parasitoids: What We Need to Further Know

Patrícia Garcia
Universidade dos Açores
Departamento de Biologia, R. Mãe de Deus, Apt 1422, 9501-855 Ponta Delgada
Centro de Investigação e Tecnologias Agrárias dos Açores (CITA-A),
Azorean Biodiversity Group, Angra do Heroísmo
Portugal

1. Introduction

In the past decade pyrethroids have emerged as a major class of highly active insecticides due to their high bio-efficacy and relatively low toxicity in comparison to organochlorine and organophosphate pesticides, which are more acutely toxic to birds and mammals than the pyrethroids. These chemicals have been largely used for the control agricultural pests, with Lepidoptera representing the highest percentage (40%) of pyrethroid targeted insects, followed by sucking insects and Coleoptera (Wirtz et al., 2009). On the other hand, insect parasitoids are of high importance in natural and agricultural ecosystems where they regulate or influence the population density of their hosts, being therefore worldwide used in the control of several agricultural pests, particularly Lepidoptera and Hemiptera.

Parasitoids can be released in the field either by inundative (for an immediate and non sustaining reduction of the target host population) or inoculative (where the objective is to release the natural enemies early in the season and so is the progeny who will have a later effect on the target host population) approaches. In the first approach, parasitoids are positioned as a fast-acting replacement of insecticides, while in the second parasitoids are considered as one aspect of integrated pest management (Smith, 1996). According to the review by Collier and Steenwyk (2004), biological control by augmentative releases of natural enemies is not likely to replace insecticides in the near future, in part due to the apparent lack of efficacy compared to conventional insecticide applications. Thus the integration of biological and chemical controls is an essential alternative to the conventional use of insecticides that requires knowledge of the lethal and sublethal effects that chemicals may have on the natural enemies. Actually most of the studies regarding the effects of insecticides (inclusive for pyrethroids) on parasitoids and other natural enemies have relied on the evaluation of acute toxicity by determination of a median lethal dose (LD_{50}) or concentration (LC_{50}) (Desneux et al., 2007). Though estimation of LD_{50} or LC_{50} is a simple and a fast approach to compare and evaluate the acute toxicity of insecticides to parasitoids, it overlooks sublethal effects that will interfere with the physiology and behaviour of the natural enemies. Insecticides may directly interfere with the efficacy of biological control agents by changing biological traits, such as, fecundity, fertility, longevity, developmental rates, emergence rates, or indirectly, by modifying behavioural traits that will affect the interactions between parasitoids and their hosts.

In view of the current literature, this chapter aims to cover the main sublethal effects of pyrethroids on insect parasitoids and then discuss the need to develop further research on physiological and behavioural effects that are considered important to the success of insect pest management programs. The chapter is organized as follows. The first section deals with a brief explanation of what are pyrethroids, discussing their role in agriculture nowadays. The second section is dedicated to the knowledge of the main aspects of the biology of insect parasitoids (particularly of egg and larval parasitoids) and makes a summarizing overview of their current use in biological control and Integrated Pest Management (IPM) programs. The third section reviews the lethal and sublethal effects of pyrethroids on parasitoids published in the past 10 years, highlighting the studies that cover the effects of these insecticides on physiological and behavioural traits.

2. Pyrethroids

2.1 What are pyrethroids?

Pyrethroids are neurotoxic manufactured insecticides that are very similar in structure to the pyrethrins, but are often more toxic to insects as well as to mammals, and last longer in the environment than the pyrethrins. Pyrethrins are naturally-occurring insecticidal esters of chrysanthemic acid (pyrethrins I) and pyrethric acid (pyrethrins II), originally found in the flowers of *Chrysanthemum cinerariaefolium* (Asteraceae) and *C. coccineum* (Todd et al., 2003; Davies et al., 2007). Pyrethrin-I, cinerin-I, and jasmolin-I are esters of chrysanthemic acid whereas pyrethrin-II, cinerin-II, and jasmolin-II are esters of pyrethric acid (Essig & Zhao, 2001).

The use of crushed and powdered *Chrysanthemum* plants as an insecticide by the Chinese dates back 1st century AD and by the Middle Ages were to be found in Persia. The "Persian Insect powder", produced from dried flowers of *Chrysanthemum roseum* and *C. cinerariaefolium*, became known and sold in Europe via Armenian traders in the beginning of 19th century (Housset & Dickmann, 2009). In the mid of 19th century, the commercial production of pyrethrins from *Chrysanthemum* flowers started at full-scale, and the chief active chemicals in the extract (pyrethrins I and II) are still in use nowadays in many household sprays and products to control insects on pets or livestock (Davies et al., 2007). Insects can become resistant to pyrethrins via the production of enzymatic detoxifiers, thus these active chemicals are usually commercialized in combination with the synergists piperonyl butoxide and n-octyl bicycloheptene dicarboximide that retard enzymatic degradation of pyrethrins within the insect (Reigart & Roberts, 1999). Moreover, pyrethrins degrade rapidly when exposed to natural sunlight (thus not persisting in the environment beyond a few weeks), so their use for the control of agricultural pests is limited; this issue was surmounted in the 1970s by the structural modification of pyrethrins (Todd et al., 2003), resulting in the production of synthetic pyrethroids that presently account for 15% of the world insecticide market (Wirtz et al., 2009). Currently pyrethroids are formulated as emulsifiable concentrates, wettable powders, granules, and concentrates for ultra low volume application, being worldwide used in agriculture, in homes and gardens, and for treatment of ectoparasitic diseases, through the commercialization of more than 50 distinct commercial products (Reigart & Roberts, 1999).

There are two types of synthetic pyrethroids that differ in chemical structure: Type I pyrethroids (allethrin, bifenthrin, permethrin, phenothrin(Bio), resmethrin, tefluthrin and tetramethrin) are derivatives of pyrethrins that do not include a cyano group and may elicit tremors, and type II pyrethroids (cyfluthrin, cyhalothrin, cypermethrin, deltamethrin, fenvalerate, fenpropathrin, flucythrinate, flumethrin, fluvalinate and tralomethrin) are derivatives of pyrethrins that include a cyano group and may elicit sinuous writhing and salivation (Todd et al., 2003). Synthetic pyrethroids and pyrethrins affect both the peripheral and central nervous system of insects, by acting on the voltage-gated sodium channel proteins found in nerve cell membranes. By prolonging the opening of these channels, pyrethrins and pyrethroids stimulate nerve cells to produce repetitive discharges, causing paralysis (known as insect 'knockdown') and possible insect death (Gunasekara, 2004; Davies et al., 2007). While the development of synthetic pyrethroids was done towards a selective toxicity to insects, they are also highly toxic to many aquatic organisms, inclusive fish (Todd et al., 2003). Though the systemic toxicity by inhalation and dermal absorption of pyrethroids is low for mammals, certain pyrethroids show a high neurotoxicity when administered by intravenous injection, and some are toxic when ingested. In addition to the limited dermal absorption, the low systemic toxicity is also related to the rapid biodegradation of pyrethroids by mammalian liver enzymes (Reigart & Roberts, 1999).

2.2 Current status of pyrethroids in agriculture

After more than 30 years on the market, the role of pyrethroids in modern pest control is still significant today. Synthetic pyrethroids represent the 3rd largest class of chemical insecticides after organophosphates and chloronicotinyl insecticides (CNI, or neonicotinoids), with a market value of 1,300 Mio. US$, a share of 15% of the global foliar and soil insecticide markets and 320 Mio. hectares of treated agricultural area (Wirtz et al., 2009). Pyrethroids are worldwide used in over 300 distinct crops, being their use particularly relevant for the control of agricultural pests of the following crops: fruit (apples, peaches, etc.), grain (sweet and field corn) and vegetables (soybeans, peppers, Brussel sprouts, cucumbers, broccoli, peas, cabbage, tomatoes, potatoes, etc.). Vegetable and flower crops represent >40% of pyrethroid targeted crops in the market, followed by corn and other cereals (30%) and, fruit and nuts (15%); regarding the targeted pests, these chemicals have been largely used for the of Lepidoptera (40% of pyrethroid targeted insects), followed by sucking insects (32%) and Coleoptera (20%). Though currently farmers have commercially available pyrethroids-based ready-mixtures, they still prefer to use straight pyrethroids products, due to their compatibility with other compounds which allows farmers to mix pyrethroids whenever practical. Even so the market share of pyrethroids-based ready-mixtures has grown strongly and constantly over the last years, particularly for pyrethroid-CNI mixtures (Wirtz et al., 2009). Mixing pyrethroid with neonicotinoids results in a product where the effects from active substances from both chemical classes complement each other, either regarding their mode of action (acting on the sodium channel or on the nicotinic acetylcholine receptor), via of contact (absorption/ingestion or systemic) and, their immediate or residual effects. Furthermore, given that pyrethroids are insecticides of broad spectrum and of low cost, the use of pyrethroid-CNI mixtures reduces the cost of only using neonicotinoids. The market share of pyrethroid-CNI mixtures is expected to be particularly relevant in the major emerging countries such as China, India and Brazil, where the impact of climate change and the growing population will promote the need for higher yield and quality products, while fulfilling the environmental safety requirements (Smith et al., 2008; Wirtz et al., 2009). To contribute to IPM programmes, these chemicals are required to be highly specific and selective (products must be benign to non-target species ranging from aquatic species through to beneficial insects), show benign environmental and toxicological profiles, and be biodegradable (Smith et al., 2008).

3. Insect parasitoids

3.1 Parasitoid definitions and natural history

Parasitoids are insects whose larvae develop by feeding on the bodies of other arthropods, usually insects, eventually resulting on the death of their host, and a single host provides all the food required for the development of the parasitoid into an adult. They represent about 8,5% of all the described insect species, although this value may rise up to 20% given that many workers argue that parasitoids are relatively poorly known (Godfray, 1994). Parasitoids are holometabolous insects (their life cycle is divided in 4 stages - egg, larva, pupa and adult), that can develop to adult singly from one host or gregariously on the same host (gregarious parasitoids). Gregarious parasitism may be the result of multiple eggs laid on or in a single host, or it may result from the repeated division of a single egg, being in this case termed as polyembryony (Gauld & Bolton, 1988).

After oviposition, the host is either killed or permanently paralyzed (idiobiont parasitoids), so the parasitoid larvae is limited to the host resources present at the moment of oviposition, or it is maintained alive or only temporarily paralyzed, and after recover continues to feed until it dies due to parasitoid development and feeding (koinobiont parasitoids). Parasitoids can be further classified as endoparasitoids, which develop inside the host, and ectoparasitoids, which develop outside the host body, though they are frequently attached or embedded in the host tissues. In some cases the female lays the eggs on the host's foodplant and parasitism occurs either when the host eats the parasitoid eggs or when there is a free-living first instar parasitoid larva that actively searches for the host. Adult parasitoids may feed from several energy sources, such as flowers, sap fluxes, and even from potential hosts, a behaviour designated host feeding. Parasitoids are classified as egg, larval, pupal or adult parasitoids when they parasitize respectively the egg, larva, pupa or adult host stage. Some parasitoids may lay the eggs in one host stage but their progeny only emerges until the host has entered a later stage, being therefore designated as egg-larval or larval-pupal parasitoids. Egg, pupal and adult parasitoids, as well as larval parasitoids whose sting causes permanent paralysis of the host, are usually idiobionts while the remaining usually are koinobionts. Idiobionts that attack concealed hosts are usually ectoparasitoids, while those attacking exposed hosts are in general endoparasitoids. Koinobionts are in majority endoparasitoids and usually attack exposed or weakly concealed hosts. Generalist parasitoid species are able to develop in several host species, while specialists can only develop on one or few closely related host species. Species that are parasitoids of other parasitoids are designated hyperparasitoids. Sex determination in hymenopteran parasitoids is haplodiploid, being the males generally haploid (unfertilized eggs develop to become males), while individuals developing from fertilized eggs become females; this type of reproduction is termed arrehenotoky. Some parasitoids reproduce asexually by thelythokous parthenogenesis; in these case diploid females produce a diploid female progeny by a variety of asexual mechanisms. One relevant aspect of arrenothokous parthenogenesis is the ability of females to determine the offspring sex ratio by altering the

proportion of laid fertilized eggs, in response to distinct factors, such as host size, host distribution and abundance (Gauld & Bolton, 1988; Godfray, 1994; Quicke, 1997).
Regarding parasitoid diversity, these insects are mainly from the order Hymenoptera (with about 50000 described parasitoids), followed by the order Diptera (with about 15000 described parasitoids), being the remaining species (about 3000) from other orders, such as Coleoptera. For Hymenoptera, parasitoids within the superfamily Ichneumonoidea represent half of the described parasitoids (about 25000 species) followed by the Chalcidoidea (>17500 species). Within the order Diptera, the family Tachinidae with 8200 described parasitoid species represents more than half of dipteran parasitoids (Godfray, 1994).

3.2 Parasitoids and insect pest management

Due to their life histories, parasitoids have an important role in natural and agricultural ecosystems by influencing and regulating the population density of their hosts. According to Van Lenteren (1986) more than 80% of the successful natural enemies are parasitoids, being the remainder predators (17%) and pathogens (1%). The role of parasitoids in the natural and agricultural ecosystems has enhanced many studies, generating a vast quantity of information on the behaviour and ecology of several distinct genera and species (Godfray, 1994); Van Lenteren (1986) estimated that up to that date, 5000 species have been tested for use in biological control and out of these, 270 species have led to partial (100), substantial (100) or complete control (70). Parasitoids can be released to the field by an inundative approach to achieve an immediate, non-sustaining reduction of the target pest or by an inoculative approach, in which few parasitoids are required early in the season because it is the progeny that will have a later effect on the target pest; therefore, in inoculative release the parasitoid must be faced as one aspect of IPM. The choice of the release approach is based on established criteria for pre-introductory evaluation of natural enemies. Among such criteria, synchronization with the host (both seasonal and internal), host specificity and high reproductive potential are quite relevant when using an inoculative approach, while for inundative the more relevant factors are adaptation to the climatic conditions of the environment and a good mass rearing method (Van Lenteren, 1986). According to Greathead (1986), in classical biological control (the introduction and permanent establishment of an exotic biological control agent) three taxa stand out as major sources of parasitoids: Ichneumonoidea and Chalcidoidea (Hymenoptera), and Tachinidae (Diptera). Within Ichneumonoidea, Braconidae and Ichneumonidae are families that have a broad range of hosts and life cycles. Braconidae are usually egg-larval or larval parasitoids of aphids, bark beetles, and foliage-feeding caterpillars; the genera *Apanteles* s.l., *Bracon* s.l. and *Opius* s.l. are within this family the more important groups of wasps used in biological control. Ichneumonidae wasps attack larvae or pupae of flies, caterpillars, beetles and sawflies; their eggs are usually placed inside the host larvae and develop internally, though a few are ectoparasitoids, feeding on the outside of the host. Within the superfamily Chalcidoidea, most of the species are parasitoids of other insects, attacking the egg or larval stage, though many other life cycles are known; the most relevant wasps in biological control are from the families Eulophidae, Pteromalidae, Encyrtidae and Aphelinidae (Greathead, 1986). Regarding Diptera, the family Tachinidae ranks with the more important families of Hymenoptera as a source of biological control agents; tachinid larvae are internal parasitoids of many insects, especially Lepidoptera and, their oviposition and larviposition strategies are very diverse, varying from those which lay eggs and wait for the host larvae to find them, to other species which retain the egg until it is about to hatch before inserting it into the body of the host (Greathead, 1986).

In biological control, using egg parasitoids over larval for the control of pests has the advantage of preventing the hatch of the host prior it causes major damages; furthermore, egg parasitoids are also easier to mass rear given that these wasps usually are less host-specific than larval parasitoids. The most successful families of egg parasitoids in biological control are Scelionidae, Mymaridae and Trichogrammatidae (Greathead, 1986).

Among Trichogrammatidae, the genus *Trichogramma* (Hymenoptera: Trichogrammatidae) as been considered for the control of lepidopterous pests for more than 100 years (Smith, 1996). The genus *Trichogramma* is worldwide distributed and consists of about 145 described species. These minute wasps (0,2-1,5 mm) have a preference for Lepidoptera, but there also records of parasitism from eggs of Coleoptera, Diptera, Hemiptera, Homoptera, Hymenoptera (Symphyta) and Neuroptera (Pinto & Stouthamer, 1994). Thelytoky is described for 10 *Trichogramma* species and is associated with an infection of the wasps with *Wolbachia* bacteria (Stouthamer et al., 1993). Arrhenotokous lines of these species can be started either by treating thelytokous females with antibiotics (tetracycline, rifampicin or sulfamethoxazole) or by rearing the parasitoids at temperatures above 28°C (Pintureau et al., 1993; Stouthamer & Werren, 1993). A species with both modes of reproduction allows the choice of which form to use in a biological control release: for inundative release thelytoky seems to be in advantage, given that the production costs per female are lowered; for inoculative releases, Stouthamer (1993) suggested that the thelytokous strains should be used when hosts are very patchily distributed because of their superior colonization characteristics. On the other hand, the sexual strains will likely be able to adapt faster to changing circumstances so they should preferentially be used for inoculative releases, where parasitoids are expected to be and reproduce in the fields for a larger period.

Research on the use of *Trichogramma* in biological control initiated in the second decade of the 20th century and a method to mass produce these wasps using *Sitotroga cerealella* Olivier (Lepidoptera: Gelechiidae) was developed in 1930 in the USA (Hassan, 1993); yet, the mass rearing and release of these wasps only fully began in Europe and North America in the 1970s (Smith, 1996). Since 1975, several species of *Trichogramma* have been used to control pests of over 20 different crops (e.g., cotton, sugar-beet, cabbage, corn, rice, sugar-cane, soybean), in more than 50 countries from all over the world being reported to be used commercially on more than 32 million ha each year (Li-Ying, 1994; Smith, 1996). Although there are almost 20 years apart between the reviews made by Stinner (1977) and by Smith (1996), both authors emphasise on integrating *Trichogramma* with other control options, particularly in combination with selective insecticides.

Aphid parasitoids (Hymenoptera, Braconidae) are also well recognized worldwide for their importance in controlling aphid pests within a range of cropping system, with more than 400 different species documented. The female parasitoid oviposits into the host but the aphid is only killed prior to parasitoid pupation, being only kept the skin of the host, termed as mummy. The parasitoid then completes the development to adult, which emerges from the host (Longley, 1999).

The immature stages (egg, larva and pupa) of both egg and aphid parasitoids can be often protected from insecticide applications since parasitoids are housed within the host egg chorion or host skin during their development, being therefore key stages for IPM programs.

4. Sublethal effects

Integrated pest management combines chemical, biological, and cultural control to provide targeted and efficient pest management solutions that can be tailored to specific climates and habitats in order to maximize a program's efficacy (Gentz et al., 2010). However, and in addition to the development of pest resistance (Denholm et al., 2001), the extensive use of insecticides in crop systems surely has effects on non-target organisms. Among these effects, the reduction of the efficiency of biological control agents, such as parasitoids, has been highlighted by several authors (Cross et al., 1999; Sterk et al., 1999; Brunner et al., 2001; Hewa-Kapuge et al., 2003). Since insecticides may cause the death of the biological control agents (lethal effects) or change several other traits of their biology (either physiological or behavioural) without killing the individuals (sublethal effects), the success of IPM programs depends, in part, on the optimal use of selective insecticides that are less harmful to natural enemies. Thus, in addition to direct mortality induced by insecticides, we should also have knowledge of the effects on physiological and behavioural traits of the natural enemies for a complete analysis of the insecticides impact (Desneux et al., 2007; Stark et al., 2007a).

The main routes of pyrethroid exposure for adult parasitoids or for ectoparasitoids are uptake after direct exposure to spray droplets, uptake of residues by contact with contaminated surfaces, e.g. soil or vegetation, and oral uptake from contaminated food sources (Langley & Stark, 1996), e.g. when parasitizing and/or feeding on insecticide-treated host. Endoparasitoids are mainly exposed to pyrethroids trough the uptake of residues from insecticide-treated host upon the emergence of the adult, or by contact with the active ingredient after its penetration trough the host skin or chorion, being therefore more often protected than adults or ectoparasitoids. Pyrethroids are considered harmful (mortality >99%) to insect parasitoids, particularly to adults; nevertheless, when testing the less susceptible life stages of endoparasitoids (e.g., larval or pupal stages), there is reduction in the mortality, lowering the evaluation category to slightly harmful (30-70% of mortality) (Sterk et al., 1999). According to the review made by Longley (1999), organophosphate insecticides prove to be generally more toxic to developing aphid parasitoids within mummified hosts than pyrethroid compounds (pyrethroids were included by the authors in a rank that gives mortalities between 21-30%, while organophosphates ranked for mortalities between 51-60%). However, even if immature stages of insect parasitoids might be less susceptible to pyrethroids, these insecticides most probably have effects on the wasp's longevity, fecundity, developmental rates, sex ratios, mobility, etc., and therefore such traits should also to be considered when assessing the side effects of insecticides on beneficial organisms. Even such approaches show the consequences of any detrimental side effects of pyrethroids that otherwise go undetected in the standard direct mortality assessments, they are laborious to perform and so are frequently ignored by researchers.

A revision of the sublethal effects of pyrethroids on parasitoids published in the past 10 years is made in this section, highlighting the studies that cover the effects of these insecticides on physiological and behavioural traits. Information regarding the main sublethal effects analysed in these studies is summarized in Table 1.

4.1 Physiological traits

The effects of insecticides caused by direct contact with the toxin are manifested as short-term mortality or relatively long-term sublethal effects. The majority of the studies that evaluate the side effects of pyrethroids on natural enemies report primarily to short-term mortality by directly analyzing the mortality of adults and immature stages and, indirectly trough the evaluation of the emergence of parasitoids from the host (e.g., Tillman & Mulrooney, 2000; Kok & Acosta-Martinez, 2001; Raposo et al., 2003; Symington, 2003; Youssef et al, 2004; Prabhaker et al., 2007; Wang et al., 2008). However, in the last years the number of studies that additionally report to the sublethal effects of pyrethroids has increased, highlighting effects on the larval and pupal development, longevity, fecundity, sex-ratio, oviposition behaviour, mobility/orientation, etc. Some of the studies mentioned herein have also analysed the effects of pyrethroids on the emergence rates of parasitoids, but this trait will not be particularly discussed in this chapter since it is an indirect via of quantifying wasp mortality within the host.

Development

Insecticides may interfere with the development rate of parasitoid preimaginal stages, which can have a large impact on a natural enemy's intrinsic rate of increase (*r*m) and phenological synchrony with the host or prey (Desneux et al. 2007). Egg-adult development time of *Apanteles galleriae* (Hymenoptera: Braconidae) reared on *Achoria grisella* (Lepidoptera: Pyralidae) larvae exposed to different doses of cypermethrin was increased by more than 50% than those parasitoids that developed on untreated hosts (Ergin et al., 2007). According to these authors, the delay in the parasitoid development may be due to changes in the hormonal milieu of the host (since the larval parasitoid synchronizes the development with the host by making use of host hormones) or to insufficient food supply from the host (due to the cypermethrin-induced decline in diet quality and to the potency of cypermethrin as an antifeedant). Conversely, Symington (2003) observed that the pupation and adult emergence of the larval endoparasitoid, *Orgilus lepidus* (Hymenoptera: Braconidae), was hastened (approximately 24h and 48h, respectively) when parasitoids were developing in permethrin-treated host larvae. Vieira et al. (2001) did not observe changes in the developmental rate of *Trichogramma cordubensis* (Hymenoptera: Trichogrammatidae) treated with pyrethroids (deltamethrin and lambda-cyhalothrin) at different preimaginal stages; in this case, the chorion of the host egg most probably provided a protection for the immature stages of the parasitoids, and thus the wasp's preimaginal development was not affected by both pyrethroids.

Longevity

Reductions in longevity have been generally observed in parasitoids that had been exposed to insecticides when developing inside of hosts (Longley, 1999; Desneux et al., 2006ab; Ergin et al. 2007).

The consequences of reduced longevity on population dynamics were recently emphasized by studies assessing pesticide impacts on arthropods using life table analysis, given that a decrease in survival may result in a strong reduction on the natural enemy's intrinsic rate of increase (Stark et al., 2007b).

Effects on parasitoid longevity may be considered a sublethal effect or latent mortality, depending on the biology of the natural enemy (e.g., proovigenic parasitoids are more likely to reproduce before their premature death than synovigenic wasps, given that proovigenic wasps have a full complement of mature eggs at emergence). According to Bayram et al. (2010), the longevity of *Trichogramma busseolae* (Hymenoptera: Trichogrammatidae) females was significantly reduced by deltamethrin or cyfluthrin, but the authors refer that such reduction may not affect the reproductive efficacy of the parasitoids under field conditions

Parasitoid species	Pyrethroid active ingredient	Exposure (individuals exposed)	Sublethal effect	Reference
Microplitis croceipes (Hym.: Braconidae)	esfenvalerate	Contact with odours (adults)	B (food searching*)	Alyokhin et al. (2010)
Telenomus busseolae (Hym.: Scelionidae)	cyfluthrin deltamethrin	Contact with residues (adults)	F*, L* B (walking and response to host cues*) OF (emergence rates, sex ratio)	Bayram et al. (2010)
Cotesia vestalis (Hym.: Braconidae)	permethrin	Spraying host plants (adults)	B (flight response and foraging behaviour)	Kawazu et al. (2010)
Trichogramma cordubensis (Hym.: Trichogrammatidae)	deltamethrin lambda-cyhalothrin	Spraying (diapausing prepupae)	L*, F*	Garcia et al. (2009)
Trichogramma pretiosum (Hym.: Trichogrammatidae)	betacyflurin esfenvalerate	Spraying host eggs (adults)	F* OF (fecundity*)	Vianna et al. (2009)
Apanteles galleriae (Hym.: Braconidae)	cypermethrin	Via the host (host diet)	L*, D* OF (sex ratio; body size)	Ergin et al. (2007)
Trichogramma pretiosum (Hym.: Trichogrammatidae)	alpha-cypermethrin zeta-cypermethrin deltamethrin	Contact with residues (adults)	F*	Bastos et al. (2006)
Diaeretiella rapae (Hym.: Braconidae)	deltamethrin	Spraying (aphid mummies) Contact with residues (adults)	L* B (response to host cues)	Desneux et al. (2006a)
Aphidius ervi (Hym.: Braconidae)	deltamethrin	Spraying/topical (aphid mummies) Contact with residues (adults)	L* B (response to host cues)	Desneux et al. (2006b)
Trichogramma cordubensis (Hym.: Trichogrammatidae)	deltamethrin	Spraying (prepupa within the host)	F, EG OF (emergence rates*)	Garcia et al. (2006)
Leptopilina heterotoma (Hym.: Figitidae)	deltamethrin	Contact with residues (adults)	B (walking* and response to host cues*)	Delpuech et al. (2005)
Trissolcus grandis (Hym.: Scelionidae)	deltamethrin	Spraying (larvae within the host)	L, F OF (emergence rates, sex ratio)	Saber et al. (2005)
Aphidius ervi (Hym.: Braconidae)	lambda-cyhalothrin	Contact with residues (adults)	B (response to host cues)	Desneux et al. (2004a)
Diaeretiella rapae (Hym.: Braconidae) *Aphidius matricariae (Hym.: Braconidae)*	deltamethrin	Contact with residues (adults)	B (oviposition behaviour)	Desneux et al. (2004b)
Aphidius ervi (Hym.: Braconidae)	lambda-cyhalothrin	Contact with residues (adults)	B (response to host cues*) B (oviposition experience and behaviour*)	Desneux et al. (2003)
Trissolcus basalis (Hym.: Scelionidae)	deltamethrin	Contact with residues (adults)	B (walking* and response to host cues*)	Salerno et al. (2002)
Trichogramma cacoeciae (Hym.: Trichogrammatidae)	deltamethrin	Contact with residues (adults)	F*	Youssef et al. (2004)
Orgilus lepidus (Hym.: Braconidae)	permethrin	Spraying (immature stages within the host)	D*	Symington (2003)
Trichogramma brassicae (Hym.: Trichogrammatidae)	deltamethrin	Contact with residues (adults)	B (sex pheromonal communication*)	Delpuech et al. (2001)
Trichogramma cordubensis (Hym.: Trichogrammatidae)	deltamethrin lambda-cyhalothrin	Spraying (immature stages within the host)	L, D	Vieira at al. (2001)

Table 1. Summary of literature published from 2000-2010 (data from ISI Web of Knowledge database) regarding the sublethal effects of pyrethroids on parasitoids; L= Longevity, F= Fecundity, EG= egg maturation, D= Development, B= Behaviour (specified), OF= offspring traits (specified); *Significant effects.

because the parasitoid lays most of its eggs within 72 h of emergence (longevity of pyrethroid-treated wasps averaged 11 days). On the other hand, Desneux et al. (2006ab) confirmed a high risk of deltamethrin treatments for aphid parasitoids [*Diaeretiella rapae* and *Aphidius ervi* (Hymenoptera: Braconidae)] that may recolonize the treated crop when emerging from the mummies, since deltamethrin induced a decrease in the longevity of emerged individuals and of the adults that walked on treated leaves, and thus, potential fecundity will also be limited. In the case of *A. ervi*, the authors have calculated a Population Survival Index (that takes into account the percentage of aphid of mummified parasitoids, the percentage of parasitoid emergence from the mummies, the percentage of emerged adults that survived more than 48h and the percentage of adults that survived to residual exposure) which foresee that only 5% of the parasitoid population would be able to recolonize a deltamethrin-treated field (at the recommended rate) by the "vertical recruitment" process (Desneux et al., 2006b); for *D. rapae* this index dropped the prediction to 3% (Desneux et al., 2006a).

Fecundity, Egg maturation and Sex-ratio

Reductions in fecundity due to pyrethroids may be due to both physiological (e.g., by interfering with oogenesis) and behavioural (e.g., by interfering with host acceptance) factors. Although several authors have shown that pyrethroids reduce the fecundity of parasitoids (Youssef et al., 2004; Bastos et al., 2006; Vianna et al., 2009; Bayram et al., 2010), the physiological and behavioural causes of such reduction are not usually discussed. Bastos et al. (2006) showed that although the pupal stage of *Trichogramma pretiosum* (Hymenoptera: Trichogrammatidae) developing in *Sitotroga cerealella* (Lepidoptera: Gelechiidae) eggs was less susceptible to alpha-cypermethrin and deltamethrin than parasitoids developing in *Ephestia kuehniella* (Lepidoptera: Pyralidae), the fecundity is highly reduced by these pyrethroids regardless of the host species. These authors hypothesized that the differences in the structure and composition of host egg chorions may have provided a higher protection for the parasitoids developing inside *S. cerealella* eggs, but variations in egg structure, however, would not account for the difference in the ability of the adults to parasitize eggs. Bayram and co-authors (2010) showed that *T. busseolae* females exposed to cyfluthrin parasitized significantly fewer eggs than untreated females, whereas deltamethrin did not significantly affect the wasp's fecundity; nevertheless, neither the emergence rates nor the sex ratio of the offspring was affected by both insecticides. These authors hypothesized that the reduction in fecundity was probably due to perturbations in behaviour (cyfluthrin also negatively affected the parasitoid behavioural responses to host cues) rather than the reduction of egg load. The influence of deltamethrin on the reproduction of *T. cordubensis* was investigated by Garcia et al. (2006) when studying egg maturation and daily fecundity of deltamethrin-treated wasps (prepupal stage) and emergence rates of their offspring. In this study, the authors demonstrated that although daily fecundity and egg maturation patterns were not modified by the pyrethroid, the offspring emergence rates of wasps that started the vitllelogenesis upon emergence from the host were most probably reduced by deltamethrin residues resting on the chorion of the host egg. Vianna et al. (2009) observed that the fecundity of *T. pretiosum* was significantly reduced by pyrethroids (betacyflurin and esfenvalerate) in two consecutive generations, despite the fact that only the first one contacted with pyrethroid-treated host eggs. Even though Saber et al. (2005) had observed a significant reduction in the adult emergence of *Trissolcus grandis* (Hymenoptera: Scelionidae) from host eggs treated with deltamethrin, the pyrethroid did not negatively affect the longevity or fecundity of emerged females.
Changes in the sex-ratio of parasitoids exposed to pyrethroids are mainly related to differential survival as a function of sex, either during development or as adults (Saber et al., 2005), or an effect during sex determination in haplodiploid species, in which the parasitoid defines the sex of the progeny when laying the eggs, a behaviour that can be disrupted by the pyrethroid. Nevertheless, when endoparasitoids are exposed to pyrethroids during the preimaginal stages it seems that the sex-ratio is not usually disrupted, as the results by Saber et al. (2005), Ergin et al. (2007) and Bayram et al. (2010) underline.

4.2 Behavioural traits

The effectiveness of a parasitoid in biological control is, to great extent, dependent on the ability of the parasitoid ability to locate, recognize and parasitize the host. Successful parasitisation has been divided into several phases, mediated by a variety of chemical and physical stimuli: host-habitat location, host location, host acceptance and host regulation (Nordlund, 1994). The recognition of chemical stimuli (e.g., host-produced cues) by foraging parasitoids plays an important role to their success on finding and parasitizing hosts. The process of odour detection of these chemical stimuli, and consequent parasitoid responses, is dependent on neural transmissions, which are expected to be affected by neurotoxic insecticides, such as pyrethroids. The recognition that the evaluation of the impacts of pyrethroids on the behavioural traits of parasitoids is also of high relevance in IPM, has conducted to an increase in the number of published studies in the past decade.

Mobility and Orientation

Pyrethroids may interfere by direct intoxication with the mobility of parasitoids, producing uncoordinated movements, trembling and even a knockdown effect; these neurotoxic insecticides may also change their mobility patterns such as, angular speed and linear speed. Disruption in the detection of habitat and host chemical cues by pyrethroids also interferes with the orientation of the parasitoids, resulting in behaviours like parasitoid arrestment or repellence (Desneux et al., 2007). Salerno et al. (2002) found significant sublethal effects of deltamethrin on the arrestment response and walking behaviour of *Trissolcus basalis* (Hymenoptera: Scelionidae) to a contact kairomone (chemical traces) from its host, *Nezara viridula* L. (Heteroptera: Pentatomidae). These authors have shown that *T. basalis* females that had been exposed to a low concentration of deltamethrin reduced their linear speed and, in the presence of host cues, females spent less time on host patches but their walking speed was not altered. In face of these results, the authors hypothesize that the sublethal effect of deltamethrin on host foraging may decrease *T. basalis* efficacy in controlling *N. viridula*. Delpuech et al. (2005) also found that deltamethrin increased the arrestment response of *Leptoplilina heterotoma* (Hymenoptera: Figitidae) to host cues and reduced their linear speed. The authors explained that the observed increase in the arrestment response is coherent with pyrethroid intoxication, given that deltamethrin induces a prolongation of nervous stimulations each time neurons are stimulated by the perception of host cues. Bayram et al. (2010) observed that cyfluthrin-treated *T. busseolae* females failed to respond to the host sex pheromone, although deltamethrin-treated females responded similarly to untreated females. These authors have also evaluated the sublethal effects of both pyrethroids on the parasitoid arrestment behaviours (residence time, linear speed) in an open arena containing abdominal scales from virgin female moths (which are a source of host-contact kairomonal cues), but no significant differences were found. According to Bayram et al. (2010), the fact deltamethrin had no effect on sex-pheromone detection by the parasitoid is probably related to the experimental procedure: all behavioural observations undertaken on *T. busseolae* were performed 12 h after the exposure, a period that may have enabled parasitoids to compensate for sublethal effects of deltamethrin. Desneux et al. (2006ab) also confirmed that the orientation behaviour *A. ervi* and *D. rapae* toward aphid-infested plants is not altered after parasitoid exposure to deltamethrin, suggesting that the greater vigour in the surviving insects may explain this lack of effect; similar results were observed by Desneux et al. (2004a) when analysing *A. ervi* responses to the odour from aphid-infested plants after parasitoid exposure to increasing doses of lambda-cyhalothrin residues. In very recent study, Kawazu et al. (2010) demonstrated that permethrin had an inhibitory effect on *Cotesia vestalis* (Hymenoptera: Braconidae) flight response, given that wasps showed significant preference for host-infested plants over insecticide-treated plants. Moreover, searching time was significantly shorter and the mortality of *C. vestalis* adults on the insecticide-treated plants significantly higher than in the control plants (treated with distilled water).

Oviposition behaviour

Once a host is detected, the female parasitoid evaluates its suitability, by checking the presence of the physiological conditions necessary for the development and growth of her progeny. Host acceptance or rejection depends on a variety of cues perceived by the female parasitoid during antennal contact with the host and/or ovipositor insertion (Godfray, 1994). All the behaviours involved in this process, from the detection of host chemical cues to oviposition, involve neural transmissions, which are targeted by neurotoxic insecticides, and thus pyrethroids have been reported to disrupt both sensory perception and motor functions of parasitoids. Such disruptions were demonstrated by Desneux et al. (2003) when studying the sublethal effects of lambda-cyhalothrin on host searching (at a low sublethal dose of exposure) of *A. ervi* females without oviposition experience and oviposition behaviours (at a lethal dose of exposure) of oviposition experienced females; however, no effect was observed on the orientation behaviour toward aphid-infested plants when females had an oviposition experience prior to the olfactometer test. Furthermore, all the behavioural effects disappeared 24 h after the end of insecticide exposure. The results obtained by these authors draw attention to the fact that orientation and oviposition behaviours of parasitoids may be disturbed by pyrethroids, depending on the dose, the parasitoid experience and the type of behaviour. Conversely, the oviposition behaviour (both frequencies and behavioural sequences) of *D. rapae* and *A. matricariae* (Hymenoptera: Braconidae) on aphid-infested plants and patch-time allocation were not disturbed following exposure to deltamethrin residues (Desneux et al., 2004b), regardless of the dose. According to these authors the unexpected lack of effects after deltamethrin exposure can have two explanations: either the deltamethrin molecules did not alter the functions necessary for host-handling behaviour, or the surviving insects were less susceptible (reflecting differences in parasitoid vigour or in genetically determined susceptibility to pesticides).

Communication

Disruption by pyrethroids of sexual communication and mate-finding in insect parasitoids may be due to changes in the capacity to create and/or perceive chemical stimulus involved in the communication process (Desneux et al. 2007)). According to Delpuech et al. (1999) deltamethrin induced an increase of the arrestment of *Trichogramma brassicae* (Hymenoptera: Trichogrammatidae) males by the sex pheromone when only males but not females were exposed a sublethal dose of the insecticide; when the pheromone was emitted by treated females, deltamethrin induced a decrease of the response of non insecticide-treated males (Delpuech et al., 1999). Afterwards, Delpuech and co-authors (2001) demonstrated that these effects are counterbalanced when both sexes are exposed to an equal sublethal dose of deltamethrin (that causes theoretically 0.1% of mortality), being the mean response to pheromone extract not significantly different from that of controls. However, the kinetics of their response was modified by the pyrethroid: for controls, the mean response to the pheromone regularly decreased during time (probably due to an increasing saturation of pheromone receptors), while the mean response of deltamethrin-treated males stabilized throughout time. As a result, the decrease of the emission of sex pheromone by treated females would be compensated by the increased receptivity of treated males.

Food searching

Volatiles present in the environment, such as insecticide odours, may interfere with the specific olfactory cues (such as odours from food sources) through repellent and/or masking action, or may enhance them by providing a contrasting background against which the informative odours become easier to distinguish. This issue was very recently investigated by Aleyokin et al. (2010) using the food-searching behaviour of *Microplitis croceipes* (Hymenoptera: Braconidae) in laboratory arenas in the presence of odours from distinct pesticides as a model system. Their results showed that the odour of esfenvalerate decreased *M. croceipes* response to food odours, being this effect less pronounced for the unfed parasitoids. In face of such behaviour, the authors suggest that provision of food may potentially be used to keep beneficial natural enemies away from insecticide-treated areas.

4.3 Other effects

Controlling diapause in egg parasitoids increases the efficiency of their long-term storage with important implications for their use in pest control programs, by allowing producers to stockpile the parasitoids for release in the growing season (Rundle et al. 2004), and therefore reduce the costs of mass rearing. However, when storing diapausing wasps one must not disregard the biological traits of post-diapausing insects, such as their susceptibility to pesticides, because poor-quality parasitoids could yield low efficiency in pest management programs. A recent study by Garcia et al. (2009) demonstrated that emergence rate, longevity and fecundity of pyrethroid (deltamethrin and lambda-cyhalothrin) treated diapausing prepupae of *T. cordubensis* generally decreased with increasing duration in cold storage. Emergence rates were particularly affected regardless of the period of cold storage, given that both pyrethroids significantly reduced the emergence rates (<25%) compared to the control (emergence varied from 83% to 89%). These results highlight the fact that the sublethal effects of pyrethroids on long-term cold stored diapausing wasps are more noteworthy that on non cold stored parasitoids and thus such effects must not be disregarded when considering the use of cold-stored wasps in IPM programs. Another important issue when analysing sublethal effects of insecticides is the variation in susceptibility that occurs between distinct populations of parasitoids. The results by Vianna et al. (2009) showed that the impact of pyrethroids on distinct populations of *T. pretiosum* can vary significantly, which indicates that the tolerance or resistance to insecticides may be related to biotic (adaptability and intrinsic capacity of the population in a specific environment) and abiotic (environmental conditions) factors of their origin place. Therefore, and as suggested by the authors, different populations of parasitoids should be used when evaluating insecticide toxicity.

5. Future prospects and needs

Although insecticides still dominate commercial agriculture, increased public awareness of their deleterious non-target effects, including those on human health, provides a strong incentive for the use of alternatives. Still, complete pesticide replacement by non-chemical methods is unlikely to happen in the near future. Therefore, compatibility with chemical control is essential for the increased adoption of alternative methods, such as the integration of biological and chemical controls. In general, field studies have shown that natural enemies, particularly adult parasitoids, are highly affected by pyrethroids, even when these insecticides are applied at a low dosage (e.g., Ruano et al., 2010). However, in the case of endoparasitoids, laboratory trials generally have shown that when pyrethroids are applied on parasitized hosts, their impact on the parasitoids is lowered (e.g., Longley, 1999; Sterk et al., 1999). Even so, their impact on sublethal effects on natural enemies, and consequently the success of biological control programs, can be as deleterious as mortality. Most of the studies for pyrethroid sublethal effects report effects in behavioural and physiological traits of parasitoids in laboratory trials, which can severely disrupt intraspecific or interspecific interactions. Such disruptions generally decrease the success of biological control or IPM programs since they usually interfere with the natural enemies' long-term survival and reproduction and, consequently, reduce their effectiveness in controlling the targeted pest. However, in some field and semi-field studies pyrethroid residues where shown to have no effect on the limitation of parasitoids when recolonizing treated fields (e.g., Desneux et al., 2005) and so, in a perspective of integrated pest management, pyrethroids and parasitoids can have additive effects in controlling the targeted pest. Increasing temporal and spatial separation between insecticide treatments and parasitoids is instrumental for their integration in pest management programs, and certainly behavioural manipulation may provide an additional tool for achieving this goal. The combination of knowledge on physiological and behavioural sublethal effects of pyrethroids, and as well of pesticides in general, provides valuable information for IPM decision-making, as highlighted by most of the studies discussed in this chapter. The integration of these effects in an analysis of population fitness, specifically focused on the population growth rate of the natural enemies, would also provide a more holistic picture of the ability of biocontrol agents to manage insect

pest populations. Therefore, predicting the overall effects of pyrethroids on natural enemies, including mortality and sublethal effects, confirmed by field data, followed with demographic and modelling analysis, is the safest via, although complex, to develop and use truly selective insecticides that cause minimal disruption to naturally occurring biological control agents.

6. References

Alyokhin, A.; Makatiani, J. & Takasu, K. (2010). Insecticide odour interference with food-searching behaviour of *Microplitis croceipes* (Hymenoptera: Braconidae) in a laboratory arena. *Biocontrol Science and Technology*, 20(3): 317-329.

Bayram, A.; Salerno, G.; Onofri, A. & Conti, E. (2010). Sub-lethal effects of two pyrethroids on biological parameters and behavioural responses to host cues in the egg parasitoid *Telenomus busseolae*. *Biological Control*, 53:153-160.

Bastos, C.S.; Almeida, R.P. & Suinaga, F.A. (2006) Selectivity of pesticides used on cotton (*Gossypium hirsutum*) to *Trichogramma pretiosum* reared on two laboratory-reared hosts. *Pest Management Science*, 62: 91-98.

Brunner, J.F.; Dunley, J.E.; Doerr, M.D. & Beers, E.H. (2001). Effect of pesticides on *Colpoclypeus florus* (Hymenoptera: Eulophidae) and *Trichogramma platneri*

(Hymenoptera: Trichogrammatidae), parasitoids of leafrollers in Washington. *Journal of Economic Entomology*, 94(5): 1073-1083.

Collier, T. & Steenwyk R.V. (2004). A critical evaluation of augmentative biological control. *Biological Control*, 31: 245-256.

Cross, J.V.; Solomon, M.G.; Babandreier, D.; Blommers, L.; Easterbrook, M.A.; Jay, C.N.; Jenser, G.; Jolly, R. L.; Kuhlmann, U.; Lilley, R.; Olivella, E.; Toepfer, S. & Vidal, S. (1999). Biocontrol of Pests of Apples and Pears in Northern and Central Europe: 2. Parasitoids. *Biocontrol Science and Technology*, 9: 277-314.

Davies, T. G. E.; Field, L. M., Usherwood, P. N. R. & Williamson, M. S. (2007). DDT, Pyrethrins, Pyrethroids and Insect Sodium Channels. *IUBMB Life*, 59(3): 151 - 162.

Delpuech, J.M.; Legallet, B.; Terrier, O. & Fouillet, P. (1999). Modifications of the sex pheromonal communication of *Trichogramma brassicae* by a sublethal dose of deltamethrin. *Chemosphere*, 38(4): 729-739.

Delpuech, J.M.; Legallet, B. & Fouillet, P. (2001). Partial compensation of the sublethal effect of deltamethrin on the sex pheromonal communication of *Trichogramma brassicae*. *Chemosphere*, 42: 985-991.

Delpuech, J.M.; Bardon, C. & Boulétreau, M. (2005). Increase of the Behavioral Response to Kairomones by the Parasitoid Wasp *Leptopilina heterotoma* Surviving Insecticides. *Archives of Environmental Contamination and Toxicology*, 49:186-191.

Denholm, I.; Pickett, J.A. & Devonshire, A.L. (2001). *Insecticide Resistance: From Mechanisms to Management*, CABI and The Royal Society publishing, ISBN 0-85199-367-2, UK.

Desneux, N.; Decourtye, A. & Delpuech, J.M. (2007). The Sublethal Effects of Pesticides on Beneficial Arthropods. *Annual Review of Entomology*, 52: 81-106.

Desneux, N.; Ramirez-Romero, R. & Kaiser, L. (2006a). Multistep Bioassay to Predict Recolonization Potential of Emerging Parasitoids after a Pesticide Treatment. *Environmental Toxicology and Chemistry*, Vol. 25 (10): 2675-2682.

Desneux, N.; Denoyelle, R. & Kaiser, L. (2006b). A multi-step bioassay to assess the effect of the deltamethrin on the parasitic wasp *Aphidius ervi*. *Chemosphere*, 65: 1697-1706.

Desneux, N.; Fauvergue, X.; Dechaume-Moncahrmont, X.; Kerhoas, L.; Ballanger, Y. & Kaiser, L. (2005). *Diaeretiella rapae* Limits *Myzus persicae* Populations After Applications of Deltamethrin in Oilseed Rape. *Journal of Economic Entomology*, 98(1): 9-17.

Desneux, N.; Rafalimanana, H. & Kaiser, L. (2004a). Dose-response relationship in lethal and behavioural effects of different insecticides on the parasitic wasp *Aphidius ervi*. *Chemosphere*, 54: 619-627.

Desneux, N. ; Wajnberg, E. ; Fauvergue, X. ; Privet, S. & Kaiser, L. (2004b). Oviposition behaviour and patch-time allocation in two aphid parasitoids exposed to deltamethrin residues. *Entomologia Experimentalis et Applicata*, 112: 227-235.

Desneux, N.; Pham-Delègue, M. & Kaiser, L. (2003). Effects of sub-lethal and lethal doses of lambda-cyhalothrin on oviposition experience and host-searching behaviour of a parasitic wasp, *Aphidius ervi*. *Pest Management Science*, 60:381-389.

Ergin, E.; Er, A.; Uçkan, F. & Rivers, D.B. (2007). Effect of Cypermethrin Exposed Hosts on Egg-Adult Development Time, Number of Offspring, Sex Ratio, Longevity, and Size of *Apanteles galleriae* Wilkinson (Hymenoptera : Braconidae). *Belgian Journal of Zoology*, 137 (1): 27-31.

Essig, K. & Zhao, Z.J. (2001). Preparation and Characterization of a Pyrethrum Extract Standard. *LCGC*, 19(7): 722-730.

Garcia, P.; Pereira, N. & Oliveira, L. (2009). Side-effects of organic and synthetic pesticides on cold-stored diapausing prepupae of *Trichogramma cordubensis*. *BioControl*, 54: 451-458.

Garcia, P.; Cabral, S.; Oliveira, L. & Rodrigues, A. (2006). Effects of deltamethrin on the reproduction of *Trichogramma cordubensis* (Hymenoptera: Trichogrammatidae). *Biocontrol Science and Technology*, 16(7): 699-708.

Gauld, I. & Bolton, B. (1988). *The Hymenoptera*, Oxford University Press, ISBN 0-19-858521, USA.

Gentz, M.C; Murdoch, G. & King, G.F. (2010). Tandem use of selective insecticides and natural enemies for effective, reduced-risk pest management. *Biological Control*, 52: 208-215.

Godfray, H.C.J. (1994). *Parasitoids: Behavioural and evolutionary ecology*, Princeton University Press, ISBN 0-691-00047-6, UK.

Greathead, D.J. (1986). Parasitoids in classical biological control, In: *Insect parasitoids*, Waage J. & Greathead D. (Ed.), 289-318, Academic Press, ISBN 0-12-728900-3, UK.

Gunasekara, A.S. (2004). *Environmental Fate of Pyrethrins*, California Environmental Protection Agency (Cal/EPA), Department of Pesticide Regulation. Sacramento, CA, U.S.

Hassan, S.A. (1993). The mass rearing and utilization of *Trichogramma* to control lepidopterous pests: achievements and outlook. *Pesticide Science*, 37: 387-391.

Hewa-Kapuge, S.; McDougall, S. & Hoffmann, A. (2003). Effects of methoxyfenozide, indoxacarb, and other insecticides on the beneficial egg parasitoid *Trichogramma* nr. *brassicae* (Hymenoptera: Trichogrammatidae) under laboratory and field conditions. *Journal of Economic Entomology*, 96(4): 1083-1090.

Housset, P. & Dickmann, R. (2009). A promise fulfilled - pyrethroid development and the benefits for agriculture and human health. *Bayer CropScience Journal*, 62/2009, 2: 135-144.

Kok, L.T. & Acosta-Martinez, J.A. (2001). Differential susceptibility of *Cotesia orobenae*, a parasitoid of the cross-striped cabbageworm,to commonly used insecticides in Cruciferae. *BioControl*, 46: 419-426.

Kawazu, K.; Shimoda, T.; Kobori1, Y.; Kugimiya, S.; Mukawa, S. & Suzuki, Y. (2010). Inhibitory effects of permethrin on flight responses, host-searching, and foraging behaviour of *Cotesia vestalis* (Hymenoptera: Braconidae), a larval parasitoid of *Plutella xylostella* (Lepidoptera: Plutellidae). *Journal of Applied Entomology*, 134: 313-322.

Langley, M. & Stark, J.D. (1996). Analytical techniques for quantifying direct, residual, and oral exposure of an Insect parasitoid to an organophosphate insecticide. *Bulletin of Environmental Contamination and Toxicology*, 57:683-690.

Li-Ying, L. (1994). Worldwide use of *Trichogramma* for biological control on different crops: A survey, In: *Biological Control with egg parasitoids*, Wajnberg E. & Hassan S. (Ed.), 37-53, CAB International, ISBN 0-851-98896-2, UK.

Longley, M. (1999). A review of pesticide effects upon immature aphid parasitoids within mummified hosts. *International Journal of Pest Management*, 45(2): 139-145.

Nordlund, D.A. (1994). Habitat location by *Trichogramma*, In: *Biological Control with egg parasitoids*, Wajnberg E. & Hassan S. (Ed.), 155-163, CAB International, ISBN 0-851-98896-2, UK.

Pinto, J.D. & Stouthamer, R. (1994). Systematics of the Trichogrammatidae, In: *Biological Control with egg parasitoids*, Wajnberg E. & Hassan S. (Ed.), 5-28, CAB International, ISBN 0851988962, UK.

Pintureau, B.; Louis, C. & Chapelle, L. (1993). Symbiose entre microorganismes et Trichogrammes (Hym. Trichogrammatidae): Intérêt pour a lutte biologique. *Bulletin de la Société Zoologique de France*, 118(2): 159-167.

Prabhaker, N.; Morse, J.G.; Castle, S.J.; Naranjo, S.E.; Henneberry, T.J. & Toscano, N.C. (2007). Toxicity of Seven Foliar Insecticides to Four Insect ParasitoidsAttacking Citrus and Cotton Pests. *Journal of Economic Entomology*, 100(4): 1053-1061.

Quicke, D.L.J. (1997). *Parasitic wasps*, Chapman & Hall, ISBN 0-412-58350-X, UK.

Raposo, F.; Oliveira, L. & Garcia, P. (2003). Susceptibility to Insecticides of *Glyptapanteles militaris* (Hymenoptera: Braconidae), a Parasitoid of *Pseudaletia unipuncta* (Lepidoptera: Noctuidae). *Biocontrol Science and Technology*, 13: 261-267-

Reigart, J.R. & Roberts, J.R. (1999). *Recognition and management of pesticide poisonings*. 5th edition. U.S. Environmental Protection Agency Publication EPA-735-R-98-003. Washington, DC, U.S.

Ruano; F.; Campos, M.; Sánchez-Raya, A.J. & Aránzazu Peña, A. (2010). Olive trees protected from the olive bark beetle, *Phloeotribus scarabaeoides* (Bernard 1788) (Coleoptera, Curculionidae, Scolytinae) with a pyrethroid insecticide: Effect on the insect community of the olive grove. *Chemosphere* 80: 35-40.

Rundle, B.J.; Thomson, L.J. & Hoffmann, A.A. (2004). Effects of cold storage on field and laboratory performance of *Trichogramma carverae* (Hymenoptera: Trichogrammatidae) and the response of three *Trichogramma* spp. (*T. carverae*, *T. nr. brassicae*, and *T. funiculatum*) to cold. *Journal of Economic Entomology*, 97: 213-221.

Saber, M.; Jalil Hejazi, M.; Kamali, K. & Moharramipour, S. (2005). Lethal and Sublethal Effects of Fenitrothion and Deltamethrin Residues on the Egg Parasitoid *Trissolcus grandis* (Hymenoptera: Scelionidae). *Journal of Economic Entomology*, 98(1): 35-40.

Salerno, G.; Colazza, S. & Conti, E. (2002). Sub-lethal effects of deltamethrin on walking behaviour and response to host kairomone of the egg parasitoid *Trissolcus basalis*. *Pest Management Science*, 58: 663-668.

Smith, S. (1996). Biological control with *Trichogramma*: Advances, Successes, and potential of their use. *Annual Review of Entomology*, 41: 375-406.

Smith, K.; Evans D.A. & El-Hiti, G.A. (2008). Role of modern chemistry in sustainable arable crop protection. *Philosophical Transactions of the Royal Society, B*, 363: 623–637.

Stark, J.D.; Vargas, R.I. & Banks, J.E. (2007a). Incorporating ecologically relevant measures of pesticide effect for estimating the compatibility of pesticides and biocontrol agents. *Journal of Economic Entomology*, 100: 1027-1032.

Stark, J.D.; Sugayama, R.L. & Kovaleski, A. (2007b). Why demographic and modeling approaches should be adopted for estimating the effects of pesticides on biocontrol agents. *BioControl*, 52: 365-374.

Sterk, G.; Hassan, S.A.; Baillod, M.; Bakker, F.; Bigler, F.; Blümel, S.; Bogenschütz, H.; Boller, E.; Bromand, B.; Brun, J.; Calis, J.N.M.; Coremans-Pelseneer, J.; Duso, C.; Garrido, A.; Grove, A.; Heimbach, U.; Hokkanen, H.; Jacas, J.; Lewis, G.; Moreth, L.; Polgar,

L.; Roversti, L.; Samsoe-Pettersen, L.; Sauphanor, B.; Schaub, L.; Stäubli, A.; Tust, J.J.; Vainio, A.; Van De Veire, M.; Viggiani, G.; Viñuela, E. & Vogt, H. (1999). Results of the seventh joint pesticide testing programme carried out by the IOBC/WPRS-Working Group "Pesticides and beneficial organisms". *BioControl*, 44: 99-117.

Stinner, R.E. (1977). Efficacy of inundative releases. *Annual Review of Entomology*, 22: 515-531.

Stouthamer, R. (1993). The use of sexual versus asexual wasps in biological control. *Entomophaga*, 38(1): 3-6.

Stouthamer, R. & Werren J.H. (1993). Microbes associated with parthenogenesis in wasps of the genus *Trichogramma*. *Journal of Invertebrate Pathology*, 61: 6-9.

Stouthamer, R.; Breeuwer, J.; Luck, R.F. & Werren, J. (1993). Molecular identification of microorganisms associated with parthenogenesis. *Nature*, 361: 66-68.

Symington, C.A. (2003). Lethal and sublethal effects of pesticides on the potato tuber moth, *Phthorimaea operculella* (Zeller) (Lepidoptera: Gelechiidae) and its parasitoid *Orgilus lepidus* Muesebeck (Hymenoptera: Braconidae). *Crop Protection*, 22: 513-519.

Tillman, P.G. & Mulrooney, J.E. (2000). Effect of Selected Insecticides on the Natural Enemies *Coleomegilla maculata* and *Hippodamia convergens* (Coleoptera: Coccinellidae), *Geocoris punctipes* (Hemiptera: Lygaeidae), and *Bracon mellitor, Cardiochiles nigriceps,* and *Cotesia marginiventris* (Hymenoptera: Braconidae) in Cotton. *Journal of Economic Entomology*, 93(6): 1638-1643.

Todd, G.D.; Wohlers, D. & Citra, M. (2003). *Toxicology Profile for Pyrethrins and pyrethroids,* Agency for Toxic Substances and Disease Registry (ATSDR), Department of Health and Human Services. Atlanta, GA, U.S.

Van Lenteren, J.C. (1986). Parasitoids in the greenhouse: Successes with seasonal inoculative release systems, In: *Insect parasitoids*, Waage J. & Greathead D. (Ed.), 341-374, Academic Press, ISBN 0-12-728900-3, UK.

Vianna, U.R.; Pratissoli, D.; Zanuncio, J.C.; Lima, E.R.; Brunner, J.; Pereira, F.F. & Serra, J.E. (2009). Insecticide toxicity to *Trichogramma pretiosum* (Hymenoptera: Trichogrammatidae) females and effect on descendant generation. *Ecotoxicology*, 18:180-186.

Vieira, A., Oliveira, L. & Garcia, P. (2001). Effects of Conventional Pesticides on the Preimaginal Developmental Stages and on Adults of *Trichogramma cordubensis* (Hymenoptera: Trichogrammatidae). *Biocontrol Science and Technology*, 11: 527- 534.

Wang, H.Y.; Yanga, Y.; Sua, J.Y.; Shena, J.L.; Gaoa, C.F. & Zhub, Y.C. (2008). Assessment of the impact of insecticides on *Anagrus nilaparvatae* (Pang et Wang) (Hymenoptera: Mymaridae), an egg parasitoid of the rice planthopper, *Nilaparvata lugens* (Hemiptera: Delphacidae). *Crop Protection*, 27: 514-522.

Wirtz, K.; Bala, S.; Amann, A. & Elbert, A. (2009). A promise extended - future role of pyrethroids in agriculture. *Bayer CropScience Journal*, 62/2009, 2: 145-158.

Youssef, A.I.; Nasr, F.N.; Stefanos, S.S.; Elkhair, S.S.A.; Shehata, W.A.; Agamy, E.; Herz, A. & Hassan, S.A. (2004). The side-effects of plant protection products used in olive cultivation on the hymenopterous egg parasitoid *Trichogramma cacoeciae* Marchal. *Journal of Applied Entomology*, 128(9-10): 593-599.

25

Pesticide Exposure and Health Related Issues in Male and Female Reproductive System

Stavros Sifakis[1], Mihalis Mparmpas[1], Offie P. Soldin[2] and Aristides Tsatsakis[3]

[1]Department of Obstetrics & Gynecology University of Crete, Heraklion, Greece

[2]Departments of Medicine, Oncology, Physiology and Obstetrics & Gynecology, PregnaTox, Georgetown University Medical Center, Washington D.C.,

[3]Laboratory of Toxicology, Department of Medicine, University of Crete, Heraklion,

[1,3]Greece

[2]USA

1. Introduction

During the last several decades there have been widespread uses of potent substances that, although effective in their intended use, have also been suspected of being harmful to reproductive health. This mixture of environmental contaminants that may adversely affect human fertility includes heavy metals (lead, mercury, arsenic), phthalates (plasticizers), bisphenol-A (building block of several plastics), polychlorinated biphenyls (lubricants), dioxins (byproducts of manufacture), pesticides and other agents. The impact of adverse effects on reproductive health include impaired gametogenesis, sperm maturation, decreased semen quality, miscarriages, ovulation and menstrual disturbances, infertility, stillbirths, developmental anomalies, cryptorchidism, hypospadias and cancer. The effects can be reversible, permanent or even transgenerational, take place in the offspring, as exposure can occur during pregnancy and intrauterine life, childhood or later. The route of exposure, dose, age, gender and genotype (susceptibility of the individual) are important factors that can determine the reproductive disorder. In this chapter we focus on the effects of exposure to pesticides during adulthood, on human male and female fertility, giving emphasis to semen quality and time to pregnancy.

2. Pesticides

Pesticides are mainly utilized in agriculture for crop protection, often replacing the natural processes on which agricultural production had previously depended. Pesticides are also applied in homes and gardens. More than 140.000 tones of pesticides are used annually in the European Union for agricultural purposes only (Ramazzini, 2009).

A pesticide is "any substance or mixture of substances intended for preventing, destroying or controlling any pest, including vectors of human or animal disease, unwanted species of plants or animals causing harm during or otherwise interfering with the production processing, storage, transport or marketing of food, agricultural commodities, wood and wood products or animal feedstuffs, or substances which may be administered to animals for the control of insects, arachnids or other pests in or on their bodies" (FAO, 2003). They fall into three major classes: insecticides, fungicides, and herbicides, classification based upon the target organism. There are also rodenticides (for control of vertebrate pests), nematicides (to kill eelworms, etc), molluscicides (to kill slugs and snails), and acaricides (to kill mites) (Cox & Surgan, 2006). Pesticides may differ according to their chemical structure, their mechanism of action and the toxicity they exhibit, but typically each pesticide consists of one (or more) active ingredient, which exerts the pesticidal activity, and an inert ingredient, which is inactive and helps in handling the active ingredient. Several studies have shown that the inert ingredient is not as inactive as it was previously believed to be (Surgan, 2005; Cox & Surgan, 2006). Over 700 active ingredients are in use worldwide as pesticides, each with distinct chemical and toxicological properties (Toppari et al., 1996). Another classification categorizes pesticides according their chemical structure (Table 1). Insecticides include organochlorines, organophosphates, and carbamates. Organochlorine hydrocarbons (DDT, heptachlor) operate by disrupting the sodium/potassium balance of the nerve fiber. They are persistent in human tissue with the potential to bioaccumulate. Organophosphate (parathion, malathion) and carbamates (carbaryl, carbofuran) are less toxic and largely replaced organochlorines. They are inhibitors of acetylcholinesterase, causing paralysis. Common herbicides include pheoxy and benzoic acid herbicides (2,4-D) and triazines (atrazine). Fungicides (vinclozolin, mancozeb) have sulfur as the most common active ingredient. Nicotinoids and pyrethroids are plant-derived pesticides (fenvalerate, pyrethrin).

In addition to the desired effects of crop protection and pest management, pesticides have some recognized adverse impacts on human health and the environment. Humans have a great risk of exposure through several pathways in occupational, agricultural and household use. Inhalation, oral, dermal and ocular, are four possible routes for pesticide exposure. Ingestion of food and water is thought to be the main routes of pesticide exposure in the general population, while dermal absorption is suspected to be the main source of occupational exposure (Toppari et al., 1996). Over 25% of fruits, vegetables, and cereals are known to contain detectable residues of at least two pesticides and more than 300 different pesticides are known to contaminate food products sold in the EU (Ramazzini, 2009). In the majority of cases, however, human exposure is unintentional and unintended (Ribas-Fitó, 2002).

Pesticides are accused of causing short-term adverse health effects. Acute health effects include stinging eyes, rashes, blisters, blindness, nausea, dizziness, diarrhea and death (Jeyaratnam, 1990; Sanborn et al., 2007). They are suspected also for a wide range of chronic effects, which can occur months or years after the exposure, such as cancers, neurological and developmental toxicity, immunotoxicity, genotoxicity, respiratory effects and disruption of the endocrine system. Pesticides may affect not only the exposed individual but also subsequent generations (Alavanja et al., 2004; Ritter et al., 2006; McCauley et al., 2006; Bassil et al., 2007).

There has been rising concern in many developed countries about the adverse effects of pesticides on human reproduction, ranging from female and male subfertility to abortion, stillbirths, birth defects and malformations (García, 2003; Weselak et al., 2007; Peiris-John & Wickremasinghe, 2008). They may cause reproductive toxicity with direct damage to the structure of the cells or as a result of biotransformation into metabolites, or interference with processes necessary for the natural homeostasis and equilibrium. They may act like hormones in the endocrine system and disrupt the function of the natural endogenous hormones, when doing so they are often called endocrine disrupting chemicals (EDC) (Lathers, 2002; Diamanti-Kandarakis et al., 2009). This group of compounds identified as EDC is heterogeneous and includes synthetic or natural chemicals.

organochlorides	DDD, DDT, DDE, chlordane, kepone, dieldrin, endosulfan, heptachlor, lindane, mirex, methoxychlor, toxaphene
organophosphorus	chlorpyrifos, glyphosate, diazinon, dimethoate, malathion, methamidophos, parathion, terbufos, tribufos, trichlorfon
carbamates	aldicarb, carbaryl, carbofuran, fenoxycarb, propoxur
pyrethroids	cypermethrin, fenvalerate, permethrin, pyrethrin, pyrethrum, resmethrin, tetramethrin
anilides/anilines	metolachlor, pretilachlor, propachlor, trifluralin
phenoxy	2,4-D, 2,4-DB , 2,4,5-T, MCPA, MCPB , fenoprop
triazines	atrazine, cyanazine, hexazinone, prometryn, propazine, simazine, terbutryn
quaternary	diquat, MPP, paraquat
ureas	chlortoluron, DCMU, metsulfuron-methyl, monolinuron
others	acetamiprid, amitraz, chlordimeform, cyromazine, diflubenzuron, nithiazine, sulfuramid, thiachloprid, xanthone

Table 1. Classification of pesticides based on their chemical structure.

3. Pesticides as Endocrine Disrupting Chemicals (EDC)

Pesticides may act as endocrine disruptors and alter the hormonal homeostasis in both males and females and lead to subfertility. The term "endocrine disruptors" was introduced into literature with an article published in 1993 (Colborn, 1993). An endocrine disruptor was defined by the U.S. Environmental Protection Agency (EPA) as "an exogenous agent that interferes with the synthesis, secretion, transport, binding, action, or elimination of natural hormones in the body that are responsible for the maintenance of homeostasis, reproduction and developmental process" (Kavlock et al., 1996). The group of EDC includes pesticides and various synthetic substances such as polychlorinated biphenils (PCB), polybrominated biphenyls (PBB), bisphenol-A (BPA), phthalates and dioxins natural compounds such as phytoestrogens (Fourth National Report on Human Exposure to Environmental Chemicals, 2009) are used as solvents, lubricants, plasticizers, cosmetics etc. They are usually small molecules (mass 1000 Daltons) and often have a phenolic moiety that probably mimics natural steroid hormones (Diamanti-Kandarakis et al., 2009). They contain chlorine or other halogens (bromine, iodine or fluorine) with strong interaction and so they resist degradation. They usually have a long half-life and accumulate in the environment, sometimes remotely from the place they were produced. Substances banned even decades ago can still be found in the environment and in living organisms. In humans and animals EDCs are stored in fatty tissue or they may be metabolized into more toxic compounds. The predominant sources of exposure are food, water and air. The routes of exposure include ingestion, inhalation and dermal absorption.

Natural hormones act in very low concentrations, similarly, can elicit adverse effects in low doses. Occasionally, the EDCs do not follow the classic dose-response effect and low doses may result in stronger effects than high doses (vom Saal et al., 2009). Humans are exposed concomitantly to a large number of compounds. These substances may enhance and have a synergistic or antagonistic effect, and as cited above, some substances can be metabolized in more toxic products (Crews et al., 2003). One of the reasons of the difficulty studying the damage after exposure to a single agent is this mixture of compounds which is accumulated in human organisms.

An issue of critical importance is the timing of exposure, since damage is age sensitive. The same dose can have different effects in fetuses, newborn, infants or adults. Given the same doses, a developing organism (embryo, neonate) whose growth is highly controlled by the endocrine system is more vulnerable to EDCs, than an adult (Guo et al., 1995; Bigsby et al., 1999; Lilienthal et al., 2006). The damage incurred by the exposure may not be immediate and may only be manifested in adulthood or during aging. The consequences may be apparent even in subsequent generations and the classic example is the cases of vaginal carcinoma in daughters of mothers who were exposed to DES during their pregnancies (Herbst et al., 1999; Anway & Skinner, 2006; Rubin, 2007). The susceptibility of an individual may vary due to genetic polymorphism and so the results of the same exposure could be different.

Initially it was thought that ECDs act via nuclear hormone receptors (e.g. estrogen, progesterone, androgen, thyroid receptors), but now it is believed that they act also via membrane, non-steroid receptors (e.g. dopamine, serotonin, nor-epinephrine receptors). Catecholamine hormones following synthesis are stored in granular vesicles intracellularly. Steroid hormones are not stored but readily synthesized following gonadotropin stimulation of the gonads and are usually found in the circulation bound by carrier proteins (only free hormones are biologically active). Once reaching their tissue targets steroid hormones exert their action by binding to different kinds of nuclear receptors. Explicit hormones bind specific receptors and individualized mechanisms follow by intracellular signalling (e.g. protein kinase-C activation or phosphatidylinositol turnover). Hormones are mostly catabolised in the liver. Consequently, EDCs can participate in most aforementioned pathways thereby changing hormone synthesis patterns, mimicing hormone function or blocking it by occupying the receptor site, modulating the number of the receptors and their affinities for specific molecules and altering hormone clearance (Gore et al., 2006; Gore, 2007; Gore, 2008). Sex hormones synthesis is regulated by the hypothalamic-pituitary-gonadal axis. LH and FSH are synthesized by the anterior pituitary under the influence of pulsatile secretion of GnRH, released by the hypothalamus.

Several pesticides have been reported to act as estrogen agonists, e.g. methoxychlor, endosulfan, toxaphene, kepone, DDT, fenarimol, alachlor, pentachrophenol, fenvalerate, chlordecone (Soto et al., 1995; Cummings & Gray, 1997; Garey & Wolff, 1998; Andersen et al., 2002; Kojima et al., 2004). On the other hand, other pesticides such as vinclozolin, p,p-DDE and o,p-DDT may have anti-androgenic activity, or both estrogenic and antiandrogenic activity (Kelce et al., 1994; Kelce et al., 1995; Kelce & Wilson, 1999). The fungicide methyl-2-benzimidazole carbamate decreases estradiol production in primary cultures of human ovarian granulosa (Can & Albertini, 1997). Treatment of rats with heptachlor suppresses progesterone and estradiol concentrations in blood (Oduma et al., 2006). Progesterone concentrations also decrease during early pregnancy in the rabbit following exposure to the pesticide DDT (Lindenau et al., 1994). DDT was found to be estrogenic, but its major metabolite DDE, has considerable antiandrogenic activity (Kelce et al., 1995). Further, atrazine seems to have estrogenic and antiandrogenic properties and was suggested to reduce testicular testosterone in male rats exposed to it (Stoker et al., 2000). Lindane intercalates into the sperm membrane and may inhibit sperm responsiveness to progesterone in vitro (Silvestroni & Palleschi, 1999). Lindane also inhibits steroidogenesis by reducing StAR (steroidogenic acute regulatory) protein-mediated cholesterol transfer (Walsh & Stocco, 2000). Neonatal exposure to either DES or flutamide also inhibited steroidogenesis and StAR protein expression in the fetal rat Leydig cell (Mikkilä et al., 2006). This protein mediates cholesterol passage through mitochondrial membranes and impaired expression results in decreased testosterone production in vitro (Manna et al., 2001). Testosterone concentration is also reduced with azole fungicides (ketoconazole) due to impaired enzymatic activity of 17α-hydroxylase and 17,20-lyase (Wang et al., 1992; Bahshwan et al., 1998; Taxvig et al., 2008). In rats, fenarimol, a different fungicide, was found to cause a dose-related decrease in fertility (Hirsch et al., 1986). In vitro studies of some pesticides such as fenarimol, prochloraz, imazalil and dicofol, indicate these pesticides inhibit the conversion of androgens to estrogens through CYP 19 aromatase inhibition (Vinggaard et al., 2000). Rats treated with mancozeb demonstrate a decrease in the number of healthy follicles and an increase in the number of atretic follicles (Mahadevaswami et al., 2000). EDCs have also been associated with breast cancer, PCO and endometriosis in women, cryptorchidism, hypospadias, testicular and prostate cancer in men, alteration in pituitary and thyroid gland functions (Crisp et al., 1998; Bretveld et al., 2006; Diamanti-Kandarakis et al., 2009).

4. Pesticides and semen quality

Approximately 6% of adult males are thought to be infertile (Purvis & Christiansen, 1992). Male factors are responsible for at least 20% of cases of infertility. Male infertility is related to impaired semen quality and may be due to a variety of causes including genetic (Klinefelter's syndrome), congenital (cryptorchidism), endocrine (hypogonadism), obstructive (vasectomy), infective (chlamydia), vascular (varicocele), neoplastic (carcinoma of the testis), lifestyle, and environmental (heat, drugs, pesticides, irradiation) factors. Others causes include sexual dysfunction related to erection and ejaculation (Purvis & Christiansen, 1992; Dohle et al., 2005). However, in many cases male infertility is regarded as idiopathic (40-75%), and no cause can be identified. Semen analysis is used to evaluate semen quality, which is taken as a surrogate measure of male infertility. The World Health Organization has provided reference values for human semen characteristics (Cooper et al., 2010).

There is evidence of regional variation in semen quality that may be an expression of gene polymorphisms, different climate, lifestyle and exposure to magnetic fields or chemical substances such as pesticides (Mallidis et al., 1991; Jørgensen et al., 2001; Swan et al., 2003; Li et al., 2010). Seasonal variation has also been detected (decrease of sperm density and total sperm count during summer) (Levine, 1999; Krause & Krause, 2002). These factors, together with the variability in techniques and methodologies used, are reasons for some controversial results in studies that analyze semen quality changes overtime. Thus, the issue of a decline in semen quality overtime is equivocal. Following the publication of the meta-analysis conducted by Carlsen et al. in 1992, demonstrating a decline in human semen quality over the last 50 years (mean sperm count from 113 millions/ml in 1940 to 66 millions/ml in 1990), numerous related studies have been published. Studies by Swan et al. had consistent results with those of Carlsen et al. (1992) and supported that historical data on sperm density, despite large random error, are reliable (Swan et al., 1997; Swan & Elkin, 1999; Swan et al., 2000). However, Olsen et al. (1995) reanalyzed the data used in a linear model to predict sperm quality deterioration in the last 50 years, advocate that the data are only robust during the last 20 years (1975-1995), in which other statistical models (quadratic, spline fit and stairstep), except the linear model, suggest constant or slightly increasing sperm counts. Studies from Italy, Denmark, Canada, Tunisia, India, Poland, Israel, Scotland, Greece and Germany suggest that there has been a decline, or sperm parameters are impaired in young populations (Adamopoulos et al., 1996; Younglai et al., 1998; Bilotta et al., 1999; Almagor et al., 2003; Vicari et al., 2003; Jørgensen et al., 2006; Sripada et al, 2007; Adiga et al., 2008; Paasch et al, 2008; Horak et al., 2008; Feki et al., 2009). Conversely, other reports from US, Japan, Korea, Sweden, Spain, Israel and Czech Republic showed no significant evidence of deterioration in sperm quality (Fisch et al., 1996; Paulsen et al., 1996; Benshushan et al., 1997; Berling & Wölner-Hanssen, 1997; Andolz et al., 1999; Seo et al., 2000; Itoh et al., 2001; Zvěrina et al., 2002). The materials and methods of the studies mentioned varied widely, as well as time period (one, two, or more decades), population sample (e.g. individuals in infertile relationship or fertile subjects who participated voluntarily) and the level of pollution in the various geographical regions. Merzenich et al. (2010) conclude that former meta-analyses of sperm count data show a global downward trend, but this conclusion should be interpreted with caution, because the included studies are of great heterogeneity. The geographic variation in semen quality may reflect different exposures to endocrine disruptors, such as pesticides (Swan, 2006). Pesticides might have the ability to interrupt male fertility at several different sites in the reproductive pathway and by one or more mechanisms, as cited previously. Thus, they can interfere with the hypothalamopituitary axis that regulates, through the production of the gonadotrophins FSH and LH, the function of Sertoli and Leydig cells, impairing spermatogenesis and steroidogenesis.

Tables 2a,b,c lists the studies published evaluating the association between exposure to pesticides and human sperm quality. Literature reviews and articles investigating chemical compounds, without including pesticides, were excluded. Studies evaluating pregnancy outcomes and no sperm quality were excluded too. Sperm quality was assessed evaluating conventional parameters (concentration, motility, morphology) or sperm DNA/chromatin integrity and aneuploidy. Sixty-three reports, satisfying these criteria were identified and included in Tables 2a,b,c. Among them six (6/63) studies evaluated the recovery of sperm quality, years after cessation of exposure to DBCP (5) and kepone (1). Thirty-six (36/63) studies examined exposure to single, specific pesticides or metabolites and included DBCP, DDT, DDE, EPB, 2,4-D, kepone (chlordecone), molinate, carbaryl, fenvalerate, ethylparathion, methamidophos, 1N (metabolite of carbaryl and naphthalene), TCPY (metabolite of chlorpyrifos and chlorpyrifos-methyl) and 3PBA, CDCCA, TDCCA (pyrethroid metabolites). The majority of the pesticides cited above are now banned or severely restricted, at least in USA or EU. Twenty-one (21/63) studies evaluated mixture of compounds such as fungicides, insecticides or herbicides with or without specifying the exact pesticides. Some of the pesticides involved in these studies are: alachlor, diazinon, acetochlor, malathion, atrazine, metolachlor, DEET (insect repellent), 2,4-D, aldicarb and cadusaphos, ethoprothos, isazophos, terbufos, pyrimiphos-ethyl (organophosphorus pesticides).

4.1 Studies with little or no evidence of an association

Eighteen studies of the 63 (18/63) found no or little evidence of association between pesticide exposure and sperm quality (Table 2a). Eight studies (8/18) involved a mixture of compounds and ten (10/18) single pesticides. Carbaryl and molinate were reported in one study each. Ten studies (10/18) were related to DDT and metabolites most of the times p,p'-DDE, the known persistent pesticide banned many years ago (not all over the world). Two of the ten studies evaluated DDT in mixtures and eight as a single pesticide. Five (5/10) of the reports that found no or little association between DDT, DDE and semen quality were carried out by the INUENDO project (including the two substudies of INUENDO by Rignell-Hydbom et al. (2004 & 2005). INUENDO (INUit-ENDOcrine) is the acronym for "Biopersistent organochlorines in diet and human fertility. Epidemiological studies in time to pregnancy and semen quality in Inuit and European populations". This EU project (2002-2005) used serum levels of CB-153 (polychlorinated biphenyl) and p,p'-DDE, the main DDT metabolite, to estimate the impact in human fertility in epidemiological studies including Inuits from Greenland and Caucasians from Poland, Sweden and Ukraine. Hauser et al. (2003) conducted two cross sectional studies and found a limited evidence of an inverse association between p,p-DDE and sperm motility as well as no strong relationships between the levels of this compound and sperm DNA damage. Furthermore, Charlier et al. (2005) estimated serum and seminal plasma concentrations of p,p-DDE in fertile and sub or infertile young men. Blood concentrations of p,p'-DDE were very low in both groups. No p,p'-DDE detected in seminal plasma of either groups. Of note, the mothers of the exposed subfertile men had serum level of p,p'-DDE significantly higher than the mothers of the control group.

Two studies did not evaluate DDT exposure separately, but it was included in a sum of other compounds. Magnusdottir et al. (2005) concluded that poor semen quality is associated with sedentary work and obesity but not with plasma levels of fourteen organochlorine pesticides including DDT and metabolites. Weiss et al. (2006) evaluating exposure, in Germany and in Tanzania, to a mixture of PCBs and pesticides, including DDT, found these pesticides had no impact on sperm quality. However high serum concentrations of DDT-DDE were associated with lower pregnancy rates in Germany.

Two out of the eighteen reports that showed little or no association between pesticide exposure and semen quality were performed by ASCLEPIOS (Larsen et al., 1998a; Härkönen et al., 1999). It was an EU project (1993-1998), that was carried out in 14 European centers and focused on occupational exposure to the fungicides styrene and inorganic lead. Questionnaire studies of time to pregnancy were combined with longitudinal and cross sectional studies of semen quality.

Tielemans et al. (1999a) conducted a case-control study and found no associations between exposure to pesticides and poor semen quality, but few subjects were exposed to pesticides (the rest of the sample was exposed to chemical pollutants different from pesticides, such as solvents and metals). Juhler et al. (1999) found similar results indicating minor association, comparing traditional and organic farmers, but for all the groups of exposure the average dietary intake of pesticides was low. Smith et al. (2004) found no significant differences in sperm aneuploidy or diploidy frequencies between men exposed to a mixture of pesticides and control groups, but the sample was rather small (n=20+20). In a cross-sectional study conducted by Multigner et al. (2008) semen characteristics were evaluated in association to exposure to a mixture of organophosphorus pesticides (cadusaphos, ethoprothos, isazophos, terbufos, pyrimiphos-ethyl and one carbamate the aldicarb). No significant difference was found between exposed and unexposed workers (in banana plantations), but exposure was assessed by a questionnaire and not by chemical analysis. Tomenson et al. (1999) conducted a longitudinal study and found no evidence that sperm and serum hormones levels were related to molinate exposure (thiocarbamate herbicide). Whorton et al. (1979) found no apparent effects on sperm count in workers exposed to carbaryl, but three subsequent relative studies indicated contrary results.

Authors & Year Country	**Pesticide**	**N**	**Study Type**
Whorton et al, 1979 (USA)	carbaryl	47 workers	CS
Larsen et al, 1998a (ASCLEPIOS) (Denmark)	Mixture	248 farmers users or not users of pesticides during a spraying season	L
Härkönen et al, 1999 (ASCLEPIOS) (Finland)	fungicides	30 healthy farmers before and after exposure	CS
Tomenson et al, 1999 (UK)	molinate (thiocarbamate herbicide)	272 workers at three US plants.	L
Tielemans et al, 1999a (The Netherlands)	Occupational exposures (solvents, metals, pesticides)	Male partners of couples having their first consultation in two infertility clinics (n=899)	CC
Juhler et al, 1999 (Denmark)	mixture	171 traditional and 85 organic farmers.	CC
Hauser et al, 2003a (USA)	DDT, DDE, PCBs	212 male partners of subfertile couples	CS
Hauser et al, 2003b (USA)	DDT, DDE, PCBs	212 male partners of subfertile couples	CS
Rignell-Hydbom et al, 2004, (Sweden)	CB-153 p,p '-DDE	195 Swedish fishermen aged 24-65	CS
Smith 2004 et al, (Canada)	mixture	20 exposed, 20 non exposed	CC
Charlier 2005 et al, (Belgium)	p,p'-DDE	73 fertile young men (controls) & 23 mothers; 82 sub or infertile young men (cases) & 19 mothers	CC
Magnusdottir 2005 et al, (Iceland)	PCBs, organochlorine pesticides, (p,p'DDE)	25 poor semen quality, 20 normal semen quality & idiopathic subfertility, 27 normal semen quality & female subfertility	CS
Rignell-Hydbom et al, 2005 (Sweden)	CB-153 p,p'-DDE	176 Swedish fishermen (low & high intake of fatty fish)	CS
Spano 2005 et al, INUENDO, (Italy)	CB-153 p,p'-DDE	193 Inuits – Greenland; 178 Swedish fishermen; 141 men from Poland; 195 men from Ukraine	CS

Stronati et al, 2006, INUENDO, (Italy)	CB-153 p,p'-DDE	200 Inuits -Greenland; 166 men from Sweden; 134 men from Poland; 152 men from Ukraine	CS
Toft et al, 2006, INUENDO, (Denmark)	CB-153 p,p'-DDE	194 men from Greenland; 185 men from Sweden; 189 men from Poland; 195 men from Ukraine	CS
Weiss et al, 2006 (Germany)	mixture of PCBs & pesticides including DDT,DDE	31 women, 16 men from Tanzania, 21 couples from Germany	CS
Multinger et al, 2007 (France)	organophosphorus pesticides and one carbamate	42 exposed & 45 non exposed in banana plantations in Guadeloupe (parallel assessment in wild rats)	CS

Table 2a. Studies evaluating the association between exposure to pesticides and human sperm quality. Studies with no or little evidence of an association; 2b: Studies with evidence of an association; 2c: Studies evaluating the recovery of sperm quality (CS: cross sectional study; CC: case-control study; L: longitudinal study; R: retrospective study; P: pilot study).

4.2 Studies with evidence of an association

Thirty-nine (39/63) studies that varied widely in materials, methods, exposure and assessment of exposure concluded that there was evidence of an association between exposure to pesticides and impaired semen quality. Twenty-six (26/39) studies evaluated exposure to single pesticides or metabolites and thirteen (13/39) studies evaluated exposure to mixtures of compounds including pesticides or mixtures of pesticides. Evidence of an association was found in studies involving:

- DDT and DDE: Only in six of the 16 studies overall involved, five as a single pesticide and one in a mixture.
- DBCP: All studies showed evidence of impairment. Thrupp in his article examined a remarkable case of massive sterilization of approximately 1500 workers exposed in banana plantations in Costa Rica (Thrupp, 1991). Slutsky et al. in his report which represents the largest cohort of DBCP exposed workers, found that after a median exposure of three years, 64.3% of these men overall, and 90.1% of men studied from the Philippines, had oligospermia or azoospermia (1999). Whorton et al. (1977), Potashnik et al. (1978), Lipshultz et al. (1980) and Egnatz et al. (1980) had the same results of a significant association. There are five studies more involving DBCP, but related to sperm recovery after cessation of exposure and are cited in the related paragraph.
- EPB: One cross-sectional and one longitudinal study with small sample size.
- Kepone: A cross-sectional study measuring blood levels found oligospermia and decreased sperm motility. One more study involving kepone was related to the recovery of sperm after exposure and is cited in the related paragraph.
- 2,4-D: One cross sectional study evaluated exposure using urine samples and observed asthenospermia, teratospermia and necrospermia. The second study evaluated this compound in a mixture of eight pesticides and found relation with poor semen quality.
- Fenvalerate: All studies conducted in China; three of them reporting impairment of conventional parameters of sperm quality and one increased percentage of sperm aneuploidy and altered morphology.

- Carbaryl: The first, a cross-sectional study, showed increase in abnormal morphology, the second study found altered seminal volume and sperm motility and the third study showed increased frequencies of aneuploidy.
- Ethylparathion and Methamidophos: Semen and urine samples were collected in order to estimate exposure to these organophosphate pesticides. There was a decrease in sperm concentration and motility but no significant difference was found in sperm morphology.
- 1N and TCPY: The first compound is a metabolite of carbaryl and naphthalene and the second a metabolite of chlorpyrifos and chlorpyrifos-methyl. This study used biological markers of exposure (urine analyses for metabolites) and several modeling approaches to test the robustness of the data. Statistically significant inverse dose- response relationships between 1N and sperm concentration and motility were found. There was a suggestive association between TCPY and sperm concentration and motility. Sperm morphology was not significantly associated with both 1N and TPCY. Only a single urine sample collected to estimate exposure and only a single semen sample collected to assess semen quality.
- 3PBA: It is the main metabolite of pyrethroids such as cypermethrin, deltamethrin, permethrin with high detection rate in the general population. The first study with rather large sample size showed suggestive association between increased urinary 3-PBA concentration (creatinine adjusted) and sperm concentration. There was an association between straight line velocity and curvilinear velocity (sperm progression and motion parameters) with urinary 3-PBA concentration (creatinine adjusted), while sperm volume, sperm number per ejaculum and sperm motility were weakly or not significantly associated. A second study that evaluated two more pyrethroid metabolites, CDCCA and TDCCA, was similar in design and found evidence for reduced semen quality and increased DNA damage related to the urine pyrethroid metabolites. Swan et al. (2003b) evaluated exposure to a mixture of eight pesticides in two different populations within Missouri and Minnesota. The small sample size limited statistical power. Exposure was assessed by urine analysis. Study concluded that Alachlor, atrazine, 2,4-D, metolachlor and a diazinon metabolite were associated with poor semen quality in Missouri. However, no significant associations were found for acetochlor, DEET and malathion dicarboxylic acid. Within Minnesota, the levels of pesticides were low for any of the pesticides, no significant associations were found too, but because of the overall results this study is classified in this category.

Dalvie et al. (2004) evaluated exposure to DDT in workers in South Africa, in relation to sperm quality and sexual function. Exposure was assessed by serum levels of o'p' and p'p' isomers of DDT, DDE, DDD. Sperm count and density were in the normal range. 84% of morphology scores were below the WHO criteria and p'p'DDT was negatively associated with sperm count (after correction for age, abstinence, physical abnormality and fever). Although no strong evidence for a DDT overall effect in sexual function and reproductive outcomes was found semen quality was impaired resulting in this study being cited in this category.

Dallinga et al. (2002) studied a group of men with poor semen quality vs. group of men with normal, based on the progressive motility of sperm. Blood samples were collected in order to determine whether differences in sperm quality were related with differences in serum concentrations of organochlorines, including DDT. No significant differences in organochlorine levels were found initially, but after adjustment for age and sperm count, sperm progressive motility were inversely related to the concentrations of metabolites in the group of men with normal semen quality. Because of this finding the study is cited in this category.

Authors & Year Country	Pesticide	N	Study Type
Whorton et al, 1977 (USA)	DBCP	25 workers	CS
Cannon et al, 1978 (USA)	kepone	133 workers	CS
Potashnik et al, 1978 (Israel)	DBCP	6 workers	CS
Lipshultz et al, 1980 (USA)	DBCP	228 workers (exposed & non exposed cohort)	CS
Egnatz et al, 1980 (USA)	DBCP	232 workers exposed, 97 workers non exposed	CS
Wyrobek et al, 1981 (USA)	carbaryl	50 men occupationally exposed, 34 unexposed newly-hired	CS
Henderson et al, 1986 (Australia)	Mixture of chemicals including pesticides	1695 men with abnormal semen quality or fertility impairment	CS
Ratcliffe et al, 1987 (USA)	EDB	46 exposed in papaya fumigation industry in Hawaii; 43 unexposed from a sugar refinery.	CS
Schrader et al, 1988 (USA)	EDB	10 exposed forestry employees, 6 unexposed (exposing time 6 wks)	L
Thrupp et al, 1991 (USA)	DBCP	exposed workers in a banana plantation in Costa Rica	L
Lerda et al, 1991 (Argentina)	2,4-D	32 farm sprayers	CS
Strohmer et al, 1993 (Austria)	mixture	101 couples seeking artificial insemination (poor semen quality), controls couples with female infertility IVF treated	CS
Slutsky 1999 et al, (USA)	DBCP	26400 workers in banana & pineapple plantation in 12 countries	R
Abell et al, 2000 (Denmark)	mixture	122 greenhouse workers	CS
Padungtod et al, 2000 (USA)	Ethylparathion Methamidophos	32 exposed workers & 43 not exposed workers in China factories	CS
Ayotte et al, 2001 (Mexico)	DDT P,P'-DDE	24 Mexican men living in endemic malaria areas not occupationally exposed	CS
Oliva et al, 2001 (Argentina)	mixture of pesticides & solvents	225 male partners from infertility clinics	CS

Dallinga et al, 2002 (The Netherlands)	Organochlorine compounds PCBs, HCB, p,p'-DDT, p,p'-DDE	34 men with poor semen quality vs. 31 men with normal (based on progressively motile sperm concentration)	CS
Tan et al, 2002 (China)	fenvalerate	32 exposed workers, 46 administrators (internal control group), 22 administrators (external control group)	CC
Swan et al, 2003b (USA)	8 metabolites of pesticides (alachlor, diazinon, acetochlor, metolachlor, 2,4-D atrazine, DEET, malathion)	Men in whom all semen parameters were low (cases) or within normal limits (controls), in Missouri & Minnesota (50 & 36 respectively)	CC
Wong et al, 2003 (The Netherl&s)	Chemicals including pesticides & other factors	73 subfertile & 92 fertile men	CC
Bian et al, 2004 (China)	fenvalerate	21 exposed workers, 23 non exposed workers (internal control group), 19 non exposed workers (external control group)	CC
Dalvie et al, 2004 (South Africa)	DDT, DDE	60 workers in South Africa	CS
Kamijima et al, 2004 (Japan)	Organophosphorus & pyrethroid insecticides	18 male sprayers, 18 age matched students or medical doctors as unexposed controls	L
Meeker et al, 2004 (USA)	1N (a metabolite of carbaryl & naphthalene), TCPY (a chlorpyrifos & chlorpyrifosmethyl metabolite)	272 men recruited through a Massachusetts Infertility clinic	CS
Pant et al, 2004 (India)	DDT, DDE, DDD, HCH	45 fertile & 45 infertile men	CS
Sánchez-Peña et al, 2004 (Mexico)	mixture of organophosphorus pesticides	33 men occupationally exposed (initially 227)	CS
Xia 2004 (China)	Fenvalerate	12 exposed workers, 12 internal control group, 18 external control group	CC
Xia et al, 2005 (China)	carbaryl	16 exposed workers, 12 internal control group, 18 external control group	CC
Tan et al, 2005 (China)	carbaryl	31 exposed workers, 46 internal control group, 22 external control group	CC

De Jager et al, 2006 (Canada)	DDT p,p'-DDE	116 men aged 27 years, non occupationally exposed living in malaria, endemic areas in Mexico	CS
Lifeng et al, 2006 (China)	fenvalerate	32 exposed workers, 46 internal control group, 22 external control group	CC
Yucra et al, 2006 (Peru)	Organophosphate pesticides	31 pesticide sprayers, 80 not exposed men	CS
Aneck-Hahn et al, 2007 (South Africa)	DDT p,p'-DDE	311 healthy men 18-40 years in an endemic malaria area in which DDT is sprayed (non occupational exposure)	CS
Perry et al, 2007 (USA)	organophosphorus & pyrethroid insecticides	18 males of reproductive age in China	P
Tuc et al, 2007 (Thail&)	Mixture	1036 rice farmers: 156 of these with abnormal semen & 314 with normal semen (as controls)	CC
Meeker et al, 2008 (USA)	Metabolites of pyrethroid insecticides: 3PBA, CDCCA, TDCCA	207 men recruited from an infertility clinic	R CC
Recio-Vega et al, 2008 (Mexico)	Organophosphorus pesticides	52 men in three occupational exposure levels	L
Xia et al, 2008 (China)	3-PBA (urinary metabolite of pyrethroids	376 men with non obstructive infertility	R CC

Table 2b. Studies evaluating the association between exposure to pesticides and human sperm quality. Studies with evidence of an association (CS: cross sectional study; CC: case-control study; L: longitudinal study; R: retrospective study; P: pilot study.)

4.3 Studies evaluating the recovery of sperm quality

There are six (6/63) studies, evaluating the recovery of sperm quality, years after cessation of exposure to DBCP and kepone (five and one study respectively). In 1986 Eaton et al. conducted a follow-up study among 44 male agricultural workers who were exposed to DBCP years ago. These workers were found to be azoospermic or oligospermic due to DBCP exposure in a previous study and their sperm was reevaluated 5 to 8 years after exposure was terminated. Only 2 of the 8 originally azoospermic workers produced sperm during the follow up and only one had normal sperm count. There was no increase in sperm production for the oligospermic men. Authors suggested a permanent destruction of germinal epithelium. Potashnik et al. conducted two follow-up studies evaluating 15 DBCP production factory workers (Potashnik & Yanai-Inbar, 1987; Potashnik & Porath, 1995). The reports reassessed their testicular function and reproductive performance, 8 and 17 years after cessation of exposure. The first study showed recovery of spermatogenesis in 4 oligo- and 3 azoospermic men, while testosterone levels of all patients were normal at all times.

The second follow up showed a significant increase in plasma FSH and LH levels in the severely affected men and no increase in the rate of spontaneous abortions or congenital malformations among pregnancies conceived during or after exposure. Recovery was evident in three of the nine azoospermic men and in three of the six oligospermic men. Olsen et al. (1990) conducted a follow-up study among azoospermic and oligospermic workers who had a maximum of 18 months of DBCP exposure. After an 11-year period, 73% of the previously azoospermic showed recovery of spermatogenesis; 13 of the men had normospermic levels and normospermic levels were found among all of the previously oligospermic men (17/17). Another study was undertaken by Lantz et al. (1981), among 14 oligospermic workers who had a maximum of 30 months of DBCP exposure . Follow up (18-21 months) showed an increase in sperm count suggesting that there is a recovery after a short term exposure. In a cross-sectional study conducted in 1982 Guzelian treated oligospermic patients who had high serum levels of chlordecone (0.6-32.0 µg/g) with cholestyramine, an anion-exchange resin. Cholestyramine binds chlordecone and increases its fecal excretion by seven-fold, resulting in reduction of chlordecone blood levels. The author found sperm motility restoration after treatment and suggested this indicated reversibility of the results.

Authors & Year Country	Pesticide	N	Study Type
Lantz et al, 1981 (USA)	DBCP	14 oligospermic workers	L
Guzelian et al, 1982 (USA)	kepone	13 workers highly exposed	CS
Eaton et al, 1986 (USA)	DBCP	44 workers 7 years after termination of exposure	L
Potashnik et al, 1987 (Israel)	DBCP	15 workers with DBCP induced azoospermia & oligospermia, 8 years after exposure.	L
Olsen et al, 1990 (USA)	DBCP	26 azoospermic and 17 oligospermic men with a maximum of 18 months of exposure	L
Potashnik et al, 1995 (Israel)	DBCP	15 workers with DBCP induced azoospermia & oligospermia, 17 years after exposure.	L

Table 2c. Studies evaluating the association between exposure to pesticides and human sperm quality. Studies evaluating the recovery of sperm quality. (CS: cross sectional study; L: longitudinal study)

4.4 Comments

The studies evaluating pesticide exposure and sperm quality were usually cross sectional. This design is relatively inexpensive and can be conducted over a short period of time. However, inherent to the study design, it is difficult to separate cause from effect, because the measurement of exposure and the possible health impairment are conducted the same simultaneously. Thus if the effect takes place later on in life, it may be misevaluated and critical windows of exposure (intrauterine, neonatal and early adulthood life) are not considered adequately.

The studies involving sperm collection have low participation rate and this, in combination with small sample size, decreases statistical power. Men providing sperm for analysis are usually self-selected volunteers that may have concerns about their fertility and this may influence the results of semen characteristics. Some embarrassment associated with sperm collection results in men who are considered fertile to be less likely to participate in such studies (selection bias). Exposure assessment in half of the reports was performed without using specific chemical analyses. Modern studies provide more accurate estimations of exposure measuring the exact levels of contaminants in humans, although there is a certain limitation regarding non-persistent pesticides that are metabolized and excreted relatively fast. Very recent studies, using biological monitoring of exposure, evaluate the susceptibility of individuals after pesticide exposure and demonstrate that genetic polymorphism modifies exposure effects on semen quality (Pérez-Herrera et al., 2008). Many of the studies that showed no or little association of exposure to pesticides and semen quality come from the north of Europe. Maybe the cumulative exposure to pollutants in this geographic area is limited compared to others. Following the evaluation of the above epidemiological studies, we conclude that overall there is suggestive evidence of an association between pesticide exposure and semen quality.

5. Pesticides and time to pregnancy

Baird et al. (1986) proposed "time to pregnancy" (TTP) as a simple measure in epidemiological studies to investigate the effects of environmental exposures on reproduction. Time to pregnancy is defined as the time interval (number of menstrual cycles, expressed in months) between the start of unprotected intercourse and a clinically recognizable pregnancy. Fecundability is defined as the probability of conception for each menstrual cycle and varies among sexually active couples not using any contraception method. The fecundability of a couple is estimated by the inverse of time to pregnancy (time to pregnancy = 1/fecundability). TTP does not consider the exact biological pathways (including gametogenesis, fertilisation, transport and implantation of the zygote, early survival of the conception product) or mechanisms involved in human fertility. Semen quality impairment, discussed earlier, is not a direct parameter of fecundity and altered semen quality does not necessarily entail changes in TTP (Joffe, 2000). TTP appears to be a non expensive, easy-to-determine indicator (obtained by questionnaires and interviews) and a sensitive measure of fecundability in either sex (Joffe, 1997).

The evaluation of pesticide effects on fecundability using TTP could be considerably susceptible to bias as highlighted by the following discussion. TTP studies are usually retrospective and based on questionnaires and interviews, susceptible to selection, response and recall bias. Individuals that have encountered problems of infertility or subjects occupationally exposed may not have the same response to participate in related studies, compared to fertile or unexposed population. Questionnaires may concern pregnancies that occurred several years back and require the recall of events that are not well remembered. Low participation rates and small sample size may limit the statistical power of the study. Time-trend bias may be prominent if the exposure under study has changed over time. TTP studies concern planned pregnancies and fertile couples with unwanted, mistimed or because of contraception failures pregnancies are more likely to be excluded (planning bias). Because of censoring prolonged time to pregnancy period (12 months for the majority of the studies), subfertile couples who need periods longer than 12 months to have conceive are underestimated and infertile couples are excluded (sterility bias). On the other hand couples who are in search of assisted reproduction programmes or benefit from a therapeutic effect of a treatment, after a long waiting period, are not included. It often difficult to recognize early fetal loss, therefore, early spontaneous abortions may lead to an overestimation of TTP. Studies that are not limited to first pregnancy may be biased by the "reproductively unhealthy worker effect": fertile women might be more likely to have children, spend their time taking care of them, or having part time jobs and as a result be less occupationally exposed. Most of the TTP studies do not assess exposure by directly measuring the concentrations of pollutants in human tissues or liquids. Usually, exposure is evaluated by the same questionnaires and interviews used to estimate TTP (self-reported data), which is simple and not expensive, but inaccurate, even if the questionnaires are detailed (specific pesticides mentioned, exact hours of spraying, quantity of pesticides applied, methods of application, protective equipment used, etc). Biological monitoring seems to be a better method to evaluate exposure, despite some limitations. The measurement of non persistent pesticides concentrations in humans cannot always reflect exposure in the time to pregnancy period accurately, as they are excreted in a relatively short period of time and many pesticides cannot be measured in biological matrices (Bradman & Whyatt, 2005). Important potential confounders that have to be taken into account include: age, ethnicity, parity, previous contraceptive use, medical conditions, and frequency of intercourse, BMI, breastfeeding, smoking, caffeine and alcohol consumption.

Table 3 shows the 29 studies that evaluate exposure to pesticides and time to pregnancy. Among these studies, three studies estimate fertilization rates and time to cleavage in women attending IVF programmes, and three studies evaluate fecundity relative to dietary intake of contaminated fish. The results are conflicting, perhaps because of the heterogeneity of study design and misclassification of exposure. The only pesticides evaluated as single pesticides and not in mixtures were glyphosate (one study), and DDT and metabolites (four studies). Some of the pesticides included in mixtures were: abamectine, imidacloprid, methiocarb, pirimicarb, deltamethrin, acephate, methomyl, cyromazine, propoxur, hexaflumuron, dichlorvos (insecticides), metalaxyl, captan, procymidon, pyrazophos, toclofosmethyl, zineb, benomyl (fungicides), acrinathrin, propargite (acaricides). DDT, DDE, HCB (hexachlorobenzene), trichlorobenzene, tetrachlorobenzene, lindane, heptachlor, aldrin, chlordane, oxychlordane, endosulfan, methoxychlor, mirex were the organochlorine pesticides most commonly used.

Sanin et al. (2009) found that reduced fecundability in some regions was not associated with the use of glyphosate. DDT and metabolites were involved in four studies as a single pesticide. Toft et al. (2005), INUENDO group, showed that fecundability is inversely related with serum concentration of p,p'DDE. Cocco et al. (2005) concluded that the fecundity ratio among spouses of DDT applicators compare to the unexposed was decreased, but the low statistical power of the study did not allow definitive conclusions. Cohn et al. (2003) measured serum levels of DDT and DDE in mothers and evaluated TTP in their daughters. Daughters' probability of pregnancy fell by 32% per 10μg/L p,p'-DDT in maternal serum, but increased 16% per 10μg/L p,p'-DDE. Axmon et al. (2006), in a big study of INUENDO, evaluating exposure to CB-153 (biphenyl) and p,p'-DDE found no effect on TTP of either male or female exposure in Sweden, Poland and Ukraine. In Greenland there seemed to be an association, but it was not possible to determine if it was due to CB-153 or p,p'-DDE.

Four studies evaluated DDT and its metabolites in mixtures. One found decreased fecundity related to exposure (Gerhard et al., 1999) and three of the studies indicated no significant association (Law et al., 2005; Axmon et al., 2006; Harley et al., 2008). Two studies examined DDE exposure in women undergoing IVF with contrary results (Jarrell et al., 1993; Younglai et al., 2002). Three studies evaluated fecundity in relation to dietary intake of contaminated fish (Courval et al., 1999; Axmon et al., 2000; Buck et al., 2000). The first suggests a modest association for men only. The second resulted in a negative association between exposure to persistent organochlorine compounds and fertility among heavy smokers and the third concluded that maternal consumption of contaminated fish may reduce fecundability. Three studies overall, evaluated the association between pesticide exposure and fertilization rates, in women attending IVF. Tielemans et al. (1999b) found an association, but exposure was assessed by questionnaires and interviews. Jarrell et al. (1993) had contrary results, but there were trace amounts of contaminants in the follicular fluid. The study conducted by Younglai et al. (2002) did not show any significant association, except for p,p'-DDE, which was associated with failed fertilization.

Curtis et al. (1999) found no significant or consistent pattern of associations between time to pregnancy and pesticides. However 6 of 13 pesticide categories were associated with decreased fecundability, during the exposure periods in which women and most of the men participated in pesticide-related activities. In a prospective study, Sallmen et al. (2003) illustrated a significant association for pyrethroids and suggestive association for carbamates and organophosphates. Restricting analyses for the first pregnancy only, a prolonged TTP was found among male and female greenhouse workers (Bretveld et al., 2006; Bretveld et al., 2008a & 2008b). According to Harley et al. (2008) prolonged TTP was related to maternal occupational pesticide exposure, home pesticide use and residence within 200 feet of an agriculture field. There was no relation with paternal occupational exposure (assessed by maternal interviews) and DDT and DDE (levels measured in maternal serum). Cole et al. (2006) using biological monitoring of exposure found that factors associated with prolonged TTP in multivariate analysis were high caffeine consumption and maternal mercury serum level. Among several pesticides analyzed, only higher maternal benzene hexachloride levels in bivariate analysis were related with prolonged TTP. However, the sample size of this study was small.

De Cock et al. (1994), Fuortes et al. (1997), Abell et al. (2000), Petrelli et al. (2001) and Idrovo et al. (2005) showed an association between impaired fecundity and exposure to pesticides, while Heacock et al. (1998), Larsen et al. (1998b), Thonneau et al. (1999), Law et al. (2005) and Lauria et al. (2006) found insignificant effects. Overall 11 studies showed an association between pesticide exposure and prolonged TTP or decreased fecundity, 7 studies found a rather inconclusive or insignificant association and 11 studies had results that are more complex to interpret and difficult to be categorised in the two groups cited above. Twenty (20/29) studies used retrospective design and only nine (9/29) used biological monitoring of exposure. Even among the studies that used biological assessment of exposure, results are conflicting. Analyses of pollutants in human tissue and liquids give important and accurate information about the cumulative exposure of an individual but there is a lack of information about timing of exposure, which may be critical for the manifestation of a reproductive impairment, such as subfecundity. Moreover, the synergistic effect of pollutants and individual susceptibility to harmful agents are factors that are difficult to estimate.

Authors & Year Country	**Pesticide**	**N**	**Stuy Type**
Jarrell 1993 et al, (Canada)	chlordane, HCB, DDE, epoxide-oxychlordane, PCB	74 women undergoing IVF	CS
De Cock 1994 et al, (The Netherlands)	mixture	43 fruit growers, (91 pregnancies)	CS
Fuortes et al, 1997 (USA)	Agriculture related exposures including pesticides	281 infertile women (cases), 216 postpartum women (controls)	CC
Heacock et al, 1998 (Canada)	Chlorophenate. fungicides	26487 sawmill workers, 23829 exposed, 2658 unexposed	R
Larsen et al, 1998b ASCLEPIOS, (Denmark)	mixture	450 traditional farmers spraying pesticides, 72 traditional not spraying, 94 organic not exposed	R
Courval et al, 1999 (USA)	mixture, including pesticides (dietary intake of sport fish)	626 couples (anglers)	R
Curtis et al, 1999 (USA)	mixture	1048 farm occupants, 2012 pregnancies	R
Gerhard et al, 1999 (Germany)	chlorinated hydrocarbons (CHC) (including pesticides)	489 infertile women	CS
Thonneau et al, 1999 ASCLEPIOS (France)	mixture	142 rural exposed workers and 220 not exposed (France); 326 exposed and 123 not exposed farmers (Denmark); 121 greenhouse workers exposed (Denmark)	R
Tielemans et al, 1999b (The Netherlands)	mixture	836 couples who sought in IVF treatment; 20 men exposed - 816 reference group	CC
Abell et al, 2000b (Denmark)	mixture used in greenhouses	492 pregnancies of women employed when they stopped contraception to have a child (starting time)	R
Axmon et al, 2000 (Sweden)	Organochlorine compounds dietary intake of fatty fish	Fishermen's wives from the Swedish east (n=399) and west coasts (n=936)	R
Buck et al, 2000 (USA)	mixture, including pesticides (dietary intake of sport fish)	895 women after years of fish consumption from lake Ontario	R
Petrelli et al, 2001 (Italy)	mixture of pesticides	127 greenhouse workers 173 clerical workers	R
Younglai et al, 2002 (Canada)	mixture of compounds including pesticides	21 couples attending IVF programme	CS
Cohn et al, 2003 (USA)	p,p'DDT, p,p'DDE	preserved maternal serum (n=289)	R

Sallmen et al, 2003 (Finland)	mixture	578 couples, Finnish green house employers and employees	P
Toft et al, 2005 INUENDO (Denmark)	CB-153 p,p'DDE	Poland: 472 women, 198 male spouses; Ukraine: 640 women, 208 male spouses; Greenland: 598 women, 201 male spouses; Sweden: 559 women, 191male spouses	R CS
Cocco et al, 2005 (Italy)	DDT	spouses of 105 men exposed	R
Idrovo et al, 2005 (Colombia)	mixture of pesticides including thiocarbamates (propineb, mancozeb)	2085 women working in cut flower production	CS R
Law et al, 2005 (USA)	PCBs, DDE	390 pregnant women enrolled at 12 study centers in US between 1959-1965 (before PCBs or DDE was banned in US)	R
Axmon et al, 2006 INUENDO (Sweden)	CB 153 p,p'-DDE	778 men, 1505 women (couples from Greenland,Poland, Ukraine & a cohort of Swedish wives)	R
Bretveld et al, 2006 (The Netherlands)	mixture of pesticides	398 female greenhouse workers - 524 referents	R
Cole et al, 2006 (Canada)	Mixture (metals, PCBs, organochlorine pesticides)	41 couples having their first pregnancy	R
Lauria et al, 2006 (Italy)	mixture	713 female greenhouse workers	R
Bretveld et al, 2008a (The Netherlands)	mixture	694 greenhouse workers exposed, 613 unexposed reference group	R
Bretveld et al, 2008b (The Netherlands)	mixture	101 couples with female greenhouse workers, 957 couples with male greenhouse workers; 1408 referents	R
Harley et al, 2008 (USA)	mixture of pesticides, DDT, DDE	402 pregnant women living in a farmworker community; in 289 of them DDT and DDE serum levels measured.	CS
Sanin et al, 2009 (Mexico)	Glyphosate (herbicide)	2592 fertile Colombian women from 5 regions with different exposure	R

Table 3. Studies evaluating exposure to pesticides and time to pregnancy. In italics: Studies that reveal no association. (CS: cross sectional study; CC: case-control study; L: longitudinal study; R: retrospective study; P: prospective study.)

6. Conclusive remarks

It is challenging to estimate if exposure to pesticides has harmful effects on human fertility, because there are many factors that can influence, limit or bias the results of the studies related to fertility. There are two main issues that have to be considered: the methods used to estimate fertility and accuracy of exposure assessment. Related epidemiological studies use basically two methods of fertility evaluation, the semen quality analysis, which is a surrogate and is indirect, and TTP assessment, which is considerably susceptible to bias. As cited before, fecundity is the important parameter that we have to study and that is the capability of producing offspring and not the actual production (fertility). Fertility rates are increasing in developed countries but this is not a proof that fecundity is increasing too. In actuality, human fecundity seems to be decreasing. Fertility rates are influenced by many factors such as socioeconomic status, immigration, health condition and medical service, the practice of contraception and abortion, the sexual behaviour and frequency of intercourse, the methods of assisted reproduction. Given the influence of age on female fecundity, the effects of the baby boom generation, the evolution of the two-child family and the trend for marriage and childbearing later on in life, a decline in fecundity would be expected.

The decline in semen quality overtime is still equivocal, perhaps because the standard assessments (count, motility and morphology) are insufficient. It is interesting that documented changes in semen quality are not always correlated with parallel changes in TTP. In the future, consideration of additional parameters of sperm quality that provide information about the sperm-egg binding event would be helpful. Modern techniques employed to evaluate DNA damage in spermatozoa such as single cell gel electrophoresis (COMET assay), Terminal transferase dUTP Nick End Labelling (TUNEL), sperm chromatin structure assay (SCSA), in situ nick translation (ISNT) as well as and sperm aneuploidy (FISH) have to be incorporated in future studies.

Pesticides are endocrine disrupting chemicals warranting research to elucidate their mechanisms of action. Timing of exposure is critical, as the same dosage seems to have diverse effects in individuals of different age and exposure in critical windows of development (intrauterine life, neonatal period, childhood) seems to be more harmful. The classical dosage-effect rule of toxicology is contradicted by some EDCs and low doses may damage more than high doses. Individuals are often exposed to a mixture of toxicants, not a single agent, and the effects may be synergistic or antagonistic. The results of an exposure may be apparent immediately but can also become apparent on at a later stage and sometimes in subsequent generations, with varying susceptibility that may depend on genetic polymorphism. Often, studies requiring semen collection have low participation rates and thus the men providing sperm for analysis are usually self selected volunteers that may have concerns about their own fertility and may not be a true reflection of the general population. Geographical and seasonal variation of semen parameters make results difficult to compare. Most of the epidemiological studies reporting chemical exposure and semen quality damage are cross-sectional. In this design, the measurement of exposure and the possible health impairment are conducted simultaneously and, because of the factors cited above, results can be easily biased. A need for a longitudinal design in studies of semen quality is apparent.

TTP studies are usually retrospective. TTP estimation and exposure assessment are carried out using questionnaires and interviews, a method relatively inexpensive, but not precise. Unplanned pregnancies are not included in TTP studies. Therefore, fertile couples, because of mistimed, unwanted or contraceptive failure pregnancies are excluded. A percentage of early fetal losses is not recognised and not included in these studies. Subfertile couples are underestimated because of censoring prolonged time to pregnancy to a period of around 12 months. Fertile women might be more likely to have children and be less busy in occupational activities and therefore less occupationally exposed (reproductively unhealthy worker effect). Restricting TTP studies for women attending only their first pregnancy is required. A prospective design that recruits couples early at the beginning of their attempt at a pregnancy, with more than one chemical analyses of specific pollutants per individual, during study period, might be less susceptible to bias and give more concrete results. In the majority of the studies cited above, exposure was evaluated by a simple questionnaire. There are few studies that directly measure the concentration of a toxicant in human tissue or liquids. Precise measurements of follicular fluid are more difficult to obtain. Biological monitoring is more precise, although with certain limits for non persistent pesticides and for the estimation of the primary source of exposure, since cumulative levels of pollutants are measured.

In evaluating epidemiological studies of exposure to pesticides and human semen quality, there is suggestive evidence of an association between exposure and semen quality impairment. The evaluation of the scientific literature related to human pesticide exposure and TTP results are conflicting. However, research results in animal and in vitro studies are clear and suggest that pesticides can adversely affect fertility. Unfortunately, these results cannot be taken as proof for humans strong enough to stop or modulate pesticide production. In the future, pesticides might be less toxic and it will therefore be even more difficult to estimate their adverse effects in human fertility. The evidence for impaired fecundity is strong in both in vivo and in vitro studies and investigators have to cooperate in order to optimise the research and raise public awareness and concern. Therefore, scientists need to emphasize the need for stricter rules dealing with chemical safety.

7. References

Abell A, Ernst E, Bonde JP. Semen quality and sexual hormones in greenhouse workers. Scand J Work Environ Health 2000a;26:492-500.

Abell A, Juul S, Bonde JP. Time to pregnancy among female greenhouse workers. Scand J Work Environ Health 2000b;26:131-6.

Adamopoulos DA, Pappa A, Nicopoulou S, Andreou E, Karamertzanis M, Michopoulos J, et al. Seminal volume and total sperm number trends in men attending subfertility clinics in the greater Athens area during the period 1977-1993. Hum Reprod 1996;11:1936-41.

Adiga SK, Jayaraman V, Kalthur G, Upadhya D, Kumar P. Declining semen quality among south Indian infertile men: A retrospective study. J Hum Reprod Sci 2008;1:15-8.

Alavanja MC, Hoppin JA, Kamel F. Health effects of chronic pesticide exposure: cancer and neurotoxicity. Ann Rev Public Health 2004;25:155–97.

Almagor M, Ivnitzki I, Yaffe H, Baras M. Changes in semen quality in Jerusalem between 1990 and 2000: a cross-sectional and longitudinal study. Arch Androl 2003;49:139-44.

Andersen HR, Vinggaard AM, Rasmussen TH, Gjermandsen IM, Bonefeld-Jørgensen EC. Effects of currently used pesticides in assays for estrogenicity, androgenicity, and aromatase activity in vitro. Toxicol Appl Pharmacol 2002;179:1-12.

Andolz P, Bielsa MA, Vila J. Evolution of semen quality in North-eastern Spain: a study in 22,759 infertile men over a 36 year period. Hum Reprod 1999;14:731-5.

Aneck-Hahn NH, Schulenburg GW, Bornman MS, Farias P, de Jager C. Impaired semen quality associated with environmental DDT exposure in young men living in a malaria area in the Limpopo Province, South Africa. J Androl 2007;28:423-34.

Anway MD, Skinner MK. Epigenetic transgenerational actions of endocrine disruptors. Endocrinology 2006;147(6 Suppl):S43-9.

Axmon A, Rylander L, Strömberg U, Hagmar L. Time to pregnancy and infertility among women with a high intake of fish contaminated with persistent organochlorine compounds. Scand J Work Environ Health 2000;26:199-206.

Axmon A, Thulstrup AM, Rignell-Hydbom A, Pedersen HS, Zvyezday V, Ludwicki JK, et al. INUENDO. Time to pregnancy as a function of male and female serum concentrations of 2,2'4,4'5,5'-hexachlorobiphenyl (CB-153) and 1,1-dichloro-2,2-bis (p-chlorophenyl)-ethylene (p,p'-DDE). Hum Reprod 2006;21:657-65.

Ayotte P, Giroux S, Dewailly E, Hernández Avila M, Farias P, Danis R, et al. DDT spraying for malaria control and reproductive function in Mexican men. Epidemiology 2001;12:366-7.

Bahshwan SA, Owen CP, Nicholls PJ, Smith HJ, Ahmadi M. Some 1-((benzofuran-2-yl)methyl) imidazoles as inhibitors of 17 alpha-hydroxylase: 17, 20-lyase (P450 17) and their specificity patterns. J Pharm Pharmacol 1998;50:1109-16.

Baird DD, Wilcox AJ, Weinberg CR. Use of time to pregnancy to study environmental exposures. Am J Epidemiol 1986;124:470-80.

Bassil KL, Vakil C, Sanborn M, Cole DC, Kaur JS, Kerr KJ. Cancer health effects of pesticides: systematic review. Can Fam Physician. 2007;53:1704-11.

Benshushan A, Shoshani O, Paltiel O, Schenker JG, Lewin A. Is there really a decrease in sperm parameters among healthy young men? A survey of sperm donations during 15 years. J Assist Reprod Genet 1997;14:347-53.

Berling S, Wölner-Hanssen P. No evidence of deteriorating semen quality among men in infertile relationships during the last decade: a study of males from Southern Sweden. Hum Reprod 1997;12:1002-5.

Bian Q, Xu LC, Wang SL, Xia YK, Tan LF, Chen JF, et al. Study on the relation between occupational fenvalerate exposure and spermatozoa DNA damage of pesticide factory workers. Occup Environ Med 2004;61:999-1005.

Bigsby R, Chapin RE, Daston GP, Davis BJ, Gorski J, Gray LE, et al. Evaluating the effects of endocrine disruptors on endocrine function during development. Environ Health Perspect 1999;107(Suppl 4):613–8.

Bilotta P, Guglielmo R, Steffè M. Analysis of decline in seminal fluid in the Italian population during the past 15 years. Minerva Ginecol 1999;51:223-31.

Bradman A, Whyatt RM. Characterizing exposures to non persistent pesticides during pregnancy and early childhood in the National Children's Study: a review of monitoring and measurement methodologies. Environ Health Perspect 2005;113:1092-9.

Bretveld RW, Thomas CM, Scheepers PT, Zielhuis GA, Roeleveld N. Pesticide exposure: the hormonal function of the female reproductive system disrupted? Reprod Biol Endocrinol 2006;4:30.

Bretveld RW, Hooiveld M, Zielhuis GA, Pellegrino A, van Rooij IA, Roeleveld N. Reproductive disorders among male and female greenhouse workers. Reprod Toxicol 2008a;25:107-14.

Bretveld R, Kik S, Hooiveld M, van Rooij I, Zielhuis G, Roeleveld N. Time-to-pregnancy among male greenhouse workers. Occup Environ Med 2008b;65:185-90.

Buck GM, Vena JE, Schisterman EF, Dmochowski J, Mendola P, Sever LE, et al. Parental consumption of contaminated sport fish from Lake Ontario and predicted fecundability. Epidemiology 2000;11:388-93.

Can A, Albertini DF. Stage specific effects of carbendazim (MBC) on meiotic cell cycle progression in mouse oocytes. Mol Reprod Dev 1997;46:351-62.

Cannon SB, Veazey JM Jr, Jackson RS, Burse VW, Hayes C, Straub WE, et al. Epidemic kepone poisoning in chemical workers. Am J Epidemiol 1978;107:529-37.

Carlsen E, Giwercman A, Keiding N, Skakkebaek NE. Evidence for decreasing quality of semen during past 50 years. BMJ 1992;305:609-13.

Charlier CJ, Foidart JM. Comparative study of dichlorodiphenyldichloroethylene in blood and semen of two young male populations: lack of relationship to infertility, but evidence of high exposure of the mothers. Reprod Toxicol 2005;20:215-20.

Cocco P, Fadda D, Ibba A, Melis M, Tocco MG, Atzeri S, et al. Reproductive outcomes in DDT applicators. Environ Res. 2005;98:120-6.

Cohn BA, Cirillo PM, Wolff MS, Schwingl PJ, Cohen RD, Sholtz RI, et al. DDT and DDE exposure in mothers and time to pregnancy in daughters. Lancet 2003;361:2205-6.

Colborn T, vom Saal FS, Soto AM. Developmental effects of endocrine-disrupting chemicals in wildlife and humans. Environ Health Perspect1993;101:378–84

Cole DC, Wainman B, Sanin LH, Weber JP, Muggah H, Ibrahim S. Environmental contaminant levels and fecundability among non-smoking couples. Reprod Toxicol 2006;22:13-9.

Cooper TG, Noonan E, von Eckardstein S, Auger J, Baker HW, Behre HM, et al. World Health Organization reference values for human semen characteristics. Hum Reprod Update 2010;16:231-45

Courval JM, De Hoog JV, Stein AD, Tay EM, He J, Humphrey HE, et al. Sport-caught fish consumption and conception delay in licensed Michigan anglers. Environ Res 1999;80:183-8.

Cox C, Surgan M. Unidentified inert ingredients in pesticides: implications for human and environmental health. Environ Health Perspect 2006;114:1803-6.

Crews D, Putz O, Thomas P, Hayes T, Howdeshell K. Animal models for the study of the effects of mixtures, low doses, and the embryonic environment on the action of endocrine disrupting chemicals. Pure and Applied Chemistry, SCOPE/IUPAC Project Implications of Endocrine Active Substances for Humans and Wildlife. 2003;75:2305-20

Crisp TM, Clegg ED, Cooper RL, Wood WP, Anderson DG, Baetcke KP, et al. Environmental endocrine disruption: an effects assessment and analysis. Environ Health Perspect 1998;106(Suppl 1):11-56.

Cummings AM, Gray LE Jr. Methoxychlor affects the decidual cell response of the uterus but not other progestational parameters in female rats. Toxicol Appl Pharmacol 1987;90:330-6.

Curtis KM, Savitz DA, Weinberg CR, Arbuckle TE. The effect of pesticide exposure on time to pregnancy. Epidemiology 1999;10:112-7.

Dallinga JW, Moonen EJ, Dumoulin JC, Evers JL, Geraedts JP, Kleinjans JC. Decreased human semen quality and organochlorine compounds in blood. Hum Reprod 2002;17:1973-9.

Dalvie MA, Myers JE, Thompson ML, Robins TG, Dyer S, Riebow J, et al. The long-term effects of DDT exposure on semen, fertility, and sexual function of malaria vector-control workers in Limpopo Province, South Africa. Environ Res 2004;96:1-8.

De Cock J, Westveer K, Heederik D, te Velde E, van Kooij R. Time to pregnancy and occupational exposure to pesticides in fruit growers in The Netherlands. Occup Environ Med 1994;51:693-9.

De Jager C, Farias P, Barraza-Villarreal A, Avila MH, Ayotte P, Dewailly E, et al. Reduced seminal parameters associated with environmental DDT exposure and p,p'-DDE concentrations in men in Chiapas, Mexico: a cross-sectional study. J Androl 2006;27:16-27.

Diamanti-Kandarakis E, Bourguignon JP, Giudice LC, Hauser R, Prins GS, Soto AM, et al. Endocrine-disrupting chemicals: an Endocrine Society scientific statement. Endocr Rev 2009;30:293-342.

Dohle GR, Colpi GM, Hargreave TB, Papp GK, Jungwirth A, Weidner W. EAU Working Group on Male Infertility EAU guidelines on male infertility. Eur Urol 2005;48:703-11

Eaton M, Schenker M, Whorton MD, Samuels S, Perkins C, Overstreet J. Seven-year follow-up of workers exposed to 1,2-dibromo-3-chloropropane. J Occup Med 1986;28:1145-50.

Egnatz DG, Ott MG, Townsend JC, Olson RD, Johns DB. DBCP and testicular effects in chemical workers: an epidemiological survey in Midland, Michigan. J Occup Med 1980;22:727-32.

Feki NC, Abid N, Rebai A, Sellami A, Ayed BB, Guermazi M, et al. Semen quality decline among men in infertile relationships: experience over 12 years in the South of Tunisia. J Androl 2009;30:541-7.

Fisch H, Goluboff ET, Olson JH, Feldshuh J, Broder SJ, Barad DH. Semen analyses in 1,283 men from the United States over a 25-year period: no decline in quality. Fertil Steril 1996;65:1009-14.

Food and Agriculture Organization of the United Nations. FAO international code of conduct on the distribution and use of pesticides. Rome: Food and Agriculture Organization of the United Nations; 2003.

Fourth National Report on Human Exposure to Environmental Chemicals. Centers for Disease Control and Prevention, U.S. Department of Health and Human Services, 2009

Fuortes L, Clark MK, Kirchner HL, Smith EM. Association between female infertility and agricultural work history. Am J Ind Med 1997;31:445-51.

García AM. Pesticide exposure and women's health. Am J Ind Med 2003;44:584-94.

Garey J, Wolff MS. Estrogenic and antiprogestagenic activities of pyrethroid insecticides. Biochem Biophys Res Commun 1998;251:855-9.

Gerhard I, Monga B, Krähe J, Runnebaum B. Chlorinated hydrocarbons in infertile women. Environ Res 1999;80:299-310.

Gore AC, Heindel JJ, Zoeller RT. Endocrine disruption for endocrinologists (and others). Endocrinology 2006;147(Suppl 6):S1-3.

Gore A. Endocrine-disrupting chemicals: from basic research to clinical practice. In: Cohn PM, ed. Contemporary Endocrinology. Totowa, NJ: Humana Press 2007.

Gore AC. Neuroendocrine systems as targets for environmental endocrine-disrupting chemicals. Fertil Steril 2008;89(Suppl 2):e101-2.

Guo YL, Lambert GH, Hsu CC. Growth abnormalities in the population exposed in utero and early postnatally to polychlorinated biphenyls and dibenzofurans. Environ Health Perspect 1995;103(Suppl 6):117–22.

Guzelian PS. Chlordecone poisoning: a case study in approaches for detoxification of humans exposed to environmental chemicals. Drug Metab Rev 1982;13:663-79.

Härkönen K, Viitanen T, Larsen SB, Bonde JP, Lähdetie J. Aneuploidy in sperm and exposure to fungicides and lifestyle factors. ASCLEPIOS. A European Concerted Action on Occupational Hazards to Male Reproductive Capability. Environ Mol Mutagen 1999;34:39-46.

Harley KG, Marks AR, Bradman A, Barr DB, Eskenazi B. DDT exposure, work in agriculture, and time to pregnancy among farmworkers in California. J Occup Environ Med 2008;50:1335-42.

Hauser R, Chen Z, Pothier L, Ryan L, Altshul L. The relationship between human semen parameters and environmental exposure to polychlorinated biphenyls and p,p'-DDE. Environ Health Perspect 2003a;111:1505-11.

Hauser R, Singh NP, Chen Z, Pothier L, Altshul L. Lack of an association between environmental exposure to polychlorinated biphenyls and p,p'-DDE and DNA damage in human sperm measured using the neutral comet assay. Hum Reprod 2003b;18:2525-33

Heacock H, Hogg R, Marion SA, Hershler R, Teschke K, Dimich-Ward H, et al. Fertility among a cohort of male sawmill workers exposed to chlorophenate fungicides. Epidemiology 1998;9:56-60.

Henderson J, Rennie GC, Baker HW. Association between occupational group and sperm concentration in infertile men. Clin Reprod Fertil 1986;4:275-81.

Herbst AL, Ulfelder H, Poskanzer DC, Longo LD. Adenocarcinoma of the vagina. Association of maternal stilbestrol therapy with tumor appearance in young women. 1971. Am J Obstet Gynecol 1999;181:1574-5.

Hirsch KS, Adams ER, Hoffman DG, Markham JK, Owen NV. Studies to elucidate the mechanism of fenarimol-induced infertility in male rats. Toxicol Appl Pharmacol 1986;86:391-9.

Horak S, Kaminska J, Olejek A. The fertilization potential of donor semen between 1982 and 2004 in the industrial area of Upper Silesia (Poland). Ann Agric Environ Med 2008;15:113-8.

Idrovo AJ, Sanìn LH, Cole D, Chavarro J, Cáceres H, Narváez J, et al. Time to first pregnancy among women working in agricultural production. Int Arch Occup Environ Health 2005;78:493-500.

Itoh N, Kayama F, Tatsuki TJ, Tsukamoto T. Have sperm counts deteriorated over the past 20 years in healthy, young Japanese men? Results from the Sapporo area. J Androl 2001;22:40-4.

Jarrell JF, Villeneuve D, Franklin C, Bartlett S, Wrixon W, Kohut J, et al. Contamination of human ovarian follicular fluid and serum by chlorinated organic compounds in three Canadian cities. CMAJ 1993;148:1321-7.

Jeyaratnam J. Acute pesticide poisoning: a major global health problem. World Health Stat Q 1990;43:139–44.

Joffe M. Time to pregnancy: a measure of reproductive function in either sex. Asclepios Project. Occup Environ Med 1997;54:289-95.

Joffe M. Time trends in biological fertility in Britain. Lancet 2000;355:1961-5.

Jørgensen N, Andersen AG, Eustache F, Irvine DS, Suominen J, Petersen JH, et al. Regional differences in semen quality in Europe. Hum Reprod 2001;16:1012-9.

Jørgensen N, Asklund C, Carlsen E, Skakkebaek NE. Coordinated European investigations of semen quality: results from studies of Scandinavian young men is a matter of concern. Int J Androl 2006;29:54-61.

Juhler RK, Larsen SB, Meyer O, Jensen ND, Spanò M, Giwercman A, et al. Human semen quality in relation to dietary pesticide exposure and organic diet. Arch Environ Contam Toxicol 1999;37:415-23.

Kamijima M, Hibi H, Gotoh M, Taki K, Saito I, Wang H, et al. A survey of semen indices in insecticide sprayers. J Occup Health 2004;46:109-18.

Kavlock RJ, Daston GP, DeRosa C, Fenner-Crisp P, Gray LE, Kaattari S, et al. Research needs for the risk assessment of health and environmental effects of endocrine disruptors: a report of the U.S. EPA-sponsored workshop. Environ Health Perspect 1996;104 Suppl 4:715-40.

Kelce WR, Monosson E, Gamcsik MP, Laws SC, Gray LE Jr. Environmental hormone disruptors: evidence that vinclozolin developmental toxicity is mediated by antiandrogenic metabolites. Toxicol Appl Pharmacol 1994;126:276-85.

Kelce WR, Stone CR, Laws SC, Gray LE, Kemppainen JA, Wilson EM. Persistent DDT metabolite p,p'-DDE is a potent androgen receptor antagonist. Nature 1995;375:581-5.

Kelce WR, Wilson EM. Environmental antiandrogens: developmental effects, molecular mechanisms, and clinical implications. J Mol Med 1997;75:198-207.

Kojima H, Katsura E, Takeuchi S, Niiyama K, Kobayashi K. Screening for estrogen and androgen receptor activities in 200 pesticides by in vitro reporter gene assays using Chinese hamster ovary cells. Environ Health Perspect 2004;112:524-31.

Krause A, Krause W. Seasonal variations in human seminal parameters. Eur J Obstet Gynecol Reprod Biol 2002;101:175-8.

Lantz GD, Cunningham GR, Huckins C, Lipshultz LI. Recovery from severe oligospermia after exposure to dibromochloropropane. Fertil Steril 1981;35:46-53.

Larsen SB, Giwercman A, Spanò M, Bonde JP. A longitudinal study of semen quality in pesticide spraying Danish farmers. The ASCLEPIOS Study Group. Reprod Toxicol 1998a;12:581-9.

Larsen SB, Joffe M, Bonde JP. Time to pregnancy and exposure to pesticides in Danish farmers. ASCLEPIOS Study Group. Occup Environ Med 1998b;55:278-83.

Lathers CM. Endocrine disruptors: a new scientific role for clinical pharmacologists? Impact on human health, wildlife, and the environment. J Clin Pharmacol 2002;42;7-23.

Lauria L, Settimi L, Spinelli A, Figà-Talamanca I. Exposure to pesticides and time to pregnancy among female greenhouse workers. Reprod Toxicol 2006;22:425-30.

Law DC, Klebanoff MA, Brock JW, Dunson DB, Longnecker MP. Maternal serum levels of polychlorinated biphenyls and 1,1-dichloro-2,2-bis(p-chlorophenyl)ethylene (DDE) and time to pregnancy. Am J Epidemiol 2005;162:523-32.

Lerda D, Rizzi R. Study of reproductive function in persons occupationally exposed to 2,4-dichlorophenoxyacetic acid (2,4-D). Mutat Res 1991;262:47-50.

Levine RJ. Seasonal variation of semen quality and fertility. Scand J Work Environ Health 1999;25(Suppl 1):34-7

Li DK, Yan B, Li Z, Gao E, Miao M, Gong D, et al. Exposure to magnetic fields and the risk of poor sperm quality. Reprod Toxicol 2010;29:86-92.

Lifeng T, Shoulin W, Junmin J, Xuezhao S, Yannan L, Qianli W, et al. Effects of fenvalerate exposure on semen quality among occupational workers. Contraception 2006;73:92-6.

Lilienthal H, Hack A, Roth-Härer A, Grande SW, Talsness CE. Effects of developmental exposure to 2,2,4,4 ,5-pentabromodiphenyl ether (PBDE-99) on sex steroids, sexual development, and sexually dimorphic behavior in rats. Environ Health Perspect 2006;114:194–201.

Lindenau A, Fischer B, Seiler P, Beier HM. Effects of persistent chlorinated hydrocarbons on reproductive tissues in female rabbits. Hum Reprod 1994;9:772-80.

Lipshultz LI, Ross CE, Whorton D, Milby T, Smith R, Joyner RE. Dibromochloropropane and its effect on testicular function in man. J Urol 1980;124:464-8.

Magnusdottir EV, Thorsteinsson T, Thorsteinsdottir S, Heimisdottir M, Olafsdottir K. Persistent organochlorines, sedentary occupation, obesity and human male subfertility. Hum Reprod 2005;20:208-15.

Mahadevaswami MP, Jadaramkunti UC, Hiremath MB, Kaliwal BB. Effect of mancozeb on ovarian compensatory hypertrophy and biochemical constituents in hemicastrated albino rat. Reprod Toxicol 2000;14:127-34.

Mallidis C, Howard EJ, Baker HW. Variation of semen quality in normal men. Int J Androl 1991;14:99-107.

Manna PR, Kero J, Tena-Sempere M, Pakarinen P, Stocco DM, Huhtaniemi IT. Assessment of mechanisms of thyroid hormone action in mouse Leydig cells: regulation of the steroidogenic acute regulatory protein, steroidogenesis, and luteinizing hormone receptor function. Endocrinology 2001;142:319-31.

McCauley LA, Anger WK, Keifer M, Langley R, Robson MG, Rohlman D. Studying health outcomes in farmworker populations exposed to pesticides. Environmental Health Perspectives 2006;114:953–60.

Meeker JD, Ryan L, Barr DB, Herrick RF, Bennett DH, Bravo R, et al. The relationship of urinary metabolites of carbaryl/naphthalene and chlorpyrifos with human semen quality. Environ Health Perspect 2004;112:1665-70.

Meeker JD, Barr DB, Hauser R. Human semen quality and sperm DNA damage in relation to urinary metabolites of pyrethroid insecticides. Hum Reprod 2008;23:1932-40.

Merzenich H, Zeeb H, Blettner M. Decreasing sperm quality: a global problem? BMC Public Health 2010;10:24

Mikkilä TF, Toppari J, Paranko J. Effects of neonatal exposure to 4-tert-octylphenol, diethylstilbestrol, and flutamide on steroidogenesis in infantile rat testis. Toxicol Sci 2006;91:456-66.

Multigner L, Kadhel P, Pascal M, Huc-Terki F, Kercret H, Massart C, et al. Parallel assessment of male reproductive function in workers and wild rats exposed to pesticides in banana plantations in Guadeloupe. Environ Health 2008;7:40.

Oduma JA, Oduor Okelo D, Odongo H, Makawiti DW. The pesticide heptachlor affects steroid hormone secretion in isolated follicular and luteal cells of rat. Comp Biochem Physiol C Toxicol Pharmacol 2006;144:76-84.

Oliva A, Spira A, Multigner L. Contribution of environmental factors to the risk of male infertility. Hum Reprod 2001;16:1768-76.

Olsen GW, Lanham JM, Bodner KM, Hylton DB, Bond GG. Determinants of spermatogenesis recovery among workers exposed to 1,2-dibromo-3-chloropropane. J Occup Med 1990;32:979-84.

Olsen GW, Bodner KM, Ramlow JM, Ross CE, Lipshultz LI. Have sperm counts been reduced 50 percent in 50 years? A statistical model revisited. Fertil Steril 1995;63:887-93.

Paasch U, Salzbrunn A, Glander HJ, Plambeck K, Salzbrunn H, Grunewald S, et al. Semen quality in sub-fertile range for a significant proportion of young men from the general German population: a co-ordinated, controlled study of 791 men from Hamburg and Leipzig. Int J Androl 2008;31:93-102.

Padungtod C, Savitz DA, Overstreet JW, Christiani DC, Ryan LM, Xu X. Occupational pesticide exposure and semen quality among Chinese workers. J Occup Environ Med 2000;42:982-92.

Pant N, Mathur N, Banerjee AK, Srivastava SP, Saxena DK. Correlation of chlorinated pesticides concentration in semen with seminal vesicle and prostatic markers. Reprod Toxicol 2004;19:209-14.

Paulsen CA, Berman NG, Wang C. Data from men in greater Seattle area reveals no downward trend in semen quality: further evidence that deterioration of semen quality is not geographically uniform. Fertil Steril 1996;65:1015-20.

Peiris-John RJ, Wickremasinghe R. Impact of low-level exposure to organophosphates on human reproduction and survival. Trans R Soc Trop Med Hyg 2008;102:239-45.

Pérez-Herrera N, Polanco-Minaya H, Salazar-Arredondo E, Solís-Heredia MJ, Hernández-Ochoa I, Rojas-García E et al. PON1Q192R genetic polymorphism modifies organophosphorous pesticide effects on semen quality and DNA integrity in agricultural workers from southern Mexico. Toxicol Appl Pharmacol 2008;230:261-8.

Perry MJ, Venners SA, Barr DB, Xu X. Environmental pyrethroid and organophosphorus insecticide exposures and sperm concentration. Reprod Toxicol 2007;23:113-8.

Petrelli G, Figà-Talamanca I. Reduction in fertility in male greenhouse workers exposed to pesticides. Eur J Epidemiol 2001;17:675-7.

Potashnik G, Ben-Aderet N, Israeli R, Yanai-Inbar I, Sober I. Suppressive effect of 1,2-dibromo-3-chloropropane on human spermatogenesis. Fertil Steril 1978;30:444-7.

Potashnik G, Yanai-Inbar I. Dibromochloropropane (DBCP): an 8-year reevaluation of testicular function and reproductive performance. Fertil Steril 1987;47:317-23.

Potashnik G, Porath A. Dibromochloropropane (DBCP): a 17-year reassessment of testicular function and reproductive performance. J Occup Environ Med 1995;37:1287-92.

Purvis K, Christiansen E. Male infertility: current concepts. Ann Med 1992;24:259-72.

Ramazzini C. Collegium ramazzini statement on the control of pesticides in the European Union: A call for action to protect human health. Am J Ind Med 2009;52:176-7.

Ratcliffe JM, Schrader SM, Steenland K, Clapp DE, Turner T, Hornung RW. Semen quality in papaya workers with long term exposure to ethylene dibromide. Br J Ind Med 1987;44:317-26.

Recio-Vega R, Ocampo-Gómez G, Borja-Aburto VH, Moran-Martínez J, Cebrian-Garcia ME. Organophosphorus pesticide exposure decreases sperm quality: association between sperm parameters and urinary pesticide levels. J Appl Toxicol 2008;28:674-80.

Ribas-Fitó N. Silent Invaders: Pesticides, Livelihoods and Women's Health. London: ZED Books, 2002. Int J Epidemiol 2006;504-5.

Rignell-Hydbom A, Rylander L, Giwercman A, Jönsson BA, Nilsson-Ehle P, Hagmar L. Exposure to CB-153 and p,p'-DDE and male reproductive function. Hum Reprod 2004;19:2066-75.

Rignell-Hydbom A, Rylander L, Giwercman A, Jönsson BA, Lindh C, Eleuteri P, et al. Exposure to PCBs and p,p'-DDE and human sperm chromatin integrity. Environ Health Perspect 2005;113:175-9.

Ritter L, Goushleff NC, Arbuckle T, Cole D, Raizenne M. Addressing the linkage between exposure to pesticides and human health effects—research trends and priorities for research. J Toxicol Environ Health B Crit Rev 2006;9:441-56.

Rubin MM. Antenatal exposure to DES: lessons learned...future concerns. Obstet Gynecol Surv 2007;62:548-55.

Sallmén M, Liesivuori J, Taskinen H, Lindbohm ML, Anttila A, Aalto L, et al. Time to pregnancy among the wives of Finnish greenhouse workers. Scand J Work Environ Health 2003;29:85-93.

Sanborn M, Kerr KJ, Sanin LH, Cole DC, Bassil KL, Vakil C. Non-cancer health effects of pesticides: systematic review and implications for family doctors. Can Fam Physician 2007; 53:1712-20.

Sánchez-Peña LC, Reyes BE, López-Carrillo L, Recio R, Morán-Martínez J, Cebrián ME, et al. Organophosphorous pesticide exposure alters sperm chromatin structure in Mexican agricultural workers. Toxicol Appl Pharmacol 2004;196:108-13.

Sanin LH, Carrasquilla G, Solomon KR, Cole DC, Marshall EJ. Regional differences in time to pregnancy among fertile women from five Colombian regions with different use of glyphosate. J Toxicol Environ Health A 2009;72:949-60.

Schrader SM, Turner TW, Ratcliffe JM. The effects of ethylene dibromide on semen quality: a comparison of short-term and chronic exposure. Reprod Toxicol 1988;2:191-8.

Seo JT, Rha KH, Park YS, Lee MS. Semen quality over a 10-year period in 22,249 men in Korea. Int J Androl 2000;23:194-8.

Silvestroni L, Palleschi S. Effects of organochlorine xenobiotics on human spermatozoa. Chemosphere 1999;39:1249-52.

Slutsky M, Levin JL, Levy BS. Azoospermia and oligospermia among a large cohort of DBCP applicators in 12 countries. Int J Occup Environ Health 1999;5:116-22.

Smith JL, Garry VF, Rademaker AW, Martin RH. Human sperm aneuploidy after exposure to pesticides. Mol Reprod Dev 2004;67:353-9.

Soto AM, Sonnenschein C, Chung KL, Fernandez MF, Olea N, Serrano FO. The E-SCREEN assay as a tool to identify estrogens: an update on estrogenic environmental pollutants. Environ Health Perspect 1995;103(Suppl 7):113-22.

Spanò M, Toft G, Hagmar L, Eleuteri P, Rescia M, Rignell-Hydbom A, et al. Exposure to PCB and p, p'-DDE in European and Inuit populations: impact on human sperm chromatin integrity. Hum Reprod 2005;20:3488-99.

Sripada S, Fonseca S, Lee A, Harrild K, Giannaris D, Mathers E, et al. Trends in semen parameters in the northeast of Scotland. J Androl 2007;28:313-9.

Stoker TE, Laws SC, Guidici DL, Cooper RL. The effect of atrazine on puberty in male wistar rats: an evaluation in the protocol for the assessment of pubertal development and thyroid function. Toxicol Sci 2000;58:50-9.

Strohmer H, Boldizsar A, Plöckinger B, Feldner-Busztin M, Feichtinger W. Agricultural work and male infertility. Am J Ind Med 1993;24:587-92.

Stronati A, Manicardi GC, Cecati M, Bordicchia M, Ferrante L, Spanò M, et al. Relationships between sperm DNA fragmentation, sperm apoptotic markers and serum levels of CB-153 and p,p'-DDE in European and Inuit populations. Reproduction 2006;132:949-58.

Surgan MH. Toxicity tests: "inert" and active ingredients. Environ Health Perspect 2005;113:A657

Swan SH, Elkin EP, Fenster L. Have sperm densities declined? A reanalysis of global trend data. Environ Health Perspect 1997;105:1228-32.

Swan SH, Elkin EP. Declining semen quality: can the past inform the present? Bioessays 1999;21:614-21.

Swan SH, Elkin EP, Fenster L. The question of declining sperm density revisited: an analysis of 101 studies published 1934-1996. Environ Health Perspect 2000;108:961-6.

Swan SH, Brazil C, Drobnis EZ, Liu F, Kruse RL, Hatch M, et al. Geographic differences in semen quality of fertile U.S. males. Environ Health Perspect 2003a;111:414-20.

Swan SH, Kruse RL, Liu F, Barr DB, Drobnis EZ, Redmon JB, et al. Semen quality in relation to biomarkers of pesticide exposure. Environ Health Perspect 2003b;111:1478-84.

Swan SH. Semen quality in fertile US men in relation to geographical area and pesticide exposure. Int J Androl 2006;29:62-8

Tan LF, Wang SL, Sun XZ, Li YN, Wang QL, Ji JM, et al. Effects of fenvalerate exposure on the semen quality of occupational workers. Zhonghua Nan Ke Xue 2002;8:273-6.

Tan LF, Sun XZ, Li YN, Ji JM, Wang QL, Chen LS, et al. Effects of carbaryl production exposure on the sperm and semen quality of occupational male workers. Zhonghua Lao Dong Wei Sheng Zhi Ye Bing Za Zhi 2005;23:87-90.

Taxvig C, Vinggaard AM, Hass U, Axelstad M, Metzdorff S, Nellemann C. Endocrine-disrupting properties in vivo of widely used azole fungicides. Int J Androl 2008;31:170-7.

Thonneau P, Abell A, Larsen SB, Bonde JP, Joffe M, Clavert A, et al. Effects of pesticide exposure on time to pregnancy: results of a multicenter study in France and Denmark. ASCLEPIOS Study Group. Am J Epidemiol 1999;150:157-63.

Thrupp LA. Sterilization of workers from pesticide exposure: the causes and consequences of DBCP-induced damage in Costa Rica and beyond. Int J Health Serv 1991;21:731-57.

Tielemans E, Burdorf A, te Velde ER, Weber RF, van Kooij RJ, Veulemans H, et al. Occupationally related exposures and reduced semen quality: a case-control study. Fertil Steril 1999a;71:690-6.

Tielemans E, van Kooij R, te Velde ER, Burdorf A, Heederik D. Pesticide exposure and decreased fertilisation rates in vitro. Lancet 1999b;354:484-5.

Toft G, Axmon A, Giwercman A, Thulstrup AM, Rignell-Hydbom A, Pedersen HS, et al. Fertility in four regions spanning large contrasts in serum levels of widespread persistent organochlorines: a cross-sectional study. Environ Health 2005;4:26

Toft G, Rignell-Hydbom A, Tyrkiel E, Shvets M, Giwercman A, Lindh CH, et al. Semen quality and exposure to persistent organochlorine pollutants. Epidemiology 2006;17:450-8.

Tomenson JA, Taves DR, Cockett AT, McCusker J, Barraj L, Francis M, et al. An assessment of fertility in male workers exposed to molinate. J Occup Environ Med 1999;41:771-87.

Toppari J, Larsen JC, Christiansen P, Giwercman A, Grandjean P, Guillette LJ Jr, et al. Male reproductive health and environmental xenoestrogens. Environ Health Perspect 1996;104(Suppl. 4):741-803

Tuc VP, Wangsuphachart V, Tasanapradit P, Fungladda W, Van Trong P, Nhung NT. Impacts of pesticide use on semen characteristics among rice farmers in Kienxuong District, Thaibinh Province, Vietnam. Southeast Asian J Trop Med Public Health 2007;38:569-75.

Vicari E, Conticello A, Battiato C, La Vignera S. Sperm characteristics in fertile men and healthy men of the south-east Sicily from year 1982 to 1999. Arch Ital Urol Androl 2003;75:28-34.

Vinggaard AM, Hnida C, Breinholt V, Larsen JC. Screening of selected pesticides for inhibition of CYP19 aromatase activity in vitro. Toxicol In Vitro 2000;14:227-34.

Von Saal FS, Akingbemi BT, Belcher SM, Birnbaum LS, Crain DA, Eriksen M, et al. Chapel Hill bisphenol A expert panel consensus statement: integration of mechanisms, effects in animals and potential to impact human health at current levels of exposure. Reprod Toxicol 2007;24:131-8.

Walsh LP, Stocco DM. Effects of lindane on steroidogenesis and steroidogenic acute regulatory protein expression. Biol Reprod 2000;63:1024-33.

Wang JM, Wu XL, You W, Ling LX, Wu J, Zhang GY. Pharmacokinetic and pharmacodynamic studies of the effect of ketoconazole on reproductive function in male rats. Int J Androl 1992;15:376-84.

Weiss JM, Bauer O, Blüthgen A, Ludwig AK, Vollersen E, Kaisi M, et al. Distribution of persistent organochlorine contaminants in infertile patients from Tanzania and Germany. J Assist Reprod Genet 2006;23:393-9.

Weselak M, Arbuckle TE, Foster W. Pesticide exposures and developmental outcomes: The epidemiological evidence. J Toxicol Environ Health B Crit Rev 2007;10:41-80.

Whorton D, Krauss RM, Marshall S, Milby TH. Infertility in male pesticide workers. Lancet 1977;2:1259-61.

Whorton MD, Milby TH, Stubbs HA, Avashia BH, Hull EQ. Testicular function among carbaryl-exposed exployees. J Toxicol Environ Health 1979;5:929-41.

Wong WY, Zielhuis GA, Thomas CM, Merkus HM, Steegers-Theunissen RP. New evidence of the influence of exogenous and endogenous factors on sperm count in man. Eur J Obstet Gynecol Reprod Biol 2003;110:49-54.

Wyrobek AJ, Watchmaker G, Gordon L, Wong K, Moore D 2nd, Whorton D. Sperm shape abnormalities in carbaryl-exposed employees. Environ Health Perspect 1981;40:255-65.

Xia Y, Bian Q, Xu L, Cheng S, Song L, Liu J, et al. Genotoxic effects on human spermatozoa among pesticide factory workers exposed to fenvalerate. Toxicology 2004;203:49-60.

Xia Y, Cheng S, Bian Q, Xu L, Collins MD, Chang HC, et al. Genotoxic effects on spermatozoa of carbaryl-exposed workers. Toxicol Sci 2005;85:615-23.

Xia Y, Han Y, Wu B, Wang S, Gu A, Lu N, et al. The relation between urinary metabolite of pyrethroid insecticides and semen quality in humans. Fertil Steril 2008;89:1743-50.

Younglai EV, Collins JA, Foster WG. Canadian semen quality: an analysis of sperm density among eleven academic fertility centers. Fertil Steril 1998;70:76-80.

Younglai EV, Foster WG, Hughes EG, Trim K, Jarrell JF. Levels of environmental contaminants in human follicular fluid, serum, and seminal plasma of couples undergoing in vitro fertilization. Arch Environ Contam Toxicol 2002;43:121-6.

Yucra S, Rubio J, Gasco M, Gonzales C, Steenland K, Gonzales GF. Semen quality and reproductive sex hormone levels in Peruvian pesticide sprayers. Int J Occup Environ Health 2006;12:355-61.

Zvěrina J, Urbánek V, Cirýn J. (Spermiologic parameters in Czech men from 1950 to 1984: analysis of data from the archives of the Sexology Institute of the First Medical School of Charles University and the General Medical School Hospital in Prague) Sb Lek 2002; 103:35-47.

26

Organophosphorous Pesticides Exacerbate the Demographic Consequences of Intersexual Selection in Fish

Irene BarbosaValero[1], Omar Arellano-Aguilar[2] and Constantino Macías Garcia[1,3]
[1]*Universidad Nacional Autónoma de México-UNAM, Departamento de Ecología Evolutiva*
[2]*Universidad Nacional Autónoma de México-UNAM, Departamento de Medicina Genómica y Toxicología Ambiental*
México

1. Introduction

In the last 150 years, the human population has increased very rapidly in part due to the generalised use of agrochemicals, which enabled the development of intensive farming of food crops. Large monocultures are particularly vulnerable to pests, thus pesticides became widely used since the 1930´s to aid producing larger crops. However, within a few decades the collateral negative effects of pesticides on the environment became apparent (Nimmo & McEwen, 1994). The first paper on adverse effects of pesticides was published in the early 1940´s, but it was not until the 1960s that the Rachel Carsons review showed strong evidence of the environmental risk from the use of pesticides, particularly organochlorines such as the DDT (Nimmo & McEwen, 1994). This led to the search for alternatives, and in recent decades the organophosphorous (OPs) pesticides have become more used due to their short persistence in the environment and their high toxicity. These are lipophilic substances that penetrate through skin, and depending on their chemical structure are divided into phosphorotiates (e.g. methyl parathion, parathion, diazinon) and phosphoroditioates (e.g. malathion, dimetoate and azinfosmethyl). The biotransformation and activation pathways of OPs have been extensively reviewed (Jokanović, 2001). Mainly, OPs interfere with transmission of nerve impulses as they inhibit the activity of serine esterases, particularly the catalytic activity of acetilcholinesterase (AChE). The chemical reactions occur largely in the liver, where the OP thioeter group is oxidized to the *oxon* metabolite (P=S→P=O) by the action of citochrome P450 monooxygenases. After this biotransformation the OPs can bioaccumulate, or they can become immediately toxic. Because of their chemical affinity, OPs bioaccumulate in adipose tissues, muscle, glands, gonads and organs such as the liver, kidneys and brain (Vittozzi et al., 2001). Fatty tissue stores both the original form and the *oxon* metabolites. This process reduces the concentrations of free OPs in blood, yet during periods of starvation or stress, when stored fat is mobilized to supply metabolic energy, the OPs are released, leading to re-intoxication (Jokanović, 2001). Toxicity itself can occur in two ways: inhibition of

[3] Author for correspondence

acetylcholinesterase, and citotoxic effects on immune cells (Vittozzi et al., 2001). Inhibition of AChE occurs through the interaction between P=O and the serine hydroxil group at the enzyme's active site. As AChE, is responsible for the hydrolysis of the neurotransmitter acetylcholine (which stimulates postsynaptic response in the nervous systems and in the neuromuscular junction; Jokanović, 2001), the enzymes inactivation disrupts the action of the nervous and neuromotor systems. Citotoxicity is the result of oxidative desulphuration of OPs in which free radicals or reactive oxygen species (ROS) are released. Li (2007) proposed this mechanism because OPs affect directly the immune system by citotoxicity on T-lymphocytes. Moreover, the high concentration of free radicals and ROS produces protein and lipid oxidation, resulting in damage to cell membranes or even DNA.

Organophosphorous pesticides cause several physiological effects -characterised by a diversity of symptoms- depending on time exposure and chemical concentration. Pesticides act at molecular level within the organisms, yet these interactions can lead to a cascade of effects, from the biochemical through the organism, the population, and up to the ecosystem levels (Walker et al., 2006; Fig. 1.). The physiological effects of OPs have been severally described. They can cause neurotoxic damage by AChE inhibition as shown in the mosquitofish (*Gambusia affinis*; Boone & Chambers, 1996), induce chromosome aberrations such as those observed in the green chromide (*Etroplus suratensis*), a cichlid fish from India and Sri Lanka (Das & John, 1999), and compromise growth and neural development in mammals (e.g. Abu-Qare et al., 2000). Citotoxic effects and damage to organs such as liver, intestine and gills have also been reported (Fanta et al., 2003), and can lead to reduced behavioural performance. In developing organism, exposure can result in a failure to attain normal weight and in reduced nestling survival (as in nestling European starlings *Sturnus vulgaris*; Parker & Goldstein, 2000).

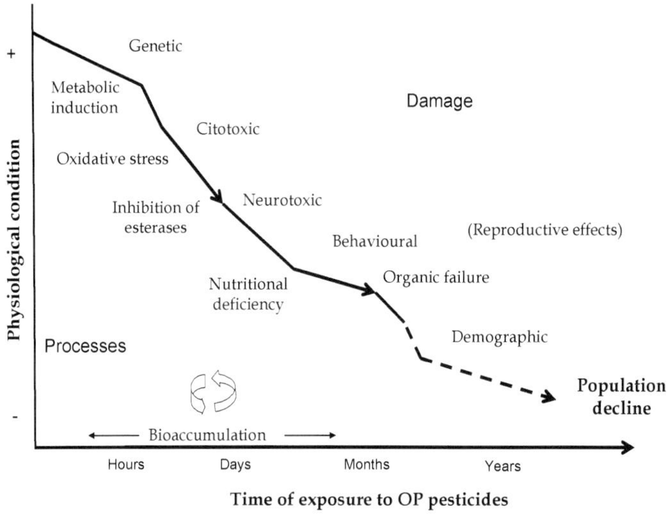

Fig. 1. Acute or chronic exposure to organophosphorous pesticides induces metabolic changes that produce a cascade of effects in the complete organism; depending on the duration of the exposure and on the pesticide concentration, the physiological condition of the exposed organisms decreases and can lead to population decline.

Perhaps the most visible effects of sub-lethal intoxication with OPs are, at the individual level, changes in the normal behaviour (see Welsh & Hanselka, 1972 and Castillo et al., 2002 for early and recent reports respectively); and, at the population level, demographic decline such as that reported by Sarma et al. (2001) for a rotifer population whose growth rate declined following continuous OP exposure. Although both the general mechanisms of action of OPs and their acute effects are well known, little has been documented on the consequences of sub-lethal concentrations and of chronic exposure. By definition, these processes would have less obvious effects than chronic/acute exposure and may only become apparent in the long run at the demographic level. Here we show evidence of such subtle effects; we deal particularly with a behavioural-ecological process -sexual selection- which is likely to be disturbed by sub-lethal exposure. In species in which male mating success depends on the production and display of costly ornaments and reproductive behaviours, the exposure to insecticides, organophosphorous or otherwise, may lead to scarcity of attractive males and thus to a reduction of the population size.

1.1 Sub-lethal exposure and behaviour

Behaviour is an aspect of the phenotype which integrates processes occurring at various levels, from cell receptors to neuromotor mechanisms, into coordinated responses whose ultimate function is to promote the individual's fitness. It is this integrative nature that has led several authors to propose that behaviour can be used to detect environmental changes in nature, such as those promoted by human activities. For instance, Little (1990) indicated that behaviour provides a "unique toxicological perspective", because it bridges the gap between the biochemical approach traditionally favoured in toxicology, and the ecological consequences of pollution. This view was more recently repeated by Walker et al. (2006) who mentioned that behaviour represents the diversity of biochemical and physiological processes. Indeed, being a biologically meaningful response, behaviour provides ecological realism to our assessments of anthropogenic impact in nature.

The term "behaviour" encompasses a very large amount of diverse phenomena. It includes 1) events which are mostly reactive, such as tropisms (the translation of one organism towards an attractant like sugar or light), predator avoidance, breathing (e.g. the opercular movements of fish, whose rhythm is partly determined by oxygen concentration in the water), panting and other simple thermoregulatory activities, etc. It also includes 2) more pro-active activities such as foraging (active search for food using endogenous rules, rather than merely responding to external stimuli), building of nests and other structures (an activity that also requires endogenous drives) or courtship displays, whose orientation and timing may be determined by external stimuli, but which follow endogenous rules that are in many cases completely unrelated to the immediate environmental conditions (see Baerends 1976; Lenher, 1996). The first, more "automatic" reactive behaviours are normally simple in structure, relatively invariant within species or populations, and easy to record. Also, the performance of those simple behaviours is often rather directly related to the organism's condition (e.g. a physiologically impaired animal would display more sluggish escape reactions than a healthy individual). It is therefore clear why behavioural toxicology relies heavily behaviours belonging to this category, particularly when dealing with acute/chronic exposure.

By contrast, behaviours in the second category are usually complex in nature, quite variable between individuals and populations, and consequently difficult to record in a standardised manner. Nevertheless, it is these behaviours that are often more closely related to important components of the species' life histories. During acute exposure pro-active behaviours are unlikely to express themselves, and thus would be of no use to the toxicological research. Yet organisms exposed to sub-lethal intoxication would still display e.g. pro-active foraging or courtship, albeit not with the vigour, frequency or efficacy as non-exposed organisms. Other, less obvious but equally insidious effects of pollutants on proactive behaviours may include interference with intra-specific recognition (e.g. Fisher et al., 2006) or abnormal sexual performance through disrupting hormonal balance (Toft et al., 2003; see Markman et al., 2008 for an example where male performance is enhanced). It is in this context that recording the expression of such behaviours can alert of potentially negative consequences of exposure to natural populations. We shall review in the following sections the effects of pollutants on behavioural performance and their implications at a demographic level.

The chemical industry has produced more than 500 compounds known to cause behavioural toxicity, which are classified as narcotics, excitatory agents, biochemical disruptors, cholinesterase inhibitors, reactive chemicals, central nervous system seizure agents and endocrine disruptors (details in Barron, 2002). The effects that have been associated with these chemicals include hyperactivity syndrome, paralysis, loss of coordination, tremors, ataxia, irritant responses, lethargy, etc., (Barron, 2002). Most of these effects have been observed following acute exposure or at high concentrations, and as mentioned above, all involve simple, automatic behaviours. To date, only a few studies have systematically investigated the behavioural consequences of exposure to sub-lethal concentrations. This is probably because the dominant paradigm in toxicology has been the dose-effect curve and the LC_{50} value, calculated as the concentration at which 50% of exposed individuals die. It is obvious that studying the consequences of exposure to LC_{50} provides an objective, repeatable experimental condition that can be compared across studies and has a clear definition. Yet before dying, the organisms experience different states of intoxication, by definition all deemed "sub-lethal", which differ in their symptoms and in their biological significance. Walker (1998) conceptualized the dose-effect curve as a function with hypothetical limits that represent biological responses. Between the hypothetical limits "c" (where compensatory mechanism can arrest the onset of overt disease) and "r" (beyond which repair mechanisms cannot longer prevent pathological damage), the behavioural symptoms enlisted above can be observed and are linked with impaired physiology and pathological damage (Fig. 2). However, there are less obvious behavioural modifications that are not necessarily directly linked to neurotoxicity. For example, as part of the M74 syndrome characterised by high mortality of salmon (*Salmo salar*) fry in the Baltic Sea (recorded between 1974 and 1993 and possibly brought about by exposure to organochlorine pesticides), both female and male adult fish showed abnormal swimming behaviour and general lethargy. During the early stages, fish showed anoxia, low levels of carotenoids and antioxidants such as α-tocopherol and ubiquinone (Börjenson & Norrgren, 1997). In addition, Baltic salmon presented induction of catalytic activity of cytochrome P450 enzymes and low levels of thiamine, which produced oxidative stress and "deficient behaviour". Although in cases such as that of the M74 syndrome the link between behavioural and physiological impairment is evident, it is still debated whether animal behaviour is a useful parameter to incorporate in toxicological tests. The main arguments against are that field observations are difficult to quantify and that the best studies are focused on behaviours that are of little ecological relevance (Walker et al., 2006). The first criticism is easily rebutted, as it can be overcome with improved recording methods (including new remote-sensing technologies). The second argument highlights the need to identify those relevant behaviours which are likely to be affected by sub-lethal exposure and to influence fitness. As we saw above, complex, pro-active behaviours such as foraging and courtship are good candidates to unmask biologically relevant effects of sub-lethal exposure in natural populations.

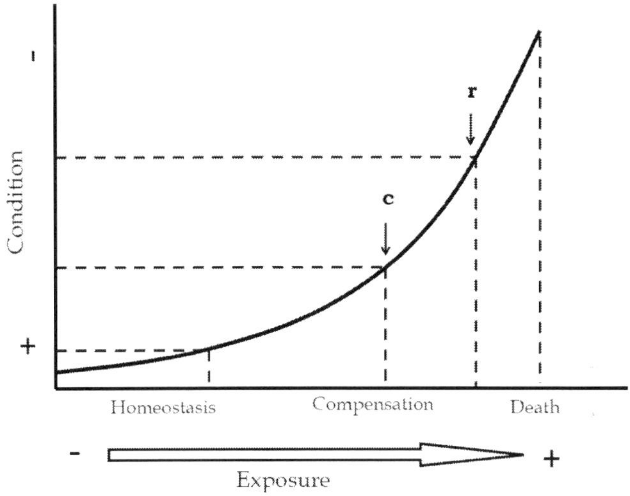

Fig. 2. The majority of behavioural changes registered during chemical exposure occur between "c" and "r", which define the conditions where disease can be fought and pathological damage averted by compensatory and repair mechanisms (modified from Walker, 1998).

1.2 Sub-lethal exposure of OPs and intersexual selection

There is a recent trend to look into the effects of pollutant exposure on animal behaviour. Particular attention have received organophosphorous pesticides because these are neurotoxic and lipophilic chemicals which inhibit acethylcholinesterase (AChE) catalytic activity and detrimentally affect the expression of behaviour (see above). Their reputedly short persistence in nature and their high toxicity were the reasons why OPs substituted in the last decades the organochlorine pesticides formerly used in agriculture. We have learned that OPs can persist at low concentrations for more than 200 days in freshwater ecosystems (Bonderenko & Gan, 2004). This is a short persistence in nature compared to the decades of the DDT half life, but even if present for only a few days, OPs become available for- and are incorporated (bioaccumulate) into aquatic fauna (De la Vega-Salazar et al., 1997), often with adverse consequences.

One obvious way in which behaviour is influenced by pesticides is through their teratological effects during embryogenesis. For instance, malformations produced by sub-lethal exposure to OP during embryonic development also cause deficient swimming (Bonfanti et al., 2004; Arellano-Aguilar & Macías Garcia, 2009). Pesticides can also have direct adverse effects on swimming behaviour when these interact with AChE activity, as is the case of OPs, or when they interfere with other and physiological processes (see Jones & Reynolds, 1997). For instance, Boone & Chambers (1996) demonstrated that more than 70%

of AChE activity is inhibited in the mosquitofish (*Gambusia affinis*) between 4 and 12 hr after sub-lethal exposure to OP pesticides such as methyl parathion (8 μg/ml concentration), chlorpyrifos (0.1 μg/ml) and parathion (0.1 μg/ml), and the physiological damage was accompanied by abnormal swimming behaviour. In fact, behavioural damage can become evident as quickly as the earliest symptoms of physiological damage. For instance, in the armoured catfish *Corydoras paleatus*, gill damage is evident one hour after exposure to 0.083 μg/g of methyl parathion, coinciding with the onset of a decrease in swimming activity that reaches between 30% and 40% three days later. Damage to the intestine and liver tissues becomes apparent only after four hours of exposure (Fanta et al., 2003). It is clear that a reduced motor performance can influence survival, yet in most cases it happens only after long-term exposure or following acute intoxication. More subtle behavioural effects are likely to follow sub-lethal exposure. We expect that organisms exposed to low concentrations of OPs will be more vulnerable to predation and their foraging behaviour will be sub-optimal. Another hypothesis is that OPs should interfere with mating behaviour. Surprisingly, in spite of having been proposed more than a decade ago (Jones & Reynolds, 1997), this possibility has seldom been investigated. In the following sections we present some recent results of our research on the links between exposure to insecticides and sexual performance. First we review the basic concepts relating sexual displays, ornamentation, male condition and female mate choice.

Charles Darwin (1871) developed the concept of intersexual selection (normally female choice) to explain the existence of structures in the males that appear to defy natural selection. Three mechanisms have been proposed to explain why ornaments can evolve through female choice: direct benefits, indirect benefits, and exploitation of female sensory biases (see review in Andersson 1994). Males may advertise with the quality of their ornaments their capacity to provide direct benefits to females such as food, nutrients, parental care or good territory. The term "indirect benefits" refers to two different processes: runaway sexual selection (Fisher, 1930), and handicap (Zahavi, 1975) or "good genes" (Andersson in 1994). The handicap mechanism proposes that phenotypic quality is to some extent based on genetic variants which correlate with the expression of secondary sexual characters. The runaway model proposes that the genes responsible for the production of the ornaments and those responsible for the female preferences are jointly inherited, although their phenotypic effects are sex-linked. According to this model it is not necessary that the genes responsible for the ornaments affect survival or fecundity (Futuyma, 1998). In the case of "indicator mechanism," the attractive male traits reflect genetic/health condition (condition-dependent indicator models) and are preferred by the females because they provide genetic benefits for her offspring. Thus, male traits correlate positively with male fitness (Andersson & Simmons, 2006). Finally, the sensory bias mechanism proposes that females are particularly responsive to certain attributes of the environment, for instance food items, which are imitated by the male ornaments (Andersson & Simmons, 2006). Of course, these mechanisms are not mutually exclusive.

The environment normally plays a significant role in the outcome of mate choice. This is in part because ornaments can evolve to signal male genetic or environmental quality (or both; Cotton et al., 2004). For instance large body size in both females and males is favoured by sexual selection because large traits generally correlate with fecundity or fertility (Curtis & Stoddard, 2003; Basolo, 2004). Yet, body size often varies with environmental factors such as temperature, food availability, social interactions, geographic distribution (Yom-Tov & Geffen, 2006), thus adaptive mate choice tracking changes in body size would uncover the environmental factors responsible for the size change, such as pollution with pesticides.

The consequences of pollution in water bodies are diverse, but can be grouped into two categories: 1) changes in abiotic parameters related to water quality such as transparency, dissolved oxygen and nutrients, and 2) presence of toxic substances. For example, eutrophication produces an increase in turbidity in which visual components of mate choice are affected. Seehausen et al. (1997) demonstrated that eutrophication has provoked biodiversity loss amongst native cichlid species in Lake Victoria because a reduced transparency has affected sexual selection based on coloration, which is responsible for maintaining reproductive isolation among sympatric species. Eutrophication also reduces the concentration of oxygen, and Reynolds & Jones (1999) observed that low oxygen condition produces a reversal of female mating preference in common gobies (*Pomatoschistus microps*). At normal concentrations of dissolved oxygen, goby females prefer males that already have eggs under their parental care over those without eggs, but female choice is reversed at 35% dissolved oxygen, presumably because large egg masses demand more oxygen than small clutches. Chemical compounds can also interfere directly with female choice in relation to their mechanism of action. For example, endocrine disrupting chemicals such as 17β-estradiol (E_2), 17α-ethinylestradiol, 4-*tert*-octylphenol (4OP), phthalates and alkylphenols interact with estrogenic receptors in the brain, causing reproductive behavioural changes, which would not be necessarily related to female choice. Bjerselius et al. (2001) observed a reduction in *following* behaviour in male goldfish (*Carassius auratus*) after exposure to 1μg E_2/L and 10μg E_2/L. An opposite effect was found by Toft & Baatrup (2003), who reported an increase in courtship behaviour in guppies (*Poecilia reticulata*) exposed to 100μg 4OP/L.

In the case of neurotoxic compounds such as OPs, several studies have reported effects on swimming behaviour (see above), however Arellano-Aguilar & Macías Garcia (2008) reported for the first time the non-neurotoxic effects of OP pesticides on intersexual selection in the Amarillo fish (*Girardinichthys multiradiatus*). This report showed that sub-lethal exposure to methyl parathion during embryonic development affects the condition of adults and their attractiveness to choosy females; particularly male ornaments (size and colour) and courtship display (details of experiment below). Following Arellano-Aguilar & Macías Garcia (2008) report, Cothran et al. (2010) could not find evidence that another OP, malathion, at sub-lethal concentration affects female mate choice in American amphipods (*Hyalella sp.*). However, there are good reasons to expect that any chemical compound that interfere with sexual signal expression would negatively influence mate choice. Accordingly, Secondi et al. (2009) found that realistic nitrate concentrations affect the sexual trait size in newts (*Triturus helveticus*), reducing the sexual attractiveness of males. Female newts showed a clear preference for unexposed males (Secondi et al, 2009), a behavioural response similar to that observed in the experiment with Amarillo fish. Clearly more research is needed to understand the ecological and evolutionary implications of the interaction between low concentrations of pollutants and sexual selection.

1.3 Demographic consequences

The demographic consequences for a population may not be apparent at the time of exposure, but become evident in the long run, even after several generations have passed. Pollutants may perturb the population dynamics or even irreversibly impair its survival. Walker et al. (2006) mentioned that in chronic pollution, populations of some species might decline and eventually become extinct, or they may recover, persist or even increases. To improve our capability to predict the possible outcome of pollution it is necessary to know the dynamic of the focal populations, through gathering accurate estimates of some demographic parameters such as sex ratio, age structure, age-specific fecundity and mortality, and rate of increase or growth rate. These parameters can be used to make reasonable predictions of the populations' growth or decline, and to determine the impact of particular life stages on the demographic behaviour of the population. They are obtained by directly counting (or sampling) the individuals and tallying their survival, mortality and fecundity. From those data indirect measures such as the sex ratio, the proportions of new born, young and adults and the population age structure are then calculated and used to estimate growth rates (Skalski et al., 2005).

One of the simplest ways to verify if a population is increasing or decreasing is subtracting the number of the individuals at a posterior time (N_{t+1}) against the current population size (N_t). The difference is the number of individuals added or lost in the next generation. It is to be expected that the number of individuals some time in the future (N_{t+1}) is affected by the population size (N), the number of births and of deaths in the current generation. It would be also necessary to determine whether there are immigrants to add and/or emigrants to subtract from the current population size, but it is often assumed for simplicity that populations are closed and that no immigration or emigration takes place.

In continuous population growth, where births (B) and deaths (D) are constantly occurring, the rate of change in population size during a given period (*dt*) is:

$$dN_t/dt=B-D \tag{1}$$

The instantaneous birth rate (*b*), or the number of births per individual per unit time is calculated by the total number of births divided by the total numbers of individuals, $b=N/B$. Similarly, the instantaneous death rate (*d*) is the division of the total number of deaths between the individuals, $d= N/D$. From these equations;

$$dN_t/dt=(b-d)N \tag{2}$$

Thus (*b-d*) represents the intrinsic or instantaneous rate of increase (*r*), which is the *per capita* rate of change, or growth rate, of the population per unit time. The value of *r* is important as it reflects the vitality and the potential of the population to increase. If the intrinsic rate of increase $r>0$, the population increases, if $r=0$, the population size remains stationary and $r<0$ indicates that the population is decreasing.

$$dN_t/dt= rN \tag{3}$$

A modification of the above is necessary when birth and death rates are not continuous. In many species births are discrete events that occur one or more times a year, at specific times or seasons (e.g. in the spring). In those species deaths may also occur at discrete periods in the year. In these cases *r* should be replaced by the finite rate of change (λ) which represents the proportional change in abundance from time *t* (Nt) to time t_{+1} (Nt_{+1}). The finite rate of increase (λ) is the growth rate per period of time (usually a year) and it is calculated as follows:

$$\lambda = Nt_{+1}/Nt \tag{4}$$

Just as before, $\lambda>1$ indicates that the population is increasing; $\lambda=1$ indicates a stationary population and $\lambda<1$ indicates a decreasing population. For example, $\lambda=1.17$ represents a 17%

of increase in the population between t and $t+1$.There are several methods to calculate r and λ, including the Lotka equation based on fecundity rate, and population projection Leslie matrices based on survival rates (Leslie 1945; Skalski et al., 2005).

These parameters (r and λ) are used commonly to describe the population growth and to compare growth rates among populations of the same or different species.

$$\lambda = e^r \text{, or equivalently, } r = \ln \lambda \tag{5}$$

To characterise demographic parameters such as absolute growth rates (r and λ) is not enough to ensure the conservation of a population. This also requires knowledge of the genetic status of the population, including how much genetic diversity it harbours, what is the degree of inbreeding amongst its members, whether it is subject to genetic drift, its mutation load and its effective population size (related to the mating system). The effective population size (N_e) is a crucial attribute: it is a genetic measure of the number of individuals needed to maintain the same genetic diversity between generations. As it is directly related to the rate at which genetic diversity is lost, N_e also measures the rate at which genetic drift affects the population (Rieman & Allendorf, 2001). In general, quantifying the amount of genetic diversity within a population or species is of the utmost importance in conservation genetics (Crandall et al., 1999), and no other genetic parameter affects the population demography as much as N_e, which is why the concept of effective population size has played an important role in the conservation genetics of endangered species.

Populations often have unequal numbers of males and females mating and contributing to the next generation. But, regardless of the total number of individuals, both sexes contribute an equal number of genes (Allendorf & Luikart, 2007). The effective size of a population with unequal numbers of males and females mating randomly is calculated as:

$$Ne = 4Nm\ Nf /\ Nm + Nf \tag{6}$$

Where Nm and Nf are the quantity of breeding males and females, respectively. Therefore, the effective population size (N_e) is slightly smaller than the absolute population size (N; Kimura & Crow, 1963). On the other hand, if there is sexual selection acting on the population, where mating is biased in favour of only a fraction of the members of one sex, one of the consequences is an even greater reduction of the number of reproductive adults that contribute to the next generation.

Although sexual selection reduces the probability that deleterious mutations get fixed (because fit individuals are chosen by females attracted by their ornaments), it also affects the population genetic structure by reducing the level of heterozygosity (Kimura & Crow, 1963; Futuyma, 1998). Strong directional selection means that some alleles become fixed and in a short time the population becomes monomorphic as it loses the additive genetic variance underlying the expression of traits (including those used in mate choice; Merilä & Sheldon, 1999). In those circumstances genetic diversity would be strongly influenced by inbreeding, genetic drift and genetic bottlenecks (Price, 2005), and populations would be therefore more vulnerable to mutation loads and demographic stochastic fluctuations (Kimura et al., 1963). For instance, Bailey et al. (2007) explored the genetic variability of wild and captive populations of two fish species with intense sexual selection. They found that low effective population size contributes to increasing the risk of local extinction. The effect is more severe when anthropogenic factors such as pollution, introduction of exotic species or overfishing are also present because both the population size and *Ne* are affected. In the next section we describe a project designed to understand the mechanism that may reduce *Ne* in a species under intersexual selection when it is also exposed to pollutants. Ecotoxicological studies have focused on demographic changes, using population growth rate as the key measure of the possible long-term effects of pollution. We attempted to gather evidence to support our view that the concept of effective population size (N_e) plays an even more important role in the management and conservation of threatened species.

2. A case of study

2.1 The organophosphate methyl parathion and the fish *Girardinichthys multiradiatus*

We shall illustrate how the exposure to an orgonophosphate -methyl parathion (MeP)- can exacerbate the demographic consequences of intersexual selection in the Mexican fish *Girardinichthys multiradiatus*. First we will briefly describe the peculiar attributes of the focal species and the pollutant involved. Then we will describe the different levels at which MeP adversely affects this fish. We will demonstrate how pesticides can affect the individual's phenotype, specifically the male ornaments, and how this consequently affects their attractiveness. We will then look into the effects at the population level; we will show the immediate demographic consequences and those that take place over successive generations of MeP exposure. Our data are mostly the result of experiments in which fish were exposed to different concentrations of MeP administered at different life stages. The study was carried out in a paradigm of artificial selection in a captive environment.

Our model is the topminnow *Girardinichthys multiradiatus*. This is a small fish that belongs to the Mexican Goodeidae, a group of viviparous fish in which sexual dimorphism is the norm. Males have large and more colourful fins than females and exhibit a complex courtship display (Macías Garcia, 1994). Females mating preferences are based on visual attributes like sexually dimorphic median fins, dimorphic colours and colour patterns (especially the yellow colour of the fins) and conspicuous courtship dances (Macías Garcia and Saborío, 2004). Fish in this species can detect UV light, and body reflectance in this wavelength is used also in female mate choice (Macías Garcia and Burt de Perera, 2002). *Girardinichthys multiradiatus* inhabits shallow areas in lakes, wetlands, streams and dams in the upper Lerma basin on the Mexican Central Plateau (Macías Garcia, 1994; Uribe, 2005). Its geographic distribution is restricted to about 11 sites (including streams and rivers) in the region of the Country where economic and industrial development is growing at a faster rate, besides being occupied in ca. 55% by agricultural activities (SNIARN 2006). Both their restricted distribution and reproductive biology -including viviparity and stringent female mate preferences- predispose the Mexican Goodeidae (or Goodeinae) to suffer from environmental fragmentation and other anthropogenic influences. In fact the principal factors of population decline in this family are desiccation and habitat destruction. Additionally, eutrophication, introduction of exotic species and pollution are also responsible for the ongoing increase in the rate of local extinction (De la Vega-Salazar & Macías Garcia, 2005). The Goodeinae encompass some 16 genera with ca. 42 species, of which 22 are in some category of risk (8 endangered, 9 threatened, 3 in extreme danger of extinction and 2 extinct in nature; Contreras-Banderas, 2005; De la Vega-Salazar et al., 2005). The International Union for the Conservation of Nature - IUCN Red list (1998) considers *G. multiradiatus* as vulnerable.

Through their natural range Goodeine fish are exposed to several pesticides, including the insecticide methyl parathion (MeP). This is an organophosphate largely used on a number of economically important agricultural crops such as cotton, soybean, corn, wheat, rice and bean (Garcia et al., 2003). The MeP used as pesticide reaches the ground and through percolation gets into the water bodies. MeP has a combination of chemical and physical properties that make it soluble in most organic solvents and also in some inorganic compounds, which is why this pollutant can easily enter into the aquatic systems. *Girardinichthys multiradiatus* is exposed to MeP in the field, where pregnant females show a negligible concentration in their tissues, about one order of magnitude below that in the tissues of the embryos they carry (De la Vega-Salazar et al., 1997).

2.2 What are the effects of early sub-lethal exposure of MeP in *G. multiradiatus*?

Two experiments were carried out to assess whether *G. multiradiatus* is negatively affected when exposed to sub-lethal concentrations of MeP; both involved contaminating a group of individuals with MeP and comparing the effects with a control group. The experimental design was planned *a priori* to reduce as much as possible the sources of variation and to increase the power of the analysis. Fish were maintained in aquaria at 25±2°C and under a 12h dark/12h light photoperiod. In the first experiment (part two) we used a paternal half-sib protocol because we presumed that secondary sexual traits are heritable in this species. In the first experiment we used the second generation of fish collected in the wild (at San Juanico; 19°55′ N, 99° 47′ W), which had been kept in outdoor ponds at the Institute of Ecology-UNAM, Mexico (details in Arellano-Aguilar & Macías Garcia, 2009).
First we calculated the LC_{50} and the bioaccumulation curve by feeding the fish 0.1 g of spiked food per day. Individuals at the start of the experiment were of the same age, weight and length. We used a gradient of MeP concentrations (0.001-0.165 µg MeP/g dry weight of food) for 30 days, and compared the effects with control fish kept under the same conditions and fed un-spiked food. Fish were measured (mm) and weighed (g) at the beginning and at the end of the experiment, and growth was calculated as the percentage of initial body length ([final length-initial length/initial length] x 100). As reported originally (Arellano-Aguilar & Macías Garcia, 2009) the accumulation of MeP in adult *G. multiradiatus* follows a dose-dependent relationship. Fish exposed to concentrations of 0.008, 0.02 and 0.04 µg of MeP for 30 days showed a reduction in size and weight. In addition, MeP was lethal at 0.165 µg/g. Bioaccumulation of MeP was measured in fish from each concentration using gas chromatography. This allowed us to determine the amount of MeP that should be added to the food given to pregnant females in order to mimic the concentrations reported in wild fish (De la Vega-Salazar et al., 1997). This procedure did in turn enable us to evaluate the consequences of pre-natal exposure to realistic concentrations on adult reproductive performance. As expected from sub-lethal concentrations, pregnant females exposed to 0.005, 0.01 and 0.1 µg/g did not show any negative effect, but they delivered varying proportions of malformed offspring. The main teratogenic effect was spinal cord malformation (scoliosis), which could be a consequence of inhibition of AChE during embryonic development. Around 60% of the total offspring exposed via mother to 0.005 µg/g MeP was apparently normal and survived to adulthood.

2.3 Are there consequences of early MeP in adult fish?

We examined the morphology and behaviour of adult fish which had been prenatally exposed to realistic concentrations of MeP. Adult males were presented with one matched-pair of non-pregnant females each. After 10 days, females were isolated in independent tanks, where they remained until parturition. One female from each pair was fed food containing a concentration of 0.005µg MeP/g (dry weight) and the other female was feed uncontaminated food. Shortly after birth each newborn brood was transferred to a MeP-free 40 l aquarium until sexes could be distinguished, when each individual was isolated in cylindrical compartments and placed in single-sex aquaria. Survival of newborn offspring prenatally exposed to 0.005µg MeP/g was comparable to that of control fish up to adulthood (90 days of age), yet morphology and behavioural performance were compromised. Dimorphic fins were less developed, colour duller, and reproductive behaviour (frequency and duration of courtship displays) less energetic in exposed than in control fish. The results (reported in Arellano-Aguilar & Macías Garcia, 2008) showed also that exposed fish were smaller than controls, and that the development of their ornamental fins did not resemble that of their fathers, whereas it was significantly heritable in their control half sibs (h_2=2m= 1.13±0.096).

2.4 Early exposure to sub-lethal dosages of MeP impairs male ornamentation; does it affect male attractiveness?

Having established that embryonic exposure to low concentrations of MeP negatively affects male ornamentation and courtship behaviour, we evaluated whether this would have reproductive consequences. Test were conducted in aquaria divided into three compartments, the central one being occupied by a female and the lateral compartments being used to place one control and one exposed male in each. Both experimental and control females were tested in their preferences. We recorded frequency and duration of female visit and frequency of copulation attempts for 20 min, and found that both types of females had significant preferences for control males, thus we conclude that exposed males were less attractive to female than their control half-siblings (Arellano-Aguilar & Macías Garcia, 2008).

2.5 Sub-lethal exposure to MeP affects the individual; are there population consequences over the long term?

The following experiment was designed to find out whether the exposure to 0.005µg of MeP had consequences at the population level and through generations. We used this concentration because it appears, from our experiments summarised above, that it has no direct effects on the individuals, but when pregnant females are exposed to this concentration, their offspring grow to be less attractive and court less vigorously than control fish. We collected pregnant females from a location in the upper Lerma basin (Salazar, 19° 18.33´ N, 99° 23.35´ W), and all the experiments were conducted on their progeny in the laboratory. Each generation was kept in aquaria only until the females gave birth, when all adults were removed and their descendants left to constitute the next generation. In this way we equalled one generation with one cohort. For example, the offspring born from females collected in the field (parental generation, or P), were separated from their mothers and were considered generation F1. We did not use the females from the field as the parental generation in order to avoid the potential effects of unrecorded variables. Members of F1 were distributed into a control and an experimental (exposed) group. Once they reached adulthood (at 90 days of age) experimental fish were exposed to food laced with 0.005µg of MeP/g of dry weight.

First we verified the results of the original experiments. As in the previous report (and in Arellano-Aguilar & Macías Garcia, 2008) we also measured morphological characters and

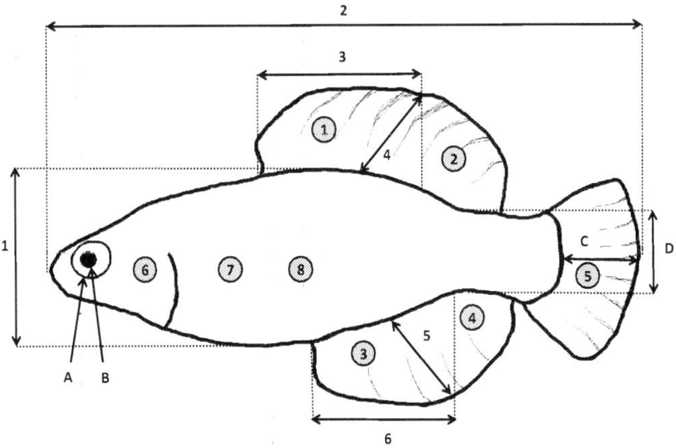

Fig. 3. Morphological measures of *Girardinichthys multiradiatus* and, within circles, the places where colour (reflectance) was measured by spectrophotometry. Numbers denote the measures that have been systematically found to influence female preferences, and letters indicate measures not known to influence female mate choice.

the attractiveness of the males, with the difference that here we did not use a half-sibling protocol, as we were interested in assessing the effects at the population level. Since the half-sibling protocol reduces the genetic variance between individuals, it is reasonable to expect differences between the two experiments. To confirm whether males exposed prenatally to sub-lethal concentrations of MeP grow to be morphologically unattractive adults, we followed the protocol used by González Zuarth & Macías Garcia (2006), and measured in males and females six attributes historically associated with female mate choice (1 to 6 in Fig. 3), and four not associated with it (A to D). The measures were then entered into a principal components analysis (PCA). This protocol was repeated for three generations.
The first component of the PCA (PC1; 74.18% variance explained) gave large loadings to measures related to fish size, whereas the second component (PC2; 20.84% variance explained) reflected sexual dimorphism. Together these components explained most of the morphological variance in our experiments (95.02%; Fig. 3). An ANOVA revealed that fish varied significantly in their PC1 scores ($F_{1,219}$=53.82, p=0.000). The difference between sexes in the PC1 was also significant ($F_{1,219}$=48.23, p=0.000), with females being normally larger as in most viviparous fish. A decrease in size (PC1) among the three generations was significant in females ($F_{1,143}$=57.05, p=0.000) and males ($F_{1,75}$=9.05, p<0.001), and there was also a significant interaction between generation and sex (generation X sex; $F_{1,219}$=5.33, p=0.005; Fig. 4) implying that females (whose size is tightly related to fecundity) become smaller in captivity. There was also a significant effect of generation in the PC2 of females ($F_{1,143}$=8.30, p<0.001), whose morphology became increasingly different from that of males.
More direct comparisons of size showed that the relative depth (the ratio of measures 1/2) decreased across generations in both females ($F_{1,143}$=26.80, p=0.000) and males ($F_{1,75}$=6.77, p<0.005), but the relative fin size decreased significantly only in males ($F_{1,75}$=20.03, p=0.000), where it is beyond the ecological optimum by the action of female mate choice (Macías Garcia et al., 1994). Unexpectedly, tail size decreased across generations in both females

($F_{1,143}$=35.14, p=0.000) and males ($F_{1,75}$=17.62, p<0.001). There was no significant difference in the eyes' measurements. All the differences over the three generations were due to a significant reduction in size. In addition, experimental females in the second generation were significantly smaller than control females ($F_{1,31}$=4.70, p=0.038; Fig. 4).

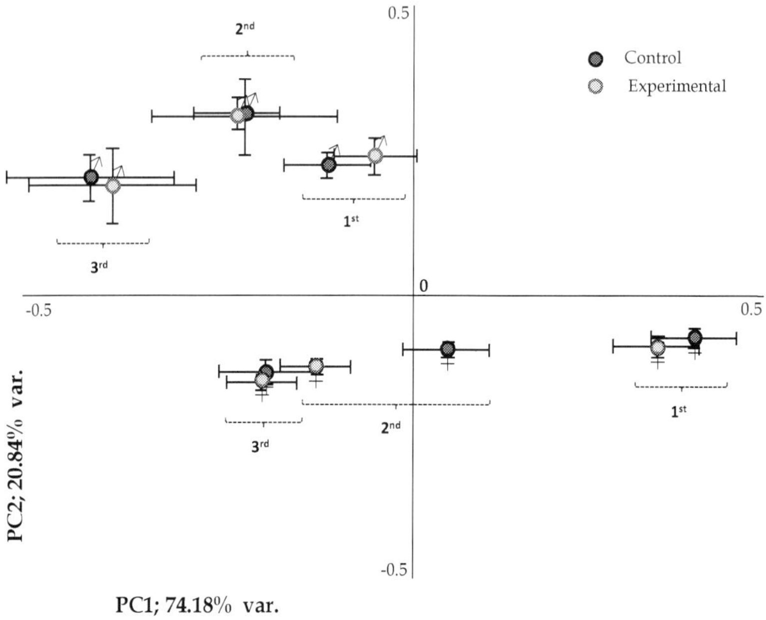

Fig. 4. Principal Component Analysis (PCA) of body measures. The first principal component captured most of the variance in size, and the second reflects sexual dimorphism. There was a decrease in size across generations in both sexes, and it was significantly large for females (the larger sex, where fecundity is related to size) in the second generation.

Fish colour was measured with a spectrophotometer and assessed with appropriate software (Spectra Suite). We analysed the UV (360-400 µm), yellow (500-630 µm) and total chroma (360-740 µm) reflectance measured at eight different regions of the body and fins (see Fig. 3). The areas measured and the regions of the spectrum selected were chosen following previous studies showing their relevance for sexual selection. The UV reflectance decreased significantly on the third generation ($F_{1,\ 227}$= 189.75, p= 0.000), in both sexes ($F_{1,\ 227}$= 24.72, p= 0.000), and was also lower in the experimental group ($F_{1,\ 227}$= 4.01, p= 0.046) than in control fish. The carotenoid (yellow) colour of the dimorphic fins showed an unexpected significant increase among generations; fin yellow reflectance was higher in the third generation in both males and females ($F_{1,\ 227}$= 208.57, p= 0.000). Still, prenatal exposure to MeP also affected dimorphic yellow chroma, with the fins of experimental males showing a reduced reflectance in the yellow region than those of control males ($F_{1,\ 227}$= 5.03, p= 0.026). Finally, total chroma decreased among generations ($F_{1,227}$= 16.91, p= 0.000) but was significant higher in control ($F_{1,227}$= 6.52, p= 0.001) than in experimental males.

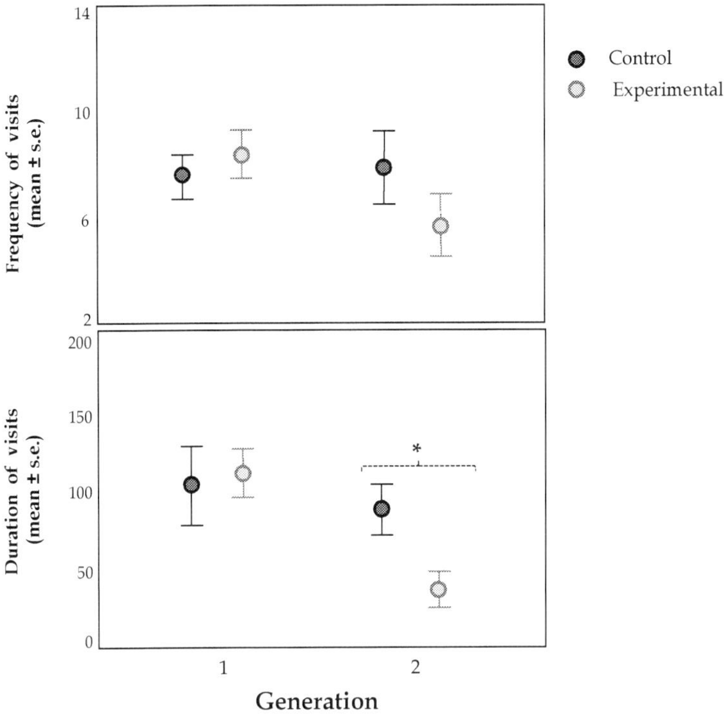

Fig. 5. Females in the first and second generation made as many visits to experimental and to control males, but their visits to the former were shorter in the second generation.

Attractiveness was as assessed by measuring female preference in dichotomous tests as before. We found (Fig. 5) that in the first generation there were no differences in the frequency ($t_{1,71}$=-1.01, p>0.05) or the duration ($t_{1,71}$=-0.30, p>0.05) of the visits by the females (control or experimental) to the males; females visited for as long, and as frequently, the experimental and the control males. The results in the second generation were different. Females again made the same number of visits to both types of male ($t_{1,34}$=3.90, p>0.05), but their visits to the control males were on average longer ($t_{1,196.12}$=2.01, p=0.047), and thus spent significantly more time next to control than to experimental males ($t_{1,34}$= 3.82, p=0.002). This seems to reflect a cumulative effect of the pesticide. The females did not discriminate between males in the first generation, but could distinguish in the second generation and preferred to remain associated with control males. It is interesting to remember that there were no differences in morphology between treatments in this generation, thus we propose that the female preference for control males in the second generation was due to differences in behaviour, in particular to courtship displays.

We followed two strategies to analyse the demographic data from this experiment. First we analysed the original data, and then we estimated population dynamics projections using a random re-sampling procedure. We did find a decrease in the population size over the generations, seemingly due selection to adapt to captive conditions. This pattern was confirmed by our re-sampling analysis, which seems to suggest, nevertheless, that the experimental populations decrease more rapidly than the control ones (Fig. 6). However, there were some demographic patterns that differ between treatments. The number of offspring born was similar in control and in experimental tanks in every generation ($F=0.74$, $p= 0.39$), but experimental offspring had a higher early mortality rate (< 1 month old) than control fish ($F=6.63$, $p= 0.011$). We also found that the frequency of fish with malformations (such as scoliosis, lordosis, etc.) in the experimental group was significantly higher than in the control group (Fig. 7; $F=5.52$, $p=0.020$).

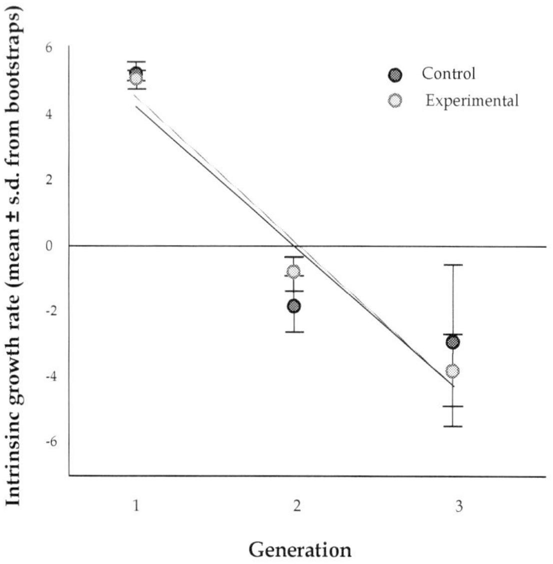

Fig. 6. The intrinsic growth rate of captive populations of *Girardinichthys multiradiatus* (confidence intervals from 1000 bootstrapped simulations) decrease from between generations due to selection to the aquarium conditions, but also in response to intra-ovary exposure to MeP. A similar pattern was observed in fecundity, which decrease between generations, was lower for exposed fish (and decreased faster for experimental fish).

3. Conclusions

It is uncertain wheter alternative technologies will ever be developed to help producing large crops without the need of pesticides. Genetic manipulation may hold a key, yet for the moment we are forced to use pesticides and to try to minimise their collateral, unintended effects. One traditional approach to assessing the risks inherent to pesticide use has involved measuring the consequences of exposure to lethal concentrations -at least to LC_{50} - and has largely negleted the effects of lower concentrations. Indeed, it is impractical to evaluate

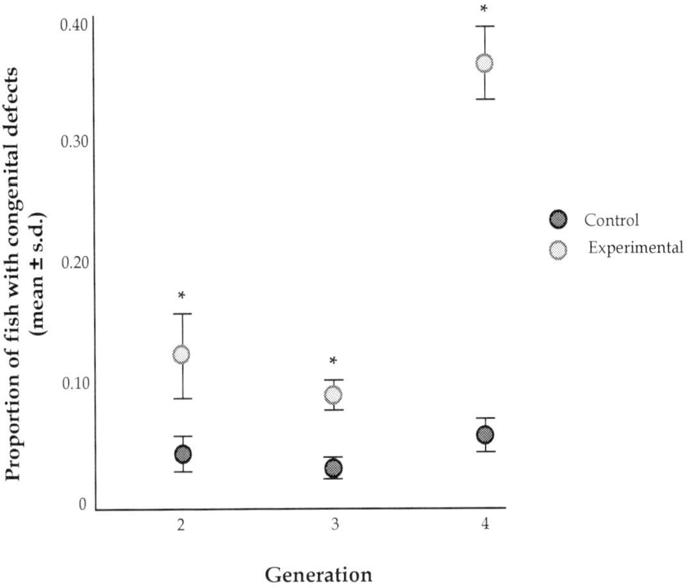

Fig. 7. The proportion of offspring born with deformities (mostly scoliosis) was significant larger (*) in fish exposed to MeP before birth than in controls. In the latter the incidence of those malformations remained below 10% in all generations, whereas in experimental fish it increased to more than one third in the last generation.

natural populations exposed to sub-lethal concentrations of pesticides, not least because effects of such exposure may be subtle, and subtle effects require large sample sizes to be detected. We propose that bioassays relying on ecologically relevant complex behaviours may facilitate the task. Unlike the relatively simple, reactive behaviours normally included in toxicological tests, the complex sequences of behaviour involved in courtship or foraging activities constitute the integrative outcome of many physiological and developmental processes, and are thus more likely to provide evidence of toxicological damage. Their study, however, requires consideration of Behavioural-Ecological theory, and the use of tools from Population Ecology and Genetics. Our initial results studying the effect of methyl-parathion on a sexually dimorphic fish are encouraging. We have shown that pre-natal exposure to a sub-lethal concentration of the insecticide increases the probability that offspring are born with defects. Surviving offspring grow to become unattractive adults, the males are shunned by their females and the females' fecundity is diminished. The effects accumulate through generations and together can compromise the survival of populations. This is likely to be a general effect in species where sexual selection is intense, and demonstrates the need to incorporate Behavioural Ecology and the study of sub-lethal effects in the growing field of Ecotoxicology.

4. Acknowledgements

Founding for this project was provided by the University's research program PAPIIT (grant IN221506) to CMG. IB and OAA were recipients of scholarships the Mexican Science Council (CONACyT). We thank E. Ávila Luna for logistical support, and M. E. Gonzalez, M. Cervantes and N. Villasana from the Animal Health Verification Center (CENAPA) for help with biochemical essays.

5. References

Abu-Qare, A.; Abdel-Rahman, A.; Kishk, A. & Abou-Donia, B (2000). Placental transfer and pharmacokinetics of a single dermal dose of [^{14}C] methyl parathion in rats, *Toxicological Sciences*, 53, 1, 5-12, ISSN 1096-6080

Allendorf, F. & Luikart, G. (2007). *Conservation and the genetics of populations*, Blackwell Publishing, ISBN 4051-2145-9, UK

Andersson, M. (1994). *Sexual Selection*. Princeton University Press, ISBN 0691 03344 7, Princeton, New Jersey, USA

Andersson, M. & Simmons, L. (2006). Sexual selection and mate choice, *Trends in Ecology and Evolution*, 21, 6, 296-302, ISSN 0169-5347

Arellano-Aguilar, O. & Macías Garcia C. (2008). Exposure to pesticides impairs the expression of fish ornaments reducing the availability of attractive males, *Proceedings of the Royal Society*, 275, 1640, 1343-1351, ISSN 0962-8452

Arellano-Aguilar, O. & Macías Garcia C. (2009). Effects of methyl parathion exposure on development and reproduction in the viviparous fish *Girardinichthys multiradiatus*, *Environmental Toxicology*, 24, 2, 178-186, ISSN 1520-4081

Baerends, G. (1976). The functional organization of behaviour, *Animal Behaviour*, 24, 726-738, , ISSN 0003-3472

Bailey, N.; Macías Garcia, C. & Ritchie, M. (2007) Beyond the point of no return? A comparison of genetic diversity in captive and wild populations of two nearly extinct species of Goodeid fish reveals that one is inbred in the wild, *Heredity*, 98, 6, 360-367, ISSN 0018-067X

Barron, M. (2002). Environmental contaminants altering behaviour, In: *Behavioural ecotoxicology*, Giacomo Dell'Omo (Ed.), 167-186, John Wiley & Sons Ltd., ISBN 0-471-96852-8, Chichester, West Sussex, UK

Basolo, A. (2004). Variation between and within the sexes in body size preferences, *Animal Behaviour*, 68, 1, 75-82, ISSN 0003-3472

Bjerselius, R.; Lundstedt-Enkel, K.; Olsén, H.; Mayer, I. & Dimberg, K. (2001) Male goldfish reproductive behaviour and physiology are severely affected by exogenous exposure to 17β-estradiol, *Aquatic Toxicology*, 53, 2, 139-152, ISSN 0166-445X

Bondarenko, S. & Gan, J. (2004). Degradation and sorption of selected organophosphate and carbamate insecticides in urban stream sediments, *Environmental Toxicology and Chemistry*, 23, 8, 1809-1814, ISSN 0730-7268

Bonfanti, P.; Colombo, A.; Orsi, F.; Nizzetto, I.; Andrioletti, M.; Bacchetta, R.; Mantecca, P.; Fascio, U.; Vailati, G. & Vismara, C. (2004). Comparative teratogenicity of chlorpyrifos and malathion on *Xenopus laevis* development, *Aquatic Toxicology*, 70, 3, 189-200, ISSN 0166-445X

Boone, S. & Chambers, J. (1996). Time course of inhibition of cholinesterase and aliesterase activities, and nonprotein sulfhydryl levels following exposure to organophosphorus insecticides in Mosquitofish (*Gambusia affinis*), *Fundamental and Applied Toxicology*, 29, 2, 202-207, ISSN 1096-6080

Börjeson, H., & Norrgren, L. (1997). M74 syndrome: a review of potential etiological factors, In: *Chemically induced alterations in functional development and reproduction of fishes*, Rosalind M. Rolland, R. M.; Gilberston, M.; & E. Peterson, R. E. (Eds.), (153-166), SETAC Technical Publications Series, ISBN 1-880611-19-8, Pensacola, FL., USA.

Castillo, C.; Montante, M.; Dufour, L.: Martinez, M.; Jimenez-Capdeville, M. (2002). Behavioural effects of exposure to endosulfan and methyl parathion in adult rats, *Neurotoxicology and Teratology*, 24, 6, 797-804, ISSN 0892-0362

Contreras-Balderas, S. (2005). Conservation Status of Mexican Freshwater Viviparous Fishes, In: *Viviparous Fishes*, Uribe, M. & Grier, H. (Eds.), 415-423, New Life Publications, ISBN 0-9645058-5-1, Mexico

Cothran, R.; Greco, R. & Relyea, R. (2010). No evidence that a common pesticide impairs female mate choice in a freshwater amphipod, *Environmental Toxicology*, 35, 3, 310-314, ISSN 1520-4081

Cotton, S.; Fowler, K. & Pomiankowski, A. (2004). Do sexual ornaments demonstrate heightened condition-dependent expression as predicted by the handicap hypothesis?, *Proceedings of the Royal Society*, 271, 1541, 771-783, ISSN 0962-8452

Crandall, K.; Posada, D. & Vasco, D. (1999). Effective population sizes; missing measures and missing concepts, *Animal Conservation*, 2, 4, 317-319, ISSN 1367-9430

Curtis, C. & Stoddard, P. (2003). Mate preference in female electric fish, *Brachyhypopomus pinnicaudatus*, *Animal Behaviour*, 66, 2, 329-336, ISSN 0003-3472

Darwin, C. (1871). *The descent of man and selection in relation to sex*, John Murray, Albemmarle Street, London UK

Das, P. & John, G. (1999). Induction of sister chromatid exchanges and chromosome aberrations in vivo in Etroplus suratensis (Bloch) following exposure to organophosphorus pesticide, *Toxicology Letters*, 104, 1-2, 111-116, ISSN 0378-4274

De la Vega Salazar, M.; Martínez-Tabche, L. & Macías Garcia, C. (1997). Bioaccumulation of methyl parathion and its toxicology in several species of the freswater community at Ignacio Ramirez dam in México, *Ecotoxicology and Environmental Safety*, 38, 53-62, ISSN 0147-6513

De la Vega-Salazar, M. & Macías Garcia, C. (2005). Principal Factors in the Decline of the Mexican Endemic Livebearing Fishes (Goodeinae: Goodeidae), In: *Viviparous Fishes*, Uribe, M. & Grier, H. (Eds.), 505-513, New Life Publications, ISBN 0-9645058-5-1, Mexico

Fanta, E.; Rios, S'A. F.; Romão, S.; Vianna, C.C.A. & Freiberger, S. (2003). Histopathology of the fish *Corydoras paleatus* contaminated with sublethal levels of organophosphorus in water and food, *Ecotoxicology and Environmental Safety*, 54, 2, 119-130, ISSN 0147-6513

Fisher, H.; Wong, B. & Rosenthal, G. (2006). Alteration of the chemical environment disrupts communication in a freswater fish. *Proceedings of the Royal Society B*, 273, 1951, 1187-1193, ISSN 0962-8452

Fisher, R. (1930). *The genetical theory of natural selection*. Oxford University Press, ISBN 0 19 850440 3, Great Clarendon Street, Oxford NY, USA

Futuyma, J.D. (1998). The evolution of behavior, In: *Evolutionary Biology*, Douglas J. Futuyma (Ed.), 586-601, Sinauer Associates, Inc., ISBN 0-87893-189-9, Sunderland, Massachusetts, USA.

Garcia, S.; Abu-Qare, A.; Meeker-O´Connell, W.; Borton, A.; Abou-Donia, M. (2003). Methyl parathion: a review of health effects, *Journal of Toxicology and Environmental Health*, Part B, 6, 2, 185-210, ISSN 1660-4601

González Zuarth, C. & Macías Garcia, C. (2006). Phenotypic differentiation and pre-mating isolation between allopatric populations of *Girardinichthys multiradiatus*. *Proceedings of the Royal Society*, 273,1584, 301-307, ISSN 0962-8452

IUCN. 2010. Red list of threatened species. http://www.iucn.org

Jokanović, M. (2001). Biotransformation of organophosphorus compounds, *Toxicology*, 166, 3, 139-160, ISSN 0300-483X

Jones, J. & Reynolds, J. (1997). Effects of pollution on reproductive behaviour of fishes, *Reviews in Fish Biology and Fisheries*, 7, 4, 463-491, ISSN 0960-3166

Kimura, M. & Crow, J. (1963). The measurement of effective population number, *Evolution*, 17, 3, 279-288, ISSN 0014-3820

Kimura, M.; Maruyama, T. Crow, F.J. (1963). The mutation load in small populations, *Genetics*, 48, 10, 1303-1312, ISSN 0016-6731

Lehner, N.P. (1996). A conceptual model of animal behavior, In: *Hanbook of ethological methods Second Edition*, Philip N. Lehner (Ed.), 19-23, Cambridge University Press, ISBN 0-521-55405-5, New York, NY., USA.

Leslie, P. (1945). On the use of matrices in certain population mathematics, *Biometrika*, 183-212, ISSN 0006-3444

Li, Q. (2007). New mechanism of organophosphorus pesticide-induced immunotoxicity, *Journal of Nippon Medical School*, 74, 2, 92-105, ISSN 1345-4676

Little, E. (1990). Behavioral toxicology: stimulating challenges for a growing discipline, *Environmental Toxicology and Chemistry*, 9, 1, 1-2, ISSN 0730-7268

Macías Garcia, C. & Burt de Perera, T. (2002). Ultraviolet-based female preferences in a viviparous fish, *Behavioral Ecology and Sociobiology*, 52, 1, 1-6, ISSN 0340-5443

Macías Garcia, C. & Saborío, E. (2004). Sperm competition in a viviparous fish, *Environmental Biology of Fishes*, 70, 211-217, ISSN 0378-1909

Macías Garcia, C. (1994). Social behaviour and operational sex ratios in the viviparous fish *Girardinichthys multiradiatus*, *Copeia*, 4, 919-925, ISSN 0045-8511

Macías Garcia, C; Jimenez, C. & Contreras, B. (1994). Correlational evidence of a sexually-selected handicap, *Behavioral Ecology and Sociobiology*, 35, 253-259, ISSN 0340-5443

Markman, S.; Leitner, S.; Catchpole, C. & Müller, C. (2008). Pollutants increase song complexity and the volume of the brain area HVC in a songbird, PLoS ONE, 3, 2, ISSN 1932-6203

Merilä, J. & Sheldon, B. (1999). Genetic architecture of fitness and nonfitness traits: empirical patterns and development of ideas, *Heredity*, 83, 2, 103-109, ISSN 0018-067X

Nimmo, R. & McEwen C. (1994). Pesticides, In: *Handbook or ecotoxicology*, Peter Calow (Ed.), 619-667, Blackwell Science Ltd., ISBN 0-632-03573-0, Main Street, Malden, USA

Parker, M. & Goldstein, M. (2000). Differential toxicities of organophosphate and carbamate insecticides in the nestling European starling (*Sturnus vulgaris*), *Archives of Environmental and Contamination Toxicology*, 39, 2, 233-242, ISSN 0090-4341

Price, D. (2005). The genetic considerations in the conservations of freshwater viviparous fish, In: *Viviparous Fishes*, Uribe, M. & Grier, H. (Eds.), 351-364, New Life Publications, ISBN 0-9645058-5-1, Mexico

Reynolds, J. & Jones, J. (1999). Female preference for preferred male is reversed under low oxygen conditions in the common goby (*Pomatoschistus microps*), *Behavioural Ecology*, 10, 2, 149-154, ISSN 1045-2249

Rieman, B. & Allendorf, F. (2001). Effective population size and genetic conservation criteria for bull trout, *North American Journal of Fisheries Management*, 21, 4, 756-764, ISSN 0275-5947

Sarma, S.; Nandini, S.; Gama-Flores, J. & Fernández-Araiza, M. (2001). Population growth of *Euchlanis dilatata* (Rotifera): combined effects of methyl parathion and food (*Chlorella vulgaris*), *Journal of Environmental Science Health, B*, 36, 1, 43-54, ISSN 0360-1234

Secondi, J.; Hinot, E.; Djalout, Z.; Sourice, S. & Jadas-Hécart, A. (2009). Realistic nitrate concentration alters the expression of sexual traits and olfactory male attractiveness in newts, *Functional Ecology*, 23, 4, 800-808, ISSN 0269-8463

Seehausen, O.; van Alphen, J. & Witte, F. (1997). Cichlid fish diversity threatened by eutrophication that curbs sexual selection. *Science* 277, 5333, 1808-1811, ISSN 0193-4511

Skalski, J.; Ryding, K. & Millspaugh, J. (2005). *Wildlife demography: analysis of sex, age, and count data*. Elsevier Academic Press, ISBN-13: 978-0-12-088773-6, USA

SNIARN (2006). El medio ambiente en México 2005: en resumen. Secretaria del Medio Ambiente y Recursos Naturales, ISBN 968-817-777-6, México

Toft, G. & Baatrup, E. (2003). Altered sexual characteristics in guppies (*Poecilia reticulata*) exposed to 17β-estradiol and 4-tert-octylphenol during sexual development, *Ecotoxicology and Environmental Safety*, 56, 2, 228-237, ISSN 0147-6513

Toft, G.; Edwards, T.; Baatrup, E. & Guillette, L. Jr. (2003). Disturbed sexual characteristics in male mosquitofish (*Gambusia holbrooki*) from a lake contaminated with endocrine disruptors, *Environmental Health Perspectives*, 111, 5, 695-701, ISSN 0091-6765

Uribe, M. & Grier, H. (2005). *Viviparous Fishes*. New Life Publications, ISBN 0-9645058-5-1, Mexico

Vittozzi, L.; Fabrizi, L.; Consigliko, E. & Testai, E. (2001). Mechanistic aspects of organophosphorothionate toxicity in fish and human, *Environment International*, 26, 3, 125-129, ISSN 0160-4120

Walker, C. (1998). Biomarker strategies to evaluate the environmental effects of chemicals, *Environmental Health Perspectives*, 106, 2, 613-620, ISSN 0091-6765

Walker, C.; Hopkin, S.; Sibly, R. & Peakall, D. (2006). Biomarkers, In: *Principles of ecotoxicology*, C. Walker, S. Hopkin, R. Sibly, D. Peakall (Eds.), 159-160, Taylor & Francis Group, ISBN 978-0-8493-3635-5, Boca Raton, FL., USA.

Welsh, J.M. & Hanselka, W.C. (1972). Toxicity and sublethal effects of methyl parathion on behavior of Siamese fighthing fish (*Betta splendens*), *The Texas Journal of Science*, 23, 4, 519-529, ISSN 0040-4403

Yom-Tov, Y. & Geffen, E. (2006). Geographic variation in body size: the effects of ambient temperature and precipitation., *Oecologia*, 148, 2, 213-218, ISSN 0029-8549

Zahavi, A. (1975). Mate selection- a selection for a handicap, *Journal of Theoretical Biology*, 53, 205-214, ISSN 0022-5193

Part 3

Pesticides Degradation and Disposal

27

Pesticides in the Environment: Impacts and Their Biodegradation as a Strategy for Residues Treatment

Ma. Laura Ortiz-Hernández, Enrique Sánchez-Salinas, Angeluz Olvera-Velona and Jorge Luis Folch-Mallol
Universidad Autónoma del Estado de Morelos
México

1. Introduction

A vast number of pollutants and waste materials including heavy metals are disposed into the environment per annum. Approximately 6 x 10^6 chemical compounds have been synthesized, with 1,000 new chemicals being synthesized annually. Almost 60,000 to 95,000 chemicals are in commercial use. According to Third World Network reports, more than one billion pounds (450 million kilograms) of toxins are released globally in air and water. The contaminants causing ecological problems leading to imbalance in nature is of global concern. The environmentalists around the world are trying to overcome this by several means. However, although they are raising their voices at international platforms regarding the depletion of natural resources; little attention is given to their words and many substances are still used without caring of the adverse consequences (Shukla et al., 2010).

Among these chemicals we can find pesticides, which are defined as any substance or mixture of substances which are used to control destructive pests such as insects, plant disease organisms and weeds, including many other living organisms such as nematodes, arthropods other than insects, and vertebrates that endanger our food supply, health, or comfort. In particular, the term pesticide refers to chemical substances that alter biological processes of living organisms deemed to be pests, whether these are insects, mould or fungi, weeds or noxious plants. Pesticides are widely used in most areas of crop production to minimize infestations by pests and thus protect crops from potential yield losses and reduction of product quality (Damalas, 2009). These pests potentially cause damage or interfere in any other way in the production, elaboration, storage, transport, or commercialization of food, agricultural products and wood products or animal food. Pesticides may be given to animals to prevent insects, arachnids or other plagues in or over their bodies (FAO, 2002).

Certainly, pesticides have improved longevity and quality of life, chiefly in the area of public health. Insect control programs have saved millions of lives by combating diseases such as malaria, yellow fever and typhus. In addition, the use of pesticides constitutes an important aspect of modern agriculture, as they are absolutely necessary for economical pest management (Gouma, 2009). The most promising opportunity for maximizing benefits and minimizing risks is to invest time, money, and effort into developing a diverse toolbox of pest control strategies that include safe products and practices that integrate chemical approaches into an overall and ecologically based framework which will optimize sustainable production, environmental quality, and human health (Damalas, 2009).

However, the application of pesticides may cause adverse effects among the different forms of life and among the ecosystems; this will depend on the sensibility grade of the organisms and the pesticide (CICOPLAFEST, 2004). Approximately 90% of agricultural pesticide application never reaches its target organisms but is, instead, dispersed through the air, soil, and water. As a result, they are routinely detected in air, surface and ground water, sediment, soil, vegetable, and to some extent in foods. In addition, many soil-applied pesticides are also intentionally introduced into the soil environment for the control of soil borne pests and pathogens, which results in the accumulation of their residues and metabolites in soil at unacceptably high levels (Gamón et al., 2003; Shalaby & Abdou, 2010).

The inadequate application practice is one of the most important ways of pollution, which has a profound impact not only on the soils of the areas in which they are applied. Pesticides are washed into aquatic ecosystems by water runoff and soil erosion. Pesticides also can drift during application and contaminate aquatic systems. Some soluble pesticides are easily leached into streams and lakes. Wild birds and mammals are damaged and destroyed by pesticides and these animals make excellent "indicator species". Deleterious effects on wildlife include death from the direct exposure to pesticides or secondary poisonings from consuming contaminated food; reduced survival, growth, and reproductive rates from exposure to sub-lethal dosages; and habitat reduction through the elimination of food resources and refuges. Pesticides easily find their way into soils, where they may be toxic to arthropods, earthworms, fungi, bacteria, and protozoa. Small organisms are vital to ecosystems because they dominate both the structure and function of ecosystems. Like pest populations, beneficial natural enemies and biodiversity (predators and parasites) are adversely affected by pesticides. Fungicides also can contribute to pest outbreaks when they reduce fungal pathogens that are naturally parasitic on many insects. When outbreaks of secondary pests occur because their natural enemies are destroyed by pesticides, additional and sometimes more expensive pesticide treatments have to be made in efforts to sustain crop yields. This raises the overall costs and contributes to pesticide-related problems. In addition to destroying natural enemy populations, the extensive use of pesticides has often resulted in the development and evolution of pesticide resistance in insect pests, plant pathogens and weeds (Pimentel, 2005; Aktar et al., 2009).

Besides, pesticide application generates social conflicts because of the elevated number of workers who are intoxicated by these products, with a high mortality rate, as well as for the suspicion of adverse effects on the health of surrounding communities, flora and fauna (Lichtinger *et al.* 2001). These toxic compounds have been implicated in various disorders and diseases including cancer, adverse reproductive outcomes, peripheral neuropathies, neurobehavioral disorders, impaired immune functions and allergic sensitization reactions, particularly of the skin, cumulative inhibition of cholinesterase activity because of long-term low doses of exposure (Al-Qurainy & Abdel-Megeed, 2009).

The metabolic fate of pesticides is dependent on abiotic environmental conditions (temperature, moisture, soil pH, etc.), microbial community or plant species (or both), pesticides characteristics (hydrophilicity, level of solubility) and biological and chemical reactions. Abiotic degradation is due to chemical and physical transformations of the pesticide by processes such as photolysis, hydrolysis, oxidation, reduction and rearrangements. Furthermore, pesticides may be biologically unavailable because of compartmentalization, which occurs as a result of pesticide adsorption to soil and soil colloids without altering the chemical structure of the original molecule. However, enzymatic transformation, which is mainly the result of biotic processes mediated by plants and microorganisms, is by far the major route of detoxification (Van Eerd et al., 2003).

Another important concern is that the millions of tons of pesticides applied annually in a worldwide level generate liquid and solid wastes. Besides, containers many times are placed directly and without control on the soil and water mainly, polluting and affecting the food chains (Ortiz-Hernández et al., 1997). Among the main consequences derived from the soil pollution, we find the loss of fertility, which directly or indirectly allow the survival of the flora and fauna, given the tight interrelationships among the different elements, which constitute the ecosystems.

On the other hand, there are more than half a million tons of obsolete, unused, forbidden or outdated pesticides, in several developing and transitional countries, which endanger the environment and health of millions of people. In the absence of a clear obsolete pesticides management strategy, over the years, significant amounts of obsolete pesticides have been stockpiled in developing countries (Dasgupta et al., 2010).

An obsolete pesticide may be recognized as one that is undesirable or impossible to use and has to be eliminated, these include (Martinez, 2004; Karstensen et al., 2006; Shah & Devkota, 2009; Dasgupta et al., 2010):

- Technical pesticides and formulations passing the expiration date (generally two years after their manufacture).
- Pesticides whose use has been forbidden or strictly restricted.
- Damaged products:
  - Those who suffered physical or chemical changes which make them phytotoxic for the crops, or with non acceptable dangerousness for human and environment health.
  - Those who suffered a loss of biological efficiency.
  - Those who present changes in their physical properties which make them incompatible with the habitual equipments of application.
- Pesticides which are undesired by their owners, even if they are in good conditions for their use.
- Products without identification.
- Products which are polluted with other substances.

It is also included:

a. Pesticide wastes generated on fire and other accidents.
b. Materials that are strongly polluted with pesticides.
c. Wastes that are generated by the fabrication or formulation of pesticides.

Because of their characteristics, obsolete pesticides are a dangerous waste, that is why they should be managed as such. Obsolete pesticides have accumulated in almost every developing country or economy in transition over the past several decades (Dasgupta et al., 2010). It is estimated that in Africa and Middle East there are more than 100,000 tons of these products, in Asia almost 200,000 and a similar quantity in East Europe and the old Soviet Union. Nowadays the FAO is elaborating the inventories of Latin America (Farrera, 2004; Karstensen et al., 2006; Ortiz-Hernández & Sánchez-Salinas, 2010).

In México, there is knowledge of the existence of obsolete pesticide products, both liquid and solid. A total of 551 records of obsolete pesticide products have been registered, distributed in 29 of 33 states of Mexico, achieving a total of 26,725.02 liters, 147,274 kg and 500 m^3 of highly polluted soils. In addition there are 28 reports of pesticide-contaminated sites in 15 states of the Mexican Republic (Giner de los Ríos, 2007). Besides, some data indicate that the total of empty pesticide containers can be about 7,000 tons annually (Albert, 2005).

Many works point to health or environmental problems from accidental or deliberate exposure to pesticides, particularly pesticides with high mammalian toxicity or those that persist in the environment. These risks should not be ignored, and efforts must be made to minimize them through rigorous regulation and proper training for users, because we should not overlook the positive impacts of pesticide use. When pesticides are used rationally and carefully, in conjunction with other technologies in integrated pest management systems, it is more likely that their use will be justifiable (Cooper & Dobson 2007).

2. The need for pesticides treatment and final disposition: biotechnological strategies

The damages caused to the environment and health, such as the existence of obsolete pesticides, make necessary the development of technologies that guarantee their elimination in a safe, efficient and economic way. Among the existent technologies there are those that apply physical treatments, such as adsorption and percolator filters; chemical treatments such as the advanced oxidation or inverse osmosis, and incineration, a treatment option usually not available in developing countries (Karstensen et al., 2006). However, a treatment that promises to be efficient, economic and safe is the biological treatment, because several reactions catalyzed by enzymes of specific microorganisms take place. This kind of treatment has been approached from a biotechnological point of view in order to be able to have a methodology that is safer and more economic than the conventional treatments, as well as avoiding additional damages to the environment. Biological processes have been used to give treatment to wastes and polluted sites with pesticides (Araya & Lakhi, 2004). Among them, the microbial metabolism is the primary force of transformation or degradation. In many cases it has been reported that the microorganisms are very important in the degradation of xenobiotic compounds. Biological treatment can be applied to compounds whose chemical structure in the nature is infrequently or inexistent because they are synthesized artificially (Ortiz-Hernández, 2002). The importance of the microorganisms lays on their great diversity and metabolic plasticity, which allows them to use diverse ecological niches. Many microorganisms may live in a large diversity of media because of their remarkable capacity of mutation and adaptation; besides, they seem to have a great potential to acquire capacities of degradation when they are exposed to xenobiotics. Additionally, isolated microorganisms with the capacity to degrade xenobiotic compounds have the potential to be used for the bioremediation of other compounds that do not have any known microbial system for their degradation (Singh & Walker, 2006). Biodegradation of these pesticides provides a cheap and efficient solution for their final disposal or for treatment of agricultural soils, contaminated water or polluted ecosystems. In 1973, the first bacteria with the capability of degrading organophosphorus compounds were described (Sethunathan, 1973; Sethunathan & Yoshida, 1973, Siddaramappa et al., 1973). Since then, a number of different genera have been identified, and the enzymes involved in pesticide degradation have been studied.

3. Obtaining isolates that degrade pesticides

The role of microorganisms in the dissipation of pesticides, especially in the soil, has long been recognized. Relative to the extended evolutionary period of microorganisms in nature, agriculture has only been around for about 10,000 years, and the introduction of organic pesticides is only a half-century old. Therefore on an evolutionary scale, the time for microbial adaptation for degrading the influx of new xenobiotic compounds is exceptionally short, and it is an ongoing process. There is much evidence from observational and molecular research that indicate microbial adaptation for the mineralization of pesticide has occurred since their first introduction into agriculture in the mid-1950´s.

The use of pesticides over time, has resulted in several microorganisms which are able to degrade xenobiotic organic compounds, including pesticides, using different strategies and enzymatic pathways. The first organophosphate pesticide-degrading bacterial strain was isolated from a paddy field in the Philippines in 1973. Since then, several phylogenetically distinct bacteria that can degrade pesticides by co-metabolism, or use them as a source of carbon, phosphorus or nitrogen, have been isolated from different parts of the world (Table 1).

Because pesticides are mainly applied to agricultural crops, soil is the medium that mostly gets these chemicals. Once in the soil, pesticides follow different ways such as degradation, volatilization, sorption, or surface transport to other places. Studies have shown that biodegradation is a process that majority occurs in pesticide dissipation, which is due to the adaptation of microorganisms after having been in contact with soils with extensive exposure to pesticides. Some findings indicate that in most cases the half-lives of pesticides in soils with a history of pesticide application are considerably shorter than in those in which no application has been performed. Besides, the pesticides that were considered non-biodegradable become biodegradable after a number of years. Therefore, the soil is a major source to isolate microorganisms that degrade pesticides (Table 1), which is an opportunity for use in the treatment of waste or obsolete pesticides.

Other sources of microorganisms with the ability to degrade pesticides are: pesticide industry's effluent- sediment, sewage sludge, activated sludge, wastewater, natural waters, sediments, areas surrounding the manufacture of pesticides, and even some live organisms. In general, microorganisms that have been identified as pesticide degraders have been isolated from a wide variety of sites contaminated with some kind of pesticide.

For the isolation of soil microorganisms, the most used procedure is as follows. Microbial population in a sample of soil is cultivated in constant agitation in flasks of different capacities with mineral salts medium using a particular pesticide (or mixture of pesticides) as the only source of carbon. This procedure is repeated several times, increasing the concentration of pesticides to ensure the adaptation of microorganisms to the conditions of the culture in the laboratory as well as the growth of those that used the pesticide as their only source of carbon. After the adaptation time, several bacterial consortiums are obtained. These consortiums do not need an additional source of carbon, since the pesticide they were adapted to is enough. At present, in different laboratories around the world, there are collections of microorganisms characterized by their identification, growth and degradation of pesticides. The isolation and characterization of microorganisms that are able to degrade pesticides give the possibility to count with new tools to restore polluted environments or to treat wastes before the final disposition.

Microorganism	Pesticide	Place of isolation	Reference
Bacteria			
Ochrobactrum sp.	Methyl parathion	Soil	Qiu et al., 2006
Arthrobacter sp.	Endosulfan	Soil	Weir et al., 2006
Sphingomonas spp.	Isoproturon	Soil	Bending & Rodríguez, 2007
Burkholderia sp.	Fenitrothion	Soil	Hong et al., 2007
Sphingomonas sp.	Chlorpyrifos	Wastewater	Li et al., 2007
Enterobacter spp.	Chlorpyrifos	Soil	Singh et al., 2004
Acinetobacter radioresistens	Methyl parathion		Liu et al., 2007
Ochrobactrum sp., *Castellaniella* sp., *Variovorax* sp., *Pseudomonas* sp., *Psychrobacter* sp.	Igepal CO-210 Igepal CO-520	Sewage sludge	DiGioia et al., 2008
Pseudomonas frederiksbergensis	Dimetoate, Malathion	Soil	Al-Qurainy & Abdel-Megeed, 2009
Bacillus pumilus	Chlorpyrifos	Soil	Anwar et al., 2009
Bacillus sp.	Mesotrione	Soil	Battison et al., 2009
Serratia liquefaciens, *Serratia marcescens*, *Pseudomonas* sp.	Diazinon	Soil	Cycón et al., 2009
Enterobacter aerogenes	Bifenthrin Fenpropathrin Cypermetrine	Sewage sludge	Liao et al., 2009
Pseudomonas putida, *Burkholderia gladioli.*	Prophenofos	Soil	Malghani et al., 2009
Stenotrophomonas sp.	DDT	Soil	Mwangi et al., 2010
Providencia stuartii	Chlorpyrifos	Soil	Rani et al., 2009
Pseudomonas putida	Propiconazole	Tea rhizosphere	Sarkar et al., 2009
Micrococcus sp.	Diuron	Diuron storage	Sharma et al., 2010
Sphingobium sp.	Methyl parathion Fenpropathrin	Sewage sludge	Yuanfan et al., 2010

Table 1. Some organisms isolated that degrade different pesticides.

Fungus			
Aspergillus niger	Endosulfan	Soil	Bhalerao & Puranik, 2007
Ganoderma australe	Lindane	*Pinus pinea* stump	Rigas et al., 2007
Trichosporon sp.	Chlorpyrifos	Sewage sludge	Xu et al., 2007
Verticillium sp. DSP	Chlorpyrifos	Soil	Fang et al., 2008
T. versicolor (R26)	Atrazine	Soil	Bastos & Magan, 2009
Aspergillus sydowii, Bionectria sp., *Penicillium miczynskii, Trichoderma* sp.	DDD	Marine sponge	Ortega et al. 2010
Algae			
Chlorophyceae sp., *Scenedesmus* spp., *Chlamydomonas* sp., *Stichococcus* sp., *Chlorella* sp. Cyanobacteria *Nostoc* spp.	Fenamiphos	Soil	Cáceres et al., 2008
Anabaena sp.	Fenamiphos	Water	Cáceres et al., 2008
Yeast			
Lipomyces kononenkoae	Picloram	Soil	Sadowsky et al., 2009

Table 1. (Continued)

4. Microbial degradation of pesticides: genes and enzymes

Microbial metabolism has proved to be very versatile and diverse. This characteristic has allowed many different bacterial and fungal genera to evolve activities capable of xenobiotic degradation and offers an important source of alternatives for bioremediation.

Bioremediation of pesticides provides a cheap and efficient solution for their final disposal or for treatment of agricultural soils, contaminated water or polluted ecosystems. Microbial degradation has advantages because a large variety of compounds can be degraded completely under mild conditions compared with degradation using physical and chemical means.

Most of the research regarding pesticide degradation by microorganisms has been performed mainly with bacteria; a few studies have focused on fungi, actinomycetes, cyanobacteria, etc. This obeys mainly to the fact that bacteria are easy to culture in simple media and grow faster than other microbes; besides, bacteria are more susceptible to genetic modifications, which give them an extra potential to increase their degradation capabilities.

Bacterial genetics and molecular biology tools have contributed widely to the understanding of the degradation processes and to the isolation and characterization of genes involved in pesticide degradation. It is important to note, however, that in nature pesticide mineralization is accomplished by microbial communities rather than isolated species and that in many instances co-metabolic pathways are used.

Since the chemical structure of pesticides is variable, individual reactions of degradation-detoxification pathways are versatile and include oxidation, reduction, hydrolysis, and conjugation. These reactions are achieved through a number of different enzymes such as dehydrogenases (Bourquin, 1977; Singh & Singh, 2005), dioxigenases (Nadeau et al., 1994; Van Eerd et al., 2003), cytochrome p450 (Castro et al., 1985; Jauregui et al., 2003), ligninases (Pizzul et al., 2009) and, in the case of organohalogenate compounds, dehalogenases (Franken et al., 1991; Sharma et al., 2006). Conjugation with glutathione is commonly used as a detoxification mechanism, especially in plants and insects, although this mechanism has also been reported in bacteria (Vuilleumier, 2001; Wei et al., 2001; Chaudhry et al., 2002).

Metabolism of pesticides may involve a three-phase process. In Phase I metabolism, the initial properties of a parent compound are transformed through oxidation, reduction, or hydrolysis to generally produce a more water-soluble and usually a less toxic product than the parent. The second phase involves conjugation of a pesticide or pesticide metabolite to a sugar or amino acid, which increases the water solubility and reduces toxicity compared with the parent pesticide. The third phase involves conversion of Phase II metabolites into secondary conjugates, which are also non-toxic. In these processes fungi and bacteria are involved producing intracellular or extra cellular enzymes including hydrolytic enzymes, peroxidases, oxygenases, etc (Van Eerd et al., 2003).

Here we will revise some aspects of microbial degradation of the most important groups of pesticides.

4.1 Organochlorines

Although organochlorine pesticides are less used every day and many countries have banned their application, in some countries they still represent a problem of disposal, since this group of pesticides is the most persistent. There are two major pathways through which microorganisms degrade organochlorine compounds: reductive dechlorination, a process that takes place under anaerobic conditions, and dehydrochlorination, occurring in the presence of oxygen. Several bacterial genera have been proven to undertake these reactions including *Klebsiella* (Kwon et al., 2005), *Alcaligenes* (Don & Pemberton, 1981) *Staphylococcus*, (Sonkong et al., 2008), and *Pseudomonas* (Barragan-Huerta et al., 2007).

Several reports have documented the capability of different genera of fungi to degrade organochlorines. Among these, basidiomycetes seem to be more resistant to these compounds (Gomes Machado et al., 2005; Rigas et al., 2005). Recently a strain of *Trichoderma harzianum* has also been shown to degrade organochlorines through an oxidative system (Katayama & Matsumura, 2009).

4.2 Organophosphates

This group of pesticides have been widely used as pesticides or chemical warfare agents because of their high toxicity towards insects, mammals and other animals. Their mechanism of action involves the irreversible inhibition of acetylcholine esterase, a key enzyme of the central nervous system, thus affecting non-target organisms as well (Singh & Walker 2006). Despite this, they are still used worldwide in large quantities as pesticides. Other organophosphorus compounds, such as nerve gases are still stored in important amounts and have to be destroyed. The basic structure of organophosphorus pesticides consists of ester or thiol derivatives of phosphoric, phosphonic or phosphoramidic acids.

The main degradation pathway starts with the hydrolysis of the P-O alkyl or P-O aryl bonds, which diminishes as much as a 100 times the toxicity of these compounds. Bacterial enzymes have been found to achieve such detoxifying reactions (Singh et al., 1999; Yañez-Ocampo et al., 2009; Yañez-Ocampo 2009). This reaction is performed by esterases or phosphotriesterases that have been described for a number of different genera of bacteria and fungi. Among these, several different genes have been described (Singh & Walker, 2006).

A large group of bacterial genera has been reported to degrade organophosphates compounds. The studied and reported enzymes are related to the phosphotriesterase, which is capable of hydrolyzing organophosphate pesticides in the central atom of pesticides' phosphorus. Hydrolysis is fundamental for the complete degradation of the molecule. Phosphotriesterase activity is the first and most important step in detoxification. Several genes encoding for different phosphotriesterase activities have been described from a number of organisms (including fungi), the most studied being the *opd* and *opaA* genes that code for organophosphorus hydrolase (OPH) and organophosphorus acid anhydrolase (OPAA), respectively (Singh & Walker 2006). *opd* genes have been described in *Flavobacterium* and *Pseudomonas* species and are plasmid borne (Serdar et al., 1982; Somara & Siddavattam, 1995), while a similar gene, *opdA*, is present in *Agrobacterium radiobacter*'s chromosome (Horne et al., 2002). *opaA* genes were found in *Alteromonas* species and are chromosomally located (Cheng et al., 1996; Cheng et al., 1997). Another gene coding for organophosphates hydrolase, *mpd*, described originally in *Plesiomonas* sp. (Zhongli et al., 2001), has also been found in other genera like *Achromobacter, Pseudaminobacter, Ochrobactrum* and *Brucella* and is located in the chromosome (Zhang et al., 2005). Other important bacterial genera able to degrade organophosphates include *Burkholderia* (Zhang, et al., 2006) and *Hyphomicrobium* (Wang et al., 2010).

Some fungal species have been reported to degrade organophosphates. Amitai et al., (1998), reported that laccase, a broad spectrum phenol oxidase from the white rot fungus *Pleurotus ostreatus*, could hydrolyze P-S bonds, which are resistant to bacterial phopshotriesterase hydrolysis; other organophosphate compounds such as nerve gases could also be hydrolyzed by this enzyme (Amitai et al., 1998). It remains to be explored if laccases from other species can also hydrolyze these compounds, since their main mechanism is an oxidative one.

Ascomycetes such as *Aspergillus* and *Penicillium* have also been shown to produce organophosphate degrading enzymes other than laccases (Liu et al., 2001; Liu et al., 2004). The *Penicillium* enzyme (P-OPH) is more versatile since it can hydrolyze P-S and P-O linkages whereas the *Aspergillus* A-OPH only splits P-S bonds and the bacterial enzymes can only hydrolyze P-O bonds.

4.3 Carbamates

Carbamates are used to control insects and nematodes in soils. Their toxicity to mammals is very high although they are not as persistent as organochlorines. A number of bacterial genera have been identified as carbamate degraders (Parekh et al., 1995). Degradation of the pesticide occurs mainly through the hydrolysis of the methylcarbamate linkage by an enzyme called carbofuran hydrolase, codified by the *mcd* gene, which was located on a plasmid first described in *Achromobacter* sp. (Tomasek & Karns, 1989). Further studies showed that a wide variety of bacteria could degrade carbamates using carbofuran hydrolase. Among other genera *Pseudomonas, Mesorhizobium, Ralstonia, Rhodococcus, Ochrobactrum,* and *Bacillus* are the most notorious (Desaint et al., 2000).

Fungal degradation of carbamates has also been reported. Of special interest is the report a novel hydrolase from *Aspergillus niger* capable of hydrolysing several N-methylcarbamate insecticides (Qing et al., 2006).

4.4 Pyrethroids

Pyrethroids insecticides are a class of lipophilic esters, with an alcohol and an acid moiety. Although less toxic and persistent than other groups of insecticides, they can still represent a problem. Pyrethroids display high affinity to Na+-channels and its binding to these channels causes a prolonged channel opening that may result in a complete depolarisation of the cell membrane thus blocking neuronal activity. There are two main routes of degradation, photo- and biodegradation, which are often superimposed. Pyrethroids developed for use in agriculture are much more photostable than the natural pyrethrins but they are still sensitive to sunlight, which provokes isomerisation or ester cleavage. The basic pathway of pyrethroid degradation by microbes proceeds through the hydrolysis of the main ester linkage. This reaction can be carried out by carboxyl esterases or phosphotriesterases.

Bacterial degradation of pyrethroids has been documented. A report by Grant et al. (2002) describes the ability to degrade synthetic pyrethroids by *Serratia* and *Pseudomonas* isolates. *Bacillus, Achromobacter* and *Pseudomonas fluorescens* have also been studied in regard to pyrethroid degradation (Maloney et al., 1988). Another interesting report shows the degradation of allethrin, a recalcitrant pyrethroid used in mosquito mats by an *Acidomonas* sp. strain (Paingankar et al., 2005).

Fungal degradation of pyrethroids seems promising. Ascomycetes (*Trichoderma, Aspergillus*) as well as basidiomycetes (*Phanerochaete*) have been reported to degrade pyrethroids through the cleavage of ester bond (Saikia and Gopal, 2004). A novel pyrethroid hydrolase from cell extracts of *Aspergillus niger* has also been characterized (Liang et al., 2005).

5. Perspectives in biodegradation of pesticides

The knowledge described so far about the genes and enzymes necessary for pesticide degradation and the description of the pathways leading to its mineralization could allow genetic manipulation of microbes to enhance the bioremediation processes for these compounds. Molecular Biology tools provide the development of novel experimental approaches to find and identify novel pesticide-degrading genes. The bottleneck for finding new tools for pesticide bioremediation is that most microorganisms cannot be studied due to culturing limitations. There are estimations that around 99% of microorganisms are uncultivable using standard cultivation methods and therefore not accessible for finding useful genes or enzymes (Lorenz & Eck, 2005; Xu, 2006). To cope with this limitation, the recent development of metagenomic technologies has provided insights about the microbial genetic information available in environmental samples, independent of cultivability. Metagenome is the total biotic genome directly isolated from natural environments, and the power of metagenomics is the access, without prior sequence information, to the so far uncultured majority of microorganisms.

Sequence information from metagenomes can provide two types of approaches for the bioremediation of pesticides. On one instance, biodiversity of contaminated environments can be monitored in order to assess the presence of pesticide degrading bacteria and fungi for biostimulation strategies. The most used molecular markers for this kind of studies are the ribosomal RNAs, although other markers have also been used (Woese, 1990; Hibbett et al., 2007). Sequence analysis of these molecules permits phylogenetic reconstructions that describe the structure and composition of the microbial population in a determined habitat thus identifying potential microbial species for bioremediation (for example presence of bacteria from genera known to degrade pesticides).

The other approach consists of searching directly for genes or enzymatic activities involved in pesticide degradation. Several successful cases have been reported for esterase activities capable of hydrolyzing organophosphates and pyrethroids (Li et al., 2008; Zhang et al., 2009). In these reports the screenings were performed using substrates that upon hydrolysis produce a distinctive color, thus detecting hydrolases not previously described. The bottleneck of this approach, however, is that the genes must be expressed in a heterologous host (normally *Escherichia coli*), thus missing proteins that need complex posttranslational modifications, eukaryotic genes (that posses introns and promoters that are not recognized in bacteria) or distantly related genes whose promoters will not function in *E. coli*. Other hosts such as yeasts, or the use of expression vectors might be used to overcome some of these limitations.

Another possible strategy is to design degenerate oligonucleotides based on known gene sequences to amplify the desired gene using the metagenome as template DNA (Rose et al., 2003). This strategy will yield genes related to those that have been already described but overcomes the expression step limitation and thus is especially suitable to look for eukaryotic genes (although for further heterologous expression it will be still necessary to obtain the cDNA and use expression vectors).

6. The immobilization of microorganisms for massive pesticides degradation

Cell immobilization has been employed for biological removal of pesticides due to the possibility of maintaining catalytic activity over long periods of time (Richins et al., 2000; Chen & Georgiou 2002; Martin et al., 2000). Cell immobilization consists of restricting cellular mobility within a defined space, thereby retaining catalytic activity.

Whole cell immobilization has been shown to have remarkable advantages over conventional biological systems using free cells, such as high cell density, avoidance of cell washout even at high dilution rates, easy separation of cells from the reaction system, repeated use of cells, and better protection of cells from harsh environments. With these advantages, immobilization of microorganisms has been applied in many areas including wastewater treatment and remediation of toxic chemicals. Comparing immobilized cell systems with conventional free cell systems, the productivity obtained with immobilized cells is considerably higher. One obvious reason for this is the high cell density maintained in the reaction system. Some research has suggested that this higher productivity results from cellular or genetic modifications induced by immobilization. Evidences indicating that the immobilized cells are much more tolerant to perturbations in the reaction environment and less susceptible to toxic substances make immobilized cell systems particularly attractive for treatment of toxic substances like pesticides (Ha, 2005).

Immobilization of microorganisms has been applied in many areas including wastewater treatment and remediation of toxic chemicals from this technique generally provides several advantages over cultures using suspended cells that include greater cellular content in the support, enhanced cellular viability (weeks or months) and greater tolerance to high concentrations of pollutants. However, the main limitations to this method are low oxygen diffusion and interference by the materials used as the support (Martin et al., 2000; Georgiou et al., 2005). Encapsulation and biofilm formation methods are commonly used in environmental contexts. Alginate, *k*-carragenin and polyvinyl alcohol are used as supports for immobilization; ceramics, diatomaceous earth and porous rocks are used for biofilm formation (Karamanev et al., 1998; Davey & Toole 2000; Watnick & Kolter 2000).

In order to have a strategy for the treatment of pesticide wastes, we are also making efforts to build bioreactors with bacteria and/or immobilized yeast cells.

As was mentioned above, there are different materials for the immobilization of cells, but we have looked for some materials that might be economic and with different advantages to design a reactor that can be used on big scale. Tezontle is a volcanic rock that is highly porous. It provides a great contact surface and it can be sterilized and reused. The presence of micropores allows the establishment of bacterial micro colonies. The immobilization method with this material is based on the colonization of the tezontle micropores through the formation of a biofilm.

For the formation of the biofilm, tezontle rocks are crushed to obtain particles of about 3 mm. Tezontle is autoclaved in an intermittent way, at 121°C for 20 minutes, leaving it rest for 24 hours. Then, the microorganisms are planted to allow the formation of the biofilm. Subsequently, a current with the pesticides wastes is passed through to allow the contact with the immobilized microorganisms, so this way the biodegradation can be executed. This strategy has been really efficient and is a tool that can be used for the degradation of pesticides wastes (Yañez-Ocampo et al., 2009 & Yañez-Ocampo 2009).

7. Ecotoxicology studies

Ecotoxicology is a toxicology discipline that was proposed by Truhaut in 1969 and studies the adverse effects of toxic substances on ecosystems. This is done by the analysis of the exposure routes, the entry to the organism and the harmful effects on individuals, populations and communities, their way of action, such as the prevention or combat of their harmful effects. A toxic substance is that one that after its penetration in the organism causes immediately, or later on, the suppression of any function of the organism or death.

Soil pollution by pesticides represents a worldwide scale problem, that is why it is necessary to deal with it with an Ecotoxicological point of view, this discipline represents an useful tool to study the destiny and effects of pesticides in the environmental compartments (soil, water, air) and it has the objective to explain the cause and anticipate possible risks of pollution of environmental compartments, and the toxic effects (mortality, immobility, growth and enzymatic inhibition, among others) that pesticides may cause. Different markers have been used (lethality, enzymatic activity ACE, loss of weight and behavior) to prove the pesticides toxicity in laboratory conditions. For their study all the levels of biological organization can be used, from molecules, tissues, organisms and communities, with the purpose of evaluate the pollutant effects. The annelids (oligochaeta), which include the earthworms, are key species in the earth ecosystems that is why they have been widely used as biomarkers in enzymatic studies (acetylcholinesterase) and in studies about the behavior depending of the specific composition of the communities, the competition among species and the digging of galleries (Lavelle *et al.* 1997, Capowiez 2000).

In our working group, the earthworms have been used to measure the effects of pesticides when they are in the soil, as well as the effects of the products that are obtained after the pesticide degradation (Olvera-Velona et al., 2008a).

8. Behaviour of pesticides in agricultural soils

The study of pesticides behavior in the soil is really interesting, because it is a heterogeneous, complex and dynamic system, in which different reactions (chemical and biochemical) take place and also it plays a role as receptor of polluting substances. The liquid and gaseous phases in the soil are the main way of transport of the soluble compounds. The solid phase is the main site where accumulation and transformation (chemical or biochemical) of contaminating substances occurs. The pesticides can be directly incorporated into the soil by surface application on the crops, injection, or inadequate aspersion techniques or indirectly through plant leaves.
When the pesticide enters in contact with the soil, sorption is the first process, including adsorption/desorption phenomena. The first one permits fixation of the compounds to the soil particles; the last one releases the pesticide into soil solutions. The sorption process is related very with the persistence and pesticide degradation, because the physicochemical and biological characteristics of soils play a key role (Madrigal-Monárrez et al., 2008).

9. Pesticides sorption in soils

The extensive use of pesticides in agriculture has resulted in the widespread distribution in the environment. Organophosphate insecticides are increasingly used in agriculture as a substitute for organochlorine and carbamate insecticides, because of their higher efficiency and lower persistence. However, it is necessary to evaluate the risks and toxicity of these compounds and their degradation products into the groundwater, and their movement through the food chain (Pehkonen & Zhang, 2002).
The pesticide fate in soils strongly depends on adsorption-desorption phenomena. These processes influence the composition of the soil solution, and knowledge and understanding of them is important to accurately predict the mobility and bioavailability of these chemicals in soils; this would permit to limit their impact on non-targeted organisms and ecosystems. Pesticide availability in soil can be evaluated using indirect and direct methods. The most common indirect method to characterize pesticide availability is by using a partition coefficient (K_d). The pesticides with high soil-liquid partition coefficients (K_d values) are more strongly adsorbed to soils and transported in the soil solution. In contrast, pesticides with low K_d values are only weakly adsorbed to soils and are more likely to leach, as they are more water-soluble and hence exhibit greater mobility in the soil solution. However, sorption-desorption processes are complex and cannot be described by a single value, assuming instantaneous totally reversible sorption (Pignatello, 2000). Desorption processes control the release rate of pesticide into the soil solution, and thus have a major control of availability and bioavailability to soil organisms (Weber et al., 1993). Increasing pesticide contact time or ageing affects sorption and desorption processes, and generally desorption decreases with the residence time in soil. As a consequence, the availability of pesticide for transport or uptake by soil organisms decreases with time (Gevao et al., 2001; Mamy & Barriuso, 2007).
The sorption and degradation are influenced by physicochemical properties of the soil [such as pH and organic carbon (OC) content], biological properties (activity and distribution of microorganisms), and environmental conditions that control soil temperature and moisture content. Both route and rate of degradation also depend on properties of the chemicals. For ionized molecules, mineralogical composition and soil pH are key parameters, whereas

sorption of neutral compounds is mostly governed by the soil organic matter (Cooke et al., 2003).

These generally accepted rules have been obtained in studies considering large number of soil types and origins. Nevertheless, the role of soil characteristics is less well documented for tropical soils, compared with temperate soils (Zheng et al., 1994; Zheng & Cooper, 1996; Wauchope et al., 2002; Oliver et al., 2005).

Several pedological factors are known to regulate sorption and desorption processes (Weber et al., 2004). For molecules, mineralogical composition and pH are key parameters. The relations between sorption and soil type characteristics strongly depend on molecular properties, such as electric state and polarity (Barriuso & Calvet, 1991). The combination of effects of soil organic matter and of mineral constituents, such as smectite-type clays and amorphous clays-allophanes, usually increases the sorption of neutral molecules. They have a higher cation exchange capacity than other soil colloids, such as clays, and thus play an important role in adsorption reactions and determining fate of pesticides in the environment (Cooke et al., 2003).

The studies about pesticide fate in soil are complex due to physical, chemical, and biological interactions occurring simultaneously. In this way, it has been suggested an integral view relating sorption and transformation (degradation or mineralization), considering the physicochemical properties of pesticides and soil. There are experimental tools, such as soil microcosms (experimental devices with nominal variables and standard conditions), that help to explain the fractioning (or mass balance) of pesticides into the liquid or solid soil phases.

Our group has performed work in assays carried out in microcosms with parathion (^{14}C-U-labeled), in two crop soils (andisol and vertisol), after an incubation time of 32 days (Olvera-Velona et al., 2008b). Some of the results are outlined below.

Soluble fraction. This pesticide fraction is easily available for microorganisms; and thus, it can be easily biodegraded. This is the fraction with the highest mobility and can be lixiviated into the aquifer mantles. The aqueous solubility of pesticides is the most related property with mobility and lixiviation. Pesticides with aqueous solubility higher than 30 mg/L represent a potential lixiviation risk (Xing & Pignatello, 1996). Adsorption coefficient (K_d) is another property related with the soluble fraction of pesticide in the soil. K_d values below 30 L/Kg indicate lesser affinity to the soil particles and higher affinity for the liquid phase (in solution) of the soil. Microcosms tests also supplied us information about parathion fractioning. The soluble fraction was 2 and 3% at 32 days, in vertisol and andisol soils, respectively.

The K_d values for parathion were 74.9±0.6 L/Kg and 38.6±0.9 L/Kg in andisol and vertisol respectively, and when related with solubility (11 mg/L), we can infer that parathion has higher affinity for the solid phase in soil and lixiviation risk is low. The result indicates that only a small fraction (2-3%) can be found in the liquid soil phase in soluble form.

Adsorbed fraction. This is the pesticide fraction that is potentially available, because it is retained in the solid particles of soil. Surely, this is the pesticide fraction with highest interaction with soil. This fraction can be studied by measuring the adsorption coefficient (K_d); with the K_d values we can estimate the amount of pesticide adsorbed by the solid phase of soil. K_d values above 200 L/Kg are related to high affinity with soil particles. Spark & Swift (2002) reported that organic matter is one of the most related parameters with adsorption process. Weber et al., (2004) reported that soil pedological factors (organic matter and clay minerals) increase pesticide adsorption. Adsorbed fractions from 50 to 46%, in andisol and vertisol soils were found, respectively. The fraction percentages reflected the affinity for soil organic matter. However, when the organic matter content is below 2% (in the case of vertisol), an affinity for soil inorganic constituents (as smectites and allophan) was found (Worall et al., 1996).

Strongly bound fraction. This fraction directs to the pesticide stabilization, and represents the less available or biodegradable forms. This fraction also is known as bound residues. Barriuso et al, (1994) suggested the formation of covalent binding with humic substances; these interactions confer stability to pesticide. The residue formation can be considered an alternative to reduce the lixiviation risk; however, also the availability (and bioavailability) is reduced. Studies carried out by Mordaunt et al., (2005) suggested that, through time and agricultural practices, these residues can be reactivated, pass to the soil solution phase, and be again available. The parathion bound residue formation was 45 and 35% in andisol and vertisol, respectively. These percentages indicated us that a significant proportion of parathion was found inactive in the soil particles (at 32 days).

Mineralized fraction. This is the fraction of pesticide that has been transformed by microbes into components, such as CO_2, H_2O and NH_{4+}. Therefore, the mineralization is the suitable process for soil depuration. The metabolic activity of microbial populations and the co-metabolism are key determinants for the process. The microbial adaptation and enzyme production are important factors to catalyze the degradation reactions. The microbial consortia can carry out the degradation or mineralization with high efficiency, because the primary consumers start the degradation process and the secondary ones utilize the metabolic products from the formers and take them to degradation. This can facilitate the growth of primary consumers, by supply of metabolic products (e. g. growth factors), and by elimination of toxic compounds through co-metabolism. The mineralized fraction of parathion was 8 and 9% in andisol and vertisol soils, respectively. We consider that parathion mineralization in both soils was low. By relating this fraction with the soluble one (available), which was 2 and 3% in andisol and vertisol, respectively, results are logic. However, we also need consider the pesticide properties such as molecular structure, concentration, and bioavailability, because they are closely related with the mineralization process.

10. Conclusion

The use of chemical pesticides has brought benefits such as the increment of agricultural production, soil productivity and products quality, which is reflected in economical benefits, vector disease control and in general, in public health. However, due to only 10 per cent of applied pesticides reach to the target organism, a high percentage is deposited on non target areas (soil, water, sediments) and impacts to non target organism such wild life, besides affecting public health. Due to the extensive pesticides use, currently there are polluted sites with these compounds (mainly soils), besides of the production of great amounts of pesticide wastes, stored obsolete pesticides and empty containers. For this reason, it is necessary to generate strategies for waste treatment and/or for the bioremediation of polluted sites. The biological treatment is an important technology from an economical and environmental point of view. Currently, the use of native or genetically modified organism to degrade or remove pesticides has emerged as a powerful technology for in situ remediation There are reports of different organisms (bacteria, algae, yeasts, fungus and plants), characterized in relation to their genome and the enzymes that they produce, that can be used for waste treatment or bioremediation of soil and water. In this last process, it is necessary to considerate the soil composition, which is the responsible of the final destination of these compounds. Another important aspect is the ecotoxicology study, which will help to prevent the native organisms of soil and water.

11. References

Aktar W., Sengupta D. & Chowdhury A. (2009). Impact of pesticide use in agriculture: their benefits and hazards. *Interdisciplinary Toxicology*. 2(1), 1-12.

Albert L. A. (2005). Panorama de los plaguicidas en México. *Revista de Toxicología en Línea*. 17 pp.

Al-Qurainy F., Abdel-Megeed A. (2009). Phytoremediation and detoxification of two organophosphorous pesticides residues in Riyadh area. *World Applied Sciences Journal*. 6 (7), 987-998.

Amitai G., Adani R., Sod-Moriah G., Rabinovitz I., Vincze A., Leader H., Chefetz B., Leiovitz-Persky L., Friesem D. & Hadar Y. (1998). Oxidative biodegradation of phosphorothiolates by fungal laccase. *FEBS Letter*. 438, 195–200.

Anwar S., Liaquat F., Khan Q.M., Khalid Z.M. & Iqbal S. (2009). Biodegradation of chlorpyrifos and its hydrolysis product 3,5,6-trichloro-2-pyridinol by *Bacillus pumilus* strain C2A1. *Journal of Hazardous Materials* 168, 400–405.

Araya M. & Lakhi A. (2004). Response to consecutive nematicide application using the same product in *Musa* AAA cv. Grande Naine originated from in vitro propagative material and cultivated on a virgin soil. *Nematologia Brasileira*. 28(1), 55-61.

Barragán-Huerta B.E., Costa-Pérez C., Peralta-Cruz J., Barrera-Cortés J., Esparza-García F. & Rodríguez-Vázquez R. (2007). Biodegradation of organochlorine pesticides by bacteria grown in microniches of the porous structure of green bean coffee. *International Biodeterioration and Biodegradation*. 59(3), 239-244.

Barriuso E. & Calvet R. (1991). Soil type and herbicide adsorption. *Int. J. Environ. An. Ch.* 46, 117–128.

Barriuso E., Laird D.A., Koskinen W., Dowdy R.H. (1994). Atrazine desorption from smectites. *Soil Sci. Soc. Am. J.* 58, 1632–1638.

Bastos A.C. & Magan N. (2009). Trametes versicolor: Potential for atrazine bioremediation in calcareous clay soil, under low water availability conditions. *International Biodeterioration and Biodegradation*. 63, 389-394.

Batisson I., Crouzet O., Besse-Hoggan P., Sancelme M., Mangot J.F., Mallet C., Bohatier J. (2009). Isolation and characterization of mesotrione-degrading Bacillus sp. from soil. *Environmental Pollution*. 157, 1195-1201.

Bending G. D. & Rodriguez-Cruz M. S. (2007). Microbial aspects of the interaction between soil depth and biodegradation of the herbicide isoproturon. *Chemosphere*. 66, 664–671.

Bhalerao T. S., Puranik P.R. (2007). Biodegradation of organochlorine pesticide, endosulfan, by a fungal soil isolate, *Aspergillus niger*. *International Biodeterioration and Biodegradation*. 59, 315–321.

Bourquin A. W. (1977). Degradation of malathion by salt-marsh microorganisms. *Appl. Environ. Microbiol.* 33, 356-362.

Cáceres P.T., Megharaj M. & Naidu R. (2008). Biodegradation of the Pesticide Fenamiphos by Ten Different Species of Green Algae and Cyanobacteria. *Curr Microbiol*. 57, 643-646.

Capowiez Y. (2000). Differences in burrowing behaviour and spatial interaction between the two earthworm species *Aporrectodea nocturna* and *Allolobophora clorótica*. *Biol. Fertil. Soils*. 30, 341-346.

Castro C. E., Wade R. S. & Balser N. O. (1985). Biodehalogenation: reactions of cytochrome P-450 with polyhalomethanes. *Biochemistry*. 24, 204-210.

Chaudhry Q., Schröder P., Werck-Reichhart D., Grajek W. & Marecik R. (2002). Prospects and limitations of phytoremediation for the removal of persistent pesticides in the environment. *Environmental Science and Pollution Research*. 9(1), 4-17.

Chen W. & Georgiou G. (2002). Cell-surface display of heterologous proteins: from high throughput screening to environmental applications, *Biotechnol. Bioeng*. 5, 496–503.

Cheng T. C., Harvey S. P. & Chen G. L. (1996) Cloning and expression of a gene encoding a bacterial enzyme for decontamination of organophosphorus nerve agents and nucleotide sequence of the enzyme. *Appl Environ Microbiol*. 62, 1636-1641.

Cheng T., Liu L., Wang B., Wu J., DeFrank J. J., Anderson D. M., Rastogi V. K. & Hamilton A. B. (1997). Nucleotide sequence of a gene encoding an organophosphorus nerve agent degrading enzyme from *Alteromonas haloplanktis*. *J. Ind. Microbiol. Biotechnol*. 18, 49-55.

Chino-Flores C. (2007) *Clonación y expresión de fosfotriesterasas (FTE´s) de un consorcio bacteriano aislado del suelo*. Tesis de Maestría. Facultad de Ciencias Biológicas. Universidad Autónoma del Estado de Morelos. UAEM. México. 58 pp.

CICOPLAFEST (2004). *Catalogo Oficial de Plaguicidas*. Comisión Intersecretarial para el Control de Proceso y Uso de plaguicidas, Fertilizantes y Sustancias Tóxicas. SAGARPA, SEDESOL, México, D. F. 481 pp.

Cooke C. M., Shaw G. & Collins C. D. (2003). Determination of solid-liquid partition coefficients (K_d) for the herbicides isoproturon and triflurelin in five UK agricultural soils, *Environmental Pollution*. 132, 541-552.

Cooper J. & Dobson H. 2007. The benefits of pesticides to mankind and the environment. *Crop Protection*. 26(9), 1337-1348.

Cycon M., Wójcik M. & Piotrowska-Seget Z. (2009). Biodegradation of the organophosphorus insecticide diazinon by *Serratia* sp. and *Pseudomonas* sp. and their use in bioremediation of contaminated soil. *Chemosphere*. 76, 494-501.

Damalas C. A. (2009). Understanding benefits and risks of pesticide use. *Scientific Research and Essay*. 4(10), 945-949.

Dasgupta S., Meisner C. & Wheeler D. (2010). Stockpiles of obsolete pesticides and cleanup priorities: A methodology and application for Tunisia. *Journal of Environmental Management*. 91(4), 824-830.

Davey M. E. & Toole G.O. (2000). Microbial biofilms: from ecology to molecular genetics. *Microbiol. Mol. Biol. Rev*. 64, 847–867.

Desaint S., Hartmann A., Parekh N. R. & Fournier J. C. (2000). Genetic diversity of carbofuran-degrading soil bacteria. *FEMS Microbiology Ecology*. 34, 173-180

DiGioia D., Michelles A., Pierini M., Bogialli S., Fava F. & Barberio C. (2008). Selection and characterization of aerobic bacteria capable of degrading commercial mixtures of low-ethoxylated nonylphenols. *J. Appl. Microbiol*. 104, 231-242.

Don R. H & Pemberton J. M. (1981). Properties of six pesticide degradation plasmids isolated from *Alcaligenes paradoxus* and *Alcaligenes eutrophus*. *J Bacteriol*. 145(2), 681-686.

Fanga H., Xianga Y.Q., Haoa Y.J., Chua X.Q., Pana X.D.,Yub J.Q. & Yua Y.L. (2008).Fungal degradation of chlorpyrifos by Verticillium sp. DSP in pure cultures and its use in bioremediation of contaminated soil and pakchoi. *International Biodeterioration y Biodegradation*. 61, 294-303.

FAO (2002). *Código internacional de conducta para la distribución y utilización de plaguicidas*. Adoptado por el 123° Período de Sesiones del Consejo de la FAO. Roma, pp 31.

Farrera P. R. (2004). *Acerca de los plaguicidas y su uso en la agricultura*. Revista Digital Ceniap Hoy, Número 6, septiembre-diciembre. Maracay, Venezuela.

Franken S. M., Rozeboom H. J., Kalk K. H. & Dijkstra D. W. (1991). Crystal structure of haloalkane dehalogenase: an enzyme to detoxify halogenated alkanes, *EMBO J*. 10, 1297- 1302.

Gamón M., E Sáez E., Gil J. & Boluda R. (2003). Direct and indirect exogenous contamination by pesticides of rice-farming soils in a Mediterranean wetland. *Arch. Environ. Contam. Toxicol*. 44, 141–151.

Georgiou D., Hatiras J. & Aivasidis A. (2005). Microbial immobilization in a two-stage fixed-bed-reactor pilot plant for on-site anaerobic decolorization of textile wastewater. *Enz. Microb. Technol*. 37, 597–605.

Gevao B., Mordaunt C., Semple K. T., Piearce T. G. & Jones K. C. (2001). Bioavailability of nonextractable (bound) pesticide residues to earthworms, *Environ. Sci. Technol*. 35, 501–507.

Giner de los Ríos D. C. F. (2007). *Estudio: Precisión del Inventario de Plaguicidas Obsoletos y Sitios Contaminados con Éstos*. Donative TF-053710. Actividades de habilitación para ayudar a México a cumplir con el Convenio de Estocolmo sobre Contaminantes Orgánicos Persistentes. 149 pp.

Gomes-Machado, K.M., Matheus D. R., Rosim-Monteiro, R. T &. Ramos-Bononi, V. L. (2005). Biodegradation of pentachorophenol by tropical basidiomycetes in soils contaminated with industrial residues. *World Journal of Microbiology and Biotechnology*. 21(3), 297-301.

Gouma S. (2009). *Biodegradation of mixtures of pesticides by bacteria and white rot fungi*. Ph.d. Thesis. School of Health Cranfield University. 416 p.

Grant R.J., Daniell T.J. & Betts W.B. (2002). Isolation and identification of synthetic pyrethroid-degrading bacteria. *Journal of Applied Microbiology*. 92, 534-540.

Ha J. (2005). *Bioremediation of the organophosphate pesticide, coumaphos, using microorganisms immobilized in calcium-alginate gel beads*. Thesis. Texas A&M University. 181 pp.

Hibbett D.S., Binder M., Bischoff J.F., Blackwell M., Cannon P.F., Eriksson O.E., Huhndorf S., James T., Kirk P.M., Lucking R., Thorsten Lumbsch H., Lutzoni F., Matheny P.B., McLaughlin D.J., Powell M.J., Redhead S., Schoch C.L., Spatafora J.W., Stalpers J.A., Vilgalys R., Aime M.C., Aptroot A., Bauer R., Begerow D., Benny G.L., Castlebury L.A., Crous P.W., Dai Y.C., Gams W., Geiser D.M., Griffith G.W., Gueidan C., Hawksworth D.L., Hestmark G., Hosaka K., Humber R.A., Hyde K.D., Ironside J.E., Koljalg U., Kurtzman C.P., Larsson K.H., Lichtwardt R., Longcore J., Miadlikowska J., Miller A., Moncalvo J.M., Mozley-Standridge S., Oberwinkler F., Parmasto E., Reeb V., Rogers J.D., Roux C., Ryvarden L., Sampaio J.P., Schussler A., Sugiyama J., Thorn R.G., Tibell L., Untereiner W.A., Walker C., Wang Z., Weir A., Weiss M., White M.M., Winka K., Yao Y.J. & Zhang N. (2007). A higher-level phylogenetic classification of the Fungi. *Mycological Research*. 111(5), 509-547.

Hong Q., Zhang Z., Hong Y. & Li S. (2007). A microcosm study on bioremediation of fenitrothion-contaminated soil using *Burkholderia* sp. FDS-1. *International Biodeterioration & Biodegradation*. 59(1), 55-61.

Horne I., Harcourt R.L., Sutherland T.D., Russell R.J. & Oakeshott J.G. (2002). Isolation of a *Pseudomonas monteilli* strain with a novel phosphotriesterase. *FEMS Microbiol Lett.* 206(1), 51-5.

Horne I., Sutherland T. D., Harcourt R. L., Russell R. J. & Oakeshott J. G. (2002). Identification of an *opd* (organophosphate degradation) gene in an *Agrobacterium* isolate. *Appl Environ Microbiol.* 68, 3371-3376.

Jauregui J. Valderrama B. Albores A. & Vazquez-Duhalt R. (2003). Microsomal transformation of organophosphorus pesticides by white rot fungi. *Biodegradation.* 14, 397–406.

Karamanev G., Chavarie C., & Samson R. (1998). Soil immobilization: new concept for biotreatment of soil contaminants, *Biotechnol. Bioeng*. 4, 471–476.

Karstensen K. H., Nguyen K. K., Le B. T., Pham H. V., Nguyen D. T., Doan T. T., Nguyen H. H, Tao M. Q., Luong D. H. & Doan H. T. (2006). Environmentally sound destruction of obsolete pesticides in developing countries using cement kilns. *Environmental Science & Policy*. 9 (6), 577-586.

Katayama A. & Matsumura F. (2009). Degradation of organochlorine pesticides, particularly endosulfan, by *Trichoderma harzianum*. *Environmental Toxicology and Chemistry*. 12(6), 1059–1065.

Kwon G. S., Sohn H. Y., Shin K. S., Kim E. & Seo B. I. (2005). Biodegradation of the organochlorine insecticide, endosulfan, and the toxic metabolite, endosulfan sulfate, by *Klebsiella oxytoca* KE-8. *Appl Microbiol Biotechnol*. 67(6), 845-850.

Lavelle P. (1997). Faunal activities and soil processes: Adaptative strategies that determine ecosystem function. *Advances in Ecological Research*. 27, 93-132.

Li G., Wang K. & Liu Y.H. (2008). Molecular cloning and characterization of a novel pyrethroid-hydrolyzing esterase originating from the Metagenome *Microbial Cell Factories*. 7, 38-48.

Li X., He J. & Li S. (2007). Isolation of a chlorpyrifos-degrading bacterium, *Sphingomonas* sp. strain Dsp-2, and cloning of the *mpd* gene. *Research in Microbiology*. 158(2), 143-149.

Liang W. Q., Wang Z. Y., Li H., Wu P. C., Hu J. M, Li N. L., Cao X., & Liu Y. H. (2005). Purification and Characterization of a Novel Pyrethroid Hydrolase from *Aspergillus niger* ZD11. *J. Agric. Food Chem*. 53 (19), 7415–7420.

Liao M., Zhang H. J. & Xie X.M. (2009). Isolation and identification of degradation bacteria *Enterobacter aerogenes* for pyrethriods pesticide residues and its degradation characteristics. *Environmental Science*. 30(8), 2445-2451.

Lichtinger W. V., Arriaga, B. R. E., Bolaños-Cacho, R. J. A. (2001). *Bases de la política para la prevención de la contaminación del suelo y su remediación*. SEMARNAT. México, D.F. pp. 5-65.

Liu F.Y, Hong M.Z, Liu D.M, Li Y.W, Shou P.S, Yan H, Shi G.Q (2007). Biodegradation of methyl parathion by Acinetobacter radioresistens USTB-04. *Journal of Environmental Sciences*. 19(10), 1257-1260.

Liu Y.H., Liu H., Chen Z. H., Lian J., Huang X. & Chung Y. C. (2004). Purification and characterization of a novel organophosphorus pesticide hydrolase from *Penicillium lilacinum* BP303. *Enzyme Microbial Technol*. 34, 297–303.

Liu Y.H., Chung Y.C. & Xiong Y. (2001). Purification and characterization of a dimethoate-degrading enzyme of *Aspergillus niger* ZHY256, isolated from sewage. *Appl Environ Microbiol.* 67, 3746–3749.

Lorenz P. & Eck J. (2005). Metagenomics and industrial applications. *Nat Rev Microbiol*, 3, 510-516.

Madrigal-Monárrez I., Benoit P., Barriuso E., Réal B., Dutertre A, Moquet M., Trejo-Hernández M. & Ortiz-Hernández M. L. (2008). Caracterización de la sorción y desorción de herbicidas en suelos con diferentes tipos de ocupación. *Revista Ingeniería e Investigación*. 28(3), 96-104.

Malghani S., Chatterjee N., Yu H. X. & Luo Z (2009). Isolation and identification of profenofos degrading bacteria. *Brazilian Journal of Microbiology*. 40, 893-900.

Maloney S. E., Maule, A. & Smith, A. R. W. (1988). Microbial Transformation of the Pyrethroid Insecticides: Permethrin, Deltamethrin, Fastac, Fenvalerate, and Fluvalinate. *Appl Environ Microbiol.* 54 (11), 2874-2876.

Mamy L. & Barriuso E. (2007). Desorption and time-dependent sorption of herbicides in soils. *Eur. J. Soil Sci.* 58, 174–187.

Martin M., Mengs G., Plaza E., Garbi C., Sánchez A., Gibello A., Gutierrez F. & Ferrer E. (2000). Propachlor removal by *Pseudomonas* strain GCH1 in a immobilized-cell system. *Appl. Environ. Microbiol.* 66, 1190–1194.

Martínez J. (2004). *Practical Guideline on Environmentally Sound Management of Obsolete Pesticides in the Latin America and Caribbean Countries.* Basel Convention Coordinating Centre for Latin America and the Caribbean, Montevideo, Uruguay. 66 pp.

Mordaunt J. C., Gevao B., Jones K.C. & Semple K. T. (2005). Formation of non-extractable pesticide residues: observations on compound differences, measurement and regulatory issues, *Environmental Pollution*. 133, 25-34.

Mwangi K., Boga H. I, Muigai A. W, Kiiyukia C. & Tsanuo M. K. (2010). Degradation of dichlorodiphenyltrichloroethane (DDT) by bacterial isolates from cultivated and uncultivated soil. *African Journal of Microbiology Research*. 4 (3): 185-196.

Nadeau L. J., Fu-Min M., Breen A. & Sayler, G. S. (1994) Aerobic degradation of (l,l, 1 trichloro-2,2-bis (4-chlorophenyl) ethane) DDT by *Alcaligenes eutrophus* A5. *Appl. Environ. Microbiol.* 60, 51-55.

Oliver D. P., Kookana R. S. & Quintana B. (2005). Sorption of pesticides in tropical and temperate soils from, Autralia and Philippines, *J. Agr. Food Chem.* 53, 6420–6425.

Olvera-Velona A., Capowiez Y., Mascle O., Ortiz-Hernández L. & Benoit P. (2008a). Assessment of the toxicity of ethyl-parathion to earthworms (*Aporrectodea caliginosa*) using behavioral, physiological and biochemical markers. *Aplied Soil Ecology*. 40, 476-483.

Olvera-Velona A., Benoit P., Barriuso E. &, Ortiz-Hernández L. (2008b). Sorption and desorption of organophosphate pesticides, parathion and cadusafos, on tropical agricultural soils. *Agron. Sustain. Dev.* 28, 231-2.

Ortega S. N., Nitschke M., Mouad A.M., Landgraf M.D., Rezende O. M.O., Seleghim R. M.H., Sette L.D. & Porto A. L. (2010). Isolation of Brazilian marine fungi capable of growing on DDD pesticide. *Biodegradation*. In Press.

Ortiz-Hernández M. L. & Sánchez-Salinas E. (2010). Biodegradation of the organophosphate pesticide tetrachlorvinphos by bacteria isolated from agricultural soils in México. *Rev. Int. Contam. Ambient.* 26(1) 27-38.

Ortiz-Hernández M.L. (2002). *Biodegradación de plaguicidas organofosforados por nuevas bacterias aisladas del suelo.* Thesis. Biotechnology PhD. Universidad Autónoma del Estado de Morelos. 130 pp. México.

Ortiz-Hernández M., Sánchez-Salinas E., Vázquez-Duhalt R. & Quintero- Ramírez, R. (1997). Plaguicidas Organofosforados y Ambiente. *Biotecnología.* 3(2), 129-151.

Paingankar M., Jain M. & Deobagkar D. (2005). Biodegradation of allethrin, a pyrethroid insecticide, by an *Acidomonas* sp. *Biotechnology Letters.* 27, 1909-1913.

Parekh, N.R., Hartmann, A., Charnay, M.-P. & Fournier, J.-C. (1995). Diversity of carbofuran-degrading soil bacteria and detection of plasmid-encoded sequences homologous to the *mcd* gene. *FEMS Microbiol. Ecol.* 17, 149-160.

Pehkonen S..O. & Zhang Q. (2002). The degradation of organophosphorus pesticides in natural waters: a critical review, *Crit. Rev. Envi. Sci. Tec.* 32, 17-72.

Pignatello J..J. (2000). The measurement and interpretation of sorption and desorption rates for organic compounds in soil media, *Adv. Agron.* 69, 1-73.

Pimentel D. 2005. Environmental and economic costs of the application of pesticides primarily in the United States. *Environment, Development and Sustainability,* 7, 229-252.

Pizzul L., Castillo M. P. & Stenström J. (2009). Degradation of glyphosate and other pesticides by ligninolytic enzymes. *Biodegradation.* 20, 751–759.

Qing Z., Yang L. & Huan L. Y. (2006). Purification and characterization of a novel carbaryl hydrolase from *Aspergillus niger* PY168. *FEMS Microbiology Letters.* 228(1), 39-44.

Rani M. S., Lakshmi K. V., Devi P. S, Madhuri R. J., Aruna S., Jyothi K., Narasimha G. & Venkateswarlu K. (2009). Isolation and characterization of a chlorpyrifosdegrading bacterium from agricultural soil and its growth response. *African Journal of Microbiology Research.* 2, 026-031.

Richins R., Mulchandani A. & Chen W. (2000). Expression, immobilization, and enzymatic characterization of cellulose-binding domain-organophosphorus hydrolase fusion enzymes, *Biotechnol. Bioeng.* 69, 591–596.

Rigas F., Papadopoulou K., Dritsa V. & Doulia D. (2007). Bioremediation of a soil contaminated by lindane utilizing the fungus *Ganoderma australe* via response surface methodology. *Journal of Hazardous Materials.* 140, 325-332.

Rigas F., Dritsa V., Marchant R., Papadopoulou K., Avramides E.J. & Hatzianestis I. (2005). Biodegradation of lindane by *Pleurotus ostreatus* via central composite design. *Environment International.* 31(2), 191-196.

Rose T..M., Henikoff J. G. & Henikoff, S. (2003). CODEHOP (COnsensus-DEgenerate Hybrid Oligonucleotide Primer) PCR primer design. *Nucleic Acids Res.* 31(13), 3763–3766.

Sadowsky J. M, Koskinen C.W., Bischoff M., Barber L.B., Becker M.J. & Turco F.R. (2009). Rapid and Complete Degradation of the Herbicide Picloram by Lipomyces kononenkoae. *J. Agric. Food Chem.* 57, 4878-4882.

Saikia N. & Gopal M. (2004). Biodegradation of b-Cyfluthrin by Fungi. *J. Agric. Food Chem.* 52, 1220-1223.

Sarkar S., Seenivasan S. & Premkumar R. (2009). Biodegradation of propiconazole by *Pseudomonas putida* isolated from tea rhizosphere. *Plant Soil Environ.* 55(5), 196-201.

Serdar C. M., Gibson D. T., Munnecke D. M. & Lancaster J. H. (1982). Plasmid Involvement in Parathion Hydrolysis by *Pseudomonas diminuta. Appl Environ Microbiol.* 44, 246-249.

Sethunathan N. (1973). Degradation of parathion in flooded rice soils, *J. Agric. Food Chem.*, 21, 602-604

Sethunathan N., & Yoshida T. (1973). A *Flavobacterium* sp. that degrades diazinon and parathion. *Can J Microbiol* 19, 873-875.

Shah B. P. & Devkota B. (2009). Obsolete Pesticides: Their Environmental and Human Health Hazards. *The Journal of Agriculture and Environment.* 10, 51-56.

Shalaby Sh. E.M.& Abdou G. Y. (2010). The influence of soil microorganisms and bio- or -organic fertilizers on dissipation of some pesticides in soil and potato tube. *Journal of Plant Protection Research.* 50 (1). 86-92.

Sharma P., Chopra A., Cameotra S. S. & Suri C. R. (2010). Efficient biotransformation of herbicide diuron by bacterial strain Micrococcus sp. PS-1. *Biodegradation.* In press.

Sharma P., Raina V., Kumari R., Malhotra S., Dogra C., Kumari H., Kohler H. P. E., Buser H.R., Holliger C. & La R. (2006). Haloalkane Dehalogenase LinB Is Responsible for b- and d-Hexachlorocyclohexane Transformation in *Sphingobium indicum* B90A. *Appl. Environ. Microbiol.* 72 (9), 5720-5727.

Shukla K.P., Nand Kumar Singh N. K. & Sharma S. (2010). Bioremediation: Developments, Current Practices and Perspectives. *Genetic Engineering and Biotechnology Journal.* In press.

Siddaramappa R., Rajaram K. P. & Sethunathan N. (1973). Degradation of parathion by bacteria isolated from flooded soil. *Appl Microbiol.* 26, 846-849.

Singh B. K., Kuhad R. C., Singh A., Lal R. & Tripathi K. K. (1999). Biochemical and molecular basis of pesticide degradation by microorganisms. *Crit Rev Biotechnol.* 19, 197-225

Singh B. K., Walker A., Alun J., Morgan W. & Wright D. J. (2004) Biodegradation of Chlorpyrifos by *Enterobacter* Strain B-14 and Its Use in Bioremediation of Contaminated Soils. *Appl. Environ. Microbiol.* 70, 4855–4863.

Singh J. & Singh D. K. (2005). Dehydrogenase and phosphomonoesterase activities in groundnut (*Arachis hypogaea* L.) field after diazinon, imidacloprid and lindane treatments. *Chemosphere.* 60 (1), 32-42.

Singh B. & Walker A. (2006). Microbial degradation of organophosphorus compounds. *FEMS Microbiol Rev.* 30, 428–471.

Somara S. & Siddavattam D. (1995). Plasmid mediated organophosphate pesticide degradation by *Flavobacterium balustinum. Biochem Mol Biol Int.* 36, 627-631

Sonkong K., Prasertsan P. & Sobhon V. (2008). Screening and identification of p,p'-DDT degrading soil isolates. *Songklanakarin J. Sci. Technol.* 30 (Suppl. 1), 103-110.

Spark K.M. & Swift R.S. (2002). Effect of soil composition and dissolved organic matter on pesticide sorption, *Science of the Total Environment.* 298, 147-161.

Tomasek P.H. & Karns J.S. (1989) Cloning of a carbofuran hydrolase gene from Achromobacter sp. strain WM111 and its expression in Gram-negative bacteria. *J. Bacteriol.* 171, 4038-4044.

Van Eerd L. L., Hoagland R. E., Zablotowicz R. M., & Hall J. C. (2003). Pesticide metabolism in plants and microorganisms. *Weed Science.* 51, (4) 472–495.

Vuilleumier S. 2001. *Bacterial glutathione S-transferases and the detoxification of xenobiotics: dehalogenation through glutathione conjugation and beyond*. Pages 240–252 in J. C. Hall, R. E. Hoagland, and R. M. Zablotowicz, eds. *Pesticide Biotransformation in Plants and Microorganisms: Similarities and Divergences*. ACS Symposium Series 777. Washington, DC: American Chemical Society. ISBN 0-8412-3704-2.

Wang L., Wen Y., Guo X., Wang G., Li S. & Jiang J. (2010) Degradation of methamidophos by Hyphomicrobium species MAP-1 and the biochemical degradation pathway. *Biodegradation*, 21 (4), 513-523

Watnick P. & Kolter R. (2000). Biofilm, city of microbes, J. Bacteriol. 182, 2675–2679.

Wauchope R.D., Yeh S., Linders J.B.H.J., Kloskowski R., Tanaka K., Rubin B., Katayama A., Kordel W., Gerstl Z., Lane M. & Unsworth J.B. (2002). Pesticide soil sorption parameters: theory, measurement, uses, limitations and reliability. *Pest Manag. Sci.* 58, 419–445.

Weber J. B., Best J. A. & Gonese J. U. (1993). *Bioavailability and bioactivity of sorbed organic chemical*, in: Linn D.M. (Ed.), *Sorption and degradation of pesticides and organic chemical in soil*, SSSA special publication No. 32, Madison, Wisconsin, pp. 153-193.

Weber J.B., Wilkerson G.G. & Reinhardt F.C. (2004). Calculating pesticides sorption coefficients (K_d) using selected soil properties, *Chemosphere*. 55, 157–166.

Wei S. H., Clark A. G. & Syvanen M. (2001). Identification and cloning of a key insecticide-metabolizing glutathione S-transferase (MdGST-6A) from a hyper insecticide-resistant strain of the housefly *Musca domestica*. *Insect Biochemistry and Molecular Biology*. 31(12), 1145-1153.

Weir K.M., Sutherland T.D., Horne I., Russell R. J. & Oakeshott J.G. (2006). A single moonoxygenase, *ese*, is involved in the metabolism of the organochlorides endosulfan and endosulphate in an *Arthrobacter* sp. *Appl. Environ. Microbiol.* 72, 3524-3530.

Woese C. R., Kandler O. & Wheelis M. L. (1990). Towards a natural system of organisms: proposal for the domains Archea, Bacteria, and Eucarya. *Proc. Natl. Acad.Sci USA*. 87, 457-459.

Worrall F., Parker A., Rae J.E. & Johnson A.C. (1996) Equilibrium adsorption of isoproturon on soil and pure clays, *Eur. J. Soil Sci.* 47, 265–272.

Qiu X. H., Bai W. Q., Zhong Q. Z., Li M., He F.Q. & Li B.T. (2006). Isolation and characterization of a bacterial strain of the genus *Ochrobactrum* with methyl parathion mineralizing activity. *Journal of Applied Microbiology*. 101, 1364-5072.

Xing B & Pignatello J.J. (1996). Time-dependent isotherm shape of organic compounds in soil organic matter: Implications for sorption mechanism, *Environmental Toxicology and Chemistry* 15, 1282-1288.

Xu G., Li Y., Zheng W., Peng X., Li W. & Yan Y. (2007). Mineralization of chlorpyrifos by co-culture of Serratia and Trichosporon spp. *Biotechnol Lett*. 29, 1469-1473.

Xu, J. (2006). Microbial ecology in the age of genomics and metagenomics: concepts, tools, and recent advances. *Molecular Ecology*. 15(7), 1713-1731.

Yañez-Ocampo G. (2009). *Tratamiento de plaguicidas organofosforados a través de la inmovilización de bacterias aisladas de sitios contaminados*. Thesis Biotechnology PhD. Universidad Autónoma del Estado de Morelos. México.

Yañez-Ocampo G., Penninckx M., Jiménez-Tobon G.A, Sánchez-Salinas E. & Ortiz-Hernández M.L. (2009). Removal of two organophosphate pesticides employing a bacteria consortium immobilized in either alginate or tezontle. *Journal of Hazardous Materials*. 168, 1554–1561.

Yuanfan H, Jin Z, Qing H, Qian W, Jiandong J, Shunpeng L (2010). Characterization of a fenpropathrin-degrading strain and construction of a genetically engineered microorganism for simultaneous degradation of methyl parathion and fenpropathrin. *Journal of Environmental Management*. 91: 2295-2300.

Zhang R., Cui Z. Jiang J. He J., Gu X. & Li S. (2005) Diversity of organophosphorus pesticide-degrading bacteria in a polluted soil and conservation of their organophosphorus hydrolase genes. *Can J Microbiol*. 51, 337-343.

Zhang Z. Hong Q., Xu J., Zhang X. & Li S. (2006). Isolation of Fenitrothion-degrading Strain *Burkholderia* sp. FDS-1 and Cloning of mpd Gene. *Biodegradation*. 17(3), 275-283.

Zhang T., Han W.J. & Liu Z.P. (2009). Gene cloning and characterization of a novel esterase from activated sludge metagenome *Microbial Cell Factories*. 8, 67-75.

Zheng S.Q., Cooper J.F., Palcy L., Coste C.M. & Marnotte P. (1994). Mobility and dissipation of cadusafos in banana fields in Martinique, *Sci. Total Environ*. 156, 1-9.

Zheng S.Q. & Cooper J.F. (1996). Adsorption, desorption, and degradation of three pesticides in different soils, *Arch. Environ. Con. Tox*. 30, 15–20.

Zhongli C., Shunpeng L. & Guoping F. (2001). Isolation of methyl parathion-degrading strain M6 and cloning of the methyl parathion hydrolase gene. *Appl Environ Microbiol*. 67, 4922-5.

28

Fate of Pesticide Residues on Raw Agricultural Crops after Postharvest Storage and Food Processing to Edible Portions

Elpiniki G. Amvrazi
University of Thessaly
Greece

1. Introduction

Most analyses of pesticide residues in foods are being performed in Raw Agricultural Commodities (RAC) for a variety of purposes, which include regulatory monitoring, import/export certification, risk assessment, field-application trials, organic food verification, and marketing to consumers. The levels of the positive detections in these analyses are generally being estimated on the basis of established Maximum Residue Limits (MRL's) which are set using field trial data for a particular pesticide to arrive at the highest residue levels expected under use according to Good Agricultural Practice (GAP). MRL's are a credible and useful means of enforcing acceptable pesticide use, and satisfy most of the above mentioned purposes of monitoring pesticide residues in the different food of plant origin. However, MRL's use, proved to be inadequate as a guide to pesticide residue consumption through nutrition in health risk assessment studies from residues in food of plant origin and this is mainly because a wide range of RAC's are processed before they are consumed.

Storage and other post-harvest practices prior the further management of the product, as well as household and industrial food preparation processes may alter pesticide residues as compared with raw crops via chemical and biochemical reactions (hydrolysis, oxidation, microbial degradation etc.) and physicochemical processes (volatilization, absorption etc.). Although these processes usually are leading to reduction of any residues left on crops at harvest (Kaushik et al. 2009; Holland et al., 1994), in special cases residues may concentrate in the final product (e.g in the production of dry fruits and unrefined vegetable oil) (Amvrazi & Albanis 2008; Guardia Ruibio et al., 2006; Lentza-Rizos & Avramides, 2006; Lentza-Rizos & Kokkinaki, 2002; Cabras et al., 2000; Cabras et al., 1998; Cabras et al., 1997a; Holland et al., 1994; Cabras et al., 1993; Leandri et al., 1993; Ferriera & Tainha, 1983) and/or be formed in more toxic by-products or metabolites of the pesticide parent compound on raw crop (Holland et al., 1994; WHO 1988). These considerations suggest that effects of post-harvest practices and food processing should be taken into account on the fate of a pesticide residue during dietary exposure assessments so as to ensure consumer safety from pesticide residues and allow a more realistic calculation of the dietary burden of livestock.

Food processing studies and their effects on pesticide residues are also very important for the monitoring of the cases that the final residue concentration is exceeding MRL in RAC.

Although the established MRLs for processed and ready to eat food are limited to present and concern mainly the processed commodities where the residue could be found concentrated, regulatory attention has been focused on this issue the last decade and a current confrontation practice is the perspective of the processing factors (P_f's) establishment to be used with already specified MRL's of pesticides in RAC's (Codex Alimentarius Commission, 2007). P_f is defined as the ratio of the residue level in processed commodity to the residue level in RAC or in the commodity to be processed.
The aim of this chapter is to review the main types of post-harvest and processing practices that typically RAC's are being subjected prior their consumption and point out the effects that these processes cause on the fate of pesticide residues that may persist on and in RAC's after harvest. The main chemical, biochemical and physical phenomena in which pesticide residues take part in during these processes and the parameters that may influence these phenomena would further be discussed, through the latest published data on this topic.

2. Mechanisms and factors affecting the fate of pesticide residues during storage and food processing

The most important mechanisms that may lead in of a possible residue alteration during storage and other household and industrial food preparation processes of RAC are summarized as follows:

1. *Dissolution*. The dissolution of a pesticide residue may take place during the washing of RAC, grape vinification, tea preparation and boiling and theoretically is related with the water solubility of the pesticide residue. Other factors such as the type of formulation applied, temperature, initial concentration of the pesticide residue on RAC, pesticide Kow, ionic strength and pH of the aqueous media and the nature of RAC may further affect the pesticide dissolution mechanism in a certain food preparation process.
2. *Heat degradation*. Frequently a number of pesticides undergo degradation, polymerization and other reactions by heating. The reactions proceed at a greater rate in the liquid phase rather than in solid state. Thermal unstable compounds (e.g. atrazine, alachlor, aldicarb, captan, daminozide, dimethoate, dodine, lenacil, phorate, and others) may significant reduced by the formation of their degraded products during cooking and oven-drying of the processed RAC. Heat may also affect almost all pesticide eliminating mechanisms occur during RAC processing (dissolution, hydrolysis, penetration, volatilization, microbial degradation, metabolism and enzymatic transformation).
3. *Hydrolysis*. Hydrolysis is one of the basic processes of most pesticides elimination through the different storage and food preparation processes of RAC. Most pesticide residues may by hydrolyzed during typical RAC processing depending on the water added or the moisture on RAC, pH, temperature and pesticide's concentration. Pesticide compounds with functionalities such as carbamate, amide, urea, thiocarbonyl and imino group are more readily hydrolyzed in the presence of trace amounts of acid and/or base during the different processes of food preparation.
4. *Metabolism (enzymatic transformation); Microbial degradation*. The enzymatic transformation of pesticides is mainly the result of biotic processes mediated by plants and microorganisms. Penetrated pesticides may be further metabolized during the storage of fresh fruits and vegetables. Microbial degradation is the breakdown of pesticides by microorganisms. It occurs when fungi, bacteria, and other microorganisms consume pesticides along with other substances. Microbial degradation is usually

mediated by enzymes. Microbial activity usually is greatest at 10-45°C and is enhanced by moisture, air and neutral pH. Thus pesticides microbial degradation may take place during the processes where these conditions exist as it happens in grain storage. Other processes of RAC where the metabolism of pesticides may occur are those include a fermentation step (bread making, alcoholic beverage production, vinegar production etc).

5. *Oxidation.* Organic pesticide residues on RAC could be oxidized rather slowly with oxygen or air during storage depending on the of the pesticide residue and on the nature of RAC. In general the degradation of pesticides through chemical oxidation is affected by all factors that may result in the formation of .OH radicals. Food exposure to air, high storage temperatures, UV irradiation, and the presence of pro-oxidant compounds in the media where the pesticide residue occurs are several factors that may enhance the oxidation process of the persistent pesticide residue. In conclusion pesticides chemical oxidation mechanism may takes place during long term storage at elevated temperatures (i.e. room temperature), during the washing of RAC with oxidizing agents such as the ozone and H_2O_2 (or combination) and or UV irradiation of the ready to eat food for preservation purposes.
6. *Penetration.* Penetration is the most common physicochemical process that may take place on the surface of RACs from the time of the pesticide application to the end of the storage of RAC. The degree of a pesticide residue penetration on a RAC significantly affects its fate during RAC storage, washing, peeling, and drying. The main factors that may affect the penetration of a pesticide residue are the characteristics of the pesticide (Kow, the molecular weight, the probable systemic action, and the formulation of applied pesticide) in relation with the nature of RAC. The initial concentration of the pesticide residue on RAC and the processing temperature could also affect significant the penetration mechanism.
7. *Photodegradation.* All pesticides are susceptible to photodegradation to some degree. The intensity and spectrum of sunlight, length of exposure, and properties of the pesticide affect the rate of pesticides photodegradation. After harvest, photodegradation is not suspended to take place on pesticide residues persist on RAC in common food preparation processes and food storage. However significant pesticide elimination through photo degradation may occur during food processing like sun drying of fruits and UV irradiation that may be used for post harvest preservation processes.
8. *Physical change of the concentration of the pesticide residue due to a possible change of the RAC's weight.* Typical examples of that type of changes are: (i) the RAC's growth (namely plant growth) during the storage of immature fresh fruits that usually lead to lower concentration in the stored RAC, and (ii) the loss of water during the storage, cooking and/or drying of RAC that usually leads in higher concentrations of the pesticide residue in the stored or processed RAC. A phase separation from RAC in food processing may also lead in a physical change of the pesticide residue depending on the partition coefficient of the pesticide in the edible part of RAC and the mass of the separated phase in the final product. A typical example of these types of changes is the concentration of fat soluble pesticides and the respective elimination of water soluble pesticides in unrefined oil production.
9. *Volatilization and co-distillation.* Non systemic pesticide residues on RAC with relative high vapour pressures and low Kow may significant be eliminated through

volatilization from the stored or prepared food after drying or cooking. During storage, the higher the temperature is, the higher the pesticide's volatilization from the furface of the RAC and the greater the air humidity is, the less pesticide volatilizes whereas the nature of the RAC (e.g. water content, % w/w) may also play a significant role on these processes of pesticides elimination from RAC.

3. Fate of pesticide residues during post harvest storage practices of RACs

Proper storage of unprocessed food of plant origin includes maintaining at appropriate temperatures, away from light, under the proper percentage of humidity and in such a way to keep it clean and safe prior the further management of the product. Temperature control is the single most important factor in maintaining quality after harvest and the most common postharvest practice of plant commodities storage is cold storage.

Cold storage is taking place at different temperature and humidity, and lasts for different time intervals for each RAC depending from the nature of the commodity. Keeping products at their lowest safe temperature (0-4 °C for temperate crops, 4-8 °C and above 8 °C for chilling sensitive crops) will increase storage life by lowering respiration rate, decreasing sensitivity to ethylene gas and reducing water loss. However, temperature may additionally control the following elements of a pesticide residue dissipation and/or alterations during storage: (a) pesticide volatilization, (b) pesticide penetration, (c) pesticide metabolism and/or enzymatic degradation through the control of crop respiration and the consequent control of crop metabolism, (d) pesticide degradation through the retarding of certain microbial growth rate and activity, and (e) pesticide concentration changes per kg of RAC due the reduction of moisture loss, and/or loss of dry weight from respiration, or due to crop growth that usually leads to a reduction of pesticide concentration on raw crop. Furthermore, photodegradation is not expected to reduce further the applied pesticide after harvest during the proper storage of raw crop away from light whereas hydrolysis does not occur readily on plant surfaces during storage and is primarily confined to absorbed materials. In consequence, pesticide residues during deep frozen storage (-10 to -20 °C) are expected to be stable or decay rather slowly. However, the higher the storage temperature and time of storage, the higher the pesticides dissipation, degradation or concentration in processed commodity will be whereas the initial pesticide concentration and its physicochemical properties may affect significant its stability through storage processes.

Residues on Fruits and Vegetables. Results of studies performed with fortified fresh fruits and vegetables with different pesticides in order to study the main effects on pesticide residues during storage indicated that the main mechanisms of pesticide residues reduction through storage are the volatilization and the slow acidic hydrolysis (since most fresh fruits have a pH=3-4). Both processes are being controlled significantly by the penetration mechanism.

Thus, although many pesticides studied (azinphos methyl, kresoxim methyl, tolylfluanid, methamidophos, fenthion) in cold storage of different RAC (apples, lemons, grapes) exhibited slow dissipation rates ($t_{1/2}$: 42-267days), or found to be stable during storage (pyrethroids and carboximides in apples) (Athanasopoulos et al., 2005; Kyriakidis et al., 2005; Athanasopoulos et al., 2003; Rasmusssen et al., 2003; Athanasopoulos & Pappas, 2000), the highly volatiles dichlorvos and diazinon dissipated rapidly (70.8% and 64.8% respectively) from cucumber stored at +4 °C for 6 days (Cengiz et al., 2006) and the same results have been reported for dichlorvos from asparagus and kiwifruit stored at +1 °C for (Holland 1994).

Rasmusssen et al. (2003) reported that the organophosphates chlorpyrifos, diazinon and fenitrothion were significantly reduced (from 25% for chlorpyrifos to 49% for fenitrothion) on apples during storage at 4 °C for 79 days, whereas quinalphos was not. Quinalphos exhibits a systemic effect and penetrates from the skin into the outer flesh of the fruits such retarding dissipation during storage through the processes taking place on fruit surface. The same low dissipation through storage (~19% and 38% reduction on tomatoes after storage at +4 °C for 7 and 14 days respectively in polyethylene bags) could be observed for procymidone (Cengiz et al., 2007) that exhibits also a systemic action as compared with other pesticide dissipation with no systemic action during tomatoes storage (54 and 64% reduction for captan in the same study and > 49% reduction of iprodione and thiacloprid in another study performed by Omirou et al. 2009 at different storage conditions; +15 °C for 7 and 12 days). However, the reduction of pirimicarb (52.7%), cyprodinil (33.83%), tebuconazole (19.34%) with systemic action as compared with the non systemic insecticides fludioxonil (28.8%), pyriproxyfen (8.26%, not significant) and buprofezin (2.55%, not significant) did not present similar differences on their reduction during cold storage of peppers (Fenoll et al., 2009) suggesting that octanol/water partition coefficients were also well correlated with penetration mechanism of pesticides in peppers (Kow: pirimicarb < tebuconzole ~ cyprodinil < fludioxonil < buprofezin < pyriproxyfen). Furthermore, it should be noted that pirimicarb and cyprodinil are unstable in basic pH as is pH of peppers (>5). Similar effects of the commodity pH on pesticides dissipation during storage have been reported by Athanasopoulos & Pappas (2000) when studied dissipation rates of azinphos methyl in apples and lemons at their typical storage conditions and a significant lower dissipation rate was observed for azinphos methyl on lemons than in apples. The differences were attributed, to the differences in acidity among the two commodities studied. The differences on the reduction of pesticides during storage of different RACs are also very pronounced in current literature and except of the observed different dissipation rates among the different commodities, significant differences were observed and among the different varieties of the same commodity studied (Rasmusssen et al., 2003).

Grain storage. A separate, *big* issue on the fate of pesticide residues on RAC during storage concerns the pesticide residues on grains. Although grains storage is being performed at ambient temperature, in bulk silos for long term (3 - 36 months), the dissipation of pesticides is a slow process as compared with the cold storage of fruits and vegetables mainly due to pesticide retention on the seed coat and/or the high degree of pesticides penetration from the seed coat to the bran and germ which contain high levels of triglycerides. Furthermore, during grain storage insecticides may be applied post-harvest to reduce losses from storage pests (Holland, 1994), thus making cereal grains a potent source of pesticide residues through diet exposure.

The most widely studied insecticides in grain storage are the organophosphates malathion, pirimiphos methyl, chlorpyrifos methyl and their dissipation after 5-8 months of storage could ranged from 50-86% depending of pesticide Kow, type of applied formulation and storage temperature and humidity (Uygun et al., 2009; Uygun et al., 2008; Uygun et al., 2007; Balinova, 2006; Uygun et al., 2005). The mechanism proposed was that insecticides adsorbed to the grain are desorbed by water and become available for degradation by storage fungi, enzymes, metal ions and other active molecules (Holland 1994). Different classes of insecticides have also been studied and their dissipation during grain storage was found higher for natural pyrethrin residues (a complete disappearance was observed by Caboni et al. 2007 after 8 months of storage at ambient temperature) and lower for organochlorines and synthetic pyrethroids that are very stable under the reported main mechanisms of dissipation through typical storage conditions (non volatiles, with high Kow values and stable in hydrolytic processes).

4. Fate of pesticide residues during common domestic and industrial processing of RACs

The processing of food commodities generally implies the transformation of the perishable raw commodity to a value added product that has greater shelf life and is closer to being table ready. In this section the fate of pesticide residues during the most common food processing techniques applied to RAC in both household and industrial processing (washing; removal of the outer parts of the RAC such as peeling, husking, hulling, shelling, and trimming; comminution such as blending, chopping and mincing; cooking, and juicing) and the most important industrial processing of RAC in terms of frequent consumption of the processed food and/or the high probability of pesticide residues concentration in the processed ready to eat food (grain milling, oil production, alcoholic beverages production and drying), are being described.

4.1 Washing

Washing of RAC is the preliminary step in both household and commercial food preparation and the effect of washing on the fate of the pesticide residues on RAC has been well studied and recently well reviewed (Kaushic & Naik, 2009; Zabik et al., 2000; Krol et al., 2000; Holland, 1994). The most interesting conclusions of these studies are that the rinsability of a pesticide is not always correlated with its water solubility (Cengiz et al., 2007; Boulaid et al., 2005; Angioni et al., 2004; Krol et al., 2000) and that different pesticides may be rinsed from processed units of RAC by different washing procedures (Angioni et al., 2004; Pugliese et al., 2004; Lentza-Rizos & Kokkinaki, 2002; Cabras at al., 1998b).
The removal of pesticides with the washing of RAC may be performed not only through the dissolution of pesticide residues in the washing water or the rinsing with chemical baths (detergents, alkaline, acid, hypochlorite, metabisulfite salt, ozonated water etc) (Holland, 1994) but also through the removal of dust or soil particles previously absorbed residues from the outer layer of RAC (Guardia Rubio et al., 2007; Guardia Rubio et al., 2006; Angioni et al., 2004; Cabras et al., 1997). Penetration is again the most dynamic process that may control the fate of a pesticide residue on RAC during washing. The systemic insecticide dimethoate with water solubility equal to 23300mg/L and quinalphos that exhibits systemic action and its water solubility is 18 mg/L, were not reduced during the washing of olives (Cabras et al., 1997). Furthermore in many studies washing was not related with the water solubility of the pesticide residue but with Kow such reinforcing the view that partition coefficients between cuticle and washing water correlate well with pesticides Kow (Baur et al., 1997). In consequence, the use of an appropriate detergent that has the ability to solubilise waxes may dissipate the residue present in the fruit's epicuticular wax layer (Angioni et al., 2004). Other washing agents or dipping treatments may also lead in the selective removal of the pesticide residues with systemic action through similar mechanisms (Cabras et al., 1998b; Femenia et al., 1998). The residue still present in the fruit after washing can be ascribed to the pesticide that has penetrated into the cuticle. The formation of possible toxic by-products during washing has also been studied by several authors (Ou-Yang wt al., 2004; Pugliese et al., 2004; Cabrera et al., 2000; Zhang & Pehkonen, 1999). According to these results, typical washing with tap water is not expected to form toxic metabolites whereas the use of high levels of sodium hypochlorite, hydrogen peroxide, and

potassium permanganate in washing water as well as ozonated water could form oxons from the organophosphorus pesticides by chemical oxidation (Ou-Yang wt al., 2004; Pugliese et al., 2004; Cabrera et al., 2000; Zhang & Pehkonen, 1999).

4.2 Peeling (husking, hulling, shelling, trimming)

The removal of the outer part of RAC by peeling, husking, hulling, shelling, or trimming is the most effective food preparation process for pesticide residues removal from RAC. Numerous studies report the elimination of different pesticide residues on different RAC through peeling to range from 70 to 100% (Cengiz et al., 2007; Boulaid et al., 2005; Fernández-Cruz et al., 2004; Rasmusssen et al., 2003; Burchat et al., 1998; Clavijo et al., 1996; Celik et al., 1995; Holland, 1994; Rouchaud et al., 1991).

The systemic action of a pesticide residue in this case is not always correlated with decreased reduction of pesticide residues through peeling. Thus, although, residues of the systemic organophosphorus phorate were only reduced by 50% through peeling of potatoes (JMPR, 1992) and similarly disyston residues in potatoes were only reduced by 35% after peeling (Holland 1994), quinalphos residues on apples and procymidone residues on tomatoes were reduced by ≥73% (Cengiz et al., 2007). Furthermore, the reported data of the reduction of dichlorvos and diazinon on cucumber by 57.2% and 67.3% respectively (Cengiz et al., 2006), the low reduction of pyridaben (~70%) in tomatoes (Boulaid et al., 2005) and the lower reduction of the metabolites of fenitrothion (80-92% for fenitrothion, 78-99% for fenitrothion oxon and 73-78% for 3-methyl-4-nitrophenol) in kakis (Fernández-Cruz et al., 2004) and endosulfan sulphate (24%) in apples (Rasmusssen et al., 2003) consist that the main factor that may affect negatively that satisfactory removal of pesticide residues from RAC is penetration into the flesh of the processed RAC that is controlled by the physicochemical properties of the pesticide residue in relation with the nature of the processed agricultural commodity.

4.3 Comminution

Comminution of RAC through chopping, blending, crushing and similar processes usually do not affect pesticide residues in RAC since most pesticides are relatively stable in acidic plant tissue homogenates for the moderate periods of time involved in food preparation. However, comminution leads to release of enzymes and acids which may increase the rate of hydrolytic and other degradative processes on residues that were previously isolated by cuticular layers. Special concern on the fate of residues should be paid on acid sensitive pesticide compounds (e.g. EBDC, carbosulfan, benfuracarb, pymetrozine, dioxacarb, thiodicarb and others) that readily are hydrolyzed in the presence of trace amounts of acids and the most toxic metabolites formed should be studied. A typical example is the rapid degradation of ethylene bis dithiocarbamate (EBDC) fungicide residues to the formation of the toxic ETU, carbon disulphide and ethylenediamine in slightly acidic media, similar to the pH of the tomato homogenates (4.0–4.2) (Kontou et al., 2004; Holland, 1994; Howard & Yip, 1971). Pesticides degradation processes during storage of blended RACs could be enhanced significant with the increase of time storage and temperature and more toxic metabolites in these cases should be taken into consideration in health risk assessments.

4.4 Juicing

The residue levels in juices from fruits or vegetables are generally reduced by 70-100% (Rasmussen et al., 2003; Zabik et al., 2000; Abou-Arab, 1999; Will & Krüger ,1999; Buchat et al., 1998; Holland, 1994) and their reduction depends on the partitioning properties of the pesticide between the fruit skin, pulp and the juice (which generally contain some solids). The pulp or pomace by-products, which often include the skin, retain a substantial proportion of lipophilic residues.

Rasmussen et al. (2003) reported that only 2-9% of different pesticide residues (chlorpyrifos, cypermethrin, deltamethrin, endosulfan, fenitrothion, fenpropathrin, iprodione, kresoxim-methyl, λ-cyhalothrin, vinclozoline) on fortified apples were transferred in apple juice whereas in apple pulp residues in mg/kg were detected 2.0-3.5 higher than in unprocessed apples (due to mass concentration). In the same study quinalphos (systemic action), endosulfan sulphate and tolylfluanid that were detected in higher amounts in the flesh of apples were transferred also in higher amounts in the produced apple juice (≤19%, 13% and 23% respectively). No correlation with the water solubility of the pesticides studied could be observed on their fate during juicing in results reported by Abou-Arab (1999) where the reduction of HCB, lindane, p,p-DDT, dimethoate, profenofos and pirimiphos-methyl, residues upon tomato juicing, ranged from 72.7% to 77.6%. However, in other studies pesticides with the highest water solubility were present in relatively higher amounts in the juiced carrots, tomatoes and strawberries (Will & Krüger, 1999; Buchat et al., 1998).

Clarification by filtration or centrifugation in juice processing may further eliminate pesticide residues retained in suspended particles (Liapis et al., 1995; Miliadis et al., 1995) and juice concentration by vacuum may also concentrate the pesticides transferred in juice (Zabik et al., 2000).

4.5 Cooking (cooking, boiling, frying)

Pesticide residues might be vaporized, hydrolysed and/or thermal degraded during cooking. However, the processes and conditions used in food cooking are highly varied. The details of time, temperature, degree of moisture loss, whether the system is open or closed and whether water is added or not in the process are important for the estimation of the fate of a residue level. In general, rates of degradation and volatilization of residues are increased by the heat involved in cooking or pasteurization and rates of hydrolysis may also be increased by the water addition and the increase of temperature.

In different studies of organophosphorus (OP) insecticides (fenitrothion, fenitrothion oxon, 3-methyl-4-nitrophenol) on RACs (cauliflower, kaki fruits) it was reported that OPs are stable through heating without the addition of water for 10-15 min (Fernández-Cruz et al., 2006; Fernández-Cruz et al., 2004) and unstable to heating in aqueous solution. Coulibaly & Smith (1993) studied famphur, fenthion, parathion, stirofos, chlorpyrifos, and ronnel in aqueous solutions that were unheated, heated to 70 °C for 1 or 2 h, and heated to 80 °C for 1 h. Stirofos and famphur were largely unaffected by heating whereas fenthion, parathion, chlorpyrifos and ronnel were hydrolyzed by 53.2%-80% in unheated water and further heating did not result in further degradation. Other studies confirm the reduction of OPs (32% reduction of fenitrothion and 89.5% reduction of triazophos) after boiling (Rasmussen et al., 2003 ; Holden et al., 2001), but also reported chlorpyrifos and acephate reduction (50.5-68%) after cooking of peppers, asparagus and peaches without the addition of water (at 100-110°C for 20-80 min) (Chavarri et al., 2005). The latter differences might be due to higher temperatures and times used in cooking processes since drying at elevated temperatures and/or times also lead in higher reduction of pesticides (including OPs) in processed commodities. Furthermore, Nagayama (1996) reported that during the cooking process, some residual pesticides were translocated into the cooking water from the raw materials according to the water solubility expression, and the pesticide remained in the processed food according to the Kow expression. These relationships were shown by simple equations. The inclination of the regression expression was similar with the same cooking process and increased with cooking time.

Other pesticides such as pyrifenox, pyridaben, and tralomethrin were not reduced after cooking applied to the tomatoes without the addition of water although residues were concentrated in the final product due to water loss through cooking by a factor of 1.9-3.0 (Boulaid et al., 2005). Captan was almost completely eliminated after cooking cauliflower without water for 15 min and after the processing of apple to sterilized puree (125 °C for 20 min at pH 4) by exhibiting its tendency to thermal degradation.

Interestingly, boiling did not reduce chlorpyrifos, cypermethrin, deltamethrin, diazinon, endosulfan (alpha, beta and endosulfan sulphate), fenpropathrin, iprodione, kresoxim methyl, λ-cyhalothrin, quinalphos and vinclozoline residues on apples (Rasmussen et al., 2003). The fact that in the same study fenitrothion and tolylfluanid were reduced by boiling was not attributed to the acid hydrolysis of pesticides but to the selective degradation of fenitrothion and tolylfluanid residues related to interaction with thiol-containing compounds present in apples. Fenitrothion was the only organophosphorus pesticide in that study that contained a methoxy group enhancing enzymatic degradation to ethoxy containing organophosphors pesticides whereas the possible proposed degradation of tolylfluanid was the cleavage of the N-S bound after tolylfluanid interaction with thiol compounds.

Randhawa et al. (2007) studied the fate of chlorpyrifos and its degradation product 3,5,6-trichloro-2-pyridinol during boiling of different vegetables with water and the decrease of chlorpyrifos ranged from 12% to 48%. The effect was more obvious in spinach (38%) followed by cauliflower (29%). 3,5,6-trichloro-2-pyridinol was substantially increased during the course of cooking consisting that metabolites of toxicological concern should also be measured in cooked fruits and vegetables (and in the boiled water that might be used in further cook). In particular, pesticides such as carbamates, amides, ureas, thiocarbonyl and those containing imino groups may be readily hydrolyzed in the presence of trace amounts of acid and/or base during heating or boiling. On this topic, a substantially amount of work (Chavarri et al., 2005; Kontou et al., 2004a; Kontou et al., 2004b; Knio et al., 2000) has been published on the fate of pesticide residues of dithiocarbamates and the more toxic ETU that during cooking concluding that during cooking dithiocarbamates may be conversed to ETU by a factor of 30-48%. Although, ETU residues during boiling may pass into the boiling water, a significant amount remains in the processed food.

4.6 Grain milling

Cleaning, conditioning and grinding are the three basic steps in grain milling process that break its grain in three main parts: wheat germ, bran and endosperm. In most studies on the distribution of pesticide residues (chlorpyrifos methyl, pirimiphos methyl, malathion, isomalathion, fenitrothion) during grain milling, the pesticide residues levels in bran are consistently higher than in wheat, usually by a factor of about 2–6 (Balinova et al., 2006; Uygun et al., 2005). Furthermore it has been reported that a considerable part of the insecticides is distributed in the semolina fractions. The residues of chlorpyrifos-methyl and pirimiphos-methyl determined in semolina were only slightly lower than the residues in bran and were, respectively, 2.0–3.6 and 1.3–3.2 times as high as those on the whole grain.

Uygun et al. (2008) reported that the carryover percentage of malathion from wheat to semolina was 16-28%, of fenitrothion 17-22%, of chlorpyrifos methyl 7-8% and of pirimiphos methyl 23-28%. As already reported for grain storage, pesticides are retained on the seed coat and tent to concentrate to the bran and germ which contain high levels of triglycerides and in consequence the lipophilcity of the pesticide residue on the processed grain may forecast its fate in the processed food product during grain milling. In conclusion Kow may adequate explain the different reductions of different pesticides that have been reported in wheat through milling; the reduction of malathion and fenitrothion was about 95%-100%, (Uygun et al., 2005), the reduction of deltamethrin was about 57.6% (Marei et al., 1995) in wheat through milling to flour.

4.7 Oil Production

Vegetable oils and fats are extracted from a variety of fruits, seeds, and nuts. Depending on the RAC, the preparation of raw materials may include husking, washing, crushing or other conditioning. The oil extraction processes are generally mechanical (boiling and/or centrifugation for fruits, pressing for seeds and nuts) or involve the use of solvents such as hexane. After boiling, the liquid oil is skimmed; after pressing, the oil is filtered; and after solvent extraction, the crude oil is separated and the solvent is evaporated and recovered. The solid by-products of oil processing usually are conditioned (for example, dried) and are reprocessed to yield other products such as for animal feed, soil amendments, food additives and soaps. The produced crude oil before its consumption is often being subjected to a refining process to remove undesirable compounds that contribute undesirable colour, flavours and aromas which are disagreeable to consumers, affect the stability of the product and/or are toxic in nature. Oil refining includes degumming, neutralization, bleaching and deodorization.

Although most pesticide residues are not affected significant by the washing procedure of fatty RAC, residues in oil-seeds following husking are very low or non-detectable as it has been reported in relative studies from field applications (Holland, 1994). However, the following oil extraction processes by means of either mechanical or chemical (solvent extraction) have high theoretical concentration factors (from 2.3 for coconut oil to 1000 for citrus oil) (EPA, 1996) and the fate of pesticide residues during edible oil production is of great importance.

4.7.1 Fate of pesticides in unrefined, edible vegetable oil

Olive oil is the more important vegetable oil in European Market and the more characteristic unrefined vegetable oil that could be characterized as *"virgin"* according to the process of its extraction. The extraction process of unrefined olive oil involves the washing and milling of the fruit, the malaxation of the produced olive paste by slow mixing at a constant temperature (usually bellow 30°C) for 30-90 min and the separation of oil by a press or a decanter (centrifugation system). Depending on the processed olives (variety and degree of maturation) and the decanting extraction technology used, additional water may be added during malaxation and centrifugation processes to better separate oil and increase oil yields. The theoretical processing factors of virgin olive oil by this process range widely (usually range from 4-6) depending on the variety of the olive fruits (oil and water content) and the decanting extraction technology used for oil extraction (Amvrazi & Albanis, 2009; EC, 2005). However, water soluble pesticides such as acephate, dimethoate, methamidophos, omethoate and phosphamidon pass into the aqueous phase during the extraction of oil from olives (Amvrazi & Albanis 2008; Cabras et al., 2000; Cabras et al., 1997; Letza-Rizos & Avramides, 1995; Leandri et al., 1993; Ferreira & Tainhan, 1983) and only a small percent is transferred into the oil (e.g. 6.3-8.8% for dimethoate) depending on the water content during the extraction of the olives (Amvrazi & Albanis, 2008). Other pesticides with lower water solubilities (azinphos methyl, buprofezin, chlorpyrifos, fenthion, deltamethrin, diazinon, endosulfan, quinalphos, λ-cyhalothrin, methidathion, parathion methyl) were found to concentrate in the oil with a concentration factor of 2-7 (Amvrazi & Albanis 2008; Cabras et al., 2000; Cabras et al., 1997; Letza-Rizos & Avramides, 1995; Leandri et al., 1993; Ferreira & Tainhan, 1983) depending on the Kow of the pesticide residue, the oil yield of the extraction procedure and the stability of the pesticide towards volatilization, and hydrolytic and other degradative processes that might take place during the malaxation stage of the procedure. The formation of fenthion sulfoxide during olive oil production was calculated ~5% of the initial fenthion and was correlated with the water addition during oil production process. However, the formation of endosulfan sulphate was not related with the water in oil extraction process (Amvrazi & Albanis, 2008).

4.7.2 Fate of pesticides in refined, edible vegetable oil

During refining process of crude oils most organochlorine and organophosphorous pesticides in edible vegetable oils can be reduced considerably (concentration factors <0.1), but pyrethorid pesticides remain to a certain extent (Fukazawa et al., 1999; Hilbert et al., 1998; Zayed et al., 1998; Miyahara & Saito, 1993; Wolff, 1974). From the four main processes of crude oil refining (degumming, neutralization, bleaching and deodorization), deodorization is the process that eliminates pesticide residues the most in the final product.

Although neutralization with alkali is not expected to affect organochlorines and pyrethroids in the refining process (Fukazawa et al., 1999; Hilbert et al., 1998), it has been reported that organophosphates may be reduced by 28-50% during alkali refining of fortified olive oil (Morchio et al., 1992).

The bleaching step is common for both physical and alkali refining and concerns the decolouration and the removal of undesirable compounds through absorption. The usual material is fuller's earth or carbon and charcoals materials in combination with fuller's earth. For optimum adsorption of most undesirable components to be achieved, the reaction time usually ranges from 15 - 30 min at 90ºC. In the field of pesticides, it has been reported that bleaching by activated earth is not affecting significant most pesticide residues transferred in crude oil from RAC. However, it has been reported that endrin (Vioque et al., 1973) and simazine (Ruiz Méndez et al., 2005) have been completely eliminated in bleached oil. Zayed et al. (1998) reported carbofuran elimination at 20 % during bleaching with Fuller's earth at 80-100ºC for 10 min and Morchio et al. (1992) reported that during docolorization, the percentage reduction of different organophosphates (dimethoate, diazinon, cis- and trans phosphamidon, parathion methyl, malathion, fenthion, and methidathion) varied from 95% (diazinon) to 30% (methidathion).

Deodorizing of oils and fats consists mainly of steam distillation under vacuum. During deodorization residual free fatty acids, aldehydes and ketones that constitute the volatile, taste and odour components are being removed from crude oil. Pesticides as well as other contaminants such as the lower molecular weight, poly aromatic hydrocarbons (PAHs) could be significant eliminated at this point (Ruiz Méndez et al., 2005; Cejpek et al., 1998; Hilbert et al., 1998; Morchio et al., 1992; Vioque et al., 1973). However, although Morchio et al. (1992) reported the complete elimination (95%) of organophosphates in refined olive oil, Hilbert et al. (1998) found that the deodorizing step decreased the amount of the most volatile organochlorine contaminants (α-hexachlorocyclohexane, lindane, hexachlorobenzene) to below the detection limit (5μg/Kg) whereas concentrations of the less volatile organochlorine pesticides (dieldrin, p,p′-Dichlordiphenyl dichloro ethylene, p,p′-1,1-dichloro-2,2-bis(p-chlorophenyl) ethene) and polychlorinated biphenyl were reduced to ~50% of the initial concentration in the crude fish oil. Ruiz Méndez et al. (2005) also reported the complete elimination of simazine, endosulfan, oxyfluorfen and diflufenican during deodorizing step of the physical refining process of olive oil at 260°C. During the refining process of soybean oil to which had been added 9 pyrethroid pesticides (pyrethrins, cyhalothrin, deltamethrin, permethrin, cypermethrin, cyfluthrin, fluvalinate, fenvalerate and flucythrinate, 5 mg/kg each) all pesticides remained for the most part in the oil during degumming, alkali-refining and bleaching. During deodorization at 260 °C, pyrethrins remarkably decreased, while the amounts of cyhalothrin, deltamethrin, permethrin, cypermethrin and cyfluthrin were reduced by ~50%. Only slight decrease was noted in fluvalinate, fenvalerate and flucythrinate under the same conditions.

4.8 Drying

Drying of RAC could be performed by the sun or a food dryer or in an oven. The different drying processes have different effects on pesticide residues on RAC since sun light may additionally photodegrade pesticide residues. Although the loss of water leads to increased theoretical processing factors during drying the respective factors of pesticide residues in dried food are generally lower.

Thus, although dried apricots have a theoretical processing factor ~5-6, specific pesticide residues (bitertanol, diazinon, procymidone, iprodione, omethoate, ziram) in dried apricots have been detected at lower levels as compared with raw fruits (Cabras et al., 1998b; Cabras 1997a). However under the same conditions of drying, fenitrothion disappeared completely, dimethoate did not change; phosalone was tripled whereas omethoate and ziram almost doubled (Cabras et al., 1997a). No changes among residues of bitertanol, diazinon, procymidone and iprodione were observed among pesticides studied with the two different drying procedures (oven and sundrying) tested, except of phosalone that doubled after sun drying (tripled by oven drying) in the processed apricots. Similar results have been reported for resin processing. Although the theoretical processing factor of resins is ~4, pesticides processing factors ranged from 0.08-1.7 for benalaxyl, dimethoate, iprodione, metalaxyl, phosalone, procymidone, vinclozoline, and cypermethrin (Lentza-Rizos & Kokkinaki, 2001; Cabras et al., 1998b). In a further study concerning the mechanisms occur during oven drying, the decrease of dimethoate was attributed to heat, of benalaxyl, procymidone and phosalone to co-distillation and of iprodione and metalaxyl to the combined action of heat and co-distillation (Cabras et al., 1998b). Furthermore, the type and the variety of the processed RAC may play a significant role on the fate of pesticides during drying; the higher the surface to-weight ratio, the more effective the loss.

4.9 Alcoholic beverages production

4.9.1 Wine making

The winemaking process begins with the pressing of the grapes by forming a biphasic system made up of the must (an acid aqueous liquid phase with pH 2.7-3.7) and the pomace (a solid phase which contain cake and lees). The following step is fermentation and this process can be carried out either with or without grape skins. In the former case (with maceration) the wine will be made with all of the residues on the grapes; in the latter case (without maceration) the process will include only the residues that have passed in the must. The grape pomace (cake and lees) is the main by-product from winemaking and traditionally has been used to produce pomace brandy (or marc brandy), and grapeseed oil. Today, it is mostly used as fodder or fertilizer.

The fate of pesticide residues on grapes during winemaking has been widely studied. In most reported data, pesticide residues present on grapes (i.e diniconazole, famoxadone, fenbuconazole, flufenoxuron, flusilazol, lufenuron, teflubenzuron, trifloxystrobin) remain adsorbed in the cake and lees (which are by-products of wine-making) at relevant levels, and are transferred to the wine in low percentages after fermentation depending, mainly on the initial partition of a pesticide residue between the must and the cake and lees (Likas, 2009; Likas & Tsiropoulos, 2009; Calhelha et al., 2006; De Melo Abreau et al. 2006; Tsiropoulos & Likas, 2005; Tsiropoulos et al., 1999). Azoxystrobin, benalaxyl, benomyl, dimethoate, fenhexamide, iprodione, metalaxyl-m, methidation, procymidone, pyrimethanil, spiroxamine and tebufenozide are several pesticide residues that have been reported to be transferred from the grapes to the wine at relevant amounts (20-30% for all pesticides but benomyl that transferred 100%) (Likas, 2009; Cabras & Angioni, 2000).

However, the partition coefficient of a pesticide residue between the solid pomace phase and the liquid phase of must that mainly depend on the Kow and water solubility of the pesticide is not the only parameter that defines pesticides fate during wine-making. Dichlofluanid (Calhelha et al., 2006; Cabras & Angioni, 2000), chlozolinate (Gennari et al., 1992), folpet, captan (Angioni et al. 2003), thiazinan, mepanipyrim and chlorpyrifos (Cabras & Angioni, 2000) are some of the pesticide residues that their dissipation in the final produced wine has been attributed the pesticide degradation from the time of pressing and during fermentation. Yeasts that usually are being used in vinification processes have shown the ability to degrade some pesticides belonging to the pyrethroid and thiophosphate classes (chlorpyrifos-methyl, fenitrothion, parathion, quinalphos). Moreover, yeasts adsorb some pesticides, thus contributing to their removal from the wine at the end of fermentation (Cabras & Angioni, 2000).

4.9.2 Pomace brandy preparation (distillation)

Pomace brandy is the liquor distilled from cake, lees (grape pomace) and/or wine. Since, grape pomace that contains most of the pesticide residues present on grapes, is the main raw material for pomace brandy production, residues are expected to be transferred in the final alcoholic beverage. However, although, the theoretical concentration factors of pesticide residues on grapes during the pomace brandy preparation process have been calculated to range from 10 to 574 (Cabras et al., 1997b), pesticide residues do not concentrate in the final edible product, mainly because of their moderate volatility that disables their transfer through distillation. In relative reported studies performed in laboratory scale processes no detectable residues of fosetyl aluminium, fenarimol, methidathion, myclobutanil, iprodione, metalaxyl, triadimefon, parathion methyl and dimethoate were detected in the final-distilled spirits of wine, whereas vinclozolin, fenthion, quinalphos and benalaxyl where transferred to the final spirits at very low percentages (0.1-2% from lees and 5-13% from wine). That remarkable decrease in residues during the distillation process could depend on fact that very low amounts of residue are transported by alcoholic vapors while higher amounts are transported by water vapors (Cabras & Angioni, 2000; Cabras et al., 1997b).

4.9.3 Beer production (malting)

In general during beer production, pesticide residues that might be present on barley and hops are transferred in beer according to their Kow. Water soluble compounds are more likely to be transferred into the beer whereas the more lipophillic are retained in the lipophillic trub and/or are metabolized by the biotic metabolism of the added yeast and/or degraded by abiotic processes from the relative reductive environment, during fermentation. In practice, a combination of the malting process, the high dilution with water and the filtering processes generally result in non detectable residues in beer.

Malting of barley resulted in loss of about 80% of fenitrothion residues (Holland 1994), 58% of tebuconazole, 48% of fenarimol, 22-23% of Z- and E-dimethomorph, and almost diminished residues of chlorfenapyr, quinozyfen and pyridaben (Hengel & Shibamoto 2002). Synthetic pyrethroid residues underwent similar high losses during malting (Holland 1994). However, in the final produced beer dimethomorph residues were detected at about 0.31% of original levels on hops (tebuconazole, fenarimol, chlorfenapyr, quinozyfen and pyridaben were not detected), glyphosate residue levels in beer were about 4% of original levels in the barley (JMPR, 1987), monocrotophos at lower than 1% of the original concentration (JMPR, 1972), and myclobutanil at lower than 0.009 (JMPR 1998).

5. Conclusions

Literature review presented in this chapter demonstrated large differences among the different pesticide residues persist in or on RAC and among the different food preparation processes studied. The largest reductions in residue levels of almost all common classes of pesticides were noticed through peeling (70-100%), juicing (73-91%), and alcoholic beverages production (most pesticides were reduced at 70-100%) and the lowest through washing (22-60%) with tap water. Grain milling and drying processes could reduce pesticides at relevant percentages (58-100% and 57.5-98% respectively) and attention should be paid for pesticides with high Kow in grain milling and for thermally stable pesticides during oven drying. Furthermore, special concern on residues levels in the final product should be taken in unrefined oil production since pesticides reduction is low for most fat soluble pesticides (35 - 78%) and processing factors range from 2 to 7. Similar considerations should be taken on pesticides residues in ready to eat food after cooking since their reduction after cooking may vary widely among the different pesticides and among the different cooking procedures (especially with or without water addition through heating) studied. The study on published data presented in this chapter pointed the selective reduction of pesticides tentative to hydrolysis (acid catalyzed mainly), heat degradation and enzymatic degradation mechanisms through cooking.

Current literature on metabolites that may be formed during storage and food processing is limited and more toxic metabolites than parent compounds with critical physicochemical processes according to the mechanisms described in this chapter in the different food preparation process should be studied for their fate in the final edible products (e.g. acid sensitive pesticides as ETDC in tomatoes after comminution and mild cooking; oxon metabolites of organophosphorous pesticides after washing of RAC with ozonated water and juicing; oxon metabolites in vegetable soups and others). This future scope is very important and of great priority to have a complete conclusion summary of the pesticides fate after the different handling of food processing technology. Results would further enable research on health risk characterization from realistic dietary exposures to pesticide residues; enable the settlement of MRLs or processing factors for pesticide residues in food products, and further enable the optimization of food technology processes with regard to pesticide residues dissipation.

6. References

Abou-Arab, A.A.K. (1999). Behavior of pesticides in tomatoes during commercial and home preparation. *Food Chemistry*, 65 (June 1999) 509–514, 0308-8146.

Amvrazi E. & Albanis, T. (2008). Multiclass pesticide determination in olives and their processing factors in olive oil: Comparison of different olive oil extraction systems, *Journal of Agricultural and Food Chemistry*, 56 (June 2008) 5700–5709, 0021-8561.

Amvrazi, E. & Albanis, T. (2009). Pesticide residue assessment in different types of olive oil and preliminary exposure assessment of Greek consumers to the pesticide residues detected. *Food Chemistry*, 113 (March 2009) 253-261, 0308-8146.

Angioni, A.; Garau, V.L.; Aguilera Del Real, A.; Melis, M.; Minelli, E.V.; Tuberoso, C.; Cabras, P. (2003). GC-ITMS determination and degradation of captan during winemaking, *Journal of Agricultural and Food Chemistry*, 51 (October 2003) 6761–6766, 0021-8561.

Angioni, A.; Schirra, M.; Garau, V.L.; Melis, M.; Tuberoso, C.I.G.; Cabras, P. (2004). Residues of azoxystrobin, fenhexamid and pyrimethanil in strawberry following field treatments and the effect of domestic washing. *Food Additives and Contaminants*, 21 (November 2004) 1065–1070, 1944-0049.

Athanasopoulos, P.E. & Pappas, C. (2000). Effects of fruit acidity and storage conditions on the rate of degradation of azinphos methyl on apples and lemons. *Food Chemistry*, 69, 1 (April 2000) 69-72, 0308-8146.

Athanasopoulos, P.E.; Pappas, C.J.; Kyriakidis, N.V. (2003). Decomposition of myclobutanil and triadimefon in grapes on the vines and during refrigerated storage. *Food Chemistry*, 82, 3 (August 2003) 367–371, 0308-8146.

Athanasopoulos, P.E.; Pappas, C.J.; Kyriakidis, N.V.; Thanos, A. (2005). Degradation of methamidophos on soultanina grapes on the vines and during refrigerated storage. *Food Chemistry*, 91, 2 (June 2005) 235–240, 0308-8146.

Balinova, A.; Mladenova, R.; D. Obretenchev, D. (2006). Effect of grain storage and processing on chlorpyrifos-methyl and pirimiphos-methyl residues in post-harvest-treated wheat with regard to baby food safety requirements. *Food Additives and Contaminants*, 23, (April 2006) 391–397, 1944-0049.

Baur, P.; Buchholz, A.; Schönherr, J. (1997). Diffusion in plant cuticles as affected by temperature and size of organic solutes: similarity and diversity among species. *Plant, Cell and Environment*, 20, (August 1997) 982–994, 0140-7791.

Boulaid, M.; Aguilera, A.; Camacho, F.; Soussi, M.; Valverde, A. (2005). Effect of Household Processing and Unit-to-Unit Variability of Pyrifenox, Pyridaben, and Tralomethrin Residues in Tomatoes. *Journal of Agricultural and Food Chemistry*, 53 (April 2005) 4054-4058, 0021-8561.

Burchat, C.S.; Ripley, B.D.; Leishmann, P.D., Ritcey; G.M.; Kakuda, Y.; Stephenson, G.R. (1998). The distribution of nine pesticides between the juice and pulp of carrots and tomatoes after home processing. *Food Additives and Contaminants*, 1998, 15, 61–71, 1944-0049.

Caboni, P.; Minello, E.V.; Cabras, M.; Angioni, A.; Sarais, G.; Dedola, F.; Cabras, P. (2007). Degradation of pyrethrin residues on stored durum wheat after postharvest treatment. *Journal of Agricultural and Food Chemistry*, 55 (January 2007) 832-835, 0021-8561.

Cabras, P. & Angioni, A. (2000). Pesticide residues in grapes, wine, and their processing products. *Journal of Agricultural and Food Chemistry*, 48 (March 2000) 967-973, 0021-8561.

Cabras, P.; Angioni, A.; Vicenzo L.; Garau, V.L.; Minelli, E.V.; Melis, M.; Pirisi, F.M. (1997b). Pesticides in the distilled spirits of wine and its byproducts. *Journal of Agricultural and Food Chemistry*, 45 (June 1997) 2248-2251, 0021-8561.

Cabras, P.; Angioni, A.; Garau, V.L.; Melis, M.; Pirisi, F.M.; Cabitza, F.; Cubeddu, M. (1998b). Pesticide residues on field-sprayed apricots and in apricot drying processes. *Journal of Agricultural and Food Chemistry*, 46 (May 1998) 2306-2308, 0021-8561.

Cabras, P.; Angioni, A.; Garau, V.L.; Melis, M.; Pirisi, F.M.; Karim, M.; Minelli, E.V. (1997a). Persistence of insecticide residues in olives and olive oil. *Journal of Agricultural and Food Chemistry*, 45 (June 1997) 2244-2247, 0021-8561.

Cabras, P.; Angioni, A.; Garau, V.L.; Pirisi, F.M.; Brandolini, V.; Cabitza, F.; Cubeddu, M. (1998a). Pesticide residues in prune processing. *Journal of Agricultural and Food Chemistry*, 46 (July 1998) 3772-3774, 0021-8561.

Cabras, P.; Angioni, A.; Garau, V.L.; Pirisi, F.M.; Cabitza, F.; Pala, M. (2000). Acephate and buprofezin residues in olives and olive oil, *Food Additives and Contaminants*, 17 (May 2000) 855-858, 1944-0049.

Cabras, P.; Garau, V.L.; Melis, M.; Pirisi, F.M.; Spanedda L. (1993). Persistence and fate of Fenthion in olives and olive by products, *Journal of Agricultural and Food Chemistry*, 41 (December 1993) 2431-2433, 0021-8561.

Cabrera, H.A.P.; Menezes, H.C.; Oliveira, J.V.; Batista, R.F.S. Evaluation of Residual Levels of Benomyl, Methyl Parathion, Diuron, and Vamidothion in Pineapple Pulp and Bagasse (Smooth Cayenne). *Journal of Agricultural and Food Chemistry*, 48 (October 2000) 5750-5753, 0021-8561.

Calhelha, R.C.; Andrade, J.V.; Ferreira, I.C.; Estevinho, L.M. (2006). Toxicity effects of fungicide residues on the wine-producing process. *Food Microbiology*, 23 (June 2006) 393–398, 0740-0020.

Celik, S.; Kunc, S.; Asan, T. (1995). Degradation of some pesticides in the field and effect of processing. *Analyst*, 120 (June 1995) 1739–1743, 0003-2654.

Cengiz, M.F.; Certel, M.; Göçmen, H. (2006). Residue contents of DDVP (Dichlorvos) and diazinon applied on cucumbers grown in greenhouses and their reduction by duration of a pre-harvest interval and post-harvest culinary applications. *Food Chemistry*, 98 (2006) 127–135, 0308-8146.

Cengiz, M.F.; Certel, M.; Karakas, B.; Göçmen, H. (2007). Residue contents of captan and procymidone applied on tomatoes grown in greenhouses and their reduction by duration of a pre-harvest interval and post-harvest culinary applications. *Food Chemistry*, 100 (2007) 1611 - 1619, 0308-8146.

Chavarri, M.J.; Herrera, A.; Ariño, A. (2005). The decrease in pesticides in fruit and vegetables during commercial processing. *International Journal of Food Science and Technology*, 40 (February 2005) 205-211, 0950-5423.

Clavijo, M.P.; Medina, M.P.; Asensio, J.S.; Bernal, J.G. (1996). Decay study of pesticide residues in apple samples. *Journal of Chromatography A*, 740 (July 1996) 146-150, 0021-9673.

Codex Alimentarius Commission. (2007). Establishment of MRLs and/or processing factors for processed and ready-to-eat foods, CX/PR 07/ 39/8 April 2007.

Coulibaly, K.; Smith, J.S. (1993). Thermostability of organophosphate pesticides and some of their major metabolites in water and beef muscle. *Journal of Agricultural and Food Chemistry*, 41 (October 1993) 1719-1723, 0021-8561.

De Melo Abreau, S.; Caboni, P.; Pirisi, F.M.; Cabras, P.; Alves, A.; Garau, V.L. (2006). Residues of the fungicide famoxadone in grapes and its fate during wine production. *Food Additives and Contaminants*, 23 (March 2006) 289-294, 1944-0049.

EC (European Communities) (2005). Council Directive 1711/2005 of 19 October 2005. Relating to the fixing the olive yields and oil yields for the 2004/05 marketing year L274/57, 20.10.2005. Off. J. Eur. Communities 2005.

EPA (1996). Environmental Protection Agency, Prevention, Pesticides and Toxic Substances (7101)-Test Guidelines, OPPTS 860.1520, Processed Food/Feed—Residue Chemistry; *EPA 712-C-96-184*, August, 1996.

Femenia, A.; Sánchez, E.S.; Simal, S.; Rossello, C. (1998). Effects of drying pretreatments on the cell wall composition of grape tissues. *Journal of Agricultural and Food Chemistry*, 46, (January 1998) 271-276, 0021-8561.

Fenoll, J.; Ruiz, E.; Hellín, P.; Lacasa, A.; Flores, P. (2009). Dissipation rates of insecticides and fungicides in peppers grown in greenhouse and under cold storage conditions. *Food Chemistry*, 113 (March 2009) 727-732, 0308-8146.

Fernández-Cruz, M.L.; Barreda, M.; Villarroya, M.; Peruga, A.; Llanos, S.; García-Baudín, J.M. (2006). Captan and fenitrothion dissipation in field-treated cauliflowers and effect of household processing. *Pest Management Science*, 62 (May 2006) 637-645, 1526-498X.

Fernández-Cruz, M.L.; Villarroya, M.; Llanos, S.; Alonso-Prados, J.L.; García-Baudín, J.M. (2004). Field incurred fenitrothion residues in kakis: comparison of individual fruits, composite samples, peeled and cooked fruits. *Journal of Agricultural and Food Chemistry*, 52 (January 2004) 860-863, 0021-8561.

Ferreira, J.R.; Tainha, A.M. (1983). Organophosphorous insecticide residues in olives and olive oil. *Pesticides Science*, 14 (April 1983) 167-172, 0031-613X.

Fukazawa, T.; Aoyama, M.; Tsutsumi, T., Tokairin, S.; Ehara, H.; Maruyama, T.; Niiya, I. (1999). Behavior of Pyrethroid Pesticides toward Edible Oils during Refining. *Journal of Japan Oil Chemists' Society*, 48, 9 (1999) 885-889, 1341-8347.

Gennari, M.; Négre, M.; Gerbi, V.; Raimondo, E.; Minati, J.L.; Gandini, A. (1992). Chlozolinate fate during vinification process. *Journal of Agricultural and Food Chemistry*, 40 (May 1992) 898-900, 0021-8561.

Guardia Rubio, M.; Ruiz Medina, A.; Molina Diaz, A.; Cañada Ayora, M.J. (2006). Influence of harvesting method and washing on the presence of pesticide residues in olives and olive oil, *Journal of Agricultural and Food Chemistry*, 54 (October 2006) 8538-8544, 0021-8561.

Guardia Rubio, M.; Ruiz Medina, A.; Molina Diaz, A.; Cañada Ayora, M.J. (2007). Multiresidue analysis of three groups of pesticides in washing waters from olive processing by solid-phase extraction-gas chromatograhy with electron capture and thermionic specific detection. *Microchemical Journal*, 85 (April 2007) 257-264, 0026-265X.

Hengel, M.J. & Shibamoto, T. Method development and fate determination of pesticide-treated hops and their subsequent usage in the production of beer. *Journal of Agricultural and Food Chemistry*, 50 (May 2002) 3412-3418, 0021-8561.

Hilbert, G.; Lillemark, L.; Balchen, S.; Højskov, C.S. (1998). Reduction of organochlorine contaminants from fish oil during refining. *Chemosphere*, 37 (September 1998) 1241-1252, 0045-6535.

Holden, A.J.; Chen, L.; Shaw, I.C. (2001). Thermal stability of organophosphorus pesticide triazophos and its relevance in the assessment of risk to the consumer of triazophos residues in food. *Journal of Agricultural and Food Chemistry*, 49 (December 2001) 103-106, 0021-8561.

Holland, P.T.; Hamilton, D.; Ohlin, B.; Skidmore, M.W. (1994). Effects of storage and processing on pesticide residues in plant products, *Pure and Applied Chemistry*, 66 (July 1994), 335-356, 0033-4545.

Howard, S.F. & Yip, G. (1971). Stability of metallic ethylenebisdithiocarbamates in chopped kale. *Journal of the Association of Official Analytical Chemists*, 54 (1971) 1371-1372, 0004-5756.

JMPR (1992). refers to the FAO Plant Production and Production Paper series "Pesticide Residues in Food-Evaluations Part 1" published annually by FAO, Rome. Each issue summarises residue data submitted to the previous JMPR meeting and is organised by pesticide in alphabetical order.

Kaushik, G.; Satya, S.; Naik, S.N. (2009). Food processing a tool to pesticide residue dissipation - A review, *Food Research International*, 42 (January 2009) 26–40, 0963-9969.

Knio, K.M.; Dagher, S.; Saad, A. (2000). The fate and persistence of zineb, maneb, and ethylenethiourea on fresh and processed tomatoes. *Food Additives and Contaminants*, 2000, 17 (5), 393-398, 1944-0049.

Kontou, S.; Tsipi, D.; Tzia, C. (2004a) Stability of the dithiocarbamate pesticide maneb in tomato homogenates during cold storage and thermal processing. *Food Additives and Contaminants*, 21 (November 2004) 1083–1089, 1944-0049.

Kontou, S.; Tsipi, D.; Tzia, C. (2004b). Kinetics of Maneb Degradation during Thermal Treatment of Tomatoes. *Journal of Agricultural and Food Chemistry*, 52 (February 2004) 1212-1219, 0021-8561.

Krol, W.J.; Arsenault, T.L.; Pylypiw, H.M. Jr.; Mattina M.J.I. (2000). Reduction of Pesticide Residues on Produce by Rinsing. *Journal of Agricultural and Food Chemistry*, 48 (August 2000) 4666-4670, 0021-8561.

Kyriakidis, N.V.; Pappas, C.; Athanasopoulos, P. (2005). Degradation of fenthion and fenthion sulfoxide on grapes on the vines and during refrigerated storage. *Food Chemistry*, 91 (June 2005) 241–245, 0308-8146.

Leandri, A.; Pompi, V.; Pucci, C.; Spanedda, A.F. (1993). Residues on olives, oil and processing wastewaters of pesticides used for control of Dacus oleae (Gmel.) (Dipt.,

Tephritidae), *Anz. Schädlingskde., Pflanzenschutz, Umweltschutz*, 66 (1993) 48-51, 0340-7330.

Lentza-Rizos, C. & Avramides, E.J. (1995), Pesticide Residues in Olive Oil. *Reviews of Environmental Contamination and Toxicology*, 141 (1995) 111-134, 0179-5953. ISBN: 978-0-387-94453-1

Lentza-Rizos, C. & Avramides, E.J.; Kokkinaki, K. (2006). Residues of Azoxystrobin from grapes to raisins, *Journal of Agricultural and Food Chemistry*, 54 (December 2006) 138-141, 0021-8561.

Lentza-Rizos, Ch. & Kokkinaki, K. (2002). Residues of cypermethrin in field-treated grapes and raisins produced after various treatments, *Food Additives and Contaminants*, 19 (July 2002) 1162–1168, 1944-0049.

Liapis, K.S.; Miliadis, G.E.; Aplada-Sarlis, P. Dicofol residues on field sprayed apricots and in apricot juice. *Bulletin of Environmental Contamination and Toxicology*, 54 (August 1995) 579-583, 0007-4861.

Likas, D. (2009). Analytical method development for the determination of selected pesticide residues in grapes, vine leaves, wine and soil. Residues dissipation study in vine and during vinification process. Doctoral Thesis, University of Thessaly, Volos, 2009.

Likas, D.T. & Tsiropoulos, N.G. (2009). Residue screening in apple, grape and wine food samples for seven new pesticides using HPLC with UV detection. An application to trifloxystrobin dissipation in grape and wine. *International Journal of Environmental Analytical Chemistry*, 89 (July 2009) 857–869, 0306-7319.

Marei, A.E.; Khattab, M.M.; Mansee, A.H.; Youssef, M.M.; Montasser, M.R. (1995). Analysis and dissipation of deltamethrin in stored wheat and milled fractions. *Alexandria Journal of Agricultural Research*, 16 (1995) 275–291, 0044-7250.

Miliadis, G. E.; Aplada-Sarlis, P.; Liapis, K. S. (1995). Disappearance of tetradifon from field sprayed apricots and the apricot juice produced from them. *Journal of Agricultural and Food Chemistry*, 43 (June 1995) 1698-1700, 0021-8561.

Miyahara, M. & Saito, Y. (1993). Pesticide removal efficiencies of soybean oil refining processes. *Journal of Agricultural and Food Chemistry*, 41 (May 1993) 731-734, 0021-8561.

Nagayama, T. (1996). Behavior of residual organophosphorus pesticides in foodstuffs during leaching or cooking. *Journal of Agricultural and Food Chemistry*, 44 (August 1996) 2388-2393, 0021-8561.

Ou-Yang, X.K.; Liu, S.M.; Ying, M. (2004). Study on the mechanism of ozone reaction with parathion-methyl. *Safety and Environmental Engineering*, 11 (2004) 38–41, 1671-1556.

Pugliese, P.; Moltó,J.C.; Damiani, P.; Marín, R.; Cossignani, L.; Mañes, J. (2004). Gas chromatographic evaluation of pesticide residue contents in nectarines after non-toxic washing treatments. *Journal of Chromatography A*, 1050 (August 2004) 185–191, 0021-9673.

Randhawa, M.A.; F. Muhammad Anjum, F.M.; Ahmed, A.; Randhawa, M.S. (2007). Field incurred chlorpyrifos and 3,5,6-trichloro-2-pyridinol residues in fresh and processed vegetables. *Food Chemistry*, 103 (2007) 1016 - 1023, 0308-8146.

Rasmusssen, R.R.; Poulsen, M.E.; Hansen, H.C.B. (2003). Distribution of multiple pesticide residues in apple segments after home processing. *Food Additives and Contaminants*, 20 (November 2003) 1044–1063, 1944-0049.

Rouchaud, J.; Gustin, F.; Creemers, P.; Goffings, G.; Herrgods, M. (1991). Fate of the fungicide tolylfluanid in pear cold stored in controlled or non controlled atmosphere. *Bulletin of Environmental Contamination and Toxicology*, 46 (1991) 499–506, 0007-4861.

Ruiz Méndez, M.V.; Pérez de la Rosa, I.; Jiménez Márquez, A.; Uceda Ojeda, M. (2005). Elimination of pesticides in olive oil by refining using bleaching and deodorization. *Food Additives and Contaminants*, 22 (January 2005) 23-30, 1944-0049.

Tsiropoulos, N.G.; Aplada-Sarlis, P.G.; Miliadis, G.E. (1999). Evaluation of teflubenzuron residue levels in grapes exposed to field treatments and in the must and wine produced from them. *Journal of Agricultural and Food Chemistry*, 47 (October 1999) 4583-4586, 0021-8561.

Tsiropoulos, N.G.; Miliadis, G.E.; Likas, D.T.; Liapis, K. (2005). Residues of spiroxamine in grapes following field application and their fate from vine to wine. *Journal of Agricultural and Food Chemistry*, 53 (December 2005) 10091-10096, 0021-8561.

Uygun, U.; Koksel, H.; Atli, A. (2005). Residue levels of malathion and its metabolites and fenitrothion in post-harvest treated wheat during storage, milling and baking. *Food Chemistry*, 92 (October 2005) 643–647, 0308-8146.

Uygun, U.; Özkara, R.; Özbey, A.; Koksel, H. (2007). Residue levels of malathion and fenitrothion and their metabolites in postharvest treated barley during storage and malting. *Food Chemistry*, 100 (2007) 1165 - 1169, 0308-8146.

Uygun, U.; Senoz, B.; Koksel, H. (2008). Dissipation of organophosphorus pesticides in wheat during pasta processing. *Food Chemistry*, 109, 2 (July 2008) 355 - 360, 0308-8146.

Uygun, U.; Senoz, B.; Öztürk, S.; Koksel, H. (2009). Degradation of organophosphorus pesticides in wheat during cookie processing. *Food Chemistry*, 117 (November 2009) 261–264, 0308-8146.

Vioque, A.; Albi, T.; Nosti, M. (1973). Residues of pesticides in edible fats. II. Elimination of chlorinated insecticides during refining. *Grasas-y-Aceites*, 25 (1973) 20-26, 0017-3495.

WHO. (1988). Environmental Health Criteria No. 78, Dithiocarbamate pesticides, ethylenethiourea and propylene thiourea: A general introduction. *World Health Organisation*, Geneva 1988.

Will, F. & Krüger, E. (1999). Fungicide Residues in Strawberry Processing. *Journal of Agricultural and Food Chemistry*, 47 (February 1999) 858-861, 0021-8561.

Wolff, J.P. (1974). Changes in pesticides during industrial processing and refining of fats. *Review Français de Crops Gras*, 21 (1974) 161-164, 0035-300.

Zabik, M.J.; El-Hadidi, M.F.A. J.; Cash, N.; Zabik, M.E.; Jones, A.L. (2000). Reduction of azinphos-methyl, chlorpyrifos, esfenvalerate, and methomyl residues in processed apples. *Journal of Agricultural and Food Chemistry*, 48 (August 2000) 4199-4203, 0021-8561.

Zayed, S.M.A.D.; Farghaly, M.; Mahdy, F. (1998). Effect of commercial processing procedures on carbofuran residues in soybean oil. *Food Chemistry*, 62 (July 1998) 265-268.

Zhang, Q. & Pehkonen, S.O. (1999). Oxidation of Diazinon by Aqueous Chlorine: Kinetics, Mechanisms, and Product Studies. *Journal of Agricultural and Food Chemistry*, 47 (March 1999) 1760-1766, 0021-8561.

Analytical Methods for Performing Pesticide Degradation Studies in Environmental Samples

Zenilda L. Cardeal[1*], Amauri G. Souza[1] and Leiliane C.A. Amorim[2]
[1]*Departamento de Química, ICEx, Universidade Federal de Minas Gerais, Belo Horizonte, MG*
[2]*Departamento de Análises Clínicas e Toxicológicas – Faculdade de Farmácia, Universidade Federal de Minas Gerais, Belo Horizonte, MG*
Brazil

1. Introduction

Pesticides are groups of artificially synthesized substances, toxic and non-biodegradable in the environment, that persist after application and are subject to some chemical processes of degradation, hydrolysis, oxidation, and photolysis by the ecosystem (Ormad et al., 1997; Ariaz-Estevez et al., 2008). According to Law 7802 of 11 July 1989, pesticides and similar substances are defined as those products and agents of physical, chemical or biological processes intended for use in the production, storage and processing of agricultural products, in pastures, in the protection of native or implanted forests and other ecosystems, and also in urban, aqueous and industrial environments, whose purpose is to change the composition of the flora or fauna, to preserve them from the harmful action of living organisms considered to be harmful (Mahalashmi et al., 2007). This group of substances can be classified according to the purpose for which they are intended, the mode or period of action, or the chemical function such as: insecticide (insects), fungicide (fungi), rodenticide (rodents), moluscicide (snails), defoliant (leaf harvesting), dissecting (foliage). They are extensively used as insecticides, herbicides and nematicides, and they are included in the classes of organochlorides, organophosphates, and pyrethroids. More than 500 different formulations of pesticides have been used in the environment, largely in agricultural activities, for many decades. Pesticides are widely studied as environmental contaminants because of their extensive use in the control of pests affecting agricultural crops, homes, and gardens. Because of their chemical characteristics, they represent a type of pollutant that shows variable persistence and biochemical and photochemical degradation (Bandala et al., 2007).
Some studies show that, less than 1% of the total quantity of pesticides used in agriculture reaches its target. The remainder contaminates soil and other environmental compartments, air, and surface and groundwater. The fact that they are not biodegradable, together with their continued use, makes them a significant problem and a critical issue, with potentially damaging and unforeseen consequences for the future (Kapustka et al., 1996; Bandala et al., 2002; Acero et al., 2008; Veiga et al., 2006). In 1995, US $1.6 million in actual pesticides were sold in Brazil. That amount increased to $2.5 million in 2007 (Andreu & Picó, 2004). The use of pesticides in the world has increased five fold in the last 30 years (Nawab et al., 2003).

According to the World Health Organization (WHO), approximately three million people are intoxicated each year as a result of the use of pesticides. In addition to the toxicity to humans, the presence of these products in the environment poses a risk to water quality and the ecosystem (Veiga et al., 2006). When applied to the soil, they can reach other levels as a result of mobility, sorption, volatilization, erosion, and leaching, thereby contaminating various environments (Andreu & Picó, 2004; Nawab et al., 2003), as is shown in Fig. (1).

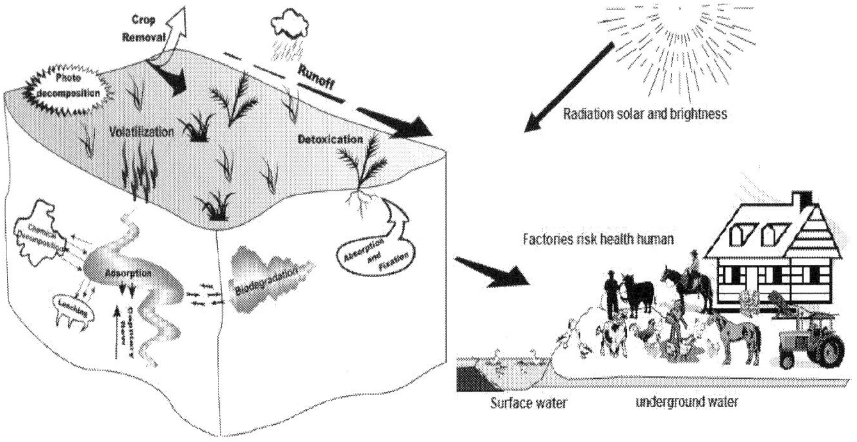

Fig. 1. Pesticide contaminants in the environment (Strandberg et al., 1998; Bavcon et al., 2002, adapted)

Other important aspects to be considered are the products of pesticide transformation (TP). There is great interest in studies on the formation of pesticide sub-products in the environment, since they can present a greater risk to the ecosystem than the original pesticides (Nawab et al., 2003; Sinclair & Boxall, 2003; Pozo et al., 2001; Sabik et al., 2000). On the other hand, some pesticide TP may present lower toxicity than the original substances from which they are formed (Borga et al., 1998). Thus, the results of pesticide use have not been completely elucidated because most studies are focused principally on the primary residues and not on their transformation products (Strandberg et al., 1998).

Some pesticides are considered to be persistent organic pollutants (POP) in the environment (Bavcon et al., 2002). These POP possess long half-lives and can accumulate in the environment and in organisms, being transferred throughout the food chain until they reach human beings (Kodaka et al., 2003). On the other hand, many pesticides can be degraded. The degradation processes generate a large number of sub-products in low concentrations that are considered to be beneficial for the systems for treatment and disinfection of crops, soils and groundwater, but hinder chemical analysis. Even in low concentrations, these residues can be prejudicial to human health and the environment because of the accumulative effect (Kapustka et al., 1996; Bandala et al., 2002). Pesticides are degradable by microbiological or chemical processes. Chemical degradation occurs by photolysis, oxidation and reduction reactions (Sassman et al., 2004), while biological degradation occurs through the action of environmental microorganisms (Nawab et al., 2003; Sassman et al., 2004; Ghardiri & Rose, 2001). The microorganisms are usually distributed in the first few centimeters of the earth's surface, where the largest quantity of the organic compounds that serve as food, including the pesticides, exist (Navarro et al., 2004; Li et al., 2002; Monkiedje et al., 2003).

The extensive presence of pesticides in the water and soil has stimulated interest in finding solutions for the treatment and/or removal of residues from the environment. Many techniques, such as adsorption, filters, biological treatment, and degradation by advanced oxidative processes (AOP) that utilize TiO_2/UV as a catalyst, photo-Fenton reagents (FR) (Legrini et al., 1993), and ozonation processes with O_3, O_3/UV, and O_3/H_2O_2 (Masten & Davies, 1994), are presently being studied for removal of pesticides.

Several analytical methods have been developed for the identification and quantification of pesticide residues and their by-products. Among the most widely used techniques are gas chromatography (GC) and liquid chromatography (LC), both in combination with mass spectrometer (MS) detector (Arnold et al., 1995). In conventional GC-MS methods for the determination of pesticides in environmental waters, solid phase extraction (SPE) and liquid-liquid extraction (LLE) are used for sample preparation. However, these methods require large volumes of organic solvents (Louter et al., 1996; Hyotylainen et al., 1998; Jongenotter et al., 1998). Therefore, several other extraction methods have been proposed, among them the technique of solid phase microextraction (SPME) (Louch et al., 1992; Zhang et al., 1994; Doong & Liao, 2001), which will be emphasized in this chapter. This review describes the current studies on degradation of pesticides and the aspects involved in the analysis of pesticides and their sub-products, sample preparation, methods of determination, and analytical techniques.

2. Environmental risk and the use of pesticides

Environmental pollution by pesticides depends on several variables, including the type and quantity of the products employed. The factors that influence the transport of these compounds to surface waters are their physical-chemical and biological properties and their capacity for degradation. These factors include solubility, vapor pressure, the partition coefficient of organic carbon with water (Koc), and the octanol-water partition coefficient (Kow). Compounds with a high solubility and low adsorption have resulted in major contamination of surface and groundwater. In addition, the pH and soil temperature, climatic conditions, landscape characteristics, including topography and drainage, and appropriate practices, also influence the pesticide content of water. Pesticides with a high solubility in water can easily contaminate surface and ground water. The increase of organic matter in the soil increases the capacity for adsorption of pesticides by reducing leaching. For example, cationic pesticides are strongly adsorbed by electrostatic interactions with negative charges in the soil surface. The weak bases with pKa values close to the soil pH are strongly adsorbed. The acid pesticides are ionized, forming negative charges, and are not adsorbed by minerals in the soil because the negative charges cause repulsion in the interactions with soil organic matter (Jinno et al., 1996; Guo et al., 2000).

The process of pesticide degradation plays an important role in the removal of their residues from the aquatic environment to mitigate the problem of pollution. This process is governed by biotic and abiotic factors (including enzymatic catalysis by microorganisms) that can cause very complex reactions involving a variety of interactions of pesticides with microorganisms that are soil constituents. The adsorption also plays a key role in the dynamics of transmission, persistence, transformation, and bioaccumulation of pesticides

(Jonge et al., 1996; Gao et al., 1998). It depends on the characteristics of these compounds and the organic matter in the soil (Spark & Swift, 2002; Coquet, 2003; Ahmad et al., 2006). The pesticides that bind covalently to humic soil matter have functions similar to the components of humus, and the processes of interaction with organic matter are generally governed by oxidative reactions. Microorganisms have always been reported as mediators in both the interaction of pesticides with the soil and their degradation (Gevao & Semple, 2000; Dubus et al., 2001). The adsorption of weakly acid organic pesticides in the soil depends on the composition of the soil and its pH (Clause & Fabricius, 2002; Boivin et al., 2004; Wauchope et al., 2002), and may sometimes favor the leaching by surface and groundwater.

Some studies have shown major interactions between the adsorption and degradation of pesticides (Guo et al., 2000; Gevao et al. 2000), since the chemical adsorption reduces the access of microorganisms, thereby limiting the degradation and transport of these compounds (Selim et al., 1999; Koskinen et al., 2001; Moyer et al., 1972). In studies assessing the rates of microbial degradation of these substances, the authors concluded that, under certain conditions of temperature and pH, the pesticides may deteriorate even before being adsorbed by the soil (Park et al., 2003; Bolan et al., 1996; Dyson et al., 2002; Elliot et al., 2000; Roulier & Jarvis, 2003). In a study that involved the evaluation of the degradation of 2,4-D in ten types of natural soils, a high degree of adsorption was observed to be associated with high microbial activity (Bolan et al., 1996). The penetration of pesticides into the soil and groundwater can be realized by galeries and by inflitration (Worrall et al., 2007). Infiltration is more common and represents a potent source of environmental pollution of groundwater. Some of the properties that favor the environmental contamination by adapted pesticides are presented in Table 1 (Strandberg et al., 1998).

Parameter	Value
Water solubility	> 30mg/L
K_{oc}*	< 5, usually < 1
K_{ow}*	< 300
Speciation	Negatively, fully or partially charged at ambient pH
Hydrolysis half-life	> 25 weeks
Photolysis half-life	> 1 week
Field dissipation half-life	> 3 weeks

*K_{oc}= organic carbon content/ water partition coefficient

*K_{ow}= octanol-water partition coefficient

Table 1. Pesticide Properties Indicating their High Potential for Groundwater Contamination.

3. Degradation process

The degradation processes described in the literature demonstrate a great efficiency in the decontamination of systems contaminated by pesticides. Several factors influence the rate of degradation, such as the chemical structure of the pollutants, pH, iron concentration, hydrogen peroxide and the organic load. Because of the great potential of contamination by pesticide residues and the variation in the time necessary for natural degradation, it is necessary to discover those processes that accelerate the decontamination of the affected environment. Thus, several degradation processes such as photocatalytic degradation, advanced oxidative processes, phytoremediation, bioremediation, ozonation and photo-Fenton reactions have been proposed. All these systems are considered to be efficient for pesticide degradation.

Microorganisms are considered to be the principal agents for the degradation of pesticides in bioremediation processes. Considering that the earth is the home for an uncountable number of microorganism species, the pesticides applied to these soils probably suffer an accelerated degradation by these organisms. Bhalero et al (Wyss et al., 2006), were able to remove the pesticide endosulfan from the environment. They isolated 16 microorganismos from the soil for this purpose. *Aspergilus* of the funghii kingdom totally removed endosulfan after incubating the mushroom for 12 days with the pesticide. The levels evaluated were between 35.0 and 350.0 $mg.L^{-1}$. The results demonstrated that *Aspergilus* is a potent and easily acquired bioremediating agent that could be used to remove other pollutants from water and even soils.

Wyss et al., 2006, isolated and characterized the bacterium *Pseudomonas sp* for the hydrolysis of atrazine. This microorganism used atrazine as a nitrogen, citrate and carbon source, and for the production of electron donor molecules under aerobic conditions. The degradation of approximately 100.0 $mg.L^{-1}$ of atrazine occurred. The *Pseudomonas sp* was first cultivated in flasks under aerobic conditions previous to the experiments. The authors observed that concentrations of atrazine above 100.0 $mg.L^{-1}$ were toxic to *Pseudomonas sp.* Giacomazzi & Cochet, 2004, studied the transformation of the herbicide diuron in water by microorganisms present in the soil. The reaction was catalyzed by OH^- and H^+ with organic and inorganic matter from soil dissolved in the aqueous phase. The proposed system presented good results for the chemical degradation of diuron. The authors suggest that microorganisms could be used to promote the biological treatment of polluted sewage water.

Phytoremediation is a degradation process that uses plants to degrade or assimilate contaminants through natural processes. The success of this process is largely the result of the photosynthetic activity and growth rate of plants. Olette et al., 2008, investigated the potential of the phytoremediation process to remove pesticides in water bodies. The ability of three aquatic plants - *Lemnos minor, Elodea canadensis* and *Cabomba aquatico* to degrade pesticides by exposing the plants to five concentrations varying from 0 to 1.0 $mg.L^{-1}$. *L. minor* was observed to be the most efficient for the capture of pesticides, followed by *E. canadensis* and finally *C. aquatico* The percentage removal of pesticides ranged from 2.5% to 50% during four days of incubation and reached 100% after seven days of incubation.

In another study, Bouldin et al., 2006, investigated the removal of atrazine and lambda-cyhalothrin from water and sediment by hydroponic plants. They used *Juncus effusus* and *Luwigia peploides*. In this study, 98% degradation in just 48 hours of exposure of the pesticide to plants was observed. The results obtained by Bouldin and Olleta demonstrate the potential of efficient phytoremediation processes for removal of pesticide residues from the environment. Xia & Xiangjuan, 2006, investigated the disappearance of the persistent pesticide ethion from water by the process of phytoremediation. They used the plant *Eichlornia crassips*, which was able to degrade 98% of the ethion. The authors concluded that this plant could be used as an efficient, economical, and ecological alternative to hasten the removal and degradation of ethion in groundwater, industrial waste and other systems, together with other types of pesticides. Amaya-Chavez et al., 2006, studied the efficiency of removal of methyl parathion from water and sediments by the plant *Typhya latifolia*. This plant was able to reduce the content of pesticides in the concentration range of 0 to 200.0 mg.L^{-1}, reaching 100% degradation during 10 days of exposure. Thus, *Typhya latifolia* may be a good candidate for phytoremediation of systems contaminated with methyl parathion.

Photocatalytic oxidation is a very advanced process used to remove and degrade pesticide residues from various environments such as soil, water and food. In recent years, several studies and reviews on advanced oxidative processes (AOP) (Wu et al., 2007; Lasa et al., 2005; Carp et al., 2004), which use UV light. Titanium dioxide is used extensively in most of these studies. It is one of the most extensively used processes among those mentioned because it is a readily available reagent, chemically robust and durable (Legrini et al., 1993; Kabra et al., 2004; Leyva et al., 1998; Oh et al., 2004; Moctezuma et al., Chen & Ray, 1998; Canle-Lopez et al., 2005). Moctezuma et al., 2007, studied the photocatalytic degradation of methyl parathion pesticides, using TiO_2 in aqueous suspension. The final products were phosphoric acid and CO_2. Furthermore, Mahalakshmi et al., 2007 demonstrated that the combination of TiO_2 and ZnO_2 was very effective in the photocatalytic mineralization reactions of carbofuran in water samples under solar radiation. They also assessed the total quantity of organic carbon (TOC) to confirm the extent and effectiveness of the mineralization process used in this study. Four intermediate products of the carbofuran pesticide were formed after only six hours of reaction. Various pathways for degradation of this compound were proposed.

Shemer & Linden, 2006, studied the degradation and the degradation products of diazinon using photocatalyzed reactions with the application of UV light and UV/H_2O_2 catalyst. The addition of H_2O_2 was very effective in this process, increasing the rate of removal of diazinon from aqueous samples. Upon photolysis, the hydrogen peroxide leads to the formation of strongly oxidant species such as the hydroxyl radical, which accelerates the degradation process. Chiron et al., 2000, used AOP to remove diazinon from aquatic environments. Similar to other studies, the hydrogen peroxide was very effective and efficient in reducing the burden of diazinon in the environment.

Chiron et al., 1997, studied the degradation of endosulfan in water samples using photocatalyzed reactions with ($FeCl_3/H_2O_2$) and (TiO_2/H_2O_2). The degradation was studied with a concentration of 35.0 g.L^{-1}, and the results showed that both catalysts were effective for the removal of pesticides and can be applied to other situations and environmental compartments. Arbeli & Fuentes, 2007, investigated the process of complete mineralization in the accelerated degradation of methyl parathion, parathion, diazinon and cypermethrin pesticides using ozonation reactions. The complete mineralization was assessed by intermediate products and the formation of CO_2. Ozonation has become a safe and promising process for the removal of pesticides from water samples, plant surfaces, and domestic waste.

3.1 Kinetic study of degradation

The time necessary for degradation of pesticides is important to assess whether pollutants are persistent in the environment. The disappearance of these compounds is related to several factors such as pH, temperature, light, oxygen, and quantity of organic matter, which alter the kinetics of degradation. The kinetic studies to assess degradation of organic pollutants can be performed and assessed by the rate of mineralization (Dyson et al., 2002). The rate of mineralization is determined by monitoring of inorganic compounds such as CO_2, Cl^-, SO_4^{2-}, NO_3^- and PO_4^{3-}. The most common method of analysis to assess

mineralization is the determination of total organic carbon (TOC). It is a rapid measurement of high reliability and sensitivity to the level of $pg.L^{-1}$.
Several authors have dedicated themselves to the study of the degradation kinetics to evaluate the persistence of the pesticides in the environmental. Penuela & Barcelo, 1998, described the use of TOC to monitor the products of catalytic degradation of triazophos with TiO_2 catalyst. The results showed that the triazophos absorbs light in the presence of TiO_2 and is degraded in 4.5 hours. The rate of carbon dioxide formation is an indication of the disappearance of the pesticide. Prevot & Pramauro, 1999, also observed the complete disappearance of 2,3,6-trichlorobenzoic acid after 60 minutes of irradiation in the presence of TiO_2. The degradation of this substance increased considerably at pH = 3, while the rates were lower at other pH values. Blanco et al., 1996, studied the degradation kinetics of pesticides in water samples by ultraviolet light. They observed that the degradation became effective after irradiation for a period longer than five hours. In another study, Konstantinou et al., 2001, investigated the photochemical degradation of herbicides (atrazine, prometryn and propazine) in different types of natural waters (river, sea and lake) and soils. The monitoring showed that the compounds exhibited degradation kinetics with very different periods of irradiation, being greater than 10 hours for the three types of matrices. Hermam et al., 1999, investigated the degradation kinetics of methyl pirimiphos in water with different photocatalytic irradiation times. The results showed that over 98% of the pesticide disappeared after 10 minutes of irradiation with $1.5x10^{17}$ photons per second. Complete (100%) degradation of the pesticide occurred upon irradiation with $1.4x10^{16}$ photons per second over a longer period (30 minutes). Shankar et al., 2004, studied the influence of pH on the degradation kinetics of pesticides in aqueous solutions. The results indicated that the degradation was rapid in solutions at pH between 2 and 7, reaching 100% degradation. At pH's above 7, the degradation rates were below 60%. A similar study was conducted by Kuo, 2002, who investigated the kinetics of degradation of the carbofuran pesticide between pH 4 and 7, using TiO_2 as the catalyst. The results showed that the degradation was slow at pH's above 7. Uygun, 1997, studied the rate of degradation of pesticides in stored carrots. The studies were performed under sunlight and ultraviolet light of 280-300 nm. The degradation of pesticides occurred after 3 to 5 hours of solar irradiation, while the degradation by UV radiation required 10 hours. Rafqah et al., 2005, investigated the rate of degradation of sulfuron methyl by a photocatalytic process in the presence of two different titanium dioxides, Degussa P_{25} and Millennium PC_{500}. The times necessary for the disappearance of the pesticide in the presence of Degussa P_{25} and Millennium PC_{500} were 20 and 80 minutes, respectively, the efficiency of TiO_2 (P_{25}) being the greater. Wang & Lemley, 2001, studied the rate of degradation of the pesticide 2,4-D in natural waters by treatment with Fenton, combined with the photocatalytic degradation of pesticides with TiO_2. Rapid degradation by the Fenton reaction occurred in the presence of hydrogen peroxide; otherwise, the process was very slow.

4. Environmental analysis of the pesticides and the sub-products

The intensive use of pesticides and the consequent contamination of surface water, soil, flora, fauna and groundwater by their residues, including the products and degradation sub-products, require the use of reliable, sensitive methods for the analysis of contamination by these pesticides. Several analytical methods were developed for identification and quantification of pesticides and their degradation products in the environment. Among the main analytical techniques used are high performance liquid chromatography (HPLC) and gas chromatography (GC), since they possess the selectivity, sensitivity and high capacity necessary for identification and quantification of pesticides and their degradation by-products (Sabik et al., 2000; Wang & Lemley, 2005; Gennaro et al., 2001; Sasano et al., 2000; Pereira et al., 1990; Faria et al., 2007; Sandra et al., 1995; Carreteur et al., 1996).

4.1 Sample preparation

The methods for extraction of pesticides and clean-up of environmental samples are extremely important for their quantitative determination in the matrices of interest. The extraction techniques used to concentrate the analytes include liquid-liquid extraction (LLE), solid phase extraction (SPE) and solid phase microextraction (SPME). The SPME technique is promising and has the advantage of not using a solvent for extraction.

The LLE extraction is based on the partition of the sample between two immiscible phases (organic and aqueous) and was used in the analysis of pesticides in water and food samples. It was used for many years as an official technique of the U.S. Environmental Protection Agency (U.S. EPA) (Watts et al., 1989). It is a classical technique based on repeated extraction of 1.0 or 0.5 L of sample with organic solvents using a separatory funnel. Its main advantage is the ability to extract a wide range of compounds with a wide range of polarities.

Solid phase extraction (SPE) is a technique commonly used as a method for pretreatment of samples in trace analysis of micro-contaminants in aqueous samples. It was introduced in mid 1970 (Pichon, 1998), and it was marketed in 1978 as an alternative to LLE. The analytes contained in an aqueous matrix are extracted together with the interfering compounds after passage through a sorbent cartridge. A selective organic solvent is commonly used to extract the analytes of interest. The selection of the SPE method depends on the physico-chemical properties of the pesticides and their concentrations, so as to process the ideal volume of solvent (Thurman et al., 1990). Sasan et al., 2000, determined the pesticides and herbicides in aqueous samples using SPE. The efficiency of the extraction method for the identification of 30 compounds was demonstrated. Using the same extraction method, Carreteur et al., 1996, analyzed pesticides in samples of sea water using SPE. The technique was very efficient for the extraction of analytes at $ng.L^{-1}$ concentrations, allowing rapid preparation of the sample at the local of collection with good performance. Other studies have focused on the triazines and their degradation products extracted by this technique (Pichon et al., 1994; Dugay et al., 1998). Penuela et al., 1996, performed the analysis of endosulfan in water samples using SPE with C18 cartridges, which permitted the study of its isomers at low concentrations. In another study, Penuela et al., 1998, used SPE to monitor the kinetics of degradation of Alachlor in water. The method was efficient for the extraction of pesticides, including for various degradation products. The pre-concentration with SPE in studies of photodegradation is a good technology, making it possible to measure the organic pollutants and identify the degradation products in low concentrations. Faria et al., 2007, proposed a new method for extraction of pesticides. To do so, they associated the SPE with a polymer supported on silica. The new system presented a high potential for extraction at gL^{-1}concentrations. However, alternative methods reduce or eliminate the use of solvents in the preparation of samples for chromatographic analysis (Chiron et al., 2000), this being one of the limitations of this technique compared to more traditional methods.

Sabik et al., 2000, used SPE and SPME to analyze triazines in water samples. They discussed the advantage of SPME over SPE, since it dispenses the use of solvent. However, a large number of pesticides, including triazines and their degradation products, are easily determined and monitored efficiently at trace concentrations ($ng.L^{-1}$) in water samples with SPE.

Chiron et al., 2000, highlighted SPME, a widely used extraction method, because it does not require the use of solvents and is simple, fast and efficient. Jinno et al., 1996, analyzed residues of some pesticides in plant samples using SPME with a PDMS fiber. The results showed that it is a technique capable of extracting traces of pesticides from aqueous solutions and from the surface of plants under domestic conditions. Sabik et al., 2000, measured pesticides in surface water and groundwater using SPME as the extraction method; Polydimethylsiloxane/Divinylbenzene (PDMS/DVB), Carbowax/Divinylbenzene (CW/DVB) and Polyacrylate (PA) fibers were tested. The CW-DVB fiber in saline solutions was more efficient for atrazine and its degradation products and was suitable for extraction of more polar compounds.

There exists a discussion about the difficulty of quantitative analysis using SPME, conversely, the studies shown below presented good results for quantitative determinations using SPME. Miege & Dugay, 1998, analyzed pesticides in aqueous samples using SPME. The most adequate fiber was PDMS (100 mm), which was able to extract organochlorine pesticides of low solubility and some organophosphates. The PA fiber was shown to be ideal for the extraction of pesticides containing nitrogen and phosphorus. The data indicated that the measurable limits were from 5.0 to 100.0 ng.L^{-1}. Navalon et al., 2002, developed a method for determination of the herbicide oxidiazon in water, soil, wine and human urine samples, using the Headspace and PDMS fiber. The best responses were obtained by extraction for 25 minutes at a temperature of 100 °C.

Carvalho et al., 2008, determined organochlorine pesticides in sediments and studied their strong interaction with organic matter using SPME in headspace mode (HS-SPME) with a PDMS fiber. The method presented detection limits of 0.005 to 0.11 ng.g^{-1} of sediment and a good linearity in the 6.0- to 1000.0-ng.g^{-1} range. Xiang et al., 2008, investigated SPME using a new plythiophene (PTH) fiber to determine organochlorine pesticides in water. The results proved the ability of the new fiber to extract these compounds from aqueous samples. The detection limit was 0.5 to 10.0 ng.L^{-1}, and the calibration curve was linear in the appropriate range, 10.0 to 100.0 ng.L^{-1} (R^2 > 0.982). The method had the advantage that the fiber is robust, with durability greater than that of the PDMS fiber used by Carvalho et al, 2008.

Gupta et al., 2008, proposed an improved SPME method using a Divinylbenzene/Carboxen/Polydimethylsiloxane (DVB/CAR/PDMS) fiber with a PTFE tube for the analysis of organophosphate pesticides. The proposed combination presented good linearity in the range from 0:03 to 150.0 µg.L^{-1}, with 78% recovery and detection limits between 6.1 to 21.8 ng.L^{-1}. The method is efficient, but requires a longer time to process the samples than the conventional SPME method. Seeking innovation, Djozan & Ebrahimi, 2008, covered the fiber with methacrylic acid and ethylene glycol by means of copolymerization for the analysis of atrazine and triazines in water by SPME. The new fiber was highly efficient, even for the analysis of rice and garlic. Similarly, Zeng et al., 2008, used ceramic material and carbon of high thermal stability as a support for SPME fibers for the analysis of organophosphorus pesticides in water samples. The method presented an LOD of 5.2 to 34.6 ng.L^{-1} and a good linearity between 0.05 to 200.0 ng.mL^{-1}.

Silva & Cardeal, 1999, developed a method for the determination of organophosphorus pesticides in water samples using SPME. A good linearity of the method was obtained in the 0.20 to 20.0 ng.L^{-1} range, with correlation coefficients above 0.999. The accuracy of the method was 5.7 to 10.2% for all the pesticides evaluated, resulting in lower limits of detection, from 0008 to 0020 ng.L^{-1}. Capobianco & Cardeal, 2005, proposed a method for the analysis of organophosphorus pesticides (co-ral, DDVP, Di-syston, phorate, phosdrin and malathion) in freshwater fish, water and food samples by applying the SPME technique. The correlation coefficients for the curve obtained were between 0.997 and 0.999, with a relative standard deviation of 4.40 to 15.13%. The detection limits ranged from 0.05 $\mu g.L^{-1}$ to 8.37 $\mu g.L^{-1}$, and the measurable limits were 0.09 $\mu g.L^{-1}$ to 8.70 $\mu g.L^{-1}$.

Among the techniques mentioned, SPME extraction preserves all the advantages of SPE, such as simplicity, low cost, and easy automation, and eliminates the disadvantages such as clogging of the cartridge and use of solvents. All the studies for the analysis of pesticides by SPME furnished good results. This fact indicates that this approach is recommended for the analysis of trace organic compounds in environmental samples. Other studies (Fernandez-Alvarez, 2008), have proposed changes in methods and alternatives for the analysis of other undesirable compounds in ecosystems.

Basher et al., 2007, compared the efficiency of extraction of organophosphorus pesticides from groundwater by SPME and by liquid-phase microextraction (LPME). The SPME method was more effective at higher concentrations (LOD between 3.1-120.5 $ng.L^{-1}$), while the LPME method was more effective at low concentrations (LOD between 0.3-11.4 $ng.L^{-1}$). Lambropoulou & Albanis, 2005, successfully developed a LPME method to determine traces of some insecticides (dichlorvos, mevinphos, ethoprophos, carbofuran,chlorpyrifos methyl, phenthoate, methidathion and carbofenothion) in water samples. Their method exhibited good linearity. The detection limits were in the 0.001 to 0.072-$\mu g.L^{-1}$ range with relative recoveries from 80 to 104%. In another work, also using LPME, Xiong & Hun, 2008, analyzed organosulfur pesticides (malation, chlorpyrifos, buprofezin, triazophos, carbofsulfan and pyridaben) in environmental samples of water. The method presented good linearity (0.80 to 850 $\mu g.L^{-1}$) and a correlation coefficient of 0.9901 to 0.9988, with limits of detection of 0.21 to 3.05 $\mu g.L^{-1}$. Thus, LPME is a promising technique for environmental analysis at trace levels.

4.2 Analytical techniques

The analytical techniques for the determination of the pesticides in environmental samples request appropriate sensibility and precision. Gas chromatography coupled to mass spectrometry is among the most frequently used tools. This analytical tool has promoted the quantification of pesticides and their sub-products in the ppt level.

Gennaro et al., 2001, investigated the process of degradation of the pesticide carbofuran using photocatalyzed reactions. The progress of the degradation of the pesticide was monitored using the techniques of HPLC and GC, both being coupled to mass spectrometer (MS) detectors. The intermediate products from carbofuran could only be identified by the GC/MS method because it presented greater selectivity, specificity, and capacity for identification.

Sasano et al., 2000, proposed the analysis of pesticides in water samples using the GC/MS technique The authors used an automated system that consisted of pneumatic valves that introduced the sample directly into the sample injector chamber, with rapid evaporation of the solvent. They determined 29 pesticides and herbicides. The system achieved a recovery of over 75% and a relative standard deviation (n = 6) of 10%. Sabik at al., 2000, described the determination of herbicides of triazines group and their degradation products in water samples using the GC/MS, HPLC-UV and HPLC/MS techniques. The technique was relevant because it became possible to propose reaction pathways for the degradation of atrazine.

Nélieu et al., 2000, analyzed the atrazine degradation products by LC and HPLC using UV and MS detectors. The results indicated that mass spectrometry is a more efficient detector. Some of the degradation products were not identified by HPLC/MS or HPLC/MS/MS. Sandra et al., 1995, investigated the degradation of the organophosphate methyl pirimiphos in water by artificial light. The analysis of the intermediate degradation products was achieved by GC/MS, HPLC-UV, HPLC/MS analysis and total organic carbon (TOC) techniques. The best results were obtained by GC/MS, and it was possible to identify all the degradation products of methyl pirimiphos. The formation of inorganic ions was verified through TOC. Prevot et al., 1999 and Ravelo-Perez et al., 2008, investigated the photocatalytic degradation of acid solutions of 2,3,6-trichlorobenzoic acid (2,3,6-TBA) in the presence of TiO_2. Several aromatic intermediates were detected by GC/MS with HS-SPME extraction. These intermediates have indicated the occurrence of hydroxylation, dechlorination, decarboxylation and oxidation-reduction reactions. Pereira at al., 1990, identified a number of pesticides including atrazine, alachlor, and metoxichlor and their degradation products in surface water using gas chromatography and mass spectrometry with an ion trap analyzer. The method resulted in the determination of approximately 1.0 $ng.L^{-1}$, with detection limits of up to 60.0 $pg.L^{-1}$.

Faria et al., 2007, proposed a method of analysis for six pesticides (imazethapyr, imazaquin, metsulfuron-methyl, bentazone, chlorimuron-ethyl and tebuconazole), by GC and HPLC with mass spectrometer detectors. The method involved the immobilization of the pesticides on a silica support immersed in water. The proposed system presented a good potential for extraction and concentration of pesticides in aqueous samples. It had the advantage of being a low-cost extraction method.

Faria et al., 2007 and Sandra et al., 1995, developed a GC/MS method for the analysis of two metabolites of the herbicide chlorotriazine and 11 pesticides in water samples. The recovery index for all the analytes studied was 105-116% for concentrations in the range of 0.5 -1.0 $ng.mL^{-1}$, with detection limits of 0.05 $ng.mL^{-1}$.

In another study, Charreteur et al., 1996, determined pesticides in water samples using sequential GC/MS with Ion trap analyzers, in electron impact (EI) and chemical ionization (CI) modes. In this study, the detection limits for both types of ionization were 0.2 to 5.0 $ng.L^{-1}$. The selectivity increased significantly with the use of EI-MS/MS, compared to CI. Steen et al., 1997, analyzed triazines using the working conditions of Charreter. Magnuson et al., 2000, also determined the degradation products of triazines by GC/MS, but noted a problem of coeluition of the two main products, deethylatrazine (DEA) and deisopropylatrazine (DIA). The strategy to avoid coeluition was to combine the products with two different compounds, octadecyl and ammelide, so as to increase the polarity of the molecule.

Uygun, 1997, investigated the degradation of chlorvenvifós by GC/MS, HPLC/UV and gel permeation chromatography (GPC) techniques. The main degradation product in all the samples was trichloroacetophenone, and the GC/MS technique was the only one able to identify this compound. The results also indicated that the degradation of chlorvenvifós is significantly lower at temperatures below 5 ^{o}C.

Penuela & Barcelo, 1998, studied the degradation of endosulfan in water samples using the GC/MS and GC/ECD techniques. The endosulfan sulfate was determined in only a few samples because the proposed method indicated that the degree of recovery was low (70-86%) and the limit of detection was high (0.5-540.0 $ng.L^{-1}$) for the levels under study.

Penuela & Barcelo, 1996, studied the degradation of Alachlor in water samples by GC/ECD and GC/MS. Three products were identified (2-hydroxy-2,6'-diethyl-N-methylacetanilide, 8-ethyl-1-methoxymethyl-4-methyl-2-oxo-1,2,3,4-tetraquionoline and hydroxyalachlor). The pesticide was highly stable when exposed to irradiation by natural light during a period of

20 h. However, in the presence of 15.0 mg.L^{-1} of $FeCl_3$ catalyst, the method became very effective for the destruction of Alachlor.

Bandala et al., 2007, studied the degradation of aldrin, and its products were analyzed by GC/MS. Three degradation products were identified: dieldrin, chlordane and 12-hydroxydieldrin. In the presence of H_2O_2, 93% degradation of aldrin was achieved.

A review of methods for the determination of pesticides and their degradation products in environmental samples using different types of detectors was performed. The methods most used in the analysis were GC with an electron capture detector (ECD), nitrogen and phosphorus detector (NPD) or mass spectrometer (MS) detector and the technique of liquid chromatography with UV and MS detectors. The best detector among the studies discussed was the mass spectrometer. A summary of some studies described in the literature regarding the pesticides examined, types of samples and techniques used are presented in Table 2.

Kouloumbos et al., 2003, investigated the products of photocatalytic degradation of diazinon in aqueous suspensions using GC/MS/MS and HPLC/MS/MS. The photocalytic degradation of diazinon catalysed by titanium dioxide was observed to proceed essentially through a hydroxylation mechanism. The results show that the combination of GC/MS/MS with EI, positive and negative ion CI, and LC/MS/MS with electrospray ionization represent a powerful analytical approach for the confirmation of the structure of photocatalytic intermediates.

Kowal et al., 2009, developed an ultraperformant liquid chromatography-tandem spectrometry (UPLC/MS) method for the analysis of metabolites of the pesticide N,N-dimethylsulfamide (DMS) in aqueous matrices. More than 600 samples of drinking water, surface water, and groundwater have been examined successfully using this method. The method furnished a relative standard deviation of 15% (n=10) and a limit of detection of 10.0 ng.L^{-1}.

Hernández et al., 2008, investigated the metabolites of pesticides in food and water by liquid chromatography with time-of-flight mas spectrometry (HPLC/TOFMS). This technique has been successfully applied in multi-residue target analysis and has allowed the safe identification of metabolites in samples, as well as their quantification.

In other study, Jeannot & Sauvard, 1999, determined pesticides in water samples using HPLC/MS/MS-APCI in positive mode. The method showed good linearity from 0.05 to 10.0 ng.L^{-1}, correlation coefficients from 0.9993 to 1.0 and detection limits from 0.02 to 0.1 μg.L^{-1}. The study furnished the identification and quantification of pesticides and their conversion products in drinking water.

Comprehensive two-dimensional gas chromatography (GCxGC) is a relatively new technique, developed in the nineties, and has great power of separation for complex samples, such as multi-residue analyses of pesticides. Banerjee et al., 2008, optimized a method for multi-residue analysis of pesticides in grapes using comprehensive GCxGC/TOFMS and GC/TOFMS. The method resolved the co-elution problems observed in full scan, one-dimensional analysis and promoted chromatographic separation of 51 pesticides within a 24-min run time with mass-spectrometric confirmation. The detection limits for GC/TOFMS were 2.0 to 19.0 ng.g^{-1} and detection limits for GCxGC/TOFMS were 0.2 to 3.0 ng.g^{-1}. Multiresidue analysis by GCxGC/TOFMS presented distinct advantages over the GC/TOFMS analysis. The technique shows promise and good separation of all co-eluted as well as closely eluted compounds with high sensitivity in the analysis of the pesticides studied.

Khummueng et al., 2006, determined residues of nine fungicides in vegetable samples using GCxGC/NPD. The concentrations ranged from 1.0 to 1000.0 $\mu g.L^{-1}$. Excellent linearity was observed for these standards in the range from 0.001 to 25.0 $mg.L^{-1}$. The limit of detection (LOD) and limit of quantification were less than 74 and 246 $ng.L^{-1}$. Degradation of one fungicide (ioprodine) was readily identified by the characteristic band in the 2D plot between the parent and the decomposition product. The study showed that GCxGC/NPD has a potential for the routine analysis of fungicides in food and vegetables samples, providing a low LOD and LOQ and a good repeatability and reproducibility of peak response. Dalluge et al., 2002, also determined 58 pesticides in food extracts using GCxGC/TOFMS. All the pesticides of interest could be identified using their full-scan mass spectra. This determination of pesticides in vegetable extracts serves as an example. It was demonstrated that GCxGC improves the separation dramatically and is very suitable for the analysis of complex food samples. The authors mention that the analytes of interest can be better separated from one another when using GCxGC, but, more importantly, they are also separated from matrix compounds, which tend to seriously interfere in the 1D-GC/MS procedure. Consequently, the quality of the TOF/MS mass spectra obtained by GCxGC is much better than those obtained with 1D-GC, as was illustrated in this study for serious pesticides. Zrostlíková et al., 2003, determined trace level residuesof 20 pesticides in complex food matrices using GCxGC/TOFMS. The repeatability of retention time as R.S.D. ranged from 0.28 to 0.56% and 0.29 to 0.78% in the first and second dimensions, respectively. Good linearity (R^2 = 0.9982-0.9996) was achieved in the concentration range of 5-500 $ng.mL^{-1}$ for standards in ethyl acetate. In this study and that of Dalluge, GCxGC/TOFMS was demonstrated to be a powerful tool for solving the problems with reliable confirmation of pesticide residues at the very low concentration levels required for the analysis of some types of samples such as baby food.

Gilbert-López et al., 2010, studied 105 pesticide residues in olive oil using fast liquid chromatography-electrospray time-of-flight mass spectrometry (LC/TOFMS) with two sample treatment methods: matrix solid-phase dispersion (MSPD) and liquid-liquid extraction. Data obtained shows that higher recoveries were obtained with LLE. The limits of detection obtained using both sample treatment methods were lower than 10 $\mu g.kg^{-1}$ for more than 85 analytes. The HPLC/MS technique provided good precision and accuracy without requiring expensive instrumentation for the sample treatment step, and it consumed relatively low amounts of solvent and generated little waste material.

Cus et al., 2010, identified the presence of 117 pesticide residues in vinification process of different grape varieties. Pesticides were determined by two different multi-residue methods. Seventy-one pesticides and dithiocarbamates were determinate by GC/MS. Another 45 pesticides were determined by HPLC/MS/MS. The LOD was 0.01 $mg.L^{-1}$. Dithiocarbamates and some pesticides that are rapidly degraded, such as chlorothalonl, were not detected in this study. However, the study demonstrated the persistence of many pesticides during the vinification process. The range of concentrations of pesticides detected was 0.01-0.013 $mg.L^{-1}$.

Ayala et al., 2010, used high performance liquid chromatography with mass spectrometry (HPLC/ESI/MS) to study the degradation of bromoxynil and trifularina through an ozonation process. This mass spectrometry method provides necessary information about the degradation products during the process that can lead to a proposal of a transformation route for the pesticides studied. In another study, Ayala et al., 2010, evaluated the degradation of bromoxinyl and trifularin in natural waters by photoradiation with a UV

Pesticides	Degradation products	Analytical Technique	Method Extraction	References
Aldrin	Dieldrin, chlordane and 1,2-hydroxy-dieldrin	GC/MS e GC/ECD	LLE	Ormad et al., 1997
Carbofuram	2,3-dichloro-2,2-dimethylbenzofuran-7-carbonato , carbofurafenol and benzofuran	HPLC,UV-vis and GC/MS	LLE	Mahalakshmi et al., 2007
36 kinds of different pesticides studied	Their respective degradation products	GC/ECD,GC/MS, GC/NPD, LC/MS and LC/UV	LSE/SPME	Bandala et al., 1998
Propan, terbutiron, propiclor, chlortoluron, thiran, ácido fenoxiacético, 2,4,5-triclorofenoxido, uracil, 5 bromuracil and bromotimol	Propham, propachlor, 3-3(3-hydroxy-4-4methylphenyl)-1,1-dimethylurea, chloro-4-methylphenyl, urea, 3-3chloro-4-methylphenylamine e chlortoluron	GC/MS	SPME	Veiga et al., 2006
Mechlorprop , dichlorprop, 2,4-D, 4-chloro-2-methylphenol, 2,4-dichlorophenol , bentazon, bromoxynil, yonyxil, dicamba dinoseb and DNOC	4-chloro-2-methyl phenol CMP) and 2,4-dichlorophenol, (DCP)	GC/MS	SPME	Ghadiri et al., 2001
29 kinds of pesticides and herbicides	-	GC/MS	SPE	Olette et al., 2008
Triazines	Chloromethoxy , methylthiotriazines, metribuzin, metamitron, triazinones and hexazinonetriazinone and hexazinone	GC/MS, LC/MS, GC/MS, LC-UV	SPME	Bouldin et al., 2006
Pirimifós metal	4-hydroxy derivative, phosphorothioic acid,	GC/MS, GC/MS-MS	LLE/SPME	Herrmann et al., 1999

	O.O.S-trimethyl ester			
Atrazine	Diethylatrazine and di-isopropylatrazine	GC/MS/MS	SPME	Shankar et al., 2004
Pirimiphos-methyl	Phosphate, sulfate and nitrate anions	GC, HPLC, GC/MS, TOC and LC/MS	SPE	Kuo et al., 2002
2,3,6-Trichlorobenzo ic	Dichlorobenzene isomers, 1,2,5-trichlorobenzene, 2,3,5-trichlorophenol, 2,3,6 – trichlorophenol, 2,3,6-trichlorohydroquinone2, 3,6-	GC/MS	LLE	Uygun,1997
Chlorfenviphos	1-(2,4-dichlorophenyl)ethan-1-ol, 2,4-dichlorobenzoic acid and 2,2-dichloro1-(2,4-dichlorophenyl) vinyl alcohol	GC/MS	LLE	Rafqah et al., 2005
Endosulfan	Endosulfan diol, endosulfan ether, endosulfan lactone, endosulfan hydroxyether and endosulfan dialdhyde	GC/MS	SPE	Wang & Lemley, 2001
Alachlor	2-hydroxy-2,6-diethyl-N-methylacetanilide, 8-ethyl-1-methoxymethyl-4-methyl-2-oxo-1,2,3,4-tetraquionoline and hydroxyalachlor	GC/MS	SPE	Gennaro et al., 2001
Parathion, atrazine e alachloro	Atrazine amide, deethylatrazine, simazine amide, deisopropylatrazine, hidroxyatrazine amide, chlorodiamino-s-triazine and deisopropylatrazine amide	GC/MS, LC/MS	LLE	Sabik et al., 2000
Triazines, phenoxyacids and organophosph orus compounds	Atrazine amide, deethylatrazine, deisopropylatrazine amide deisopropylatrazine simazine amide, chlorodiamino-s-triazine, ammelie and hydroxyatrazine amide	GC/MS, LC/MS/MS	LSE/LLE	Picho et al.,1998

Diazinon	2-isopropil-6-metil-pirimidin-4-ol (IMP) and 2-isopropil-6-metilpirimidin-4-il fosfato (dizoxon)	UV-vis, GC/MS	LSE/SPME	Chiron et al., 1993

Table 2. Studies of the Determination of Pesticides and their Degradation in Environmental Samples

lamp and with a combination of UV and hydrogen peroxide (H_2O_2). The concentration of these compounds was monitored by high performance liquid chromatography with a UV detector using the wavelengths of 237 and 220 nm to quantify trifularina and bromoxinyl, respectively. The study showed that the UV/H_2O_2 process enhanced the oxidation rate in comparison to direct photolysis.

Chung et al., 2010, validated a method of multi-residue analysis by liquid chromatography with a sequential mass spectrometer detector (HPLC/MS/MS). Ninety-eight pesticides, including organophosphorus and carbamates, were determined in various types of food, edible oil, meat, egg, cheese, chocolate, coffee, rice, tree nuts, citric fruits and vegetables. The method was suitable for the practical determination of multi-class pesticides in food. It presented good linearity in the range of 1 to 20 $\mu g.L^{-1}$. The coefficients of determination were greater than 0.995 for all the compounds studied. The limit of detection was of 2 $\mu g.kg^{-1}$, and the limit of quantification (LOQ) was 10 $\mu g.kg^{-1}$. The average recoveries, measured at 10 $\mu g.kg^{-1}$, were in the 70-120% range for all of the compounds tested.

5. Conclusion

Degradation methods have been extensively used and proposed as an alternative for the complete destruction of pesticide residues, or as a means of obtaining less harmful compounds in different environmental matrices and especially in water. The main methods proposed include photoirradiation, advanced oxidative processes (AOP), phytoremediation and bioremediation. AOP allied with irradiation was efficient in the elimination of harmful pesticide residues and has also been used to study the degradation products to determine the kinetics of formation and disappearance of more toxic products and to establish the routes of degradation and their relative importance.

Chromatographic techniques such as gas chromatography with mass spectrometer detector are proposed for the determination of pesticide residues and their degradation products because of the selectivity, sensitivity and relative speed of analysis. However, liquid chromatographic techniques with mass spectrometer detectors are more suitable for the analysis of more polar compounds because the samples are analyzed directly without the need for derivatization.

Commonly used extraction processes involving analytical methods such as LLE, SPE or SPME are associated with the chromatographic techniques. Considering the use of solvents and the formation of residues by LLE and SPE methods, methods using SPME are considered more advantageous. Moreover, several studies have shown that the SPME technique is selective, rapid and of low cost.

6. References

Acero, J.L.; Benitez, F.J; Real, F.J.; Gonzalés, M. (2008). Chlorination of organophosphorus pesticides in natural waters. *J. Hazard. Mater., 153(1-2)*, 320-328.

Ahmad, R.; Nelson, P.N.; Kookana, R.S. (2006). The molecular composition of soil organic matter as determined by 13C NMR and elemental analyses and correlation with pesticide sorption. *Eur. J. Soil Sci., 57(6),* 883–893.

Amaya-Chavés, A.; Martinéz, T.L.; López, L. E.; Galard, M.M. (2006). Methyl parathion toxicity to and removal efficiency by Typha latifólia in water and artificial sediments. *Chemosphere, 63(7),* 1124-1129.

Andreu, V.; Picó, Y. (2004). Determination of pesticides and their degradation products in soil: critical review andcomparison of methods. *Trac-Trends Anal. Chem., 23(10-11),* 772-789.

Arbeli, Z.; Fuentes, C. L. (2007). Accelerated biodegradation of pesticides: An overview of the phenomenon, its basis and possible solutions; and a discussion on the tropical dimension. *Crop Prot., 26(12),* 1733-1746.

Arias-Estevez, M.; Lopez-Periago, E.; Martinez-Carballo, M.; Simal-Gandara, J.; Mejuto, J.C.; Garcia-Rio, L. (2008). The mobility and degradation of pesticides in soils and the pollution of groundwater resources. *Agric. Ecosyst. Environ., 123(4),* 247-260.

Arnold, S.M.; Tallat, R.E.; Hickey, W.J.; Harris, R.F. (1995). Identification of Fenton's reagent-generated atrazine degradation products by high-performance liquid chromatography and megaflow electrospray ionization tandem mass spectrometry. *J. Mass Spectrom., 30(3),* 452-460.

Ayala, C. P. El-Din, M. G. Smith, D. W. (2010). Kinetics and mechanism of the degradation of two pesticides in aqueous solutions by ozonation. *Chemosphere*, 78 (2010) 557-562.

Ayala, P. C. El-Din, M. G. Smith, D. W. (2010). Degradation of bromoxynil and trifularin in natural water by direct photolysis and UV plus H_2O_2 advanced oxidation process. *Water Research*, 44, 2221-2228.

Bandala, E.R.; Gelover, S.; Leal, M.T.; Bulnes, C.A.; Jimenez, A.; Estrada, C.A. (2007). Solar photocatalytic degradation of Aldrin. *Catal. Today, 76(2-4),* 189-199.

Bandala, E.R.; Octaviano, J.A.; Albiter, V.; Torres, L.G.; Wickramanayake, G.B.; Hinchee, R.E. (2002). Solar photocatalytic degradation of Aldrin. *Catal. Today, 76(2-4),* 189-199

Banerjee, K.; Patil, S. H.; Dasgupta, S.; Oulkar, D. P.; Patil, S. B.; Savant, R. Adsule, P. (2008). Optimization of separation and detection condicitions for the multiresidue analysis of pesticides in grapes by comprehensive two-dimensional gas chromatography-time-of-flight mass spectrometry. *J. Chromatogr. A, 1190(1-2),* 350-357.

Basheer, C.; Alnedhary, A. A.; Rao, B.S.M.; Lee, H. K. (2007). Determination of organophosphorous pesticides in wastewater samples using binary-solvent liquid-phase microextraction and solid-phase microextraction: A comparative study. *Anal. Chim. Acta, 605(2),* 147-152.

Bavcon, M.; Trebese, P.; Zupancic-Kralj, L. (2002). Investigations of the determination and transformations of diazinon and malathion under environmental conditions using gas chromatography coupled with a flame ionization detector. *Chemosphere, 50(5),* 595-601.

Bhalerao, T. S.; Puranik, P. R. (2007). Biodegradation of organochlorine pesticide, endosulfan, by a fungal soil isolate, Aspergillus niger. *Int. Biodeterior. Biodegrad., 59(4),* 315-321.

Blanco, J.; Minero, C.; Pelizetti, E.; Malato, S. (1996). Large solar plant photocatalyticac water decontamination degradation of atrazine. *Solar Energy*. 56(5), 411-419.

Boivin, A.; Cherrier, R.; Perrin-Ganier, C.; Schiavon, M. (2004).Time effect on bentazone sorption and degradation in soil. *Pest Manag. Sci., 60(8),* 809–814.

Bolan, N.S.; Baskaran, S. (1996). Biodegradation of 2,4-D herbicide as affected by its adsorption–desorption behaviour and microbial activity of soils. *Aust. J. Soil Res., 34(6),* 1041–1053.

Borga, K.; Gabrielsen, G.W.; Shaare, J.U. (2001). Biomagnification of organochlorines along a Barents Sea food chain. *Environ. Pollut. 113(2),* 187-198.

Bouldin, J. L.; Farris, J.L.; Moore, M.T.; Smith, S.J.; Cooper, C.M. (2006). Hydroponic uptake of atrazine and lambda-cyhalothrin in Juncus effusus and Ludwigia peploides. *Chemosphere, 65(6),* 1049-1057.

Canle-Lopez, M.; Santaballa, J.A.; Vulliet, E. (2005). On the mechanism of TiO_2 - photocatalyzed degradation of aniline derivatives. *J. Photochem. Photobiol. A: Chem., 175(2-3),* 192–200.

Capobiango, H.L.V.; Cardeal, Z.L. (2005). A solid-phase microextraction method for the chromatographic determination of organosphosphorus pesticides in fish, water, potatoes, guava and coffee. *J. Braz. Chem. Soc., 16(5),* 907-914.

Carp, O.; Huisman, C.L.; Reller, A. (2004). Photoinduced reactivity of titanium dioxide. *Prog. Solid State Chem., 32(1-2),* 33–177.

Carreteur, C.; Kerbaol, N.; Peron, J.J. (1996). Contribution to the analysis of triazines in water by gas chromatography and ion trap tandem mass spectrometry. *Analusis, 24(8),* 336-343.

Carvalho, P. N.; Rodrigues, P. N.; Alves, F.; Evangelista, R.; Basto, M.C.P.; Vasconcelos, M.T.S.D. (2008). An expeditious method for the determination of organochlorine pesticides residues in estuarine sediments using microwave assisted pre-extraction and automated headspace solid-phase microextraction coupled to gas chromatography-mass spectrometry. *Talanta, 76(5),* 1124-1129.

Chen, D.; Ray. A.K. (1998). Photodegradation kinetics of 4-nitrophenol in TiO_2 suspension. *Water Res., 32(11),* 3223–3234.

Chiron, S.; Fernandez-Alba, A.; Barcelo, D. (1993). Comparison of on-line solid-phase disk extraction to liquid-liquid extraction for monitoring selected pesticides in environmental waters. *Environ. Sci. Technol.,* 27, 2352-2359.

Chiron, S.; Fernandez-Alba, A.; Rodriguez, A.; Garcia-Calvo, E. (2000). Pesticide chemical oxidation: state-of-the-art. *Wat. Research,* 343 (2), 366-377.

Chung, S. W.C. Chan, B. T.P. (2010). Validation and use of fast sample preparation method and liquid chromatography-tandem mass spectrometry in analysis of ultra-trace levels of 98 organophophorus pesticide and carbamate residues in a total diet study involving diversified food types. *J. Chromatogr. A,* 1217, 4815-4824.

Clausen, L.; Fabricius, I. (2002). Atrazine, isoproturon, mecoprop, 2,4-D, and bentazone adsorption onto iron oxides. *J. Environ. Qual., 30 (3),* 858–869.

Coquet, Y. (2003). Sorption of pesticide atrazine, isoproturon, and metamitron in the Vadose Zone. *Vadose Zone J., 2 (1),* 40–51.

Cus, F. Cesnik, H. B. Bolta, S. V. Gregorcic, A. (2010). Pesticide residues in grapes and during vinification process. *Food Control,* 21, 1512-1518.

Dalluge, J.; Rijn, M. V.; Beens, J.; Vreuls, J. J.; Brinkman, U.A. T. (2002). Comprehensive two-dimensional gas chromatography with time-of-flight mass spectrometry detection applied to the determination of pesticides in food extracts. *J. Chromatogr.* A, *965 (1-2)*, 207-217.

Djozan, D.; Ebrahimi, B. (2008). Preparation of new solid phase micro extraction fiber on the basis of atrazine-molecular imprinted polymer: Application for GC and GC-MS screening of triazine herbicides in water, rice and onion. *Anal. Chim. Acta, 616*, 152-159.

Doong, R.A.; Liao, P.L. (2001). Determination of organochlorine pesticides and their metabolites in soil samples using headspace solid-phase microextraction. *J. Chromatogr. A, 918 (1)*, 177-188.

Dubus, I.G.; Barriuso, E.; Calvet, R. (2001). Sorption of weak organic acids in soils: clofencet 2,4-D and salicylic acid. *Chemosphere, 45 (6–7)*, 767–774.

Dugay, J.; Miege, C.; Hennion, M.C. (1998). Effect of the various parameters governing solid-phase microextraction for the trace-determination of pesticides in water. *J. Chromatogr. A, 795 (1)*, 27- 42.

Dyson, J.S.; Beulke, S.; Brown, C.D.; Lane, M.C.G. (2002). Adsorption and degradation of the weak acid mesotrione in soil and environmental fate implications. *J. Environ. Qual., 31 (2)*, 613–618.

Elliott, J.A.; Cesna, A.J.; Best, K.B.; Nicholaichuk, W.; Tollefson, L.C. (2000). Leaching rates and preferential flow of selected herbicides through tilled and untilled soil. *J. Environ. Qual., 29 (5)*, 1650–1656.

Faria, A. M.; Maldaner, L.; Santana, C.C.; Jardim, I.C.S.F.; Collins, C.H. (2007). Poly(methyltetradecylsiloxane) immobilized onto silica for extraction of multiclass pesticides from surface waters. *Anal. Chim. Acta, 582 (1)*, 34-40.

Fernandez-Alvarez, M.; Llompart, M.; Lamas, J. P.; Lores, M.; Garcia-Jares, C.; Cela, R.; Dagnac, T. (2008). Development of a solid-phase microextraction gas chromatography with microelectron-capture detection method for a multiresidue analysis of pesticides in bovine milk. *Anal. Chim. Acta, 617 (1-2)*, 37-50.

Gao, J.P.; Spitzuer, J.; Kettrup, A. (1998). Sorption of pesticides in the sediment of the Teufelsweiher pond (Southern Germany), I: equilibrium assessments, effect of organic carbon content and pH. *Water Res. 32 (5)*, 1662-1672.

Gennaro, M.C.; E. Marengo, E.; Gianotti, V.; Maurino, V. (2001). New strategies for the determination of phenylurea pesticides by gas chromatography with hot splitless inlet systems. *J. Chromatogr. A, 910 (1)*, 79-86.

Gevao, B.; Semple, K.T.; Jones, K.C. (2000). Bound pesticide residues in soils: a review. *Environ. Pollut. 108 (1)*, 3–14.

Ghardiri, H. ; Rose, C.W. (2001). Degradation of endosulphan in clay soil from cotton farms of western Queensland. *J. Environ. Manag. 62 (2)*, 155-69.

Giacomazzi, S.; Cochet, N. (2004). Environmental impact of diuron transformation: a review. *Chemosphere, 56 (11)*, 1021-1032.

Gilbert-López, B.; García-Reyesa, J. F.; Fernández-Alba, A.R.; Molina-Díaz, A. (2010). Evaluation of two sample treatment methodologies for large-scale pesticide residue analysis in olive oil by fast liquid chromatography–electrospray mass spectrometry, *J. Chromatogr. A*, 1217 , 3736-3747.

Guo, L.; Jury, W.A.; Wagenet, R.J.; Flury, M. (2000). Dependence of pesticide degradation on sorption: nonequelibrium model and application to soil reators. *J. Contam. Hydrol. 43 (1)*, 45-62.

Gupta, M.; Pillai, A.K.K.V.; Jain, A.; Verma, K.K. (2008). Coupled in-tube and on-fibre solid-phase microextractions for cleanup and preconcentration of organic micropollutants from aqueous samples and analysis by gas chromatography-mass spectrometry. *Anal. Chim. Acta, 618 (1),* 61-69.

Hernnández, F.; Sancho, J.V.; Ibáñez, M. Grimalt, S. (2008). Investigation of pesticide metabolites in food and water by LC-TOF-MS. *Trac-Trends Anal. Chem. 27 (10),* 862-872.

Herrmann, J.M.; Guillard, C.; Arguello, M.; Guera, A.; Tejedor, A.; L. Piedra, L.; Fernández-Alba, A. (1999). Photocatalytic degradation of pesticide pirimiphos-methyl. Determination of the reaction pathway and identification of intermediate products by various analytical methods. *Catal. Today, 54 (2-3),* 353-367.

Hyotylainen, T.; Jauho, K.; Riekkola, M.(1998). Analysis of pesticides in red wines by online coupled reversed-phase liquid chromatography-gas chromatography with vaporizer/precolumn solvent split/gas discharge interface. *J. Chromatogr. A, 813 (1),* 113-119.

Jeannot, R.; Sauvard, E. (1999). High-performance liquid chromatography coupled with mass spectrometry applied to analyses of pesticides in water. Results obtained in HPLC-MS-MS-APCI in positive mode. *Analusis, 27(3),* 271-280.

Jinno, K.; Muramatsu, T.; Kiso, Y.; Magdic, S.; Pawliszyn, J. (1996). Analysis of pesticides in environmental water samples by solid-phase micro-extraction high performance liquid chromatography. *J. Chromatogr. A, 754 (22),* 137-144.

Jonge, R.J.; Breure, A.M.; Andel, J.G. (1996). Reversibility of adsorption of aromatic compounds onto powdered activated carbon (PAC). *Water Res., 30 (4),* 883–892.

Jongenotter, G.A.; Kerkhoff, M.A.T.; van der Knaap, H.C.M.; Vandeginste, B.G.M. (1999). Automated online GPC-GC-FPD involving co-solvent trapping and the on-column interface for the determination of organophosphorus pesticides in olive oils. *J. High Resol. Chromatogr, 22 (1),* 17-23.

Kabra, K.; Chaudhary, R.; Sawheney, R.L.(2004). Treatment of Hazardous Organic and Inorganic Compounds through Aqueous-Phase Photocatalysis: A Review. *Ind. Eng. Chem. Res, 43 (24),* 7683–7696.

Kapustka, L.A.; Williams, B.A; Fairbrother, A. (1996). Evaluating risk predictions at population and community levels in pesticide registration - hypotheses to be tested. *Environ. Toxicol. Chem. 15 (4),* 427-31.

Khummueng, W.; Trenerry, C.; Rose, G.; Marriott, P. J. (2006). Application of comprehensive two-dimensional gas chromatography with nitrogen-selective detection for the analysis of fungicide residues in vegetable samples. *J. Chromatogr. A, 1131 (1-2),* 203-214.

Kodaka, R.; Sugano, T.; Katagi, T.; Takimoto, Y. (2003). Clay-catalyzed nitration of a carbamate fungicide diethofencarb. *J. Agric. Food Chem., 51 (26),* 7730-7737.

Konstantinou, I.K.; Zarkadis, A.T.; Albanis, T.A. (2001). Photodegradation of selected herbicides in various natural waters and soils under environmental condictions. *J. Environ. Qual. 30 (1),* 121-130.

Koskinen, W.C.; Cox, L.; Yen, P.Y. (2001). Changes in sorption/bioavailability of imidacloprid metabolites in soil with incubation time. *Biol. Fertil. Soils, 33 (6),* 546–550.

Kouloumbos, V. N.; Tsipi, D. F. (2003). Identification of photocatalytic degradation products of diazinon in TiO_2 aqueous suspensions using GC-MS-MS and LC-MS with quadrupole time-of-line mass spectrometry. *J. Am. Soc. Mass Spectrom., 14*, 803-817

Kowal, S.; Balsaa, P.; Werres, F. (2009). Determination of the polar pesticide degradation product N,N-dimethylsulfamide in aqueous matrices by UPLC/MS/MS. *Anal. Bioanal. Chem. 395 (6)*, 1787-1794.

Kuo, W.S. (2002). Photocatalytic oxidation of pesticide rinsate. *J. Environ. Sci. Health, Part B: Pesticides, Food Contam. Agric. Wastes, 37 (1)*, 65–74.

Lambropoulou, D.; Albanis, T. A. (2005). Application of hollow fiber liquid phase microextraction for the determination of insecticides in water. *J. Chromatogr. A, 1072 (1)*, 55-61.

Lasa, H.; Serrano, B.; Salaices, M. (2005). Photocatalytic Reaction Engineering, Springer. *Germany*, 265.

Legrini, O.; Oliveros, E.; Braun, A.M. (1993). Photochemical processes for water treatment. *Chem. Rev., 93(2)*, 671-98.

Leyva, E.; Moctezuma, E.; Ruiz, M.G.; L. M.; Torres-Martínez, L. M. (1998). Photodegradation of phenol and 4-chlorophenol by $BaO-Li_2O-TiO_2$ catalysts. *Catal. Today, 40 (4)*, 367–376.

Li, Z.Y.; Zhang, Z.C.; Zhou, Q.L.; Gao, R.Y.; Wang, Q.S. (2002). Fast and precise determination of phenthoate and its enantiomeric ratio in soil by the matrix solid-phase dispersion method and liquid chromatography. *J. Chromatogr. A, 977(1)*, 17-25.

Louch, D.; Motlagh, S.; Pawliszyn, J. (1992). Dynamics of organic compound extraction from water using liquid-coated fused silica fibres. *Anal. Chem. 64 (10)*, 1187-1199.

Louter, A.J.H.; van Beekvelt, C.A.; Montanes, P.Cid.; Slobodnik, J.; Vreuls, J.J.; Brinkman, U.A. (1996). Analysis of microcontaminants in aqueous samples by fully automated online solid-phase extraction-gas chromatography-mass selective detection. *J. Chromatogr. A, 725 (1)*, 67-83.

Magnuson, M.L.; Speth, T.F.; Kelty, C.A. (2000). Determination of interfering triazine degradation products by gas chromatography-ion trap mass spectrometry. *J. Chromatogr. A, 868 (1)*, 115-119.

Mahalakshmi, M.; Arabindoo, B.; Palanichamy, A.; Murugesan, V. (2007). Photocatalytic degradation of carbofuran using semiconductor oxides. *J. Hazard. Mater., 143 (1-2)*, 240-245.

Masten, S.J.; Davies, S.H.R. (1994). The Use of Ozonization to Degrade Organic Contaminants in Wastewaters. *Environ. Sci. Techonol., 28(4)*, 180A-185A.

Miege, C.; J. Dugay. (1998). Solide-phase microextration and gas chromatography for rapid analysis of pesticides. *Analusis, 26(6)*, M137-M143.

Moctezuma, E.; Leyva, E.; Monreal, E.; Villegas, N.; Infante, D. (1999). Photocatalytic degradation of the herbicide "paraquat". *Chemosphere, 39(3)*, 511–517.

Moctezuma, E.; Leyva, E.; Palestino, G.; Lasa, H. (2007). Photocatalytic degradation of methyl parathion: Reaction pathways and intermediate reaction products. *J. Photochem. Photobiol. A: Chem., 186(1)*, 71-84.

Monkiedje, A.; Spiteller, M.; Bester, K. (2003). Degradation of racemic and enantiopure metalaxyl in tropical and temperate soils. *Environ. Sci. Technol., 37(4)*, 707-712.

Moyer, J.R.; Hance, R.J.; McKone, C.E. (1972). The effects of adsorbents on the rate of degradation of herbicides incubated with soil. *Soil Biol. Biochem., 4(3),* 307–311.

Navalon, A.; Prieto, A.; Araújo, L.; Vilchez, L.J. (2002). Determination of oxadiazon residues by headspace solid-phase microextraction and gás chromatography-mass spectrometry. *J. Chromatogr. A, 946(1-2),* 239-245.

Navarro, S.; Vela, N.; García, C.; Navarro, G. (2004). Persistence of Simazine and Terbuthylazine in a Semiarid Soil after Organic Amendment with Urban Sewage Sludge. *J. Agric. Food Chem., 51(25),* 7359-65.

Nawab, A.; Aleem, A.; Malik, A. (2003). Determination of organochlorine pesticides in agricultural soil with special reference to γ-HCH degradation by Pseudomonas strains. *Biores. Technol., 88(1),* 41-46.

Nelieu, S.; Stobiecki, M.; Einhorn, J. (2000). Tandem solid-phase extraction of atrazine ozonation products in water. *J. Chromatogr. A, 866(2),* 195-201.

Oh, Y.; Li, X.; Cubbage, J.W.; Jenks, W. S. (2004). Mechanisms of catalyst action in the TiO2-mediated photocatalytic degradation and cis-trans isomerization of maleic and fumaric acid. *Appl. Catal. B: Environ., 54(2),* 105–114.

Olette, R.; Couderchet, M.; Biagianti, S.; Eullaffroy, P. (2008). Toxicity and removal of pesticides by selected aquatic plants. *Chemosphere, 70(8),* 1414-1421.

Ormad, P.; Cortes, S.; Puig, A.; Ovelleiro, J.L. (1997). Degradation of organochloride compounds by O_3 and O_3/H_2O_2. *Water Res., 31,* 2387-2391.

Park, J.H.; Feng, Y.C.; Ji, P.S.; Voice, T.C.; S.A. Boyd, S.A. (2003). Assessment of bioavailability of soil-sorbed atrazine. *Appl. Environ. Microbiol., 69(6),* 3288–3298.

Penuela, G.A.; Barceló, D. (1998). Application of C_{18} disks followed by gas chromatography techniques to degradation kinetics, stability and monitoring of endosulfan in water. *J. Chromatogr. A, 795(1),* 93-104.

Penuela, G.A.; Barceló, D. (1996). Comparative degradation kinetics of alachlor in water by photocatalysis with $FeCl_3$, TiO_2 and photolysis, studied by solid-phase disk extraction followed by gas chromatographic techniques. *J. Chromatogr. A, 754(1-2),* 187-195.

Pereira, W.E.; Rostad, C.E.; Leiker, T.J. (1990). Determination of trace levels of herbicides and their degradation products in surface and ground waters by gas chromatography/ion-trap mass spectrometry. *Anal. Chim. Acta, 228(1),* 69-75.

Pichon, V. (1998). Multiresidue solid-phase extraction for trace-analysis of pesticides and their metabolites in environmental water. *Analusis, 26(6),* M91-M98.

Pichon, V.; Hennion, M.C. (1994). Determination of pesticides in environmental waters by automated online trace-enrichment and liquid chromatography. *J. Chromatogr. A, 665(2),* 269-81.

Pozo, O. ; Pitarch, E. ; Sancho, J.V. ; Hernandez, F. (2001). Determination of the herbicide 4-chloro-2-methylphenoxyacetic acid and its main metabolite, 4-chloro-2-methylphenol in water and soil by liquid chromatography-electrospray tandem mass spectrometry. *J. Chromatogr. A., 923(1-2),* 75-85.

Prevot, A.B.; Pramauro, E. (1999). Analytical monitoring of photocatalytic treatments. Degradation of 2,3,6-trichlorobenzoic acid in aqueous TiO_2 dispersions. *Talanta, 48(4),* 847-857.

Rafqah, S.; Wong-Wah-Chung, P.; Aamili, A.; Sarakha, M. (2005). Degradation of metsulfuron methyl by heterogeneous photocatalysis on TiO2 in aqueous suspensions Kinetic and analytical studies. *J. Mol. Catal. A: Chem., 237(1-2)*, 50-59.

Ravelo-Pérez, R. M. L.; Borges, J. H.; Miquel, T. M.B.; Delgado, M. A. R. (2008). Pesticides analysis in tomatoes by solid-phase microextration and micellar electrokinetic chromatography. *J. Chromatogr. A, 1185(1)*, 151-154.

Roulier, S.; Jarvis, N.(2003). Modeling macropore flow effects on pesticide leaching: Inverse parameter estimation using microlysimeters. *J. Environ. Qual., 32(6)*, 2341–2353.

Sabik, H.; Jeannot, R.; Rondeau, B. (2000). Multiresidue methods using solid-phase extraction techniques for monitoring priority pesticides, including triazines and degradation products, in ground and surface waters. *J. Chromatogr. A., 885(1-20)*, 217-236.

Sandra, P.; Beltran, J.; David, F. (1995). Enhanced selectivity in the determination of triazines in environmental samples by benchtop CG/MS/MS. *J. High Resol. Chromatogr., 18(9)*, 545-50.

Sasano, R.; Hamada, T.; Kurano, M.; Furuno, M. (2000). On-line coupling of solid-phase extraction to gas chromatography with fast solvent vaporization and concentration in an open injector liner. Analysis of pesticides in aqueous samples. *J. Chromatogr. A, 896(1-2)*, 41-49.

Sassman, S.A.; Lee, L.S.; Bischoff, M.; Turco, R.F. (2004). Assessing N,N'-Dibutylurea (DBU) Formation in Soils after Application of n-Butylisocyanate and Benlate Fungicides. *J. Agric. Food Chem., 52(4)*, 747-754.

Selim, H.M.; Ma, L.; Zhu, H. (1999). Predicting solute transport in soils: second-order two site models. *Soil Sci. Soc. Am. J., 63(4)*, 768–777.

Shankar, M.V.; Anandan, S.; Venkatachalam, N.; Arabindoo, B.; Murugesan, V. (2004). Novel thin-film reactor for photocatalytic degradation of pesticides in aqueous Solutions. *J. Chem. Technol. Biotechnol., 79(11)*, 1279–1285.

Shemer, H.; Linden, K.G.(2006). Degradation and by-product formation of diazinon in water during UV and UV/H2O2 treatment. *J. Hazard. Mater., 136(*3), 553-559.

Silva, F.C.; Cardeal, Z.L. (1999). Determination of organophosphorus pesticides in water using SPME-GC-MS. *Quím. Nova, 22(2)*, 197-200.

Sinclair, C.J. ; Boxall, A.B. (2003). Assessing the ecotoxicity of pesticide transformation products. *Environ. Sci. Technol., 37(20)*, 4617-4625.

Spark, K.M.; Swift, R.S. (2002). Effect of soil composition and dissolved organic matter on pesticide sorption. *Sci. Environ., 298 (1–3)*, 147–161.

Steen, R.J.C.A.; Freriks, I.L.; Cofino, W.P.; Brinkman, U.A. (1997). Large-volume injection in gas chromatography-ion trap tandem mass spectrometry for the determination of pesticides in the marine environment at the low ng/l level. *Anal. Chim. Acta, 353(2-3)*, 153-163.

Strandberg, B.; Strandberg, L.; Bergqvist, P.; Falandysz, J.; Rappe,C. (1998). Concentrations and biomagnification of 17 chlordane compounds and other organochlorines in harbor porpoise (Phocoena phocoena) and herring from the southern Baltic Sea. *Chemosphere, 37(9/12)*, 2513-2523.

Thurman, E.M.; Meyer, M.; Pomes, M.; Perry, C.A.; Schwab, P. (1990). Enzyme-Linked-Immunosorbent-Assay Compared with Gas-Chromatography Mass-Spectrometry for the Determination of Triazine Herbicides in water. *Anal. Chem., 62(18),* 2043-2048.

Uygun, U.(1997). Degradation of chlorfenvinphos in carrots during storage. *Food Chemistry, 60(4),* 479-487.

Veiga, M.M.; Silva, D.M.; Veiga, L.B. E. (2006). Pesticide pollution in water systems in a small rural community in Southeast Brazil. *Cadernos de saúde publica / Ministério da Saúde, Fundação Oswaldo Cruz, Escola Nacional de Saúde Publica, 22(11),* 2391-2399.

Wang, Q.; Lemley, A. T. (2001). Kinetic Model and Optimization of 2,4-D degradation by nodic Fenton Treatment. *Environ. Sci. Technol. 35(22),* 4509-4514.

Watts, C.D.; Clark, L.; Hennings, S.; Moore, K.; Parker, C.; Crathorne, B.; Angeletti, G. (1989). Pesticides: analytical requirements for compliance with EEC directives. *Water Pollut. Res.,* 16–34.

Wauchope, R.D.; Yeh, S.; Linders, J.B.H.; Kloskowski, R.; Tanaka, K.; Katayama, A.; Kördel, W.; Gerstl, Z.; Lane, M.; Unsworth, J. (2002). Pesticide soil sorption parameters: theory, measurement, uses, limitations and reliability. *Pest Manag. Sci., 58(5),* 419-445.

Worrall, F.; Besien, T.; Kolpin, D.W. (2002). Groundwater vulnerability: interactions of chemical and site properties. *Sci. Total Environ. 299(1-3),* 131–143.

Wu, J.G.; Luan, T.G.; Lan, C.Y.; Lo, T.W.H.; Chan, G.Y.S. (2007). Removal of residual pesticides on vegetable using ozonated water. *Food Control. 18(5),* 466-472.

Wyss, I.; Boucher, J.; Montero, A.; Marison, I. (2006). Micro-encapsulated organic phase for enhanced bioremediation of hydrophobic organic pollutants. *Enzyme Microb. Technol., 40(1),* 25-31.

Xia, H.; Xiangjuan, M. (2006). Phytorremediation of ethion by water hyacinth (Eichhornia crassipes) from water. *Biores. Technol., 97(8),* 1050-1054.

Xiang. L.; Li, C.; Chen, J.; Li, C.; Sun, C. (2008). Polytiophene as a novel fibre coating for solid-phase microextration. *J. Chromatogr. A, 1198-1199,* 7-13.

Xiong, J.; Hu, B. (2008). Comparison of hollow fiber liquid phase microextraction and dispersive liquid-liquid microextraction for the determination of organosulfur pesticides in environmental and beverage samples by gas chromatography with flame photometric detection. *J. Chromatogr. A, 1193(1-2),* 7-18.

Zeng, J.; Yu, B.; Chen, W.; Lin, Z.; Zhang, L.; Lin, Z.; Chen, X.; Wang, X. (2008). Application of ceramic/carbon composite as a novel coating for solid-phase microextration. *J. Chromatogr. A, 1188(1),* 26-33.

Zhang, Z.; Yang, M.J.; Pawliszyn, J. (1994). Solid-phase microextraction. A solvent-Free Alternative for Sample preparation. *Anal. Chem., 66(17),* 844A-854A.

Zrostlíková, J.; Hajslová, J.; Cajka, T. (2003). Evaluation of two-dimensional gas chromatography-time-of-flight mass spectrometry for the determination of multiple pesticide residues in fruit. *J. Chromatogr. A, 1019(1-2),* 173-186.

30

Micropollutant Degradation Mechanism

Brigita Tepuš[1], Irena Petrinić[2] and Marjana Simonič[2]
[1]*Municipal Enterprise Ptuj*, [2]*University of Maribor, Faculty of Chemisty and Chemical Engineering Slovenia*

1. Introduction

The organic pollution is a major concern during the treatment of drinking-water as organic micro-pollutants might show disruptive and toxic properties. Organic micro-pollutants are found in surface and groundwaters at different concentrations, mostly between 0,1 and 100 μg/L (Panno&Kelly, 2004).

Pesticides are known contaminants of concern. 363 kt of pesticides were used between 1980 and 1990 in the USA. From among triazine pesticides, atrazine and its metabolites, deethylatrazine and deisopropylatrazine, can still be found in drinking-water supplies throughout the EU, due to their usage as maize and sugar beet pesticide. They are slowly biodegradable microbiologically (Reid et al, 2003). They have to be removed from drinking-water sources because they are classified as possible human carcinogens (Legube et al, 2004). Atrazine, with the chemical name 2-chloro-4-(ethylamino)-6-(isopropylamino)-s-triazine ($C_8H_{14}ClN_5$, M_{CIET} = 215,7 g/mol) is soluble in water at 30 mg/L and half live in soil for atrazine is 15–100 days (Ralebitso et al, 2002). Atrazine is classified as a class C carcinogen. Chromosom damage to chinese hamster egg cells were observed if they were exposed to 0,005–0,080 μmol/L of atrazine, within two days. Two well-known atrazine metabolites, deethylatrazine and deisopropylatrazine, were found to be potentially carcinogenic, therefore the admissible levels for each pesticide individually in water are set at 0,1 ug/L, and the sum should not exceed 0,5 μg/L in EU (Thurman et. al, 1994). US EPA (US Environmental Protection Agency) set the total admissible levels for atrazine, deethylatrazine and deisopropylatrazine in groundwater at 3 μg/L (Richards et al, 1995). A study by US EPA in 2003 showed that triazines - atrazine, simazine and propazine - as well as metabolites - deethylatrazine and deisopropylatrazine in deethyl- deisopropylatrazine - have the same mechanism concerning endocrine disruptions. Anumnerated compounds act the same way on human bodies, therefore, US EPA introduced the sum of all chloro–s–triazines. Atrazine removal from drinking water sources is impossible using chlorination, aeration, filtration or coagulation. Quite effective technologies include activated carbon, ozonation, membrane separatoin, and biofiltration. The most efective are RO and NF membranes (Jiang&Adams, 2006).

During a study of atrazine degradation within concentrations ranging from 5 to 1700 ng/L, the only metabolite found was deethylatrazine within a concentration range from 10–850 ng/L (Garmouna et al., 1997).

A study of atrazine monitoring in Slovenia from 1993 to 1996 showed that atrazine concentrations were up to 7,23 μg/L. Then they started to decrease to 0,06 μg/L due to the elimination of Primextra and Atrapine T in 1994 (Pintar & Lobnik, 2001). In March, 1999 atrazine use was prohibited (Official Gazette RS, 1999).
Evidently, organic micro-pollutants represent only a minor fraction of organic pollution. The major fraction of organic pollution is attributed to natural organic matter (NOM). NOM is a heterogeneous mixture of undefined structurally complex organic compounds derived from plants, animals, microorganisms, and their waste and metabolic products. Therefore, NOM inevitably occurs in all natural water sources and, like micro-pollutants, must be removed from water sources. NOM interferes with the performances of several unit processes. NOM could be responsible for high coagulant demand, rapid clogging of filters by biofilm growth on media, rapid saturation of activated carbon beds, thereby increasing the regeneration frequency, high disinfectant demand, inhibiting the impact of disinfectants, and the rapid decay of ozone. NOM is a major membrane foulant and may inhibit the removal of organic micro-pollutants by activated carbon. NOM should be carefully considered when choosing the optimal process and design for organic removal (Haarhoff, 2010).

2. Ozone reaction in water

Ozone in ground and surface water reacts with dissolved organic substances (DOC) and micropollutants. Ozonation decreases the formation of disinfection by-products, such as tri-halometanes (THM) and haloacetic acid (HAA). NOM influences ozone decay. Ozonation is one of the better known technologies for atrazine removal (Von Gunten, 2003). Ozone is unstable in water and its half-life ranges from a few seconds to a few hours, depending on pH, NOM, and water alkalynity. Ozone decomposition constitutes the first step of a complicated mechanism for indirect reactions, which are accelerated by initiators such as OH-ions. The resulting radicals react instantly (k = 108-1010 L mol^{-1} s^{-1}) and non-selectively with pollutants. The radical pathway is influenced by the type of dissolved substances in the water. This mechanism, consisting of three different steps is widely used: the initiation step - formation of superoxide anion radical ($O_2^{\cdot -}$), the propagation step - formation of hydroxyl radicals and re-initiation of the chain reaction, and the termination step - inhibitors (scavengers) stop the re-formation of the superoxide anion radical. The direct reaction of organic compounds with ozone is a selective process and has a slow reaction rate constant. It takes place when the radical mechanism is inhibited the oxydation of ozone with NOM is a typically second-order reaction, but with a first-order reaction with respect to ozone and to the organic compound following eqs. 1 and 2. Second- order reactions are typical for reactions of organic compounds and hydroxyl radicals following eq. 3 (Von Gunten, 2003).

$$\gamma_{NOM} + \gamma_{O3} \rightarrow \text{products} \tag{1}$$

$$(-d\gamma_{NOM}/d\,t) = k_r \cdot \gamma_{NOM} \cdot \gamma_{O3} \tag{2}$$

$$(-d\gamma_{NOM}/d\,t) = k_{O3} \cdot \gamma_{NOM} \cdot \gamma_{O3} + k_{OH} \cdot \gamma_{NOM} \cdot \gamma_{OH^{\cdot}} \tag{3}$$

γ_{NOM} – NOM, mg/L
γ_{O3} – ozone, mg/L
$\gamma_{OH^{\cdot}}$ – hydroxyl radicals, mg/L
t – time, t
k_r – reaction rate constant, mol/(L. s)

k_{O3} – reaction rate constant for ozone, mol/(L.s)
k_{OH} – reaction rate constant for hydroxyl radicals, mol/(L.s)
Various reaction orders for ozone degradation from 0 to 2 have been reported (Hermannowicz, 1999). The order of reaction depends on the reaction time. Ozone reacts with organic pollutants and, as by-products, various metabolites are formed which can affect ozone decomposition. Some accelerate while others inhibit decomposition. Decomposition change can be noticed in the change of reaction order. Ozone degradation follows first reaction order in batch experiments. If the reaction time is prolonged and the concentrations of ozone are negligible in comparison with initial ozone concentractions, the reaction order changes. In reported experiment, the first-order reaction coeficients are calculated as being higher compared with batch experiments. Various data concerning first order kinetics are found in literature: 0,031–0,23/min for millipore water, up to 0,27–11,3 1/min for continuous systems and 0.16 to 0.361/min for batch systems in surface water. Batch experiments are conducted on the laboratory scale, where different conditions appear, compared with real water samples. Ozone may destroy organic components in water and, consequently, its concentration decreases.
It has been reported that concentrations of atrazine could be lowered by the formation of OH radicals, at pH 7.60, because of initiation with OH^- ions in water (Gottschalk, 2000).
In aqueous solutions, ozone may react with various dissolved compounds in one of two ways: either direct reaction of the molecular ozone or indirect reaction through the formation of secondary oxidants (radical species: hydroxyl radicals) during ozone decomposition in water. These different reaction pathways lead to different oxidation products, and are controlled by different types of kinetics.

Indirect reactions

Any indirect reaction of ozone with pollutants generates radicals, such as hydroxyl radicls (OH^0) which can then accelerate ozone decomposition. They react un-selective and rapidly with $k_r = (10^8–10^{10})$ L/(mol.s) (Gottschalk et al., 2000). Radical mechansms are complex and depend on various factors. Major reactions are presented in eqs. 4 to 14, based on 2 models. Mechanisms consist, basically, of 3 stages: reaction initiation, radical chain-reaction, and reaction termination.
Reaction initiation between ozone and hydroxyl ions leads to the formation of superoxide anionic radical $O_2^{\cdot -}$ and hydrogen peroxide radical $HO_2^{\cdot}$ (see eqs. 4, 5).

$$O_3 + OH^- \rightarrow O_2^{\cdot -} + HO_2^{\cdot}, \qquad k_r = 70 \text{ L/mol.s} \tag{4}$$

$$HO_2^{\cdot} \leftrightarrow O_2^{\cdot -} + H^+, \qquad pK_a = 4{,}8 \tag{5}$$

Radical chain-reaction: according to the reaction between ozone and superoxide anionic radical $O_2^{\cdot -}$ ozonide anionic radical $O_3^{\cdot -}$ is formed, which immediately decomposes to OH-radicals, following eqs. 6 to 8.

$$O_3 + O_2^{\cdot -} \rightarrow O_3^{\cdot -} + H^+, \qquad k_r = 1{,}6\ .10^9 \text{ L/(mol.s)} \tag{6}$$

$$HO_3^{\cdot -} \rightarrow O_3^{\cdot -} + H^+, \qquad pK_a = 6{,}2 \tag{7}$$

$$HO_3^{\cdot} \rightarrow OH^{\cdot} + O_2, \qquad k_r = 1{,}1.\ 10^8 \text{ L/(mol.s)} \tag{8}$$

$OH\cdot$ can react eather way (eqs. 2.9 and 2.10):

$$OH\cdot + O_3 \rightarrow HO_4\cdot, \qquad k_r = 2{,}0 \,.\, 10^9\ L/(mol.s) \qquad (9)$$

$$HO_4\cdot \rightarrow O_2 + HO_2\cdot, \qquad k_r = 2{,}8 \,.\, 10^4\,/s \qquad (10)$$

Following eq. 10 oxygen and $HO_2\cdot$ are formed, and the reaction can begin again. The promoter is a compound which enables the transformation of $OH\cdot$ into superoxide radical $O_2\cdot^-/HO_2\cdot$, and catalyses the chain-reaction.
Organic molecules R are promoters. They contain functional groups and react with $OH\cdot$. Organic radicals $R\cdot$ are generated.

$$H_2R + OH\cdot \rightarrow HR\cdot + H_2O \qquad (11)$$

Organic peroxide radical $ROO\cdot$ is generated in the presence of oxygen and can react to form $O_2\cdot^-/HO_2\cdot$ during chain-reaction:

$$HR\cdot + O_2 \rightarrow HRO_2\cdot \qquad (12)$$

$$HRO_2\cdot \rightarrow R + HO_2\cdot \qquad (13)$$

$$HRO_2\cdot \rightarrow RO + OH\cdot \qquad (14)$$

Reaction termination

Compounds react with $OH\cdot$ to produce radicals $O_2\cdot^-/HO_2\cdot$. These are called inhibitors and terminate the reaction (eqs. 15 and 16). Known inhibitors are carbonate ions (k = 4,2 10^8 L/(mol s)) and hydrogencarbonate ions (k_r = 1,5 10^7 L/(mol s)), PO_4^{3-} , humic acids, and tertial butil alchohole (t-BuOH). There is a second method for terminating reaction when two radicals react to form oxygen and water (eq. 17).

$$OH\cdot + CO_3^{2-} \rightarrow OH^- + CO_3\cdot^-, \qquad k_r = (4{,}2\ 10^8)\ L/(mol\ s) \qquad (15)$$

$$OH\cdot + HCO_3^- \rightarrow OH^- + HCO_3\cdot, \qquad k_r = (1{,}5\ 10^7)\ L/(mol\ s) \qquad (16)$$

$$OH\cdot + HO_2\cdot \rightarrow O_2 + H_2O, \qquad k_r = (3{,}7\ 10^{10})\ L/(mol\ s) \qquad (17)$$

Two OH-radicals are formed per three molecules of ozone.
Other mechanisms are also possible. Some aromatic comounds with buffers decompose ozone within a system: aromatic ring ⇒ olefine ⇒ H_2O_2 ⇒ HO_2^-. The aromatic ring reacts with hydroxyl radical or ozone, olefine is generated, and a chain is formed with two bonds (C–C=C–C=C–C). Olefine immediately reacts with ozone to form H_2O_2. A part of this molecule dissociates to HO_2^-, which accelerates ozone degradation. t–butanole inhibits the aroomatic ring decay, and no H_2O_2 is formed. Some aromatic compaunds do not react with ozone but they do react with OH-radical. This second pathway is faster than the mechanism explained above (Pi, 2005).

Direct reactions

The direct reaction of organic compounds with ozone is a selective process and the reaction rate constant k_r = 1,0.10^3 L/(mol.s) is low (Gottschalk et al., 2000). It takes place when the radical mechanism is inhibited. Ozone reacts slowly with different organic compounds, whilest it reacts quickly with electron donors such as the hydroxyl group in phenole. Direct ozonation prevails if the radical mechanism is inhibited and if water contains terminating compounds. The direct mechanism is more important than the radical within pH ranges below 4 and vice-versa at higher pH above 10, while both mechansms are important within neutral range. Inorganic copmpounds, such as iron, manganesse, nitrite, cianide, bromide can be oxydized during ozonation. Fe^{2+} forms $Fe(OH)_3$, Mn^{2+} forms MnO_2, NO_2^- forms NO_3^- ions. The most problematic is the oxydation of bromide to bromate, which is a carcinogen. Certain chlorine by-products can be formed, such as HOCl, OCl^-, ClO_2^- v ClO_3^-. However, more organics can be directly oxiydized by ozone.

2.1 Micropollutants' oxidation with ozone

Atrazine can be degraded by mechanisms involving dealkylation, deamination, dehalogenation, and hydroxylation. Atrazine degradation is pH and temperature dependent. The atrazine degradation efficiency was 17 % at pH 3,3, and was much higher up to 71 % at pH = 9,7. If a higher pH value is applied more polar metabolites are formed and more atrazine is degraded (Kearney, 1988). Laboratory scale experiments showed that alkyl groups are oxydized, while amino alkyl groups are oxydized into acetamide. Ozonation of the N-ethyl group is five-times faster compared with ozonation of the N-isopropyl group (Hapeman, 1994). Two metabolites with imino-groups are formed. The N-ethyl-group is more reactive compared with the N-isopropyl group: 19-times by ozonation and 4-times by radical attack, therefore acetamide or imine is predominantly formed and does not react with ozone. The N-isopropyl group forms a free amino-group following dealkylation. The major reaction products released during atrazine ozonation according to Acero (Acero et al., 2000), which are:

- Atrazine: 2-chloro-4-(ethylamino)-6-(isopropylamino)-s-triazine, (CIET, 67 %)
- 4–acetamido–2–chloro–6–isopropylamino–s–triazine (CDIT, 24 %),
- deisopropylatrazine: 2-amino-4-chloro-6-(ethylamino)-s-triazine (CEAT, 5 %),
- deethylatrazine: 2-amino-4-chloro-6-(isopropylamino)-s-triazine (CIAT, 4 %; four primary metabolites),
- 4–acetamido–6–amino–2–chloro–s–triazine (CDAT),
- 6–amino–2–chloro–4–ethylimino–s–triazine (CIAT–imine),
- deethyldeisopropylatrazine (DEDIA; three secundary metabolites).

Degradation of 2–kloro–4–ethylimino–6–isopropylamino–s–triazine using ozone is slow and the end products are unknown.

4–acetamido–2–chloro–6–isopropylamino–s–triazine is degraded by ozone to 4–acetamido–6–amino–2–chloro–s–triazine (100 %). Ozonation of deethylatrazine leads to deethyldeisopropylatrazine formation (100 %). Ozone attacks isopropyl groups and leads to dealkyilation of isopropyl group. Due to the ozonation of deisopropylatrazine 6–amino–2-chloro–4–ethylimino–s–triazine (66 %) and 4–acetamido–6–amino–2–chloro–s–triazine (34 %) are formed.

The dealkylation of 6–amino–2–chloro–4–ethylimino–s–triazine is very difficult to carry out. 4–acetamido–6–amino–2–chloro–s–triazine and deethyldeisopropylatrazine are the end products. 2–chloro–4–ethylimino–6–isopropylamino–s–triazine (50%) is the major metabolite. Three products formed during the radical reactions plus a small portion of undefined products:

- 4–acetamido–2–chloro–6–isopropylamino–s–triazine,

- 4–acetamido–6–amino–2–chloro–s–triazine
- 6–amino–2–chloro–4–ethylimino–s–triazine.

Oxydation of 4–acetamido–2–chloro–6–isopropylamino–s–triazine leads to

- 4–acetamido–2–hydroxi–4–isopropylamino–s–triazine (ODIT, 10 %) and
- 4–acetamido–6–amino–2–chloro–s–triazine (90 %) formation.

While oxydation of deisopropylatrazine forms:

- 4–acetamido–6–amino–2–chloro–s–triazine (30 %) and
- 6–amino–2–chloro–4–ethylimino–s–triazine (70 %).

Deethyldeisopropylatrazine and 4–acetamido–6–amino–2 –chloro–s–triazine are the end products. Hydrolisis of acetamide following dealkilation forms acetic acid. 4–acetamido–2–chloro–6–isopropylamino–s–triazine hydrolyses at pH 6 to 8 to deethyldeisopropylatrazine. The imino group hydrolyses to acetaldehide. Deethyilatrazine is also end-product (Acero et al, 2000).

Drinking water with atrazine at 190 m^3/d flow was filtered and ozonized. The treated water contained increased concentrations of: deisopropylatrazine, deethylatrazine, deethyldeisopropylatrazine, hydroxydeethylatrazine 4-acetamido-2-chloro-6-ethylamino-s-triazine, hydroxydeisopropylatrazine, and other metabolites. (Verstraeten et al., 2002)

The ozone doze was 1,5 mg/L, and the ozonation time 20 min. After ozonation, the water was filtered. 2,2 mg/L of chlor 0,6 mg/L of fluor and 0,38 mg/L NH_4^+ were added before distribution to the city collection reservoir. The rate constants of N–ethyl group ozonation were double that of N–izopropyl group ozonation. Atrazine oxydation pathway was via N–acethyl group and included N–isopropyl group in smaller portion. Imine and amide were formed. The major reaction products released were (Verstraeten et al., 2002):

- deethylatrazine,
- deethyldeisopropylatrazine,
- 4-acetamido-2-chloro-6-etilamino-s-triazine (CDET),
- deisopropylatrazine,

while hydroxyatrazine (OIET; 2-(ethylamino)-4-hydroxy-6-(isopropylamino)-s-triazine) was not considered due to negligible dehalogenation. The pH of the water was above 8,5, more deethylatrazine was formed in comparison with deisopropylatrazine. A similar process could be done with hydroxyatrazine where dealkyl products being formed (hydroxydeethylatrazine (OIAT; 2-amino-4-hydroxy-6-(isopropylamino)-s-triazine), and hydroxydeisopropylatrazine (OEAT; 2-amino-4-(ethylamino)-6-hydroxy -s-triazine (Verstraeten et al, 2002).

Ozonation alone is sometimes insufficient for successful oxidation of micropollutants. Ozone might be combined with other processes, such as UV-light, hydrogen peroxide, Fenton. The range of processes, termed advanced oxidation processes AOP' dosplayed great potential for treating organic micropollutants. Metabolites are formed during AOP, which could be even more toxic as target micropollutants. Therefore, ozonation is combined using filtration or adsorption.

3. Advanced oxidation processes

The range of AOP processes have been developed over the last 40 years in order to degrade organic micro-pollutants. AOPs are based on the generation of powerful oxidizing agents, especially hydroxy-radicals, which destroy micro-pollutants. The best-known is the direct photolysis of hydrogen peroxide and UV - direct photolysis or photo-induced processes, such as Photo-Fenton oxidation. Enhanced degradation of atrazine by ozonation is achieved by combining ozonation with other processes, such as H_2O_2/ozone within the pH range above 7 (Meunier et al., 2006, Huang et al, 2004), and UV/ozone (Panno, 2004). Other combined processes are also known from the literature, such as UV/TiO_2 (Van Gunten, 2003), and UV/Fenton reagent (Hermanowicz et al, 1999). Chan & Chu (Chan & Chu, 2005) reported on the dependence of atrazine removal from concentration of dissolved iron ions, using Fenton reagent. Park (Park et al., 2004) proved the dependence of ozone decomposition on the pH of goethite surfaces during para-chlorobenzoic acid degradation. Ni (Ni et al., 2002) used different metal ions for 2-dichlorophenol degradation with ozone. $TiO_2/h\nu/O_2$ is a powerful combination for pesticides degradation (Andreozzi et al, 1999). Compared to OH· (1.8-2.7 V depending on the pH), $SO_4^{\cdot-}$ demonstrated higher standard reduction potential (2.5-3.1 V) at neutral pH. At acidic pH, they both demonstrated similar reduction potential, but $SO_4^{\cdot-}$, in general, was more selective for oxidizing organics than that of hydroxyl radicals. There were few studies on the generation mechanism of $SO_4^{\cdot-}$ by cobalt- catalyzed decomposition in the homogeneous system. This $SO_4^{\cdot-}$ radicals were very effective in oxidizing and transforming organic compounds like atrazine (Chan & Chu, 2009).

3.1 Ozone/Hydrogen peroxide

H_2O_2 reacts with ozone as anion HO_2^-(Gottschalk et al., 2000). The reaction constant of the system H_2O_2/O_3 depends on the initial oxydant's concentration. This reaction follows eqs. 18 and 21. The reaction of ozone and undisociiated H_2O_2 is also possible (see eq. 20), but the degradation rate is low. Initiation is also possible following eq.21 (Sunder & Hempel, 1997).

$$H_2O_2 \leftrightarrow HO_2^- + H^+, \qquad pK_a = 11,8 \tag{18}$$

Iniciation of reaction:

$$HO_2^- + O_3 \rightarrow HO_2^{\cdot} + O_3^{\cdot-}, \qquad k_r = 2,2\,.10^6 \text{ L/(mol.s)} \tag{19}$$

$$H_2O_2 + O_3 \rightarrow H_2O + O_2, \qquad k_r < 10^{-2} \text{ L/(mol.s)} \tag{20}$$

$$H^+ + O_3 \rightarrow HO_3^{\cdot} \rightarrow HO^{\cdot} + O_2, \qquad k_r = 1,1.\,10^5\text{/s} \tag{21}$$

During the chain-reaction phase hydroxyl radicals are transformed into peroxy radicals (eqs. 22 to 24). pK_a for $O_2^{\cdot-}/HO_2^{\cdot}$ is 4,8 (see eq. 5), and radical chain-reaction ends following eq 6. Compounds in water can act as promoters or scavengers, such as organic pollutants. The radical chain-reaction goes as follows (Sunder & Hempel, 1997):

$$HO^{\cdot} + O_3 \rightarrow HO_2^{\cdot} + O_2, \qquad k_r = 1,1\,.10^8 \text{ L/(mol.s)} \tag{22}$$

$$H_2O_2 + HO^{\cdot} \rightarrow H_2O + HO_2^{\cdot}, \qquad k_r = 2,7\,.10^7 \text{ L/(mol.s)} \tag{23}$$

$$HO^{\cdot+} + HO_2^{\cdot} \rightarrow H_2O + O_2^{\cdot-}, \qquad k_r = 7,5\,.10^9 \text{ L/(mol.s)} \tag{24}$$

An electron pathway from H_2O_2 to HO_2^- or bimolecular degradation is possible. During the ozonation of perchloroethene inorganic carbon acts as a hydroxyl radical scavenger and is formed due to the addition of H_2O_2. Increased concentration of inorganic carbon in water caused lower pollutant concentration. Free radicals are un-available due to the reaction with ozone, the ozone degradtaion decreases, and the concentration of inorganic carbon affects ozone stabilization. This effect is impossible at high pH. Most of the inorganic carbon is in hydrogencarbonate ion form at pH = 7. The inorganic carbon concentration was between 50 and 200 mg/L and consequently alkalynity was high. Hydrogencarbonate ions react with hydroxyl radicals at a lower constant rate of 8,5. 10^6 L/(mol. s) (Acero & Von Gunten, 1998) compared with atrazine and nitrobenzene with k = 3 .10^9 L/(mol.s) (Hoigne, 1997). Due to a higher concentration of hydrogencarbonate compared with organic pollutants it can be presumed that hydrogencarbonate consumes hydroxyl radicals. The inhibitory effect of hydrogen-carbonate is formed originating from the reaction with hydroxyl and hydrogencarbonate radicals which act selectively. They also express a lower reaction constant compared with hydroxyl radicals for the oxidation of organic pollutants (Hoigne, 1998). The reaction of hydrogencarbonate and hydroxyl radicals leads to intermediate formation, which can enable the radicals to form (Legube et al., 2004).

Termination phase:

The reaction between hydroxyl radicals and inorganic carbon produces carbonate radical $CO_3^{\cdot -}$, and the reaction mechanism is insufficiently established in the literature (eqs. 25 to 27).

$$HO^{\cdot} + HCO_3^- \rightarrow H_2O + CO_3^{\cdot -},$$
$$k_r = 1{,}5\ .10^7\ L/(mol\ s)\ (Rosenfeldt\ et\ al.,\ 2006),\ 8{,}5.\ 10^6\ L/(mol.s)\ (Ma,\ 2000) \quad (25)$$

$$HO^{\cdot} + CO_3^{2-} \rightarrow HO^- + CO_3^{\cdot -},$$
$$k_r = 4{,}2\ 10^8\ L/(mol\ s)\ (Rosenfeldt\ et\ al.,\ 2006),\ 3{,}9\ .10^8\ L/(mol.s)\ (Ma,\ 2000) \quad (26)$$

$$HO^{\cdot} + pollutant \rightarrow H_2O + CO_2 + products \quad (27)$$

The reaction constant for HO_2^- was determined at 2,2 .10^6 L/(mol.s) and for HO^- at 70 L/(mol.s). The latter is negligible in comparison with the first.

The constant for atrazine degradation in the presence of hydroxyl radicals was determined at:

- k_r = 3 .10^9 L/(mol.s) at pH = 2 (Accero et al, 2000),
- k_r = 2,4. 10^9 L/(mol.s) at pH = 2 (DeLaat et al, 1994),
- k_r = 2,6 .10^9 L/(mol.s) at pH = 3.6 (photo-Fenton) (Haag & Yao, 1992).

The direct constant of atrazine decomposition was calculated at:

- k_r = 7,90 ± 0,62 L/(mol.s) at pH = 3 and 25 ±0,2 °C (Camel & Bermont, 1998), which is very low.

Two end products are formed if atrazine is exposed to the H_2O_2/O_3 process,: 2,4–diamino–6–hydroxy–s–triazine at pH = 8 and deethyldeisopropylatrazine. The share of both compounds depends on hydroxyl radical concentration (Nelieu et al., 2000).

The H_2O_2/O_3 process is advisable for oxidation in water if the ozone molecule is relatively stable (Acero & von Gunten, 2001). Such waters do not contain many organic substances but the alkalinity is high. During ground-water ozonation, the ozone reacts quickly during the initial phase, then the first-order reaction takes place. Sometimes we can not differentiate between these two phases due to the very rapid end of the first one. During the H_2O_2/O_3 process only one phase was obseved for both types of drinking water. The ozone degradation rate constant was 10- times lower compared with the H_2O_2/O_3 process, in ground-water, while twice as low in surface water. The quantity of hydroxyl radicals was measured by p–chlorobenzoic acid transformation. It reacted with hydroxyl radicals at k_r = 5,2. 10^9 L/(mol. s) and not with ozone, due to the very low constant k_r = 0,15 L/(mol.s). One hydroxyl radical per three molecules of ozone is formed during ozonation while one hydroxyl radical per one molecule of ozone is formed during the H_2O_2/O_3 process in pure water, which means 2/3 of the hydroxyl radicals during H_2O_2/O_3 are formed due to ozonation during the initiation phase. Hydroxyl radicals' formation per ozone molecule is 0,5 in real water samples. The formation of hydroxyl radicals increased from 23 to 54 % during ozonation and the H_2O_2/O_3 process in ground-water. In the surface-water, the formation of hydroxyl radicals was comparable with the ozonation and H_2O_2/O_3 processes. The surface-water contained more NOM, which acted as a hydroxyl radicals' promoters; one hydroxyl radical formed per one molecule of ozone (theoretical value for $OH\cdot/O_3$ was the same as for O_3/H_2O_2). The addition of H_2O_2 did not significantly accelerate the ozone degradation. The alkalinity of the surface-water was higher than that of the ground-water, therefore, the p–chlorobenzoic acid degradation rate was higher in groundwater. Namely, the inorganic carbon species are scavengers of hydroxyl radicals. Atrazine was decomposed in 300 min during ozone oxidation, and in 80 mins during the H_2O_2/O_3 process (Acero & von Gunten, 2001).

3.2 Ozone/UV

The degradation of organic compounds takes place by photolysis (Beltran, 1996). Ozone in water is decomposed into H_2O_2 (eq. 28). Ultraviolet lights should expose photolysis at 254 nm.

$$O_3 + H_2O \rightarrow H_2O_2 + O_2 \tag{28}$$

Oxidation of the compound can be achieved by each oxidant: UV-radiation, ozone, and H_2O_2. Direct photolysis of UV-light absorption can take place. Direct oxydation by H_2O_2 is impossible under normal conditions at pH = 5–10 and room temperature. The extinction coeficient ε for ozone at 254 nm is higher (ε = 3300 L/(mol.cm)) comapred with H_2O_2 (ε = 18,6 L/(mol.cm)). The rate of ozone degradation is 1000–higher compared with H_2O_2 degradation (Gottschalk et al., 2000).

During ozonation, the ozonide radical ion $O_3\cdot^-$ is formed from $O\cdot^-$ and O_2. If $O\cdot^-$ are absent, $O_3\cdot^-$ degrade to $O\cdot^-$ and O_2, and reacts with $O\cdot^-$. $O_3\cdot^-$ degradation is very sensitive to $O\cdot^-$ presence. $O_3\cdot^-$ are generated with photolysis alkaline water (pH > 12,7) and H_2O_2 or $S_2O_8{}^{2-}$ addition. The first order reaction takes place with H_2O_2, whilest then using $S_2O_8{}^{2-}$ a more complex degradation process take palce and more intermediates are formed (Gonzales&Martire, 1997).

The fluorescence method was introduced to analyze hydroxyl radical levels during indirect ozone process, and O_3 /UV processes. It was observed that the amount of hydroxyl radical exposure during the O_3 /UV process was much higher than in the indirect ozone process. If the alkalinity in water is high, the inhibition is significant and the linear correlation between alkalinity and hydroxyl radical exposure was revealed which might have insight into the effect of alkalinity on the inhibition of hydroxyl radicals. Consequently, more reduction of TOC and DBP during the O_3 /UV process would be observed.

The chlorine demand increases with decreasing pH and increasing alkalinity. It could be concluded that hydroxyl radical can more strongly destroy the organic precursors resulting in reducing chlorine consumption than an ozone molecule (Gonzales & Martire, 1997).

3.3 UV/Hydrogen peroxide

Degradation of organic compounds can take place if the light energy is adsorbed by the molecule to produce an electronically-excited molecular state, and chemical transformation is competitive with deactivation process. UV/ H_2O_2 generally involves the generation of hydroxyl radicals' formation. The photolysis of H_2O_2 yields two hydroxyl radicals formed per photon, absorbed by 254 nm during a direct process (eq. 29). HO_2^- also absorbs energy by 254 nm and is in acid-base equilibrium with H_2O_2 (eqs. 30 and 31) (Gottschalk et al., 2000).

$$H_2O_2 \rightarrow 2OH\cdot, \qquad \varepsilon = 18{,}6\ L/(mol.s) \tag{29}$$

$$HO_2^- \rightarrow OH\cdot + O\cdot^-, \qquad \varepsilon = 240\ L/(mol.s) \tag{30}$$

$$HO_2^- + O\cdot^- \rightarrow OH^- + O_2\cdot^-, \qquad k_r = 4.\ 10^8\ L/(mol.s) \tag{31}$$

Prado (Prado & Esplugas, 1999) studied atrazine degradation using UV light at different pH values 4,74, 6,85 and 11,71. The best results were achieved at pH 11,71 over 50 min. In a second set of experiments, atrazine was treated with H_2O_2. 4 % of atrazine was oxidized at pH = 4,8, 9 % at pH = 6,8, and total atrazine degradation was achieved within 240 min at pH = 11,4. Enough radicals were available at a high pH without UV-irradiation. Atrazine was oxidized with ozone and 50 % of atrazine decomposed within 90 min at pH = 4,74, while 70 % at pH = 6,88 and again total degradation was achieved at high pH 11,55 within 30 mins. It can be conceluded from the results, that direct ozonation is slower compared with the radical. During indirect reaction a lot of hydroxyl radicals were formed at high pH values. The UV/H_2O_2 process was studied for atrazine removal. Within the neutral pH range only 15 min was needed for atrazine removal, whilest when using UV 50 min was necessary. The atrazine disappeared at pH = 4,74 in 25 min. The process is slower compared with the use of UV at only pH = 11,55. Other reactions take place at higher pH values and the hydroxyl radical formation is deactivated. At O_3/UV process, H_2O_2 decomposes to hydroxyl radicals just like the H_2O_2/UV process. Atrazine decomposes within 80 min at pH = 4,65, within 40 mins at pH = 7,2, and within 30 mins at pH = 11,23. During the O_3/H_2O_2 process, 70 % of atrazine decomposes within 90 mins at pH = 4,0, total degradation takes place within 40 mins at pH = 6,92, and within 30 mins at pH = 10,1. In this study H_2O_2 concentration was 10–times higher than stechioometric concentration. The O_3/H_2O_2/UV process showed that atrazine decomposes within 30 mins at pH = 4,31 t = 30 min, within 15 mins at pH = 6,7, and within 100 mins at pH = 11,03. The best results were achieved using UV/H_2O_2 at pH = 6,8 followed by O_3/H_2O_2/UV within the neutral pH range. The half-life times for atrazine degradation are listed in Table 1. The concentration of atrazine was 6,95. 10^{-5} mol/L at the temperature 20–23 °C. (Prado & Esplugas, 1999).

Atrazine was degraded in ultra-pure water. The best results were achieved using O_3/UV which is in accordance with other experiments. The water type has a major influence on removal efficiency (Beltran et al., 2000).

pH	(4,3–4,8)	(6,7–7,2)	(10,1–11,7)
process	t/min	t/min	t/min
UV	9,7	8,2	7,1
O_3	81	38	3,6
H_2O_2	/	/	32
H_2O_2/UV	4,7	2,2	25,4
O_3/UV	14	9	8,5
O_3/ H_2O_2	52	10	16
O_3/ H_2O_2/UV	4,9	3,1	20

Table 1. The half-life time of atrazine

3.4 Ozone/Metal catalyst

Catalytic ozonation is a new technology developed over recent years. It was discovered, that the reaction rate increased when Pb^{2+}, Cu^{2+}, Zn^{2+}, Fe^{2+}, Ti^{2+}, and Mn^{2+} ion were applied during the ozonation of 2-dichlorophenol (Ni et al., 2002).
The best results were achieved using Mn^{2+} (k_r = 227 L/(mol.min)) followed by Fe $^{2+}$ (k_r = 143 L/(mol.min), , Ti^{2+} with k_r = 139 L/(mol.min), Zn (k_r = 107 L/(mol.min), Cu k_r = 89 L/(mol.min) and Pb k_r = 81 L/(mol.min). The reaction rate was three time higher at pH = 3 and 1 mg/L Mn^{2+} in comprarison with treatment without Mn. With an initial Mn-concentration of 0 to 2 ppm, after gas exposure for 20 min the removal rate can be increased from 38% to 93%. The TOC removal rate increased from 13 % to 38 % over 60 min. The reaction rate improved greatly at an initial pH = 3. Linear correlation was established between ozone degradation with a metal catalyst, and the oxydation abilities of pollutants. At high pH values atrazine degradation is high due to high concentrations of OH-radicals (Ni et al., 2002). Manganesse is non-toxic to humans. The maximum allowable concentration (MAC) of 50 μg/L was set for Mn^{2+}. Linear correlation was established between Mn^{2+} concentration and concentration of undecomposed atrazine in dependence of time. The catalytical ability of Mn^{2+} activation was higher than that of the Mn^{4+} ions created after the reaction of $KMnO_4$ and $MnSO_4$. Comercially-available MnO_2 proves the non-degradation ability of atrazine. The authors explained that fact by the formation of hydroxyl radicals if Mn^{2+} is combined with ozone generation. If the Mn^{2+} concentration was higher, more ozone reacted with the organic pollutant and less ozone remained in the solution. Even the flow-rate of ozone from the air-gas into the solution was higher with higher manganesse ions in the solution. If the Mn^{2+} concentration was 1.5 g/L 65 % of ozone flows from air and only 35 % without Mn^{2+} (Ma & Graham, 1997). The humic acid influence on atrazine degradation was studied by the same authors. At a low quantity of humic acids at 1 mg/L the degradation process was accelerated with the presence of Mn^{2+} or MnO_2, due to the humic acids acting as initiators and promoters of radical reaction. At higher humic acid concentrations (2,4 and 6 mg/L) the process of atrazine degradation was inhibited by the humic acids despite different additions of Mn^{2+} or MnO_2, due to the reaction of humic acid with hydroxyl radicals. MnO_2 exhibits certain adsorption properties for certain pollutants, therefore the theory of atrazine adsorption onto MnO_2 was experimentally proven. The results show that only 10 % of atrazine adsorbed onto MnO_2 which is low, regardless on the concentration of humic acid in the solution (Ma & Graham, 1999). In the same reasearch, the

effect of hydrogencarbonate and t-butanole on atrazine ozonation was studied and using catalyst Mn^{2+}. Mn^{2+} improved the atrazine degradation. Polar low molecular weight metabolites were formed. Increased hydrogencarbonate ion concentrations or t-butanole decreased the atrazine degradation rate, because they acted as OH-radical scavengers. The worst results were achieved with t-butanole, which reacts with hydroxyl radicals at a rate of $5 \cdot 10^8$ L/(mol.s), and is faster compared with hydrogencarbonate ions. Inert intermediates are formed during t-butanole reaction with hydroxyl radicals, which terminate the reaction, and with t-butanole acting as the inhibitor. Ozone degradation is slower even at low t-butanole concentrations of 5 mg/L (Ma & Graham, 2000).
p–chlorobenzoic acid was exposed to ozonation using goethite (FeOOH), with particle sizes 0,3–0,6 mm, and a specific area of 147 m^2/g. The ozone rate constants were double using 2 g/L FeOOH compared with ozonation alone (Park, 2004).
Simazine, atrazine (ethyl and isopropyl groups), and terbuthylazide (ethyl and t-buthyl groups) at pH = 3 were oxidized by ozone, as well as $Ce(NH_4)_2(NO_3)_6$ in acetonitrile. Dealkyilation of the N-ethyl group took place, whilest it was negligible for atrazine. Dealkylation of t-buthyl group did not appear, therefore N-deethylation dominated. The ratio N–deethylation/N–deisopropylation was determined at 11,5 using $Ce(NH_4)_2(NO_3)_6$ whilest it was 5,3 using ozone. Oxidative N-dealkylation of small linear alkyl groups is faster if alkyl chained groups are larger at chemical oxidants, and in enzimatic systems. Ozonation of atrazine and terbuthylazide generated traces of amides during oxidation of carbon next to nitrogen on N-ethyl group (Bolzacchini et al., 1994).
The atrazine ring opens using Al_2O_3 from 240 to 450 °C. Triazine ring hydrolises to amonium and carbon dioxide. Those groups which are bonded to the triazine ring reacted with hydrogen to form small molecules (Zhan et al., 1996).
Kinetic decomposition of ozone and atrazine (and its metabolites), were studied using ozonation and catalytic ozonation. Three different types of Pt-catalyst were applied by studying the atrazine decay-rate. It was found that the addition of Pt-catalyst improves the atrazine decay rate at higher pH. The improvement was more significant using $Dohr_1$-Pt catalyst. After 30 min of catalytic ozonation, up to 93% of atrazine was removed, whilest only 33% of atrazine was removed after 30 min of ozonation without the catalyst. HPLC analyses showed that atrazine did not decompose to form deethylatrazine, but some other substances which could not be detected using our analytical methods. Pt-catalysts increased the ozone decomposition rate. The determined rate constants using ozonation with Pt-catalyst were twice the values of ozonation without catalyst. The ozone decomposition was generated in the bulk solution and on the surface of the Pt-catalyst. The highest rate of decomposition in the bulk solution was achieved within the pH range above 7, and was equal to 0.0262 1/min; on the catalyst surface the highest decomposition rate of 0.0120 1/min.g L was achieved in a neutral pH. The decomposition of ozone is proportional to the Pt catalyst mass (Tepuš & Simonič, 2007; Tepuš & Simonič, 2008).

3.5 UV/Metal catalyst

Pesticides could be degraded using high pressure mercury or xennon light due to different photochemicall processes. Long reaction times and highly energetic photones are needed. Pesticides are often incompletely removed. The major reactions using UV-light are dehalogenation, substitution of chloride atom by hydroxil groups, and radical formation advanced oxidation processes with UV-light are promising processes for pesticide degradation. Ther is a difference between:

- homogenous (H_2O_2/UV, Fe^{3+}/UV) and
- heterogenous photocatalitic processes (TiO_2/UV, ZnO_2/UV).

Photocatalitic advanced oxidation processes are light-induced reactions based on hydroxil radicals' formation in combination with oxidants, such as TiO_2, ZnO_2, Fenton. Titan dioxide (TiO_2) is a more frequently used photocatalyst for pesticide degradation. Light is absorbed at λ < 385 nm (while H_2O_2/UV needs λ between 210 and 230 nm). Positive sites in the net are generated acting as strong oxidants, or hydroxil radicals are formed. TiO_2 uses sun-light as an energy source. If the catalyst is bonded to the surface, the efficiency is lower in comparison with those processes where TiO_2 ions are mixed into the solution. If TiO_2 is bonded to the activated carbon, the pesticide reaction rate might increase and the generation of atrazine metabolites decrease. For 90 % atrazine degradation 9,1 min was necessary during the TiO_2/UV process, 9,2 min for ozonation, and below 0,5 min using Fe/UV (Chiron, 2000).

Atrazine degradation using UV light and TiO_2 catalyst was studied (Pelizzetti, 1992). Hydrolisys of 2–chloro substituent on a ring took place, oxidation of alkyln group, and dealkylation and deamination of the chain. Finally, the amino groups were replaced by hydroxil ones. A series of intermediates were analysed, while the cianuric acid did not decompose. Atrazine decomposition was fast, whilest the cianuric acid rate was slow due to the amino group bonding to the triazine ring replacement by hydroxil ones. Inorganic compounds, such as peroxodisulphate accelerate cianuric acid formation. Many products are generated with high hydrofilicity, and are less toxic.

95 % of atrazine was degraded photocatalytically on immobilised TiO_2 at pH = 7,1 with the initial concentration of atrazine at 1mg/L, over 24 h. When real water was used as matrix, the atrazine degradation rate was reduced at a factor of 3. Orto–phosphate and carbonate ions slightly improved the process, whilest other inorganic species did not influence the reaction rate. Atrazine in destilated water decomposes into deethylatrazine, deisopropylatrazine, hydroxiatrazine, deethyldeisopropylatrazine, deethylhydroxiatrazine (OIAT), deisopropylhydroxiatrazine (OEAT), deethyldeisopropylhydroxiatrazine (OAAT). The latter decomposes into cianuric acid (OOOT). In drinking-water atrazine decomposes into deethylatrazine, deisopropylatrazine and deethyldeisopropylatrazine. Photochemical degradation yields dehalogenated products, while photocatalytical degradation yields dealkilated products (Ziegmann et al., 2006).

3.6 Hydrogen peroxide/metal catalyst

Fenton reagent (H_2O_2 and Fe^{2+}) enables hydroxil radicals formation following eqs. 32 and 33.

$$Fe^{2+} + H_2O_2 \rightarrow Fe^{3+} + HO\cdot + HO^- \quad (32)$$

$$HO\cdot + Fe^{2+} \rightarrow HO^- + Fe^{3+} \quad (33)$$

Oxidation of alkylamino groups and/or dealkylation take palce. The exact mechanism is unknown, as yet.

Atrazine was degraded into:

- 4–acetamido–2–chloro–6–isopropylamino–s–triazine,
- deethylatrazine,
- 4–acetamido–2–chloro–6–ethylamino–s–triazine (CDET),

- deisopropylatrazine,
- 4–acetamido–2–hydroxi–6–isopropilamino–s–triazine, due to dehalogenation, oxidation of alkyl groups. In less than 30 s out of atrazine:
- deethyldeisopropylatrazine (23 %) and
- 4–acetamido–6–amino–2–chloro–s–triazine (28 %) were formed.

Higher Fenton reagent concentrations enables the formation of 4–acetamido–6–amino–2–chloro–s–triazine, and deethyldeisopropylatrazine. The end-product was 2, 4–diamino–6–hydroxy–s–triazine. The efficiency of atrazine degradation was 99 % at pH = 3, and only 37 % at pH = 9 (Arnold et al., 1995).

The next study suggest two phases of atrazine decomposition: 'faster and slower 'following second order kinetics (Chan & Chu, 2003), more metabolites were analysed:

- 2–chloro–4–(1–carboxylethanolamino)–6–isopropylamino–s–triazine (CIET-carboxylethanolamino),
- 2-hydroxy-4-acetamido-6-ethylamino-s-triazine (ODET),
- 6-hydroy-4-ethylamino-2-amino-s-triazine,
- 4-hydroxy-6-isopropylamino-2-amino-s-triazine (Chan & Chu, 2005)

Fenton reagent consists of an iron salt which is usually Fe- sulphate. In this study, iron hydride within an anaerobic environment was used, due to the fact that oxygen leads to organic radicals' formation and peroxyl radicals, which affect the Fe^{2+} and hydroxyl radical concentrations, but not the reaction rate of atrazine degradation. Therefore, secondary reactions do not affect rate reaction. The constant reaction rate using iron hidride was ten times higher at pH = 3 than at pH = 8 (Barreiro et al., 2007).

Reaction rate constants were determined at 0,24–2,83. 10^3 1/s for atrazine and (1,57–12,75).10^5 1/s for H_2O_2 using Fe^{3+}/H_2O_2 for 80 % atrazine removal in 1 h. Higher H_2O_2 led to higher rate constants til the certain value and after that they decrease (Gallard & De Laat, 2000).

Continuously electrogenerating of H_2O_2 from the electro-reduction of dissolved O_2 and combination of Fe^{3+} and Cu^{2+} leads to the optimum degradation rate for which complete disaperance of atrazine is achivied at 22 min. However, Cu^{2+} concentrations higher than 10 mM inhibit H_2O_2 generation and consequently atrazine degradation rate because of copper deposition on the carbon- felt cathode surface. In this study degradation of cyanuronic acid, the ulitmate product of atrazine was observed, which is very rare (Balci et al., 2009).

4. Adsorption media for atrazine removal

4.1 Adsorption isotherms

Adsorption isotherms are developed by exposing a given amount of adsorbate in a fixed volume of liquid to varying amounts of activated carbon. The adsorbent phase after equlibrium is calculated using eq. 34:

$$c_e = ((\gamma_0 - \gamma_e)\ V)/m \qquad (34)$$

c_e – adsorbent phase concentration after equlibrium , mg/g
γ_0 – initial concentration of adsorbate, mg/L
γ_e – final equlibrium concentration of adsorbate, mg/L
V – volume of water in reactor, L
m – mass of adsorbent, g

Various adsorption isotherms were developed, but the Freundlich isotherm is used more commonly, followed by the Langmuir isotherm.
The Freundlich isotherm is defined as follows (Metcalf & Eddy, 2003):

$$c_e = k_f \, \gamma_e^{1/n} \tag{35}$$

k_f – Freundlichov capacity factor (mg/g) $(L/mg)^{1/n}$
$1/n$ – Freundlich intensity parameter,
The constants in Freundlich isotherm can be determined by plotting log c_e versus log γ_e. Eq. 35 can be rewritten as eq. 36:

$$\log c_e = \log k_f + 1/n \log \gamma_e \tag{36}$$

Langmuir isotherm is defined as (eq. 37):

$$c_e = (a \, b \, \gamma_e)/(1 + b \, \gamma_e) \tag{37}$$

Where
a – empirical constant, mg/g
b – empirical constant, cm^3/mg
The Langmuir isotherm was developed by assuming that a fixed number of accessible sites are available on the adsorption surface and that adsorption is reversible. Equilibrium is reached when the rate of adsorption of molecules onto activated carbon is the same as the rate of the molecules desorption. The Langmuir isotherm can be rearanged to eq. 38:

$$\gamma_e / c_e = 1/(a \, b) + \gamma_e / a \tag{38}$$

4.2 Pesticide removal achievements by adsorption

Different adsorption media could be used for atrazine removal such as activated carbon, zeolite, resins, and others (Nyex 100).
Adsorption resins are similar to ion-exchange resins. They express high porosity, include different exchange groups or none, and are utilised for anionic and weak ionic compounds adsorption. Resins could be divided into three groups regarding polarity:

- ion adsorption resins which are strongly base, as used for organic adsorption
- phenole adsorption resins which are weak base amino and phenole groups, used for coloured articles' removal in the food industry;
- inert adsorption resins macroporous copolimers of styrene and divinylbenzene with a high net-degree and high ratio between area and volume; used for weak ionised substances.

Atrazine, simazine and propazine as well as deethylatrazine, deisopropylatrazine and deethyldeisopropylatrazine were efficiently removed when using Calgon WPH and Norit HDB activated carbons. Freundlich constants were calculated, as presented in Table 2, for atrazine and metabolites. Calgon WPH was more efficinet for atrazine removal (Jiang & Adams, 2006).
Non of the metabolites are formed if atrazine is adsorbed onto activated carbon. The procedure is simple, also for deethylatrazine and in deisopropylatrazine removal. The water solubilty of atrazine is 33 mg/L, for Deethylatrazine and deisopropylatrazine it is higher at 380 mg/L and 210 mg/L at 25 °C, respectively. deethylatrazine and deisopropylatrazine expose a lower capacity for activated carbon compared with atrazine, due to the rule

Pesticide	Sample	$k_f/((mg/g)(L/mg)^{1/n}$	$1/n$
Atrazine	millipore	13,518[a] 6,15[b]	0,491[a] 0,44[b]
Deethylatrazine	millipore	1,793[a] 5,09[b]	0,294[a] 0,56[b]
Deisopropylatrazine	millipore	1,829[a] 6,13[b]	0,308[a] 0,65[b]
Atrazine	groundwater	2,211[a]	0,358[a]
Deethylatrazine	groundwater	0,651[a]	0,832[a]
Deisopropylatrazine	groundwater	1,385[a]	0,621[a]
Atrazine	millipore	10,654[c]	0,221[c]
Deethylatrazine	millipore	1,659[c]	0,219[c]
Deisopropylatrazine	millipore	1,837[c]	0,377[c]
Atrazine	groundwater	0,885[c]	0,973[c]
Deethylatrazine	groundwater	0,000[c]	7,516[c]
Deisopropylatrazine	groundwater	1,076[c]	0,420[c]

Table 2. Freundlih constants for different carbon types ([a]Calgon WPH, [c]Norit HDB, pH = 7, room temperature (Jiang & Adams, 2006), [b]Calgon WPL, pH = 6, ϑ = 21 °C (Adams & Watson, 1996))

that substances with higher solubility have lower adsorption capability when binding to activated carbon. Due to this rule, it can be expected that the adsorption capacities of other s–triazine metabolites are lower due to their high solubilities in water. Higher adsoprtion capacities were determined at lower pH = 6 compared with higher pH values (e.g.pH = 8). pH change within the neutral region does not affect the solubility of atrazine due to pK_a of atrazine = 1,7. Adsorption is a reliable treatment method for pollutant removal until certain value. The costs rise very quickly if it is necessary to remove the pollutant below this mentioned value (Adams & Watson, 1996).

Picabiol and WCM 106 activated carbon gave beter results concerning atrazine removal compared with WCM 106, due to a higher specific area. The efficiency was improved by combinig atrazine adsorption with pre-ozonation (Pryor, 1999). NOM has a huge influence on atrazine adsorption. It was discovered out that 3,4–0,4 mg/g lower adsorption capacity within 62 days is achieved due to the high NOM content in water. High DOC also inteferes with atrazine adsorption on granular activated carbon. Up to two thirds lower adsorption capacities were determined (Lebeau et al., 1999).

Organic zeolites were found to exhibit an adsortion capacity for organic pollutants. Clay with negative charge and zeolites have an affinity to cationic exchange. In contrast to clay, zeolites with grain sizes around 1 milimetre or more might be used as filter media for inorganic substances' removal, such as ammonia and heavy metals. If the functional groups on a zeolite surface are replaced by high-molecular weight quarter amine, they could be applied for neionic organic contaminants' removal from water. The capacities for atrazine bonding to stearyle-dimethylbenzi ammonium chloride modified zeolite surfaces was 0,43 mg/g, following the Langmuir model (Lemić et al., 2006) .

In study in which activated caron, carbonaceous resin and high- silica zeolites were studied to evaluate their effectivenes activated carbon was the most effective and zeolites were less effective because zeolites contain pores of uniform size and shape, and pesticides must matching pore size/shape requirements (Rossner et al., 2009).

Nyex 100 is an adsorption media containing non-porous particles of carbon, and expresses high conductivity. Adsorption and electrochemical regeneration are rapid due to hindered intramolecular diffusion. Nyex 100 is a cost-effective carbon dust material with particle diameters from 10 to 600 μm, and an average diameter of 124 μm. Its specific area is low at 2,75 m^2/g whilest activated carbon has 2000 m^2/g. The Freundlich adsorption isotherm for atrazine was calculated at 0,279 $(mg/g)(L/mg)^{1/n}$, and $1/n$ = 0,550 at pH = 3, at room temperature 17–26 °C. This is lower compared with activated carbon. If Nyex 100 is electrochemically regenerated using cathode and anode, and salt. However, the adsorption isotherm was similar to when using fresh Nyex 100 (Brown et al., 2004).

Adsorption isotherms were determined using Filtrasorb 400 (Chemviron Carbon) and two resins: Dowex Optipore L 493 (Dow Chemical Company) and Lewatit VP OC 1064 MD PH. The Freundlich equation was employed. Lewatit VP OC 1064 MD PH was the best adsorbent for atrazine, followed by Filtrasorb 400, and Dowex Optipore L 493 resin with only half the Lewatit VP OC 1064 MD PH capacity. Filtrasorb 400 was determined to be the better solution for deethylatrazine removal with a third higher adsorption capacity than Dowex Optipore L 493 (Tepuš et al., 2009).

5. Membrane technologies

Over recent years, membranes have become fully or partially integrated into all facilities that produce **drinking water** (Duranceau, 2000). This is due to the fact that membrane processes can resolve technically complex and, at times, conflicting requirements relating to compliance with multi-contaminant regulations (Taylor & Hong, 2000). With the tightening of regulations in the future, the need for membrane technology such as reverse osmosis (RO) and nanofiltration (NF) will increase significantly. However, wider use of reverse osmosis membrane technology in the drinking water industry has been hampered greatly by membrane fouling (Hong & Elimelech, 1997). The extent and rate of membrane fouling are largely affected by membrane surface characteristics (Elimelech et al., 1997; Vrijenhoek & Hong, 2001).

Because it is generally accepted that, besides the operation values (flux, pressure), membrane performance in RO and/or NF processes is influenced by membrane porosity and by physicochemical interaction in a system's membrane-water-solute(s) (Kosutic & Kunst, 2002). It has been discovered, that rejection of the model solution by very tight RO membranes is dominantly affected by the membrane porosity parameters (pore size distribution and effective number of pores), whilst, the rejection of charge ions and organics by NF membrane is expected to be influenced more by the physicochemical parameters (charge, hydrophobicity).

Therefore, in NF retention properties are very important: the possibility of retaining relatively small organic molecules and multivalent ions from aqueous solution is crucial for most applications. NF and RO offer very good removal possibilities for most **organic micro-pollutants**, since the molecular weights of these pollutants are often around 200-300 g/mol, and the molecular weight cut-off (MWCO) values of NF membranes are also often within this region (for RO membranes, the MWCO values are even lower). However, removal of some organic micropollutants is still incomplete and traces may still be detected in the permeate of NF and RO installations (Bellona et al., 2004).

Considering that the molecular weights of almost all pesticides range from 200 to 400 Da, NF membranes are potentially useful for pesticide removal. Since NF membranes can simultaneously remove both hardness and pesticides, their application to the treatment of drinking water has been increased (Reinhard et al., 1986; Baier et al., 1987; Duranceau et al., 1992; Hofman et al., 1993; Berg et al., 1997; Hofman et al., 1997; Van der Bruggen et al., 1998).

General

Membrane separation is addressed as a pressure-driven process. Pressure driven processes are commonly divided into four overlapping categories of increasing selectivity: microfiltration (MF), ultrafiltration (UF), nanofiltration (NF), and hyperfiltration or reverse osmosis (RO). MF can be used to remove bacteria and suspended solids with pore sizes of 0.1 to micron. UF will remove colloids, viruses and certain proteins with pore sizes of 0.0003 to 0.1 microns. NF relies on physical rejection based on molecular size and charge. Pore sizes are within the range 0.001 to 0.003 microns. RO has a pore size of about 0.0005 microns and can be used for desalination (Mulder, 1991).

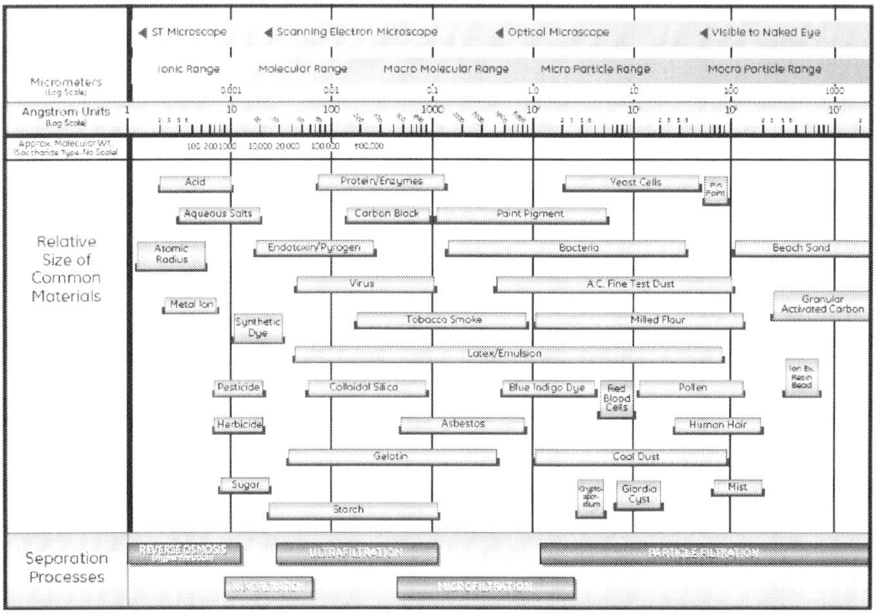

Fig. 1. Filtration and Separation Spectrum (Aim Filtration Systems, Aug. 2010).

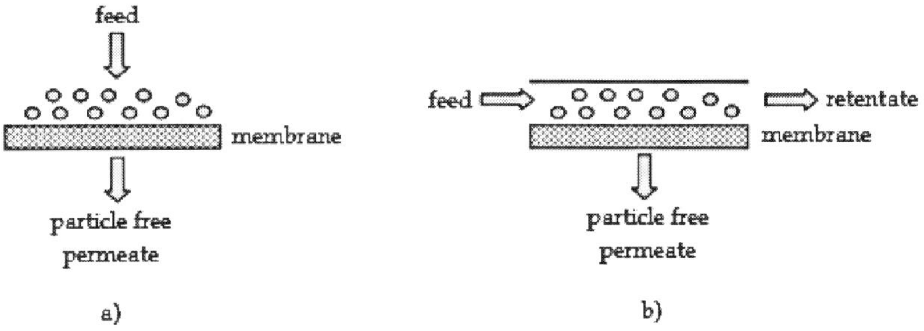

Fig. 2. Comparison between: (a) dead-end, (b) cross-flow configuration (Saxena et al., 2009).

During membrane filtration, there are two major filtration modes, dead-end filtration and cross-flow filtration. In the cross-flow mode, the fluid to be filtered flows parallel to the membrane surface and permeates through the membrane due to pressure difference. The cross-flow reduces the formation of the filter cake to keep it at a low level (Negaresh, 2007).

Membrane Materials

The membrane material refers to the substance from which the membrane itself is made. Normally, the membrane material is manufactured from a synthetic polymer, although other forms, including ceramic and metallic "membranes," may be available (Allgeier, 2005). MF and UF membranes may be constructed from a wide variety of materials, including cellulose acetate (CA), polyvinylidene fluoride (PVDF), polyacrylonitrile (PAN), polypropylene (PP), polysulfone (PS), polyethersulfone (PES), or other polymers. Each of these materials has different properties with respect to surface charge, degree of hydrophobicity, pH and oxidant tolerance, strength, and flexibility.

NF and RO membranes are generally manufactured from cellulose acetate or polyamide materials (and their respective derivatives), and there are various advantages and disadvantages associated with each. While cellulose membranes are susceptible to biodegradation and must be operated within a relatively narrow pH range of about 4 to 8, they do have some resistance to continuous low-level oxidant exposure. Polyamide (PA) membranes, by contrast, can be used under a wide-range of pH conditions and are not subject to biodegradation. Although PA membranes have very limited tolerance for the presence of strong oxidants, they are compatible with weaker oxidants such as chloramines. PA membranes require significantly less pressure to operate and have become the predominant material used for NF and RO applications (Allgeier, 2005).

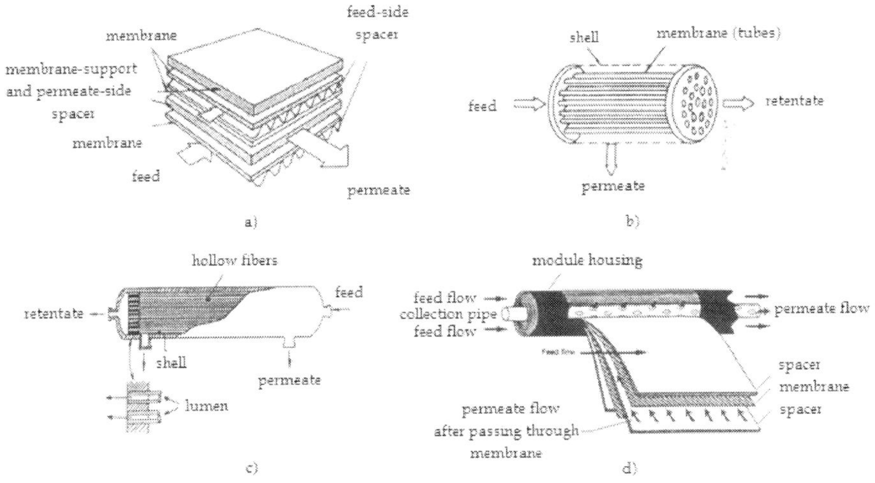

Fig. 3. Membrane module: a) Plate-and-frame membrane module b) Tubular membrane module c) Hollow fibre module with opened-end design d) Spiral Wound Membrane Module (Mulder, 1991).

	Reverse Osmosis	Nanofiltration	Ultrafiltration	Microfiltration
Membrane	Asymmetrical	Asymmetrical	Asymmetrical	Symmetrical Asymmetrical
Thickness **Thin film**	150 μm 1 μm	150 μm 1 μm	150 -250 μm 1 μm	10 - 150 μm
Pore size	<0.002 μm	<0.002 μm	0.2 - 0.02 μm	4 - 0.02 μm
Rejection of	HMW, LMWC, sodium chloride, glucose, amino acids	HMWC mono-,di- and oligosaccharides, polyvalent neg. ions	Macro molecules, Proteins, Polysaccharides, vira	Particles, Clay, Bacteria
Membrane material(s)	CA Thin film	CA Thin film	Ceramic, PSO, PVDF, CA Thin film	Ceramic, PP, PSO, PVDF
Membrane Module	Tubular, Spiral wound, Plate-and-frame	Tubular, Spiral wound, Plate-and-frame	Tubular, Hollow fiber, Spiral wound, Plate-and-frame	Tubular, Hollow fiber
Operating pressure	15-150 bar	5-35 bar	1-10 bar	<2 bar

Table 3. Comparing four membrane processes (Wagner, 2001).

Membrane Modules

The feasibility of a membrane process depends on the design of membrane module since the active separation membrane area is directly influenced by the membrane modules configuration. Plate-and-frame and tubular membrane module are two of the earliest module designs based on simple filtration technology. Both systems are still available today but, due to their relatively high cost and inefficiency, they have been mainly substituted by hollow fiber and spiral wound membranes (Cheryan, 1998).

Nanofiltration principle and mechanism

Among all the separation operations in the liquid phase using membranes, nanofiltration (NF) is the latest one to be developed. NF is a process located between UF and RO. Some authors refer to NF as charged UF (Simpson et al., 1987), softening, low pressure RO (Rohe et al., 1990). NF is generally expected to remove 60 to 80% of hardness, >90% of colour, and all turbidity. The process has the advantage of low operating pressures compared to RO, and a high rejection of organics compared to UF. Monovalent salt is not retained to a significant extent; however this is not normally required in the water treatment of surface water.

Rejection Mechanisms

Due to its small pore size, the observed mass transfer mechanism for NF is **diffusion** and **convection**. In addition, the active normal layer normally consists of negatively-charged chemical groups, thus mass transfer via migration of ions in an electrical field must also be considered (Tsuru T. et al., 1991). The transport mechanism is normally explained in terms of charge and or size effects (Peeters J. M. M, 1999). Transport of uncharged solutes takes place by convection due to a pressure difference and by diffusion due to the concentration gradient across the membrane. A sieving mechanism is responsible for the retention of the uncharged solutes. For charged components an electrostatic interaction takes place between the component and the membrane as the most nanofiltration membrane is charged (mostly negatively). The effect of membrane charge on the transport of charged components has already been described by Donnan at the beginning of the 20th century.

Equilibrium/Fixed charge effects

For charged solutes two additional mechanisms can be recognised:

1. Donnan exclusion:

When a charged membrane is placed in a salt solution, equilibrium occurs between the membrane and the solution. Because of the presence of the fixed membrane charge, the ionic concentration in the membrane is not equal to those in a solution. The counter-ion (opposite sign of the charge to the fixed charge in the membrane) concentration is higher in the membrane phase than in the bulk solution, while the co-ions (same sign of charge at the fixed membrane charge) concentration is lower in the membrane phase. A potential difference at the inter-phase, called the Donnan potential, is created to counteract the transport of counter ions to the solution phase and the co-ions in the membrane phase. When a pressure gradient across the membrane is applied, water is transported though the membrane. The effect of the Donnan potential is then to repeal the co-ions from the membrane. Because of the electro-neutrality requirements the counter ion is also rejected and salt retention occurs.

For every charge that passes through the membrane, an opposite charge must also pass to maintain charge-neutrality. This phenomenon is complicated since different ions have different diffusivities. So alone each ion would move through the membrane at a different speed. When several different ions are passing through the membrane together some are slowed down and some are sped up in order to maintain charge neutrality.

2. Dielectric exclusion:

Dielectric exclusion, which does not generally play a role in ultrafiltration and microfiltration but is of major importance in electrodialysis (Bontha & Pintauro, 1994). Due to the charge of the membrane and the dipole momentum of water, water molecules will show a polarisation in the pore. This polarisation results in a decrease in the dielectric constant inside the pore, thereby making it less favourable for a charged-solute to enter. However, even in a situation that the dielectric constant inside the pore is equal to the one of water, a change in electrostatic free-energy of the ion occurs when the ion is transferred from the bulk into the pore. This also results in exclusion. The relative importance of two mechanisms in NF is still a point of debate within the scientific community (Hagmeyer & Gimbel, 1998, Yaroshchuk, 2000). Most of literature on NF uses Donnan exclusion as the distribution mechanism (Tsuru et al., 1991, Wang et al., 1995, Bowen & Mukhtar, 1996).

The principal transport mechanisms of NF are depicted in Figure 4.

Macoun (Macoun, 1998) summarised NF rejection mechanisms as follows:

- **Wetted Surface** – water associates with the membrane through hydrogen bonding and those molecules which form hydrogen bonds with the membrane can be transported,
- **Preferential Sorption/Capillary Rejection** – the membrane is heterogeneous and microporous, electrostatic repulsion is based on different electrostatic constants in solution and membrane,
- **Solution Diffusion** – membrane is homogeneous and non-porous, solute and solvent dissolve in the active layer and diffusion determines transport,
- **Charged Capillary** – the electric double layer in pores determines rejection, ions of same charge as membrane are attracted and counter-ions are rejected due to the streaming potential,
- **Finely Porous** – membrane is a dense material punctured by pores, transport is determined by partitioning between bulk and pore fluid.

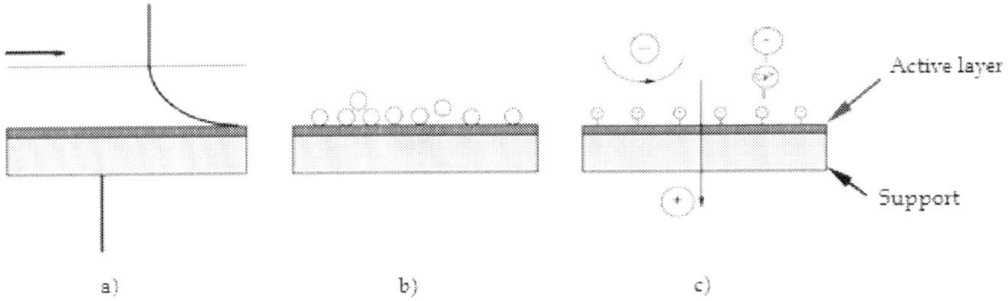

Fig.4. Transport phenomena in NF, (a) concentration polarisation (b) sieving (c) charge effects (e.g. charge repulsion or electrical double layer formation).

Filtration Models

The Extended-Nernst Planck Equation (equation (39)) is a means of describing NF behaviour. The extended Nernst Planck equation, proposed by Deen (Deen et al., 1980), includes the Donnan expression, which describes the partitioning of solutes between solution and membrane. The model can be used to calculate an effective pore size (which does not necessarily mean that pores exist), and to determine thickness and effective charge of the membrane. This information can then be used to predict the separation of mixtures (Bowen & Mukhtar, 1996). No assumptions regarding membrane morphology are required (Peeters, 1997). The terms represent transport due to diffusion, electrical field gradient and convection, respectively. J_{Si} is the flux of an ion i, $D_{I,P}$ is the ion diffusivity in the membane, R the gas constant, F the Faraday constant, Ψ the electrical potential, and $K_{I,c}$ the convective hindrance factor in the membrane.

$$J_{si} = -D_{i,p}\frac{dc_i}{dx} - \frac{z_i c_i D_{i,p}}{R \cdot T} \cdot F \cdot \frac{d\Psi}{dx} + K_{i,c} c_i J \tag{39}$$

The equation predicts solute rejection as a function of feed concentration, ion charge, convection across the membrane, and solute diffusion (Braghetta, 1995). The model has proven to be successful for modelling the solute transport in simple electrolyte solutions, although its applicability in the presence of organics is questionable.

Wang et al (Wang et al., 1995b) developed the model further to account for the transport phenomena of organic electrolytes, thus combining electrostatic and steric hindrance effects. The steric hindrance pore model suggested by Nakao et al. (Nakao et al., 1982) was incorporated into the modified Nernst Planck equation. For mixed solutions, hindered diffusivity becomes more significant. The rejection depends on electrolyte concentration and the membrane charge increases with salt concentration. This indicates co-ion adsorption on the membrane and, in fact, the effective membrane charge was described as a Freundlich isotherm being function of bulk concentration by Bowen and Mukhtar (Bowen & Mukhtar, 1996).

The Fine Porous Model, as presented by Xu and Spencer (Xu & Spencer, 1997), describes the equilibrium and non-equilibrium factors of rejection. Only coupling between solvent and solute is taken into account, and no solute-solute coupling is permitted. Equilibrium parameters dominated separation, and these are described by the reflection coefficient σ in equation (40), where k_M is the solute mass transfer coefficient in the membrane.

$$R = 1 - \left[1 + \left(\frac{\sigma}{1-\sigma}\right) \cdot \left(1 - e^{-\frac{J}{k_M}}\right) \cdot e^{-\frac{J}{k_S}}\right]^{-1} \qquad (40)$$

The Hindrance Pore Model was introduced by Wang et al. (Wang et al, 1995). This model also allows the calculation of an effective pore radius and the ratio of membrane porosity to membrane thickness. As can be seen with the various models, determination of an effective pore size has become an issue. This is due to the fact that NF pores are too small to be measured directly by various methods, as in MF or UF.

Micropollutants removal using NF

Viable technologies to remove micropollutants, such as pesticides and alkyl phthalates and NOM from water of impaired quality are high-pressure membrane processes such as nanofiltration (NF) or reverse osmosis (RO). In past research, it has been demonstrated that some micropollutants such as pesticides (e.g., atrazine) can be effective by NF membranes (Kiso et al., 2001; Kiso et al., 2000; Cho et al., 1999; Kiso et al., 2001; Kiso et al., 2002).

Pesticide rejection by NF and RO membranes is thought to be influenced by compound physical-chemical properties (e.g., molecular size, solubility, diffusivity, polarity, hydrophobicity, and charge), membrane properties (e.g., permeability, pore size, hydrophobicity, and charge), and membrane operating conditions (e.g., flux, transmembrane pressure, and recovery). Several studies have reported that the molecular size of the molecule was the most important structural property for retention (Van der Bruggen et al., 1999; Ozaki & Li, 2002). In addition to steric hindrance, Kiso et al. (Kiso et al. 2000; Kiso et al. 2001; Kiso et al. 2001) determined the hydrophobicity of compounds quantified as n-octanol/water partition coefficient (Kow), as another key parameter for rejection. Studies conducted by Van der Bruggen et al. (Van der Bruggen et al., 1998) using NF membranes indicated that a higher dipole moment resulted in a lower retention and that the retention of a compound with a high dipole moment was lower than that expected when based on molecular size. Most of these studies used surrogate compounds (e.g., alcohols) or pesticides; in many cases, higher than relevant concentrations were employed. There is still a lack of understanding about whether DBPs/EDCs/PhACs (disinfection byproducts/endocrine disruptors/Pharmaceuticals) can be sufficiently removed by NF and RO membranes. (Kimura et al., 2003)

Studies above pesticide removal have mostly focused on the removal mechanisms between pesticides and membranes. Van der Bruggen et al. (Van der Bruggen et al., 1998; Van der Bruggen et al., 1999) demonstrated that molecular weight and size were the most critical mechanisms for pesticide removal using different kinds of NF membranes. Kiso et al. (Kiso et al., 2000; Kiso et al., 2001; Kiso et al., 2001; Kiso et al., 2002) studied the rejection of alkyl phthalates, nonphenylic pesticides, and aromatic pesticides by flat-sheet and hollow fine fibre types membranes. Both RO and NF were used in their studies. The results also showed that molecular weight, size, and hydrophobicity were all significant. However, the combined effect of the flux, recovery, molecular weight and size were seldom discussed together, although flux and recovery are two of the critical operational parameters for NF membranes.

A single-element Filmtec NF70 nanofilter was operated for six l-month periods in which each of the pesticides was studied (Duranceau et al., 1992). The results showed that rejection of these six pesticides was dependent on pesticide molecular weight. EDB (molecular weight 190) completely passed the NF70 for all test conditions. DBCP (molecular weight 236) was partially rejected and indicated diffusion control mass transport. All other pesticides having molecular weights greater than 278 were completely rejected by the membrane. Variations in recovery and feed-stream velocity had no effect on pesticide rejection by the membrane with the exception of Dibromochloropropane, which did show an increase in permeate concentration w ith increasing recovery.

The removal of simazine, atrazine, diuron, bentazone, DNOC, and dinoseb has also been investigated using four different nanolilters - Fluid Systems 4~21PZ, Filmtec NF70, Hydranautics PVD 1, and a Toray SU6 10 on a pilot scale in the Netherlands by KIWA (Hofman et al., 1993). Atrazine was as consistently rejected as any pesticide, which was due to steric effects; diuron was the most poorly-rejected pesticide. These results showed that pesticide rejection varied by membrane and did not always increase with pesticide molecular weight. Lower rejection of diuron with the NF70 membrane might have been due to the surface interaction of the membrane film with the diuron.

Another pilot plant was operated in Germany for a 5-month period, in order to study the rejection of simazine, atrazine, diuron, terbutylazine by a Hydranautics PVD- 1, polyvinyl alcohol membrane. For 75% recovery, simazine, atrazine, terbutylazine were rejected over 90% and diuron was rejected for about 85%. Diuron was again the lowest rejected pesticide in this study. When the recovery increased to 80%, all the pesticide rejection was decreased by about 5%. This result can be explained by the higher concentration in the feed-side, which results in high permeate concentration in the diffusion-controlled membrane system.

The rejection properties of pesticides and alkyl phthalates were examined using flat-sheet-type NF membranes (Kiso et al., 2000; Kiso et al., 2000; Kiso et al., 2001) and the following results obtained: (1) higher desalting NF membranes rejected almost all solutes at more than 95%, (2) some compounds were rejected effectively even by lower desalting membranes, (3) the rejection properties were influenced, not only by steric hindrance, but also by an affinity to the membrane. The rejection properties of a hollow-fiber membrane (HNF- 1) for non-phenilic pesticides were also investigated in our previous work (Kiso et al., 2002) where the rejection properties were discussed on the basis of short-term (5 h) of membrane separation experiments. The fact that the pesticides were adsorbed on the membrane suggested that it is necessary to conduct the experiments for longer periods, in order to evaluate the effects of the adsorption. In addition, it was found that aromatic pesticides were adsorbed more than non-aromatic pesticides. (Yung et al., 2005)

Van der Bruggen et al. (Van der Bruggen et al., 2006) attempted to develop a semi-quantitative method for estimating rejection of organic micropollutants by NF. This model provides an approximation of rejection by taking into account compound molecular weight, hydrophobicity, and charge combined with the membrane's molecular weight cut-off (MWCO) and surface charge. Further development of this type of model is needed as molecular parameters including, among others, dipole moment and effective hydrated radius, along with membrane parameters such as pore size distribution, hydrophobicity and charge are not excluded. It is also important that operational parameters are considered, such as recovery and cross-flow velocity.

In addition, an increase in compound rejection may result from the binding of EDCs and PhACs to NOM due to hydrogen bonding, forming NOM-compound complexes that are larger, have an increased negative-charge, and/or a higher affinity for adsorption to the membrane when compared to the compound alone (Plakas et al., 2006; Zhang et al., 2004; Devitt et al., 1998). The presence of cations can also influence the membrane charge and the interaction of compounds and humic acids with each other and the membrane surface (Cho et al., 2000; Jucker & Clark, 1994). For example, Devitt et al. (Devitt et al., 1998) investigated the rejection of atrazine by NF and UF membranes and observed that atrazine-NOM association decreased in the presence of cations (principally calcium). Plakas et al. (Plakas et al., 2006) studied the removal of atrazine, isoproturon and prometryn by NF and found that the presence of calcium ions alone has a positive effect on pesticide retention but can interfere with the pesticide-NOM complex, thus reducing overall retention. (Comerton et al., 2008)

6. Conclusion

Atrazine is still one of the most commonly used herbicides in the world and is used on most corn, sugarcane and sorghum acreage in the United States. It is used to stop pre- and post-emergence broadleaf and grassy weeds, and is generally applied in the spring. Thus, atrazine concentrations are greatest in streams during the spring, when most fish in North America are attempting to reproduce. Investigations and evaluations of the potential risks posed by atrazine, particularly in wild populations of fish from streams in agricultural areas with high use of this herbicide are still the important issue worldwide.

7. References

Acero J. L., von Gunten (1998). Influence of carbonate on advanced oxidation process for drinking water treatment, *Proceedings of ozonation and AOPs in water treatment: Application and research*, Poitiers, France, 23-25 Septemebr, 13.1-13.9.

Acero J. L. & Von Gunten U. (2001). Characterization of oxidation processes: ozonation and the AOP O_3/H_2O_2, *Journal Awwa*, 90-100.

Acero J. L., Stemmler K. & von Gunten U. (2000). Degradation kinetics of atrazine and its degaradtion products with ozone and OH radicals: a predictive tool for drinking water treatment, *Environ. Sci. Technol.*, 34, 591-597.

Adams C. D. & Watson T. L. (1996). Treatability of s-triazine herbicide metabolites using powdered activated carbon, *Journal of environmental engineering*, 327-330.

AIM Filtration Systems, *Filtration and Separation*, <http://www.filters-cartridges.com/filtration-spectrum.html>, August 2010.

Allgeier S. (2005). *Membrane Filtration Guidence Manual,* Edited by Malcolm Pirnie, Inc., Separation Processes, Inc., and The Cadmus Group, Inc., US EPA's Office of Water.

Andreozzi R., Caprio V., Insola A., Marotta R. (1999). Advanced oxidation processes (AOP) for water purification and recovery, *Catalysis Today,* 53, 51-59.

Arnold S. M., Hickey W. J. & Harris R. F. (1995). Degradation of atrazine by Fenton s reagent: condition optimization and product quantification, *Environ. Sci. Technol.,* 29, 2083-2089.

Baier J.H., Lyskins Jr. B.W., Fronk C.A., Kramer S.J. (1987). Using reverse osmosis to remove agricultural chemicals from ground water. *Journal American Water works Association,* 79, 55.

Balci B., Oturan N., Cherrier R., Oturan M. A. (2009). Degradation of atrazine in aqueous medium by electrocatalytical generated hydroxyl radicals. A kinetic and mechanistic study, *Wat. Res.,* 43, 1924-1934.

Barreiro J. C., Capelato M. D., Martin-Neto L., Hansen H. C. B. (2007). Oxidative decomposition of atrazine by a Fenton-like reaction in a H_2O_2/ferrihydride system, *War. Res.,* 41, 55-62.

Bellona C., Drewes J.E., Xu P., Amy G. (2004). Factors affecting the rejection of organic solutes during NF/RO treatment - a literature review. *Water Research,* 38, 2795.

Beltran F. J. (1996). Theoretical aspects of the kinetics of competitive first reactions of ozone in the O_3/H_2O_2 and O_3/UV oxidation processes, *Ozone Science & Engineering,* 13-38.

Beltran F. J., Gonzalez M., Acedo b. & Rivas F. J. (2000). Kinetic modelling of aqueous atrazine ozonation processes in a continuous flow bubble contactor, *Journal of Hazardous Materials,* B(80), 189-206.

Berg P., Hagmeyer G., Gimbel R. (1997). Removal of pesticides and other micropollutants by nanofiltration. *Desalination,* 113, 205–208.

Bolzacchini E., Brambilla A., Orlandi M., Polesello S. & Rindone B. (1994). Oxidative pathways in the degradation of triazine herbicides: a mehanistic approach, *Wat. Sci. Tech.,* 30(7), 129-136.

Bontha J.R. & Pintauro P.N. (1994). Water orientation and ion solvation effects during multicomponent salt partitioning in a nafion cation exchange membrane. *Chem. Eng. Sci.,* 49, 3835-3851.

Boussahel R., Bouland S., Moussaoui K.M., Montiel A. (2000). Removal of pesticide residues in water using the nanofiltration process. *Desalination,* 132, 205–209.

Bowen W.R., Mukhtar H. (1996). Characterisation and prediction of separation performance of nanofiltration membranes. *Journal of Membrane Science,* 112, 263-274.

Braghetta A. (1995). *The influence of solution chemistry and operating conditions on nanofiltration of charged and uncharged organic macromolecules.* Dissertation, University of North Carolina, Chapel Hill.

Brown N. W., Roberts E.P.L., Chasiotis A., Cherdron T. & Sanghrajka N., (2004). Atrazine removal using adsrobption and electrochemical regeneration, *Wat. Res.,* 38, 3067-3074.

Chan K. H. & Chu W. (2003). The system design of atrazine oxidation by catalytic oxidation process through a kinetic approach, *Wat. Sci.,* 37, 3997-4003.

Chan K. H. & Chu W. (2005). Model application and mechanism study on the degradation of atrazine by Fenton s system, *Journal of Hazardous Material,* B118, 227-237.

Chan K. H. & Chu W. (2009). Degradation of atrazine by cobalt-mediated activation of peroxymonosulfate: Different cobalt conteranions in homogenous process and cobalt oxide catalysts in photolytic heterogenous process, *Wat. Res.*, 43, 2513-2521.

Cheryan M. (1998). *Ultrafiltration and microfiltration handbook*. University of Illinois, USA. Technomic publishing company, Inc., USA.

Chiron S., Fernandez- Alba A., Rodriguez A. & Garcia-Calvo E. (2000). Pesticide chemical oxidation: state-of-the-art, *Wat. Res.*, 34 (2), 366-377.

Cho J.W., Amy G., Pellegrino J. (1999). Membrane filtration of natural organic matter: initial comparison of rejection and flux decline characteristics with ultrafiltration and nanofiltration membranes. *Water Research*, 33, 2517–2526.

Cho J., Amy G., Pellegrino J. (2000). Membrane filtration of natural organic matter: factors and mechanisms affecting rejection and flux decline with charged ultrafiltration (UF) membrane. *J. Membr. Sci.*, 164, 89.

Comerton A.M., Andrewsa R.C., Bagley D.M., Hao C. (2008). The rejection of endocrine disrupting and pharmaceutically active compounds by NF and RO membranes as a function of compound and water matrix properties. *Journal of Membrane Science*, 313, 323–335.

Deen W.M., Satvat B., Jamieson J.M. (1980). Theoretical model for glomerular filtration of charged solutes. *Am. J. Physiol.*, 238, F126-139.

De Laat J., Chramosta N., Dore M., Suty H. & Pouillot M. (1994). Constantes Cinétiques de Réaction des Radicaux Hydroxyles sur Quelques Sous-Produits d'Oxydation de l'Atrazine par O_3 ou par O_3/H_2O_2, *Environ. Technology*, 15, 419-428.

Devitt E.C., Ducellier F., P. Cote, Wiesner M.R. (1998). Effects of natural organic matter and the raw water matrix on the rejection of atrazine by pressuredriven membranes. *Wat. Res.*, 32 (9), 2563.

Duranceau S. (2000). Future of membranes. *Journal American Water works Association*, 92, 70–71.

Duranceau S.J., Taylor J.S., Mulford L.A. (1992). SOC removal in a membrane softening process. *Journal American Water works Association*, 84, 68-78.

Elimelech M., Zhu X.H., Childress A.E., Hong S. (1997). Role of surface morphology in colloidal fouling of cellulose acetate and composite polyamide RO membranes. *Journal of Membrane Science*, 127 (1), 101–109.

Gallard H. & De Laat J. (2000). Kinetic modelling of Fe^{3+}/H_2O_2 oxidation reactions in dilute aqueous solution using atrazine as a model organic compound, *Wat. Res.*, 34 (12), 3107-3116.

Gonzalez M. C. & Martire D. O. (1997). Kinetics of O^{0-} in O_3^{0-} in alkaline aqueous solutions, *Wat. Sci. Tech.*, 35(4), 49-55.

Garmouna M., Blanchard M., Chesterikoff A., Ansart P. & Chevreuil M. (1997). Sesonal trasnport of herbicides (triazines and phenylureas) in a small stream draining an agricultural basin: Melarchez (France), *Wat. Res.*, 31(6), 1489-1503.

Gottschalk, C., Libra J. A. & Saupe A. (2000). *Ozonation of Water and Waste Water*, Wiley-VHC: Federal Republic of Germany, 11-20, 120-121.).

Haag W. R.& Yao C. C. D. (1992). Rate constants for reaction of hydroxyl radicals with several drinking water contaminants, *Environ. Sci. Technol.*, 26, 1005-1013.

Haarhoff J., Kubare M., Mamba B., Krause R., Nkambule T., Matsebula B. & Menge J. (2010). NOM characterization and removal at six Southern African water treatment plants *Drink. Water Eng. Sci.*, 3, 53–61.

Hagmeyer G. & Gimbel R. (1999). Modelling the rejection of nanofiltration membranes using zeta potential measurements. *Sep. Purif. Technol*, 15, 19–30.

Hapeman C. J. (1994). *Oxidation of s-triazinepesticides*. In Teller, D. W., Pohland, F. G. (Eds.), Emerging Technologies in Hazardous Waste Management IV, ACS Syposium Series, American Chemical Society, 223-233.

Hermanowicz S. W., Bellamy W. D. & Fung L. C. (1999). Variability of ozone reaction kinetics in batch and continuous flow reactors, *Wat. Res.*, 33 (9), 2130-2138.

Hoigne J. (1997). Inter-calibration of OH radicals sources and water quality parameter, *Wat. Sci. Tech.*, 35 (4), 1-8.

Hoigne J. (1998). *Chemistry of aqueous ozone and transformation of pollutans by ozonation and advanced oxidation processes*, Part C, Quality and treatment of drinking water II, The handbook of environmental chemistry, Vol 5, ed. J. Hrubec, Springer- Verlag, Berlin Heidelberg, 83-141.

Hong S.K., Elimelech M. (1997). Chemical and physical aspects of natural organic matter (NOM) fouling of nanofiltration membranes. *Journal of Membrane Science*, 132, 159–181.

Hofman J.A.M., Noij Th.H.M., Schippers J.C. (1993). Removal of pesticides and other organic micropollutants with membrane filtration. *Water Supply*, 11, 101.

Hofman J.A.M.H., Beerendonk E.F., Folmet H.C., Kruithof J.C. (1997). Removal of pesticides and other micropollutants with cellulose-acetate, polyamide and ultra-low pressure reverse osmosis membranes. *Desalination*, 113, 209.

Hofman J.A.M.H., Noij Th.H.M., Kruithof J.C., Schippers J.C. (1993). *Removal of pesticides by nanofiltration*. Proc. AWWA Membrane Technology Conference, Reno, NV.

Huang W. J., Peng M. Y., Chen L. Y. (2004). Removal of bromate and assimilable organic carbon from drinking water using granular activated carbon, *Water science and Technology*, 50(8), 73-80.

Jiang H. &Adams C. (2006). Treatability of chloro-s-triazines by conventional drinking water treatment technologies, *Wat. Res.*, 40, 1657-1667.

Jucker C., Clark M.M. (1994). Adsorption of aquatic humic substances on hydrophobic ultrafiltration membranes. *J. Membr. Sci.*, 97, 37.

Jung Y.J., Kiso Y., Othman R.A.A.B., Ikeda A., Nishimura K., Min K.S., Kumano K., Ariji A. (2005). Rejection properties of aromatic pesticides with a hollow-fiber NF membrane. *Desalination*, 180, 63-71.

Kearney P. C., Muldoon M. T., Somich C. J., Ruth J. M. & Voaden D. J. (1988). Biodegradation of ozonated atrazine as a wastewater disposal system, *J. Agric. Food Chem.*, 36, 1301-1306.

Kimura K., Amy G., Drewes J., Watanabe Y. (2003). Adsorption of hydrophobic compounds onto NF/RO membranes: an artifact leading to overestimation of rejection. *Journal of Membrane Science*, 221, 89–101.

Kimura K., Amy G., Drewes J., Heberer T., Kim T., Watanabe Y. (2003). Rejection of organic micropollutants (disinfection by-products, endocrine disrupting compounds, and pharmaceutically active compounds) by NF/RO membranes. *Journal of Membrane Science*, 227, 113- 121.

Kimura K., Toshima S., Amy G., Watanabe Y. (2004). Rejection of neutral endocrine disrupting compounds (EDCs) and pharmaceutical active compounds (PhACs) by RO membranes. *Journal of Membrane Science*, 245, 71–78.

Kiso Y., Nishimura Y., Kitao T., Nishimura K. (2000). Rejection properties of non-phenylic pesticides with nanofiltration membranes. *Journal of Membrane Science*, 171, 229–237.

Kiso Y., Kon T., Kitao T., Nishimura K. (2001). Rejection properties of alkyl phthalates with nanofiltration membranes. *Journal of Membrane Science*, 182, 205–214.

Kiso Y., Mizuno A., Othman R., Jung Y.J., Kumano A., Ariji A. (2002). Rejection properties of pesticides with a hollow fiber NF membrane (HNF-1). *Desalination*, 143, 147–157.

Kiso Y., Sugiura Y., Kitao T., Nishimura K. (2001). Effects of hydrophobicity and molecular size on rejection of aromatic pesticides with nanofiltration membranes. *J. Membr. Sci.*, 192, 1-10.

Kosutic K, Kunst B. Removal of organics from aqueous solutions by commercial RO and NF membranes of characterized porosities. *Desalination*, 142 (1), 47-56

Lebeau T., Lelievre C., Wolbert D., Laplanche A. (1997). Effect of natural organic matter loading on the atrazine adsorption capacity of an aging powdered activated carbon slurry, *Wat. Res.*, 33(7), 1695-1705.

Legube B., Parinet B., Gelinet K., Berne F. & Croue J. P. (2004). Modeling of bromate formation by ozonation of surface waters in drinking water treatment, *Wat. Res.*, 38, 2185-2195.

Lemić J., Kovačević D., Tomašević-Čanović Magdalena, Kovačević D., Stanić T. & Pfend R. (2006). Removal of atrazine, lindane and diazinone from water by organo-zeolites, *Wat. Res.*, 40, 1097-1085.

Ma J. & N. J. D. Graham, (1997). Preliminary investigation of manganese catalyzed ozonation for the destruction of atrazine, *Ozone Science & Engineering*, 19, 227- 240.

Ma. J. & Graham N. J. D. (1999). Degradation of atrazine by manganese- catalysed ozonation: influence of humic substances, *Wat. Res.*, 33 (3), 785-793.

Ma. J. & Graham N. J. D. (2000). Degradation of atrazine by manganese-catalysed ozonation-influence of radical scavengers, *Wat. Res.*, 34 (15), 3822-3828.

Macoun R.G. (1998). *The mechanisms of ionic rejection in nanofiltration*. PhD thesis, UNSW, Chemical Engineering.

Mallevialle J., Odendaal P.E., Wiesner M.R. (1996). *Water treatment membrane processes*, McGraw-Hill.

Metcalf & Eddy. (2003). *Wastewater engineering treatment and reuse, Fourth edition*, McGraw-Hill Higher education, Chapter 11.

Meunier l., Canonica S. & Von Gunten U. (2006). Implication of sequential use of UV and ozone for drinking water quality, *Wat. Res.*, 40, 1864-1876.

Mulder M. (1991). *Basic Principles of Membrane Technology*, Kluwer Academic Publishers, Dordrecht.

Nakao S., Kimura S. (1982). Models of membrane transport phenomena and their application for ultrafiltration data. *J. Chem. Eng. Japan*, 15, 200-205.

Negaresh E. (2007). *Particle and Macromolecule fouling in Submerged Membrane*. Master Thesis, Chemical Sciences & Engineering, Faculty of Engineering, UNSW.

Nelieu S., Kerhoas L., Einhorn J. (2000). Degradation of atrazine into ammeline by combined ozone/hydrogen peroxide treatment in water, *Environ. Sci. Technol.*, 34, 430–437.

Nghiem L.D., Schafer A.I., Elimelech M. (2004). Removal of natural hormones by nanofiltration membranes: measurement, modeling, and mechanisms. *Environ. Sci. Technol.*, 38, 1888–1896.

Nghiem L.D., Manis A., Soldenhoff K., Schafer A.I. (2004). Estrogenic hormone removal from wastewater using NF/RO membranes. *Journal of Membrane Science*, 242, 37–45.

Ni C. H., Chen J. N. & Yang P. Y. (2002). Catalytic ozonation of 2-dichlorophenol by metallic ions, *Water science and Technology*, 47, 77- 82.

Official Gazette RS N. 13, (1999). Odločba o prepovedi oziroma omejitvi prometa in uporabe strupenih substanc in iz njih izdelanih preparatov, ki se uporabljajo kot fitofarmacevtska sredstva, Ljubljana, Slovenia.

Ozaki H., Li H. (2002). Rejection of organic compounds by ultra-low pressure reverse osmosis membrane. *Wat. Res.*, 36, 123–130.

Panno S. V.& Kelly W. R. (2004). Nitrate and herbicide loading in two groundwater basins of Illinois sinkhole plain, *Journal of Hydrology*, 290, 229-242.

Park J. S., Choi H. & Cho J. (2004). Kinetic decomposititon of ozone and para-chlorobenzoic acid (pCBA) during catalytic ozonation, *Wat. Res.*, 38, 2285-2292.

Peeters J. M. M. et al. (1999). Streaming potential measurements as a characterization method for nanofiltration membranes. *Colloids and Surface: A.*, 150, 247–259.

Peeters J.M.M. (1997). *Characterization of nanofiltration membranes*. Dissertation, University of Enschede, Netherlands.

Pelizzeti E., Carlin V., Minero C., Pramauro E. & Vincenti M. (1992). Degradation pathways of atrazine under solar light and in the presence of TiO_2 colloidal particles, *The Science of Total Environment*, 123/124, 161-169.

Pi Y., Schumacher J. & Jekel M. (2005). Decomosition of aqueous ozone in the presence of aromatic organic solutes, *Wat. Res.*, 39, 83-88.

Pintar M. & Lobnik F. (2001). *Dinamika pojavljanja pesticidov in nitratov v podtalnici pod intenzivnimi kmetijskimi površinami*, Proceedings (Zbornik Seminarja): Zdravstvena ustreznost pitne vode in varnost oskrbe, Otočec ob Krki.

Plakas K.V., Karabelas A.J., Wintgens T., Melin T. (2006). A study of selected herbicides retention by nanofiltration membranes—the role of organic fouling. *J. Membr. Sci.*, 284, 291.

Prado J. & Esplugas S. (1999). Comparison of different advance oxidation processes involving ozona to eliminate atrazine, *Ozone Science & Engineering*, 21, 39-52.

Pryor M. J., Nozaic D., Freese S. D. & Rajagopaul R. (1999). The use of granular activated carbon for the treatment of impounded surface water, *Wat. Sci. Tech*, 39 (10-11), 197-200.

Reid D. C., Dwards A. C., Cooper D., Wilson E. & Mcgaw B. A. (2003). *The quality of drinking water from private water supplies in Aberdeenshire*, UK, 37, 245-254.

Reinhard M., Goodman N.L., McCarty P.L., Argo D.G. (1986). Removing trace organics by reverse osmosis using cellulose acetate and polyamide membranes. *Journal American Water works Association*, 78, 163.

Richards R. P., Baker D. B., Christensen B. R. & Tierney D. P. (1995). Atrazine exposures through drinking water: Exposure assessments for Ohio, Illinois, and Iowa, *Environ. Sci. Technol.*, 29, 406-412.

Rohe D. L., Blanton T. C., Mariñas B. J. (1990). Drinking Water Treatment by Nanofiltration. *National Conference on Environmental Engineering*, 707-716.

Rosenfeldt E. J., Linden K. G., Canonica S., Von Gunten U. (2006). Comparison of the efficiency of OH radical formation during ozonation and the advanced oxidation processes O_3/H_2O_2 and UV/H_2O_2, *Wat. Res.*, 40, 3695–3704.

Rossner A., Snyder S. A., Knappe D. R. U. (2009). Removal of emerging contaminates of concern by alternative adsorbents, *Wat. Res.*, 43, 3787-3796.

Saxena A., Tripathi B.P., Kumar M., Shahi V.K. (2009). Membrane-based techniques for the separation and purification of proteins: An overview. *Advances in Colloid and Interface Science*, 145, 1–22.

Schafer A. I., Nghiem L. D., Waite T. D. (2003). Removal of the natural hormone estrone from aqueous solutions using nanofiltration and reverse osmosis. *Environ. Sci. Technol.*, 37, 182–188.

Simpson A. E., Kerr C. A., Buckley C. A. (1987). The effect of pH on the nanofiltration of the carbonate system in solution. *Desalination*, 64, 305-319.

Sunder M. & Hempel D. C. (1997). Oxidation of tri- and perchloroethene in aqueous solution with ozone and hydrogen peroxide in a tube reactor, *Wat. Res*, 31 (1), 33-40.

Taylor J. S., Hong S. K. (2000). Potable water quality and membrane technology. *Laboratory medicine*, 31, 10, 563-568.

Tepuš B. & Simonič M. (2007). The effect of platinum catalyst on decomposition of ozone and atrazine removal. *Journal of advanced oxidation technologies*, 10, 1, 202-208.

Tepuš B. & Simonič M. (2008). Kinetic Studies of Catalytic ozonation of atrazine, *Chroatica Chemica Acta*, 81, 4, 673-679.

Tepuš B., Petrinić I., Simonič M. (2009). Comparison between nitrate and pesticide removal from ground water using adsorbents and NF and RO membranes, *Journal of Hazardous Materials*, 170, 1210-1217.

Thurman E. M., Meyer M. T., Mills M. S., Zimmerman L. R. & Perry C. A. (1994). Formation and transport of deethylatrazine and deisopropylatrazine in surface water, *Environ. Sci. Technol.*, 28, 2267-2277.

Tsuriu T. et al. (1991). Negative rejection of anions in the loose reverse osmosis separation of mono- and divalent ion mixtures. *Desalination*, 81, 219-231.

Tsuru T. et al. (1991). Calculation of ion rejection by extended Nernst-Planck equation with charged reverse osmosis membranes for single and mixed electrolyte solutions. *J. Chem. Eng. Of Japan*, 24, 511-517.

Van der Bruggen B., Schaep J., Maes W., Wilms D., Vandecasteele C. (1998). Nanofiltration as a treatment method for the removal of pesticides from ground waters. *Desalination*, 117, 139-147.

Van der Bruggen B., Schaep J., Wilms D., Vandecasteele C. (1999). Influence of molecular size, polarity and charge on the retention of organic molecules by nanofiltration. *J. Membr. Sci.*, 156, 29-41.

Van der Bruggen B., Verliefde A., Braeken L., Cornelissen E. R., Moons K., Verberk J. Q. J. C., van Dijk H. J. C., Amy G. (2006). Assessment of a semiquantitative method for estimation of the rejection of organic compounds in aqueous solution in nanofiltration. *J. Chem. Technol. Biotechnol.*, 81, 1166.

Verstraeten I. M., Thurman E. M., Lindsey M. E., Lee E. C. & Smith R. D. (2002). Changes in concentrations of triazine and acetamide herbicides by bank filtration, ozonation and chlorination in public water supply, *Journal of hidrology*, 26, 190-208.

Von Gunten U., Ozonation of drinking water: Part I: Oxidation kinetics and product formation, *Wat. Res.* 37, 2003, 1443-1467.

Vrijenhoek E.M., Hong S., Elimelech M. (2001). Influence of membrane surface properties on initial rate of colloidal fouling of reverse osmosis and nanofiltration membranes. *Journal of Membrane Science*, 188, 115–128.

Wagner J. (2001). *Membrane Filtration Handbook: Practical Tips and Hints*, second edition.

Wang, G.-S. & Alben, K. T. (1988). Effect of preadsorbed background organic matter on granular activated carbon adsorption of atrazin, *The science of the Total Environment*, 224, 221-226

Wang X.L. et al. (1995). Electrolyte transport through nanofiltration membranes by the space charge model and the comparison with the Teorell Meyer Sievers model. *J. Membr. Sci.*, 103, 117-133.

Wang X.-L., Tsuru T., Togoh M., Nakao S.-i., Kimura S. (1995). Evaluation of pore structure and electrical properties of nanofiltration membranes. *Journal of Chemical Engineering of Japan*, 28, 2, 186-192.

Wang X.-L., Tsuru T., Togoh M., Nakao S.-i., Kimura S. (1995). Transport of organic electrolytes with electrostatic and steric-hindrance effects through nanofiltration membranes. *Journal of Chemical Engineering of Japan*, 28, 4, 372-380.

Xu X., Spencer H.G. (1997). Transport of electrolytes through a weak acid nanofiltration membrane: effects of flux and crossflow velocity interpreted using a fine-porous membrane model. *Desalination*, 113, 85-93.

Yaroshchuk A.E. (2000). Dielectric exclusion of ions from membranes. *Adv. Coll. Int. Sci.*, 85, 193-230.

Yoon Y., Westerhoff P., Yoon J., Snyder S.A. (2004). Removal of 17bestradiol and fluoranthene by nanofiltration and ultrafiltration. *J. Environ. Eng.-ASCE*, 130, 1460–1467.

Zhan Z., Müllner M. & Lercher J. A. (1996). Catalytic hydrolysis of s-triazine compounds over Al_2O_3, *Catalysis Today*, 27, 167-173.

Zhang Y., Van den Bruggen B., Chen G. X., Braeken L. & Vandecasteele C. (2004). Removal of pesticides by nanofiltration: effect of the water matrix, *Separation and Purification Technology* 38, 163–172.

Zhang Y., Ban der Bruggen B., Chen G.X., Braeken L., Vandecasteele C. (2004). Removal of pesticides by nanofiltration: effect of water matrix. *Sep. Purif. Technol.*, 38, 163.

Zhu X., Elimelech M. (1995). Fouling of reverse osmosis membranes by aluminium oxide colloids. *Journal of Environmental Engineering*, 121, 884-892.

Ziegmann M., Doll T. & Frimmel F. H. (2006). Matrix effects on the photocatalytical degradation of dichloroacetic acid and atrazine in water, *Acta Hydrochim. Hydrobiol.*, 34, 146-154.

31

Bacterial-Degradation of Pesticides Residue in Vegetables During Fermentation

Aslan Azizi
Agricultural Engineering Research Institute (AERI), Karaj
Iran

1. Introduction

An application of large quantity of agricultural pesticides in rural area is a common practice in order to increase the productivity and yield, to protect the agricultural crop from pests and prevent products lost due to insect and bacterial contamination is a common practice. Resistance and mutation of some pests to chemicals are the causes of using larger quantity of pesticide in the developing countries. According to the rate of degradation of chemicals, pesticides can be categorized as sensitive or tolerant to decomposition. Their destruction might be occurred under exposing to the normal atmospheric conditions or by biological activity of the soil microorganisms such as *Pseudomonas, Flavobacterium, Alcaligenes, Rhadococcus, Gliocladium, Trichoderma* and *Penicillium*. These microorganisms use the pesticides as their carbon and energy sources (Aislabie & Lloyd-Jones, 1995). Because agricultural pesticides are mostly artificial synthetic compounds without any identical in the nature, they are substantially tolerant towards degradation in natural conditions. In many cases, stability of these pesticides to the biological destruction arises from their insolubility in water, as the microorganisms are incapable of decaying such materials. Malathion, carbamate, pyrethriod, diazinon, dichloropicolinic acid and phenylalkanbic are sensitive pesticides to the hydrolytic activity of microorganism enzymes. Extracellular enzymes of the bacteria are capable of cleavage broad range of chemical pesticides. Apart from the natural structure of the pesticides, their volatility and adsorption ability to the soil compounds are also important factors affect sensitivity to the biological cleavage. These factors themselves are dependent on temperature, light, soil moisture and pH. The more fugacity of the pesticides, the more transfer of them to the atmosphere. Higher moisture of the soil ease degradation rate of the water soluble pesticides by the microorganisms, while reduce their volatility. Some of the pesticides such as diazinon are very sensitive to the low pH range and their degradation at this range dramatically occur (Muller & Korte, 1975; Freed et al., 1979). Because organophosphorous compounds are decomposed faster and easier compared to the organochlorine compounds, their application have been increasing day by day. Consumption of fruits and vegetables containing organochlorine components residue causes undesirable health disorders especially on the nerve system (Racke & Frank, 1997; Pehkonen & Zhang, 2002; Lal & Saxena, 1982). This danger is more acute in Iran because of the improper attitude that excessive application of the pesticides leads to the more efficient deterioration of the pests.

Because of the important impact of the microorganisms in the degradation of the pesticides, numerous researches have been done regarding qualitative and quantitative aspects of this phenomenon. Navab *et al.* (2003) studied the effects of isolated *Pseudomonas* spp. from soil on the DDT, DDD, DDF and HCH under the laboratory conditions. The bacteria were able to partially degrade the pesticides. It has been reported that *Flavobacterium* and *Spingomonaspaucimobil* spp. decayed some types of the pesticides during 48 h of fermentation process (Dimitrios et al.). Comprehensive research done by Peric *et al.* (1981) about the biological decomposability of the pesticides revealed that DDT (mainly) and HCH (partially) degraded by the species of *Debariomyces, Micrococcus* and *Lactobacillus. Lactobacilli* showed the lowest effect. Peric *et al.* (1981) also perceived that adding mentioned microbial mix to the fermented sausage led to the significant decrease in HCH concentration. Similar degradation of DDT and DDE in the Roquefort blue cheese by the using different species of gram positive *Lactobacilli, Streptococci* and yeasts were reported by Ledford and Chen (1969) and Mirna and Coretti (1979). Therefore, isolation, identification and screening of the microorganisms which are capable of pesticides residue degradation in food materials are important issues. No research has been done about the isolation and identification of the indigenous pesticides decomposing microflora in vegetable materials. Therefore, this study intends to investigate the effects of isolated indigenous microflora from Iranian vegetable source on the degradation of containing pesticides residue during the fermentation process.

2. Experimental

In order to produce vegetable with desirable pesticides residue, broadcasting precise quantity pesticide and conducting systematic experiments, experimental farm with a surface area of 1000 m^2 was prepared. The land preparation stages namely: land excavation, land leveling, land grading, primary and secondary tillage operations, sowing, irrigation and cultural practices were carried out.

Types of pesticides used in cultivated vegetables: Diazinon and malathion, which are the most consuming agricultural pesticides in Iran were selected for this study (the pesticides were obtained from Plant Production Department, Ministry of Agriculture, Tehran, Iran). The pesticides were sprayed on the vegetables in a concentration of 0.002 g L^{-1}. The cultivated vegetables were tomato (Super Queen), celery (Tail Utah), green bean (Sun Ray), pea (Green Arrow), cabbage (Space Star) and cauliflower (Globe Master). The original seeds were obtained from seed and plant production institute of Iran.

Preparation of vegetable samples: Cultivated vegetables were harvested at maturation stage and were immediately transferred to the laboratory. The external waste and damaged leaves of vegetables were removed then, piled, trimmed, washed and cut to the desired sizes. Cut vegetables were mixed together and filled into glass jars and covered with 2% (w/v) hot brine (95°C). The aim of adding hot brine was to destroy the heat labile anaerobic nonspore forming microorganisms such as coliforms, improve colour and texture of final product, accelerate the acidification rate and improve nutritional property of the final product. Finally, about 4 mL of vinegar was added to the top of samples in order to prevent activity of unwanted microflora. The mouth of the jars were covered with nylon film having low permeability to water vapour and oxygen, tied with thread and kept at room temperature.

Fermentation process was immediately started. In order to isolate and identify microorganism involved in fermentation of mixed vegetable in different stages, sampling was conducted every 12 h. Fermentation was continued till the pH of the product reached about 4.0. Fermentation process was stopped by opening the jars. Vegetables were removed from jars washed and stored at -4°C until pesticide detection experiments were done.

Culture media used for the isolation, screening and enumeration of the microorganisms: MRS broth/agar and LSDM broth/agar media (Merck, Darmstadp, Germany) were used for the isolation, partial identification, screening and enumeration of lactic acid bacteria. Media compounded, heated in a flask and boiled in a thermostatically controlled heater till cleared. Then pH of the medium was adjusted to 6.20. The media solutions were distributed in 250 mL flasks. The flasks were autoclaved at 121°C for 15 min and kept in refrigerator until used. In order to evaluate fermentation rate of the different carbohydrates, MRS broth without glucose and beef extract containing 0.05% chlorophenol red was used as a base medium. The ingredients of the medium were separately prepared from Merck (Darmstadp, Germany) and mixed carefully under controlled conditions. All the carbohydrates were sterilized using membrane filtration and added to the basal media to have final concentration of 1% in media. After inoculation of the microflora to the media, incubation process was carried out at 37°C for 7 d and colour variations within this period were studied.

Isolation and identification of microorganisms involved in fermentation of vegetables: Fermentation of vegetables started after few hours, subsequent to sealing the jars. Lactic acid bacteria were isolated from fermenting vegetables using MRS agar and LSDM at 12, 24 and 48 h of fermentation. The contents of the jars were mixed thoroughly and 10 mL of brine was withdrawn under sterile condition using a syringe. Fermenting brine 1 mL was serially diluted in saline and was plated on MRS agar and LSDM agar. The streaked plates were incubated at 30°C for 72 h. At the end of the incubation period, bacteria colonies were counted. Individual colonies were isolated based on morphology, gram reaction, cell morphology, catalase production, presence of spores and aerobic and anaerobic growth. Isolated cultures were purified by repeated streaking on MRS agar and isolation. The microflora at the end of 12, 24 and 48 h were isolated and examined for gram reaction, morphology, catalase production, presence of spore and growth under aerobic and anaerobic conditions. The general key used for the identification of gram positive bacteria was done according to the procedure given in Bergey's manuals determinative bacteriology (Sneath et al., 1986). The general morphological and biochemical characteristics of lactic acid bacteria were determined according to the procedure of (Sharpe et al., 1979). In the course of identification of microorganism, to identify microorganism capable of standing presence of two pesticides malathion and diazinon in MRS and LSDM media, the amount of carbohydrate in the media was reduced and same quantity of malation and diazonin was added. They were inoculated with the desired purified cultures. Finally, the isolates were washed under aseptic conditions and centrifuged at 3000 g for 10 min to separate the cells from culture medium. Recovered cells were washed with sterile distilled water several times and kept in a 1% sterilized brine solution for further studies.

Identification procedure was adapted based on the classical biochemical and morphological tests. Biochemical tests included fermentation patterns of the sugars (arabinose, fructose, esculin, glucose, galactose, lactose, mannose, maltose, manitol, rhamnose, raffinose, ribose, salicin, sucrose, sorbitol and xylose), capability of hydrolyzing casein and gelatin, indol production, ammonia from argenin, catalaze and pseudocatalaze tests, gas formation from consumption of glucose, VP and MR tests, being homofermentative or heterofermentative, reaction to the molecular oxygen, organic acid production under aerobiosis and anaerobiosis, growth at 15, 30 and 45°C, survival at 60°C after 0.5 h, reaction to the 0.1% MB solution, growth at pH 4.0 and 9.6, growth at the 4 and 10% brine solutions and cells motility. Morphological tests consisted of considering cell appearance, cell arrangement, spore production and gram staining (Sharpe et al., 1979; Anderson, 1984).

Detecting pesticides residue in the vegetable samples: Liquid-liquid extraction method with the solvents of acetone and dichloromethane were used for extraction of the malathion and diazinon from vegetable samples. Quantification analysis of the pesticides was done by applying GC method (Shimadzu 2100, Japan) with the NPD detector and DB5 column 20.

Replications: Experiments were performed three times in duplicate and the mean of the results were considered as final data.

3. Results and discussion

Enumeration of vegetable microflora during the fermentation process: Extra care must be taken when comparing the results; *in vitro* studies are not always real situation in food products. This is due to the fact that the biodegradation process may be affected by a number of factors such as the interaction between microorganisms, the microbial concentration of the medium, whether the medium is liquid or solid and the microbial growth conditions of temperature and pH.

Table-1 indicates total counts of lactic acid bacteria in MRS and LSDM agar along with the pH drop kinetic, in 12 h of vegetable fermentation. The population of microorganisms increased during 24 h of fermentation reaching pH 4.2, then, sudden decrease of population was observed. According to the Table, the population of *Lactobacillus* and *Streptococcus genera* (which were enumerated by LSDM media) was considerably higher than other genera of lactic acid bacteria. It can be attributed to the naturally higher number of bacteria belong to the two mentioned genera in the vegetable mix.

Fermentation time	Bacterial counts (log cfu/g)		pH of the vegetable mix
(h)	MRS	LSDM	5.8
12	9.30	8.18	4.2
24	10.81	9.09	3.7
48	8.04	8.84	3.6

Table 1. Total counts of microorganisms in MRS and LSDM during fermentation of vegetables.

Identification of isolated bacteria: According to the results obtained from classical tests (Tables 2-5), the isolated microorganisms totally belonged to the lactic acid bacteria group and consisted all four genera of these bacteria including *Lactobacillus, Streptococcus, Pediococcus* and *Leuconostoc*. According to Table-2, isolates 12.2, 12.3, 12.11, 12.14 and 24.5 were from *Lactobacillus genus* including *L. delbrueckii* ssp. *Lactis* and *L. plantarum*. Apart from the *Lactobacilli*, 11 types of cocci were isolated (Table-2). These bacteria belonged to the genera *Streptococcus* species *S. lactis* (isolates 48.2 and 48.3) and *S. rafinolactis* (isolate 48.4), *Pediococcus* species *P. pentosasus* (isolate 12.7), *P. acidilactici* (isolate 12.13) and *P. damnosus* (isolate 12.11) and *Leuconostoc* species *L. mesentroides* (isolate 12.4), *L. cremoris* (isolate 48.6) and *L. oenus* (isolate 48.7). Some of the unidentified species concerned to the genus *Leuconostoc* were also recognized.

Isolate No.	Colony morphology Gram +v/-v	B/C	Arrangement	Gas from glucose	Catalaze	Pseudo catalaze	Motility	Growth pH 4.0	Growth pH 9.6	Salt 4 %	Salt 10 %	Survival at 60 °C for 0.5 h	Vp test	MR test
At the end of 12 h									+		-	+	+	+
1	+	B/Sh	Cl	-	-	-			+	+	-	-	-	+
2	+	B/Sh	Cl	-	-	-			-		-	-		+
3	+	B/Sh	2's	+	-	-		+		+	-	+	-	+
4	+	C/Tiny	1's	+	-	-		+			-	-	-	+
5	+	B	2's	+	-	-		+	+		-	-		+
6	+	B	2's	+	-	-		+	+	+	-	+		+
7	+	C	Ch, 2's	-	-	-		-	+	-	-	-	+	+
8	+	B	2's	+	-	-			+		-		-	-
9	+	B	2's	+	-	--					-		-	+
10	+	B	2's	-	-	-					-		-	+
11	+	C/B	2's, Cl's	-	-	-		-	+	+	-	+		+
12	+	C/B	2's	-	-	-	+	+	+		--			+
13	+	C	2's	-	-	-	+	+	+	+	-	+		+
14	-	C/B	2's	-	--	---			+	+	-	+		+
At the end of 24 h					-									
1	+	B	1's, 2's	-	-	-		-	+		-		-	
2	+	C	Cl's, 4, 1's	-	-	-	+		+		-		-	+
3	+	C	Cl's, 4's	-	--	-	+	+	+		-		+	+
4	+	C/B	1.2's, Ch	-	-	-	-	-	+		-		-	+
5	+	C/B	1's, Ch	-	-	-	-	+	+		-		-	-
6	+	C/B	2's	+	-	-			+		-		-	+
7	+	C	Cl's, Ch	+	+	+			Nd		-		-	+
At the end of 48 h														
1	+	C/B	2's	+	-	+		-	-		-		-	+
2	+	C/B	2's	-	-	+		-	-		-		-	-
3	+	C	1.2's	-	-	-		-	-		-		--	-
4	+	C	1.2's	-	-	-	+	+	+		-		-	-
5	+	C	1.2's	-	+	+	+	+	+		-		-	+
6	+	C	1.2's	+	-	+		-	-		-		-	+
7	+	C	1.2's	+	-	+					-		-	+

B: Bacilli; C: cocci; Cis: Clusters; Sh: Short; 1's: Single; 2's: Two cells together; Ch: Chain; 1.2's: Single and two cells; 4's: Four cells together.

Table 2. Characteristics of isolates from the brine of fermented vegetables at different intervals

Characteristics	A	B	C	D	E	F	G
1. Gram reaction	+	+	+	+	+	+	+
2. Arrangement	-	-	-	-	-	--	-
3. Gas from glucose	--	-	-	--	--	--	-
4. Catalase	-	-	-	-	-	-	-
5. Pseudo catalase	--	-	-	-	-	-	-
6. Motility	-	-	-	-	-	-	-
7. pH 4	-	-	-	+	+	+	+
8. pH 9.6	+	+	+	+	+	+	+
9. Salt 4 %	+	+	+	+	+	+	+
10. Salt 10 %	-	-	-	-	-	-	-
11. Growth at 25 °C	-	-	-	+	+	+	+
30 °C	+	+	+	+	+	+	+
45 °C	+	+	+	+	+	+	+
12. Vp test	+	+		-	-	-	-
13. MR test	+	+	+	+	+	+	+
14. Survival at 60 °C for 0.5 h	+	+	+	+	+	+	+
15. Reaction to 0.1 % MB	+	+	+	+	+	+	+
16. NH_3 from arginine	+	+	+	+	+	+	+
17. Indol produced	-	-	-	-	-	-	-
18. Casein hydrolyzed	-	-	-	-	-	-	-
19. Gelatin hydrolyzed	-	---	-	-	-	-	-
20. Arabinose	+	+	+	+	+	+	+
21. Esculin	+	+	+	+	+	+	+
22. Fructose	+	+	+	+	+	+	+
23. Galactose	+	+	+	+	+	+	+
24. Glucose							
25. Lactose	-	+	+	+	+	+	+
26. Maltose	+	+	+	+	+	+	+
27. Mannitol	-	-	-	+	+	+	+
28. Mannose	+	+	+	+	+	+	+
29. Raffinose	-	-	-	+	+	+	+
30. Rhamnose	-	-	-	-	-	-	-
31. Ribose	-	-	-	+	+	+	+
32. Salicin	+	+	+	+	+	+	+
33. Sorbitol	-	-	-	+	+	+	+
34. Sucrose	+	+	+	+	+	+	+
35. Xylose	-	-	-	+	+	+	+

A = Isolate 12-2; B = Isolate 12-3; C = *L. delbrukii* Sub. Sp. lactic; D = Isolate 12-11; E = Isolate 12-14; F = Isolate 24.5; G = *L. plantarum*

Table 3. Physiological and biochemical characteristics of isolates from fermenting brine

Isolate No.	Identification
Obligate homofermentative *Lactobacilli*	
12-2	*L. delbrueckii* sub. Sp. Lactis
12-3	*L. delbrueckii* sub. Sp. Lactis
Facultative heater fermentative *Lactobacilli*	
12-11	*L. plantarum*
12-14	*L. plantarum*
12-12	*L. plantarum*
12-13	*L. plantarum*

Table 4. Characterization of Lactobacillus isolates

Characteristic	Isolate No. 48-2, 45-3	*S. lactis*	Isolate No. 12, No. 11	*Leuconostuc mesentroides*	Isolate 48.6	*Leuconostuc cremoris*	Isolate No. 25-7	*Leuconostuc oenos*	Isolate No. 12-7	*Pediococcus pentosaceus*	Isolate No. 12-3	*P. acidilactici*	Isolate No. 24-2, 24-3	*P. damnosus*
Gram reaction	+	+	+	+	+	+	+	+	+	+	+	+	+	+
Gas from glucose	–	–	+	d	+	–	+	d	–	–	–	–	–	–
Catalase	–	–	–	–	–	–	–	–	–	–	–	–	–	–
Pseudo catalase	–	–	–	–	+		+		–	+	–	+	–	
Motility	–	–	–	–	–	–	–	–	–	–	–	–	–	–
Growth at pH 4	–	–	+	–	–	–	–	+	–	+	+	+	+	+
At pH 9.6	–	–	+	+	–		ND	d	+	d	+	d	–	–
Growth in salt 4 %	–	+	+	+	–	–	+	ND	+	+	+	+	+	–
In salt 10 %	–	–	–	–	–	d	–	ND	d	d	–	–	–	–
Growth at 25 °C	–	+	+	+	–		–		+	+	+	+	+	–
at 37 °C	+	+	+	d	+	–	+	d	+	+	+	+	+	–
at 45 °C	–	–	+		+		+		–		–		+	
Vp test	–		–		–		–		+		+		–	
MR test	–		–		+		–		+	+	+			
Survival at 60 °C for 0.5 h	+	+	–	–	+		–							
Reaction to 0.1 % MB		+	+		–		–							
NH_3 from arginine	+	+	–	–	–		+		+	+	+	+	–	–
Indol produced			–	–	–	–	–	–	–	–	–	–	–	–
Casein hydrolyzed	+	+												
Gelatin hydrolyzed														
Arabinose	–	–	+	+	–	–	–	d	+	+	+	d	–	–
Esculin	–	d	+	d	–	–	+	+		+	+	+	+	+
Galactose	+		+		+		+		d					
Fructose	+		+		–		+		+					
Lactose	+	+	+	d	+	+	–	–	–	d	+	d	–	–
Maltose	+	+	+	+	–	d	+	–	+	+	+	–	+	d
Mannitol	+		+		–		+	–	–		–	–	–	–
Mannose	+		+		–		+	d						
Raffinose	–		+		–		–							
Rhamnose ribose	–								+					
Ribose	+		+		–		–		+		+	d	–	–
Salicin	–	–	+	d	+	–	+	d	+		+	+	+	+
Sorbitol	–													
Sucrose	–	–	+	+	–	–	–	–	+	–	–	–	–	d
Xylose	–		–		–		+		+		+	+	–	

Table 5. Physiological and biochemical characteristics of isolates from fermenting brine

Degradation of the pesticides by the lactic acid bacteria: Table 6 shows the effect of indigenous lactic acid bacteria in vegetable mix on the degradation of malathion and diazinon after 48 h of fermentation process. According to the Table-6, the initial concentrations of malathion and diazinon in the vegetable (unprocessed sample) were 3.5 and 0.6 mg kg^{-1}, respectively. This fact implies more penetration of the first pesticide into the plant tissues during the pesticide spraying stage. After 48 h of fermentation, the concentration of malathion considerably decreased and reached to 0.5 mg kg^{-1}, whereas, diazinon concentration only decreased about 0.1 mg kg^{-1}. The remarkably degradation of the malathion during the fermentation could be attributed (Freed et al.,1979) to its instability at low pH ranges, regardless of bacterial decomposition.

Sample name	Diazenon (mg/kg)	Malation (mg/kg)
Control	Not detected	Not detected
Un processed sample	0.6	3.5
Processed sample	0.5	0.5

Table 6. Dagradation of Malation and Diazinon by lactic acid bacteria in 48 h.

4. Conclusion

According to the results obtained from this research, indigenous microflora of Iranian vegetables which are substantially consisted of different species of lactic acid bacteria, were capable of degrading malathion and diazinon, the two common pesticides used in Iran. Regardless of enhancing hygienic value of the vegetable products, fermentation led to the formation of a novel fermented vegetable with well organoleptic characteristics. Among the isolated lactic acid bacteria, *L. plantarum* was a probiotic bacterium. Because the mix microflora was able to grow fast up to about pH 4.2 (Table-1), special attention should be made on this point whether mentioned bacterium is tolerant towards low pH ranges. This fact would be very important because probiotics are generally sensitive to low pH and high acidic media. The lactic acid microflora showed synergistic relationships among the species, because single cultures were not able to reduce the medium pH compared with the mix cultures, when inoculated to the vegetable, separately (data not shown). Adapted microflora after 3 times of transfer in MRS broth containing malathion and diazinon instead of glucose, made the tolerant microorganisms which were highly capable of growing and decreasing the pH of the media. Therefore, producing such a mix culture in the form of lyophilized starter culture for production of fermented vegetable products could be an important objective. Determination of optimum fermentation time is also an important issue, because along with the increase of fermentation time, the amount of pesticides as well as the viable counts of the bacteria decreases. The second fact is not favourable. Because the number of viable cells remarkably decreases during the period of 24 to 48 h after the start of fermentation (Table-1), it is recommended that the optimum fermentation time to be identified during 24 to 48 h. Complementary research should be done about the degrading effects of fermentation on the other consuming chemical pesticides. Moreover, it might be interesting to determine the contribution of each species in the degradation of the pesticides, during the fermentation period. Finally, the components produced from degraded pesticides must be identified and evaluated from safety as well as sensory points of view.

5. References

Aislabie, J. & Lloyd-Jones, G. (1995). A review of bacterial degradation of pesticides, In: *Australian Journal of Soil Research* 33, pp. 925-942.

Anderson R., (1998), Biogenic amines in lactic acid-fermented vegetables. *Lebensm-Wiss, U-Technol.*, 21, 68-69.

AOAC, (1995), Association of Official Analytical Chemists, Official Methods of Analytical, Association Official Analytical Chemist International, Arlington, VA, USA, end. 16.

Briggs M., (1953), An improved medium for lactobacilli, In: *J. Dairy Res.* 20, pp. 36-40.

Dannenberg A, & Pehkonen S.O. (1998) Investigation of the heterogeneously catalyzed hydrolysis of organophosphorous pesticides. *J. Agric Food Chem* 46: pp. 325-334.

Dimitrios, G.; Karpouzas, A.; Fotopoulou, O.; Menkissoglu-Spiroudi, U. & Brajesh, K., (2005). Non-specific biodegradation of the organophosphorus pesticides, cadusafos and ethoprophos, by two bacterial isolates. In: *FEMS Microbiol. Ecol.*, 53, pp. 369-378.

Freed, VH,; Chiou, CT, & Schmedding, DW. (1979) Degradation of selected organophosphorous pesticides in water and soil. In: *J. Agric. Food Chem.* 27, pp. 706-708.

Lal, R. & Saxena, D.M., (1982). Accumulation, metabolism and effects of organochlorine insecticides on microorganisms. In: *Microbiol. Rev.*, 46, pp. 95-127.

Ledford, RA. & Chen, JH. (1969). Degradation of DDT and DDE by cheese microorganisms. In: J Food Sci 34, pp. 386-388

Mirna, A. & Coretti K., (1979). Influence of processing on the degradation of pesticides in meat products. In: *Meat Sci.*, Volume 3, Issue 2, April 1979, pp. 97-108

Muller, W.P. & F. Korte, (1975). Microbial degradation of benzo-(a)-pyrene, monolinuron and dieldrin in waste composting. In: *Chemosphere*, 4, pp. 195-198

Muller, W.P. & Korte, F., in eds.: F. Coulston, & F. Korte, (1976), Ecological Chemical Evaluation of Waste Treatment Procedures, Environmental Quality and Safety: In: *Global Aspects of Chemistry, Toxicology and Technology as Applied to the Environment*, New York, Academic Press, pp. 215-236.

Nawab, A.; Alim, A. & Malik, A., (2003). Determination of organochlorine pesticides in agricultural soil with special reference to γ-HCH degradation by *Pseudomonas* strains, In: *Biores. Technol.* 88, pp. 41-46

Pehkonen, S.O. & Zhang Q. (2002). The degradation of organophosphorus pesticides in natural waters, In: *Crit. Rev. Env. Sci. Tec.* 32, pp. 17-72

Peric, M., Raseate, J. & V. Visaki, (1981). Degradation Organochlorione Pesticide on Influenced by Micrococci Isolated from Fermented Sausages. In: *European Meeting of Meat Research Workers, (25th Symposium) Belgrade, Yugoslavia.*

Racke, K. D., & Frink C. R., (1989). Fate of Organic Contaminants During Sewage Sludge Composting. *Bulletin of Environmental Contamination and Toxicology*, Vol. 42, No. 4, pp. 526-533

Racke, K.D., in eds.: Chambers, J.E. & Levi, P.E. (1992). Degradation of Organ Phosphorus Insecticides in Environmental Matrices, In: *Organophosphates, Chemistry, Fate and Effects*, Academic Press, San Diego, pp. 47-73.

Sharpe, M.E., in eds.:. Skinner, F.A. & Lovelock, D.W., (1979). Identification of the Lactic Bacteria, *Identification Methods for Microbiologists*, Academic Press, London, pp. 233-235.

Sneath, P.H.A.; Mair, N.S; Sharpe, M.E. & G.G. Holt, (1986). Bergey's Manual of Systematic Bacteriology, Williams and Wilkins, Baltimore, M.D., Vol. 2.

Vandervoort, C., et al. Fate of Selected Pesticides Applied to Turfgrass: Effects of Composting on Residues. *Bulletin of Environmental Contamination and Toxicology*, Vol. 58, No. 1, January 1997, pp. 38-45.

32

Interpretation and Modelling of Environmental Behaviour of Diverse Pesticides by Revealing Photodecomposition Mechanisms

Attila Kiss and Diána Virág
Eszterházy Károly College, Institute for Food Sciences
Hungary

1. Introduction

Photodecomposition of several organic contaminants involving distinctive pesticides is a topic of high interest for researchers in recent years. Several efforts have recently been made aiming at reduction of potential risks attributed to pesticides, however comprehensive assessment of potential hazards, including characterisation of photodegradates, has been marginally performed (Grover and Cessna, 1991; Burrows et al. 2002). Investigating pesticide degradation occurring in the environment is of high interest as both parent compounds and decomposition products can be hazardous because of their toxicity. Photochemical degradation of pesticides is the breakdown of pesticides by light, particularly sunlight. Photochemical degradation of pesticides can be important in the decontamination of natural water or contaminated soils (Aaron et al., 2001; Coly et al., 1994). Essence of previous studies concerning the examined pesticide is presented as follows.

Atrazine (2-chloro-4-ethylamine-6-isopropylamino-S-triazine) was one of the most popular selective triazine herbicide until it has been classified as a Restricted Use Pesticide (RUP) due to its potential for groundwater contamination (Ware, 1986). It is a photosynthetic inhibitor and used to control broadleaf and grassy weeds. The degradation pathway of the most frequently used triazine pesticide, atrazine, was investigated in aqueous phase by sonolysis, ozonation, photolysis at 254 nm and photocatalysis in the presence of TiO_2 (Bianchi et al., 2006). Dealkylation and dechlorination was induced by ozonation and photocatalysis, while direct photolysis at 254 nm promoted very efficient dechlorination. Determination of products obtained by photolysis of atrazine in soil samples was performed by Durand and Barcelo (1990, 1991). Hydroxy, 2-H and 2-methoxy derivatives of the parent compound were identified. TiO_2 catalysed mineralization of atrazine was also studied along with other triazines (Konstantinou et al. 2001; Parra et al. 2004; Lányi and Dinya, 2005; Pérez et al. 2006). Products formed by oxidation, ring opening, decarboxylation, isomerization, hydroxilation and cyclization were found to be the main degradates. By the loss all of the side chains formation of cyanuric acid and 2-amino-4,6-dihydroxy-1,3,5-s-triazine were observed. The structure of each photolytic degradates (in the presence of porphyrin and phthalocyanine complexes as catalysts) was given by Hequet et al. (2000, 2001).

Simazine (6-chloro-N2,N4-diethyl-1,3,5-triazine-2,4-diamine), a wide-spread representative of s-triazine type pesticide, is a selective herbicide with photosynthetic inhibiting effect. It is used to control broad-leaved weeds and annual grasses. A comparative study between fragmentation processes taking place in mass spectrometry using an electron ionisation source and photodegradation processes has been carried out for atrazine, simazine and trietazine. The same kinds of fragmentations were observed for the three compounds: C–N bond cleavage in the lateral chains, C–Cl bond scission and heteroatomic ring cleavage. The photochemical degradation and the kinetics of the degradation processes of s-triazine herbicides (atrazine, propazine, and prometryne) has been investigated in case of several types of natural waters and soils (Konstantinou et al. 2001). The photolytic behaviour of triazine herbicides (atrazine, simazine, trietazine, prometon, prometryn) in the presence of TiO_2 as a special photocatalyst has already been studied (Pelizzetti et al., 1990, Hequet et al., 2001). All the herbicides degraded rapidly, full mineralization was not observed. Cyanuric acid was found to be the common final photoproduct of all herbicides. Triazine-derivatives are considered to be the representatives of pesticides of the most wide-spread practical application; therefore it is of crucial importance to evaluate their fate in the environment (Vidal et al. 1999). It was shown that some triazine herbicides undergo photodegradation to form deaminated derivatives (Mansour et al. 1993). The photodegradation products of some commonly used N-containing herbicides were detected however entire mechanisms have not been revealed (Lányi and Dinya, 2005). High-pressure mercury vapour lamp (254 nm, 125 W) and GC/MS technique were used during the examinations. Decomposition products steming plausibly from loss of side-chains and substitution with OH-group were detected. Different metabolites formed having mixed side-chains, and the presence of dimer products were also observed.

Acetochlor (2-chloro-N-(ethoxymethyl)-N-(2-ethyl-6-methylphenyl)acetamide), as a member of the chloroacetanilide class of broad leaf herbicides, is one of the most widely used herbicide. It is a growth inhibitor and applied as preemergence for control of annual grasses.

Chloracetanilide herbicides have been investigated the terms of revealing stability, water solubility and toxicity of degradates (Belfroid et al., 1998). Brekken and Brezonik (1998) studied the reaction between acetochlor and HO-, assuming that the primary source of HO is nitrate photolysis. According to their experimental data, the direct photolysis would be much slower than HO -mediated degradation. In case of acetochlor, serious efforts have been made in order to identify biodegradation products of the pesticide, however no specific reaction pathways have been mapped (Coleman et al., 2000; Thurman et al., 2002; Zheng et al., 2003).

Chlorpyrifos (O,O-diethyl O-(3,5,6-trichloro-2-pyridyl)phosphorothioate) is an organophosphate insecticide, acaricide and miticide used to control foliage and soil-borne insect pests on a variety of food and feed crops. The photodegradation of chlorpyrifos by simulated sunlight in water/methanol has been studied by Barcelo et al. (1993) and 3,5,6-trichloro-2-pyridinol was identified as the only degradation product. A method was developed to determine the rate of reaction of chlorpyrifos with HO radicals in the gas phase at high temperatures during photodecomposition (Hebert et al., 2000). Kamiya and Kameyama (1998) studied the effects of humic materials and metal ions on the photochemical degradation of various organophosphorus pesticides (including chlorpyrifos) (Kamiya et al., 2001).

Carbendazim (methyl benzimidazol-2-ylcarbamate) is a benzimidazole carbamate fungicide with systemic activity and broad effect spectrum. It inhibits fungal mitotic microtubule formation. The visible-light-promoted photodegradation of carbendazim was studied in water or water–methanol solution under various conditions (in the presence of air and a photosensitizer xanthene dye or pigment riboflavin, at various pH values (Escalada et al., 2006, Panades et al., 2000, Mazellier et al., 2002). It was established that the rate of photodegradation increased with pH and oxygen concentrations. The aqueous photodegradation of carbendazim was studied by Panades et al. (2000). The kinetics of the photodecomposition was determined using HPLC–DAD and the identification of photoproducts was carried out with HPLC–MS by Boudina et al., (2003). Three products were detected after the UV irradiation. One of them, 2,4-amino-benzimidazol has already been identified in a previous paper (Mallat et al, 1997, Tomlin, 1994). A plausible pathway for the photolytic degradation of carbendazim in pure water was proposed as well, however our studies pointed out marked differences when comparing the two different mechanisms.

Diuron (N-(3,4-dichlorophenyl)-N,N-dimethyl urea) is a substituted urea type non-selective herbicide with photosynthesis inhibition effect (Wessels and Veen, 1956). Its principal breakdown product 3,4-dichloroaniline is more toxic than diuron itself. It is persistent and contaminates marine waters, groundwater, sediment and soil (Giacomazzi and Cochet, 2004). It is considered as a Priority Hazardous Substance by the European Commission (Mehlhorn, 2001).

EPTC (*S*-ethyl dipropylthiocarbamate) is a selective, growth inhibitor herbicide of thiocarbamate-type. The UV-photodegradation of EPTC in hexane has been studied by Marco and Hayes (1979) as well as Abu-Qare et al. (2002). Formation of several photoproducts and the cleavage of C-S and C-N bonds were observed but no reaction pathway was revealed. The kinetics of photodegradation of EPTC was studied by Dinya and Lányi (2005.).

In summary it might be stated that huge efforts have been made to characterise both the chemical and environmental behaviour of pesticides, however the precise reaction mechanisms of photodegradation have not been revealed, and in some cases controversial results have been obtained. Hence identification of major degradates might be considered as a major task in order to gain insight into the photodegradation and environmental behaviour of the selected and investigated pesticides.

The abovementioned information shows that numerous photodegradation studies have been performed especially concerning s-triazines. However, with the exception of atrazine, detailed reaction mechanisms of the concerned pesticides have not been identified. In some cases, specific degradation products were detected without the aim of mapping the entire pathway of photodegradation. Thus our work contributes to a more extensive and comprehensive knowledge on pesticide photodecomposition with regard to both reaction mechanism and chemical characteristics of degradation products. The pathway of photolytic degradation of seven pesticides having diverse chemical structure and practical application is mapped by GC/MS identification of degradation products.

2. Specified results

2.1 Photodegradation of ATRAZINE

The five-hour-long UV-irradiation of atrazine [2-chloro-4-ethylamine-6-isopropylamino-S-triazine] resulted in the cleavage of the chloro-group and the hydroxylation of the ring. As a consequence [1,3,5-triazin-2-hydroxy-4-(ethylamino)-6-(isopropyl)amino] could be detected. [1,3,5-triazin-2-hydroxy-4,6-(diethylamino)] was generated by the scission of a methyl-

group. [1,3,5-triazin-2-hydroxy-4,6-diamine] might be identified as the end product of the photodegradation by consecutive de-ethylation processes.
Novelty: In contrast to other studies 20-hour-long UV-irradiation did not lead to the formation of cyanuric acid as the end-product. Under natural conditions this compound is not likely to be formed; it could only be obtained by $TiO_2/Na_2S_2O_8$ catalyzed degradation (Lányi and Dinya, 2005, Pérez et al. 2006). In comparison with the results of previous studies it might be established that dechlorination, hydroxylation and scission of methyl- and ethyl groups are regarded as major steps of the photodegradation of atrazine. Ring opening, dimerisation and decarboxylation were not detected during our studies.

2.2 Photodegradation of SIMAZINE

The degradation of simazine [2,6-di(ethylamino)-4-chloro-1,3,5-triazine] effected by UV-photons can take place via two parallel reaction pathways. Major steps of the photodecomposition were found to be as follows: cleavage of a chloro-group and its partial substitution to OH-group, loss of methyl and ethyl groups, and scissoring of OH-group. The most stabile degradation product of simazine is [2,4-di(ethylamino)-1,3,5-triazine].
Novelty: In contrast with the finding of Pelizzetti, and identification of cyanuric acid, symmetrical [2,4-diamino-1,3,5-triazine] was obtained by us as the end-product of the degradation of the triazine-type pesticide. Studies aiming at investigating the photolytic degradation of simazine so far have only pointed out the fact of degradation and the factors influencing it (Mansour et al. 1993, Hequet et al., 2001, Bianchi et al., 2006). Previously efforts were made to propose a reaction pathway for the decomposition (Pelizzetti et al., 1990), and to detect the loss of side-chains and their substitution with OH-group (Lányi et al., 2005), however identification of all the major degradates ([2,4-di(ethylamino)-hydroxy-1,3,5-triazine], [2-amino-4-chloro-6-methylamino-1,3,5-triazine], [2-ethylamino-4-methylamino-1,3,5-triazine], [2-amino-4-methylamino-1,3,5-triazine]) and revealing the complete decomposition pathway might be regarded as new findings.

2.3 Photodegradation of ACETOCHLOR

In case of the photodegradation of acetochlor [2-chloro-N-(ethoxymethyl)-N-(2-ethyl-6-methylphenyl)acetamide] there are alternative reaction pathways according to our findings. Several degradation products could be detected after some hours of irradiation: [2-chloro-N-hydroximethyl-N-(2-ethyl-6-methylphenyl)acetamide], [N-hydroximethyl-N-(2-ethyl-6-methylphenyl)acetamide], [N-methyl-N-(2-ethyl-6-methylphenyl)acetamide]. Major steps of photodecomposition are as follows: cleavage of ester-bond of N-ethoxy-methyl group, breaking off the chloro- and the hydroxyl-groups, resulting in the formation of [2-ethyl-6-methyl-N-methyl-aniline]. This last degradation product might be formed from the parent compound in a direct way as well. Alternatively, the cleavage of chloro-, methyl- and ethoxy-groups of the parent compound and the production of a formanilid-derivative [N-methyl-N-(2-ethyl-6-methylphenyl)formamide] might also be observed. Cleavage of methyl-, ethyl and amino-groups, through the formation of [2-ethyl-6-methyl-aniline] led to the formation of [3-ethyl-toluene] as an end product.
Novelty: Three main degradation products of **acteochlor** we detected were also determined by other studies aiming at modelling biodegradation of acetochlor, but the total degradation pathway of acetochlor has not been revealed so far. The biological degradation of acetochlor led to the formation of metabolites being not analogous to intermediers detected during our investigations (Coleman et al., 2000; Zheng et al., 2003). The determination of all 9

degradation products and mapping the entire degradation pathway by our experiments contributes to the entire understanding of acetochlor's environmental behaviour.

2.4 Photodegradation of chlorpyrifos

The photodegradation of chlorpyrifos [O,O-diethyl-O-(3,5,6-trichloro-2-pyridyl)phosphorothioate] may occur via two reaction patterns. It might be initiated by the cleavage of either an ethyl-group or a chloro-group resulting in the formation of [O-ethyl-O-(3,5,6-trichloro-2-pyridil)-hydrogene- phosphorothioate] and [O,O-diethyl-O-(3,5-dichloro-2-pyridil)phosphorothioate]. The following step of the decomposition is the scission of chloro and ethyl groups ([O-ethyl-O-(3,5-dichloro-2-pyridil)-hydrogene-phosphorothioate]). Breaking away of another chloro-group leads to the formation of [O-ethyl-O-(5-chloro-2-pyridil)-hydrogene- phosphorothioate]. Loss of the other ethyl-group transforms the previous compound into the end-product [O-(5-chloro-2-pyridil)-dihydrogene-phosphorothioate], whose presence is shown in the GC-chromatogram and the mass spectrums.

Novelty: Neither the detected and identified UV-induced degradation products of chlorpyrifos, nor its degradation pathway has not been described and observed so far. These degradates are not analogous with the results of Barcelo (1993) and Xu (2008): only 3,5,6-trichloro-2-pyridinol was found as the product of the biological degradation of chlorpyrifos by a newly isolated *Paracoccus* sp. strain TRP.

2.5 Photodegradation of carbendazim

The first step of the degradation of carbendazim [methyl-benzimidazole-2-ylcarbamate] was the loss of the methyl group and the formation of [benzimidazole-2-ylcarbamic-acide], which product showed small stability, and subsequent to the cleavage of a hydroxyl- and a carbonyl group the origination of [benzimidazole-2-ylcarbamate] was observed. This product was transformed into [2-amino-benzimidazole] which was converted to the rather stabile benzimidazole by deamination as a result of further UV-treatment. Several hours of following UV-irradiation led to opening of the imidazole-ring and to the production of [2-methyl-amino-aniline]. Subsequently, by the cleavage of the N-methyl bond the end-product of the photodegradation, [1,2-diaminobenzene] was acquired.

Novelty: Formation of 2,4-amino-benzimidazole, benzene, phenol or aniline, or any dimmer products was not observed in our studies, contrary to the findings of Mallat et al., Mazellier et al. and Tomlin.

2.6 Photodegradation of DIURON

The photocatalytic degradation of diuron [N-(3,4-dichlorophenyl)-N,N-dimethyl urea] was easy to be initiated. The main degradation products could already be detected just after 2 hours of UV-irradiation. By the scission of the N-dimethyl group [N-4,5-dichloro-phenyl-isocyanate] was formed. Dechlorination processes were also observed generating [N-dichloro-phenyl-isocyanate] and [benzene-isocyanate]. An alternative reaction pathway may also occur: dechlorination of the parent compound led to the formation of [N,N-dimethyl-N'-phenylurea]. The loss of N-dimethyl group and the incorporation of a propyl-group yielded [N-(4-isopropyl-phenyl)-formamide] as an intermedier of the photolytic decomposition. Decarbonylation was induced by further UV-irradiation and [aniline] was detected as the end product of the photodegradation of diuron. [2-methyl-propinic acid] was produced as a secondary- or by-product of the degradation processes.

Novelty: The degradation mechanism of **diuron** was not thoroughly investigated yet prior to our studies. The main steps of phototransformation were found to be the sequential cleavage of the chloro-group and the N-dimethyl-group, as well as hydroxylation and dehydroxylation processes.

2.7 Photodegradation of EPTC

Photochemical decomposition of EPTC [S-ethyl-dipropyl-thiocarbamate] occurs rapidly as the appearance of the first degradation product could already be detected subsequent to twenty-minute-long UV-irradiation. During the photochemical degradation of EPTC [N,N-dipropyl-formamide] and [N,N-diethyl-propionamide] are formed at the first stage of the transformation via two alternative decomposition routes. Both the cleavage of S-ethyl-group and the demethylation of N-propyl groups are possible. In accordance with the given reaction pathways it might be established that both the consecutive losses of alkyl-groups and the cleavage of the amide bond lead to the degradation end-product: [diethyl-amine]. Some other degradates were also identified, such as: [tripropylamine], [dipropylamine], [dipropyl-ethyl-amine].

Novelty: When comparing findings of our studies with previous research on revealing products of biological degradation of EPTC (Abu-Qare and Duncan, 2002) it might be established that photodecomposition and biological degradation do not result in the formation of analogous products, as nor EPTC-sulfon or EPTC-sulphoxide were detected throughout our examinations. Breaking off of S- and N-alkyl groups resulted in the same degradates as identified previously, however ketoformyl and ketocarbonyl derivatives might only be observed in TiO_2 catalytized photodegradation processes (Lányi and Dinya, 2005).

3. Summary

Our study is focused on revealing specific details of photolytic degradation of pesticides whose decomposition mechanisms have not been mapped entirely yet. Importance of the study is enhanced by the fact that scarce information is available regarding their natural degradation processes, as well as the quality, structure and biological impact of the decomposition products.

In order to clarify the photolytic decomposition behaviour of selected pesticides and to be capable of right interpretation of the complete mechanisms, our major goal was to isolate and identify the degradation products by applying GC-MS and HPLC-MS techniques. Subsequently the most plausible degradation pathways were established, and the rate of the photodecomposition was also followed. Frequently applied pesticides of distinctive chemical structure and physical behaviour have been selected for our studies. Our study presents the total mechanisms of the photoinduced degradation of various pesticides in case of acetochlor, simazine, chlorpyrifos, carbendazim (Kiss and Virág, 2009/a) and diuron, atrazine, and EPTC (Kiss and Virág, 2009/b).

4. References

Aaron, J.J., and M.A. Oturan. 2001. New photochemical and electrochemical methods for the degradation of pesticides in aqueous media. Environmental applications.Turk. J. Chem. 25:509-520.

Abu-Qare, A. W., H. J. Duncan, Photodegradation of the herbicide EPTC and the safener dichlormid, alone and in combination. Chemosph. 46 (2002) 1183-1189.

Barcelo, D., G. Durand, and N.D. Bertrand. 1993. Photodegradation of the organophosphorus pesticides chlorpyrifos, fenamiphos and vamidothion in water. Toxicol. Environ. Chem. 38:183–199.

Belfroid A.C., M. van Drunen, M.A. Beek, S.M. Schrap, C.A.M. van Gestel, and B. van Hattum. 1998. Relative risks of transformation products of pesticides for aquatic ecosystems.Sci. Total. Environ. 222:167 –183.

Brekken, J.F., and P.L. Brezonik. 1998. Indirect photolysis of acetochlor: Rate constant of a nitrate-mediated hydroxyl radical reaction. Chemosphere 36:2699–2704.

Bianchi, C. L., C. Pirola, V. Ragaini, and E. Selli. 2006. Mechanism and efficiency of atrazine degradation under combined oxidation processes. Applied Catalysis B: Environmental 64:131–138.

Boudina, A., C. Emmelin, A. Baaliouamer, M. F. Grenier-Loustalot, and J. M. Chovelon. 2003. Photochemical behaviour of carbendazim in aqueous solution. Chemosph. 50:649-655

Burrows, H.D., L.M. Canle, J.A. Santaballa, S. Steenken, Reaction pathways and mechanisms of photodegradation of pesticides. J. Photochem. Photobiol. B: 67 (2002) 71–108.

Coleman, S., R. Linderman, E. Hodgson, and R.L Rose. 2000. Comparative metabolism of chloroacetamide herbicides and selected metabolites in human and rat liver microsomes. Environ. Health. Perspect. 108: 1151-1157.

Coly, A., and Aaron, J. J. 1994. Fluorimetric determination of aromatic persticides in technical formulations. Effects of solvent and of ultraviolet photolysis. Talanta 41:1475-1480.

Durand, G., D. Barceló, Determination of chlorotriazines and their photolysis products by liquid chromatography with photodiode-array and thermospray mass spectrometric detection. J. Chromatography A. 502 (1990) 275-286.

Durand, G., D. Barceló, Confirmation of chlorotriazine pesticides, their degradation products and organophosphorus pesticides in soil samples using gas chromatography-mass spectrometry with electron impact and positive- and negative-ion chemical ionization. Anal. Chim. Acta, 243 (1991) 259-271.

Escalada, J. P., A. Pajares, J. Gianotti, W.A. Massad, S. Bertolotti, F. Amat-Guerri, and N.A. García. 2006. Dye-sensitized photodegradation of the fungicide carbendazim and related benzimidazoles. Chemosphere 65:237-244.

Giacomazzi, S., N. Cochet, Environmental impact of diuron transformation: a review Chemosph.56 (2004) 1021-1032.

Grover, R., A.J. Cessna, (Eds.), Environmental Chemistry of Herbicides, CRC Press, Boca Raton, 1991.

Hebert, V.R., C. Hoonhout, and G.C. Miller. 2000. Use of stable tracer studies to evaluate pesticide photolysis at elevated temperatures. J. Agric. Food Chem. 48:1916–1921.

Hequet, V., P. Cloirec, C. Gonzalez, B. Meunier, Photocatalytic degradation of atrazine by porphyrin and phthalocyanine complexes. Chemosph. 41 (2000) 379-386.

Hequet, V., C. Gonzalez, P. Cloirec, Photochemical processes for atrazine degradation: methodological approach. Water Research, 35 (2001) 4253-4260.

Kamiya, M., and K. Kameyama. 1998. Photochemical effects of humic substances on the degradation of organophosphorus pesticides. Chemosphere 36:2337–2344.

Kamiya, M., K. Kameyama, S. Ishiwata. 2001. Effects of cyclodextrins on photodegradation of organophosphorus pesticides in humic water. Chemosphere 42:251–255.

Kiss, A., D. Virág, Photostability and photodegradation pathways of distinctive pesticides. J. Environ. Qual. 38 (2009/a) 1.

Kiss, A., D. Virág, Interpretation and modelling of environmental behaviour of diverse pesticides by revealing photodecomposition mechanisms. Microchemical Journal. (2009/b) 92(2)119-122

Konstantinou, I.K., A.K. Zarkadis, T.A. Albanis, Photodegradation of selected herbicides in various natural waters and soils under environmental conditions. J. Environm. Qual. 30 (2001) 121-130.

Lányi, K., Z. Dinya, Photodegradation study for assessing the environmental fate of some triazine-, urea- and thiolcarbamate-type herbicides. Microchem. J. 80 (2005) 79-87.

Mallat, E., D. Barcelo, and R. Tauler 1997. Degradation study of benomyl and carbendazim in water by liquid chromatography and multivariate curve resolution methods. Chromatographia 46:342–350.

Mansour, M. Ed. 1993. Fate and Prediction of Enviromental Chemicals in Soils, Plants and Aquatic Systems. Lewis, Boca Raton, Ann Arbor, USA.

Marco, A.C.D., E.R. Hayes, Photodegradation of thiolcarbamate herbicides. Chemosph 1 (1979) 321–326.

Mazellier, P. É. Leroy, and B. Legube. 2002. Photochemical behavior of the fungicide carbendazim in dilute aqueous solution. J. Photochem. Photobiol. A: Chemistry; 153:221-227.

Mehlhorn, B., The EU-list of priority substances - a new strategy against water pollution. (No. 2477/2001/EC) Annex X of Water Framework Directive (2000/60/EC). 2001, Umweltbundesamt, Berlin.

Panadés, R., A. Ibarz, and S. Esplugas. 2000. Photodecomposition of carbendazim in aqueous solutions. Water Res 34:2951-2954.

Parra, S., S.E. Stanca, I.K. Guasaquillo, R. Thampi, Photocatalytic degradation of atrazine using suspended and supported TiO_2. Appl. Cat. B: Environm. 51 (2004) 107–116.

Pelizzetti, E., V. Maurino, C. Minero, V. Carlin, E. Pramauro, O. Zerbinat, and M.L. Tosato. 1990. Photocatalytic degradation of atrazine and other s-triazine herbicides. Environ. Sci. Technol. 24:1559–1565.

Pérez, M.H., G. Peñuela, M.I. Maldonado, O. Malato, P. Fernández-Ibáñez, I. Oller, W. Gernjak, S. Malato, Degradation of pesticides in water using solar advanced oxidation processes. Appl. Cat. B: Environm. 64 (2006) 272-281.

Thurman, E.M., I. Ferrer, and R. Parry 2002. Accurate mass analysis of ethanesulfonic acid degradates of acetochlor and alachlor using high-performance liquid chromatography and time-of-flight mass spectrometry. J. Chromatogr. A 957:3-9.

Tomlin. 1994. The Pesticide Manual., Tomlin Crop Protection Publications, Cambridge United Kingdom.

Zheng H.H., and C.M. Ye. 2003. Photodegradation of acetochlor in water and UV photoproducts identified by mass spectrometry. J. Environ. Sci.-China 15:783-790.

Vidal, A., Z. Dinya, and F. Mogyoródi. 1999. Photocatalytic Degradation of Thiocarbamate Herbicide Active Ingredients in Water. Appl. Catal. B: Environ. 21:259-267.

Ware, G.W., Fundamentals of Pesticides: A Self-Instruction Guide, 2nd Edition. Thomson Publications (Eds.), Fresno, CA, 1986, pp. 8-3.

Wessels, J.S.C., R. van der Veen, The action of some derivatives of phenylurethan and of 3,phenyl-1,1-dimethylurea on Hill reaction. Biochem. Biophys. Acta 19 (1956) 548–549.

Xu, G., W. Zheng, Y. Li, S. Wang, J. Zhang and Y. Yan. 2008. Biodegradation of chlorpyrifos and 3,5,6-trichloro-2-pyridinol by a newly isolated Paracoccus sp. strain TRP. *International Biodeterioration & Biodegradation In Press, Corrected Proof*

33

Degradation of Organochlorine and Organophosphorus Pesticides by Photocatalysis: Chlorpiryfos and Endosulfan Case Study

Rosalina González Forero
La Salle University
Bogotá, Colombia

1. Introduction

In Colombia, the use of pesticides has increased vastly during the last two decades representing a large problem for the aquatic ecosystems and the water purification (Duarte 2009, Cáceres 2008 and and Cárdenas, 2006). Colombia possesses a large quantity of hydro resources, several of which have been transformed into reservoirs for energy generation and for use in drinking water plants. Most of the reservoirs are used for reception from agricultural regions, and therefore, the water arrives with large quantities of residual pesticides.

On the other hand, the amount of obsolete pesticides confiscated by the environmental authorities in Colombia is increasing year per year (Duarte 2009, Cáceres 2008 and ICA, 2006), this situation has produced a serious problems because there are not enough suitable sites for storage operations. Many stocks of these pesticides are being abandoned, burned, or added to water surfaces or stored in places without technical parameters for the collection of hazardous substances. For this reasons, technology is required to be able to destroy the pesticides and be sure that there are no residual organic compounds or minerals present therefore, hydro resources. One of these technologies is The AOP's (Advanced Oxidation Processes), and more specific The *Photocatalysis*, which uses ultraviolet radiation, oxygen and in some cases titanium dioxide like catalyst was selected. This presents numerous advantages in comparison with the conventional technologies: It achieves the destruction of a large number of organic pollutants and it also transforms them into innocuous products. It doesn't generate undesirable secondary compounds. It can be used for the treatment of water, air and contaminated ground; and it allows the employment of solar radiation as an energy source, adding to this process a significant environmental value.

The greatest amounts of pesticides confiscated by the environmental authorities in Colombia and the largest stocks of obsolete pesticides stored throughout the country are chlorinated organic substances, whose active ingredient is Endosulfan. (Gonzalez 2009 and

Carol Ximena Duarte, Víctor Fabian Forero, Carmen Emilce Cáceres, Katherine Marcela Gómez performed the experiments presented in this chapter at The Environmental Engineering Laboratory at La Salle University. All the experiments were funded by La Salle University

Cárdenas 2006), for this reason this substance was selected for this study. The second substance selected was Attamix Sb because it is used massively and it is part of the phosphorous organic substances and the comparison between these two substances was interesting to test the technology.

This technology can be studied by two most common cases:

1, Homogeneous Photocatalysis: UV/H_2O_2 System: As described by Gonzalez and Braun (1996), the combination of ultraviolet radiation and hydrogen peroxide causes the breakage of the O-O bond by the action of radiation, to form two hydroxyl radicals. As Eq.(1)

$$H_2O_2 + hv \rightarrow 2HO \quad (1)$$

Then the OH • radical attacks the hydrogen peroxide, resulting in the following sequence of reactions. Eq.(2,3,4). When organic compounds are present in the environment, the oxidation reactions start with the different radicals formed. (Fung, 1999)

$$H_2O_2 + OH \rightarrow HO_2 + H_2O \quad (2)$$

$$H_2O_2 + HO_2 \rightarrow HO + O_2 + H_2O \quad (3)$$

$$2HO_2 \rightarrow H_2O_2 + O_2 \quad (4)$$

The use of UV/peroxide offers great advantages: it is a very accessible, commercially available oxidant; it is thermally stable, and can be stored safely in the work place. Since it has an infinite solubility in water, there is no mass transfer problems associated with gases, as in the case of O_3. The capital investment is minimal and the operation is simple. The disadvantage is that it has low absorption at 254 nm, and high concentrations of H_2O_2 are required to maintain this process.

Heterogeneous Photocatalysis: This process is based on direct or indirect absorption of radiant energy (visible or UV) from a solid. This solid is known as photocatalyst, and usually a broadband semiconductor. In this process, the destruction or removal reactions of contaminants occur in the interfacial region between the solid and the solution, without the catalyst undergoes chemical changes.

In this research Dioxide of Titanium (TiO_2) was used as a catalyst, because, it has a high chemical stability and it works in a wide range of pH, being capable to produce electronic transitions by absorption of light in the near ultraviolet (UV-A) ((PSA-CIEMAT, 2001)

2. Materials and methods

2.1 Pesticides

2.1.1 Attamix S.B.

It has as active ingredient Chlorpyrifos, also known as Dursban 23, which belongs to the family of pesticides, insecticides and in turn to organophosphates. Chlorpyrifos is a crystal white solid with strong aroma, not very soluble in water, so it is usually mixed with oily liquids before being applied to crops or animals. In the United States, Chlorpyrifos was used for residential use until 2000 when it was restricted by the United States Environmental Protection Agency U.S. EPA. In 1999 in Colombia were sold around 2.1 million liters and 1.6 tons of Chlorpyrifos, which is the organophosphate with the biggest percentage of the total sales of this group of insecticides.

The following shows the structural form of chlorpyrifos as well as physical and chemical characteristics of the pesticide Attamix. The next figure shows the Structural form of chlorpyrifos:

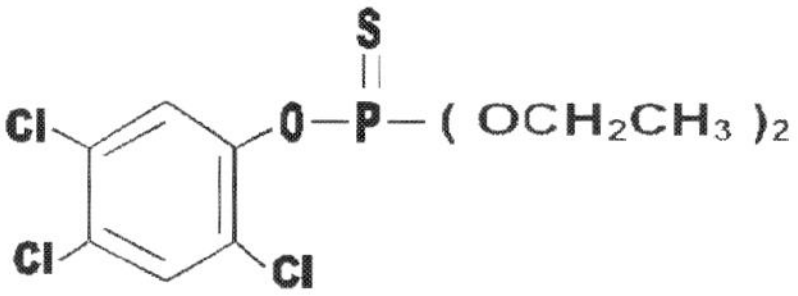

Fig. 1. Structural form of Chlorpyrifos. Organophosphorus pesticides (I): general aspects and toxicokinetics (Obiols, 2000)

Chlorpyrifos is used to control insect pests because inhibits acetylcholinesterase, is moderately toxic and its chronic exposure has been linked to neurological effects, developmental disorders, and autoimmune disorders. (Muller, 2000)

2.1.2 Endosulfan

The pesticide used in this study was obtained at ICA (Agriculture Colombian Institute) in Villavicencio City Branch, because in that place there were many containers confiscated by this institute, because they are the environmental authority in this zone of the country. The pesticide selected was Thiodan 35EC (I.A. Endosulfan). In the next pictures there are the containers used.

Copyright Forero,2008

Copyright Cardenas,2006

Fig. 2. Thiodan 35EC used in the trials and Thiodan Confiscated by the authorities.

Endosulfan is an organochlorine compound that is used as an insecticide and acaricide. This colourless solid has emerged as a highly controversial agrochemical (ABC, 2009) due to its acute toxicity, potential for bioaccumulation, and role as an endocrine disruptor. Banned in more than 62 countries, including the European Union, several Asian and West African nations (Weekly Times, 2009) and soon in the United States, (Cone, 2010 and EPA 2010) it is still used extensively in many other countries including India, Brazil and Australia. Because of its threats to the environment, a global ban on the use and manufacture of Endosulfan is being considered under the Stockholm Convention (Earth Negotiations Bulletin, 2008).
Endosulfan is acutely neurotoxic to insects and mammals, including humans. The US EPA classifies it as Category I: "Highly Acutely Toxic" based on a LD_{50} value of 30 mg/kg for female rats (EPA, 2002) while the World Health Organization (1984) classifies it as Class II "Moderately Hazardous" based on a rat LD_{50} of 80 mg/kg. It is involved in the transfer of nerve impulses; symptoms of acute poisoning include hyperactivity, tremors, convulsions, lack of coordination, staggering, difficulty breathing, nausea and vomiting, diarrhea, and in

severe cases, unconsciousness (EPA, 2000), Doses as low as 35 mg/kg have been documented to cause death in humans, (International Programme on Chemical Safety, 2000) and many cases of sub-lethal poisoning have resulted in permanent brain damage (EPA, 2000). Farm workers with chronic Endosulfan exposure are at risk of rashes and skin irritation (EPA 2002). EPA's acute reference dose for dietary exposure to Endosulfan is 0.015 mg/kg for adults and 0.0015 mg/kg for children. For chronic dietary exposure, the EPA references doses are 0.006 mg/(kg.day) and 0.0006 mg/(kg.day) for adults and children, respectively (EPA, 2002).

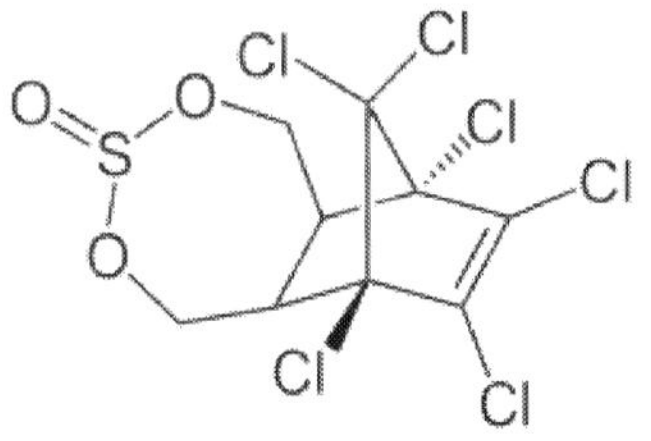

Fig. 3. Structural form of Endosulfan

2.2 Photochemical system

2.2.1 Reactor

The system used in the experiments was an annular type photochemical reactor (See Figures 4, 5 and 6). It consists of a cylindrical chamber, closed at each end, equipped with an inlet and an outlet for water, and a glass tube running through it from end to end by a central axis.

This tube protects the low pressure mercury ultraviolet lamp, connected to its corresponding electrical system.

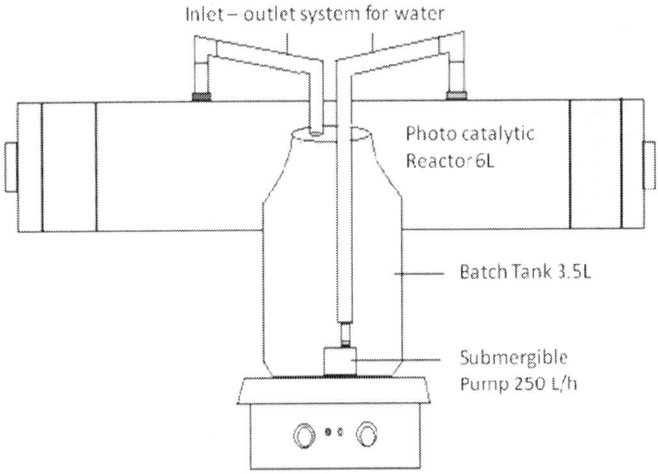

Fig. 4. Diagram of the Photocatalytic treatment units used at the laboratory

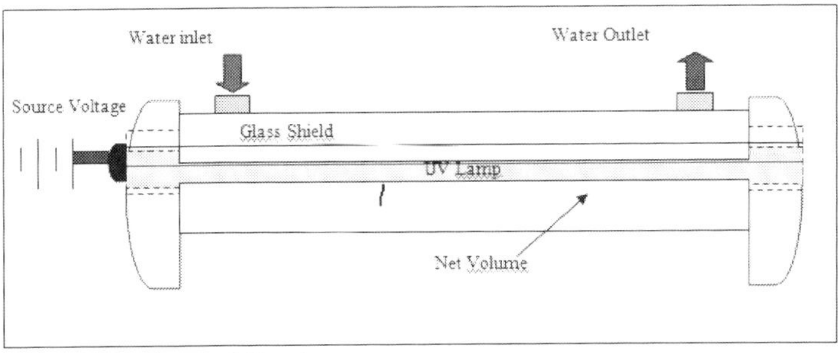

Fig. 5. Annular Photochemical Reactor

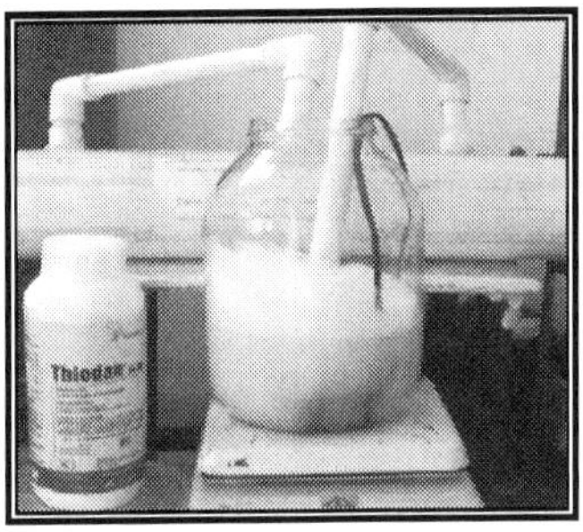

Copyright Forero,2008

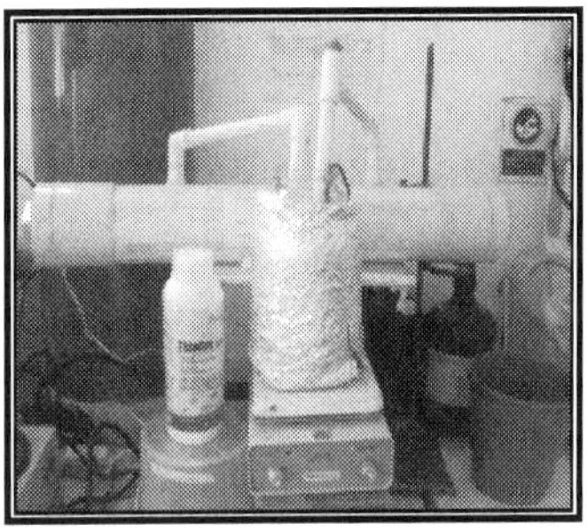

Copyright Forero,2008

Fig. 6. Photocatalytic System

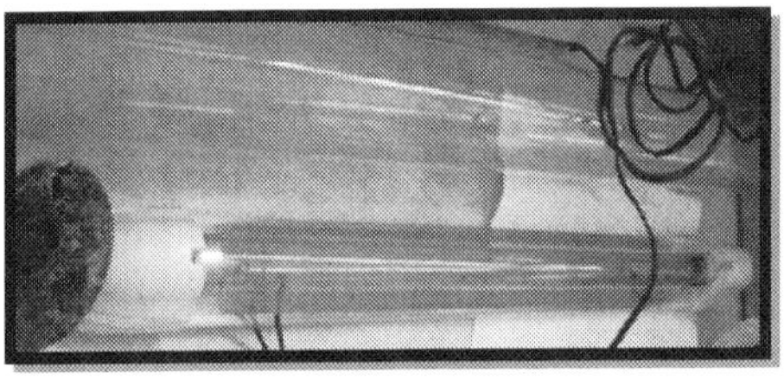

Copyright Caceres, 2008

Copyright Forero,2008

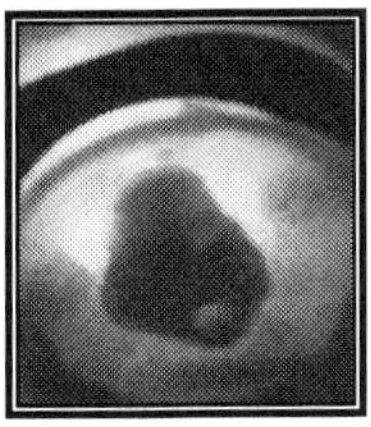

Copyright Forero, 2008

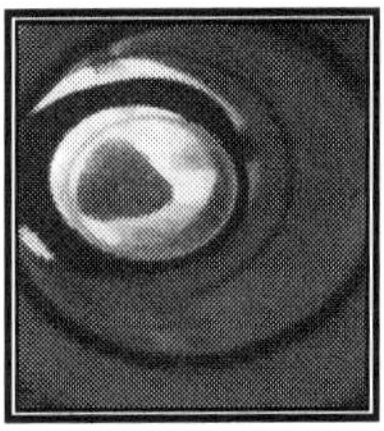

Copyright Forero, 2008

Fig. 7. UV Lamp

2.2.2 UV Lamp

The lamp used was a germicide UV lamp of mercury at low pressure. To protect the lamp from the water was used an isolation of Duran™ glass completely close in one side, the other side had a movable piece to do the maintenance. The wavelength used was 254 nm. The figure 7 shows the lamp.

2.2.3 Batch tank

To carry out the recirculation and homogenization of the chemicals, one glass container was used, because glass material is less reactive to chemicals than other materials. It was 25.8 cm of length and 50 cm of diameter.
The next figure shows the batch tank:

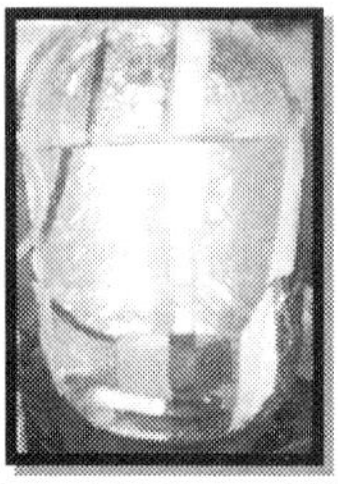

Copyright Forero,2008

Copyright Forero,2008

Fig. 8. Batch Tank

2.2.4 Recirculation system

The water recirculation was done by a submergible pump of 250L/h, with an input of 60Hz 3 Watts and Voltage of AC 110V -120V

2.2.5 Homogenization system

To obtain the homogeneity of the Photocatalytic system it was used a magnetic stir.

2.3 Chemicals

2.3.1 Catalyst

It was used Titanium Oxide TiO_2 Anatase type, because it has good photo adsorption properties. The concentrations used were between 100 and 500 mg/L and how this chemical is used in a solid form it was design a package to prevent the increment in the turbidity and the decrement of efficiency of the system. This package was elaborated with filter paper with 100 mm of pore diameter and 65 g/m^2 of surface area.

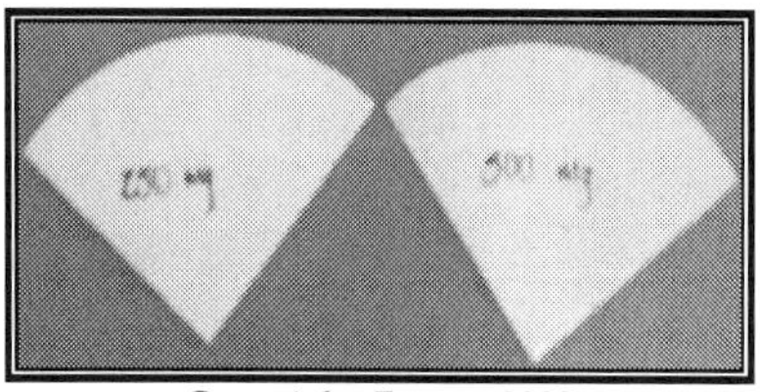

Copyright Forero, 2008

Fig. 9. Catalyst Package

2.3.2 Hydrogen Peroxide

In this kind of studies is common the used of Hydrogen Peroxide as strong oxidant. It was used 30% v/v and the concentrations in the system were from 1 to 10mL/L.

2.4 Other characteristics

2.4.1 Residence time

Is the time in which the pesticide will expose to radiation in its way inside the reactor, it was determined 120 min per trial because previous studies (Barrios 2005) suggested that.

2.4.2 Temperature

This process is not sensible to the temperature because the activation energy (kT= 0.026 eV), is very low compared with the activation energy of TiO_2 (3.2eV) and its contribution to generation e-h+ is very low. For this reason heat was not added to the system. The temperature of the system was 22°C. (Herrmann 1991)

2.5 Test performed

2.5.1 Chlorpyrifos

In the next table there are the concentration of Hydrogen Peroxide and Titanium Dioxide used in the degradation of Chlorpyrifos. The trials had an irradiation time of 120 min and 22°C, the 15th trial was divided in three parts 40, 80 and 120 min to check the influence of the irradiation time in the efficiency.

Concentrations Trial	Hydrogen Peroxide (H_2O_2) (mL/L)	Titanium Dioxide (TiO_2) (mg/L)
1	2.5	250
2	1	100
3	1.5	100
4	2	100
5	2.5	200
6	1	200
7	1.5	200
8	2	200
9	2.5	200
10	1	250
11	2	250
12	3	100
13	5	100
14	7	100
15A	10	100
15B	10	100
15C	10	100

Table 1. Concentration of Hydrogen Peroxide and Titanium Dioxide used in the degradation of Chlorpyrifos

In each trial was used 20 g of Attamix and 3L of the solution was added to the batch tank. The catalyst was adhered to the reactor and the Hydrogen Peroxide volume was added then. Was used and homogenization time of 10 min and the lamp was turned on. After 120 min an aliquot was extracted from the batch tank and analyzed by gas chromatography.

2.5.1.1 Sample Extraction

Each sample obtained from the reactor was extracted by the EPA 3510 Method using Dichloromethane (CH_2Cl_2). The solution pesticide-solvent was concentrated in Randall equipment to a 3 mL volume.

Fig. 10. Randall equipment

2.5.1.2 Gas chromatography

It was used a Thermo Finnigan model Trace GC Ultra Chromatograph, with a column RTX5 30m long and internal diameter 0.32mm and 0.5µm of phase, detector FID and an auto injector. The makeup gas was Nitrogen at 35 mL/min and the carrier gas Hydrogen at 30 mL/min. The injection volume was 2 µL and splitless mode. The standard was a Restek 18000235 and the temperature program was:

	Flow°C/min	Temperature °C	Heat Time min
Initial		60	1
Ramp 1	7	210	10
Ramp 2	5	250	0

Table 2. Temperature Program

2.5.2 Ensosulfan

To perform the degradation of Endosulfan, the tests were different than Chlorpyrifos tests, because this compound is totally prohibited in Colombia and the authorities find this chemical in different presentations and with different level of contamination. Below there are the trials done:

Hydrogen Peroxide concentrations: 1ml/L, 2ml/L

Titanium Dioxide concentrations: 250mg/L, 500mg/L

Reaction Time: 30, 60, 120 min

Pesticide concentrations: 1000, 750, 500, 250 y 100 mg/L

The system - reactor conditions were the same as with Chlorpyrifos, but the differences are that also for Endosulfan was used Homogeneous Photocatalysis, it means without catalyst and the extraction was done using Hexane.

The same trials were done by Homogeneous to compare with heterogeneous.

3. Results

3.1 Chlorpyrifos

The results were expressed by the Elimination Percent and it is 100 minus the relationship between the difference of the initial and final concentration over the initial concentration. In the next table are the concentrations and elimination percent of each trial. How it was used 20gr of pesticide, the initial concentration was 416 mg/L of Chlorpyrifos (the active ingredient).

TiO_2 Concentration (mg/L)	Hydrogen Peroxide (mL/L)	Trial	Pesticide Concentration (mg/L)	Elimination Percent (%)
100	1	2	0.646	84.5
100	1.5	3	0.633	84.81
100	2	4	0.612	85.31
100	2.5	5	0.569	86.35
200	1	6	0.751	81.98
200	1.5	7	0.702	83.15
200	2	8	0.644	84.55
200	2.5	9	0.638	84.69
250	1	10	0.765	81.64
250	2	11	0.731	82.46
100	3	12	0.562	86.51
100	5	13	0.553	86.73
100	7	14	0.523	87.45
100	10	15A	0.5	88
100	10	15B	0.497	88.08
100	10	15C	0.456	89.06

Table 3. Elimination percent for each trial performed

The table before shows the elimination percent increases if the volume of Hydrogen Peroxide increases. For 100 and 200 mg/L of Titanium Dioxide The figure 11 shows the elimination percent.

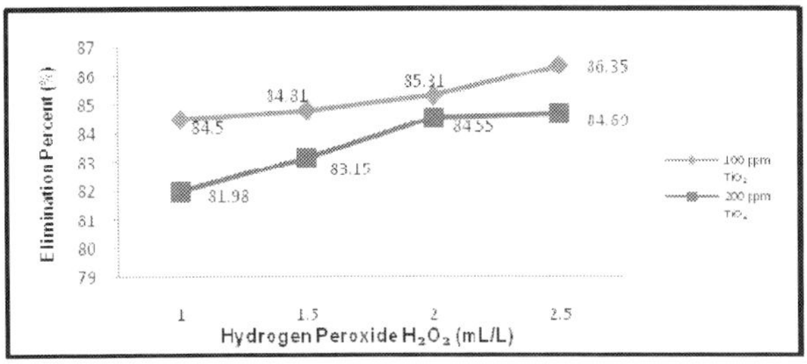

Fig. 11. Elimination Percent of Chlorpyrifos for 100 and 200 mg/L of TiO_2

The trials with 250 mg/L of Titanium Dioxide showed the lowest elimination percent, because the turbidity presented by the catalyst. The best elimination percent was 89.06% in the 15th trial with a concentration of Titanium Dioxide of 100 mg/L using Hydrogen Peroxide of 10mL/L with 120 min of irradiation UV at 254 nm. At 120 min of reaction time all the experiments showed more than 80% of elimination percent when the pesticide Attamix is degradated with the Heterogeneous Photocatalysis. With the better trial (15) were performed two additional test at 40 and 80 min and the best elimination was at 120 min, but the difference between them are so low (88 - 89%).

This part of the investigation concluded that the pesticide (Attamix SB) can be degraded efficiently using Heterogeneous photocatalysis.

3.2 Endosulfan

The results showed that two species were presented in the pesticide α and β endosulfan.

3.2.1 Homogeneous photocatalysis

The next figure shows the trials performed and the next table presents the elimination percents obtained:

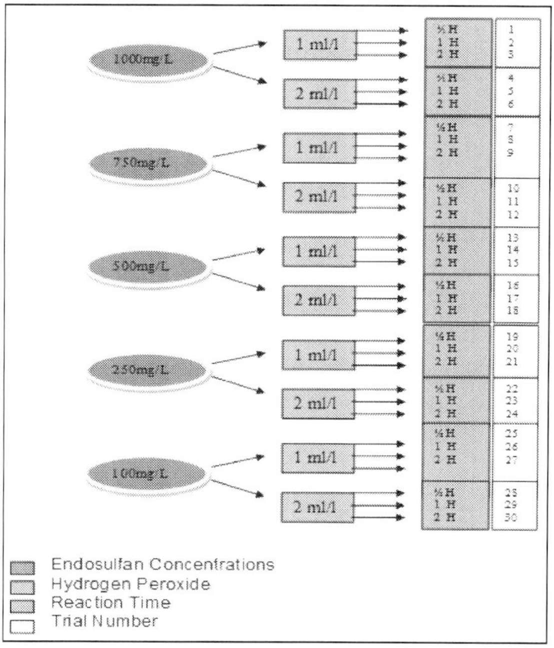

Fig. 12. Homogeneous Photocatalysis Trials

Pesticide Concentration	Trial	Elimination % *α - Endosulfan*	Elimination % *β - Endosulfan*
1000ppm	1	8,87	8,10
	2	12,76	11,48
	3	13,15	12,33
	4	8,88	8,59
	5	12,37	11,66
	6	13,58	12,50
750ppm	7	10,78	8,80
	8	11,12	10,93
	9	11,55	12,22
	10	10,90	7,75
	11	13,43	10,58
	12	13,66	13,05
500ppm	13	10,12	9,35
	14	12,32	10,87
	15	13,92	13,48
	16	10,94	11,65
	17	12,42	12,82
	18	13,72	13,36
250ppm	19	12,72	12,43
	20	16,08	14,93
	21	17,42	17,31
	22	12,80	12,66
	23	16,22	15,41
	24	17,44	17,87
100ppm	25	17,79	17,59
	26	18,37	18,65
	27	19,71	19,69
	28	18,07	17,61
	29	19,08	18,82
	30	19,80	19,70

Table 4. Elimination Percent of α and β Endosulfan using Homogeneous Photocatalysis

For the homogeneous photocatalysis system UV/H_2O_2, it was obtained the higher elimination percent for α y β Endosulfan of 19.7% y 19.8%, with the lowest concentration of pesticide 100 mg/L, with the highest reaction time 120 min and the highest volume of Hydrogen Peroxide 2 ml/L .

3.2.2 Heterogeneous photocatalysis

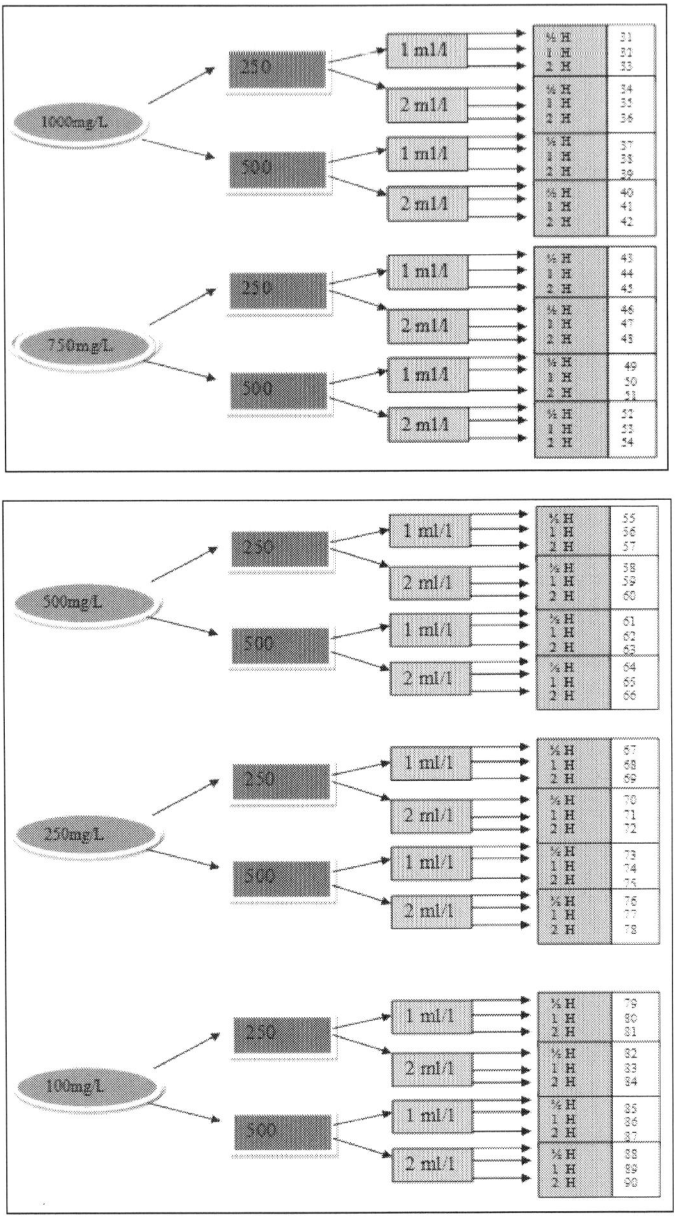

Fig. 13. Heterogeneous Photocatalysis Trials

Pesticide Concentration	Trial	Elimination % *a - Endosulfan*	Elimination % *β - Endosulfan*
1000ppm	31	62,49	64,72
	32	63,29	64,44
	33	63,28	65,28
	34	63,05	65,54
	35	63,08	66,09
	36	63,71	66,84
	37	61,72	61,00
	38	62,13	60,99
	39	61,92	61,35
	40	62,11	64,68
	41	62,32	64,89
	42	62,12	65,32
750ppm	43	62,68	67,93
	44	62,70	67,93
	45	62,81	67,98
	46	62,95	69,92
	47	63,19	69,89
	48	63,41	69,87
	49	62,14	66,22
	50	61,56	65,54
	51	62,36	64,87
	52	61,39	67,20
	53	61,15	67,58
	54	62,30	67,38
500ppm	55	72,66	66,22
	56	72,39	65,54
	57	75,31	64,87
	58	73,26	70,79
	59	73,89	70,82
	60	75,86	70,61
	61	71,13	59,22

USEPA. (2000). Action to Terminate Endosulfan *United States Environmental Protection Agency*. http://www.epa.gov/pesticides/reregistration/endosulfan/endosulfan-cancl-fs.html.

US EPA. (2002). *Reregistration Eligibility Decision for Endosulfan*, November 2002.

Weekly Times. (2009). Australia should ban endosulfan: Greens".. January 8, 2009. http://www.weeklytimesnow.com.au/article/2009/01/08/40315_horticulture.html.

World Health Organization (1984).Environmental Health Criteria 40.

Advanced Oxidation Processes (AOPs) for Removal of Pesticides from Aqueous Media

Marco A. Quiroz[1], Erick R. Bandala[1] and Carlos A. Martínez-Huitle[2]
[1]*Universidad de las Américas Puebla. Grupo de Investigación en Energía y Ambiente. Sta. Catarina Mártir, Cholula 72820 Puebla*
[2]*Universidade Federal do Rio Grande do Norte, CCET – Department of Chemistry, Lagoa Nova - CEP 59.072-970 – Natal, RN*
[1]*México*
[2]*Brazil*

1. Introduction

Advanced oxidation processes (AOPs) are technologies with significant importance in environmental restoration applications (Anipsitakis and Dionysiou, 2003; Bandala et al., 2007). The AOPs concept was established by Glaze et al., (Huang et al., 1993, Glaze, 1987; Glaze et al., 1987) who defined AOPs as processes involving the generation of highly reactive oxidizing species able to attack and degrade organic substances (Bolton, 2001). Nowaday AOPs are considered high efficiency physical-chemical processes due to their thermodynamic viability and capable to produce deep changes in the chemical structure of the contaminants (Domenech et al., 2004) via the participation of free radicals (Domenech et al., 2004). These species, mainly hydroxyl radicals ($HO^{\bullet}$), are of particular interest because their high oxidation capability (Andreozzi et al., 1999; Goswami and Blake, 1996; Huston and Pignatello, 1999; Legrini et al., 1993; Rajeshwar, 1996). However, other studies have suggested that, besides hydroxyl radicals, AOPs can also generate other oxidizing species (Anipsitakis and Dionysiou, 2003; 2004). Generated radicals are able to oxidize organic pollutants mainly by hydrogen abstraction (eq. 1) or by electrophylic addition to double bonds to generate organic free radicals ($R^{\bullet}$) which can react with oxygen molecules forming peroxyradicals and initiate oxidative degradation chain reactions that may lead to the complete mineralization of the organics, as proposed in eq. (1) (Blanco, 2003).

$$RH + HO^{\bullet}\left(or\ SO_4^{\bullet -}\right) \rightarrow HR^{\bullet} + H_2O \qquad (1)$$

Free radicals in AOPs, may be produced by photochemical and non-photochemicals procedures. Table 1 list some of the most frequently reported AOPs for application in water restoration.

Among the different approaches for pollutants removal from water, some of them are recognized as mainly efficient for pesticide degradation. Ozonation and ozone related processes (O_3/H_2O_2, UV/O_3), heterogeneous photocatalysis (TiO_2/UV), homogeneous photocatalysis (Fenton and Fenton-like processes) and electrochemical oxidation are considered as the most efficient for pesticide degradation in water (Somich et al., 1990; Scott

Non-photochemical AOPs	Photochemical AOPs
Alkaline media ozonation	Fenton and Fenton-like reactions
O_3/H_2O_2	Heterogeneous photocatalysis
Fenton reaction	UV/H_2O_2
Electrochemical oxidation	UV/O_3
Hydrodynamic/ultrasonic cavitation	
Sub/super critical water	

Table 1. Frequently reported AOPs

and Ollis, 1995; Zepp et al., 1994; Legrini et al. 1993, Bandala et al., 2002a; Arancibia et al., 2002; Masten and Davies, 1994; Chiron et al., 2000; Ikehata and El-Din, 2005; Ikehata and El-Din, 2006, Bandala and Estrada, 2007, Martínez-Huitle et al., 2008). Several different successful laboratory scale applications have been reported for many of these methodologies (Malato et al., 2004; Blanco et al.,2007), however only few full scale development are currently reported, pendent task mostly depending on deep knowledge and analysis of current results and the generation of new approaches to the engineering of the processes (Malato et al., 1999; 2000).

2. Pesticide degradation using photocatalysis

2.1 Heterogeneous photocatalysis (HP)

Photocatalysis have been defined by Kisch (1989) as the acceleration of a photoreaction by a catalyzer. To take place, homogeneous photocatalysis require that the catalyzer (usually a semiconductor) absorbs an energy quantum. After energy absorption, the absorber specie (C) generates energy carriers (e^- and h^+) and excited electrons are transferred to the oxidant (Ox_1). At the same time, the catalyzer accepts electrons from the reducer (Red_2) which fill the holes generated in valence band of the semiconductor. Electron flux in both directions is null and the catalyzer remains unaltered as proposed in reaction sequence (2) (Malato, 1999):

$$\begin{aligned} &C \xrightarrow{h\nu} C\left(e^- + h^+\right) \\ &h^+ + Red_2 \rightarrow Ox_2 \\ &e^- + Ox_1 \rightarrow Red_1 \end{aligned} \tag{2}$$

The heterogeneous photocatalytic degradation concept involves the use of a solid semiconductor (i.e. TiO_2, ZnO, others) to generate a colloidal suspension stable under radiation for stimulate a reaction in the solid/liquid (or solid/gas) interface. When the semiconductor is in contact with a solution containing a redox pair, charge transference occurs along the interface to balance chemical potentials between the two faces. Metallic oxides and sulfurs are among the most used semiconductor materials available for photocatalytic purposes. Norwadays, titanium dioxide (TiO_2) is the most frequently used semiconductors for heterogeneous photocatalytic processes anytime it has demonstrated to be the most active (Blake, 2000; Blanco et al., 2007). Table 2 depicts some of the semiconductor materials used in photocatalytic reactions along with their band gap energy required for catalyzer activation and the maximal wavelength required for activation.

Degradation of organic pollutants by HP is among the most successful applications of the AOPs as suggested by the wide variety of research groups, installations, references and patents for use of this technology for removing toxic substances in water (Ajona and Vidal,

Material	Band gap energy (eV)	Activation wavelength (nm)
$BaTiO_3$	3.3	375
CdO	2.1	590
CdS	2.5	497
CdSe	1.7	730
Fe_2O_3	2.2	565
GaAs	1.4	887
GaP	2.3	540
SnO_2	3.9	318
$SrTiO_3$	3.4	365
TiO_2	3.2	387
WO_3	2.8	443
ZnO	3.2	390
ZnS	3.7	336

Table 2. Band gap energy and activation wavelength for some semiconductors (Malato, 1999).

2000; Blake, 2000; Bandala and Estrada, 2007). But, the use of heterogeneous photocatalysis for restoration of water contaminated with pesticides has been shown as one of the best fields for application of this technology. It is proposed as an ideal methodology because it can be used for low concentration effluents or complex multicomponent commercial suspensions. Its success application has being recognized by GEF as a promising innovative technology for the destruction and decontamination of Persistent Organic Pollutants (POPs) in developing countries (McDowall et al., 2004). The number of tested pesticides for heterogeneous photocatalytic degradations is wide. Among them, chlorinated, phosphorated, carbamic, thiocarbamic and triazine type pesticides are the most frequently reported. Table 3 shows an actualized reference collection of works published for pesticide degradation using TiO_2 mediated photocatalytic degradation in recent years. This Table shows the importance on the treatment of this type of pollutans, due to the extensive use.

Pesticide	**References**	**Pesticide**	**References**	**Pesticide**	**References**
Aldrin	Bandala et al., 2002; Ormad et al., 2010	DMMP	O´Shea,1997.	Permethrin	Chiaranzelli et al., 1995
Acrinatrin	Malato et al., 2000a; Malato et al., 2004	3,4-DPA	Pathirana,1997	Phorate	Chen et al., 1996: Hisanaga et al., 1990
Alachlor	Chiron et al., 1997; Moza et al.,1992;Muszkat et al., 1992, 1995; Wong and Chu, 2003a,b; Hincapié et al., 2005; Ormad et al., 2010; Farre et al., 2005	Endosulfan and derivatives	Ormad et al., 2010	Pyrimethanil	Oller et al., 2006
Aldicarb	Parreño et al., 1994.	Endrin	Ormad et al., 2010	Pirimiphos-methyl	Herrmann et al., 1999
Ametryn	Ormad et al., 2010	EPTC	Mogyoródi et al., 1993. Vidal et al., 1991	Procimidona	Hustert and Moza, 1997

Pesticide	References	Pesticide	References	Pesticide	References
Asulam	Tanaka et al., 1992.	Fenitrothion	Chiron et al., 1997, Tanaka et al., 1992; Herrmann,1999, Hasegawa,1998; Tapalov et al., 2003; Mahmoodi et al., 2008	Prometon	Borio et al., 1998;Herrmann, 1999c;Pelizzetti, 1990b, 1993; Ormad et al., 2010
Atrazine	Parra et al., 2004; Clestur et al., 1993; Lackhoff and Niessner, 2002; McMurray et al., 2006; Campanella and Vitalliano, 2006; Zhang et al., 2006; Bellobono,1995 Chiron et al., 2000; Herrmann,1999; Minero et al., 1996b;Muszkat etal.,1992,1995; Pelizzetti, 1987,1990a,1990b,1991,1992,1993. Sullivan et al., 1994; Texier,1999a, 1999b; Parra et al., 2004; Ormad et al., 2010; Farre et al., 2005	Fenobucarb	Hasegawa, 1998.	Prometryn	Muszkat et al., 1992, 1995; Pelizzetti, 1990b, 1993; Borio et al., 1998; Evgenidou et al., 2007; Ormad et al., 2010
Azinfphos-methyl	Domínguez,1998; Calza et al., 2008	Fenuron	Richard and Bengana, 1996.	Propachlor	Muszkat et al., 1995; Konstantinou et al., 2002; Muneer et al., 2005
Bendiocarb	Hasegawa, 1998.	Imidachloprid	Chiron et al., 1997; Texier et al., 1999a; 1999b; Agüera et al., 1998; Fernández et al., 1999; Sharma et al., 2009	Propanil	Sturini at al., 1997; Konstantinou et al., 2001
Carbaryl	Arancibia et al., 2002; Gelover et al., 2004	HCH and derivatives	Ormad et al., 2010	Propazine	Muszkat et al., 1992, 1995; Pelizzetti, 1992; Ormad et al., 2010
Carbetamid	Brun, 1995; Percherancier et al., 1995.	Heptachlor	Ormad et al., 2010	Propetryne	Herrmann, 1999
Chlorfenvinfos	Farre et al., 2005; Ormad et al., 2010	Iprobenfos	Hasegawa, 1998.	Propoxur	Lu et al., 1995, 1999
Chlorpyrifos	Ormad et al., 2010	Isoprothiolane	Hasegawa et al., 1998.	Propyzamide	Chiarenzelli et al., 1995; Hasegawa, 1998; Torimoto et al., 1996.

Pesticide	References	Pesticide	References	Pesticide	References
Carbofuran	Kuo and Lin, 2000; Mahalakshmi et al., 2006; Mansour, 1997; Tennakone, 1997	Isoproturon	Amorisco et al., 2005. Mansour, 1997; Farre et al., 2005; Sharma et al., 2008a,b,c,d; 2009; Ormad et al., 2010; Haque and Muneer, 2003	Simazine	Hasegawa, 1998.;Pelizzetti et al., 1990b;1992;1993; Ormad et al., 2010
Cyanobenzoate	Muszkat et al., 1995	Lindane	Chiron et al., 1997; Herrmann, 1999c; Sabin , 1992; Guillard et al., 1995; Vidal, 1998; Zaleska et al., 2000.	2,4,5-T	Barbeni et al., 1987; Chiron et al., 1997; Ollis et al., 1991a; Pelizzetti, 1993; Kamble et al., 2006
Cycloate	Vidal et al., 1999; Mogyoródi et al., 1993; Vidal, 1991	Malathion	Muszkat et al., 1995; Mak and Huang, 1992; 1993; Doong and Chang 1997	2,3,6-TBA	Bianco-Prevot et al., 1999
Chloroxynil	Muszkat et al., 1992	Manuron	Herrmann et al., 1999	Terbutylazina	Mansour et al., 1997
Chlorpyriphos	Picaht et al., 2007;Chiarenzelli et al., 1995.	MCC	Tanaka et al., 1999	Terbutryn	Muszkat et al., 1992; Ormad et al., 2010
Chlorsulfuron	Fresno et al., 2005; Maurino et al., 1999	Metamidophos	Doong and Chang, 1997; Hisanaga et al., 1990; Malato et al., 1999	Tetrachlorophenol	Pelizzetti, 1985
2,4-D	Terashima et al., 2006: Sanjay et al., 2004; Singh and Muneer, 2004; Chiron et al., 1997; 2000; D´Oliveira,1993a; Herrmann et al., 1998;1999; Lu et al., 1995,1997; Martin et al., 1995; Müller, 1998;Pichat et al.,1993a;1993b; Trillas et al., 1995;Sun and Pignatello, 1995; Kamble et al., 2006	Metamitron	Mansour ,1997.	Tetrachlorvinphos	Herrmann,1999; Kerzhentsev et al., 1996
DBS	Domínguez et al., 1998;	Metolachlor	Sakkas et al., 2004; Chiron et al., 1997; Ormad et al., 2010	Tetradifon	Chiron et al., 1997; Ormad et al., 2010
DCB	Muszkat et al., 1992.	Metobromuron	Amine et al., 2005; Muszkat et al., 1992.	Thiram	Hasegawa, 1998.
DDT and DDT derivatives	Borello et al., 1989; Chiron et al., 1997; Herrmann,1999c; Pelizzetti, 1985; 1993;Sabin, 1992; Zaleska et al., 2000; Ormad et al., 2010	Methoxychlor	Ormad et al., 2010	Tifensulfuron-methyl	Maurino et al., 1999.

Pesticide	References	Pesticide	References	Pesticide	References
DEMP	O´Shea, 1997a	MIPC	Tanaka et al., 1999	Thiobencarb	Nishida and Ohgaki, 1994.
DEP	Tanaka et al., 1992; Hisanaga et al., 1990; Muneer et al., 1998.	MMPU	Muszkat et al., 1992.	Thiocarbaryl	Nishida and Ohgaki, 1994.
Diazinon	Sakkai et al., 2005; Mahmoodi et al., 2007;Doong and Chang, 1997;Hasegawa, 1998.; Mak,1992; Mansour,1997; Kouloumbos et al., 2003;	Molinate	Mogyoródi et al, 1993; Konstantinou et al., 2001; Ormad et al., 2010;	Triadimefon	Chiarenzelli et al., 1995.
Dichloran	Chiarenzelli et al., 1995.	Monocrotophos	Shankar et al., 2004;Chen et al., 1996	Trifluralin	Ormad et al., 2010
Dichloroaniline	Muszkat et al., 1995.	Monuron	Augliaro, 1993; Pramauro et al., 1993	Trichlorophenol	Barbeni et al., 1986;D´Oliveira et al., 1993; Jardim et al., 1997; Ollis et al., 1991a; Pelizzetti, 1985; 1993; Tseng and Huang, 1991; Tanaka et al., 1992
Dichlorophenol	Bhatkhade et al., 2004; Kim and Choi, 2005; Boyarri et al., 2005; Texier et al., 1999a,b;Jardim et al., 1997; Manilal, 1992	MPMC	Tanaka et al., 1999.	Trichlopyr	Poulius, 1998; Qamar et al., 2006
Dichloropyridine	Kyriacou et al., 1997	MTMC	Tanaka , 1999.	Trietazine	Muszkat et al., 1992, 1995
Dichlorvos	Hasegawa et al., 1998; Chen et al., 1996; Chen, 1997; Lu, 1993, 1995; Mak et al., 1992, Mak and Hung, 1993; Hisanaga et al., 1990; Evgenidou et al., 2005;2006; Oancea and Oncescu, 2008	Oxamil	Texier 1999a.; Malato et al., 2000b; Oller et al., 2006	Vernolate	Mogyoródi et al., 1993; Vidal, 1991.
Dicofol	Chiron et al., 1997; Ormad et al., 2010	Paraquat	Florencio et al., 2004;Moctezuma, 1999.	Vinclozoline	Hustert and Moza, 1997.
Dieldrin	Ormad et al., 2010	Parathion	Zoh et al., 2005, 2006;Chen et al., 1996; Chiron et al.,1997; Herrmann, 1999; Sakkas et al., 2002	Pendimetalin	Mansour, 1997; Moza et al., 1992
Diquat	Florencio et al., 2004; Kinkennon et al., 1995.	Paration-metil	Sakellarides et al., 2004; Zoh et al., 2005,2006;Chiron et	XMC	Tanaka et al. , 1999

Pesticide	References	Pesticide	References	Pesticide	References
			al., 1997, 2000; Evgenidou et al., 2007b; Ormad et al., 2010		
Dimethoate	Oller et al., 2006;Domínguez et al., 1998;Evgenidou et al., 2006; Oller et al., 2006; Ormad et al., 2010	PCDD	Barbeni et al., 1986; Pelizzetti, 1985		
Diuron	Canle et al., 2005; Kinkennon et al.,1995;Muneer et al., 1998; Farre et al., 2005;Katsumata et al., 2009;Macounova et al., 2003;Ormad et al., 2010	PCDF	Barbeni et al., 1986; Pelizzetti, 1985.		

Table 3. References on heterogeneous photocatalytic degradation of pesticides in water using TiO_2.

2.2 Kinetics and reaction mechanisms

For an extended period of time different works analyzing heterogeneous photocatalysis mechanisms have proposed hypotheses on the generation of photoproduced holes (h^+) and surface trapped hydroxyl radicals ($HO^\bullet$) (Romero et al., 1999). Initial steps involved in band-gap irradiation of TiO_2 particles (or any other semiconductor) have been studied in detail by lasser-flash photolysis measurements (Bahnemann et al., 1997; Serpone, 1996). It is well stablished that TiO_2 illumination with radiation of the proper wavelength ($\geq E_g$) generates electron/hole pair which can recombine or dissociate (both reactions are in competition) to produce, in the latter case, a conduction band electron and a valence band hole which are able to migrate to the particle surface. Once in the surface, both charge carriers will be able to interacting with adsorbed electron acceptors and oxidize electron donors. In the heterogenous process in aqueous face, oxygen is often present as electron acceptor and HO^- and H_2O are available as electron donors to yield hydroxyl radicals. It is well documented that these trapping reactions occurs in less than 30 ps (Colombo et al., 1995; Skinner et al., 1995; Serpone et al., 1995).

Considering the importance of mass transference in the process, initial practical approaches to quantitative description of HP kinetics has been commonly carried out using a Langmuir-Hinshelwood (L-H) kinetics model (Al-Ekabi et al., 1988, 1989). This mathematical model assumes that the reaction occurs on the catalyst surface. According to L-H model, the reaction rate (r) is proportional to the fraction of particle surface covered by the pollutant (θ_x). Mathematically,

$$r=-\frac{dC}{dt}=k_r\theta_x=\frac{k_rKC}{1+KC+K_sC_s} \quad (3)$$

where k_r is the reaction rate constant, K is the pollutant adsorption constant, C is the pollutant concentration at any time, K_s is the solvent adsorption constant and C_s is its concentration. During eighties, many authors presented their data using L-H kinetic approach (Chen et al., 1983; Herrmann et al., 1983; Matthews, 1988; Nguyen and Ollis, 1984; Ollis, 1984; Pruden and Ollis, 1983). Nevertheless, despite L-H approach fits properly experimental data, it does not consider the interaction of the radiation field (Bandala et al., 2004; Arancibia et al., 2002).

Other kinetic studies on heterogeneous photocatalysis suggest that reaction rate increases with catalyst concentration to get a maximum value for catalyst concentration between 0.2 and 1 g/L, depending on the compound and the reactor used. Over these concentrations, reaction rate remains unaffected or decreases when catalyst concentration increases (Jimenez et al., 2000; Arancibia et al., 2002; Curco et al., 1996; Gimenez et al., 1999). An interesting problem is the relation between catalyst concentration, reaction rate, radiation absorption and process improvement, because, several studies have suggested important associations depending on the catalyst radiation absorbed (Schiavello et al., 1999; Brandi et al., 1999, Arancibia et al., 2002; Bandala et al., 2004). From these results, several models, most of them based on complex mathematical or static computational approaches, have been developed and proposed in order to predict radiation absorption and scattering as function of catalyst concentration, optical path and catalyst type and its relation to pseudokinetic constants experimentally obtained (Bandala et al., 2004; Arancibia et al., 2002; Curco et al., 2002). Based on the radiation absorbed by the catalyst, some authors, as Cassano's group considered the most representative in the field, have considered that the vital point in this process resides on the *a priori* design of photochemical reactors, that improve of HP reactions and the generation of intrinsic reaction kinetic that may lead to process scaling-up (Alfano et al., 2000; Cassano and Alfano, 2000; Romero et al., 1999; Brandi et al., 2000).

Besides reactor design, heterogeneous photocatalytic degradation reaction can be enhanced by the use of higher active catalyst or inorganic oxidizing species. In the first case, activation of TiO_2 under visible light is a desirable technological approach. In order to utilize visible light for TiO_2 excitation, several dye-synthesized and ion-doped TiO_2 have been developed achieving higher performances in their use for photocatalyzed degradation of different organic substrates (Bae and Choi, 2003; Lin et al., 2006; Xu et al., 2002; Iwasaki et al., 2000; Asah et al., 2001; Irie et al., 2003; Burda et al., 2003) using the band gap narrowing effect produced. However, only few recent reports deals with application of visible light activated TiO_2 to photoassisted pesticide degradation (Senthilnathan and Philip, 2010; Sojicetal, 2010).

2.3 Effect of oxidizing species on the reaction rate

According to reaction sequence 2, production of charge carriers is a fundamental step in degradation processes using HP. Once generated, these species may lead to hydroxyl radicals generation (and the subsequent organic matter degradation) or can recombine to generate the initial state and energy emission. This latter reaction, known as recombination, is a practical problem when using TiO_2 catalyst and it is extremely efficient (reaction rate = 10^{-9} s) when no proper electron acceptor is present in the reaction media (Malato et al., 1998; Hoffman et al., 1995). This side process is energy-wasting and limiting to get high quantum yield (i.e. number of primary chemical reactions per photon absorbed). In most of the cases, dissolved oxygen is used as electron scavenger in these processes and several works have dealed on its efficiency as oxidant agent to complete organic matter mineralization (Li Puma et al., 1993; Martin et al., 1995; Mills et al., 1993; Ollis et al., 1991; Pelizzetti and Minero, 1993). Nevertheless, it has been demonstrated that only low mineralization is reached when dissolved oxygen is used as oxidant agent in, for example, the photoassisted degradation of pesticides (Mills and Morris, 1993; Serra et al., 1994; Minero et al., 1996). Several previous studies have investigated the role of alternative electron acceptors such as peroxide compounds (Wang and Hong., 1999; Wong and Chu, 2003; Dionysiou et al., 2004). Among them, hydrogen peroxide has been identified as widely used to improve photocatalytic processes. This simple peroxide is considered as environmentally friendly and of great interest for "green" chemistry and engineering applications (Ghosh et al., 2001). Hydrogen peroxide has been applied to enhance the rates of TiO_2 photocatalytic reactions (Madden et al., 1997; Pacheco et al., 1993; Malato et al., 1998; Wang and Hong, 1999; Doong and Chang, 1997; Wong and Chu, 2003) using UV radiation (Mengyue et al., 1995; Haarstrick et al., 1996; Pacheco et al., 1993; Malato et al., 1998). The improvement of photocatalytic rates using H_2O_2 has been attributed to many factors, mainly: hydrogen peroxide is better electro acceptor than oxygen (Ollis et al., 1991; Madden et al., 1997; Malato et al., 1998; Peterson et al., 1991; Cornish et al., 2000; Ohno et al., 2001), its potential for reduction is 0.72 V while this value for oxygen reduction is - 0.13 V (Cornish et al., 2000), it is considered able to favor photocatalytic mechanisms by the removal of photogenerated electrons in the conduction band (Dionysios et al., 2004). Nevertheless it has been well documented that, at high concentrations of H_2O_2, it can compete for adsorption with organic matter (Dionysios et al., 2004; Bandala et al., 2002; Sauer et al., 2002; Cornish et al., 2000). Besides hydrogen peroxide, other oxidant agents have been tested for improve photocatalytic reactions (Martin et al., 1995; Pelizzetti et al., 1991; Al-Ekabi et al., 1992; Kenneke et al., 1993). For example, peroxidisulphate ($S_2O_8^{2-}$) has been indicated as an important oxidant, allowing drastical improvements in the TiO_2 photocatalyzed mineralization of pesticides and pesticide mixtures by Malato et al. (1998; 1999; 2000) and they think its use is justified when pesticide mineralization is the major concern.

2.4 Material science implications: slurries or immobilized photocatalyst

Generation of catalyst sludges is among main disadvantages for HP processes in water treatment. This kind of treatment, currently available at pilot-plant level, uses suspended TiO_2 in photoreactors where the semiconductor is recovered after the treatment (Malato et al., 2000; 2002). According to various lab scale research reports (Bideau et al., 1995; Matthews and McEvoy, 1992; Sabate et al., 1992; Chester et al., 1993), the use of TiO_2 in suspensions is more efficient than on its immobilized form. Nevertheless, this latter form posses specific advantages, such as cost reductions, material losses decrease and skipping recovery steps in the process, which make desirable the generation of immobilized titania photocatalyst with higher efficiency as compared with those reported to date (Balasubramanian et al., 2004; Gelover et al., 2004).

Several supporting materials, from sand to quartz optical fiber, have been reported so far for TiO_2 immobilization. In the same way, a wide number of methods for catalyst fixation as reviewed by Pozzo et al., (1997). In last years, the use of *in situ* catalyst generation method seems to be the most promising technology for catalyst immobilization (Rachel et al., 20002; Guillard et al., 2002; Gelover et al., 2004). Other authors (Guillard et al., 2003; Gelover et al., 2004) has demonstrated that, by the use of these *in situ* catalyst generation method, fixed form of titanium dioxide generated present equal efficiency as Degussa P-25 (considered as the most efficient form of titanium dioxide) suspended catalyst for pesticide degradation. However, more scientific research is necessary about the development of this promising idea before it can be considered for future design of efficient photocatalytic plants.

2.5 Homogeneous photocatalysis

Homogeneous photocatalysis refers to those photocatalytic processes in which the catalyst is dissolved in water during the redox process. In general, homogeneous processes can be represented as depicted in reaction sequence (4) (Domenech et al., 2004):

$$\begin{aligned} & C \xrightarrow{h\nu} C^* \\ & C^* + R \rightarrow R^* + C \\ & R^* \rightarrow P \end{aligned} \tag{4}$$

Similarly to heterogeneous photocatalysis, homogeneous processes are based in the generation of hydroxyl radicals but, in difference, some other highly oxidant species can be generated and be responsible of organic contaminant degradation (Anipsitakis and Dionysiou, 2004; Yamazaki and Piette, 1991; Sawyer et al., 1996). Since the well known Fenton's experiments in the latest XIX century, it is documented that hydrogen peroxide/ferrous salts solutions are capable to oxidize organic compounds (Fenton, 1894). Fenton reagent has been reported of high efficiency degrading aliphatic hydrocarbons, halogenated aromatics, polychlorinated byphenils, nitroaromatics, azo-dyes and pesticides (Bigda, 1995) as shown in Table 4.

3. Fenton-like reactions

Besides Fenton reaction, several Fenton-based procedures have been developed, being these reactions, inspired on the Fenton reaction chemistry (so-called Fenton-like processes). It has been demonstrated that, in many of the cases, Fenton-like processes are more efficient than Fenton reaction to water treatment and will, probably, be the next step in the scaling-up of AOPs application to pesticide treatment in water.

When Fenton reaction involves ultraviolet radiation, visible light or both, the reaction is known as the photo-Fenton process. Compared with dark Fenton reaction, photo-Fenton process has numerous advantages such as the increase of degradation rate, minimize in sludge generation and the use of solar energy, among others (Malato et al., 2002; De Laat and Le Troung, 2006; Chacón et al., 2006, Orozco et al., 2008). Photo-Fenton process is among the most efficient methods to generate hydroxyl radicals (Bauer et al., 1999). Even higher than other very well studied and widely applied AOPs such as TiO_2/UV and H_2O_2/UV as shown in comparative studies using 4-chlorophenol as model wastewater contaminant (Krutzler and Bauer, 1999). Many parameters, such as initial concentration of ferric salt and hydrogen peroxide, the ratio of $[H_2O_2]_0/[Fe(II)]_0$, pH, light intensity and temperature influence on the efficiency of photo-Fenton process (Bandala et al., 2007; Lee and Yoon, 2004) are determinants in the efficiency.

Except for Fenton reagent, the potential of generating highly reactive radical species using transition metals coupled with electron acceptors have not been explored completely for water treatment (Anipsitakis and Dionysiou, 2004). Recently, Anipsitakis and Dionysiou (2004) have carried out experiments in order to identify radical generation by the interaction of transition metals with common oxidants. They tested 14 different combinations of metals and oxidant and found that cobalt (II)/potassium peroximonosulfate (Co/PMS) system posses very attractive characteristics for water decontamination (Anipsitakis and Dionysiou, 2004). This homogeneous system have been shown to be able for generate sulfate radicals and demonstrate greater efficiencies when compared with Fenton reagent for the treatment of water containing organic pollutants (Anipsitakis and Dionysiou, 2004).

Pesticide	References	Pesticide	References	Pesticide	References
Abamectin	Fallaman et al., 1999			Metoxichlor	Houston and Pignatello, 1999; Pignatello and Sun, 1995
Acephate	Yu, 2002	Dichlorophenol	Wadley and Waite, 2002; Aaron and Oturan, 2001; Detomaso et al., 2003; Momani et al., 2006; Momani, 2006; Bayarri et al., 2007	Metolachlor	Malato et al., 2002, 2003
Acrinatrin	Fallaman et al., 1999	Dimethoate	Nikolaki et al., 2005; Oller et al., 2005	Metomyl	Wang et al., 2004; Scherer et al., 2004; Muszkat et al., 2002
Alachlor	Houston and Pignatello, 1999; Laperlot et al., 2006; Farre et al., 2007,2005; Wang and Lemley, 2001; Hincapie et al., 2005; Perez et al., 2006	Diuron	Malato et al., 2002; Lapertot et al., 2006; Farre et al., 2007; Hincapié et al., 2005; Perez et al., 2006; Farre et al., 2005; Malato et al., 2003; Edelahi et al., 2004	Metribuzin	Yu, 2002
Aldicarb	Houston and Pignatello, 1999	DMDT	Barbusinski and Filipek, 2001	Metamidophos	Fallaman et al., 1999
Atrazine	Bandala et al., 2007; Houston and Pignatello, 1999; Sun and Pignatello, 1993; Laperlot et al., 2006; Wang et al., 2003; Farre et al., 2007; Ostra et al., 2007; Pignatello, 1993; Arnold et al., 1995; Adams et al., 1990; Ijpelaar et al., 2000; Hincapie et al., 2005; Perez et al., 2006; Farre et al., 2005	3,4-DPA	Saltmiras and Lemley, 2000	Parathion-ethyl	Oturan, 2003
Azinphos-methyl	Houston and Pignatello, 1999	Ethylene thiourea	Fallaman et al., 1999	Paration-methyl	Pignatello and Sun, 1995; Roe and Lemley, 1997; Gutierrez et al., 2007
Bromacil	Muszkat et al., 2002	Endosulfan	Yu, 2002	Pentachlorophenol	Farre et al., 2007
Carbaryl	Kong and Lemley, 2006; Wang et al., 2003	Edifenphos	Barbusinski and Filipek, 2001; Yu, 2002; Badaway et al., 2006	Pyrimethanil	Fallaman et al., 1999

Carptan	Houston and Pignatello, 1999	Fenitrothion	Fallaman et al., 1999; Malato et al., 2002; Malato et al., 2003	Pichloram	Houston and Pignatello, 1999
Carbofuran	Houston and Pignatello, 1999; Wang et al., 2003	Formetanate	Houston and Pignatello, 1999	Profenophos	Badawy et al., 2006
Chlorfenvinphos	Farre et al., 2005; Hincapié et al., 2005; Lapertot et al., 2006; Barbusinski and Filipec, 2001	Glyphosate	Barbusinski and Filipek, 2001	Propamocarb	Fallaman et al., 1999
Chlorophenol	Krutzler et al., 1999; Detomaso et al., 2003	HCH	Yu, 2002	Simazine	Houston and Pignatello, 1999; Adams et al., 1990
Chlorotalonil	Gutierrez et al., 2007	Fenthion	Fallaman et al., 1999; Malato et al., 2002; Malato et al., 2003	2,4,5-T	Pignatello, 1993; Aarón and Oturan, 2001; Pignatello, 1992; Sun and Pignatello, 1993
Chlorpyriphos	Yu, 2002	Imidachloprid	Lapertot et al., 2006; Farre et al., 2007; Hincapie et al., 2005; Farre et al., 2005	Tebuconazole	Faxeira et al., 2005
4-chloro-phenoxia-cetic acid	Sedlak et al., 1992	Isoproturon	Fallman et al., 1999	Tamaron	Faxeira et al., 2005
2,4-D	Oturan et al., 1999; Pignatello, 1992; Sun and Pignatello, 1993; Bandala et al., 2007; Sun and Pignatello, 1992; Sun and Pignatello 1993a,b; Wang et al., 2003; Wang and Lemley, 2001; Aaron and Oturan, 2001; Kong and Lemley, 2006	Luteron	Roe and Lemley, 1997; Oturan, 2003; Houston and Pignatello, 1999	Treflan MTF	Saltmiras and Lemley, 2001
Dicamba	Houston and Pignatello, 1999	Malathion	Houston and Pignatello, 1999	Terbutryn	Adams et al., 1990
Disulfoton	Houston and Pignatello, 1999	MCPP	Aaron and Oturan, 2001	Tetraethyl pyrophosphate	Oturan, 2003
DDT	Barbusinsky and Filipek, 2001; Bousahel et al., 2007	Metamidophos	Gutierrez et al., 2007; Yu, 2002; Fallaman et al., 1999; Faxeira et al., 2005	Trichlorophenol	Aaron and Oturan, 2001
Diazinon	Wang et al., 2003; Badaway et al., 2006; Yu, 2002			Trifluralin	Wang et al., 2003

Table 4. References on homogeneous photocatalytic degradation of pesticides in water

3.1 Effect of metal counterion

An interesting effect that should be take account when applying homogeneous photocatalysis is salt counterion. Inorganic anions (Cl^- , SO_4^{2-} , HPO_4^{2-})in wastewater or added as reagents have a significant effect on the reaction rate in the case of Fenton process. These effects are a) complexation with Fe(II) or Fe(III), affecting iron species reactivity and distribution; b) Precipitation reactions leading to a decrease of the active dissolved Fe(III); c) Scavenging of hydroxyl radicals and d) oxidation reactions involving these inorganic radicals. It have been well documented that cloride ions shows inhibitory effect for oxidation reactions, using both Fe(II) and Fe(III), of phenols (Tang and Huang, 1996), dichlorvos (Lu et al., 1997), atrazine (De Laat et al., 2004) and azo-dyes (Orozco et al., 2008). On the other hand, the effect of inorganic salt counterion in cobalt-mediated Fenton-like processes is not completely clear. It has been shown that precence of chloride ions produce highly chlorinated intermediates during oxidation process probably due to chloride radicals generation (Anipsitakis et al., 2006). The presence of sulfate or nitrate ions did not show any effect on reaction rate. The effect of organic counterion for cobalt salts can be related with the availability to cobalt (II) re-generation during oxidation processes and enhancing of reaction rate by radiation. Currently, we are testing the effect of several organic cobalt salts in the degradation rate of the herbidice 2,4-D and observed that the counterion effect is very important on the global reaction rate.

4. Radiation source

In homogeneous and heterogeneous photocatalysis, radiation is identified as a very important supply to the overall process. Two main radiation sources have been used to promote these processes: artificial radiation and solar radiation. The use of artificial radiation (generally a high pressure mercury or xenon arc lamp) sources has been widely applied for pesticide degradation by mean of different photochemical processes, among them homogeneous or heterogeneous photocatalysis (Chiron et al., 2000). In recent years, application of photocatalytic processes using solar radiation has increased as a cost-effective alternative for these technologies. It is interesting note that, actual industrial/commercial applications developed recently are related to solar enhanced processes (Blanco and Malato, 2003).

Different to solar thermal processes, where large amounts of radiation of any wavelength is collected, in solar photocatalytic processes only high-energy radiation is able to be used to promote photochemical reactions (i.e. λ<600 nm). This selective wavelength range produce that only very specific solar collection geometries can be useful to be applied for solar driven photocatalytic reactions. Several different solar collector geometries have been tested for application to solar photocatalytic processes (both, homogeneous and heterogeneous) and a wide number of works dealing with the comparison between all these experimental results have been reported (Bandala and Estrada, 2007). From all these information, the actual consense is that low concentration collectors seems to be the best technological option instead of earlier high concentration designs (Blanco et al., 2007; Bandala and Estrada, 2007). In particular, compound parabolic concentrators (CPCs) have been identified as very promising technological approach to industrial application of solar photocatalysis. CPCs combine the characteristics and advantages of high range concentrators and static flat systems. Among their main advantages are use of global solar radiation, absence of tracking systems, low evaporation of volatile compounds, low cost and high optical and quantum efficiencies conditions. Some authors have reported the comparison some solar collection geometries and found that V trough concentator is able to perform solar photocatalytic processes in practically equivalent conditions than widely reported CPCs (Bandala and Estrada, 2007). This solar collection geometry have not being tested enough for solar chemistry applications but, as far as we can see, could be an interesting alternative anytime the actual solar collection geometry design is simpler than CPCs, optical and quantum yields are similar and cost could be considerably lower.

5. Coupled advanced technologies for pesticide degradation

Despite AOPs are cost effective processes for water and mainly wastewater treatment, one of their main problems is their cost when compared with other conventional treatment processes such as biological treatment (Sarria et al., 2003). The treatment of water containing non-biodegradable toxic organic compounds is an environmentally complex issue in several industries such as pulp and paper, textile and petroleum industries. Considering the toxic nature of pesticides, it is clear that these kinds of xenobiotics are, in many cases, low biodegradable and, in most cases, highly refractory organic compounds. Due to this reasons, coupling AOPs and biological processes should be a good alternative to minimize the costs of treatment of water or wastewater containing this kind of pollutants. The strategy of combining chemical and biological processes to degrade contaminants in water has been proposed since middle of 90's (Scott and Ollis, 1995; 1997). Since then several works on the biological treatment of wastewater deal with the combined operation of chemical and biological oxidations (Scott and Ollis; 1995; Beltran et al; 1997; Benitez et al., 2001). Felsot *et al.* (2003), among other authors, have suggested that the combination of physical or chemical methods with biological treatment is likely a feasible option for the treatment of pesticide wastewater. In all these works is demonstrated the beneficial use of chemical oxidation process as a pretreatment or post-treatment of a biological process (Beltran, 2004, Lapertot et al., 2007).

Usually, when coupling chemical and biological processes the aim of the chemical oxidation is not to mineralize the organic contaminants but produce the conversion of high toxic, refractory parent components into biodegradable intermediates capable to be completely removed by biological processes (Esplugas et al., 2004). The possibility of minimal use of the oxidant agent, usually the most expensive component of the chemical process, followed by a low cost biological process (i.e. activated sludge, biofilm reactors) can help to improve the cost efficiency of a high effective process.

The effectivity of the coupled process is usually recorded using time evolution of coarse concentration variables such as total organic carbon (TOC), chemical oxygen demand (COD), biochemical oxygen demand (BOD) or some of their relationships (Esplugas et al., 2004; Sarria 2003; Pulgarin et al., 1999).

Relatively few works on the application of this kind of coupled methodologies are available in literature. Most of them corresponds to ozonation processes (Marco et al., 1997; Helble et al., 1999; Yeber et al., 1999; Beltran et al., 1999; Benitez et al., 2001; Ledakowicz et al., 2001), H_2O_2/UV (Adams and Kuzhikanni, 2000; Ledakowicz et al., 2001), TiO_2/UV oxidation (Li and Zhang, 1996; Li and Zhao, 1997; Chum and Yizgohon, 1999; Hess et al., 1998; Parra et al., 2002), Fenton and Fenton-like (Pulgarin et al., 1999; Chamarro et al., 2001; Sarria et al., 2003; Rodriguez et al., 2002; Sarria et al., 2001; Sarria et al., 2002) and wet oxidation (Donlagic and Levec, 1998). Table 5 shows some examples of physical-chemical/biological processes, including the treated pesticide, both process and the correspondent reference.

Pesticide(s)	Biological process	Physical-chemical process	References
EPTC, molinate, propazine, atrazine, simazine, prometryn, ametryn, simetryn, pyrazon, tris MEA,	Attached biomass (biofilter)	ozonation	Mezzanote *et al.* (2005)
Tetraconazole, metribuzin	suspended cells	anodic Fenton	Scherer *et al.* (2004)
Atrazine	biomineralization	ozonation	Scherer et al. (2004).
Atrazine/sotriazine	microorganisms	chemical	Scherer et al. (2004).
Atrazine	*Klebsiella terragena* DRS-1	ozone	Ikehata and El-Din, 2005.
Eyanazine, atrazine, metachlor and paraquat	microorganisms	ozone	Ikehata and El-Din, (2005)
Simazine	various microorganisms	oxzone, UV, photolisis or O_3/UV	Ikehata and El-Din, (2005)
Eyanuric acid, amino-S-triazines, chloro-amino-strazines, chloroethyl-S-triazines	microbial culture	ozone	Ikehata and El-Din, (2005).

Table 5. Some examples of coupled biological-physical-chemical process reported in literature

It is clear that pesticide removal from water should be one of the main applications of this coupled methodology. Nevertheless few reports are available in literature dealing with the use of this approach to pesticide remotion (Parra et al., 2000; 2002; Sarria et al., 2002; Lapertot et al., 2007; Al-Momani et al., 2006; Contreras et al., 2003). They had found that most of the tested pesticide effluents, readily determined as non-biodegradable by the Zahn-Wellens test, increased in their biodegradability once the photoassisted process was applied. The actual behavior of toxicity of isoproturon effluent, for example, showed an increase in this parameter during the first reaction minutes followed of sharp decrease. Authors suggest (Parra et al., 2000) that this behavior could be due to formation of intermediate compounds with higher toxicity than the parent pesticide and its further oxidation. For some other cases, effluent biodegradability was not completely reached after photoassisted process. For example, in the case of metobromuron the BOD/COD ratio went from 0.0 (stated as completely non-biodegradable) to 0.1, too low if compared with the BOD/COD ratio considered for municipal biodegradable wastewater, 0.4 (Parra et al., 2000).

As another example, Table 5 shows the partial contribution of the pre- and post treatment using ozone, over the entire coupled ozonation-biological process applied to in streams containing different pesticides, at different concentrations. As observed, preozonation of the stream can be very advantageous for the coupled process, contributing with a 56-98% of the overall pesticide removal. Biological process can contribute (in this specific case) with 1.8-41% of the removal, and finally, post-ozonation process can polish the stream, with additional removals of 0-3%.

6. Pesticide degradation by advanced electrochemical oxidation processes

6.1 General aspects

In the last years, there has been great interest in the development of effective methods of pollutans removal from aqueous solutions based on direct and indirect electrochemical techniques. The most useful direct electrochemical method is anodic oxidation (Kaba et al., 1990; Kotz et al., 1991; Stucki et al., 1991; Comninellis and Pulgarin, 1991, 1993; Murphy et al., 1992; Comninellis and Nerini, 1995; Feng et al., 1995; Johnson et al., 1999; Gandini et al., 2000; Rodrigo et al., 2001; Rodgers and Bunce, 2001; Wu and Zhou, 2001) where organic compounds are essentially degraded by reaction with adsorbed hydroxyl radicals at the anode surface, which are generated from water oxidation:

$$H_2O \rightarrow \bullet OH_{ads} + H^+ + e^- \quad (5)$$

Since the participation of $\bullet OH_{ads}$ radicals in the reaction is the key factor to degrade the pollutan, then the generation efficiency of them should be tightly related to the nature of the anodic material. Thus, although the traditional Pt anodes has been used for this purpose (Kaba et al., 1990; Kotz et al., 1991; Stucki et al., 1991; Comninellis and Pulgarin, 1991, 1993; Murphy et al., 1992; Comninellis and Nerini, 1995), it are less efficient that the oxide-base electrodes such as PbO_2 (Kaba et al., 1990; Feng et al., 1995; Wu and Zhou, 2001), doped PbO_2 (Feng et al., 1995), doped SnO_2 (Kotz et al., 1991; Stucki et al., 1991; Comninellis and Pulgarin, 1991, 1993; Johnson et al., 1999), IrO_2 (Comninellis and Nerini, 1995; Rodgers and Bunce, 2001) or more recently to the boron-doped diamond thin-layer anode, BDD (Gandini et al., 2000; Rodrigo et al., 2001).
On the other hand, the indirect electrochemical methods involves the previous formation of oxidizing agents such as H_2O_2 (Hsiao and Nobe, 1993; Do 1993, 1994; Ponce de Leon and Pletcher, 1995; Brilla et al., 1996; Brillas et al., 1998; Alvarez-Gallegos and Pletcher, 1999; Harrington and Pletcher, 1999; Oturan et al., 1999; Brillas et al., 2000; Oturan et al., 2000, Oturan, 2000; Oturan et al., 2001):

$$O_2 + 2H^+ + 2e^- \rightarrow H_2O_2 \quad (6)$$

or the well known Fenton's reagent (H_2O_2/Fe^{2+}) (Hsiao and Nobe, 1993; Do 1993, 1994; Ponce de Leon and Pletcher, 1995; Alvarez-Gallegos and Pletcher, 1999; Oturan et al., 1999; Oturan et al., 2000, Oturan, 2000; Oturan et al., 2001):

$$Fe^{2+} + H_2O_2 \rightarrow Fe(OH)^{2+} + {}^{\cdot}OH \quad (7)$$

The combination of chemical and electrochemical procedures has also been reported as a good alternative to water treatment. The electro-Fenton and photoelectro-Fenton methods can be considered as advanced electrochemical oxidation processes, AEOPs (Brilla et al., 1996; Brillas et al., 1998; Brillas et al., 2000; Boye, et al., 2002).

6.2 Mechanism of the electrochemical pollutan oxidation

Principal advantages of the electrooxidation method are the case of operations, a wide range of treatment conditions and eliminations of the need to generate, dispense and store treatment reagents, but more important is their capability to induce a very deep oxidation that can result in a virtually complete mineralization of the pollutan (Comninellis C. 1994; Houk et al., 1998; Feng and Li, 2003). It has been shown that the electrode material plays a key role on the evolution of the oxidation process (Martínez-Huitle and Ferro, 2006; Martínez-Huitle et al, 2004; Belhadj and Savall, 1998) and consequently on the by-products of oxidation. According to the mechanism involved in the pollutant oxidation (Martínez-Huitle and Ferro, 2006), the electrode materials have been classified in two main groups: *active* and *non-active* electrode material (Martínez-Huitle and Ferro, 2006; Martínez-Huitle et al, 2004).

The proposed model assumes that the initial reaction in both kind of anodes (generically denoted as M) corresponds to the oxidation of water molecules leading to the formation of physisorbed hydroxyl radical ($M(^{\bullet}OH)$): $M + H_2O \rightarrow M(^{\bullet}OH) + H^+ + e^-$. Both the electrochemical and chemical reactivity of heterogeneous $M(^{\bullet}OH)$ are dependent on the nature of the electrode material. The surface of *active* anodes interacts strongly with $^{\bullet}OH$ radicals and then (Martínez-Huitle and Ferro, 2006; Martínez-Huitle et al, 2004; Quiroz et al., 2005; Quiroz et al., 2006), a so-called higher oxide or superoxide (MO) may be formed. This may occur when higher oxidation states are available for a metal oxide anode, above the standard potential for oxygen evolution (E° = 1.23 V vs. SHE): $M(^{\bullet}OH) \rightarrow MO + H^+ + e^-$. The redox couple MO/M acts as a mediator in the oxidation of organics by $MO + R \rightarrow M + RO$; which competes with the side reaction of oxygen evolution via chemical decomposition of the higher oxide species: $MO \rightarrow M + \frac{1}{2} O_2$.

In contrast, the surface of a *non-active* anode interacts so weakly with $^{\bullet}OH$ radicals that allows the direct reaction of organics with $M(^{\bullet}OH)$ to give fully oxidized reaction products such as CO_2 and H_2O ($M(^{\bullet}OH) + R \rightarrow M + m\, CO_2 + n\, H_2O + H^+ + e^-$) where R is an organic compound with m carbon atoms and $2n$ hydrogen atoms, without any heteroatom, which needs ($2m + n$) oxygen atoms to be totally mineralized to CO_2 and H_2O. This reaction also competes with the side reaction of $M(^{\bullet}OH)$ like direct oxidation to O_2 ($M(^{\bullet}OH) \rightarrow M + \frac{1}{2} O_2 + H^+ + e^-$) or indirect consumption through dimerization to hydrogen peroxide by $2\, M(^{\bullet}OH) \rightarrow 2\, M + H_2O_2$. A *non-active* electrode does not participate in the direct anodic reaction of organics and does not provide any catalytic active site for their adsorption from the aqueous medium (Martínez-Huitle and Ferro, 2006; Quiroz et al., 2006). It only acts as an inert substrate and as a sink for the removal of electrons. In principle, only outer-sphere reactions and water oxidation are possible with this kind of anode. Hydroxyl radical produced from water discharge is subsequently involved in the oxidation process of organics. The model presupposes that the electrochemical activity (related to the overvoltage for O_2 evolution) and chemical reactivity (related to the rate of organics oxidation) of physisorbed $M(^{\bullet}OH)$ are strongly linked to the strength of the $M-^{\bullet}OH$ interaction. As a general rule, the weaker the interaction, the lower the anode reactivity for organics oxidation with faster chemical reaction with $M(^{\bullet}OH)$. The BDD anode is the best non-active electrode verifying this behavior (Martínez-Huitle and Ferro 2006; Belhadj and Savall 1998; Quiroz et al., 2006; Marcelli et al., 2003), then being proposed as the preferable anode for treating organics by electrochemical oxidation.

On the basis of this model, metal oxides such as IrO_2 and RuO_2 (Martínez-Huitle and Ferro 2006; Da Pozzo et al, 2005) known as *active* electrodes, achieving an incomplete oxidation of organic pollutants; whereas *non-active* oxides, such as Ti/SnO_2 and Pb/PbO_2 and their doped analogues are capable to oxidized organics to CO_2 (Martínez-Huitle and Ferro 2006;

Quiroz et al., 2005; Panizza et al., 2001). Within this last group of electrode materials, boron doped diamond (Si/BDD) electrodes have received great attention due to the wide range of their electrochemical properties (Quiroz et al., 2006; Marcelli et al., 2003).

6.3 Application of the direct electrochemical oxidation to removal pesticides from aqueous media

There is a scarce range of studies concerned with direct electrochemical oxidation for removal pesticides from aqueous media. Several reasons can be wielded to explain this little attention given to the study of their degradation, but all seems to indicated that this lack of attention is the risk to form degradation products of pesticides even more toxic than the parent compound that forms the pesticide. This assumption it is addmited if we takes into account the experimental conditions by which various electrooxidation pesticides processes quoted in literature has been performed. However, other important factor of pesticides to be condidered is their unique chemical structure which can associate functional groups with diferent susceptibility to the oxidation. This last characteristic make difficult to determine the degree of pesticide degradation and their corresponding oxidation pathway.
In spite of to be a known fact that the best anode materials to degrade pollutan organic compounds are those based in metallic oxides, the use of Pt electrodes has still been the preferable choise as anode material to degrade pesticides by direct electrochemical oxidation.

6.4 Organophosphates

This is the type of pesticides more reported being the more commonly quoted in literature methidathion, methylparathion, monochrotophos, phosphamidon, demeton-S-methyl, methamidophos, fenthion, and diazinon.
(a) Methylparathion ($C_{10}H_{14}NO_5PS$)
Methylparathion is a sintetic insecticide widely used in farm crops but with a strict control by the Environmental Protection Agency (EPA). The EPA allows 0.002 mg of methylparathion per liter of drinking water, which made justifiable the application of AOPs methods for their destruction from residual waters of agricultural nature. Arapoglou et al. (2003) reported by the first time the application of a direct electrochemical oxidation for the treatment of organophosphoric pesticides. Their electrochemical system was a Ti/Pt anode and a stainless steel 304 as cathode in a brine solution (H_2O + NaCl) under an applied currect of 36 A. After 2h of electrolysis a high reduction of COD and BOD_5 of the oxidized methylparathion as well as a low kWh/COD_r ration were reported. No degradation by-products of this organophosphoric pesticide were identified in any of these experiments.
Vlyssides et al. (2004) reported the electrochemical degradation of methylparathion by using Ti/Pt as anode in an aqueous medium of sodium chloride as electrolyte at 45°C and an applied current density of 560 mA/cm^2. It was shown that an 8% w/w aqueous suspension of methylparathion and 20 g/L of sodium chloride can be electrolyzed in 2 h of reaction time. Methylparathion is quickly degraded, but a complete mineralization was not observed. Several degradation by-products and intermediates of methylparathion produced by electrochemical oxidation were reported. Formation of paraoxon, p-nitrophenol, benzoquinone, and hydroquinone were identified as primary intermediates of methylparathion degradation. The formation of these type of intermediates originates the

formation of carboxylic acids such as oxalic, formic, and acetic acids as final products of the degradation process. Inorganic species were also identified between them nitrate, sulfate, phosphate, as well some oxides such as nitrogen oxides, sulfur dioxide and carbon dioxide. The full chemical analysis of liquid phase as well as of gas phase allows to the author to propose a degradation pathways for methylparathion electrochemical oxidation.

(b) Methidathion ($C_6H_{11}N_2O_4PS_3$)

Hachami et al. (2008) investigated the degradation of 1.4 mM of methidathion in in aqueous solution by anodic oxidation using a boron-doped diamond (BDD) anode. They observed an important reduction of chemical oxygen demand (COD) in the presence of 2-3 % of NaCl, as well as in the pH of electrolyzed solution. From these results the authors has suggested a pseudo first-order kinetics for the COD reduction of methidathion with a rate constant dependent on the applied current and on the electrolysis temperature: k = 0.0073 s^{-1} at 20 mA and 0.0146 s^{-1} at 60 mA, while k = 0.0131 s^{-1} at 298 K and 0.0077 at 363 K. It was concluded that applied current increases the rate of electrochemical oxidation but decreases it with the increases in temperature. The obtained activation energy (- 10.75 kJ) is in agree with the stablished conclusions. No attempt was made to identify the degradation products of methidathion although was suggested that mechanism of electrochemical mineralization can involve some mediators like chlorinated species or other radicals.

(c) Monochrotophos ($C_7H_{14}NO_5P$)

Yatmaz and Uzman (2009) investigated the direct electrochemical oxidation for removal of monochrotophos on Ti electrodes in aqueous solution of sodium salts (chloride or sulphate) as a function of applied current density and initial concentrations of pesticide. At 50 A/m^2 the monochrotophos degradation efficiencies were increased from 40 to 62% with the increase of initial concentration from 50 to 300 mg/L in the first five minutes of electrolysis after which the degradation reaction was stopped The increase in current density from 50 to 100 A/m^2 has a negligible effect on the degradation parameters owing to a poor generation of ·OH radicals on this type of anodes. The use of high concentration of NaCl electrolyte solution increases the electrochemical oxidation efficiency but increases also the risk to formation of chlorined compounds as residuals of degradation. In general, this electrochemical arrangement based on use of Ti as anodes for direct oxidation of monochrotophos was not an efficient method for removal this organophosphorous pesticide from aqueous media.

(d) Phosphamidon ($C_{10}H_{19}ClNO_5P$)

Phosphamidon is also an organophosphate insecticide, considered as an obsolete pesticide but whose disposal provokes serious environmental problems. It is soluble in water and stable in neutral and acid media and for this reason easy to find in aquatic media. This organophosphoric pesticide has been treated by direct electrochemical oxidation using Ti/Pt as anodes. Vlyssides et al. (2005) has reported experimental results from a laboratory scale pilot plant where the achieved reduction was nearly 26%.

Vlyssides et al. (2005) has also reported the electrochemical oxidation of the phosphorothioate pesticides Demeton-S-methyl ($C_6H_{15}O_3PS_2$), Methamidophos ($C_2H_8NO_2PS$), Fenthion ($C_{10}H_{15}O_3PS_2$), and Diazinon ($C_{12}H_{21}N_2O_3PS$). These pesticides were treated by an electrolysis system using Ti/Pt anode and a stainless steel 304 as cathode and also in a laboratory scale pilot plant. They reported that for Fenthion the achieved reduction was over 60%, while for Demeton-S-methyl, Methamidophos and Diazinon was more than 50%.

(e) Methamidophos ($C_2H_8NO_2PS$)

The anodic oxidation of methamidophos was studied by Martínez-Huitle at al. (2008) in a sodium sulphate aqueous solution on Pb/PbO_2, Ti/SnO_2, and Si/BDD (boron doped diamond) electrodes at 30°C. Under galvanostatic conditions, it was observed that the performance of the electrode material is influenced by pH and current density as it was shown by HPLC and ATR-FTIR analyses of methamidophos and its oxidation products along the electrolysis. It was found that methamidophos degradation using Pb/PbO_2 in acid media (pH 2.0 and 5.6) generates formaldehyde as the main product of reaction giving evidence of an indirect mineralization mechanism. Under the same conditions, Ti/SnO_2 showed poor formaldehyde production compared to the Pb/PbO_2 electrode. On Si/BDD electrodes formaldehyde production was not observed, instead the ATR-FTIR results showed the formation of phosphate as the reaction progressed suggesting a complete methamidophos mineralization on this electrode. In addition, HPLC results showed that the electrode efficiency is also dependant on the applied current density. This current density influence is remarkably clear on the Si/BDD electrodes where was evident that the most efficient current density towards a complete methamidophos mineralization was reached with the application of 50 mA/cm^2.

(f) Other pesticides

Until now, electrochemical methods of direct oxidation have seldom been applied to the degradation of other pesticides different to the organophosphorus. However, the electrochemical oxidation of some thiocarbamate (R_1,R_2,$NCOSR_3$, where R's are alkyl, cicloalkyl or aryl groups) herbicides in aqueous NaCl solutions has been investigated (Mogyoródy 2006), as well the oxidation of thiram ($C_6H_{12}N_2S_4$) (Priyantha and Weliwegamage 2008), an organo-sulfur fungicide, and also of the atrazine ($C_8H_{14}ClN_5$) herbicide (Malpass et al. 2006; Mamián et al. 2009). In addition, the electrochemical combustion of mecoprop ($C_{10}H_{11}ClO_3$) (Flox et al. 2006), carbaryl ($C_{12}H_{11}NO_2$) (Miwa et al. 2006, Malpass et al. 2009), and propham ($C_{10}H_{13}NO_2$) (Ozcan et al. 2008) herbicides has also been reported recently.

(g) In conclusion, the application of electrochemical methods by direct oxidation in pesticide removal has scarcely been explored. The complex nature of the molecular structure of pesticides, highly heteroatomic, is a restritive factor to stablish the chemical composition characteristics of solution due to the solubility problems and/or generation of dangerous intermediates. Thus, for instance, pesticides containing N atoms can to form chloramines if the aqueous solution has NaCl as electrolyte (Mogyoródy 2006a, 2006b). However, it is important to point out that presence of NaCl in solution can also confer to the electrodes an enhanced activity. In this case the Cl^- species at the electrode surface act as intermediates in the electron transfer between the pesticide molecule and the electrode (Miwa et al. 2006). The anode material is other important restrictive factor which determinates reaction parameters such as current efficiency, selectivity and product composition. Several works have reported the use of Ti or Pt electrodes (Mamián et al. 2009; Yatmaz and Uzman 2009; Mogyoródy 2006a, 2006b; Vlyssides et al. 2005a, 2005b, 2004; Arapoglou et al. 2003; Pulgarin and Kiwi 1996) with results little adequated for consider its as anodic material for removal of pesticides from aquatic media. The formation of complex mixture of oxidation by-products in solution, no detoxification of solution, or desactivation phenomena of anodes are some of limitations of these type of electrodes for their use in the electrochemical method of direct oxidation. Better results has beeb achieved by using metallic oxides such as

SnO_2, PbO_2, or RuO_2 (Martínez-Huitle et al. 2008; Mogyoródy 2006a, 2006b; Pulgarin and Kiwi 1996), dimensionally stable anodes (Miwa et al. 2006; Malpass G.R.P. et al. 2006, 2009), and more recently boron-doped diamond surfaces (Gao et al. 2009; Ozcan et al. 2008; Flox et al. 2006; Hachami et al. 2008; Martínez-Huitle et al. 2008).

7. References

Aaron, J.J. & Oturan, M.A. (2001). New photochemical and electrochemical methods for the degradation of pesticides in aqueous media. Environmental applications. *Turkish Journal of Chemistry* 25, 509-520. ISSN 1300-0527.

Adams, N.; Levi, P. & Hadgson, E. (1990). In vitro studies of the metabolism of atrazine, simazine and terbutryn in several vertebrate studies. *Journal of Agricultural and Food Chemistry* 38, 1411-1417. ISSN 0021-8561.

Adams, C.D. & Kuzhikanni, J.J. (2000). Effect of UV/H_2O_2 preoxidation on the aerobic biodegradability of quaternary amine surfactants. *Water Research* 34(2), 668-672. ISSN 0043-1354.

Agüera, A.; Almansa, E.; Malato, S.; Maldonado, I. & Fernandez-Alba, A. (1998). Evaluation of photocatalytic degradation trough advanced oxidation processes: An overview. *Analusis* 26, 245-251. ISSN 0365-4877.

Ajona, J. & Vidal, A. (2000). The use of CPC collectors for detoxification of contaminated water: design, construction and preliminary results. *Solar Energy* 68(1), 109-120. ISSN 0038-092X.

Al-Ekabi, H. & Serpone, N. (1988). Kinetic studies in heterogenoeous photocatalysis. Photocatalytic degrdation of chlorinated phenols in aereated aqueous solutions over TiO_2 suported on a glass matrix. *Journal of Physical Chemistry* 92, 5726-5731. ISSN 1089-5674.

Al-Ekabi, H.; Serpone, N.; Pelizzetti, E.; Minero, C.; Fox, M.A. & Barton R. (1989). Kinetics studies in heterogeneous photocatalysis. TiO_2-mediated degradation of 4-chlorophenol alone and in three-component mixture of 4-chlorophenol, 2,4-dichlorophenol and 2,4,5-trichlorphenol in air-equilibrated aqueous media. *Langmuir* 5, 250-255. ISSN 0734-7463.

Alfano, O.M.; Bahnemann, D.; Cassano, A.E.; Dillert, R. & Goslich, R. (2000). Photocatalysis in water environments using artificial and solar light. *Catalysis Today* 58, 199-230. ISSN 0920-5861.

Al-Momani, F.; Sans, C.; Contreras, S. & Esplugas, S. (2006). Degradation of 2,4-dichlorophenol by combining photo-assited Fenton reaction and biological treatment. *Water and Environmental Research* 78(6), 590-597. ISSN 1061-4303.

Alvarez-Gallegos, A.; Pletcher, D. (1999). The removal of low level organics via hydrogen peroxide formed in a reticulated vitreouscarbon cathode cell. Part 2: The removal of phenols and related compounds. *Electrochimica Acta* 44, 2483. ISSN 0013-4686.

Amorisco, A.; Losito, I.; Palmisano, F. & Zambonin, P.G. (2005). Photocatalytic degradation of the herbicide isoproturon: characterization of by-products by liquid chromatography with electrospray ionization tandem mass spectrometry. *Rapid communications in Mass Spectrometry* 19(11), 1507-1516. ISSN 0951-4198.

Amine, A.; Boulkamh, A. & Richard, C. (2005). Phototransformation of metobromuron in the presence of TiO_2. *Applied Catalysis B: Environmental* 59(3-4), 147-154. ISSN 0926-3373.

Andreozzi, R.; Caprio, V.; Insola, A. & Martota, R. (1999). Advanced oxidation proceses (AOP) for water purification and recovery. *Catalysis Today* 53, 51-59. ISSN 0920-5861.

Anipsitakis, G. & Dionysiou, D.D. (2003). Degradation of organic contaminants in water with sulfate radicals generated by the conjunction of peroximonosufate with cobalt. *Environmental Science and Technology* 37, 4790-4797. ISSN 0013-936X.

Anipsitakis, G. & Dionysiou, D.D. (2004). Radical generation by the interaction of transition metals with common oxidants. *Environmental Science and Technology* 38, 3705-3712. ISSN 0013-936X.

Anipsitakis, G.P. & Dionysiou, D.D. (2004). Transition metal/UV-based advanced oxidation technologies for water decontamination. *Applied Catalysis B: Environmental* 54, 155-163. ISSN 0926-3373.

Anipsitakis, G.; Dionysiou, D.D. & Gonzalez, M.A. (2006). Cobalt mediated activation of peroxymonosulfate and sulfate radical attack on phenolic compounds. Implication of chloride ions. *Environmental Science and Technology* 40, 1000-1007. ISSN 0013-936X.

Arancibia, C.; Bandala, E.R. & Estrada, C. (2002). Radiation absorption and rate constants for carbaryl photocatalytic degradation in a solar collector. *Catalysis Today* 76(2), 149-159. ISSN 0920-5861.

Arapoglou, D.; Vlyssides, A.; Israilides, C.; Zorpas, A. & Karlis P. (2003). Detoxification of methyl-parathion pesticide in aqueous solutions by electrochemical oxidation. *Journal of Hazardous Materials* B98, 191-199. ISSN 0304-3894.

Arnold, S.M.; Tallat, R.E.; Hickey, W.J. & Harris, R.F. (1995). Identification of Fenton's reagent generated atrazine degradation products by high performance liquid chromatography and mega flow electrospray ionisation tandem mass spectrometry. *Journal of Mass Spectrometry* 30(3), 452-460. ISSN 1075-5174.

Asah, R.; Morikawa, T.; Ohwaki, T.; Aoki, L. & Taga, Y. (2001). Visible-light photocatalyst in nitrogen-doped titanium dioxides. *Science* 293 (5528), 269-271. ISSN 0036-8075.

Augliaro, V. (1993). Kinetics of heterogeneous photocatalytic decomposition of monuron over anatase titanium dioxide powder. *Research in Chemical Intermediates* 19(9), 839-853. ISSN 0922-6168.

Bae, E. & Choi, W. (2003). Highly enhanced photoreductive degradation of perchlorinated compounds on dye-synthetized metal/TiO_2 under visible light. *Environmental Science and Technology* 37, 147-152. ISSN 0013-936X.

Badaway, M.I.; Ghaly, M.Y. & Gad-Allah, T.A. (2006). Advanced oxidation process for the removal of organophophorus pesticides fro water. *Desalination* 194(1-3), 166-175. ISSN 0011-9164.

Bandala, E.R.; Gelover, S.; Leal, T.; Arancibia, C.; Jiménez, A. & Estrada C. (2002). Solar photocatalytic degradation of aldrin. *Catalysis Today* 76(2), 189-199. ISSN 0920-5861.

Bandala, E.R.; Arancibia, C.; Orozco, S. & Estrada, C. (2004). Solar photoreactors comparison based on oxalic acid photocatalytic degradation. *Solar energy* 77, 509-512. ISSN 0038-092X.

Bandala, E.R.; Pelaez, M.A.; Dionysiou, D.D.; Gelover, S.; García, J. & Macias, D. (2007a). Degradation of 2,4-dichlorophenoxyacetic acid (2,4-D) using cobalt-peroxymonosulfate in Fenton-like processes. *Journal of Photochemistry and Photobiology A: Chemistry* 186, 357-363. ISSN 1010-6030.

Bandala, E.R.; Domínguez, Z.; Rivas, F. & Gelover, S. (2007b). Degradation of atrazine using solar driven Fenton-like advanced oxidation technologies. *Journal of Environmental Science and Health Part B* 42, 21-26. ISSN 0360-1234.

Bandala, E.R. & Estrada, C. (2007). Comparison of solar collection geometries for application to photocatalytic degradation of organic contaminants. *Journal of Solar Energy Engineering* 129, 22-26. ISSN 0199-6231.

Bahnemann, D.W.; Hilgendorff, M. & Memming, R. (1997). Charge carriers dynamic at TiO_2 particles: Reactivity of free and trapped holes. *Journal of Physical Chemistry B:* 101(21), 4265-4275. ISSN 1089-5674.

Balasubramanian, G.; Dionysiou, D.D.; Suidan, M.T.; Baudin, I. & Laine, J.M. (2004). Evaluating the activities of immobilized TiO_2 powder films for the photocatalytic degradation of organic contaminants in water. *Applied Catalysis B: Environmental* 47, 73-84. ISSN 0926-3373.

Barbeni, M.; Pramauro, E. & Pelizzetti, E. (1986). Photochemical degradation of chlorinated dioxins, biphenils phenols and benzene on semiconductor dispersions. *Chemosphere* 15 (9), 1913-1916. ISSN 0045-6535.

Basbusinski, K. & Filipek, K. (2001). Use of Fenton's reagent for removal of pesticides from industrial wastewaters. *Polish Journal of Environmental Studies* 10(4), 207-212. ISSN 1230-1485.

Bauer, R.; Waldner, G.; Fallman, H.; Hager, S.; Klare, M.; Krutzler, T.; Malato S. & Maletzky, P. (1999). The photo-Fenton reaction and the TiO_2/UV process for waste water treatment- novel developments. *Catalysis Today* 53, 131-144.ISSN 0920-5861.

Bayarri, B.; Gonzalez, O.; Maldonado, M.I.; Gimenez, J. & Esplugas, S. (2007). Comparative study of 2,4-dichlorophenol degradation with different advanced oxidation processes. *Journal of solar Energy Engineering* 129(1), 60-67. ISSN 0920-5861.

Beltran, F.; Garcia, J. & Alvarez, P. (1999). Integration of continuous biological and chemical (ozone) treatment of domestic wastewater II: Ozonation followd by biological oxidation. *Journal of Chemical Technology and Biotechnology* 74(9), 884-890. ISSN 1097-4660.

Beltran, F.J. (2004). *Ozone reaction kinetics for Water and wastewater systems*. Lewis Publishers. New York, USA.

Bellobono, I.R. (1995). Pre-industrial experience in advanced oxidation and integral photodegradation of organics in potable waters and wastewaters by Photoperm membranes inmovilizing TiO_2 and promoting photocatalysis. *Journal of Membrane Science* 102, 139-147. ISSN 0376-7388.

Belhadj T. N., A. Savall, Mechanistic aspects of phenol electrochemical degradation by oxidation on a Ta/PbO_2 anode, *J. Electrochem. Soc.* 145 (1998) 3427-3434. ISSN 0013-4651.

Benitez, F.J.; Acero, J.L.; Gonzalez, T. & García, J. (2001). Ozonation and biodegradation process in batch reactors treating black table olive washing wastewater. *Industrial Engineering and Chemistry Research* 40(14), 3144-3151. ISSN 1520-5045.

Bhatkhade, D.; Sawant, S.; Schouten, J. & Pangrkar, V.G. (2004). Photocatalytic degradation of chlorobencene using solar and artificial UV radiation. *Journal of Chemical Technology and Biotechnology* 79(4), 354-360. ISSN 1097-4660.

Bianco-Prevot, A.; Pramauro, E. & de la Guardia, M. (1999). Photocatalytic degradation of carbaryl in aqueous TiO_2 suspensions containing surfactants. *Chemosphere* 39 (3), 493-502. ISSN 0045-6535.

Bideau, M.; Claudel, B.; Dubein, C.; Faure, I. & Kazouan, H. (1995). The immobilization of titanium dioxide in the photocatalytic oxidation of spent waters. *Journal of Photochemistry and Photobiology A: Chemistry* 91(2), 137-144. ISSN 1010-6030.

Bigda R.J. (1995). Consider Fenton's chemistry for wastewater treatment. Chemical Engineering Progress 91(2), 62-66. ISSN 0360-7275.

Blake, D.M. (2000). Bibliography of work on the photocatalytic removal of hazardous compounds from water and air. Update number 3 to January 1999. National Technical Information Service. US Dept. of Commerce. Springfield, USA.

Blanco, J. & Malato, S. (2003). *Solar Detoxification*. UNESCO Publishing, ISBN 92-3-103916-4. France.

Blanco, J. (2003). *Development of CPC solar collectors for application to photochemical degradation of persistent pollutants in water*. Editorial CIEMAT. Madrid, Spain. (In Spanish).

Blanco, J.; Fernandez, P. & Malato, S. (2007). Solar photocatalytic detoxification and disinfection of water: Recent overview. *Journal of Solar Energy Engineering* 129, 4-15. ISSN 0199-6231.

Bolton, J.R. (2001). *Ultraviolet applications Handbook*. Bolton Photosciences Inc. Ontario, Canada.

Borello, R.; Minero, C.; Pramauro, E.; Pelizzetti, E.; Serpone, N. & Hidaka, H. 1989. Photocatalytic degradation of DDT mediated in aqueous semiconductor slurries by simulated sunlight. *Environmental Toxicology and Chemistry* 8(11), 997-1002. ISSN 0730-7268.

Borio, O.; Gawlik, B.M.; Bellobono, I.R. & Muntau, H. (1998). Photooxidation of prometryn and prometon in aqueous solution by hydrogen peroxide on photocatalytic membranes immobilizing titanium dioxide. *Chemosphere* 37(5), 975-989. ISSN 0045-6535.

Bousahel, R.; Harik, D.; Mammar, M. & Lamara, S. (2007). Degradation of obsolete DDT by Fenton oxidation with zero-valent iron. *Desalination* 206, 369-372. ISSN 011-6194.

Boyarri, B.; Gimenez, J.; Curco, D. & Esplugas, S. (2005). Photocatalytic degradation of 2,4-dichlorophenol by TiO_2/UV: kinetics, actimometries and models. *Catalysis Today* 101(3-4), 227-236. ISSN 0920-5861.

Boye, B.; Dieng, M.M. & Brillas, E. (2002). Degradation of herbicide 4-chlorophenioxyacetic acid by advanced electrochemical oxidation methods. *Environmental Science and Technology* 36, 3030-3035. ISSN 0013-936X.

Brandi, R.J.; Alfano, O.M. & Cassano, A.E. (1999). Collection efficiencies of UV radiation in solar photocatalytic reactors. Comparison of flat plate and parabolic trough reactors with a rigorous mathematical model. *Journal of Advanced Oxidation Technologies* 4(1), 76-84. ISSN 1203-8407.

Brandi, R.; Alfano, O.M. & Cassano, A.E. (2000). Evaluation of radiation absorption in slurry photocatalytic reactors 1. Assessment of methods in use and new proposals. *Environmental Science and Technology* 34(12), 2631-2639. ISSN 0013-936X.

Bigda, R.J. (1995). Consider Fenton's chemistry for wastewater treatment. *Chemical Engineering Progress* 91(12), 62-66. ISSN 0360-7475.

Brillas, E.; Mur, E. & Casado, J. (1996). Iron (II) catalized mineralization of aniline using a carbon-PTBFE O_2-fed cathode. *Journal of Electrochemical Society* 143(3), L49-L53. ISSN 0013-4651.

Brillas, E.; Calpe, J.C. & Casado, J. (2000). Mineralization of 2,4-D by advanced electrochemical oxidation processes. *Water Research* 34, 2253. ISSN 0043-1354.

Brillas, E.; Sauleda, R.; Casado, J. (1998). Degradation of 4-chlorophenol by anodic oxidation, electro-fenton, photoelectro-fenton, and peroxi-coagulation processes. Journal of the Electrochemical Society 145 (3), 759-765. ISSN 0013-4651.

Brun, H. (1995). Photocatalytic degradation of carbethamyn pesticide. *Environmental Technology* 16, 395-400. ISSN 0959-330.

Burda, C.; Lou, Y.; Chen, X.; Samia, A.C.S.; Stout, J. & Gole, J.L. (2003). Enhanced nitrogen doping in TiO_2 nanoparticles. *Nano Letters* 3, 1049-1054. ISSN 1530-6984.

Calza, P.; Massolino, C. & Pelizzetti, E. (2008). Light induced transformations of selected organophosphorus pesticides on titanium dioxide: Pathways and by-products evaluation using LC-MS technique. *Journal of Photochemistry and Photobiology A: Chemistry* 199(1), 42-49. ISSN 1010-6030.

Campanella, L. & Vitaliano, R. (2006). Atrazine toxicity reduction following H_2O_2/TiO_2 photocatalized reaction and comparison with H_2O_2-photolytic reaction. *Annali di Chimica* 97(1-2), 123-134. ISSN 1612-8877.

Canle, M.; Fernandez, M.I.; Rodriguez, S.; Santaballa, J.A.; Steenken, S. & Vullier, E. (2005). Mechanisms of direct and TiO_2-photocatalyzed UV degradation of phenyl urea herbicides. European Journal of Chemical *Physics and Physical Chemistry* 6(6), 2064-2074. ISSN 1439-4235.

Cassano, A.E. & Alfano, O.M. (2000). Reaction engineering of suspended solid heterogeneous photocatalytic reactors. *Catalysis today* 58, 167-197. ISSN 0920-5861.

Chacón, J.M.; Leal, M.T.; Sanchez, M.; Bandala, E.R. (2006). Solar photocatalytic degradation of azo-dyes by photo-Fenton process. *Dyes and Pigments* 69, 144-150. ISSN 0143-7208.

Chamarro, E.; Marco, A. & Esplugas, S. (2001). Use of Fenton reagent to improve organic chemical biodegradability. *Water Research* 35(4), 1047-1051. ISSN 0043-1354.

Chum, H. & Yizgohon, W. (1999). Decolourization and biodegradability of photocatalytic treated azo dyes and wool textile wastewater. *Chemosphere* 39 (12), 2107-2155. ISSN 0045-6535.

Chen, H.; Chun, L. & Ollis, D.F. (1983). Heterogeneous photocatalysis: degradation of diluted solutions of dichloromethane (CH_2Cl_2), chloroform ($CHCl_3$) and carbon tetrachloride (CCl_4) with illuminated TiO_2 photocatalyst. *Journal of Catalysis* 82, 418-423. ISSN 0021-9517.

Chen, S.F.; Mengyue, Z. & Yaowu, T.J. (1996). Photocatalytic degradation of pesticide dicofol. *Microchemistry* 54, 54-58. ISSN 0026-265X.

Chen, S.F. (1997). Intermediates isolation and structural characterization from dicofol photocatalytic degradation. *Journal of Environmental Sciences* 9(3), 278-282.ISSN 0976-4402.

Chester, G.; Anderson, M. & Read, H. (1993). A jacketed annular membrane photocatalytic reactor for water treatment: degradation of formic acid and atrazine. *Journal of Photochemistry and Photobiology A: Chemistry* 71(3), 291-297. ISSN 1010-6030.

Chiarenzelli, J.R.; Scrudato, R.J.; Rafferty, D.E.; Wunderlich, M.L.; Roberts R.N.; Pagano, J.J. & Yates, M. (1995). Photocatalytic degradation of simulated pesticides rinsates in water and water soil matrices. *Chemosphere* 30 (1), 173-185. ISSN 0045-6535.

Chiron, S.; Fernández-Alba, A.R. & Rodríguez, A. (1997). Pesticide chemical oxidation processes: an analytical approach. *Trends in Analytical Chemistry* 16 (9), 518-525. ISSN 0165-9936.

Chiron, S.; Fernandez, A.; Rodríguez, A. & Garcia-Calvo, E. (2000). Pesticide chemical oxidation: State-of-the-art. *Water Research* 34(2), 366-377. ISSN 0043-1354.

Clestur, G.; Anderson, M.; Read, H. & Esplugas, S. (1993). A jacketed annular membrane photocatalytic reactor for wastewater treatment degradation of formic acid and atrazine. *Journal of Photochemistry and Photobiology A; Chemistry* 71(3), 291-297. ISSN 1010-6030.

Colombo, D.P.; Russel, K.A.; Saeh, J.; Skinner, D.E. & Bowman, R.M. (1995). Femtosecond study of the intensity dependence of electron-hole dynamics on TiO_2 nanoclusters. *Chemical Physics Letter* 232(3), 207-214. ISSN 0009-2614.

Comninellis, C. & Pulgarin, C. (1991). Anodic oxidation of phenol for waste water treatment. *Journal of Applied Electrochemistry* 21(8), 703-703. ISSN 0021-891X.

Comninellis, C. & Pulgarin, C. (1993). Electrochemical oxidation of phenol for waste water treatment using SO_2 anodes. *Journal of Applied Electrochemistry* 23(2), 108-112. ISSN 0021-891X.

Comninellis, C. (1994) Electrocatalysis in the electrochemical conversion/combustion of organic pollutants for waste water treatment. *Electrochimica Acta* 39(11-12), 1857-1862. ISSN 0013-4686.

Comninellis, C. & Nerini, A. (1995). Anodic oxidation of phenol in the presence of NaCl for wastewater treatment. *Journal of Applied Electrochemistry* 25(1), 23-28. ISSN 0021-891X.

Contreras, S.; Rodriguez, M.; Al-Momani, F.; Sans, C. & Esplugas, S. (2003). Contribution of the ozonation pre-treatment to the biodegradation of aquous solutions of 2,-dichlorphenol. *Water Research* 37(13), 3167-3171. ISSN 0043-1354.

Cornish, B.J.P.A.; Lawton, L.A. & Robertson, P.K.J. (2000). Hydrogen peroxide enhanced photocatalytic oxidation of microcystin-LR using titanium dioxide. *Applied Catalysis B: Environmental* 25(1), 59-67. ISSN 0926-3373.

Curco, D.; Malato, S.; Blanco, J. & Jimenez, J. (1996). Photocatalysis and radiation absorption in a solar plant. *Solar Energy Materials and Solar Cells* 44(22), 197-217. ISSN 0927-0248.

Curco, D.; Gimenez, J.; Addardak, A.; Cervera, S. & Esplugas, S. (2002). Effects of radiation absortion and catalyst concentration on the photocatalytic degradation of pollutants. *Catalysis Today* 75, 177-188. ISSN 0920-5861.

Dionysiou, D.D.; Suidan, M.T.; Baudin, I. & Laine, J.M. (2004). Effect of hydrogen peroxide on the destruction of organic contaminants-synergism and inhibition in a continuous-mode photocatalytic reactor. *Applied Catalysis B: Environmental* 50, 259-269. ISSN 0926-3373.

Da Pozzo, A.; Merli, C.; Sirés, I.; Garrido, J.; Rodríguez, M. & Brillas, E. (2005). Removal of the herbicide amitrole from water by anodic oxidation and electro-Fenton. *Environmental Chemistry Letters* 3, 7-11. ISSN 1610-3653.

De Laat, J.; Lee, T.G. & Legube, B. (2004). A comparative study of the effects of chloride, sulfate and nitrate ions on the rates of decomposition of H_2O_2 and organic compounds by $Fe(II)/H_2O_2$ and $Fe(III)/H_2O_2$. *Chemosphere* 55(5), 715-723. ISSN 0045-6535.

De Laat, J. & Lee, T.G. (2006). Effects of chloride ions on the iron (III)-catalyzed decomposition of hydrogen on the effciency of the Fenton-like oxidation proceses. *Applied Catalysis B: Environmental* 66, 137-146. ISSN 0926-3373.

Detomaso, A.; Lopez, A.; Lovecchio, G.; Mascolo, G. & Curci, R. (2003). Practical applications of the Fenton reaction to the removal of chlorinated aromatic pollutants. Oxidative degradation of 2,4-dichlorophenol. *Environmental Science and Pollution Research International* 10(6), 379-384. ISSN 0944-1344.

D´Oliveira, J.C.; Minero, C.; Pelizzetti, E. & Pichat, P. (1993). Photodegradation of dichlorophenols and trichlorophenols in TiO_2 aqueous suspensions: kinetic effects of positions of the CI atoms and identification of the intermediates. *Journal of Environmental Science and Health A:* 28(4), 941-962. ISSN 1093-4529.

Do, J. S. & Chen, C. P. (1993). In situ oxidative degradation of formaldehyde with electrogenerated hydrogen peroxide. *Journal of the Electrochemical Society* 140(6), 1632-1637. ISSN 0013-4651.

Do, J. S. & Chen, C. P. (1994). In situ oxidative degradation of formaldehyde with electrogenerated hydrogen peroxide electrogenerated on the modified graphites. *Journal of Applied Electrochemistry* 24(9), 936-942. ISSN 0021-891X.

Domenech, X.; Jardim, W.F. & Litter, M.I. (2004). Advanced oxidation processes for contaminant removal. In: *Contaminants removal by heterogeneous photocatalysis,* Blesa M.A., Sánchez B. (Ed). Editorial CIEMAT, Madrid, Spain. (In Spanish).

Domínguez, C.; García, J.; Pedraz, M.A.; Torres, A. & Galán, M.A. (1998). Photocatalytic oxidation of organic pollutants in water. *Catalysis Today* 40(1), 85-101. ISSN 0920-5861.

Donlagic, J. & Levec, J. (1998). Comparison of catalysed and non-catalyzed oxidation of azo dye and effect on biodegradability. *Environmental Science and Technology* 32, 1294-1302. ISSN 0013-936X.

Doong, R. & Chang, W. (1997). Photoassisted titanium dioxide mediated degradation of organophosphorus pesticides by hydrogen peroxide. *Journal of Photochemistry and Photobiology A: Chemistry* 107(1), 239-244. ISSN 1010-6030.

Edelahi, M.C.; Oturan, N.; Oturan, M.A.; Padellec, Y.; Bermond, A. & El Kacemi, K. (2004). Degradation of diuron by the electro-Fenton process. *Environmental Chemistry Letters* 1(4), 233-236. ISSN 1610-3661.

Esplugas, S.; Contreras, S. & Ollis, D.F. (2004). Engineering aspects of the integration of chemical and biological oxidation: simple mechanistic models for the oxidation treatment. *Journal of Environmental Engineering* 130(9), 967-974. ISSN 0733-9372.

Evgenidou, E.; Fytianos, K. & Poulios, I. (2005). Semiconductor-sensitized photodegradation of dichlorvos in water using TiO_2 and ZnO as catalysts. *Applied Catalysis B: Environmental* 59(1-2), 81-89. ISSN 0926-3373.

Evgenidou, E.; Konstantinou, I.; Fytianos, K. & Albanis, T. (2006). Study of the removal of dichlorvos and dimetoate in a titanium dioxide mediate photocatalytic process through the examination of intermediates and the reaction mechanisms. *Journal of Hazardous Materials* 137(2), 1056-1064. ISSN 0304-3894.

Evgenidou, E.; Bizani, E.; Christophoridis, C. & Fytianos, K. (2007a). Heterogeneous photocatalytic degradation of prometryn in aqueous solutions under UV-Vis irradiation. *Chemosphere* 68(10), 1877-1882. ISSN 0045-6535.

Evgenidou, E.; Konstantinou, I.K.; Fytianos, K.; Poulios, I. & Albanis, T. (2007b). Photocatalytic oxidation of methyl parathion over TiO_2 and ZnO suspensions. *Catalysis Today* 124(3-4), 156-162. ISSN 0920-5861.

Fallaman, H.; Krutzler, T.; Bauer, R.; Malato, S. & Blanco, J. (1999). Applicability of the photo-Fenton method for treating water containing pesticides. *Catalysis Today* 54, 309-319. ISSN 0920-5861.

Farre, M.J.; Franch, M.I.; Malato, S; Ayllon, J.A.; Peral, J. & Domenech, X. (2005a). Degradation of some biorecalcitrant pesticides by homogeneous and heterogeneous photocatalytic ozonation. *Chemosphere* 58(8), 1127-1133. ISSN 0045-6535.

Farre, M.J.; Franch, M.I.; Malato, S.; Ayllon, J.A.; Peral, J. & Domenech, X. (2005b). Degradation of some biorecalcitrant pesticides by homogeneous and heterogeneous photocatalytic ozonation. *Chemosphere* 58(8), 1127-1133. ISSN 0045-6535.

Farre, M.J.; Franch, M.I.; Ayllon, J.A.; Peral, J. & Domenech, X. (2007). Biodegradability of treated aqueous solutions of biorecalcitrant pesticides by means of photocatalytic ozonation. *Desailinization* 211, 22-33. ISSN 0011-9164.

Faxeira, A.C.; Mendez, L.; Stollar, G.; Guardani, R. & Oller, C.A. (2005). Photo-Fenton remediation of wastewater containing agrochemicals. *Brazilian Archives of Biology and Technology* 48, 207-218. ISSN 1516-8913.

Feng, J.; Houk, L.L.; Johnson, D.C.; Lowery, S.N. & Carey, J.J. (1995). Electrocatalysis of anodic oxygen-transfer reactions: The electrochemical incineration of benzoquinone. Journal of the Electrochemical Society 142(11), 3626-3632. ISSN 0013-4651.

Feng, Y.J. & Li, X.Y. (2003). Electro-catalytic oxidation of phenol on several metal-oxide electrodes in aqueous solutions. Water Research 37, 2399-2407. ISSN 0043-1354.

Felsot, A.; Racke, K.D. & Hamilton, D.J. (2003). Disposal and degradation of pesticide waste. *Reviews of Environmental Contamination and Toxicology* 177, 123-200. ISSN 0179-5953.

Fenton, H.J. (1894). Oxidation of tartaric acid in precence of iron. *Journal of Chemical Society* 65, 899-910. ISSN 0022-4936.

Fernández, P.; Malato, S. & De Las Nieves, F. (1999). Relationship between TiO_2 particle size and reactor diameter in solar photodegradation efficiency. *Catalysis Today* 54, 195-204. ISSN 0920-5861.

Florencio, M.N.; Pires, E.; Castro, A.L.; Nunes, M.R.; Borges, C. & Costa, F.M. (2004). Photodegradation of diquat and paraquat in aqueous solutions by titaniuk dioxide: Evolution of degradation reactions and characterization of intermediates. *Chemosphere* 55(3), 345-355. ISSN 0045-6535.

Flox, C.; Cabot, P.L.; Centelas, F.; Garrido, J.A.; Rodríguez, R.M.; Arias, C. & Brillas, E. (2006). Electrochemical combustion of herbicide mecoprop in aqueous medium using a flow reactor with a boron-doped diamond anode. *Chemosphere* 64(6), 892-902. ISSN 0045-6535.

Fresno, F.; Guillard, C.; Coronado, J.M.; Chovelon, S.M.; Tudela, D.; Soria, J. & Herrmann, J.M. (2005). Photocatalytic degradation of sulfonylurea herbicides over pure and tin-doped TiO_2 photocatalyst. *Journal of Photochemistry and Photobiology A: Chemistry* 173(1), 13-20. ISSN 1010-6030.

Gandini, D.; Mahé, E.; Michaud, P.A.; Haenni, W.; Perrer, A. & Comninellis, C. (2000). Oxidation of carboxilyc acids at boron-doped diamond electrodes for waste water treatment. *Journal of Applied Electrochemistry* 30(12), 1345-1350.

Gao, J.; Zhao, G.; Shi, W. & Li, D. (2009). Microwave activated electrochemical degradation of 2,4-dichlorophenoxyacetic acid at boron-doped diamond electrode. *Chemosphere* 75(4), 519-525. ISSN 0045-6535.

Gelover, S.; Mondragon, P. & Jimenez, A. (2004). Titanium dioxide sol-gel deposited over glass and its applications as photocatalyst for water decontamination. *Journal of Photochemistry and Photobiology A: Chemistry* 165, 241-246. ISSN 0021-891X.

Ghosh, A.; Gupta, S.S.; Bartos, M.J.; Hangun, Y.; Voucolo, L.D.; Steinhoff, B.A.; Noser, C.A.; Horner, D.; Mayer, S.; Inhehees, K.; Horwitz, C.P.; Spatz, J.; Raybov, A.D.; Modal, S. & Collins, T.J. (2001). Green chemistry. Sustaining a high-technology civilization. *Pure and Applied Chemistry* 73(1), 113-118. ISSN 1365-3075.

Gimenez, J.; Curco, D. & Queral, M.A. (1999). Photocatalytic treatment of phenol and 2,4-dichlorophenol in a solar plant in the way to scaling-up. *Catalysis Today* 54, 229-243. ISSN 0920-5861.

Glaze, W.H. (1987). Drinking water treatment with ozone. *Environmental Science and Technology* 21, 224-230. ISSN 0013-936X.

Glaze, W.H.; Kwang, J.W. & Chapin, D.H. (1987). Chemistry of water treatment processes involving ozone, hydrogen peroxide and ultraviolet radiation. *Ozone Science and Engineering* 9, 335-352. ISSN 0191-9512.

Goswami, D.Y. & Blake, D.M. (1996). Cleanup with sunshine. *Mechanical Engineering* 118(8), 56-59. ISSN 16878132.

Guillard, C.; Pichat, P.; Huber, G. & Hoang, C. (1996). The GC-MS analysis of organic intermediates from the *TiO_2-UV photocatalytic treatment of water contaminated with lindane*. Journal of Advanced Oxidation Technologies 1, 53-60. ISSN 1203-8407.

Guillard, C.; Beaugiraud, B.; Dutriez, C.; Herrmann, J.M.; Jaffrazic, H.; Jaffrazic-Renault, N. and Lacroix, M. (2002). Physicochemical properties and photocatalytic activities of TiO_2 films by sol-gel methods. *Applied Catalysis B: Environmental* 39(4), 331-342. ISSN 0926-3373.

Guillard, C.; Disdier, J.; Monnet, C.; Dussaud, J.; Malato, S.; Blanco, J.; Maldonado, M.I. & Herrmann, J.M. (2003). Solar efficiency of a new deposited titania catalyst: chlorophenol, pesticide and dye removal applications. *Applied Catalysis B: Environmental* 46, 319-332. ISSN 0926-3373.

Gutierrez, F.; Santiesteban, A.; Cruz, L. & Bello, R. (2007). Removal of chlorotalonil, methyl parathion and methamidophos from water by the Fenton reaction. *Environmental Technology* 28(3), 267-272. ISSN 0959-3330.

Haarstrick, A.; Kut, O.M. & Heinzle, E. (1996). TiO_2-assisted degradation of environmentally relevant organic compounds in wastewater using a novel fluidised bed photoreactor. *Environmental Science and Technology* 30(3), 817-824. ISSN 0013-936X.

Hachami, F.; Salghi, R.; Mihit, M.; Bazzi, L.; Serrano, K.; Hormatallah, A. & Hilali, M. (2008). Electrochemical destruction of methidathion by anodic oxidation using a boron-doped diamond electrode. *International Scientific Journal for Alternative Energy and Ecology* 6 (62), 35-40. ISSN 1757-3971.

Harrington, T. & Pletcher, D. (1999). The removal of low levels of organics fromaqueous solutions using Fe(II) and hydrogen peroxide forme in situ at gas diffusion electrodes. *Journal of the Electrochemical Society* 146(8), 2983-2989. ISSN 0013-4651.

Hasegawa, K. (1998). Photocatalytic degradation of tetradifon pesticide. *Denki Kagaku* 66(6), 625-634. ISSN 1344-3542.

Haque, M.M. & Muneer, M. (2003). Heterogeneous photocatalysed degradation of a herbicide derivative, isoproturon in aqueous suspensionsof titanium dioxide. *Journal of Environmental Management* 69(2), 169-176. ISSN 0301-4797.

Helble, A.; Schlayer, W.; Liechti, P.; Jenny, R. & Möbius, C. (1999). Advanced effluent treatment in the pulp and paper industry with a combined process of ozonation and fixed bed biofilm reactors. *Water Science and Technology* 40(11-12), 343-350. ISSN 0273-1223.

Herrmann, J.M.; Mozzanega, M.N. & Pichat, P. (1983). Oxidation of oxalic acid in aqueous suspensions of semiconductors illuminated with UV or visible light. *Journal of Photochemistry* 22, 333-343. ISSN 1010-3060.

Herrmann, J.M.; Disdier, J.; Pichat, P.; Malato, S. & Blanco, J. (1998). TiO_2 based solar photocatalytic detoxification of water containing organic pollutants. Case studies of 2,4-dichlorophenoxyacetic acid (2,4-D) and benzofuran. *Applied Catalysis B: Environmental* 17, 15-23. ISSN 0926-3373.

Herrmann, J. M. (1999). Water treatment by heterogeneous catalysis. *Catalysis Science and Service*, 1 (Environmental Catalysis), 171-194. ISSN 0920-5861.

Hess, T.F.; Lewis, T.A.; Crawford, R.L.; Katamneni, S.; Wells, J.H. & Watts, R.J. (1998). Combined photocaalytic and fungal treatment for destruction of 2,4,6-trinitrotoluene (TNT). *Water Research* 32(5), 1481-1491. ISSN 0043-1354.

Hincapie, M.; Maldonado, M.I.; Olleri, I.; Gernjak, W.; Sánchez, J.A.; Ballesteros M. & Malato, S. (2005). Solar photocatalytic degradation and detoxification of UE priority substances. *Catalysis Today* 101(3-4), 203-210. ISSN 0920-5861.

Hisanaga, T.; Harada, K. & Tanaka, K. (1990). Photocatalytic degradation of organichloride compounds in suspended TiO_2. *Journal of Photochemistry and Photobiology A: Chemistry* 54, 113-118. ISSN 1010-3060.

Hoffmann, M.R.; Martin, S.T.; Choi, W. & Bahnemann, D.W. (1995). Environmental applications of semiconductor photocatalysis. *Chemical Reviews* 95, 69-96. ISSN 1520-6890.

Houk, L.L.; Johnson, S.K.; Feng, J.; Houk,R.S. & Johnson,D.C. (1998) Electrochemical incineration of benzoquinone in aqueous media using a quaternary metal oxide electrode in the absence of a soluble supporting electrolyte. *Journal of Applied Electrochemistry* 28 1167-1177. ISSN 0021-891X.

Houston, P.L. & Pignatello, J. (1999). Degradation of selected pesticide active ingredients and commercial formulations in water by the photo-assisted Fenton reaction. *Water Research* 33(5), 1238-1246. ISSN 0043-1354.

Hsiao, Y.L. & Nobe, K. (1993). Hydroxylation of chlorobencene and phenol in a packed flow reactor with electrogenerated Fenton's reagent. *Journal of Applied Electrochemistry* 23(9), 943-946. ISSN 0021-891X.

Huang, C.P.; Dong, Ch. & Tang, Z. (1993). Advanced chemical oxidation: its present role and potential future in hazardous waste treatment. *Waste Management* 13, 361-377. ISSN 0956-053X.

Hustert, K. & Moza, P.N. (1997). Photochemical degradation of dicarboximide fungicides in the presence of soil constituent. *Chemosphere* 35(1-2), 33-37. ISSN 0045-6535.

Ijpelaar, G.F.; Meijers, R.T.; Hopman, R. & Krivithot, J.C. (2000). Oxidation of herbicides in ground water by the Fenton process: A realistic alternative for O_3/H_2O_2. *Ozone: Science and Engineering* 22(6), 607-616. ISSN 0191-9512.

Ikehata, K. & El-Din, M.G. (2005a). Aqueous pesticide degradation by ozonation and ozone-based advanced oxidation processes: A review (Part I). *Ozone: Science and Engineering* 27, 83-114. ISSN 0191-9512.

Ikehata, K. & El-Din, M.G. (2005b). Aqueous pesticide degradation by ozonation and ozone-based advanced oxidation proceses: A review (Part II). *Ozone Science and Engineering* 27(3), 173-202. ISSN 0191-9512.

Ikehata, K. & El-Din, M.G. (2006). Aqueous pesticide degradation by hydrogen peroxide/ultraviolet irradiation and Fenton type advanced oxidation processes: a review. *Journal of Environmental Engineering and Sciences* 5(2), 81-135. ISSN 1496-9018.

Iwasaki, K.; Hara, M.; Kawada, H.; Tada, H. & Ito, S. (2000). Cobalt iron-doped TiO2 photocatalyst response to visible light. *Journal of Colloids and Interface Science* 224(1), 202-204. ISSN 0021-9797.

Jardim, W.F.; Morales, S.G.; Morales, M. & Takiyama, K. (1997). Photocatalytic degradation of aromatic chlorinated compounds using TiO_2: toxicity of intermediates. *Water Research* 31(7), 1728-1732. ISSN 0043-1354.

Jimenez, A.; Estrada, C.; Cota, A.D. & Roman, A. (2000). Photocatalytic degradation of DBSNa using solar energy. *Solar Energy Materials and solar Cells* 60(1), 85-95. ISSN 0927-0248.

Johnson, S.K.; Houk, L.L.; Feng, J.; Houk, R.S. & Johnson, D.C. (1999). Electrochemical incineration of 4-chlorophenol and the identification of products and intermediates by mass spectrometry. *Environmental Science & Technology* 33 (15), 2638-2644. ISSN 0013-936X.

Kaba, L.; Hitchens, G.D. & Bockris, J.M. (1990). Electrochemical incineration of wastes. *Journal of the Electrochemical Society* 137(5), 1341-1345. ISSN 0013-4651.

Kamble, S.P.; Sawant, S.B. & Pangarkar, V.G. (2006). Photocatalytic mineralization phenoxiacetic acids using concentrated solar radiation and titanium dioxide in slurry photoreactor. *Chemical Engineering Research and Design* 84(5), 355-362. ISSN 0263-8762.

Katsumata, H.; Sada, M.; Nakaoka, Y.; Kaneco, S.; Suzuki, T. & Ohta, K. (2009). Photocatalytic degradation of diuron in aqueous solutions of platinized TiO_2. *Journal of Hazardous Materials* 171(1-3), 1081-1087. ISSN 0304-3894.

Kerzhentsev, M.; Guillard, C.; Herrmann, J.M. & Pichat, P. (1996). Photocatalytic pollutant removal in water at room temperature: case study of the total degradation of the insecticide fenitrotion. *Catalysis Today* 27, 215-220. ISSN 0920-5861.

Kim, S. & Choi, W. (2005). Visible-light-induced photocatalytic degradation of 4-chlorophenol and phenolic compounds in aqueous suspensions of pure titania: Demonstrating the existence of a surface-complex-mediated path. *Journal Physical Chemistry* 109(11), 5143-5149. ISSN 1089-5647.

Kinkennon, A.E.; Green, D.B. & Hutchinson, B. (1995). UV degradation of pyrymidinic herbicides using powder titanium dioxide suspensions. *Chemosphere* 31(7), 3663-3671. ISSN 0045-6535.

Kish, H. (1989). What is photocatalysis? In: Serpone N., Pelizzetti E. (Eds.) *Photocatalysis fundamentals and applications*. John Wiley and Sons. ISBN 0471626031. New York, USA.

Kong, L. & Lemley, A.T. (2006a). Modeling evaluation of carbaryl degradation in a continuously stirred tank reactor by anodic Fenton treatment. *Journal of Agricultural and Food Chemistry* 54(26), 10061-10069. ISSN 0021-8561.

Kong, L. & Lemley, A.T. (2006b). Kinetic modelling of 2,4-D degradation in soil slurry by anodic Fenton treatment. *Journal of Agricultural and Food Chemistry* 54(11), 3945-3950. ISSN 0021-8561.

Kouloumbos, V.N.; Tsipi, D.F.; Hiskia, A.E.; Nikolic, D. and van Breemen, R.B. (2003). Identification of photocatalytic degradation products of diazinon in TiO_2 aqueous suspensions using GC/MS/MS and LC/MS with quadrupole time-of-flight mass spectrometry. *Journal of the American Society for Mass Spectrometry* 14(8), 803-817. ISSN 1044-0305.

Konstantinou, I.K.; Sakkas, V.A. & Albanis, T.A. (2001). Photocatalytic degradation of the herbicides propanil and molinate over aqueous TiO_2 suspensions: identification of intermediates and the reaction patway. *Applied Catalysis B: Environmental* 34(3), 227-239. ISSN 0926-3373.

Konstantinou, I.K.; Sakkas, V.A. & Albanis, T.A. (2002). Photocatalytic degradation of propachlor in aqueous TiO_2 suspensions. Determination of the reaction patways and identification of intermediate products by various analytical methods. *Water Research* 36(11), 2733-2742. ISSN 0043-1354.

Kotz, R.; Stucki, S. & Carcer, B. (1991). Electrochemical waste water treatment using high overvoltage anodes. Part I: Physical and electrochemical properties of SnO_2 anodes. *Journal of Applied Electrochemistry* 21(1), 14-20. ISSN 0021-891X.

Kruztler, T.; Fallaman, H.; Maletzky, P.; Bauer, R.; Malato, S. & Blanco, J. (1999a). Solar driven degradation of 4-chlorophenol. *Catalysis Today* 54, 321-327. ISSN 0920-5861.

Krutzler, T. & Bauer, R. (1999b). Optimization of a photo-Fenton prototype reactor. *Chemosphere* 38(11), 2517-2532. ISSN 0045-6535.

Kuo, W.S. & Lin, Y.T. (2000). Photocatalytic oxidation of xenobiotics in water with immobilized TiO_2 on agitator. *Journal of Environmental Sciences and Health B* 35(1), 61-75. ISSN 0360-1234.

Kyriacou, G.; Tzoanas, K. & Poulius, I. (1997). Photocatalytic degradation of pesticides. *Journal of Environmental Science and Health A* 32 (4), 963-977. ISSN 1093-4529.

Lackhoff, M. & Niessner, R. (2002). Photocatalytic atrazine degradation by synthetic minerals, atmospheric aerosols and soil particles. *Environmental Science and Technology* 36(2-4), 5342-5347. ISSN 0013-936X.

Laperlot, M., Pulgarin, C.; Fernandez, P.; Maldonado, M.I.; Perez, L.; Oller, I.; Gernjak, W. & Malato, S. (2006). Enhancing biodegradability of priority substances (pesticides) by solar photo-Fenton. *Water Research* 40(5), 1086-1094. ISSN 0043-1354.

Lapertot, M.; Ebrahimi, S.; Dazio, S.; Rubinelli, A. & Pulgarin, C. (2007). Photo-Fenton and biological integrated process for degradation of a mixture of pesticides. *Journal of Photochemistry and Photobiology A: Chemistry* 186(1), 34-40. ISSN 1010-6030.

Ledakowicz, S.; Solecka, M. & Zylla, R. (2001). Biodegradation, decoourization and detoxification of textile wastewater enhanced by advanced oxidation processes. *Journal of Biotechnology* 89(2-3), 175-184. ISSN 0168-1656.

Lee, C. & Yoon, J. (2004). Teperature dependence of hydroxyl radical formation in the $hv/Fe^{3+}/H_2O_2$ and Fe^{3+}/H_2O_2 systems. *Chemosphere* 56, 923-934. ISSN 0045-6535.

Legrini, O.; Oliveros, E. & Braun, A.M. (1993a). Photochemical process for water treatment. *Chemical Reviews* 93, 671-698. ISSN 1520-6890.

Legrini, O.; Oliveros, E. & Braun, M. (1993b). Photochemical processes for water treatment. *Chemical Reviews* 93, 671-689. ISSN 1520-6890.

Lin, Y.M.; Tseng, Y.H.; Huang, J.H.; Chao, C.C.; Chen, C.C. & Wang, I. (2006). Photoctalytic activity for degradation of nitrogen oxides over visible light responsive titania-based photocatalyst. *Environmental Science and Technology* 40, 1616-1621. ISSN 0013-936X.

Li Puma, G.L. & Yue, P.L. (1993). Photodegradation of pentachlorophenol. In: *Photocatalytic purification and treatment of water and air,* Ollis D.F., Al-Ekabi H. (Ed.) Elsevier. ISBN 0444898557. Amsterdam.

Li, X. & Zhang, M. (1996). Decolorization and biodegradability of dyeing wastewater treated by a TiO_2-sensitized photo-oxidation process. *Water Science and Technology* 34(9), 49-55. ISSN 0273-1223.

Li, X. & Zhao, Y. (1997). On-site treatment of dyeing wastewater by a biophotoreactor system. *Water Science and Technology* 36(2-3), 165-142. ISSN 0273-1223.

Lu, M. C. (1993). Herbicides degradation using advanced oxidation technologies. *Journal of Photochemistry and Photobiology A: Chemistry* 76, 103-110. ISSN 1010-6030.

Lu, M.C.; Roam, G.D.; Chem, J.N. & Huang, C.P. (1995). Photocatalytic degradation of herbicides: intermediates identification. *Chemical Engineering Communiations* 139, 1-13. ISSN 0098-6445.

Lu, M.C. & Chen, J.N. (1997). Pretreatment of pesticide wastewater by photocatalytic oxidation. *Water Science and Technology* 36 (2-3), 117-122. ISSN 0273-1223.

Lu, M.C.; Chen, J.N. & Chang, C.P. (1997). Effect of inorganic ions on the oxidation of dichlorvos insecticide with Fenton's reaction. *Chemosphere* 35, 2285-2293. ISSN 0045-6535.

Lu, M.C.; Chen, J.N. & Tu, M.F. (1999). Photocatalytic oxidation of herbidie Diquat. Journal of *Environmental Science Health B:* 34 (5), 859-872. ISSN 0360-1234.

Mahalakshmi, M.; Arabindoo, B.; Palanichamy, M. & Murugesan, V. (2006). Photodegradtion of carbofuran using semiconductor oxides. *Journal of Hazardous Materials* 134(1-2), 240-245. ISSN 0304-3894.

Mahmoodi, N.M.; Arami, M.; Limaec, N.N. & Gharanjiq, K. (2006). Photocatalytic degradation of agricultural N-heterocyclic organic pollutants using immobilized nanoparticles of titania. *Journal of Hazardous Materials* 145(1-2), 65-71. ISSN 0304-3894.

Mahmoodi, N.M.; Arami, M.; Gharanjig, K.; Nourmohammadian, F. & Bidokhti, A.Y. (2008). Purification of water containing agricultural organophosphorus pollutant using nanophotocatalysts: Laboratory studies and numerical modeling. *Desalination* 230(1-3), 183-192. ISSN 0011-9164.

Mak, M.K.S. & Hung, T. (1992). Pesticide wastewater treatment using photocatalytic processes. *Toxicology and Environmental Chemistry* 36, 155-168. ISSN 0277-2248.

Mak, M.K.S. & Hung, S.T. (1993). Toxicity reduction of water polluted with pesticides by advanced oxidation technologies. *Environmental Science and Technology* 14, 265-269. ISSN 0013-936X.

Malato, S.; Blanco, J.; Richter, C.; Braun, B. & Maldonado, M.I. (1998). Enhancement of the rate of solar photocatalytic mineralization of organic pollutants by inorganic oxidizing species. *Applied Catalysis B: Environmental* 17, 347-356. ISSN 0926-3373.

Malato, S. (1999). *Solar photocatalytic decomposition of pentachlorophenol dissolved in water.* Editorial CIEMAT, Madrid, Spain.

Malato, S.; Blanco, J.; Richter, C.; Milow, B. & Maldonado, M.I. (1999a). Solar photocatalytic mineralization of commercial pesticides: methamidophos. *Chemosphere* 38(5), 1145-1156. ISSN 0045-6535.

Malato, S.; Blanco, J.; Richter, C.; Milow, B. & Maldonado, M.I. (1999b). Pre-industrial experience in solar photocatalytic mineralization of real wastewaters. Application to pesticide container recycling. *Water Science and Technology* 40(4-5), 123-130. ISSN 0098-6445.

Malato, S.; Blanco, J.; Richter, C. & Maldonado, M.I. (2000a). Optimization of pre-industrial solar photocatalytic mineralization of comercial pesticides. Application to pesticide container recycling. *Applied Catalysis B: Environmental* 25, 31-38. ISSN 0926-3373.

Malato, S.; Blanco, J.; Richter, C.; Fernández, P. & Maldonado, M.I. (2000b). Solar photocatalytic mineralization of commercial pesticides: Oxamyl. *Solar Energy Materials and Solar Cells* 64, 1-14. ISSN 0927-0248.

Malato, S.; Blanco, J.; Fernández, A.R. & Agüera, A. (2000c). Solar photocatalytic mineralization of commercial pesticides: acrinathrin. *Chemosphere* 40(4), 403-409. ISSN 0045-6535.

Malato, S.; Blanco, J.; Vidal, A. & Richter, C. (2002a). Photocatalysis with solar energy at a pilot plant-scale: an overview. *Applied Catalysis B: Environmental* 37, 1-15. ISSN 0926-3373.

Malato, S.; Blanco, J.; Caceres, J.; Fernandez, A.R.; Aguera, A. & Rodriguez, A. (2002b). Photocatalytic treatment of water soluble pesticides by photo-Fenton and TiO_2 using solar energy. *Catalysis Today* 76(2), 209-220. ISSN 0920-5861.

Malato, S.; Blanco, J.; Vidal, A.; Alarcon, D.; Maldonado, M.I.; Caceres, J. & Gernjak, W. (2003). Photocatalytic treatment of diuron by solar photocatalysis: Evaluation of main intermediates and toxicity. *Solar Energy* 75(4), 329-336. ISSN 0927-0248.

Malato, S.; Blanco, J.; Estrada, C.; Bandala, E.R. & Peñuela, G. (2004). Pesticide degradation. In: *Contaminants removal by heterogeneous photocatalysis*. Blesa, M.A.; Sánchez B. (Ed) Editorial CIEMAT, Madrid, Spain. (In Spanish).

Malpass, G.R.P.; Miwa, D.W.; Machado, S.A.S.; Olivi, P. & Motheo A.J. (2006). Oxidation of the pesticide atrazine at DSA® electrodes. *Journal of Hazardous Materials* 137(1), 565-572. ISSN 0263-8762.

Malpass, G.R.P.; Miwa, D.W.; Miwa, A.C.; Machado, S.A.S. & Motheo, A.J. (2009). Study of photo-assisted electrochemical degradation of carbaryl at dimensionally stable anodes (DSA). *Journal of Hazardous Materials* 167 (1-3), 224-229. ISSN 0263-8762.

Mamián, M.; Torres, W. & Larmat, F.E. (2009). Electrochemical degradation of atrazine in aqueous solution at a platinum electrode. *Portugaliae Electrochimica Acta* 27(3), 371-379. ISSN 0872-1904.

Mangyue, Z.; Shifu, C. & Yaowu, T. (1995). Photocatalytic degradation of organophosphorus pesticides using thin films TiO_2. *Journal of Chemical Technology and Biotechnology* 64(4), 339-344. ISSN 0268-2575.

Manilal, V.B. (1992). Photocatalytic treatment of toxic organics in wastewater: toxicity of photodegradation products. *Water Research* 26 (8), 1035-1038. ISSN 0043-1354.

Mansour, M.; Feich, E.A.; Behecti, A. & Scheunert, I. (1997). Experimental approaches to studying the photostability of selected pesticides in water and soil. *Chemosphere* 35(1-2), 39-50. ISSN 0045-6535.

Marco, A.; Esplugas, S. & Saum, G. (1997). How and why combine chemical and biological proceses for wastewater treatment. *Water Science and Technology* 35(4), 321-327. ISSN 0273-1223.

Marselli, B.; García-Gomez, J.; Michaud, P.A.; Rodrigo, M. & Comninellis, C. (2003). Electrogeneration of hydroxyl radicals on boron-doped diamond electrodes. *Journal of the Electrochemical Soc*iety 150, D79-D83. ISSN 0013-4651.

Martin, S.T.; Lee, A.T. & Hoffmann, M.R. (1995a). Photocatalytic degradation of pesticide-acaricides in aqueous suspensions of TiO_2. *Environmental Science and Technology* 29, 2567-2573. ISSN 0013-936X.

Martin, S.T.; Lee, A.T. & Hoffmann, M.R. (1995b). Chemical mechanisms of inorganic oxidants in the TiO_2/UV process: Increased rates of degradation of chlorinated hydrocarbons. *Environmental Science and Technology* 29(10), 2567-2573. ISSN 0013-936X.

Martínez-Huitle, C.A.; Quiroz, M.A. Comninellis, C. Ferro, S. & De Battisti, A. (2004). Electrochemical incineration of chloranilic acid using Ti/IrO_2, Pb/PbO_2 and Si/BDD electrodes. *Electrochimica Acta* 50(4), 949-956. ISSN 0013-4686.

Martinez-Huitle, C.A. & Ferro, S. (2006) Electrochemical oxidation of organic pollutants for the wastewater treatment: direct and indirect processes. *Chemical Society Reviews* 35, 1324-1340. ISSN 0306-0012.

Martínez-Huitle, C.A.; De Battisti, A.; Ferro, S.; Reyna, S.; Cerro, M. & Quiroz, M.A. (2008). Removal of the methamidophos pesticide from aqueous solution by electrooxidation using Pb/PbO_2, Ti/SnO_2 and Si/BDD electrodes. *Environmental Science and Technology* 42, 6929-6935. ISSN 0013-936X.

Masten, S.J. & Davies, H.R. (1994). The use of ozonation to degrade organic contaminants in wastewaters. *Environmental Science and Technology* 28(4), 180-185. ISSN 0013-936X.

Matthews, R.W. (1988). Kinetics of photocatalytic oxidation of organic solutes over titanium dioxide. *Journal of Catalysis* 111, 264-272.ISSN 0021-9517.

Matthews, R. & McEvoy, S. (1992). Photocatalytic degradation of phenol in the presence of near UV illuminated titanium dioxide. *Journal of Photochemistry and Photobiology A: Chemistry* 64(2), 231-246. ISSN 1010-6030.

Maurino, V.; Minero, C.; Pelizzetti, E. & Vincenti, M. (1999). Photocatalytic transformation of sulfonylurea herbicides over irradiated titanium dioxide particles. *Colloids and Surfaces A: Physicochemistry and Engineering Aspects* 151, 329-338. ISSN 0927-7757.

McDowall, R.; Boyle, C. & Graham, B. (2004). *Review of emerging, innovative technologies for the destruction and decontamination of POPs and the identification of promising technologies for use in developing countries*. Technical Report GF/8000-02-02-2205. The Scientific and Technical Advisory Panel of the GEF, United Nations Environmental Programme.

McMurray, T.A.; Dunlop, S.M. & Byrne, S.A. (2006). The photocatalytic degradation of atrazine on nanoparticulate TiO_2 films. *Journal of Photochemistry and Photobiology A: Chemistry* 182 (1), 43-51. ISSN 1010-6030.

Mezzanote, V.; Canziani, R.; Sardi, E. & Spada, L. (2005). Renoval of pesticidas by cmbined ozonation/attached biomasa process sequence. *Ozone: Science and Engineering* 27, 327-331. ISSN 0191-9512.

Mills, A.; Davies, R.H. & Worsley, D. (1993). Water purification by semiconductor photocatalysis. *Chemical Society Reviews* 22, 417-425. ISSN 0306-0012.

Mills, A. & Morris, S. (1993). Photomineralization of 4-chlorophenol sensitized by titanium dioxide: a study of the initial kinetics of carbon dioxide photogeneration. *Journal of Photochemistry and Photobiology A: Chemistry* 71(1), 75-83. ISSN 0045-6535.

Minero, C.; Pelizzetti, E.; Malato, S. & Blanco, J. (1996). Large solar plant photocatalytic water decontamination: Effect of operational parameters. *Solar Energy* 56 (5), 421-428.ISSN 0038-092X.

Miwa, D.W.; Malpass, G.R.P; Machado, S.A.S & Motheo, A.J. (2006). Electrochemical degradation of carbaryl on oxide electrodes. *Water Research* 40(17), 3281 - 3289. ISSN 0043-1354.

Moctezuma, E. (1999). Mineralization of pesticides in polluted water. *Chemosphere* 39 (3), 511-517. ISSN 0045-6535.

Mogyoródi, F.; Vidal, A.; Romero, M. & Sánchez, B. (1993). Photolytic and photocatalytic degradation of thiocarbamate pesticides in water. *ISES Solar World Congress*. August, 1993. Budapest, Hungry.

Mogyoródy, F. (2006a). Reaction pathways in the electrochemical degradation of thiocarbamate herbicides in NaCl solution. *Journal of Applied Electrochemistry* 36(6), 635-642. ISSN 0021-891X.

Mogyoródy, F. (2006b). Electrochemical degradation of thiocarbamates in NaCl solution. *Journal of Applied Electrochemistry* 36(7), 773-781. ISSN 0021-891X.

Momani, F.A.; Sans, C.; Contreras, S. & Esplugas, S. (2006a). Degradation of 2,4-dichlorophenol by combining Photoassisted Fenton reaction and biological treatment water. *Environmental Research* 78(6), 590-597. ISSN 0013-9351.

Momani, F.A. (2006b). Biodegradability enhancement of 2,4-dichlorophenol aqueous solutions by means of photo-Fenton reaction. *Environmental Engineering Science* 23(4), 722-733. ISSN 1092-8758.

Moza, P.N.; Hustert, K.; Pal, S. & Sukul, P. (1992). Photocatalytic degradation of pendimethanil and alachlor. *Chemosphere*, 25 (11), 1675-1682. ISSN 0045-6535.

Müller, T.S. (1998). Formation of toxic intermediates upon the photocatalytic degradation of pesticides. *Chemosphere* 26(9), 2043-2055. ISSN 0045-6535.

Muneer, M.; Theurich, J. & Bahnemann, D. (1998). Titanium dioxide mediated photocatalytic degradation of two major organic polutants: 1,2-diethyl phtalate and Diuron. *Journal of the Electrochemical Society* 98(5), 174-187. ISSN 0013-4651.

Muneer, M.; Qamar, M.; Saquib, M. & Bahnemann, D.W. (2005). Heterogeneous photocatalysed reaction of three selected pesticide derivatives propham, propachlor and tebuthiuron in aquous suspension of titanium dioxide. *Chemosphere* 61(4), 457-468. ISSN 0045-6535.

Murphy, O.J.; Hitchens, G.D.; Kaba, L. & Verotsko, C.E. (1992). Direct electrochemical oxidation of organics for wastewater treatment. *Water Research* 26(4), 443-451. ISSN 0043-1354.

Muszkat, L.; Bir, L. & Feigelson, L. (1992). Solar photodegradation of xenobiotic contaminants in polluted well water. *Journal of Photochemistry and Photobiology A: Chemistry* 65, 409-417. ISSN 1010-6030.

Muszkat, L.; Bi, L. & Feigelson, L. (1995). Solar photocatalytic mineralization of pesticides in polluted waters. *Journal of Photochemistry and Photobiology A: Chemistry* 87, 85-88. ISSN 1010-6030.

Muszkat, L.; Feigelson, L.; Bir, L. & Muszkat, K.A. (2002). Photocatalytic degradation of pesticides and bio-molecules in water. *Pesticide Management Science* 58(11), 1143-1148. ISSN 1526-0874.

Nguyen, T. & Ollis, D.F. (1984). Complete heterogeneously photocatalyzed transformation of 1,1- and 1,2-dibromoethane to CO_2 and HBr. *Journal of Physical Chemistry* 88, 3386-3388. ISSN 1520-5207.

Nikolaki, M.D.; Oreopoulou, A.G. & Philippopoulos, C.J. (2005). Photo-Fenton assited reaction of dimethoate in aqueous solutions. *Journal of Environmental Science and Health B*. 40(2), 233-246. ISSN 0360-1234.

Nishida, K. & Ohgaki, S. (1994). Sulfurated pesticide oxidation using titanium dioxide suspensions. *Water Science and Technology* 30(9), 39-46. ISSN 0273-1223.

Oancea, P. & Oncescu, T. (2008). The photocatalytic degradation of dichlorvos under solar irradiation. *Journal of Photochemistry and Photobiology A: Chemistry* 199(1), 8-13. ISSN 1010-6030.

Ohno, T.; Masaki, Y.; Hirayama, S. & Matsumara, M. (2001). TiO_2-photocatalyzed epoxidation of 1-decene by H_2O_2 under visible light. *Journal of Catalysis* 204(1), 163-168. ISSN 0021-9517.

Oller, J.; Gernjak, W.; Maldonado, M.I.; Fernandez, P.; Blanco, J.; Sanchez, J.A. & Malato, S. (2005). Photocatalytic treatment of dimethoate by solar photocatalysis at pilot plant scale. *Environmental Chemistry Letters* 3(5), 118-121. ISSN 1610-3653.

Oller, I.; Gernjak, W.; Perez, L.; Sanchez, J.A. & Malato, S. (2006a). Solar photocatalytic degradation of some hazardous water-soluble pesticides at pilot-plant scale. *Journal of Hazardous Materials* 138(3), 507-517. ISSN 0304-3894

Oller, I.; Gernjak, W.; Maldonado, M.I.; Perez-Estrada, L.A.; Sanchez, J.A. & Malato, S. (2006b). Solar photocatalytic degradation of some hazardous water-soluble pesticides at pilot-plant scale. Journal of Hazardous Materials 138(3), 507-517. ISSN 0304-3894.

Ollis, D.F. (1984). Heterogeneous photoassisted catalysis: Conversions of perchloroethyhlene, dichloroethane, chloroacetics acids and chlorobenzenes. *Journal of Catalysis* 88, 89-96.ISSN 0021-9517.

Ollis, D.F.; Pelizzetti, E. & Serpone, N. (1991a). Destruction of water contaminants. *Environmental Science and Technology* 25 (9), 1523-1528. ISSN 0013-936X.

Ollis, D.F.; Pelizzetti, E. & Serpone, N. (1991b). Photocatalyzed destruction of water contaminants. *Environmental Science and Technology* 25(9), 1522-1529. ISSN 0013-936X.

Ormad, M.P.; Miguel, N.; Lanao, M.; Mosteo, R. & Ovelleiro, J.L. (2010). Effect of application of ozonde combined with hydrogen peroxide and titanium dioxide in the removal of pesticides fro water. *Ozone: Science and Engineering* 32(1), 25-32.ISSN 0191-9512.

Orozco, S.; Bandala, E.R.; Arancibia, C.A.; Serrano, B.; Suarez, R. & Hernandez, I. (2008). Effect of iron salt on the color removal of water containing the azo-dye reactive blue using photo-assisted Fe(II)/H_2O_2 and Fe(III)/H_2O_2 systems. *Journal of Photochemistry and Photobiology A: Chemistry* 198, 144-149. ISSN 1010-6030.

O´Shea, K.E. (1997). TiO_2 photocatalytic degradation of dimethyl- and diethyl-methylphosphonate, effect of catalyst and environmental factors. *Journal of Photochemistry and Photobiology A: Chemistry* 107, 221-226. ISSN 1010-6030.

Ostra, M.; Ubide, C. & Zuriarrain, J. (2007). Interference modeling, experimental design and pre-concentration steps in validation of the Fenton's reagent for pesticide determination. *Analitica Chimica Acta* 548(1), 228-235. ISSN 0003-2670.

Oturan, M.A.; Aaron, J.J.; Aturan, N. & Pinson, J. (1999). Degradation of chlorophenoxyacid herbicides in aqueous media using a novel electrochemical meted. *Pesticide Science* 55(5), 558-562. ISSN 1526-4998.

Oturan, M.A.; Aaron, J.J.; Oturan, N. & Pinson, J. (1999). Degradation of chlorophenoxyacid herbicides in aqueous media, using a novel electrochemical method. *Pesticide Science* 55(5), 558-562. ISSN 1526-4998.

Oturan, M.A.; Peiroten, J.; Chartrin, P. & Acher, A.J. (2000). Complete destruction of p-nitrophenol in aqueous medium by electro-Fenton method. *Environmental Science and Technology* 34(16), 3474-3479. ISSN 0013-936X.

Oturan, M.A. (2000). An ecologically effective water treatment technique using electrochemically generated hydroxyl radicals for in situ destruction of organic pollutants: Application to herbicide 2,4-D. *Journal of Applied Electrochemistry* 30(4), 475-482. ISSN 0021-891X.

Oturan, M.A.; Oturan, N.; Lahitte, C. & Trevin, S. (2001). Production of hydroxyl radicals by electrochemically assisted Fenton's reagent: Application to the mineralization of an organic micropollutant, pentachlorophenol. *Journal of Electroanalytical Chemistry* 507(1-2), 96-102. ISSN 0022-0728.

Oturan, M.A. (2003). Removal of organophosphorous pesticides from water by electrogenerated Fenton's reagent. *Environmental Chemistry Letters* 1(3), 165-168. ISSN 1610-3661.

Ozcan, A.; Sahin, Y.; Koparal, A.S. & Oturan, M.A. (2008) Propham mineralization in aqueous medium by anodic oxidation using boron-doped diamond anode: influence of experimental parameters on degradation kinetics and mineralization efficiency. *Water Research* 42(12), 2889-2898. ISSN 0043-1354.

Pacheco, J.E.; Prairie, M.R. & Yellowhorse, L. (1993). Photocatalytic destruction of chlorinated solvents in water with solar energy. *Journal of Solar Energy Engineering* 115(3), 123-129. ISSN 0199-6231.

Parra, S.; Sarria, V.; Malato, S.; Peringer, P. & Pulgarin, C. (2000). Photochemical versus coupled photochemical-biological flow system for the treatment of two biorecalcitrant herbicides: metobromuron and isoproturon. *Applied Catalysis B: Environmental* 27(3), 153-168. ISSN 0926-3373.

Parra, S.; Malato, S. & Pulgarin, C. (2002). New integrated photocatalytic-biological flow system using supported TiO_2 and fixed bacteria for the mineralization of isoproturon. *Applied Catalysis B: Environmental* 36(2), 131-144. ISSN 0926-3373.

Parra, S.; Stanea, S.E.; Guasaquillo, I. & Thampi, K.R. (2004). Photocatalytic degradation of atrazine using suspended and supported TiO_2. *Applied Catalysis B: Environmental* 51(2), 107-116.ISSN 0926-3373.

Parreño, R.; Morales-Rubio, A. & De la Guardia, M. (1994). On-line catalytic photodegradation of aldicarb. *Journal of Flow Injection Analysis* 11(1), 79-93. ISSN 0911-775X.

Pathirana, H.M.K.K. (1997). TiO_2 oxidation of 3,4-DPA in water using UV irradiation. *Journal of Photochemistry and Photobiology A: Chemistry* 102, 273-277. ISSN 1010-6030.

Pelizzetti, E. (1985). Sunlight photodegradation of haloaromatic pollutants catalyzed by semiconductor particulate materials. *La Chimica e L´Industria* 67(11), 623-625. ISSN 0009-4315.

Pelizzetti, E. (1987). Using sunlight to fight pollution. *Chimica e L'Industria* 69 (10), 88-89. ISSN 0009-4315.

Pelizzetti, E. (1990a). Photodegradation of xenobiotic contaminant in polluted well water. Soil Science 150(2), 523-526. ISSN 0038-075X.

Pelizzetti, E. (1990b). Photocatalytic degradation of atrazine and other s-triazine herbicides. Environmental Science and Technology 22, 1559-1565. ISSN 0013-936X.

Pelizzetti, E.; Carlin, V.; Minero, C. & Grätzel, C. K. (1991). Enhancement of the rate of photocatalytic degradation on TiO_2 of 2-chlorophenol, 2,7-dichlorodibenzodioxin and atrazine by inroganic oxidizing species. *New Journal of Chemistry* 15 (5), 351-359. ISSN 1144-0546.

Pelizzetti, E. (1992). Identification of photocatalytic degradation pathways of 2-Cl-s-triazine herbicides and detection of their decomposition intermediates. *Chemosphere* 24 (7), 891-910. ISSN 0045-6535.

Pelizzetti, E.; Minero, C. & Pramauro, E. (1993a). Photocatalytic Process for Destruction of Organic Water Contaminants. In: *Chemical Reactor Technologies for Environ Safe Reactors and Products*. De Lasa H.I. (Ed) pp. 577-608. Kluwer Academic Pub., ISBN: 0792320328, Amsterdam.

Pelizzetti, E. & Minero, C. (1993). Mechanisms of the photo-oxidative degradation or organic pollutants over TiO_2 particles. *Electrochimica Acta* 38(1), 47-55. ISSN 0013-4686.

Percherancier, J.P.; Chapelon, R. & Pouyet, B. (1995). Semiconductor sensitized photodegradation of pesticides in water. The case of carbetamide. *Journal of Photochemistry and Photobiology A: Chemistry* 87, 261-266. ISSN 1010-6030.

Perez, M.H.; Penuela, G.; Maldonado, M.I.; Malato, O.; Fernández, P.; Oller, I., Gernjak W. & Malato, S. (2006). Degradation of pesticides in water using solar advanced oxidation proceses. *Applied Catalysis B: Environmental* 64 (3-4), 272-281. ISSN 0926-3373.

Panizza, P.A.; Michaud, G.; Cerisola, C. & Comninellis, C. (2001) Anodic oxidation of 2-naphtol at boron doped diamond electrodes. *Journal of Electroanalitical Chemistry* 10, 1-9. ISSN 0022-0728.

Peterson, M.W.; Turner, J.A. & Nozik, A.J. (1991). Mechanistic studies of the photocatalytic behavior of TiO_2. Particles in a photoelectrochemical slurry cell and the relevance to photodetoxification reactions. *Journal of Physical Chemistry B* 95(1), 221-225. ISSN 1089-5647.

Pichat, P.; D´Oliveira, J.C.; Maffre, J.F. & Mas, D. (1993a). Destruction of 2,4-dichlorophenoxyacetic acid (2,4-D) in water by TiO_2-UV, H_2O_2-UV or direct photolysis. In: *Photocatalytic Purification and Treatment of Water and Air*. Al-Hekabi, H.; Ollis, F. (Ed) Elsevier Sci. Publish. pp. 683-688. London, U.K. ISBN-10: 0444898557

Pichat, P.; Guillard, C.; Maillard, C.; Amalric, L. & D'Oliveira, J.C. (1993b). TiO_2 photocatalytic destruction of water aromatic pollutants: intermediates; properties-degradability correlation; effects of inorganic ions and TiO_2 surface area; comparison with H_2O_2 processes. In: *Photocatalytic Purification and Treatment of Water and Air*. Al-Hekabi, H.; Ollis, F. (Ed). Elsevier Sci. Publish. B.V., pp. 207-223. London, U.K. ISBN-10: 0444898557.

Pichat, P.; Vanner, S.; Dussaud, J. & Rubio, J.P. (2007). Field solar photocatalytic purification of pesticides-containing rinse waters from tractors cisterns used for grapevine treatment. *Solar Energy* 77(5), 533-542. ISSN 0038-092X.

Pignatello, J.J. (1992). Dark and photoassited Fe^{3+}-catalyzed degradation of chlorophenoxy herbicides by hydrogen peroxide. *Environmental Science and Technology* 26, 944-951. ISSN 0013-936X.

Pignatello, J.J. (1993). Degradation of pesticides by ferric reagents and peroxide in the presence of light. US Patent No. 5232484. Int. Classification A01N3738, 4370, C02F172.

Pignatello, J.J. & Sun, Y. (1995). Complete oxidation of metolachlor and methyl parathion in water by the photoassisted Fenton reaction. *Water Research* 29(8), 1837-1844. ISSN 0043-1354.

Ponce de Leon, C. & Pletcher, D. (1995). Removal of formaldehyde from aqueous solutions via oxygen reduction using a reticulated vitreous carbon cathode cell. *Journal of Applied Electrochemistry* 25(4), 307-314. ISSN 0021-891X.

Poulios, I. (1998). Photocatalytic decomposition of triclopyr over aqueous semiconductor suspensions. *Journal of Photochemistry and Photobiology A: Chemistry* 115, 175-183. ISSN 1010-6030.

Pramauro, E.; Vincent, M.; Augliaro, V. & Palmisano, L. (1993). Photocatalytic degradation of monuron in aqueous TiO_2 dispersions. *Environmental Science and Technology* 27, 1790-1795. ISSN 0013-936X.

Priyantha, N. & Weliwagamage, S. (2008). Interaction of thiram with glassy carbon electrode surfaces under applied potential conditions. *International Journal of Electrochemical Science* 3, 125-135. ISSN 1452-3981.

Pruden, A.L. & Ollis, D.F. (1983). Photoassisted heterogeneous catalysis: the degradation of trichloroethylene in water. *Journal of Catalysis* 82, 404-417. ISSN 0021-9517.

Pulgarin, C. & Kiwi, J. (1996). Overview on Photocatalytic and Electrocatalytic pretreatment of Industrial Non-Biodegradable Pollutants and Pesticides. *Chimie* 50, 50-55. ISSN 1631-0748.

Pulgarin, C.; Invernizzi, M.; Parra, S.; Sarriá, V.; Polania, R. & Peringer, P. (1999). Strategy for the coupling of photochemical and biological flow reactors useful in mineralization of biorecalcitrant industrial pollutants. *Catalysis today* 54(2), 341-352. ISSN 0920-5861.

Qamar, M.; Muneer, M. & Bahnemann, D. (2006). Heterogeneous photocatalysed degradation of two selected pesticide derivatives trichlopyr and daminozid in aquous suspensions of titanium dioxide. *Journal of Environmental Management* 180(2), 99-106. ISSN 0301-4797.

Quiroz, M.A.; Reyna, S.; Martínez-Huitle, C.A.; Ferro, S. & De Battisti, A. (2005) Electrocatalytic oxidation of *p*-nitrophenol from aqueous solutions at Pb/PbO_2 anodes. *Applied Catalysis B* 59, 259-266. ISSN 0926-3373.

Quiroz, M.A.; Ferro, S.; Martínez-Huitle, C.A. & Meas Vong, Y. (2006). Boron doped diamond electrode for the wastewater treatment. *Journal of the Brazilian Chemical Society* 17, 227-236. ISSN 0103-5053.

Rachel, A.; Subrahmanyam, M. & Boule, P. (2002). Comparison of photocatalytic efficiencies of TiO_2 in suspended and immobilized form for the photocatalytic degradation of nitrobenzenesulfonic acids. *Applied Catalysis B: Environmental* 37(4), 301-308. ISSN 0926-3373.

Rajeshwar, K. (1996). Photochemical strategies for abating environmental pollution. *Chemistry and Industry* 12, 454-458. ISSN 0009-3068.

Richard, C. & Bengana, S. (1996). pH effect in the photocatalytic transformation of a phenyl-urea herbicide. *Chemosphere* 33 (4), 635-641. ISSN 0045-6535.

Rodrigo, M.A.; Michaud, P.A.; Duo, I.; Panizza, M.; Cerisola, G. & Comninellis, C. (2001). Oxidation of 4-chlorophenol at boron-doped diamond electrode for wastewater treatment. *Journal of Electrochemical Society* 148(5), D60-D64. ISSN 0013-4651.

Rodgers, J.D. & Bunce, N.J. (2001). Electrochemical treatment of 2,4,5-trinitrotoluene and related compounds. *Environmental Science and Technology* 35(2), 406-410. ISSN 0013-936X.

Rodriguez, M.; Sarria, V.; Espulgas, S. & Pulgarin, C. (2002). Photo-Fenton treatment of a biorecalcitrant wastewater generated in textile activities: biodegradability of photo-treated solution. *Journal of Photochemistry and Photobiology A: Chemistry* 151(1), 129-135. ISSN 1010-6030.

Roe, B.A. & Lemley, A.T. (1997). Treatment of two insecticides in an electrochemical Fenton system. *Journal of Environmental Science and Health B* 32(2), 261-281. ISSN 1532-4109.

Romero, M.; Blanco, J.; Sánchez, B.; Vidal, A.; Malato, S.; Cardona, A.I. & Garcia, E. (1999a). Solar photocatalytic degradation of water and air pollutants: challenges and perspectives. *Solar Energy* 66(2), 169-182. ISSN 0038-092X.

Romero, R.L.; Alfano, O.M. & Cassano, A.E. (1999b). UV radiation distribution in an annular photocatalytic reactor using two different types of titanium dioxide catalyst. *Journal of Advanced Oxidation Technologies* 4, 27-34. ISSN 1203-8407.

Sabate, J.; Anderson, M.; Aguado, M.; Gimenez, J. & Cervera, S. (1992). Comparison of TiO_2 powder suspensions and TiO_2 ceramic membranes supported on glass as photocatalytic system in the reduction of chromium (VI). *Journal of Molecular Catalysis* 71(1), 57-68. ISSN 1381-1169.

Sabin, F. (1992). Photo-oxidation of organic compounds in the presence of titanium dioxide: determination of the efficiency. *Journal of Photochemistry and Photobiology A: Chemistry* 63, 99-106. ISSN 1010-6030.

Sakkas, V.A.; Arabatzis, M.; Konstantinou, I.K.; Dimou, A.D.; Albanis, T.A. & Falaras, P. (2004). Metolachlor photocatalytic degradation using TiO_2 photocatalysts. *Applied Catalysis B: Environmental* 49(3), 195-205. ISSN 0926-3373.

Sakellarides, T.; Sakkas, V.A.; Lambropoulou, D.A. & Albanis, T.A. (2004). Application of solid phase microextraction for photocatalytic studies of fenitrothion and methyl-parathion in aqueous TiO_2 suspensions. International *Journal of Environmental Analytical Chemistry* 84(1-3), 161-172. ISSN 0306-7319.

Sakkai, V.A.; Dimou, A.; Pitarakis, K.; Mantis, G. & Albanis, T. (2005). TiO_2 photocatalyzed degradation of diazinon in aqueous medium. *Environmental Chemistry Letters* 3(2), 1610-1636. ISSN 1610-3661.

Saltmiras, D.A. & Lemley, A.T. (2000). Degradation of ethylene thiourea (ETU) with three Fenton treatment. *Journal of Agricultural and Food Chemistry* 48(12), 6149-6157. ISSN 0021-6531.

Sanjay, P.K.; Deosarkar, P.; Sudhir, B.S.; Jacob, A.M. & Vishwas, G.P. (2004). Photocatalytic degradation of 2,4-dichlorophenoxiacetic acid using concentrate solar radiation: Batch and continuous operation. *Industrial and Engineering Chemistry Research* 43(26), 8178-8187. ISSN 0888-5885.

Sarria, V.; Parra, S.; Invernizzi, M.; Peringer, P. & Pulgarin, C. (2001). Photochemical-biological treatment of a real industrial biorecalcitrant wastewater containing 5-amino-6-methyl-2-benzimidazolone. *Water Science and Technology* 44(5), 93-101. ISSN 0273-1223.

Sarria, V.; Parra, S.; Adler, N.; Peringer, P.; Benitez, N. & Pulgarin, C. (2002). Recent developments in the coupling of photoassisted and aerobic biological proceses for the treatment of biorecalcitrant compounds. *Catalysis Today* 76(2), 301-315. ISSN 0920-5861.

Sarria, V.; Deront, M.; Peringer, P. & Pulgarin, C. (2003a). Degradation of a biorecalcitrant dye precursor present in industrial wastewater by a new integrated iron (III) Photoassisted-biological treatment. *Applied Catalysis B: Environmental* 40(3), 231-246. ISSN 0926-3373.

Sarria, V.; Kenfack, S.; Guillod, O. & Pulgarin, C. (2003b). An innovative coupled solar-biological system at field pilot scale for the treatment of biorecalcitrant pollutants. *Journal of Photochemistry and Photobiology A: Chemistry* 139 (1), 89-99. ISSN 1010-6030.

Sauer, T.; Cesconet, G.; Jose, H.J. & Moreira, R.F.P.M. (2002). Kinetics of photocatalytic degradation of reactive dyes in a TiO_2 slurry reactor. *Journal of Photochemistry and Photobiology A: Chemistry* 149(1-3), 147-154. ISSN 1010-6030.

Sawyer, D.T.; Sobkowiak, A. & Matsushita, T. (1996). Metal MLX; M=Fe, Cu, Co, Mn/hydroperoxide-induced activation of dioxygen for the oxygenation of hydrocarbons: oxygenated Fenton chemistry. *Accounts of Chemical Research* 29(9), 409-416. ISSN 1520-4898.

Scherer, E.M.; Wang, Q.; Hay, A.G. & Lemley, A.T. (2004a). The binary treatment of aqueous metribuzin using anodic fenton treatment and biodegradation. *Archives of Environmental Contamination and Toxicology* 47(2), 154-161. ISSN 0090-4341.

Scherer, E.M.; Wang, Q.Q.; Hay, A.G. & Lemley, A.T. (2004b). The binary treatment of aqueous metribuzyn using anodic Fenton treatment and biodegradation. *Archives on Environmental Contamination and Toxicology* 47, 154-161. ISSN 0090-4341.

Scott, J.P. & Ollis, D.F. (1995). Integration of chemical and biological oxidation processes for water treatment: review and recommendations. *Environmental Progress* 14(2), 88-103. ISSN 1547-5921.

Scott, J.P. & Ollis, D.F. (1997). Integration of chemical and biological oxidation processes for water treatment II. Recent developments. *Journal of Advanced Oxidation Technologies* 2, 374-380. ISSN 1203-8407.

Sedlak, P.; Brodilove, J. & Lunak, S. (1992). Photochemical degradation of pesticides, photocatalytic effect of Fe(III) ions on methylene blue sensitised interaction of 4-chlorophenoxiacetic acid with H_2O_2. *Bulletin of Environmental Contamination and Toxicology* 49(6), 834-838. ISSN 0007-4861.

Senthilnathan, J. & Pjilip, L. (2010). Photocatalytic degradation of lindane under UV and visible light using N-doped TiO_2. *Chemical Engineering Journal* 161 (1-2), 83-92. ISSN 1385-8947.

Serra, F.; Trillas, M.; Garcia, J. & Domenech, J. (1994). Titanium dioxide-photocatalyzed oxidation of 2,4-dichlorophenol. *Journal of Environmental Science and Health A: Environmental* 29(7), 1409-1421. ISSN 1093-4529.

Schiavello, M.; Augugliaro, V.; Loddo, V.; Lopez, M.J. & Palmisano, L. (1999). Quantum yield of heterogenous photocatalytic systems: further applications of an experimental method for determining the absorbed photon flow. *Research on Chemical Intermediates* 25(2), 213-227. ISSN 0922-6168.

Scott, J.P. & Ollis, D.F. (1995). Integration of chemical and biological oxidation processes for water treatment review and recommendations. *Environmental Progress* 14, 88-103. ISSN 1547-5921.

Serpone, N.; Lawless, D.; Khairutdinov, R. & Pelizzetti E. (1995). Subnanosecond relaxation dynamics in TiO_2 colloidal sols (Particle size= 1.0-13.4 nm). Relevance to heterogeneous photocatalysis. *Journal of Physical Chemistry B* 99, 16655-16661. ISSN 1089-5647.

Serpone, N. (1996). Relative photonic efficiencies and quantum yields in heterogeneous photocatalysis. *EPA Newsletter* 58, 54-81. ISSN 0966-0429.

Shankar, M.V.; Cheralathan, K.K.; Arabindoo, B.; Palamichamy, M. & Murugesan, V. (2004). Enhanced photocatalytic activity for the destruction of monocrotophos pesticide by TiO_2. *Journal of Molecular Catalysis A: Chemical* 223(1), 195-200. ISSN 1381-1169.

Sharma, M.V.P.; Kumari, V.D. & Subramanyam, M. (2008a). Photocatalytic degradation of isoproturon herbicide over TiO_2/Al-MCM-41 composite system using solar light. *Chemosphere* 72(4), 644-651. ISSN 0045-6535.

Sharma, M.V.P.; Durgakumari, V. & Subramanyam, M. (2008b). Solar photocatalytic degradation of isoproturon over TiO_2/H-MOR composite system. *Journal of Hazardous Materials* 160(2-3), 568-675. ISSN 0304- 3894.

Sharma, M.V.P.; Lalitha, K.D.; Kumari, V. & Subramanyam, M. (2008c). Solar photocatalytic mineralization of isoproturon over TiO_2 /HY composite systems. *Solar Energy Materials and Solar Cells* 92(3), 332-342.ISSN 0927-0248.

Sharma, M.V.P.; Sadanandam, G.; Ratnamala, A.; Kumari V.D. & Subramanyam, M. (2009). An efficient and novel porous nanosilica supported TiO2 photocatalyst for pesticide degradation using solar light. *Journal of Hazardous Materials* 171 (1-3), 626-633. ISSN 0304-3894.

Sharma, M.V.P.; Kumari, V.D. & Subramanyam, M. (2008d). TiO_2 supported over SBA-15: An efficient photocatalyst for the pesticide degradation using solar light. *Chemosphere* 73(9), 1562-1569.ISSN 0045-6535.

Singh, H.K. & Muneer, M. (2004). Photodegradation of herbicide 2,4-dichlorophenoxiacetic acid in aqueous suspensions of titanium dioxide. *Research on Chemical Intermediates* 30(3), 1568-1575. ISSN 0922-6168.

Sojic, D.V.; Despotovic, V.N.; Abazovic, N.D.; Comor, M.I. & Abramovic, B.F. 2010. Photocatalytic degradation of selected herbicides in aqueous suspensios of doped titania under visible light irradiation. *Journal of Hazardous Materials* 179(1-3), 49-56. ISSN 0304-3894.

Somich, C.J.; Muldoon, M.T. & Kearney, P.C. (1990). On site treatment of pesticide waste and rinsate using ozone and biologically active soil. *Environmenal Science and Technology* 24, 745-749. ISSN 0013-936X.

Skinner, D.E.; Colombo, D.P.; Cavaleri, J.A. & Bowman, R.M. (1995). Femtosecond investigation of electron trapping in semiconductor nanoclusters. *Journal of Physical Chemistry* 99, 7853-7856. ISSN 1089-5647.

Stucki, S.; Kotz, R.; Carcer, B. & Suter, W. (1991). Electrochemical waste water treatment using high overvoltage anodes Part II: Anode performance and applications. *Journal of Applied Electrochemistry* 21(2), 99-104. ISSN 0021-891X.

Sturini, M.; Fasani, E.; Prandi, C. & Albini, A. (1997). Titanium dioxide- photocatalyzed degradation of some anilides. *Chemosphere* 35(5), 931-937. ISSN 0045-6535.

Sullivan, J.M. (1994). TiO_2 catalyzed photo oxidation of atrazine in dilute aqueous solutions under solar irradiation: process development. *Proceedings of the ASME International Solar Energy Conference*. San Francisco, USA. March, 1994.

Sun, Y. & Pignatello, J.J. (1992). Chemical treatment of pesticide wastes. Evaluation of Fe(III) chelates for catalytic hydrogen peroxide oxidation of 2,4-D at circumneutral pH. *Journal of Agricultural and Food Chemistry* 40, 322-327. ISSN 0021-8561.

Sun, Y. & Pignatello, J.J. (1993a). Organic intermediates in the degradation of 2,4-dichlorophenoxyacetic acid by Fe^{3+}/H_2O_2 and $Fe^{3+}/H_2O_2/UV$. *Journal of Agricultural and Food Chemistry* 41, 1139-1142. ISSN 0021-8561.

Sun, Y. & Pignatello, J.J. (1993b). Photochemical reactions involved in the total mineralization of 2,4-D by $Fe^{3+}/H_2O_2/UV$. *Environmental Science and Technology* 27, 304-310. ISSN 0013-936X.

Sun, Y. & Pignatello, J.J. (1993c). Activation of hydrogen peroxide by iron (III) chelates for abiotic degradation of herbicides and insecticides in water. *Journal of Agricultural and Food Chemistry* 41, 308-312. ISSN 0021-8561.

Sun, Y. & Pignatello, J. (1995). Evidence for surface dual hole-radical mechanism in the TiO_2 photocatalytic oxidation of 2,4-dichlorophenoxyacetic acid. *Journal of Environmental Science and Technology* 29, 2065-2072. ISSN 1735-2630.

Tahar, N. B. & Savall, A. (1998) Mechanistic aspects of phenol electrochemical degradation by oxidation on a Ta/PbO_2 anode, *Journal of the Electrochemical Society* 145, 3427-3434. ISSN 0013-4651.

Tanaka, K.; Abe, K.; Sheng, C. Y. & Hisanaga, T. (1992). Efficient photocatalytic degradation of chloral hydrate in aqueous semiconductor suspension. *Environmental Science and Technology* 26 (12), 2534-2536. ISSN 0013-936X.

Tang, W.Z. & Huang, C.P. (1996). 2,4-dichlorophenol oxidation kinetics by Fenton's reagent. *Environmental Technology* 17, 1371-1378. ISSN 0959-3330.

Tapalov, A.; Molnar, V.; Abramovic, B.; Korom, S. & Pericin, D. (2003). Photocatalytic removal of the insecticide fenitrothion from water sensitized with TiO_2. *Journal of Photochemistry and Photobiology A: Chemistry* 160(3), 195-201. ISSN 1010-6030.

Tennakone, K. (1997). Carbamic pesticide photocatalytic oxidation. *Water Research* 31(8), 1909-1912. ISSN 0043-1354.

Terashima, Y.; Ozaki, H.; Giri, R.R.; Tano, T.; Nakatsuji, S.; Takanami R. & Taniquchi, S. (2006). Photocatalytic oxidation of low concnetration 2,4-dichlorophenoxiacetic acid solutions with new TiO_2 fibber catalyst in a continuous flow reactor. *Water Science and Technolgy* 54(8), 55-63. ISSN 0273-1223.

Texier, I. (1999a). Organophosphorous pesticides mineralization using powdered titanium dioxide and UV radiation. *Journal of Physics* 9(3), 289-294. ISSN 0953-8984.

Texier, I. (1999b). Oxidation by-products identification of pesticide TiO_2 photocatalysis. *Journal of Physics* 96, 430-436. ISSN 0953-8984.

Torimoto, T.; Ito, S.; Kuwabata, S. & Yoneyama, H. (1996). Effects of absorbents used as supports for titanium dioxide loading on photocatalytic degradation of polypyazimide. *Environmental Science and Technology* 30, 1275-1281. ISSN 0013-936X.

Trillas, M.; Peral, J. & Domenech, X. (1995). Kinetics constant in photocatalytic process. *Applied Catalysis B: Environmental* 5, 377-387. ISSN 0926-3373.

Tseng, J.M. & Huang, C.P. (1991). Removal of chlorophenols from water by photocatalytic oxidation. *Water Science and Technology* 23, 377-387. ISSN 0273-1223.

Vidal, A. (1991). Degradación fotocatalítica contaminantes en agua: Catalizadores soportados sobre matrices. IER-CIEMAT. Proy. M8H01. December, 1991.

Vidal, A. (1998). Developments in solar photocatalysis for water purification. *Chemosphere* 36(12), 2593-25606. ISSN 0045-6535.

Vidal, A.; Dinya, Z. & Mogyorod, F. (1999). Photocatalytic degradation of thiocarbamate herbicide active ingredients in water. *Applied Catalysis B: Environmental* 21(4), 259-267. ISSN 0926-3373.

Vlyssides, A.; Barampouti, E.M.; Mai, S.; Arapoglou, D. & Kotronarou, A. (2004). Degradation of methylparathion in aqueous solution by electrochemical oxidation. Environmental Science and Technology 38, 6125-6131. ISSN 0013-936X.

Vlyssides, A.; Arapoglou, D.; Mai, S. & Barampouti, E.M. (2005a). Electrochemical detoxification of four phosphorothioate obsolete pesticides stocks. *Chemosphere* 58(4), 439-447. ISSN 0045-6535.

Vlyssides, A.; Arapoglou, D.; Mai, S.; Barampouti E.M. (2005b). Electrochemical oxidation of two organophosphoric obsolete pesticide stocks. *International Journal of Environmental and Pollution* 23(3), 289-299. ISSN 0957-4352.

Wadley, S. & Waite, T.D. (2002). Photo-Fenton oxidation of pesticides. *Water Supply* 2(5-6), 249-256. ISSN 1606-9749.

Wang, Y. & Hong, C.S. (1999). Effect of hydrogen peroxide, periodate and persulfate on photocatalysis of 2-chlorobiphenil in aqueous TiO_2 suspensions. *Water Research* 33, 2031-2036. ISSN 0043-1354.

Wang, L. & Lemley, A.T. (2001). Kinetic model and optimisation of 2,4-D degradation by anodic Fenton treatment. *Environmental Science and Technology* 35(22), 4509-4514. ISSN 0013-936X.

Wang, Q.; Lemley, A.T. & Saltmiras, D.A. (2003). Anodic Fenton degradation of pesticides. In: *Pesticide decontamination and detoxification*. Gan, J.J.; Zhu, P.C.; Aust, S.D. & Lemley, T.A. (Ed.) Oxford University Press. London, UK. ISBN 0841238472.

Wang, Q. & Lemley, A.T. (2004). Kinetic effect of humic acid on alachlor degradation by anodic Fenton treatment. *Journal of Environmental Quality* 33, 2343-2352. ISSN 0045-2425.

Wang, Q.; Schere, E.M. & Lemley, A.T. (2004). Metribuzin degradation by membrane anodic Fenton treatment and its interaction with Ferric ions. *Environmental Science and Technology* 38(4), 1221-1227. ISSN 0013-936X.

Wong, C.C. & Chu, W. (2003). The hydrogen peroxide-assisted photocatalytic degradation of alachlor in TiO_2 suspensions. *Environmental Science and Technology* 37(10), 2310-2316. ISSN 0013-396X.

Wong, C.C. & Chu, W. (2003). The hydrogen peroxide-assisted photocatalytic degradation of alachlor in TiO_2 suspensions. *Environmental Science and Technology* 37(10), 2310-2316. ISSN 0013-936X.

Wong, C.C. & Chu, W. (2003). The direct photolysis and photodegradation of alachlor at different TiO_2 and UV sources. *Chemosphere* 50(8), 981-987. ISSN 0045-6335.

Wu, Z.C. & Zhou, M.H. (2001). Partial degradation of phenol by advanced electrochemical oxidation process. *Environmental Science and Technology* 35(13), 2698-2703. ISSN 0013-936X.

Xu, A.W.; Gau Y. & Liu, H.Q. (2002). the preparation, characterization and their photocatalytic activities of rar-earth-doped TiO_2 nanoparticles. *Journal of Catalysis* 207(2), 151-157. ISSN 0045-6535.

Yamazaki, I. & Piette, L.H. (1991). EPR spin-trapping study on the oxidizing species formed in the reaction of the ferrous ion with hydrogen peroxide. *Journal of the American Chemical Society* 113, 7588-7593. ISSN 0002-7836.

Yatmaz, H.C. & Uzman, Y. (2009). Degradation of Pesticide Monochrotophos from Aqueous Solutions by Electrochemical Methods. International. *Journal of Electrochemical Sciences* 4, 614-626. ISSN 1452-3981.

Yeber, C.; Rodriguez, J.; Baeza, J.; Freer, J.; Zaror, C.; Duran, N. & Mansilla, H.D. (1999). Toxicity abatement and biodegradability enhancement of pulp and paper mill bleanching effluent by advanced chemical oxidation. *Water Science and Technology* 40(11-12), 337-342. ISSN 0273-1223.

Yu, J.J. (2002). Removal of organophosphate pesticides from wastewater by supercritical carbon dioxide extraction. *Water Research* 36(4), 1095-1101. ISSN 0043-1354.

Zaleska, A.; Hupka, J.; Wiergowsky, M. & Biziuk, M. (2000). Photocatalytic degradation of lindane, p,p'-DDT and methoxychlor in an aqueous environment. *Journal of Photochemistry and Photobiology A: Chemistry* 135(2-3), 213-220. ISSN 1010-6030.

Zepp, G.R.; Helz, D.G. & Corsby, D.G. 1994. *Aquatic surface photochemistry*. Lewis Publishers, Boca Raton, US.ISBN 0873718712.

Zhanqi, G.; Shaoqui, Y.; Na, T. & Cheng, S. (2007). Microwave assisted rapid and complete degradation of atrazine using TiO_2 nanotube photocatalyst suspensions. *Journal of Hazardous Materials* 145(3), 424-430. ISSN 0304-3894.

Zoh, K.D.; Kim, T.S.; Kim, J.G. & Choi, K.H. (2005). Degradation of parathion and the reduction of acute toxicity in TiO_2 photocatalysis. *Water Science and Technology* 52(8), 45-52. ISSN 0273-1223.

Zoh, K.D.; Kim, T.S.; Kim, J.G.; Choi, K.H. & Yi, S.M. (2006). Parathion degradation and toxicity reduction in solar photocatalysis and photolysis. *Water Science and Technology* 53(3), 1-8. ISSN 0273-1223.

35

Low-Cost Sorbent for Removing Pesticides during Water

Katarzyna Ignatowicz
Białystok University of Technology
Poland

1. Introduction

Pesticides are toxic chemicals for fighting against various diseases and pests. These compounds are carcinogenic, teratogenic, embryotoxic, and mutagenic. Expired or not used, they become very dangerous wastes, that when improperly stored, penetrate in uncontrolled way to the natural environment making the threat for all living forms. Therefore, the Sophia Declaration was created to underline the negative influences of durable organic pollutants, useless plant protection means, and other substances unsafe for human, environment, animals, and natural resources conditions such as ground waters or soil, and to emphasize the economic results related. The Declaration stresses the accelerating activities to remove above mentioned pollutants and appeals to governments and local organizations that removal of useless dangerous substances was a priority. It also turns to European Union and other sponsors to support domestic initiatives that introduce strategies of removal the durable organic pollutants, useless plant protection means, and other dangerous substances.

The past years remained dozens of thousands of tons of pesticide wastes that have been stored since 50's of the 20th century. There is also possibility to worsen the construction condition along with the occurrence of corroding the concrete bunkers and wells, the outdated pesticides are deposited, and in consequence, a toxic leakage. The leakage can be transported by underground water and then in a form of so-called underground inflow, it is caught by a network of surface waters. Therefore, there is a necessity to search for the solutions to reduce the pesticide migration in an environment as well as to introduce new concepts. Thus, it seems to be purposeful to undertake studies upon application of sorption process using selected adsorbents as a screen for pesticides in order to reduce their migration from other graveyards, stores and contaminated soils and concrete. The manuscript presented the study upon the possibility to removal of pesticide from aqueous solutions by using low-cost sorbent.

2. Material and methods

2.1 Sorbates and sorbent

On a basis of literature data and own studies, chloroorganic pesticides that most often occurred near the graveyards at the highest concentrations were selected as representative sorbats. (Ignatowicz, 2008; 2009) Individual pure active substances (DDT, DDE, DDD) were applied. Technical grade DDT of 99,8±0,1% purity, DDE 99,8±0,2% purity and DDD of 98,5±0,3% purity obtained from Institute of Industrial Organic Chemistry Analytical Department in Poland were used as an adsorbates. A sample solution of pesticide has been prepared by dissolving 1 g of pesticide in 10 ml of methanol and then diluted to 1 L with doubly distilled deionised water. The concentrations of prepared solutions were applied: 5 mg pesticide per litre.

Fig. 1. Compost prisms.

Properties of compost						
Manurial (mg/kg$_{dm}$)						
Ca	Mg	Nog	N-NH_4^+	Pog	C	K
20.2	4.2	8.1	0.2	6.1	265.8	2.2
Metal (mg/kg$_{dm}$)						
Pb	Cu	Cd	Cr	Ni	Zn	Hg
5.4	25.1	0.3	4.3	3.5	123	0.2
Permissible standard						
500	800	10	500	100	2500	5
Other (%)						
pH	Hydration		Dry mass		Organic matter	
5.88	73.7		26.3		70.3	

Table 1. The characteristic of compost.

Sewage sludge achieved directly from dairy treatment plant "Mlekovita" in Wysokie Mazowieckie (Dąbrowski, 2006) with sawdust addition composted under natural conditions in Rudka Forest Inspectorate was used as natural adsorbents. (Dąbrowski, 2005) „Mlekovita" is the largest dairy plant in Podlasie region, and one of the largest in Poland.

Individual biological sewage treatment station generates almost 1 100 tons of dry sludge that can be re-used. The characteristics of the compost are given in Table 1. The used waste products comply with the requirements of the ordinance of the Minister of the Environment concerning agricultural usage of municipal sewage sludge (Dz.U. z 2002 r., nr 134, poz. 1140).

2.2 Sorption procedure

Studies under static conditions were performed in accordance to methodology applied in Belgium, Germany, France, Italy, England, USA, Poland and other. (Spadotto, 2003; Ignatowicz, 2009; Hamadi, 2004; Kumar, 2003; Witbowo, 2007; Yuh-Shan, 2006; Mashayekhi, 2006; Tsui, 2007) They were aimed at plotting the adsorption isotherms due to which it is possible to compare the sorption capacities of different adsorbats on different adsorbents. Selected adsorbent, previously degassed, washed with distilled water and dried, was ground in spherical mortar and dried in electric drier at 150 °C for 3 hours till constant weight. Such prepared sorbent served for weighing following samples: 0.001, 0.002; 0.005; 0.01; 0.025 g per 100 ml solution. Representative samples of adsorbent were added into the conical flasks with glass stopper and containing working solution of the pesticide (5 mg L^{-1}). Flasks were shaken in electric oven at constant oscillation amplitude (9) for 24 hours, and then remained for 48 hrs to reach a complete adsorption equilibrium. After that, samples were subjected to double filtration on soft filter paper. Then, pesticide concentration in a filtrate according to obligatory methods was determined using gas chromatograph AGILENT6890. (Balinova, 1996; Siepak, 2001, 2009; Hussen, 2007; Munoz, 2009) Freundlich's, Langmuir's, Temkin, Jovanovic, BET's and Huttig's isotherms (Atkins, 2006) were plotted on a base of achieved results applying Statistica software in order to analyze the processes.

2.3 Analytical procedure

Pesticide concentrations were determined in collected samples in accordance to obligatory methodology using gas chromatograph AGILENT6890 equipped with ECD and NPD detector. The injector temperature was 210 °C and the flow rate of helium was 1.0 mL min^{-1}. The column DB (35m length 0.32 mm i.d. 0.5 µm film thickness) temperature was set at 120 °C for 2 min and increased at a rate of 13 °C min^{-1} to 190 °C. The temperature was finally increased to 295 °C and maintained isothermally for 20 min.

Moreover, after sample digestion according to EPA 3015 procedure using microwave digester Mars 5, also metals concentrations were determined by means of ICP-AES technique, except of mercury determined by means of CV-AAS technique. (Balinova, 1996; Siepak, 2001, 2009; Hussen, 2007; Munoz, 2009) Chromatogram of chloroorganics pesticide present Figure 2.

3. Sorption isotherms

Among the several existing isotherms, the sorption datas were subjected to four commonly used isotherms models (Tab. 2), namely Langmuir (1918), Freundlich (1894), BET (1938), Huttig (1948), Jovanovic (1969) and Temkin (1963), to evaluate the maximum saturation capacity of adsorbent.(Atkins, 2006)

```
Data File C:\HPCHEM\1\DATA\30112007\010F1001.D                Sample Name: W/99/07

    woda

    Injection Date   : 2007-11-30 18:49:19          Seq. Line :  10
    Sample Name      : W/99/07                       Location : Vial 10
    Acq. Operator    : Ewa Rutkowska                      Inj :   1
    Acq. Instrument  : Instrument 1                Inj Volume : 2 µl
    Method           : C:\HPCHEM\1\METHODS\NP-EC.M
    Last changed     : 2007-11-30 13:04:44  by Ewa Rutkowska
```

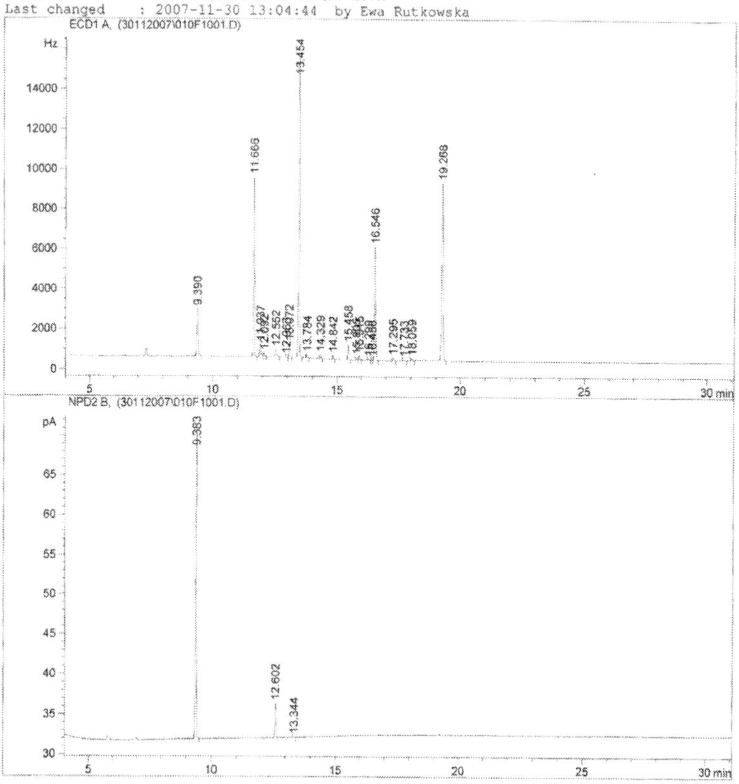


```
                      External Standard Report

    Sorted By              :      Signal
    Multiplier             :      1.0000
    Dilution               :      1.0000
    Use Multiplier & Dilution Factor with ISTDs

Instrument 1 2007-12-03 11:14:18  Ewa Rutkowska                     Page 1 of 2
```

Fig. 2. The chromatogram of chloroorganic pesticide.

3.1 Freundlich model

The first mathematical fit to an isotherm was published by Freundlich and is a purely empirical formula for microporous and heterogeneous adsorbates:

$$q_e = K_F c^{1/n} \quad (1)$$

where q_e ($mg.g^{-1}$) is the amount of chloro - pesticide adsorbed on the adsorbent surface at equilibrium), c ($mg.L^{-1}$) the pesticide concentration in aqueous solutions at equilibrium, K_F ($mg.g^{-1}$) - constant - is the maximum multilayer adsorption capacity and $1/n$ is a characteristic constant which measures the adsorption intensity (k and n are empirical constants for each adsorbent-adsorbate pair at a given temperature). As the temperature increases, the constants k and n change to reflect the empirical observation that the quantity adsorbed rises more slowly and higher pressures are required to saturate the surface. The linear form of the Freundlich isotherm is shown in Table 2.

3.2 Langmuir model

Langmuir isotherm is a semi-empirical isotherm derived from a proposed kinetic mechanism. Langmuir isotherm is a model for monolayer localized physical adsorption on homogeneous surface; may be extended with heterogeneity effects, lateral interactions and multilayer effects. It is based on four assumptions:

- The surface of the adsorbent is uniform, that is, all the adsorption sites are equivalent,
- Adsorbed molecules do not interact,
- All adsorption occurs through the same mechanism,
- At the maximum adsorption, only a monolayer is formed: molecules of adsorbate do not deposit on other, already adsorbed, molecules of adsorbate, only on the free surface of the adsorbent.

The Langmuir equation may be written as:

$$q_e = q_m K C_e / 1 + K C_e \quad (2)$$

where q_e ($mg.g^{-1}$) is the amount of pesticide adsorbed on the adsorbent surface at equilibrium, C_e ($mg.L^{-1}$) the pesticide concentration in aqueous solutions at equilibrium, q_m ($mg.g^{-1}$) is the maximum monolayer adsorption capacity, K (L mg^{-1}) is the constant related to the free energy of adsorption. Eq. (1) can be linearized to five different linear forms as shown in Table 2.

3.3 BET model

Often molecules do form multilayer, that is, some are adsorbed on already adsorbed molecules and the Langmuir isotherm is not valid. In 1938 Stephan Brunauer, Paul Emmett, and Edward Teller developed a model isotherm that takes that possibility into account. The Langmuir isotherm is usually better for chemisorption and the BET isotherm works better for physisorption for non-microporous surfaces. The BET equation may be written as:

$$A = ac / (1+c)(1+Kc) \quad (3)$$

where A ($mg.g^{-1}$) is the amount of pesticide adsorbed on the adsorbent surface at equilibrium, c ($mg.L^{-1}$) the pesticide concentration in aqueous solutions at equilibrium, a ($mg.g^{-1}$) is the maximum multilayer adsorption capacity, K ($L\ mg^{-1}$) is the constant related to the free energy of adsorption.

3.4 Jovanovic model

The Jovanovic model keeps the same assumptions contained in the Langmuir model, only considering, in addition the possibility of some mechanical contacts between the adsorbing and desorbing molecules. Moreover, a different extension of Jovanovic model for heterogeneous surface, the Jovanovic-Freundlich model, has been applied as a semiempirical model to the adsorption data of imprinted polymers. The Jovanovic-Freundlich model for single component adsorption was derived from a different relationship relating the surface coverage and the bulk concentration of the adsorbate. The model reduces to the Jovanovic model when the surface is homogeneous. It reduces to the monolayer isotherm at high concentration but does not obey the Henry's law concentrations. The energy distribution corresponding to this model for Jovanovic local behavior is a quasi-Gaussian function skewed in the direction of high adsorption energies. The Jovanovic equation may be written as:

$$q_e=q\ (1-\exp(-Kc)) \qquad (4)$$

where q_e ($mg.g^{-1}$) is the amount of pesticide adsorbed on the adsorbent surface at equilibrium, c ($mg.L^{-1}$) the pesticide concentration in aqueous solutions at equilibrium, q ($mg.g^{-1}$) is the maximum multilayer adsorption capacity, K ($L\ mg^{-1}$) is the constant related to the free energy of adsorption.

3.5 Temkin model

The Temkin [22] isotherm equation assumes that the heat of adsorption of all the molecules in the layer decreases linearly with coverage due to adsorbent - adsorbate interactions, and that the adsorption is characterized by a uniform distribution of the binding energies, up to some maximum binding energy. Temkin model is given by:

$$\theta = R\ T/\ \Delta Q\ \ln K_0 C_e \qquad (5)$$

where θ is the fractional coverage, R the universal gas constant ($kJ\ mol^{-1}\ K^{-1}$), T the temperature (K), $\Delta Q = (-\Delta H)$ the variation of adsorption energy ($kJ\ mol^{-1}$), and K_0 is the Temkin equilibrium constant ($L\ mg^{-1}$). If the adsorption obeys Temkin equation, the variation of adsorption energy and the Temkin equilibrium constant can be calculated from the slope and the intercept of the plot θ versus $\ln C_e$ (Table 2).

3.6 Huttig model

Huttig's isother is a model for adsorption on homogeneous surface with multilayer effects. The Huttig equation may be written as:

$$A= ac\ (1+c)\ /\ 1-Kc \qquad (6)$$

where A ($mg.g^{-1}$) is the amount of pesticide adsorbed on the adsorbent surface at equilibrium, c ($mg.L^{-1}$) the pesticide concentration in aqueous solutions at equilibrium, a ($mg.g^{-1}$) is the maximum multilayer adsorption capacity, K ($L\ mg^{-1}$) is the constant related to the free energy of adsorption.

Isotherm	Type of relation of physicochemical quantities	Formula	Linear formula	Formula constant	Method of estimation
Monolayer					
Freundlich	Adsorption on heterogeneous surface with monolayer effect; micro-porous solids	$q_e = K_F c_e^{1/n}$	$\ln qe = \ln K_F + \frac{1}{n} \ln c_e$	$A=kc^{1/n}$ n, k	Gauss-Newton
Langmuir	Adsorption on homogeneous surface with monolayer effect (lateral inter-actions and multilayer effect may be easily incorporated)	$q_e = \frac{qbc}{1+bc}$	$q_e = -\frac{1}{b}\frac{q}{c} + q$	A=ac/1+kc a, k	Gauss-Newton
Temkin	Adsorption on heterogeneous surface with monolayer effect	$\Theta = \frac{RT}{\Delta Q} \ln k c_e$ $q_e = q_m + K \lg c$	$\Theta = \frac{RT}{\Delta Q} \ln k + \frac{RT}{\Delta Q} \ln c_e$	A = a + k lgc a, k	Gauss-Newton
Jovanovic	Adsorption of organic sub-stance from di-lute solutions on heterogeneous surface with monolayer effect (vertical inter-actions and multilayer effect may be easily incorporated)	$\Theta = 1 - \exp(-kc)$		A= a(1-exp(-kc) a,k	Gauss-Newton
Multilayer					
Huttig	Adsorption on homogeneous surface with multilayer effect	$\Theta = (1+c)\frac{Kc}{1+Kc}$		A=(1+c)ca/(1-kc) a, k	Gauss-Newton
BET	Adsorption on homogeneous surface with multilayer effect	$\Theta = \frac{1}{1-c}\left(\frac{Kc}{1+(k-1)c}\right)$	$\frac{c}{q_e(1-c)} = \frac{1}{q_m k} + \frac{k-1}{q_m k} C$	A=ac/(1+c)(1+kc) a, k	Gauss-Newton

Table 2. The model sorption isotherm characteristic.

3. Results and discussion

Achieved study results are presented in Figures 3-7 and Table 3. Characteristics of applied adsorbent (Table 1) indicate that meet requirements of compost for natural applications. The parameters calculated according to the isotherm models studied are listed in Table 3. The adsorption process is described using Freundlich's, Langmuir's, Huttig's, Tiemkin, Jovanovic and BET's formulae:

$$\text{Freundlich } A=kc^{1/n}$$

$$\text{Langmuir } A=a_m kc/1+kc$$

$$\text{BET } A=ac/(1+c)(1+kc)$$

$$\text{Huttig } A=(1+c)ca/(1-kc)$$

$$\text{Temkin } A=a+k\,\lg c$$

$$\text{Jovanovic } A=a(1-\exp(-kc))$$

Following curves were achieved $A_F = 417.28\ c^{0.273}$ for compost at correlation coefficient of R=0.91 (Fig. 3); $A_L = 3149.6\ c/1\text{-}1056.4\ c$ at correlation coefficient of R=0.55 (Fig. 4); and $A_{BET} = 5298.18\ c/(1+c)(1\text{-}4.836c)$ at correlation coefficient of R=0.76 (Fig. 5); and $A_H = (1+c)c51159.3/(1\text{-}249.86c)$ at R=0.86 (Fig. 6), and $A_T = 285.97 + 71.6\ \lg c$ at R=0.88 (Fig.7), and $A_J = 226.07\ (1\text{-}\exp(\text{-}63.98c))$ at R=0.82 (Fig.8).
Constants k and 1/n for Freundlich model were estimated by means of the least squares by Gauss-Newton method applying Statistica software, and then the errors for these constants were evaluated (Fig. 3). The other isotherms were calculated applying Statistica software by

Izoterm	a	b	R
Freundlich	417.283	0.273	0.91
Langmuir	3149.6	-1056.4	0.55
Temkin	285.967	71.603	0.88
Huttig	51159.3	-249.86	0.86
BET	5298.18	-4.8358	0.76
Jovanovic	226.07	63.988	0.82

Table 3. The coefficients of the adsorption isotherms (level of confidence 95%, α=0,05).

means of the least squares by Gauss-Newton method and achieved constants a and k are presented in Figure 3-9. Figures 3-9 present adsorption isotherms for studied pesticides on applied low-cost natural adsorbents as a function of adsorbate amount adsorbed by adsorbent weight unit (x/m) vs. concentration of the pesticide at the equilibrium (c_0).

The correlation coefficients (R) were employed to ascertain the fit of all isotherms with the experimental data. From Table 3, the coefficients R values were found higher for the Freundlich, Temkin, Huttig and Jovanovic models than for the Langmuir and BET models. This indicates that the Freundlich, Temkin, Huttig and Jovanovic isotherm are clearly the best fitting isotherm model for experimental data. According to the two used models (BET and Langmuir), important gaps in the adsorption capacities of the low-cost adsorbents have been noted. Based on the values fit a BET line to experimental data the parameters q and k are set. In general, the parameter of monolayer capacity adsorption q is consistent with other methods (the differences rarely exceed 20%). Adsorption equilibrium constant k determined by this method can be even negative, which is physically meaningless and can be regarded as an artifact of the method of matching. In fact, such differences may arise in the case of highly heterogeneous adsorbents or microporous. This confirms the theory of heterogeneous surfaces of compost (Atkins, 2006).

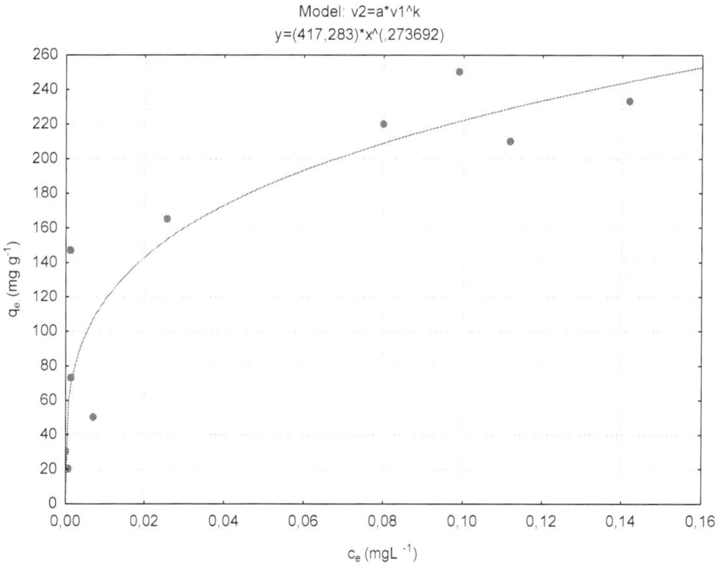

Fig. 3. Freundlich isotherm obtained using the nonlinear method for the sorption of pesticide on compost.

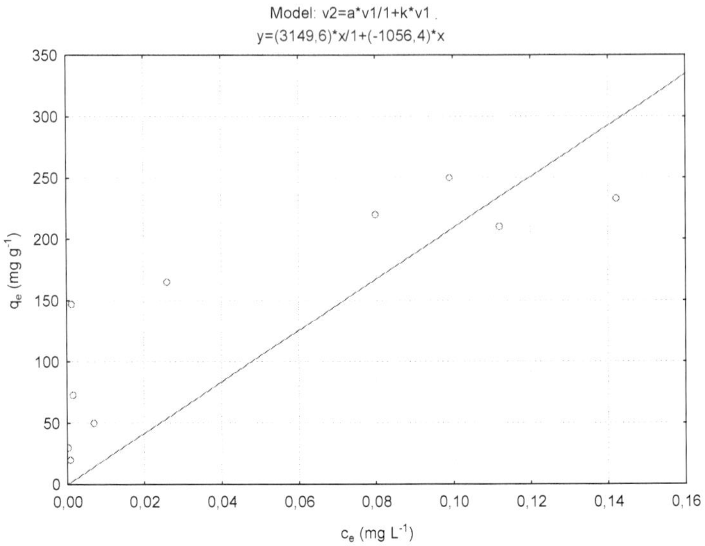

Fig. 4. Langmuir isotherm obtained using the nonlinear method for the sorption of pesticide on compost.

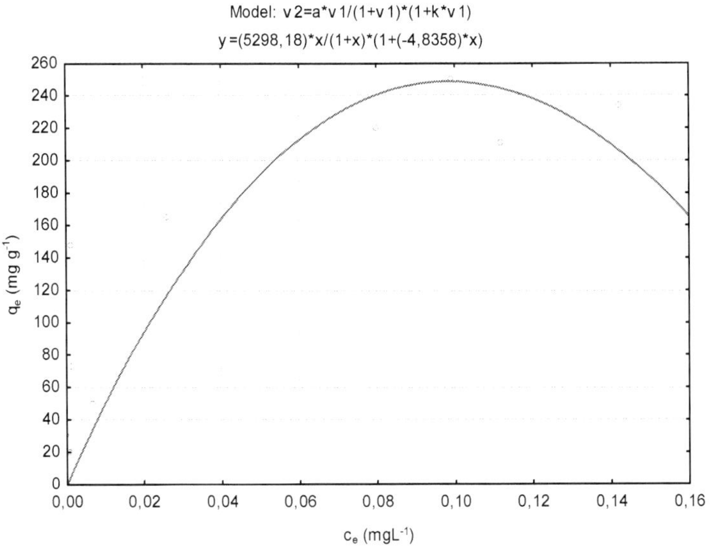

Fig. 5. BET isotherm obtained using the nonlinear method for the sorption of pesticide on compost.

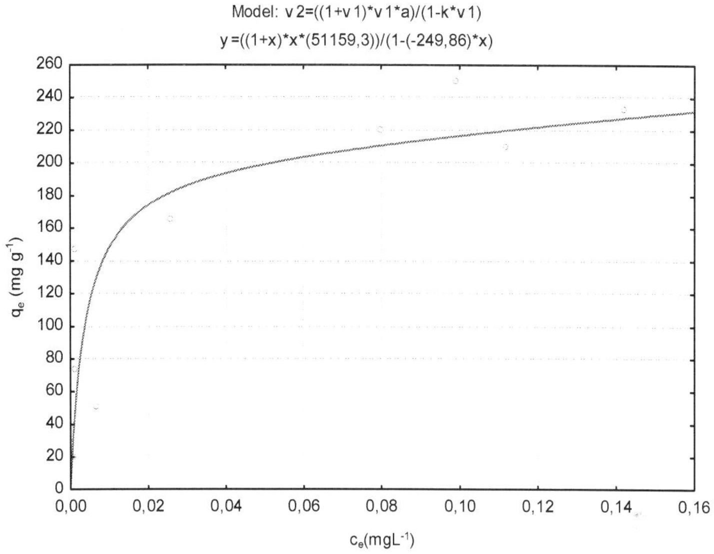

Fig. 6. Huttig isotherm obtained using the nonlinear method for the sorption of pesticide on compost.

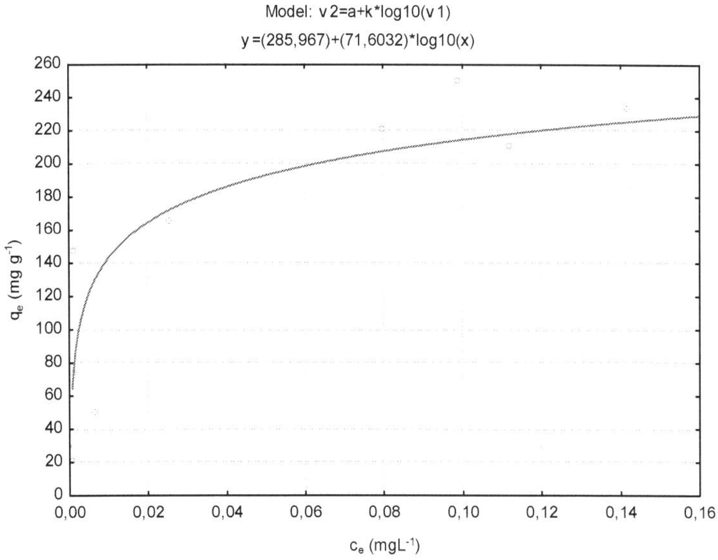

Fig. 7. Temkin isotherm obtained using the nonlinear method for the sorption of pesticide on compost.

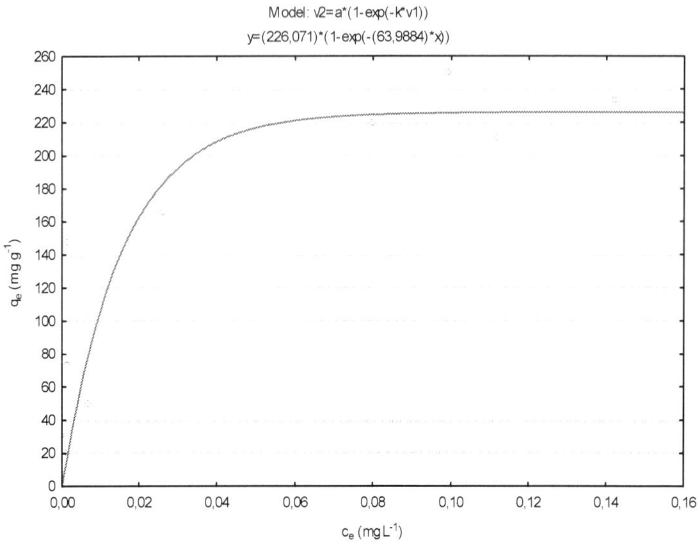

Fig. 8. Jovanovic isotherm obtained using the nonlinear method for the sorption of pesticide on compost.

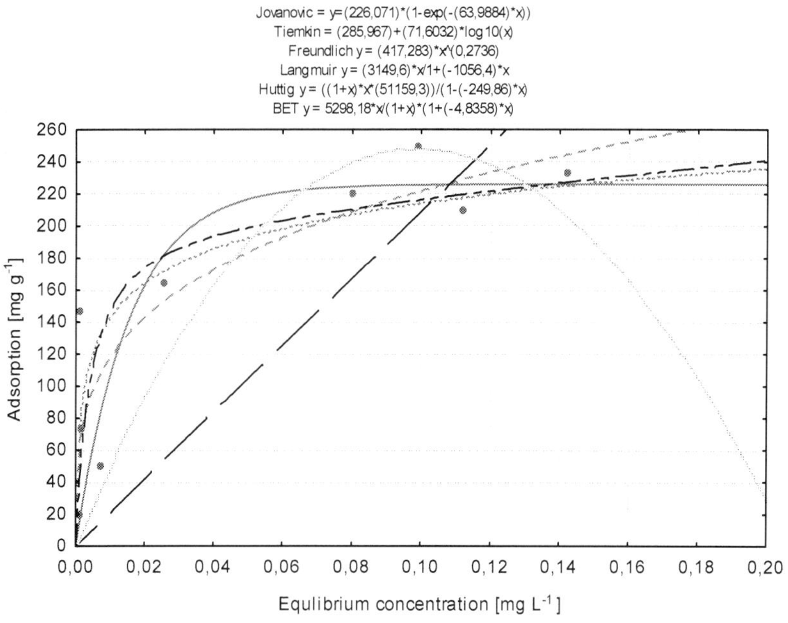

Fig. 9. The comparison of model's isotherms of pesticide on compost.

The nature of the studied pesticides (a chlorinated and hydrophobic molecule) suggests that its adsorption is of the hydrophobic type directly bound to the specific surface of the adsorbent particles. The isotherms study was performed according to the protocol described in the experimental chapter. The obtained results lead, firstly, to plot a, the amount of pesticide adsorbed on the adsorbent surface at equilibrium ($mg.g^{-1}$) against c_e, the pesticide concentration in the aqueous solution at equilibrium in order to classify the isotherms according to the classification of Giles et al., 1960 (Atkins, 2006). This one includes four main groups: L, H, S and C. The experimental adsorption isotherms of pesticides in aqueous solutions on the studied materials are presented in Fig. 3-9. The same group of isotherms according to Giles' classification (L) was achieved for all pesticides.(Atkins, 2006) The L shape of the adsorption isotherms means that there is no strong competition between solvent and the adsorbate to occupy the adsorbent surface sites. In this case, the longitudinal axes of the adsorbed molecules are parallel to the adsorbent surface (molecules adsorbed flat on the surface). From Fig. 3, the values of experimental maximum adsorption capacity (q_m experimental) for the three chloroorganic compounds on compost is about 260 $mg\ g^{-1}$.
According to Hamdaoui and Nafrechoux (2007) the pesticide molecules bind to the adsorbent through only one grouping and the adsorption becomes progressively easier as the absorbed quantity increases. Thus, the first fixed molecules facilitate the adsorption of the following molecules because of the lateral attraction. In the shape of these isotherms indicates that the chloroorganic pesticide is adsorbed as a monolayer and that there is no strong competition between the pesticide molecules and water to occupy the adsorption surface sites. In this case, the longitudinal axes of the adsorbed molecules are parallel to the adsorbent surface. This type of isotherm is relative to microporous adsorbents with a diameter lower than 25 Å, the adsorbent being saturated at the moment of the monolayer replenishment. (Atkins, 2006; Ignatowicz, 2009) There would be weak interactions therefore on these adsorbent surfaces because the number of layers cannot increase freely.
The equilibrium data were further analyzed using the Freundlich equation using the same set experimental data, by plotting ln qe versus Ce. The calculated Freundlich isotherm constants and the corresponding coefficient of correlation values were shown in Table 3. The coefficients of correlation are high ($R \geq 0.91$) showing a good linearity. The magnitude of the exponent n ($n=3.66$) gives an indication on the favorability of adsorption. It is generally stated that values of n in the range 2–10 represent good, 1–2 moderately difficult, and less than 1 poor adsorption characteristics (Hamdaoui, 2007).
Knowledge on 1/n parameter value in Freundlich's formula allows for assessing the adsorption intensity of a given substance from water phase on adsorbent; value of k constant determines the sorption capacity of an adsorbent at balance concentration in a solution. Higher k value corresponds to higher sorption capacity. In own studies, higher value of k coefficient was achieved for compost, which proves its usefulness in application as sorption screen around the pesticide graveyard. Constants 1/n in Freundlich's formula is directional coefficients of isotherms equal to the tangent of line inclination angle in logarithmic coordinates. Therefore, the higher 1/n value, the more intensive adsorption process. (Ignatowicz, 2009; Pagnanelli, 2009; Akhtar, 2009)
The parameters of Temkin model as well as the correlation coefficients are given in Table 3. The very higher values of the coefficient of correlation show a good linearity whatever the

maximum adsorption capacity used for the calculation of surface coverage. The variation of adsorption energy, $\Delta Q = (-\Delta H)$, is positive for all the studied compounds, which indicates that the adsorption reaction is exothermic. In order to seek a systematic for the observed changes between the variations of adsorption energy of the tested chloroorganic compounds, it seems that the chloro group has a negative increment (exothermic effect).

4. Conclusions

The present study indicates the suitability of the low-cost adsorbent - compost for removal of graveyard's chloroorganic pesticides from aqueous solutions. The adsorption process is described using Freundlich, Temkin, Huttig and Jovanovic formulae. The Freundlich and Temkin models were fitting better the experimental datas. According to the two used models, important gaps in the adsorption capacities of the low-cost adsorbents have been noted.

The same group of isotherms according to Giles' classification (L) was achieved for chloroorganic pesticide. The pesticide molecules bind to the adsorbent through only one grouping and the adsorption becomes progressively easier as the absorbed quantity increases. Thus, the first fixed molecules facilitate the adsorption of the following molecules because of the lateral attraction. The shape of these isotherms indicates that the chloroorganic pesticide is adsorbed as a monolayer and that there is no strong competition between the pesticide molecules and water to occupy the adsorption surface sites. In this case, the longitudinal axes of the adsorbed molecules are parallel to the adsorbent surface. This type of isotherm is relative to microporous adsorbents with a diameter lower than 25 Å, the adsorbent being saturated at the moment of the monolayer replenishment.

5. Acknowledgments

Financial support for this research was provided by Ministry of Science and Higher Education within the project W/WBiIŚ/5/10 and N305 070 32/2535.

6. References

Akhtar M., Iqbal S., Bhanger M.I., Moazzam M. (2009). Utilization of organic by-products for the removal of organophosphorous pesticide from aqueous media. *J. Hazard. Mater.* 162, 703-707.

Anderson P., Jack R., Burke Ch., Cowles J., Moran B. (2004). *Surface Water Monitoring Program for Pesticides in Salmonid-Bearing Streams April to December 2003*, WSDA, Washington State, Department of Agriculture, Ecology Publication 04-03-048.

Atkins P.W. (2006). *Physical chemistry*, 8th ed. Oxford University Press

Balinova A. (1996). Strategies for chromatographic analysis of pesticide residues in water. *J. Chromatogr. A* 754, 125-135

Hamadi N.K., Swaminathan S.,. Chen X.D. (2004). Modeling of adsorption isotherms of phenol and chlorophenols onto granular activated carbon Part I. Two-parameter models and equations allowing determination of thermodynamic parameters. *J. Hazard. Mater.* B112, 133-141.

Hamdaoui O., Naffrechoux E. (2007). Modeling of adsorption isotherms of phenol and chlorophenols onto granular activated carbon - Part I. Two-parameter models and equations allowing determination of thermodynamic parameters. *J. Hazard. Mater.* 147, 381-394.

Hussen A., R. Westbom, N. Megersa, L. Mathiasson, E. Bjorklund J Chromatogr. A 1152 (2007) 247.

Ignatowicz K. (2004). Sorption process for migration reduction of pesticides from tombs. *Archives of Environmental Protection,* 30, 51-61.

Ignatowicz K. (2008). Zastosowanie sorpcji na odpadowych materiałach naturalnych do ograniczenia migracji pestycydów z mogilników. *Przemysł Chemiczny,* 87/5, 464-467.

Ignatowicz K. (2009). Selection of sorbent for removing pesticides during water treatment. *J. Hazard. Mater.* 169, 953-957.

Ignatowicz K. (2009). Sewage sludge compost applying for migration reduction of pesticide from graveyards. *Pol. J. Environ. Stud.* 6, 49-55

Kumar A., Kumar S., Kumar S. (2003). Adsorptionof resorcinol and catechol on granular activated carbon: Equilibrium and kinetics. *Carbon* 41, 3015-3025

Mashayekhi H., Veneman P., Xing B. (2006). Phenanthrene sorption by compost humic acid. *Biol. Rertil. Soils* 42, 426-431

Munoz-Arnanza J., Boschb C., Fernandezb P., Grimaltb J.O., Jimeneza B. (2009). Optimization of a heart-cutting multidimensional gas chromatography-based method for the assessment of enantiomeric fractions of o,p-DDT in environmental samples. *J Chromatogr. A* 1216, 6141-6145

Ochsner T.E., Stephens B.M., Koskinen W.C., Kookana R.S. (2006). Sorption of a hydrophilic pesticide. *Soil Sci. Soc. Am.* 70, 1991-1997

Pagnanelli F., Viggi C.C., Mainelli S., Toro L. (2009). Assessment of solid reactive mixtures for the development of biological permeable reactive barriers *J. Hazard. Mater.* 170, 998-1005.

Siepak J., Zerbe J., Kabacinski M. (1993). Wst pna ocena degradacji studni na obszarach wiejskich województwa poznańskiego. *Przyroda i Człowiek,* 4, 117-127.

Siepak J. (2001). Slurry sampling for electrothermal atomic absorption spectrometric determination of chromium, nickel, lead and cadmium in sewage sludge. *Analytica Chimica Acta* 437, 11-16

Siepak J., Zioła-Frankowska A., Frankowski M. (2009). Development of a New analytical metod for online simultaneous qualitative determination of aluminium by HPLC-FAAS. *Talanta* 78, 623-630

Spadotto C.A., Hornsby A.G. (2003). Soil sorption of acidic pesticides: modeling pH effects. *J. Environ. Qual.* 32, 949-956

Tsui L., Roy W.R. (2007). Effect of compost age and composition on the atrazine removal from solution. *J. Hazard. Mater.* B139, 79-85.

Witbowo N., Setyadhi L., Witbowo D., Setiawan J., Ismadji S. (2007). Adsorption of benzene and toluene from aqueous solutions onto activated carbon and its acid and heat treated forms: Influence of surface chemistry on adsorption. *J. Hazard. Mater.* 146, 237-242.

Yuh-Shan Ho. (2006). Isotherms for the sorption of lead onto peat: comparison of linear and non-linear methods. *Pol. J. Environ. Stud.* 15, 81-86

Inffluence of the Activated Carbon Nature and the Aqueous Matrix on the Pesticides Adsorption

Natividad Miguel, María P. Ormad, Rosa Mosteo,
Jorge Rodríguez and José L. Ovelleiro
Department of Chemical Engineering and Environmental Technologies
University of Zaragoza. Pedro Cerbuna, 12. 50009. Zaragoza
Spain

1. Introduction

Water flowing through surface and/or subsoil acquires a chemical composition due to its dissolution effect on soluble minerals of rocks and organic compounds deriving from the degradation of organic matter. This natural composition of water is basically altered by four pollution sources: domestic wastewaters, industrial wastewaters, uncontrolled wastewaters and run-off pollution in agricultural areas. The latter can result in the presence of pesticides in natural waters, because these compounds can pass through the soil and subsoil and pollute surface and groundwaters which are supplies sources for water intended for human consumption.

Pesticides are a group of artificially synthesized substances used to fight pests and improve agricultural production. However, they are generally toxic for living organisms and are very difficult to degrade, being toxic agents with persistent and bioaccumulative effects. In spite of their benefits in the agriculture, they have undesirable effects due to its toxicity, carcinogenesis and mutagenesis (Becker & Wilson, 1980; Kouras et al., 1998).

In Europe, pesticides are considered Hazardous Pollutants in accordance with current legislation relating to water (Directive 2000/60/CE; Directive 2006/11/CE). In the Ebro River Basin (Spain), these substances are controlled via a Pesticides Control Network, which systematically analyzes 44 organic pesticides in surface waters. These pesticides were selected in accordance with their appearance in lists of hazardous substances and/or their high level of use in Spanish agriculture (Claver et al., 2006). Among these substances, there are a big variety of pesticides, such as triazines, urea derivates, drins, etc.

Although the concentration of these substances detected in natural waters is generally very low, the maximum permissible concentration in human drinking waters in Spain is often exceeded (ROYAL DECREE 140/2003), which establishes a limit of 0.5 $\mu g\ L^{-1}$ as the total amount of pesticides and 0.1 $\mu g\ L^{-1}$ for any single pesticide. Consequently, the treatment used to produce drinking water must guarantee the removal of these types of substances or at least reduce their concentration below the limits established in current legislation.

Systems of drinking water production consist of different stages depending on the initial water quality. Actually, a lot of drinking water plants use an adsorption stage onto activated

carbon in their treatments systems (Ormad et al., 2008). The aim of this stage is the removal of organic compounds, micropollutants such as volatile organic compounds, pesticides, PCBs and phenolic compounds, and some heavy metals with trace concentrations in water. In general, any organic compound with a molecular weight greater than 45 can be adsorbed onto activated carbon (Kenneth et al., 1992). For this purpose, the activated carbon can be used with two shapes: granular activated carbon (GAC) or powered activated carbon (PAC). The adsorption is a superficial phenomenon which is influenced by a lot of factors. These factors are related both the activated carbon (specific surface, particle and pore size, pore distribution, etc.), both the environmental conditions (pH, temperature, chemical composition of the solution, etc.) (Smith & Weber, 1985; Kilduff et al., 1998; Matsui et al., 2003; Gun'ko et al., 2008). The activated carbon is produced from different carbonic substances, of vegetal or mineral origin, by an activation process by which particles with specific surfaces about 10^3 m^2 g^{-1} are achieved. The final properties of activated carbon depend on both the used raw, both the activation method applied (Gun'ko et al., 2008).
The aim of this work is to study the effectiveness of the adsorption process with activated carbon in order to remove the 44 organic pesticides detected systematically in the Ebro river. Moreover, the influence of the activated carbon nature and the aqueous matrix in the treatment is studied. For this, two types of PAC are used: one of mineral origin and other of vegetal one; and the treatment is carried out with two types of solutions, in distilled and natural water. Finally, the influence of the contact time in the removal of the studied pesticides is studied.

2. Materials and methods

2.1 Samples

The adsorption treatment onto PAC is applied on solutions of pesticides dissolved in distilled water (pH = 5.5, dissolved organic carbon (DOC) = 0 mg C L^{-1}) and in natural water coming from Ebro river (Spain) (pH = 8, DOC = 3 mg C L^{-1}). Each sample is fortified with 500 ng L^{-1} of each studied pesticide in order to ensure their presence and to study their possible removal (Miguel et al., 2008). The studied pesticides and their classification according their biological activity, chemical nature and toxicity are shown in table 1.

PESTICIDE	BIOLOGICAL ACTIVITY	CHEMICAL NATURE	TOXICITY
Alachlor	Herbicide	Organic-chlorinated	Moderately toxic
Aldrin	Insecticide	Organic-chlorinated	Very toxic
Ametryn	Herbicide	Heterociclic compound	Moderately toxic
Atrazine	Herbicide	Heterociclic compound	Moderately toxic
Chlorpiryfos	Insecticide	Organic-phosphorated	Very toxic
Chlorfenvinfos	Insecticide	Organic-phosphorated	Extremely toxic
pp'-DDD	Insecticide	Organic-chlorinated	Moderately toxic
pp'-DDE	Insecticide	Organic-chlorinated	Moderately toxic
op'-DDT	Insecticide	Organic-chlorinated	Very toxic
pp'-DDT	Insecticide	Organic-chlorinated	Very toxic
Desethylatrazine	Herbicide	Heterociclic compound	Moderately toxic
3,4-Dichloroaniline	Herbicide	Organic-chlorinated	Moderately toxic

Table 1. Studied pesticides and their characteristics

PESTICIDE	BIOLOGICAL ACTIVITY	CHEMICAL NATURE	TOXICITY
4,4′-Dichlorobenzophenone	Acaricide	Organic-chlorinated	Moderately toxic
Dicofol	Acaricide	Organic-chlorinated	Moderately toxic
Dieldrin	Insecticide	Organic-chlorinated	Extremely toxic
Dimethoate	Insecticide	Organic-chlorinated	Very toxic
Diuron	Herbicide	Urea derivate	Moderately toxic
α-Endosulphan	Insecticide	Organic-chlorinated	Very toxic
Endosulphan-sulphate	Insecticide	Organic-chlorinated	Very toxic
Endrin	Insecticide	Organic-chlorinated	Extremely toxic
α-HCH	Insecticide	Organic-chlorinated	Very toxic
β-HCH	Insecticide	Organic-chlorinated	Moderately toxic
χ HCH	Insecticide	Organic-chlorinated	Very toxic
δ-HCH	Insecticide	Organic-chlorinated	Moderately toxic
Heptachlor	Insecticide	Organic-chlorinated	Very toxic
Heptachlor epoxide A	Insecticide	Organic-chlorinated	Very toxic
Heptachlor epoxide B	Insecticide	Organic-chlorinated	Very toxic
Hexachlorobenzene	Fungicide	Organic-chlorinated	Very toxic
Isodrin	Insecticide	Organic-chlorinated	Extremely toxic
4-Isopropylaniline	Herbicide	Heterociclic compound	Moderately toxic
Isoproturon	Herbicide	Urea derivate	Moderately toxic
Metholachlor	Herbicide	Organic-chlorinated	Moderately toxic
Methoxychlor	Insecticide	Organic-chlorinated	Slightly toxic
Molinate	Herbicide	Carbamate	Very toxic
Parathion methyl	Acaricide/ Insecticide	Organic-phosphorated	Extremely toxic
Parathion ethyl	Acaricide/ Insecticide	Organic-phosphorated	Extremely toxic
Prometon	Herbicide	Heterociclic compound	Moderately toxic
Prometryn	Herbicide	Heterociclic compound	Slightly toxic
Propazine	Herbicide	Heterociclic compound	Slightly toxic
Simazine	Herbicide	Heterociclic compound	Moderately toxic
Terbuthylazine	Herbicide	Heterociclic compound	Moderately toxic
Terbutryn	Herbicide	Heterociclic compound	Moderately toxic
Tetradiphon	Insecticide/ Acaricide	Organic-chlorinated	Slightly toxic
Trifluralyn	Herbicide	Heterociclic compound	Slightly toxic

Table 1. Studied pesticides and their characteristics (continuation)

2.2 Analytical methodology

The EPA Method 525.2 is used in order to determine the pesticides in samples (EPA Method 525.2). This methods is based on the analysis by gas chromatography together with mass spectrometry (GC/MS), with a previous solid-liquid extraction. An Autotrace Workstation (Zymark) automatic extractor was used for the extraction. The chromatographic conditions and equipment used are shown in table 2 and results of the methodology validation in table 3. The results were obtained using the Xcalibur POLARIS 1.2 version program (ThermoQuest).

Gas Chromatographer TRACE GC 2000 (TermoFinnigan)	
Column	DB5-MS (J&W, 30 m, 0,25 mm, 0,25 μm)
Temperature program	90 °C (1 min) – 20 °C min^{-1} – 180 °C (1 min) – 2 °C min^{-1} – 240 °C (1 min) – 20 °C min^{-1} – 310 °C (10 min)
Injector temperature	250°C
Injection volume	1 μL, splitless 0.8 min
Carrier gas	He (N55), 1mL min^{-1}
Mass Spectrometer POLARIS (ThermoFinnigan)	
Ionization energy	70 eV
Acquisition mode	Full scan
Mass array	50-450 amu
Screener speed	1 scan s^{-1}
Acquisition time	32.5 min

Table 2. Pesticides analysis conditions

Pesticide	Quantification limit (μgL^{-1})		Calibration interval (μgL^{-1})	Validity interval (μgL^{-1})	Recovery interval (%)	
	Instrumental step	Full method			Instrumental step	Full method
Isoproturon	20	0.030	20-500	0.030-300	75-130	63-110
Diuron	20	0.030	20-500	0.030-300	82-128	70-123
3.4-Dichloroaniline	20	0.030	20-500	0.030-300	88-130	47-106
4-Isopropylaniline	20	0.030	20-500	0.030-300	80-130	60-125
Desethylatrazine	20	0.030	20-500	0.030-300	76-130	80-129
Trifluralyn	20	0.015	20-500	0.030-300	70-130	70-127
Dimethoate	20	0.030	50-500	0.030-300	66-124	54-137
Simazine	50	0.030	20-500	0.030-600	75-135	64-127
Prometon	20	0.030	20-500	0.030-300	76-124	0-125
Atrazine	200	0.100	200-5000	0.100-300	78-130	75-127
Propazine	20	0.015	20-500	0.015-300	86-130	73-127
Terbuthylazine	20	0.015	20-500	0.015-300	79-130	83-128
Parathion methyl	50	0.030	50-500	0.030-300	78-139	72-130
Parathion ethyl	20	0.030	20-500	0.030-300	74-122	64-128
Alachlor	20	0.015	20-500	0.015-300	75-125	70-124
Ametryn	20	0.030	20-500	0.030-300	78-130	0-116
Prometryn	20	0.030	20-500	0.030-300	80-120	17-116
Terbutryn	20	0.030	20-500	0.030-300	80-120	13-114
Chlorpyrifos	20	0.015	20-500	0.015-300	75-120	73-116
Chlorfenvinfos	20	0.015	20-500	0.015-300	76-130	70-126
HCHs	20	0.015	20-500	0.015-300	84-124	70-120
Hexachlorobenzene	20	0.030	20-500	0.030-300	70-130	74-136
Heptachlor	20	0.015	20-500	0.015-300	75-130	58-113

Table 3. Results of the validation of the pesticides analysis methodology

Pesticide	Quantification limit (μgL^{-1})		Calibration interval (μgL^{-1})	Validity interval (μgL^{-1})	Recovery interval (%)	
	Instrumental step	Full method			Instrumental step	Full method
Heptachlor epoxide A	20	0.015	20-500	0.015-300	85-125	62-112
Heptachlor epoxide B	20	0.015	20-500	0.015-300	84-130	58-113
Aldrin	20	0.015	20-500	0.015-300	85-125	64-126
4,4′-Dichlorobenzophenone	20	0.015	20-500	0.015-300	75-120	68-126
Isodrin	20	0.015	20-500	0.015-300	85-125	66-120
α-Endosulphan	20	0.015	20-500	0.015-300	70-125	70-93
pp′-DDE	20	0.015	20-500	0.015-300	89-122	64-107
Dieldrin	20	0.015	20-500	0.015-300	70-125	62-120
Endrin	20	0.015	20-500	0.015-300	80-125	74-122
pp′-DDD + op′-DDT	40	0.030	40-1000	0.030-600	79-125	66-139
Endosulphan-sulphate	20	0.015	20-500	0.015-300	83-125	73-126
pp′-DDT	20	0.030	20-500	0.030-300	76-130	50-120
Dicofol	50	0.030	50-500	0.030-300	80-148	63-136
Methoxychlor	20	0.015	20-500	0.015-300	77-126	75-130
Metholachlor	20	0.015	20-500	0.015-300	76-115	73-128
Molinate	20	0.015	20-500	0.015-300	91-130	75-113
Tetradifon	20	0.015	20-500	0.015-300	85-130	70-116

Table 3. Results of the validation of the pesticides analysis methodology (continuation)

2.3 Activated carbon characterization

The characterization of used PACs is carried out with analysis by screener electronic microscopy (SEM) and by the method Brunauer-Emmet-Teller (BET). This characterization is carried out before and after each applied treatment.

The BET method is used in order to determine the specific surface by the measurement of the gas adsorption at low termperature. The equipment used is a Micromeritic Instruments Co., Pulse Chemisorb 2700.

The SEM analysis is carried out with a screener electronic microscope (JEOL JSM 6400) which can generates images of secondary and retrodispersed electrons accelerated with tensions between 0.2 and 40 KV. This microscope allow observations up to 3.5 nm of resolution and has coupled a computerized system of dispersed X rays energy (INCA 300 X-Sight, Oxford Instruments) with a resolution of 133 eV to 5.9 KeV. Moreover, it has coupled a computerized system to register and analizar difraction diagrams of retrodispersed electrons (Electron Back Scatter Diffraction).

2.4 Experimental procedure

The treatment of PAC adsorption is applied with distilled and natural water fortified with studied pesticides and the PAC concentration used is 10 mg L^{-1} (in a similar way than in drinking water plants in Spain). The PAC is put into de sample and it is softly stirred during 10 min. in a Jar-Test (SBS).

Two types of PAC are used: BM8 (CHIEMIVALL), of mineral origin and coming from bitominous mineral; and VPlus (CHIEMIVALL), of vegetal origin and coming from wood. Two PACs are activated with vapour and obey with the norm EN-12903 to their application in drinking water. Their characteristics are shown in the table 4.

Specifications	BM8	VPlus
Iodine number (mg g^{-1})	800	950
Ashes content (%)	< 10	3
pH of the aqueous extract	9.0 - 10.0	9.0 - 10.0
Humidity (%)	< 8	5
Particle size	< 0.044 mm (90%)	< 0.044 mm (90%)
Origin	Mineral	Vegetal

Table 4. Characteristics of PAC BM8 and VPlus (CHIEMIVALL)

The concentrations of each pesticide before and after applying the treatment are analysed by GC/MS and the removal yields are calculated according to the equation 1:

$$\eta_{removal\ i} = \left[\frac{C_{i\ initial-solution} - C_{i\ final-solution}}{C_{i\ initial-solution}}\right] * 100 \qquad (1)$$

where $\eta_{removal\ i}$ is the removal yield of each pesticide, $Ci_{initial-solution}$ is the concentration of each pesticide in the sample before the treatment, $Ci_{final-solution}$ is the concentration of each pesticide in the sample after the treatment, and *i* each studied pesticide.

Moreover, the average removal yield of all pesticides is calculated according to the equation 2:

$$\eta_{average-removal} = \left[\frac{\sum_i \eta_{removal_i}}{n}\right] \qquad (2)$$

where $\eta_{average-removal}$ is the average removal yield of all studied pesticides, $\eta_{removal\ i}$ is the removal yield of each pesticide and *n* is the number of pesticides (n=44).

The study of the influence of the treatment time is carried out by experiments at different contact times: 10 min., 30 min., 1 h., 2 h. and 4 h..

3. Results

3.1 Pesticides removal

Pesticides removal percentages achieved after the application of the PAC adsorption treatment, with two types of used PACs and in two studied aqueous matrix, are shown in the table 5.

Average removal percentages achieved in the solution of distilled water are: 34% with mineral PAC and 46% with vegetal PAC. All pesticides in distilled water are better remove by vegetal PAC adsorption, except for the endosulphan-sulphate which is lightly better removed with mineral PAC. It is due to the fact that, in general, the adsorption equilibrium is reached faster with vegetal PAC tan with mineral (Matsui et al., 2003).

With respect to the pesticides solution in natural water, the average removal percentages achieved are: 33% with mineral PAC and 26% with vegetal PAC. In this case, all pesticides are better removed with mineral PAC, with the exception of triazines (simazine, atrazine, propazine, terbuthylazine, prometon, ametryn, prometryn, terbutryn and desethylatrazine) which are removed equal or lightly better with vegetal PAC. It is due to the influence of the organic matter contained in natural water and pore size of the used PACs. Although *a priori* the adsorption onto vegetal PAC is faster, this PAC has a higher pore size, therefore the competition for its occupation between the organic matter of natural water and pesticides increases, decreasing in this way the adsorption (Matsui et al., 2003).

In general, it can be observed that the aqueous matrix doesn't influence in the adsorption treatment with mineral PAC for the group of studied pesticides since the average removal percentages achieved are similar to solutions in distilled and natural water. It is due to the fact that the pore size of the mineral PAC is smaller than the particle size of the organic matter and then their interaction and influence is prevented. However, the adsorption treatment with vegetal PAC is influenced by the used aqueous matrix since to the pesticides solution in distilled water the removal percentages achieved are higher than to the pesticides solution in natural water. This is because the pore size of vegetal PAC is greater and the presence of organic matter in natural water competes with pesticides for the occupation of PAC pores or blocks them getting worse the adsorption capacity (Gillogly et al., 1998; Knappe et al., 1999; Pelekani et al., 1999; Newcombe et al., 2002; Matsui et al., 2003). Moerover, the quantity of pollutants that can be adsorbed in a PAC is lower as the pH of the solution is greater. For this, the adsorption is more effective in distilled water than in natural water (Hu et al., 1998).

Therefore, according to achieved results, the adsorption with vegetal PAC is more effective in order to remove the studied pesticides when they are in distilled water and the adsorption with mineral PAC is more effective, when pesticides are in natural water.

Triazines (simazine, atrazine, propazine, terbuthylazine, prometon, ametryn, prometryn, terbutryn and desethylatrazine) are heterociclic compounds used as herbicides. The adsorption of these substances is more effective with mineral PAC both in natural and distilled water. In distilled water the removal is about 40% and about 20% in natural water.

With respect to organic-phosphorated insecticides (parathion methyl, parathion ethyl, chlorpiryfos, chlorfenvinfos and dimethoate), they are removed in a similar way in distilled and natural water, although their removal in distilled water is more effective using vegetal PAC (45%) and in natural water using mineral PAC (45%).

Hexachlorocilohexanes pesticides (HCHs), which are organic-chlorinated insecticides, are removed in a similar way with mineral PAC both in distilled and natural water. However, the removal of theses pesticides is greater in distilled water with vegetal PAC (40-45%) than with mineral PAC in natural water (30%).

Hexachlorobezene, organic-chlorinated compound used as insecticide, is removed in a similar way with two PACs in distilled water, about 65%. However, it is removed 60% with mineral PAC and 50% with vegetal PAC in natural water.

PESTICIDE	η removal (%) DISTILLED WARTER		η removal (%) NATURAL WATER	
	MINERAL PAC	VEGETAL PAC	MINERAL PAC	VEGETAL PAC
Simazine	30	40	15	25
Atrazine	30	50	15	15
Propazine	30	50	15	20
Terbuthylazine	30	50	25	25
Prometon	20	40	10	10
Ametryn	30	50	25	30
Prometryn	35	50	20	20
Terbutryn	30	45	25	25
Desethylatrazine	25	25	25	25
Parathion methyl	35	55	45	40
Parathion ethyl	40	55	55	45
Chlorpiryfos	35	45	45	30
Chlorfenvinfos	25	30	45	35
Dimethoate	30	40	45	35
α-HCH	30	45	30	30
β-HCH	25	40	30	20
χ-HCH	25	40	30	20
δ-HCH	30	45	30	20
Hexachlorobenzene	65	65	60	50
Heptachlor	20	30	30	15
Heptachlor epoxide A	25	40	25	15
Heptachlor epoxide B	30	40	25	15
α-endosulphan	30	20	20	10
Endosulphan sulphate	40	30	20	10
Endrin	20	35	25	15
Dieldrin	40	50	25	10
Isodrin	30	45	30	15
Aldrin	20	35	25	10
pp´-DDE	20	30	15	10
pp´-DDD+op´-DDT	20	35	30	20
pp´-DDT	20	35	20	10
Isoproturon	15	40	10	15
4-isopropylaniline	90	100	60	60
Diuron	100	100	100	100
3,4-dichloroaniline	80	85	70	65
Molinate	30	50	15	10
Trifluralyn	20	30	35	20
Alachlor	30	45	35	25
Metholachlor	30	45	35	30

Table 5. Removal yields of studied pesticides

PESTICIDE	η removal (%) DISTILLED WARTER		η removal (%) NATURAL WATER	
	MINERAL PAC	VEGETAL PAC	MINERAL PAC	VEGETAL PAC
Methoxychlor	20	40	35	20
Tetradiphon	40	50	35	25
Dicofol	25	50	40	35
4,4´-dichlorobenzophenone	65	75	55	50
$\eta_{average\text{-}removal}$ (%)	**34**	**46**	**33**	**26**

Table 5. Removal yields of studied pesticides (continuation)

Heptachlor, organic-chlorinated insecticides, are removed equal in distilled and natural water with mineral PAC, about 30%. Their removal increases when these pesticides are in distilled water with vegetal PAC achieving removal percentages about 40%.

With respect to drins (aldrin, isodrin, dieldrin and endrin), all of them organic-chlorinated compounds used as insecticides, the vegetal PAC is more effective when they are in distilled water (40%) and the mineral PAC is more effective in order to remove them in natural water (25%).

Regarding to endosulphans, organic-chlorinated insecticides too, they are better removed with mineral PAC both in distilled and natural water.

With respect to DDTs, organic-chlorinated insecticides, the mineral PAC is more effective in order to remove them in natural water achieving removal percentages about 20%, and the vegetal PAC is more effective in distilled water, increasing their removal up to 35%.

Finally, for the rest of studied pesticides, they are better removed with vegetal PAC in distilled water and with mineral PAC in natural water. It's worth noting that diuron, urea derivate pesticide, which is complete removed with two PACs and in both solutions, and anilines (4-isopropylaniline and 3,4-dichloroaniline) for which the removal percentages achieved are very high in all cases.

Since in natural water the more effective PAC is the mineral PAC in order to adsorb the studied pesticides, the study of the influence of the contact time is carried out with mineral PAC and solutions of pesticides in natural water at different contact times (10 min., 30 min., 1 h., 2 h. and 4 h.). Average removal yields of the group studied pesticides at different treatment times are shown in figure 1.

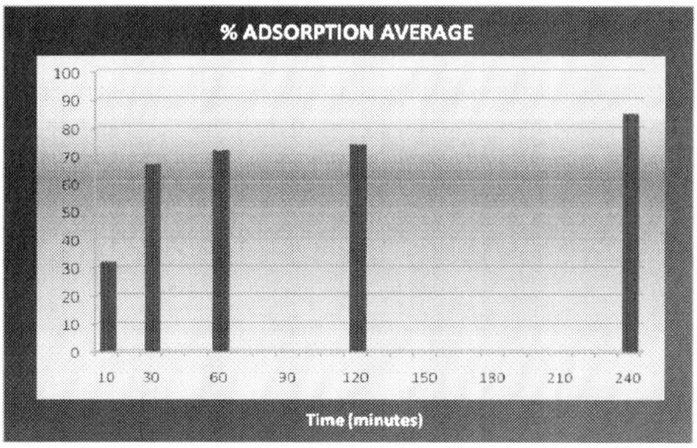

Fig. 1. Average yields of pesticides adsorption with mineral PAC in natural water

Average yields of the pesticides adsorption at different contact times are: 32% in 10 min., 67% in 30 min., 71% in 1 h., 74% in 2 h. and 85% in 4 h. In accordance with these results, it can be observed that the adsorption of pesticides is produced mainly in the first 30 minutes. Since this moment and up to 2 hours of treatment the adsorption continues but lesser. Finally, between 2 and 4 hours of treatment the adsorption lightly increases and the group of studied pesticides is adsorbed 85%.

The evolution of the pesticides adsorption in the time is the typically observed in multi-component mixtures: the main adsorption happens in the first minutes of the treatment, after the adsorption is lower, and finally the adsorption lightly increases (Noll et al., 1992). In the concrete case of the studied pesticides, the main adsorption occurs in the first 30 minutes of treatment, time which coincides with the recommendations given by the WEF-ASCE (WEF-ASCE, 1998) regarding to the contact time in the activated carbon adsorption.

Results about the adsorption of different groups of studied pesticides in natural water with mineral PAC at different contact times are shown in figures 2 to 12.

Triazines adsorption is shown in figure 2.

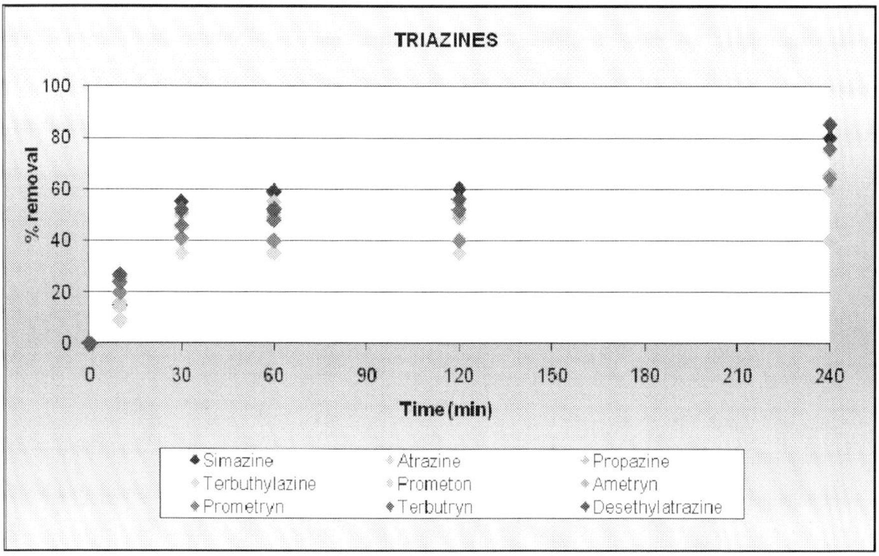

Fig. 2. Triazines adsorption

In accordance with achieved results, and as it is showed in the figure 2, at 10 minutes of contact time the triazines adsorption is between 10 and 25%. This percentage considerably increases up to 35-55% in 30 minutes of treatment. Next, until 2 hours of treatment adsorption isn't observed, and it increases up to 60-85% at 4 hours of treatment for all triazines, with the exception of atrazine which is adsorbed 40% in 4 hours.

With respect to organic-phosphorated pesticides, in the figure 3 it can be observed that the removal at 10 minutes of treatment is about 50%. The adsorption is complete at 30 minutes for all of these pesticides except to chlorpiryfos which is adsorbed 80% in 30 minutes and it requires 4 hours of treatment to remove it completely.

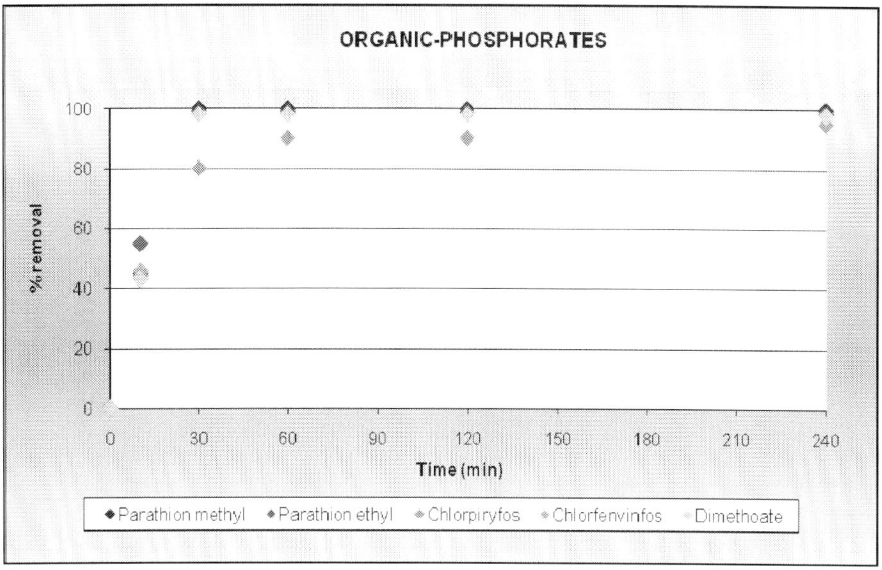

Fig. 3. Organic-phosphorated pesticides adsorption

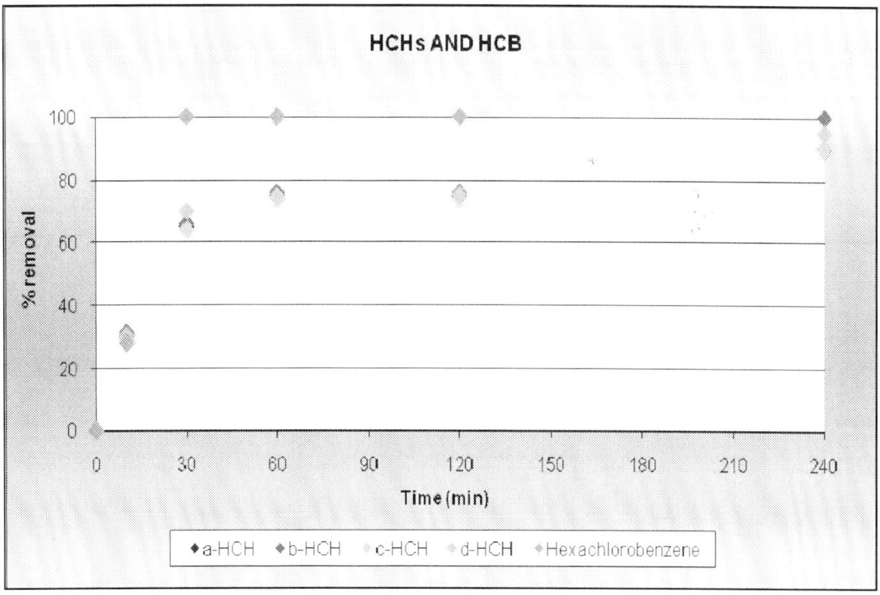

Fig. 4. HCHs and HCB adsorption

Regarding to HCHs (figure 4), it can be observed the same behavior. The adsorption yield at 10 minutes is 30%. This percentage notably increases up to 65% in 30 minutes of treatment and next, practically there isn't adsorption until 2 hours. The adsorption increases up to the complete removal of HCHs at 4 hours of treatment.

The adsorption of hexachlorobenzene (HCB, figure 4) is faster since at 10 minutes of treatment the adsorption is 60% and it is complete removed at 30 minutes.

With respect to heptachlors (figure 5) it can be observed that three heptachlor are adsorbed about 25% in 10 minutes. The adsorption increases, mainly for heptachlor, until 30 minutes. At this time the removal of heptachlor is 90% and of heptachlors epoxide is 60%. Until 2 hours the adsorption doesn't increase and it lightly increases at 4 hours. The complete removal of heptachlor is achieved at 2 hours and after 4 hours of treatment the removal of heptachlors epoxide is about 70%.

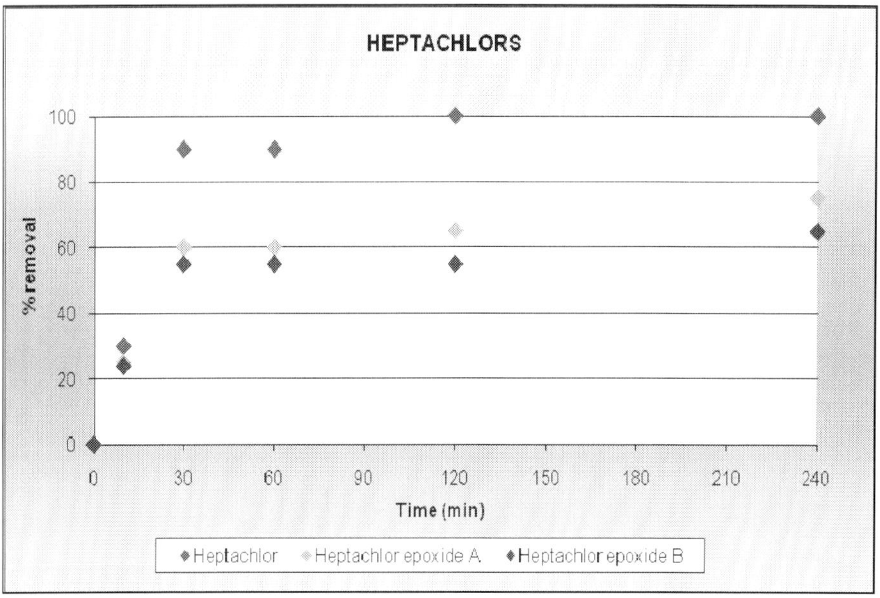

Fig. 5. Heptachlors adsorption

The behavior of two studied endosulphans is completely different, as can be observed in figure 6. Two of them are adsorbed 20% at 10 minutes, but α-endosulphan is complete removed at 30 minutes meanwhile endosulphan sulphate is adsorbed 70% after 4 hours of treatment.

With respect to drins (figure 7), all of them are removed about 30% in 10 minutes of treatment. At this time, the aldrin and isodrin (isomers between them) adsorption increases with the time, achieving adsorption yields about 85-100% in 4 hours. In the case of endrin and dieldrin (isomers between them), the adsorption is lower and after 4 hours their removal yields are about 60%.

In accordance with results obtained for DDTs (figure 8), their adsorptions are similar until 1 hour of treatment (50-60%). After and until 4 hours, the adsorption of pp'-DDE and pp'-DDD+op'-DDT lightly increases up to 65-85%. On the contrary, the adsorption of pp'-DDT notably increases at 2 hours or treatment achieving its complete removal.

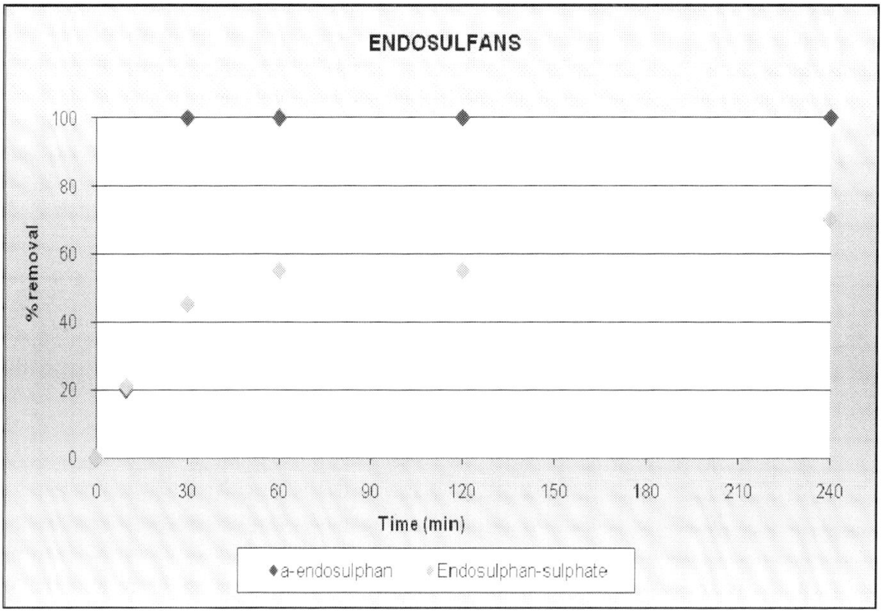

Fig. 6. Endosulphans adsorption

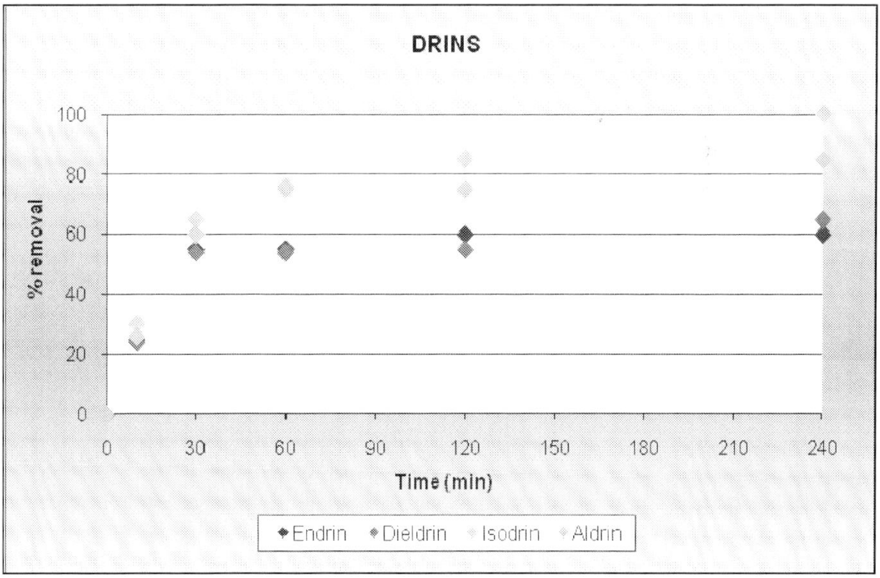

Fig. 7. Drins adsorption

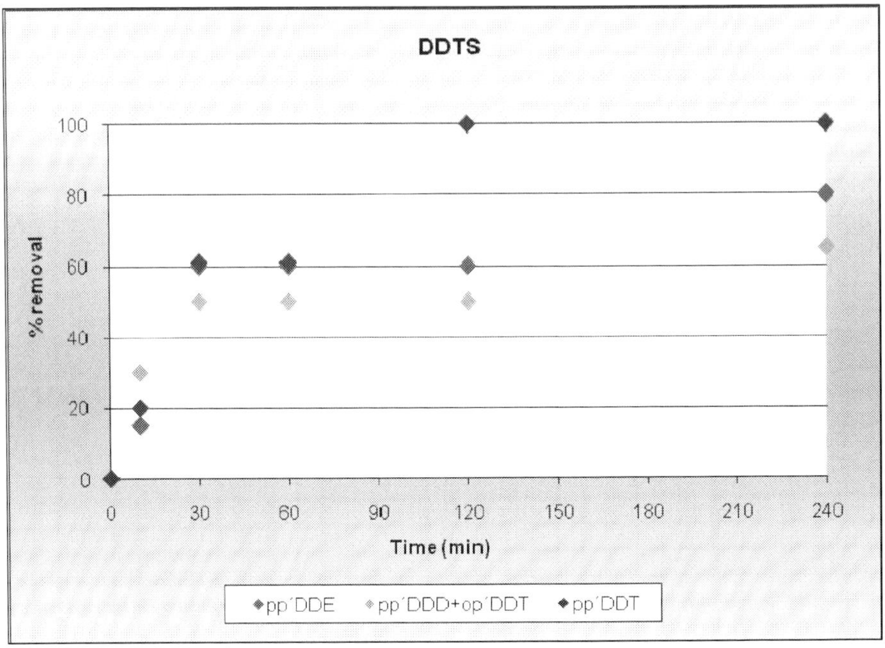

Fig. 8. DDTs adsorption

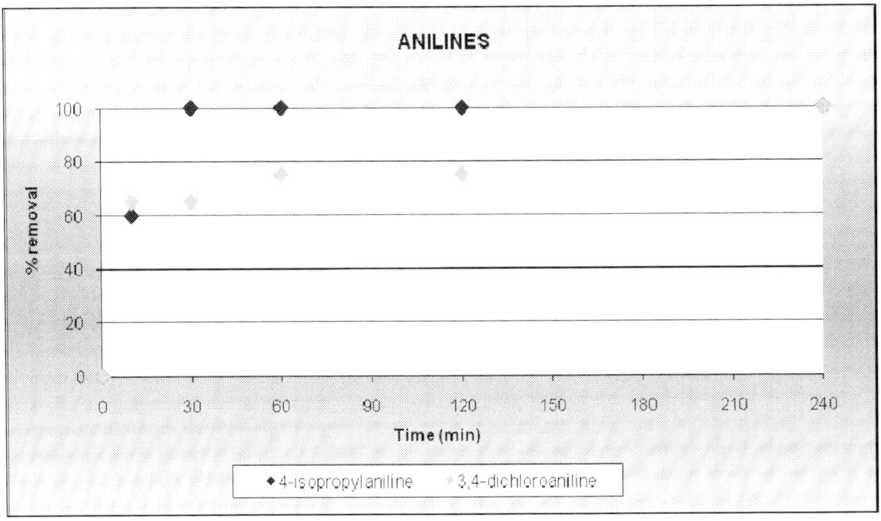

Fig. 9. Anilines adsorption

With respect to anilines (figure 9), it can be observed their different behavior to the adsorption versus the time. Two anilines are removed about 70% at 10 minutes of treatment. However, 4-isopropylaniline is complete adsorbed at 30 minutes meanwhile 3,4-dichloroaniline requires 4 hours of treatment for its total adsorption.

Regarding urea derivates pesticides (figure 10), their behavior is complete different too. At 10 minutes the diuron removal is very high, 65%. Its removal increases up to 95% after 4 hours of treatment. On the contrary, isoproturon is adsorbed 10% in 10 minutes although after 4 hours the same adsorption than diuron is achieved (95%).

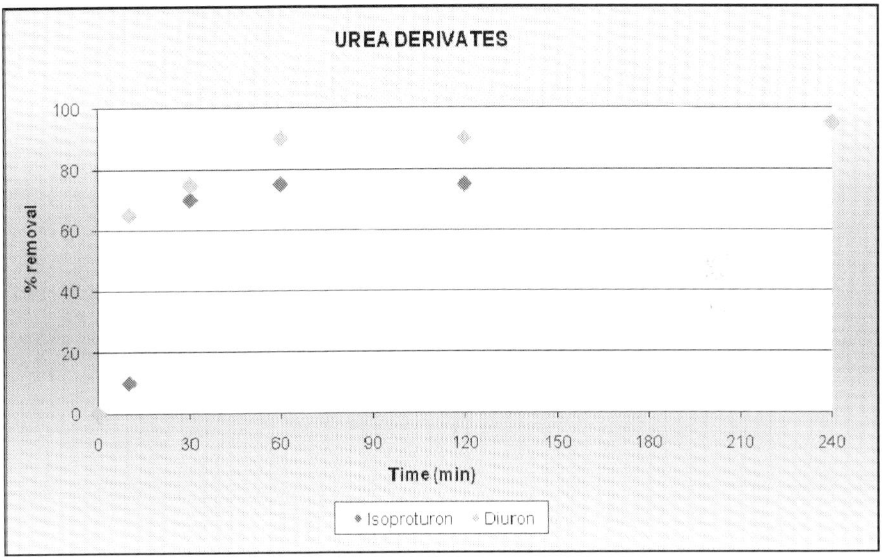

Fig. 10. Urea derivates adsorption

In accordance with the results obtained to anilides (figure 11), it can be observed the same behavior for alachlor and metholaclor, and different one for methoxychlor. The three pesticides are adsorbed 35% in 10 minutes. However, alachlor and metholachlor increase their adsorption gradually with the time, achieving a removal about 70% in 4 hours of treatment, while the methoxychlor adsorption is 90% in 30 minutes and total in 4 hours.

With respect to the rest of studied pesticides (figure 12), dicofol and 4,4'-dichlorobenzophenone are quickly adsorbed achieving their complete adsorption in 30 minutes of treatment. Molinate and trifluralyn are mainly adsorbed in the first 30 minutes of treatment and after 4 hours, the adsorption is total, like a lot of studied pesticides. Finally, the tetradiphon adsorption is progressive in the time and its complete adsorption is achieved in 2 hours of treatment.

As it has been mentioned before, the adsorption of studied pesticides shows the typical tendency of a multicomponent mixture. Moreover, the main adsorption of the group of pesticides happens in the first 30 minutes of treatment. In this time, the removal of each individual studied pesticide presents a pseudo-first order kinetic, that is corresponded with the next equation:

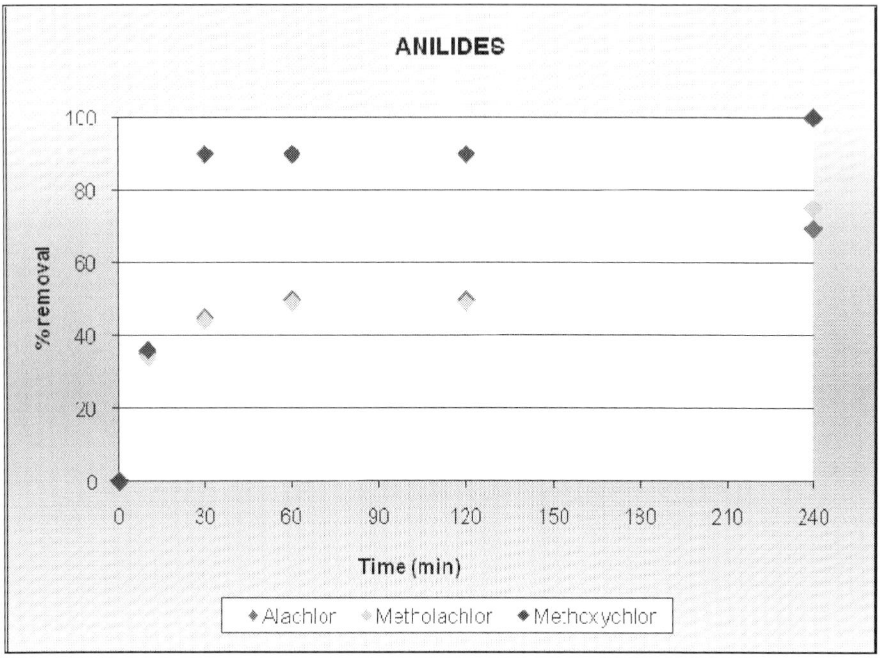

Fig. 11. Anilides adsorption

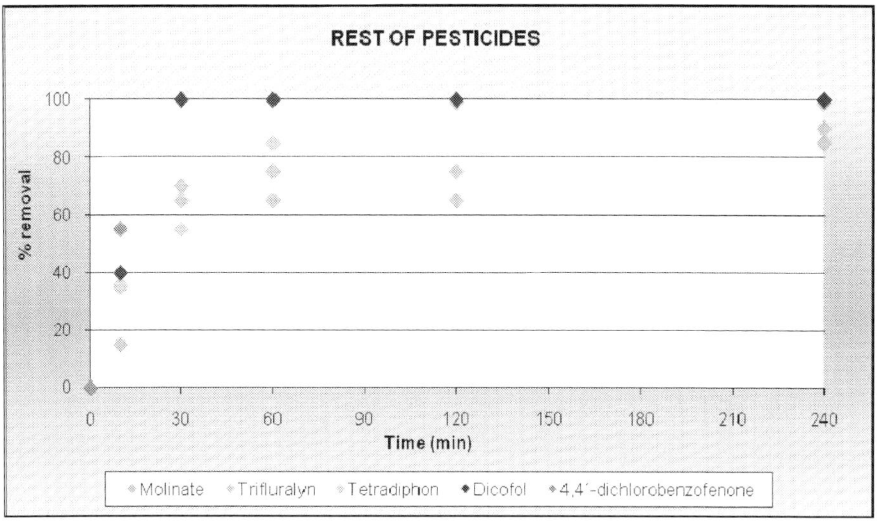

Fig. 12. Rest of studied pesticides adsorption

$$\log\frac{C}{C_0}=k*t \tag{3}$$

Where C is the pesticide adsorbed concentration, C_0 is the pesticide initial concentration, t is the time of treatment and k is a constant which represents the adsorption rate of each pesticide. Rate constants, k, of each studied pesticide and regression coefficients, R^2, obtained for each lineal equation are shown in table 6.

As it can be observed, with a contact time of 30 minutes, the most of studied pesticides follows a kinetic of pseudo-first order, just like in studies carried out by other authors (Ayranci and Hoda, 2005; Salman and Hameed, 2010). Moreover, rate constants obtained haven't any direct relation neither solubility nor molecular weight of studied pesticides (Ayranci and Hoda, 2005).

With respect to individual pesticides, parathion methyl and ethyl, chlorfenvinfos, dimethoate, hexachlorobenzene, α-endosulphan, 4-isopropylaniline, dicofol and 4,4'-dichlorobenzophenone are the pesticides that are adsorbed faster.

Three times slower than these pesticides, are adsorbed heptaclor and methoxyclor, and about a half of these, chlorpiryfos, molinate, HCHs, diuron and isoproturon are adsorbed.

PESTICIDE	k (h^{-1})	R^2	PESTICIDE	k (h^{-1})	R^2
Simazine	0,712	0,985	Heptachlor epoxide B	0,687	0,998
Atrazine	0,369	0,996	α-endosulphan	5,703	0,920
Propazine	0,525	0,997	Endosulphan sulphate	0,517	0,997
Terbuthylazine	0,589	0,989	Endrin	0,687	0,998
Prometon	0,462	0,984	Dieldrin	0,687	0,998
Ametryn	0,589	0,989	Isodrin	0,908	1,000
Prometryn	0,432	0,983	Aldrin	0,797	1,000
Terbutryn	0,496	0,965	pp´-DDE	0,822	0,977
Desethylatrazine	0,595	0,989	pp´-DDD+op´-DDT	0,575	0,952
Parathion methyl	5,649	0,960	pp´-DDT	0,810	0,993
Parathion ethyl	5,691	0,947	Isoproturon	1,102	0,944
Chlorpiryfos	1,381	0,997	4-isopropylaniline	5,634	0,967
Chlorfenvinfos	5,661	0,948	Diuron*	1,075	0,771
Dimethoate	5,858	0,946	3,4-dichloroaniline*	0,769	0,557
α-HCH	0,908	1,000	Molinate	0,947	0,970
β-HCH	0,908	1,000	Trifluralyn	1,037	0,998
χ-HCH	0,908	1,000	Alachlor*	0,471*	0,807
δ-HCH	1,052	0,999	Metholachlor*	0,479*	0,815
Hexachlorobenzene	5,618	0,967	Methoxychlor	2,061	0,980
Heptachlor	2,076	0,970	Tetradiphon	0,659	0,939
Heptachlor epoxide A	0,797	1,000	Dicofol	5,812	0,941
			4,4´-dichlorobenzophenone	5,671	0,960

*In the case of diuron, 3,4-dichloroaniline, alachlor and metholachlor, the regression coefficiente obtained are lower than for the rest of pesticides. It is due to the fact that the main adsorption of these pesticides happens in the first 10 minutes of treatment

Table 6. Rate constants and regression coefficients of studied pesticides

Finally, pesticides wich are adsorbed slowest onto activated carbon are triazines, heptachlors epoxide, endosulphan-sulphate, drins, DDTs, 3,4-dichloroaniline, alachlor, metholachlor and tetradiphon.

3.2 Results of activated carbon characterization

Results corresponding to the chemical analysis by SEM of used PACs (vegetal and mineral), before and after applying treatments in distilled and natural water, are shown in table 7.

In accordance with the comparison of the SEM analysis before and after the treatments, the main variations produced are the following:

- Oxygen percentage lightly increases after the treatment in natural water for both PACs, vegetal and mineral.
- A little quantity of calcium appears after the treament in natural water for both PACs.
- For the rest of elements, the variations produced after applying treatments aren't significants.

Results corresponding to the characterization of vegetal and mineral PAC after the treatment in distilled and natural water by the BET method are shown in table 8.

In accordance with the results obtained about BET surfaces, it can be said that:

- The treatment in distilled water doesn't cause the reduction of the surface are in both used PACs.
- The treatment in natural water causes a reduction of the surface are in both PACs, about 10%. This decrease is due to the adsorption of organic matter contained in natural water onto activated carbon.

VEGETAL PAC			
ELEMENT	% WEIGHT BEFORE THE TREATMENT	% WEIGHT AFTER TREATMENT IN DISTILLED WATER	% WEIGHT AFTER TREATMENT IN NATURAL WATER
O	1.41	1.57	4.44
Si	-	0.30	0.49
Ca	-	-	0.41
Al	-	-	-
S	-	-	-
Fe	-	-	-
MINERAL PAC			
ELEMENT	% WEIGHT BEFORE THE TREATMENT	% WEIGHT AFTER TREATMENT IN DISTILLED WATER	% WEIGHT AFTER TREATMENT IN NATURAL WATER
O	2.93	3.01	3.47
Si	0.69	1.03	1.03
Ca	-	-	0.26
Al	0.35	0.33	0.38
S	0.24	0.37	0.16
Fe	0.24	0.32	0.37

Table 7. Results about the chemical analysis by SEM of used PACs

PAC	BET SURFACE ($m^2 g^{-1}$)		
	Before the treatments	After the treatments in distilled water	After the treatment in natural water
VEGETAL	555.2±10.9	563.2±12.0	497.6±9.9
MINERAL	745.4±14.5	746.7±14.3	676.9±13.0

Table 8. BET surfaces of used PACs

4. Conclusions

In accordance with the study of the activated carbon nature and aqueous matrix influence it can be said that:

- Removal percentages of studied pesticides achieved in a contact time of 10 minutes are: 34% with mineral PAC and 46% with vegetal PAC to the solution in distilled water; 33% with mineral PAC and 26% with vegetal PAC to the solution in natural water.
- In general, the adsorption with vegetal PAC is more effective in order to remove the studied pesticides in distilled water, and the adsorption with mineral PAC is more effective in order to remove the studied pesticides in natural water.
- In the solution of distilled water, all pesticides are better removed by the treatment with vegetal PAC, with the exception of endosulphan-sulphate which is lightly better removed with mineral PAC.
- With respect to the solution in natural water, all studied pesticides have greater removal percentages by the adsorption with mineral PAC, with the exception of some triazines which are lightly better removed with vegetal PAC.
- In general, the adsorption treatment with mineral PAC isn't influenced by the aqueous matrix since the average removal percentages obtained are similar both distilled and natural water. However, the use of vegetal PAC is influenced by the aqueous matrix since the average removal percentages achieved in natural water are considerably lower that the percentages achieved in distilled water.
- The treatment of activated carbon adsorption in distilled water doesn't cause the surface are reduction in any used PAC. However, the treatment in natural water causes a reduction about 10% of two PACs. This reduction is due to the adsorption of organic matter of the natural water onto the activated carbon.

According to the study of the influence of the contact time it can be said that:

- The adsorption with mineral PAC in natural water achieves a partial removal of the studied pesticides.
- Average adsorption percentages of studied pesticides in natural water in different contact times using mineral PAC are: 32% in 10 minutes, 67% in 30 minutes, 71% in 1 hour, 74% in 2 hours and 85% in 4 hours.
- The main adsorption of studied pesticides is achieved in the first 30 minutes of treatmen with mineral PAC. This time coincides with the recommendations given by the WEF-ASCE (WEF-ASCE, 1998) regarding to the contact time in the activated carbon adsorption.

WEF-ASCE. (1998) Design of Municipal Wastewater Treatment Plants. WEF Manual of Practice 8, ASCE Manual and Report on Engineering Practice 76. ISBN 9780784403426.

37

Adsorption Properties of Sediments for Pesticides: Investigation by Supercritical Fluid Extraction and Gas Chromatography Mass Spectrometry

Hiroaki Chikushi[1,2], Natsuko Yoshida[1] and Kei Toda[1]
[1]*Department of Chemistry, Kumamoto University, 2-39-1, Kurokami, Kumamoto 860-8555*
[2]*Institute of Environmental Analysis, Nodaichi-Denshi Co., 335 Yoyashu, Kumamoto 860-0823*
Japan

1. Introduction

Use of pesticides is necessary in the recent agricultural production for removal of weeds and pests which interfere growth of crops. Though use amount and use methods are determined by standards for the each pesticide, residual pesticides in environments and foods are paid attentions much from the view point of food safety. Therefore, analytical methods to evaluate pesticide residue contained in foods are investigated well (Okihashi et al. 2005). On the other hand, pesticides spread in agricultural fields and golf links are not only taken by plants but also drawn to environment as waste water with rain water and supplied waters, and it is issued that the pesticides pass through sediments to reach to ground waters (Sudo et al., 2002). Japanese Ministry of Environment determined pesticide control law and standard indexes for golf link pesticides to monitor and control the pesticides emitted to environment (Ministry of Environment, Government of Japan, 2001). Determination methods for the pesticides contained in several kinds of samples are usually complicated and difficult to determine accurately without sophisticated skill (Ormad et al., 1996; Zhou et al., 2009). In addition, the determination methods are different with different target analyte and samples, and it takes long time to determine several kinds of pesticides. In 2005, the positive list for residual pesticides was presented and multiple analysis methods were shown such as gas chromatography (GC) analysis after supercritical fluid extraction (SFE) (Okihashi et al., 2005; Tobino et al., 2007). For each pesticide, on the other hand, liquid-liquid extraction (LLE) (Ormad et al., 1996) and solid phase extraction (SPE) (Zhou et al., 2009) are popular for the determination. However, they are time consuming and require good experimental skill and also consume much organic solvent not to be good for environmental charging. Accordingly, we tried to apply SFE method to environmental analysis which is already used for food analysis as Tobino et al. presented before (2007). The SFE is effective extraction utilizing supercritical fluid characteristics such as high permeability of the fluid and high solubility of organic solutes to the fluid. Such SFE has been applied to determination of dioxines contained in sediments (Miyawaki et al., 2003). Glazkov et al.

temperature was initially 50°C for 2 min, ramped to be 120°C at the rate of 20°C/min and upto 270°C at 5°C/min. The each pestcide was analyzed at the indicated M/z number in single ion monitoring (SIM) mode.

Fig. 2. Supercritical fluid extraction instrument used in this invesitigation

3. Results and discussion

3.1 Water analysis with SFE

Firstly we investigated how to analyze pesticides contained in water. This is important because conventional water analysis is complicated. We use SFE for sediment analysis, so if the same SFE method can be applied to water analysis, a series of measurement would be quite convenient. Therefore, our strategy for water pesticide analysis was adsorption of solid adsorbent and extraction from adsorbent by SFE. This running cost is much cheaper than solid phase extraction using commercial cartridges and not affected by extraction condition such as sample flow rate. There is no need of solvent exchange of sample in the pretreatment which is necessary in conventional solid phase extraction with cartridges.

Adsorption properties of the each adsorbent were tested and the results obtained by 15 min adsorption were plotted against octanol/water partition coefficient (log K_{ow}). Interestingly, most of the adsorbents showed relationship between the adsorption property (recovery of pesticide) and log Kw in this condition, especially in cases of XAD and silica gel. On the other hand small relationship was observed in cases of ODS and Tenax GR. With these adsorbents, recoveries of the pesticides were mostly quantitative and it can be said that the pesticide molecular sizes were suitable for the adsorption on those. These results agreed

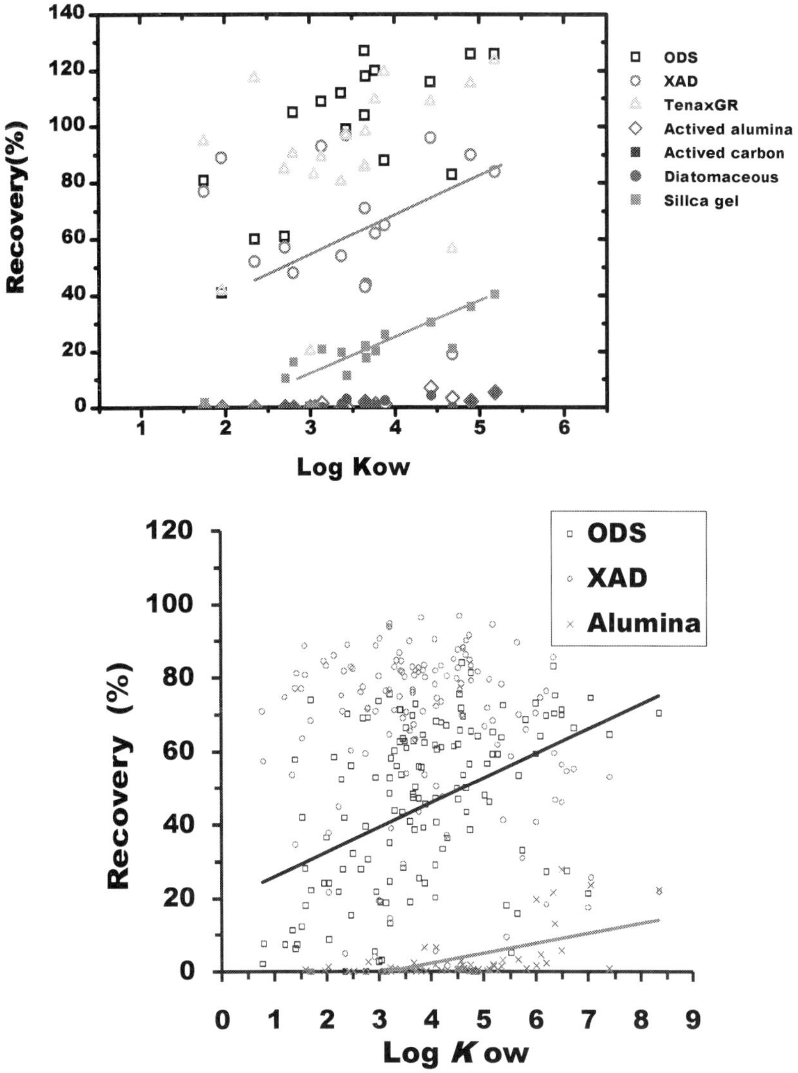

Fig. 3. Recoveries of pesticides obtained in the 15-min adsorption of 20 µg/L pesticide test plotted against their octanol/water partition coefficient (top) and same examination for 100 µg/L of 170 kinds of pesticides examined with XAD, ODS and alumina (bottom)

with the reported results by Junk and Richard (1988) where they evaluated for ODS. From the data shown in Fig. 3 and other data, it was decided that ODS, XAD and Tenax GR were suitable for the pesticide adsorption. However, Tenax GR was not used in the further experiments due to high cost of the adsorbent.

Next, effect of adsorption time was tested. Adsorption time where the adsorbent in sample water was stirred was changed 5, 10, 15, 30 and 60 min and recoveries were plotted against the adsorption time. In case of ODS, the amount was saturated at 15 min and did not increase in the longer adsorption. Meanwhile, adsorption equilibrium was completed at 30 min in case of XAD. It seemed that adsorption onto ODS occur only on the adsorbent surface while pesticide molecules enter inside of XAD particles. That is why equilibrium times were different with these adsorbents. Average particle sizes were ODS 0.04 mm and XAD 0.6 mm. It means that relative surface area to their weight for ODS was 15 times greater than that of XAD. This was also the reason of fast equilibrium in case of ODS. According to these results, adsorption times were decided as 15 min for ODS and 30 min for XAD.

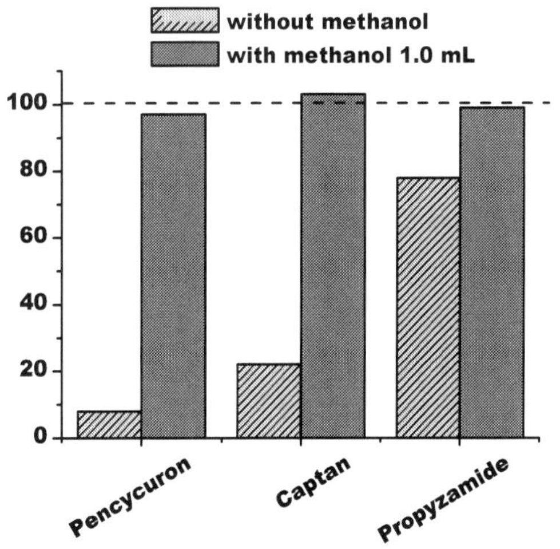

Fig. 4. Effect of methanol as modifier. Methanol 1 mL was added in the SFE cartridge as modifier. Recoveries for pencycuron, captan and propyzamide obtained with and without methanol were shown. SFE was performed at 50°C and 3000 psi.

After adsorption, the adsorbent particles were collected, dried overnight and then the pesticides were extracted from the adsorbent by SFE. Conditions of SFE were examined such as extraction temperature and extraction pressure. Also effect of modifier was tested. Modifier was necessary to extract effectively from the adsorbent. As shown in Fig. 4, good recoveries were obtained with methanol addition as modifier.

Effects of SFE pressure and temperature were also examined. In case of ODS, the results were mostly independent on the conditions and good recoveries were obtained from 1000 to 3000 psi; a little drop in recovery was observed at 4000 psi. On the other hand, effect of pressure was critical in case of XAD. Probably supercritical fluid must enter into the inside of XAD particle and more pressure was required in case of XAD. But good recoveries were obtained at 3000 psi or more. We decided to use 3000 psi for the both adsorbents. Effect of extraction temperature was not big as shown in Fig. 5. But recoveries and deviation were the best at 50°C. In conclusion, SFE procedure was the best at 50°C, 3000 psi with methanol as modifier.

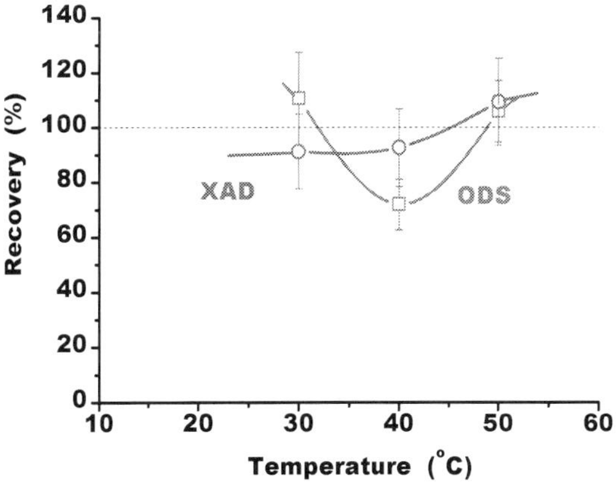

Fig. 5. Effect of SFE temperature. Methanol 1 mL was added in the SFE cartridge. SFE was carried out at 3000 psi.

The optimal conditions were obtained and recoveries of the 26 kinds of pesticides were determined accordingly for pure water as shown in Fig. 6. Peak was not obtained for isoprodion and bensride in GC-MS chromatogram due to difficulty of GC measurement. For most of other compounds, recoveries were between 80 and 120% which was said as quantitative range. Even tested concentrations were low (20 µg/L), good recoveries were obtained in this method. Mostly guideline level for the pesticide varies 30 to 800 g/L. In addition, it can be seen that recoveries with this method were obviously better than those obtained with SPE. The proposed method is simple and easy to operate and good recoveries can be obtained.

Next, linearity of calibration curve and limit of detections in this method were examined. Calibration curves for 0, 1, 5, 10, 20, 40 (or 50), 100 µg/L of pesticides were evaluated for XAD and ODS adsorbents. Series of adsorption - SFE - GC-MS were repeated three times for each solution. For most of pesticides, good linearity was obtained (R^2 0.869 ~ 0.999 with XAD adsorption, R^2 0.963 ~ 0.9989 with ODS adsorption).

Further applications were examined; namely tea bag extraction. In this proposal, XAD was placed in a thin paper bag which was used for tea serving. The tea bag containing XAD resins was placed in a stirred water sample, and then the XAD resins were treated as same as mentioned before. Good results were obtained as well as the batch-wise collection as the chromatogram in Fig. 7. The XAD particles were relatively large and could be kept in the tea bag whereas ODS came out to the solutions in the adsorbing process. This method is applicable to on site adsorption. The tea bag containing XAD resins can be placed in stream of river water or waste water for a day or a week and taken back to the laboratory to determine. This method is expected to be useful for simple water screenings after supplying the pesticides to golf links and agricultural fields (Chikushi et al., 2009).

3.2 Sediment analysis with SFE

The five kinds of sediments obtained mainly around Kumamoto were separately placed in individual glass column after appropriate pretreatment as mentioned in experimental. Into these sediment columns, 25 mL of 40 µg/L pesticide aqueous solution was added and gradually passed through the sediment by gravity. After all pesticide water passed the

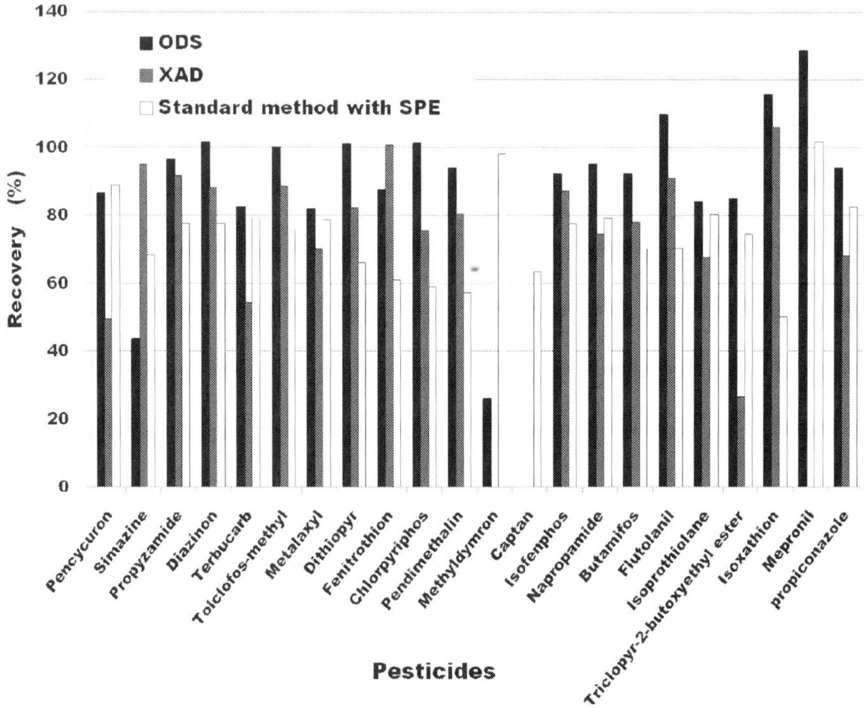

Fig. 6. Recoveries for water pesticides obtained by this method with XAD and ODS adsorbents and standard method with SPE extraction. Pure waters (50 mL) were spiked to contain 20 µg/L of each pesticide.

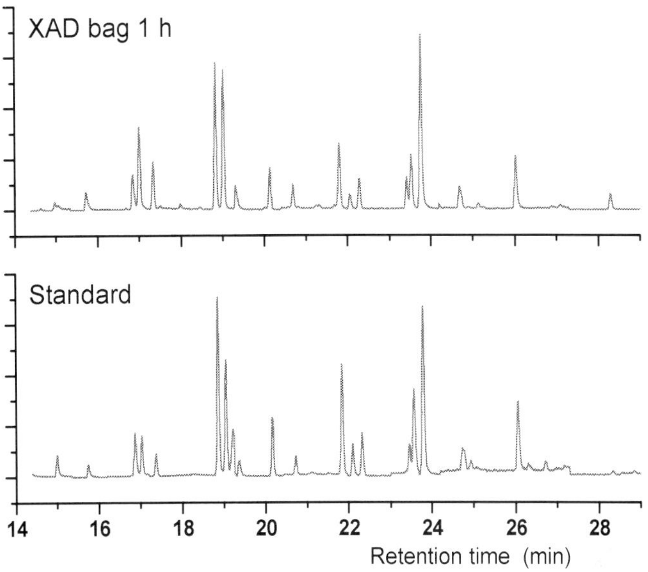

Fig. 7. GC-MS chromatograms obtained with bag sampling. The top panel chromatogram was obtained by bag sampling where XAD resin was placed in a tea bag and put in 50 mL of 20 µg/L aqueous sample water. The bottom is for comparison: 1-mg/L standard (acetone medium) was injected directly without any treatment.

sediment, pure water 25 mL was flowed through the column. All eluents were collected under the column. Pesticides adsorbed in sediments and eluted waters were measured with direct SFE and XAD adsorption and SFE as described before, respectively. Adsorption efficiencies were obtained from the pesticide amounts remaining in sediments and eluted with water and calculated results are shown in Table 1.

As shown in Table 1, good adsorptions were observed with Ariake Sea tidal flat sediment, kuroboku soil, limonite and mudstone in this order. The pesticides were not trapped in akahoya ash and sandstone.

3.3 Properties of sediment and pesticide adsorption

First, sediment determination method with SFE-GC-MS was evaluated. 1 mL of 1-mg/L pesticide mixture was added to sediment, stored in decicator overnight and the pesticides were extracted with SFE to measure by GC-MS. Results are shown in Fig. 8. For most of pesticides, reasonable values (100 +/- 20%) were obtained with this method.

Generally, adsorptions of non-ionic hydrophobic compounds such as these pesticides are related to organic matter contents of sediments. Characteristics of sediment vary in dependent with place, depth, surrounding conditions. Physical and chemical properties of sediments were investigated to make clear the relationship between the sediment characteristics and adsorption property.

Organic contents (humus) was extremely high in Kuroboku soil (25.1%) compared with the other's (0.45 ~ 2.84%). Organic contents are thought to help adsorption of pesticides with

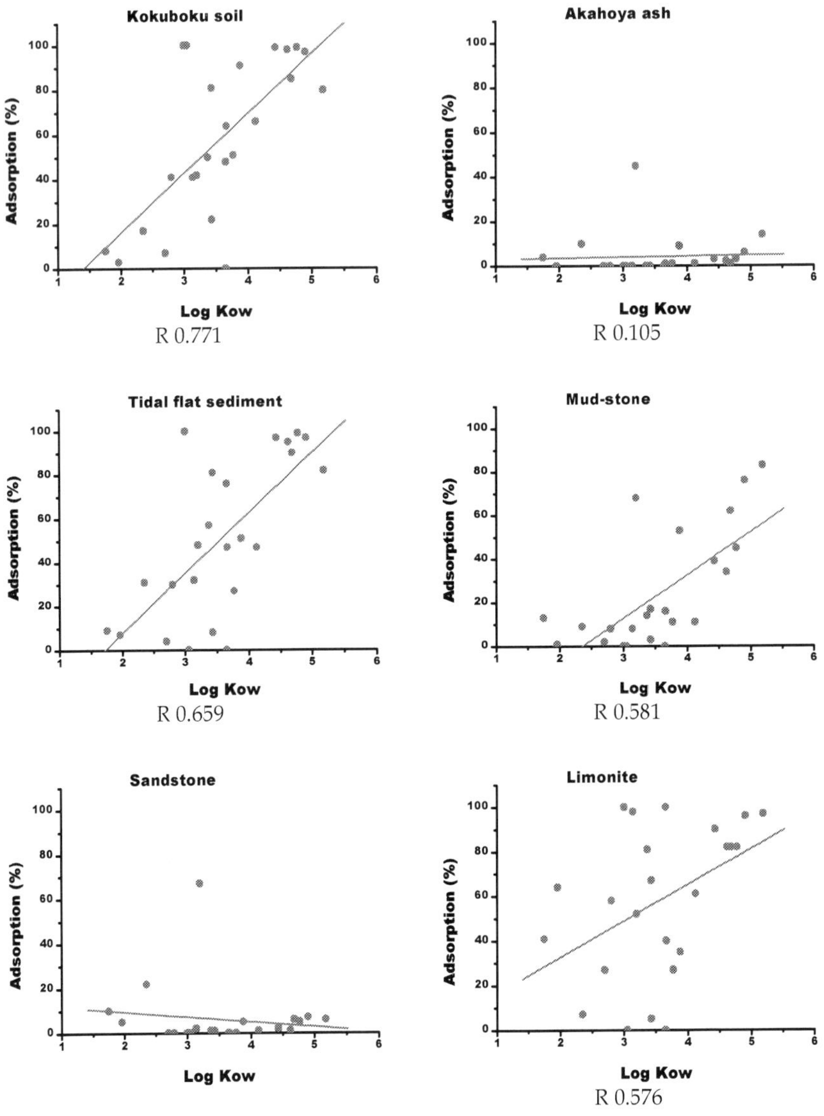

Fig. 9. Adsorption of pesticides against octanol/water partition coefficient

Fig. 10. SEM image of limonite observed at amplitude of x 10,000

Adsorptions are plotted against pesticide solubility to water as shown in Fig. 11. Adsorption decreased with increase in water solubility in cases of Kuroboku, Mudstone and Tidal sediments. This indicates that adsorption mechanism was same in those sediments. On the other hand, adsorptions of pesticides on Limonite were independent on the solubility. Limonite had unique adsorption characteristics for organic compounds. Kah and Brown (2007) reported that adsorption of pesticides, especially which had ionic property, was dependent of sediment pH. Pateiro-Moure et al. (2009) reported that pesticides adsorption was related with organic matter and ferric oxide. It is said that adsorption ability is not dependent of only humus in the both reports. Among the sediments tested in this work, Limonite adsorbed the pesticides well even though it had only little humus.

4. Conclusions

We have developed a simple analytical method for pesticides contained water samples by stirring of adsorbent particles in a small beaker followed by supercritical fluid extraction and gas chromatography mass spectrometry. This method provides not only labour less procedure for the measurement but also use of less organic solvent for green analysis. This is suited for screening of samples. The supercritical fluid extraction was also applied to analysis of adsorbed pesticides in several kinds of sediments. By combining water and sediment analysis both with SFE, adsorption properties of sediments were successfully investigated. More hydrophobic (large K_{ow}) pesticides adsorb more strongly to the sediments. Humus is important factor in the pesticide adsorption. Kuroboku and Tidal sediment contain high humus and adsorb pesticides well. However, Limonite which contains less humus adsorbs pesticides in good efficiency as well. This is unique property of Limonite probably due to microporous ferric oxide as a good adsorber.

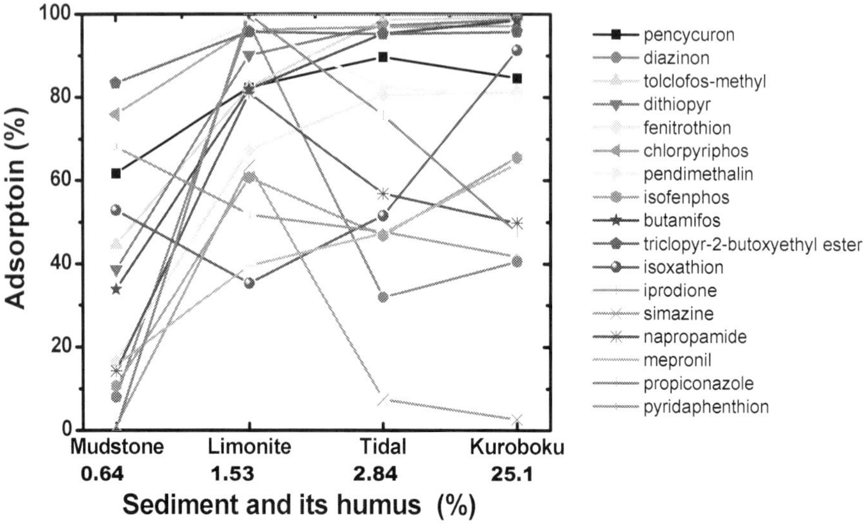

Fig. 11. Adsorption property of sediments (difference with sediment humus)

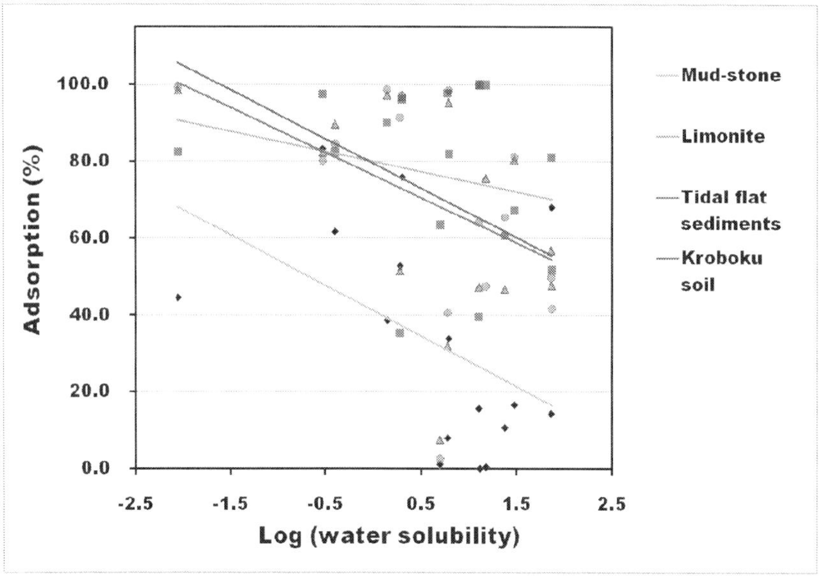

Fig. 12. Adsorption properties of pesticides plotted against their water solubility

5. References

Chikushi, H.; Hirota, K.; Yoshida, N.; Edamura, T. & Toda, K. (2009). Simple water analysis of golf link pesticides by means of batch-wise adsorption and supercritical fluid extraction. *Talanta*, Vol. 80, No. 2, 738 - 743.

Daikuhara, G. & Sakamoto, Y. (1911). Study on cause and characteristics of sediment acidifying and distribution of acid sediments (in Japanese). *Nouji Kenkyu Shiken Houkoku*, Vol. 37, 66 - 72.

Glazkov, I.N.; Revelsky, I.A.; Efimov, I.P. & Zoltov, Y.A. (1999). Supercritical fluid extraction of water samples containing ultratrace amounts of organic micropollutants. *J. Microcolumn Separations*, Vol. 11, No. 10, 729 - 736.

Janda, V.; Mikešová, M & Vejrosta, J. (1996). Direct supercritical fluid extraction of water-based matrices. *J. Chromatogr. A*, Vol. 733, No. 1-2, 35 - 40.

Junk, G.A. & Richard, J.J. (1988). Organics in water: solid phase extraction on a small scale. *Anal. Chem.*, Vol. 60, No. 5, 451 - 454.

Kah, M. & Brown, S.D. (2007). Prediction of the adsorption of ionizable pesticides in soil. *J. Agric. Food Chem.*, Vol. 55, No. 6, 2312 - 2322.

Kononova, M.M. (1966). *Soil organic matter, 2nd English ed.*, 378 - 382, Pergamon Press., Oxford.

Ministry of Environment, Government of Japan, (2001). Standard Method of Waste Water Analysis(Multi-Analysis of Pesticides), in Provisional Guideline for Preventing Water Pollution with Golf Link Chemicals (in Japanese), pp.93-96.
(http://www.env.go.jp/water/dojo/noyaku/law_data/f402kansuido0077.pdf).

Miyawaki, S.; Kawashima, F. & Honda, K. (2003). Influence of the moisture content on supercritical fluid extraction and soxhlet extraction of dioxins from soil. *Bunseki Kagaku*, Vol. 52, No. 11, 989 - 995.

Okihashi,M.; Kitagawa,Y.; Akutsu,K.; Obana,H. & Tanaka, Y. (2005). Rapid method for the determination of 180 pesticide residues in foods by gas chromatography/mass Spectrometry and flame photometric detection. *J. Pestic. Sci.*, Vol. 30, No. 4, 368-377.

Ormad, P.; Cortés, S.; Sarasa, J.; Martín, A.; Puig, A. & Ovellerio, J.L. (1996). Characterization of organic micropollutants in the River Cinca (Spain) by using granular activated carbon and liquid-liquid Extraction. *J. Chromatogr. A*, Vol. 733, No. 1-2, 159 - 169.

Pateiro-Moure, M.; Perez-Novo, C.; Arias-Estevez, M.; Rial-Otero, R. & Simal-Gandara, J. (2009). Effect of organic matter and iron oxides on quaternary herbicide sorption-desorption in vineyard-devoted soils. *J. Colloid Interface Sci.*, Vol. 333, No. 2, 431 - 438.

Schollenberger, C.J. & Simon, R.H. (1945). Determination of exchange capacity and exchangeable bases in soils. *Soil Sci.*, Vol. 59, No. 1, 13 - 24.

Sudo, M.; Kunimatsu, T. & Okubo, T. (2002). Concentration and loading of pesticide residues in Lake Biwa basin (Japan). *Water Res.* Vol. 36, No. 1, 315 - 329.

Tobino, T. (2007). *J.Pesitic.Sci.*,Vol. 32, No. 3, 322-327.

ion approaches any surfaces, it encounters many attractive forces that eventually will cause its physical adherence or its bonding to the surface of condensed phase. This process in which molecules or ions become associated with solid phases is generally called sorption. Therefore, sorption is a generic term that designates all the relevant processes, which are involved when a pesticide comes into contact with solid matrices, without reference to a specific mechanism. The surface of the material where the sorption occurs is called substrate or sorbent and the chemical that interacts with the sorbent is the sorbate.

Chemical characteristics	Examples	Effects
Ionic		
Cationic	Diquat, paraquat	Herbicide
Basic	Atrazine, simazine	Herbicide
Acidic	Dicamba, MCPA,	Herbicide
Miscelleneous	Bromacil	Herbicide
Nonionic		
Chlorinated hydrocarbons	Lindane, DDT, Toxaphene	Insecticide
Organophosphorous	Methyl parathion, diazinon	Insecticide, fungicide
Dinitroanilides	Trifluralin, oryzalin	Herbicide
Carbanilates	Chlorpropham, barban	Herbicide
Benzonitriles	Dichlobenil	Herbicide
Acetamides	Allidochlor	Herbicide
Carbothioates	Molinate	Herbicide
Thiocarbamates	Triallate, cycloate	Herbicide, fungicide
Anilides	Alachlor,propanil	Herbicide, fungicide
Ureas	Linuron, diuron	Herbicide, algicide
Methylcarbamates	Carbaryl, terbutol	Insecticide

Table 1. Classification of pesticides according to chemical properties (modified from Gevao et al., 2000)

In the environment, the natural sorbents, called geosorbents, are soils, aquifer solids, suspended particulate matter or sediments. Geosorbents, shown in Figure 1, are very heterogeneous at various particle, aggregate and sample scales which vary temporally as well as spatially (Luthy et al., 1997, Bronick & Lal, 2005, Ehlers & Loibner, 2006). If the process of sorption occurs onto a two-dimensional surface it is called adsorption. It is always an exothermic process where geosorbent-adsorbate bonds (see section 4) are usually stronger than the bonds between adsorbed chemicals (Somorjai & Li, 2010). Those interactions conduct to the formation of a layer of adsorbate retained at the surface of the geosorbent where it is difficult to remove the sorbate. If the molecule, instead of residing at the surface of the geosorbent, penetrates into the three-dimensional matrix or into the pores in the solid, the process is called absorption.

3. Geosorbent composition

Structural and molecular compositions of geosorbents have a direct control on the sorption and retention of pesticides. As shown in Figure 1, geosorbents are primary formed by a complex assemblage of inorganic minerals and natural organic matter. These two types of material are free or associated to form macro- and microaggregates with size distribution and inter- and intra-aggregate pore continuity (Luthy et al., 1997, Six et al., 2004, Bronick & Lal, 2005, Jasinka et al., 2006, Kögel-Knabner et al., 2008). The macropores (pore size larger than 50 nm), mesopores (pore size between 2 and 50 nm) and micropores (pore size smaller than 2 nm) limit more or less the advective and diffusive transport of the contaminant into the geosorbents. The transport of pollutants at the surface of a geosorbent or inside its core allows them to reach potential sorption domains or sites of the most reactive surfaces. The pores increase the surface where pesticides can interact with the geosorbents. For pesticides with low polarity and slight solubility in water, e.i. in the mg liter^{-1} range or less, organic matter surface of the geosorbent is the principal domain where sorption occurs (Wauchope et al., 2002). However, when organic carbon content of the geosorbent becomes relatively low (<0.1 %,) other reactive surfaces can favor the sorption of pesticides (Wauchope et al., 2002, Ehlers & Loibner, 2006, Kah & Brown, 2006). Among the inorganic surfaces present in the geosorbents, the clay minerals and sesquioxides (oxides, hydroxides and oxyhydroxides of Al and Fe) have been reported to significantly contribute to the sorption of pesticides (Luthy et al., 1997, Wauchope et al., 2002, Ehlers & Loibner, 2006, Kah & Brown, 2006). Therefore, the sorption capacity of a geosorbent for pesticides is dependant of the clay minerals, Fe- and Al-oxide contents as well as organic matter content and its composition.

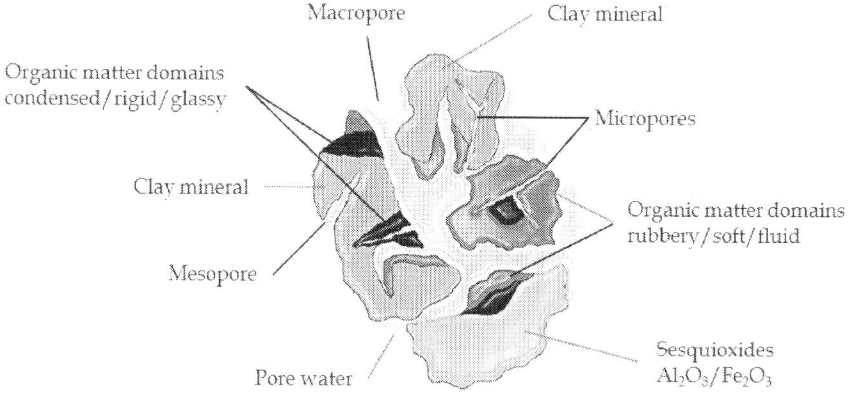

Fig. 1. Simple schematic of a geosorbent particule aggregate showing inorganic and organic materials and different porosities.

4. Interactions between chemicals and geosorbents

The type of interaction between a chemical and a geosorbent reactive surface depends on the nature of the organic and inorganic matter as well as chemical moieties present in the

pesticide in a soil we must use an empirical approach. Under laboratory controlled conditions it is possible to estimate the extent of all the interactions committed in the sorption or desorption process by the measurement of a sorption coefficient, K^S_d, or a desorption coefficient, K^D_d. These sorption coefficients are specific to a pesticide/geosorbent system and are useful for comparative and modeling purposes.

Delle Site (2001) identified and reviewed fifteen experimental methods and seven prediction methods for the determination and estimation of the sorption coefficients of organic compounds in natural sorbent/water systems. The most common method to study the sorption/desorption process is however the batch equilibrium method (EPA, 1999, OECD, 2000, Wauchope et al., 2002). In this method, a known mass of geosorbent (soil, sediment or suspended matter) is introduced into a vial (a batch reactor) and is conditioned in a known volume of appropriate liquid matrix (distilled water or water containing electrolytes (NaCl, $CaCl_2$, or sea salts)). A known volume of solution (in the appropriate matrix) of the pesticide at known concentration is added to the slurry. A minimum headspace is left to avoid losses of solute in vapour phase. The vial is then mixed gently for a specified period of time suitable to reach equilibrium, typically from 2 to 48h, 24h being usual (Wauchope et al., 2002). At the end of the agitation period, the solid is separated from the solution and the concentration of the pesticide remaining in solution is measured. The amount of pesticide sorbed on the geosorbent sample is obtained by the difference between the amount of pesticide, introduce in the batch reactor and the amount remaining in the solution at the end of experiment. The amount of sorbed substance can also be measured by the direct extraction of the geosorbent. This procedure is however more tedious than analysing the aqueous phase. Desorption studies could also be operated after the adsorption step. In the desorption assay, a specific volume of supernatant is removed for analysis, and is replaced by the same volume of the appropriate liquid matrix. The sample is shaken, centrifuged and the supernatant analyzed to complete the first cycle. At least two further decant/refill desorption cycles should be completed to generate acceptable desorption data (Bowman, 1979, Agriculture Canada, 1987, OECD, 2000).

Batch sorption measurements present few disadvantage (Delle Site, 2001, EPA, 1999, Strandberg & Fortkamp, 2005, Soubaneh et al., 2008). Losses of chemicals by volatilization and by biological or chemical degradation are possible, and the length of the experiment could be insufficient to reach equilibrium; also the complete separation between geosorbent and the water phase is difficult. Another important point reported for selected inorganic substances (EPA, 2004) is that a batch sorption test does not necessarily reproduce the chemical reaction conditions that take place in the real environment. For instance, in a soil column, water percolates through it at a finite rate and both reaction time and degree of mixing between water and soil can be much less than those occurring in a laboratory batch test. Consequently, K^S_d, values obtained from batch experiments can be different than those in a real system. Another disadvantage is that they do not accurately simulate desorption of the contaminants from a contaminated soil because it is often hypothesized that sorption and desorption reactions are reversible and K^S_d and K^D_d values are the same (EPA, 2004). This assumption is contrary to many observations that show that the sorption/desorption equilibrium may not be fully reversible, a phenomenon called hysteresis (Delle Site, 2001 and references therein, Wauchope et al., 2002).

Even though the drawbacks of the batch sorption technique are multiple, the method presents numerous advantages (Strandberg & Fortkamp, 2005). The assay is inexpensive, there is no requirement for complex or extra equipment in the lab, the methodology is quite simple and the tests can be done in few days. Therefore, it allows studying the sorption of harmful substances under a wide variety of environmental conditions relatively quickly. Last but not least, this method is so commonly used that it facilitates the comparison with results from other studies.

6. Sorption equilibrium models

The equilibrium exchange of pesticides between the aqueous phase and a geosorbent, such as soils, aquifer solids, suspended particulate matter or sediments, is quantified by the geosorbent/solution sorption distribution coefficient, $K^S{}_d$, which is defined as the ratio between the concentration of the pesticide in the geosorbent phase (C_s) and its concentration in the solution (C_e). As shown in equation 1, the sorption distribution coefficient corresponds also to the ratio of the amount of pesticide (Q_s) fixed on the solid phase (M) to the amount of pesticide freely dissolved (Q_e) in the volume of aqueous phase (V).

$$K_d^s = \frac{C_s}{C_e} = \frac{Q_s/M}{Q_e/V} \tag{1}$$

In many cases, the sorption distribution coefficient of non-polar pesticides and the fraction (%) of organic carbon content (f_{oc}) in geosorbents are correlated (Weber et al., 2000, Wauchope et al., 2002). This correlation means that pesticides have a preferential affinity to bind with organic matter instead of minerals present in the geosorbents. Thus, a carbon-normalised sorption coefficient, K_{oc}, allows comparison among pesticide sorption affinities for geosorbents with different amount of organic matter. The K_{oc} parameter can be calculated from equation 2. K_{oc} values are universally used as a measure of the relative

$$K_{OC} = \frac{K_d^S}{f_{OC}} \tag{2}$$

potential mobility of hydrophobic pesticides in soils (Wauchope et al., 2002). High values of K_{oc} suggest that pesticides could be considered as immobile in soils or sediments (McCall et al., 1980) and the low values caution a high leaching risk.

It seems easy to measure $K^S{}_d$ from equation 1. In fact, the determination of the quantity of pesticide in the aqueous or solid phase of batch sorption experiments may be associated with difficulties if the pesticide is liquid-liquid extracted with an organic solvent. For instance, pesticide can be lost during the evaporation treatments used to concentrate the analyte. In addition, loss of pesticide may also occur by its adsorption onto the glass walls, stir bar or Teflon caps of the material used. This phenomenon is expected since many other nonpolar organic compounds in aqueous solutions have strong affinities for glass and Teflon-coating parts (Baltussen et al., 1999; Ackerman & Hurtubise, 2000). Loss of pesticide may also happen by degradation in the reactor as reported for few toxaphene congeners (Lacayor et al., 2004). As clearly shown by Soubaneh et al. (2008), the evaluation and correction of these potential losses will improve the determination of the distribution coefficient values.

It is important to recall that empirical K_d measured in the batch sorption experiments represent an integration of all sorptive processes occurring in the batch sorption reactor as illustrated in Figure 2, where the toxaphene pesticide is used as a model compound. In the

values of n are generally comprised between 0.7 and 1.2 (von Oepen et al., 1991, Wauchope et al., 2002). Value of n = 1 occurs when all the sorption domains behave in a similar way. When n= 1, K_f is equivalent to K_d irrespectively of the concentration C_e and the system performs like the linear model. If n ≠ 1, K_f and K_d cannot be compared because each constant has its own unit that are different.
Sorption isotherm data could also be fitted to the Langmuir model given by equation 5, with the assumption that geosorbents have a finite number of sorption sites of uniform energy.

$$C_S = \frac{C_{s,max}K_L C_e}{1+K_L C_e} \tag{5}$$

In equation 5, $C_{s,max}$ designates the maximal sorption capacity and K_L is the Langmuir sorption coefficient related to the energy of sorption. C_s and C_e have the same meaning as previously. Please note that if K_LC_e is high (e.g., at relatively high C_e), Cs approaches $C_{s,max}$, and gives the maximum sorbate capacity of the geosorbent while, for sufficiently small values of C_e (K_LC_e near zero), C_s is linearly related to C_e and $C_{s,max}K_L$ equals K_d the sorption coefficient.
The examination of the form of the isotherm curves for sorption and desorption for the same chemical in a pesticide/geosorbent system is interesting. Under apparent equilibrium conditions, different curves can be obtained for sorption and desorption. In this phenomenon, called hystereris, the distribution coefficient $K^D{}_d$ for desorption can be greater than the $K^S{}_d$ measured for the sorption at a constant C_e concentration (Huang et al, 2003). Experimental artifacts in the desorption experiment may result in the hysteresis (Bowman & Sans, 1985, Huang et al., 1998, Delle Siete, 2001, Calvet et al., 2005). Huang et al. (1998) have suggested, under rigorously controlled experiments, to characterize this hysteresis phenomenon by an hysteresis index defined in equation 6 where C_s^S and C_s^D are the sediment sorbate concentrations for the sorption and single-cycle desorption experiments, respectively. Parameters T and C_e refer to the conditions of constant temperature and constant aqueous phase concentration.

$$\text{Hysteresis Index (HI)} = \left.\frac{C_s^D - C_s^S}{C_s^S}\right|_{T,Ce} \tag{6}$$

The higher the value of HI is, the stronger the sediment will sequestrate the chemical. A zero HI value indicates that the hysteresis is insignificant and the sorption is reversible.

7. Conclusion

It is of paramount importance to understand the behaviour and fate of excess pesticides released into the environment which is strongly influenced by the interactions between the pesticides and natural solids called geosorbents. The chemical composition of the pesticides and the physical and chemical complexities of the geosorbents modulate the diversity and the intensities of interactions between the pesticides and geosorbents resulting in various destinies for pesticides in the fields. Even though our understanding of the pesticide/geosorbent interactions is mostly qualitative, and that it is difficult to examine the sorption processes at the molecular scale, we have some simple experimental approaches to quantitatively characterize the sorption phenomenon. The simple batch sorption experiment allows the measurement of a sorption distribution coefficient under near realistic field conditions at a specific concentration of pesticide. However, this datum is of limited interest considering its high specificity. For non-polar pesticides, if the coefficient is normalized for the organic content of the geosorbent, this allows discriminating the affinities of pesticides for various soils or sediments and can be used to evaluate the potential mobility of pesticides in the environment. Batch sorption experiments at different concentrations of a given pesticide provide isotherm curves that are more realistic of the variation in the distribution coefficients in the fields. Close examination of the sorption and desorption isotherms permits some insights into the mechanism of sorption and can also provide a useful hysteresis index to discuss the reversibility of the sorption and the persistence of the pesticides in the environment.

8. References

Ackerman, A.H. & Hurtubise, R.J. (2000). The effects of adsorption of solutes on glassware and teflon in the calculation of partition coefficients for solid-phase microextraction with 1PS paper. *Talanta*, 52: 853–861

Agriculture Canada, (1987). Guideline for determining environmental chemistry and fate of pesticides in Canada. T-1-255, Food Production and Inspection Branch, Ottawa

Ainsworth, C.C.; Fredrickson, J.K. & Smith, S.C. (1993). Effects of sorption on the degradation of aromatic acids and bases. In: *Sorption and degradation of pesticides and organic chemicals in soils*, Linn D.M.; Carski, T.H.; Brusseau, M.L. & Chang, F.H. (Eds.) 125-144, SSSA Special publication N0 32. SSSA and ASA, Madison, USA

Anonynous, (november 2010). Charge-transfer complex available from http://en.wikipedia.org/wiki/Charge-transfer_complex

Arias-Estevez, M.; Lòpez-Periago, E.; Martinez-Carballo, E.; Simal-Gandara, J. & Mejuto, J.-C. (2008). The mobility and degradation of pesticides in soils and the pollution of groundwater resources. *Agricultural, Ecosystems and Environment*, 123: 247-260

Azizian, S. (2004). Kinetic models of sorption: a theoretical analysis. *Journal of Colloid and Interface Science*, 276 : 47–52

Baltussen, E.; Sandra, P.; David, F.; Janssen, H.-G. & Cramers, C. (1999). Study into the equilibrium mechanism between water and poly(dimethylsiloxane) for very apolar solutes: adsorption or sorption? *Analytical Chemistry*, 71, 5213–5216

Barriuso, E.; Laird, D.A.; Koskinen, W.C. & Dowdy, R.H. (1994). Atrazine desorption from smectites. *Soil Science Society of America Journal*, 58:1632–1638.

Bowman, B.T., 1979. Method for repeated additions for generating pesticide adsorption-desorption isotherm data. *Canadian Journal of Soil Science*, 59: 435–437.

Bowman, B.T. & Sans, W.W. (1985). Partioning behavior of insecticides in soil-water systems: II. Desorption hysteresis effects. *Journal of environmental Quality*, 14:270-273

Bronick, C.J. & Lal, R. (2005) Soil structure and management: a review. *Geoderma*, 124: 3-22

Calvet, R. (1989). Adsorption of organic chemicals in soils. *Environmental Health Perspective*, 83:145–177

Calvet, R.; Barriso, E.; Bedos, C.; Benoit, P.; Charnay, M.-P. & Coquet Y. (2005). *Les pesticides dans le sol.* Editions France Agricole. ISBN 2-85557-119-7, Paris

Celano, G. ; Smejkalova, D. ; Spaccini, R. & Piccolo A. (2008). Interactions of three s-triazines with humic acids of different structure. *Journal of Agricultural and Food Chemistry* 56 : 7360-7366

Celis, R.; Cox, L.; Hermosin, M.C. & Cornejo, J. (1997). Sorption of thiazafluron by iron- and humic acid-coated montmorillonite. *Journal Environmental Quality*, 26 :472-479

Chefetz, B. & Xing, B. (2009). Relative role of aliphatic and aromatic moieties as sorption domains for organic compounds: a review. *Environmental Science and Technology*, 43:1680-1688

Chiou, C.T.; Peters, L.J. & Freed V.H.A. (1979). A physical concept of soil-water equilibria for nonionic compounds. *Science*, 206 (4420) 831-832

Clausen, L.; Fabricius, I. & Madsen, L. (2001) Adsorption of Pesticides onto Quartz, Calcite, Kaolinite, and a-Alumina. *Journal Environmental Quality*, 30:846–857

Cornejo, J.; Celis, R.; Pavlovic, I & Ulibarri, M.A. (2008). Interactions of pesticides with clays and layered double hydroxides: a review. *Clay Minerals*, 43 : 155–175

Cornelissen, G. ; Gustafsson, O .; Bucheli, T.D. ; Jonker, M.O. ; Koelmans, A. & Vannoort, P.M. (2005). Extensive sorption of organic compounds to black carbon, coal, and kerogen in sediments and soils: Mechanisms and consequences for distribution, bioaccumulation, and biodegradation. *Environmental Science and Technology*, 39:6881-6895

Cox, L.; Hermosın, J. & Cornejo J. (1993). Adsorption of methomyl by soils of southern Spain and soils components. *Chemosphere*, 27 : 837-849

Delle Site, A. (2001). Factors affecting sorption of organic compounds in natural sorbent/water systems and sorption coefficients for selected pollutants. A review. *Journal of Physical and Chemistry Reference data*, 30:187-438

de Oliviera M.F.; Johnson, C.; Premachandra, G.S.; Teppen, B.J.; Laird H.D.; Zhu, D. & Boyd S. (2005). Spectroscopic Study of Carbaryl Sorption on Smectite from Aqueous Suspension. *Environmental Science and Technology*, 39: 9123–9129

Dubus, I.G.; Barriuso, E. & Calvet, R. (2001). Sorption of weak organic acids in soils : clofencet, 2,4-D and salicilic acid. *Chemosphere*, 45 :767-774

Ehlers, G.C.A. & Loibner, A.P. (2006). Linking organic pollutant (bio)availability with geosorbent properties and biomimetic methodology: A review of geosorbent characterization and (bio)availability prediction. *Environmental Pollution*, 141:494-512

EPA (U.S. Environmental Protection Agency), 1999. Understanding variation in partition coefficient, Kd, values. Volume I. The Kd model, methods of measurement, and application of chemical reaction codes. EPA 402-R-99-004A, U.S. Environmental Protection Agency, Cincinnati, Ohio

EPA (U.S. Environmental Protection Agency), 2004. Understanding variation in partition coefficient, Kd, values. Volume III. Review of Geochemistry and Available Kd Values for Americium, Arsenic, Curium, Iodine, Neptunium, Radium, and Technetium, U.S. Environmental Protection Agency, Cincinnati, Ohio

Ferreira, J.A. Martin-Neto, L, Vaz, C.M.P. & Regitano, J.B. (2002) Sorption interactions between imazaquin and a humic acid extracted from a typical Brazilian oxisol. *Journal of Environmental Quality*, 31:1665–1670

Gevao, B.; Semple, K.T. & Jones K.C. (2000). Bound pesticide residues in soils: a review. *Environmental Pollution*, 108 : 3-14

Herwig, U,; Klumpp. E.; & Narres, H.D. (2001). Physicochemical interactions between atrazine and clay minerals. *Applied Clay Science,* 18:211–222

Hillel, D. (2004). *Introduction to environmental soil physics.* Elsevier, ISBN 0-12-348655-6, Amsterdam.

Hiradate, S.; Furubayashi, A.; Uchida, N. & Fujii, Y. (2007). Adsorption of 2,4-dichlorophenoxyacetic acid by an Andosol. *Journal of Environmental Quality,* 36:101–109

Ho, Y.S.; Ng, J.C.Y. & McKay, G. (2000). Kinetics of pollutant sorption by biosorbents : Review. *Separation and Purification methods,* 29 : 189-232

Huang, W.; Yu, H. & Weber Jr., W.J. (1998). Hysteresis in the sorption and desorption of hydrophobic organic compounds by soils and sediments. 1. A Comparative analysis of experimental protocols. *Journal of Contaminant Hydrology,* 31:129-148

Huang W.; Peng, P.; Yu, Z. & Fu, J. (2003) Effects of organic matter heterogeneity on sorption and desorption of organic contaminants by soils and sediments. *Applied Geochemistry,* 18:955-972

Hyun, S.; Lee, L.S. & Rao P.S.C. (2003). Significance of anion exchange in pentachlorophenol sorption by variable-charge soils *Journal of Environmental Quality,* 32:966–976

Hyun, S. & Lee, L.S. (2005). Quantifying the contribution of different sorption mechanisms for 2,4-dichlorophenoxyacetic acid sorption by variable-charge soils.*Environmental Science and Technology,* 39:2522-2528

Jasinka, E.; Wetzel, H.; Baumgartl, T. & Horn R. (2006) Heterogeneity of physico-chemical properties in structured soils and its consequences. *Pedosphere,* 16(3): 284-296

Johnson, W. G.; Lavy, T.L. & Gbur, E.E. (1995). Sorption, mobility and degradation of trichlopyr and 2,4-D on 4 soils. *Weed Science,* 43:678–684

Kah, M. & Brown, C.D. (2006). Adsorption of ionisable pesticides in soils, *Reviews of Environmental Contamination & Toxicology, 188:149–217*

Karickhoff, S.W. (1984) Organic pollutant sorption in aquatic systems. *Journal of Hydraulic Engineering,* 110 (6) 707-735

Keiluweit, M. & Kleber, M. (2009). Molecular-level interactions in soils and sediments: The role of aromatic π-systems. *Environmental Science and Technology,* 43: 3421-3429

Khan, S.U. (1982). Distribution and characteristics of bound residues of prometryn in an organic soil. *Journal of Agricultural and Food Chemistry,* 30:175–179

Kögel-Knabner, I.; Guggenberger, G.; Kleber, M.; Kandeler, E.; Kalbitz, K.; Scheu, S.; Eusterhues, K. & Leinweber, P. (2008). Organo-mineral associations in temperate soils: Integrating biology, mineralogy, and organic matter chemistry. *Journal of Plant Nutrition and Soil Science,* 171 : 61-82

Konstantinou, I.K. & Albanis T.A. (2000) Adsorption-Desorption Studies of Selected Herbicides in Soil-Fly Ash Mixtures, *Journal of Agricultural and Food Chemistry,* 48:4780-4790

Kovaios, I.D.; Paraskeva, C.A.; Koutsoukos, P.G. & Payatakes A.Ch. (2006). Adsorption of atrazine on soils : Model study. *Journal of Colloid and Interface Science,* 299 :88–94

Lacayor, M.; van Bavel, B. & Mattiasson, B. (2004). Degradation of toxaphene in water during anaerobic and aerobic conditions. *Environmental Pollution,* 130, 437–443.

Lagaly, G. (2001). Pesticide–clay interactions and formulations. *Applied Clay Science,* 18: 205–209

Vereecken, H. (2005). Mobility and leaching of glyphosate: a review. *Pest Management Science*, 61:1139-1151

von Oepen, B.; Kördel, W. & Klein, W. (1991). Sorption of nonpolar compounds to soils: Processes, measurements and experience with the application of the modified OECD-Guideline 106. *Chemosphere*, 22:285-304

von Oepen, B.; Kördel, W.; Klein, W. & Schuurmann G. (1991). Predicitve QSPR models for estimating soil sorption coefficients: Potential and limitations based on dominating processes. *Science of the Total Environment*, 109:343-354

Watson, J.R.; Posner, A.M. & Quirk, J.P. (1973). Adsorption of the herbicide 2,4-D on goethite. *Journal of Soil Science*, 24:503–511

Wauchope, R. D.; Yeh, S.; Linders, J.B.H.J.; Kloskowski, R.; Tanaka, K.; Rubin, B.; Katayama, A.; Kördel, W.; Gerstl, Z.; Lane M. & Unsworth J.B. (2002). Pesticide soil sorption parameters: theory, measurement, uses, limitations and reliability. *Pest Management Science*, 58:419-445

Weber, J.B. & Weed, S.B. (1968). Adsorption and desorption of diquat, paraquat and prometone by montmorillonitic and kaolinitic clay minerals. *Soil Science Society of America Proceedings*, 32, 485487

Weber, J.B.; Shae P.H. & Weed, S.B. (1986). Fluridone retention and release in soils. *Soil Science Society America Journal*, 50: 582-588

Weber, J.B.; Wilkerson, G.G.; Linker, H.M.; Wilcut, J.W., Leidy, R.B.; Senseman, S.; Witt, W.W.; Barret, M.; Vencill, W.K.; Shaw, D.R.; Mueller, T.C.; Miller, D.K.; Brecke, B.J.; Talbert, R.E. & Peeper, T.F. (2000). A proposal to standardize soil/solution herbicide distribution coefficients. *Weed Science*, 48:75-88

Weber, Jr., W.J., McGinley, P.M. & Katz, L.E. (1991). Sorption phenomena in subsurface systems: Concepts, models and effects on contaminant fate and transport. *Water Research*, 25: 499-528

Welhouse, G. J. & Bleam, W. F. (1993). Atrazine hydrogen-bonding potentials. *Environmental Science and Technology*, 27:494–500

Worrall, F.; Parker, A.; Rae, A, & Johnson, A.C. (1996). Equilibrium adsorption of isoproturon on soil and pure clays. *European Journal of Soil Science*, 41:265-212

Yang, G-P. ; Zhao, Y-H. ; Lu, X-L. & Gao X.-C. (2005). Adsorption of methomyl on marine sediments. *Colloids and Surfaces A: Physicochem. Eng. Aspects*, 264 : 179–186

Yu, Y. & Zhou, Q.-X. (2005). Adsorption characteristics of pesticides methamidophos and glyphosate by two soils. *Chemosphere*, 58:811-816

Zhang, P. C. & Sparks, D.L. (1989). Kinetics and mechanisms of molybdate adsorption/ desorption at the goethite/water interface using pressure-jump relaxation. *Soil Science Society American Journal*, 53:1028-1034

39

Pesticides as a Waste Problem with Examples from Norway

Ketil Haarstad
Bioforsk Soil and Environment
Norway

1. Introduction

Pesticides can produce waste problems when disposing off the surplus from usage, containers and other residues, but also during the production and distribution. Obsolete pesticides such as the chlorinated γ-HCH (lindane) and DDT have created huge problems also because the use of toxic raw materials, such as chlorophenols and dioxins, and the production of hazardous and useless by-products form the production, such as e.g. α- and β- HCH that constitutes about 85% of the production output. The cleanup of production sites and dumps has created major problems - the work is almost completed in Western Europe - but many locations remain polluted, especially in Central- and Eastern Europe, and possibly also in other sites around the globe. Sites polluted with obsolete pesticides are mainly threatening rural areas. According to the World Health Organization (WHO) half of the deaths due to cancer in 2050 will be caused by polluted food, soil, water and environment, mainly from pesticides. The economical burden will also be heavy. The scale of the problem with obsolete pesticides is estimated to more than 250 000 tons of pesticides in Easter Europe and Central Asia (Vijgen & Egenhofer, 2009).

Landfilling is the most common waste treatment worldwide. In The European Union (EU) several waste directives have improved the quality of the waste going to landfills, both containing hazardous waste and ordinary household or municipal solid waste (MSW), but as we show here there is a potential for pollution from pesticides both from old and new landfills. As outlined in the Waste Framework Directive from the EU pesticides are one of the key parameters that needs to be monitored. Based on the results from Norwegian investigations on pesticides from greenhouse production an increased focus is put on this problem in Europe.

Large quantities of numerous chemicals have the potential to pollute the waste that is landfilled. Annually close to 500 kg is landfilled for each person in the developed world, producing on the average more than 30 tons of waste over a lifetime. Before a more managed waste disposal started in the nineties many landfills were established and scattered almost everywhere, with the risk of polluting the environment and threatening human health. Landfills create polluted wastewater, or leachate, ca. 50-200 mm annually in dry areas, 400-800 mm on the average, and >1000 mm under wet conditions. The content of organic matter in the waste is often the most important factor influencing the quality of the leachate, in addition to the local geology, hydrology and climate. Landfills usually create large emissions of organic matter and nitrogen, see Figure 1, measured as chemical oxygen demand (COD), biochemical oxygen demand (BOD) and ammonia-nitrogen (NH_4 N). Nitrogen is usually the parameter with the longest half-life. Also the leachate contains suspended solids including organic and inorganic colloids, estimated to more than 12 kg/day and 2.5 kg/day, respectively, from an average landfill. These particles also have a strong potential to erode and transport pesticides from the waste, especially the hydrophobic ones with low water solubility and often high toxicity. The COD and BOD contain mainly organic acids in leachate from fresh waste. Later the BOD is reduced, and the COD will consist mainly of "hard" COD due to non-degradable humic substances.

Pesticides are labelled as hazardous and are generally not allowed to be landfilled. Pesticides that are resisting anaerobic degradation and have a high water solubility and a low acid partition coefficient, $pK_a<7$, are expected to be more readily leached from the waste. Herbicides generally have a high solubility, in the order of several hundreds or thousands of mg/l, fungicides have intermediate solubility, up to a few mg/l, while insecticides are usually non-soluble in water, with water solubility less than 10 µg/l. The solubility is, however, depending on factors such as the content of dissolved organic carbon, e.g. for DDT (Haarstad & Fresvig, 2000). Landfill leachate has high concentrations of suspended and soluble organic matter and thus has a potential of transporting relatively large amounts of hydrophobic compounds. Leachate is generally anaerobic and methanogenic with a high pH, or acidic and low in pH if the waste is fresh, or a combination of these. The ecotoxicity of the pesticides are usually inversely related to their solubility, but the solubility is often much greater than the ecotoxicological limit value (the predicted no effect concentration or PNEC).

The production and use of pesticides clearly are able to produce waste problems that can give environmental and health problems. It is necessary to impose strict regulations both on the production, sale, use and waste handling of these compounds, all based on intensive monitoring.

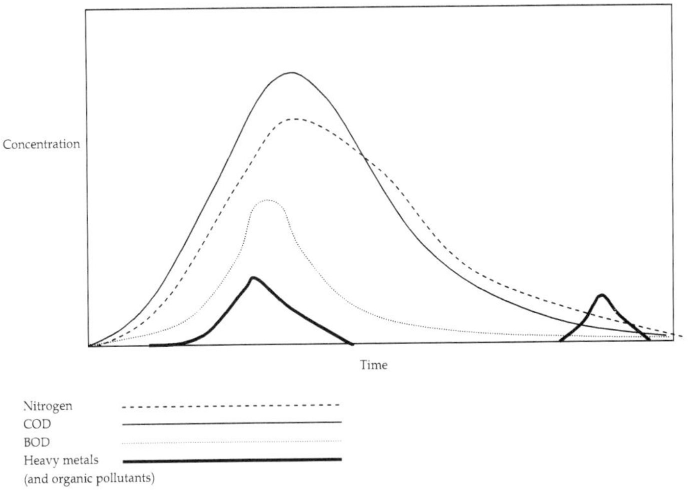

Fig. 1. Trends in concentrations (mg/l or µg/l) in leachate emissions over a lifetime of a landfill.

2. Pesticides in ordinary landfills

There are limited reports on the concentrations of pesticides in MSW leachate. A review listed 24 compounds, including 3 metabolites, found in concentrations ranging from 0.025 to 26 µg/l (chlorprofam) - most of the findings were however from one sampling (Kjeldsen et al., 2008). Another study found phenoxy acids in concentrations up to 65 µg/l in leachate-contaminated ground water (Tuxen et al., 2003). A report showed the occurrence of mecoprop and isoproturon in UK leachates, of which the former was readily removed by aerobic treatment (Robinson & Barr, 1999). In a study pesticides were analyzed in 28 grab samples of untreated and 12 samples of treated leachate from eight municipal solid waste (MSW) landfills, over a period of 10 years, with the purpose to screen for compounds and to evaluate the removal in different treatment systems (Haarstad & Mæhlum, 2008). Also 1 grab sample of leachate sediments is included. The most frequently detected group of compounds are the phenoxy acids, and they also occur in highest concentrations, up to 230 µg/l for mecoprop, in this study. Also three fungicides and one insecticide were detected, but in much lower concentrations. All samples exceeded the maximum limit value (MLV) for the sum concentration of pesticides in drinking water (0.5 µg/l), and six compounds exceeded the PNEC. Reverse osmosis showed good removal of phenoxy acids, while sequential batch reactor aerobic treatment, as well as aerated lagoons in combination with wetlands, groundwater infiltration and reactor treatment showed slightly lower removal.

Table 1 shows that most pesticides are detected in leachate treated in wetlands, mostly with pre-areation (Haarstad & Mæhlum, 2008), where the concentrations can reach 50 ppb for the most water soluble compounds. A total of 13 compounds were detected.

	Reverse osmosis	Aeration/wetlands	SBR
Pesticides in treated leachate	phenoxy acids	phenoxy acids	phenoxy acids
		chlorfenvinphos	
		isoptroutron	
		azoxystrobin	
		clopyralid	
		mecoprop	
Concentrations (µg/l)	0.01-0.08	0.16-50	0.03-1.1

Table 1. Pesticides detected in MSW leachate (from Haarstad & Mæhlum, 2008).

3. Pesticides in green waste from agriculture

Pesticides in agricultural waste products are also a problem. A landfill consisting mainly of organic waste from a tree nursery and containing an estimated 900 kg of DDT has been monitored since 1994. Downstream ground water was sampled from four wells. More than 10 years of monitoring of two of the wells is presented in Table 2, in addition to sampling of the waste. A total of seven pesticides were detected in the ground water (Haarstad, 2008). In addition to DDT, there were two other insecticides and four fungicides occurring in the ground water downstream of the landfill. The maximum concentration of pesticides was 3.76 µg/L of which 2.70 µg/L was permethrin in October 2000. The maximum concentration of DDT in the ground water was 0.52 µg/L, indicating that the leachate DDT concentrations probably exceed the water solubility of the compound. The investigation shows that the practice of establishing local landfills of waste from nurseries is environmentally unsafe. In 2002, the landfill was covered with clayey soil and vegetated. It seems that this has stabilized the pesticide concentrations in the ground water and removed the occurrence of extreme values. However, the vegetated soil layer has not been able to prevent the leakage of pesticides to the ground water, still present 5 years later.

	Well P2								Well P3					
Sampling date	Σ*	D	F	P	L	PE	I	T	Σ	D	L	P	PE	F
04.11.1994									0.12	0.12				
10.04.1995									0.15	0.15				
16.04.1996									0.07	0.07				
27.05.1998	0.32		0.07	0.25			0		0.3	0.22	0.08			
01.12.1998	0.28		0.15				0.13		0.89	0.47	0.33			0.09
11.08.1999									0.34	0	0.34			
10.12.1999	0.14			0.14					0.07	0.07				
16.10.2000	0.77		0.16			0.61			3.76	0.52	0.54		2.7	
26.11.2001	0.25	0.1	0.06	0.06	0.03				0.63	0.14	0.47			0.02
26.05.2003	0.14	0.03	0.11						0.76	0.11	0.65			
03.10.2003									0.94	0.08	0.86			
05.11.2004	0.11		0.02	0.09					0.27	0.04	0.18	0.04		0.01
26.06.2007	0.29							0.29	0.16	0.02	0.12	0.02		

From Haarstad 2008. Blank=below the limit of detection. Σ =sum pesticides. D=sum DDT, F=fenpropimorph, P=propiconazole, L=lindane, PE=permethrin, I=iprodione, T=trifloxystrobin

Table 2. Pesticides from green waste (µg/l)

4. Pesticides in products, waste and wastewater from greenhouse production

In a study of a greenhouse waste tip a total of 8 pesticides were detected with concentrations from 10 - 170 μg/kg. In the runoff receiving leachate from the tip concentrations of pesticides varied between 0.03 - 1. 2 μg/l.

In a study of runoff from greenhouses water samples downstream 9 large facilities focusing on flower production were analyzed for 56 pesticides. The samples included concentrated runoff from the production of tomatoes and cucumbers. Pesticides were detected at all locations and in 90% of the samples, with a total of 18 compounds, of which were 9 fungicides, 5 herbicides and 4 insecticides in concentrations exceeding the PNEC value. Detected fungicides were cyprodinil, propikonazol, iprodion, azoksystrobin, prokloraz and vinklozolin. Insecticides with detections above the PNEC were pirimikarb, diazinon and klorfenvinfos. The fungicides pyrimetanil, iprodion and imazalil had concentrations >1μg/l in the concentrated runoff from vegetable greenhouses. The findings where the concentrations were above the PNEC values came from two locations with the production of flowers, in addition to water from the floor of one of the greenhouses. Generally the most likely pesticides to leak from greenhouse productions are pyrimethanil, cyprodinil, propiconazole, iprodione, azoxystrobin, imazil, prochloraz and pirimicarb, based on the facts that they are common in use in such productions, results from other monitoring programs, timing of their detections downstream greenhouses and a correlation between findings and high levels of nutrients in the samples.

Repeated sampling of imported and domestic flowers, pot plants and pot soil shows the presence of occasional high levels of residual pesticides. In the pot plants 22 compounds were detected with concentrations varying from 0.02 to 8.8 mg/kg, while in pot soil 7 compounds were detected with concentrations from 0.05 to 2.2 mg/kg. Most detection was made in flowers, where 31 compounds were detected, with concentrations from 0.01 to 5.3 mg/kg. For pot plants 11 fungicides and 11 insecticides were detected, for pot soil 4 insecticides and 3 fungicides, and for flowers 14 insecticides and 18 fungicides.

5. References

Haarstad, K. 2008. Long-Term leakage of DDT and Other Pesticides from a Tree Nursery Landfill. Ground Water Monitoring & Remediation, 28, no. 4, 107-111.

Haarstad, K., Mæhlum, T. 2008. Pesticides in Norwegian Landfill leachate. The Open Environmental & Biological Monitoring Journal, 1, 8-15.

Haarstad, K. and Fresvig, M. 2000. The Influence of Organic Matter and pH on DDT Aqueous Solubility. Journal of Soil Contamination, 9, 4, 347-358, July 2000

Kjeldsen P, Barlaz MA, Rooker AP, Baun A, Ledin A, Christensen, TH. Present and long-term composition of MSW landfill leachate: A review. Crit Rev Environ Sci Technol 2002; 32: 297-336.

Robinson, HD, Barr, MJ. Aerobical biological treatment of landfill leachate. Waste Man Res 1999; 17: 478-86.

Roger Roseth. 2010. Pesticides in waste from the production and import of flowers. (In Norwegian with English summary). Bioforsk-report 67/2010 28 pp.

Roger Roseth, Gro Hege Ludvigsen, Roald Aasen. 2007. Preliminary project-pesticides in runoff from greenhouses. (In Norwegian with English summary). Bioforsk-report 162/2007. 26 pp.

Roseth, R., Haarstad, K. 2010. Pesticide runoff from greenhouse production. Water Sci. & Tech., 61.6, 1373-1381.

Tuxen N, Ejlskov P, Albrechtsen HJ, Reitzel LA, Pedersen JK, Bjerg PL. Application of natural attenuation to ground water contaminated by phenoxy acid herbicides at an old landfill in Sjølund, Denmark. Ground Water Monitoring & Remediation 2003; 23(4): 48-58.

Vijgen, J., Egenhofer, C. 2009. Obsolete (lethal) Pesticides, a ticking time bomb and why we have to act now. Tauw Group, The Netherlands, c/o Elmevej 14, DK-2840 Holte. 40 pp.